Compendium of Inflammatory Diseases

Michael J. Parnham
Editor-in-Chief

Compendium of Inflammatory Diseases

Volume 2

I–Z

With 218 Figures and 116 Tables

Editor-in-Chief
Michael J. Parnham
Fraunhofer IME-TMP &
JW Goethe University Frankfurt
Frankfurt am Main, Germany

ISBN 978-3-7643-8530-9   ISBN 978-3-7643-8550-7 (eBook)
ISBN 978-3-7643-8677-1 (print and electronic bundle)
DOI 10.1007/978-3-7643-8550-7

Library of Congress Control Number: 2016946035

© Springer International Publishing AG 2016
This work is subject to copyright. All rights are reserved by the Publisher, whether the whole or part of the material is concerned, specifically the rights of translation, reprinting, reuse of illustrations, recitation, broadcasting, reproduction on microfilms or in any other physical way, and transmission or information storage and retrieval, electronic adaptation, computer software, or by similar or dissimilar methodology now known or hereafter developed.
The use of general descriptive names, registered names, trademarks, service marks, etc. in this publication does not imply, even in the absence of a specific statement, that such names are exempt from the relevant protective laws and regulations and therefore free for general use.
The publisher, the authors and the editors are safe to assume that the advice and information in this book are believed to be true and accurate at the date of publication. Neither the publisher nor the authors or the editors give a warranty, express or implied, with respect to the material contained herein or for any errors or omissions that may have been made.

Printed on acid-free paper

This Springer imprint is published by Springer Nature
The registered company is Springer International Publishing AG
The registered company address is: Gewerbestrasse 11, 6330 Cham, Switzerland

Preface

Inflammation has become one of the most exciting and rewarding areas of medical research. Recent years have seen a revolution in our understanding of how blood and tissue cells interact and of the intracellular mechanisms controlling their activation. This has revealed the underlying inflammatory pathology of a wide variety of diseases and provided multiple new targets for drug therapy. As a consequence, there has been a burst in the development of new anti-inflammatory and immunomodulatory drugs for these diseases. In parallel, the literature on inflammatory diseases, the mechanisms and mediators involved, and approaches to drug therapy has expanded dramatically so that it is very difficult to obtain an overview of the field.

The *Compendium of Inflammatory Diseases* is an attempt to rectify this situation, providing in-depth information for students, teachers, researchers, and clinicians working in academia, hospital, and industry. The reference work provides insight into the cellular and humoral components and processes of inflammation, its treatment, and its clinical expression in a collection of articles each written by experts in their fields. The simple A–Z format provides easy access to relevant information. The compendium is being published simultaneously in print and electronic forms. Extensive cross references to entries in the compendium and in other Springer Reference works enable efficient and user-friendly searches.

Because of the constant change in our understanding of inflammation, the print version provides a selective view of the current status, and in such a large reference work omissions are inevitable. However, a significant aspect of the format is that a living reference or dynamic online version is available, which can be updated by authors and editors to keep up with new developments. Since new authors can also submit contributions to the online edition, this version will continue to grow. The DOI for the living reference (currently, *Encyclopedia of Inflammatory Diseases*) is http://link.springer.com/referencework/10.1007/978-3-0348-0620-6. As a living project, the compendium will serve as a reliable data pool for all those working in inflammation research, becoming an indispensible source of information for academia, clinical practitioners, and industry.

Work on the compendium was started in 2004 and has involved a large number of people over the years. I thank, in particular, the associate editors of the various parts of the compendium: Anne-Marie Irani, Jason McDougall, and Marc Feldmann for inflammatory diseases; Sylvie Chollet-Martin for inflammatory cells and processes; Heiko Mühl and Heinfried Radeke for inflammatory mediators; and Garry Graham and Kevin Pile for the pharmacology of inflammation. Thanks for your patience, dedication, and hard work. I also acknowledge the assistance of John Hamilton, Alan Lewis, Luke O'Neill, and Peter Lipsky who originally helped to put a list of potential articles together and Vincent Lagente and Valerie Quesniaux who gave support along the way. I am also extremely grateful to all the authors who gave their time and energy to the writing of the approximately 120 chapters. A special word of thanks to the people at Springer who encouraged, cajoled, and supported the collation and processing of the manuscripts: Detlef Klüber who initiated and promoted the project together with Sandra Fabiani and Susanne Friedrichsen; Lydia Müller, Kerstin Kindler, and Somodatta Roy who did the hands-on processing work in the early phases of the project; and Sunali Mull and Keerthi Sudevan who pulled the strings together to bring the project to its first print milestone.

Frankfurt am Main   Michael J. Parnham
April 2016

Article Note

The copyright holder on page IV has been updated. An Erratum is available at DOI 10.1007/978-3-7643-8550-7_601.

About the Editor-in-Chief

Michael J. Parnham B.Sc., Ph.D., CBiol MSB, FBPhS Fraunhofer Institute for Molecular Biology and Applied Ecology IME, Project Group Translational Medicine and Pharmacology TMP and JW Goethe University Frankfurt, Frankfurt am Main, Germany

Michael Parnham studied Pharmacology at Chelsea (now King's) College, London, including a year working on prostaglandins and pain with Sergio Ferreira at The Royal College of Surgeons and completed doctoral studies at Bristol University in 1973. Postdoctoral research took him to Erasmus University Rotterdam, The Netherlands, to work with Ivan Bonta on prostaglandins and chronic inflammation. This led to the award of the Goslings Prize for research on antirheumatic therapy and the cofounding with Bonta and Kay Brune of the European Workshop on Inflammation, of which Dr. Parnham was secretary for many years. In 1980, he moved to Germany for a decade in antiinflammatory research with A. Nattermann/Rhône-Poulenc, becoming Head of General Biology. In 1990, he and his colleagues were awarded the Prix Galien Germany for the development of ebselen, a novel organoselenium approach to anti-inflammatory therapy. In the same year, Dr. Parnham established a consultancy and became a lecturer in pharmacology and toxicology at Goethe University, Frankfurt. Appointed Professor in Frankfurt in 1998, Dr. Parnham also moved to Zagreb, Croatia, where he established new pharmacology and toxicology labs in the Research Institute of PLIVA Pharmaceuticals. He and his team characterized the immunomodulatory effects of macrolide antibiotics and developed two new chemical agents and two biogenerics to clinical trials, receiving two awards from PLIVA. Subsequently, after ownership changed in 2006, he was Director of Preclinical Discovery at GlaxoSmithKline in Zagreb. Currently, he is Head of Preclinical Research at Fraunhofer IME-TMP in Frankfurt, Germany, carrying out applied research in neuroinflammation, autoimmunity, sepsis, and pain. He is editor of the monograph series *Progress in Inflammation Research* and coeditor of *Milestones in Drug Therapy* and of the textbook *Principles of Immunopharmacology*. From 1982 to 2013, Prof. Parnham was Editor, then Managing Editor of the peer-reviewed journal *Inflammation Research*. In 2015, he received the Lifetime Achievement Award of the International Association of Inflammation Societies.

Section Editors

Section: Inflammatory Diseases

Anne-Marie Irani Division of Allergy and Immunology, Children's Hospital of Richmond at VCU, Virginia Commonwealth University, Richmond, VA, USA

Sir Marc Feldmann Emeritus Professor, Kennedy Institute of Rheumatology, NDORMS, University of Oxford, Botnar Research Centre, Headington, Oxford, UK

Jason J. McDougall Departments of Pharmacology and Anaesthesia, Pain Management and Perioperative Medicine, Dalhousie University, Halifax, NS, Canada

Section: Mediators of Inflammation

Michael Parnham Goethe University Frankfurt and Fraunhofer Institute for Molecular Biology and Applied Ecology IME, Project Group Translational Medicine and Pharmacology TMP, Frankfurt am Main, Germany

Heiko Mühl Pharmazentrum Frankfurt, University Hospital Frankfurt, Goethe-University Frankfurt am Main, Frankfurt, Germany

Heinfried H. Radeke Immune Pharmacology/ZAFES, Pharmazentrum Frankfurt, Bldg. Institute of General Pharmacology and Toxicology, Clinic of the Goethe University Frankfurt Main, Frankfurt Main, Germany

Section: Pharmacology of Inflammation

Kevin Pile Campbelltown Hospital, Sydney, Australia
Western Sydney University, Sydney, Australia

Garry Graham Honorary Professor, Department of Pharmacology, School of Medical Sciences, University of NSW, Sydney, NSW, Australia

Department of Clinical Pharmacology and Toxicology, St Vincent's Hospital, Sydney NSW 2010, Australia

Section: Inflammatory Processes and Cells

Sylvie Chollet-Martin Professor of Immunology, Faculté de Pharmacie de Paris-Sud 11, Département d'Immunologie et INSERM UMR 996, Châtenay-Malabry, France

Service d'Immunologie, Hôpital Bichat, Paris, France

Contributors

Behdad Afzali Molecular Immunology and Inflammation Branch, National Institute of Arthritis and Musculoskeletal and Skin Diseases, National Institutes of Health, Bethesda, MD, USA

Division of Transplant Immunology and Mucosal Biology, MRC Centre for Transplantation, King's College London, London, UK

Nadine Ajzenberg Department of Haematology AP-HP, Bichat-Claude Bernard Hospital, University Paris Diderot, Paris, France

Francesco Annunziato Department of Experimental and Clinical Medicine and Denothe Center, University of Florence, Florence, Italy

Victor Appay Sorbonne University, UPMC University Paris 06, UMR-S CR7, Paris, France

Jesús Arenas Department of Biology, Section Molecular Microbiology, Utrecht University, Utrecht, The Netherlands

Natalia Arenas-Ramirez Department of Immunology, University Hospital Zurich, Zurich, Switzerland

Amon Asgharpour Division of Gastroenterology, Hepatology and Nutrition, Department of Internal Medicine, VCU Medical Center, Richmond, VA, USA

Joseph Avruch Diabetes Unit, Medical Services and Department of Molecular Biology, Massachusetts General Hospital, Boston, MA, USA

The Department of Medicine, Harvard Medical School, Simches Research Center, Massachusetts General Hospital, Boston, MA, USA

Gholamreza Azizi Department of Laboratory Medicine, Imam Hassan Mojtaba Hospital, Alborz University of Medical Sciences, Karaj, Iran

Research Center for Immunodeficiencies, Pediatrics Center of Excellence, Children's Medical Center, Tehran University of Medical Sciences, Tehran, Iran

Magnus Bäck Center for Molecular Medicine, Department of Medicine, Karolinska Institutet, Stockholm, Sweden

Lorena Barrientos Faculté de Pharmacie, Université Paris Sud, INSERM UMR 996, Paris, France

Marcel Batten Immunology Division, Garvan Institute of Medical Research, and St. Vincent's Clinical School, University of New South Wales, Sydney, NSW, Australia

Michael Bauer Department for Anesthesiology and Intensive Care, Jena University Hospital, Jena, Germany

Center for Sepsis Control and Care, Jena University Hospital, Jena, Germany

Robert P. Baughman Department of Internal Medicine, Interstitial Lung Disease and Sarcoidosis Clinic, University of Cincinnati Medical Center, Cincinnati, OH, USA

Stephen Bickston Division of Gastroenterology, Hepatology and Nutrition, Department of Internal Medicine, VCU Medical Center, Richmond, VA, USA

Jacques Bienvenu Université de Lyon, INSERM U1111, Lyon, France

Armelle Biola-Vidamment Faculté de Pharmacie, Université Paris-Sud, INSERM UMR 996, Châtenay-Malabry, France

Joel A. Block Division of Rheumatology, Department of Internal Medicine, Rush University Medical Center, Chicago, IL, USA

Jennifer V. Bodkin Cardiovascular Division, King's College London, London, UK

Tobias Bopp Institute for Immunology, University Medical Center of the Johannes Gutenberg University Mainz, Mainz, Germany

Markus Bosmann Center for Thrombosis and Hemostasis, University Medical Center, Mainz, Germany

Paola Bossù Department of Clinical and Behavioral Neurology, IRCCS Fondazione Santa Lucia, Rome, Italy

Barbara Bottazzi Humanitas Clinical Research Center, Rozzano (Milano), Italy

Joëlle Botti INSERM U984, Paris–Sud 11 University, Châtenay-Malabry, France

Medical School, Paris 7 University, Denis Diderot, Paris, France

Yoram Bouhnik INSERM, UMR773, Team "Physiopathology of Inflammatory Bowel Diseases", Centre de Recherche Bichat Beaujon, Paris, France

Service de Gastroentérologie, MICI et Assistance Nutritive, PMAD Hôpital Beaujon, Clichy Cedex, France

University Paris–Diderot Sorbonne Paris–Cité, Paris, France

1Pôle des Maladies de l'Appareil Digestif (PMAD), Service de Gastroentérologie e, Hôpital Beaujon, Clichy, France

Gerd Bouma Immunology and the Department of Gastroenterology, VU University Medical Center, Amsterdam, The Netherlands

Onur Boyman Department of Immunology, University Hospital Zurich, Zurich, Switzerland

Allergy Unit, Department of Dermatology, University Hospital Zurich, Zurich, Switzerland

Samuel N. Breit St Vincent's Centre for Applied Medical Research, St Vincent's Hospital and University of New South Wales, Sydney, NSW, Australia

Elizabeth Brint Department of Pathology, Clinical Sciences Building, Cork University Hospital, University College Cork, National University of Ireland, Cork, Ireland

David A. Brown St Vincent's Centre for Applied Medical Research, St Vincent's Hospital and University of New South Wales, Sydney, NSW, Australia

Pierre Bruhns Institut Pasteur, Department of Immunology, Unit of Antibodies in Therapy and Pathology, INSERM, U1222, Paris, France

Anne B. Chang Centre for Children's Health Research, Children's Health Queensland, Brisbane, QLD, Australia

Department of Respiratory and Sleep Medicine, Lady Cilento Children's Hospital, Brisbane, QLD, Australia

Nicolas Charles Faculté de Médecine Site Bichat, INSERM UMR-S 699, Université Paris Diderot, Paris, France

Jianfeng Cheng Division of Gastroenterology, Hepatology and Nutrition, Department of Internal Medicine, VCU Medical Center, Richmond, VA, USA

Cécile Chenivesse Assistance Publique – Hôpitaux de Paris, Groupe Hospitalier Pitié–Salpêtrière Charles Foix, Service de Pneumologie et Réanimation Médicale, Paris, France

Neurophysiologie Respiratoire Expérimentale et Clinique, Sorbonne Universités, UPMC Univ Paris 06, INSERM, UMR_S 1158, Paris, France

Bradley E. Chipps Capital Allergy and Respiratory Disease Center, Sacramento, CA, USA

Ka-Yee Grace Choi Manitoba Centre for Proteomics and Systems Biology, Departments of Internal Medicine and Immunology, University of Manitoba, Winnipeg, MB, Canada

Sylvie Chollet-Martin Faculté de Pharmacie, Université Paris Sud, INSERM UMR 996, Châtenay-Malabry, France

Urs Christen Pharmazentrum Frankfurt/ZAFES, Goethe University Hospital Frankfurt, Klinikum der Johann Wolfgang Goethe Universität, Frankfurt am Main, Germany

Gustavo Citera Section of Rheumatology, Instituto de Rehabilitación Psicofísica, Buenos Aires, Argentina

Leslie Cleland Rheumatology Unit, Royal Adelaide Hospital, Adelaide, SA, Australia

Patrice Codogno INSERM U984, Paris-Sud 11 University, Châtenay-Malabry, France

Sina M. Coldewey Department for Anesthesiology and Intensive Care, Jena University Hospital, Jena, Germany

Center for Sepsis Control and Care, Jena University Hospital, Jena, Germany

Isabelle Coornaert Laboratory of Physiopharmacology, University of Antwerp, Campus Drie Eiken, Antwerp, Belgium

Lorenzo Cosmi Department of Experimental and Clinical Medicine and Denothe Center, University of Florence, Florence, Italy

Celine Cougoule Département Mécanismes Moléculaires des Infections Mycobactériennes, CNRS, Institute of Pharmacology and Structural Biology, Toulouse, France

Luc Cynober Department of Experimental, Metabolic and Clinical Biology, Paris Descartes University, Paris, France

Service de Biochimie, Hôpital Cochin, Paris, France

Marc Daëron Institut Pasteur, Paris, France

Centre d'Immunologie de Marseille–Luminy, Marseille, France

Pham My-Chan Dang INSERM U1149, CNRS ERL8252, Centre de Recherche sur l'Inflammation, Paris, France

Richard O. Day Department of Pharmacology, School of Medical Sciences, University of New South Wales, Sydney, NSW, Australia

Department of Clinical Pharmacology and Toxicology, St Vincent's Hospital, Sydney, NSW, Australia

Jean-Pascal De Bandt Department of Experimental, Metabolic and Clinical Biology, Paris Descartes University, Paris, France

Service de Biochimie, Hôpital Cochin, Paris, France

Marie de Bourayne INSERM UMR-996, Faculté de Pharmacie, UniverSud, Université Paris-Sud, Châtenay-Malabry, France

Luc de Chaisemartin INSERM UMRS 996, Universté Paris Sud, Chatenay-Malabry, France

Guido R. Y. De Meyer Laboratory of Physiopharmacology, University of Antwerp, Campus Drie Eiken, Antwerp, Belgium

Veronica De Rosa Laboratorio di Immunologia, Istituto di Endocrinologia e Oncologia Sperimentale, Consiglio Nazionale delle Ricerche (IEOS-CNR), Università di Napoli, Napoli, Italy

Unità di NeuroImmunologia, IRCCS Fondazione Santa Lucia, Rome, Italy

Alexiane Decout Département "Mécanismes moléculaires des infections mycobactériennes", Institut de Pharmacologie et de Biologie Structurale, UMR 5089 CNRS; IPBS/Université Paul Sabatier, Toulouse, France

Université de Toulouse; UPS; IPBS, Toulouse, France

Nathalie Dehne Faculty of Medicine, Institute of Biochemistry I, Goethe University Frankfurt, Frankfurt, Germany

Suelen Detoni Department of Pharmacology, Federal University of Santa Catarina, Florianópolis, SC, Brazil

Tessa Dieckman Department of Molecular Cell Biology, VU University Medical Center, Amsterdam, The Netherlands

Immunology and the Department of Gastroenterology, VU University Medical Center, Amsterdam, The Netherlands

Charles A. Dinarello Department of Medicine, University of Colorado Denver, Aurora, CO, USA

Department of Medicine, University Medical Center Nijmegen, Nijmegen, The Netherlands

Ahmet Eken Center for Immunity and Immunotherapies, Seattle Children's Research Institute, Seattle, WA, USA

Faculty of Medicine, Erciyes University, Medical Biology, Kayseri, Turkey

Zeina El Ali Faculté de Médecine, Institut Mondor de Recherche Biomédicale (IMRB), Université Paris-Est (UPEC), INSERM UMR-955, Team 12, Créteil, France

Jamel El-Benna INSERM U1149, CNRS ERL8252, Centre de Recherche sur l'Inflammation, Paris, France

Carole Elbim Sorbonne University, UPMC University Paris 06, UMR-S CR7, Paris, France

INSERM, Centre d'Immunologie et des Maladies Infectieuses, UMR-S CR7, INSERM U1135, Paris, France

Madeleine Ennis Centre for Infection and Immunity, School of Medicine, Dentistry and Biomedical Sciences, The Queen's University of Belfast, Belfast, UK

Vesna Erakovic Haber Fidelta d.o. o., Zagreb, Croatia

Marie-Alix Espinasse Faculté de Pharmacie, Université Paris-Sud, INSERM UMR 996, Châtenay-Malabry, France

Gerard Espinosa Department of Autoimmune Diseases, Institut Clínic de Medicina i Dermatologia, Universitat de Barcelona, Hospital Clinic, Barcelona, Catalonia, Spain

Elizabeth S. Fernandes Cardiovascular Division, King's College London, London, UK

Programa de Pós-Graduação, Universidade Ceuma, São Luís, MA, Brazil

Gwen S. Fernandes Academic Rheumatology, School of Medicine, University of Nottingham, Nottingham, UK

Arthritis Research UK Centre for Sport, Exercise and Osteoarthritis, University of Nottingham, Nottingham, UK

Philana Fernandes Cork Cancer Research Centre, 5th Floor Biosciences Institute Rm 5.27, University College Cork, National University of Ireland, Cork, Ireland

Juliano Ferreira Department of Pharmacology, Federal University of Santa Catarina, Florianópolis, SC, Brazil

Thiago A. F. Ferro Programa de Pós-Graduação, Universidade Ceuma, São Luís, MA, Brazil

Hannah Flaßkamp Department of Biomedical Sciences, School of Human Sciences, University of Osnabrück, Osnabrück, Germany

Wieslawa Agnieszka Fogel Department of Hormone Biochemistry, Faculty of Military Medicine, Medical University of Lodz, Lodz, Poland

Ulf Forssmann Genmab A/S, Copenhagen K, Denmark

Center of Pharmacology and Toxicology, Hannover Medical School, Hannover, Germany

Gary L. Francis Department of Pediatrics, Division of Pediatric Endocrinology and Metabolism, Children's Hospital of Richmond at Virginia Commonwealth University, Richmond, VA, USA

Antony B. Friedman Department of Gastroenterology, The Alfred Hospital and Monash University, Melbourne, VIC, Australia

Cecilia Garlanda Humanitas Clinical Research Center, Rozzano (Milano), Italy

Romain Génard INSERM UMR-996, Faculté de Pharmacie, UniverSud, Université Paris-Sud, Châtenay-Malabry, France

Bernhard F. Gibbs Medway School of Pharmacy, University of Kent, The Universities of Greenwich and Kent at Medway, Chatham, Kent, UK

Peter R. Gibson Department of Gastroenterology, The Alfred Hospital and Monash University, Melbourne, VIC, Australia

Caitlin M. Gillis Institut Pasteur, Department of Immunology, Unit of Antibodies in Therapy and Pathology, INSERM, U1222, Paris, France

Université Pierre et Marie Curie, Paris, France

Christian A. Gleissner Department of Cardiology, University Hospital, Heidelberg, Germany

Erik Oliver Glocker Department of Medical Microbiology and Hygiene, University Hospital Freiburg, Freiburg, Germany

Kamar Godder Blood and Marrow Transplant Program, Kidz Medical Services at Nicklaus Children's hospital, Miami, FL, USA

Ana Gomez-Larrauri Department of Biochemistry and Molecular Biology, Faculty of Science and Technology, University of the Basque Country (UPV/EHU), Bilbao, Spain

Antonio Gómez-Muñoz Department of Biochemistry and Molecular Biology, Faculty of Science and Technology, University of the Basque Country (UPV/EHU), Bilbao, Spain

Aurélie Gouel-Chéron Institut Pasteur, Department of Immunology, Unit of Antibodies in Therapy and Pathology, INSERM, U1222, Paris, France

Université Pierre et Marie Curie, Paris, France

Département d'anesthésie-réanimation, Hôpital Bichat-Claude-Bernard, Hôpitaux de Paris, Université Paris-VII, Paris, France

Vikas Goyal Centre for Children's Health Research, Children's Health Queensland, Brisbane, QLD, Australia

Garry G. Graham Department of Pharmacology, School of Medical Sciences, University of New South Wales, Sydney, NSW, Australia

Department of Clinical Pharmacology and Toxicology, St Vincent's Hospital, Sydney, NSW, Australia

Daniel Neil Granger Department of Molecular and Cellular Physiology, LSU Health Sciences Center, Shreveport, LA, USA

Lovorka Grgurevic Department of Anatomy, Laboratory for Mineralized Tissues, Center for Translational and Clinical Research, University of Zagreb School of Medicine, Zagreb, Croatia

Shipra Gupta SNBL USA, Ltd., Everett, WA, USA

Dávid Győri Department of Physiology, Semmelweis University, Budapest, Hungary

Ahmed Hamaï INSERM U984, Paris-Sud 11 University, Châtenay-Malabry, France

C. E. M. Hollak Department of Endocrinology and Metabolism, Academic Medical Center (AMC), Amsterdam, The Netherlands

Morley D. Hollenberg Inflammation Research Network-Snyder Institute for Chronic Disease, Department of Physiology and Pharmacology and Department of Medicine, Cumming School of Medicine, University of Calgary, Calgary, AB, Canada

Richard Horuk Department of Pharmacology, UC Davis, Davis, CA, USA

Jan Hošek Department of Molecular Biology and Pharmaceutical Biotechnology, Faculty of Pharmacy, University of Veterinary and Pharmaceutical Sciences Brno, Brno, Czech Republic

Kai-Sheng Hsieh Department of Pediatrics, Kaohsiung Chang Gung Memorial Hospital, Chang Gung University College of Medicine, Kaohsiung city, Taiwan

Christoph Hudemann Institute of Laboratory Medicine and Pathobiochemistry, Molecular Diagnostics, Philipps University Marburg, University Hospital GmbH and Marburg GmBH, Marburg, Germany

Marie-Geneviève Huisse Department of Haematology AP-HP, Bichat-Claude Bernard Hospital, University Paris Diderot, Paris, France

Margarita Hurtado-Nedelec AP-HP, Centre Hospitalier Universitaire Xavier Bichat, UF Dysfonctionnements Immunitaires, Paris, France

Antonio Inforzato Humanitas Clinical Research Center, Rozzano (Milano), Italy

Makoto Inoue Department of Immunology, Duke University School of Medicine, Durham, NC, USA

Yoriko Inoue Respiratory Disease Center, Yokohama City University Medical Center, Yokohama City, Japan

Estela Jacinto Department of Biochemistry and Molecular Biology, Rutgers-Robert Wood Johnson Medical School, Piscataway, NJ, USA

Chris John Jackson Sutton Arthritis Research Laboratory, Kolling Institute of Medical Research, The University of Sydney, St Leonards, NSW, Australia

Adam Jaffe Department of Respiratory Medicine, Sydney Children's Hospital, Randwick, NSW, Australia

School of Women's and Children's Health, UNSW Medicine, University of New South Wales, Sydney, NSW, Australia

Sébastien Jaillon Humanitas Clinical Research Center, Rozzano (Milano), Italy

Tomas Jakobsson Department of Laboratory Medicine, Division of Clinical Chemistry, Karolinska Institutet C1:62, Karolinska University Hospital Huddinge, Stockholm, Sweden

Sangmin Jeong College of Veterinary School, Konkuk University, Seoul, South Korea

Susan John Department of Immunobiology, Kings College London, London, UK

Michaela Jung Faculty of Medicine, Institute of Biochemistry I, Goethe University Frankfurt, Frankfurt, Germany

Masashi Kanayama Department of Immunology, Duke University School of Medicine, Durham, NC, USA

Watcharoot Kanchongkittiphon Division of Allergy and Immunology, Department of Medicine, Boston Children's Hospital, Boston, MA, USA

Harvard Medical School, Boston, MA, USA

Department of Pediatrics, Ramathibodi Hospital, Mahidol University, Bangkok, Thailand

Matthew Kaspar Division of Gastroenterology, Hepatology and Nutrition, Department of Internal Medicine, VCU Medical Center, Richmond, VA, USA

Marija Kastelan Department of Dermatovenerology, Clinical Hospital Centre Rijeka, Medical Faculty, University of Rijeka, Rijeka, Croatia

Department of Dermatovenerology, School of Medicine, University of Rijeka, Rijeka, Croatia

Constance H. Katelaris Department of Immunology, Campbelltown Hospital, Sydney, NSW, Australia

Joshua L. Kennedy Arkansas Children's Hospital Research Institute, University of Arkansas for Medical Sciences, Little Rock, AR, USA

Saadia Kerdine-Römer INSERM UMR-996, Faculté de Pharmacie, UniverSud, Université Paris-Sud, Châtenay-Malabry, France

Claus Kerkhoff Department of Biomedical Sciences, School of Human Sciences, University of Osnabrück, Osnabrück, Germany

Soohyun Kim Laboratory of Cytokine Immunology, Department of Biomedical Science and Technology, Konkuk University, Seoul, South Korea

College of Veterinary School, Konkuk University, Seoul, South Korea

Tadamitsu Kishimoto Laboratory of Immune Regulation, World Premier International Immunology Frontier Research Center, Osaka University, Suita, Osaka, Japan

Matthias Klein Institute for Immunology, University Medical Center of the Johannes Gutenberg University Mainz, Mainz, Germany

Rufino J. Klug Programa de Pós-Graduação, Universidade Ceuma, São Luís, MA, Brazil

Ioannis Kourtzelis Department of Clinical Pathobiochemistry and Institute for Clinical Chemistry and Laboratory Medicine, Faculty of Medicine, Technische Universität Dresden, Dresden, Germany

Andrew Kovalenko Department of Biological Chemistry, Weizmann Institute of Science, Rehovot, Israel

Georg Kraal Department of Molecular Cell Biology, VU University Medical Center, Amsterdam, The Netherlands

Peter Kraiczy Institute of Medical Microbiology and Infection Control, University Hospital Frankfurt, Frankfurt/Main, Germany

Santhosh Kumar Allergy and Immunology, Virginia Commonwealth University Health System, Richmond, VA, USA

Ho-Chang Kuo Kawasaki Disease Center and Department of Pediatrics, Division of Allergy, Immunology and Rheumatology, Kaohsiung Chang Gung Memorial Hospital, Chang Gung University College of Medicine, Kaohsiung city, Taiwan

John M. Kyriakis Mercury Therapeutics, Inc., Woburn, MA, USA

Trang Le Department of Pediatrics, Division of Pediatric Endocrinology and Metabolism, Children's Hospital of Richmond at Virginia Commonwealth University, Richmond, VA, USA

Francesco Liotta Department of Experimental and Clinical Medicine and Denothe Center, University of Florence, Florence, Italy

M. Letizia Lo Faro Oxford Transplant Centre, University of Oxford, Nuffield Department of Surgical Sciences, Churchill Hospital, Oxford, UK

Stephen M. Mahler Australian Institute for Bioengineering and Nanotechnology, University of Queensland, Brisbane, QLD, Australia

Anne-Marie Malfait Division of Rheumatology, Department of Internal Medicine, Rush University Medical Center, Chicago, IL, USA

Isabelle Maridonneau-Parini Institut de Pharmacologie et de Biologie Structurale - CNRS, UMR 5089, Toulouse, France

Viviana Marin-Esteban Faculté de Pharmacie, Université Paris Sud, INSERM UMR 996, Châtenay-Malabry, France

Stefan F. Martin Allergy Research Group, Department of Dermatology, Medical Center - University of Freiburg, Freiburg, Germany

Wim Martinet Laboratory of Physiopharmacology, University of Antwerp, Campus Drie Eiken, Antwerp, Belgium

Gita V. Massey Division of Hematology and Oncology, Department of Pediatrics, Virginia Commonwealth University Medical Center, Richmond, VA, USA

Giuseppe Matarese Dipartimento di Medicina e Chirurgia, Università di Salerno, Salerno, Italy

IRCCS Multimedica, Milan, Italy

Shiro Matsubara Department of Neurology, Tokyo Metropolitan Neurological Hospital, Fuchu, Tokyo, Japan

Sandra Mazzoli STDs Center Santa Maria Annunziata Hospital, Florence, Tuscany, Italy

Michael H. Mellon Department of Allergy and Immunology, Southern California Permanente Medical Group, San Diego, CA, USA

Christina Mertens Faculty of Medicine, Institute of Biochemistry I, Goethe University Frankfurt, Frankfurt, Germany

Nikita Minhas Sutton Arthritis Research Laboratory, Kolling Institute of Medical Research, The University of Sydney, St Leonards, NSW, Australia

Abbas Mirshafiey Department of Immunology, School of Public Health, Tehran University of Medical Sciences, Tehran, Iran

Ioannis Mitroulis Department of Clinical Pathobiochemistry and Institute for Clinical Chemistry and Laboratory Medicine, Faculty of Medicine, Technische Universität Dresden, Dresden, Germany

Attila Mócsai Department of Physiology, Semmelweis University, Budapest, Hungary

Barbara Moepps Institute of Pharmacology and Toxicology, University of Ulm, Medical Center, Ulm, Germany

Philippe Montravers Département d'anesthésie-réanimation, Hôpital Bichat-Claude-Bernard, Hôpitaux de Paris, Université Paris-VII, Paris, France

Neeloffer Mookherjee Manitoba Centre for Proteomics and Systems Biology, Departments of Internal Medicine and Immunology, University of Manitoba, Winnipeg, MB, Canada

Javier Mora Faculty of Medicine, Institute of Biochemistry I, Goethe University Frankfurt, Frankfurt, Germany

Eric Morand Monash Medical Centre, Clayton, VIC, Australia

Christine V. Möser Pharmazentrum Frankfurt/ZAFES, Klinikum der Johann Wolfgang Goethe-Universität Frankfurt, Frankfurt am Main, Germany

William B. Moskowitz Pediatric Cardiac Catheterization Laboratory, The Children's Hospital of Richmond at VCU, Richmond, VA, USA

The Children's Hospital of Richmond at VCU, Richmond, VA, USA

George F. Moxley Department of Internal Medicine, Division of Rheumatology, Allergy and Immunology, VCU School of Medicine, Richmond, VA, USA

Heiko Mühl Department of General Pharmacology and Toxicology, Pharmazentrum Frankfurt/ZAFES, University Hospital Goethe University Frankfurt, Frankfurt am Main, Germany

Ari Murad Department of Immunology, Campbelltown Hospital, Sydney, NSW, Australia

Paul H. Naccache Department of Microbiology-Infectiology and Immunology, Faculty of Medicine, Centre de recherche du CHU de Québec, Université Laval, Québec, QC, Canada

Devi Ngo Inflammation, Walter and Eliza Hall Institute of Medical Research, Parkville, VIC, Australia

Ellen Niederberger Pharmazentrum Frankfurt/ZAFES, Klinikum der Johann Wolfgang Goethe-Universität Frankfurt, Frankfurt am Main, Germany

JW Goethe University Frankfurt, Frankfurt, Germany

Jérôme Nigou Département "Mécanismes moléculaires des infections mycobactériennes", Institut de Pharmacologie et de Biologie Structurale, UMR 5089 CNRS; IPBS/Université Paul Sabatier, Toulouse, France

Université de Toulouse; UPS; IPBS, Toulouse, France

Daniela Novick Department of Molecular Genetics, Weizmann Institute of Science, Rehovot, Israel

Rolf M. Nüsing Department of Clinical Pharmacology, Johann Wolfgang Goethe-University, Frankfurt, Germany

Eric Ogier-Denis INSERM, UMR773, Team "Physiopathology of Inflammatory Bowel Diseases", Centre de Recherche Bichat Beaujon, Paris, France

Service de Gastroentérologie, MICI et Assistance Nutritive, PMAD Hôpital Beaujon, Clichy Cedex, France

University Paris–Diderot Sorbonne Paris–Cité, Paris, France

Sara M. Oliveira Department of Biochemistry and Molecular Biology, Federal University of Santa Maria, Santa Maria, RS, Brazil

Marta Ordoñez Department of Biochemistry and Molecular Biology, Faculty of Science and Technology, University of the Basque Country (UPV/EHU), Bilbao, Spain

Mohamed Oukka Center for Immunity and Immunotherapies, Seattle Children's Research Institute, Seattle, WA, USA

Department of Immunology, University of Washington, Seattle, WA, USA

Department of Pediatrics, University of Washington, Seattle, WA, USA

Alberto Ouro Department of Biochemistry and Molecular Biology, Faculty of Science and Technology, University of the Basque Country (UPV/EHU), Bilbao, Spain

Christian Pagnoux Division of Rheumatology, Vasculitis Clinic, Rebecca McDonald Centre for Arthritis and Autoimmune Diseases, Mount Sinai Hospital, University of Toronto, Toronto, ON, Canada

Teresa Faria Pais Instituto Gulbenkian de Ciência, Oeiras, Portugal

Ilaria Palladino Department of Clinical and Behavioral Neurology, IRCCS Fondazione Santa Lucia, Rome, Italy

Marc Pallardy Faculté de Pharmacie, Université Paris-Sud, INSERM UMR 996, Châtenay-Malabry, France

Michael J. Parnham Fraunhofer Institute for Molecular Biology and Applied Ecology IME, Project Group for Translational Medicine and Pharmacology TMP, Frankfurt am Main, Germany

D. P. Patel University of Louiseville, Louiseville, KY, USA

Stanford Peng University of Washington School of Medicine, Seattle, WA, USA

Melinda M. C. Penn Division of Pediatric Endocrinology and Metabolism, Center for Endocrinology, Diabetes and Metabolism, VCU Medical Center, Richmond, VA, USA

Aurélie Pépin Faculté de Pharmacie, Université Paris-Sud, INSERM UMR 996, Châtenay-Malabry, France

Wanda Phipatanakul Division of Allergy and Immunology, Department of Medicine, Boston Children's Hospital, Boston, MA, USA

Harvard Medical School, Boston, MA, USA

Kevin D. Pile Campbelltown Hospital, School of Medicine, University of Western Sydney, Campbelltown, NSW, Australia

Hadeesha Piyadasa Manitoba Centre for Proteomics and Systems Biology, Departments of Internal Medicine and Immunology, University of Manitoba, Winnipeg, MB, Canada

Gabor Pozsgai Department of Pharmacology and Pharmacotherapy, Faculty of Medicine, University of Pécs, Pécs, Hungary

Bernadette Prentice Department of Paediatric Respiratory and Sleep Medicine, Sydney Children's Hospital, Randwick, NSW, Australia

Natalia Presa Department of Biochemistry and Molecular Biology, Faculty of Science and Technology, University of the Basque Country (UPV/EHU), Bilbao, Spain

Claudio Procaccini Laboratorio di Immunologia, Istituto di Endocrinologia e Oncologia Sperimentale, Consiglio Nazionale delle Ricerche (IEOS-CNR), Università di Napoli, Napoli, Italy

Yvonne Radon Department of Biomedical Sciences, School of Human Sciences, University of Osnabrück, Osnabrück, Germany

Laila Rahbar Arthritis and Rheumatic Diseases PC, Richmond, VA, USA

Anil Kumar Ramaswamy Division of Endocrinology and Metabolism, Center for Endocrinology, Diabetes and Metabolism, Virginia Commonwealth University, Richmond, VA, USA

Pediatric Endocrinology and Diabetes Associates, Richmond, VA, USA

Tangada Sudha Rao Department of Rheumatology, McGuire VA Medical Center, Richmond, VA, USA

Gabriele Reichmann Department of Immunology, Paul-Ehrlich-Institut, Langen, Germany

Harald Renz Institute of Laboratory Medicine and Pathobiochemistry, Molecular Diagnostics, Philipps University Marburg, University Hospital GmbH and Marburg GmBH, Marburg, Germany

Theresa K. Resch Junior Research Group "Novel Vaccination Strategies and Early Immune Responses", Paul-Ehrlich-Institut, Langen, Germany

Rudolf Richter Department of Internal Medicine, Clinic of Immunology and Rheumatology, Research Group of Experimental and Clinical Peptide Chemistry, Lower Saxony Institute of Peptide Research, Hannover Medical School, Hannover, Germany

Blood Donation Service of the German Red Cross, Institute of Transfusion Medicine and Immune Hematology, Frankfurt, Germany

Io-Guané Rivera Department of Biochemistry and Molecular Biology, Faculty of Science and Technology, University of the Basque Country (UPV/EHU), Bilbao, Spain

W. N. Roberts University of Louiseville, Louiseville, KY, USA

Stephen F. Rodrigues Department of Clinical and Toxicological Analyses, School of Pharmaceutical Sciences, University of Sao Paulo, Sao Paulo, Brazil

Lynn Roth Laboratory of Physiopharmacology, University of Antwerp, Campus Drie Eiken, Antwerp, Belgium

Paul Rouzaire Service d'Immunologie Biologique, CHU de Clermont-Ferrand, Université d'Auvergne, ERTICa EA4677, Clermont-Ferrand, France

Brent J. Ryan Department of Physiology, Anatomy and Genetics, University of Oxford, Oxford, UK

Christian D. Sadik Department of Dermatology, Allergy, and Venereology, University of Lübeck, Lübeck, Germany

Jean Sainte-Laudy Laboratoire d'Immunologie, CHU Limoges, Limoges, France

Claire Sand Cardiovascular Division, King's College London, London, UK

Marianna Sari Otolaryngology Unit, Medical and Surgical Specialties, University of Padua, Padua, Italy

Julio Scharfstein Immunobiology, Institute of Biophysics Carlos Chagas Filho, Federal University of Rio de Janeiro, Rio de Janeiro, Brazil

Center of Health Sciences (CCS), Cidade Universitária, Rio de Janeiro, Brazil

Edgar Schmitt Institute for Immunology, University Medical Center of the Johannes Gutenberg University Mainz, Mainz, Germany

Emilce E. Schneeberger Section of Rheumatology, Instituto de Rehabilitación Psicofísica, Buenos Aires, Argentina

Hélène Schoemans Department of Hematology, Bone Marrow Transplantation Unit, University Hospitals Leuven, Leuven, Belgium

Masaharu Shinkai Respiratory Disease Center, Yokohama City University Medical Center, Yokohama City, Japan

Mari L. Shinohara Department of Immunology, Duke University School of Medicine, Durham, NC, USA

Department of Molecular Genetics and Microbiology, Duke University School of Medicine, Durham, NC, USA

F. Siddique Department of Rheumatology, Hunter Holmes McGuire VA Medical Center, Richmond, VA, USA

Sandro Silva-Gomes Département "Mécanismes moléculaires des infections mycobactériennes", Institut de Pharmacologie et de Biologie Structurale, UMR 5089 CNRS; IPBS/Université Paul Sabatier, Toulouse, France

Université de Toulouse; UPS; IPBS, Toulouse, France

Akhilesh K. Singh Center for Immunity and Immunotherapies, Seattle Children's Research Institute, Seattle, WA, USA

Chrysanthi Skevaki Institute of Laboratory Medicine and Pathobiochemistry, Molecular Diagnostics, Philipps University Marburg, University Hospital GmbH and Marburg GmBH, Marburg, Germany

Karel Šmejkal Department of Natural Drugs, Faculty of Pharmacy, University of Veterinary and Pharmaceutical Sciences Brno, Brno, Czech Republic

Fernando A. Sommefleck Section of Rheumatology, Instituto de Rehabilitación Psicofísica, Buenos Aires, Argentina

Miles P. Sparrow Department of Gastroenterology, The Alfred Hospital and Monash University, Melbourne, VIC, Australia

Sarah Spiegel Department of Biochemistry and Molecular Biology, VCU Massey Cancer Center, Virginia Commonwealth University School of Medicine, Richmond, VA, USA

Martin Stacey Faculty of Biological Sciences, School of Molecular and Cellular Biology, University of Leeds, Leeds, UK

Lisa Stamp University of Otago, Christchurch, New Zealand

John W. Steinke Carter Center for Immunology Research, Asthma and Allergic Disease Center, University of Virginia Health System, Charlottesville, VA, USA

Kazuaki Takabe Department of Surgery, VCU Massey Cancer Center, Virginia Commonwealth University School of Medicine, Richmond, VA, USA

Toshio Tanaka Department of Clinical Application of Biologics, Osaka University Graduate School of Medicine, Suita, Osaka, Japan

Jean-Baptiste Telliez Inflammation and Immunology, Pfizer Inc., Cambridge, MA, USA

Marcus Thelen Institute for Research in Biomedicine, Bellinzona, Switzerland

Paul S. Thomas Inflammation and Infection Research Centre, Faculty of Medicine, University of New South Wales, Sydney, NSW, Australia

Department of Respiratory Medicine, Prince of Wales Hospital, Sydney, NSW, Australia

Department of Respiratory Medicine, Prince of Wales Clinical School, University of New South Wales, Randwick, NSW, Australia

Xavier Treton INSERM, UMR773, Team "Physiopathology of Inflammatory Bowel Diseases", Centre de Recherche Bichat Beaujon, Paris, France

Service de Gastroentérologie, MICI et Assistance Nutritive, PMAD Hôpital Beaujon, Clichy Cedex, France

University Paris–Diderot Sorbonne Paris–Cité, Paris, France

Eckardt Treuter Department of Biosciences and Nutrition, Karolinska Institutet Huddinge, Stockholm, Sweden

Gabriela Trevisan Graduate Program in Health Sciences, University of the Extreme South of Santa Catarina, Criciúma, SC, Brazil

Miguel Trueba Department of Biochemistry and Molecular Biology, Faculty of Science and Technology, University of the Basque Country (UPV/EHU), Bilbao, Spain

Anne Tsicopoulos Institut National de la Santé et de la Recherche Médicale U1019, Lille, France

Institut Pasteur de Lille, Center for Infection and Immunity of Lille, Lille, France

CNRS UMR 8204, Lille, France

Univ Lille Nord de France, Lille, France

Clinique des Maladies Respiratoires et Centre Hospitalier Régional et Universitaire de Lille, Lille, France

Ana M. Valdes Academic Rheumatology, School of Medicine, University of Nottingham, Nottingham, UK

Bart M. Vanaudenaerde Department of Clinical and Experimental Medicine, Division of Respiratory Diseases, Lung Transplant Unit, KULeuven and University Hospitals Leuven, Leuven, Belgium

Frank L. van de Veerdonk Department of Internal Medicine, Radboud University Nijmegen Medical Centre, GA, Nijmegen, The Netherlands

Department of Medicine, University of Colorado Denver, Aurora, CO, USA

Bieke Van der Veken Laboratory of Physiopharmacology, University of Antwerp, Campus Drie Eiken, Antwerp, Belgium

Ruurd van der Zee Division of Immunology, Department of Infectious Diseases and Immunology, Faculty of Veterinary Medicine, Utrecht University, Utrecht, The Netherlands

Willem van Eden Division of Immunology, Department of Infectious Diseases and Immunology, Faculty of Veterinary Medicine, Utrecht University, Utrecht, The Netherlands

Nicolas Venteclef Centre de recherche des Cordeliers, INSERM UMRS 1138, Université Pierre et Marie Curie, Paris, France

Geert M. Verleden Department of Clinical and Experimental Medicine, Division of Respiratory Diseases, Lung Transplant Unit, KULeuven and University Hospitals Leuven, Leuven, Belgium

Stijn E. Verleden Department of Clinical and Experimental Medicine, Division of Respiratory Diseases, Lung Transplant Unit, KULeuven and University Hospitals Leuven, Leuven, Belgium

Sébastien Viel Université de Lyon, INSERM U1111, Lyon, France

Dipti Vijayan Immunology Division, Garvan Institute of Medical Research, and St. Vincent's Clinical School, University of New South Wales, Sydney, NSW, Australia

Alexandra Villa-Forte Center for Vasculitis Care and Research, Department of Rheumatic and Immunologic Diseases, Cleveland Clinic, Cleveland, OH, USA

Joana Vitte Laboratoire d'Immunologie, Assistance Publique Hôpitaux de Marseille, INSERM UMR 1067, Aix-Marseille University, Marseille, France

Cosmin Sebastian Voican INSERM, UMR773, Team "Physiopathology of Inflammatory Bowel Diseases", Centre de Recherche Bichat Beaujon, Paris, France

Service de Gastroentérologie, MICI et Assistance Nutritive, PMAD Hôpital Beaujon, Clichy Cedex, France

University Paris–Diderot Sorbonne Paris–Cité, Paris, France

Andreas von Knethen Institute of Biochemistry I, Faculty of Medicine, Goethe-University Frankfurt, Frankfurt/Main, Germany

Robin Vos Department of Clinical and Experimental Medicine, Division of Respiratory Diseases, Lung Transplant Unit, KULeuven and University Hospitals Leuven, Leuven, Belgium

Slobodan Vukicevic Department of Anatomy, Laboratory for Mineralized Tissues, Center for Translational and Clinical Research, University of Zagreb School of Medicine, Zagreb, Croatia

Zoe Waibler Junior Research Group "Novel Vaccination Strategies and Early Immune Responses", Paul-Ehrlich-Institut, Langen, Germany

David Wallach Department of Biological Chemistry, Weizmann Institute of Science, Rehovot, Israel

Thierry Walzer Université de Lyon, INSERM U1111, Lyon, France

Brant R. Ward Allergy and Immunology, Virginia Commonwealth University School of Medicine, Richmond, VA, USA

Lehn K. Weaver Division of Pediatric Rheumatology, The Children's Hospital of Philadelphia, Philadelphia, PA, USA

Andreas Weigert Faculty of Medicine, Institute of Biochemistry I, Goethe University Frankfurt, Frankfurt, Germany

Pamela F. Weiss Division of Pediatric Rheumatology, The Children's Hospital of Philadelphia, Philadelphia, PA, USA

Guy Werlen Department of Biochemistry and Molecular Biology, Rutgers-Robert Wood Johnson Medical School, Piscataway, NJ, USA

Matthew Whiteman Inflammation Research Group, University of Exeter Medical School, Exeter, UK

Christina M. Wiedl Division of Hematology and Oncology, Department of Pediatrics, Virginia Commonwealth University Medical Centre, Richmond, VA, USA

Kenneth Williams Department of Pharmacology, School of Medical Sciences, University of New South Wales, Sydney, NSW, Australia

Paul G. Winyard Inflammation Research Group, University of Exeter Medical School, Exeter, UK

Miriam Wittmann Leeds Institute of Rheumatic and Musculoskeletal Medicine, University of Leeds and NIHR Leeds Musculoskeletal Biomedical Research Unit Leeds, Chapel Allerton Hospital, Leeds, UK

Centre for Skin Sciences, University of Bradford, Bradford, UK

Meilang Xue Sutton Arthritis Research Laboratory, Kolling Institute of Medical Research, The University of Sydney, St Leonards, NSW, Australia

Akimitsu Yamada Division of Surgical Oncology, Department of Surgery, Virginia Commonwealth University School of Medicine and Massey Cancer Center, Richmond, VA, USA

Maria Elisabetta Zannin Ophthalmology Unit, Department of Pediatrics, University of Padua, Padua, Italy

Francesco Zulian Rheumatology Unit, Department of Pediatrics, University of Padua, Padua, Italy

Antonie Zwiers Department of Molecular Cell Biology, VU University Medical Center, Amsterdam, The Netherlands

Idiopathic Thrombocytopenic Purpura

Gita V. Massey
Division of Hematology and Oncology, Department of Pediatrics, Virginia Commonwealth University Medical Center, Richmond, VA, USA

Synonyms

Immune thrombocytopenia; Immune-mediated thrombocytopenic purpura; ITP

Definition

Idiopathic thrombocytopenic purpura (ITP) is generally believed to be an autoimmune process in which various pathologic immune mechanisms lead to the accelerated destruction of platelets and/or inhibition of their production. The end result is thrombocytopenia (defined as platelets $<100 \times 10^9/l$) which leads to clinical symptoms of bleeding (Rodeghiero et al. 2009). It is a very heterogeneous disorder arising through diverse mechanisms which are modified by environmental and genetic influences. ITP is considered to be primary in the absence of any known comorbidity and secondary when accompanied by other inflammatory, infectious, or immune pathologies (Stasi 2012).

Epidemiology and Genetics

ITP occurs in about 100/1 million persons/year (Frederiksen and Schmidt 1999). Approximately 50 % of the diagnoses occur in children at a peak age of 5 years. In most children, the disease is of acute onset with spontaneous recovery of platelet counts within 6 months. Approximately 20 % of cases are secondary and have a more chronic course. Most of these cases occur in adults. Disorders that are known to be associated with secondary ITP include infections such as hepatitis C, HIV, and *H. pylori*; immunodeficiencies such as common variable immunodeficiency (CVID) and autoimmune lymphoproliferative syndrome (ALPS); and autoimmune disorders such as systemic lupus erythematosus (SLE) and antiphospholipid syndrome (APS) (Cines and Blanchette 2002).

To date there is no known common genetic predisposition to ITP, although increased incidence in monozygotic twins (Laster et al. 1982) and in some families has been described (Lippman et al. 1982). HLA antigen alleles DRw2 and DRB1*0410 may predispose to ITP. Certain alleles (HLA-DR4 and HLA-DRB1*1501) have been associated with unfavorable responses to steroids and splenectomy, respectively. DRB1*0410 has been associated with a favorable response to steroids (Cines and Blanchette 2002).

Pathophysiology

ITP is caused by decreased life span of platelets as well as decreased platelet production. Platelet kinetic studies done with indium-111-labeled autologous platelets have not only shown markedly decreased life span of platelets, but also inappropriately low platelet turnover and production (Stratton et al. 1989; Louwes et al. 1999). Sites of destruction have been identified as the spleen primarily and secondarily the liver. Abnormal platelet turnover occurs through a diversity of immune mechanisms. As more inciting events and immune defects are identified, the proportion of patients thought to have primary ITP has decreased.

B Cells and Autoantibodies

The classic pathogenetic pathway has been thought to be autoantibody production by B cells to a platelet antigen(s). Often multiple antigens are involved. Examples include surface glycoproteins IIb/IIIa, Ib/IX, Ia/IIa, IV, and V (McMillan 2000). Antibody-coated platelets then undergo accelerated clearance via Fc receptors in macrophages through a process known as opsonization. Platelet destruction may also lead to new neoantigens which can lead to additional antibody formation. Antibodies may also impede megakaryocyte development, induce apoptosis of platelet precursors in the bone marrow, impede platelet release from the bone marrow, and promote intramedullary phagocytosis (McMillan et al. 2004). In addition, platelet function can also be impaired. Platelet antibodies are detected in approximately 50–60 % of ITP patients; however, this is neither a sensitive nor specific test in diagnosing ITP (Stasi et al. 2008). Remission in ITP can also occur despite continued presence of platelet antibodies. Many antigenic targets are not yet known, and often additional immune mechanisms contribute to the evolution of ITP. However, it is generally accepted that B cell production of autoantibodies is what leads to the approximate 80 % response rate of ITP to immune globulin (IgG) infusions and/or splenectomy (Cines et al. 2009).

T Cell-Dependent Responses

Increased numbers of HLA-DR-positive T cells and elevated sIL2 receptors as well as changes in cytokine profiles in patients with ITP suggest activation of precursor helper T cells and type 1 helper T cells (Semple et al. 1996). This is consistent with a pro-inflammatory response that is also seen in other organ-specific autoimmune diseases. These platelet-reactive T cells stimulate antibody synthesis by B cells after exposure to fragments of platelet antigens such as GPIIb/IIIa. The continued activation of these T cells in ITP patients suggests a role for altered or impaired peripheral tolerance mechanisms in the pathophysiology of ITP (Cines and Blanchette 2002; Stasi 2012).

Cytotoxic T cells may also play a role in the development of ITP. Killer cell immunoglobulin-like receptors (KIRs) have been shown to be upregulated in patients with ITP (Olsson et al. 2003). The function of KIRs is to downregulate cytotoxic T lymphocyte and natural killer cell responses by binding to major histocompatibility complex (MHC) class I receptors and thus preventing the lysis of target cells.

Finally $CD4^+$, $CD25^+$ regulatory T cells (Tregs) from ITP patients have decreased suppressive activity when compared to normal controls (Liu et al. 2007; Olsson et al. 2005).

Antigen-Presenting Cells (APCs) and Macrophages

Antigen-presenting cells are responsible for presenting specific peptide sequences to activated T helper cells which then stimulate production of autoreactive IgG by B cells against the protein antigen. The role of dendritic cells and macrophages in the pathophysiology of ITP, antigen presentation, and loss of tolerance is not yet clear. Preliminary reports indicate a bigger role for macrophages than dendritic cells (Catani et al. 2006; Kuwana et al. 2009).

Mesenchymal Stem Cells

Mesenchymal stem cells or pluripotent stem cells have the capacity to differentiate into multiple cell types. In addition, they also serve as immunomodulators (mostly immunosuppressive) and can

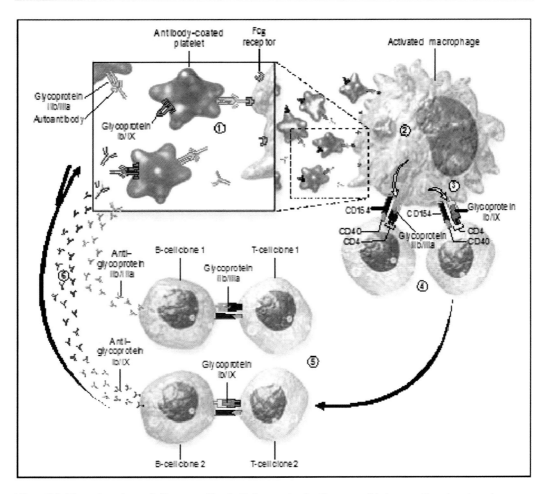

Idiopathic Thrombocytopenic Purpura, Fig. 1 Pathogenesis of epitope spread in immune thrombocytopenic purpura

prevent T cell responses. ITP patients have decreased proliferative capacity of their mesenchymal stem cells, thus leading to less inhibition of T cell activation. The exact mechanisms of this dysfunction are not yet known; however, innovative therapies such as donor-derived mesenchymal cell infusions in chronic ITP patients have met with some success (Perez-Simon et al. 2009; Fang et al. 2012).

The factors that initiate autoantibody production are unknown (Fig. 1). Most patients have antibodies against several platelet-surface glycoproteins at the time the disease becomes clinically evident. Here, glycoprotein IIb/IIIa is recognized by autoantibody (orange, inset), whereas antibodies that recognize the glycoprotein Ib/IX complex have not been generated at this stage (1). Antibody-coated platelets bind to antigen-presenting cells (macrophages or dendritic cells) through Fcg receptors and are then internalized and degraded (2). Antigen-presenting cells not only degrade glycoprotein IIb/IIIa (light-blue oval), thereby amplifying the initial immune response, but also may generate cryptic epitopes from other platelet glycoproteins (light-blue cylinder) (3). Activated antigen-presenting cells (4) express these novel peptides on the cell surface along with costimulatory help (represented in part by the interaction between CD154 and CD40) and the relevant cytokines that facilitate the proliferation of the initiating CD4-positive T cell clones (T cell clone 1) and those with additional specificities (T cell clone 2) (5). B cell immunoglobulin receptors that recognize additional platelet antigens (B cell clone 2) are thereby also induced to proliferate and synthesize anti-glycoprotein Ib/IX antibodies (green) in addition to amplifying the production of anti-glycoprotein IIb/IIIa antibodies (orange) by B cell clone 1 (6).

Immune Tolerance Checkpoint Defects

B and T cells are both involved in immune tolerance. In general three immune tolerance defects in ITP are proposed (Table 1). These include central tolerance defects (those which arise in early stages of lymphocyte development or in the bone marrow), differentiation blocks (leading to skewed peripheral B cell subsets), and peripheral tolerance defects (those arising in the setting of immune stimulation such as viral infections or immunizations) (Cines et al. 2009). Examples of central tolerance defects include inherited conditions such as ALPS and some subsets of autoimmune diseases; examples of differentiation blocks include CVID and immunosuppressive therapy, and examples of peripheral tolerance defects include infections such as HIV, hepatitis C, and varicella and vaccinations such as MMR (Cines et al. 2009). In general immune tolerance checkpoint defects are thought to correlate with clinical course of ITP and response to therapy. The peripheral tolerance defects tend to be the most platelet specific as well as the most responsive to immunotherapy or antigen clearance. The central defects are more commonly associated with more than one cell line being affected and are more resistant to immunotherapy and splenectomy. In the latter, reconstitution of the immune system is also more rapid, given the self-renewal capacity of early lymphocytes; thus recurrences of ITP are also more common with central tolerance defects than peripheral tolerance defects. Better understanding of the immune pathophysiology undoubtedly leads to better understanding of the etiology of ITP and thus to better targeted therapies (Cines et al. 2009).

Molecular Mimicry

Certain infections (hepatitis C, HIV, and H. pylori) are known to be associated with ITP.

Idiopathic Thrombocytopenic Purpura, Table 1 Types of ITP classified by pathophysiology (Cines et al. 2009; Stasi 2012)

Type of ITP	Mechanism	Tolerance defect
Neonatal ITP	Passive transfer of Ab	None
Acute childhood ITP	Molecular mimicry B cells T cells	Peripheral
Vaccine-related ITP	Unclear	Peripheral
Infectious ITP (HIV, Hep C, H. pylori)	Molecular mimicry B cells Viral suppression of bone marrow	Peripheral
Neoplastic ITP (CLL, lymphomas)	B cells T cells	Differentiation
Common variable immune deficiency ITP	B cells T cells Unclear	Differentiation
Autoimmune lymphoproliferative syndrome ITP	B cells T cells	Central
Antiphospholipid syndrome ITP	B cells T cells	Central
Systemic lupus erythematosus ITP	B cells T cells Antigen-presenting cells Mesenchymal stem cells Marrow suppression	Central
Post-transplant ITP	B cells T cells Mesenchymal cells Antigen-presenting cells Marrow suppression	Central

Although multiple pathophysiologic mechanisms are at play, an important one is molecular mimicry, in which the antiviral/microbial antibodies may also recognize the restricted peptide sequences (i.e., GPIIIa) targeted by autoimmune antiplatelet antibodies (Stasi et al. 2009).

Passive Transfer of Antibodies (Ab)

Neonates may also present with ITP. A common mechanism in this situation is placental transfer of IgG antibodies from the mother to infant. Maternal antiplatelet antibodies may either be the result of ITP in the mother (autoimmune) or the result of platelet-surface antigen incompatibility between the mother and father, leading to maternal production of antibodies during pregnancy (alloimmune).

Clinical Presentation and Course

Clinical manifestations of ITP are also variable. In general, bleeding symptoms are known to roughly correlate with platelet count at diagnosis:

Platelet count ($\times 10^9$/l)	Bleeding symptoms
50–150	None
30–50	Bruising/bleeding with trauma
10–30	Petechiae and ecchymosis
<10	Danger of internal bleeding such as intracranial hemorrhage

Most of the bleeding symptoms involve mucosal surfaces, such as the buccal mucosa, epistaxis, and menorrhagia. Bleeding may also be exacerbated by vascular inflammation and, in some cases, by DIC or a microangiopathic picture (most commonly associated with infections). Paradoxically there may be an increased thrombotic risk in certain cases. The most notable is ITP associated with antiphospholipid syndrome. Recent evidence also indicates increased fatigue with decreased platelet counts, although the etiology of the latter remains unclear (Newton et al. 2011). The prevalence of hypothyroidism in patients with ITP is also greater than in the general population (Ioachimescu et al. 2007).

From a clinical standpoint, adult ITP is two to three times more common in women. At presentation, ITP is still classified as primary (i.e., isolated low platelet count unrelated to other conditions) or secondary (thrombocytopenia is related to the underlying condition such as infection, vaccine, autoimmune disorder, or immunodeficiency). As more knowledge is gained about the pathophysiology of ITP, less and less cases can be classified as primary. Even the common childhood acute ITP is thought to be secondary to viral infection and subsequent immune dysregulation.

Historically ITP has also been categorized as acute (<6 months duration) or chronic (>12 months duration). More recently the disease phases have been modified and standardized to include "newly diagnosed ITP" (<3 months duration), "persistent ITP" (3–12 months duration), "chronic ITP" (>12 or 18 months duration), and "refractory ITP" (no initial response or relapse after splenectomy) (Stasi 2012).

Complete response is defined as no bleeding symptoms and platelets >100 $\times$ 10^9/l; response is defined as platelets between 30–100 $\times$ 10^9/l with a doubling from baseline platelet count; no response is defined as platelets <30 $\times$ 10^9/l or less than a doubling of the baseline count (Stasi 2012).

The diagnosis of ITP is typically a diagnosis of exclusion. There are no specific or sensitive laboratory tests available; this is not surprising, given the diversity of pathophysiologic mechanisms. A careful history and physical examination are required along with supportive laboratory testing based on clinical findings.

The history should concentrate on the acuity of onset of bleeding symptoms, type and extent of bleeding symptoms, associated symptoms (i.e., infectious symptoms, signs of autoimmune disease), immunization history, and medication history. A positive family history for ITP should lead to the consideration of other diagnoses such as inherited platelet disorders, as familial ITP is exceedingly rare.

Except for the presence of bruises or petechiae, the physical examination is often benign. Young children especially often appear completely well. The presence of fevers, weight loss, hepatosplenomegaly, lymphadenopathy, diffuse rashes, and joint pain should again lead to consideration of other etiologies for thrombocytopenia.

The most helpful laboratory test is the complete blood count and examination of the peripheral blood smear. Other cell lines (leukocytes and erythrocytes) are usually not abnormal. Platelets are usually large (increased mean platelet volume) with no other morphologic abnormalities. A bone marrow aspirate/biopsy is not usually clinically indicated but should be considered if there are additional cytopenias, no response to therapy, and other atypical morphologic features (i.e., erythrocyte macrocytosis) seen on the blood smear. In some cases bone marrow aspirates are performed in children prior to starting steroids, as the latter may partially treat an infiltrative process such as acute lymphoblastic leukemia.

Detection of platelet-associated antibodies (a direct assay for measuring platelet-bound antibodies) is only 49–66 % sensitive and 78–92 % specific for ITP with an overall positive predictive value of approximately 80 %. In addition, platelet antibodies are detected in only 50–60 % of ITP patients (Brighton et al. 1996).

Therapy

The first major question to answer is whether any therapy is needed at all. Therapy with the specific aim of increasing the platelet count should take into consideration the presence and severity of clinical bleeding and/or the risk of a life-threatening hemorrhage such as an intracranial bleed. To answer this question, the practitioner needs to consider the platelet count, the age of the patient, other comorbidities, previous history, and risk for trauma. Increased risk of morbidity from a serious bleed is associated with older age, lower platelet count, and increased number of comorbidities as well as previous bleeding events (Cines and Bussell 2005).

In pediatrics, the decision to treat is often based on the fear of intracranial hemorrhage. The incidence of this event is 0.2–1 % and almost always occurs with platelet counts of less than $20 \times 10^9/l$ (usually less than $10 \times 10^9/l$). Most often the bleed will occur within 4 weeks of diagnosis (usually within the first week) (Lilleyman 1994). The event is so rare and the outcomes with observation alone so favorable that, to date, there is no convincing evidence that therapy prevents intracranial hemorrhage.

The second question to answer is what to treat with. The mechanism of action of the therapy should take into consideration the pathophysiology of the ITP (Fig. 2).

Mechanism of action	Therapeutic modality
Platelet/antibody clearance by Fc receptors	Steroids Immune globulin Anti-D immune globulin Splenectomy
Impaired bone marrow production of megakaryocytes and intramedullary destruction of platelets and platelet precursors	TPO receptor agonists
B cell modulation and anti-idiotypic antibodies	Anti-CD20 Immune globulin
T cell modulation	Steroids Cytoxan Cyclosporine Anti-CD154
Removal of antibodies	Plasmapheresis

Thus therapy should be individualized and predicated on the severity of thrombocytopenia and bleeding and/or risk of bleeding events. Obviously medications that interfere with platelet function (i.e., aspirin) should be discontinued. If treatment is indicated, the first line of therapy is corticosteroids (usually prednisone). These can be dosed in two ways: continuous daily oral prednisone (1.0–1.5 mg/kg/day) with a rapid taper in 3–4 weeks, or higher-dose pulsed methylprednisolone or dexamethasone (40 mg daily for 4 days) (Cheng et al. 2003). Response rates of 50–75 % are reported and outcomes are better in patients with acute onset of ITP (George et al. 1996).

In children, because of the concerns about steroid side effects and potential "masking" of an early leukemia, immune globulin and anti-D immune globulin are often used first. Both may have side effects of headache, fever, and aseptic meningitis (which could mimic symptoms of intracranial hemorrhage). The latter is effective in only Rh + individuals, and its use may be accompanied by a drop in hemoglobin of 1–2 g as well as the possibility of severe intravascular hemolysis. The former may cause acute anaphylaxis (especially in patients with IgA deficiency) and contribute to renal and pulmonary insufficiency. Both modalities, however, shorten the duration of time for which platelets remain less

Idiopathic Thrombocytopenic Purpura, Fig. 2 Mechanisms of action of therapies for immune thrombocytopenic purpura. Many drugs used in the initial treatment of immune thrombocytopenic purpura impair the clearance of antibody-coated platelets (*1*) by the Fcg receptors expressed on tissue macrophages (*inset*). Splenectomy works at least in part by this mechanism but may also impair the T cell–B cell interactions involved in the synthesis of antibody in some patients. Corticosteroids may also increase platelet production by impairing the ability of macrophages within the bone marrow to destroy platelets, and thromboprotein stimulates megakaryocyte progenitors (*2*). Many nonspecific immunosuppressants, such as azathioprine and cyclosporine, act at the level of the T cell (*3*). A monoclonal antibody against CD154 that is now in clinical trials targets a costimulatory molecule needed for the optimization of the T cell–macrophage and T cell–B cell interactions involved in antibody production and class switching (*4*). Intravenous immunoglobulin may contain anti-idiotypic antibodies that impede antibody production. A monoclonal antibody that recognizes CD20 expressed on B cells (*5*) is also under study. Plasmapheresis may transiently remove antibody from the plasma (*6*), and platelet transfusions are used in emergencies to treat bleeding (*7*). The effect of staphylococcal protein A on the antibody repertoire is under study

than 20×10^9/l, with short-term responses of 80 % reported (Blanchette et al. 1994; Gaines 2000).

Other drugs (danazol, dapsone, azathioprine, mercaptopurine, vinca alkaloids, cyclosporin A, and mycophenolate) have all been used, but with inconsistent responses (Stasi and Provan 2004).

Second-line therapy is currently a bit more controversial. Historically, splenectomy has been the treatment of choice. Short-term response rates of 80 % are reported with long-term responses of 60–70 % 5–10 years out from surgery. However, it has been difficult to predict who will respond to splenectomy; the operative procedure is invasive, carries surgical risks of bleeding and infection, and is irreversible. Laparoscopic splenectomy has recently decreased operative mortality rates from 1 % to 0.2 % (Kojouri et al. 2004). In addition to the subsequent risk of sepsis with encapsulated bacteria, other long-term complications such as increased risk for thrombosis, atherosclerosis, and pulmonary hypertension have also been reported (Rodeghiero and Ruggeri 2012). Splenectomy is also usually contraindicated in the very young (<5 yo), the elderly (>70 yo) and those with other medical comorbidities or secondary ITP.

More recently anti-CD20 and thrombopoietin (TPO) receptor agonists have been used with increasing frequency as second-line therapeutic modalities. Anti-CD20 gives initial response rates of 50–60 %; however sustained responses 3–5 years out are much lower (20 %) (Stasi 2012; Arnold et al. 2007). Again, there are no good predictors of response. Adverse effects include chills, fever, dyspnea, and sometimes neutropenia. Anti-CD20 is also associated with reactivation of certain viral infections (i.e., Hepatitis B), leads to decreased IgG levels, and has been reported to cause serum sickness as well as multifocal leukoencephalopathy.

TPO receptor agonists have also demonstrated 60–80 % response rates with sustained responses of 70–90 %. Two TPO receptor agonists are now licensed for adult patients with ITP. They are romiplostim (given subcutaneously) and eltrombopag (given orally). The length of therapy has not yet been well established, and adverse effects include reticulin fibrosis of bone marrow with possible development of myelodysplastic syndrome as well as rebound thrombocytopenia once the therapy is stopped (Kuter et al. 2010; Cheng et al. 2011).

Newer experimental therapies include the use of campath-1H, protein A immunoadsorption columns, peripheral blood stem cell transplantation, and infusion of adipose tissue-derived mesenchymal cells from haploidentical family donors (Stasi 2012).

Finally mention should be made of urgent therapy of ITP in the setting of a life-threatening bleed. In this situation, again splenectomy is advised in addition to high-dose methylprednisolone (30 mg/kg for 2–3 days), immune globulin infusion (1 g/kg 2–3 days), and platelet transfusion (usually two to three times the usual dose in order to override the presence of antibodies). Adjunct therapies may include the use of antifibrinolytic agents as well as recombinant Factor VIIa (Cines and Bussell 2005).

Outcome

Overall the outcome of ITP is favorable, and in most cases (especially in childhood) it is a self-remitting, transient, benign disorder with >70 % resolution of thrombocytopenia within 6 months of diagnosis irrespective of therapy (Cines and Blanchette 2002). Adults with ITP do have a higher rate of mortality – 53.0 per 1000 person years compared with 29.1 per 1000 person years for normal controls. Predictors of increased morbidity include age (>60 years) and increased chronicity of platelet count <30 × 10^9/l (Norgaard et al. 2011). The major cause of mortality remains intracranial hemorrhage. The risk for this rare event is highest in the elderly, patients with previous history of significant bleeding, and patients with suboptimal response to therapy (Lee and Kim 1998; Cohen et al. 2000).

Cross-References

▶ Autoinflammatory Syndromes
▶ Corticosteroids
▶ Immunoglobulin Receptors and Inflammation

References

Arnold, D. M., Dentali, F., Crowther, M. A., et al. (2007). Systematic review: Efficacy and safety of rituximab for adults with idiopathic thrombocytopenic purpura. *Annals of Internal Medicine, 146*(1), 25–33.

Blanchette, V., Imbach, P., Andrew, M., et al. (1994). Randomised trial of intravenous immunoglobulin G, intravenous anti-D, and oral prednisone in childhood acute immune thrombocytopenic purpura. *Lancet, 344*, 703–707.

Brighton, T. A., Evans, S., Castaldi, P. A., Chesterman, C. N., & Chong, B. H. (1996). Prospective evaluation of the clinical usefulness of an antigen-specific assay (MAIPA) in idiopathic thrombocytopenic purpura and other immune thrombocytopenias. *Blood, 88*, 194–201.

Catani, L., Fagioli, M. E., Tazzari, P. L., et al. (2006). Dendritic cells of immune thrombocytopenic purpura (ITP) show increased capacity to present apoptotic platelets to T lymphocytes. *Experimental Hematology, 34*(7), 879–887.

Cheng, Y., Wong, R. S. M., Soo, Y. O. Y., et al. (2003). Initial treatment of immune thrombocytopenic purpura with high-dose dexamethasone. *New England Journal of Medicine, 349*, 831–836.

Cheng, G., Saleh, M. N., Marcher, C., et al. (2011). Eltrombopag for management of chronic immune thrombocytopenia (RAISE): A 6-month, randomised, phase 3 study. *Lancet, 377*(9763), 393–402.

Cines, D. B., & Blanchette, V. S. (2002). Immune thrombocytopenic purpura. *New England Journal of Medicine, 346*(13), 995–1008.

Cines, D. B., & Bussell, J. B. (2005). How I treat idiopathic thrombocytopenic purpura (ITP). *Blood, 106*, 2244–2251.

Cines, D. B., Bussel, J. B., Liebman, H. A., & Luning Prak, E. T. (2009). The ITP syndrome: Pathogenic and clinical diversity. *Blood, 113*, 6511–6521.

Cohen, Y. C., Djulbegovic, B., Shamai-Lubovitz, O., & Mozes, B. (2000). The bleeding risk and natural history of idiopathic thrombocytopenic purpura in patients with persistent low platelet counts. *Archives of Internal Medicine, 160*, 1630–1638.

Fang, B., Mai, L., Li, N., & Song, Y. (2012). Favorable response of chronic refractory immune thrombocytopenic purpura to mesenchymal stem cells. *Stem Cells and Development, 21*(3), 497–502.

Frederiksen, H., & Schmidt, K. (1999). The incidence of idiopathic thrombocytopenic purpura in adults increases with age. *Blood, 94*, 909–913.

Gaines, A.R. (2000). Acute onset hemoglobinemia and/or hemoglobinuria and sequelae following Rh(o)(D) immune globulin intravenous administration in immune thrombocytopenic purpura patients. *Blood, 95*(8), 2523–2529.

George, J. N., Woolf, S. H., Raskob, G. E., et al. (1996). Idiopathic thrombocytopenic purpura: A practical guideline developed by explicit methods for the American Society of Hematology. *Blood, 88*, 3–40.

Ioachimescu, A. G., Makdissi, A., Lichtin, A., & Zimmerman, R. S. (2007). Thyroid disease in patients with idiopathic thrombocytopenia: A cohort study. *Thyroid, 17*(11), 1137–1142.

Kojouri, K., Vesely, S. K., Terrell, D. R., & George, J. N. (2004). Splenectomy for adult patients with idiopathic thrombocytopenic purpura: A systematic review to assess long-term platelet count responses, prediction of response, and surgical complications. *Blood, 104*, 2623–2634.

Kuter, D. J., Rummel, M., Boccia, R., et al. (2010). Romiplostim or standard of care in patients with immune thrombocytopenia. *New England Journal of Medicine, 363*(20), 1889–1899.

Kuwana, M., Okazaki, Y., & Ikeda, Y. (2009). Splenic macrophages maintain the anti-platelet autoimmune response via uptake of opsonized platelets in patients with immune thrombocytopenic purpura. *Journal of Thrombosis and Haemostasis, 7*(2), 322–329.

Laster, A. J., Conley, C. L., Kickler, T. S., Dorsch, C. A., & Bias, W. B. (1982). Chronic immune thrombocytopenic purpura in monozygotic twins: Genetic factors predisposing to immune thrombocytopenic purpura. *New England Journal of Medicine, 307*, 1495–1498.

Lee, M. S., & Kim, W. C. (1998). Intracranial hemorrhage associated with idiopathic thrombocytopenic purpura: A report of seven patients and a meta-analysis. *Neurology, 50*, 1160–1163.

Lilleyman, J. S. (1994). Intracranial haemorrhage in idiopathic thrombocytopenic purpura. *Archives of Disease in Childhood, 71*, 251–253.

Lippman, S. M., Arnett, F. C., Conley, C. L., Ness, P. M., Meyers, D. A., & Bias, W. B. (1982). Genetic factors predisposing to autoimmune diseases: Autoimmune hemolytic anemia, chronic thrombocytopenic purpura, and systemic lupus erythematosus. *American Journal of Medicine, 73*, 827–840.

Liu, B., Zhao, H., Poon, M. C., et al. (2007). Abnormality of CD4(+)CD25(+) regulatory T cells in idiopathic thrombocytopenic purpura. *European Journal of Haematology, 78*(2), 139–143.

Louwes, H., Zeinali Lathori, O. A., Vellenga, E., & de Wolf, J. T. (1999). Platelet kinetic studies in patients with idiopathic thrombocytopenic purpura. *American Journal of Medicine, 106*(4), 430–434.

McMillan, R. (2000). Autoantibodies and autoantigens in chronic immune thrombocytopenic purpura. *Seminars in Hematology, 37*, 239–248.

McMillan, R., Wang, L., Tomer, A., Nichol, J., & Pistillo, J. (2004). Suppression of in vitro megakaryocyte production by antiplatelet autoantibodies from adult patients with chronic ITP. *Blood, 103*, 1364–1369.

Newton, J. L., Reese, J. A., Watson, S. I., Vesely, S. K., Bolton-Maggs, P. H., George, J. N., et al. (2011). Fatigue in adult patients with primary immune thrombocytopenia. *European Journal of Haematology, 86*(5), 420–429.

Norgaard, M., Jensen, A. O., Engebjerg, M. C., et al. (2011). Long-term clinical outcomes of patients with primary chronic immune thrombocytopenic purpura: A Danish population-based cohort study. *Blood, 117*(13), 3514–3520.

Olsson, B., Andersson, P. O., Jernas, M., et al. (2003). T-cell mediated cytotoxicity toward platelets in chronic idiopathic thrombocytopenic purpura. *Nature Medicine, 9*(9), 1123–1124.

Olsson, B., Andersson, P. O., Jacobsson, S., Carlsson, L., & Wadenvik, H. (2005). Disturbed apoptosis of T-cells in patients with active idiopathic thrombocytopenic purpura. *Thrombosis and Haemostasis, 93*(1), 139–144.

Perez-Simon, J. A., Tabera, S., Sarasquete, M. E., et al. (2009). Mesenchymal stem cells are functionally abnormal in patients with immune thrombocytopenic purpura. *Cytotherapy, 11*(6), 698–705.

Rodeghiero, F., & Ruggeri, M. (2012). Short- and long-term risks of splenectomy for benign haematological disorders: Should we revisit the indications? *British Journal of Haematology, 158*(1), 16–29.

Rodeghiero, F., Stasi, R., Gernsheimer, T., et al. (2009). Standardization of terminology, definitions and outcome criteria in immune thrombocytopenic purpura of adults and children: Report from an international working group. *Blood, 113*(11), 2386–2393.

Semple, J. W., Milev, Y., Cosgrave, D., et al. (1996). Differences in serum cytokine levels in acute and chronic autoimmune thrombocytopenic purpura: Relationship to platelet phenotype and antiplatelet T-cell reactivity. *Blood, 87*, 4245–4254.

Stasi, R. (2012). Immune thrombocytopenia: Pathophysiologic and clinical update. *Seminars in Thrombosis and Hemostasis, 38*, 454–462.

Stasi, R., & Provan, D. (2004). Management of immune thrombocytopenic purpura in adults. *Mayo Clinic Proceedings, 79*(4), 504–522.

Stasi, R., Evangelista, M. L., Stipa, E., Buccisano, F., Venditti, A., & Amadori, S. (2008). Idiopathic thrombocytopenic purpura: Current concepts in pathophysiology and management. *Thrombosis and Haemostasis, 99*, 4–13.

Stasi, R., Willis, F., Shannon, M. S., & Gordon-Smith, E. C. (2009). Infectious causes of chronic immune thrombocytopenia. *Hematology/Oncology Clinics of North America, 23*(6), 1275–1297.

Stratton, J. R., Ballem, P. J., Gernsheimer, T., Cerqueira, M., & Slichter, S. J. (1989). Platelet destruction in autoimmune thrombocytopenic purpura: Kinetics and clearance of indium-111-labeled autologous platelets. *Journal of Nuclear Medicine, 30*(5), 629–637.

IkappaB

Ellen Niederberger
Pharmazentrum frankfurt/ZAFES, Klinikum der Johann Wolfgang Goethe-Universität Frankfurt, Frankfurt am Main, Germany
JW Goethe University Frankfurt, Frankfurt, Germany

Synonyms

I(kappa)B; I-kappa-B; **IkappaBalpha**, MAD3; **IkappaBbeta**, thyroid hormone receptor (TR)-interacting protein 9; **IkappaBdelta**, T-cell activation NFKB-like protein; **IkappaBepsilon**, Slc35b2; IkB; IKB; Inap; INAP; Inhibitor of kappa B; Inhibitor of nuclear factor of kappa light chain gene enhancer in B cells; Mail; MAIL; Major histocompatibility complex enhancer-binding protein alpha; Member B2, **IkappaBzeta**, IL-1-inducible nuclear ankyrin repeat protein; Molecule possessing ankyrin repeats induced by lipopolysaccharide; NF-kappa-B inhibitor; NF-kappaB1; NF-kappaB2; Nfkbi; Nuclear factor of kappa light chain gene enhancer in B-cell inhibitor; **p100**, NF-KB2; **p105**, NF-KB1; p50/p105; RL/IF-1; Solute carrier family 35; Ta-nfkbh; TA-NFKBH; Thyroid receptor-interacting protein 9; TR-interacting protein 9; TRIP9; TRIP-9

Definition

The transcription factor nuclear factor kappa B (NFκB) plays an important role in immune responses, inflammatory diseases, and cell death. It is composed of homo- and heterodimers of different Rel family proteins (p65, RelB, c-Rel, p52, and p50) which share a Rel homology domain (RHD) which is crucial for dimerization. NFκB activation is tightly regulated by interaction with inhibitory IκB proteins, nuclear translocation, and DNA binding. The best studied classical IκB function is described for the canonical NFκB activation cascade. In this pathway, the majority of Rel/NFκB dimers – the most common form is the p50/p65 heterodimer – are sequestered in the cytoplasm of most unstimulated cells by binding to cytosolic IκBs. These IκBs act as inhibitors of NFκB activity by trapping it in cytoplasmic complexes and thereby preventing nuclear translocation and NFκB-DNA binding, respectively. Upon activation by different stimuli, IκB is degraded and releases NFκB to allow its nuclear translocation and transcription of target genes (reviewed in Hayden and Ghosh (2008), Karin et al. (2004)).

In addition to this prototypical action of cytoplasmic NFκB binding and inhibiting its function, it is meanwhile clear that a number of different

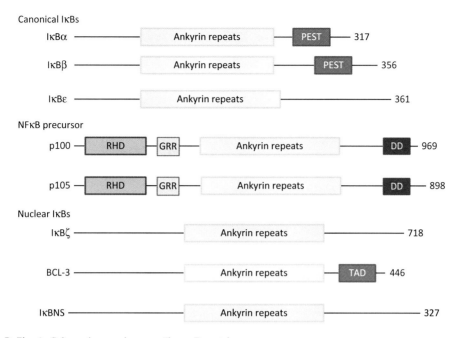

IkappaB, Fig. 1 Schematic overview over IkappaB proteins

IκBs exist and that IκB functions are much more diverse and not completely specific for NFκB subunits. Thus, IκBs might also contribute to cross talk between NFκB and other signaling cascades. Interestingly, the family of IκB proteins is slightly larger and more diverse than the NFκB proteins (Whiteside and Israel 1997).

Biosynthesis and Release

IκB proteins were first detected during purification of NFκB where it came out that treatment of cytosolic extracts with dissociating agents such as sodium deoxycholate or formamide stimulated DNA binding of cytoplasmic NFκB (Baeuerle and Baltimore 1988). At that time point, only two isoforms of IκBs, IκBα and IκBβ, were identified, and it could be shown that purified fractions of these proteins were able to effectively and specifically inhibit NFκB-DNA binding (Zabel and Baeuerle 1990). In the meantime, in the mouse and the human genome, the IκB protein family comprises at least eight members (IκBα, IκBβ, IκBε, IκBζ, BCL-3, IκBNS, p105, and p100) which are all characterized by a number of ankyrin repeat domains (ARDs) that are responsible for the interaction with NFκB (Fig. 1) (Hayden and Ghosh 2008).

IκBα, IκBβ, and IκBε are referred to as typical, canonical IκB proteins and IκBζ, BCL-3, and IκBNS as atypical nuclear IκB proteins. p105 and p100 are NFκB precursor proteins which function as IκBs before processing. The ARDs of all IκB proteins are localized in their C-terminus where they mediate interaction with the NFκB RHDs and modulate the function of the nuclear localization signals (NLSs) in several NFκB dimers. The ankyrin repeats are usually formed of homologous 33-residue long sequences with similar folds which are arranged in a pile in the ARD to form an ordered protein structure (Li et al. 2006). In addition to carrying ankyrin repeats, N- and C-terminal domains of IκBs can fulfill important functions which are described in more detail below.

While the abovementioned IκBs are mostly well studied, other IκB-like proteins are cell type specific, and their function is not completely clarified at date. One example is IκBγ which is generated by proteasomal degradation of the C-terminal part of the murine p105 precursor protein in B lymphocytes, whereas NFκB p50 is generated from the N-terminal part of p105 (Gerondakis et al. 1993). IκBγ is an inhibitor of the NFκB subunits p50, p65, and c-Rel but appears to be extremely unstable in living cells. Another less well-studied IκB isoform is IκBR (IκB related protein) (Ray et al. 1995), which has first been cloned from human lung alveolar epithelial cells and differs in its structure from the other IκBs. Its functional role in vivo is not clarified so far, and it is not really clear if it really functions as IκB protein. IκBL (Albertella and Campbell 1994) and IκBn (Yamauchi et al. 2010) are also reported as IκB proteins; however, there are further studies necessary to clarify their role in NFκB signaling.

Typical, Traditional IκBs (Canonical IκBs)

The first IκBs identified were the two proteins called IκBα and IκBβ which both show a similar protein structure with a central ARD, a signal-responsive region (SRR, "degron") at the amino-terminus, and a Thr-rich proline, glutamic acid, serine, and threonine (PEST) region on the carboxy-terminus. The PEST region of IκBα and IκBβ can be phosphorylated in vivo by casein kinase II and is required for inhibition of NFκB-DNA binding activity by stabilizing the IκB molecule as well as promoting interaction with NFκB dimers and their subsequent removal from DNA. Deletion of the PEST domain revealed that this site is able to protect IκBα from stimulus-induced degradation during NFκB activation (McKinsey et al. 1997). In 1997, a third isoform was detected and named IκBε (Li and Nabel 1997). IκBα and IκBβ are expressed in almost all tissues, while IκBε is expressed mainly in hematopoietic cells. The SRR in the N-terminal region of IκBα, IκBβ, and IκBε expresses signal-specific phosphorylation sites with two conserved serine residues which need to be phosphorylated for degradation of IκB and activation of NFκB. An acidic region in the C-terminal domain is involved in the IκB-mediated prevention of NFκB-DNA binding and dissociation of DNA-NFκB dimers. In cells, a variety of stimuli such as cytokines and bacterial lipopolysaccharides (LPSs) are able to induce IκB phosphorylation at these serine residues by specific IκB kinases (IKKs). Subsequently, IκB is ubiquitinated and then degraded by the 26S proteasome complex. The mechanism of NFκB inhibition is relatively similar for all of the typical IκBs, and therefore they can compensate at least partially for each other. However, it is not possible to exchange them completely which has been shown by experiments with mouse embryonic fibroblast with a knockout of different typical IκBs (Hoffmann et al. 2002). These studies showed that each typical IκB is unique in its function which is most likely due to differences in their kinetics of degradation and resynthesis, their cell type specificity, and the substrate specificity of IKKs. Furthermore, a knockout of all three typical IκB isoforms in cell culture indicated that p65 NFκB is not completely dependent on IκB-mediated cytoplasmatic retention since cells showed a relatively normal p65 distribution and an increased basal NFκB-dependent gene expression.

NFκB Precursors

p105 and p100

A further class of IκBs consists of the NFκB precursor proteins p105 and p100 which are processed into the mature NFκB subunits p50 and p52. The p50 and p52 RHDs are located at the amino-terminal end of the precursor proteins which is adjacent to a glycine-rich polypeptide region. An ARD with seven ankyrin repeats is located carboxy-terminal to this region which also contains a pair of serine residues (Moorthy et al. 2007).

p100 is known as an important part of the noncanonical NFκB activation cascade which is driven via NFκB-inducing kinase (NIK) and phosphorylation of p100 by IKKα. The phosphorylation promotes polyubiquitination and proteasomal processing to generate mature p52 (Xiao et al. 2004).

p105 processing occurs constitutively and independent of ubiquitination but also upon stimulation of cells leading to phosphorylation of p105 at two C-terminal serine residues and subsequent proteasomal degradation (Lin et al. 1998).

Novel, Atypical IκB-Like Proteins

A third class of IκBs comprises IκBζ (also known as MAIL (molecule possessing ankyrin repeats induced by LPS) and INAP (interleukin-1 inducible ankyrin repeat protein)), BCL-3, and IκBNS which are all so-called novel, atypical or nuclear IκB proteins. They differ from the typical IκBs in their sequence and domain structure, their regulation, and also their different functions (Moorthy et al. 2007).

BCL-3

In accordance with other IκBs, BCL-3 carries a central ARD with seven consensus ankyrin repeats. The structure amino-terminal of this ARD is not completely defined so far. However, unique for IκBs, there is a well-defined transactivation domain (TAD) in this region but no consensus sequences for IKK phosphorylation and destruction boxes, respectively. A serine-rich segment and a further TAD were detected carboxyl-terminal to the ARD of BCL-3. It has been suggested that both amino- and carboxy-terminal sequences are required for full transcriptional activity of BCL-3 indicating their cooperative mode of action. BCL-3 itself is regulated by NFκB as well as other transcription factors such as AP1 or Stat3, respectively. BCL-3 induction has been shown after stimulating macrophages with IL-10 which is known to suppress NFκB-DNA binding. Further interactions have been found with glycogen synthase kinase 3β (GSK-3β) which constitutively phosphorylates the carboxy-terminus of BCL-3 (Viatour et al. 2004). BCL-3 is subsequently ubiquitinated and degraded by proteolysis which modulates the expression of BCL-3-dependent genes.

IκBζ

IκBζ is transcriptionally upregulated by bacterial lipopolysaccharide, which also strongly activates NFκB (Yamazaki et al. 2001). It consists of a nuclear localization signal (NLS) and an ARD with seven ankyrin repeats. The expression of a TAD has also been discussed; however, this does not appear well defined. Interestingly, IκBζ has a long amino-terminal region which does not show sequence homology to any known protein. A carboxyl-terminal sequence does not exist. Similar to BCL-3, IκBζ expression is regulated by specific NFκB pathways, and the protein is localized in the nucleus by the NLS as confirmed by transfection experiments.

Biological Activities

Different IκB isoforms seem to be responsible for inhibition of distinct NFκB/Rel protein dimers, e.g., IκBα and IκBβ strongly inhibit c-Rel- and p65-containing complexes, whereas IκBζ and BCL-3 bind to homodimers of p50 or p52. IκBε has been described as a specific inhibitor of only c-Rel, p65, and their respective homodimers. The p52 precursor p100 is often associated with RelB; however, the NFκB precursors p100 and p105 are not that specific and able to bind to p50, p52, p65, and c-Rel.

Typical, Traditional IκBs (Canonical IκBs)

In the late 1980s, traditional IκBs were originally described to function as proteins which retain NFκB in the cytoplasm and inhibit it from nuclear translocation and DNA binding by masking the NLS of the NFκB subunits (Baeuerle and Baltimore 1988).

IκBα

IκBα is the most common and most extensively studied member of the IκB family. It may bind to a number of NFκB dimers, but the p50/p65 dimer in the canonical NFκB activation cascade appears to be particularly important for IκBα function. Binding of IκBα keeps NFκB in a complex which can be released after phosphorylation of IκBα at two conserved serine residues (Ser32 and Ser36) by the IκB kinases IKKα and IKKβ. The phosphorylation subsequently induces polyubiquitination through type E3 ubiquitin-protein ligase and degradation of IκBα by the 26S proteasome. This procedure releases NFκB and allows its nuclear translocation and DNA binding. NFκB activation in turn stimulates de novo synthesis of IκB proteins and constitutes a negative feedback loop and termination of NFκB signaling (Sun et al. 1993). The kinetics of this feedback loop is strongly involved in regulating the duration of the NFκB response.

IκBβ

IκBβ shows similar NFκB binding specificity as IκBα but at lower binding affinity to both the p50/p65 homo- and heterodimers in comparison to IκBα. In vitro, it has been shown that recombinant IκBβ only binds to p50/c-Rel hetero- or c-Rel homodimers with high affinity when its PEST domain is phosphorylated which is not a prerequisite for IκBα binding (McKinsey et al. 1997). In addition, IκBβ is phosphorylated at Ser19 and Ser23 also by IKKα and IKKβ, however not to the same extent as described for IκBα. Furthermore, it has been described that p50/p65/IκBα complexes constantly shuttle between the cytoplasm and nucleus, while p65/p65/IκBβ complexes are exclusively localized in the cytoplasm of unstimulated cells (Malek et al. 2001). After stimulation with LPS, IκBβ is slowly degraded and accumulates in the nucleus in a hypophosphorylated form which does not mask the p65 NLS and is able to bind to the DNA together with p65 and c-Rel (Suyang et al. 1996). The IκBβ/NFκB/DNA complexes are resistant to IκBα and therefore prolong the NFκB response indicating that IκBβ might be able to promote both NFκB activating and inhibiting properties (Rao et al. 2010).

IκBε

IκBε was first described in 1997 as a third canonical IκB member which is structurally related to IκBα and IκBβ (Li and Nabel 1997). It also contains an ARD with six ankyrin repeats; however, the amino-terminus is longer than that of IκBα and IκBβ, and IκBε lacks the carboxyl-terminal PEST region. Furthermore, IκBε shows a more limited tissue-specific expression and is mainly expressed in hematopoietic cells. IκBε is degraded after phosphorylation of its serine residues Ser157 and Ser161 in an IKK-dependent manner but shows delayed kinetics in comparison to IκBα. It preferentially binds to p65/p65 homodimers and c-Rel/p65 heterodimers and might be responsible for the late NFκB gene activation (Whiteside et al. 1997).

NFκB Precursors

p105 and p100

The physiological relevance of the IκB function of the NFκB precursor proteins is not completely elucidated so far. Both p105 and p100 act as NFκB inhibitors which have been described for their exclusive expression in the cytoplasm. This localization led to the suggestion that they may maintain a reservoir of inactive NFκB subunits irrespective of NFκB activation by proteolysis of canonical IκBs.

The ankyrin repeats of p105 preferentially bind to p50, p65, and c-Rel and trap them in a cytoplasmatic complex. The main function of constitutive p105 degradation is suggested as providing a pool of p50 molecules for the formation of p50-p50 homodimers. Furthermore, p105 has also been described as typical IκB by binding to other NFκB dimers and being processed by phosphorylation and degradation. In this case, activation of p105 is mediated by canonical NFκB signaling where IKKβ phosphorylates p105 C-terminal serine residues. In addition to its role in the NFκB cascade, p105 is also involved in other signaling pathways such as MAP kinase pathways (Hayden and Ghosh 2012).

p100 and p52 can bind to other IκB proteins, but only p52 is able to bind to the DNA. p100 preferentially binds to RelB in the cytoplasm which fits well with the assembly of p52/RelB in the alternative NFκB pathway. However, it is also able to bind to p65 homodimers and other NFκB subunits involved in canonical NFκB signaling (Hayden and Ghosh 2008). It is inducibly phosphorylated at several N-terminal serine residues by IKKα and NIK which, similar to typical IκBs, leads to polyubiquitination and proteasomal degradation of the C-terminal IκB-like part of the protein thus generating p52. Induction of p100 by NFκB is able to shift p65-containing NFκB dimers to RelB/p52 dimers which are resistant to IκBα and might therefore control late transcriptional responses (Saccani et al. 2003).

Novel, Atypical IκB-Like Proteins

IκBζ, BCL-3, and IκBNS not only have, in contrast to the typical IκBs, been described as NFκB inhibitors but might also function as activators depending on the cell types and the conditions applied. Since atypical IκBs are upregulated upon NFκB activation, it has been suggested that they might be responsible for secondary or late responses to a variety of NFκB-activating stimuli. Among each other, they are similar in their binding specificity to NFκB subunits, their subcellular localization, and their function as transcriptional coactivators (Schuster et al. 2013). The atypical IκBs are mostly expressed in low levels in resting cells and increase their expression after stimulation with NFκB activators.

BCL-3

BCL-3 was originally discovered as a gene translocation into an immunoglobulin locus. It has been suggested that it is a proto-oncogene with overexpression in B-cell chronic lymphocytic leukemia (Ohno et al. 1990). In contrast to the canonical IκBs and the NFκB precursors, it is localized in the nucleus where it interacts with p50 and p52 homodimers but is also able to regulate a number of non-NFκB transcription factors. It has been speculated that it has both repressive and activating functions on transcription since in vitro studies showed that BCL-3 overexpression resulted in NF κB-mediated gene expression as well as suppression. These effects have at least partially been attributed to its interaction with histone acetyltransferases and histone deacetylases, respectively. BCL-3 phosphorylation crucially influences its binding properties to p50/p52 homodimers. Its nuclear localization depends on ubiquitination and allows its transactivating functions.

IκBζ

IκBζ shows specificity toward the NFκB p50 homodimer. TLR and IL-R signaling appears to activate IκBζ leading to coactivation of enhanced IL-6 expression with delayed expression kinetics indicating that IκBζ is required for late NFκB-dependent gene expression. On the other hand, TNFα signaling cascades are inhibited by IκBζ indicating its dual functionality (Yamamoto et al. 2004).

IκBNS

Another atypical IκB protein is IκBNS which has been identified as an inducible IκB in T cells undergoing selection (Fiorini et al. 2002); however, it is also expressed in a number of other cell types. It can be induced by IL-10 and might therefore control inhibition of proinflammatory protein expression. IκBNS preferentially binds to and stabilizes p50 homodimers. Overexpression of IκBNS alleviated phorbol myristate acetate (PMA)-/ionophore-induced activation of NFκB and NFκB-DNA binding indicating that it functions as a negative regulator of NFκB responses. Accordingly, IκBNS-deficient cells showed a prolonged NFκB activity after stimulation with LPS. In general, depending on the cell type, IκBNS has been suggested to play an important role in the negative regulation of TLR-mediated immune responses and, on the other hand, as a positive regulator of NFκB-induced IL-2 production.

Pathophysiological Activities

The transcription factor NFκB plays important roles in inflammatory and immune responses as well as cell proliferation, apoptosis, and cell cycle progression. Therefore, dysregulations of NFκB signaling have been associated with a number of diseases such as cancer, autoimmune diseases, metabolic disorders, and pain. Since IkappaB proteins are central elements in NFκB signaling, they are involved in similar pathologies. In particular, IκBα, which is crucial in the canonical NFκB cascade, is strongly involved in inflammatory responses. Mice with a knockout of IκBα die shortly after birth which further supports its important physiological functions (Klement et al. 1996). However, the specific role of single different IkappaBs is not completely clarified so far. Several deletion studies only delivered data concerning the function of individual IkappaBs, e.g., a loss of IκBε has been associated with defects in hematopoietic cells which, however, can be largely compensated by IκBα activity (Memet et al. 1999). Mice with a deletion of IκBβ show a decreased transcription of proinflammatory genes associated with resistance to LPS-induced septic shock and collagen-induced arthritis supporting the assumption that it might act as a positive and negative regulator of NFκB (Rao et al. 2010; Scheibel et al. 2010). IκBζ knockout mice reveal a proinflammatory phenotype and atopic dermatitis with increased cytokine production (Shiina et al. 2004). BCL-3 knockout mice display severe impairments in their immune system indicating its important role in immune and anti-inflammatory gene responses (Schwarz et al. 1997). Furthermore, it has been suggested that IkappaB proteins might exert several pathophysiological activities which are independent of NFκB signaling, but this is, at date, also not completely clear. BCL-3 has been associated with cancer development probably due to regulation of cyclin D1 and p53 (Kashatus et al. 2006; Westerheide et al. 2001). IκBα together with the tumor suppressor p53 has been described to interact with cyclin-dependent kinase 4 and is also able to inhibit the HIV Rev protein indicating functions in cancer and viral diseases independently of NFκB (Li et al. 2003; Wu et al. 1997).

Modulation by Drugs

There are a number of established inhibitors which are able to inhibit NFκB activity by modulating IkappaB phosphorylation, ubiquitination, and/or proteasomal degradation. The dibenzylbutyrolactone lignan arctigenin is a potent inhibitor of LPS-induced IkappaBalpha phosphorylation and nuclear translocation of p65 (Cho et al. 2002). BAY 11-7083 and BAY 11-7085 decrease the expression of proinflammatory mediators in cell culture by irreversibly inhibiting TNFα-induced phosphorylation of IκBα. However, constitutive IκBα phosphorylation remained unaltered. BAY 11-7085 was also investigated in vivo and showed anti-inflammatory activity in the rat carrageenan paw edema assay and rat adjuvant arthritis model (Pierce et al. 1997).

The naturally occurring antibacterial peptide PR39 selectively binds to the alpha 7 subunit of the 26S proteasome thereby impairing IκBα degradation by the ubiquitin-proteasome pathway without disturbing total proteasome activity. IκBα phosphorylation and ubiquitination are not affected by PR39 treatment (Gao et al. 2000). Another proteasome inhibitor, MG-132, is also able to inhibit NFκB activation by interaction with the p105 NFκB precursor (Palombella et al. 1994). Further proteasome inhibitors which are able to interfere with IkappaB degradation are bortezomib and PR171, respectively.

The metal chelator pyrrolidine dithiocarbamate (PDTC) inhibits the release of the inhibitory subunit IκB from the latent cytoplasmic form of NFκB in stimulated cells by unspecific inhibition of IκB-ubiquitin ligase (Hayakawa et al. 2003; Iseki et al. 2000). Another small molecule inhibitor, Ro106-9920, selectively blocks LPS- and TNFα-induced IκBα ubiquitination, degradation, and NFκB activation (Swinney et al. 2002).

A further possibility of IκBα modulation is constituted by its overexpression via an IκBα protein mutated at the serine residues Ser32 and Ser36 (IκB super-repressor) (DiDonato et al. 1996).

Cross-References

▶ Bacterial Lipopolysaccharide
▶ Coxibs

References

Albertella, M. R., & Campbell, R. D. (1994). Characterization of a novel gene in the human major histocompatibility complex that encodes a potential new member of the I kappa B family of proteins. *Human Molecular Genetics, 3*(5), 793–799.

Baeuerle, P. A., & Baltimore, D. (1988). I kappa B: A specific inhibitor of the NF-kappa B transcription factor. *Science, 242*(4878), 540–546.

Cho, M. K., Park, J. W., Jang, Y. P., Kim, Y. C., & Kim, S. G. (2002). Potent inhibition of lipopolysaccharide-inducible nitric oxide synthase expression by dibenzylbutyrolactone lignans through inhibition of I-kappaBalpha phosphorylation and of p65 nuclear translocation in macrophages. *International Immunopharmacology, 2*(1), 105–116. [pii] S1567-5769(01)00153-9.

DiDonato, J., Mercurio, F., Rosette, C., Wu-Li, J., Suyang, H., Ghosh, S., et al. (1996). Mapping of the inducible IkappaB phosphorylation sites that signal its ubiquitination and degradation. *Molecular and Cellular Biology, 16*(4), 1295–1304.

Fiorini, E., Schmitz, I., Marissen, W. E., Osborn, S. L., Touma, M., Sasada, T., et al. (2002). Peptide-induced negative selection of thymocytes activates transcription of an NF-kappa B inhibitor. *Molecular Cell, 9*(3), 637–648. [pii] S1097276502004690.

Gao, Y., Lecker, S., Post, M. J., Hietaranta, A. J., Li, J., Volk, R., et al. (2000). Inhibition of ubiquitin-proteasome pathway-mediated I kappa B alpha degradation by a naturally occurring antibacterial peptide. *Journal of Clinical Investigation, 106*(3), 439–448. doi:10.1172/JCI9826.

Gerondakis, S., Morrice, N., Richardson, I. B., Wettenhall, R., Fecondo, J., & Grumont, R. J. (1993). The activity of a 70 kilodalton I kappa B molecule identical to the carboxyl terminus of the p105 NF-kappa B precursor is modulated by protein kinase A. *Cell Growth and Differentiation, 4*(8), 617–627.

Hayakawa, M., Miyashita, H., Sakamoto, I., Kitagawa, M., Tanaka, H., Yasuda, H., et al. (2003). Evidence that reactive oxygen species do not mediate NF-kappaB activation. *EMBO Journal, 22*(13), 3356–3366.

Hayden, M. S., & Ghosh, S. (2008). Shared principles in NF-kappaB signaling. *Cell, 132*(3), 344–362. doi:10.1016/j.cell.2008.01.020.

Hayden, M. S., & Ghosh, S. (2012). NF-kappaB, the first quarter-century: Remarkable progress and outstanding questions. *Genes and Development, 26*(3), 203–234. doi:10.1101/gad.183434.111.

Hoffmann, A., Levchenko, A., Scott, M. L., & Baltimore, D. (2002). The IkappaB-NF-kappaB signaling module: Temporal control and selective gene activation. *Science, 298*(5596), 1241–1245. doi:10.1126/science.1071914. [pii] 298/5596/1241.

Iseki, A., Kambe, F., Okumura, K., Niwata, S., Yamamoto, R., Hayakawa, T., et al. (2000). Pyrrolidine dithiocarbamate inhibits TNF-alpha-dependent activation of NF-kappaB by increasing intracellular copper level in human aortic smooth muscle cells. *Biochemical and Biophysical Research Communications, 276*(1), 88–92.

Karin, M., Yamamoto, Y., & Wang, Q. M. (2004). The IKK NF-kappa B system: A treasure trove for drug development. *Nature Reviews Drug Discovery, 3*(1), 17–26.

Kashatus, D., Cogswell, P., & Baldwin, A. S. (2006). Expression of the BCL-3 proto-oncogene suppresses p53 activation. *Genes and Development, 20*(2), 225–235. doi:10.1101/gad.1352206.

Klement, J. F., Rice, N. R., Car, B. D., Abbondanzo, S. J., Powers, G. D., Bhatt, P. H., et al. (1996). IkappaBalpha deficiency results in a sustained NF-kappaB response and severe widespread dermatitis in mice. *Molecular and Cellular Biology, 16*(5), 2341–2349.

Li, Z., & Nabel, G. J. (1997). A new member of the I kappaB protein family, I kappaB epsilon, inhibits RelA (p65)-mediated NF-kappaB transcription. *Molecular and Cellular Biology, 17*(10), 6184–6190.

Li, J., Joo, S. H., & Tsai, M. D. (2003). An NF-kappaB-specific inhibitor, IkappaBalpha, binds to and inhibits cyclin-dependent kinase 4. *Biochemistry, 42*(46), 13476–13483. doi:10.1021/bi035390r.

Li, J., Mahajan, A., & Tsai, M. D. (2006). Ankyrin repeat: A unique motif mediating protein-protein interactions. *Biochemistry, 45*(51), 15168–15178. doi:10.1021/bi062188q.

Lin, L., DeMartino, G. N., & Greene, W. C. (1998). Cotranslational biogenesis of NF-kappaB p50 by the 26S proteasome. *Cell, 92*(6), 819–828. [pii] S0092-8674(00)81409-9.

Malek, S., Chen, Y., Huxford, T., & Ghosh, G. (2001). IkappaBbeta, but not IkappaBalpha, functions as a classical cytoplasmic inhibitor of NF-kappaB dimers by masking both NF-kappaB nuclear localization sequences in resting cells. *Journal of Biological Chemistry, 276*(48), 45225–45235. doi:10.1074/jbc.M105865200.

McKinsey, T. A., Chu, Z. L., & Ballard, D. W. (1997). Phosphorylation of the PEST domain of IkappaBbeta regulates the function of NF-kappaB/IkappaBbeta complexes. *Journal of Biological Chemistry, 272*(36), 22377–22380.

Memet, S., Laouini, D., Epinat, J. C., Whiteside, S. T., Goudeau, B., Philpott, D., et al. (1999). IkappaBepsilon-deficient mice: Reduction of one T cell precursor subspecies and enhanced Ig isotype switching and cytokine synthesis. *Journal of Immunology, 163*(11), 5994–6005.

Moorthy, A. K., Huxford, T., & Ghosh, G. (2007). Structural aspects of NF-κB and IκB proteins. In S. Ghosh (Ed.), *Handbook of transcription factor NF-kappaB*. Boca Raton: CRC Press.

Ohno, H., Takimoto, G., & McKeithan, T. W. (1990). The candidate proto-oncogene BCL-3 is related to genes implicated in cell lineage determination and cell cycle control. *Cell, 60*(6), 991–997. [pii] 0092-8674(90)90347-H.

Palombella, V. J., Rando, O. J., Goldberg, A. L., & Maniatis, T. (1994). The ubiquitin-proteasome pathway is required for processing the NF-kappa B1 precursor protein and the activation of NF-kappa B. *Cell, 78*(5), 773–785. [pii] S0092-8674(94)90482-0.

Pierce, J. W., Schoenleber, R., Jesmok, G., Best, J., Moore, S. A., Collins, T., et al. (1997). Novel inhibitors of cytokine-induced IkappaBalpha phosphorylation and endothelial cell adhesion molecule expression show anti-inflammatory effects in vivo. *Journal of Biological Chemistry, 272*(34), 21096–21103.

Rao, P., Hayden, M. S., Long, M., Scott, M. L., West, A. P., Zhang, D., et al. (2010). IkappaBbeta acts to inhibit and activate gene expression during the inflammatory response. *Nature, 466*(7310), 1115–1119. doi:10.1038/nature09283.

Ray, P., Zhang, D. H., Elias, J. A., & Ray, A. (1995). Cloning of a differentially expressed I kappa B-related protein. *Journal of Biological Chemistry, 270*(18), 10680–10685.

Saccani, S., Pantano, S., & Natoli, G. (2003). Modulation of NF-kappaB activity by exchange of dimers. *Molecular Cell, 11*(6), 1563–1574.

Scheibel, M., Klein, B., Merkle, H., Schulz, M., Fritsch, R., Greten, F. R., et al. (2010). IkappaBbeta is an essential co-activator for LPS-induced IL-1beta transcription in vivo. *Journal of Experimental Medicine, 207*(12), 2621–2630. doi:10.1084/jem.20100864.

Schuster, M., Annemann, M., Plaza-Sirvent, C., & Schmitz, I. (2013). Atypical IkappaB proteins – Nuclear modulators of NF-kappaB signaling. *Cell Communication Signal, 11*(1), 23. doi:10.1186/1478-811X-11-23.

Schwarz, E. M., Krimpenfort, P., Berns, A., & Verma, I. M. (1997). Immunological defects in mice with a targeted disruption in BCL-3. *Genes and Development, 11*(2), 187–197.

Shiina, T., Konno, A., Oonuma, T., Kitamura, H., Imaoka, K., Takeda, N., et al. (2004). Targeted disruption of MAIL, a nuclear IkappaB protein, leads to severe atopic dermatitis-like disease. *Journal of Biological Chemistry, 279*(53), 55493–55498. doi:10.1074/jbc.M409770200.

Sun, S. C., Ganchi, P. A., Ballard, D. W., & Greene, W. C. (1993). NF-kappa B controls expression of inhibitor I kappa B alpha: Evidence for an inducible autoregulatory pathway. *Science, 259*(5103), 1912–1915.

Suyang, H., Phillips, R., Douglas, I., & Ghosh, S. (1996). Role of unphosphorylated, newly synthesized I kappa B beta in persistent activation of NF-kappa B. *Molecular and Cellular Biology, 16*(10), 5444–5449.

Swinney, D. C., Xu, Y. Z., Scarafia, L. E., Lee, I., Mak, A. Y., Gan, Q. F., et al. (2002). A small molecule ubiquitination inhibitor blocks NF-kappa B-dependent cytokine expression in cells and rats. *Journal of Biological Chemistry, 277*(26), 23573–23581. doi:10.1074/jbc.M200842200. [pii] M200842200.

Viatour, P., Dejardin, E., Warnier, M., Lair, F., Claudio, E., Bureau, F., et al. (2004). GSK3-mediated BCL-3 phosphorylation modulates its degradation and its oncogenicity. *Molecular Cell, 16*(1), 35–45. doi:10.1016/j.molcel.2004.09.004.

Westerheide, S. D., Mayo, M. W., Anest, V., Hanson, J. L., & Baldwin, A. S., Jr. (2001). The putative oncoprotein BCL-3 induces cyclin D1 to stimulate G(1) transition. *Molecular and Cellular Biology, 21*(24), 8428–8436. doi:10.1128/MCB.21.24.8428-8436.2001.

Whiteside, S. T., & Israel, A. (1997). I kappa B proteins: Structure, function and regulation. *Seminars in Cancer Biology, 8*(2), 75–82. doi:10.1006/scbi.1997.0058.

Whiteside, S. T., Epinat, J. C., Rice, N. R., & Israel, A. (1997). I kappa B epsilon, a novel member of the I kappa B family, controls RelA and cRel NF-kappa B activity. *EMBO Journal, 16*(6), 1413–1426. doi:10.1093/emboj/16.6.1413.

Wu, B. Y., Woffendin, C., MacLachlan, I., & Nabel, G. J. (1997). Distinct domains of IkappaB-alpha inhibit human immunodeficiency virus type 1 replication through NF-kappaB and Rev. *Journal of Virology, 71*(4), 3161–3167.

Xiao, G., Fong, A., & Sun, S. C. (2004). Induction of p100 processing by NF-kappaB-inducing kinase involves docking IkappaB kinase alpha (IKKalpha) to p100 and IKKalpha-mediated phosphorylation. *Journal of Biological Chemistry, 279*(29), 30099–30105. doi:10.1074/jbc.M401428200.

Yamamoto, M., Yamazaki, S., Uematsu, S., Sato, S., Hemmi, H., Hoshino, K., et al. (2004). Regulation of Toll/IL-1-receptor-mediated gene expression by the inducible nuclear protein IkappaBzeta. *Nature, 430*(6996), 218–222. doi:10.1038/nature02738.

Yamauchi, S., Ito, H., & Miyajima, A. (2010). IkappaBeta, a nuclear IkappaB protein, positively regulates the NF-kappaB-mediated expression of proinflammatory cytokines. *Proceedings of the National Academy of Sciences of the United States of America, 107*(26), 11924–11929. doi:10.1073/pnas.0913179107. [pii] 0913179107.

Yamazaki, S., Muta, T., & Takeshige, K. (2001). A novel IkappaB protein, IkappaB-zeta, induced by proinflammatory stimuli, negatively regulates nuclear factor-kappaB in the nuclei. *Journal of Biological Chemistry, 276*(29), 27657–27662. doi:10.1074/jbc.M103426200. [pii] M103426200.

Zabel, U., & Baeuerle, P. A. (1990). Purified human I kappa B can rapidly dissociate the complex of the NF-kappa B transcription factor with its cognate DNA. *Cell, 61*(2), 255–265.

Immunoglobulin Receptors and Inflammation

Marc Daëron
Institut Pasteur, Paris, France
Centre d'Immunologie de Marseille–Luminy, Marseille, France

Synonyms

Adaptive immunity; Allergy; Antibodies; Cell activation; Fc receptors; Immunoregulation; Immunotherapy; Inflammation; Signal transduction

Definitions

Inflammation

Inflammation is a physiological defense response. It associates destruction and repair mechanisms that are instrumental in wound healing and protection against infection. It involves the many myeloid cells that contribute to the effector phase of innate and adaptive immune responses. Normally, inflammation is tightly controlled and remains unapparent. It can be pathogenic if it escapes from regulation. Autoimmunity and allergy are examples of pathological inflammation generated by not properly controlled adaptive immune responses. Receptors for the Fc portion of antibodies (FcR) play prominent roles in both the induction and the control of these inflammatory diseases.

Innate and Adaptive Immunity

The innate immune system is made of numerous differentiated cells of several types, mostly of the myeloid lineage. They express pattern-recognition receptors which enable them to interact with structures borne or secreted by microorganisms in innate immunity. They have no antigen receptors, but they express FcR. When binding to FcR, specific antibodies enroll effector cells of innate immunity in adaptive immunity. It follows that innate and adaptive immunity uses the same effector cells.

The adaptive immune system is essentially made of limited numbers of lymphoid cells equipped with antigen receptors. Lymphocytes need to proliferate and to differentiate into effector cells of different types before they can act on specific antigens. Adaptive immune responses generate effector cells and molecules. Cells include T cells, endowed with various effector functions and capable of secreting numerous cytokines upon cognate interactions with antigen-presenting cells. Molecules include antibodies that recognize specifically the antigen against which they were raised and diffuse in the whole body. At least quantitatively, antibodies are the major effector molecules of adaptive immunity.

Fc Receptors

Antibodies, however, have no biological activity by themselves. Their Fab portions can bind to antigens, but except in rare instances, binding to antigen has little or no biological consequences. For antibodies to affect antigens, they not only need to bind to antigens through their Fab portions but also to interact through their Fc portion with effector systems. Among these are the cells that express FcR. When binding to FcR, antibodies provide myeloid cells with B cell receptor-like immunoreceptors and a bona fide immunological specificity (Daëron 1997). This specificity is an intrinsic property neither of the cell nor of FcR but of antibodies.

FcR are immunoreceptors of the third type. While B cell receptors (BCR) "recognize" antigen as native molecules and T cell receptors (TCR) as peptides associated with major histocompatibility complex molecules, FcR "recognize" antigen as immune complexes. BCR, TCR, and FcR are receptors for the three forms under which any given antigen can interact with and deliver signals to cells of the immune system.

Structure, Functions, and Physiopathological Relevance

Fc Receptors as Binding Sites for Antibodies

There are FcR for IgA (FcαR), IgG (FcγR and FcRn), and IgE (FcεR). There are also FcR for polymeric IgM and IgA (polyIgR). Many FcR names include a Roman number and a capital letter. Roman numbers refer to the binding affinity for immunoglobulins. FcRI and IV are high-affinity receptors; FcRII and III are low-affinity receptors. Capital letters refer to the genes that encode FcR. Human FcR are often referred to using the CD nomenclature. CD16 designate FcγRIII, CD32 FcγRII, CD64 FcγRI, CD23 FcεRII, and CD89 FcαRI.

Most FcR belong to the immunoglobulin superfamily (IgSF). A minority (FcγRIIA in humans; FcγRIIB and polyIgR in mice and humans) are single-chain receptors. Other FcR are multichain receptors composed of one immunoglobulin-binding subunit (FcRα) and one (FcRγ) or two (FcRγ and FcRβ) common subunits. FcRγ is a homodimer shared by all multichain receptors. FcRβ is a four-transmembrane domain polypetide that associates with multichain FcR expressed by mast cells and basophils. FcRn is a unique MHC class I-like molecule that binds the Fc portion of IgG instead of peptides. FcεRII are C-type lectins expressed as homotrimeric molecules.

Most multichain FcR must associate with FcRγ in order to reach the plasma membrane. The expression of these receptors therefore depends on the tissue distribution of FcRγ, and FcRγ-deficient mice have no activating FcR (Takai et al. 1994). Mouse FcεRI, but not human FcepsilonRI, also need to associate with FcRβ. As FcRβ is expressed by mast cells and basophils only in both species, the expression of FcepsilonRI is restricted to these cells in mice, but not in humans. FcRn do not associate with FcRγ but with β-2 microglobulin, like other MHC-I molecules, and this association is mandatory for FcRn to be expressed.

Antibodies bind to FcR with a variable affinity (Hulett and Hogarth 1994). The binding of antibodies to FcR is reversible, and it obeys the mass action law. The affinity of FcR is characterized by an affinity constant Ka which is the quotient of an association constant (k_{on}) divided by a dissociation constant (k_{off}). A proportion of high-affinity FcR, which can bind monomeric immunoglobulins in the absence of antigen, are occupied in vivo, whereas low-affinity FcR, which can bind antibodies as multivalent immune complexes only, are not in spite of the high concentration of circulating immunoglobulins. Occupied high-affinity FcR, however, can be freed as bound antibodies dissociate. The dissociation constant therefore critically determines the availability of high-affinity FcR. The Ka of FcεRI is especially high not because the association constant is high but because the dissociation constant is extremely low. The result is that, in spite of their extremely low concentration in plasma, IgE remain bound to FcεRI for extended periods of time where they are ready to trigger mast cells and basophil activation when allergen comes. One point to keep in mind is that even though their affinity for is too low for monomeric immunoglobulins to bind, low-affinity FcR bind immune complexes with a high avidity. The affinity with which IgG bind to FcγR further depends on the glycosylation of their Fc portion.

High-affinity FcR include IgA (FcαRI, in humans, pIgR in humans and mice), IgE (FcεRI, in humans and mice), and IgG receptors (FcγRI and FcRn, in humans and mice, and FcγRIV, in mice only). Low-affinity FcR include IgE (FcεRII, in humans and mice) and IgG receptors (FcγRII and III, in humans and mice). Humans have three FcγRII (FcγRIIA, B and C) and two FcγRIII (FcγRIIIA and B), whereas mice have one receptor of each type (FcγRIIB and FcγRIIIA) only. The diversity of human FcγRII and III is further increased by polymorphisms of selected residues in their extracellular domains.

Fc Receptors as Initiators of Adaptive Immune Responses

Antigen presentation to naive T cells is the first step of adaptive immune responses. Antibodies were found to function as potent adjuvants in a FcR-dependent fashion.

The uptake of antigen-IgG antibody complexes via dendritic cell FcR may indeed enhance both cross presentation and MHC class II presentation (Heyman 1990). The engagement of dendritic cell FcγR, by antigen-antibody complexes, induces the activation and maturation of dendritic cells.

FcγRIIB expressed by follicular dendritic cells can trap immune complexes in secondary lymphoid tissues, and these complexes are periodically arranged like epitopes on T-independent antigens. Antigens in immune complexes bound by follicular dendritic cells are thus much more potent inducers of antibody responses than free antigen in vitro (Tew et al. 2001) and in vivo (Wu et al. 2008). Supporting this point of view, the blockade of FcγRIIB expressed by follicular dendritic cells inhibited the ability of immune complexes to induce T-independent responses.

IgE antibodies were also found to be potent adjuvants for adaptive immune responses. IgE-induced enhancement affects all classes of antibodies. Although FcεRII are expressed by both types of cells, B cells but not follicular dendritic cells are involved. This explains that enhancement is antigen specific: only FcεRII-expressing B cells that possess the specific BCR receive cognate T cell help leading to antibody production (Hjelm et al. 2006).

Fc Receptors as Inducers of Inflammation

FcR signal when they are aggregated by antibodies and plurivalent antigens. Although the result is the same, the sequence of events that lead to receptor aggregation is different for high-affinity and low-affinity FcR. Monomeric antibodies bind first to high-affinity FcR that are aggregated when a plurivalent antigen binds to receptor-bound antibodies. Antibodies bind first to antigen, generating immune complexes that can bind to and, therefore, simultaneously aggregate low-affinity FcR. Most FcR activate cellular responses.

Activating FcR contain *immunoreceptor tyrosine-based activation motifs* (ITAM). ITAM consist of two YxxL motifs separated by six to eight variable amino acids. Activating FcR are FcαRI, FcεRI, FcγRI, FcγRIIA, FcγRIIC, FcγRIIIA, and FcγRIV. FcγRIIA and FcγRIIC are the only single-chain receptors that possess an ITAM. FcαRI, FcεRI, FcγRI, FcγRIIIA, and FcγRIV associate with the common FcR subunit FcRγ. FcRγ contains two ITAM; FcRβ contains one ITAM. Upon receptor aggregation, ITAM are phosphorylated by src family tyrosine kinases, which initiates the constitution of dynamic intracellular signalosomes in which activation signals are dominant over inhibition signals.

Activating FcR are expressed by myeloid cells and by lymphoid cells with no classical antigen receptor (i.e., NK cells and intraepithelial γ/δ T cells of the intestine). They are not expressed by mature T and B lymphocytes. Activating FcR therefore do not interfere with lymphocyte activation triggered by clonally expressed antigen receptors.

The biological functions of FcR depend on the functional repertoire of individual FcR-expressing cells. Some cells can release granules that contain cytotoxic mediators and other cells granules that contain vasoactive or proinflammatory mediators. Many cells can synthetize cytokines, chemokines, or growth factors of different types. FcR therefore are involved in a variety of biological functions. Activating FcR can trigger the transcytosis of immunoglobulins, the endocytosis of soluble immune complexes, the phagocytosis of particulate complexes, the exocytosis of granular mediators, the production of lipid-derived proinflammatory mediators, or the secretion of newly transcribed cytokines, chemokines, and growth factors. Activating FcR do not induce specific biological responses but cell type-dependent responses that can be induced by other receptors in the same cell.

Biological functions, however, are not ensured by single cells but by cell populations that are either present or recruited by chemokines and/or that proliferate locally. The composition of such populations depends on time and location. Biological functions in which FcR are involved are therefore a resultant of the biological properties of the many cells that are engaged in the reaction at a given place and at a given time. They are pleiotropic and dynamic.

FcR as Inducers of Acute Allergic Inflammation

It is well established that FcεRI initiate allergic reactions when receptor-bound IgE are aggregated by a plurivalent allergen on mast cells or basophils. FcεRI-deficient mice indeed fail to develop IgE-induced passive systemic anaphylaxis (PSA) or as passive cutaneous anaphylaxis (PCA) (Dombrowicz et al. 1993), and mast cells are critical for both reactions. Human FcεRI can also trigger PCA and PSA (Dombrowicz et al. 1996) induced by human IgE in transgenic mice. FcεRI expressed by eosinophils, monocytes, alveolar macrophages, neutrophils, and platelets in patients with high IgE levels may increase allergic symptoms.

Although it has been known since the 1950s, IgG antibodies, especially of the IgG1 subclass, were recently rediscovered to induce PSA in mouse models. FcγRIIIA were demonstrated to be the main receptors involved in experimental anaphylaxis using FcgammaRIIIA-deficient mice. FcγRIV, however, which bind IgG2 only, were found to contribute to active systemic anaphylaxis (ASA), together with FcγRIIIA (Jönsson et al. 2011). The expression of FcγRIV is restricted to neutrophils and monocytes/macrophages, and both cell types were found to contribute to systemic anaphylaxis. FcγRIIIA, however, play a predominant role in anaphylaxis because they bind not only IgG2 but also IgG1 antibodies that are produced in much higher amounts than IgG2, and because they are expressed by mast cells and basophils (Mancardi et al. 2011). FcγRIIIA-expressing mast cells are indeed mandatory for PCA (Hazenbos et al. 1996).

Unlike the well-established role of FcγR in experimental anaphylaxis, the role of activating FcγR in human allergies is unclear. Recent studies nevertheless demonstrated the ability of human FcγR to induce allergic reactions using transgenic mice (Jönsson et al. 2012). Both human FcγRI and FcγRIIA triggered IgG-induced PSA and ASA. FcγRIIA expressed by mast cells were also responsible for IgG-induced PCA. Interestingly, the transfer of human neutrophils expressing FcγRIIA could restore anaphylaxis in resistant mice, suggesting that not only human FcγRIIA, but also human neutrophils, can trigger IgG-induced anaphylaxis (Jönsson et al. 2011).

FcR as Inducers of Autoimmune Inflammation

Most pathogenic autoantibodies are of the IgG class, and all activating FcγR contribute to the development of most antibody-dependent autoimmune diseases. Activating FcR are expressed by mouse monocytes/macrophages and by neutrophils. Macrophages and neutrophils are mandatory for autoimmune diseases such as arthritis, lung inflammation, and thrombocytopenia.

Antiplatelet-induced idiopathic thrombocytopenic purpura (ITP) was prevented in FcRγ-deficient mice. FcγRI, FcγRIIIA, and FcγRIV were demonstrated to contribute to platelet depletion. Each of these receptors was sufficient to induce ITP, suggesting a redundant role of activating FcγR (Nimmerjahn and Ravetch 2005).

The involvement of FcγRI, FcγRIIIA, and FcγRIV was also reported in systemic lupus erythematosus, experimental hemolytic anemia, glomerulonephritis, and arthritis. The role of human FcR in experimental autoimmune diseases has been investigated in transgenic mice. Human FcγRIIA induced ITP or arthritis. The expression of human FcγRI in mice lacking activating FcR restored arthritis symptoms. The contribution of IgA and human FcαRI in glomerulonephritis is well established.

FcR are also involved in autoimmune disorders of the central nervous system. Anti-myelin antibodies found in multiple sclerosis and antidopaminergic neurons antibodies found in Parkinson disease are thought to induce inflammation by activating FcR-expressing phagocytic cells. FcRγ-deficient mice displayed less lesions in a model of Parkinson disease and a reduced mortality in a model of ischemic stroke (Komine-Kobayashi et al. 2004).

Fc Receptors as Regulators of Inflammation

Inhibitory FcR contain one *immunoreceptor tyrosine-based inhibition motif* (ITIM). ITIM consist of a single YxxL motif preceded by a loosely conserved often hydrophobic residue at position Y-2. Inhibitory FcR are the members of one family of low-affinity receptors for IgG, referred to as FcγRIIB (Daëron 1997; Ravetch and Bolland 2001). FcγRIIB are expressed by most myeloid cells and by B lymphocytes. NK cells and T cells, which do not express FcγRIIB, express other inhibitory receptors involved in cell-cell interactions. FcγRIIB have a more restricted tissue distribution in humans than in mice.

FcγRIIB are encoded by a single gene that generates two (FcγRIIB1 and FcγRIIB2 in humans) or three (FcγRIIB1, FcγRIIB1', and FcγRIIB2 in mice) isoforms of membrane receptors, by alternative splicing of sequences encoded by the first intracytoplasmic exon. The inhibitory properties of FcγRIIB depend on an ITIM (Daëron et al. 1995), encoded by the third intracytoplasmic exon of the *fcgr2b* gene, present in the intracytoplasmic domain of all murine and human FcγRIIB isoforms. Unlike activating receptors, FcγRIIB trigger no intracellular signal upon aggregation. They trigger negative signals when they are co-aggregated with activating receptors by immune complexes (Daëron et al. 1995). Under these conditions, the ITIM of FcγRIIB is phosphorylated by the same src family tyrosine kinase that phosphorylates ITAM in activating receptors. Phosphorylated FcγRIIB recruit inhibitory molecules that are brought into signalosomes. This renders inhibition signals dominant over activation signals (Daëron and Lesourne 2006).

The aggregation of identical FcR only (homoaggregation) is a rare situation. Even when cells express one type of FcR only (e.g., FcγRIIB in murine B cells or FcγRIIIA in murine NK cells), immune complexes can co-engage FcR with other immunoreceptors (BCR in B cells or NKR on NK cells). Several FcR are co-aggregated when IgG immune complexes interact with cells that co-express several FcγR or with cells that co-express FcR for several classes of antibodies. Heteroaggregation, i.e., the co-aggregation of different types of FcR or the co-aggregation of FcR with other immunoreceptors, is actually a rule, rather than an exception.

The enhancing effect of IgG antibodies in antigen presentation is under the control of FcγRIIB. The co-engagement of FcγRIIB with activating FcR, suppresses immune complex-induced dendritic cell maturation and dampens antigen presentation (Kalergis and Ravetch 2002). Noticeably, FcγRIIB expressed by follicular dendritic cells can be engaged by the Fc portion of antibodies present in immune complexes and prevent them from binding to B cell FcγRIIB. In the absence of FcγRIIB on follicular dendritic cells, immune complexes can co-engage the BCR with FcγRIIB on B cells and inhibit B cell activation.

FcγRIIB are potent inhibitors of acute allergic reactions. FcγRIIB profoundly inhibit IgG1-induced anaphylaxis. FcγRIIB-deficient mice indeed display markedly increased anaphylaxis compared to wild-type mice (Ujike et al. 1999). Importantly, human basophils were recently found to express more FcγRIIB than any other blood cells in normal donors and to control both IgG- and IgE-induced basophil activations. As a consequence, basophils failed to be activated by IgG immune complexes, and IgG immune complexes that co-engaged FcγR with FcεRI on basophils inhibited IgE-dependent basophil activation in all normal donors tested (Cassard et al. 2012).

FcγRIIB-deficient C57BL/6 mice spontaneously develop lupus-like autoimmune diseases when aging. Anti-DNA and anti-chromatin antibodies are found in these mice, which usually succumb at 8 months of age, due to fatal autoimmune glomerulonephritis. Likewise, FcγRIIB-deficient mice had an enhanced disease susceptibility, whereas mice lacking IgG activating FcR were protected in a mouse model of multiple sclerosis. FcγRIIB therefore prevent the outcome of autoimmunity by contributing to peripheral tolerance (Bolland and Ravetch 2000).

Fc Receptors as Therapeutic Targets in Inflammation

Intravenous immunoglobulins (IVIG) have been successfully used in the treatment of autoimmune disorders. IVIG consist of pools of IgG from thousands of normal donors. Initially conceived as a substitutive treatment of immunodeficiencies, IVIG proved efficient in other pathologic conditions. High doses of IVIG have indeed anti-inflammatory effects, and they became an efficient treatment of several autoimmune diseases such as arthritis, idiopathic thrombocytopenic purpura, or systemic lupus erythematosus. This effect can be reproduced with IVIG Fc fragments, suggesting that FcR contribute to the anti-inflammatory effect of IVIG. The therapeutic effect of IVIG can be enhanced by increasing their concentration in sialic acid-rich immunoglobulins. The mechanism underlying this phenomenon remains unclear. Divergent hypotheses point out a role for FcRn, activating FcR or FcγRIIB, which can be saturated, blocked, or upregulated, respectively, by IVIG (Nimmerjahn and Ravetch 2007).

One clinical trial used blocking anti-FcγRI antibody as a treatment of idiopathic thrombocytopenia. This antibody markedly reduced symptoms (Ericson et al. 1996). The disease was however not abolished, suggesting that FcR other than FcγRI participate to pathogenesis. Supporting this conclusion, the administration of anti-FcγRIIIA antibodies to patients had some therapeutic effect.

Therapeutic approaches to allergic disease mainly focus on IgE and/or FcεRI. Omalizumab is a humanized monoclonal antibody directed against the FcεRI-binding site of IgE. It was developed to prevent mast cell and basophil sensitization by IgE. Omalizumab was however found to form IgE-anti-IgE complexes that are rapidly degraded via binding to FcγR, so that serum IgE becomes undetectable (Djukanovic et al. 2004). As the half-life of FcεRI is markedly reduced in the absence of receptor-bound IgE, basophils and mast cells have a reduced FcεRI expression. Omalizumab efficiently reduces the severity of symptoms in allergic diseases such as asthma, seasonal rhinitis, and chronic urticaria. Other experimental therapeutic approaches have been developed, aiming at co-engaging FcεRI or FcεRI-bound IgE with mast cell or basophil FcγRII (Zhu et al. 2002; Tam et al. 2004).

Conclusion

FcR are critical molecules of the immune system as they mediate most biological activities of antibodies. As they are ubiquitously expressed and as antibodies circulate in the blood stream, FcR are involved in a wide array of biological activities in physiology and pathology. These activities concur with homeostasis and protective responses including physiological inflammation. FcR, however, can also generate pathological inflammation. Activating FcR indeed triggers the release of potentially harmful – in some cases, life-threatening – inflammatory mediators and induces destructive cytotoxic mechanisms. The activating properties of FcR are however (or therefore?) tightly controlled by regulatory mechanisms that depend on inhibitory FcR. As a consequence of these antagonistic effects of FcR, physiological inflammation remains inapparent, and immune responses are normally nonpathogenic. These facts taken together identify FcR as valuable therapeutic tools or targets in inflammatory diseases.

Cross-References

▶ Allergic Disorders
▶ Anaphylaxis (Immediate Hypersensitivity): From Old to New Mechanisms
▶ Asthma
▶ Basophils
▶ Dendritic Cells
▶ Immunoglobulin Receptors and Inflammation
▶ Mast Cells
▶ Rheumatoid Arthritis

References

Bolland, S., & Ravetch, J. V. (2000). Spontaneous autoimmune disease in Fc(gamma)RIIB-deficient mice results from strain-specific epistasis. *Immunity, 13*, 277–285.

Cassard, L., Jonsson, F., Arnaud, S., & Daëron, M. (2012). Fcgamma receptors inhibit mouse and human basophil activation. *Journal of Immunology, 189*, 2995–3006.

Daëron, M. (1997). Fc receptor biology. *Annual Review of Immunology, 15*, 203–234.

Daëron, M., & Lesourne, R. (2006). Negative signaling in Fc receptor complexes. *Advances in Immunology, 89*, 39–86.

Daëron, M., Latour, S., Malbec, O., Espinosa, E., Pina, P., Pasmans, S., et al. (1995). The same tyrosine-based inhibition motif, in the intracytoplasmic domain of FcgRIIB, regulates negatively BCR-, TCR-, and FcR-dependent cell activation. *Immunity, 3*, 635–646.

Djukanovic, R., Wilson, S. J., Kraft, M., Jarjour, N. N., Steel, M., Chung, K. F., et al. (2004). Effects of treatment with anti-immunoglobulin E antibody omalizumab on airway inflammation in allergic asthma. *American Journal of Respiratory and Critical Care Medicine, 170*, 583–593.

Dombrowicz, D., Flamand, V., Brigman, K. K., Koller, B. H., & Kinet, J. P. (1993). Abolition of anaphylaxis by targeted disruption of the high affinity immunoglobulin E receptor alpha chain gene. *Cell, 75*, 969–976.

Dombrowicz, D., Brini, A. T., Flamand, V., Hicks, E., Snouwaert, J. N., Kinet, J. P., et al. (1996). Anaphylaxis mediated through a humanized high affinity IgE receptor. *Journal of Immunology, 157*, 1645–1651.

Ericson, S. G., Coleman, K. D., Wardwell, K., Baker, S., Fanger, M. W., Guyre, P. M., et al. (1996). Monoclonal antibody 197 (anti-Fc gamma RI) infusion in a patient with immune thrombocytopenia purpura (ITP) results in down-modulation of Fc gamma RI on circulating monocytes. *British Journal of Haematology, 92*, 718–724.

Hazenbos, W. L., Gessner, J. E., Hofhuis, F. M., Kuipers, H., Meyer, D., Heijnen, I. A., et al. (1996). Impaired IgG-dependent anaphylaxis and Arthus reaction in Fc gamma RIII (CD16) deficient mice. *Immunity, 5*, 181–188.

Heyman, B. (1990). The immune complex: Possible ways of regulating the antibody response. *Immunology Today, 11*, 310–313.

Hjelm, F., Carlsson, F., Getahun, A., & Heyman, B. (2006). Antibody-mediated regulation of the immune response. *Scandinavian Journal of Immunology, 64*, 177–184.

Hulett, M. D., & Hogarth, P. M. (1994). Molecular basis of Fc receptor function. *Advances in Immunology, 57*, 1–127.

Jonsson, F., Mancardi, D. A., Kita, Y., Karasuyama, H., Iannascoli, B., Van Rooijen, N., et al. (2011). Mouse and human neutrophils induce anaphylaxis. *Journal of Clinical Investigation, 121*, 1484–1496.

Jonsson, F., Mancardi, D. A., Zhao, W., Kita, Y., Iannascoli, B., Khun, H., et al. (2012). Human FcgammaRIIA induces anaphylactic and allergic reactions. *Blood, 119*, 2533–2544.

Kalergis, A. M., & Ravetch, J. V. (2002). Inducing tumor immunity through the selective engagement of activating Fcgamma receptors on dendritic cells. *Journal of Experimental Medicine, 195*, 1653–1659.

Komine-Kobayashi, M., Chou, N., Mochizuki, H., Nakao, A., Mizuno, Y., & Urabe, T. (2004). Dual role of Fcgamma receptor in transient focal cerebral ischemia in mice. *Stroke, 35*, 958–963.

Mancardi, D. A., Jonsson, F., Iannascoli, B., Khun, H., Van Rooijen, N., Huerre, M., et al. (2011). Cutting edge: The murine high-affinity IgG receptor FcgammaRIV is sufficient for autoantibody-induced arthritis. *Journal of Immunology, 186*, 1899–1903.

Nimmerjahn, F., & Ravetch, J. V. (2005). Divergent immunoglobulin g subclass activity through selective Fc receptor binding. *Science, 310*, 1510–1512.

Nimmerjahn, F., & Ravetch, J. V. (2007). Fc-receptors as regulators of immunity. *Advances in Immunology, 96*, 179–204.

Ravetch, J. V., & Bolland, S. (2001). IgG Fc receptors. *Annual Review of Immunology, 19*, 275–290.

Takai, T., Li, M., Sylvestre, D., Clynes, R., & Ravetch, J. V. (1994). FcR gamma chain deletion results in pleiotrophic effector cell defects. *Cell, 76*, 519–529.

Tam, S. W., Demissie, S., Thomas, D., & Daëron, M. (2004). A bispecific antibody against human IgE and human FcgammaRII that inhibits antigen-induced histamine release by human mast cells and basophils. *Allergy, 59*, 772–780.

Tew, J. G., Wu, J., Fakher, M., Szakal, A. K., & Qin, D. (2001). Follicular dendritic cells: Beyond the necessity of T-cell help. *Trends in Immunology, 22*, 361–367.

Ujike, A., Ishikawa, Y., Ono, M., Yuasa, T., Yoshino, T., Fukumoto, M., et al. (1999). Modulation of immunoglobulin (Ig)E-mediated systemic anaphylaxis by low-affinity Fc receptors for IgG. *Journal of Experimental Medicine, 189*, 1573–1579.

Wu, Y., Sukumar, S., EL Shikh, M. E., Best, A. M., Szakal, A. K., & Tew, J. G. (2008). Immune complex-bearing follicular dendritic cells deliver a late antigenic signal that promotes somatic hypermutation. *Journal of Immunology, 180*, 281–290.

Zhu, D., Kepley, C. L., Zhang, M., Zhang, K., & Saxon, A. (2002). A novel human immunoglobulin Fc gamma Fc epsilon bifunctional fusion protein inhibits Fc epsilon RI-mediated degranulation. *Nature Medicine, 8*, 518–521.

Inflammasomes

Makoto Inoue[1], Masashi Kanayama[1] and Mari L. Shinohara[1,2]
[1]Department of Immunology, Duke University School of Medicine, Durham, NC, USA
[2]Department of Molecular Genetics and Microbiology, Duke University School of Medicine, Durham, NC, USA

Definition

Inflammasomes are multiprotein oligomer complexes that control activation of the proteolytic enzyme caspase-1. If the inflammasome includes NLRP3 as a component, it is called the NLRP3 inflammasome. In addition to NLRP3, other proteins, such as NLRC4, AIM2, or NLRP6, form other types of inflammasome.

Biosynthesis and Release

Components of Inflammasomes

Inflammasomes are cytosolic sensors, which mature and secrete the pro-inflammatory cytokines, interleukin (IL)-1β, and IL-18. Inflammasomes are expressed mainly in phagocytes, such as macrophages and dendritic cells (DCs), and form a multiprotein complex. Each inflammasome is termed by a protein included in the complex, such as NLRP3 [pyrin domain (PYD) containing 3] (also termed cryopyrin, CIAS1, NALP3), NLRP1 [NLR family, PYD containing 1], NLRC4 [NLR family, caspase recruitment domain (CARD) domain containing 4] (also termed IRAF), NLRP6 [NLR family, PYD containing 6] (also termed Pypaf5), NLRP12 [NLR family, PYD containing 12], and AIM2 (absent in melanoma 2). Except for AIM2, these proteins belong to the NLR family. Each inflammasome consists of distinct components,

---

Makoto Inouea and Masashi Kanayama equally contributed first authors.

but all of the inflammasomes include pro-caspase-1. Active caspase-1 processes the maturation of IL-1β and IL-18 and elicits rapid release of these inflammatory cytokines by cell death, termed pyroptosis, which is characterized by cytoplasmic swelling and early plasma membrane rupture.

NLRP3 is the most comprehensively studied NLR family to form an inflammasome complex with ASC [apoptosis-associated speck-like protein containing a CARD] and pro-caspase-1. Human CARDINAL [CARD inhibitor of NFkB-activating ligands], which does not have a mouse homolog, is known to be involved in the human NLRP3 inflammasome. However, a functional role of CARDINAL in the NLRP3 inflammasome is not clear, because CARDINAL is dispensable for IL-1β production in human cells (Allen et al. 2009). NLRP3 is auto-repressed by an internal interaction between the NACHT [neuronal apoptosis inhibitory protein (*NA*IP), MHC class II transcription activator (*C*IITA), incompatibility locus protein from *Podospora anserina* (*H*ET-E), and telomerase-associated protein (*T*P1)] domain and leucine-rich repeat (LRR) when it is not activated. When activated, NLRP3 changes conformation and interacts with ASC through its PYD, followed by further interaction between ASC and pro-caspase-1 through CARD. Oligomerization of the inflammasome heterotrimer unit leads to pro-caspase-1 self-cleavage to generate activated caspase-1. Pro-caspase-1 in other inflammasomes is similarly activated through the oligomerization of inflammasome units.

NLRP1 protein differs structurally from other inflammasome NLRs and associates with ASC, pro-caspase-1, and pro-caspase-5. Human NLRP1 protein contains five domains, PYD, NACHT, LRR, FIND, and CARD, while mouse NLRP1 lacks PYD (Moayeri et al. 2012). Thus, human NLRP1 is different from rodent NLRP1 in possessing an ASC-interacting pyrin domain.

NLRC4 contains CARD, NACHT, and LRR domains, and it associates with NAIPs (NAIP2, 5, 6), which are receptors for bacteria, to recruit pro-caspase-1 for the formation of the NLRC4 inflammasome. Although ASC seems not to be structurally required for the NLRC4 inflammasome formation, some studies have demonstrated that ASC may facilitate NLRC4-mediated caspase-1 recruitment.

Both NLRP6 and NLRP12 proteins are structurally similar to NLRP3 protein and are associated with ASC and pro-caspase-1 to form inflammasomes to generate mature IL-1β and IL-18. In addition, NLRP6 and NLRP12 proteins can additionally trigger the activation and attenuation of NFκB signaling, respectively (Grenier et al. 2002; Lich et al. 2007).

AIM2 consists of PYD and HIN200 [hemopoietic interferon (IFN)-inducible nuclear proteins with a 200 amino acid] domain, which detects cytosolic DNA, and requires ASC to recruit pro-caspase-1 and the formation of the AIM2 inflammasome.

Activation of Inflammasomes

Inflammasomes are activated by pathogen-associated molecular patterns (PAMPs), as well as damage-associated molecular patterns (DAMPs). NLRP3 inflammasome senses various pathogens and damage-associated molecules, including viruses (influenza A virus, Sendai virus, adenovirus), bacteria (*Staphylococcus aureus*, *Listeria monocytogenes*, *Salmonella typhimurium*), fungi (*Candida albicans*), extracellular ATP, amyloid β, uric acid, monosodium urate, islet amyloid polypeptide (IAPP), oxidized low-density lipoprotein (oxLDL), hyaluronan, as well as various environmental irritants, such as silica, asbestos, and alum. Studies showed that these stimuli activate the NLRP3 inflammasome through production of reactive oxygen species (ROS) by mitochondria, release of mitochondrial DNA (mtDNA), leakage of cathepsins by disrupted lysosomal membrane, or the efflux of potassium by the loss of cell membrane integrity (Levine and Elazar 2011; Zhou et al. 2011). Microtubules mediate the assembly of the NLRP3 inflammasome and are thus required for NLRP3 activation (Misawa et al. 2013). Mitochondrial antiviral-signaling protein (MAVS) recruits NLRP3 to the mitochondrial surface for NLRP3 inflammasome activation (Subramanian et al. 2013).

NLRP1 inflammasome is known to be activated by peptidoglycan component muramyl dipeptide (MDP) and the anthrax lethal factor of *Bacillus anthracis*. The anthrax lethal factor activates capsase-1 and induces NLRP1-dependent cell death, but how it triggers the NLRP1 inflammasome is unclear.

NLRC4 inflammasome is activated by the gram-negative bacteria S*almonella typhimurium*, *Legionella pneumophila*, *Shigella flexneri*, and *Pseudomonas aeruginosa*. These bacteria have either a type III secretion system (T3SS) or type IV secretion system (T4SS). Flagellin from several bacterial binds to NAIP5/6, which subsequently binds NLRC4 to recruit pro-caspase-1 and inflammasome formation, while PrgJ binds to NAIP2.

Specific ligands for activation of NLRP6 and NLRP12 are currently unknown.

Negative Regulation of Inflammasome

Inflammatory responses need to be well controlled, because excessive inflammation is harmful. In fact, several inhibitory pathways against inflammasome activity are reported. For example, NLRP3 inflammasome activation is inhibited by type I IFNs (IFN-I), autophagy, nitric oxide (NO), T cells, and components of bacterial and viral pathogens.

IFN-I inhibits NLRP3 inflammasome activity through IFN-I receptor (IFNAR) (Guarda et al. 2011; Inoue et al. 2012b). To achieve inhibition of NLRP3 inflammasome activity, IFNAR signaling upregulates suppressor of cytokine signaling-1 (SOCS-1), leading to ubiquitination and degradation of active Rac1. Consequently, the downregulation of active Rac-1 decreases ROS generation by mitochondria and downregulates NLRP3 inflammasome activity (Inoue et al. 2012b). Autophagy also inhibits NLRP3 inflammasome activity via elimination of damaged mitochondria, which is a source of ROS and mtDNA (Nakahira et al. 2011). NO is known to induce the S-nitrosylation of NLRP3 and prevents the interaction of the ASC adaptor complex to NLRP3 (Mishra et al. 2013). On the other hand, T cells can suppress IL-1β production in macrophage and DCs by downregulating NLRP3 inflammasome activity via cell-cell contact and function of CD40L (CD154), OX40L (CD252), and RANKL [receptor activator of nuclear factor-κB ligand] (CD254) on T cells, although the molecular mechanism by which T cells suppress NLRP3 inflammasome activity is still unclear (Guarda et al. 2009). T-cell-derived IFNγ is known to suppress NLRP3 inflammasome-mediated IL-1β production via NO production (Mishra et al. 2013). The components of bacteria and viruses are also known to inhibit inflammasome (NLRP3, NLRP1, and AIM2)-mediated IL-1β production via several pathways, such as preventing inflammasome formation through decoy proteins for NLR, ASC, or pro-caspase-1, suppression of caspase-1 enzyme activity, and inhibition of NF-κB translocation and transcription of the gene encoding IL-1β (see the "Modulation by Drugs" section).

Biological Activities

Caspase-1 Activation

Formation of an inflammasome complex leads to self-cleavage of inactive 45 kDa pro-caspase-1 to active caspase-1, which consists of two polypeptides of 20 and 10 kDa (p20 and p10) with a 1:1 ratio. p20 and p10 peptides are generated from a single pro-caspase-1 protein by cleaving at Asp^{119}-Asn^{120}, Asp^{297}-Ser^{298}, and Asp^{316}-Asn^{317} (Yamin et al. 1996). In addition, it has been reported that caspase-11 is also required for caspase-1 activation in macrophages during infections with *Escherichia coli*, *Citrobacter rodentium*, or *Vibrio cholera* (Kayagaki et al. 2011).

Maturation of IL-1β and IL-18

Active caspase-1 induces the maturation of IL-1β and IL-18 by proteolysis. Active and mature IL-1β (17.5 kDa) is formed by a cleavage of inactive 31-kDa precursor at Asp^{116}-Ala^{117}. IL-18 is synthesized as an inactive 24-kDa precursor, which is structurally similar to IL-1β. Active 18-kDa peptide is generated by proteolytic cleavage by caspase-1.

Induction of Pyroptosis

Activation of inflammasome results in pyroptosis, a form of caspase-1-dependent cell death, which is distinct from apoptosis and necrosis. Caspase-1 activation leads to rapid formation of plasma membrane pores, which allow osmotic cell lysis. Pyroptosis is considered to be a mechanism to release IL-1β and IL-18 to the extracellular space, although the exact mechanism of secretion remains unknown. In addition, pyroptosis enhances elimination of intracellular *Salmonella* by releasing the bacterium from macrophages and induces neutrophil-mediated killing of the bacterium (Miao et al. 2010a). However, pyroptosis exacerbates tissue injury. For example, pyroptosis mediated by NLRP1 and NLRP3 inflammasomes causes lethal toxin-induced lung injury (Kovarova et al. 2012) and hepatic fibrosis (Wree et al. 2013), respectively.

Protection from Microbial Infections

Inflammasomes are activated during infection with various microorganisms and enhances antimicrobial immunity to protect hosts. For example, mutant mice lacking NLRP3 inflammasome components (NLRP3, ASC, or caspase-1) show increased susceptibility to bacterial, viral, parasitic, and fungal infections (Table 1).

In bacterial infections, the inflammasome is activated by bacterial products such as bacterial DNA and toxins (Table 1). Because *Casp1* and *Casp11* loci are very close in the genome, "*Casp1*$^{-/-}$ mice" in most published studies prior to 2011 were actually double knockout of *Casp1* and *Casp11* genes. A study showed that *Casp11* single-knockout mice have increased susceptibility to *S. typhimurium* (Broz et al. 2012), suggesting the previously unknown host protective role of caspase-11. *E. coli* lipopolysaccharide (LPS) or *S. typhimurium* LPS activates caspase-11, independently of the LPS receptor toll-like receptor (TLR) 4 (Kayagaki et al. 2011).

In viral infections, viral nucleotides are capable of inducing activation of NLRP3 and AIM2 inflammasomes. In influenza A virus infection, the pH gradient caused by viral M2 protein, a proton-selective ion channel critical for the replication and pathogenesis of influenza A virus, is required for activation of the NLRP3 inflammasome (Ichinohe et al. 2010) (Table 1).

The NLRP3 inflammasome plays a role in host protection also in parasitic infections. In *Leishmania amazonensis* infection, deficiency of *Nlrp3*, *Pycard* (*Asc*), or *Casp1* results in an increase in parasite load in vivo. IL-1β processed by the NLRP3 inflammasome enhances NO production in macrophages and inhibits *Leishmania* multiplication (Lima-Junior et al. 2013) (Table 1).

In fungal infections, signal pathways through TLR2 and dectin-1, a β-glucan receptor, are required for inflammasome activation (Table 1). Recognition of *Candida albicans* by TLR2 and dectin-1 triggers the activation of NLRP3, NLRC4, and a non-canonical caspase-8 inflammasomes. Inflammasome-mediated IL-1β production increases T helper 17 (Th17) cells and enhances neutrophil recruitment to clear fungi. The protective roles of NLRP3 and NLRC4 inflammasomes were demonstrated by in vivo experiments using $Tlr2^{-/-}$, $Clec7a^{-/-}$ (dectin-1-deficient), $Nlrp3^{-/-}$, $Nlrc4^{-/-}$, $Pycard^{-/-}$, $Casp1^{-/-}$, and $Il1r1^{-/-}$ mice in disseminated and oral candidiasis (Table 1).

On the other hand, in some cases, excessive activation of inflammasomes leads to hyperinflammation and collateral damage to host tissues. The NLRP3 inflammasome is responsible for the release of high-mobility group box 1 (HMGB1), which is a factor that exacerbates endotoxemia. In *Plasmodium* (malaria) infection, $Nlrp3^{-/-}$ mice showed reduced fever and decreased susceptibility, compared to wild-type mice (Shio et al. 2009) (Table 1), suggesting a benefit from the absence of inflammasome-mediated inflammation. In influenza A virus infection, inhibition of NLRP3 inflammasome activation by the mitophagy-mediated removal of mitochondria prevents tissue damage (Lupfer et al. 2013) (Table 1). Thus, inflammasomes can be a double-edged sword for hosts in microbial infections as well.

Pathophysiological Relevance

Inflammasomes appear to correlate with the pathology of autoinflammatory, autoimmune, and sterile inflammatory diseases. Table 2 lists

Inflammasomes, Table 1 The role of inflammasomes in bacterial, viral, parasitic, and fungal infections

Pathogen	Activator	Induced inflammasome	Relationship to elimination of pathogens in vivo and survival of host	References
Bacteria				
Bacillus anthracis	Lethal toxin	NALP1b/ NLRP1b	Mice expressing a lethal toxin-sensitive allele of Nlrp1b showed increased resistance to *B. anthracis* infection	(Boyden et al. 2006; Terra et al. 2010; Nour et al. 2009)
Burkholderia pseudomallei	BsaK	NLRP3, NLRC4	Deficiency of *Nlrp3, Nlrc4, Pycard, Casp1, Il18* enhances mortality against *B. pseudomallei* infection	(Miao et al. 2010b; Ceballos-Olvera et al. 2011)
Escherichia coli	mRNA, heat-labile enterotoxin	NLRP3, caspase-11	*Pycard$^{-/-}$, Nlrp3$^{-/-}$* mice show increased bacterial burden in spleen	(Rathinam et al. 2012; Sander et al. 2011; Brereton et al. 2011; Kayagaki et al. 2011)
Francisella tularensis	Bacterial DNA	AIM2	*Aim2$^{-/-}$* mice show higher mortality in *F. tularensis* infection	(Broz et al. 2010a; Rathinam et al. 2010; Fernandes-Alnemri et al. 2010)
Group B streptococcus	β-hemolysin	NLRP3	Mice lacking caspase-1, ASC, and NLP3 are highly susceptible to GBS infection	(Costa et al. 2012)
Legionella pneumophila	flagellin	NLRC4/ NAIP5	Mice lacking NLRC4 and caspase-1 increases bacterial burden	(Zhao et al. 2011; Akhter et al. 2009; von Moltke et al. 2012; Zamboni et al. 2006; Ren et al. 2006; Amer et al. 2006; Lightfield et al. 2008; Lightfield et al. 2011; Kofoed et al. 2011)
Listeria monocytogenes	LLO, flagellin, bacterial DNA	NLRP3, NLRC4, AIM2	Defect of NLRP3 inflammasome activation by guanylate-binding protein 5 (GBP5) deficiency increases bacterial burden and weight loss	(Franchi et al. 2007; Meixenberger et al. 2010; Mariathasan et al. 2006; Warren et al. 2008; Wu et al. 2010; Tsuchiya et al. 2010; Sauer et al. 2011; Kim et al. 2010; Shenoy et al. 2012)
Mycobacterium tuberculosis	DNA, ESX-1, ESAT-6	NLRP3, NLRC4, AIM2	*Aim2* deficiency increases susceptibility against *M. tuberculosis* infection	(Saiga et al. 2012; Mishra et al. 2010; Mayer-Barber et al. 2010; Master et al. 2008; Koo et al. 2008)

(*continued*)

Inflammasomes, Table 1 (continued)

Pathogen	Activator	Induced inflammasome	Relationship to elimination of pathogens in vivo and survival of host	References
Salmonella typhimurium	PrgJ, T3SS, flagellin	NLRC4, NLRP3, caspase-11	Deficiency of *Casp1*, *Nlrc4* and double knockout of *Nlrp3* and *Nlrc4* increases bacterial burden	(Miao et al. 2010b; Zhao et al. 2011; Kofoed et al. 2011; Miao et al. 2006; Franchi et al. 2006; Broz et al. 2010b; Franchi et al. 2012; Qu et al. 2012)
			Nlrc4 deficiency increases mortality against orogastric infection of *S. typhimurium*	
			Caspase-11 increases susceptibility to *Salmonella* infection in the absence of caspase-1	
Staphylococcus aureus	α-hemolysin	NLRP3	*Nlrp3*$^{-/-}$ mice show lower mortality when heat-killed *S. aureus* and alpha-hemolysin are intratracheally injected	(Miller et al. 2007; Kebaier et al. 2012; Munoz-Planillo et al. 2009)
Streptococcus pneumoniae	Pneumolysin	NLRP3	*Nlrp3* deficiency increases bacterial burden	(Witzenrath et al. 2011; McNeela et al. 2010)
Vibrio cholera	Cholera toxin B, HlyA, MARTXvc	NLRP3	Not determined	(Kayagaki et al. 2011; Toma et al. 2010)
Yersinia pestis	T3SS, YopJ	NLRP3, NLRC4, NLRP12	Deficiency of *Il1beta*, *Il1r1*, *Il18*, *Nlrp3*, and *Nlrp12* increases susceptibility in *Y. pestis* infection	(Brodsky et al. 2010; Zheng et al. 2011; Vladimer et al. 2012)
			Deficiency of *Casp1* and *Pycard* increases bacterial burden	
Virus				
Adenovirus	Not determined	NLRP3	Not determined	(Muruve et al. 2008; Barlan et al. 2011)
Cytomegalovirus	Viral double-strand DNA	AIM2	Deficiency of *Pycard* and *Aim2* increases virus titer in the spleen	(Rathinam et al. 2010)

Herpes virus	Not determined	NLRP3, AIM2	Not determined	(Nour et al. 2011; Johnson et al. 2013)
Influenza A virus	M2 channel, viral RNA, PB1-F2	NLRP3	Deficiency of *Nlrp3*, *Pycard*, and *Casp1* increases susceptibility against influenza A virus infection	(Ichinohe et al. 2010; Thomas et al. 2009; Allen et al. 2009; Lupfer et al. 2013; McAuley et al. 2013)
Sendai virus	Not determined	NLRP3	Not determined	(Kanneganti et al. 2006)
Vaccinia virus	Viral double-strand DNA	AIM2	Not determined	(Rathinam et al. 2010; Hornung et al. 2009)
Parasite				
Leishmania amazonensis	Not determined	NLRP3	The deficiency of *Nlrp3*, *Pycard*, and *Casp1* results in the increase of parasite load	(Lima-Junior et al. 2013)
Plasmodium chabaudi	Hemozoin	NLRP3	Deficiency of *Nlrp3* and *Il1b* decreases the susceptibility in *Plasmodium* infection	(Shio et al. 2009)
Fungi				
Aspergillus fumigatus	β-glucan	NLRP3, caspase-8	Not determined	(Said-Sadier et al. 2010; Gringhuis et al. 2012)
Candida albicans	β-glucan	NLRP3, NLRC4	Lower susceptibility in *Dectin-1*$^{-/-}$, *Nlrp3*$^{-/-}$, *Asc*$^{-/-}$, *Casp1*$^{-/-}$, and *Il1r1*$^{-/-}$ mice	(Gringhuis et al. 2012; Gross et al. 2009; Hise et al. 2009; Tomalka et al. 2011; Joly et al. 2009)
Cryptococcus neoformans	Not determined	NLRP3	Deficiency of *Nlrp3* and *Pycard* increases the susceptibility in *C. neoformans* infection	(Lei et al. 2013)

Inflammasomes, Table 2 Involvement of inflammasomes in autoinflammatory, autoimmune, and inflammatory diseases

Disease	Inflammasome	Human	References	Animals	References
Autoinflammatory disease					
MWS	NLRP3	*Nlrp3* mutation	(Arostegui et al. 2004)	Systemic inflammation in *Nlrp3* mutant Tg mice	(Brydges et al. 2009)
FCAS	NLRP3	*Nlrp3* mutation	(Dode et al. 2002)	Not determined	
CINCA	NLRP3	*Nlrp3* mutation	(Feldmann et al. 2002)	Not determined	
NOMID	NLRP3	*Nlrp3* mutation	(Hoffman et al. 2001)	Systemic inflammation in *Nlrp3* mutant Tg mice	(Meng et al. 2009)
Autoimmune disease					
SLE	NLRP3, AIM2	*Nlrp3* SNPs	(Pontillo et al. 2012)	AIM2 expression up	(Zhang et al. 2013)
		Aim2 expression up	(Zhang et al. 2013)	NLRP3 activation up	(Kahlenberg et al. 2013)
Type 1 diabetes	NLRP1, NLRP3	*Nlrp1, Nlrp3* SNPs	(Magitta et al. 2009; Pontillo et al. 2010)	Not determined	
Celiac disease	NLRP1, NLRP3	*Nlrp1, Nlrp3* SNPs	(Pontillo et al. 2010)	Not determined	
Rheumatoid arthritis (RA)	NLRP1, NLRP3	*Nlrp1, Nlrp3* SNPs	(Sui et al. 2012; Kastbom et al. 2008)	Not determined	
		Nlrp3 expression up	(Mathews et al. 2013)		
Psoriasis	NLRP3, AIM2	*Nlrp3* SNPs	(Carlstrom et al. 2012; de Koning et al. 2012)	Not determined	
		Aim2 expression up	(Dombrowski et al. 2011)		
Vitiligo	NLRP1	*Nlrp1* SNPs	(Jin et al. 2007)	Not determined	
Addison's disease	NLRP1	*Nlrp1* SNPs	(Magitta et al. 2009)	Not determined	
Type 2 diabetes	NLRP3	*Nlrp3* expression up	(Lee et al. 2013)	Improvement of glucose tolerance and insulin sensitivity	(Zhou et al. 2010; Vandanmagsar et al. 2011; Youm et al. 2011)
Obesity					
Multiple sclerosis (MS)	NLRP1, NLRP3	Not determined		Lack of EAE disease in $Nlrp3^{-/-}$ mice	(Gris et al. 2010; Jha et al. 2010; Inoue et al. 2012a, b)
				Nlrp1 expression up in EAE	(Soulika et al. 2009)

(*continued*)

Inflammasomes, Table 2 (continued)

Disease	Inflammasome	Human	References	Animals	References
Inflammatory disease					
Crohn's disease	NLRP3, NLRP6	*Nlrp3* SNPs	(Villani et al. 2009)	Increase in chemical-induced colitis in *Nlrp3$^{-/-}$* mice	(Allen et al. 2010; Hirota et al. 2011; Zaki et al. 2010)
				Exacerbation of chemical colitis in *Nlrp6$^{-/-}$* mice	(Elinav et al. 2011)
Alzheimer's disease	NLRP3	Not determined		Reduction of spatial memory in APP/PS1/ *Nlrp3$^{-/-}$* mice	(Heneka et al. 2013)
Atherosclerosis	NLRP3	NLRP3 expression up	(Zheng et al. 2013)	Reduction of crystal-induced atherosclerosis in *Nlrp3$^{-/-}$* mice	(Duewell et al. 2010)

such diseases that were studied in the context of inflammasomes both in animals and humans, and details are described below.

Autoinflammatory Diseases

Mutations in the human *Nlrp3*, encoding NLRP3, were found to be associated with cryopyrin-associated periodic syndromes (CAPSs). The mutations generate constitutively active forms of NLRP3, which lead to increased activation of the NLRP3 inflammasome and IL-1β production. Over 50 different mutations in the *Nlrp3* locus have been identified in CAPS. Several CAPS subtypes include Muckle-Wells syndrome (MWS), familial cold-induced autoinflammatory syndrome (FCAS), chronic infantile neurological cutaneous and articular (CINCA) syndrome, and neonatal-onset multisystem inflammatory disease (NOMID) syndrome. Mutant mice expressing the *Nlrp3* gene mutations, associated with MWS or FCAS, also showed hyperactivation of the NLRP3 inflammasome and IL-1β production.

Autoimmune Diseases

Single-nucleotide polymorphism (SNP) analysis of the *Nlrp1* locus suggested that the NLRP1 inflammasome is associated with autoimmune diseases, such as vitiligo, rheumatoid arthritis (RA), and Addison's disease. However, animal studies for these diseases have yet to be performed to evaluate how the NLRP1 inflammasome is involved in the diseases. SNPs within the *Nlrp3* locus predispose to various human autoimmune diseases, such as systemic lupus erythematosus (SLE), type-1 diabetes, celiac disease, RA, and psoriasis. For example, RA patients showed significantly elevated gene expression of NLRP3-inflammasome components (Mathews et al. 2013). On the other hand, a strong upregulation of epidermal AIM2 protein expression is observed in psoriasis and venous ulcera (de Koning et al. 2012). Cytosolic DNA is an important psoriasis-associated molecule and triggers AIM2 inflammasome activation in psoriasis. Although association of AIM2 with SLE is not identified by SNP analysis, AIM2 expression is correlated with the severity of disease in SLE patients and in mice with lupus (Zhang et al. 2013). In addition, IL-1β and IL-18 are known to be involved in SLE and lupus. Involvement of the NLRP3 inflammasome in type-2 diabetes and obesity is suggested by multiple studies. For example, IAPP, a hallmark of type-2 diabetes, is also known to activate the NLRP3 inflammasome (Masters et al. 2010). These studies strongly implicate the link between inflammasome-mediated inflammation and metabolic disorders.

Association of the NLRP3 inflammasome with multiple sclerosis (MS) has not been identified by SNP analysis to date, but a number of reports suggest the involvement of the NLRP3 inflammasome in the development of MS. Increased levels of caspase-1, IL-1β, IL-18, and activators of the NLRP3 inflammasome (extracellular ATP, uric acid, cathepsin B) have been reported in MS patients. In addition, signaling from P2X7R, which detects extracellular ATP, is enhanced in normally appearing axonal tracts of the CNS in MS patients. The critical role of the NLRP3 inflammasome in experimental autoimmune encephalomyelitis (EAE), an animal model for MS, has become clear (reviewed in Inoue and Shinohara (2013a, b)). $Nlrp3^{-/-}$ mice were characterized as being resistant to EAE and exhibited a reduction in both Th1 and Th17 cells in the peripheral lymphoid tissues, as well as in the CNS (Gris et al. 2010; Inoue et al. 2012a). Although attenuated Th17 differentiation was considered to be a major underlying mechanism for the resistance of $Nlrp3^{-/-}$ mice against EAE due to the lack of production of IL-1β, an inducer of Th17 responses, it appears that the inability to migrate T cells and antigen-presenting cells (APCs) to the CNS is a major cause of resistance to EAE in $Nlrp3^{-/-}$ mice (Inoue et al. 2012a). It is of note that there is an NLRP3 inflammasome-independent subset in EAE that is induced by aggressive disease induction methods (Inoue et al. 2012b). Thus, it is possible that the NLRP3 inflammasome is not involved in the development of all kinds of MS, which is a multifactorial and heterogeneous disease. On the other hand, NLRP1 is reported in the context of EAE for its intra-axonal accumulation (Soulika et al. 2009), but involvement of the NLRP1 inflammasome in EAE is not yet known.

Inflammatory Diseases

SNPs within the *Nlrp3* locus predispose to various human inflammatory diseases, such as Crohn's disease and ulcerative colitis. $Nlrp3^{-/-}$ mice are more susceptible to colitis induced by dextran sulfate sodium (DSS) or acute 2,4,6-trinitrobenzene sulfonate (TNBS) (Zaki et al. 2010), suggesting a crucial role of the NLRP3 inflammasome in regulating gut homeostasis.

The NLRP3 inflammasome is also involved in the pathology of neurodegenerative disease such as Alzheimer's disease. Amyloid-β, a trigger of Alzheimer's disease, activates the NLRP3 inflammasome (Halle et al. 2008). The pathogenic role of the NLRP3 inflammasome in Alzheimer's disease was demonstrated in a mouse model using the APP/PS1 strain (Heneka et al. 2013).

Expression of NLRP3 is enhanced in patients with atherosclerosis. Oxidized oxLDL, a hallmark molecule of atherosclerosis, activates the NLRP3 inflammasome. Cholesterol crystals are present in early diet-induced atherosclerotic lesions and activate the NLRP3 inflammasome. In addition, the NLRP3 inflammasome mediates crystal-induced peritoneal inflammation and atherosclerosis using $Nlrp3^{-/-}$ mice. Intracellular conversion of soluble oxLDL and amyloid-β into crystals was shown to disrupt lysosomes and activate the NLRP3 inflammasome (Sheedy et al. 2013).

Modulation by Drugs

Targeting Upstream of Inflammasomes

P2X7R is a receptor for extracellular ATP, which stimulates the NLRP3 inflammasome. Cathepsin B leaking out from lysosomes during "frustrated phagocytosis" contributes to ROS generation and activates the NLRP3 inflammasome. Antagonists against P2X7R (such as A 438079 and JNJ-47965567) and cathepsin B (such as CA-074Me and Ac-LVK-CHO) were tested at the preclinical level, and the therapeutic efficacy of these antagonists in inflammation and tissue injury has been demonstrated (Feng et al. 2013; Hoque et al. 2012). Currently, several ROS inhibitors, such as YCG063 and N-acetyl-L-cysteine (NAC), are available. NAC is widely used in medical treatment for many diseases such as chronic bronchitis. Because ROS strongly induces inflammasome activation, eutralization of ROS by agents such as NAC may serve to suppress inflammasome-mediated inflammation.

Targeting Inflammasomes

Preventing inflammasome formation may be one possible approach to target inflammasomes. Assembly of inflammasomes is based upon interactions through CARD and PYD; therefore, small single-domain proteins such as CARD-only proteins (COPs) and pyrin-only proteins (POPs) interrupt the interactions. POPs inhibit the interaction between ASC and NLRPs, and COPs impair CARD interactions of pro-caspase-1 with ASC. As we described above, a number of bacterial and viral proteins are known to inhibit inflammasome activation. Recombinant proteins of these microbial products may serve as therapeutic drugs for inflammasome-mediated diseases. In poxvirus infection, viral pyrin-only proteins (vPOPs) act as a decoy to interfere with the interaction between ASC and NLR family proteins and thereby impair inflammasome activity. Other microbial inhibitory proteins which interfere with inflammasome activity include myxoma virus Serp2 and M013, Shope fibroma virus gp013L, cowpox virus CrmA, vaccinia virus serpin B13R, influenza virus NS1, baculovirus p35, *P. aeruginosa* ExoU and ExoS, *Y. enterocolitica* YopE and YopT, *M. tuberculosis* Zmp1, and *F. tularensis* MviN and RipA.

On the other hand, small chemical agents targeting caspase-1, including VX-765 and VX-740 (Pralnacasan), have been used in clinical trials for RA and psoriasis. VX-765 is currently in a phase II trial. Because VX-740 causes liver abnormalities in animal toxicological studies, the clinical trial was discontinued.

Targeting Downstream of Inflammasomes

Targeting IL-1β has been practiced in treating autoimmune diseases. Such drugs include those that block IL-1β/IL-1 signaling, such as anakinra (recombinant human IL-1R antagonist), rilonacept (soluble decoy IL-1R), and canakinumab (human anti-IL-1β antibody). The efficacy of IL-1β blockers in the therapy of autoimmune diseases has already been confirmed. In addition, the efficacy of these agents in CAPS, such as FCAS, MWS, and CINCA, has also been reported.

Cross-References

▶ Autoinflammatory Syndromes
▶ Autophagy and Inflammation
▶ Genetic Susceptibility to Inflammatory Diseases
▶ Obesity and Inflammation

References

Akhter, A., et al. (2009). Caspase-7 activation by the Nlrc4/Ipaf inflammasome restricts *Legionella pneumophila* infection. *PLoS Pathogens, 5*, e1000361.

Allen, I. C., Scull, M. A., Moore, C. B., Holl, E. K., McElvania-TeKippe, E., Taxman, D. J., et al. (2009). The NLRP3 inflammasome mediates in vivo innate immunity to influenza A virus through recognition of viral RNA. *Immunity, 30*, 556–565.

Allen, I. C., et al. (2010). The NLRP3 inflammasome functions as a negative regulator of tumorigenesis during colitis-associated cancer. *The Journal of Experimental Medicine, 207*, 1045.

Amer, A., et al. (2006). Regulation of Legionella phagosome maturation and infection through flagellin and host Ipaf. *The Journal of Biological Chemistry, 281*, 35217.

Arostegui, J. I., et al. (2004). Clinical and genetic heterogeneity among Spanish patients with recurrent autoinflammatory syndromes associated with the CIAS1/PYPAF1/NALP3 gene. *Arthritis and Rheumatism, 50*, 4045.

Barlan, A. U., et al. (2011). Adenovirus membrane penetration activates the NLRP3 inflammasome. *Journal of Virology, 85*, 146.

Boyden, E. D., et al. (2006). Nalp1b controls mouse macrophage susceptibility to anthrax lethal toxin. *Nature Genetics, 38*, 240.

Brereton, C. F., et al. (2011). Escherichia coli heat-labile enterotoxin promotes protective Th17 responses against infection by driving innate IL-1 and IL-23 production. *Journal of Immunology, 186*, 5896.

Brodsky, I. E., et al. (2010). A Yersinia effector protein promotes virulence by preventing inflammasome recognition of the type III secretion system. *Cell Host & Microbe, 7*, 376.

Broz, P., et al. (2010a). Differential requirement for Caspase-1 autoproteolysis in pathogen-induced cell death and cytokine processing. *Cell Host & Microbe, 8*, 471.

Broz, P., et al. (2010b). Redundant roles for inflammasome receptors NLRP3 and NLRC4 in host defense against Salmonella. *The Journal of Experimental Medicine, 207*, 1745.

Broz, P., Ruby, T., Belhocine, K., Bouley, D. M., Kayagaki, N., Dixit, V. M., et al. (2012). Caspase-11 increases susceptibility to Salmonella infection in the absence of caspase-1. *Nature, 490*, 288–291.

Brydges, S. D., et al. (2009). Animal models of inflammasomopathies reveal roles for innate but not adaptive immunity. *Immunity, 30*, 875.

Carlstrom, M., et al. (2012). Genetic support for the role of the NLRP3 inflammasome in psoriasis susceptibility. *Experimental Dermatology, 21*, 932.

Ceballos-Olvera, I., et al. (2011). Inflammasome-dependent pyroptosis and IL-18 protect against *Burkholderia pseudomallei* lung infection while IL-1β Is deleterious. *PLoS Pathogens, 7*, e1002452.

Costa, A., et al. (2012). Activation of the NLRP3 inflammasome by group B streptococci. *Journal of Immunology, 188*, 1953.

de Koning, H. D., Bergboer, J. G., van den Bogaard, E. H., van Vlijmen-Willems, I. M., Rodijk-Olthuis, D., Simon, A., Zeeuwen, P. L., & Schalkwijk, J. (2012). Strong induction of AIM2 expression in human epidermis in acute and chronic inflammatory skin conditions. *Experimental Dermatology, 21*, 961–964.

Dode, C., et al. (2002). New mutations of *CIAS1* that are responsible for Muckle–Wells syndrome and familial cold urticaria: A novel mutation underlies both syndromes. *American Journal of Human Genetics, 70*, 1498.

Dombrowski, Y., et al. (2011). Cytosolic DNA triggers inflammasome activation in keratinocytes in psoriatic lesions. *Science Translational Medicine, 3*, 82ra38.

Duewell, P., et al. (2010). NLRP3 inflamasomes are required for atherogenesis and activated by cholesterol crystals that form early in disease. *Nature, 464*, 1357.

Elinav, E., et al. (2011). NLRP6 inflammasome is a regulator of colonic microbial ecology and risk for colitis. *Cell, 145*, 745.

Feldmann, J., et al. (2002). Chronic infantile neurological cutaneous and articular syndrome is caused by mutations in *CIAS1*, a gene highly expressed in polymorphonuclear cells and chondrocytes. *American Journal of Human Genetics, 71*, 198.

Feng, Y., Ni, L., & Wang, Q. (2013). Administration of cathepsin B inhibitor CA-074Me reduces inflammation and apoptosis in polymyositis. *Journal of Dermatological Science, 72*, 158–167.

Fernandes-Alnemri, T., et al. (2010). The AIM2 inflammasome is critical for innate immunity against *Francisella tularensis*. *Nature Immunology, 11*, 385.

Franchi, L., et al. (2006). Cytosolic flagellin requires Ipaf for activation of caspase-1 and interleukin 1beta in salmonella-infected macrophages. *Nature Immunology, 7*, 576.

Franchi, L., et al. (2007). Differential requirement of P2X7 receptor and intracellular K+ for caspase-1 activation induced by intracellular and extracellular bacteria. *The Journal of Biological Chemistry, 282*, 18810.

Franchi, L., et al. (2012). NLRC4-driven interleukin-1β production discriminates between pathogenic and commensal bacteria and promotes host intestinal defense. *Nature Immunology, 13*, 449.

Grenier, J. M., Wang, L., Manji, G. A., Huang, W. J., Al-Garawi, A., Kelly, R., et al. (2002). Functional screening of five PYPAF family members identifies PYPAF5 as a novel regulator of NF-kappaB and caspase-1. *FEBS Letters, 530*, 73–78.

Gringhuis, S. I., et al. (2012). Dectin-1 is an extracellular pathogen sensor for the induction and processing of IL-1β via a noncanonical caspase-8 inflammasome. *Nature Immunology, 13*, 246.

Gris, D., Ye, Z., Iocca, H. A., Wen, H., Craven, R. R., Gris, P., et al. (2010). NLRP3 plays a critical role in the development of experimental autoimmune encephalomyelitis by mediating Th1 and Th17 responses. *Journal of Immunology, 185*, 974–981.

Gross, O., et al. (2009). Syk kinase signalling couples to the Nlrp3 inflammasome for anti-fungal host defence. *Nature, 459*, 433.

Guarda, G., Dostert, C., Staehli, F., Cabalzar, K., Castillo, R., Tardivel, A., et al. (2009). T cells dampen innate immune responses through inhibition of NLRP1 and NLRP3 inflammasomes. *Nature, 460*, 269–273.

Guarda, G., Braun, M., Staehli, F., Tardivel, A., Mattmann, C., Forster, I., et al. (2011). Type I interferon inhibits interleukin-1 production and inflammasome activation. *Immunity, 34*, 213–223.

Halle, A., Hornung, V., Petzold, G. C., Stewart, C. R., Monks, B. G., Reinheckel, T., et al. (2008). The NALP3 inflammasome is involved in the innate immune response to amyloid-beta. *Nature Immunology, 9*, 857–865.

Heneka, M. T., Kummer, M. P., Stutz, A., Delekate, A., Schwartz, S., Vieira-Saecker, A., et al. (2013). NLRP3 is activated in Alzheimer's disease and contributes to pathology in APP/PS1 mice. *Nature, 493*, 674–678.

Hirota, S. A., et al. (2011). The NLRP3 inflammasome plays key role in the regulation of intestinal homeostasis. *Inflammatory Bowel Diseases, 17*, 1359.

Hise, A. G., et al. (2009). An essential role for the NLRP3 inflammasome in host defense against the human fungal pathogen Candida albicans. *Cell Host & Microbe, 5*, 487.

Hoffman, H. M., et al. (2001). Mutation of a new gene encoding a putative pyrin-like protein causes familial cold autoinflammatory syndrome and Muckle–Wells syndrome. *Nature Genetics, 29*, 301.

Hoque, R., Sohail, M. A., Salhanick, S., Malik, A. F., Ghani, A., Robson, S. C., & Mehal, W. Z. (2012). P2X7 receptor-mediated purinergic signaling promotes liver injury in acetaminophen hepatotoxicity in mice. *American Journal of Physiology. Gastrointestinal and Liver Physiology, 302*, G1171–G1179.

Hornung, V., et al. (2009). AIM2 recognizes cytosolic dsDNA and forms a caspase-1 activating inflammasome with ASC. *Nature, 458*, 514.

Ichinohe, T., Pang, I. K., & Iwasaki, A. (2010). Influenza virus activates inflammasomes via its intracellular M2 ion channel. *Nature Immunology, 11*, 404–410.

Inoue, M., & Shinohara, M. L. (2013a). NLRP3 inflammasome and MS/EAE. *Autoimmune Diseases, 2013*, 859145.

Inoue, M., & Shinohara, M. L. (2013b). The role of interferon-beta in the treatment of multiple sclerosis and experimental autoimmune encephalomyelitis – In the perspective of inflammasomes. *Immunology, 139*, 11–18.

Inoue, M., Williams, K. L., Gunn, M. D., & Shinohara, M. L. (2012a). NLRP3 inflammasome induces chemotactic immune cell migration to the CNS in experimental autoimmune encephalomyelitis. *Proceedings of the National Academy of Sciences of the United States of America, 109*, 10480–10485.

Inoue, M., Williams, K. L., Oliver, T., Vandenabeele, P., Rajan, J. V., Miao, E. A., et al. (2012b). Interferon-beta therapy against EAE is effective only when development of the disease depends on the NLRP3 inflammasome. *Science Signaling, 5*, ra38.

Jha, S., et al. (2010). The inflammasome sensor, NLRP3, regulates CNS inflammation and demyelination via caspase-1 and interleukin-18. *Journal of Neuroscience, 30*, 15811.

Jin, Y., et al. (2007). NALP1 in vitiligo-associated multiple autoimmune disease. *The New England Journal of Medicine, 356*, 1216.

Johnson, K. E., et al. (2013). Herpes simplex virus 1 infection induces activation and subsequent inhibition of the IFI16 and NLRP3 inflammasomes. *Journal of Virology, 87*, 5005.

Joly, S., et al. (2009). Cutting edge: Candida albicans hyphae formation triggers activation of the Nlrp3 inflammasome. *Journal of Immunology, 183*, 3578.

Kahlenberg, J. M., et al. (2013). Neutrophil extracellular trap-associated protein activation of the NLRP3 inflammasome is enhanced in lupus macrophages. *Journal of Immunology, 190*, 1217.

Kanneganti, T. D., et al. (2006). Critical role for Cryopyrin/Nalp3 in activation of caspase-1 in response to viral infection and double-stranded RNA. *The Journal of Biological Chemistry, 281*, 36560.

Kastbom, A., et al. (2008). Genetic variation in proteins of the cryopyrin inflammasome influences susceptibility and severity of rheumatoid arthritis (the Swedish TIRA project). *Rheumatology (Oxford), 47*, 415.

Kayagaki, N., Warming, S., Lamkanfi, M., Vande Walle, L., Louie, S., Dong, J., et al. (2011). Non-canonical inflammasome activation targets caspase-11. *Nature, 479*, 117–121.

Kebaier, C., et al. (2012). *Staphylococcus aureus* α-hemolysin mediates virulence in a murine model of severe pneumonia through activation of the NLRP3 inflammasome. *The Journal of Infectious Diseases, 205*, 807.

Kim, S., et al. (2010). Listeria monocytogenes is sensed by the NLRP3 and AIM2 inflammasome. *European Journal of Immunology, 40*, 1545.

Kofoed, E. M., et al. (2011). Innate immune recognition of bacterial ligands by NAIPs dictates inflammasome specificity. *Nature, 477*, 592.

Koo, I. C., et al. (2008). ESX-1-dependent cytolysis in lysosome secretion and inflammasome activation during mycobacterial infection. *Cellular Microbiology, 10*, 1866.

Kovarova, M., Hesker, P. R., Jania, L., Nguyen, M., Snouwaert, J. N., Xiang, Z., et al. (2012). NLRP1-dependent pyroptosis leads to acute lung injury and morbidity in mice. *Journal of Immunology, 189*, 2006–2016.

Lee, H. M., et al. (2013). Upregulated NLRP3 inflammasome activation in patients with type 2 diabetes. *Diabetes, 62*, 194.

Lei, G., et al. (2013). Biofilm from a clinical strain of *Cryptococcus neoformans* activates the NLRP3 inflammasome. *Cell Research, 23*, 965.

Levine, B., & Elazar, Z. (2011). Development. Inheriting maternal mtDNA. *Science, 334*, 1069–1070.

Lich, J. D., Williams, K. L., Moore, C. B., Arthur, J. C., Davis, B. K., Taxman, D. J., et al. (2007). Monarch-1 suppresses non-canonical NF-kappaB activation and p52-dependent chemokine expression in monocytes. *Journal of Immunology, 178*, 1256–1260.

Lightfield, K. L., et al. (2008). Critical function for Naip5 in inflammasome activation by a conserved carboxy-terminal domain of flagellin. *Nature Immunology, 9*, 1171.

Lightfield, K. L., et al. (2011). Differential requirements for NAIP5 in activation of the NLRC4 inflammasome. *Infection and Immunity, 79*, 1606.

Lima-Junior, D. S., Costa, D. L., Carregaro, V., Cunha, L. D., Silva, A. L., Mineo, T. W., et al. (2013). Inflammasome-derived IL-1beta production induces nitric oxide-mediated resistance to Leishmania. *Nature Medicine, 19*, 909–915.

Lupfer, C., Thomas, P. G., Anand, P. K., Vogel, P., Milasta, S., Martinez, J., et al. (2013). Receptor interacting protein kinase 2-mediated mitophagy regulates inflammasome activation during virus infection. *Nature Immunology, 14*, 480–488.

Magitta, N. F., et al. (2009). A coding polymorphism in NALP1 confers risk for autoimmune Addison's disease and type 1 diabetes. *Genes and Immunity, 10*, 120.

Mariathasan, S., et al. (2006). Cryopyrin activates the inflammasome in response to toxins and ATP. *Nature, 440*, 228.

Master, S. S., et al. (2008). Mycobacterium tuberculosis prevents inflammasome activation. *Cell Host & Microbe, 3*, 224.

Masters, S. L., Dunne, A., Subramanian, S. L., Hull, R. L., Tannahill, G. M., Sharp, F. A., et al. (2010). Activation of the NLRP3 inflammasome by islet amyloid polypeptide provides a mechanism for enhanced IL-1beta in type 2 diabetes. *Nature Immunology, 11*, 897–904.

Mathews, R. J., Robinson, J. I., Battellino, M., Wong, C., Taylor, J. C., Eyre, S., et al. (2013). Evidence of NLRP3-inflammasome activation in rheumatoid arthritis (RA); genetic variants within the NLRP3-inflammasome complex in relation to susceptibility to RA and response to anti-TNF treatment. *Annals of the Rheumatic Diseases, 73*, 1202–1210.

Mayer-Barber, K. D., et al. (2010). Caspase-1 independent IL-1beta production is critical for host resistance to mycobacterium tuberculosis and does not require TLR signaling in vivo. *Journal of Immunology, 184*, 3326.

McAuley, J. L., et al. (2013). Activation of the NLRP3 inflammasome by IAV virulence protein PB1-F2 contributes to severe pathophysiology and disease. *PLoS Pathogens, 9*, e1003392.

McNeela, E. A., et al. (2010). Pneumolysin activates the NLRP3 inflammasome and promotes proinflammatory cytokines independently of TLR4. *PLoS Pathogens, 6*, e1001191.

Meixenberger, K., et al. (2010). Listeria monocytogenes-infected human peripheral blood mononuclear cells produce IL-1beta, depending on listeriolysin O and NLRP3. *Journal of Immunology, 184*, 922.

Meng, G., et al. (2009). A NLRP3 mutation causing inflammasome hyperactivation potentiates Th17-dominant immune responses. *Immunity, 30*, 860.

Miao, E. A., et al. (2006). Cytoplasmic flagellin activates caspase-1 and secretion of interleukin 1beta via Ipaf. *Nature Immunology, 7*, 569.

Miao, E. A., Leaf, I. A., Treuting, P. M., Mao, D. P., Dors, M., Sarkar, A., et al. (2010a). Caspase-1-induced pyroptosis is an innate immune effector mechanism against intracellular bacteria. *Nature Immunology, 11*, 1136–1142.

Miao, E. A., et al. (2010b). Innate immune detection of the type III secretion apparatus through the NLRC4 inflammasome. *Proceedings of the National Academy of Sciences of the United States of America, 107*, 3076.

Miller, L. S., et al. (2007). Inflammasome-mediated production of IL-1beta is required for neutrophil recruitment against Staphylococcus aureus in vivo. *Journal of Immunology, 179*, 6933.

Misawa, T., Takahama, M., Kozaki, T., Lee, H., Zou, J., Saitoh, T., et al. (2013). Microtubule-driven spatial arrangement of mitochondria promotes activation of the NLRP3 inflammasome. *Nature Immunology, 14*, 454–460.

Mishra, B. B., et al. (2010). Mycobacterium tuberculosis protein ESAT-6 is a potent activator of the NLRP3/ASC inflammasome. *Cellular Microbiology, 12*, 1046.

Mishra, B. B., Rathinam, V. A., Martens, G. W., Martinot, A. J., Kornfeld, H., Fitzgerald, K. A., et al. (2013). Nitric oxide controls the immunopathology of tuberculosis by inhibiting NLRP3 inflammasome-dependent processing of IL-1beta. *Nature Immunology, 14*, 52–60.

Moayeri, M., Sastalla, I., & Leppla, S. H. (2012). Anthrax and the inflammasome. *Microbes and Infection/Institut Pasteur, 14*, 392–400.

Munoz-Planillo, R., et al. (2009). A critical role for hemolysins and bacterial lipoproteins in Staphylococcus aureus-induced activation of the Nlrp3 inflammasome. *Journal of Immunology, 183*, 3942.

Muruve, D. A., et al. (2008). The inflammasome recognizes cytosolic microbial and host DNA and triggers an innate immune response. *Nature, 452*, 103.

Nakahira, K., Haspel, J. A., Rathinam, V. A., Lee, S. J., Dolinay, T., Lam, H. C., et al. (2011). Autophagy proteins regulate innate immune responses by inhibiting the release of mitochondrial DNA mediated by the NALP3 inflammasome. *Nature Immunology, 12*, 222–230.

Nour, A. M., et al. (2009). Anthrax lethal toxin triggers the formation of a membrane-associated inflammasome complex in murine macrophages. *Infection and Immunity, 77*, 1262.

Nour, A. M., et al. (2011). Varicella-zoster virus infection triggers formation of an interleukin-1β (IL-1β)-processing inflammasome complex. *The Journal of Biological Chemistry, 286*, 17921.

Pontillo, A., et al. (2010). Two SNPs in NLRP3 gene are involved in the predisposition to type-1 diabetes and celiac disease in a pediatric population from northeast Brazil. *Autoimmunity, 43*, 583.

Pontillo, A., et al. (2012). Polimorphisms in inflammasome genes are involved in the predisposition to systemic lupus erythematosus. *Autoimmunity, 45*, 271.

Qu, Y., et al. (2012). Phosphorylation of NLRC4 is critical for inflammasome activation. *Nature, 490*, 539.

Rathinam, V. A., et al. (2010). The AIM2 inflammasome is essential for host-defense against cytosolic bacteria and DNA viruses. *Nature Immunology, 11*, 395.

Rathinam, V. A., et al. (2012). TRIF licenses caspase-11-dependent NLRP3 inflammasome activation by gram-negative bacteria. *Cell, 150*, 606.

Ren, T., et al. (2006). Flagellin-deficient Legionella mutants evade caspase-1- and Naip5-mediated macrophage immunity. *PLoS Pathogens, 2*, e18.

Said-Sadier, N., et al. (2010). Aspergillus fumigatus stimulates the NLRP3 inflammasome through a pathway requiring ROS production and the Syk tyrosine kinase. *PLoS ONE, 5*, e10008.

Saiga, H., et al. (2012). Critical role of AIM2 in Mycobacterium tuberculosis infection. *International Immunology, 24*, 637.

Sander, L. E., et al. (2011). Sensing prokaryotic mRNA signifies microbial viability and promotes immunity. *Nature, 474*, 385.

Sauer, J. D., et al. (2011). Listeria monocytogenes engineered to activate the Nlrc4 inflammasome are severely attenuated and are poor inducers of protective immunity. *Proceedings of the National Academy of Sciences of the United States of America, 108*, 12419.

Sheedy, F. J., Grebe, A., Rayner, K. J., Kalantari, P., Ramkhelawon, B., Carpenter, S. B., et al. (2013). CD36 coordinates NLRP3 inflammasome activation by facilitating intracellular nucleation of soluble ligands into particulate ligands in sterile inflammation. *Nature Immunology, 14*, 812–820.

Shenoy, A. R., et al. (2012). GBP5 promotes NLRP3 inflammasome assembly and immunity in mammals. *Science, 336*, 481.

Shio, M. T., Eisenbarth, S. C., Savaria, M., Vinet, A. F., Bellemare, M. J., Harder, K. W., et al. (2009). Malarial hemozoin activates the NLRP3 inflammasome through Lyn and Syk kinases. *PLoS Pathogens, 5*, e1000559.

Soulika, A. M., Lee, E., McCauley, E., Miers, L., Bannerman, P., & Pleasure, D. (2009). Initiation and progression of axonopathy in experimental autoimmune encephalomyelitis. *The Journal of Neuroscience: The Official Journal of the Society for Neuroscience, 29*, 14965–14979.

Subramanian, N., Natarajan, K., Clatworthy, M. R., Wang, Z., & Germain, R. N. (2013). The adaptor MAVS promotes NLRP3 mitochondrial localization and inflammasome activation. *Cell, 153*, 348–361.

Sui, J., et al. (2012). NLRP1 gene polymorphism influences gene transcription and is a risk factor for rheumatoid arthritis in Han Chinese. *Arthritis and Rheumatism, 64*, 647.

Terra, J. K., et al. (2010). Cutting edge: Resistance to Bacillus anthracis infection mediated by a lethal toxin sensitive allele of Nalp1b/Nlrp1b. *Journal of Immunology, 184*, 17.

Thomas, P. G., et al. (2009). NLRP3 (NALP3/CIAS1/Cryopyrin) mediates key innate and healing responses to influenza A virus via the regulation of caspase-1. *Immunity, 30*, 566.

Toma, C., et al. (2010). Pathogenic Vibrio activate NLRP3 inflammasome via cytotoxins and TLR/nucleotide-binding oligomerization domain-mediated NF-kappa B signaling. *Journal of Immunology, 184*, 5287.

Tomalka, J., et al. (2011). A novel role for the NLRC4 inflammasome in mucosal defenses against the fungal pathogen Candida albicans. *PLoS Pathogens, 7*, e1002379.

Tsuchiya, K., et al. (2010). Involvement of absent in melanoma 2 in inflammasome activation in macrophages infected with Listeria monocytogenes. *Journal of Immunology, 185*, 1186.

Vandanmagsar, B., et al. (2011). The NALP3/NLRP3 inflammasome instigates obesity-induced autoinflammation and insulin resistance. *Nature Medicine, 17*, 179.

Villani, A. C., et al. (2009). Genetic variation in the familial Mediterranean fever gene (*MEFV*) and risk for Crohn's disease and ulcerative colitis. *PLoS ONE, 4*, e7154.

Vladimer, G. I., et al. (2012). The NLRP12 inflammasome recognizes *Yersinia pestis*. *Immunity, 37*, 96.

von Moltke, J., et al. (2012). Rapid induction of inflammatory lipid mediators by the inflammasome in vivo. *Nature, 490*, 107.

Warren, S. E., et al. (2008). Multiple Nod-like receptors activate caspase 1 during Listeria monocytogenes infection. *Journal of Immunology, 180*, 7558.

Witzenrath, M., et al. (2011). The NLRP3 inflammasome is differentially activated by pneumolysin variants and contributes to host defense in pneumococcal pneumonia. *Journal of Immunology, 187*, 434.

Wree, A., Eguchi, A., McGeough, M. D., Pena, C.A., Johnson, C. D., Canbay, A., Hoffman, H. M., & Feldstein, A. E. (2013). NLRP3 inflammasome activation results in hepatocyte pyroptosis, liver inflammation and fibrosis. *Hepatology, 59*, 898–910.

Wu, J., et al. (2010). Involvement of the AIM2, NLRC4, and NLRP3 inflammasomes in caspase-1 activation by Listeria monocytogenes. *Journal of Clinical Immunology, 30*, 693.

Yamin, T. T., Ayala, J. M., & Miller, D. K. (1996). Activation of the native 45-kDa precursor form of interleukin-1-converting enzyme. *The Journal of Biological Chemistry, 271*, 13273–13282.

Youm, Y. H., et al. (2011). Elimination of the NLRP3-ASC inflammasome protects against chronic obesity-induced pancreatic damage. *Endocrinology, 152*, 4039.

Zaki, M. H., Boyd, K. L., Vogel, P., Kastan, M. B., Lamkanfi, M., & Kanneganti, T. D. (2010). The NLRP3 inflammasome protects against loss of epithelial integrity and mortality during experimental colitis. *Immunity, 32*, 379–391.

Zamboni, D. S., et al. (2006). The Birc1e cytosolic pattern-recognition receptor contributes to the detection and control of Legionella pneumophila infection. *Nature Immunology, 7*, 318.

Zhang, W., Cai, Y., Xu, W., Yin, Z., Gao, X., & Xiong, S. (2013). AIM2 facilitates the apoptotic DNA-induced systemic lupus erythematosus via arbitrating macrophage functional maturation. *Journal of Clinical Immunology, 33*, 925–937.

Zhao, Y., et al. (2011). The NLRC4 inflammasome receptors for bacterial flagellin and type III secretion apparatus. *Nature, 477*, 596.

Zheng, Y., et al. (2011). A *Yersinia* effector with enhanced inhibitory activity on the NF-kB pathway activates the NLRP3/ASC/caspase-1 inflammasome in macrophages. *PLoS Pathogens, 7*, e1002026.

Zheng, F., et al. (2013). *Heart Lung and Circulation*.

Zhou, R., et al. (2010). Thioredoxin-interacting protein links oxidative stress to inflammasome activation. *Nature Immunology, 11*, 136.

Zhou, R., Yazdi, A. S., Menu, P., & Tschopp, J. (2011). A role for mitochondria in NLRP3 inflammasome activation. *Nature, 469*, 221–225.

Inflammatory Bowel Disease

Amon Asgharpour, Jianfeng Cheng, Matthew Kaspar and Stephen Bickston
Division of Gastroenterology, Hepatology and Nutrition, Department of Internal Medicine, VCU Medical Center, Richmond, VA, USA

Synonyms

Crohn's disease; Ulcerative colitis

Definition

Inflammatory bowel disease (IBD) predominantly consists of two chronic, often relapsing, immunologically mediated gastrointestinal disorders: ulcerative colitis (UC) and Crohn's disease (CD). The two diseases share many clinical, pathogenetic, and epidemiologic characteristics, suggesting that their underlying causes may be similar. The chronic inflammation associated with IBD is related to a sustained immune response. It has yet to be determined whether this is an appropriate response to an unknown pathogen or an inappropriate response to normal gut contents.

Crohn's disease primarily affects the terminal ileum and colon, but can involve any portion of the gastrointestinal tract from the mouth to anus. It is often characterized by discontinuous lesions ("skip lesions") with inflammation that can involve the full thickness of the affected portion of bowel for mucosa to serosa.

Ulcerative colitis begins in the rectum and can extend proximally (longitudinally) to affect other portions of the colon. UC can be classified by extent of disease as proctitis, left-sided colitis, or pancolitis (when more than the left colon is affected). The inflammatory lesions seen are in a continuous pattern and only affect the mucosal layers of the colon without deeper involvement.

In about 85 % of patients, the abnormalities can be distinguished between the two diseases based on clinical, endoscopic, radiologic, and histologic features. Reassignment of a diagnosis of CD or UC may be as high as 9 % in the first 2 years after diagnosis.

Epidemiology and Genetics

Approximately 1.4 million people are affected by IBD in the USA and 2.2 million in Europe. Incidence (frequency of new case over a certain time) and prevalence (total number of cases of a disease in the population at a given time) of IBD vary with geographic location and race. The incidence rates for UC in North America range from 6.0 to 15.6 cases per 100,000 person-years, and the prevalence ranges from 38 to 246 cases per 100,000 persons. In Europe, incidence rates range from 1.5 to 20.3 cases per 100,000 person-years, with prevalence of 21 to 243 cases per 100,000 persons (Loftus 2004). Those of Ashkenazi Jewish descent have incidence rates that are severalfold higher than in the non-Jewish population across various geographic regions.

The peak incidence of UC occurs in the second and third decades of life with a second smaller peak in the ages 60–70. Studies have shown no gender difference in the occurrence of UC, and a female-to-male ratio of nearly 1:1 applies to all age groups.

For CD, the incidence rate of 3.1 to 14.6 cases per 100,000 person-years in North America, the prevalence of CD in the USA is estimated to be 201 per 100,000 adults. Although the age of diagnosis can range from early childhood throughout an entire lifespan, the peak age for CD diagnosis is 15–30s. There is a small female predominance in CD, with female-to-male ratio of 1.3:1. IBD is generally less common in the developing world than in developed nations. It has been more common in the Western countries. Across Asia the incidence and prevalence of IBD has increased rapidly over the last two to four decades. The incidence of Crohn's disease in Japan now equals that of Israel.

Multiple genetic factors contribute to IBD pathogenesis. Genome-wide association studies (GWAS) have identified approximately 70 genetic loci that confer susceptibility to CD and over 30 loci that are associated with UC (Vora et al. 2012). IBD as classified into UC and CD appears to be polygenic. It is possible that subsets may prove to be monogenic. Gene loci specifically associated with CD include NOD2 (nucleotide oligomerization domain (NOD)-like receptors, also known as *CARD15* (*ca*spase-*r*ecruitment *d*omain 15)), autophagy genes *ATG16L1* (which encodes autophagy related 16-like protein 1), and IRGM (which encodes immunity-related GTPase family M), whereas genes related to regulatory pathways (*IL10* and *ARPC2*), intestinal epithelial cell (IEC) function (e.g., *ECM1*), and an E3 ubiquitin ligase (e.g., *HERC2*) appear to be specific for UC (Zhernakova et al. 2009). There is a considerable overlap in genetic risk factors between both forms of IBD, such as genes in the IL-23 pathway including IL23R (which encodes a critical subunit for IL-23 receptor), IL12B (which encodes the P40 subunit of IL-R and IL-23), and STAT3 (which encodes signal transducer and activator of transcription 3, with roles in IL23R signaling)

(Ghishan 2012). Studies of twins suggest that monozygotic twins (identical twins) have a greater tendency to develop IBD than do dizygotic twins (fraternal twins). Genetic composition is a more powerful determinant for CD than for UC: The concordance rate among identical twins is 50–75 % with an approximately 800-fold increased relative risk for developing CD disease compared to the general population but only 10–20 % for UC.

Pathophysiology

The pathogenesis of IBD remains one of the most complex and elusive areas in gastroenterology. In the last two decades, many studies in genetics, immunology, and the microbiota have made significant progress in understanding possible underlying mechanisms; however, there are many questions that remain to be answered. Collectively, current understandings of the pathogenesis of IBD can be categorized into host (gene, immune system, gut epithelial cells, etc.), environment (commensal microbiome, infections, etc.), and the interaction between these two. Although different, there are significant overlaps in pathogenesis of CD and UC. Many of these key factors in pathogenesis have been the targets of therapeutic interventions.

Immune cells including macrophages, dendritic cells, neutrophils, natural killer cells, and B and T cells (especially certain helper T cells such as Th1, Th2, and Th17) are present at the lamina propria in IBD patients. When exposed to an antigen, the antigen-presenting cells (APC), such as macrophages, process it and present the antigen to CD4+ T cells. Macrophages also release *TNF*, *IFNγ*, and *IL-12* which differentiate some activated T cells into Th1 cells that subsequently produces *IL-2* and *IFNγ*; some of the activated T cells become Th2 which releases *IL-10*, some into Th3 which produces *TGFβ* and T regulatory cell (Tr1). Both *IL-10* and *TGFβ* are anti-inflammatory. Nuclear factor kappa B (*NFκB*) in the cytoplasm of mononuclear cells is also a key factor in regulating cytokines and immune response system. Studies show that patients with Crohn's disease have increased Th17 cytokine *IL-17* and Th1 cytokines, while patients with ulcerative colitis have elevated Th17 and Th2 cytokines. The balance of pro- and anti-inflammatory states is disturbed in IBD patients (Duchmann et al. 1999; Fuss et al. 2004).

In both good health and in disease, the gut flora is in close contact with the mucosa lining of the gastrointestinal tract. In health, the immune system tolerates them without causing pathologic inflammation. It is thought that controlled sampling of the normal flora is an important part of the immune tolerating mechanism. The epithelial system is critical in this regard. In IBD, studies have revealed that tight junctions among the epithelial cells are compromised and pericellular space and permeability increased. This leads to loss of barrier function, which may be a cause and result of aberrant inflammation among IBD patients (Turner 2006; Wang et al. 2006; Dahan et al. 2008). Organic cation transporter (*OCTN*) and two types of membrane associate guanylate kinases (*MAGUK*) are considered to be involved in protecting the integrity of the epithelial cell barrier. Goblet cells and Paneth cells are also important in this system: goblet cells produce mucus and factors that are essential in defense, repair, and inflammation regulation, and Paneth cells make α-defensins which have antimicrobial property. Both the production and composition of mucin are altered in IBD patients, which could contribute to the pathogenesis of IBD by facilitating exposure of the epithelial cell layer to pathogenic bacteria.

Microbiome studies show that the bacterial profile not only differs between IBD patients and healthy controls but also between UC and CD (Eckburg and Relman 2007; Frank et al. 2007). There are a decreased number of commensal organisms including *Lachnospiraceae* and *Bacteroidetes* and an increase in the number of *Proteobacteria* in IBD patients.

Novel research in the microbiome supports the notion that the innate and adaptive immune system can be modulated and primed by the intestinal microbiome. The priming of the host's immune response can allow for rapid activation and enhanced killing when exposed to pathogenic organisms (see Fig. 1).

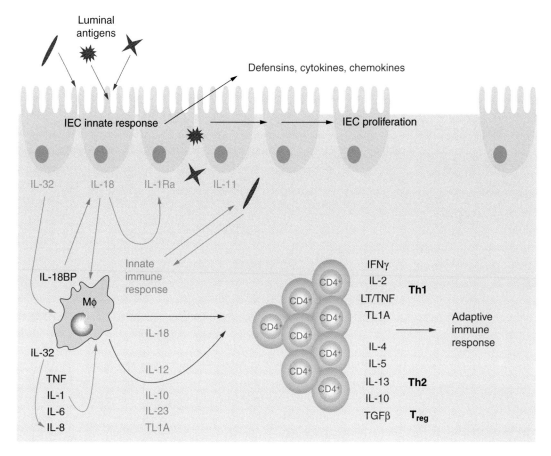

Inflammatory Bowel Disease, Fig. 1 Novel cytokine paradigm in IBD. Based on emerging evidence that innate immune responses may play a primary role in the initiation phases of IBD and that adaptive immune responses may occur secondary to innate mucosal immune dysregulation, the classic Th1/Th2 cytokine paradigm may be outdated. Classifying immune mediators as innate or adaptive cytokines may be more applicable to IBD pathogenesis. Mucosal innate cytokines (*blue*) can be classified as those primarily produced by intestinal epithelial cells and macrophages and are important in promoting innate immune responses (*orange arrows*). In health, their main function is that of protection (i.e., maintenance and repair of the epithelial barrier); however, in disease states they can exacerbate and prolong inflammation. Categorizing mucosal adaptive cytokines can be more complex because they can be produced early on, primarily by macrophages and dendritic cells (early adaptive cytokines are shown in *green*), whereas cytokines produced later follow the classic Th1/Th2 as well as T regulatory cell (Treg) cytokine profiles (*red*). The end result, however, is the orchestration of mucosal adaptive immune responses (*gray arrows*), which in health coordinate normal Th1 and Th2 effector functions, whereas aberrant or uncontrolled Th1/Th2 immune responses (or dysregulated Treg function) can lead to a chronic state of inflammation (Pizarro and Cominelli: Cytokine therapy for Crohn's disease: advances in translational research. Annual Review of Medicine, 2007; 58:433–44. Reproduced with permission)

Clinical Presentation

Ulcerative Colitis

Patients with UC can present with a variety of symptoms, depending on the severity, extent, and the duration of the disease. In general, there is a correlation between the severity of the symptoms and the severity of the disease. At the time of initial presentation, approximately 45 % of patients with UC have disease limited to the rectosigmoid (proctitis), 35 % have disease extending beyond the sigmoid but not involving the entire colon, and 20 % of patients have pancolitis (Langholz et al. 1994). Common symptoms are rectal bleeding, diarrhea, abdominal pain, passage of mucus, tenesmus, and urgency.

Patients with proctitis often present with fecal urgency (a sensation of incomplete fecal evacuation), fecal incontinence, passage of blood, and mucus, while patients with extensive colitis usually present with diarrhea (with more blood loss), fever, weight loss, and cramping abdominal pain. In more severe active disease, a patient can have anorexia, nausea, and vomiting. The most extreme presentation of UC is toxic megacolon, a severe and life-threatening condition which was reported at a rate of up to 7.5 % of UC patients on admission (Latella et al. 2002; Langholz et al. 1994). Clinical systemic toxicity (such as fever, tachycardia, leukocytosis, or anemia) and colonic dilatation (diameter >6 cm) indicate the presence of toxic megacolon. Signs of peritonitis indicate colon perforation.

Crohn's Disease

The presentation of Crohn's disease can vary considerably, depending on the location of the disease, the severity of the inflammation, and specific intestinal complications such as fistula or stricture formation. Diarrhea is the most common complaint in CD. Abdominal pain is a more frequent and persistent symptom than in UC. Another common presentation of Crohn's disease is perianal disease such as skin lesions, anal canal lesions, and perianal fistulae. The transmural nature of Crohn's disease makes fistula formation a frequent presentation. Fistulae can manifest as different forms: perianal fistula (arising from the anal gland); enteroenteric, enterocolonic, and colocolonic fistulae (from one segment of the gastrointestinal tract to another); rectovaginal/enterovaginal fistula (from the intestine to the vagina); enterovesicular/colovesicular fistula (from the intestine to the bladder); and enterocutaneous fistula (from the intestine to the anterior abdominal wall). If a forming sinus tract does not adhere to adjacent tissue, an abscess or perforation can form from the tract rather than a fistula. These patients can have a varying presentation including fever, anorexia, abdominal pain, nausea, vomiting, and/or peritoneal signs. Longstanding inflammation can cause strictures, another characteristic presentation in Crohn's disease, that can present as a bowel obstruction.

Weight loss, malnutrition, and anemia are often seen in Crohn's disease. If the degree of blood loss is severe enough, patients can have profound anemia and symptoms of fatigue and dyspnea. Anemia can also be seen with vitamin B-12 deficiency secondary to ileal disease, surgical resection, and bacterial overgrowth.

Extraintestinal Manifestations in IBD

It is reported that extraintestinal manifestations occur in 21–47 % of patients with IBD (Bernstein et al. 2005). In a large series, extraintestinal manifestations are found to occur more often in Crohn's disease than in UC and are more common among patients with colonic involvement than in patients with no colonic inflammation (Sands and Siegel 2010).

Musculoskeletal disease is the most common extraintestinal manifestation, including peripheral and axial arthropathy. Peripheral arthropathy comprises pauciarticular arthropathy (involving <5 large joints, usually parallels underlying bowel disease activity) and polyarticular arthropathy (involving ≥5 small joints, independent of bowel disease activity). Axial arthropathy is less common than peripheral arthropathy and does not parallel bowel disease activity. It includes spondylitis (HLA-B27 positive) and isolated sacroiliitis (HLA-B27 negative). IBD patients also have decreased bone mineral density at initial presentation that can be worsened by the use of steroid therapy to treat an IBD flare. These patients also tend to have decreased physical activity, which worsens bone mineral density. Decreased bone mineral density can lead to osteoporosis and bone fracture.

Pyoderma gangrenosum (PG) and erythema nodosum (EN) are the most common dermatologic manifestations associated with IBD. PG (an idiopathic non-tender skin lesion) occurs commonly in the lower extremities (usually at the site of trauma) and follows a course independent of bowel activity. EN (tender, red-purple nodules) most commonly appears on the pretibial region and parallels the bowel disease activity.

Ocular manifestations of IBD include episcleritis, scleritis, and uveitis. Episcleritis and scleritis are generally mild and present with red eyes, burning, and itching, which parallel bowel disease activity. Uveitis presents with photophobia, blurred vision, and headache, which is a true emergency potentially leading to blindness if untreated. Uveitis is independent of bowel disease activity.

Primary sclerosing cholangitis (PSC) is the most common hepatobiliary manifestations in IBD, occurring from 1.4 % to 7.5 %, while approximately 70–80 % of patients with PSC have concomitant IBD, much more UC than CD (Broome´ and Bergquist 2006). PSC-IBD patients have increased risk for developing colorectal cancer and cholangiocarcinoma. PSC has no treatment and may require liver transplantation.

IBD is associated with hypercoagulability that can lead to both venous thromboembolic disease and arterial thrombus. The most common manifestations are deep vein thrombosis and pulmonary embolism. Hypercoagulability in these patients is multifactorial in causation. The inflammatory state itself physiologically leads to microvascular coagulation. In addition, thrombocytosis, increased levels of fibrinogen and factors V and VIII, decreased levels of protein C and S, and circulating immune complexes are often seen in severe cases of ulcerative colitis and Crohn's disease.

Therapy

Ulcerative Colitis

The goals of UC therapy are to induce and maintain remission and further improve the patient's quality of life. The strategy is based on the severity and extent of the disease. For mild to moderate disease, first-line therapy is 5-aminosalicylic acid (mesalamine), including oral and topical formulation (enema/suppository). Patients with moderate disease unresponsive to mesalamine, glucocorticoids, azathioprine, or 6-MP are recommended to receive induction therapy. For severe disease, IV glucocorticoids, cyclosporine, or infliximab is used as induction therapy. However, glucocorticoids are only for induction and not for maintenance. Maintenance therapy for UC includes mesalamine (topical or oral), azathioprine or 6-MP, and infliximab.

Aminosalicylates

Sulfasalazine is a compound that is split by the colonic flora into sulfapyridine (inactive carrier) and 5-aminosalicylic acid (5-ASA, the active constituent) and therefore works only in the colon. It is effective in mild to moderate UC with 39–62 % induction remission rate at dose of 3–6 g/day (Sutherland et al. 2000). Side effects are common (up to 45 % patients) and majority are caused by sulfapyridine moiety, including dose-related such as nausea, vomiting, dyspepsia, back pain, and headache and dose-unrelated such as fever, rash, agranulocytosis, hepatitis, pancreatitis, interstitial nephritis, and pneumonitis. Sulfasalazine competitively inhibits intestinal absorption of folic acid and causes folate deficiency; therefore, at least 1 mg per day of folic acid supplementation is recommended. Sulfasalazine can also cause reversible oligospermia and sperm dysmotility in male patients. To balance the efficacy and adverse side effect, a dose of 2 g/day is recommended to maintain remission. Similar to sulfasalazine, olsalazine (1.5–3 g/day) and balsalazide (6.75 g/day, equivalent to mesalamine 2.4 g/day) also work only in the colon. Side effect of diarrhea from olsalazine often limits its use (Pruitt et al. 2002).

To avoid the systemic side effect from sulfapyridine, various formulations (either encapsulated by a time-release or PH-dependent coating) were developed to deliver 5-ASA to specific sites of the gastrointestinal site (Table 1). Pentasa®, Asacol®, and Lialda® are three commonly used mesalamine medications that work in both the distal ileum and colon. In particular, Pentasa releases 5-ASA from the duodenum throughout the small bowel and the colon via a time-release capsule. It has been shown that mesalamine 2.4–4.8 g/day has efficacy for mild UC and higher doses, 4.8 g/day, are more efficacious in moderately active disease (Hanauer et al. 2007). Mesalamine adds little for severely active disease.

Topical forms include 5-ASA enema (Rowasa® 1–4 g nightly, being able to deliver up to the level of the splenic flexure) and suppository (Canasa® 1–1.5 g nightly, delivers up to 15–20 cm from the anal verge), which are effective in induction and maintenance of remission.

Inflammatory Bowel Disease, Table 1 Oral 5-aminosalicylic acid preparations and sites of delivery in the gastrointestinal tract

Drug	Formulation	Site of delivery
Prodrugs		
Sulfasalazine	Sulfapyridine + 5-ASA	Colon
Olsalazine	5-ASA dimer	Colon
Balsalazide	4-aminobenzoyl β-alanine + 5-ASA	Colon
Mesalamine Preparations		
Asacol, Claversal, Salofalk	pH sensitive, resin-coated; delayed release	Distal ileum, colon
Rowasa	Enema	Distal colon
Canasa	Suppository	Rectum
Pentasa	Ethylcellulose-coated microgranules; controlled release	Duodenum to colon
Lialda	pH sensitive, multi-matrix and polymethacrylate coated; delayed and slow release	Distal ileum, colon

From Osterman MT and Lichtenstein GR. Ulcerative colitis, Sleisenger & Fordiran's Gastrointestinal and liver disease 9th edition, vol 2 page 1994 with permission

Corticosteroids

Glucocorticoids are effective for induction in moderate to severe UC patients. Usually oral prednisone at 40 mg/day is used for moderate UC and gradually tapered in accordance to clinical response. Parenteral administration of hydrocortisone (100 mg IV Q8 h) or methylprednisolone (16–20 mg IV Q8 h) is reserved for severe or fulminant UC and switched to oral prednisone when bloody diarrhea and appetite improve. Glucocorticoids have no effect in maintenance of remission in patients with UC. For steroid-dependent patients, immunomodulatory agents or biologic agents should be considered. Budesonide is a glucocorticoid that has a different structure from prednisone, which results in significantly less systemic toxicity compared with traditional glucocorticoids. Controlled studies have not shown that oral budesonide is effective for active UC (Lofberg et al. 1996).

Immunomodulators

Azathioprine and 6-mercaptopurine (6-MP) are the most widely used immunomodulators in UC patients, which are effective in both induction and maintenance remission. Thiopurine methyltransferase (TPMT) genotype or phenotype should be tested before initiating treatment, as this enzyme is responsible for most of the drug's biotransformation. For patients with homozygous wild-type TPMT or normal TPMT enzyme, azathioprine can be started at 2.5 mg/kg/day and 6-MP at 1.5 mg/day. For patients with heterozygous TPMT mutation or intermediate enzyme activity, both azathioprine and 6-MP dosage should be decreased to 50 %. Azathioprine and 6-MP have a delayed onset of action with mean time to clinical response of 3–4 months. Common side effects of azathioprine and 6-MP include bone marrow suppression (especially leukopenia), nausea, vomiting, pancreatitis, allergic reaction, and infections. Routine monitoring of complete blood count with differential and liver function testing is recommended.

Cyclosporine A (CSA), a potent inhibitor of cell-mediated immunity, can be used in severe, steroid-refractory UC patients for a short duration. At 2 mg/kg/day continuous intravenous dosing, it is used as a bridging therapy to control active disease in patients with steroid-refractory UC while waiting for the onset of azathioprine/6-MP or surgical resection. There is a growing body of data suggesting that anti-TNF therapy is as good or better in such patients. Since CSA has not been useful in maintaining remission of UC, its role for severe UC is shrinking.

Nutritional Support

Unlike in CD, bowel rest and total parenteral nutrition have not been shown to be effective as primary treatment in UC. However, for patients with severe colitis having anorexia, hypoalbuminemia, or malnutrition, parenteral nutrition can provide benefits.

Biological Therapy: Anti-TNF Agents

Two anti-TNF agents (infliximab and adalimumab) were proven to be effective in moderate to severe UC for both induction and maintenance of remission. Infliximab (Remicade) can be given at 5–10 mg/kg infusions at 0, 2, and 6 weeks as induction therapy and then every 8 weeks as maintenance therapy. Adalimumab (Humira) is given at an initial dose of 160 mg subcutaneously as induction followed by a 80 mg dose 2 weeks after and then 40 mg every other week for maintenance. Side effects include injection site and infusion reactions, delayed hypersensitivity reactions (secondary to the development of autoantibodies), headaches, rash, demyelinating disorders, pancytopenia, and T-cell lymphoma. Since anti-TNF agents can cause activation of tuberculosis, all patients should have PPD, chest X-ray, and hepatitis B screening prior to starting these medications.

Surgical Therapy

Surgical resection of the colon and rectum is curative for UC. Emergency surgery is indicated in severe fulminant colitis, toxic megacolon refractory to medical therapy, perforation, and peritonitis. Elective surgery is indicated in medically refractory disease, intractable disease with impaired quality of life, and intolerance of side effects from medical therapy. The surgical options include subtotal colectomy with ileostomy, colectomy with ileorectal anastomosis, proctocolectomy with continent ileostomy, and restorative proctocolectomy with ileal pouch-anal anastomosis (IPAA).

Crohn's Disease

Crohn's disease is a chronic debilitating disease that cannot be cured by either medical or surgical therapy. The treatment goal is to induce and maintain remission, decrease complications, and improve patient's quality of life.

Unlike UC, 5-ASA provides minimal benefit in CD. Sulfasalazine 4–6 g/day may be beneficial to mild to moderate Crohn's disease with colonic involvement, while other mesalamine products are of uncertain benefit.

Antibiotics

Several antibiotics, such as metronidazole, ciprofloxacin, and rifaximin have therapeutic efficacy in CD, especially in cases of pyogenic complications, perineal diseases, enterocutaneous fistulae, or active colonic disease.

Glucocorticoids

Glucocorticoids play a central role in the short-term control of symptoms of CD and not effective for maintenance of remission. Glucocorticoid formulation dosage options include hydrocortisone (100 mg IV every 8 h), prednisolone (30 mg IV twice daily), or methylprednisolone (16–20 mg IV every 8 h). Budesonide possesses glucocorticoid receptor affinity and limited systemic side effect by enhanced first-pass metabolism and may be considered as first-line therapy for patients with active ileal, ileocecal, or right colonic disease. For steroid-dependent or refractory patients, immunomodulators or biologic agents should be considered. Clinical factors associated with glucocorticoid dependence include smoking, younger age at onset, colonic location, and non-fibrostenotic disease (Franchimont et al. 1998).

Immunomodulators

Azathioprine and 6-MP are used in CD for maintenance of remission, for patients with active CD in whom first-line therapy fails, for steroid-dependent or refractory cases, and for postsurgical prophylaxis of CD. The onset of action of this medication is delayed, 3–4 months, and many patients will need tapering glucocorticoids as bridging. For patients with normal TPMT enzyme level, it is recommended that azathioprine is used in doses of 2.0–2.5 mg/kg/day, and 6-MP is given in doses of 1.0–1.5 mg/kg/day. For patients with heterozygous TPMT mutation or intermediate enzyme activity, both azathioprine and 6-MP dosage should be decreased to 50 %. For patients with homozygous mutation in TPMT, azathioprine/6-MP use should be avoided due to severe bone marrow suppression secondary to high levels of 6-thioguanine (metabolites from these medications). Most authorities agree that azathioprine maintenance therapy should be continued longer than 3.5 years. The decision to withdraw thiopurine therapy should only be undertaken after discussion between doctor and patient of possible risks and benefits (Sands and Siegel 2010).

Methotrexate is another immunomodulator to be used in chronically active CD with steroid dependence or refractory to steroid therapy, and it is also beneficial in maintaining remission. The suggested dosage is 25 mg subcutaneously weekly. Common side effects include stomatitis, nausea, diarrhea, hair loss, mild leukopenia, and serum aminotransferase elevations. Methotrexate can interact with sulfa medications and azathioprine/6-MP to cause severe leukopenia, and these combinations should be avoided. Methotrexate is strongly teratogenic, and women of childbearing age must use highly effective contraception while on methotrexate.

In contrast to UC, cyclosporine has not been shown effective in CD.

Biologic Therapies: Anti-TNF Agents

All three anti-TNF agents (infliximab, adalimumab, and certolizumab pegol) are proven to be effective in induction and maintenance remission for CD. Proper selection of patients is very important in starting the anti-TNF agents since it is unlikely to have benefit for patients without objective findings of inflammation or with fibrostenotic disease. Infliximab (Remicade) can be given at 5–10 mg/kg infusion at 0, 2, and 6 weeks as induction therapy and then every 8 weeks as maintenance therapy. Adalimumab (Humira) is given at an initial dose of 160 mg subcutaneously as induction followed by a 80 mg dose 2 weeks after and then 40 mg every other week for maintenance. Certolizumab pegol (CIMZIA) is also given subcutaneously at dosage of 400 mg at 0, 2, and 4 weeks as induction and then every 4 weeks as maintenance. Since anti-TNF agents can cause activation of tuberculosis, all patients should have PPD, chest X-ray, and hepatitis B screening prior to starting these medications. For patients with pyogenic complications of CD or any underlying serious infection, adequate drainage and treatment with antibiotics should be done first before starting or continuing anti-TNF agents. Smoking cessation education is very important for CD patients since smoking can both exacerbate CD and decrease response to anti-TNF agents. Side effects include injection site and infusion reactions, delayed hypersensitivity reactions (secondary to the development of autoantibodies), headaches, rash, demyelinating disorders, pancytopenia, and T-cell lymphoma.

Leukocyte trafficking Agents

Natalizumab is an α_4-integrin antagonist that inhibits leukocyte adhesion and migration into inflamed tissue and is proven to be effective for induction and maintenance of moderate to severe CD. It is administered intravenously at 300 mg every 4 weeks. Since it can cause fatal complication of progressive multifocal leukoencephalopathy (PML), it is recommended to check for JC virus before initiating the medication.

Another agent, vedolizumab, acts selectively an antagonist to $\alpha 4\beta 7$-integrin. As $\alpha 4\beta 7$ is found only in gut (and not CNS) tissues, vedolizumab does not appear to carry the same risk for progressive multifocal leukoencephalopathy as natalizumab. Vedolizumab is currently approved by the FDA for treatment of moderate to severe UC and Crohn's disease. It is administered as an IV infusion on an induction and maintenance regimen. A third drug, etrolizumab, is currently being investigated as a therapy for IBD. As a selective antagonist to $\beta 7$-integrin, it appears to confer a similar efficacy and side effect profile to vedolizumab on phase II trials. Its efficacy in moderate-to-severe ulcerative colitis is currently being evaluated in a phase III clinical trial.

Other Selected Therapies Under Investigation

A variety of novel therapeutic targets have been identified and are currently under investigation for treatment of IBD.

In the IL-12/23 pathway, ustekinumab (originally approved for treatment of plaque psoriasis and psoriatic arthritis) suppresses expression of both IL-12 and IL-23 by selectively inhibiting their common p40 subunit. Ustekinumab is administered as a subcutaneous injection and is currently undergoing phase III trials for the treatment of Crohn's disease.

A number of oral agents are also being developed. Tofacitinib (originally used in treatment of rheumatoid arthritis), an inhibitor of the JAK kinase system, has shown promise in phase II trials and is currently being evaluated in a phase

III trial for treatment of ulcerative colitis. Ozanimod (originally approved for treatment of multiple sclerosis), an agonist of the sphingosine-1-phosphate receptor subtypes 1 and 5 ($S1P_{1R}$), functions to "trap" lymphocytes in the lymph nodes and as such prevent their migration to sites of inflammation. A placebo-controlled phase trial showed marginal improvement in rate of clinical response over placebo in patients with moderate-to-severe ulcerative colitis. In Crohn's disease, mongersen, an oral SMAD7 antisense oligonucleotide, has been shown in a phase 2 trial to be effective for induction of remission in patients with active Crohn's disease.

Nutritional Support

Total parenteral nutrition and bowel rest are proven to be effective in patients with severe malnutrition before surgery or severe CD patients. For patients with ileal disease or resection, it is important to have parenteral supplementation of vitamin B12. The addition of cholestyramine (1–4 g/day) or colesevelam (625–3800 mg/day) may be required to control bile salt diarrhea. Bone loss is a complication anticipated in all patients. Vitamin D and calcium supplementation is required to preserve bone density.

Surgical Therapy

In contrast to UC, surgical therapy is used to control symptoms and treat complications and cannot cure Crohn's disease. Indications for surgery include complications such as intra-abdominal abscess, medically intractable fistula, fibrotic stricture with obstructive symptoms, toxic megacolon, hemorrhage, and cancer. The surgical principle in CD is to preserve intestinal length and function because there is a high probability of recurrence post segmental resection. At some point in the course of their disease, 75 % of patients will require surgery for complications related to CD.

Outcome

CD is a chronic disease with a highly variable course. It has been shown that 22 % of patients remain in remission, 25 % experience chronically active symptoms, and 53 % have a course that fluctuates between active and inactive disease. There appears to be a moderate increase in mortality compared to healthy controls, an excess of 17 deaths per 100,000. The morbidity of this disease can greatly impact patient quality of life. CD patients are at increased risk for developing colorectal cancer and should be screened with colonoscopy.

UC, like CD, has a highly variable disease course. The majority of patients with UC have recurrent flares and durations of disease remission. UC has a slightly higher mortality rate compared to healthy controls. UC patients are also at increased risk of colorectal cancer and should be screened with colonoscopy. The morbidity of this disease can greatly impact patient quality of life. The mortality rates of both UC and CD have been decreasing over time. There is still more to learn about these disease processes, and further investigation will provide future therapy.

Cross-References

▶ Corticosteroids
▶ Inflammatory Bowel Disease Models in Animals
▶ Lymphocyte Homing and Trafficking
▶ Mechanisms of Macrophage Migration in 3-Dimensional Environments
▶ Tumor Necrosis Factor (TNF) Inhibitors

References

Amiot, A., & Peyrin-biroulet, L. (2015). Current, new and future biological agents on the horizon for the treatment of inflammatory bowel diseases. *Therapeutic Advances in Gastroenterology, 8*(2), 66–82.

Bernstein, C. N., Wajda, A., & Blanchard, J. F. (2005). The clustering of other chronic inflammatory diseases in inflammatory bowel disease: A population-based study. *Gastroenterology, 129*, 827–836.

Broome´, U., & Bergquist, A. (2006). Primary sclerosing cholangitis, inflammatory bowel disease, and colon cancer. *Seminars in Liver Disease, 26*, 31–40.

Dahan, S., Roda, G., Pinn, D., et al. (2008). Epithelial: Lamina propria lymphocyte interactions promote epithelial cell differentiation. *Gastroenterology, 134*, 192–203.

Duchmann, R., May, E., Heike, M., et al. (1999). T cell specificity and cross reactivity towards enterobacteria, bacteroides, bifidobacterium, and antigens from resident intestinal flora in humans. *Gut, 44*, 812.

Eckburg, P. B., & Relman, D. A. (2007). The role of microbes in Crohn's disease. *Clinical Infectious Diseases, 44*, 256–262.

Franchimont, D. P., Louis, E., Croes, F., & Belaiche, J. (1998). Clinical pattern of corticosteroid dependent Crohn's disease. *European Journal of Gastroenterology and Hepatology, 10*, 821–825.

Frank, D. N., St Amand, A. L., Feldman, R. A., Boedeker, E. C., Harpaz, N., & Pace, N. R. (2007). Molecular-phylogenetic characterization of microbial community imbalances in human inflammatory bowel diseases. *Proceedings of the National Academy of Sciences of the United States of America, 104*, 13780–13785.

Fuss, I. J., Heller, F., Boirivant, M., et al. (2004). Nonclassical CD1d-restricted NK T cells that produce IL-13 characterize an atypical Th2 response in ulcerative colitis. *Journal of Clinical Investigation, 113*, 1490.

Ghishan, F. (2012). Inflammatory bowel disease. In A. Elzouki (Ed.), *Textbook of clinical pediatrics: SpringerReference* (www.springerreference.com). Berlin/Heidelberg: Springer . doi:10.1007/SpringerReference_323865 2012-05-03 10:47:29 UTC.

Hanauer, S. B., Sandborn, W. J., Dallaire, C., et al. (2007). Delayed-release oral mesalamine 4.8 g/day (800 mg tablets) compared to 2.4 g/day (400 mg tablets) for the treatment of mildly to moderately active ulcerative colitis: The ASCEND I trial. *Canadian Journal of Gastroenterology, 21*, 827.

Khanna, R., Preiss, J. C., Macdonald, J. K., & Timmer, A. (2015). Anti-IL-12/23p40 antibodies for induction of remission in Crohn's disease. *Cochrane Database of Systematic Reviews, 5*, CD007572.

Langholz, E., Munkholm, P., Davidsen, M., et al. (1994). Course of ulcerative colitis: Analysis of changes in disease activity over years. *Gastroenterology, 107*, 3.

Latella, G., Vernia, P., Viscido, A., et al. (2002). GI distension in severe ulcerative colitis. *American Journal of Gastroenterology, 97*, 1169–1175.

Lofberg, R., Danielsson, A., Suhr, O., et al. (1996). Oral budesonide versus prednisolone in patients with active extensive and left-sided ulcerative colitis. *Gastroenterology, 110*, 1713.

Loftus, E. V., Jr. (2004). Clinical epidemiology of inflammatory bowel disease: Incidence, prevalence, and environmental influences. *Gastroenterology, 126*, 1504.

Monteleone, G., Neurath, M. F., Ardizzone, S., et al. (2015). Mongersen, an oral SMAD7 antisense oligonucleotide, and Crohn's disease. *The New England Journal of Medicine, 372*(12), 1104–1113.

Pruitt, R., Hanson, J., Safdi, M., et al. (2002). Balsalazide is superior to mesalamine in the time to improvement of signs and symptoms of acute mild-to-moderate ulcerative colitis. *American Journal of Gastroenterology, 97*, 3078.

Sandborn, W. J., Ghosh, S., Panes, J., Vranic, I., Su, C., Rousell, S., et al. (2012). *The New England Journal of Medicine, 367*(7), 616–624.

Sands, B. E., & Siegel, C. A. (2010). Crohn's disease. In M. Feldman et al. (Eds.), *Sleisenger and fordtran's gastrointestinal and liver disease* (9th ed., vol. 2, pp. 1941–1973). Philadelphia: Saunders Elsevier.

Sutherland, L., Roth, D., & Beck, P., et al. (2000). Oral 5-aminosalicylic acid for inducing remission in ulcerative colitis. *Cochrane Database of Systematic Reviews, 2* CD000543.

Turner, J. R. (2006). Molecular basis of epithelial barrier regulation: From basic mechanisms to clinical application. *American Journal of Pathology, 169*, 1901–1909.

Vora, P., Shih, D. Q., McGovern, D. P., & Targan, S. R. (2012). Current concepts on the immunopathogenesis of inflammatory bowel disease. *Frontiers Bioscience* (Elite Ed), *4*, 1451–1477.

Wang, F., Schwarz, B. T., Graham, W. V., et al. (2006). IFN-gamma-induced TNFR2 expression is required for TNF-dependent intestinal epithelial barrier dysfunction. *Gastroenterology, 131*, 1153–1163.

Zhernakova, A., van Diemen, C. C., & Wijmenga, C. (2009). Detecting shared pathogenesis from the shared genetics of immune-related diseases. *Nature Reviews. Genetics, 10*, 43–55.

Inflammatory Bowel Disease Models in Animals

Cosmin Sebastian Voican[1,2,3], Xavier Treton[1,2,3], Eric Ogier-Denis[1,2,3] and Yoram Bouhnik[1,2,3,4]

[1]INSERM, UMR773, Team "Physiopathology of Inflammatory Bowel Diseases", Centre de Recherche Bichat Beaujon, Paris, France

[2]Service de Gastroentérologie, MICI et Assistance Nutritive, PMAD Hôpital Beaujon, Clichy Cedex, France

[3]University Paris–Diderot Sorbonne Paris–Cité, Paris, France

[4]1Pôle des Maladies de l'Appareil Digestif (PMAD), Service de Gastroentérologie e, Hôpital Beaujon, Clichy, France

Synonyms

Adaptive immunity; Colitis; Ileitis; Inflammation; Innate immunity; Intestinal injury; Knockout mice; Microbiota; Transgenic mice

Definitions

Inflammatory Bowel Diseases

Inflammatory bowel diseases (IBD) are idiopathic chronic disorders of the intestine that are mediated by environmental factors in genetically susceptible individuals. The complexity of IBD pathogenesis is well suggested by the recent genome-wide association studies that led to the discovery of more than 100 susceptibility genes (Anderson et al. 2011; Franke et al. 2010). In spite of major therapeutic advance during the past 15 years, many patients with IBD still undergo surgery due to complications of the disease and lack of treatment response. Therefore, understanding the complex pathophysiology of IBD is critical to develop new therapeutic avenues.

Animal Models of Inflammatory Bowel Disease

The IBD animal models have significantly contributed to the dissection of mechanism and assessment of new strategies for IBD treatment. More than 70 animal models have been established to study IBD which are classified into four major groups: chemically induced model, spontaneous model, cell transfer model, and genetically engineered model. However, an important limitation in illuminating the mechanism of IBD is the availability of an animal model that perfectly reproduces the spectrum of human IBD. This chapter aims to describe the main animal models of IBD and their implication in understanding IBD pathophysiology.

Structure, Functions, and Physiopathological Relevance

Chemically Induced Models of IBD

The first model of IBD was designed by Kirsner et al. in 1957 who induced colic inflammation by sensitization of rats to crystalline egg albumin and rectal instillation of dilute formalin (Kirsner and Elchlepp 1957). Other types of chemically induced IBD have since been developed (Table 1). These models have the advantage of causing inflammation with a normal immune system.

Oxazolone Model

Oxazolone is a chemical allergen, firstly used for studying delayed-type hypersensitivity response in skin, which can bind to proteins gaining antigenic properties. Intrarectal administration of 1 % oxazolone (resolved in ethanol) in C57BL/6 mice presensitized with 3 % oxazolone by skin painting 5 days before rectal challenge induces colitis (Table 1) (Heller et al. 2002). This colitis is characterized by superficial lesions with hemorrhagic inflammation and severe submucosal edema, and it is mediated by IL4- and IL13-producing NKT cells. Oxazolone-modified self-proteins and bacterial antigens may activate NKT cells in colic mucosa inducing IL13 production, which may further mediate mucosal damage. This hypothesis is supported by the observations that oxazolone-induced colitis is prevented by IL13 blockade or NKT cell inhibition. Human ulcerative colitis (UC) colic mucosa also contains an increased number of IL13-secreting $CD161^+$ NKT cells (Fuss et al. 2004). Thus, oxazolone-induced colitis is a good model of Th2 cytokine mediated IBD.

Dextran Sulfate Sodium (DSS) Model

Administration of 1–5 % DSS, a polymer of sulfated polysaccharide, in drinking water can induce acute or chronic intestinal injury, according to the type of regime. Acute colitis is induced by continuous administration of DSS for a short period (4–9 days) and is characterized by bloody diarrhea, intestinal inflammation, ulcerations, and weight loss (Table 1). Intestinal recovery upon DSS withdrawal allows the study of mechanisms involved in the resolution of gut inflammation (epithelial healing, development of regulatory immune cells) (Cadwell et al. 2008; Pickert et al. 2009). Chronic colitis may be induced by cyclical administration of DSS and consists of mononuclear leukocyte infiltration, crypt architectural disarray, and transmural inflammation, features of Crohn's disease (CD) (Table 1) (Melgar et al. 2005). Furthermore, dysplasia frequently occurs in DSS-induced chronic colitis which may allow the study of colitis-associated carcinogenesis (De Robertis et al. 2011).

The mechanisms of DSS-induced intestinal damage are unclear, but direct hyperosmotic

Inflammatory Bowel Disease Models in Animals, Table 1 Chemically induced and spontaneous models of IBD

Class	Animal model IBD	Lesions	Contribution	Advantage	Disadvantage
Chemically induced models	Oxazolone model	Superficial colic inflammation	Role of NKT cells in human UC	Model of UC	Short duration of mucosal lesion
		Edema	Role of Th2 cytokines (IL13 and IL4) in human UC	Model of Th2 cytokine-induced colitis	Probably reflects early UC injury
		Ulceration of the epithelial cell layer			
		Neutrophil infiltration			
	DSS model	Acute: mucin depletion, epithelial degeneration and necrosis, neutrophils infiltration, crypt abscesses	Role of innate immune system	Easy to use	Variability of inflammation severity from mouse to mouse and region to region
		Chronic: mononuclear leukocyte infiltration, crypt architectural disarray, mucosal lymphocytosis, transmural inflammation	Role of epithelial barrier disarrays	Time and cost saving	Difficulties in assessing intestinal inflammation
			Epithelial repair mechanisms	Frequently used in genetically engineered mice to assess the role of different mediators of inflammation	
			Mechanisms of inflammation-associated colon cancer		
	TNBS model	Bowel wall thickening	Role of NOD2 in colitis	Time and cost saving	Highly dependent on mouse strains
		Extensive ulcerations	Role of Th1 cells in mucosal immune response	Model of CD	A high number of animals is required in each treatment group
		Transmural leukocyte infiltration	Therapeutic effect of anti-IL12p40 antibodies in colitis	Model of Th1 cell-induced colitis	
		Crypt distortion			
Spontaneous models	SAMP1/Yit mouse model	Mostly on terminal ileum	Insights into the pathogenic mechanisms of CD	Very good model of CD	
		Mild to moderate ileitis	Therapeutic effect of anti-TNFα antibodies in CD	Chronic ileitis with 100 % penetrance	
		Discontinuous transmural intestinal inflammation		Th1- and Th2-mediated ileal inflammation	
		Neutrophil and lymphocyte infiltration			
		Granuloma formation			
		Crypt abscesses		Suitable for testing novel therapies	

injury to epithelial cells seems to play a major role. Breakdown of epithelial barrier function allows the passage of bacterial antigens into the mucosa triggering the inflammatory response through the release of different pro-inflammatory mediators (TNFα, IL1β, IL12). T cell- and B cell-deficient mice (Rag1 −/− mice) can also develop acute colitis as a result of DSS exposure suggesting that the adaptive immune system does not play a major role in this model (Dieleman et al. 1994). The severity of inflammation considerably varies from mouse to mouse as well as from region to region in an individual colon, being influenced by factors such as genetic background, protocol of DSS administration, or DSS molecular weight. Although a very useful model to study some aspects of intestinal inflammation, attention should be made when results are interpreted (Table 1).

Trinitrobenzene Sulfonic Acid (TNBS) Model
Intrarectal administration of TNBS in 50 % ethanol solution can induce colitis in susceptible strains of mice, which resembles CD (Table 1) (Scheiffele and Fuss 2002). Thus, SJL/J, Balb/c, and C3HeJ mice show a high susceptibility to TNBS-induced intestinal inflammation, while C57BL/6 mice are resistant. It is believed that ethanol-induced disruption of mucosal epithelial cell barrier function facilitates the passage of TNBS into the intestinal mucosa. Similar to oxazolone, TNBS haptenates self-proteins and bacterial proteins triggering a Th1 cell-mediated immune response. The importance of intestinal bacteria is suggested by the observation that mice with TNBS-induced colitis are responsive to their own flora. Furthermore, transfer of T cells extracted from lamina propria of mice with TNBS colitis to naïve mice induces colitis in the absence of TNBS (Neurath et al. 1996). Th1 cells play a central role in TNBS colitis. IL12, a cytokine known to mediate Th1 cell differentiation from naïve cells, is upregulated in the inflamed tissue of TNBS colitis mice. Moreover, anti-IL12p40 antibody therapy can prevent or improve TNBS-induced colitis. In humans, mononuclear antibodies against anti-IL12 may induce clinical response and remission in patients with CD (Mannon et al. 2004). Overall, TNBS model of colitis is a good tool to study mechanisms of gut inflammation related to CD and in particular Th1-mediated inflammatory response.

Spontaneous Models of IBD

In some animal models of IBD, chronic gut inflammation may spontaneously develop in the absence of chemical manipulations. On such model is SAMP1/Yit mouse which develops a spontaneous and chronic ileitis with 100 % penetrance by week 30, mostly located in the terminal ileum that closely resembles human CD (Table 1) (Pizarro et al. 2011). Different from other models, intestinal inflammation can occur even in the absence of enteric bacteria. Th1 cells (high IFNγ and TNFα production) are involved in the development of ileal inflammation in SAMP1/Yit mice, and anti-TNFα therapy significantly alleviates this process (Marini et al. 2003). Th2 response is also responsible for ileitis, as IL13 and IL5 are significantly increased in the inflamed ileum of SAMP1/Yit mice (Takedatsu et al. 2004). Furthermore, neutralization of IL5 attenuates ileitis in the same mouse model (Takedatsu et al. 2004).

Cell-Transfer Models of IBD

The most widely used model of IBD in this group is $CD45RB^{high}$ transfer model (Table 1). $CD4^+CD45RB^{high}$ T cells, that are naïve $CD4^+$ T cells, isolated via fluorescence-activated cell sorting (FACS) from the spleen of wild-type mice transferred to immunodeficient $RAG1/2^{-/-}$ or SCID-recipient mice induce transmural colitis at 5–8 weeks after cell transfer (Powrie et al. 1994; Read et al. 2000). In mice, naïve $CD4^+$ T cells express a high level of CD45RB, whereas $CD4^+$ T cells that have encountered antigen (memory T cells) express a lower level of CD45RB. It was shown that cotransfer of $CD4^+CD45RB^{low}$ T cell subset protects against colitis development (Morrissey and Charrier 1994). Further studies demonstrated that $CD25^+FoxP3+$ regulatory T cell (Treg) subpopulation accounts for the protective role against colitis since depletion of $CD25^+$ cells from the $CD4^+CD45RB^{low}$ cells abrogates their potential to prevent colitis (Read et al. 2000). IL10 and TGFβ seem to be central anti-inflammatory factors in $CD45RB^{low}$ model of colitis. Anti-IL10 administration blocked Treg capacity to prevent CD4$^+$CD45RB^{high} T cell-induced colitis in immunodeficient mice (Asseman et al. 1999). Thus, this model majorly contributed to the understanding of Treg role in controlling intestinal inflammation.

Th1 pro-inflammatory response mediates gut inflammation in CD45RB^{low} model of colitis via production of TNFα and IFNγ (Powrie et al. 1994). Furthermore, the role of IL23 and its receptor in the pathogenesis of CD45RB^{low} model was also documented (Ahern et al. 2010). The mechanism of IL23-induced gut inflammation is probably mediated by the ability of this cytokine to promote T cell proliferation, maintain Th17 cell differentiation, and inhibit Treg differentiation (Ahern et al. 2010).

Genetically Engineered Models of IBD

Genetically engineered animals have opened new opportunities in the study of mechanisms of disease. They offered more accurate models of disease and allowed development of efficient therapies. Conventional or conditional transgenic (Tg) mice overexpress a gene of interest in all or a specific cell type, respectively. Oppositely, conventional or conditional knockout (KO) mice lack a gene of interest in all or a specific cell type, respectively. Another model is the knockin (KI) group of mice that are genetically engineered to carry a mutation in the gene of interest. The most relevant genetically engineered models of IBD are discussed below.

IL10 KO Mouse Model

IL10 is a potent suppressor cytokine that inhibits pro-inflammatory response from innate and adaptive immune system, preventing tissue damage caused by exacerbated inflammation. The protective role of Treg against development of bowel inflammation is mainly mediated by IL10 production (Asseman et al. 1999). Thus, recent data suggest that IL10 is involved in the immune tolerance to gut microbiota. The transfer of commensal bacteria, *Bacteroides fragilis*, in the intestine of germ-free mice can induce Treg cell that protect from colitis via an IL10-mediated mechanism (Round and Mazmanian 2010).

Genetically engineered mice lacking 1L10 gene spontaneously develop colitis after 3 month of age (Table 1) (Kuhn et al. 1993). Gut microbiota plays a major role in this model of colitis as intestinal inflammation does not occur in mice maintained in germ-free conditions (Matharu et al. 2009). The colitis of IL10 KO mice is mediated by CD4$^+$T cells. IL12 induces Th1 cell differentiation. The protective role of anti-IL12p40 antibodies, that blocks both IL12 and IL23 cytokines, has already been proven in IL10 KO mice colitis (Davidson et al. 1998). However, IL12p35 (IL12)-deficient IL10 double KO mice develop colitis at an early age (Yen et al. 2006). Furthermore, deficiency of INFγ (a product of Th1 cells) was associated with exacerbation of colitis in IL10 KO mice (Sheikh et al. 2010). Oppositely, IL12p19 (IL23)-deficient IL10 double KO mice do not show colitis even at 12 months of age (Yen et al. 2006). IL23 cytokine is a major factor for maintenance of Th17 differentiation. Therefore, IL10 KO mice colitis seems to be mediated by Th17 cytokines, and Th1 cytokines may exhibit a protective effect. Recent data suggest that disbalance of Th1 and Th17 can lead to inflammatory disease (Pan et al. 2012) which may explain the opposite role of the two pro-inflammatory pathways in IL10 KO mouse model of colitis.

STAT3 KO Mouse Model

Signal transducer and activator of transcription (STAT) 3 is a transcription factor implicated in the regulation of both innate and adaptive immune response. The function of STAT3 greatly depends on cell type. In macrophages and neutrophils, STAT3 has anti-inflammatory effects mediating the signals of IL10. Oppositely, STAT3 pathway activation by IL6 is critical for differentiation of CD$^+$T cells into Th17 cells with pro-inflammatory properties. Therefore, STAT3 deficiency in CD4$^+$T cells improves colitis in CD45RB^{high} model (Durant et al. 2010). STAT3 is activated in the intestinal epithelial cells upon mucosa damage and drive tissue regeneration (Pickert et al. 2009). Mice specifically lacking STAT3 in the intestinal epithelial cells are more susceptible to DSS-induced colitis (Pickert et al. 2009).

In humans, STAT3 has recently been identified as a susceptibility gene of both CD and UC (Anderson et al. 2011; Franke et al. 2010).

Conditional KO mice with macrophage and neutrophil deletion of STAT3 spontaneously develop colitis at the age of 20 weeks (Table 1) (Takeda et al. 1999). This colitis is associated with an increased Th1 cytokine production by macrophages which are not responsive to anti-inflammatory effects of IL10 (Takeda et al. 1999). The absence of colitis in STAT3 KO mice lacking T and B lymphocytes suggests that interaction between innate and adaptive immune cells is essential in this model of colitis (Reindl et al. 2007).

TGFβ KO Mouse Model

Transforming growth factor (TGF) β is a multifunctional cytokine that holds a role in immune homeostasis and tolerance, modulation of inflammation, and wound repair. TGFβ is important to maintain Treg suppressor function and differentiation; blocking TGFβ activity prevents the Treg from suppressing immunity (Nakamura et al. 2004).

TGFβ KO mice develop necrotic inflammation in the intestine and other organs, being able to survive only until 3–4 weeks of age (Shull et al. 1992). To overcome the problem of early death of TGFβ KO mice, conditional transgenic mice specifically overexpressing the dominant negative TGFβ receptor type II in CD^+ T cells have been created (Gorelik and Flavell 2000). TGFβ receptor type II overexpression can block TGFβ signaling pathway by interfering with the assembly of TGFβ-binding receptor complex. This type of mice develops inflammation in the colon, stomach, duodenum, as well as other organs at about 3–4 months of age (Gorelik and Flavell 2000). Gut inflammation related to TGFβ signal alteration seems to be mediated by both Th1 and Th2 pathways, and B cells play a protective role in this model of colitis. Oppositely, a suppression of Th17 response is seen.

TCRα KO Mouse Model

T cell receptor (TCR) plays a central role in the specific immune response (adaptive immune response) to foreign antigens while maintaining self-tolerance. TCR, composed of TCRα and TCRβ chains, is required for the recognition of antigens bound to the molecules of the major histocompatibility complex expressed on antigen-presenting cells and consequent activation of T cell-mediated adaptive immune response.

Treg cell development in the thymus requires TCR gene rearrangement at the TCRα locus. Furthermore, TCR is essential for the suppressive function of Treg cells (Kim et al. 2009). TCRα KO mice develop Th2-mediated colitis at 6 month of age in a proportion of approximately 60 % (Mombaerts et al. 1993) (Table 1). TCR-deficient mice produce T cells expressing only TCRβ chains (TCRα-$β^+$T cells). The TCR-α-$β^+$T cells are immunologically functional and can induce colitis through production of IL4 (Iijima et al. 1999). Anti-IL4-neutralizing antibody treated TCRα KO mice, and TCRα KO/IL4 KO mice exhibit no or only mild colitis (Iijima et al. 1999). Intestinal flora plays a major role in this model, as colitis does not develop in TCRα KO mice maintained in germ-free conditions (Dianda et al. 1997). Similarly to human UC, resection of cecal patches (equivalent to human appendectomy) or carbon monoxide exposure (mimic of smoking) suppresses colitis in TCRα KO mice (Sheikh et al. 2011).

CD40L Tg Mouse Model

CD40 ligand (CD40L or CD154) is mainly expressed on T cells and interacts with CD40 receptor expressed on antigen-presenting cells such as macrophages, dendritic cells, B cells, or endothelial cells. CD40L/CD40 interaction triggers stimulatory signals in both antigen-presenting cells and T cells. Inflammatory infiltrate containing CD40L-expressing T cells have been detected in lamina propria of patients with inflammatory bowel disease and can contribute to intestinal inflammation (Battaglia et al. 1999). Mice overexpressing CD40L develop colitis between 3 and 6 weeks of age (Table 1) (Clegg et al. 1997). Increased mononuclear cell infiltrate is also detected in other organs of CD40L Tg mice, such as the lung, liver, and pancreas.

Furthermore, overexpression of CD40L on B cells is associated with development of colitis and ileitis (Kawamura et al. 2004).

STAT4 Tg Mouse Model

STAT4 is a transcription factor that mediates IL12-induced development of Th1 cells. Furthermore, STAT4 has an inhibitory effect on Treg cell differentiation. Thus, STAT4 promotes a Th1 pro-inflammatory response that plays a capital role in human CD (Simpson et al. 1998). Mice overexpressing STAT4 transcription factor develop severe transmural colitis (Table 1) within 7–14 days after immunization with dinitrophenyl keyhole limpet hemocyanin (DNP-KLH) (Wirtz et al. 1999). Colonic CD^+T cells produce high levels of Th1 cytokines (TNFα, IFNγ) and adoptive transfer of CD^+T cells from the colon of STAT4 Tg mice induces colitis in SCID mice (Wirtz et al. 1999).

TNFSF15 Tg Mouse Model

TNFSF15 (or TL1A) is a TNF-like factor that binds to death receptor (DR) 3 primarily expressed on activated lymphocytes. TNFSF15-DR3 interaction promotes effecter T cell proliferation at the site of inflammation and induces activation signals probably through activation of NFκB pathway (Pobezinskaya et al. 2011). Furthermore, TNFSF15 can act as a costimulator of T cells enhancing both Th1 and Th17 responses (Meylan et al. 2008). In humans, TNFSF15 has been identified as a susceptibility gene for CD and UC (Anderson et al. 2011; Franke et al. 2010). TNFSF15 is produced by lamina propria macrophages and promotes Th cell production of IFNγ and IL17 in patients with CD (Kamada et al. 2010). Transgenic mice overexpressing TNFSF15 in T lymphocytes spontaneously develop small intestine inflammation (Table 1) at 6 weeks of age in a proportion of 100 % (Meylan et al. 2011). Moreover, TNFSF15 overexpression on dendritic cells is also associated with development of ileitis, but pathological changes were milder. Interestingly, inflammation does not develop in the colon. IL13 is highly upregulated in the small intestine of TNFSF15 Tg mice, and treatment with anti-IL13-neutralizing antibodies effectively alleviates intestinal inflammation (Meylan et al. 2011). IL17-pathway is also involved in this model of IBD but to a lesser extent.

TNF (ARE) Mouse Model

Adenylate-uridylate-rich elements (AREs) are regulatory sequences localized in the untranslated region of mRNAs that strongly accelerates degradation of mammalian mRNAs. Deletion of ARE in the untranslated region of TNFα mRNA in mice is associated with an enhanced constitutive and inducible production of TNFα due to an increased mRNA stability. TNFα overexpression leads to spontaneous intestinal inflammation and chronic polyarthritis by 4 weeks of age in this mouse model (Table 1) (Kontoyiannis et al. 2002). Development of intestinal inflammation is mediated by T cells and Th1 pro-inflammatory cytokines (Kontoyiannis et al. 2002). Similarly to CD, mucosal inflammation is mainly located in the terminal ileum.

SHIP KO Mouse Model

Src homology 2 (SH2)-containing inositol-5-phosphatase (SHIP) protein becomes activated in immune cells as a response to various growth factors as well as TCR or B cell receptor stimuli (Kerr 2008). SHIP signaling pathway is critical in limiting the number and function of regulatory immune cells in peripheral lymph nodes and spleen, by reducing survival of Treg cells or by preventing FoxP3 expression in naïve T cells. Therefore, SHIP deficiency is associated with an expansion of immunoregulatory cells in peripheral lymphoid tissues (Paraiso et al. 2007). However, regulation of the immune system is altered in this mouse model. Furthermore, granulocytes of SHIP KO mice are less susceptible to apoptosis, and an increased granulocyte-monocyte infiltration is seen in different organs such as the gut, liver, kidneys, or lymph nodes.

SHIP KO mice develop transmural segmental ileitis at 6–8 weeks of age, and ileal granulomas are detected in almost one third of mice (Kerr et al. 2011) (Table 1). Adoptive transfer of splenocytes from SHIP KO mice can induce ileitis in wild-type mice (Kerr et al. 2011). Interestingly, ileitis is not induced either by transfer of CD3$^+$T cells or NK cells, suggesting that granulocyte-monocyte lineage cells may play a major role in this animal model of colitis.

Integrin αV and β8 KO Mouse Models

Integrins are a major family of cell-surface-adhesion receptors that are heterodimers of noncovalently associated α and β subunits. αV subunit associates with different β subunits (β1, β3, β5, β6, β8) to form integrins with key role in cell adhesion, migration, and survival. β8 subunit is associated with αV subunit to form αV β8 integrin that is involved in the activation of TGFβ.

Conditional KO mice with αV subunit deletion in endothelial and hematopoietic cells develop chronic and progressive inflammation in the colon by 20 weeks of age, but not in the small intestine (Table 1) (Lacy-Hulbert et al. 2007). Other organs such as the liver, respiratory tract, and peritoneum are also affected. Colic injury is associated with enhanced expression of IFNγ, IL6, TNFα, as well as IL4. A reduced infiltration of Treg cells is also observed in the colon of αV KO mice and may be related to the loss of αV-mediated TGFβ activation. Indeed, generation of Treg in the intestinal mucosa and induction of tolerance require expression of αV subunit by dendritic cells (Paidassi et al. 2011). Furthermore, macrophages and dendritic cells from αV KO mice have impaired capacity to phagocytize apoptotic cells which are a major source of self-antigens for the stimulation of adaptive immunity.

Mice with conditional loss of β8 unit on leukocyte develop a progressive wasting disorder by 4–6 months of age and severe colitis by 10 months of age (Table 1) (Travis et al. 2007). High levels of autoantibodies directed against double-stranded DNA and ribonuclear proteins are detected in these mice. Specific deletion of β8 unit on dendritic cells induced colitis, whereas deletion of β8 unit only in CD4$^+$ cells failed to induce colitis. Thus, expression of αVβ8 integrin on dendritic cells is capital for the negative regulation of adaptive immunity. Disturbance of TGFβ activation and consequent impairment of Treg differentiation accounts for uncontrolled intestinal inflammation in this model of colitis.

NOD2 Mouse Models

NOD2 is a member of cytoplasmic NLR (nucleotide-binding domain, leucine-rich repeats) proteins which play a key role in the recognition of and host defense against pathogens or extracellular danger signals. NOD2 is mainly expressed on dendritic cells, macrophages, Paneth cells, and intestinal epithelial cells (Voss et al. 2006). It can recognize muramyl dipeptide, a bacterial composant, conferring resistance to both Gram-positive and Gram-negative bacteria. NOD2, expressed mainly on the terminal ileum, plays a major role in the regulation of commensal microbiota, and NOD2 deficiency associates dysbiosis in the terminal ileum (Rehman et al. 2011). Dysbiosis of microbiota composition has also been reported in CD patients (Podolsky 2002), and NOD2 has been identified as a CD susceptibility gene (Franke et al. 2010) strongly associated with ileal CD.

Three mouse models bearing nonfunctional NOD2 have been generated. NOD2 mice do not develop spontaneous intestinal inflammation and display normal lymphoid and myeloid cellular composition in the thymus and spleen (Kobayashi et al. 2005). Nevertheless, they show some abnormalities which may participate to the pathogenesis of CD. The first mouse model carries a NOD2^{2939insC} mutation similar to human CD (Maeda et al. 2005). Macrophages isolated from NOD2^{2939insC} mutant mice show an enhanced pro-inflammatory response upon muramyl dipeptide stimulation. Furthermore, NOD2^{2939insC} mutant mice have increased susceptibility to DSS-induced colitis. The second model is a NOD2 KO mouse strain generated by deletion of exon 1 (Watanabe et al. 2004). Macrophages isolated from this NOD2 KO mice show an enhanced production of IL12 upon TLR2 stimulation. IL12 is an inductor of Th1 cell differentiation, suggesting that NOD2 may have an inhibitor role on TLR2-mediated Th1 activation. The third model is a NOD2 KO mouse strain generated by deletion of exon 3. In the intestinal tract, the NOD2 is mainly expressed by the Paneth cells of crypts in the terminal ileum (Lala et al. 2003). NOD2 is involved in the regulation of gut microbiota by stimulating the production of antimicrobial peptides such α-defensins. NOD2-deficient mice had a reduced crypt capacity of killing Gram-positive or Gram-negative bacteria (Petnicki-Ocwieja et al. 2009). In accordance with this observation, an increased colonization by commensal bacteria (*Bacteroides*, Firmicutes, and *Bacillus*) has been detected in the terminal ileum of NOD2 deficient mice (Petnicki-Ocwieja et al. 2009). In humans, CD patients with ileal involvement showed decreased defensin production independently of the presence of NOD2 mutation (Wehkamp et al. 2005). Nevertheless, the level of defensin expression was much more reduced in patients bearing NOD2 mutation (Wehkamp et al. 2005). A recent study showed that NOD2 KO mice can develop ileitis and granuloma upon inoculation of the opportunistic bacterium, *Helicobacter hepaticus* (Biswas et al. 2010). Thus, alteration of host-microbiota interaction may account for the association of NOD2 with ileal Crohn. However, NOD2 deficiency is not sufficient to induce inflammatory bowel disease, and additional factors are required.

XBP1 KO Mouse Model

X-box-binding protein 1 (XBP1) plays a key role in unfolded protein response. The unfolded protein response is a cellular adaptive mechanism that alleviates endoplasmic reticulum stress by increasing the protein folding capacity and simultaneously reducing the influx of nascent polypeptides into the endoplasmic reticulum. Conditional KO mice with XBP1 deletion in the intestinal epithelial cells develop spontaneous small intestinal mucosa inflammation (Table 1), in association with increased endoplasmic reticulum stress (ERS) (Kaser et al. 2008). Ileal lesions are characterized by the absence of Paneth cells and an altered local antimicrobial function. Intestinal epithelial cells of XBP1 KO mice show an enhanced pro-inflammatory response to environmental stimuli. Furthermore, XBP1 KO mice are more susceptible to DSS-induced colitis (Kaser et al. 2008). Interestingly, human intestinal mucosa exhibits increased endoplasmic reticulum stress, and some XBP1 variants have been associated with both CD and UC (Kaser et al. 2008).

Autophagy-Related Mouse Models: Atg5 KO and ATG16-L1 KO

Autophagy is a cellular process involved in the degradation of cytosolic proteins, protein aggregates, damaged or excess organelles, and invasive microbes. It consists of sequestration of organelles or large unfolded proteins within membranes and fusion with lysosomes for subsequent degradation. Recent studies identified Atg16L1, a key gene for autophagy, as a susceptibility gene for CD (Anderson et al. 2011; Franke et al. 2010). Three mouse models lacking autophagy genes (Atg16L1 and Atg5) have been developed. The first mouse strain lacking functional Atg16L1 was generated through genet-trap-mediated disruption (Cadwell et al. 2008). Atg16L1-deficient mice do not spontaneously develop intestinal inflammation. However, abnormalities of Paneth cell such as altered granule exocytosis and increases in expression of genes involved in regulating injury responses are detected. Interestingly, patients carrying the Atg16L1 risk allele bear similar abnormalities of Paneth cells (Cadwell et al. 2008). The second Atg16L1-deficint mouse model carries a mutant AtgL1 gene coding for a deleted form of Atg16L1 protein (Saitoh et al. 2008). Macrophages isolated from the Atg16L1-deficint mice have an enhanced pro-inflammatory response (IL1β, IL18) upon lipopolysaccharide stimulation (Saitoh et al. 2008). Furthermore, this Atg16L1-deficint mouse strain is more susceptible to DSS-induced colitis. Nevertheless, spontaneous colitis does not develop, suggesting that additional factors are required for the induction of intestinal inflammation. Another mouse model of altered autophagy consists of transplantation of thymi from Atg5 KO mice under the renal capsule of normal adult recipients (Nedjic et al. 2008). These mice develop inflammation in the colon but also in the lung, liver, and uterus, suggesting the key role of autophagy in the adaptive immune response.

Muc2 KO Mouse Model

Intestinal goblet cells produce the mucus layer that covers intestinal epithelium and form a mechanical barrier between mucosal surface and luminal content. Muc2 mucin is a key compound of intestinal mucus responsible for its high viscosity. It forms a protease-resistant matrix that retains molecules involved in host defense. Patients with UC have a decrease in goblet cell number, Muc2 production, and mucus secretion (Hanski et al. 1999). Muc2 KO mice develop distal colitis as early as 5 weeks of age (Table 1) (Van der Sluis et al. 2006). Interestingly, colic adenoma and adenocarcinoma emerged at 6-month follow-up (Velcich et al. 2002). Muc2 deficiency increased the susceptibility to DSS-induced colonic damage even in heterozygous state. Moreover colitis was further exacerbated in Muc2/IL10 double-knockout mice (van der Sluis et al. 2008). Thus, alteration of mucus barrier function seems to be a key factor for development of colitis (Table 2).

Winnie Mice

By murine N-ethyl-N-nitrosourea mutagenesis, Heazlewood et al. (2008) identified two distinct noncomplementing missense mutations in Muc2 causing an ulcerative colitis-like phenotype. One hundred percent of mice of both strains developed mild spontaneous distal intestinal inflammation by 6 weeks (histological colitis scores versus wild-type mice, $p < 0.01$) and chronic diarrhea. Mutant mice showed aberrant Muc2 biosynthesis, less stored mucin in goblet cells, a diminished mucus barrier, and increased susceptibility to colitis induced by a luminal toxin. Enhanced local production of IL-1beta, TNFα, and IFN-gamma was seen in the distal colon, and intestinal permeability increased twofold. The number of leukocytes within mesenteric lymph nodes increased fivefold, and leukocytes cultured in vitro produced more Th1 and Th2 cytokines (IFN-gamma, TNFα, and IL13). This pathology was accompanied by accumulation of the Muc2 precursor and ultrastructural and biochemical evidence of endoplasmic reticulum (ER) stress in goblet cells, activation of the unfolded protein response, and altered intestinal expression of genes involved in ER stress, inflammation, apoptosis, and wound repair. This model reinforces the idea that abnormalities of the ER stress in goblet cells may be an initial event in UC lesions. However, such mutations in the MUC2 gene have not been showed in human ulcerative colitis.

EXCY2 Model: Double Nox1/IL10 Knock-Out Mice (IL10/Nox1dKO Mice)

In ulcerative colitis, ER stress is disturbed in the uninflamed colonic mucosa, with a specific eIF2α hypo-phosphorylation (Treton et al. 2011). To study the effect of ER stress on goblet cells, Nox1-deficient mice, which exhibit increased goblet cell number (Coant et al. 2010), were crossed with IL10KO mice, which express deregulated ER stress in epithelial intestinal cells and develop mild enterocolitis. IL10/Nox1dKO mice spontaneously developed colitis at 6/7 weeks old. Colitis features accurately imitated the entire human UC phenotype, i.e., anatomical, histological, and immunological (increase of pro-inflammatory cytokines TNF, IL17A, IL13, and $CD4^+$ T lymphocytes in the lamina propria). At a molecular level, the colonic epithelium expressed eIF2α hypo-phosphorylation before onset of histological colitis. Long-term follow-up (week 32) showed worsening of colitis associated with high-grade dysplasia or invasive colon cancer and typical sclerosing cholangitis lesions in 40 % and 25 % of cases, respectively. Treatment of IL10/Nox1dKO mice with salubrinal, a drug which acts as a specific inhibitor of eIF2α phosphatase enzymes, restored colonic ERS homeostasis and prevented colitis onset in all mice.

Conclusion

The large number of animal model of intestinal inflammation (>70) highlights the fact that there is yet no ideal model reproducing human IBD. This is partly due to the fact that systematic studies evaluating how well murine models mimic human inflammatory diseases were lacking. A recent study published in PNAS showed that, although acute inflammatory stress from different etiologies result in highly similar genomic responses in humans, the responses in corresponding mouse models correlate poorly with the human conditions (Seok et al. 2013).

Another limitation is related to the multifactorial nature of IBD, involving epithelial,

Inflammatory Bowel Disease Models in Animals, Table 2 Cell transfer and genetically engineered models of IBD

Class	Animal model IBD	Lesions	Contribution	Advantage	Disadvantage
Cell transfer models	CD45RB^{high} transfer model	Wasting disease	Role of Treg cells in the control of gut inflammation	Colitis with 100 % penetrance	Immunodeficiency of recipient mice
		Chronic progressive colitis	Protective role of IL10	Model of Th1/Th17 cytokine-induced colitis	
		Transmural inflammation	The pathogenic role of IL23 pathway in colitis	Good model for testing new therapeutic molecules	
		Crypt degeneration	Insights into the Th17 cell role in colitis		
Genetically engineered models	IL10 KO mouse model	Chronic transmural inflammation	Role of gut microbiota	Gut microbiota-induced colitis	Disturbed immunoregulatory function
		Mucosal hyperplasia	Role of Th17 pathway	Th17 cytokine-induced colitis	
		Granulomas	Role of disbalance of Th1 and Th17		
		Pseudopolyp formation			
		Aberrant expression of MHC class II on intestinal epithelia			
	TGFβ KO mouse model	Wasting disease	Role of TGFβ in the modulation of gut inflammation	Model of Th1 and Th2 cytokine-induced colitis	Suppression of Th17 response
		Moderate to severe transmural inflammation (lymphocytes, macrophages, plasma cells) in the colon			
		Crypt abscesses and distortion			
		Mild inflammation in the stomach and duodenum			
	TCRα KO mouse model	Inflammation limited primarily to colic mucosa	Role of Th2 cytokines (IL4) in human UC	Good model of UC (common etiological factors)	Dysregulation T cells
		Oligoclonal immune response against microbiota	Protective role of IL22 against colitis	Model of Th2-induced colitis	
		Anti-tropomyosin antibodies, anti-neutrophil cytoplasmic antibodies		Gut microbiota-induced colitis	
	SHIP KO mouse model	Segmental ileitis	Role of SHIP as a susceptibility gene in Crohn's disease	Model of Crohn's disease	Profound alterations in the homeostasis of immunoregulatory cells
		Transmural ileal inflammation with a predominance of neutrophil infiltration	Role of granulocyte-monocyte lineage	Gut microbiota-induced colitis	
		Ileal granulomas			
		Fistula formation			
		Crypt abscesses			

(continued)

Inflammatory Bowel Disease Models in Animals, Table 2 (continued)

Class	Animal model IBD	Lesions	Contribution	Advantage	Disadvantage
	STAT3 KO mouse model	Colic injuries	Role of myeloid cell lineage	Model of UC	Alterations in the homeostasis of myeloid cell lineage
		Reduced gland number; depletion of goblet cells	Role of innate-adaptive immune cell interaction	Model of Th1-induced colitis	
		Lamina propria infiltration of neutrophils, lymphocytes, plasma cells		Intact adaptive immune system	
		Crypt abscesses			
		Mucosa ulcerations			
	αV KO and β8 KO mouse models	Injury limited to the colon	Role of impaired phagocytosis in intestinal inflammation	Models of UC	
		Inflammatory infiltrations (eosinophils, plasma cells); ulcerations; crypt abscesses; extensive epithelial proliferation	Role of impaired generation of Treg cells	Models of Th1 and Th2 cytokine-induced colitis	
		Adenocarcinoma development			
	NOD2 mouse models	No spontaneous intestinal inflammation	Role NOD2 in host-microbiota interaction	Ileal abnormalities similar to CD (dysbiosis; altered defensins production)	No spontaneous intestinal lesions
		Ileitis with granuloma formation upon inoculation of Helicobacter hepaticus		Intact adaptive immune system	
	XBP1 KO mouse model	Inflammation of the small intestine	Role of endoplasmic reticulum stress in CD and UC	Model of intestinal inflammation induced by defects in mucosa function	
		Lamina propria neutrophil infiltrates	Role of environmental factors in intestinal inflammation	Intact adaptive immune system	
		Crypt abscesses			
		Ulcerations without granulomas			
	Atg16L1 and Atg5 mouse models	No spontaneous intestinal inflammation	Role of autophagy in the regulation of Paneth cell function	Paneth cell abnormalities similar to CD	No spontaneous intestinal lesions
		Spontaneous colic inflammation in mice with specific thymic Atg5 deificiency	Role of autophagy in the control of endotoxin-induced inflammation		
			Autophagy as a susceptibility factor for colitis		

(*continued*)

Inflammatory Bowel Disease Models in Animals, Table 2 (continued)

Class	Animal model IBD	Lesions	Contribution	Advantage	Disadvantage
	Muc2 KO mouse model	Thickening of the intestinal mucosa	Role of mucus barrier function in UC	Model of UC	
		Superficial erosions		Model of local barrier dysfunction-induced colitis	
		Inflammatory infiltrate (lymphocytes)		Intact adaptive immune system	
		Crypt hyperplasia			
		Flattening of epithelial cells			
		Distortion of the lamina propria			
		Colic adenoma and adenocarcinoma			
	CD40L Tg mouse model	Colic transmural mononuclear infiltrate	Role of CD40L-/CD40-mediated inflammation		
		Granuloma formation			
		Ulcerations			
		Glandular loss and dysplasia			
	STAT4 Tg mouse model	Colic transmural inflammation	Role of IL12 in IBD	Th1-mediated colitis	
		Inflammatory infiltrate containing lymphocytes, granulocytes, and macrophages	Role of Th1 pro-inflammatory response in IBD		
		Destruction of crypt architecture			
	TNFSF15 Tg mouse model	Inflammation limited to the small intestine	Role of TNF super family in IBD	Model of CD	
		Ileal dilatation	Role of IL13 and IL17 in IBD	T-lymphocyte-induced intestinal inflammation	
		Lamina propria infiltration with lymphocytes, macrophages and neutrophils		IL13- and IL17-mediated intestinal inflammation	
		Hyperplasia of goblet cells			
	TNF(ARE) mouse model	Transmural intestinal inflammation mainly localized to the terminal ileum	Insides into the pathogenic action of TNFα in IBD	Good model to study pathogenic action of TNFα in IBD	
		Infiltration of lymphocytes, plasma cells, neutrophils		T-lymphocyte and Th1 cytokine-induced intestinal inflammation	
		Granuloma formation			

environmental, and immune abnormalities. If each model is used to study one aspect of IBD, it is difficult to find models that incorporate all elements of IBD pathogenesis. This limit may explain, in part, the fact that many treatments, mainly immunomodulators, which were found to be very effective in mice, failed in humans (abatacept, anti-CD3, anti-IL17). In this context, the most accomplished models are probably those involving abnormalities in both epithelial and immune compartments such as EXCY2 for ulcerative colitis. The validation of this model should include both effective and noneffective evaluated treatments in this condition to be confirmed as the ideal go/no go model. Furthermore, EXCY2 should be used for drug development both to control disease activity (as induction and/or maintenance therapy) and to prevent or treat major UC complications such as colonic dysplasia/cancer and cholangitis.

Cross-References

- ▶ Autophagy and Inflammation
- ▶ Cancer and Inflammation
- ▶ Genetic Susceptibility to Inflammatory Diseases
- ▶ Inflammatory Bowel Disease
- ▶ Lymphocyte Homing and Trafficking
- ▶ Tumor Necrosis Factor (TNF) Inhibitors

References

Ahern, P. P., Schiering, C., Buonocore, S., McGeachy, M. J., Cua, D. J., Maloy, K. J., et al. (2010). Interleukin-23 drives intestinal inflammation through direct activity on T cells. *Immunity, 33*, 279–288.

Anderson, C. A., Boucher, G., Lees, C. W., Franke, A., D'Amato, M., Taylor, K. D., et al. (2011). Meta-analysis identifies 29 additional ulcerative colitis risk loci, increasing the number of confirmed associations to 47. *Nature Genetics, 43*, 246–252.

Asseman, C., Mauze, S., Leach, M. W., Coffman, R. L., & Powrie, F. (1999). An essential role for interleukin 10 in the function of regulatory T cells that inhibit intestinal inflammation. *Journal of Experimental Medicine, 190*, 995–1004.

Battaglia, E., Biancone, L., Resegotti, A., Emanuelli, G., Fronda, G. R., & Camussi, G. (1999). Expression of CD40 and its ligand, CD40L, in intestinal lesions of Crohn's disease. *American Journal of Gastroenterology, 94*, 3279–3284.

Biswas, A., Liu, Y. J., Hao, L., Mizoguchi, A., Salzman, N. H., Bevins, C. L., et al. (2010). Induction and rescue of Nod2-dependent Th1-driven granulomatous inflammation of the ileum. *Proceedings of the National Academy of Sciences of the United States of America, 107*, 14739–14744.

Cadwell, K., Liu, J. Y., Brown, S. L., Miyoshi, H., Loh, J., Lennerz, J. K., et al. (2008). A key role for autophagy and the autophagy gene Atg16l1 in mouse and human intestinal paneth cells. *Nature, 456*, 259–263.

Clegg, C. H., Rulffes, J. T., Haugen, H. S., Hoggatt, I. H., Aruffo, A., Durham, S. K., et al. (1997). Thymus dysfunction and chronic inflammatory disease in gp39 transgenic mice. *International Immunology, 9*, 1111–1122.

Coant, N., Ben Mkaddem, S., Pedruzzi, E., Guichard, C., Treton, X., Ducroc, R., et al. (2010). NADPH oxidase 1 modulates WNT and NOTCH1 signaling to control the fate of proliferative progenitor cells in the colon. *Molecular and Cellular Biology, 30*, 2636–2650.

Davidson, N. J., Hudak, S. A., Lesley, R. E., Menon, S., Leach, M. W., & Rennick, D. M. (1998). IL-12, but not IFN-gamma, plays a major role in sustaining the chronic phase of colitis in IL-10-deficient mice. *Journal of Immunology, 161*, 3143–3149.

De Robertis, M., Massi, E., Poeta, M. L., Carotti, S., Morini, S., Cecchetelli, L., et al. (2011). The AOM/DSS murine model for the study of colon carcinogenesis: From pathways to diagnosis and therapy studies. *Journal of Carcinogenesis, 10*, 9.

Dianda, L., Hanby, A. M., Wright, N. A., Sebesteny, A., Hayday, A. C., & Owen, M. J. (1997). T cell receptor-alpha beta-deficient mice fail to develop colitis in the absence of a microbial environment. *American Journal of Pathology, 150*, 91–97.

Dieleman, L. A., Ridwan, B. U., Tennyson, G. S., Beagley, K. W., Bucy, R. P., & Elson, C. O. (1994). Dextran sulfate sodium-induced colitis occurs in severe combined immunodeficient mice. *Gastroenterology, 107*, 1643–1652.

Durant, L., Watford, W. T., Ramos, H. L., Laurence, A., Vahedi, G., Wei, L., et al. (2010). Diverse targets of the transcription factor STAT3 contribute to T cell pathogenicity and homeostasis. *Immunity, 32*, 605–615.

Franke, A., McGovern, D. P., Barrett, J. C., Wang, K., Radford-Smith, G. L., Ahmad, T., et al. (2010). Genome-wide meta-analysis increases to 71 the number of confirmed Crohn's disease susceptibility loci. *Nature Genetics, 42*, 1118–1125.

Fuss, I. J., Heller, F., Boirivant, M., Leon, F., Yoshida, M., Fichtner-Feigl, S., et al. (2004). Nonclassical CD1d-restricted NK T cells that produce IL-13 characterize an atypical Th2 response in ulcerative colitis. *Journal of Clinical Investigation, 113*, 1490–1497.

Gorelik, L., & Flavell, R. A. (2000). Abrogation of TGFbeta signaling in T cells leads to spontaneous

T cell differentiation and autoimmune disease. *Immunity, 12*, 171–181.

Hanski, C., Born, M., Foss, H. D., Marowski, B., Mansmann, U., Arasteh, K., et al. (1999). Defective post-transcriptional processing of MUC2 mucin in ulcerative colitis and in Crohn's disease increases detectability of the MUC2 protein core. *Journal of Pathology, 188*, 304–311.

Heazlewood, C. K., Cook, M. C., Eri, R., Price, G. R., Tauro, S. B., Taupin, D., et al. (2008). Aberrant mucin assembly in mice causes endoplasmic reticulum stress and spontaneous inflammation resembling ulcerative colitis. *PLoS Medicine, 5*, e54.

Heller, F., Fuss, I. J., Nieuwenhuis, E. E., Blumberg, R. S., & Strober, W. (2002). Oxazolone colitis, a Th2 colitis model resembling ulcerative colitis, is mediated by IL-13-producing NK-T cells. *Immunity, 17*, 629–638.

Iijima, H., Takahashi, I., Kishi, D., Kim, J. K., Kawano, S., Hori, M., et al. (1999). Alteration of interleukin 4 production results in the inhibition of T helper type 2 cell-dominated inflammatory bowel disease in T cell receptor alpha chain-deficient mice. *Journal of Experimental Medicine, 190*, 607–615.

Kamada, N., Hisamatsu, T., Honda, H., Kobayashi, T., Chinen, H., Takayama, T., et al. (2010). TL1A produced by lamina propria macrophages induces Th1 and Th17 immune responses in cooperation with IL-23 in patients with Crohn's disease. *Inflammatory Bowel Diseases, 16*, 568–575.

Kaser, A., Lee, A. H., Franke, A., Glickman, J. N., Zeissig, S., Tilg, H., et al. (2008). XBP1 links ER stress to intestinal inflammation and confers genetic risk for human inflammatory bowel disease. *Cell, 134*, 743–756.

Kawamura, T., Kanai, T., Dohi, T., Uraushihara, K., Totsuka, T., Iiyama, R., et al. (2004). Ectopic CD40 ligand expression on B cells triggers intestinal inflammation. *Journal of Immunology, 172*, 6388–6397.

Kerr, W. G. (2008). A role for SHIP in stem cell biology and transplantation. *Current Stem Cell Research & Therapy, 3*, 99–106.

Kerr, W. G., Park, M. Y., Maubert, M., & Engelman, R. W. (2011). SHIP deficiency causes Crohn's disease-like ileitis. *Gut, 60*, 177–188.

Kim, J. K., Klinger, M., Benjamin, J., Xiao, Y., Erle, D. J., Littman, D. R., et al. (2009). Impact of the TCR signal on regulatory T cell homeostasis, function, and trafficking. *PloS One, 4*, e6580.

Kirsner, J. B., & Elchlepp, J. (1957). The production of an experimental ulcerative colitis in rabbits. *Transactions of the Association of American Physicians, 70*, 102–119.

Kobayashi, K. S., Chamaillard, M., Ogura, Y., Henegariu, O., Inohara, N., Nunez, G., et al. (2005). Nod2-dependent regulation of innate and adaptive immunity in the intestinal tract. *Science, 307*, 731–734.

Kontoyiannis, D., Boulougouris, G., Manoloukos, M., Armaka, M., Apostolaki, M., Pizarro, T., et al. (2002). Genetic dissection of the cellular pathways and signaling mechanisms in modeled tumor necrosis factor-induced Crohn's-like inflammatory bowel disease. *Journal of Experimental Medicine, 196*, 1563–1574.

Kuhn, R., Lohler, J., Rennick, D., Rajewsky, K., & Muller, W. (1993). Interleukin-10-deficient mice develop chronic enterocolitis. *Cell, 75*, 263–274.

Lacy-Hulbert, A., Smith, A. M., Tissire, H., Barry, M., Crowley, D., Bronson, R. T., et al. (2007). Ulcerative colitis and autoimmunity induced by loss of myeloid alphav integrins. *Proceedings of the National Academy of Sciences of the United States of America, 104*, 15823–15828.

Lala, S., Ogura, Y., Osborne, C., Hor, S. Y., Bromfield, A., Davies, S., et al. (2003). Crohn's disease and the NOD2 gene: A role for paneth cells. *Gastroenterology, 125*, 47–57.

Maeda, S., Hsu, L. C., Liu, H., Bankston, L. A., Iimura, M., Kagnoff, M. F., et al. (2005). Nod2 mutation in Crohn's disease potentiates NF-kappaB activity and IL-1beta processing. *Science, 307*, 734–738.

Mannon, P. J., Fuss, I. J., Mayer, L., Elson, C. O., Sandborn, W. J., Present, D., et al. (2004). Anti-interleukin-12 antibody for active Crohn's disease. *New England Journal of Medicine, 351*, 2069–2079.

Marini, M., Bamias, G., Rivera-Nieves, J., Moskaluk, C. A., Hoang, S. B., Ross, W. G., et al. (2003). TNF-alpha neutralization ameliorates the severity of murine Crohn's-like ileitis by abrogation of intestinal epithelial cell apoptosis. *Proceedings of the National Academy of Sciences of the United States of America, 100*, 8366–8371.

Matharu, K. S., Mizoguchi, E., Cotoner, C. A., Nguyen, D. D., Mingle, B., Iweala, O. I., et al. (2009). Toll-like receptor 4-mediated regulation of spontaneous helicobacter-dependent colitis in IL-10-deficient mice. *Gastroenterology, 137*(1380–90), e1–e3.

Melgar, S., Karlsson, A., & Michaelsson, E. (2005). Acute colitis induced by dextran sulfate sodium progresses to chronicity in C57BL/6 but not in BALB/c mice: Correlation between symptoms and inflammation. *American Journal of Physiology – Gastrointestinal and Liver Physiology, 288*, G1328–G1338.

Meylan, F., Davidson, T. S., Kahle, E., Kinder, M., Acharya, K., Jankovic, D., et al. (2008). The TNF-family receptor DR3 is essential for diverse T cell-mediated inflammatory diseases. *Immunity, 29*, 79–89.

Meylan, F., Song, Y. J., Fuss, I., Villarreal, S., Kahle, E., Malm, I. J., et al. (2011). The TNF-family cytokine TL1A drives IL-13-dependent small intestinal inflammation. *Mucosal Immunology, 4*, 172–185.

Mombaerts, P., Mizoguchi, E., Grusby, M. J., Glimcher, L. H., Bhan, A. K., & Tonegawa, S. (1993). Spontaneous development of inflammatory bowel disease in T cell receptor mutant mice. *Cell, 75*, 274–282.

Morrissey, P. J., & Charrier, K. (1994). Induction of wasting disease in SCID mice by the transfer of normal $CD4^+/CD45RBhi$ T cells and the regulation of this autoreactivity by CD4$^+$/CD45RBlo T cells. *Research in Immunology, 145*, 357–362.

Nakamura, K., Kitani, A., Fuss, I., Pedersen, A., Harada, N., Nawata, H., et al. (2004). TGF-beta 1 plays an important role in the mechanism of CD4$^+$CD25$^+$ regulatory T cell activity in both humans and mice. *Journal of Immunology, 172*, 834–842.

Nedjic, J., Aichinger, M., Emmerich, J., Mizushima, N., & Klein, L. (2008). Autophagy in thymic epithelium shapes the T-cell repertoire and is essential for tolerance. *Nature, 455*, 396–400.

Neurath, M. F., Fuss, I., Kelsall, B. L., Presky, D. H., Waegell, W., & Strober, W. (1996). Experimental granulomatous colitis in mice is abrogated by induction of TGF-beta-mediated oral tolerance. *Journal of Experimental Medicine, 183*, 2605–2616.

Paidassi, H., Acharya, M., Zhang, A., Mukhopadhyay, S., Kwon, M., Chow, C., et al. (2011). Preferential expression of integrin alphavbeta8 promotes generation of regulatory T cells by mouse CD103$^+$ dendritic cells. *Gastroenterology, 141*, 1813–1820.

Pan, B., Zeng, L., Cheng, H., Song, G., Chen, C., Zhang, Y., et al. (2012). Altered balance between Th1 and Th17 cells in circulation is an indicator for the severity of murine acute GVHD. *Immunology Letters, 142*, 48–54.

Paraiso, K. H., Ghansah, T., Costello, A., Engelman, R. W., & Kerr, W. G. (2007). Induced SHIP deficiency expands myeloid regulatory cells and abrogates graft-versus-host disease. *Journal of Immunology, 178*, 2893–2900.

Petnicki-Ocwieja, T., Hrncir, T., Liu, Y. J., Biswas, A., Hudcovic, T., Tlaskalova-Hogenova, H., et al. (2009). Nod2 is required for the regulation of commensal microbiota in the intestine. *Proceedings of the National Academy of Sciences of the United States of America, 106*, 15813–15818.

Pickert, G., Neufert, C., Leppkes, M., Zheng, Y., Wittkopf, N., Warntjen, M., et al. (2009). STAT3 links IL-22 signaling in intestinal epithelial cells to mucosal wound healing. *Journal of Experimental Medicine, 206*, 1465–1472.

Pizarro, T. T., Pastorelli, L., Bamias, G., Garg, R. R., Reuter, B. K., Mercado, J. R., et al. (2011). SAMP1/YitFc mouse strain: A spontaneous model of Crohn's disease-like ileitis. *Inflammatory Bowel Diseases, 17*, 2566–2584.

Pobezinskaya, Y. L., Choksi, S., Morgan, M. J., Cao, X., & Liu, Z. G. (2011). The adaptor protein TRADD is essential for TNF-like ligand 1A/death receptor 3 signaling. *Journal of Immunology, 186*, 5212–5216.

Podolsky, D. K. (2002). Inflammatory bowel disease. *New England Journal of Medicine, 347*, 417–429.

Powrie, F., Leach, M. W., Mauze, S., Menon, S., Caddle, L. B., & Coffman, R. L. (1994). Inhibition of Th1 responses prevents inflammatory bowel disease in scid mice reconstituted with CD45RBhi CD4$^+$ T cells. *Immunity, 1*, 553–562.

Read, S., Malmstrom, V., & Powrie, F. (2000). Cytotoxic T lymphocyte-associated antigen 4 plays an essential role in the function of CD25(+)CD4(+) regulatory cells that control intestinal inflammation. *Journal of Experimental Medicine, 192*, 295–302.

Rehman, A., Sina, C., Gavrilova, O., Hasler, R., Ott, S., Baines, J. F., et al. (2011). Nod2 is essential for temporal development of intestinal microbial communities. *Gut, 60*, 1354–1362.

Reindl, W., Weiss, S., Lehr, H. A., & Forster, I. (2007). Essential crosstalk between myeloid and lymphoid cells for development of chronic colitis in myeloid-specific signal transducer and activator of transcription 3-deficient mice. *Immunology, 120*, 19–27.

Round, J. L., & Mazmanian, S. K. (2010). Inducible Foxp3$^+$ regulatory T-cell development by a commensal bacterium of the intestinal microbiota. *Proceedings of the National Academy of Sciences of the United States of America, 107*, 12204–12209.

Saitoh, T., Fujita, N., Jang, M. H., Uematsu, S., Yang, B. G., Satoh, T., et al. (2008). Loss of the autophagy protein Atg16L1 enhances endotoxin-induced IL-1beta production. *Nature, 456*, 264–268.

Scheiffele, F. & Fuss, I. J. (2002). Induction of TNBS colitis in mice. *Current Protocols in Immunology, 15*, 15.19.1–14.

Seok, J., Warren, H. S., Cuenca, A. G., Mindrinos, M. N., Baker, H. V., Xu, W., et al. (2013). Genomic responses in mouse models poorly mimic human inflammatory diseases. *Proceedings of the National Academy of Sciences of the United States of America, 110*, 3507–3512.

Sheikh, S. Z., Matsuoka, K., Kobayashi, T., Li, F., Rubinas, T., & Plevy, S. E. (2010). Cutting edge: IFN-gamma is a negative regulator of IL-23 in murine macrophages and experimental colitis. *Journal of Immunology, 184*, 4069–4073.

Sheikh, S. Z., Hegazi, R. A., Kobayashi, T., Onyiah, J. C., Russo, S. M., Matsuoka, K., et al. (2011). An anti-inflammatory role for carbon monoxide and heme oxygenase-1 in chronic Th2-mediated murine colitis. *Journal of Immunology, 186*, 5506–5513.

Shull, M. M., Ormsby, I., Kier, A. B., Pawlowski, S., Diebold, R. J., Yin, M., et al. (1992). Targeted disruption of the mouse transforming growth factor-beta 1 gene results in multifocal inflammatory disease. *Nature, 359*, 693–699.

Simpson, S. J., Shah, S., Comiskey, M., de Jong, Y. P., Wang, B., Mizoguchi, E., et al. (1998). T cell-mediated pathology in two models of experimental colitis depends predominantly on the interleukin 12/Signal transducer and activator of transcription (Stat)-4 pathway, but is not conditional on interferon gamma expression by T cells. *Journal of Experimental Medicine, 187*, 1225–1234.

Takeda, K., Clausen, B. E., Kaisho, T., Tsujimura, T., Terada, N., Forster, I., et al. (1999). Enhanced Th1 activity and development of chronic enterocolitis in mice devoid of Stat3 in macrophages and neutrophils. *Immunity, 10*, 39–49.

Takedatsu, H., Mitsuyama, K., Matsumoto, S., Handa, K., Suzuki, A., Takedatsu, H., et al. (2004). Interleukin-5 participates in the pathogenesis of ileitis in SAMP1/Yit mice. *European Journal of Immunology, 34*, 1561–1569.

Travis, M. A., Reizis, B., Melton, A. C., Masteller, E., Tang, Q., Proctor, J. M., et al. (2007). Loss of integrin alpha(v)beta8 on dendritic cells causes autoimmunity and colitis in mice. *Nature, 449*, 361–365.

Treton, X., Pedruzzi, E., Cazals-Hatem, D., Grodet, A., Panis, Y., Groyer, A., et al. (2011). Altered endoplasmic reticulum stress affects translation in inactive colon tissue from patients with ulcerative colitis. *Gastroenterology, 141*, 1024–1035.

Van der Sluis, M., De Koning, B. A., De Bruijn, A. C., Velcich, A., Meijerink, J. P., Van Goudoever, J. B., et al. (2006). Muc2-deficient mice spontaneously develop colitis, indicating that MUC2 is critical for colonic protection. *Gastroenterology, 131*, 117–129.

van der Sluis, M., Bouma, J., Vincent, A., Velcich, A., Carraway, K. L., Buller, H. A., et al. (2008). Combined defects in epithelial and immunoregulatory factors exacerbate the pathogenesis of inflammation: Mucin 2-interleukin 10-deficient mice. *Laboratory Investigation, 88*, 634–642.

Velcich, A., Yang, W., Heyer, J., Fragale, A., Nicholas, C., Viani, S., et al. (2002). Colorectal cancer in mice genetically deficient in the mucin Muc2. *Science, 295*, 1726–1729.

Voss, E., Wehkamp, J., Wehkamp, K., Stange, E. F., Schroder, J. M., & Harder, J. (2006). NOD2/CARD15 mediates induction of the antimicrobial peptide human beta-defensin-2. *Journal of Biological Chemistry, 281*, 2005–2011.

Watanabe, T., Kitani, A., Murray, P. J., & Strober, W. (2004). NOD2 is a negative regulator of Toll-like receptor 2-mediated T helper type 1 responses. *Nature Immunology, 5*, 800–808.

Wehkamp, J., Salzman, N. H., Porter, E., Nuding, S., Weichenthal, M., Petras, R. E., et al. (2005). Reduced paneth cell alpha-defensins in ileal Crohn's disease. *Proceedings of the National Academy of Sciences of the United States of America, 102*, 18129–18134.

Wirtz, S., Finotto, S., Kanzler, S., Lohse, A. W., Blessing, M., Lehr, H. A., et al. (1999). Cutting edge: Chronic intestinal inflammation in STAT-4 transgenic mice: Characterization of disease and adoptive transfer by TNF- plus IFN-gamma-producing CD4[+] T cells that respond to bacterial antigens. *Journal of Immunology, 162*, 1884–1888.

Yen, D., Cheung, J., Scheerens, H., Poulet, F., McClanahan, T., McKenzie, B., et al. (2006). IL-23 is essential for T cell-mediated colitis and promotes inflammation via IL-17 and IL-6. *Journal of Clinical Investigation, 116*, 1310–1316.

Interferon gamma

Miriam Wittmann[1,2] and Martin Stacey[3]
[1]Leeds Institute of Rheumatic and Musculoskeletal Medicine, University of Leeds and NIHR Leeds Musculoskeletal Biomedical Research Unit Leeds, Chapel Allerton Hospital, Leeds, UK
[2]Centre for Skin Sciences, University of Bradford, Bradford, UK
[3]Faculty of Biological Sciences, School of Molecular and Cellular Biology, University of Leeds, Leeds, UK

Synonyms

IFNγ; Immune interferon; Type II IFN

Definition

Interferon-gamma (IFNγ) is one of the most important and potent mediators of inflammatory responses and elicits varied biological responses by regulating up to 2000 genes. It is thus involved in many pathophysiological processes including host defense, autoimmunity, and cancer. IFNγ is a key regulator of inflammatory responses and has strong effects on immune cells as well as tissue cells. IFNγ is distinct from the "classical" alpha and beta IFNs (type I IFNs) and is also referred to as type II or immune IFN. It is structurally unrelated and thus does not show homology to type I INFs, and it binds to a different receptor. IFNγ is a key regulator of immune responses and is the hallmark product of cells involved in a type I immune response (type I immune responses are thus characterized by the production of IFNγ). Main producer cells include CD4+ T helper cell type 1 (Th1 cells), innate lymphoid cells (ILC) type 1, NK cells (of note: ILC1 and NK cells are now referred to as group 1 ILCs), NKT cells, and CD8+ lymphocytes.

The human *IFNG* gene encodes for a protein which forms the biologically active homodimer by non-covalent self-assembly of two mature 143 amino acid glycopeptides of 21–24 kDa (depending on glycosylation patterns). The IFNγ receptor (IFNGR) is formed by two ligand-binding chains, IFNGR1 and IFGR2, both of which belong to the class II cytokine receptor family. The active receptor complex consists of two IFNGR1 chains and two INFGR2 chains.

Biosynthesis and Signaling

Main producer cells of IFNγ are Th1 lymphocytes, CD8 cytotoxic lymphocytes, and group 1 ILCs. Other cells, including B cells and to some extent antigen-presenting cells (APCs), can produce this cytokine as well. IFNγ production by APCs, such as dendritic cells (DCs) and monocytes/macrophages, has been discussed to be of importance in initial steps of the host defense against infection (Bogdan and Schleicher 2006). However, with regard to the source of IFNγ in early immune responses (e.g., before adaptive immune response comes into play), group 1 ILCs stimulated by IL-12 and IL-18 may be of greater importance.

As mentioned, Th1 CD4+ T cells are a major source of IFNγ. Initiation of IFNγ production requires T cell receptor (TCR) activation by antigens, superantigens, or mitogens. In addition, activation of T cells by costimulatory signals including surface receptors (e.g., CD54–LFA1, CD28–CD80/CD86, CD40–CD40L interaction) and a range of soluble mediators can largely enhance IFNγ production. Importantly, IL-12 and the master Th1 transcription factor T-bet (encoded by *Tbx21*) (Lazarevic et al. 2013) but also IFNγ itself, signal transducer, and activator of transcription (STAT)4 and STAT1 play an important role in polarizing naïve T cells along the Th1 lineage (Szabo et al. 2003). IL-12, produced by APC, is indispensable for robust Th1 polarization. Once undergone lineage commitment, Th1 as well as cytotoxic T cells (Tc1) produce IFNγ in response to further stimulation present in the tissue environment.

In humans, the high-affinity IL-12Rβ2 is upregulated under the influence of type I IFNs which can thus increase the sensitivity of human lymphocytes to IL-12. IL-12 (IL-12 family) and IL-18 (IL-1 family member) are important soluble mediators which work alone or in synergy to enhance and sustain the amount of IFNγ produced. Other IL-12 family members IL-23 and IL-27 and IL-1 family members in particular IL-1α/β can also enhance IFNγ production by T cells but seem not potently involved in the Th1 lineage polarization process. As for all CD4+ T cells, the T cell growth factor IL-2 and also IL-15 enhance their specific function and proliferation. Soluble negative regulators of IFNγ production include among others IL-4, IL-10, and TGFβ.

Signal Transduction

Signal transduction (for review, see Szabo et al. 2003; Hu and Ivashkiv 2009; Schroder et al. 2004; Gough et al. 2008). IFNGR1 is constitutively expressed, whereas the expression of IFNGR2 seems tightly regulated depending on cell activation status. Consequently, expression levels of IFNGR2 rather than IFNGR1 can limit cellular responsiveness to IFNγ (Regis et al. 2006). IFNγ only associates with IFNRG2 when the IFNGR1 chain is present. The IFNGR1 chain contains binding motifs for the Janus tyrosine kinase (JAK)1 and STAT1. The cytoplasmic region of IFNGR2 contains a binding motif for recruitment of Jak2. Although IFNγ primarily signals through the JAK-STAT pathway, the complex activation of and cooperation with other signaling pathways including MAP kinases, PI3-K, CaMKII, and NF-κB is well described (for review, see Gough et al. 2008) and differs depending on cell type and microenvironment. With regard to STAT1 activation, binding of IFNγ induces JAK2 autophosphorylation which allows JAK1 transphosphorylation. Active JAK kinases can phosphorylate the cytoplasmic domains of the IFNGR1 creating docking sites for the STAT1 protein. The STAT1 pair recruited to the receptor is then phosphorylated at Y701. Phosphorylation induces the STAT1 homodimer (also referred to as gamma IFN activation factor) to dissociate from the receptor and the active homodimer then translocates to the nucleus. Regulation of transcription is activated by binding of the STAT1 homodimer to IFNγ-activated sequences (GAS). STAT1 target genes include chemokines (CXCL9, CXCL10, CXCL11, CCL5), adhesion molecules (ICAM-1), and transcription factors (IFN regulatory factor (IRF)1, IRF8). Transcription of, e.g., IRF1 occurs within 15–30 min of IFNγ stimulation. IRFs bind to conserved consensus sites in the IFN-inducible genes named IFN-stimulated response elements (ISRE). IRF1 and IRF-8 are important enhancers of STAT1 initiated expression of IFNγ-responsive genes. With IFNγ being such a potent, mainly pro-inflammatory cytokine, tight control is in place at the level of transcription, translation, and protein activity. Negative feedback to the activated JAK-STAT1 signaling pathway is provided by the silencer of cytokine signaling (SOCS)1 which is one of the IFNγ-induced genes. A kinase inhibitory region allows SOCS1 to directly inhibit JAK1 signaling, but it has other modes of action including targeting its binding partners for proteasomal degradation.

Epigenetic Regulation

Epigenetic regulation (for review, see Aune et al. 2013; Pollard et al. 2013). Epigenetic regulation in form of complex patterns of histone modifications at the *IFNG* locus occurs in the context of Th1 lineage commitment and production of IFN. Lineage commitment in human cell is not as strong as described for the mouse system (e.g., human Th cells show more plasticity); nevertheless, cells undergo significant epigenetic changes enabling them to maintain their lineage characteristics. It seems that STAT4 and T-bet are both important for epigenetic modifications seen at the *IFNG* locus which are sustained in committed, memory CD4+ and CD8+ T cells (Schoenborn and Wilson 2007). Higher order chromatin remodeling influenced by T-bet places the *IFNG* locus in a distinct chromosomal "neighborhood" allowing for altered gene interactions.

In contrast to CD4 and CD8 T cells, NK and NKT cells are competent to produce IFNγ without prior polarization processes. For their response to extracellular stimulation with rapid IFNγ production, the transcription factors Eomes and T-bet are of importance.

Recent findings suggest that long non-coding RNAs (lncRNA) play a significant role in IFNγ transcription. One recently identified example is the lncRNA *TMEVPG1* (aka *NeST*) which acts as an enhancer of IFNγ transcription in CD4+ and CD8+ T cells; STAT4/T-bet support the expression of this particular lncRNA.

Biological Activities

IFNγ regulates a broad range of biological responses and affect almost all cells and tissues (for further review, see Hu and Ivashkiv 2009; Schroder et al. 2004; Pollard et al. 2013; Billiau and Matthys 2009; Ikeda et al. 2002; Kelchtermans et al. 2008; Muhl and Pfeilschifter 2003; Young and Bream 2007; de Bruin et al. 2014). Hallmark functions of IFNγ include its activities against intracellular pathogens such as mycobacteria and viruses, its effect on class I and class II antigen presentation pathways, and its antiproliferative and tumor surveillance properties. It is one of the key mediators of adaptive immune responses. IFNγ is important for full activation of monocytes/macrophages. It is an essential component of macrophage-activating factor (MAF) (Nathan et al. 1983), an obsolete term used to describe the ability of cell culture supernatants to augment macrophage functions and in particular bactericidal activity, intracellular killing of microorganisms, oxidative metabolism, MHC II expression, or enhanced tumor cell killing. As mentioned, IFNγ can regulate many genes which are referred to as IFN-regulated genes (IRGs). For antiviral properties of IFNγ, MxA, PKR, and GBP-1 are among important molecules, and Mad1 and c-myc are involved in its antiproliferative actions. IFNγ has proapoptotic properties in which IRF1, caspase 1, cathepsin D, TNFα, and Fas/FasL are involved. The antimicrobial effector functions are partly mediated by its action on inducible NO synthetase (iNOS) and reactive oxygen species and complement. IFNγ induces the expression of high-affinity IgG receptors

(CD16, CD32, CD64) on neutrophils and cells of the monocyte-macrophage lineage. It regulates leukocyte trafficking by acting on chemokine release and adhesion molecules which include CXCL10, CXCL9, CXCL11, ICAM1, VCAM1, and many others.

IFNγ influences the class I (presentation of intracellular antigens via MHC class I) and class II (presentation of extracellular antigens via MHC class II) antigen presentation pathways (further detailed in (Schroder et al. 2004)). Upregulation of MHC class I on the cell surface by IFNγ is seen in almost all nucleated cells of our body. Intracellular-derived antigens are presented via MHC I and these include proteins derived from or altered by pathogens such as viruses or mycobacteria. Increased MHC class I expression ascertains that CD8+ cytotoxic T cells (CTL) can recognize and respond to these antigens. The actions of IFNγ on the MHC presentation pathway ultimately result in an increased number and repertoire of antigen epitopes presented to CD8+ cells (Schroder et al. 2004; Heink et al. 2005). IFNγ is also an important effector molecule for CTL. Activated CTLs express CXCR3 and migrate along CXCR3 ligand (CXCL9/10/11) gradients produced by epithelial cells or monocytes/macrophages under the influence of IFNγ to infected/damaged tissues.

IFNγ shares the action on the MHC I pathway with other IFNs. But it is mainly IFNγ which can significantly enhance the competence of the MHC class II antigen presentation pathway. IFNγ upregulates MHC class II cell surface expression on APCs. In addition, it can also lead to the expression of MHC class II in other cells types such as endothelial and epithelial cells (nonprofessional APC). By influencing multiple steps in the MHC II antigen presentation pathway, it again leads to an increased quantity and variety of peptides presented to CD4+ T cells. Taken together, IFNγ augments the presentation of intracellular (MHC I pathway) and extracellular (MHC II pathway) derived antigens in terms of quantity and quality and thus enhances immune surveillance with regard to intra- and extracellular microbial pathogens but also tumor surveillance.

A function of IFNγ on CD4+ Th1 cell polarization and T-bet expression has been mentioned above. It also plays an important role in regulating other CD4+ T cell populations. The well-described inhibitory effect on Th2 cells is in part mediated by IFNγ and Th2 polarized cells are particularly susceptible to the antiproliferative effect of IFNγ by virtue of high IFNGR2 expression. IFNγ also counteracts the development of Th17 cells. Indeed, a gain-of-function mutation in STAT1 was shown to be associated with higher degree of fungal infection which is interpreted as a lack of Th17 immunity. IFNγ thus further skews the immune response toward a Th1 phenotype.

IFNγ and IL-12 are involved in a positive feedback regulation. IL-12 induces IFNγ production, which is much enhanced when acting in synergy with IL-18, and IFNγ is an important factor for IL-12 production by APC. This amplification may be important in stabilization of a robust Th1 response needed, e.g., for efficient defense against and clearance of pathogens causing tuberculosis, leishmaniasis, or leprosy.

As mentioned, IFNγ has strong macrophage/APC activating potential and this activation is crucial for host defense against intracellular pathogens. Activated macrophages also play an important role in tissue repair and remodeling following inflammation, infection, and tissue damage. The ability of MAF (derived from conditioned cell culture supernatant) to activate macrophages has been mentioned above. Only a proportion of properties of MAF can be reproduced by IFNγ only. In particular intracellular killing against mycobacteria or listeria requires the presence of other cytokines such as TNFα. This highlights the ability of IFNγ to act as a priming signal (Hu and Ivashkiv 2009). IFNγ works in synergy with a great number of other mediators and TLR ligands. One of the best studied synergistic actions is between IFNγ and TNFα or CD40L. One example (among many) of their synergistic action is IL-12 production. Monocytes/macrophages need 2 signals to become fully competent IL-12 producers; they depend on a priming signal which increases their susceptibility to a second signal which can be a TLR ligand, TNFα, or cell contact-dependent costimulatory signals such as CD40L. The most significant priming signal known is IFNγ. Thus exposure to

IFNγ is a prerequisite for optimal macrophage/DC activation and IL-12 production. The synergistic effect (e.g., the biological response is much higher than just adding the effects of both signals) depends on IFNγ being present before the action of the second signal.

The immunoregulatory function of IFNγ on B cells is evidenced by its action of differentiation and Ig isotype class switch (Szabo et al. 2003). IFNγ inhibits the production of IgG1 and IgE induced by the presence of IL-4 and promotes class switch to IgG2a, IgG2b, and IgG3. It induces the expression of MHC class II on B cells and acts on B-cell maturation/differentiation.

Much as CTLs, NK cells are both targets of IFNγ action and producers of IFNγ. IFNγ can upregulate NK cell activity. NK cells produce IFNγ in response to various exogenous and endogenous stimuli including IL-12 and IL-18. IFNγ affects hematopoiesis (for review, see de Bruin et al. 2014) and facilitates the expansion of myeloid cells. It induces monopoiesis at the cost of neutrophil and eosinophil formation. Prolonged high levels of IFNγ have been associated with bone marrow failure and anemia, both of which are seen in chronic inflammatory conditions. SOCS molecules may play an important role in the negative effects of IFNγ on differentiation of hematopoietic lineages.

Effect on Nonimmune Cells

IFNγ influences diverse biological responses of tissue-resident cells. It is known to act on osteoclasts and can inhibit their bone resorption activity. The antiproliferative property of IFNγ is also seen on the level of smooth muscle cells and mesenchymal and endothelial cells. It inhibits collagen production by myofibroblasts and these actions may result in a function as endogenous inhibitor for vascular overreactions in the context of injuries. It functions as an inhibitor of capillary growth and this action may also add to its antitumor activities. Both epithelial and endothelial cells are very responsive to IFNγ. This cytokine increases adhesion molecule and chemokine/cytokine production by endothelial cells and thus plays an important role in immune cell trafficking to infected/injured tissue. However these properties of IFNγ may also add to its pro-atherogenic role (Sikorski et al. 2012). IFNγ is one of the strongest activator of epithelial cells, in particular skin keratinocytes. They upregulate expression and production of many surface molecules and soluble mediators and among the upregulated chemokines are CXCL9 and CXCL10. These chemokines in turn attract CXCR3+ cells which are mainly IFNγ producers and further enhance an IFNγ dominated inflammatory response. In both epithelial and endothelial tissues, IFNγ can also impair barrier functions/increase permeability. In the skin this seems mediated by downregulation of tight junctions and increased keratinocyte apoptosis. Thus excessive IFNγ in epithelial organs can have negative effects on tissue integrity and barrier function.

Anti-inflammatory Properties of IFNγ

Anti-inflammatory properties of IFNγ (Kelchtermans et al. 2008; Muhl and Pfeilschifter 2003). All powerful stimulatory molecules are subjected to a tight control at several levels from gene expression to protein secretion and function. IFNγ itself induces the expression of regulatory molecules and their secretion typically follows a different (delayed) kinetic compared to pro-inflammatory mediators. In this context, inflammation-limiting mediators include IL-18 binding protein, IL-1 receptor antagonist, and indoleamine 2,3 dioxygenase (IDO). IFNγ is known to inhibit IL-8 release and it can limit inflammation by acting on regulatory T cells (Treg) and by inducing apoptosis. On the cellular level, IFNγ stimulation strongly induces SOCS1 which inhibits the IFNγ signaling JAK/STAT1-dependent pathway. Besides the inhibitory activity in JAK/STAT signaling, SOCS molecules have also effects on MAP-kinase activation and are involved in the regulation of TIRAP, IRAK, and NF-κB stability. The regulating and tolerogenic properties of IFNγ are of pathophysiological importance, and in a number of animal models for autoimmune diseases, there is evidence of anti-inflammatory properties. Diverse mechanisms seem to be involved in the IFNγ-mediated, disease-protective function including inhibition of

Th17 cells, increase of Treg activity, inhibition of osteoclastogenesis, induction of IDO, and increased apoptosis as well as inhibition of proliferation (Kelchtermans et al. 2008). IDO contributes to create a tolerogenic environment both by direct suppression of T cell responses and enhancement of local Treg cell-mediated immunosuppression. While IFNγ has an inhibitory effect on both Th17 and Th2 populations, it does not inhibit Treg development. Treg themselves are capable of producing IFNγ, and this may be important for activation of APC in their microenvironment to allow antigen specific Treg responses.

Pathophysiological Relevance

In clinical setting IFNγ is rather used in diagnostic tests than in therapeutic applications. IFNγ release assays are used for screening of latent TB and experimentally as outcome parameter in drug allergy and contact allergy assays.

Pathologies Associated with Lack of INFγ Activity

Mice deficient in IFNγ, IFNGR1/2, or STAT1 have severely impaired immune responses as demonstrated by an increased susceptibility to microbial pathogens and certain viruses. Human deficiency of IFNγ is associated with severe infection. Patients with loss-of-function mutations in the INFGR1 or IFNGR2 chain show severe susceptibility to otherwise weakly virulent mycobacteria, often resulting in childhood fatality. In addition to recurrent infections, infants with deficient production of IFNγ exhibit decreased neutrophil mobility and NK cell activity. Mendelian susceptibility to mycobacterial disease (MSMD) is a rare inherited condition (reviewed in (Bustamante et al. 2014)) characterized by selective predisposition to clinical disease caused by weakly virulent mycobacteria, such as BCG vaccines and environmental mycobacteria, in otherwise healthy patients. Since 1996 nine MSMD-causing genes have been discovered. The products of all these genes are involved in IFN-γ-dependent immunity. These disorders impair the production (IL12B, IL12RB1, IRF8, ISG15, NEMO) or the response to IFNγ (IFNGR1, IFNGR2, STATt1, IRF8, CYBB).

In mouse models, neutralization of IFNγ or knockout of its receptors resulted in compromised tumor rejection. While atopic disease (asthma, atopic dermatitis, hay fever) shows increased Th2-associated immunity, the notion that addition of exogenous IFNγ could help prevent or attenuate these conditions has not proved correct. For atopic dermatitis, it is well described that IFNγ- and Th1-mediated immunity plays an important role in the chronic phase of the disease.

Pathologies Associated with Increased or Prolonged IFNγ Activity

Aside from functions in host defense, IFNγ may also contribute to autoimmune pathology. The signature of IFNs is measurable in a number of autoimmune diseases (Kelchtermans et al. 2008), and among the diseases in which IFNs are important effectors are lupus erythematosus, Sjögren's syndrome, polymyositis, dermatomyositis, and systemic sclerosis. In these conditions disease activity correlates with what is known as type I IFN signature. While the traditional concept is that this "signature" is caused by type I IFN, due to significant overlap between the genes induced by type I and II IFNs, it is difficult to ascertain which IFN is the major contributor and there is good evidence for IFNγ to also play a role in autoimmune diseases (Hu and Ivashkiv 2009). IFNγ can certainly play a role in end-organ damage due to infiltration of IFNγ-secreting T cells resulting in macrophage activation, inflammation, and tissue damage. Other chronic inflammatory diseases where IFNγ is considered of pathophysiological significance include Crohn's disease, septic shock, Guillain-Barré syndrome, and artherosclerosis (Sikorski et al. 2012). As mentioned above, chronically high levels of this IFNγ are considered to contribute to bone marrow suppression and anemia.

Gain-of-function mutations in human STAT1 have been linked to chronic mucocutaneous candidiasis due to impaired IL-17 immunity (Liu et al. 2011). Although other cytokines including type I IFNs and IL-27 also signal via STAT1 activation, IFNγ certainly plays an important role in this clinical manifestation. The inhibitory effect of IFNγ on Th17 cells seems of pathophysiological relevance in this condition.

Clinical symptoms of allergic contact dermatitis, a classical example of delayed-type hypersensitivity, are certainly mediated to a large extent by an IFNγ-mediated response. The associated inflammation results in eczema, a superficial inflammation of the skin. Interestingly, alopecia areata, considered as autoimmune in origin, can be treated with both immunosuppression (glucocorticoids, calcineurin antagonists) and induction of inflammation by using a strong contact allergen. Indeed, the latter seems to be a rather successful approach. Whether this points to both a pro- and anti-inflammatory role of IFNγ depending on context remains to be elucidated in more detail.

Modulation by Drugs

In contrast to type I IFN, recombinant IFN or specific neutralizing antibodies are currently not used in clinic. In spite of the antiproliferative activities of IFNγ, trials aiming for more successful treatment of malignancies have been disappointing. Treatment with IFNγ in patients with sepsis and underlying impaired monocyte function has been shown to improve clinical outcome in some of these patients. However, IFNγ has not entered clinical practice as therapeutic agent.

Some approaches to inhibit the action of IFNγ in inflammatory and autoimmune conditions are being pursued. Fontolizumab is a humanized monoclonal antibody against IFNγ which was well tolerated and showed some efficacy in patients with Crohn's disease. However a phase II clinical trial investigating its use in rheumatoid arthritis was terminated as it failed to meet the endpoint. Another human monoclonal antibody against IFNγ is being tested in lupus erythematosus conditions (AMG811). Overall, levels of IFNγ in inflammatory conditions are inhibited by a range of immunosuppressants including calcineurin inhibitors (such as cyclosporine A, tacrolimus), mycophenolate, methotrexate, and glucocorticosteroids; however the cytokine reduction is rather a result of impaired lymphocyte function than a specific approach to inhibit IFNγ. Emerging lncRNA (e.g., tmevpg1) which regulate IFNγ activities may be interesting for targeting IFNγ expression in T cells in the future and may show a better effect/side effect balance as compared to neutralizing IFNγ directly.

Cross-References

- ▶ Antiviral responses
- ▶ Costimulatory Receptors
- ▶ CXCR3 and its Ligands
- ▶ Dendritic cells
- ▶ Janus kinases (Jaks)/STAT pathway
- ▶ Interleukin 2
- ▶ Interleukin 4 and the related cytokines (Interleukin 5 and Interleukin 13)
- ▶ Interleukin 10
- ▶ Interleukin 12
- ▶ Interleukin 17
- ▶ Interleukin 18
- ▶ Interleukin-18 Binding Protein
- ▶ Interleukin 23
- ▶ Interleukin 27
- ▶ Leukocyte recruitment
- ▶ MAP kinase pathways
- ▶ NFkappaB
- ▶ Osteoclasts in Inflammation
- ▶ TH17 response
- ▶ Toll-like receptors
- ▶ Tumor Necrosis Factor Alpha (TNFalpha)
- ▶ Type I Interferons

References

Aune, T. M., Collins, P. L., Collier, S. P., Henderson, M. A., & Chang, S. (2013). Epigenetic activation and silencing of the gene that encodes IFN-gamma. *Frontiers in Immunology, 4*, 112.

Billiau, A., & Matthys, P. (2009). Interferon-gamma: A historical perspective. *Cytokine & Growth Factor Reviews, 20*, 97–113.

Bogdan, C., & Schleicher, U. (2006). Production of interferon-gamma by myeloid cells – fact or fancy? *Trends in Immunology, 27*, 282–290.

Bustamante, J., Boisson-Dupuis, S., Abel, L., & Casanova, J. L. (2014). Mendelian susceptibility to mycobacterial disease: Genetic, immunological, and clinical features of inborn errors of IFN-gamma immunity. *Seminars in Immunology, 26*, 454–470.

de Bruin, A. M., Voermans, C., & Nolte, M. A. (2014). Impact of interferon-gamma on hematopoiesis. *Blood, 124*, 2479–2486.

Gough, D. J., Levy, D. E., Johnstone, R. W., & Clarke, C. J. (2008). IFNgamma signaling-does it mean JAK-STAT? *Cytokine & Growth Factor Reviews, 19*, 383–394.

Heink, S., Ludwig, D., Kloetzel, P. M., & Kruger, E. (2005). IFN-gamma-induced immune adaptation of the proteasome system is an accelerated and transient response. *Proceedings of the National Academy of Sciences of the United States of America, 102*, 9241–9246.

Hu, X., & Ivashkiv, L. B. (2009). Cross-regulation of signaling pathways by interferon-gamma: Implications for immune responses and autoimmune diseases. *Immunity, 31*, 539–550.

Ikeda, H., Old, L. J., & Schreiber, R. D. (2002). The roles of IFN gamma in protection against tumor development and cancer immunoediting. *Cytokine & Growth Factor Reviews, 13*, 95–109.

Kelchtermans, H., Billiau, A., & Matthys, P. (2008). How interferon-gamma keeps autoimmune diseases in check. *Trends in Immunology, 29*, 479–486.

Lazarevic, V., Glimcher, L. H., & Lord, G. M. (2013). T-bet: A bridge between innate and adaptive immunity. *Nature Reviews Immunology, 13*, 777–789.

Liu, L., et al. (2011). Gain-of-function human STAT1 mutations impair IL-17 immunity and underlie chronic mucocutaneous candidiasis. *The Journal of Experimental Medicine, 208*, 1635–1648.

Muhl, H., & Pfeilschifter, J. (2003). Anti-inflammatory properties of pro-inflammatory interferon-gamma. *International Immunopharmacology, 3*, 1247–1255.

Nathan, C. F., Murray, H. W., Wiebe, M. E., & Rubin, B. Y. (1983). Identification of interferon-gamma as the lymphokine that activates human macrophage oxidative metabolism and antimicrobial activity. *The Journal of Experimental Medicine, 158*, 670–689.

Pollard, K. M., Cauvi, D. M., Toomey, C. B., Morris, K. V., & Kono, D. H. (2013). Interferon-gamma and systemic autoimmunity. *Discovery Medicine, 16*, 123–131.

Regis, G., Conti, L., Boselli, D., & Novelli, F. (2006). IFNgammaR2 trafficking tunes IFNgamma-STAT1 signaling in T lymphocytes. *Trends in Immunology, 27*, 96–101.

Schoenborn, J. R., & Wilson, C. B. (2007). Regulation of interferon-gamma during innate and adaptive immune responses. *Advances in Immunology, 96*, 41–101.

Schroder, K., Hertzog, P. J., Ravasi, T., & Hume, D. A. (2004). Interferon-gamma: An overview of signals, mechanisms and functions. *Journal of Leukocyte Biology, 75*, 163–189.

Sikorski, K., et al. (2012). STAT1 as a central mediator of IFNgamma and TLR4 signal integration in vascular dysfunction. *JAK-STAT, 1*, 241–249.

Szabo, S. J., Sullivan, B. M., Peng, S. L., & Glimcher, L. H. (2003). Molecular mechanisms regulating Th1 immune responses. *Annual Review of Immunology, 21*, 713–758.

Young, H. A., & Bream, J. H. (2007). IFN-gamma: Recent advances in understanding regulation of expression, biological functions, and clinical applications. *Current Topics in Microbiology and Immunology, 316*, 97–117.

Interleukin-1 (IL-1) Inhibitors: Anakinra, Rilonacept, and Canakinumab

Kevin D. Pile[1], Garry G. Graham[2,3] and Stephen M. Mahler[4]
[1]Campbelltown Hospital, School of Medicine, University of Western Sydney, Campbelltown, NSW, Australia
[2]Department of Pharmacology, School of Medical Sciences, University of New South Wales, Sydney, NSW, Australia
[3]Department of Clinical Pharmacology and Toxicology, St Vincent's Hospital, Sydney, NSW, Australia
[4]Australian Institute for Bioengineering and Nanotechnology, University of Queensland, Brisbane, QLD, Australia

Synonyms

Anti-IL-1; IL-1 antagonists; IL-1 blockers; Rilonacept and canakinumab have been termed IL-1 trap and ACZ885, respectively

Definition

Inhibitors of interleukin-1 (IL-1) are proteins which decrease the actions of the inflammatory cytokine, IL-1. There are two general mechanisms of IL-1 inhibitors: binding to the IL-1 receptor (anakinra) or binding directly to IL-1 (rilonacept and canakinumab). They are members of a general class termed biological disease-modifying antirheumatic drugs (bDMARDs). They are not classified as slow-acting antirheumatic drugs (SAARDs) because they are considered to have more specific and rapidly developing actions. Clinically, the major IL-1 inhibitor is anakinra.

Chemical Structures and Properties

Three types of inhibitors of interleukin-1 have been approved or trialed in the clinic (Moll and Kuemmerle-Deschner 2013):

- Anakinra is a protein with a relatively small molecular mass (17,258 D). Anakinra is very similar to the human IL-1 receptor antagonist (IL-1ra), differing in that anakinra is not glycosylated and, also, that anakinra contains a terminal methionine residue which is necessary for its biological production. Anakinra is produced from cultures of genetically modified *Escherichia coli*, using recombinant DNA technology.
- Rilonacept is a soluble decoy receptor composed of a large dimeric fusion protein (molecular mass 251,000 D) consisting of the ligand-binding domains of the extracellular portions of the human IL-1 receptor (IL-1R1) and IL-1 receptor accessory protein (IL-1RAcP) linked in line to the Fc region of human IgG1. Rilonacept is produced by recombinant DNA technology in Chinese hamster ovary (CHO) cells.
- Canakinumab is a human antihuman-IL-1β monoclonal antibody that belongs to the IgG1/κ isotype subclass. It is comprised of two 447- (or 448-) residue heavy chains and two 214-residue light chains, with a molecular mass of 145,157 D when deglycosylated. Both heavy chains of canakinumab contain oligosaccharide chains linked to the protein backbone at asparagine 298 (Asn 298). Canakinumab is produced from murine Sp2/0 cells in a serum-free cell culture medium.

Metabolism, Pharmacokinetics, and Dosage

Having protein structures, the interleukin inhibitors are not absorbed orally and are administered by subcutaneous injection. The three interleukin inhibitors have differing metabolic and pharmacokinetic profiles:

- Anakinra has limited distribution in the body because of its protein structure, but, like other proteins with molecular masses less than about 30,000 D, anakinra is readily filtered in the glomerulus and then hydrolyzed within the kidney (Meibohm and Zhou 2012). The naturally occurring IL-1 inhibitor is eliminated similarly. The half-life of anakinra is short at about 5 h in patients with normal renal function. The half-life is approximately doubled in patients with end-stage renal failure, while its apparent clearance (CL/F) is reduced by 75 %. Consequently, the dose should be reduced in patients with significant renal impairment (Yang et al. 2003). Anakinra is normally administered once daily (100 mg by subcutaneous injection). This dosage interval is much shorter than rilonacept and canakinumab because of the much shorter half-life of anakinra. Anakinra is not significantly dialyzable despite its relatively low molecular mass (Yang et al. 2003).
- Rilonacept is absorbed slowly and achieves peak plasma concentrations at about 3 days after its subcutaneous dosage. It has a half-life of about 7.5 days (Radin et al. 2010) which allows its once a week dosage (160 mg by subcutaneous injection). Like other proteins of very large molecular mass, rilonacept is expected to be cleared by the mononuclear phagocytes in the reticuloendothelial system. Rilonacept is not dialyzable because of its large molecular mass, and, consequently, its dosage does not require change in dialyzed patients with end-stage renal disease (Radin et al. 2010).
- Canakinumab has a bioavailability of about 70 % after subcutaneous injection and a half-life of about 21–28 days which leads to once a month or once every 2 months dosage. This long half-life is typical of antibodies. Presumably, it is cleared by the mononuclear phagocytes in the reticuloendothelial system. Canakinumab is

typically used at 150 mg subcutaneously (Moll and Kuemmerle-Deschner 2013). With a molecular mass of 145,200 D, canakinumab should also not be cleared by the kidney or be dialyzable.

Pharmacological Effects

IL-1 promotes cytokine release and release of prostaglandins and metalloproteinases, activates endothelial cell adhesion molecules, and reduces glycosaminoglycan synthesis resulting in increased cartilage breakdown and activation of osteoclasts with bone and cartilage damage.

There are two IL-1 proteins, IL-1α and IL-1β, which are synthesized through the activities of two related genes. Both bind to the same receptor, but IL-1β is the more significant in inflammation, and its precursor is synthesized in monocytes, B cells, fibroblasts, and chondrocytes of the rheumatoid synovium. IL-1β production is stimulated by the polyprotein particle, the inflammasome, which occurs in many cells and is a major stimulant of acute gout. Constituents of the inflammasome include caspase 1 and cryopyrin. Caspase 1 is a peptidase which converts the IL-1β precursor to IL-1β and is activated from procaspase by the assembly of proteins to form the inflammasome. Cryopyrin is a regulator of inflammation and is encoded by the NLRP3 gene (see section "Cryopyrin-Associated Periodic Syndromes (CAPS)" below).

As is the case with other cytokines, the significance of IL-1 in rheumatological diseases is shown in mice. Notably, mice deficient in the naturally occurring IL-1 receptor antagonist (IL-1ra) develop a spontaneous erosive arthritis.

The molecular interactions of the three IL-1 inhibitors differ:

- Anakinra binds reversibly to IL-1 receptors but does not have agonist activity. Thus, anakinra is a competitive inhibitor of both IL-1α and IL-1β with the native receptor and blocks the biological actions of these cytokines.
- Rilonacept binds directly to IL-1α and IL-1β. Thus, it blocks the activity of both IL-1α and IL-1β by direct binding of these two cytokines.
- Canakinumab binds only IL-1β and, thus, selectively decreases the biological activity of this cytokine.

Clinical Indications and Efficacy

Rheumatoid Arthritis

Anakinra has moderate efficacy in treating rheumatoid arthritis in adults where it appears to have, on average, less activity than the TNF inhibitors, although head to head comparisons are lacking (Mertens and Singh 2009). It may be more effective than other TNF inhibitors in the treatment of both systemic juvenile idiopathic arthritis and adult-onset Still's disease, where IL-1 may be a more critical cytokine than TNF.

Gout

The IL-1 inhibitors decrease the symptoms of acute gout but are considered as fourth-line drugs behind nonsteroidal anti-inflammatory drugs, colchicine, and corticosteroids (Tran et al. 2013). They may be useful for gout which is difficult to treat, particularly due to flares which may occur during initiation of urate-lowering therapy with drugs such as allopurinol and febuxostat. Anakinra and canakinumab reduce the pain and inflammation of acute gout, while rilonacept reduces the rate of gout recurrence during initiation of urate-lowering therapy. Anakinra is particularly useful for acute gout because its short half-life is sufficient since most attacks of acute gout are self-limiting. At this stage, rilonacept and canakinumab are not approved by the FDA. In contrast to anakinra, we suggest that their long half-lives, particularly canakinumab, are not necessary for the generally short-term treatment of acute gout.

The IL-1 inhibitors have also been shown to suppress the inflammation and pain of acute pseudogout (chondrocalcinosis), a disease due to the deposition of calcium pyrophosphate in joints. This disease which is very common in the elderly does not, however, always require drug treatment.

Cryopyrin-Associated Periodic Syndromes (CAPS)

The three IL-1 inhibitors are useful for the treatment of cryopyrin-associated periodic syndromes (CAPS) which are a group of inherited inflammatory diseases of varying intensities: familial cold auto-inflammatory syndrome, Muckle–Wells syndrome, and neonatal-onset multisystem inflammatory disease (NOMID) (Dinarello et al. 2012). At this stage, canakinumab has been approved by drug agencies in the USA and Europe for the treatment of Muckle–Wells syndrome, and it is likely that both rilonacept and canakinumab will be recommended officially for the therapy of the several CAPS.

Schnitzler syndrome, a rare disease with some aspects similar to those of CAPS, also responds to anakinra and rilonacept (Krause et al. 2012). CAPS and its closely related diseases are termed IL-1-mediated inflammasomopathies (Moll and Kuemmerle-Deschner 2013).

A major cause of CAPS is an autosomal dominant mutation in the NLRP3 gene that encodes cryopyrin (or NALP3). Cryopyrin is a protein that normally limits the activation of caspase 1, the enzyme which hydrolyzes the precursor of IL-1 β to active IL-1 β (Dinarello et al. 2012), eventually leading to increased secretion of IL-1β. Not surprisingly, IL-1β inhibitors are active in the treatment of the CAPS diseases (Moll and Kuemmerle-Deschner 2013). By comparison, blockers of tumor necrosis factor are not useful for CAPS.

Familial Mediterranean Fever

The IL-1 inhibitors are also beneficial for the treatment of this complex disease which has variable episodic symptoms including fever, abdominal pain, joint pain, and erythematous skin lesions. The disease is described as partially IL-1 related. The present use of the IL-1 inhibitors is in familial Mediterranean fever patients who are intolerant or inadequately responsive to colchicine.

Atherosclerosis

There has been much interest in the possible beneficial effects of IL-1 inhibitors in atherosclerosis and other cardiovascular diseases which may have an inflammatory component as "atherosclerotic plaques express IL-1β, and IL-1 β expression appears to correlate with the progression of atherosclerotic plaques" (Abbate et al. 2012). Consequently, canakinumab is being trialed in atherosclerosis, to determine whether it reduces myocardial infarction, stroke, or cardiovascular death (Dinarello et al. 2012). In a small clinical trial, anakinra has been shown to improve cardiac relaxation or stiffness in patients with preserved ejection fraction at rest (Van Tassell et al. 2010).

Dry Eye Disease

The topical administration of anakinra has led to fewer symptoms in dry eye disease, at least in some patients (Amparo et al. 2013).

Adverse Effects

Local reactions at the sites of injection have been noted with all three IL-1 inhibitors. The three IL-1 inhibitors are remarkably safe, but their major safety consideration is the potential for infection. Anakinra has been trialed and used for a longer time than rilonacept or canakinumab, and the limitations of the use of anakinra in infections are clearer than with the other two IL-1 inhibitors. From the results of trials on anakinra, it is recommended that its treatment should not be commenced if the patients have an active infection and should be stopped if a serious infection or sepsis develops. Furthermore, anakinra should be used with care if the patient has a history of severe infections and used only with great care if the patient has had tuberculosis. The use of rilonacept and canakinumab should be considered similarly with regard to infections (Lyseng-Williamson 2013).

Interactions

- In the treatment of rheumatoid arthritis, anakinra should be administered with methotrexate in order to increase the response.
- Anakinra should not be administered with TNF inhibitors because of the risk of infections.

Cross-References

► Corticosteroids
► Disease-Modifying Antirheumatic Drugs: Overview
► Gout
► Inflammatory Bowel Disease
► Interleukin-1 (IL-1) Inhibitors: Anakinra, Rilonacept, and Canakinumab
► Methotrexate
► Rheumatoid Arthritis
► Spondyloarthritis
► Tumor Necrosis Factor (TNF) Inhibitors

References

Abbate, A., Van Tassell, B. W., & Biondi-Zoccai, G. G. L. (2012). Blocking interleukin-1 as a novel therapeutic strategy for secondary prevention of cardiovascular events. *BioDrugs, 26*(4), 217–233.

Amparo, F., Dastjerdi, M. H., Okanobo, A., Ferrari, G., Smaga, L., Hamrah, P., et al. (2013). Topical interleukin 1 receptor antagonist for treatment of dry eye disease: A randomized clinical trial. *JAMA Ophthalmology, 131*(6), 715–723.

Dinarello, C. A., Simon, A., & van der Meer, J. W. (2012). Treating inflammation by blocking interleukin-1 in a broad spectrum of diseases. *Nature, 11*, 633–652.

Krause, K., Weller, K., Stefaniak, R., Wittkowski, H., Altrichter, S., Siebenhaar, F., et al. (2012). Efficacy and safety of the interleukin-1 antagonist rilonacept in Schnitzler syndrome: An open-label study. *Allergy, 67*(7), 943–950.

Lyseng-Williamson, K. A. (2013). Canakinumab: A guide to its use in acute gouty arthritis flares. *BioDrugs, 27*(4), 401–406.

Meibohm, B., & Zhou, H. (2012). Characterizing the impact of renal impairment on the clinical pharmacology of biologics. *Journal of Clinical Pharmacology, 52*, 54S–62S.

Mertens, M., & Singh, J. A. (2009). Anakinra for rheumatoid arthritis: A systematic review. *Journal of Rheumatology, 36*(11), 1118–1125.

Moll, M., & Kuemmerle-Deschner, J. B. (2013). Inflammasome and cytokine blocking strategies in autoinflammatory disorders. *Clinical Immunology, 147*(3), 242–275.

Radin, A., Marbury, T., Osgood, G., & Belomestnov, P. (2010). Safety and pharmacokinetics of subcutaneously administered rilonacept in patients with well-controlled end-stage renal disease (ESRD). *Journal of Clinical Pharmacology, 50*(7), 835–841.

Tran, T. H., Pham, J. T., Shafeeq, H., Manigault, K. R., & Arya, V. (2013). Role of interleukin-1 inhibitors in the management of gout. *Pharmacotherapy, 33*(7), 744–753.

Van Tassell, B. W., Varma, A., Salloum, F. N., Das, A., Seropian, I. M., Toldo, S., et al. (2010). Interleukin-1 trap attenuates cardiac remodeling after experimental acute myocardial infarction in mice. *Journal of Cardiovascular Pharmacology, 55*(2), 117–122.

Yang, B. B., Baughman, S., & Sulivan, J. T. (2003). Pharmacokinetics of anakinra in subjects with different levels of renal function. *Clinical Pharmacology & Therapeutics, 74*, 85–94.

Interleukin 2

Natalia Arenas-Ramirez[1] and Onur Boyman[1,2]
[1]Department of Immunology, University Hospital Zurich, Zurich, Switzerland
[2]Allergy Unit, Department of Dermatology, University Hospital Zurich, Zurich, Switzerland

Synonyms

Common gamma chain cytokine; Effector T cell; IL-2 receptor x (IL-2Rα), CD25, Tac antigen; IL-2Rβ, CD122; IL-2Rγ, common gamma chain, γc, CD132; Immune regulation; Lymphokine-activated killer cell; Natural killer cell; Regulatory T cell; T cell growth factor

Definition

In 1976, Gallo and colleagues showed that supernatants of activated T cells contained a factor that induced the proliferation of antigen-activated T cells in vitro (Morgan et al. 1976). This mitogen, then called T-cell growth factor, was subsequently purified, cloned, and named interleukin 2 (IL-2).

IL-2 is a 15-kDa four-α-helix-bundle cytokine that contains a single disulfide bond, which is essential for its biological activity (Malek 2008). IL-2 plays an important role in immune homeostasis and activation as well as in immunity (Boyman and Sprent 2012). It is a member of a cytokine family that includes IL-2, IL-4, IL-7, IL-9, IL-15, and IL-21, which all bind to the common gamma chain ($γ_c$) cytokine receptor (Liao et al. 2013).

Interleukin 2, Fig. 1 Interleukin 2 (IL-2) is secreted (indicated by *blue arrows*) mainly by CD4$^+$ T cells and at a lower extent by CD8$^+$ T, natural killer (*NK*), and NKT cells. IL-2 can bind to and be consumed (*red arrows*) by cells expressing the dimeric low-affinity IL-2 receptor, such as cytotoxic CD8$^+$ T cells and NK cells, or the trimeric high-affinity receptor present at steady-state conditions on CD4$^+$ Foxp3$^+$ regulatory T (T$_{Reg}$) cells and, upon antigen-mediated stimulation, on activated CD4$^+$ (CD4$_{act}$) and CD8$^+$ (CD8$_{act}$) T cells

Based on its in vitro properties, IL-2 was initially thought to be a crucial factor of T-cell proliferation, expansion, and immunity (Smith 1988). This idea changed when IL-2- and IL-2 receptor (IL-2R)-deficient mice were generated, which were found to harbor above-normal counts of activated T cells and to develop autoimmune disorders (Sadlack et al. 1993). The subsequent finding that IL-2 was crucial for the generation and homeostasis of CD4$^+$ IL-2Rα (CD25)$^+$ regulatory T (T$_{Reg}$) cells addressed the autoimmunity associated with IL-2 and IL-2R deficiency (Setoguchi et al. 2005).

IL-2 is not only important for the maintenance of T$_{Reg}$ cells but also for the differentiation of CD4$^+$ T cells into defined effector T-cell subsets following antigen-mediated activation (Boyman and Sprent 2012). On the other hand, IL-2 signals stimulate effector CD8$^+$ T-cell generation and differentiation into memory cells. These points are important to consider when designing therapeutic strategies targeting IL-2.

Biosynthesis and Release

IL-2 is produced mainly by activated CD4$^+$ T cells. In wild-type mice, the most prominent source of IL-2 is CD4$^+$ T cells expressing low to intermediate levels of CD25 (Setoguchi et al. 2005). Other cells are also able to produce IL-2, albeit at lower levels than CD4$^+$ T cells, including CD8$^+$ T cells, natural killer (NK) cells, and NKT cells (Fig. 1) (Malek 2008). Moreover, activated dendritic cells (DC) and mast cells have been reported to secrete IL-2, although the biological relevance of IL-2 produced by these cells remains elusive (Boyman and Sprent 2012). IL-2 production by naïve T cells is rapidly stimulated after engagement of the T-cell receptor (TCR) and co-stimulatory molecules such as CD28 with their counterparts on antigen-presenting cells.

On resting $CD4^+$ T_{Reg} cells and recently activated T cells, IL-2 binds to and signals through a trimeric IL-2R complex consisting of IL-2Rα (CD25, previously called the Tac antigen), IL-2Rβ (CD122), and γ_c (CD132) (Fig. 1). Signaling is mediated by CD122 and γ_c, whereas CD25 does not contribute to IL-2 signaling but increases the affinity of the IL-2R for IL-2 by about ten- to 100-fold (Taniguchi and Minami 1993). Structural data of human IL-2 complexed with the trimeric IL-2R suggested that IL-2 initially binds to CD25 (with a Kd $\sim 10^{-8}$ M) conferring a slight structural change to IL-2 thus allowing the recruitment of CD122 and subsequently γ_c (Malek and Castro 2010). However, immune cells expressing only a dimeric IL-2R composed of CD122 and γ_c are also able to bind IL-2 and receive a signal, although IL-2 binding on a per receptor complex basis is lower for dimeric (Kd $\sim 10^{-9}$ M) compared to trimeric IL-2R (Kd $\sim 10^{-11}$ M) (Taniguchi and Minami 1993). In mice, dimeric IL-2Rs are found at a high density on antigen-experienced (memory) $CD8^+$ T cells and NK cells, at intermediate levels on naïve $CD8^+$ T cells and memory $CD4^+$ T cells, and at a low density on naïve $CD4^+$ T cells (Boyman and Sprent 2012). Expression levels of dimeric CD122-γ_c IL-2R levels on human lymphocyte subsets are less well characterized. In addition to $CD4^+$ T_{Reg} cells, conventional T cells, and NK cells, IL-2 has been reported to also stimulate activated B cells and DC precursors (Boyman and Sprent 2012; Liao et al. 2013).

Upon binding to its receptor, IL-2 and the three IL-2R subunits are internalized, whereby IL-2, CD122, and γ_c are targeted to degradation, whereas CD25 can be recycled to the cell surface (Malek and Castro 2010). IL-2R signal transduction occurs via the Janus kinase (JAK)−signal transducer and activator of transcription (STAT), the phosphoinositide 3-kinase (PI3K)−AKT, and the mitogen-activated protein kinase (MAPK) pathways (Boyman and Sprent 2012; Liao et al. 2013). TCR signaling promotes the binding of transcription and constitutive factors to specific sites contributing to the transcription of the IL-2 gene (Rao et al. 1997). IL-2 production is regulated by an autoregulatory feedback loop. IL-2 inhibits its own production depending on the activation levels of STAT-5 and the transcription factor B-lymphocyte-induced maturation protein 1 (BLIMP1), the latter of which leads to IL-2 gene suppression (Malek 2008). Upon TCR engagement of a naïve T cell, IL-2 is produced and CD25 is upregulated on the cell surface (Smith 1988), leading to the binding of IL-2 to trimeric IL-2Rs, followed by STAT-5 activation and BLIMP1 induction (Malek 2008). However, the mechanism of BLIMP1- and STAT-5-mediated repression of the IL-2 gene remains unclear. Moreover, constant T-cell stimulation by antigen can induce the expression of the death receptor FAS (CD95) and its ligand (CD95L) on T cells, while IL-2 signaling suppresses inhibitors of apoptosis, such as c-FLIP, in these cells, thereby promoting the apoptosis of T cells through activation-induced cell death (Lenardo 1991).

T_{Reg} cells at steady state and activated $CD4^+$ and $CD8^+$ T cells during immune responses are the main consumers of IL-2 (Fig. 1). This correlates well with the high systemic IL-2 levels in mice depleted of these cells by the injection of an anti-CD25 monoclonal antibody or in CD25-deficient mice. Notably, adoptive transfer of wild-type $CD25^+$ T_{Reg} cells to CD25-deficient mice leads to normalization of systemic IL-2 levels in these mice, showing the importance of T_{Reg} cells in systemic IL-2 homeostasis. Furthermore, $CD25^+$ nonimmune cells appear to contribute as well to the control of systemic IL-2 levels (Letourneau et al. 2009).

Biological Activities

IL-2 is crucial for the homeostasis of thymus-derived ("natural") T_{Reg} cells, as evidenced by a decrease of T_{Reg} cell counts in the absence of IL-2 signaling in the thymus and periphery (Malek and Castro 2010). Moreover, treatment of wild-type mice with neutralizing anti-IL-2 antibodies significantly reduces the number of thymic and peripheral T_{Reg} cells (Malek and Castro 2010; Setoguchi et al. 2005). These findings indicate that IL-2 signals are needed for thymic development and peripheral homeostasis of T_{Reg} cells. Moreover, IL-2 signals upregulate the expression levels of CD25 and forkhead box P3 (FoxP3) on T_{Reg} cells, thereby maintaining the suppressive capacity of these cells (Boyman and Sprent 2012).

Naïve T cell can develop into in vitro-induced or peripherally derived ("induced") FoxP3$^+$ T_{Reg} cells upon activation of their TCR and in the presence of IL-2 and TGF-β. Induction of FoxP3 by naïve T cells depends on the presence of these two cytokines, although expression levels of FoxP3 can also be modulated by other γ_c cytokines (Malek and Castro 2010).

The increased susceptibility of mice to autoimmune and inflammatory disorders in the absence of IL-2 signaling correlates with low numbers of T_{Reg} cells and high counts of T-helper (Th) 17 cells (Boyman and Sprent 2012). Previously it has been shown that IL-2 inhibits the expression of IL-6Rβ (gp130) in non-polarized CD4$^+$ T cells, leading to a decrease of IL-6-mediated STAT-3 activation needed for the development of RORγt$^+$ Th17 cells. Moreover, IL-2-mediated activation of STAT-5 inhibits the binding of STAT-3 to the IL-17 gene locus. In already-polarized Th17 cells, the presence of T_{Reg} cells leads to the consumption of IL-2 thereby favoring the survival and function of Th17 cells (Boyman and Sprent 2012).

The generation of Th1 and Th2 cells from Th0 cells is also shaped by IL-2 signals. Polarization toward Th1 cells depends on the transcription factor T-bet and IL-12 signaling, both of which is enhanced in the presence of IL-2. Likewise, in Th2 cells, IL-2 induces the expression of IL-4Rα on CD4$^+$ T cells and maintains the accessibility of the IL-4 and IL-13 gene loci (Liao et al. 2013).

Different studies have shown that the primary expansion of the CD8$^+$ T cells' population is two- to threefold lower in IL-2$^{-/-}$ mice compared to wild-type animals. Experiments using bone marrow chimeric mice, carrying a mixture of CD25$^{-/-}$ and wild-type CD8$^+$ T cells, have provided important information concerning the requirement of IL-2 by CD8$^+$ T cells in a competitive environment following an acute infection (Williams and Bevan 2007). The expansion of specific CD8$^+$ T cells in the chimeric mice was two- to threefold higher for cells of wild-type origin than for their CD25$^{-/-}$ counterparts. The recall response to a secondary antigen challenge showed a lack of expansion of antigen-specific CD25$^{-/-}$ CD8$^+$ T cells, indicating that IL-2 is essential early during the primary response. These results suggest that early IL-2 signals during primary expansion are essential for optimal CD8$^+$ T-cell memory and recall responses. Although the importance of CD4$^+$ T cell help in the primary expansion of CD8$^+$ T cells was highlighted by several studies (Williams and Bevan 2007), more recent evidence suggests that activated antigen-specific CD8$^+$ T cells can produce their own IL-2, making them less dependent on paracrine IL-2 signals from CD4$^+$ T cells and DCs (Boyman and Sprent 2012). IL-2 levels also affect the differentiation of naïve CD8$^+$ T cells into either CD25^{high} short-lived effector T cells, receiving strong IL-2 signals stimulating their proliferation and subsequent apoptosis, or long-lived CD25^{low} memory precursor T cells, expressing – among others – high levels of IL-7Rα, allowing them to survive and proliferate after a secondary antigen challenge (Malek and Castro 2010).

In an analogous manner, the relative expression levels of CD25 on CD4$^+$ T cells upon TCR stimulation correlates with the percentage of cells, which differentiate into CD25^{high} short-lived effector CD4$^+$ T cells, whereas CD25^{low} effector CD4$^+$ T cells can either become long-lived central memory cells or follicular helper T cells, depending on localization and context (Boyman and Sprent 2012).

Collectively, strength and duration of IL-2 signals control CD4$^+$ and CD8$^+$ T-cell fates. Strong IL-2 signals lead to terminal differentiation of both subsets into effector T cells entering apoptosis when antigen becomes limiting. Weaker IL-2 signals promote CD4$^+$ T cells to differentiate into either follicular helper T cells or central memory T cells and facilitate the survival and function of long-lived memory CD8$^+$ T cells.

Antigen-presenting cells can also express CD25 on their surface. This is the case of murine DCs and Langerhans cell and of human DCs. It has been shown in vitro and in vivo that mature DCs can also be a source of IL-2, presumably providing some help to T cells (Boyman and Sprent 2012). Activated DCs can either bind IL-2 in a paracrine or autocrine manner. Since DCs secrete lower amounts of IL-2 than T cells, DCs might contribute to early T-cell stimulation by simply capturing T-cell-derived IL-2 via their CD25 molecules, thereby presenting IL-2 *in trans* to newly activated T cells expressing yet dimeric IL-2Rs. In support of this hypothesis, a study suggested that DCs were indeed able to transpresent IL-2 by CD25 to antigen-specific T cells (Wuest et al. 2011), analogous to IL-15 transpresentation by IL-15Rα expressed on the surface of DCs. However, the low affinity of CD25 to IL-2 somewhat questions the biological in vivo relevance of this mechanism.

Some studies have shown the expression of CD25 and CD122 on nonlymphoid cells (Letourneau et al. 2009). One relevant case is the expression of low but significant levels of trimeric IL-2Rs on the surface of pulmonary endothelial cells. Direct binding of IL-2 to these cells might lead to the vascular leak syndrome (Krieg et al. 2010), as observed with the administration of high doses of IL-2 (see below).

Pathophysiological Relevance

As mentioned before, IL-2 plays an important role for the homeostasis and suppressive activity of T_{Reg} cells as well as for the efficient stimulation of cytotoxic T cells. A balance between these subsets of cells is important in order to maintain immune tolerance and immunity. Thus, IL-2 deficiency has multiple consequences both in humans and animals.

A complete lack of IL-2 or CD25 leads to a deficiency in T_{Reg} cells and subsequent immunopathology. A few cases of human IL-2 and CD25 deficiency have been reported in the literature, leading to inflammatory and/or autoimmune phenomena along with susceptibility toward certain pathogens, including opportunistic infections, similar to the immune dysregulation polyendocrinopathy enteropathy X-linked (IPEX) syndrome (Aoki et al. 2006; Caudy et al. 2007; Pahwa et al. 1989; Sharfe et al. 1997; Weinberg and Parkman 1990). However, unlike IPEX syndrome, immunodeficiency due to a defect in IL-2 or CD25 is not X-linked but autosomal recessive, thus affecting both genders. Mice deficient in IL-2 or CD25 contain increased numbers of activated $CD44^{high}$ T cells, but severely decreased counts of $CD4^+$ $Foxp3^+$ T_{Reg} cells, and succumb to systemic autoimmune disease (Malek 2008). As discussed earlier, the lack of CD25 on $CD4^+$ and $CD8^+$ T cells impairs these cells' ability to differentiate into functional Th cell subsets and memory T cells, respectively, which affects immunity against pathogens (Boyman and Sprent 2012).

While complete IL-2 or CD25 defects are very rare in humans, partial IL-2 and T_{Reg} cell deficiencies have been reported in various immune pathologies. Some chronic-inflammatory and autoimmune diseases, including type 1 diabetes mellitus, multiple sclerosis, Graves' disease, rheumatoid arthritis, and Celiac disease, have been associated with polymorphisms in the *IL-2* or *CD25* gene (Gregersen and Olsson 2009). Experimental data in nonobese diabetic mice, which spontaneously develop autoimmune diabetes mellitus, suggested that a deficiency in IL-2 production underlies a T_{Reg} cell dysfunction leading to the development of autoimmunity (Tang et al. 2008). This finding has resulted in early phase clinical trials, including one using low-dose IL-2 administration (plus sirolimus; see below) in type 1 diabetes mellitus, which led to increased T_{Reg} cell counts but transient worsening of the disease in all subjects (Long et al. 2012). Moreover, other immunopathologies appear to feature a T_{Reg}-cell deficiency that might be amenable to IL-2 therapy, including certain forms of autoimmune vasculitis and chronic graft-versus-host disease (GVHD) (see below). While isolation, in vitro expansion, and reinfusion of T_{Reg} cells harbor several technical difficulties, the administration of IL-2 is an attractive alternative approach.

Modulation by Drugs

IL-2 levels and signaling can be therapeutically modulated using different strategies, including the provision of recombinant human IL-2 (rhIL-2), anti-CD25 monoclonal antibody (mAb), and inhibitors of IL-2 gene expression or of molecules downstream of the IL-2R.

Administration of rhIL-2 has been applied to two broad areas of medicine, with high-dose rhIL-2 used against certain metastatic cancers (Rosenberg 2012) and low-dose rhIL-2 tried in the treatment of autoimmune vasculitis and chronic GVHD (Boyman and Sprent 2012). The principle of using low- vs. high-dose IL-2 for these indications arises from the notion that $CD25^+$ $Foxp3^+$ T_{Reg} cells express "high-affinity" trimeric IL-2Rs and are thus more sensitive to low doses of IL-2, whereas $CD8^+$ T cells and NK cells express "low-affinity" dimeric IL-2Rs and therefore require the provision of high doses of IL-2 in order to be stimulated (Fig. 1). Following successful preclinical studies in different murine tumor models, high-dose rhIL-2 was tested and approved by the Food and Drug Administration (FDA) in humans for the treatment of metastatic renal cell carcinoma in 1992 and metastatic melanoma in 1998. A high-dose IL-2 treatment usually consists of intravenous infusions of 600,000–720,000 international units (IU) rhIL-2/kg body weight every 8 h for up to 12–15 doses, followed – after rest period of 5–14 days – by an additional cycle (Boyman et al. 2006b; Smith et al. 2008). Following a 6- to 12-week interval without IL-2 treatment, patients showing a clinical benefit receive further cycles of IL-2 up to a total of five courses. Based on long-term surveys, high-dose IL-2 treatment of patients with metastatic melanoma led to 27 % overall survival at 2 years, 16 % at 4 years, 10 % at 10 years, and 9–10 % at 20 years and more following IL-2 therapy (Smith et al. 2008). The drawback of high-dose IL-2 treatment lies in the occurrence of adverse effects, including cardiovascular, pulmonary, renal, hepatic, gastrointestinal, neurological, cutaneous, hematological, and systemic events (Boyman et al. 2006b). IL-2-induced toxicity is explained by a vascular leak syndrome, which correlates with the dose of IL-2 therapy. Indeed, low-dose IL-2 shows less adverse effects but elicits suboptimal antitumor immune responses. Experimental mouse data support a model where VLS is mediated by direct and indirect effects of IL-2 (Krieg et al. 2010). Firstly, high doses of IL-2 lead to the binding of IL-2 to $CD25^+$ lung endothelial cells, thus directly damaging these cells. Secondly, upon activation of T cells and NK cells by IL-2, these cells secrete vasoactive pro-inflammatory cytokines, such as tumor necrosis factor, which further contribute to endothelial cell damage.

Unlike high-dose IL-2 (where a typical patient receives about 100–200 million IU rhIL-2 daily), low-dose IL-2 leads to preferential stimulation of $CD25^+$ T cells, notably $CD4^+$ $CD25^+$ $Foxp3^+$ T_{Reg} cells. The use of low-dose rhIL-2 in chronic GVHD was assessed in a clinical trial involving 29 patients with chronic GVHD non-responsive to standard glucocorticoid therapy (Koreth et al. 2011). Administration of IL-2 leads to an increase in counts of T_{Reg} cells along with disease improvement in 12 out of 23 patients that could be evaluated. In this study, one million IU rhIL-2/m^2 body-surface area (resulting in about 1.5–2.5 million IU rhIL-2 daily) were given to patients and no major adverse effects were described. Another trial was performed in ten patients with hepatitis C virus-associated cryoglobulinemic vasculitis non-responsive to antiviral therapy and/or B-cell-targeted anti-CD20 (rituximab) treatment (Saadoun et al. 2011). The dose used in this trial was 1.5 million IU rhIL-2/day for the first 5 days, followed by three 5-day courses of 3 million IU rhIL-2/day at weeks 3, 6, and 9. Also the authors of this study observed an increase of $CD4^+$ $CD25^+$ $Foxp3^+$ T_{Reg} cells along with an increase of the minor population of $CD8^+$ $CD25^+$ $Foxp3^+$ T cells and an improvement of vasculitis in 8 out of 10 patients. No severe adverse effects were reported.

Although these results are encouraging, achieving lymphocyte-specific selectivity by the provision of low- vs. high-dose IL-2 is somewhat arbitrary and bears the risk of activating unwanted (effector) immune cells, as evidenced in a recent early phase clinical trial in type 1 diabetes mellitus (Long et al. 2012). Thus, the generation of IL-2 formulations selectively (or preferentially targeting) $CD4^+$ T_{Reg} cells vs. cytotoxic lymphocytes would be preferable. A technique allowing such selectivity of IL-2 consists in the provision of IL-2 bound to a particular anti-IL-2 mAb, thereby forming so-called IL-2/anti-IL-2 mAb complexes (briefly, IL-2 complexes). IL-2 complexes of murine IL-2 plus anti-murine IL-2 mAb S4B6 or of human IL-2 with antihuman IL-2 $mAb_{(CD122)}$ have been shown to preferentially stimulate $CD8^+$ T cells and NK cells, both of which express high levels of the dimeric $CD122$-γ_c IL-2R (Boyman et al. 2006a). Alternatively, when IL-2 is in complex with anti-murine IL-2 mAb JES6-1 or anti-human IL-2 $mAb_{(CD25)}$, IL-2 is directed selectively toward $CD25^+$ cells expressing trimeric IL-2Rs, mainly $CD4^+$ $CD25^+$ $Foxp3^+$ T_{Reg} cells (Boyman et al. 2006a). Thus, use of CD25-directed IL-2 complexes (e.g., JES6-1) allows the preferential stimulation of $CD4^+$ T_{Reg} cells for the treatment of chronic-inflammatory disorders, autoimmune disease, and graft rejection, whereas administration of CD122-directed IL-2 complexes (e.g., S4B6) leads to potent activation and expansion of $CD8^+$ T cells and NK cells, while T_{Reg} cell stimulation remains minimal, thereby favoring antitumor immune responses (Boyman and Sprent 2012; Boyman et al. 2006b). Interestingly, such therapy allows the administration of lower doses of IL-2 and disfavors the binding of IL-2 to pulmonary endothelial cells (Krieg et al. 2010). Alternatively, targeting IL-2 to the tumor site represents another strategy to reach high local concentrations of IL-2 while reducing systemic adverse effects. This can be achieved by linking IL-2 or IL-2 complexes to an antibody targeting specific tumor antigens. For IL-2, such IL-2 immunocytokines directed against the tumor neovasculature have been shown efficacious in mice and patients suffering from an acute myeloid leukemia (Gutbrodt et al. 2013).

Another way of modulating IL-2 activity is to block specific chains of the IL-2R. Examples that have been applied in the clinic are daclizumab and basiliximab, both mAbs targeting human CD25. Basiliximab and daclizumab have been used in the prevention of acute graft rejection, while daclizumab has also been tried in relapsing-remitting multiple sclerosis. The mechanism of action of daclizumab and basiliximab in the prevention of acute graft rejection appears to rely on the inhibition of $CD25^+$ anti-graft effector T cells. Conversely, the mechanism of action of daclizumab in relapsing-remitting multiple sclerosis is more complicated and has been suggested to involve the increase of regulatory NK cells as well as other effects.

Several molecules have been shown to interfere with IL-2 production or signaling. The most prominent drugs include cyclosporin (cyclosporin A, CsA), tacrolimus (FK506), and sirolimus (rapamycin), an inhibitor of the mammalian target of rapamycin (mTOR). Cyclosporin is derived from a fungus, while both tacrolimus and sirolimus are macrolides of bacterial origin. Cyclosporin and tacrolimus act by inhibiting the complex of calcineurin and nuclear factor of activated T cells (NFAT) that is necessary for IL-2 and other immune genes (Rao et al. 1997). Conversely, sirolimus does not interfere with the calcineurin-NFAT complex but blocks mTOR, thus preventing mTOR-mediated signaling upon IL-2R activation (Liao et al. 2013). More recently, tofacitinib, an inhibitor of mainly JAK3 and JAK1, has been tried in chronic-inflammatory disorders, including rheumatoid arthritis, psoriasis, and inflammatory bowel disease. By interfering with JAK3 and JAK1, tofacitinib temporarily blocks signaling of IL-2 and other γ_c cytokines (Liao et al. 2013).

Acknowledgments We thank the members of the Boyman laboratory for the critical reading of this article. This work was funded by SNF grants PP00P3-128421 and CRSII3-136203, a National Psoriasis Foundation USA grant, a Zurich Integrative Human Physiology (ZIHP) cooperative project grant, a grant of the Stiftung fuer wissenschaftliche Forschung of the University of Zurich, and a grant of the Novartis Foundation (all to OB).

References

Aoki, C. A., Roifman, C. M., Lian, Z. X., Bowlus, C. L., Norman, G. L., Shoenfeld, Y., et al. (2006). IL-2 receptor alpha deficiency and features of primary biliary cirrhosis. *Journal of Autoimmunity, 27,* 50–53.

Boyman, O., & Sprent, J. (2012). The role of interleukin-2 during homeostasis and activation of the immune system. *Nature Reviews Immunology, 12*, 180–190.

Boyman, O., Kovar, M., Rubinstein, M. P., Surh, C. D., & Sprent, J. (2006a). Selective stimulation of T cell subsets with antibody-cytokine immune complexes. *Science, 311*, 1924–1927.

Boyman, O., Surh, C. D., & Sprent, J. (2006b). Potential use of IL-2/anti-IL-2 antibody immune complexes for the treatment of cancer and autoimmune disease. *Expert Opinion on Biological Therapy, 6*, 1323–1331.

Caudy, A. A., Reddy, S. T., Chatila, T., Atkinson, J. P., & Verbsky, J. W. (2007). CD25 deficiency causes an immune dysregulation, polyendocrinopathy, enteropathy, X-linked-like syndrome, and defective IL-10 expression from CD4 lymphocytes. *The Journal of Allergy and Clinical Immunology, 119*, 482–487.

Gregersen, P. K., & Olsson, L. M. (2009). Recent advances in the genetics of autoimmune disease. *Annual Review of Immunology, 27*, 363–391.

Gutbrodt, K. L., Schliemann, C., Giovannoni, L., Frey, K., Pabst, T., Klapper, W., Berdel, W. E., et al. (2013). Antibody-based delivery of interleukin-2 to neovasculature has potent activity against acute myeloid leukemia. *Science Translational Medicine, 5*, 201ra118.

Koreth, J., Matsuoka, K., Kim, H. T., McDonough, S. M., Bindra, B., Alyea, E. P., et al. (2011). Interleukin-2 and regulatory T cells in graft-versus-host disease. *New England Journal of Medicine, 365*, 2055–2066.

Krieg, C., Letourneau, S., Pantaleo, G., & Boyman, O. (2010). Improved IL-2 immunotherapy by selective stimulation of IL-2 receptors on lymphocytes and endothelial cells. *Proceedings of the National Academy of Sciences of the United States of America, 107*, 11906–11911.

Lenardo, M. J. (1991). Interleukin-2 programs mouse alpha beta T lymphocytes for apoptosis. *Nature, 353*, 858–861.

Letourneau, S., Krieg, C., Pantaleo, G., & Boyman, O. (2009). IL-2- and CD25-dependent immunoregulatory mechanisms in the homeostasis of T-cell subsets. *Journal of Allergy and Clinical Immunology, 123*, 758–762.

Liao, W., Lin, J. X., & Leonard, W. J. (2013). Interleukin-2 at the crossroads of effector responses, tolerance, and immunotherapy. *Immunity, 38*, 13–25.

Long, S. A., Rieck, M., Sanda, S., Bollyky, J. B., Samuels, P. L., Goland, R., et al. (2012). Rapamycin/IL-2 combination therapy in patients with type 1 diabetes augments Tregs yet transiently impairs beta-cell function. *Diabetes, 61*, 2340–2348.

Malek, T. R. (2008). The biology of interleukin-2. *Annual Review of Immunology, 26*, 453–479.

Malek, T. R., & Castro, I. (2010). Interleukin-2 receptor signaling: At the interface between tolerance and immunity. *Immunity, 33*, 153–165.

Morgan, D. A., Ruscetti, F. W., & Gallo, R. (1976). Selective in vitro growth of T lymphocytes from normal human bone marrows. *Science, 193*, 1007–1008.

Pahwa, R., Chatila, T., Pahwa, S., Paradise, C., Day, N. K., Geha, R., et al. (1989). Recombinant interleukin 2 therapy in severe combined immunodeficiency disease. *Proceedings of the National Academy of Sciences of the United States of America, 86*, 5069–5073.

Rao, A., Luo, C., & Hogan, P. G. (1997). Transcription factors of the NFAT family: Regulation and function. *Annual Review of Immunology, 15*, 707–747.

Rosenberg, S. A. (2012). Raising the bar: The curative potential of human cancer immunotherapy. *Science Translational Medicine, 4*, 127ps128.

Saadoun, D., Rosenzwajg, M., Joly, F., Six, A., Carrat, F., Thibault, V., et al. (2011). Regulatory T-cell responses to low-dose interleukin-2 in HCV-induced vasculitis. *New England Journal of Medicine, 365*, 2067–2077.

Sadlack, B., Merz, H., Schorle, H., Schimpl, A., Feller, A. C., & Horak, I. (1993). Ulcerative colitis-like disease in mice with a disrupted interleukin-2 gene. *Cell, 75*, 253–261.

Setoguchi, R., Hori, S., Takahashi, T., & Sakaguchi, S. (2005). Homeostatic maintenance of natural Foxp3(+) CD25(+) CD4(+) regulatory T cells by interleukin (IL)-2 and induction of autoimmune disease by IL-2 neutralization. *The Journal of Experimental Medicine, 201*, 723–735.

Sharfe, N., Dadi, H. K., Shahar, M., & Roifman, C. M. (1997). Human immune disorder arising from mutation of the alpha chain of the interleukin-2 receptor. *Proceedings of the National Academy of Sciences of the United States of America, 94*, 3168–3171.

Smith, K. A. (1988). Interleukin-2: Inception, impact, and implications. *Science, 240*, 1169–1176.

Smith, F. O., Downey, S. G., Klapper, J. A., Yang, J. C., Sherry, R. M., Royal, R. E., et al. (2008). Treatment of metastatic melanoma using interleukin-2 alone or in conjunction with vaccines. *Clinical Cancer Research, 14*, 5610–5618.

Tang, Q., Adams, J. Y., Penaranda, C., Melli, K., Piaggio, E., Sgouroudis, E., et al. (2008). Central role of defective interleukin-2 production in the triggering of islet autoimmune destruction. *Immunity, 28*, 687–697.

Taniguchi, T., & Minami, Y. (1993). The IL-2/IL-2 receptor system: A current overview. *Cell, 73*, 5–8.

Weinberg, K., & Parkman, R. (1990). Severe combined immunodeficiency due to a specific defect in the production of interleukin-2. *New England Journal of Medicine, 322*, 1718–1723.

Williams, M. A., & Bevan, M. J. (2007). Effector and memory CTL differentiation. *Annual Review of Immunology, 25*, 171–192.

Wuest, S. C., Edwan, J. H., Martin, J. F., Han, S., Perry, J. S., Cartagena, C. M., et al. (2011). A role for interleukin-2 trans-presentation in dendritic cell-mediated T cell activation in humans, as revealed by daclizumab therapy. *Nature Medicine, 17*, 604–609.

Interleukin 4 and the Related Cytokines (Interleukin 5 and Interleukin 13)

Chrysanthi Skevaki, Christoph Hudemann and Harald Renz
Institute of Laboratory Medicine and Pathobiochemistry, Molecular Diagnostics, Philipps University Marburg, University Hospital GmbH and Marburg GmBH, Marburg, Germany

Synonyms

B151-TRF; B-cell differentiation factor-epsilon (BCDF-epsilon); B-cell growth factor-gamma (BCGF-gamma); B-cell stimulatory factor 1 (BSF-1); B-cell growth and differentiation factor (BGDF); B-cell growth factor 2 (BCGF-2); BCDF-alpha; BCDF-gamma; BCDF-mu; BCGF-1; Binetrakin; BSF-p1; EL4 B-cell growth factor (EL4-BCGF); Colony-forming unit eosinophil growth stimulating factor (CFU-Eo GSF); Dennert cell-derived BCGF (DL-BCGF); Eosinophil colony-stimulating factor (Eo-CSF); Eosinophil differentiation factor (Eo-DF or EDF); Eosinophil stimulation promoter (ESP); Hodgkin's cell growth factor (HCGF); IgA-enhancing factor (IgA-EF); IgE-enhancing factor (IgE-EF); IgG1-enhancing factor; IgG1-induction factor; *IL-4*: Ia-inducing factor (IaIF); *IL-5*: B-cell differentiation factor (BCDF); *IL-13*: NC30; Killer helper factor (KHF); Low molecular weight B-cell growth factor (LMW-BCGF); Lymphocyte stimulatory factor; Macrophage fusion factor (MFF); Mast cell growth enhancing factor (MaGEF); Mast cell growth factor 2 (MCGF-2); P600 (murine); Pitrakinra (recombinant human IL-4); T-cell growth factor 2 (TCGF-2); T-cell replacing factor 1 (TRF-1)

Definition

Interleukin 4 (IL-4) is a 129-amino acid-long (20 kDa) pleiotropic cytokine that influences T- and B-cell responses including cell proliferation, survival, and gene expression. The human interleukin 5 (IL-5) is a 115-amino acid-long (22 kDa) Th2 cytokine that is part of the hematopoietic family and can be considered as the major differentiation factor of eosinophils. The pleiotropic Th2 cytokine interleukin 13 (IL-13) has in its main form a length of 111 amino acids (12 kDa) and was first characterized as a T-cell-derived regulator of inflammatory cytokine production. Its general role lies in suppression of cell-mediated immune responses and promotion of B-cell differentiation.

Biosynthesis and Release

Murine IL-4 gene maps to chromosome 11, while the human gene along with the genes encoding for IL-3, IL-5, IL-13, granulocyte–macrophage colony-stimulating factor (GM–CSF), and CSF2 form a cytokine gene cluster on chromosome 5q, with genes for IL-4 and IL-13 particularly close to each other. It consists of four exons and has a length of approximately 10 kb. The 5′ region contains conserved lymphokine elements (CLE) which bind to transcription factors and control gene expression. IL-4 has a compact, globular fold stabilized by three disulfide bonds, which are essential for biological activity. IL-4 is mainly produced by Th2 cells but may also be released by mast cells, basophils, eosinophils, and alveolar macrophages. The IL-5 gene is located on chromosome 11 in the mouse and chromosome 5 in humans within a 500 kb cytokine gene cluster that includes interleukin 4 (IL-4), interleukin 3 (IL-3), and GM–CSF, which are often co-expressed in Th2 cells. The human gene spans over 4 kb, contains four exons, and may be regulated coordinately by long-range regulatory elements spread over 120 kb on chromosome 5q23–31. The 5′ region of the IL-5 gene contains CLEs which serve as binding sites for transcription factors. A highly glycosylated 22 kDa IL-5 monomer forms a bioactive 40–50 kDa disulfide-linked homodimer. IL-5 is mainly produced by Th2 lymphocytes following stimulation by allergens or pathogens, by mast cells upon allergen–IgE complex activation, and by eosinophils. Innate lymphocytes type II (ILC2) have also been shown to be high IL-5 producers, while IL-5 mRNA is expressed in γδT cells, NK and Natural Killer T (NKT) cells, and nonhematopoietic cells as well. The coding region for *IL-13* is located on chromosome 5 segment q23–31 and is closely related to the genes that encode for IL-3, IL-5, and IL-9 (Smirnov et al. 1995). It belongs to the class of type I cytokines that share a tertiary structure formed by 4 α-helical bundles (A, B, C, and D). This ~16 kDa glycoprotein displays only 25 % sequence homology to IL-4, yet the structural core is fully conserved and thus functionally closely related. IL-13 production is regulated by the Th2 transcription factor GATA-3; therefore, it is deemed to represent a classical Th2 cytokine. Its main source is activated CD4+ Th2 cells, but also Th1 and innate immune cells such as eosinophils, basophils, mast cells, or natural killer cells have the capacity to produce IL-13.

Biological Activities

IL-4 binds to receptor IL-4Rα which is ubiquitously expressed (on T cells, B cells, eosinophils, mononuclear phagocytes, endothelial cells, lung fibroblasts, bronchial epithelial cells, and smooth muscle cells), and this is followed by heterodimerization with a second receptor chain, either IL-2Rγc or IL-13-bound IL-13 receptor alpha 1 (IL-13Rα1) chain. These complexes (type I and II receptors) lead to activation of the Janus family of protein kinases (Jak). Specifically, IL-4Rα, γc, and IL-13Rα1 activate Jak1, Jak3, and tyrosine kinase (Tyk2), respectively. This induces a series of intracellular signaling cascades by phosphorylating specific tyrosine residues in the cytoplasmic domain of IL-4Rα. Signal transducer and activator of transcription 6 (Stat6) is subsequently recruited and phosphorylated resulting in dimerization and translocation to the nucleus. This in turn upregulates the expression of Th2 lineage-specific zinc-finger transcription factor GATA-binding protein 3 (GATA-3). The latter binds to the DNA sequence "GATA" to modulate target Th2 cytokine genes (IL-4, IL-5, IL-9, and IL-13) and promote their expression. GATA-3 expression can be induced via IL-4, IL-33, and other mediators, i.e., even in the absence of allergen-mediated IL-4 exposure. GATA-3 also promotes the expression of MAF, a transcription factor which selectively transactivates IL-4 alone. IL-2Rγc and IL-13Rα1 are expressed only on some cell types, thus leading to cell-specific downstream effects. Indeed, type I receptors (IL-4Rα/IL-2R γc/IL-4) are typically found on hematopoietic cells and are involved in Th2 development, while type II receptors (IL-4Rα/IL-13Rα1/IL-4) are found on both hematopoietic and nonhematopoietic cells. IL-4 signaling through the type II receptor on airway smooth muscle and bronchial epithelial cells can directly induce airway hyperresponsiveness (AHR) and mucus secretion, respectively. The IL-13Rα1 subunit in the type II complex may amplify IL-4Rα signaling but can also initiate independent signaling pathways involving other Stat proteins (Stat3, Stat1) and possibly playing a role in AHR. IL-4 induces isotype class switching of B cells to IgG1 and IgE synthesis and recruitment of mast cells. High-affinity IgE receptors on mast cells and low-affinity IgE receptors on B cells and mononuclear phagocytic cells are also upregulated as a result of IL-4 action. Furthermore, IL-4 stimulates differentiation of naïve CD4+ T cells into the Th2 phenotype and expression of other Th2 cytokines including IL-5 and IL-9. IL-4 also induces endothelial expression of vascular cell adhesion molecule-1 (VCAM-1), which facilitates transmigration of T cells, monocytes, eosinophils, and basophils contributing to local inflammatory responses.

IL-5 regulates its own receptor, IL-5Rα, during eosinophil ontogeny. The regulatory factor X (RFX) family of DNA-binding proteins binds to the cis element of the IL-5Rα promoter. Expression of RFX proteins is ubiquitous, yet they are suggested to act as lineage-specific activators in cooperation with other factors. Receptor binding leads to recruitment of the βc chain to form the IL-5–IL-5Rα–βc complex. The βc chain alone does not bind any cytokines, has a relatively long cytoplasmic portion and several functional domains, and is deeply involved in signal transduction. High- and low-affinity IL-5Rs are expressed on all hematopoietic and lymphoid cells and transduce multiple functions. The active form of IL-5 induces rapid tyrosine phosphorylation of βc; Src-homology 2 (SH2)/SH3-containing proteins; Bruton agammaglobulinemia tyrosine kinase (Btk) and Btk-associated molecules; JAK1; JAK2; signal transducer and activator of transcription 1 (STAT1), STAT3, and STAT5; phosphoinositide 3-kinase (PI3K); Lyn tyrosine kinase; and mitogen-activated protein kinases (MAPKs). Phosphorylation of these cellular proteins results in activation of downstream signaling molecules. IL-5 also activates v-raf-1 murine leukemia viral oncogene homologue 1 (Raf-1) and the Src-homology-2-domain-containing protein tyrosine phosphatase (SHP2) (Pazdrak et al. 1997). IL-5 induces the expression of CIS (cytokine-inducible SH2 protein) and JAB (JAK-binding protein) in eosinophils in the context of negative feedback loops (Zahn et al. 2000). In addition to the JAK2–STAT5 pathway, the Ras GTPase–extracellular signal-regulated kinase (Ras–ERK) pathway has also been implicated in signaling of IL-5 (Hall et al. 2001). IL-5 does not induce eosinophil lineage commitment but rather enhances differentiation and proliferation of eosinophil progenitors in the bone marrow. Moreover, its chemotactic activity leads to eosinophil migration from bone marrow to the blood and subsequently to sites of inflammation also via increased eosinophil adhesion to endothelial cells. In the lung, IL-5 prolongs the survival of eosinophils by preventing apoptosis (Giembycz and Lindsay 1999; Kouro and Takatsu 2009). IL-5 also plays a role in B-cell growth and differentiation in antibody-secreting plasma cells.

IL-4 and IL-13 partially share the common receptor subunit IL-4Rα (also known as CD124), which together with the IL-13-bound IL-13Rα1 receptor form a heterodimer complex comprising the dominant type II IL-4R signaling pathway. Following binding, constitutively associated JAK1 and tyrosine kinase 2 (Tyk-2) become phosphorylated and allow activation of STAT6 or STAT3 in nonhematopoietic cells (LaPorte et al. 2008), with subsequent translocation into the nucleus and induction of IL-13-dependent gene expression. Monocytes, macrophages, and fibroblasts express both type I and type II IL-4Rs, while nonhematopoietic cells such as smooth muscle cells and epithelial cells predominantly express the type II receptor (Wills-Karp and Finkelman 2008). Additionally, the high-affinity receptor IL-13Rα2 exists largely as a monomer and is sufficient for binding IL-13 alone. It mainly exists intracellularly in nonlymphoid tissue and tumor cells. Upon IFN-γ treatment, IL-13Rα2 translocates to the cell membrane, and subsequent binding of IL-13 leads to reduced signaling. In human bronchial epithelial cells, antibody treatment against IL-13Rα2 leads to an IL-13-dependent allergic airway model with an increase of PAS positive cells and total MUC5A mRNA, which underlines its role as a functional regulator of IL-13 (Tanabe et al. 2008). While IL-13Rα1 is expressed on virtually every cell except T and B cells (Hershey 2003) and does not influence the outcome and survival of several tumors, IL-13Rα2 expression was found to be high in various malignant tumor cell lines and patient tissue derived from malignant glioma, ovarian cancer, and different carcinomas. Th1 polarization of the immune system is widely believed to mediate tumor rejection; Th2 polarization on the other hand might favor tumor cell formation. Direct activation of apoptosis via a 15-LOX pathway indicates an antitumorigenic behavior, as it has been shown with defined antiproliferative effects on B-cell acute lymphoblastic leukemia, breast cancer, and renal cell carcinoma (Lebel-Binay et al. 1995). However, several studies have also shown a decreased tumor rejection by IL-13. Natural killer T cells stimulated by tumor antigens produce IL-13, which in turn leads to secretion of macrophage-specific TGF-β. $CD8^+$ T cells involved in tumor immune surveillance are then inhibited. IL-13 is therefore discussed as a biomarker for tumor formation and survival; however, a defined link between this cytokine and cell survival remains to be established. Along these lines, IL-13 inhibitors or novel IL-13 antagonists may prove to be novel tools in cancer immunotherapeutics.

Pathophysiological Relevance

IL-4 plays a critical role in the development of allergic inflammation and asthma. Asthmatic patients have high IL-4 levels in serum, bronchoalveolar lavage (BAL), and bronchial biopsies and increased IL-4 expressing T-cell counts. Along these lines, levels of IL-4Rα and Stat6, which are critical components of the IL-4 signaling pathway, have also been found elevated in sputum and bronchial biopsies of asthmatic patients. Allergen challenge of atopic asthmatic subjects results in induction of IL-4 release in BAL and peripheral blood mononuclear cell cultures. Moreover, nebulization of IL-4 in mild asthmatics induces features of asthma such as AHR and eosinophilia (Shi et al. 1998a). Genetic variants in the region of IL-4 (and IL-13) were associated with susceptibility to severe asthma, lung function, and total IgE. The IL-4/IL-13 signaling pathway is linked to the symptoms experienced by a subset of severe asthmatics with allergen-associated symptoms and high serum IgE levels (Caruso et al. 2013). In a chronic house dust mite (HDM) exposure model, IL-4-deficient mice failed to exhibit airway remodeling despite the presence of airway inflammation (Johnson et al. 2007). However, in a chronic ovalbumin exposure model, IL-4-deficient mice were similar to wild-type mice in regard to airway inflammation, epithelial hyperplasia, subepithelial fibrosis, and AHR suggesting that airway remodeling processes may be antigen type dependent (Foster et al. 2000). Earlier studies demonstrated reduced eosinophilic and peribronchial inflammation, no allergen-specific and total IgE production, and absence of AHR in a murine ovalbumin asthma model of IL-4 deficient versus wild-type animals (Brusselle et al. 1994, 1995).

The ability of IL-5 to defend against host infections with large organisms such as helminths has been a subject of investigation for over 100 years (Brown 1898). Following helminth infection, Th2 immune polarization is characterized by an increase of IL-5 and a systemic eosinophil rush that migrate to the infection site where they either contribute to tissue rearrangement processes or directly kill tissue-traversing larval stages of this type of parasite and act as IL-5-dependent T-helper cells for protective antibody production. Surprisingly, neutralizing IL-5 antibody treatment has been shown to have little effect on the survival or reproduction of nematodes and trematodes, raising the possibility of other IL-5-independent or redundant mechanisms leading to increased eosinophilia. The usage of IL-5R knockout as well as IL-5 knockout mice revealed a slightly reduced constitutive eosinophilic population, which only in very few cases resulted in stronger worm burden. Thus, until today, the precise role regarding IL-5-dependent eosinophilia in helminth infection can only be speculated, while it is likely to serve against reinfection by larval stages of these parasites (Maizels and Balic 2004). Besides worm infections, IL-5-dependent eosinophilia is also associated with atopic diseases and asthma, neoplastic disorders, and drug hypersensitivity. Asthmatic patients present increased IL-5 concentration in BAL and serum originating primarily from activated CD4+ T cells, and IL-5 inhalation even further increases eosinophil numbers worsening AHR (Shi et al. 1998b). Airway remodeling is widely used as a collective term of eosinophil-related structural rearrangements in the airways such as those causing fibroblast hypertrophy or collagen deposition. Nevertheless, IL-5 knockout mice still present a certain amount of local eosinophil infiltration, which can be attributed to the CC chemokine ligand 11 (CCL11; eotaxin) in an IL-5-independent manner. In asthmatics, eosinophilic inflammation is even further increased during asthmatic exacerbations, and corticosteroid-induced IL-5 reduction correlates with reduced inflammation and improvement of lung function. In corticosteroid-resistant patients, interleukin 33 (IL-33) causes the production of additional IL-5 which then leads to treatment-resistant severe airway inflammation.

Over the past few years, numerous transgenic mouse lines with tissue-specific targeted overexpression of IL-13 as well as IL-13-deficient mouse lines allowed a wide functional characterization that is constantly growing. While IL-4 is crucial for the initial polarization of CD4 naïve

T cells to the TH2 phenotype, IL-13 appears to be essential in the effector phase of airway pathophysiology which includes IgE synthesis, airway inflammation, mucus secretion, airway hyperresponsiveness (AHR), and fibrosis. Here, mouse and human studies have clearly shown that IL-13 induces mucus hypersecretion and goblet cell hyperplasia (Zhu et al. 1999). IL-13 blockade using an IL-13 receptor α_2-human Fc fusion protein inhibited many allergic characteristics in allergen-immunized mice, revealing dependence of allergen-induced AHR on IL-13 (Grünig et al. 1998). Local fibroblasts are stimulated directly or indirectly via the TGF-β secretion of activated macrophages leading to an increased type I collagen synthesis and deposition (Jinnin et al. 2004). Smooth muscle cells display increased proliferation and cholinergic contractions by modulated Ca^{2+} responses. Extracellular matrix metalloproteinases (MMPs) are also found to be upregulated during many pathological conditions such as allergic inflammation, tissue injury and repair, remodeling, and host defense against pathogens. Interestingly, IL-13 is a potent stimulator of MMPs in the lung, and overexpression of these proteins resulted in emphysematous alterations (Greenlee et al. 2007). IgE production is stimulated by activated mast cells and the switch of antibody-producing plasma cells from IgM to IgE. Transgenic IL-13 mice display an about 10–100-fold higher level of serum IgE independent of the presence of IL-4, which indicates that IL-13 is crucial for IgE production in these animals. Overall, IL-13 can be considered a key player in a broad range of respiratory diseases such as asthma and chronic obstructive pulmonary disease (COPD).

Modulation by Drugs

Blocking IL-4 has been attempted by means of soluble IL-4R (sIL-4R), monoclonal antibodies, or antisense technologies. All such approaches were successful in preclinical models but did not provide sufficiently convincing data in clinical studies. Soluble IL-4R (sIL-4R) is a naturally occurring secreted form of the receptor which contains the extracellular portion but lacks the transmembrane and cytoplasmic domains. Therefore, IL-4 binding to sIL-4R is neutralized avoiding cellular activation effects. Administration of sIL-4R in mice during sensitization or even allergen challenges attenuated characteristics of allergic airway inflammation. Clinical trials initially produced promising results with nebulization of recombinant human sIL-4R (Nuvance; Immunex, Seattle, WA) being safe and effective for the treatment of moderate asthma also in the long term (Borish et al. 2001). Subsequent studies did not confirm the agent's usefulness which may be attributed to the fact that it can be recruited by the IL-13/IL-13Rα1 complex and stabilize IL-13 binding (Andrews et al. 2006). Another approach has been the use of a murine humanized anti-IL-4 monoclonal antibody, pascolizumab (Biopharma, Winchester, UK). Anti-IL-4 treatment has been shown to be effective in vitro with decreased IL-4 bioactivity, attenuation of Th2 responses, and IgE production of cell lines. Similarly, results from murine asthma models showed significant inhibition of AHR and IgE levels. In vivo pharmacokinetic and chronic safety studies in cynomolgus monkeys as well as Phase I clinical trials demonstrated good tolerability. Phase II trials were terminated prematurely due to lack of evidence for clinical benefit. RNA-based gene silencing strategies may serve to attenuate IL-4 expression by means of degradation of target mRNA. Such an approach has provided promising results in animal models but has not been tested clinically yet. Nevertheless, targeting IL-4 alone may be too selective due to the redundancy of IL-4 and IL-13 biological activities. Therefore, different approaches for blocking IL-4Rα receptor have been developed. These include an IL-4 mutein (pitrakinra, Aerovance, Berkley), an IL-4Rα antibody (AMG317), and antisense IL-4Rα (AIR-645). Pitrakinra has high affinity for IL-4Rα and acts as a competitive receptor antagonist. Results from murine models and cynomolgus monkeys were positive in terms of asthma characteristics such as AHR, goblet cell hyperplasia, and eosinophilic inflammation. Subsequent clinical trials involving atopic asthmatic patients following allergen challenge showed that pitrakinra possesses good pharmacokinetic and pharmacodynamic features and is clinically effective and safe. In a multicenter Phase IIb study in subjects with moderate-to-severe asthma, pitrakinra was unable to reduce exacerbation incidence while tapering off other antiasthmatic medication except from a group of patients with eosinophilic asthma (Burmeister Getz et al. 2009). Usage of AMG317 in mouse asthma models has provided beneficiary effects, while sustained subcutaneous administration over months in Phase I and II trials resulted in reduced total serum IgE levels. In the Phase II studies (Corren et al. 2010), AMG317 failed to improve symptom scores overall. However, a subgroup of severe asthma patients with the highest baseline Asthma Control Questionnaire (ACQ) scores) improved with this treatment. Preclinical testing of antisense oligonucleotides against IL-4Rα effectively reduced several asthma hallmarks in mice. Clinical trials did not warrant further development despite good tolerance and low systemic exposure (Seguin and Ferrari 2009). Finally, an additional strategy targets the Stat6 transcription factor using a Stat6-inhibitory peptide, a small molecule inhibitor (AS1517499), an antisense RNA or small interfering RNA (siRNA). Results in murine models were promising in terms of inhibition of allergen-induced airway inflammation, goblet cell hyperplasia, mucus production, and AHR. Human lung epithelial cells previously exposed to IL-4/IL-13 have also been successfully treated with siRNA in vitro, but clinical studies are still missing. GATA-3-dependent Th2-immune polarization by IL-4 has been shown to constitute a promising therapeutic target using novel DNAzymes. In human and mouse, pathological immune mechanisms such as in allergic bronchial asthma were significantly reduced without any additional delivery systems, demonstrating the power of DNAzymes as a novel class of antisense technology (Sel et al. 2008; Turowska et al. 2013).

IL-5 specifically binds to its heterodimeric IL-5 receptor mainly present on eosinophils and B cells, the main players in allergic airway inflammation, and IL-5-based therapy might therefore attenuate key features. Human α-IL-5 monoclonal antibodies (mAB) such as mepolizumab (GlaxoSmithKline, Brentford) or reslizumab (Teva, Petah Tikva) have been considered as promising therapeutic agents for eosinophil-mediated asthma. In wild-type mice, treatment with an anti-IL-5 antibody almost completely prevented subepithelial and peribronchial fibrosis. Early patient studies did show that intravenous application resulted in a significant decrease of blood and sputum eosinophil numbers. However, most clinical endpoints like lung function or symptom scores failed to improve. In long-term studies, anti-IL-5 mAB treatment reduced the incidence of asthma exacerbations and patient-reported symptom scores (Nair et al. 2009). This implies that IL-5 is rather the result of eosinophilic inflammation closely related to asthma exacerbations. Direct targeting of IL-5-associated effector cells (eosinophils, basophils) is achieved by humanized anti-IL-5 receptor antibody therapy. Candidate antibodies have already shown promising antibody-dependent cell-mediated cytotoxicity in vitro and in vivo (mouse models). Benralizumab (MedImmune, Gaithersburg) was tested in a small cohort study of mild atopic asthmatics and resulted in reduced eosinophil counts in bone marrow and peripheral blood; yet larger studies are needed for analysis of clinical benefits in asthmatic patients (Laviolette et al. 2013). Other eosinophilic conditions, such as the hyper-eosinophilic syndrome (HES) that displays a characteristic eosinophilia in the heart, skin, gastrointestinal tract, and lung, showed benefit by mepolizumab treatment, which allowed reduction and even discontinuance of corticosteroid intake. A novel approach to suppress key features of airway inflammation is the usage of antisense technology. OVA-sensitized mice displayed decreased infiltration of inflammatory cells and AHR when treated with lentivirus-delivered siRNA targeting IL-5 directly. TPI ASM8 is a mixture of two antisense oligonucleotides (AONs) that target the human β subunit of the IL-3, IL-5, and GM-CSF receptors and the chemokine receptor CCR3. Studies with mild asthmatics displayed an attenuated broad range of inflammatory and physiological changes after inhaled allergen challenge, suggesting that TPI

ASM8 had broader effects than merely inhibition of eosinophils in asthmatics (Gauvreau et al. 2011).

As mentioned, IL-13 and its receptors represent valid therapeutic targets for airway diseases and different cancers. To date, most human trials using anti-IL-13 agents have been performed among asthmatic patients with varying degrees of success. The humanized IgG4 IL-13-antibody lebrikizumab (Roche, Basel) improved lung function in patients with moderate-to-severe asthma with a high serum periostin level (Corren et al. 2011). Also, treated patients experienced a significant reduction of the rather unspecific FeNO pointing to a reduced airway inflammation. Such findings show that in an IL-13-driven asthmatic phenotype, lebrikizumab would indeed be a suitable therapeutic strategy. An ongoing Phase II placebo-controlled study of tralokinumab (MedImmune, Gaithersburg) shows a dose-dependent increase from baseline of forced expiratory volume in 1 s (FEV$_1$) (Antohe et al. 2013). Another strategy focuses on blocking IL-13-dependent internal signaling via either direct inhibition of specific IL-13 receptors (AMG-317) or prevention of receptor dimerization (IMA-638). Both did not alter AHR or eosinophil percentages in sputum, yet a significant dose-dependent decrease in serum IgE levels and slightly attenuated asthmatic response to allergens was recorded (Corren 2013). In cancer patients, IL-13-based treatment strategies focus on selective blockage of IL-13Rα2 due to its expression in various solid tumor forms. Central memory cells engineered to express a chimeric antigen receptor (CAR) targeting IL-13Rα2 in malignant glioblastomas display transient and promising anti-glioma tumor responses. Furthermore, therapeutic vaccination of immunocompetent mice with a cDNA vaccine encoding human IL-13Rα2 caused significant antitumor response. The combination of immunotoxins with cancer vaccines based on the disturbance of IL-13-dependent pathways is promising novel treatment tools for patients with advanced cancer (Nakashima et al. 2011). To summarize, interleukin 13 is a key player in lymphocyte polarization and subsequent immune system activation. It possesses several unique effector functions clearly distinguishing its role from that of IL-4; related signaling pathways altered in numerous pathological manifestations make it an ideal candidate for future treatment development.

References

Andrews, A. L., Holloway, J. W., Holgate, S. T., & Davies, D. E. (2006). IL-4 receptor alpha is an important modulator of IL-4 and IL-13 receptor binding: Implications for the development of therapeutic targets. *Journal of Immunology, 176*(12), 7456–7461.

Antohe, I., Croitoru, R., & Antoniu, S. (2013). Tralokinumab for uncontrolled asthma. *Expert Opinion on Biological Therapy, 13*(2), 323–326. doi:10.1517/14712598.2012.748740.

Borish, L. C., Nelson, H. S., Corren, J., Bensch, G., Busse, W. W., Whitmore, J. B., et al. (2001). Efficacy of soluble IL-4 receptor for the treatment of adults with asthma. *The Journal of Allergy and Clinical Immunology, 107*(6), 963–970.

Brown, T. R. (1898). Studies on trichinosis, with especial reference to the increase of the eosinophilic cells in blood and muscle, the origin of these cells and their diagnostic importance. *The Journal of Experimental Medicine, 3*(3), 315–347.

Brusselle, G. G., Kips, J. C., Tavernier, J. H., van der Heyden, J. G., Cuvelier, C. A., Pauwels, R. A., et al. (1994). Attenuation of allergic airway inflammation in IL-4 deficient mice. *Clinical and Experimental Allergy: Journal of the British Society for Allergy and Clinical Immunology, 24*(1), 73–80.

Brusselle, G., Kips, J., Joos, G., Bluethmann, H., & Pauwels, R. (1995). Allergen-induced airway inflammation and bronchial responsiveness in wild-type and interleukin-4-deficient mice. *American Journal of Respiratory Cell and Molecular Biology, 12*(3), 254–259.

Burmeister Getz, E., Fisher, D. M., & Fuller, R. (2009). Human pharmacokinetics/pharmacodynamics of an interleukin-4 and interleukin-13 dual antagonist in asthma. *Journal of Clinical Pharmacology, 49*(9), 1025–1036.

Caruso, M., Crisafulli, E., Demma, S., Holgate, S., & Polosa, R. (2013). Disabling inflammatory pathways with biologics and resulting clinical outcomes in severe asthma. *Expert Opinion on Biological Therapy, 13*(3), 393–402.

Corren, J. (2013). Role of interleukin-13 in asthma. *Current Allergy and Asthma Reports, 13*(5), 415–420.

Corren, J., Busse, W., Meltzer, E. O., Mansfield, L., Bensch, G., Fahrenholz, J., et al. (2010). A randomized, controlled, phase 2 study of AMG 317, an IL-4Ralpha antagonist, in patients with asthma.

American Journal of Respiratory and Critical Care Medicine, 181(8), 788–796.

Corren, J., Lemanske, R. F., Hanania, N. A., Korenblat, P. E., Parsey, M. V., Arron, J. R., et al. (2011). Lebrikizumab treatment in adults with asthma. *The New England Journal of Medicine, 365*(12), 1088–1098. doi:10.1056/NEJMoa1106469.

Foster, P. S., Ming, Y., Matthei, K. I., Young, I. G., Temelkovski, J., & Kumar, R. K. (2000). Dissociation of inflammatory and epithelial responses in a murine model of chronic asthma. *Laboratory Investigation; A Journal of Technical Methods and Pathology, 80*(5), 655–662.

Gauvreau, G. M., Pageau, R., Séguin, R., Carballo, D., Gauthier, J., D'Anjou, H., et al. (2011). Dose – response effects of TPI ASM8 in asthmatics after allergen. *Allergy, 66*(9), 1242–1248.

Giembycz, M. A., & Lindsay, M. A. (1999). Pharmacology of the eosinophil. *Pharmacological Reviews, 51*(2), 213–340.

Greenlee, K. J., Werb, Z., & Kheradmand, F. (2007). Matrix metalloproteinases in lung: Multiple, multifarious, and multifaceted. *Physiological Reviews, 87*(1), 69–98.

Grünig, G., Warnock, M., Wakil, A. E., Venkayya, R., Brombacher, F., Rennick, D. M., et al. (1998). Requirement for IL-13 independently of IL-4 in experimental asthma. *Science, 282*(5397), 2261–2263.

Hall, D. J., Cui, J., Bates, M. E., Stout, B. A., Koenderman, L., Coffer, P. J., et al. (2001). Transduction of a dominant-negative H-Ras into human eosinophils attenuates extracellular signal-regulated kinase activation and interleukin-5-mediated cell viability. *Blood, 98*(7), 2014–2021.

Hershey, G. K. (2003). IL-13 receptors and signaling pathways: An evolving web. *The Journal of Allergy and Clinical Immunology, 111*(4), 677–690. quiz 691.

Jinnin, M., Ihn, H., Yamane, K., & Tamaki, K. (2004). Interleukin-13 stimulates the transcription of the human alpha2(I) collagen gene in human dermal fibroblasts. *The Journal of Biological Chemistry, 279*(40), 41783–41791.

Johnson, J. R., Swirski, F. K., Gajewska, B. U., Wiley, R. E., Fattouh, R., Pacitto, S. R., et al. (2007). Divergent immune responses to house dust mite lead to distinct structural-functional phenotypes. *American Journal of Physiology. Lung Cellular and Molecular Physiology, 293*(3), L730–L739.

Kouro, T., & Takatsu, K. (2009). IL-5- and eosinophil-mediated inflammation: From discovery to therapy. *International Immunology, 21*(12), 1303–1309.

LaPorte, S. L., Juo, Z. S., Vaclavikova, J., Colf, L. A., Qi, X., Heller, N. M., et al. (2008). Molecular and structural basis of cytokine receptor pleiotropy in the interleukin-4/13 system. *Cell, 132*(2), 259–272. doi:10.1016/j.cell.2007.12.030.

Laviolette, M., Gossage, D. L., Gauvreau, G., Leigh, R., Olivenstein, R., Katial, R., et al. (2013). Effects of benralizumab on airway eosinophils in asthmatic patients with sputum eosinophilia. *The Journal of Allergy and Clinical Immunology, 132*(5), 1086–1096.

Lebel-Binay, S., Laguerre, B., Quintin-Colonna, F., Conjeaud, H., Magazin, M., Miloux, B., et al. (1995). Experimental gene therapy of cancer using tumor cells engineered to secrete interleukin-13. *European Journal of Immunology, 25*(8), 2340–2348.

Maizels, R. M., & Balic, A. (2004). Resistance to helminth infection: The case for interleukin-5-dependent mechanisms. *The Journal of Infectious Diseases, 190*(3), 427–429.

Nair, P., Pizzichini, M. M., Kjarsgaard, M., Inman, M. D., Efthimiadis, A., Pizzichini, E., et al. (2009). Mepolizumab for prednisone-dependent asthma with sputum eosinophilia. *The New England Journal of Medicine, 360*(10), 985–993. doi:10.1056/NEJMoa0805435.

Nakashima, H., Terabe, M., Berzofsky, J. A., Husain, S. R., & Puri, R. K. (2011). A novel combination immunotherapy for cancer by IL-13Rα2-targeted DNA vaccine and immunotoxin in murine tumor models. *Journal of Immunology, 187*(10), 4935–4946. doi:10.4049/jimmunol.1102095.

Pazdrak, K., Adachi, T., & Alam, R. (1997). Src homology 2 protein tyrosine phosphatase (SHPTP2)/Src homology 2 phosphatase 2 (SHP2) tyrosine phosphatase is a positive regulator of the interleukin 5 receptor signal transduction pathways leading to the prolongation of eosinophil survival. *The Journal of Experimental Medicine, 186*(4), 561–568.

Seguin, R. M., & Ferrari, N. (2009). Emerging oligonucleotide therapies for asthma and chronic obstructive pulmonary disease. *Expert Opinion on Investigational Drugs, 18*(10), 1505–1517.

Sel, S., Wegmann, M., Dicke, T., Sel, S., Henke, W., Yildirim, A. O., et al. (2008). Effective prevention and therapy of experimental allergic asthma using a GATA-3-specific DNAzyme. *The Journal of Allergy and Clinical Immunology, 121*(4), 910–916.

Shi, H. Z., Deng, J. M., Xu, H., Nong, Z. X., Xiao, C. Q., Liu, Z. M., et al. (1998a). Effect of inhaled interleukin-4 on airway hyperreactivity in asthmatics. *American Journal of Respiratory and Critical Care Medicine, 157*(6 Pt 1), 1818–1821.

Shi, H. Z., Xiao, C. Q., Zhong, D., Qin, S. M., Liu, Y., Liang, G. R., et al. (1998b). Effect of inhaled interleukin-5 on airway hyperreactivity and eosinophilia in asthmatics. *American Journal of Respiratory and Critical Care Medicine, 157*(1), 204–209.

Smirnov, D. V., Smirnova, M. G., Korobko, V. G., & Frolova, E. I. (1995). Tandem arrangement of human genes for interleukin-4 and interleukin-13: Resemblance in their organization. *Gene, 155*(2), 277–281.

Tanabe, T., Fujimoto, K., Yasuo, M., Tsushima, K., Yoshida, K., Ise, H., et al. (2008). Modulation of mucus production by interleukin-13 receptor alpha2 in the human airway epithelium. *Clinical and Experimental Allergy, 38*(1), 122–134. Epub 2007 Nov 19.

Turowska, A., Librizzi, D., Baumgartl, N., Kuhlmann, J., Dicke, T., Merkel, O., et al. (2013). Biodistribution of the GATA-3-specific DNAzyme hgd40 after inhalative exposure in mice, rats and dogs. *Toxicology and Applied Pharmacology, 272*(2), 365–372.

Wills-Karp, M., & Finkelman, F. D. (2008). Untangling the complex web of IL-4- and IL-13-mediated signaling pathways. *Science Signaling, 1*(51), e55. doi:10.1126/scisignal.1.51.pe55.

Zahn, S., Godillot, P., Yoshimura, A., & Chaiken, I. (2000). IL-5-induced JAB-JAK2 interaction. *Cytokine, 12*(9), 1299–1306.

Zhu, Z., Homer, R. J., Wang, Z., Chen, Q., Geba, G. P., Wang, J., et al. (1999). Pulmonary expression of interleukin-13 causes inflammation, mucus hypersecretion, subepithelial fibrosis, physiologic abnormalities, and eotaxin production. *The Journal of Clinical Investigation, 103*(6), 779–788.

Interleukin 6

Tadamitsu Kishimoto[1] and Toshio Tanaka[2]
[1]Laboratory of Immune Regulation, World Premier International Immunology Frontier Research Center, Osaka University, Suita, Osaka, Japan
[2]Department of Clinical Application of Biologics, Osaka University Graduate School of Medicine, Suita, Osaka, Japan

Synonyms

26kDa protein; B cell differentiation factor (BCDF); B cell stimulatory factor 2 (BSF-2); Hepatocyte-stimulating factor (HSF); Hybridoma growth factor (HGF); Interferon-β2 (IFNβ2)

Definition

Interleukin-6 (IL-6), originally identified as a B cell stimulatory factor 2 that enhances immunoglobulin synthesis by activated B cells, is a prototypical cytokine featuring pleiotropic and redundant activity. It is a proinflammatory and immunoregulatory cytokine contributing to host defense against infections and tissue injuries, while persistent dysregulated IL-6 synthesis causes the development of various diseases.

Biosynthesis and Release

Human IL-6 is a glycoprotein with 21–26 kDa consisting of 212 amino acids including a 28-amino-acid-signal peptide (Hirano et al. 1986). In infectious inflammation, IL-6 is produced by monocytes and macrophages after the stimulation of Toll-like receptors (TLRs) with distinct pathogen-associated molecular patterns, while in noninfectious inflammations, such as burns or traumatic injuries, damage-associated molecular patterns from damaged or dying cells stimulate TLRs to produce IL-6. In addition to these cells, IL-6 can be produced by a panoply of cells: dendritic cells, T and B cells, neutrophils, mast cells, fibroblasts, synovial cells, keratinocytes, endothelial cells, stromal cells, mesangial cells, glial cells, neurons, chondrocytes, osteoblasts, smooth muscle cells, adipocytes, and other cells including tumor cells (Akira et al. 1993; Kishimoto 1987).

The synthesis of IL-6 is strictly regulated by transcriptional and posttranscriptional mechanisms (Tanaka et al. 2014). Transcriptional factors including nuclear factor kappa B (NF-kB), nuclear factor IL-6 (NF-IL6), activation protein 1, and interferon regulatory factor 1 activate the IL-6 gene, whereas the activation of receptors, such as glucocorticoid, estrogen, and aryl hydrocarbon, represses IL-6 gene expression. Some microRNAs control the activity of transcriptional factors. For instance, microRNA-155 promotes degradation of NF-IL6 by interacting with the 3′-untranslated region (3′-UTR) of NF-IL6 mRNA. Moreover, several microRNAs and RNA-binding proteins bind to 3′-UTR of IL-6 mRNA and control its stability. Recently, a regulatory RNase-1 (Regnase-1) has been found to promote IL-6 mRNA degradation (Iwasaki et al. 2011), whereas an AT-rich interactive domain-containing protein 5a (Arid5a) selectively stabilizes IL-6 but not TNF-α or IL-12 mRNA through interaction with 3′-UTR (Masuda et al. 2013). In this way, Arid5a counteracts the destabilizing function of Regnase-1, indicating that the balance between Arid5a and Regnase-1 determines IL-6 mRNA stability.

Interleukin 6, Fig. 1 Pleiotropic activity of IL-6. IL-6 promotes hepatocytes to produce acute-phase proteins such as CRP, serum amyloid A, fibrinogen, and hepcidin, whereas it reduces synthesis of albumin and cytochrome p450. In the acquired immune response, IL-6 augments immunoglobulin synthesis by directly acting on activated B cells and by promoting the development of T follicular helper T cells, which produce IL-21. IL-6 also induces Th17 differentiation from naive CD4-positive T cells and inhibits Treg differentiation. In bone marrow, IL-6 induces maturation of megakaryocytes into platelets and activation of hematopoietic stem cells. IL-6 also acts on synovial fibroblasts to produce RANKL, which promotes differentiation of osteoclasts, and VEGF, which enhances angiogenesis and vascular permeability. Furthermore, IL-6 induces dermal fibroblasts to produce collagen and the growth of cells including keratinocytes, myeloma cells, and mesangial cells. Abbreviations: *CRP* C-reactive protein, *RANKL* receptor activator of NF-κB ligand, *Tfh* follicular helper T cells, *Treg* regulatory T cells, *VEGF* vascular endothelial growth factor

Biological Activities

IL-6 is an important mediator of inflammation. As mentioned earlier, IL-6 is promptly produced in an infected or a damaged lesion and provides an emergent signal to the entire body (Fig. 1). When stimulating hepatocytes, IL-6 strongly induces a broad spectrum of acute-phase proteins such as C-reactive protein (CRP), serum amyloid A (SAA), fibrinogen, hepcidin, haptoglobin, and antichymotrypsin, whereas it reduces albumin, cytochrome p450, fibronectin, and transferrin (Heinrich et al. 1990). CRP is a well-known biomarker of inflammation, and it is measured in clinical laboratory examinations to evaluate the existence and severity of inflammation, while elevations in serum IL-6 levels have been observed before an increase in serum CRP accompanying acute infection and abdominal surgery (Heney et al. 1992; Ohzato et al. 1992). High levels of hepcidin block iron transporter ferroportin 1 on macrophages, hepatocytes, and gut epithelial cells, leading to hypoferremia and anemia associated with inflammation (Gantz and Nemeth 2011). A long-term elevation of SAA results in the development of amyloid A amyloidosis, a serious complication of chronic inflammatory diseases (Obici and Merlini 2012). The production of acute-phase proteins such as CRP, hepcidin, and SAA by hepatocytes primarily depends on IL-6, since it has been shown that IL-6 blockade treatment leads to normalization of their serum levels.

IL-6 also plays a major role in local inflammation by stimulating IL-8 production by endothelial cells, monocyte chemoattractant protein, and expression of adhesion molecules, resulting in recruitment of leukocytes to the inflamed lesion (Gabey 2006).

In the acquired immune response, IL-6 induces B cell differentiation into immunoglobulin-producing cells and promotes the growth of plasmablasts and myeloma cells. IL-6 also governs the differentiation of naive CD4-positive T cells into specific effector T cells. IL-6 together with transforming growth factor (TGF)-β preferentially promotes their differentiation into IL-17-producing cells (Th17), whereas IL-6 inhibits TGF-β-induced regulatory T cell (Treg) development. The resultant Th17/Treg imbalance leads to eradication of immunologic tolerance and is thus of pathological importance for the development of autoimmune and chronic inflammatory diseases (Kimura and Kishimoto 2010). Furthermore, IL-6 promotes T follicular helper cell development and IL-21 production, which also enhances immunoglobulin synthesis (Awasthi and Kuchroo 2009).

In hematopoiesis, IL-6 induces maturation of megakaryocytes into platelets as well as activation of hematopoietic stem cells. IL-6 production by bone marrow stromal cells induces the receptor activator of NF-kB ligand, which is an essential factor for the differentiation and activation of osteoclasts and bone resorption, leading to osteoporosis. Enhanced angiogenesis and increased vascular permeability are often observed in inflamed lesions due to the excess production of vascular endothelial growth factor, which is also induced by IL-6. The promotional activities by IL-6 of collagen production by dermal fibroblasts and of their differentiation into myofibroblasts may also play a pathological role in fibrosing disorders such as systemic sclerosis (SSc). Furthermore, it has been demonstrated that IL-6 interacts with various other cells and organ systems including the endocrine system of the hypothalamic-pituitary-adrenal axis and the neuropsychological system.

Pathophysiological Relevance

A transient expression of IL-6 contributes to host defense against environmental stress such as infections and tissue injuries by triggering a broad spectrum of biological events. Once the source of stress is removed from the host, IL-6-mediated activation of the signal transduction cascade is terminated by negative regulatory systems such as ligand-induced internalization and degradation of gp130 and recruitment of suppressors of cytokine signaling (Naka et al. 1997), while at the same time serum IL-6 and CRP levels become normalized. However, persistent dysregulated IL-6 production can lead to the development of various autoimmune and chronic inflammatory diseases and even cancers (Akira et al. 1993; Kishimoto 2005). Although the mechanism(s) through which such an abnormally persistent IL-6 production is induced in association with various diseases has not been fully clarified, it may be of some relevance that some viral infections may cause persistent IL-6 production. Viral products such as the transcription protein of the human T lymphotropic virus 1, the transactivator of the transcription of the human immunodeficiency virus 1, and the human hepatitis B virus X protein have been found to enhance IL-6 gene transcription by modulating the DNA binding activity of NF-kB and/or NF-IL6. Moreover, IL-6 mRNA stabilization is promoted by the human herpes virus 8 open reading frame 57, which competes with the binding of microRNA-608 to IL-6 mRNA. As mentioned before, Arid5a stabilizes IL-6 mRNA, and a recent study has demonstrated that the expression of Arid5a in peripheral blood CD4-positive T cells is significantly higher in untreated rheumatoid arthritis (RA) patients than in healthy volunteers (Saito et al. 2013), so that excess expression of Arid5a may be one of the mechanisms which account for abnormal persistent IL-6 synthesis in RA.

The pathological significance of IL-6 was first demonstrated in a case of cardiac myxoma, in which excess production of IL-6 from myxoma tissues was speculated to be responsible for clinical symptoms and abnormal laboratory findings such as fever, polyarthritis, anemia, CRP elevation, and hypergammaglobulinemia with positivity of the antinuclear factor (Hirano et al. 1987). Dysregulation of IL-6 production was subsequently demonstrated in the lymph nodes involved in Castleman's disease, synovial fluids of RA, myeloma cells, and various cells of other diseases and has been implicated in the onset or development of chronic autoimmune inflammatory diseases and cancers.

Numerous animal models of diseases have also confirmed the pathological role of IL-6 in disease development. It was verified in these models that IL-6 blockade by gene knockout or by administration of anti-IL-6 or anti-IL-6 receptor (IL-6R) antibody suppressed disease development either preventatively or therapeutically. For example, IL-6 blockade resulted in limited susceptibility to Castleman's disease-like symptoms in IL-6 transgenic mice and was found to be efficacious in models of RA, multiple sclerosis, noninfectious uveitis, systemic lupus erythematosus (SLE), SSc, polymyositis, pristane-induced plasmacytomas, and other diseases.

Modulation by Drugs

Because of the pathological role of IL-6 in various diseases, IL-6 blockade was anticipated to constitute a novel treatment strategy for these diseases (Kishimoto 2005, 2010; Tanaka et al. 2012).

The multiple functions of IL-6 are initiated by its binding to IL-6R, which is made up of two receptor chains, an 80 kDa transmembrane IL-6R (cluster of differentiation [CD]126) or a 50–55 kDa soluble IL-6R and a 130 kDa transmembrane gp130 signal-transducing chain (CD130) (Kishimoto et al. 1992, 1994; Fig. 2). Upon binding to either the transmembrane or the soluble IL-6R, the IL-6/receptor complex induces the homodimerization of gp130 chains leading to the downstream signaling cascade.

Several attempts were made to block IL-6 activity, and through trial and error, a humanized antihuman IL-6R monoclonal antibody (chemical name, tocilizumab; generic name, ACTEMRA outside the EU or RoACTEMRA inside the EU) was finally developed, by grafting the complementarity-determining regions of a mouse antihuman IL-6R antibody onto human IgG1 (Sato et al. 1993). Tocilizumab blocks IL-6-mediated signal transduction by inhibiting IL-6 binding to transmembrane and soluble IL-6R (Fig. 2). In clinical terms, if free tocilizumab concentration is maintained at more than 1 μg/ml, CRP remains negative, so that the serum concentration of CRP is a hallmark for determining whether IL-6 activity is completely blocked in vivo (Nishimoto et al. 2008).

The first in-man study of tocilizumab was started at Osaka University Hospital in the late 1990s, and subsequently various worldwide clinical trials have demonstrated its outstanding efficacy for RA, systemic juvenile idiopathic arthritis (sJIA), and Castleman's disease. For RA patients with a moderately to severely active disease, tocilizumab is now used as an innovative drug in more than 100 countries. As a monotherapy or in combination with disease-modifying antirheumatic drugs (DMARDs), it has significantly suppressed disease activity and radiographic progression of joint deformity and improved daily functional activity. Due to its antirheumatic potency, tocilizumab is now positioned as the first-line biologic for the treatment of patients with moderate to severe RA for whom conventional DMARDs cannot be used (Smolen et al. 2014). Moreover, tocilizumab is currently

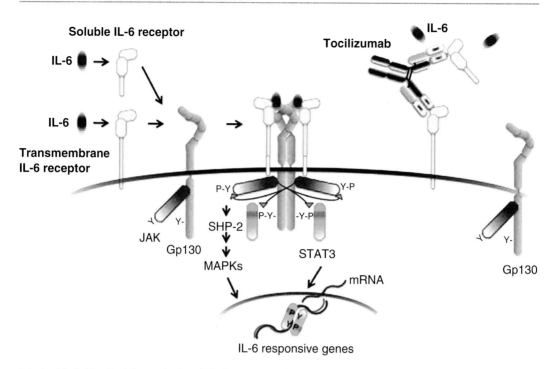

Interleukin 6, Fig. 2 A humanized anti-IL-6 receptor antibody, tocilizumab, inhibits IL-6-mediated signal transduction. IL-6-mediated signal transduction is initiated after binding of IL-6 to either the transmembrane IL-6 receptor or soluble IL-6 receptor. The resultant IL-6/IL-6 receptor complex associates with gp130 and induces homodimerization of gp130, which in turn triggers activation of JAKs and tyrosine phosphorylation of gp130. The phosphorylated gp130 recruits STAT3 via the SH2-domain, after which the activated STAT3 translocates into the nucleus and regulates transcription for various sets of IL-6 responsive genes. Tyrosine-phosphorylated gp130 also recruits SHP-2 and activates the MAP kinase pathway. Tocilizumab, a humanized anti-IL-6 receptor antibody, blocks IL-6-mediated signal transduction by inhibiting IL-6 binding to the transmembrane and soluble IL-6 receptors. Abbreviations: *JAKs* Janus kinase family tyrosine kinases, *MAPKs* mitogen-activated protein kinases, *SHP-2* SH2-domain-containing protein tyrosine phosphatase-2, *STAT3* signal transducer and activator of transcription 3

approved for Castleman's disease in Japan and India and for sJIA in Japan, India, the EU, and the USA. The striking efficacy of tocilizumab for sJIA has led to the recognition of the start of a new era in the treatment of this disease, which had been previously thought to be one of the most intractable and "unmet medical needs" diseases.

The great success of tocilizumab in the treatment of these diseases indicates that IL-6 blockade is indeed a therapeutic strategy for refractory diseases and has accelerated the development of other IL-6 inhibitors (Jones et al. 2011). Furthermore, favorable results of off-label use with tocilizumab in pilot studies, case series, and case studies strongly suggest that it will be applicable for a wide variety of diseases (Tanaka et al. 2014) (Table 1). These include systemic autoimmune diseases such as SLE, SSc, polymyositis, and large-vessel vasculitis syndrome (giant cell arteritis and Takayasu arteritis); organ-specific autoimmune diseases including neuromyelitis optica, relapsing polychondritis, acquired hemophilia A, autoimmune hemolytic anemia, and Graves' ophthalmopathy; chronic inflammatory diseases including adult-onset Still's disease, amyloid A amyloidosis, polymyalgia rheumatica, remitting seronegative symmetrical synovitis with pitting edema, Behcet's disease, noninfectious uveitis, graft-versus-host diseases, cytokine release syndrome, pulmonary arterial hypertension, IgG4-related disease, and hereditary autoinflammatory syndromes; and other diseases including atherosclerosis, type 2 diabetes mellitus, atopic dermatitis, sciatica, amyotrophic

Interleukin 6, Table 1 Clinical application of IL-6 blockade strategy for autoimmune, chronic inflammatory diseases, and cancers

I. Indicated diseases
1. Rheumatoid arthritis (in more than 100 countries)
2. Systemic juvenile idiopathic arthritis (in Japan, India, the EU, and the USA)
3. Polyarticular juvenile idiopathic arthritis (in Japan and India)
4. Castleman's disease (in Japan and India)
II. Candidate diseases
1. Autoimmune diseases
(I) Systemic lupus erythematosus
(II) Systemic sclerosis
(III) Polymyositis and dermatomyositis
(IV) Large-vessel vasculitis (Takayasu arteritis and giant cell arteritis)
(V) Sjogren's syndrome
(VI) Relapsing polychondritis
(VII) Autoimmune hemolytic anemia
(VIII) Acquired hemophilia A
(IX) Multiple sclerosis and neuromyelitis optica
(X) Cogan's syndrome
(XI) Graves' ophthalmopathy
2. Chronic inflammatory diseases
(I) Adult-onset Still's disease
(II) Amyloid A amyloidosis
(III) Crohn's disease
(IV) Polymyalgia rheumatica and remitting seronegative symmetrical synovitis with pitting edema
(V) Behcet's disease and noninfectious uveitis
(VI) Graft-versus-host diseases
(VII) Cytokine release syndrome
(VIII) Hereditary autoinflammatory syndromes and goat
(IX) IgG4-related disease
(X) Schnitzler syndrome
(XI) Atherosclerosis
(XII) Type II diabetes mellitus
(XIII) Atopic dermatitis
(IXV) Sciatica
(XV) Amyotrophic lateral sclerosis
3. Neoplastic diseases
(I) Multiple myeloma
(II) Pancreatic cancer
(III) Prostate cancer
(IV) Colon cancer
(V) Ovarian cancer
(VI) Malignant mesothelioma
(VII) Cancer-associated cachexia

The anti-IL-6 receptor antibody, tocilizumab, is currently approved as a biological drug for the treatment of rheumatoid arthritis, juvenile idiopathic arthritis, and Castleman's disease and is expected to be used for the treatment of a wide variety of autoimmune, chronic inflammatory diseases, and cancers lateral sclerosis, and some types of cancers and cancer-related cachexia. However, further clinical studies will be essential to determine the efficacy and safety of tocilizumab for such a broad range of clinical indications.

Cross-References

▶ Cytokines
▶ Rheumatoid Arthritis

References

Akira, S., Taga, T., & Kishimoto, T. (1993). Interleukin-6 in biology and medicine. *Advances in Immunology, 54*, 1–78.

Awasthi, A., & Kuchroo, V. K. (2009). The yin and yang of follicular helper T cells. *Science, 325*, 953–955.

Gabey, C. (2006). Interleukin-6 and chronic inflammation. *Arthritis Research & Therapy, 8*(Suppl 2), S3.

Gantz, T., & Nemeth, E. (2011). Hepcidin and disorders of iron metabolism. *Annual Review of Medicine, 62*, 347–360.

Heinrich, P. C., Castell, J. V., & Andus, T. (1990). Interleukin-6 and the acute phase response. *Biochemical Journal, 265*, 621–636.

Heney, D., Lewis, I. J., Evans, S. W., Banks, R., Bailey, C. C., & Whicher, J. T. (1992). Interleukin-6 and its relationship to C-reactive protein and fever in children with febrile neutropenia. *The Journal of Infectious Diseases, 165*, 886–890.

Hirano, T., Yasukawa, K., Harada, H., Taga, T., Watanabe, Y., Matsuda, T., et al. (1986). Complementary DNA for a novel human interleukin (BSF-2) that induces B lymphocytes to produce immunoglobulin. *Nature, 324*, 73–76.

Hirano, T., Taga, T., Yasukawa, K., Nakajima, K., Nakano, N., Takatsuki, F., et al. (1987). Human B-cell differentiation factor defined by an anti-peptide antibody and its possible role in autoantibody production. *Proceedings of the National Academy of Sciences of the United States of America, 84*, 228–231.

Iwasaki, H., Takeuchi, O., Teraguchi, S., Matsushita, K., Uehata, T., Kuniyoshi, K., et al. (2011). The IkB kinase complex regulates the stability of cytokine-encoding mRNA induced by TLR-IL-1R by controlling degradation of regnase-1. *Nature Immunology, 12*, 1167–1175.

Jones, S. A., Scheller, J., & Rose-John, S. (2011). Therapeutic strategies for the clinical blockade of IL-6/gp130 signaling. *The Journal of Clinical Investigation, 121*, 3375–3383.

Kimura, A., & Kishimoto, T. (2010). IL-6: Regulator of Treg/Th17 balance. *European Journal of Immunology, 40*, 1830–1835.

Kishimoto, T. (1987). The biology of interleukin-6. *Blood, 74*, 1–10.

Kishimoto, T. (2005). Interleukin-6: From basic sciences to medicine-40 years in immunology. *Annual Review of Immunology, 23*, 1–21.

Kishimoto, T. (2010). IL-6: From its discovery to clinical applications. *International Immunology, 22*, 347–352.

Kishimoto, T., Akira, S., & Taga, T. (1992). Interleukin-6 and its receptor: A paradigm for cytokines. *Science, 258*, 593–597.

Kishimoto, T., Taga, T., & Akira, S. (1994). Cytokine signal transduction. *Cell, 76*, 253–262.

Masuda, K., Ripley, B., Nishimura, R., Mino, T., Takeuchi, O., Shioi, G., et al. (2013). Arid5a controls IL-6 mRNA stability, which contributes to elevation of IL-6 level in vivo. *Proceedings of the National Academy of Sciences of the United States of America, 110*, 9409–9414.

Naka, T., Narazaki, M., Hirata, M., Matsumoto, T., Minamoto, S., Aono, A., et al. (1997). Structure and function of a new STAT-induced STAT inhibitor. *Nature, 387*, 924–929.

Nishimoto, N., Terao, K., Mima, T., Nakahara, H., Takagi, N., & Kakehi, T. (2008). Mechanisms and pathologic significances in increase in serum interleukin-6 (IL-6) and soluble IL-6 receptor after administration of an anti-IL-6 receptor antibody, tocilizumab, in patients with rheumatoid arthritis and Castleman disease. *Blood, 112*, 3959–3964.

Obici, L., & Merlini, G. (2012). AA amyloidosis: Basic knowledge, unmet needs and future treatments. *Swiss Medical Weekly, 142*, w13580.

Ohzato, H., Yoshizaki, K., Nishimoto, N., Ogata, A., Tagoh, H., Monden, M., et al. (1992). Interleukin-6 as a new indicator of inflammation status: Detection of serum levels of interleukin-6 and C-reactive protein after surgery. *Surgery, 111*, 201–209.

Saito, Y., Kagami, S., Sanayama, Y., Ikeda, K., Suto, A., Kashiwakuma, D., et al. (2014). At-rich-interactive domain-containing protein 5A functions as a negative regulator of retinoic acid receptor-related orphan nuclear receptor γt-induced Th17 cell differentiation. *Arthritis & Rheumatology, 66*, 1185–1194.

Sato, K., Tsuchiya, M., Saldanha, J., Koishihara, Y., Ohsugi, Y., Kishimoto, T., et al. (1993). Reshaping a human antibody to inhibit the interleukin 6-dependent tumor cell growth. *Cancer Research, 53*, 851–856.

Smolen, J. S., Landewe, R., Breedveld, F. C., Buch, M., Burmester, G., Dougados, M., et al. (2014). EULAR recommendations for the management of rheumatoid arthritis with synthetic and biological disease-modifying antirheumatic drugs: 2013 update. *Annals of the Rheumatic Diseases, 73*, 492–509.

Tanaka, T., Narazaki, M., & Kishimoto, T. (2012). Therapeutic targeting of the interleukin-6 receptor. *Annual Review of Pharmacology and Toxicology, 52*, 199–219.

Tanaka, T., Narazaki, M., Ogata, A., & Kishimoto, T. (2014). A new era for the treatment of inflammatory autoimmune diseases by interleukin-6 blockade strategy. *Seminars in Immunology, 26*, 88–96.

Interleukin-6 Inhibitor: Tocilizumab

Kevin D. Pile[1], Garry G. Graham[2,3] and Stephen M. Mahler[4]

[1]Campbelltown Hospital, School of Medicine, University of Western Sydney, Campbelltown, NSW, Australia

[2]Department of Pharmacology, School of Medical Sciences, University of New South Wales, Sydney, NSW, Australia

[3]Department of Clinical Pharmacology and Toxicology, St Vincent's Hospital, Sydney, NSW, Australia

[4]Australian Institute for Bioengineering and Nanotechnology, University of Queensland, Brisbane, QLD, Australia

Synonyms

Interleukin-6 antagonists; Interleukin-6 blockers

Definition

Inhibitors of interleukin-6 (IL-6) are proteins which decrease the actions of the inflammatory cytokine, IL-6. Clinically, tocilizumab is the major inhibitor of IL-6. Several other proteins with IL-6 inhibitory actions (e.g., sirukumab) are being tested in clinical trials. IL-6 inhibitors are members of a general class, biological disease-modifying anti-rheumatic drugs (bDMARDs). Unlike the conventional synthetic DMARDs (csDMARDs) which have small molecular masses, they are not classified as slow-acting anti-rheumatic drug (SAARDs) because they are considered to have more specific and rapidly developing actions.

Chemical Structures and Properties

Tocilizumab is a humanized monoclonal antibody with molecular mass of 145,000 D. Tocilizumab was developed by grafting the complementarity determining regions of a nude mouse anti-human IL-6R antibody onto human IgG1. The light chain is made up of 214 amino acids. The heavy chain is made up of 448 amino acids. The four polypeptide chains are linked intra- and inter-molecularly by disulfide bonds. Tocilizumab is produced in a Chinese hamster ovary (CHO) cell line by recombinant DNA technology.

Tocilizumab is a protein which prevents any passive diffusion through cell membranes and, like other bDMARDs of similar molecular mass, it is not dialyzable.

Metabolism, Pharmacokinetics, and Dosage

Having protein structures, the IL-6 inhibitors are not absorbed orally and are administered parenterally. Like other bDMARDs, tocilizumab is metabolized slowly by the reticuloendothelial system. The terminal half-life of tocilizumab is about 21 days which allows the dosage interval to be approximately 4 weeks (Frey et al. 2010). The recommended initial dosage is 4 mg/kg administered by intravenous infusion over one hour. This dosage can be raised to 8 mg/kg if the response is inadequate. Its subcutaneous dosage is being investigated. The maximal dosage has been set at 800 mg.

Pharmacological Activities

IL-6 has widespread effects and signals primarily through a membrane-bound complex containing IL-6 bound to IL-6 receptors (IL-6Rs) and two signal-transducing glycoprotein 130 (gp130) subunits. IL-6R is predominantly expressed on neutrophils, monocytes/macrophages, hepatocytes, and some lymphocytes. However, soluble IL-6R (sIL-6R) can be generated by proteolysis of the membrane-bound IL-6R or by alternative mRNA splicing. Released sIL-6R is able to complex with IL-6 and be transported in bodily fluids, with the IL-6/sIL-6R complex able to bind and activate the ubiquitously expressed signaling gp130 subunit which is found on a range of cells including endothelial cells and synoviocytes. Thus, IL-6 has predictable actions via IL-6R-bearing cells but has a broader potential effect on any cell expressing gp130.

IL-6 is a monocyte-derived cytokine, important in B-cell maturation (and hence antibody production). IL-6 mobilizes marginated neutrophils into the circulation. Binding of IL-6 to neutrophil membrane-bound IL-6R induces secretion of proteolytic enzymes and reactive oxygen intermediates causing cartilage degradation. During acute inflammation in rheumatoid arthritis, monocytes, macrophages, and endothelial cells release IL-6 accompanied by an increase in neutrophils in synovial fluid. Subsequently, IL-6 is thought to influence the shift from acute to chronic inflammation, marked by an increase in the recruitment of monocytes.

Tocilizumab binds to both the membrane-bound and soluble forms of the IL-6 receptor (IL-6R) while an IL-6 inhibitor under development, sirukumab, binds to and inactivates IL-6 (Xu et al. 2011). The molecular mechanism of action of the two agents is, therefore, different but both decrease the proinflammatory activities of IL-6 (Dayer and Choy 2010). Further IL-6 antagonists are being developed: sarilumab which binds to IL-6R (like tocilizumab) and two proteins (clazakizumab and olokizumab) which are antibodies to IL-6 (like sirukumab) (Semerano et al. 2014).

The pharmacological effects of and clinical findings with tocilizumab, resulting from inhibition of IL-6 activity, include:

- Decreased production of acute phase reactants, such as C-reactive protein (CRP). It should be noted that IL-6 is the principal stimulator of synthesis of CRP which is elevated in patients with rheumatoid arthritis. As a result, the fall in the plasma concentrations of CRP correlates with plasma concentrations of tocilizumab (Schmitt et al. 2011). Elevated values of erythrocyte sedimentation and serum amyloid A are also normalized (Dhillon 2014).
- Increased levels of plasma albumin (Davies and Choy 2014; Ohsugi 2007). A common feature of rheumatoid arthritis is a fall in circulating albumin.

- Increased activity of hepatic P450 enzymes which are downregulated by IL-6 (see section "Adverse Effects" below).
- Decreased production of hepcidin by hepatocytes. The production of hepcidin is stimulated by IL-6 with consequent blockade of iron absorption from the gastrointestinal tract and the release of iron from macrophages. Thus, excessive IL-6 is a major cause of anemia of chronic inflammation, an effect which is largely reversed by Il-6 blockers (Navarro-Millan et al. 2012; Semerano et al. 2014).
- Increased hemoglobins. Levels of hemoglobin are increased by IL-6 blockers with the greatest improvement being produced at low baseline hemoglobin values (Hashimoto et al. 2014; Navarro-Millan et al. 2012; Semerano et al. 2014).
- Increased plasma concentrations of IL-6. The elimination of IL-6 is primarily through uptake by receptors, and inhibition of binding of IL-6 by tocilizumab therefore leads to increased plasma concentrations of IL-6. The increased levels of IL-6 are not, however, deleterious because the effects of IL-6 are blocked by the tocilizumab (Schmitt et al. 2011). Also, the plasma concentrations of the soluble IL-6 receptor are apparently increased, probably because of binding to form a sIL-6R/tocilizumab complex (Schmitt et al. 2011). On the other hand, the concentrations of IL-6 in the synovium of patients with rheumatoid arthritis may be increased (Dhillon 2014).

Clinical Indications and Efficacy

Tocilizumab suppresses several rheumatic diseases, its activity being consistent with the significance of IL-6 in the pathophysiology of the rheumatic diseases. In rheumatoid arthritis, the number and tenderness of inflamed joints is reduced with decreased progression of joint damage (Davies and Choy 2014; Navarro-Millan et al. 2012)

Tocilizumab is currently utilized for the treatment of rheumatoid arthritis, systemic onset juvenile idiopathic arthritis, psoriatic arthritis, and Castleman's disease (a rare lymphoproliferative disease) (Davies and Choy 2014). Small studies on occasional patients with polymyalgia rheumatica, giant cell arteritis, adult-onset Still's disease, and several other rare inflammatory diseases have shown positive responses to tocilizumab. However, results in spondyloarthropathies have been disappointing (Davies and Choy 2014).

Tocilizumab is recommended primarily for the treatment of rheumatoid arthritis which has failed to respond to one or more DMARDs. It may be used as monotherapy or in conjunction with methotrexate or other nonbiological DMARDs. In rheumatoid patients who had failed to respond or been intolerant to a TNF inhibitor within the previous year, dosage with tocilizumab (8 mg/kg) plus methotrexate was superior to methotrexate alone (Maini et al. 2006). In some trials, however, tocilizumab alone has shown good activity (Dhillon 2014; Semerano et al. 2014). Tocilizumab has been trialed in Crohn's disease. Improvement was noted in about 80 % of patients but only 20 % achieved disease remission (Danese 2012).

Adverse Effects

Tocilizumab is generally tolerated well, although a number of adverse effects have been noted (Dhillon 2014) including:

- Infections. These are the most common serious adverse effects of tocilizumab. The rates of tuberculosis do not seem to be increased, but all rheumatoid arthritis patients should be screened for latent disease. Respiratory tract infections, including pneumonia, have been reported.
- Gastrointestinal. Upper gastrointestinal events suggestive of inflammation, gastritis, or ulcer are common. Tocilizumab should be used with caution in patients with a previous history of intestinal ulceration or diverticulitis.
- Neutropenia. Reduction in neutrophil count occurs more during treatment with tocilizumab than with TNF inhibitors. However, the reduction does not appear to have significant clinical consequences as the neutropenia has been

transient and, furthermore, there has been no association between low neutrophil counts and infection-related serious adverse reactions. In trials, the recommendation was not to treat with tocilizumab if neutrophil counts $<0.5 \times 10^9$/L or platelet counts $< 50 \times 10^9$/L. These limits appear reasonable for more general use of the drug.

- Liver reactions. Elevation of transaminases (AST and ALT), up to three times the upper limit of normal, occurs frequently with the highest mean increases 2 weeks after each infusion. More than threefold elevation of enzymes is uncommon (2–4 %) but remains three to five times higher than with DMARD alone. A total bilirubin increase up to three times, the upper limit of normal is also seen. It is recommended to be cautious with the use of tocilizumab in patients with active liver disease or elevated hepatic enzymes (50 % above the upper limit of normal) and not to treat patients with ALT or AST more than five times the upper normal level. Transaminases should be monitored every 1–2 months during the first 6 months of therapy and every 3 months thereafter.
- Plasma lipids. Increases in fasting plasma lipids occur early after treatment and remain elevated during therapy with approximate mean changes of: total cholesterol 0.8 mmol/L, HDL cholesterol 0.1 mmol/L, and LDL cholesterol 0.5 mmol/L. Approximately 24 % of patients receiving tocilizumab experienced sustained elevations in total cholesterol to at least 6.2 mmol/L. Lipid parameters should be measured 4–8 weeks after commencing therapy and managed according to local guidelines and taking into account the individual risk factors.

IL-6 is an important factor in the differentiation of B cells, and it was considered that the antagonist, tocilizumab, may block the efficacy of vaccinations. However, in contrast to methotrexate, influenza and pneumococcal vaccinations were still successful during tocilizumab treatment (Dhillon 2014). Oddly, the combination of tocilizumab and methotrexate still allowed a positive response to these vaccinations.

Drug Interactions

As discussed above, tocilizumab is frequently used with methotrexate as the combination, tocilizumab (8 mg/kg) plus methotrexate, is superior to methotrexate alone (Maini et al. 2006). Cytochrome P450 enzymes in the liver are downregulated by IL-6 but may be brought back to normal or above normal by tocilizumab. Consequently, tocilizumab may increase the plasma concentrations of drugs metabolized by cytochrome P450 enzymes. This may be of particular note for drugs with narrow therapeutic ranges, such as warfarin and theophylline (Navarro-Millan et al. 2012). A known example is the substantial increase in the plasma concentrations of simvastatin after dosage with tocilizumab (Schmitt et al. 2011). This effect on cytochrome P450 should be a general phenomenon with IL-6 inhibitors, and dosage of metabolized drugs should be checked when treatment with tocilizumab or any novel IL-6 inhibitor is commenced or stopped (Schmitt et al. 2011).

Cross-References

- ▶ Corticosteroids
- ▶ Interleukin 6
- ▶ Methotrexate
- ▶ Rheumatoid Arthritis
- ▶ Spondyloarthritis
- ▶ Tumor Necrosis Factor (TNF) Inhibitors

References

Danese, S. (2012). New therapies for inflammatory bowel disease: From the bench to the bedside. *Gut, 61*, 918–932.

Davies, R., & Choy, E. (2014). Clinical experience of IL-6 blockade in rheumatic diseases – Implications on IL-6 biology and disease pathogenesis. *Seminars in Immunology, 26*(1), 97–104.

Dayer, J. M., & Choy, E. (2010). Therapeutic targets in rheumatoid arthritis: The interleukin-6 receptor. *Rheumatology (Oxford), 49*, 15–24.

Dhillon, S. (2014). Intravenous tocilizumab: A review of its use in adults with rheumatoid arthritis. *BioDrugs, 28*(1), 75–106.

Frey, N., Grange, S., & Woodworth, T. (2010). Population pharmacokinetic analysis of tocilizumab in patients with rheumatoid arthritis. *Journal of Clinical Pharmacology, 50*(7), 754–766.

Hashimoto, M., Fujii, T., Hamaguchi, M., Furu, M., Ito, H., Terao, C., et al. (2014). Increase of hemoglobin levels by anti-IL-6 receptor antibody (tocilizumab) in rheumatoid arthritis. *PloS One, 9*(5), e98202.

Maini, R. N., Taylor, P. C., Szechinski, J., Pavelka, K., Broll, J., Balint, G., et al. (2006). Double-blind randomized controlled clinical trial of the interleukin-6 receptor antagonist, tocilizumab, in European patients with rheumatoid arthritis who had an incomplete response to methotrexate. *Arthritis and Rheumatism, 54*(9), 2817–2829.

Navarro-Millan, I., Singh, J. A., & Curtis, J. R. (2012). Systematic review of tocilizumab for rheumatoid arthritis: A new biologic agent targeting the interleukin-6 receptor. *Clinical Therapeutics, 34*(4), 788–802.

Ohsugi, Y. (2007). Recent advances in immunopathophysiology of interleukin-6: An innovative therapeutic drug, tocilizumab (recombinant humanized anti-human interleukin-6 receptor antibody), unveils the mysterious etiology of immune-mediated inflammatory diseases. *Biological and Pharmaceutical Bulletin, 30*(11), 2001–2006.

Schmitt, C., Kuhn, B., Zhang, X., Kivitz, A. J., & Grange, S. (2011). Disease-drug-drug interaction involving tocilizumab and simvastatin in patients with rheumatoid arthritis. *Clinical Pharmacology & Therapeutics, 89*, 735–740.

Semerano, L., Thiolat, A., Minichiello, E., Clavel, G., Bessis, N., & Boissier, M. C. (2014). Targeting IL-6 for the treatment of rheumatoid arthritis: Phase II investigational drugs. *Expert Opinion on Investigational Drugs, 23*(7), 979–999.

Xu, Z., Bouman-Thio, E., Comisar, C., Frederick, B., Van Hartingsveldt, B., Marini, J. C., et al. (2011). Pharmacokinetics, pharmacodynamics and safety of a human anti-IL-6 monoclonal antibody (sirukumab) in healthy subjects in a first-in-human study. *British Journal of Clinical Pharmacology, 72*(2), 270–282.

Interleukin 9

Edgar Schmitt, Matthias Klein and Tobias Bopp
Institute for Immunology, University Medical Center of the Johannes Gutenberg University Mainz, Mainz, Germany

Definition

Interleukin-9 (IL-9) is a 14 kd glycoprotein composed of 144 amino acids with a typical signal peptide of 18 amino acids and is mainly produced by T cells, mast cells, and granulocytes. Multiple targets of IL-9 have been described among them cells of the hematopoietic and nonhematopoietic lineage. It acts through the IL-9 receptor, which mainly activates signal transducer and activator (STAT) proteins STAT1, STAT3, and STAT5. Because of its multiple sources and multifaceted functions, IL-9 is involved in a plethora of different biological processes and diseases. Genetic studies in mice demonstrated an essential role of IL-9 in the pathophysiology of asthma, nematode infection, gut inflammation, experimental autoimmune encephalomyelitis (EAE), and melanoma.

Biosynthesis and Release

Interleukin-9 was originally characterized as a cytokine with growth-enhancing capacity on T cells and mast cells and, respectively, termed P40, T-cell growth factor (TCGF)III, or mast cell growth-enhancing activity (MEA) (reviewed in Stassen et al. (2012)). Ultimately p40, TCGFIII, and MEA were shown to represent the same cytokine, and based on its biological effects on cells of the lymphoid and myeloid lineage, the cytokine was renamed IL-9 (reviewed in Stassen et al. (2012)). Further studies demonstrated that IL-9 was secreted by T-cell lines that were generated in the presence of IL-4 and mainly produced cytokines characteristic for Th2 cells (reviewed in Schmitt et al. (2013)). Continuative analyses by Schmitt and colleagues demonstrated that stimulation of naïve $CD4^+$ T cells in the presence of IL-4 and TGF-β results in vigorous IL-9 production that is dependent on IL-2 and inhibited by IFN-γ (Schmitt et al. 2013). In 1993, Gessner and colleagues further corroborated the assumption that IL-9 is a cytokine mainly produced by Th2 cells and suggested that upon infection of mice with *Leishmania major*, Th2 cells represent the main source of IL-9 (Gessner et al. 1993). Next to *L. major*, infection with pathogens like *Trichuris muris* and *Schistosoma mansoni* provoked expression of IL-9 by T cells (reviewed in Stassen et al. (2012)). Consistent with this, depletion of $CD4^+$ T cells strongly reduced IL-9 production upon administration of cognate protein antigen (Monteyne et al. 1997). However, in the same publication, it was demonstrated that T-cell-derived IL-9 preceded IL-4 production. Using Il4-deficient mice, the authors further demonstrated that IL-9 production by CD4$^+$ T cells can be independent from this Th2-polarizing cytokine and hence challenged the concept of IL-9 being a Th2 cytokine for the first time. Recently, two independent studies demonstrated that IL-9 and IL-4 are rarely produced by the same T cell, indicating that IL-9-producing T cells represent a discrete Th cell subpopulation distinct from Th2 cells (reviewed in Schmitt et al. (2013)). Furthermore, Staudt and colleagues demonstrated that IL-9-single-producing T cells are the major source of this cytokine in a preclinical model of asthma (Staudt et al. 2010). Eventually, it was demonstrated that the usage of a combination of IL-4 and TGF-β as growth and differentiation factors led to the development of a CD4$^+$ T-cell subset that preferentially produced IL-9 and therefore was called Th9 cells (reviewed in Schmitt et al. (2013)). Yet, the precise role of Th9 cells in health and disease still remains unclear. Next to Th9 cells, other CD4$^+$ T-cell subsets including Th17 cells and Treg cells have been described to produce IL-9 (reviewed in Schmitt et al. (2013)). However, these findings are mainly based on the detection of IL-9 in the supernatant of in vitro stimulated T-cell cultures or by quantitative RT-PCR-based analysis of IL-9 mRNA expression. Since these analyses do not allow the identification of IL-9-producing cells on the single-cell level, it is legitimate to assume that detection of IL-9 upon in vitro stimulation of Th17 or Treg cells might be due to a small number of Th9 cells present in these cultures.

Using IL-9 fate reporter mice and an unusual preclinical model of asthma, namely the papain-induced airway inflammation model, Wilhelm and colleagues showed that in this model, most IL-9-producing cells are found among a population of innate lymphoid cells (ILCs) that lack lineage markers clearly delineating these cells from T cells (Wilhelm et al. 2011). According to Th2 cells, these innate lymphoid cells were termed ILC2. In contrast, Licona-Limón and colleagues generated IL-9-deficient and IL-9-fluorescent reporter mice to visualize the real-time expression of the Il9 gene (Licona Limon et al. 2013). Upon infection of those mice with the worm Nippostrongylus brasiliensis, the authors described that Th9 cells represent the major source of IL-9. Furthermore, by employing an adoptive transfer model, the authors showed that Th9 cell-derived IL-9 is crucially involved in rapid expulsion of these parasitic worms.

In these parasitic nematode infection models, mast cells appear to be the main targets of IL-9. With their ability to rapidly release different mediators upon IgE-mediated activation, mast cells critically contribute to mucus production, eosinophilia, and increased intestinal permeability ultimately leading to worm expulsion. Of note, mast cells, basophils, and eosinophils represent another important source of IL-9 in mice (Hultner et al. 2000) and humans (Gounni et al. 2000).

Taken together, it is likely that depending on the experimental mouse model and the time of analyses, IL-9 can be produced by a variety of cells. Among these, Th9 cells produce this cytokine in an antigen-dependent manner and certainly represent an important and major source of this cytokine in a variety of different diseases.

Biological Activities and Targets of IL-9

IL-9 acts on a multitude of hematopoietic and nonhematopoietic cells thereby crucially affecting the development and manifestation of various diseases that are caused by immunological disorders (Fig. 1).

Initially, IL-9 was identified and characterized as a growth factor for long-term T-cell lines but not for primary T cells. In parallel, it was found that IL-9 exhibited strong survival- and growth-enhancing activity for bone marrow-derived mast cells (BMMC) (reviewed in Stassen et al. (2012)). Subsequently, IL-9 was shown to promote the expression of various effector molecules in mast cells like FcεRI, cytokines, mast cell proteases, and members of the granzyme family (reviewed in Renauld et al. (1995)). IL-9 can also foster the

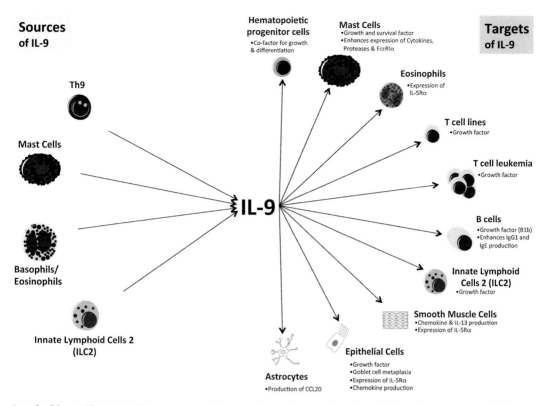

Interleukin 9, Fig. 1 Multiple sources and targets of interleukin-9 (IL-9). IL-9 is a pleiotropic cytokine that is produced by distinct hematopoietic cells and that influences directly and indirectly a multitude of different cell types which crucially affect the onset and maintenance of inflammatory immune responses.

development of hematopoietic progenitors; in particular it supports the maturation of erythroid progenitors and the generation of fetal multipotent colony-forming unit colonies. In addition, the expansion of B1b-B cells is supported by IL-9 and also IL-4-induced production of IgG1 and IgE (reviewed in Goswami and Kaplan (2011)). While primary T cells are not affected, it was found that IL-9 can serve as an autocrine growth factor for thymomas and T-cell leukemias (reviewed in Renauld et al. (1995)).

Analyses of the asthma-promoting activity of IL-9 revealed that this cytokine acts in this pathologic scenario not only on mast cells but also on nonhematopoietic cells like lung epithelial goblet cells and airway smooth muscle cells eventually leading to increased mucus production, tissue remodeling, mastocytosis, and eosinophilia in the lungs. These severe asthmatic symptoms largely result from IL-9-mediated effects, especially elevated production of IL-5, IL-13, and eotaxin1/CCL11 and increased proliferation of goblet cells and airway smooth muscle cells, as well (reviewed in Goswami and Kaplan (2011)). Finally, it was shown using an experimental EAE model that IL-9 induces the expression of CCL-20 in astrocytes thus recruiting inflammatory $CCR6^+$ Th17 cells into the CNS. This could be confirmed by blocking of IL-9 in vivo that strongly inhibited Th17 cell migration into the CNS (reviewed in Stassen et al. (2012)).

Pathophysiological Relevance of IL-9

IL-9 is a multifunctional cytokine that plays an important role concerning the development and maintenance of diverse diseases and thus represents a valuable target for innovative therapeutic strategies.

IL-9 and Tumor Diseases

Originally, IL-9 was found to be a growth factor for long-term in vitro-cultured T-cell lines and BMMC lines. Interestingly, autonomously proliferating T-cell sublines developed sporadically from these long-term IL-9-dependent T-cell lines that have the potential to form tumors in vivo. In addition, 5–10 % of IL-9 transgenic mice spontaneously developed lymphoblastic lymphomas without showing a general T-cell hyperproliferation corroborating the initial finding that IL-9 is not a growth factor for primary T cells. Furthermore, IL-9 transgenic mice were highly susceptible to mutagenic treatments and developed tumors after administration of marginal doses that were completely innocuous in wild-type control mice. The oncogenic potential of IL-9 is additionally supported by the observation that it protects T-cell lymphomas against dexamethasone-induced apoptosis. Furthermore, constitutive IL-9 production was found in human HTLV-1-transformed T cells and in Hodgkin cell lines implying a tumor-promoting role of IL-9 (reviewed in Renauld et al. (1995)). This assumption is also supported by recent findings that culture supernatants of PBMCs from adult T-cell leukemia patients with spontaneous proliferation contained high amounts of IL-9, and the proliferation of certain leukemia cells from these patients could be blocked by IL-9R monoclonal antibodies. Furthermore, IL-9 is produced by more than 80 % of ex vivo PBMC cultures from chronic adult T-cell leukemia patients suggesting a para/autocrine function of IL-9 for this type of tumor (Chen and Wang 2014). Similar conclusions can be drawn from the fact that high serum levels of IL-9 correlate with negative prognostic factors in extranodal NK/T-cell lymphomas (Zhang et al. 2014).

Completely the opposite was observed with regard to the effect of IL-9 on the growth of melanoma cells in vivo. Th9-derived IL-9 was demonstrated to inhibit the growth of subcutaneous and pulmonary melanoma. Mast cells appeared to be the effector cells in the cutaneous melanoma model, whereas in the pulmonary melanoma model, the IL-9-dependent tumor protection was based on a CCL20/CCR6-mediated tumor-specific recruitment of cytotoxic $CD8^+$ T cells and $CD8^+$ DC (reviewed in Schmitt and Bopp (2012)). Subsequently, it was shown that also IL-9-producing $CD8^+$ Tc9 cells exerted strong anti-melanoma activity that was based on the IL-9-dependent recruitment of tumor-infiltrating $Thy1^+$ $CD8^+$ T cells (reviewed in Schmitt and Bopp (2012)).

Obviously, IL-9 plays an ambiguous role in tumor immunology; on the one hand, it acts as a detrimental tumor promoter if cells express a functional IL-9 receptor, and on the other hand, it represents a potent tumor suppressor for IL-9 nonresponsive tumors by recruiting mast cells, cytotoxic T cells, and $CD8^+$ DC. Therefore, this Dr. Jekyll and Mr. Hyde character of IL-9 requires a very careful genetic and immunologic phenotyping of tumors and tumor-bearing patients before the potential application of an IL-9-based cancer immune therapy.

IL-9 and Allergic Asthma

IL-9 was initially thought to be preferentially produced by Th2 cells, and therefore the relevance of this cytokine for asthma as a classical Th2-mediated disease was carefully analyzed in preclinical and clinical studies. As outlined in detail below in the section "Modulation by Drugs," genetic linkage analyses in humans and intratracheal application of IL-9 in mice as well as lung-specific transgenic expression of mouse IL-9 clearly indicated a central role for this cytokine in the development and maintenance of allergic asthma. The asthma-promoting effect of IL-9 is based on its ability to directly induce mastocytosis, goblet cell metaplasia, and the expression of several cytokines and chemokines in lung epithelial and smooth muscle cells. IL-13 and eotaxin1/CCL11 could be identified in this context as IL-9-induced mediators of typical asthma symptoms like increased mucus production, eosinophilia, and remodeling (reviewed in Goswami and Kaplan (2011)). The characterization of Th9 cells as a discrete IL-9-producing Th subtype led to studies that demonstrated a decisive contribution of Th9-derived IL-9 for allergic airway pathologies independent from Th2 cells (Staudt et al. 2010). Recently, ILC2 cells (innate lymphoid cells) were identified to be a new source for IL-9 in the lungs (Wilhelm et al. 2011). IL-9 can also serve as a growth factor for ILC2 cells suggesting that such cells potentiate in an autocrine fashion the pathologic effect of Th9 cells during chronic allergic asthma.

In several clinical studies, it could be convincingly demonstrated that human IL-9, IL-9R, and eotaxin1/CCL11 exhibited a strongly increased expression in the lung tissue of asthmatic patients as compared to healthy controls (Yamasaki et al. 2010). In addition, in vitro stimulation of human PBMCs with various allergens resulted in the production of IL-9 predominantly in asthma patients indicating a close correlation between disease and IL-9 production (Devos et al. 2006). Hence, two phase 2 studies were conducted in which asthma patients were treated with neutralizing humanized monoclonal antibodies against IL-9 (for details see "Modulation by Drugs" section).

IL-9 and Inflammatory Gut Responses

Murine models of parasitic intestinal infections (*Trichinella spiralis, Trichuris muris*) clearly showed that IL-9 is crucially involved in a protective immune response upon worm infection. One of the main targets of IL-9 is the mucosal mast cell that are the source of different mediators which caused increased mucus production, strong eosinophilia, and intestinal hypercontractility eventually favoring the expulsion of parasites. Hence IL-9 causes an inflammatory intestinal environment largely based on the activation of mucosal mast cells in order to get rid of intestinal parasites (reviewed in Stassen et al. (2012)).

Concerning inflammatory bowl diseases, clinical analyses of ulcerative colitis (UC) patients who suffer from a sustained epithelial damage of the gut as a result of a dysregulated immune response exhibited an increased IL-9 mRNA expression in inflamed samples (Hufford and Kaplan 2014). The major source of IL-9 was found to be $CD3^+CD4^+$ T cells, and the key IL-9 transcription factors IRF4 and PU.1 were upregulated and co-localized with IL-9 in T cells from patients with severe disease. In addition, PMN cells and eosinophils from UC patients expressed high levels of IL-9 receptor, and IL-9 treatment of isolated PMC cells led in a dose-dependent manner to an increased production of pro-inflammatory IL-8. Using a mouse model of oxazolone-induced colitis, the increase of IL-9-producing T cells observed in humans could be confirmed. Furthermore it was suggested that such Th9-derived IL-9 impairs mucosal wound healing and exacerbates acute colitis. The neutralization of IL-9 five days after oxazolone treatment led to an alleviation of the disease implying that Il-9 can represent a promising target for the treatment of chronic inflammatory bowl diseases.

IL-9 in Autoimmunity

The transfer of MOG_{35-55}-specific IL-9-producing Th9 cells was shown to induce the development of experimental autoimmune encephalomyelitis (EAE). Since neutralization of IL-9 retarded the development of EAE, it was anticipated that IL-9 plays an important pathological role although this EAE model is known to require the activation of Th17 cells (Nowak et al. 2009). Subsequently, it was found that IL-9 can induce the expression of CCL20 in astrocytes and thus recruiting EAE-promoting $CCR6^+$ Th17 effector cells to the CNS (reviewed in Stassen et al. (2012)). Hence, this fatal interplay between IL-9-producing Th9 cells and Th17 cells can have a potent disease-aggravating influence on the course of a MOG_{31-55}-induced EAE.

A similar scenario was described concerning the function of IL-9 in a mouse Th17 cell-associated psoriasis model (Singh et al. 2013). Expression of IL-9 and IL-9 receptor was increased in the skin of TGF-β1 transgenic mice which exhibit a phenotype comparable to human psoriasis. Intradermal injection of IL-9 caused Th17 cell-related inflammations, and the neutralization of IL-9 reduced prototypic psoriatic inflammation and angiogenesis and delayed the development of the psoriasis-like phenotype as well. In humans comparative analyses of skin biopsies from psoriatic patients demonstrated an increased expression of the IL-9 receptor at the dermal-epidermal junction and within the basal layers as compared to healthy volunteers (Singh et al. 2013). Accordingly, polyclonally activated CD4$^+$ T cells from psoriatic patients produced significantly higher amounts of IL-9.

Thus, the analyses of both autoimmune diseases – the mouse models of EAE and psoriasis as well as the human correlates – impressively demonstrated that a dysregulated local secretion of IL-9 will lead to a strong local inflammation that cannot be curbed by immune tolerance mechanisms.

Modulation by Drugs

In humans the *Il9* locus is located in the chromosome 5q31–q33 that bears several candidate genes known to be risk factors for the development of allergic asthma. In mice, the *Il9* locus is located on the corresponding syntenic chromosome 13 and is also implicated as a candidate for bronchial responsiveness. Indeed, in humans an *Il9* allelic variation is associated with increased serum IgE levels and allergy (Doull et al. 1996). In addition, bronchial biopsy specimens from asthmatic patients demonstrated increased IL-9 and IL-9 receptor mRNA expression when compared to healthy controls (Shimbara et al. 2000). Concomitantly, forced ectopic expression by an *Il9* transgene in mice as well as intratracheal instillation of IL-9 induced asthma-like syndromes in mice, including eosinophilia, increased IgE serum levels, and airway hyperresponsiveness (reviewed in Stassen et al. (2012)). Of note, ectopic expression of IL-9 in the lungs of these mice resulted in the expression of several Th2 cytokines, including IL-13. Importantly, all asthma syndromes noted in *Il9* transgenic mice were abrogated either upon blockage of IL-13 during *Il9* transgene expression or in the progeny of those mice crossed with *Il13*-deficient mice (Steenwinckel et al. 2007; Temann et al. 2007). Further evidence for a detrimental role of IL-9 in allergic airway inflammation stems from preclinical studies in which neutralization of IL-9 during the sensitization phase ameliorated asthma symptoms, including eosinophilia, epithelial damage, and airway hyperresponsiveness (Cheng et al. 2002; Kung et al. 2001).

Hence, the close chromosomal linkage of the *Il9* locus to those loci encoding for IL-4, IL-5, and IL-13 in humans and the fact that IL-9 has long been attributed to Th2 immune responses including allergic asthma provided the rationale for the development of an IL-9-neutralizing antibody for the potential treatment of asthma (Antoniu 2010). The humanized IL-9-neutralizing antibody MEDI-528 has been reported to have an acceptable safety profile and seems to have clinical activity in alleviating asthmatic symptoms in patients suffering from mild to moderate asthma (Parker et al. 2011; White et al. 2009).

In line with the findings in humans described above, mice genetically deficient in *Il9* developed allergen-induced lung inflammation and airway hyperresponsiveness comparable to their wild-type littermates (McMillan et al. 2002). Hence, it appears likely that, in the complex pathogenesis of asthma, several compensatory mechanisms, including enhanced production of IL-13, act in the absence of IL-9 and, thus, hamper amelioration of asthma symptoms in severe asthmatics upon IL-9 neutralization (Oh et al. 2013).

Cross-References

▶ Acute Exacerbations of Airway Inflammation
▶ Allergic Disorders
▶ Anti-asthma Drugs, Overview
▶ Asthma
▶ Cancer and Inflammation
▶ Cytokines
▶ Inflammatory Bowel Disease
▶ Inflammatory Bowel Disease Models in Animals
▶ Interleukin 2
▶ Janus Kinases (Jaks)/STAT Pathway
▶ Leukocyte Recruitment
▶ Mast Cells

References

Antoniu, S. A. (2010). MEDI-528, an anti-IL-9 humanized antibody for the treatment of asthma. *Current Opinion in Molecular Therapeutics, 12*(2), 233–239.

Chen, N., & Wang, X. (2014). Role of IL-9 and STATs in hematological malignancies (review). *Oncology Letters, 7*(3), 602–610.

Cheng, G., Arima, M., Honda, K., Hirata, H., Eda, F., Yoshida, N., et al. (2002). Anti-interleukin-9 antibody treatment inhibits airway inflammation and hyperreactivity in mouse asthma model. *American Journal of Respiratory and Critical Care Medicine, 166*(3), 409–416.

Devos, S., Cormont, F., Vrtala, S., Hooghe-Peters, E., Pirson, F., & Snick, J. (2006). Allergen-induced interleukin-9 production in vitro: Correlation with atopy in human adults and comparison with interleukin-5 and interleukin-13. *Clinical and Experimental Allergy, 36*(2), 174–182.

Doull, I. J., Lawrence, S., Watson, M., Begishvili, T., Beasley, R. W., Lampe, F., et al. (1996). Allelic association of gene markers on chromosomes 5q and 11q with atopy and bronchial hyperresponsiveness. *American Journal of Respiratory and Critical Care Medicine, 153*(4 Pt 1), 1280–1284.

Gessner, A., Blum, H., & Rollinghoff, M. (1993). Differential regulation of IL-9-expression after infection with Leishmania major in susceptible and resistant mice. *Immunobiology, 189*(5), 419–435.

Goswami, R., & Kaplan, M. H. (2011). A brief history of IL-9. *Journal of Immunology (Baltimore, Md.: 1950), 186*(6), 3283–3288.

Gounni, A. S., Nutku, E., Koussih, L., Aris, F., Louahed, J., Levitt, R. C., et al. (2000). IL-9 expression by human eosinophils: Regulation by IL-1beta and TNF-alpha. *Journal of Allergy and Clinical Immunology, 106*(3), 460–466.

Hufford, M. M., & Kaplan, M. H. (2014). A gut reaction to IL-9. *Nature Immunology, 15*(7), 599–600.

Hultner, L., Koelsch, S., Stassen, M., Kaspers, U., Kremer, J. P., Mailhammer, R., et al. (2000). In activated mast cells, IL-1 up-regulates the production of several Th2-related cytokines including IL-9. *Journal of Immunology, 164*(11), 5556–5563.

Kung, T. T., Luo, B., Crawley, Y., Garlisi, C. G., Devito, K., Minnicozzi, M., et al. (2001). Effect of anti-mIL-9 antibody on the development of pulmonary inflammation and airway hyperresponsiveness in allergic mice. *American Journal of Respiratory Cell and Molecular Biology, 25*(5), 600–605.

Licona Limon, P., Henao-Mejia, J., Temann, A. U., Gagliani, N., Licona-Limón, I., Ishigame, H., et al. (2013). Th9 cells drive host immunity against gastrointestinal worm infection. *Immunity, 39*(4), 744–757.

McMillan, S. J., Bishop, B., Townsend, M. J., McKenzie, A. N., & Lloyd, C. M. (2002). The absence of interleukin 9 does not affect the development of allergen-induced pulmonary inflammation nor airway hyperreactivity. *Journal of Experimental Medicine, 195*(1), 51–57.

Monteyne, P., Renauld, J. C., Van Broeck, J., Dunne, D. W., Brombacher, F., & Coutelier, J. P. (1997). IL-4-independent regulation of in vivo IL-9 expression. *Journal of Immunology, 159*(6), 2616–2623.

Nowak, E. C., Weaver, C. T., Turner, H., Begum-Haque, S., Becher, B., Schreiner, B., et al. (2009). IL-9 as a mediator of Th17-driven inflammatory disease. *The Journal of Experimental Medicine, 206*(8), 1653–1660.

Oh, C. K., Leigh, R., McLaurin, K. K., Kim, K., Hultquist, M., & Molfino, N. A. (2013). A randomized, controlled trial to evaluate the effect of an anti-interleukin-9 monoclonal antibody in adults with uncontrolled asthma. *Respiratory Research, 14*, 93.

Parker, J. M., Oh, C. K., Laforce, C., Miller, S. D., Pearlman, D. S., Le, C., et al. (2011). Safety profile and clinical activity of multiple subcutaneous doses of MEDI-528, a humanized anti-interleukin-9 monoclonal antibody, in two randomized phase 2a studies in subjects with asthma. *BMC Pulmonary Medicine, 11*(1), 14.

Renauld, J. C., Kermouni, A., Vink, A., Louahed, J., & Van Snick, J. (1995). Interleukin-9 and its receptor: Involvement in mast cell differentiation and T cell oncogenesis. *Journal of Leukocyte Biology, 57*(3), 353–360.

Schmitt, E., & Bopp, T. (2012). Amazing IL-9: Revealing a new function for an "old" cytokine. *The Journal of Clinical Investigation, 122*(11), 3857–3859.

Schmitt, E., Klein, M., & Bopp, T. (2013). Th9 cells, new players in adaptive immunity. *Trends in Immunology, 48*(2), 115–125.

Shimbara, A., Christodoulopoulos, P., Soussi-Gounni, A., Olivenstein, R., Nakamura, Y., Levitt, R. C., et al. (2000). IL-9 and its receptor in allergic and nonallergic lung disease: Increased expression in asthma. *Journal of Allergy and Clinical Immunology, 105*(1), 108–115.

Singh, T. P., Schön, M. P., Wallbrecht, K., Gruber-Wackernagel, A., Wang, X.-J., & Wolf, P. (2013). Involvement of IL-9 in Th17-associated inflammation and angiogenesis of psoriasis. *PloS One, 8*(1), e51752.

Stassen, M., Schmitt, E., & Bopp, T. (2012). From interleukin-9 to T helper 9 cells. *Annals of the New York Academy of Sciences, 1247*(1), 56–68.

Staudt, V., Bothur, E., Klein, M., Lingnau, K., Reuter, S., Grebe, N., et al. (2010). Interferon-regulatory factor 4 is essential for the developmental program of T helper 9 cells. *Immunity, 33*(2), 192–202.

Steenwinckel, V., Louahed, J., Orabona, C., Huaux, F., Warnier, G., McKenzie, A., et al. (2007). IL-13 mediates in vivo IL-9 activities on lung epithelial cells but not on hematopoietic cells. *Journal of Immunology, 178*(5), 3244–3251.

Temann, U.-A., Laouar, Y., Eynon, E. E., Homer, R., & Flavell, R. A. (2007). IL9 leads to airway inflammation by inducing IL13 expression in airway epithelial cells. *International Immunology, 19*(1), 1–10.

White, B., Leon, F., White, W., & Robbie, G. (2009). Two first-in-human, open-label, phase I dose-escalation safety trials of MEDI-528, a monoclonal antibody against interleukin-9, in healthy adult volunteers. *Clinical Therapeutics, 31*(4), 728–740.

Wilhelm, C., Hirota, K., Stieglitz, B., Van Snick, J., Tolaini, M., Lahl, K., et al. (2011). An IL-9 fate reporter demonstrates the induction of an innate IL-9 response in lung inflammation. *Nature Immunology, 12*(11), 1071–1077.

Yamasaki, A., Saleh, A., Koussih, L., Muro, S., Halayko, A. J., & Gounni, A. S. (2010). IL-9 induces CCL11 expression via STAT3 signalling in human airway smooth muscle cells. *PloS One, 5*(2), e9178.

Zhang, J., Wang, W.-D., Geng, Q.-R., Wang, L., Chen, X.-Q., Liu, C.-C., et al. (2014). Serum levels of interleukin-9 correlate with negative prognostic factors in extranodal NK/T-cell lymphoma. *PloS One, 9*(4), e94637.

Interleukin 10

Erik Oliver Glocker
Department of Medical Microbiology and Hygiene, University Hospital Freiburg, Freiburg, Germany

Synonyms

Cytokine synthesis inhibitory factor (CSIF)

Definition

IL-10 is the most important and best characterized anti-inflammatory cytokine. It keeps inflammatory reactions in check and prevents from tissue damage. IL-10 controls the expression of proinflammatory cytokines, inhibits activation of macrophages, limits the presentation of antigens to T cells and has a direct inhibitory effect on T cells. In addition, it regulates differentiation and proliferation of macrophages, T and B cells.

Biosynthesis and Release

IL-10 is the founding member of the IL-10 cytokine family which includes IL-19, IL-20, IL-22, IL-24, and IL-26 (Fickenscher et al. 2002). The IL-10 family forms the class 2 cytokine family, with IL-10 being a unique member as it potently inhibits the production of proinflammatory cytokines.

IL-10 is a 20kDa protein encoded by 178 amino acids and forms a noncovalent homodimer of two interpenetrating polypeptide chains (Walter 2014).

IL-10 is produced by cells of the innate immune system (macrophages/monocytes, dendritic cells (DCs), mast cells etc.) and the adaptive immune system (CD4+ T cells, CD8+ T cells and B cells). Cells of the innate immune system usually produce IL-10 on activation by pathogen-derived products that had bound to pattern recognition receptors (PRR) (Moore et al. 2001).

Toll-like receptor 2 (TLR2) agonists seem to be particularly important for inducing IL-10 expression by antigen-presenting cells (APCs). Macrophages and myeloid DCs may also produce significant amounts of IL-10 after stimulation with TLR4 and TLR9 ligands, but IL-10 production following TLR3 stimulation was only observed in macrophages. Activation of macrophages results in high levels of IL-10 production, whereas myeloid DCs produce far less and plasmacytoid DCs do not produce any IL-10 (Boonstra et al. 2006).

Binding to TLRs kicks of signaling cascades including myeloid differentiation primary-response protein 88 (MyD88) and TIR-domain-containing adaptor protein inducing IFN-β (TRIF). Signaling via MyD88 then leads to the activation of mitogen-activated protein kinases (MAPKs) and nuclear factor-κB (NF-κB). The MAPK cascade consists of extracellular signal-regulated kinases (ERKs; comprising ERK1 and ERK2); JUN N-terminal kinases (JNKs; comprising JNK1 and JNK2); and p38 (Akira and Takeda 2004; Symons et al. 2006).

The strength of ERK activation correlates with IL-10 production with ERK being most activated in macrophages and only slightly activated plasmacytoid DCs.

IL-10 can also be induced independent of TLRs by stimulation of C-type lectin receptors such as Dectin-1. Dectin-1 is usually involved in the sensing of components of the cell wall of fungi; once activated and phosphorylated, Dectin-1 recruits spleen tyrosine kinase (SYK) and the production of IL-2 and IL-10 is initiated; signaling downstream of Dectin-1 also involves the ERK pathway (Saraiva and O'Garra 2010).

T cells recognize microbial peptides that have been presented on major histocompatibility complexes by antigen presenting cells. In contrast to cells of the innate immunity that can instantly produce IL-10, naive CD4+ T cells first need to differentiate into different T helper (Th) cell subsets to produce IL-10 (Saraiva and O'Garra 2010).

IL-10 production was first described in Th2 cells, but has also been shown for Th1 and IL-17 producing T helper (Th17) cells (Saraiva and O'Garra 2010; Fiorentino et al. 1989).

IL-10 producing Th1 cells have been observed in infectious diseases; they can be generated by high levels of antigen-specific or polyclonal stimulation in the presence of IL-12 and is STAT4 and ERK dependent. Notch signaling can also induce IL-10 expression by Th1 cells, a process that requires STAT4. In Th2 cells, IL-10 production is regulated by IL-4, STAT6 and GATA binding protein 3 (GATA3), in Th17 cells by STAT3- and/or STAT1. In contrast to macrophages and DCs where both ERK and p38 are equally required to induce IL-10, ERK plays a critical role for IL-10 expression in Th1, Th2, and Th17 cells (Saraiva and O'Garra 2010).

Regulatory T (Treg) cells, which are characterized by the expression of the lineage defining transcription factor forkhead box P3 (FoxP3), inhibit naive T cell proliferation in vitro independently of IL-10, but in vivo through IL-10. The signals that induce IL-10 expression by FoxP3+ Treg cells still remain elusive, although transforming growth factor β (TGFβ) has been shown to be a critical in vivo (Hawrylowicz and O'Garra 2005; Roncarolo et al. 2006).

FoxP3⁻ and IL-10 producing T cells have also been described. These cells produce IL-10, but not IL-2, IL-4 or IFNγ, and can be generated in vitro by stimulation with TGF-β, IL-10, and IFN-α or dexamethasone. Costimulation through CD2, CD46, inducible T cell costimulator (ICOS) and/or type I IFNs provide additional signals for IL-10 expression (Hawrylowicz and O'Garra 2005; Roncarolo et al. 2006).

IFN-γ potently inhibits ERK- and p38-dependent production of IL-10 and interferes with the phosphoinositide 3-kinase (PI3K)–AKT pathway what then leads to the release of glycogen synthase kinase 3 (GSK3). GSK3 acts on the transcription factors activator protein 1 (AP1) and cAMP response element-binding protein (CREB) and inhibits IL-10 production by suppressing their binding to the promoter of the IL-10 gene. IL-10 itself induces the expression of dual-specificity protein phosphatase 1 (DUSP1), which counteracts p38 phosphorylation, thereby decreasing IL-10 production (Saraiva and O'Garra 2010).

IL-10 expression in immune cells is modulated by IL-21 and IL-27: IL-21 enhances IL-10 expression by CD4+ T cells; and IL-27 enhances IL-10 expression by Th1, Th2 and Th17 cells but impairs TLR-induced IL-10 expression by monocytes. IL-21 and IL-27 both activate ERK and STAT3 (Saraiva and O'Garra 2010).

Biological Activities

IL-10 was first described in 1989 as cytokine synthesis inhibitory factor (CSIF) that was released by Th2 cells and capable of controlling the secretion of cytokines by murine Th1 cells (Fiorentino et al. 1989).

IL-10 controls the expression of cytokines, surface molecules and soluble factors in myeloid cells. IL-10 inhibits the release of several pro-inflammatory cytokines including IL-1α, IL-1β, IL-6, IL-12, IL-18 and TNF. In particular its inhibitory effect on the release of IL-1 and TNF makes IL-10 a key controller of inflammation (Moore et al. 2001).

Furthermore, IL-10 blocks the expression of CC chemokines (such as MCP1, MCP-5, MIP-1α MIP-1β and RANTES) and CXC chemokines (such as IL-8 and MIP-2).

Apart from controlling the release of pro-inflammatory cytokines and effector molecules, IL-10 also up-regulates the expression of anti-inflammatory proteins including IL-1 receptor antagonist IL-1RA and soluble TNF receptors. Destructive activities of matrix metalloproteinases (MMP) are constrained by boosting the expression of tissue inhibitor of MMPs (TIMP), TIMP1 and inhibiting the production of MMP2 and MMP9 (Lacraz et al. 1995).

IL-10 inhibits expression of MHC class II molecules, CD54, CD80 and CD86 on monocytes what significantly affects the T cell activation. It also enhances expression of Fc-γR leading to an improved capacity of monocytes/macrophages to phagocytose opsonized particles, bacteria, or fungi despite IL-10 reduced the ability of the cells to kill the ingested organisms by decreasing the generation of superoxide anion ($O2^-$) and nitric oxide (NO) (Moore et al. 2001; Bogdan et al. 1991).

In contrast to macrophages and monocytes, the impact of IL-10 on B cells is rather stimulatory. IL-10 enhances B cell survival by induction of expression of the antiapoptotic protein Bcl-2 and is a cofactor for proliferation of B cell precursors and mature B cells activated by CD40 or anti-IgM. This proliferation is further cranked up by IL-2 and IL-4; IL-10 also controls B cell differentiation and isotype switching (Moore et al. 2001).

IL-10 inhibits cytokine production and proliferation of CD4+ T cells by curbing the function of antigen presenting cells. IL-10 also directly affects the function of T cells and inhibits IL-2, TNF, and IL-5 production. In contrast, IL-10 may also induce recruitment, proliferation and cytotoxic activity of CD8+ T cells. Interestingly, activation of T cells in the presence of IL-10 can induce anergy, which cannot be reversed by IL-2 or stimulation by anti-CD3and anti-CD28. IL-10-mediated anergy is associated with the induction of Treg cells that produce high levels of IL-10 and can suppress antigen-specific responses (Roncarolo et al. 2014).

Signaling Pathway

To exert its anti-inflammatory properties, IL-10 molecules form homodimers that bind to a cell-bound receptor structure (IL-10R): this receptor is a tetrameric complex that is made up of two molecules of the IL-10R α-chain (IL-10R1) and two molecules of the accessory IL-10R β-chain (IL-10R2). IL-10R1 is an approximately 80,000 kDa protein with an extracellular domain of 227 amino acid residues, a transmembrane domain of 21 residues, and an intracellular domain (ICD) of 322 amino acids. The extracellular domain of IL-10R2 consists of 201 amino acid residues, but its intracellular domain consists of only 83 residues. In contrast to IL-10R1 that binds IL-10 with high affinity, IL-10R2 interacts with only low affinity and is furthermore shared by several other cytokines, including IL-22, IL-26, and the λ-interferons IL-28A/B and IL-29 (Walter 2014).

Upon binding of IL-10 to IL-10R1 and recruitment of the accessory IL-10R2, two members of the Janus kinase family, Janus kinase (JAK)1 and tyrosine kinase (Tyk)2 are activated and trigger phosphorylation of themselves and then of IL-10R1 at the intracellular tyrosine residues 446 and 496. This results in the generation of docking sites for signal transducer and activator of transcription-3 (STAT3). STAT3 then gets phosphorylated (at tyrosine residue 705) by JAK1 and Tyk2, dimerizes and translocates to the nucleus where it induces expression of its target genes such as the suppressor of cytokine signaling (SOCS)-3 gene (Donnelly et al. 1999; Walter 2014). Both IL-10R1 and IL-10R2 are required for proper IL-10 signaling. Loss-of-function mutations in either IL-10R1 or IL-10R2 result in a complete signaling failure that cannot be compensated by any other pathway.

Intestinal Immunity

The human intestine is colonized with a massive amount of different bacteria that permanently challenge the intestinal immune system. The intestinal immune system not only needs to be capable of distinguishing the physiological flora from invading pathogens but also to control any inflammatory events. A variety of different cells including epithelial cells, Paneth cells, goblet cells, cells of the adaptive immune system, and the powerful anti-inflammatory cytokine IL-10 are required to maintain the integrity of the gut.

Why IL-10 is this important for mammals was first demonstrated in mice: Mice without functional IL-10 ($IL10^{-/-}$) or mice lacking a component of the IL-10 receptor (the β-unit IL10R2 subunit) spontaneously developed severe and progressive enterocolitis (Kole and Maloy 2014).

CD4+/CD25+/FoxP3 positive Treg cells and FoxP3- Tr1-like cells are both important sources of IL-10 and hence key to maintain intestinal immune homeostasis. FoxP3+ Treg deficient mice develop severe multiorgan inflammation, which improves on transfer of FoxP3+ positive Treg cells and an already manifest colitis in mice can be cured. The majority of CD4+/CD25+/FoxP3+ Treg cells develop in the thymus; some Treg cells then migrate to the gastrointestinal tract and prevent there excessive immune reactions (Barnes and Powrie 2009).

IL-10 Producing B Cells

TCR$\alpha^{-/-}$ mice with TCRα–β+ CD4+ T cells spontaneously develop an inflammatory disease similar to ulcerative colitis and show a strongly activated B cell compartment with an increased B cell number in mesenteric lymph nodes (MLN) and increased serum autoantibodies against neutrophil cytoplasmic antigens (ANCA) and DNA. B cell-deficient TCR$\alpha^{-/-}$ mice develop an even earlier and more severe colitis disease than the nondeficient mice and TCR$\alpha^{-/-}$ mice, suggesting a protective role for B cells. The protective function of B cells required their provision of IL-10 and B cells were the major source of IL-10 in MLN of TCR$\alpha^{-/-}$ mice with colitis. IL-10 was absent in MLN B cells from healthy TCR$\alpha^{-/-}$ mice, indicating that its expression was induced in MLN B cells as a counter-regulatory mechanism (Hilgenberg et al. 2014).

Whether such IL-10 producing B cells could dampen autoimmune disease in humans was investigated in patients with encephalitis disseminata (ED). Indeed, B cells from ED patients secreted less IL-10 than B cells from healthy controls did, hinting at the possibility that progression of ED might be the result of an impaired capacity of B cells to produce IL-10. The capacity of B cells to produce IL-10 was restored after infection of ED patients with helminths what was then followed by fewer flare ups of the disease (Hilgenberg et al. 2014).

Pathophysiological Relevance

IL-10/IL-10 Polymorphisms and Mutations

Lack of IL-10 or IL-10R (IL-10/IL-10R deficiency) result in severe inflammatory bowel disease (Glocker and Grimbacher 2012; Glocker et al. 2011). Both deficiencies are rare congenital disorders characterized by inadequate excessive proinflammatory responses on activations of the immune system. IL-10R deficiency was first identified in 2009 in children with severe early-onset inflammatory bowel disease. Two pediatric IBD patients harbored homozygous loss-of-function mutations in *IL10RA* (resulting in amino acid exchanges Gly141Arg and Thr84Ile), two other children carried homozygous loss-of-function mutations in *IL10RB* (resulting in a premature stop codon: (Trp159X). The identified mutations blocked the essential IL-10 mediated phosphorylation of the signal transducer and activator of transcription (STAT)3 and led to a complete interruption of the anti-inflammatory responses (Glocker et al. 2011).

IL-10 deficiency may be caused by loss-of-function mutations in IL-10 itself. Affected patients are critically ill and present with an as severe early-onset enterocolitis as patients with IL-10R deficiency do. Patients with IL-10/IL-10R deficiencies typically come down within the first months of life with intractable inflammatory bowel disease. The disease is often accompanied with perianal/rectovaginal and enterocutaneous fistulae and abscesses and often requires surgery including subtotal colectomy to keep the disease under control. Due to the severe colitis and the accompanying malabsorption children fail to thrive properly.

IL-10/Il-10R deficient patients may have recurrent infections, primarily affecting the respiratory tract. Despite completely abolished IL-10-signaling, the immunological workup is largely normal or shows only slight abnormalities such as decreased CD4/CD8 ratios. Patients may reveal hypergammaglobulinemia due to the ongoing inflammation, but may also show reduced serum levels of immunoglobulins which might be the reason for an increased susceptibility to infectious diseases (Engelhardt et al. 2013).

Whether the immunological alterations and the occurrence of apparently frequent infections were caused by the underlying disease, nutritional deficiencies, or the strong immunosuppressive therapy is unclear. Of interest are EBV-associated lymphomas in patients who harbored mutations in IL-10R2, which is also part of the receptor for λ-interferons. Thus, a lack of λ-interferon signaling, which is a component of the antiviral defense may have combined with immunosuppressive therapy to predispose to lymphomas (Glocker et al. 2011; Shah et al. 2012).

IL-10/IL-10R deficiencies are life threatening and patients die without medical support. The use of anti-inflammatory drugs such as steroids, methotrexate, thalidomide, and anti-TNF-α monoclonal antibodies does not induce remission or long-term improvement. The so far only curative therapeutical approach is hematopoietic stem cell transplantation (HSCT). HSCT should be considered as a potential therapy option from early on. In several IL-10/IL-10R deficient patients, HSCT has been proven to be successful and suggests that IL-10 signaling in hematopoietic cells rather than IL-22/IL-26/IFN-λ signaling in nonhematopoietic cells is critical to induce remission (Glocker et al. 2011; Shah et al. 2012).

Patients with severe early-onset IBD and loss-of-functions mutations in IL-10 or IL-10R demonstrate the need for IL-10 to balance the intestinal ecosystem. Ahead of the identification of IL-10/IL-10R deficiencies, genome-wide association studies (GWAS) already suggested a role of IL-10 in the pathogenesis of adult IBD. A single nucleotide polymorphism (SNP rs3024505) that flanks the IL10 gene was significantly associated with UC in adult patients. A GWAS in a pediatric population by a Canadian group reported a significant allelic association between two further IL-10 SNPs (rs2222202 and rs1800871) and Crohn's disease. A meta-analysis of six GWAS confirmed an association of the IL-10 gene locus with both Crohn's disease and ulcerative colitis (Engelhardt and Grimbacher 2014).

An interesting association between the frequently described frameshift insertion 3020insC in the intracellular sensor molecule nucleotide-binding oligomerization domain containing 2 (NOD2) that is associated with Crohn's disease has been shown to inhibit transcription of the IL-10 gene what might explain the tight link to intestinal hyperinflammation in Crohn's disease (Noguchi et al. 2009).

Immunodysregulation, Polyendocrinopathy, Enteropathy, X-Linked (IPEX) Syndrome

A lack of IL-10 producing Treg cells is the hallmark of the Immunodysregulation, polyendocrinopathy, enteropathy, X-linked (IPEX) syndrome. IPEX-patients carry mutations in the *FOXP3* gene, that is located on the X-chromosome, suffer from autoimmune lymphoproliferation and multiple autoimmune disorders. If not fatal in early childhood, IPEX patients usually present with insulin-dependent diabetes mellitus, skin disease, hypo- or hyperthyreoidism and recurrent infections. The predominant clinical feature, however, is an autoimmune enteropathy that is similar to Crohn's disease, ulcerative colitis or celiac disease. The small intestine reveals villous atrophy, the large intestine shows inflammatory infiltrates with CD3+ cells and plasma cells (Gambineri et al. 2003).

Rheumatoid Arthritis and Lupus Erythematosis (SLE)

IL-10 is considered to play a key role in autoimmune diseases such as rheumatoid arthritis and systemic Lupus erythematosus (SLE). The production of IL-10 by macrophages and T cells in the synovia inhibits the release of proinflammatory cytokines by synovial cells. Even though IL-10 expression in rheumatoid arthritis has been associated with enhanced B cell activation and production of autoantibodies, IL-10 was also shown to be protective in animal models. When administered to animals before and/or after induction of disease, IL-10 reduced joint swelling, cytokine production, and cartilage degradation (Moore et al. 2001).

IL-10 secreting B cells (regulatory B cells) have been shown to prevent the development of arthritis. The adoptive transfer of B cells prevented the development of arthritis upon immunization with bovine collagen in an IL-10-dependent manner.

SLE is an utterly complex autoimmune disorder characterized by polyclonal B cell activation, the occurrence of high levels of autoantibodies including autoantibodies to nuclear antigens (ANA) that are typically directed to double-stranded DNA and nucleosomes and the deposition of glomerular immune complexes.

The role of IL-10 and its release from B cells in the pathogenesis of SLE is as complex as the disease is and is controversially discussed. B cells and macrophages from SLE patients were shown to produce high levels of IL-10, and several studies have confirmed a correlation between serum levels of IL-10 and disease activity. Treatment of SLE mice with anti-IL-10 antibodies inhibited the production of autoantibodies and immune pathology, suggesting that IL-10 stimulation of immunoglobulin production by B cells plays a major role in the pathogenesis of SLE. In a small pilot study, treatment of 6 SLE patients with a mouse anti-hIL-10 Mab achieved reduction of most symptoms and improvement of disease. In contrast, two randomized and placebo-controlled trials on the effect of B cell-depletion treatment in SLE failed to meet their primary and secondary endpoints (Hilgenberg et al. 2014).

The controversially discussed role of IL-10 in SLE is still under investigation as there are also data that hint at a suppressive and protective function. In this context, B cells from patients with SLE were shown to produce less IL-6 and IL-10 than B cells from healthy individuals did on stimulation of TLR9 (Hilgenberg et al. 2014).

Modulation by Drugs

As IL-10 is a multifunctional anti-inflammatory cytokine which targets T cells, monocytes, macrophages, and most other hematopoietic cell types, it has always been of great interest to use its immune-regulatory potential therapeutically. Parenteral application of recombinant IL-10 to patients with refractory Crohn's disease resulted in some clinical improvement but not significant remission.

Targeted enteral application of IL-10 for the treatment of inflammatory bowel disease has been carried out by genetically engineered bacteria producing IL-10 and resulted in a reasonable reduction of colitis in a murine model. The rectal application of IL-10 containing gelatine microspheres caused improvement of colitis in $IL10^{-/-}$ mice and was better than systemic treatment with IL-10 (Shah et al. 2012).

Corticosteroids such as Methyl-Prednisolon are well-known immunosuppressive drugs that exert several mechanisms of action including induction of IL-10. Inhaled steroids that are commonly used in patients with asthma have been shown to induce the tryptophan-degrading enzyme Indoleamine 2, 3-dioxygenase (IDO) in an IL-10 dependent manner. IDO is an important immunomodulator and its metabolites inhibit T cell proliferation, thereby controlling inflammation of the airways in these patients. The inducible effect on IDO was reported to be stepped up by statins such as simvastatin (Unterberger et al. 2008; Maneechotesuwan et al. 2010).

Application of intravenous immunoglobulins (IVIG) for immunotherapies has been associated with an increased IL-10 production and release by Treg cells. This is of particular interest, as IVIG preparations may be given to patients with autoimmune disorders. Heavy and light chains of IgG contain regulatory T-cell epitopes (Tregitopes) that result in activation of Tregs, increased expression of FoxP3 and induced IL-10 expression in T cells. This may explain why patients with autoimmune diseases benefit from IVIG (Cousens et al. 2013).

Cross-References

▶ Interleukin 22

References

Akira, S., & Takeda, K. (2004). Toll-like receptor signalling. *Nature Reviews Immunology, 4*, 499–511.

Barnes, M., & Powrie, F. (2009). Regulatory T cells reinforce intestinal homeostasis. *Immunity, 31*, 401–411.

Bogdan, C., Vodovotz, Y., & Nathan, C. (1991). Macrophage deactivation by interleukin 10. *Journal of Experimental Medicine, 174*, 1549–1555.

Boonstra, A., Rajsbaum, R., Holman, M., Marques, R., Asselin-Paturel, C., Pereira, J. P., et al. (2006). Macrophages and myeloid dendritic cells, but not plasmacytoid dendritic cells, produce IL-10 in response to MyD88- and TRIF-dependent TLR signals, and TLR-independent signals. *Journal of Immunology, 177*, 7551–7558.

Cousens, L. P., Tassone, R., Mazer, B. D., Ramachandiran, V., Scott, D. W., & De Groot, A. S. (2013). Tregitope update: mechanism of action parallels IVIg. *Autoimmunity Reviews, 12*, 436–443.

Donnelly, R. P., Dickensheets, H., & Finbloom, D. S. (1999). The interleukin-10 signal transduction pathway and regulation of gene expression in mononuclear phagocytes. *Journal of Interferon & Cytokine Research, 19*, 563–573.

Eberhardt, M. K., & Barry, P. A. (2014). Pathogen manipulation of cIL-10 signaling pathways: opportunities for vaccine development? *Current Topics in Microbiology and Immunology, 380*, 93–128.

Engelhardt, K. R., & Grimbacher, B. (2014). IL-10 in humans: lessons from the gut, IL-10/IL-10 receptor deficiencies, and IL-10 polymorphisms. *Current Topics in Microbiology and Immunology, 380*, 1–18.

Engelhardt, K. R., Shah, N., Faizura-Yeop, I., Kocacik Uygun, D. F., Frede, N., Muise A. M., et al. (2013). Clinical outcome in IL-10- and IL-10 receptor-deficient patients with or without hematopoietic stem cell transplantation. *Journal of Allergy and Clinical Immunology*, 131:825–830.

Fickenscher, H., Hor, S., Kupers, H., Knappe, A., Wittmann, S., & Sticht, H. (2002). The interleukin-10 family of cytokines. *Trends in Immunology, 23*, 89–96.

Fiorentino, D. F., Bond, M. W., & Mosmann, T. R. (1989). Two types of mouse T helper cell. IV. Th2 clones secrete a factor that inhibits cytokine production by Th1 clones. *Journal of Experimental Medicine, 170*, 2081–2095.

Gambineri, E., Torgerson, T. R., & Ochs, H. D. (2003). Immune dysregulation, polyendocrinopathy, enteropathy, and X-linked inheritance (IPEX), a syndrome of systemic autoimmunity caused by mutations of FOXP3, a critical regulator of T-cell homeostasis. *Current Opinion in Rheumatology, 15*, 430–435.

Glocker, E. O., Kotlarz, D., Klein, C., Shah, N., & Grimbacher, B. (2011). IL-10 and IL-10 receptor defects in humans. *Annals of the New York Academy of Sciences, 1246*, 102–107.

Glocker, E., & Grimbacher, B. (2012). Inflammatory bowel disease: is it a primary immunodeficiency? *Cellular and Molecular Life Sciences, 69*, 41–48.

Hawrylowicz, C. M., & O'Garra, A. (2005). Potential role of interleukin-10-secreting regulatory T cells in allergy and asthma. *Nature Reviews Immunology, 5*, 271–283.

Hilgenberg, E., Shen, P., Dang, V. D., Ries, S., Sakwa, I., & Fillatreau, S. (2014). Interleukin-10-producing B cells and the regulation of immunity. *Current Topics in Microbiology and Immunology, 380*, 69–92.

Kole, A., & Maloy, K. J. (2014). Control of intestinal inflammation by interleukin-10. *Current Topics in Microbiology and Immunology, 380*, 19–38.

Lacraz, S., Nicod, L. P., Chicheportiche, R., Welgus, H. G., & Dayer, J. M. (1995). IL-10 inhibits metalloproteinase and stimulates TIMP-1 production in human mononuclear phagocytes. *Journal of Clinical Investigation, 96*, 2304–2310.

Maneechotesuwan, K., Ekjiratrakul, W., Kasetsinsombat, K., Wongkajornsilp, A., & Barnes, P. J. (2010). Statins enhance the anti-inflammatory effects of inhaled corticosteroids in asthmatic patients through increased induction of indoleamine 2, 3-dioxygenase. *Journal of Allergy and Clinical Immunology, 126*, 754–762.

Moore, K. W., de Waal Malefyt, R., Coffman, R. L., & O'Garra, A. (2001). Interleukin-10 and the interleukin-10 receptor. *Annual Review of Immunology, 19*, 683–765.

Noguchi, E., Homma, Y., Kang, X., Netea, M. G., & Ma, X. (2009). A Crohn's disease-associated NOD2 mutation suppresses transcription of human IL10 by inhibiting activity of the nuclear ribonucleoprotein hnRNP-A1. *Nature Immunology, 10*, 471–479.

Roncarolo, M. G., Gregori, S., Battaglia, M., Bacchetta, R., Fleischhauer, K., & Levings, M. K. (2006). Interleukin-10-secreting type 1 regulatory T cells in rodents and humans. *Immunology Reviews, 212*, 28–50.

Roncarolo, M. G., Gregori, S., Bacchetta, R., & Battaglia, M. (2014). Tr1 cells and the counter-regulation of immunity: natural mechanisms and therapeutic applications. *Current Topics in Microbiology and Immunology, 380*, 39–68.

Saraiva, M., & O'Garra, A. (2010). The regulation of IL-10 production by immune cells. *Nature Reviews Immunology, 10*, 170–181.

Shah, N., Kammermeier, J., Elawad, M., & Glocker, E. O. (2012). Interleukin-10 and interleukin-10-receptor defects in inflammatory bowel disease. *Current Allergy and Asthma Reports, 12*, 373–379.

Symons, A., Beinke, S., & Ley, S. C. (2006). MAP kinase kinase kinases and innate immunity. *Trends in Immunology, 27*, 40–48.

Unterberger, C., Staples, K. J., Smallie, T., Williams, L., Foxwell, B., Schaefer, A., et al. (2008). Role of STAT3 in glucocorticoid-induced expression of the human IL-10 gene. *Molecular Immunology, 45*, 3230–3237.

Walter, M. R. (2014). The molecular basis of IL-10 function: from receptor structure to the onset of signaling. *Current Topics in Microbiology and Immunology, 380*, 191–212.

Interleukin 12

Tessa Dieckman[1,2], Antonie Zwiers[1], Georg Kraal[1] and Gerd Bouma[2]
[1]Department of Molecular Cell Biology, VU University Medical Center, Amsterdam, The Netherlands
[2]Immunology and the Department of Gastroenterology, VU University Medical Center, Amsterdam, The Netherlands

Synonyms

IL-12; IL-12p70; Interleukin-12

Definition

The cytokine IL-12 (IL-12p70) is a heterodimer, which consists of IL-12p35 (encoded by IL12A) and IL-12p40 (encoded by IL12B). It is produced by myeloid cells including dendritic cells and phagocytes (monocytes/macrophages and neutrophils) in response to a variety of stimuli (Trinchieri 2003). These include activation of Toll-like receptors (TLRs) and other receptors by pathogens (bacteria, fungi, intracellular parasites, and viruses), membrane-bound and soluble signals from activated T cells and natural killer (NK) cells, and components of the extracellular matrix. A variety of cell types are target for the IL-12 cytokine such as T lymphocytes, NK cells and NKT cells, hematopoietic precursor cells, and B cells. In these cells, IL-12 induces proliferation, enhancement of functional activity, such as cytotoxicity, and the production of cytokines, particularly type-1 cytokines such as IFN-γ. Because of its production which is induced after encountering bacterial and other pathogenic stimuli and its effect on T cell activation, interleukin 12 is seen as an important link between the innate and adaptive immune system, with a strong pro-inflammatory function.

IL-12 is a member of the IL-12 cytokine family, which is at the moment composed of four heterodimeric cytokines: interleukin-12 (IL-12), interleukin-23 (IL-23), interleukin-27 (IL-27), and interleukin-35 (IL-35) (Vignali and Kuchroo 2012). Both IL-12 and IL-23 are pro-inflammatory cytokines, whereby IL-23 is formed by a combination of IL-12p40 with the IL-23p19 chain (encoded by IL23A). The other two cytokines, interleukin-27 (IL-27) and interleukin-35 (IL-35), are characterized as more anti-inflammatory cytokines and consist of dimers of Epstein–Barr virus-induced molecule 3 (EBI3) and IL-27p28 (IL-27) and IL-12p35 and EBI3 forming the IL35 cytokine. The structural relationship of the cytokines of this family is reflected in a similar structural overlap in the composition of their dimeric receptors. The receptors for IL-12 and IL-23 share the interleukin-12-receptor-β1 (IL-12Rβ1) chain, in combination with a unique interleukin-12-receptor-β2 (IL-12Rβ2) and IL-23 receptor (IL23R) chain, respectively. The receptor for IL-27 consists of a combination of WSX-1 (IL-27R) and gp130, a component of several cytokine receptors, whereas IL-35 signals via a combination of the same gp130 and IL-12Rβ2. In addition, IL-35 can signal via homodimers of either gp130 or IL-12Rβ2 chains (Vignali and Kuchroo 2012).

Biosynthesis and Release

The most important cell types that are involved in the production of IL-12 are macrophages and dendritic cells (DCs). In addition neutrophilic granulocytes and B cells have been reported to produce IL-12. Both subunits of the dimeric molecule are to be produced in the same cell. Upon stimulation and induction, dendritic cells, macrophages, and monocytes produce the 40 kDa fragment of IL-12p70 in excess of the IL-12p35 35 kDa subunit. To be produced and secreted as intact dimeric molecule, the two IL-12 subunits undergo posttranslational modification and dimerization. The rate of synthesis of the intact cytokine is limited by the rate of IL-12p35 production, whereby both production and release are tightly regulated (Carra et al. 2000).

The p40 gene promoter in macrophages is normally configured in a nucleosome array. Upon activation of macrophages with TLR ligands, remodeling of the nucleosome makes the region accessible for the transcription factor CCAAT/enhancer-binding protein (C/EBP) (Albrecht et al. 2004). In addition, the p40 promoter contains several elements that are important for its expression such as NF-κB elements. A downstream C/EBP site cooperates with Rel proteins binding to the NF-κB site, and an E26 transformation-specific (ETS) consensus element upstream of the NF-κB site binds two members of the ETS family of transcriptional factors, ETS-2 and PU.1. These can complex with c-Rel and several members of the IRF family, including IRF-1, IRF-2, and IFN consensus sequence-binding protein (ICSB) or IRF-8 (Lyakh et al. 2008). Actors that are known to inhibit the production of IL-12 have been shown to affect one or more of these transcriptional factors. It has been demonstrated that the interaction of IL-12-producing cells with IL-4 and PGE2, which are both inhibitors of IL-12p40 expression, leads to enhanced GAP-12 binding to the GA-12 (GATA sequence in the IL-12 promoter) repressor site between the ETS and NF-κB sites.

Also for the IL-12p35 promoter region, multiple transcriptional motifs like Sp1, IFN-γ-response element (γ-IRE), PU.1, and C/EBP have been described. Interestingly four p35 mRNA isoforms have been described. Certain isoforms appear only after stimulation, which indicates that p35 gene expression is regulated by both transcriptional and translational mechanisms. It is assumed that the presence of several transcription initiation sites in the IL-12p35 promoter makes it possible to synthesize different instances of the p35 subunit and hence of different sets of dimers of IL-12/IL-23 depending on the cell types employing these alternative initiation sites (Lyakh et al. 2008).

The p40 and p35 subunits are processed via different pathways. Whereas during processing of p40 the signal peptide is removed together with translocation into the endoplasmic reticulum (ER), the processing of p35 does not conform to this cotranslational model of signal peptide removal. Instead, removal of the p35 signal peptide occurs via two sequential cleavages. The first cleavage takes place within the ER in the hydrophobic region of the signal peptide, which is not affected by glycosylation. A second cleavage, possibly involving a metalloprotease, removes the remaining portion of the p35 signal peptide. The secretion of p35 can be inherited by interfering with glycosylation, but p40 secretion is not affected by inhibition of glycosylation (Murphy et al. 2000). Together, diverse mechanisms for regulation of IL-12 production seem to exist with an important role for p35 processing in the control of IL-12 heterodimer production.

The generation of biologically active IL-12 is crucially dependent on signaling through Toll-like receptors (TLR) and will be augmented by the presence of secondary pro-inflammatory signaling. Within the populations of cells that produce IL-12, macrophages and dendritic cells seem to differ in their requirements for secondary signaling. Whereas for dendritic cells TLR ligation may be sufficient for IL-12 induction, in macrophages secondary, accessory signals are required. On the other hand, macrophages react to a wider variety of antigens. Secondary signals that enhance IL-12 production are IFN-γ, IL-4, IL-6, and TNF-α. The latter is able to upregulate CD40 on DCs, whereby IL-12 itself induces CD40L on adjacent T cells and NK cells. The CD40–CD40L interaction is a strong activator of IL-12 production, and through these, not necessarily antigen-specific, cell–cell interaction type I immune responses can be initiated. Since IFN-γ and TNF-α are products of type I activated T cells, these cytokines will also help to maintain the production of IL-12 (Carra et al. 2000; Lyakh et al. 2008; Vignali and Kuchroo 2012).

Several mechanisms exist to avoid IL-12 production and inappropriate immune responses and to downregulate ongoing IL-12 production. In mice a controlling role for IL-12p40 homodimers has been demonstrated. By antagonistic binding to the IL-12 receptor, such homodimers could play an inhibitory role. However, in male, the role of p40 dimers has not been demonstrated. Interestingly, a cytokine that plays an important role in the downregulation of ongoing IL-12-mediated immunity is IL-4. Although clearly involved in Th2 stimulation, this cytokine is able to promote IL-12 production in the presence of Th1 factors such as IFN-γ and TNF-α. When the production of these factors starts to wane, an opposite effect of IL-4 is seen, whereby the synthesis of IL-12 is inhibited and homeostasis will be restored. A comparable role has been attributed to tumor necrosis factor α (TNF-α) which, after release by T and NK cells upon IL-12 stimulation, can either stimulate or inhibit IL-12 production depending on the balance between signaling via TNF receptor and via CD40–CD40L.

Modulation of IL-12 production can also be exerted by the influence of the inhibitory cytokine IL-10, which affects the transcription of both p35 and p40 genes. In addition TGF-β is a negative regulator of IL-12 production (Trinchieri 2003; Vignali and Kuchroo 2012).

Biological Activity

IL-12 plays an important role in both innate immunity and acquired immunity, because it induces interferon-γ (IFN-γ) production by NK and NKT cells early in the immune response and is involved in the differentiation of Th1 cells, in which T cell subset plays a crucial role in the anti-infective and antitumor responses.

In addition to Th1 T cells, two other T cell helper subsets have been identified, Th2 and Th17, with IL-4 and IL-23 as primarily inducing cytokines, respectively (Zhu et al. 2010). The presence of the IL-12 cytokine favors IL-2 and IFN-γ production by Th1 cells and at the same time suppresses Th2 cell differentiation and functioning of its accessory interleukins IL-4, IL-5, and IL-10, as well as suppressing Th17 production by inhibiting the transcription factors, RORγt and RORα (Lyakh et al. 2008; Vignali and Kuchroo 2012).

In spite of its crucial role in Th1 immune response skewing, IL-12 does not lead to a direct proliferative effect on resting, naïve T cells, in concordance with the fact that naïve T cells express only limited amounts of the IL-12 receptor. When cells have been activated before, IL-12 can act directly, whereas with resting cells, an enhancing effect is seen when the cells are activated simultaneously with other factors like mitogens that are important to enhance the expression of the IL-12 receptor. In addition the microenvironment in which activation takes place, the mode of co-stimulation, the presence of low levels of TNF-α and IL-1, and the T cell receptor strength will influence the outcome of the Th differentiation.

The importance of IL-12 as an IFN-γ inducer lies not only in its high efficiency at low concentrations, but also in its synergy with many other activating stimuli, in particular in combination with other pro-inflammatory cytokines like IL-2, IL-25, and IL-18 (Akira 2000). Through the activation of T cells and NK cells leading to IFN-γ production, IL-12 has an important role in resistance to infections, particularly with bacteria and intracellular parasites. Its original role in chronic inflammation and organ-specific autoimmunity has become less clear, with the finding that many of the processes attributed to the influence of IL-12 were resulting from the IL-23/Th17 axis (Lyakh et al. 2008; Oppmann et al. 2000). So is IL-23 involved in tissue pathology through the induction of Th17 responses, and a role for IL-23 in chronic inflammation has been proposed based on the association of polymorphisms in the IL-23R locus with Crohn's disease, ulcerative colitis, celiac disease, multiple sclerosis, psoriasis, and ankylosing spondylitis (Zwiers et al. 2004; Strober and Fuss 2011; van Wanrooij et al. 2012).

In addition to IFN-γ, IL-12 induces T cells and NK cells to produce several other cytokines including granulocyte–macrophage colony-stimulating factor (GM-CSF) and TNF-α. The generation of cytotoxic T lymphocytes (CTLs) is enhanced by IL-12, and it augments the cytotoxic activity of CTLs and NK cells by inducing the transcription of genes that encode cytotoxic granule-associated molecules, such as perforin and granzymes. Furthermore, it upregulates the expression of adhesion molecules (Trinchieri 2003).

Effects of IL-12 on B cell proliferation and differentiation are likely indirect via the induction of Th1 that directly or via their products such as IFN-γ lead to production of particular immunoglobulin isotypes (enhanced IgG1 and decreased IgE).

Biological activities of IL-12 are mediated via binding to its receptor complex which is composed of the two subunit receptor chains IL-12Rβ1 and IL-12Rβ2, binding the IL-12 subunits IL-12p40 and IL-12p35, respectively. Genes for the receptor chains are located on chromosome 9 (β1) and 1 (β2). The receptor complex signals via tyrosine kinases of the Janus kinase (JAK) family, JAK2, and tyrosine kinase 2 (Tyk2). Receptor chain IL-12Rβ2 is the signaling chain: upon binding of the p35 and p40 subunit to the receptor complex, IL-12Rβ2 passes down its signal, which leads to transphosphorylation of JAK2 and Tyk2. Upon IL-12Rβ2 phosphorylation, the intracellular tyrosine residue 800 is activated, in turn recruiting and activating transcription factor STAT4 (signal transducer and activator of transcription 4). Subsequently, STAT4 homodimerizes will be transported into the nucleus where it modulates as transcription factor the expression of several genes, for example, IFN-γ and TNF-α. Not only STAT4 mediates cellular responses by IL-12, also other transcription factors like STAT1, STAT3, and STAT5 can be activated by IL-12 (Langrish et al. 2004).

The IL-12 receptor complex is expressed on T cells, NK cells, and DCs. Naïve T cells solely express receptor chain IL-12Rβ1 at low levels and upon activation; T cells will start to upregulate expression of both receptor chains, whereby a stronger induction of the IL-12Rβ2 expression is seen. Cytokines that lead to the upregulation of the receptor chain include IL-12 itself, IFN-γ, and type I interferon. Thus, IL-12 receptor expression is limited by IL-12Rβ2 expression. Resting NK cells and DCs already express both receptor chains.

Pathophysiological Relevance

IL-12 production is rapidly induced upon encounter of dendritic cells and macrophages with bacteria, viruses, and parasites through activation of the Toll-like receptors. In response to IL-12, NK cells are able to produce IFN-γ, which is an important activator of macrophage activity, enhancing their capacity to kill bacteria. This macrophage activation can occur independently of T cell activation and forms an important early type of innate immune response. Once T cells have been activated and the differentiation of Th1 T cells has taken place under the influence of IL-12, the cytokine is crucial for maintaining the Th1 response ensuring an effective antimicrobial resistance.

As mentioned above, the role of IL-12 in chronic inflammation has become less outspoken with the notion that in particular IL-23 is responsible for many of the phenomena that have originally been described for IL-12. This is in particular the case for induction of chronicity of inflammation seen in many types of inflammatory and autoimmune diseases and that seem to rely on the persistence of inflammatory Th17 cells induced and maintained by IL-23.

The activation of NK cells and the induction of IFN-γ make IL-12 a relevant cytokine in cancer therapy. NK cell and macrophage activation will be important for tumor eradication. Furthermore, the IFN-γ production induced by IL-12 has an anti-angiogenic effect, leading to reduction of vascular endothelial growth factor (VEGF) in tumors (Del Vechroo et al. 2007).

Modulation By Drugs

The use of therapeutic agents to interfere with the function of the IL-12 cytokine has primarily been focused on targeting the p40 subunit. Monoclonal antibodies directed against p40, e.g., ustekinumab, are now used for treatment of psoriasis and inflammatory bowel disease. Based on what is known of the IL-12–IL-23 functions and the overlapping role of the p40 chain, these approaches are targeted to interfere with IL-23 effects rather than those of IL-12. This is also the case for therapeutic reagents that react with the STAT signaling pathways downstream of the IL-12Rβ1 (Grivennikov et al. 2010).

Cross-References

▶ Allergic Disorders
▶ Antiviral Responses
▶ Atopic Dermatitis
▶ Autoinflammatory Syndromes
▶ Bacterial Lipopolysaccharide
▶ Cancer and Inflammation
▶ Cytokines
▶ Dendritic Cells
▶ Foxp3
▶ Genetic Susceptibility to Inflammatory Diseases
▶ Interferon gamma
▶ Interleukin 4 and the related cytokines (Interleukin 5 and Interleukin 13)
▶ Inflammatory Bowel Disease
▶ Inflammatory Bowel Disease Models in Animals
▶ Interleukin 6
▶ Interleukin 12

- Interleukin 17
- Interleukin 18
- Interleukin 23
- Interleukin 27
- Leukocyte Recruitment
- Lymphocyte Homing and Trafficking
- Natural Killer Cells
- Pathogen-Associated Molecular Patterns (PAMPs)
- Rheumatoid Arthritis
- TH17 Response
- Toll-Like Receptors
- Type I Interferons

References

Akira, S. (2000). The role of IL-18 in innate immunity. *Current Opinion in Immunology, 12*, 59–63.

Albrecht, I., Tapmeier, T., Zimmermann, S., Frey, M., Heeg, K., & Dalpke, A. (2004). Toll-like receptors differentially induce nucleosome remodelling at the IL-12p40 promoter. *EMBO Reports, 5*, 172–177.

Carra, G., Gerosa, F., & Trinchieri, G. (2000). Biosynthesis and posttranslational regulation of human IL-12. *Journal of Immunology, 164*, 4752–4761.

Del Vechroo, M., Bajetta, E., Canova, S., Lotze, M. T., Wesa, A., Parmiani, G., et al. (2007). Interleukin-12: Biological properties and clinical application. *Clinical Cancer Research, 13*, 4677–4685.

Grivennikov, S. I., Greten, F. R., & Karin, M. (2010). Immunity, inflammation, and cancer. *Cell, 140*, 883–899.

Langrish, C. L., McKenzie, B. S., Wilson, N. J., de Waal, M. R., Kastelein, R. A., & Cua, D. J. (2004). IL-12 and IL-23: Master regulators of innate and adaptive immunity. *Immunological Reviews, 202*, 96–105.

Lyakh, L., Trinchieri, G., Provezza, L., Carra, G., & Gerosa, F. (2008). Regulation of interleukin-12/interleukin-23 production and the T-helper 17 response in humans. *Immunological Reviews, 226*, 112–131.

Murphy, F. J., Hayes, M. P., & Burd, P. R. (2000). Disparate intracellular processing of human IL-12 preprotein subunits: Atypical processing of the P35 signal peptide. *Journal of Immunology, 164*, 839–847.

Oppmann, B., Lesley, R., Blom, B., Timans, J. C., Xu, Y., Hunte, B., et al. (2000). Novel p19 protein engages IL-12p40 to form a cytokine, IL-23, with biological activities similar as well as distinct from IL-12. *Immunity, 13*, 715–725.

Strober, W., & Fuss, I. J. (2011). Proinflammatory cytokines in the pathogenesis of inflammatory bowel diseases. *Gastroenterology, 140*, 1756–1767.

Trinchieri, G. (2003). Interleukin-12 and the regulation of innate resistance and adaptive immunity. *Nature Reviews. Immunology, 3*, 133–146.

van Wanrooij, R. L., Zwiers, A., Kraal, G., & Bouma, G. (2012). Genetic variations in interleukin-12 related genes in immune-mediated diseases. *Journal of Autoimmunity, 39*, 359–368.

Vignali, D. A., & Kuchroo, V. K. (2012). IL-12 family cytokines: Immunological playmakers. *Nature Immunology, 13*, 722–728.

Zhu, J., Yamane, H., & Paul, W. E. (2010). Differentiation of effector CD4 T cell populations. *Annual Review of Immunology, 28*, 445–489.

Zwiers, A., Seegers, D., Heijmans, R., Koch, A., Hampe, J., Nikolaus, S., et al. (2004). Definition of polymorphisms and haplotypes in the interleukin-12B gene: Association with IL-12 production but not with Crohn's disease. *Genes and Immunity, 5*, 675–677.

Interleukin-12/23 Inhibitors: Ustekinumab

Kevin D. Pile[1], Garry G. Graham[2,3] and Stephen M. Mahler[4]

[1]Campbelltown Hospital, School of Medicine, University of Western Sydney, Campbelltown, NSW, Australia

[2]Department of Pharmacology, School of Medical Sciences, University of New South Wales, Sydney, NSW, Australia

[3]Department of Clinical Pharmacology and Toxicology, St Vincent's Hospital, Sydney, NSW, Australia

[4]Australian Institute for Bioengineering and Nanotechnology, University of Queensland, Brisbane, QLD, Australia

Synonyms

Interleukin-12/23 blockers; Interleukin 12/23 inhibitors; Interleukin-12/23p40 inhibitors; Ustekinumab initially termed CNTO 1275

Definition

Inhibitors of interleukin-12/23 (IL-12/23) are compounds which bind to both IL-12 and IL-23 and inactivate them. The major IL-12/23 inhibitor is ustekinumab.

Interleukin-12/23 Inhibitors: Ustekinumab, Fig. 1 Antibody structure of ustekinumab, a humanized antibody with human IgG1 heavy (449 amino acid residues) and kappa light chain (214 amino acid residues) isotypes, respectively

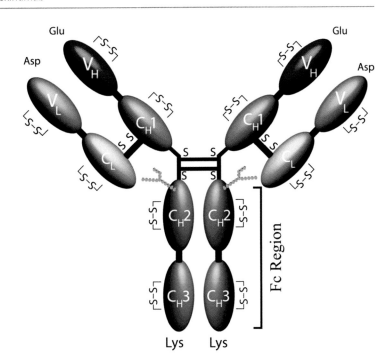

Chemical Structures and Properties

Ustekinumab is a fully humanized IgG1κ monoclonal antibody composed of an IgG1 heavy chain isotype and a kappa light chain isotype with an approximate molecular weight of 148,600 D (Fig. 1). It is expressed in Sp2/0 murine myeloma cell line using a protein-free, chemically defined cell culture medium and purified by a series of affinity and ion exchange chromatographic steps and viral inactivation steps. There are two identical heavy and light chains linked by covalent disulfide bonds (-S-S-) and non-covalent heavy-heavy and heavy-light chain interactions. Ustekinumab contains a single N-linked glycosylation site at the Asp 299 amino acid residue of each heavy chain. The major source of molecular weight and charge heterogeneity is due to posttranslational modifications caused by different glycoforms and partial loss of the C-terminal lysine.

The large molecular mass of ustekinumab prevents any passive diffusion through cell membranes and, like other pharmacologically active proteins, is not dialyzable.

Metabolism, Pharmacokinetics, and Dosage

Ustekinumab has an immunoglobulin-like structure and, like other drugs of this type, is slowly broken down by the mononuclear phagocytes in the reticuloendothelial system ($t_{1/2} \approx$ 21 days). Ustekinumab is administered subcutaneously, and, as is the case with other large proteins which are not administered intravenously, it is absorbed slowly with peak plasma concentrations achieved after about 8 days. Both factors, slow absorption and long half-life, indicate that its dosage should be infrequent and the present recommended dose is 45 mg at commencement of treatment and then at 4 weeks, but more occasional subsequent dosage (once every 12 weeks thereafter) appears to maintain benefit. Patients weighing over 100 kg are given 90 mg at the same times.

Pharmacological Effects

Ustekinumab is an antagonist of the inflammatory cytokines, IL-12 and IL-23. Ustekinumab binds to the p40 subunit of both IL-12 and IL-23 resulting in the inhibition of the inflammatory functions of both cytokines (Luo et al. 2010; Wada et al. 2012).

Ustekinumab was developed primarily for plaque psoriasis and psoriatic arthritis following the discovery of immunopathological mechanisms of psoriasis. However, recent research indicates that IL-23 alone is largely responsible for the development of psoriasis. Thus, targeting IL-23 may be a more satisfactory and specific treatment of psoriasis than inhibition of the activity of both IL-12 and IL-23 (Levin and Gottlieb 2014).

Clinical Indications and Efficacy

Ustekinumab is active in the treatment of plaque psoriasis (Croxtall 2011) where it maintains its benefit for a considerable time. Treatments for plaque psoriasis have included oral drugs such as methotrexate, cyclosporin, and retinoids. Phototherapy has also been used widely. These treatments may produce toxicity and do not always work.

Ustekinumab reduces the severity of the psoriatic arthritis including inflammation of peripheral joints and also the inflammation of whole fingers and toes (dactylitis) or where ligaments or tendons attach to bones (enthesitis) (Olivieri et al. 2014). Despite its activity in psoriatic arthritis, ustekinumab is regarded as a second-line drug for this arthritis after failure of conventional synthetic disease-modifying antirheumatic drugs (csDMARDs) such as methotrexate. At this stage, ustekinumab has not been compared with tumor necrosis factor (TNF) inhibitors in psoriatic arthritis (Weitz and Ritchlin 2014). However, largely on the basis of clinical trials in rheumatoid arthritis, TNF inhibitors are regarded as the first-line treatments for psoriatic arthritis and ustekinumab is presently reserved for patients in whom the TNF inhibitors have failed.

Ustekinumab induces a clinical response in a third of TNF resistant Crohn's disease but does not induce remission (Sandborn et al. 2012). There have been insufficient studies to date to make any definite conclusion about the activity of ustekinumab in rheumatoid arthritis. As is the case with CD20 inhibitors, ustekinumab has been tested for the treatment of multiple sclerosis but appears inactive.

Two other inhibitors of IL-12/IL-23 have been evaluated. Briakinumab (ABT-874), a fully human antibody, was active in the treatment of psoriasis but was withdrawn because of adverse effects (see "Adverse Effects" below). Briakinumab was also tested for therapy of multiple sclerosis but was inactive.

Adverse Effects

Several adverse effects occurred during treatment with briakinumab including serious infections, nonmelanoma skin cancers, and major cardiac events. It was withdrawn because of its adverse effects.

Ustekinumab has been generally well tolerated in trials up to 5 years, without reports of tuberculosis. Increased nasopharyngitis, headache, and arthralgia are reported.

Drug Interactions

No significant metabolic or pharmacokinetic interactions have been detected between ustekinumab and other drugs (Gupta et al. 2014).

Cross-References

▶ Corticosteroids
▶ Disease-Modifying Antirheumatic Drugs: Overview
▶ Methotrexate
▶ Rheumatoid Arthritis
▶ Spondyloarthritis
▶ Tumor Necrosis Factor Alpha (TNFalpha)
▶ Tumor Necrosis Factor (TNF) Inhibitors

References

Croxtall, J. D. (2011). Ustekinumab. A review of its use in the management of moderate to severe plaque psoriasis. *Drugs, 71*(13), 1733–1753.
Gupta, R., Levin, E., Wu, J. J., Koo, J., & Liao, W. (2014). An update on drug-drug interactions with biologics for

the treatment of moderate-to-severe psoriasis. *Journal of Dermatological Treatment, 25*(1), 87–89.

Levin, A. A., & Gottlieb, A. B. (2014). Specific targeting of interleukin-23p19 as effective treatment for psoriasis. *Journal of the American Academy of Dermatology, 70*, 555–561.

Luo, J., Wu, S. J., Lacy, E. R., Orlovsky, Y., Baker, A., Teplyakov, A., et al. (2010). Structural basis for the dual recognition of IL-12 and IL-23 by ustekinumab. *Journal of Molecular Biology, 402*(5), 797–812.

Olivieri, I., D'Angelo, S., Palazzi, C., & Padula, A. (2014). Advances in the management of psoriatic arthritis. *Nature Reviews. Rheumatology, 10*(9), 531–542.

Sandborn, W. J., Gasink, C., Gao, L. L., Blank, M. A., Johanns, J., Guzzo, C., et al. (2012). Ustekinumab induction and maintenance therapy in refractory Crohn's disease. *New England Journal of Medicine, 367*(16), 1519–1528.

Wada, Y., Cardinale, I., Khatcherian, A., Chu, J., Kantor, A. B., Gottlieb, A. B., et al. (2012). Apilimod inhibits the production of IL-12 and IL-23 and reduces dendritic cell infiltration in psoriasis. *PLoS One, 7*(4), e35069.

Weitz, J. E., & Ritchlin, C. T. (2014). Ustekinumab: Targeting the IL-17 pathway to improve outcomes in psoriatic arthritis. *Expert Opinion on Biological Therapy, 14*(4), 515–526.

Interleukin 17

Christian D. Sadik
Department of Dermatology, Allergy, and Venereology, University of Lübeck, Lübeck, Germany

Synonyms

IL-17/IL-17A/CTLA-8

Definition

"Interleukin-17" (IL-17A) is a commonly used colloquial term referring to the cytokine IL-17A. IL-17A is the defining cytokine of the IL-17 cytokine family, which is constituted by the six cytokines IL-17A, IL-17B, IL-17C, IL-17D, IL-17E, and Il-17F. All IL-17 cytokines consist of two single glycoprotein chains forming homo- or heterodimers linked via disulfide bonds. The complete IL-17A dimer is composited by of 155 amino acids and weighs 35 kDa. In addition to IL-17A and IL-17F homodimers, IL-17A and IL-17F chains can also combine to IL-17A/F heterodimers, which are biologically active (Wright et al. 2007).

IL-17A was the first member of the IL-17 cytokine family described. Its sequence was discovered in 1993 in hybridoma cells emerging from fusion of murine cytotoxic T cells with a rat T lymphoma cell line, which had been stimulated with combination of PMA and ionomycin (Rouvier et al. 1993). The product had first been designated as *cytotoxic T lymphocyte antigen 8* (CTLA-8) and was renamed "IL-17A" when identified as cytokine in 1995 (Yao et al. 1995a, b). Later, homologues of IL-17A, now designated as IL-17B, IL-17C, IL-17D, and IL-17E, were identified.

In humans, the genes for IL-17A and IL-17F are sited on chromosome 5q and show close structure similarities. The genes for IL-17C, IL-17D, and IL-17E are found on chromosome 16q, 13q, and 14q, respectively. The members of the IL-17 cytokine family show homology of 16–55 % to each other. Herein, IL-17A and IL-17F exhibit the highest homology (55 %) (Wright et al. 2007) and IL-17A and IL-17E the lowest (16 %). All IL-17 family cytokines feature five spatially highly conserved cysteinyl residues, which are instrumental for dimerization by forming cystine-knot structures (disulfide bonds) between the two monomer chains.

Homologues of IL-17 cytokines have been identified in a diverse array of life-forms, including sea lamprey, rainbow trout, zebrafish, and *Caenorhabditis elegans*. Additionally, IL-17A displays 58 % homology to an open reading frame in herpes virus saimiri, which is a T cell-tropic virus (Rouvier et al. 1993). The significance of this unusual homology, however, is still unclear. IL-17A is major orchestrator of host defense and, for instance, promotes neutrophil recruitment into peripheral tissues in bacterial and fungal infections.

Biosynthesis and Release

IL-17A production and release occur in T_H17 cells, $\gamma\delta$ T cells, $CD8^+$ T cells, natural killer

(NK) cells, natural killer T (NKT) cells, lymphoid tissue inducer-like (LTi) cells, macrophages, dendritic cells, neutrophils, mast cells, and Paneth cells. T_H17 cells are in general considered the most abundant source for IL-17A, although in certain (patho-)physiological settings other cell lineages may be the more important source.

Given the pivotal role of IL-17A in inflammation, much research efforts were made in the last decade to characterize T_H17 cells and to delineate the mechanisms of their differentiation. In addition to IL-17A, T_H17 cells produce a unique set of cytokines including IL-17F, IL-22, and CCL20. T_H17 cells are $CD4^+$ and differentiate upon simultaneous stimulation with TGF-β and IL-6. This cell population is further expanded by IL-23, IL-21, and IL-1β. Recently, the existence of two different T_H17 subpopulations, a pathogenic and a nonpathogenic one, has been uncovered, and it has been suggested that IL-23 is responsible for inducing the differentiation specifically of pathogenic T_H17 cells by a mechanisms including the paracrine/autocrine actions of TGF-β3 on T_H17 cells. Intracellularly, T_H17 cell differentiation is regulated by the transcription factors *signal transducer and activator of transcription 3* (STAT3), *retinoic acid-receptor-related orphan receptor γt* (RORγt), *IFN regulatory factor 4* (IRF4), *aryl hydrocarbon receptor* (ARH), *basic leucine zipper transcription factor ATF-like* (BATF), and *runt-related transcription factor 1* (Runx1). RORγt is almost specific for T_H17 cells and is the signature transcription factor of this $CD4^+$ T cell subset. The importance of STAT3 for T_H17 differentiation is underscored by the inability of individuals suffering from hyper-IgE ("Job's") syndrome, which is caused by mutations in the *stat3* gene, to produce functional T_H17 cells (Milner et al. 2008). The consequently impaired biosynthesis of sufficient amounts of IL-17A is likely partially responsible for the high susceptibility of Job's syndrome patients to recurrent infections.

The molecular mechanisms governing IL-17A production in other cell lineages are only known in less detail. Alveolar and peritoneal macrophages, for instance, release IL-17A and IL-17F in a Toll-like receptor 4- (TLR4)-dependent fashion. This effect was specific for TLR4 and was not shared by other Toll-like receptors.

Biological Activities

The IL-17 cytokine family binds to different receptor complexes of the IL-17R family, which is constituted by the five receptor subunits IL-17RA, IL-17RB, IL-17RC, IL-17RD, and IL-17RE. Receptors for IL-17 cytokine family members commonly form heteromeric receptor complexes consisting of the receptor chain IL-17RA combined with one of the receptor chain IL-17RB, IL-17RC, IL-17RD, or IL-17RE (Miossec and Kolls 2012). For signaling, both chains of the receptor complex are required. The receptors vastly differ in their structure and their subsequent cell signal from other cytokine receptor families. Their subunits commonly all possess a single transmembrane domain, an extracellular-fibronectin III-like domain as well as an intracellular SEF/IL-17R (SEFIR) domain (Gaffen 2009). The SEFIR domain shows some structural similarities to the Toll-/IL-1R (TIR) domains found Toll-like receptors.

IL-17RA is expressed ubiquitously but is found in highest quantity on hematopoietic cells. Conversely, IL-17RC is highly expressed on non-hematopoietic cells, and its expression is only low on hematopoietic cells. This pattern of distribution also determines the target cells of IL-17A and IL-17F and restricts their biological effects mainly to non-hematopoietic cells, particularly epithelial cells and fibroblasts. Homodimers and heterodimers of IL-17A and IL-17F all bind to the IL-17RA/IL-17RC receptor complex (Kuestner et al. 2007).

IL-17RA/IL-17RB binds IL-17B and IL-17E. IL-17E is considered to promote T_H2 cell immunity, among others, in fighting infections with the parasite *Trichuris muris* and the helminth *Nocardia brasiliensis*. Furthermore, it may play a role in atopic diseases, including asthma and atopic dermatitis.

Differences in the roles of IL-17A and IL-17F homodimers as well as of the IL-17A/IL-17F heterodimers have been suggest but are still poorly understood. They differ at least in their strength of signaling at the IL-17RA/IL-17RC receptor complex with IL-17F quantitatively activating the downstream of IL-17RA/IL-17RC receptor complex engagement 10–30-fold weaker than IL-17A. The strength of activation mediated by the IL-17A/IL-17F heterodimer is somewhere in-between of that of IL-17A and IL-17F. Although previously thought to exert vastly overlapping effects in vivo, later investigations suggested that IL-17A and IL-17F may also exert distinct, partially opposite effects in vivo. In DSS-induced colitis mouse models, for instance, IL-17F aggravated diseases, while IL-17A, in contrast, was protective (Yang et al. 2008).

Engagement of the IL-17RA/IL-17RC receptor complex activates recruitment of the adaptor molecule and E3 ubiquitin ligase Act1 (also named *Connection to IKK and SAPK/JNK*; CIKS) to the receptor complex. Act1 is indispensable for all actions of IL-17A (Fig. 1). Subsequently, TRAF6, another adaptor molecule and E3 ubiquitin ligase, associates with Act1 and is Lys-ubiquitinated by Act1. These two proteins are joined by *TGFβ-activated kinase* (TAK). This protein complex initiates degradation of IκBα, consequently activating NF-κB. Additionally MAP kinase pathways are activated as well as the transcription factors *CCAAT/enhancer binding proteins* C/EBPβ and C/EBPδ, which regulate, for instance, transcription of IL-6 mRNA (Fig. 1).

The signal transduction and function of IL-17B, IL-17C, and IL-17D are less well defined. IL-17E activates Act1, NFATc1, and JunB via IL-17RB, inducing IL-4 release from T_H2 cells (Angkasekwinai et al. 2007). The ligand for IL-17RD has not been defined, although it has been highly conserved with homologues found in frogs, sea lamprey, and *C. elegans* (Pancer et al. 2004). Knowledge on IL-17RE is very scarce.

IL-17A and IL-17 F often synergistically interact with TNF-α, IL-1β, and IL-22. Furthermore, IL-17A acts in concert with G-SCF to promote granulopoiesis (von Vietinghoff and Ley 2009). A broad spectrum of proinflammatory genes is subject to direct modulation by IL-17A. The first genes described modulated in their expression by IL-17A were IL-6 and IL-8/CXCL8. Other important IL-17A target genes are the neutrophil-active chemokines CXCL1, CXCL2, CXCL5, and CXCL6. Regulation of these genes supposedly underlies the pivotal neutrophil recruiting effect of IL-17A in vivo. Other IL-17A target genes are

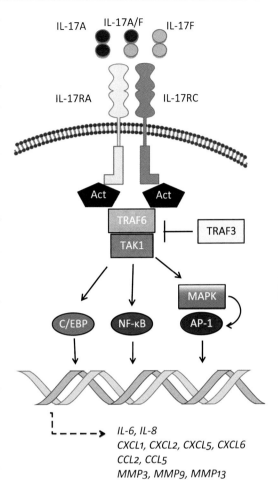

Interleukin 17, Fig. 1 Signal transduction of the IL-17RA/IL-17RC receptor complex. The receptor complex is activated by IL-17A or IL-17F homodimers or, alternatively, by IL-17A/F heterodimers. Engagement of the IL-17RA/IL-17RC receptor complex initiates recruitment of the adaptor molecules Act1 and TRAF6 as well as the *TGFβ-activated kinase (TAK)* to the receptor complex. This protein complex in turn mediates the activation of the transcription factors NF-κB, C/EBP, and AP-1, which modulate gene expression of diverse genes, particularly upregulation of proinflammatory cytokines and chemokines

the CXC chemokines CXCL9, CXCL10, and CXCL11, which in general are involved in the regulation of T cell migration. Another target is the homeostatic chemokine CXCL12. IL-17A also regulates the expression of some CC chemokines, such as CCL2, CCL5, CCL7, and CCL11. Remarkable is also the induction of CCL20 expression, the only ligand of CCR6, by IL-17A, as this chemokine is crucially involved in T_H17 recruitment into peripheral tissues. Other important genes enhanced in their expression by IL-17A are NOS2, PTGS2, and mPGES-1. This way IL-17A enhances nitric oxide and prostaglandin release. Additionally, IL-17A can upregulate several matrix metalloproteases, including MMP3, MMP9, and MMP13, the complement factors C3 and factor B, as well as the antimicrobial proteins S100A7, S100A8, and S100A9 (Benedetti and Miossec 2014).

Pathophysiological Relevance

IL-17 family cytokines, particularly, IL-17A, are considered to play a major role in host defense. At the same time particularly IL-17A has been implicated in the pathogenesis of a diverse panel of inflammatory diseases, including rheumatoid arthritis, psoriasis, psoriasis arthritis, contact hypersensitivity, atherosclerosis, asthma, chronic obstructive pulmonary disease (COPD), ulcerative colitis, Crohn's disease, multiple sclerosis, autoimmune uveitis, type I diabetes, and systemic lupus erythematosus (SLE) (Sadik et al. 2011).

The role of IL-17A has been investigated in most detail for rheumatoid arthritis. Data from of different rodent models of arthritis suggest a crucial role for IL-17A at several points of the pathogenesis of the disease (Doodes et al. 2008; Koenders et al. 2005; Lubberts et al. 2001; Nakae et al. 2003a, b; Sadik et al. 2011). IL-17A levels are elevated in the synovium of rheumatoid arthritis patients, and high levels of IL-17A in the joint correlated with a poor outcome of disease. IL-17A activates diverse cell types within the joint, including fibroblast-like synoviocytes, endothelial cells, and osteoclasts, inducing the expression of chemokines, matrix metalloproteases, and RANKL, thus promoting the recruitment of immune cells into the joint and inducing bone and cartilage erosions. The immune cells recruited into the joint in turn release additional IL-17A this way closing a vicious circle progressing arthritis into its chronic phase (Benedetti and Miossec 2014). IL-17A also contributes the progression of arthritis by promoting synovial fibroblast hyperproliferation and the formation of ectopic follicular structures in arthritic joints (Toh et al. 2010; Zhang et al. 2009). Furthermore, it has been shown that the number of T_H17 cells and consequently the production of IL-17A are modulated by the composition of the bacterial gut flora and are critical for the break of the immunotolerance and the subsequent production of autoantibodies in the K/BxN mouse model of rheumatoid arthritis (Wu et al. 2010). The relevance for the pathogenesis of disease in humans, however, has not been determined yet but suggests that IL-17A is crucial for the development of autoimmunity and is part of an intricate network between humans – their bacterial microflora – and autoimmune disease.

Multiple lines of evidence also suggest a major role for IL-17A in psoriasis vulgaris. In psoriatic skin lesions, expression of both IL-17A and IL-17F is elevated (Wilson et al. 2007), and polymorphisms in the IL-17 receptor-specific and essential signaling molecule Act1 are a risk factor for psoriasis (Wang et al. 2013). IL-17A is required for skin inflammation in several mouse models of psoriasis, including the Aldara-induced psoriasiform dermatitis model as well as in IL-23-induced psoriasiform dermatitis (Rizzo et al. 2011; van der Fits et al. 2009). Additionally, the therapeutic efficiency of IL-23 inhibition in psoriasis lends further support for a crucial role of IL-17A in the pathogenesis of psoriasis because IL-23 is a major driver of IL-17A production.

Similarly, both human and murine data strongly indicate a role for Il-17A in multiple sclerosis. T_H17 cells, for instance, are enriched in active lesions of multiple sclerosis (Tzartos et al. 2008), and disease in experimental autoimmune encephalomyelitis (EAE), an animal model for multiple sclerosis, is attenuated with deficiencies in IL-17A and its up- and downstream regulatory molecules.

On the other hand, IL-17A has been suggested to restrict tissue destruction under rare, specific conditions. Thus, IL-17A was required in the sensitization phase of mouse models of allergic asthma to establish disease but conversely counteracted disease in the effector phase reducing T_H2 cytokine release and consequently chemokine expression and eosinophil influx into the lung (Schnyder-Candrian et al. 2006). Similarly, in models of graft-vs.-host disease (GvHD) and colitis, IL-17A exerted protective effects (Yang et al. 2008; Yi et al. 2008).

Modulation by Drugs

With IL-17 family cytokines, particularly IL-17A, crucially involved in the pathogenesis of various diseases, much effort has been put into the development of therapeutics targeting IL-17A and its effects. To this end, several approaches have been developed, including neutralization of IL-17A by antibodies, direct inhibition of the IL-17RA/IL-17RC receptor complex as well as inhibition of IL-17 receptor signaling. Other strategies seek to indirectly curb IL-17A by inhibiting T_H17 cell differentiation or activity.

Up to date (March 2014), clinical trials well document significant beneficial effects of IL-17A inhibition in psoriasis, psoriasis arthritis, and ankylosing spondylitis. Likewise, a significant improvement of multiple sclerosis upon inhibition of IL-17A has also been reported in a small clinical study. Although IL-17A inhibition in general improves rheumatoid arthritis, beneficial effects of IL-17A inhibition are minor and are not superior to already established treatment strategies. In Crohn's disease, inhibition of IL-17A aggravated disease.

The most advanced drugs (as of March 2014) targeting the IL-17 cytokine family are ixekizumab (LY2439821), secukinumab (AIN457), and brodalumab (AMG 827). Ixekizumab and secukinumab are both neutralizing antibodies directed to IL-17A. They are humanized and fully human antibodies, respectively. Brodalumab, in contrast, is an antibody directed to the IL-17RA receptor chain this way inhibiting receptor activation presumably of all IL-17 cytokine family receptors. Brodalumab is therefore expected to exert a broader spectrum of pharmacological effects than ixekizumab and secukinumab.

Secukinumab was the first of the antibodies tested in humans. First testing was conducted in psoriasis patients. After, as a proof of principle, the antibody appeared effective in ameliorating psoriasis, it was subsequently tested in patients suffering from rheumatoid arthritis or from uveitis. In rheumatoid arthritis patients, it slightly reduced disease activity by at least 20 % in 46 % of all patients included in the study. Later studies also suggested effectivity of secukinumab in ankylosing spondylitis. Similarly, ixekizumab was effective in attenuating rheumatoid arthritis and psoriasis. Also, inhibiting IL-17RA proved efficient in treating inflammatory diseases. Thus, it was shown to clear psoriasis completely within 12 weeks in 72 % of patients receiving the drug.

Several key players of T_H17 cell differentiation and proliferation, including IL-6, IL-23, and IL-1, are molecular target structures of licensed, commonly used drugs, whose therapeutic effects may be consequently partially mediated by modulation of IL-17A and IL-17F levels. These drugs include the IL-6 receptor inhibitor tocilizumab (Actemra®), which is commonly used in the treatment of rheumatoid arthritis, the soluble IL-1 receptor anakinra (Kineret®) as well as the neutralizing anti-IL-1 antibody canakinumab (Ilaris®). Most important are ustekinumab (Stelara®) and briakinumab, which target the common p40 cytokine chain of IL-23 and IL-12. Ustekinumab is one of the most potent drugs used in the treatment of psoriasis as well as of psoriasis arthritis. Thus, inhibition upstream of IL-17A is already part of our therapeutic everyday armamentarium. Another potential therapeutic strategy pursuit is to block IL-17 signaling, for instance, at the level of Act1. The development of such therapeutic approaches, however, is still in an early stage.

Cross-References

▶ Antibacterial Host Defense Peptides
▶ Chemokine CCL14
▶ Chemokine CCL15
▶ Chemokine CCL18
▶ Cytokines
▶ IkappaB
▶ Interleukin 22
▶ Interleukin 6
▶ Leukocyte Recruitment
▶ Lymphocyte Homing and Trafficking
▶ Mast Cells
▶ Nuclear Receptor Signaling in the Control of Inflammation
▶ Rheumatoid Arthritis
▶ Skin Inflammation Models in Animals
▶ TGF beta Superfamily Cytokine MIC-1/GDF15 in Health and Inflammatory Diseases
▶ TH17 Response
▶ Tumor Necrosis Factor Alpha (TNFalpha)
▶ Tumor Necrosis Factor (TNF) Inhibitors

References

Angkasekwinai, P., Park, H., Wang, Y.H., Wang, Y.H., Chang, S.H., Corry, D.B., et al. (2007). Interleukin 25 promotes the initiation of proallergic type 2 responses. *Journal Experimental Medicine, 204*(7), 1509–1517. doi:10.1084/jem.20061675.

Benedetti, G., & Miossec, P. (2014). Interleukin 17 contributes to the chronicity of inflammatory diseases such as rheumatoid arthritis. *European Journal of Immunology, 44*(2), 339–347. doi:10.1002/eji.201344184.

Doodes, P. D., Cao, Y., Hamel, K. M., Wang, Y., Farkas, B., Iwakura, Y., et al. (2008). Development of proteoglycan-induced arthritis is independent of IL-17. *Journal of Immunology, 181*(1), 329–337.

Gaffen, S. L. (2009). Structure and signalling in the IL-17 receptor family. *Nature Reviews Immunology, 9*(8), 556–567. doi:10.1038/nri2586.

Koenders, M.I., Kolls, J.K., Oppers-Walgreen, B., van den Bersselaar, L., Joosten, L.A., Schurr, J.R., et al. (2005). Interleukin-17 receptor deficiency results in impaired synovial expression of interleukin-1 and matrix metalloproteinases 3, 9, and 13 and prevents cartilage destruction during chronic reactivated streptococcal cell wall-induced arthritis. *Arthritis and Rheumatology, 52*(10), 3239–3247.

Kuestner, R.E., Taft, D.W., Haran, A., Brandt, C.S., Brender, T., Lum, K., et al. (2007). Identification of the IL-17 receptor related molecule IL-17RC as the receptor for IL-17F. *Journal of Immunology, 179*(8), 5462–5473.

Lubberts, E., Joosten, L. A., Oppers, B., van den Bersselaar L., Coenen-de Roo, C. J., Kolls, J. K., et al. (2001). IL-1-independent role of IL-17 in synovial inflammation and joint destruction during collagen-induced arthritis. *Journal of Immunology, 167*(2), 1004–1013.

Milner, J. D., Brenchley, J. M., Laurence, A., Freeman, A. F., Hill, B. J., Elias, K. M., et al. (2008). Impaired T (H)17 cell differentiation in subjects with autosomal dominant hyper-IgE syndrome. *Nature, 452*(7188), 773–776. doi:10.1038/nature06764.

Miossec, P., & Kolls, J. K. (2012). Targeting IL-17 and TH17 cells in chronic inflammation. *Nature Reviews Drug Discovery, 11*(10), 763–776. doi:10.1038/nrd3794.

Nakae, S., Nambu, A., Sudo, K., & Iwakura, Y. (2003a). Suppression of immune induction of collagen-induced arthritis in IL-17-deficient mice. *Journal of Immunology, 171*(11), 6173–6177.

Nakae, S., Saijo, S., Horai, R., Sudo, K., Mori, S., & Iwakura, Y. (2003b). IL-17 production from activated T cells is required for the spontaneous development of destructive arthritis in mice deficient in IL-1 receptor antagonist. *Proceedings of the National Academy of Sciences of the United States of America, 100*(10), 5986–5990.

Pancer, Z., Mayer, W. E., Klein, J., & Cooper, M. D. (2004). Prototypic T cell receptor and CD4-like coreceptor are expressed by lymphocytes in the agnathan sea lamprey. *Proceedings of the National Academy of Sciences of the United States of America, 101*(36), 13273–13278. doi:10.1073/pnas.0405529101.

Rizzo, H. L., Kagami, S., Phillips, K. G., Kurtz, S. E., Jacques, S. L., & Blauvelt, A. (2011). IL-23-mediated psoriasis-like epidermal hyperplasia is dependent on IL-17A. *Journal of Immunology, 186*(3), 1495–1502. doi:10.4049/jimmunol.1001001.

Rouvier, E., Luciani, M. F., Mattei, M. G., Denizot, F., & Golstein, P. (1993). CTLA-8, cloned from an activated T cell, bearing AU-rich messenger RNA instability sequences, and homologous to a herpesvirus saimiri gene. *Journal of Immunology, 150*(12), 5445–5456.

Sadik, C. D., Kim, N. D., Alekseeva, E., & Luster, A. D. (2011). IL-17RA signaling amplifies antibody-induced arthritis. *PLoS One, 6*(10), e26342. doi:10.1371/journal.pone.0026342.

Schnyder-Candrian, S., Togbe, D., Couillin, I., Mercier, I., Brombacher, F., Quesniaux, V., et al. (2006). Interleukin-17 is a negative regulator of established allergic asthma. *Journal of Experimental Medicine, 203*(12), 2715–2725. doi:10.1084/jem.20061401.

Toh, M.L., Gonzales, G., Koenders, M.I., Tournadre, A., Boyle, D., Lubberts, E., et al. (2010). Role of interleukin 17 in arthritis chronicity through survival of synoviocytes via regulation of synoviolin expression. *PLoS One, 5*(10), e13416. doi:10.1371/journal.pone.0013416.

Tzartos, J. S., Friese, M. A., Craner, M. J., Palace, J., Newcombe, J., Esiri, M. M., et al. (2008). Interleukin-17 production in central nervous system-infiltrating T cells and glial cells is associated with active disease in multiple sclerosis. *American Journal of Pathology, 172*(1), 146–155. doi:10.2353/ajpath.2008.070690.

van der Fits, L., Mourits, S., Voerman, J.S., Kant, M., Boon, L., Laman, J.D., et al. (2009). Imiquimod-induced psoriasis-like skin inflammation in mice is mediated via the IL-23/IL-17 axis. *Journal of Immunology, 182*(9), 5836–5845. doi:10.4049/jimmunol.0802999.

von Vietinghoff, S., & Ley, K. (2009). IL-17A controls IL-17F production and maintains blood neutrophil counts in mice. *Journal of Immunology, 183*(2), 865–873. doi:10.4049/jimmunol.0804080.

Wang, C., Wu, L., Bulek, K., Martin, B. N., Zepp, J. A., Kang, Z., et al. (2013). The psoriasis-associated D10N variant of the adaptor Act1 with impaired regulation by the molecular chaperone hsp90. *Nat Immunol, 14*(1), 72-81. doi:10.1038/ni.2479

Wilson, N.J., Boniface, K., Chan, J.R., McKenzie, B.S., Blumenschein, W.M., Mattson, J.D., et al. (2007). Development, cytokine profile and function of human interleukin 17-producing helper T cells. *Nature Immunology, 8*(9), 950–957. doi:10.1038/ni1497.

Wright, J.F., Guo, Y., Quazi, A., Luxenberg, D.P., Bennett, F., Ross, J.F., et al. (2007). Identification of an interleukin 17F/17A heterodimer in activated human CD4+ T cells. *Journal of Biological Chemistry, 282*(18), 13447-13455. doi:10.1074/jbc.M700499200.

Wu, H.J., Ivanov, II, Darce, J., Hattori, K., Shima, T., Umesaki, Y., et al. (2010). Gut-residing segmented filamentous bacteria drive autoimmune arthritis via T helper 17 cells. *Immunity, 32*(6), 815-827. doi:10.1016/j.immuni.2010.06.001.

Yang, X.O., Chang, S.H., Park, H., Nurieva, R., Shah, B., Acero, L., et al. (2008). Regulation of inflammatory responses by IL-17F. *Journal of Experimental Medicine, 205*(5), 1063–1075. doi:10.1084/jem.20071978.

Yao, Z., Fanslow, W.C., Seldin, M.F., Rousseau, A.M., Painter, S.L., Comeau, M.R., et al. (1995a). Herpesvirus Saimiri encodes a new cytokine, IL-17, which binds to a novel cytokine receptor. *Immunity, 3*(6), 811–821.

Yao, Z., Painter, S. L., Fanslow, W. C., Ulrich, D., Macduff, B. M., Spriggs, M. K., et al. (1995b). Human IL-17: A novel cytokine derived from T cells. *Journal of Immunology, 155*(12), 5483–5486.

Yi, T., Zhao, D., Lin, C.L., Zhang, C., Chen, Y., Todorov, I., et al. (2008). Absence of donor Th17 leads to augmented Th1 differentiation and exacerbated acute graft-versus-host disease. *Blood, 112*(5), 2101–2110. doi:10.1182/blood-2007-12-126987.

Zhang, Q., Wu, J., Cao, Q., Xiao, L., Wang, L., He, D., et al. (2009). A critical role of Cyr61 in interleukin-17-dependent proliferation of fibroblast-like synoviocytes in rheumatoid arthritis. *Arthritis and Rheumatology, 60*(12), 3602–3612. doi:10.1002/art.24999.

Interleukin 18

Paola Bossù and Ilaria Palladino
Department of Clinical and Behavioral Neurology, IRCCS Fondazione Santa Lucia, Rome, Italy

Synonyms

Interferon-gamma-inducing factor (IGIF); Interleukin-1 family member 4 (IL-1F4); Interleukin-1 gamma (IL-1γ)

Definition

The pro-inflammatory cytokine interleukin (IL)-18 is a pleiotropic molecule broadly expressed throughout the body. It was cloned in 1995 from lipopolysaccharide-triggered liver cells obtained from *Propionibacterium acnes*-treated mice (Okamura et al. 1995). Initially identified as a factor able to stimulate immune cells to release interferon gamma (IFN-γ) and thus named IFN-γ-inducing factor (IGIF), IL-18 has been included into the interleukin-1 superfamily (IL-1F) on the basis of its several structural and functional homologies with IL-1β, the primary member of the family (Sims et al. 2001). IL-18 shares many biological properties with IL-1β. However, the latter, which is not constitutively expressed in healthy conditions, holds more wide inflammatory properties, with broad effects also on nonimmune cells, and it is central in fever induction. At variance, IL-18 does not induce fever, neither affects cyclooxygenase-2 nor prostaglandin E2 production, and its precursor is normally expressed in many cell types. In addition, IL-18 uniquely activates the release of IFN-γ from T lymphocytes and natural killer cells and induces FasL expression. In general, the complex biological properties of IL-18, which mainly lead to the amplification of inflammation and modulation of innate and acquired immune responses, go from host defense to tissue damage, sometimes with causative role in diseases, especially when the cytokine is released in unrestrained amounts, as in sepsis, autoimmunity, chronic inflammatory diseases, and neurodegeneration. Because of its different properties, IL-18 is under study for therapeutic advance either as a tool to promote immune response or, more frequently, as a target to reduce inflammation in diseases characterized by a prominent inflammatory component.

Biosynthesis and Release

IL-18 is a member of the IL-1 family (IL-1F), on the basis of its similarities with the master cytokine IL-1β as for molecular structure, processing mechanisms, receptor complex organization, and signal transduction pathways. In fact, despite the limited sequence identity between the two molecules (17 %), IL-18 folds similarly to IL-1β into a β-trefoil structure formed by the arrangement of antiparallel β-strands (Kato et al. 2003), but with length and conformation of segments between the strands considerably different among the two cytokines. Furthermore, likewise IL-1β, the IL-18 protein lacks classical secretory signal peptides, and it is expressed as a precursor molecule that typically requires secondary proteolytic cleavage to be secreted and activated. More specifically, the human IL-18 gene, which is different from the genes of most of the other IL-1F members that are clustered on human chromosome 2, is located on chromosome 11q22.2–q23.3 and encodes a leaderless and biologically inactive precursor with a molecular weight of 24 kDa (pro-IL-18), which is normally present in the cytosol of a large variety of cells of both immune and nonimmune type, such as macrophages, monocytes, dendritic cells, microglia, Kupffer cells, intestinal and epithelial cells, fibroblasts, keratinocytes, and osteoblasts, as well as adrenal cortex cells, astrocytes, and neurons. The IL-18 precursor molecule is cleaved into its 18 kDa biologically active mature form mainly by the intracellular enzyme called IL-1β-converting enzyme (ICE), also known as caspase-1 (Gu et al. 1997). However, in specific conditions, the pro-IL-18 processing can also occur in the absence of caspase-1, by means of other caspases or even neutrophil- and macrophage-derived serine proteases (Sugawara et al. 2001). Notably, caspase-1 itself is synthesized as an inactive pro-form (pro-caspase-1) that requires either proteolytic cleavage to become activated. Its activation occurs within particular cytoplasmic multiprotein complexes, the so-called inflammasomes, which are proteolytic platforms containing sensor proteins belonging to the family of nucleotide-binding oligomerization domain-like receptors (NLRs). Different inflammasome complexes have been so far identified, mainly defined by the NLR protein that they contain. On the whole, in response to pathogen- or damage-associated molecular patterns (PAMPs or DAMPs), such as exogenous molecules from microbes, exogenous nonmicrobial particulates, and endogenous danger signals, including ATP, amyloid-β, and sodium urate crystals, NLR proteins assemble, activate their effectors, and lead to IL-18 processing (Bauernfeind et al. 2011). Even though accumulating results in the last years suggest that inflammasome complexes play a crucial role in driving inflammatory reactions through the regulation of pro-IL-1β and pro-IL-18 activation, the exact molecular mechanisms underlying this event still need to be further elucidated and alternative pathways of IL-1β/IL-18 activation may also be relevant.

Notably, compelling evidence points to the $P2X_7$ receptor, the plasma membrane purinergic receptor for extracellular ATP, as a key player in IL-1/IL-18 maturation and externalization. This receptor, whose expression is upregulated by pro-inflammatory cytokines, is an ATP-gated ion channel able to alter the ionic environment of the cell and trigger a chain of intracellular events which culminate with initiation and amplification of the innate immune response. Specifically, in the presence of high concentrations of ATP, due for instance to release by injured or dying cells, $P2X_7$ receptor comes into play, causing lower intracellular K^+ and enabling the oligomerization of NALP3 inflammasome with stimulation of caspase-1 activity which, in turn, catalyzes conversion of pro-IL-1β and pro-IL-18 into their respective mature forms (Di Virgilio 2007).

Furthermore, with regard to the maturation and release of IL-18, it is important to remind that, differently from IL-1β that is generally absent in basal conditions and induced upon specific inflammatory stimuli only in a limited number of inflammatory cells, the IL-18 gene appears constitutively expressed in healthy cells (Puren et al. 1999). Thus, the mechanisms leading to IL-18 processing are particularly relevant in regulating its activity. Similarly to other leaderless cytokines including IL-1β, IL-18 is released by unconventional protein secretion pathways, and the activation of pro-IL-18 depends on a sequence of molecular mechanisms which can be organized according to a two-signal model, similarly to what previously described also for IL-1β activation. The first signal involves the stimulation of extracellular sensors belonging to the family of pattern-recognition receptors (PRRs), named Toll-like receptors (TLRs), which are able to detect PAMPs and to induce the NF-kB-mediated transcription of pro-cytokines, as well as modulation of NLR expression. The second signal, apparently independent from the first, is due to the stimulation of the intracellular sensor NLR, which, after being triggered by damage-associated molecular patterns, oligomerizes to make the formation and activation of a caspase-1-activating multiprotein complex, such as the NALP3 inflammasome, which in turn may lead to the production of bioactive IL-18 (Davis et al. 2011). Accordingly, LPS and several TLR agonists, particularly when used in combination with other stimuli that activate caspase-1, have been reported to induce high levels of IL-18 production.

After processing, IL-18 leaves the cell mostly in a regulated manner, and it is largely believed that its externalization may occur in the same way as for IL-1β, either via a vesicle system composed of plasma membrane microvesicles and exosomes or via a pro-inflammatory program of cell death mediated by inflammasomes, called pyroptosis. The activation of the vesicular versus the non-vesicular secretory pathway might occur on the basis of specific cell types and the environmental or experimental conditions, but the exact mechanisms of IL-18 secretion still remain a matter of active investigation (Carta et al. 2013).

Biological Activities

The biological effects of IL-18 differ from those of IL-1β in some ways, even though the two cytokines share several similarities about the structure of their receptor complex and signaling pathways, as well as their peculiar property to be functionally regulated by soluble decoy proteins.

In line with the pleiotropic nature of IL-18, its receptor complex (IL-18R) belonging to the IL-1/Toll-like receptor superfamily is expressed on several types of both malignant and normal cells, including natural killer (NK) cells, macrophages, T and B lymphocytes, microglia, astrocytes, and even neurons. More in detail, IL-18R is composed of two different proteins: the IL-18-binding chain IL-18Rα, member of IL-1 receptor-related protein (also named IL-1R Rp1or IL-1R5), and the signaling accessory chain IL-18Rβ (also named IL-18RAcP, AcP-like, or IL-1R7). Both chains are characterized by three extracellular immunoglobulin domains and one intracellular portion containing the Toll/IL-1 receptor (TIR) domain. In order to prime IL-18 signaling that takes place by transduction pathways similar to those of IL-1β, the presence on the target cell of both α- and β-chains of IL-18R is necessary (Thomassen et al. 1998). While IL-18Rα can bind the cytokine with a low affinity and is constitutively expressed on many cell types, only the copresence of the nonbinding IL-18Rβ protein allows the formation of a high-affinity complex, able to signal. Particularly, IL-18Rβ, which is generally present in a restricted number of cells and is inducible by specific conditions, is recruited into the signaling complex upon the binding of IL-18 to IL-18Rα, allowing the approximation within the heterodimer of the TIR domains and leading to the recruitment of the IL-1R-activating kinase (IRAK), via the adaptor molecule myeloid differentiation primary response protein 88 (MyD88). The autophosphorylation of IRAK then allows the activation of TNFR-associated factor-6 (TRAF-6) resulting in phosphorylation of NF-kB-induced kinase (NIK), which in turn leads to activation of two IkB kinases (IKK-1 and IKK2) that induce the phosphorylation of IkB. This latter step brings to the degradation and release of two components of

NF-kB: p50 and p65. Therefore, NFkB is made free to migrate in the nucleus to induce gene transcription. Accordingly, most of the immune and inflammatory genes induced by IL-18 are NF-κB regulated. IL-18 can also activate c-Jun N-terminal kinase (JNK) and mitogen-activated protein kinases as p38MAPK, which also translocate to the nucleus where they can phosphorylate several transcription factors like c-Jun and c-Fos (Smith 2011).

In general, the biologic activities of cytokines are controlled by endogenous inhibitory proteins. In the specific case of IL-18, like in the IL-1β system, some decoy molecules that interfere with or prevent the interaction between IL-18 and its receptor complex, eventually modulating the cytokine biological activity, have been described. An important specific inhibitor of IL-18 is the IL-18-binding protein (IL-18BP), a natural occurring, soluble 40 kDa protein, constitutively expressed at molar excess than IL-18 itself. IL-18BP can bind selectively and with high affinity the mature form of IL-18 (not pro-IL-18), preventing its interaction with IL-18R (Novick et al. 1999). Different isoforms of IL-18BP resulting in alternative splicing have been described in humans. In particular, IL-18BPa is the most secreted form of the protein with the highest binding affinity for IL-18.

Since IL-18 can modulate both type 1 (Th1) and type 2 (Th2) T helper responses, the main property of IL-18BP is to inhibit these activities. Furthermore, IL-18BP expression is activated by IFN-γ, which in turn is induced by IL-18 itself. Hence, within a pathway of pathophysiological significance, IFN-γ-mediated induction of IL-18BP plays as a negative feedback loop in IL-18-elicited immune responses, controlling excessive interleukin-18 bioactivity and probably occurring after some delay, to avoid premature termination of IL-18 activity (Paulukat et al. 2001; Hurgin et al. 2002).

One more modulator of IL-18 functions is IL-37 (IL-1F7), another cytokine processed by caspase-1 and belonging to IL-1 family. IL-37 is emerging as a potent anti-inflammatory cytokine and an efficient general inhibitor of innate immunity (Nold et al. 2010). More specifically, IL-37 can prevent the formation of the functional IL-18R and inhibit the action of IL-18 by two different possible mechanisms: either by binding with a low affinity IL-18Rα and not allowing the recruitment of IL-18Rβ or by binding IL-18BP and increasing its inhibitory activity, though other still unidentified IL-37-mediated mechanisms of IL-18 inhibition may also exist (Bufler et al. 2002). Furthermore, different truncated forms of both IL-18Rα and IL-18Rβ, which may act as soluble and negative regulators of IL-18 action, have been identified.

The biological properties of IL-18 are several and only partly overlapping with those of IL-1β. On the whole, the IL-18 activity is primarily related to host defense processes and ranges from amplification of inflammation and modulation of innate and acquired immune responses addressed to fight infection or other types of body offenses, to tissue damage when its expression is unrestricted leading to the development of chronic inflammations. The protective or detrimental outcome of IL-18 bioactivity mainly depends on the physiopathological context in which the molecule works, including the process timing and the general cell and cytokine milieu.

As an amplificatory factor of inflammation, IL-18 plays a role in activating neutrophils, in triggering the release from monocytes of cytokines associated with innate responses, as tumor necrosis factor (TNF)-α and IL-1β, in inducing chemokine production, cell adhesion molecules, and nitric oxide synthesis (Dinarello 2007). However, in contrast with IL-1β, IL-18 lacks relevant pyrogenic effects, does not induce acute phase protein synthesis, and is unable to induce cyclooxygenase-2 and prostaglandin E2 production. At the same time, IL-18 is able to induce the expression of Fas ligand with possible detrimental consequences on tissues. Furthermore, IL-18 is able to enhance inflammation through T lymphocyte activation, by amplifying Th1 response, promoting IL-17 production by γδ and CD4 T cells, amplifying both Th1- and Th2-associated cytokines, and blocking the production of the regulatory cytokine IL-10 in Th1 cultures (Blom and Poulsen 2012). All the above reported biological properties may account for the potential amplification of local inflammation as well as tissue destruction that IL-18 mediates under determinate pathological conditions.

The relevance of IL-18 as an immunoregulatory cytokine is largely derived from its major ability of inducing IFN-γ. In fact, as also observed when originally discovered, IL-18 is a potent inducer of IFN-γ production from splenocytes, T cells, B cells, NK cells, and dendritic cells, especially in collaboration with IL-12. Moreover, depending on genetic factors and cytokine milieu, IL-18 can promote Th1 or Th2 differentiation (Nakanishi et al. 2001). Thus, on the basis of the multiple inflammatory and immunoregulating activities here described, IL-18 has been proposed to play an important role in host defense against microbial and viral infection; besides it may participate in the pathogenesis of human autoimmune diseases associated with elevated production of IFN-γ and other clinical conditions characterized by an inflammatory component, as reported in greater detail below.

Noteworthy, IL-18 is expressed throughout various organ systems of the body, in cells of both hematopoietic and non-hematopoietic cell lineages, and even more intriguingly, it appears to be involved beyond the immune responses. To this regard, of particular interest is its emerging key role within the nervous system, where it may play an important activity in neuroinflammatory and neurodegenerative conditions in both immature and adult brains (Felderhoff-Mueser et al. 2005). Indeed, resident cells of the brain have been observed to express the cytokine, its receptor, and its activating enzyme caspase-1, supporting the concept that IL-18 can contribute to the homeostasis of central nervous system (CNS). In general, IL-18 modulates neuronal and glial cell function to control brain inflammatory processes, but also to mediate cellular mechanisms underlying cognition and behavior (Alboni et al. 2010). In agreement with these activities, IL-18 has been described to inhibit the induction of LTP (long-term potentiation) in the dentate gyrus, to play a role in processes related to stress and HPA axis activation; to be a critical regulator for exploratory activity, fear memory, and spatial learning; to induce and regulate sleep in animals; and to affect appetite control.

Pathophysiological Relevance

Similarly to IL-1β and likewise other cytokines that orchestrate a complex feedback network, IL-18 may have opposite function: either protective or detrimental.

In particular, as suggested by several studies based on IL-18 neutralization or knock-out animal models, under physiological conditions IL-18 is able to amplify the immune response through the triggering of the first-line inflammatory defense mechanisms and T lymphocyte stimulation, with the overall aim to maintain body homeostasis. However, when IL-18 expression is deregulated, as occurring for instance because of an unsuccessful or defective immune response, or owing to other diseases which imply an inflammatory component, or simply due to senescence, the cytokine effects can switch from beneficial to harmful. Regarding the latter condition, an important role for IL-18 has been depicted in aging, where a low-grade inflammation, including elevated levels of the cytokine and its natural inhibitor, has been included among the factors able to influence healthy versus unhealthy aging processes. Accordingly, in pathological conditions related to aging, such as atherosclerosis, IL-18 appears to have a relevant role (Dinarello 2006). Furthermore, IL-18 expression and effector functions have been found changed in a number of different pathological conditions, where the cytokine may take part to the processes which underlie or exacerbate the disease, such as in sepsis, chronic inflammatory diseases (like RA or inflammatory bowel disease), and autoimmune diseases, which are often associated with elevated production of IFN-γ and IL-18 (like lupus erythematosus, Crohn's disease, and multiple sclerosis), as well as in the metabolic syndrome, heart disease, and cancer, as recently reviewed by Dinarello et al. (2013).

Moreover, IL-18 is recently emerging as a mediator between the immune system and the nervous system, and it may play within CNS an important role in maintaining homeostasis, even acting beyond its immunomodulating properties. Accordingly, other than in the onset and progression of autoimmune CNS disease, exemplified by multiple sclerosis, IL-18 might also participate in and often correlate with clinical severity of stroke, traumatic brain injuries, neurodegeneration, and dementia, as well as psychiatric disturbances like schizophrenia (Bossù et al. 2010; Palladino et al. 2012).

Notably, the balance between IL-18 and IL-18BP appears to be skewed in most of these human illnesses, and elevated levels of free, bioactive IL-18 have been actually observed in the blood of patients suffering from diseases associated with the cytokine, pointing to a probable pathological importance of IL-18 and to a beneficial effect and potential therapeutic use of IL-18BP.

Modulation by Drugs

Given the multifaceted biological functions of IL-18, its modulation can be useful either as a tool or target for therapeutic interventions. However, the conditions in which the cytokine can be exploited for its protective abilities are much less frequent than the cases where instead its inhibition is expected to improve a specific pathological condition, especially in the case of chronic inflammatory or autoimmune diseases.

In force of its activity to induce IFN-γ and activate NK functions, IL-18 has been proposed in the past as a therapeutic option in cancer, even though contrasting roles of the cytokine as either tumor suppressive or cancer promoter have been described, depending on the specific context (Park et al. 2007). Furthermore, since IL-18 has been identified as an amplifier for Th1 and Th2 immune responses, a putative treatment based on IL-18 administration may be an attractive therapeutic instrument against bacterial complications in immunocompromised hosts (Kinoshita et al. 2013). Other studies came out consistent with a potential protective role for IL-18, such as in the homeostasis of the gastrointestinal tract and in macular degeneration, as well as in certain symptoms related to neuropsychiatric diseases.

As a therapeutic target, together with many other pro-inflammatory mediators, IL-18 may be downregulated by classical anti-inflammatory drugs, including both corticosteroids and nonsteroidal anti-inflammatory drugs (NSAID), which are broadly used as first-line anti-inflammatory treatment for a spectrum of disorders. For instance, glucocorticoids have been demonstrated to have effect on secretion of IL-18BP and to reduce IL-18 release, leading to a downregulation of IL-18/IL-18BP ratio, which may be beneficial in some autoimmune disorders. At variance with nonspecific anti-inflammatory drugs, a more focused approach based on the use of biotechnological compounds addressed to selectively control the functions of IL-18 would be a more efficient intervention for the treatment of those diseases where inflammation has detrimental consequences and a pathogenic involvement of IL-18 has been delineated. Thus, the use of IL-18BP, neutralizing monoclonal antibodies to IL-18, caspase-1 inhibitors, or blocking antibodies to the IL-18 receptor chains has been proposed to specifically counteract the cytokine function.

Interestingly, transgenic mice overexpressing IL-18BP exhibit reduced inflammation and disease severity, and a large number of encouraging preclinical data have provided evidence about the importance of reducing IL-18 activity to achieve disease improvement in animal models of autoimmune and inflammatory diseases, such as lupus erythematosus, inflammatory bowel disease, arthritis, acute renal injury, atherosclerosis, and even traumatic brain injury. Concordantly, IL-18BP is considered a promising therapeutic concept for humans suffering from autoimmune/inflammatory disturbances, and clinical trials using exogenous IL-18BP molecule as treatment for patients with psoriasis or rheumatoid arthritis have been initiated (Tak et al. 2006) and up till now are under investigation. Similarly, other IL-18-targeting drugs, including neutralizing antibodies directed against the cytokine or its receptor chains, are under investigation for their potential use in clinical trials. Preclinical studies strongly support this possibility, and as an example, neutralizing antibodies to IL-18 have been found to ameliorate ischemia-/reperfusion-induced myocardial injury in mice (Venkatachalam et al. 2009). A strong possibility to be clinically beneficial has been ascribed also to caspase-1 or inflammasome inhibitors, which by preventing the activation of both IL-1β and IL-18 may result in novel and specific therapeutic interventions for several inflammatory diseases, even though the relative contribution of each cytokine to disease pathogenesis would remain to be further determined.

Finally, if clinical trials will confirm the results of animal studies underscoring the potential benefit of IL-18 targeting in diseases characterized by exaggerate innate immune responses and inflammation, it is very likely that in the near future, new perspectives will be open for the treatment of severe and sometimes incurable CNS diseases where IL-18 appears involved, like stroke, neurodegeneration, traumatic brain injuries, and even psychiatric disturbances.

References

Alboni, S., Cervia, D., Sugama, S., & Conti, B. (2010). Interleukin 18 in the CNS. *Journal of Neuroinflammation, 7*, 9.

Bauernfeind, F., Ablasser, A., Bartok, E., Kim, S., Schmid-Burgk, J., Cavlar, T., et al. (2011). Inflammasomes: Current understanding and open questions. *Cellular and Molecular Life Sciences, 68*, 765–783.

Blom, L., & Poulsen, L. K. (2012). IL-1 family members IL-18 and IL-33 upregulate the inflammatory potential of differentiated human Th1 and Th2 cultures. *Journal of Immunology, 189*, 4331–4337.

Bossù, P., Ciaramella, A., Salani, F., Vanni, D., Palladino, I., Caltagirone, C., et al. (2010). Interleukin-18, from neuroinflammation to Alzheimer's disease. *Current Pharmaceutical Design, 16*, 4212–4223.

Bufler, P., Azam, T., Gamboni-Robertson, F., Reznikov, L. L., Kumar, S., Dinarello, C. A., et al. (2002). A complex of the IL-1 homologue IL-1F7b and IL-18-binding protein reduces IL-18 activity. *Proceedings of the National Academy of Sciences of the United States of America, 99*, 13723–13728.

Carta, S., Lavieri, R., & Rubatelli, A. (2013). Different members of the IL-1 family come out in different ways: DAMPs vs. cytokines? *Frontiers in Immunology, 4*, 123.

Davis, B. K., Wen, H., & Ting, J. P. (2011). The inflammasome NLRs in immunity, inflammation, and associated diseases. *Annual Review of Immunology, 29*, 707–735.

Di Virgilio, F. (2007). Liaisons dangereuses: P2X(7) and the inflammasome. *Trends in Pharmacological Science, 28*, 465–472.

Dinarello, C. A. (2006). Interleukin 1 and interleukin 18 as mediators of inflammation and the aging process. *The American Journal of Clinical Nutrition, 83*, 447S–455S.

Dinarello, C. A. (2007). Interleukin-18 and the pathogenesis of inflammatory diseases. *Seminars in Nephrology, 27*, 98–114.

Dinarello, C. A., Novick, D., Kim, S., & Kaplanski, G. (2013). Interleukin-18 and interleukin-18 binding protein. *Frontiers in Immunology, 4*, 289.

Felderhoff-Mueser, U., Schmidt, O. I., Oberholzer, A., Bührer, C., & Stahel, P. F. (2005). IL-18: A key player in neuroinflammation and neurodegeneration? *Trends in Neurosciences, 28*, 487–493.

Gu, Y., Kuida, K., Tsutsui, H., Ku, G., Hsiao, K., Fleming, M. A., et al. (1997). Activation of interferon-gamma inducing factor mediated by interleukin-1beta converting enzyme. *Science, 275*, 206–209.

Hurgin, V., Novick, D., & Rubinstein, M. (2002). The promoter of IL-18 binding protein: activation by an IFN gamma-induced complex of IFN regulatory factor 1 and CCAAT/enhancer binding protein beta. *Proceedings of the National Academy of Sciences of the United States of America, 99*, 16957–16962.

Kato, Z., Jee, J., Shikano, H., Mishima, M., Ohki, I., Ohnishi, H., et al. (2003). The structure and binding mode of interleukin-18. *Nature Structural & Molecular Biology, 10*, 966–971.

Kinoshita, M., Miyazaki, H., Ono, S., & Seki, S. (2013). Immunoenhancing therapy with interleukin-18 against bacterial infection in immunocompromised hosts after severe surgical stress. *Journal of Leukocyte Biology, 93*, 689–698.

Nakanishi, K., Yoshimoto, T., Tsutsui, H., & Okamura, H. (2001). Interleukin-18 is a unique cytokine that stimulates both Th1 and Th2 responses depending on its cytokine milieu. *Cytokine & Growth Factor Reviews, 12*, 53–72.

Nold, M. F., Nold-Petry, C. A., Zepp, J. A., Palmer, B. E., Bufler, P., & Dinarello, C. A. (2010). IL-37 is a fundamental inhibitor of innate immunity. *Nature Immunology, 11*, 1014–1022.

Novick, D., Kim, S. H., Fantuzzi, G., Reznikov, L. L., Dinarello, C. A., & Rubinstein, M. (1999). Interleukin-18 binding protein: A novel modulator of the Th1 cytokine response. *Immunity, 10*, 127–136.

Okamura, H., Tsutsi, H., Komatsu, T., Yutsudo, M., Hakura, A., Tanimoto, T., et al. (1995). Cloning of a new cytokine that induces IFN-gamma production by T cells. *Nature, 378*, 88–89.

Palladino, I., Salani, F., Ciaramella, A., Rubino, I. A., Caltagirone, C., Fagioli, S., et al. (2012). Elevated levels of circulating IL-18BP and perturbed regulation of IL-18 in schizophrenia. *Journal of Neuroinflammation, 22*, 206.

Park, S., Cheon, S., & Cho, D. (2007). The dual effects of interleukin-18 in tumor progression. *Cellular & Molecular Immunology, 4*, 329–335.

Paulukat, J., Bosmann, M., Nold, M., Garkisch, S., Kämpfer, H., Frank, S., et al. (2001). Expression and release of IL-18 binding protein in response to IFN-gamma. *Journal of Immunology, 167*, 7038–7043.

Puren, A. J., Fantuzzi, G., & Dinarello, C. A. (1999). Gene expression, synthesis and secretion of IL-1β and IL-18 are differentially regulated in human blood mononuclear cells and mouse spleen cells. *Proceedings of the*

National Academy of Sciences of the United States of America, 96, 2256–2261.

Sims, J. E., Nicklin, M. J., Bazan, J. F., Barton, J. L., Busfield, S. J., Ford, J. E., et al. (2001). A new nomenclature for IL-1-family genes. Trends in Immunology, 22, 536–537.

Smith, D. E. (2011). The biological paths of IL-1 family members IL-18 and IL-33. Journal of Leukocyte Biology, 89, 383–392.

Sugawara, S., Uehara, A., Nochi, T., Yamaguchi, T., Ueda, H., Sugiyama, A., et al. (2001). Neutrophil proteinase 3-mediated induction of bioactive IL-18 secretion by human oral epithelial cells. Journal of Immunology, 167, 6568–6575.

Tak, P. P., Bacchi, M., & Bertolino, M. (2006). Pharmacokinetics of IL-18 binding protein in healthy volunteers and subjects with rheumatoid arthritis or plaque psoriasis. European Journal of Drug Metabolism and Pharmacokinetics, 31, 109–116.

Thomassen, E., Bird, T. A., Renshaw, B. R., Kennedy, M. K., & Sims, J. E. (1998). Binding of interleukin-18 to the interleukin-1 receptor homologous receptor IL-1Rrp1 leads to activation of signaling pathways similar to those used by interleukin-1. Journal of Interferon & Cytokine Research, 18, 1077–1088.

Venkatachalam, K., Prabhu, S. D., Reddy, V. S., Boylston, W. H., Valente, A. J., & Chandrasekar, B. (2009). Neutralization of interleukin-18 ameliorates ischemia/reperfusion-induced myocardial injury. Journal of Biological Chemistry, 284, 7853–7865.

Interleukin-18 Binding Protein

Daniela Novick[1], Soohyun Kim[2] and Charles A. Dinarello[3,4]
[1]Department of Molecular Genetics, Weizmann Institute of Science, Rehovot, Israel
[2]Laboratory of Cytokine Immunology, Department of Biomedical Science and Technology, Konkuk University, Seoul, South Korea
[3]Department of Medicine, University of Colorado Denver, Aurora, CO, USA
[4]Department of Medicine, University Medical Center Nijmegen, Nijmegen, The Netherlands

Synonyms

IL-18 antagonist; IL-18 decoy receptor; IL-18 inhibitor; IL-18 regulator; IL-18BP

Definition

IL-18BP is a unique binding protein that deviates from the canonical definition of soluble receptors (Novick and Rubinstein 2007). It functions mainly as an IL-18 decoy protein and modulates the biological activity of this cytokine (Novick et al. 1999). The salient property of IL-18BP in immune responses is in downregulating Th1 activities by binding to mature IL-18 and thus reducing the induction of interferon-γ (IFN-γ) (Nakanishi et al. 2001). Since IL-18 also affects Th2 responses, its antagonist, the IL-18BP, is naturally involved in these reactions too (Nakanishi et al. 2001). Encoded by a separate gene, IL-18BP deviates from the classical definition of soluble receptors, namely, it does not correspond to the extracellular ligand-binding domain of the IL-18 receptor but rather belongs to a separate family of secreted proteins. IL-18BP is exceptionally high-affinity (400 pM) (Kim et al. 2000) and naturally occurring protein present in the serum of healthy humans at a 20-fold molar excess compared to the cytokine it binds, the IL-18 (Novick et al. 2001). In humans, increased disease severity can be associated with an imbalance of IL-18 to IL-18BP dictated by the levels of free IL-18 in the circulation. IL-18BP is highly regulated at the level of gene expression and protein synthesis by IFN-γ (Hurgin et al. 2002; Muhl et al. 2000) in a negative feedback loop manner.

In high concentrations IL-18BP binds the anti-inflammatory cytokine IL-37 (Bufler et al. 2002), thus reducing its ability to inhibit the induction of IFN-γ by IL-18.

Biosynthesis and Release

IL-18BP has a classic signal peptide and therefore is readily secreted. Structurally, IL-18BP has one immunoglobulin (Ig) domain, which displays some sequence homology to the third Ig domain of IL-18 receptor alpha (IL-18Rα). The human IL-18BP gene encodes four different isoforms (IL-18BPa, IL-18BPb, IL-18BPc, IL-18BPd), whereas two isoforms (IL-18BPc and IL-18BPd) have been identified in the mouse. These IL-18BP isoforms are produced by alternative mRNA splicing and differ primarily in their C-terminal region. IL-18BPa exhibits the greatest affinity for IL-18 with a dissociation constant of 400 pM. IL-18BPc shares the Ig domain with IL-18BPa, except for 29 amino acids in the C-terminal region. IL-18BPc has 10 times less binding affinity for IL-18 than IL-18BPa. Human IL-18BPa and IL-18BPc neutralize the biological activity of IL-18, while IL-18BPb and IL-18BPd lack a complete Ig domain and do not have the ability to bind or inhibit IL-18. Both mouse IL-18BP isoforms possess an identical Ig domain and are able to neutralize murine IL-18 activities. Murine IL-18BPd, which shares a common C-terminal motif with human IL-18BPa, is also able to inhibit human IL-18 (Kim et al. 2000). In the human and mouse IL-18BP is expressed constitutively in the spleen as shown by Northern blot analysis. In addition, IL-18BP mRNA is detected constitutively in the intestinal tract and in the prostate (Novick et al. 1999). Endothelial cells and monocyte/macrophages also represent important sources of IL-18BP.

Amino acid similarity exists between human IL-18BP and a gene found in various members of the poxviruses. The greatest degree of homology is to IL-18BP expressed by the *Molluscum contagiosum* gene (Xiang and Moss 2001). This infection is characterized by large numbers of viral particles in the epithelial cells of the skin but very few inflammatory or immunologically active cells in or near the lesions. Clearly, the virus fails to elicit an inflammatory or immunological response that might be explained by the ability of viral IL-18BP to reduce the activity of mammalian IL-18 in fighting this viral infection. This provides further evidence for the natural ability of IL-18BP to interfere with IL-18 activity.

Biological Activities

The biological activities of IL-18BP stem mainly from those of the cytokines it binds. IL-18BP is a naturally occurring extremely potent antagonist of IL-18 (Novick et al. 1999), but when in high concentration it binds IL-37 (Bufler et al. 2002). IL-18 is considered to be a link between innate and adaptive immune responses. As such IL-18 is particularly important for the clearance of intracellular pathogens, a process which requires the induction of IFN-γ, and of viruses, which involves induction of cytotoxic T cells and natural killer cells (Gracie et al. 2003). Thus IL-18BP plays a major role in the Th1 response. It does so via inhibiting the production of IFN-γ induced by IL-18 together with IL-12 or IL-15 (Nakamura et al. 2000), while without IL-12, IL-18 is a proinflammatory cytokine in an IFN-γ independent manner and plays a role in Th2 diseases (Nakanishi et al. 2001).

Increasing numbers of animal and clinical studies indicate a role for IL-18 in heart disease. Heart disease includes coronary vessel disease with associated myocardial infarction, post-viral myocardiopathies, autoimmune heart disease, and chronic heart failure. Reducing cytokines is tested as a possible therapy in these conditions. The myocardium of patients with ischemic heart failure expresses the alpha chain of the IL-18 receptor and has elevated levels of circulating IL-18 that is associated with death (Mallat et al. 2004). Daily administration of IL-18 into mice results in ventricular hypertrophy, increased collagen (Platis et al. 2008), and elevated left ventricular diastolic pressure (Woldbaek et al. 2005). Validation of the role of a cytokine in a disease process is best assessed by specific blockade. To test blockade of IL-18 by IL-18BP, a tissue from human atrial muscle strips obtained from patients undergoing bypass surgery was exposed to ischemia while contractile strength was measured. The addition of IL-18BP to the perfusate during and after the ischemic event resulted in improved contractile function from 35 % of control to 76 % with IL-18BP (Pomerantz et al. 2001). IL-18BP treatment also preserved intracellular tissue creatine kinase levels.

IL-18 has apparently a role also in inflammatory bowel diseases. A commonly used mouse model for colitis is dextran sodium sulfate (DSS) that damages the intestinal wall when added to the drinking water. Blocking IL-18 with either anti IL-18 antibodies or IL-18BP reduces colitis induced by antigen sensitization (Ten Hove et al. 2001).

IL-18 was shown to play a role in Th17 responses via its contribution to the production of IL-17. In a model for multiple sclerosis termed experimental autoimmune encephalomyelitis (EAE) (Sutton et al. 2006), a role for IL-18 was studied. In this model a transfer to healthy mice of myelin-derived immunogen primed dendritic cells that induce IL-17 from T cells resulted in EAE. However, the disease did not develop when the dendritic cells were exposed to a caspase-1 inhibitor (Lalor et al. 2011) implying that IL-1β and IL-18 are involved, since caspase-1 converts these cytokines into their active forms. Indeed injection of the mice with either IL-1β or IL-18 restored the disease.

There are several activities of IL-18 that are independent of IFN-γ. For example, IL-18 inhibits proteoglycan synthesis in chondrocytes (Joosten et al. 2000), cells that are essential for maintaining healthy cartilage. Indirect involvement of IL-18 was shown in multiple sclerosis, in other autoimmune diseases, as well as in the metastatic process, since IL-18 increases VCAM-1 expression in endothelial cells independent of IFN-γ (Carrascal et al. 2003).

In certain conditions IL-18 is a protective cytokine rather than inflammatory. Mice deficient in caspase-1 experience increased disease severity when subjected to DSS colitis. Administration of exogenous IL-18 restored mucosal healing in these mice (Dupaul-Chicoine et al. 2010). In addition, IL-18 deficiency or IL-18 receptor deficiency results in the development of a metabolic syndrome in mice (Henao-Mejia et al. 2012). Mice deficient in NLRP3 are more susceptible to DSS colitis, which is thought to be due to decreased levels of IL-18 (Hirota et al. 2011). Mice deficient in NLRP6 are also more vulnerable to DSS (Chen et al. 2007; Elinav et al. 2011) and the susceptibility appears to be due to lack of sufficient IL-18. A protective role for IL-18 is not limited to the gastrointestinal tract. In the eye, a condition resembling "wet macular degeneration" worsens with antibodies to IL-18 (Doyle et al. 2012).

In combination with IL-2, IL-18 enhances the production of IL-13 by cultured T lymphocytes and NK cells. IL-18 can potentially induce IgG1, IgE, and Th2 cytokines such as IL-4, IL-5, and IL-10 production in murine experimental models. Transgenic mice overexpressing IL-18 produced high levels of both Th1 and Th2 cytokines and of IgE and IgG1 (Hoshino et al. 2001).

Mice transgenic for human IL-18BP isoform a (IL-18BP-Tg) expressed high levels of bioactive human IL-18BPa in the circulation and were viable, were fertile, and had no tissue or organ abnormality. This circulating IL-18BP completely neutralized the ability of exogenously administered IL-18 to induce IFN-γ and lowered significantly serum levels of induced macrophage-inflammatory protein-2 (MIP-2) compared with non-transgenic littermates. In addition IL-18BP-Tg mice were completely protected in a model of hepatotoxicity induced by administration of concanavalin A (Fantuzzi et al. 2003). Mice deficient of IL-18BP were not reported.

In high concentrations IL-18BP binds IL-37, resulting in a reduction of its protective activity and a decrease in the anti-inflammatory activity of IL-37. Indeed this pattern has been observed in mice injected with IL-18BP (Banda et al. 2003; Bufler et al. 2002).

Pathophysiological Relevance

Cytokines are involved in many acute and chronic inflammatory diseases and autoimmune disorders. The cytokine production in these situations is deregulated, namely, its level is too high and/or persists for too long. Therefore, the specific and effective sequestration of these surplus cytokine molecules is highly desirable. Naturally occurring soluble cytokine receptors and binding proteins are very attractive and potent reagents to neutralize cytokines since they ideally fulfill the requirement of highly specific drugs, which combine high-affinity binding with very low immunogenicity. As the natural soluble receptors have a rather short protein half-life in vivo, genetically modified molecules have been generated, which increase their clearance time. IL-18BP is a unique protein that fulfills these three prerequisites with no modifications required: (a) It is naturally occurring. (b) Unlike the canonical soluble cytokine receptors, its affinity to the ligand is extremely high (400 pM), an affinity significantly higher than that of IL-18Rα to IL-18. (c) Its K_{off} rate is very low, namely, it dissociates from the complex with its ligand at a very slow rate (association rate constant is 1.38×10^{-6} M^{-1} s^{-1} and a markedly reduced dissociation rate constant of 6.43×10^{-4} s^{-1}) (Kim et al. 2000).

Serum levels of IL-18BP in healthy subjects are in the range of 2000–5000 pg/mL, while the levels of IL-18 in the same sera are much lower (80–120 pg/ml/mL) (Novick et al. 2001). In a pathological situation what apparently counts is the level of the free cytokine. Bound versus free IL-18 can be calculated based on the law of mass action, the affinity of IL-18 to IL-18BP, the concentrations of IL-18 and IL-18BP (each measured by a specific ELISA), and the fact that a single IL-18BP molecule binds a single IL-18 molecule (Novick et al. 2001). In many diseases both IL-18BP and IL-18 are high but the level of IL-18BP is not high enough to neutralize these high levels of IL-18, and therefore, the level of free IL-18 remains higher than in healthy subjects. Several human autoimmune diseases and other pathological situations are associated with elevated production of IL-18. Those include systemic lupus erythematosus, Wegener's disease, rheumatoid arthritis, systemic juvenile idiopathic arthritis, adult Still's disease, macrophage activation syndrome (MAS), metabolic syndrome, Crohn's disease, psoriasis, graft-versus-host disease, sepsis, trauma, ulcerative colitis, myocardial infarction, coronary artery disease, and acute kidney injury. Levels of IL-18, IL-18BP, and free IL-18 are presented in Table 4 of Novick et al. (2013). In acute kidney injury the IL-18 level in urine is one of the prognostic biomarkers indicating the severity of the disease (Parikh et al. 2005, 2006). MAS and adult Still's disease deserve special attention referring to extremely high levels of IL-18 which reach nanogram level (Kawashima et al. 2001; Mazodier et al. 2005). In MAS, though increased levels of IL-18BP (average of 35 ng/ml) were observed, the calculated free IL-18 remained extremely high (660 pg/ml) and unlike IL-12 significantly correlated with clinical status and the hematological markers of the disease (Mazodier et al. 2005).

A large body of evidence has accumulated during the recent years, suggesting that the inflammasome is a major player in innate immune host defense. Inflammasomes, which are potent inducers of IL-1β and IL-18 during inflammation, are large protein complexes typically consisting of a NOD-like receptor (NLR), the adapter protein ASC, and caspase-1. Inflammasome seems to affect not only infection and inflammation but also metabolic disorders, including atherosclerosis, type 2 diabetes, gout and obesity, and aerobic glycolysis (Brydges et al. 2013; van de Veerdonk et al. 2011; Wen et al. 2012). In addition, chronic inflammatory responses have been observed to be associated with various types of cancer and the link is assumed to be via the inflammasome. During malignant transformation or cancer therapy, the inflammasomes are activated in response to danger signals arising from the tumors or from therapy-induced damage to the tumor or healthy tissue (Kolb et al. 2014; Terlizzi et al. 2014).

Thus, inflammasome components, pathways, and activated cytokines may serve as targets toward therapy of the various pathological conditions.

Cross-References

▶ Cytokines
▶ Inflammasomes
▶ Inflammatory Bowel Disease
▶ Interferon gamma
▶ Interleukin 17
▶ Interleukin 18
▶ Interleukin 2
▶ Natural Killer Cells
▶ Rheumatoid Arthritis
▶ TH17 Response

References

Banda, N. K., Vondracek, A., Kraus, D., Dinarello, C. A., Kim, S. H., Bendele, A., et al. (2003). Mechanisms of inhibition of collagen-induced arthritis by murine IL-18 binding protein. *Journal of Immunology, 170*(4), 2100–2105.

Brydges, S. D., Broderick, L., McGeough, M. D., Pena, C. A., Mueller, J. L., & Hoffman, H. M. (2013).

Divergence of IL-1, IL-18, and cell death in NLRP3 inflammasomopathies. *Journal of Clinical Investigation, 123*(11), 4695–4705. doi:10.1172/JCI71543

Bufler, P., Azam, T., Gamboni-Robertson, F., Reznikov, L. L., Kumar, S., Dinarello, C. A., et al. (2002). A complex of the IL-1 homologue IL-1F7b and IL-18-binding protein reduces IL-18 activity. *Proceedings of the National Academy of Sciences of the United States of America, 99*(21), 13723–13728. doi:10.1073/pnas.212519099.

Carrascal, M. T., Mendoza, L., Valcarcel, M., Salado, C., Egilegor, E., Telleria, N., et al. (2003). Interleukin-18 binding protein reduces b16 melanoma hepatic metastasis by neutralizing adhesiveness and growth factors of sinusoidal endothelium. *Cancer Research, 63*(2), 491–497.

Chen, C. J., Kono, H., Golenbock, D., Reed, G., Akira, S., & Rock, K. L. (2007). Identification of a key pathway required for the sterile inflammatory response triggered by dying cells. *Nature Medicine, 13*(7), 851–856. doi:10.1038/nm1603.

Doyle, S. L., Campbell, M., Ozaki, E., Salomon, R. G., Mori, A., Kenna, P. F., et al. (2012). NLRP3 has a protective role in age-related macular degeneration through the induction of IL-18 by drusen components. *Nature Medicine, 18*(5), 791–798. doi: 10.1038/nm.2717.

Dupaul-Chicoine, J., Yeretssian, G., Doiron, K., Bergstrom, K. S., McIntire, C. R., LeBlanc, P. M., et al. (2010). Control of intestinal homeostasis, colitis, and colitis-associated colorectal cancer by the inflammatory caspases. *Immunity, 32*(3), 367–378. doi:10.1016/j.immuni.2010.02.012.

Elinav, E., Strowig, T., Kau, A. L., Henao-Mejia, J., Thaiss, C. A., Booth, C. J., et al. (2011). NLRP6 inflammasome regulates colonic microbial ecology and risk for colitis. *Cell, 145*(5), 745–757. doi: 10.1016/j.cell.2011.04.022.

Fantuzzi, G., Banda, N. K., Guthridge, C., Vondracek, A., Kim, S. H., Siegmund, B., et al. (2003). Generation and characterization of mice transgenic for human IL-18-binding protein isoform a. *Journal of Leukocyte Biology, 74*(5), 889–896. doi: 10.1189/jlb.0503230.

Gracie, J. A., Robertson, S. E., & McInnes, I. B. (2003). Interleukin-18. *Journal of Leukocyte Biology, 73*(2), 213–224.

Henao-Mejia, J., Elinav, E., Jin, C., Hao, L., Mehal, W. Z., Strowig, T., et al. (2012). Inflammasome-mediated dysbiosis regulates progression of NAFLD and obesity. *Nature, 482*(7384), 179–185. doi: 10.1038/nature10809.

Hirota, S. A., Ng, J., Lueng, A., Khajah, M., Parhar, K., Li, Y., et al. (2011). NLRP3 inflammasome plays a key role in the regulation of intestinal homeostasis. *Inflammatory Bowel Disease, 17*(6), 1359–1372. doi:10.1002/ibd.21478.

Hoshino, T., Kawase, Y., Okamoto, M., Yokota, K., Yoshino, K., Yamamura, K., et al. (2001). Cutting edge: IL-18-transgenic mice: in vivo evidence of a broad role for IL-18 in modulating immune function. *Journal of Immunology, 166*(12), 7014–7018.

Hurgin, V., Novick, D., & Rubinstein, M. (2002). The promoter of IL-18 binding protein: Activation by an IFN-gamma -induced complex of IFN regulatory factor 1 and CCAAT/enhancer binding protein beta. *Proceedings of the National Academy of Sciences of the United States of America, 99*(26), 16957–16962. doi:10.1073/pnas.262663399.

Joosten, L. A., van De Loo, F. A., Lubberts, E., Helsen, M. M., Netea, M. G., van Der Meer, J. W., et al. (2000). An IFN-gamma-independent proinflammatory role of IL-18 in murine streptococcal cell wall arthritis. *Journal of Immunology, 165*(11), 6553–6558.

Kawashima, M., Yamamura, M., Taniai, M., Yamauchi, H., Tanimoto, T., Kurimoto, M., et al. (2001). Levels of interleukin-18 and its binding inhibitors in the blood circulation of patients with adult-onset Still's disease. *Arthritis and Rheumatism, 44*(3), 550–560. doi: 10.1002/1529-0131(200103)44:3 < 550::AID-ANR103 > 3.0.CO;2–5.

Kim, S. H., Eisenstein, M., Reznikov, L., Fantuzzi, G., Novick, D., Rubinstein, M., et al. (2000). Structural requirements of six naturally occurring isoforms of the IL-18 binding protein to inhibit IL-18. *Proceedings of the National Academy of Sciences of the United States of America, 97*(3), 1190–1195.

Kolb, R., Liu, G. H., Janowski, A. M., Sutterwala, F. S., & Zhang, W. (2014). Inflammasomes in cancer: a double-edged sword. *Protein & Cell, 5*(1), 12–20. doi:10.1007/s13238-013-0001-4.

Lalor, S. J., Dungan, L. S., Sutton, C. E., Basdeo, S. A., Fletcher, J. M., & Mills, K. H. (2011). Caspase-1--processed cytokines IL-1beta and IL-18 promote IL-17 production by gammadelta and CD4 T cells that mediate autoimmunity. *Journal of Immunology, 186*(10), 5738–5748. doi:10.4049/jimmunol.1003597.

Mallat, Z., Heymes, C., Corbaz, A., Logeart, D., Alouani, S., Cohen-Solal, A., et al. (2004). Evidence for altered interleukin 18 (IL)-18 pathway in human heart failure. *FASEB Journal, 18*(14), 1752–1754. doi: 10.1096/fj.04-2426fje.

Mazodier, K., Marin, V., Novick, D., Farnarier, C., Robitail, S., Schleinitz, N., et al. (2005). Severe imbalance of IL-18/IL-18BP in patients with secondary hemophagocytic syndrome. *Blood, 106*(10), 3483–3489. doi: 10.1182/blood-2005-05-1980.

Muhl, H., Kampfer, H., Bosmann, M., Frank, S., Radeke, H., & Pfeilschifter, J. (2000). Interferon-gamma mediates gene expression of IL-18 binding protein in nonleukocytic cells. *Biochemical and Biophysical Research Communications, 267*(3), 960–963. doi:10.1006/bbrc.1999.2064.

Nakamura, S., Otani, T., Ijiri, Y., Motoda, R., Kurimoto, M., & Orita, K. (2000). IFN-gamma-dependent and -independent mechanisms in adverse effects caused by concomitant administration of IL-18 and IL-12. *Journal of Immunology, 164*(6), 3330–3336.

Nakanishi, K., Yoshimoto, T., Tsutsui, H., & Okamura, H. (2001). Interleukin-18 is a unique cytokine that stimulates both Th1 and Th2 responses depending on its cytokine milieu. *Cytokine and Growth Factor Reviews, 12*(1), 53–72.

Novick, D., & Rubinstein, M. (2007). The tale of soluble receptors and binding proteins: From bench to bedside. *Cytokine and Growth Factor Reviews, 18*(5–6), 525–533. doi:10.1016/j.cytogfr.2007.06.024.

Novick, D., Kim, S. H., Fantuzzi, G., Reznikov, L. L., Dinarello, C. A., & Rubinstein, M. (1999). Interleukin-18 binding protein: A novel modulator of the Th1 cytokine response. *Immunity, 10*(1), 127–136.

Novick, D., Schwartsburd, B., Pinkus, R., Suissa, D., Belzer, I., Sthoeger, Z., et al. (2001). A novel IL-18BP ELISA shows elevated serum IL-18BP in sepsis and extensive decrease of free IL-18. *Cytokine, 14*(6), 334–342. doi: 10.1006/cyto.2001.0914.

Novick, D., Kim, S., Kaplanski, G., & Dinarello, C. A. (2013). Interleukin-18, more than a Th1 cytokine. *Seminars in Immunology, 25*(6), 439–448. doi:10.1016/j.smim.2013.10.014.

Parikh, C. R., Abraham, E., Ancukiewicz, M., & Edelstein, C. L. (2005). Urine IL-18 is an early diagnostic marker for acute kidney injury and predicts mortality in the intensive care unit. *Journal of the American Society of Nephrology, 16*(10), 3046–3052. doi:10.1681/ASN.2005030236.

Parikh, C. R., Mishra, J., Thiessen-Philbrook, H., Dursun, B., Ma, Q., Kelly, C., et al. (2006). Urinary IL-18 is an early predictive biomarker of acute kidney injury after cardiac surgery. *Kidney International, 70*(1), 199–203. doi: 10.1038/sj.ki.5001527.

Platis, A., Yu, Q., Moore, D., Khojeini, E., Tsau, P., & Larson, D. (2008). The effect of daily administration of IL-18 on cardiac structure and function. *Perfusion, 23*(4), 237–242. doi:10.1177/0267659108101511.

Pomerantz, B. J., Reznikov, L. L., Harken, A. H., & Dinarello, C. A. (2001). Inhibition of caspase 1 reduces human myocardial ischemic dysfunction via inhibition of IL-18 and IL-1beta. *Proceedings of the National Academy of Sciences of the United States of America, 98*(5), 2871–2876. doi:10.1073/pnas.041611398.

Sutton, C., Brereton, C., Keogh, B., Mills, K. H., & Lavelle, E. C. (2006). A crucial role for interleukin (IL)-1 in the induction of IL-17-producing T cells that mediate autoimmune encephalomyelitis. *Journal of Experimental Medicine, 203*(7), 1685–1691. doi:10.1084/jem.20060285.

Ten Hove, T., Corbaz, A., Amitai, H., Aloni, S., Belzer, I., Graber, P., et al. (2001). Blockade of endogenous IL-18 ameliorates TNBS-induced colitis by decreasing local TNF-alpha production in mice. *Gastroenterology, 121*(6), 1372–1379.

Terlizzi, M., Casolaro, V., Pinto, A., & Sorrentino, R. (2014). Inflammasome: Cancer's friend or foe? *Pharmacology and Therapeutics.* doi:10.1016/j.pharmthera.2014.02.002.

van de Veerdonk, F. L., Netea, M. G., Dinarello, C. A., & Joosten, L. A. (2011). Inflammasome activation and IL-1beta and IL-18 processing during infection. *Trends in Immunology, 32*(3), 110–116. doi:10.1016/j.it.2011.01.003.

Wen, H., Ting, J. P., & O'Neill, L. A. (2012). A role for the NLRP3 inflammasome in metabolic diseases–did Warburg miss inflammation? *Nature Immunology, 13*(4), 352–357. doi:10.1038/ni.2228.

Woldbaek, P. R., Sande, J. B., Stromme, T. A., Lunde, P. K., Djurovic, S., Lyberg, T., et al. (2005). Daily administration of interleukin-18 causes myocardial dysfunction in healthy mice. *American Journal of Physiology - Heart and Circulatory Physiology, 289*(2), H708–714. doi: 10.1152/ajpheart.01179.2004.

Xiang, Y., & Moss, B. (2001). Correspondence of the functional epitopes of poxvirus and human interleukin-18-binding proteins. *Journal of Virology, 75*(20), 9947–9954. doi:10.1128/JVI.75.20.9947-9954.2001.

Interleukin 22

Heiko Mühl
Department of General Pharmacology and Toxicology, Pharmazentrum Frankfurt/ZAFES, University Hospital Goethe University Frankfurt, Frankfurt am Main, Germany

Synonyms

Interleukin 10-related T-cell-derived inducible factor (ILTIF); IL-D110; Zcyto18

Definition

Interleukin (IL)-22 is a member of the IL-10 cytokine family that installs biological activity by binding to the heterodimeric IL-22R1/IL-10R2 receptor complex thereby initiating STAT3 activation as dominant signaling principle. Being a cytokine principally acting on epithelial(-like) cells, IL-22 is regarded pivotal for host defense, tissue protection, and repair at biological barriers. Unleashed IL-22 biological activity, however, has been related to various pathological conditions, most notably psoriasis, arthritis, and carcinogenesis.

Biosynthesis and Release

Human IL-22 is a glycosylated protein of 179 amino acids with a regular N-terminal hydrophobic signal peptide consisting of 33 amino acids. Being a member of the IL-10 cytokine family, human IL-22 displays 25 % sequence homology with human IL-10 and exhibits prominent bundle-forming clustering of α-helices, a structural feature characteristic of IL-10 and its cytokine siblings (Dumoutier et al. 2000; Wolk and Sabat 2006; Xie et al. 2000).

In the course of appropriate immunoactivation, IL-22 is expressed and secreted by a broad array of leukocytic cell types. Initially linked to Th1 differentiation (Wolk and Sabat 2006), it became rapidly apparent that IL-22 production is likewise achieved by Th17 cells (Liang et al. 2006) and a population of $CD4^+$ skin-homing T cells recently coined Th22 (Duhen et al. 2009). Those T cells decisively fail to efficiently express interferon (IFN)-γ and IL-17. Notably, $CD8^+$ T cells can likewise acquire the potential to produce IL-22 in significant amounts. Besides aforementioned cellular components of adaptive immunity, also cells of the innate arm of immunity generate IL-22, not only in pathophysiology but interestingly enough also in the steady state (see below). Those include activated γδT cells, dendritic cells (DC)/mononuclear phagocytes, conventional natural killer (NK) cells, invariant NK T cells, and array of NK-like cells recently coined innate lymphoid cells (ILC). Most recently, even neutrophils have been added to the list of IL-22 producers (Pickert et al. 2009; Rutz et al. 2013; Sanos et al. 2009; Sonnenberg et al. 2011; Zindl et al. 2013).

As diverse the cellular sources of IL-22 turned out to be as manifold apparently are the signals initiating production of the cytokine. Along with activation of the T-cell receptor complex, cytokines that initiate T-cell differentiation are pivotal for IL-22 derived from adaptive T cells. Those are, among others, IL-12 and IL-18 for TH1 as well as IL-1 and IL-23 for TH17 differentiation. The latter two cytokines are also regarded key to IL-22 production by aforementioned innate leukocytes (Rutz et al. 2013). Besides that, activation of the toll-like receptor (TLR) system should be crucial for innate IL-22 production under conditions of host defense. At this point it should be emphasized that IL-22 production may not only be a consequence of microbial infection or autoimmunity but should be initiated/amplified in diverse pathological conditions via TLR-activating danger-associated molecular patterns (DAMPs) that are released from necrotic or necroptotic cells during tissue damage (Kulkarni et al. 2014). Notably, activation of innate signaling may also enforce IL-22 production by $CD4^+$ or $CD8^+$ T cells in specific pathophysiological settings.

On a molecular level, some transcription factors and intracellular receptor complexes have been proven to directly bind to and contribute to activation of the *IL22* promoter. Those include the retinoid orphan receptor-γt (RORγt in mice; RORC in humans), the aryl hydrocarbon receptor (AhR), the retinoic acid receptor (RAR), B-cell-activating transcription factor (BATF), signal transducer and activator of transcription (STAT)-3, nuclear factor of activated T cells (NF-AT), and cAMP response element-binding protein (CREB). In contrast, v-maf avian musculoaponeurotic fibrosarcoma oncogene homologue (c-maf) and forkhead-box-protein P3 (FOXP3) have the capability to curb *IL22* transcription (Jeron et al. 2012; Mielke et al. 2013; Rudloff et al. 2012; Rutz et al. 2013).

Biological Activities

Biological properties of IL-22 are determined by restrictive expression and decisive signaling of its heterodimeric IL-22R1/IL-10R2 receptor complex. Whereas IL-10R2 is ubiquitously expressed, IL-22R1 is predominantly detected on epithelial (-like) cell types, among others keratinocytes, hepatocytes, and lung and intestinal as well as renal proximal tubular epithelial cells. Other types of cells displaying IL-22R1 expression and thus IL-22 responsiveness are rare. Exceptions include synovial fibroblasts and myofibroblasts. Notably, leukocytic cells generally lack IL-22R1 expression and do not respond to the cytokine (Wolk et al. 2004; Wolk and Sabat 2006).

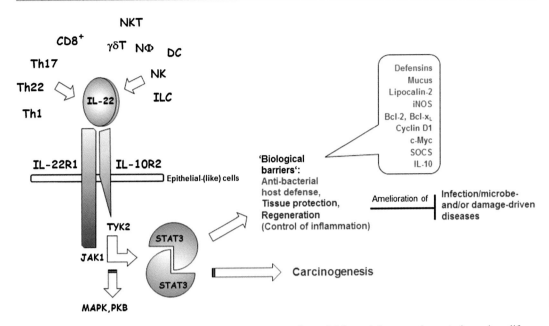

Interleukin 22, Fig. 1 Tissue protective properties of IL-22 at biological barriers. After being released from diverse leukocytic cell types, IL-22 binds to its IL-22R1/IL-10R2 heterodimeric receptor on epithelial(-like) cells which initiates signaling dominated by STAT3. As a result, a specific gene expression profile is generated favoring antibacterial host defense, anti-apoptosis, and proliferation. The latter two cellular functions are especially supported by simultaneous action of MAPK and PkB. This specific profile of activation is supposed to mediate tissue-protective properties of IL-22 at biological barriers but may likewise promote carcinogenesis

However, aberrant functional IL-22R1 on tumor cells of hematopoietic and lymphoid origin may, via autocrine IL-22, contribute to disease progression as seen in anaplastic large-cell lymphoma (Bard et al. 2008). Restrictive expression of IL-22R1 on non-leukocytic cells supports the view that leukocyte-derived IL-22 can be regarded as specialized cytokine messenger transmitting information between the leukocytic and the non-leukocytic cell compartments (Wolk et al. 2004).

Ligation of functional IL-22 receptors mediates, via recruitment of Janus kinase (JNK)-1 and tyrosine kinase (Tyk)-2, activation of the transcription factor STAT3 as principal means of signal transduction. Besides that, signaling by pro-proliferative mitogen-activated protein kinases (MAPK), specifically extracellular signal-regulated kinase (ERK)-1/2, and protein kinase B (PKB or Akt) is commonly detected in IL-22R1$^+$ cells under the influence of this cytokine (Fig. 1). In contrast, activation of pro-inflammatory nuclear factor-κB (NF-κB) is not a general feature of the IL-22/IL-22R1 axis. Cellular functions of IL-22 are primarily determined by the aforementioned profile of signal transduction (STAT3, Erk1/2, PKB), are usually linked to changes in gene expression, and overall favor antibacterial host defense, proliferation, and anti-apoptosis. Specifically, IL-22 enhances expression of antibacterial genes such as defensins, lipocalin, and inducible NO synthase, induces proliferation by action on cyclin D1 and c-myc, and promotes anti-apoptosis via B-cell lymphoma (Bcl)-2 and Bcl-X$_L$. In addition, IL-22 may also, in context-specific manner, mediate anti-inflammatory signals, possibly via upregulation of IL-10 and suppressors of cytokine signaling (Fig. 1). Those properties altogether form the basis of tissue-protective functions ascribed to IL-22 biological activity. IL-22 also activates the hepatic acute-phase response, a function likely of relevance in that context (Mühl et al. 2013; Wolk and Sabat 2006). The pro-proliferative action of IL-22, however, also has a pathogenic site that may promote processes connected to chronic inflammation (e.g., the "malignant transformation" of synovial fibroblasts in rheumatoid arthritis) and carcinogenesis. Although IL-22 signal transduction is dominated by STAT3, an intrinsic capacity to activate pro-inflammatory STAT1 has been recognized early on. This feature is strongly enhanced under the influence of type I IFN, a path that may contribute to context-specific pro-inflammatory properties of IL-22 (Mühl 2013). Furthermore, the actual function of IL-22 in chronic inflammation appears to be likewise influenced by either presence or absence of IL-17. In fact, both cytokines show a strong potential for pro-inflammatory synergy which has been confirmed in vivo using a lethal model of murine airway inflammation (Sonnenberg et al. 2010). This key observation should be highly relevant in the context of several Th17-driven diseases.

Overt IL-22 activity is curbed by IL-22 binding protein (IL-22BP), a secreted decoy receptor encoded by a separate gene (Dumoutier et al. 2001) and functionally related to IL-18BP. *IL22BP* (*IL22Ra2*) is a gene constitutively expressed by DC/mononuclear phagocytes which are regarded the major source of the protein. IL-22BP displays inverse regulation as compared to IL-22. Interestingly, production of IL-22BP is decreased in an inflammatory environment, partly via IL-18 and PGE_2 (Huber et al. 2012).

Pathophysiological Relevance

Diverse manifestations of inflammation, among others infections, autoimmunity, or conditions of prolonged tissue stress/damage, associate with augmented IL-22 production. Accordingly, patients diagnosed with bacterial infections, sepsis, rheumatoid arthritis, psoriasis, inflammatory bowel diseases, and liver cirrhosis display increased serum levels of IL-22. Likewise, animal models of aforementioned disorders confirm upregulation of IL-22 expression under those conditions and suggest an active role of the cytokine in pathophysiology. Blockage of IL-22 biological activity by use of either IL-22 knockout mice or by administration of IL-22 neutralizing antibodies or IL-22BP in fact aggravates symptoms in various models of infection/microbe- or damage-driven diseases. Selected examples are models of acute liver injury induced by concanavalin A (ConA) or by the combination of lipopolysaccharide (LPS) plus D-galactosamine or those of *influenza A* virus or *Klebsiella pneu*monia infections as well as experimental colitis induced by dextran sulfate sodium (DSS)-associated epithelial injury. In addition, endogenous IL-22 supports regeneration in experimental hepatectomy. These observations emphasize the relevance of endogenously produced IL-22 for tissue protection and regeneration in pathophysiology (Mühl et al. 2013).

Notably, IL-22 likewise fulfills key physiological functions in the steady state which is well characterized for the intestinal compartment. In the intestine, mucosal innate lymphoid cells/lymphoid tissue inducer-like cells efficiently secrete IL-22 under physiological conditions. The driving-force behind this phenomenon is exposure of those cells to commensal microbes (Sanos et al. 2009) and/or to nutritional components that may facilitate IL-22 expression, e.g., by activation of AhR signaling (Qiu et al. 2012).

As already alluded to, IL-22 displays Janus-faced immunoregulatory properties that also have the capability to promote specific aspects of inflammation (Pan et al. 2013). The best studied case in this regard is psoriasis, an autoimmune disorder of the skin (Sabat and Wolk 2011). Here, IL-22 production correlates with clinical disease activity, and blockage of IL-22 biological activity by the use of knockout mice or neutralizing antibodies reduces severity of experimental disease (van Belle et al. 2012). Main targets of IL-22 in psoriasis are certainly keratinocytes. IL-22 modulates keratinocyte differentiation and enhances their proliferation, two processes key to psoriasis pathogenesis. In addition, IL-22 upregulates keratinocyte expression of crucial disease parameters, among others the chemokine CXCL5 (epithelial-derived neutrophil-activating peptide 78) and matrix metalloproteinases (MMP)-1 and MMP-3. A further autoimmune disease in which IL-22 apparently displays a pathogenic potential is rheumatoid arthritis. Again, patients' IL-22 production correlates with disease severity. Furthermore, IL-22 knockout mice display ameliorated course of collagen-induced arthritis. IL-22 is supposed to contribute to arthritis pathogenesis by action on synovial fibroblasts (Ikeuchi et al. 2005). Likely through STAT3, IL-22 is driving proliferation of this cell type, thereby establishing a transformed-like malignant character that is crucial for disease progression. In this scenario, IL-22 likewise promotes production of pro-inflammatory CCL2 (monocyte chemotactic protein-1) and joint destructive receptor activator of NF-κB ligand (RANKL) by synovial fibroblasts. However, the role of IL-22 in murine collagen-induced arthritis is actually more complex and may depend on the particular disease phase. Specifically, IL-22 apparently is protective in the early initiation phase (prior to the onset) but pathogenic in the later phases of murine collagen-induced arthritis (Justa et al. 2014).

Because STAT3 is inherent to IL-22 signal transduction and uncontrolled STAT3 activation is evidently connected to carcinogenesis (Bournazou and Bromberg 2013), chronic overt IL-22 biological activity is supposed to support tumor formation. Interestingly, increased expression of IL-22 has been observed in different human cancer types, among others in lung and hepatocellular carcinoma as well as in gastric and colon cancer. Murine models of hepatocellular carcinoma and colon cancer in fact confirm the potential of endogenously produced IL-22 to serve pro-tumorigenic functions. Moreover, serum IL-22 has been shown to be a negative prognostic indicator in clinical hepatocellular carcinoma. Taken together, current data suggest a pro-tumorigenic role of the IL-22/STAT3 axis (Jiang et al. 2011; Huber et al. 2012; Mühl et al. 2013). This not only relates to the endogenously produced cytokine but obviously may curb the option of administering recombinant IL-22 in long-term therapeutic regimes.

Modulation by Drugs: Application of Recombinant IL-22 in Pharmacotherapy

Manipulating the IL-22/STAT3 axis may open the avenue toward novel pharmacological strategies either combating specific autoimmune inflammatory diseases and cancer (blockage of IL-22) or providing tissue-protective impulses in the context of damage-associated diseases (IL-22 administration).

To block IL-22 biological activity may turn out to be highly efficient especially in psoriasis. Rheumatoid arthritis and certain forms of cancers are further key clinical indications whereby blocking IL-22 in patients appears to be promising. As already alluded to, preclinical studies and suitable animal models suggest a significant potential for IL-22 blockage in these disorders. Various strategies for inhibiting IL-22 action are conceivable, currently being tested in clinical trials, or already in use. Those include interventions to manipulate T-cell differentiation, to neutralize IL-22 by antibodies or IL-22BP, or to block IL-22R1 by antibodies or by IL-22 muteins which tightly bind to IL-22R1 but fail to recruit IL-10R2 for signal transduction (Niv-Spector et al. 2012). Of those therapeutic strategies, suppression of TH1/TH17 cell activity is already clinically approved for the treatment of severe psoriasis. This is achieved by application of ustekinumab (Stelara®), a human monoclonal antibody directed against the shared p40 subunit of IL-12 and IL-23 (Griffiths et al. 2010) which promote TH1 and TH17 differentiation, respectively. As TH1/TH17 cells are principal sources of IL-22 in psoriasis and IL-22 is regarded pivotal for disease pathology, reduction of IL-22 production should, to a significant degree, contribute to the striking clinical success of ustekinumab in severe plaque psoriasis.

Patients suffering from infection/microbe- and/or damage-driven diseases, particularly at biological barriers such as the lung, intestine, and liver, would supposedly benefit from interventions that aim at increasing IL-22 biological activity with subsequent activation of STAT3/PKB/MAPK signaling (Fig. 1). This hypothesis is the rationale for therapeutic approaches using IL-22 for treating disease. In fact, administration of recombinant IL-22 mediates amelioration in a broad array of rodent disease models that, for example, cover allergic inflammation-, fibrosis-, and ventilator-induced injury in the lung;

alcohol-, ConA-, acetaminophen-, steatosis-, LPS/D-galactosamine-, ischemia-reperfusion-, and fibrosis-induced injury/dysfunction in the liver; DSS-induced injury and colitis; and choline-deficient/DL-ethionine diet-induced or cerulein-induced pancreatitis (Mühl et al. 2013). The point that IL-22 normally neither activates leukocytes nor affects the key pro-inflammatory transcription factor NF-κB concurs with the notion that recombinant IL-22 administration of up to 8 μg per mouse is generally not associated with acute side effects in vivo. In fact, a phase I clinical trial evaluating IL-22 in healthy human volunteers is currently ongoing. However, certain restriction likely applies. Although application of recombinant IL-22 may be suitable to treat acute disease and tissue damage subsequent to infection, trauma, or exposure to xenobiotics, chronic administration of the cytokine is likely confined by the inherent potential of IL-22 to augment some aspects of autoimmune inflammation and to promote carcinogenesis. Whatever the circumstances are, close surveillance of patients undergoing a potential recombinant IL-22 therapy is certainly advised.

Cross-References

▶ Interleukin 10
▶ Interleukin 17
▶ Janus Kinases (Jaks)/STAT Pathway
▶ TH17 Response

References

Bard, J. D., Gelebart, P., Anand, M., Amin, H. M., & Lai, R. (2008). Aberrant expression of IL-22 receptor 1 and autocrine IL-22 stimulation contribute to tumorigenicity in ALK+ anaplastic large cell lymphoma. *Leukemia, 22*, 1595–1603.

Bournazou, E., & Bromberg, J. (2013). Targeting the tumor microenvironment: JAK-STAT3 signaling. *JAK-STAT, 2*, e23828.

Duhen, T., Geiger, R., Jarrossay, D., Lanzavecchia, A., & Sallusto, F. (2009). Production of interleukin 22 but not interleukin 17 by a subset of human skin-homing memory T cells. *Nature Immunology, 10*, 857–863.

Dumoutier, L., Louahed, J., & Renauld, J. C. (2000). Cloning and characterization of IL-10- related T cell-derived inducible factor (IL-TIF), a novel cytokine structurally related to IL-10 and inducible by IL-9. *Journal of Immunology, 164*, 1814–1819.

Dumoutier, L., Lejeune, D., Colau, D., & Renauld, J. C. (2001). Cloning and characterization of IL-22 binding protein, a natural antagonist of IL-10-related T cell-derived inducible factor/IL-22. *Journal of Immunology, 166*, 7090–7095.

Griffiths, C. E., Strober, B. E., van de Kerkhof, P., Ho, V., Fidelus-Gort, R., Yeilding, N., et al. (2010). Comparison of ustekinumab and etanercept for moderate-to-severe psoriasis. *New England Journal of Medicine, 362*, 118–128.

Huber, S., Gagliani, N., Zenewicz, L. A., Huber, F. J., Bosurgi, L., Hu, B., et al. (2012). IL-22BP is regulated by the inflammasome and modulates tumorigenesis in the intestine. *Nature, 491*, 259–263.

Ikeuchi, H., Kuroiwa, T., Hiramatsu, N., Kaneko, Y., Hiromura, K., Ueki, K., et al. (2005). Expression of interleukin-22 in rheumatoid arthritis: potential role as a proinflammatory cytokine. *Arthritis and Rheumatism, 52*, 1037–1046.

Jeron, A., Hansen, W., Ewert, F., Buer, J., Geffers, R., & Bruder, D. (2012). ChIP-on-chip analysis identifies IL-22 as direct target gene of ectopically expressed FOXP3 transcription factor in human T cells. *BMC Genomics, 13*, 705.

Jiang, R., Tan, Z., Deng, L., Chen, Y., Xia, Y., Gao, Y., et al. (2011). Interleukin-22 promotes human hepatocellular carcinoma by activation of STAT3. *Hepatology, 54*, 900–909.

Justa, S., Zhou, X., & Sarkar, S. (2014). Endogenous IL-22 plays a dual role in arthritis: regulation of established arthritis via IFN-γ responses. *PLoS One, 9*, e93279.

Kulkarni, O. P., Hartter, I., Mulay, S. R., Hagemann, J., Darisipudi, M. N., Kumar, Vr. S., et al. (2014). Toll-like receptor 4-induced IL-22 accelerates kidney regeneration. *J Am Soc Nephrol, 25*, 978–989.

Liang, S. C., Tan, X. Y., Luxenberg, D. P., Karim, R., Dunussi-Joannopoulos, K., Collins, M., et al. (2006). Interleukin (IL)-22 and IL-17 are coexpressed by Th17 cells and cooperatively enhance expression of antimicrobial peptides. *Journal of Experimental Medicine, 203*, 2271–2279.

Mielke, L. A., Jones, S. A., Raverdeau, M., Higgs, R., Stefanska, A., Groom, J. R., et al. (2013). Retinoic acid expression associates with enhanced IL-22 production by γδ T cells and innate lymphoid cells and attenuation of intestinal inflammation. *Journal of Experimental Medicine, 210*, 1117–1124.

Mühl, H. (2013). Pro-inflammatory signaling by IL-10 and IL-22: bad habit stirred up by interferons? *Frontiers in Immunology, 4*, 18.

Mühl, H., Scheiermann, P., Bachmann, M., Härdle, L., Heinrichs, A., & Pfeilschifter, J. (2013). IL-22 in tissue-protective therapy. *British Journal of Pharmacology, 169*, 761–771.

Niv-Spector, L., Shpilman, M., Levi-Bober, M., Katz, M., Varol, C., Elinav, E., et al. (2012). Preparation and characterization of mouse IL-22 and its four single-amino-acid muteins that act as IL-22 receptor-1 antagonists. *Protein Engineering Design and Selection, 25*, 397–404.

Pan, H. F., Li, X. P., Zheng, S. G., & Ye, D. Q. (2013). Emerging role of interleukin-22 in autoimmune diseases. *Cytokine & Growth Factor Reviews, 24*, 51–57.

Pickert, G., Neufert, C., Leppkes, M., Zheng, Y., Wittkopf, N., Warntjen, M., et al. (2009). STAT3 links IL-22 signaling in intestinal epithelial cells to mucosal wound healing. *Journal of Experimental Medicine, 206*, 1465–1472.

Qiu, J., Heller, J. J., Guo, X., Chen, Z. M., Fish, K., Fu, Y. X., et al. (2012). The aryl hydrocarbon receptor regulates gut immunity through modulation of innate lymphoid cells. *Immunity, 36*, 92–104.

Rudloff, I., Bachmann, M., Pfeilschifter, J., & Mühl, H. (2012). Mechanisms of rapid induction of interleukin-22 in activated T cells and its modulation by cyclosporin a. *Journal of Biological Chemistry, 287*, 4531–4543.

Rutz, S., Eidenschenk, C., & Ouyang, W. (2013). IL-22, not simply a Th17 cytokine. *Immunology Reviews, 252*, 116–132.

Sabat, R., & Wolk, K. (2011). Research in practice: IL-22 and IL-20: significance for epithelial homeostasis and psoriasis pathogenesis. *Journal der Deutschen Dermatologischen Gesellschaft, 9*, 518–523.

Sanos, S. L., Bui, V. L., Mortha, A., Oberle, K., Heners, C., Johner, C., et al. (2009). RORgammat and commensal microflora are required for the differentiation of mucosal interleukin 22-producing NKp46+ cells. *Nature Immunology, 10*, 83–91.

Sonnenberg, G. F., Nair, M. G., Kirn, T. J., Zaph, C., Fouser, L. A., & Artis, D. (2010). Pathological versus protective functions of IL-22 in airway inflammation are regulated by IL-17A. *Journal of Experimental Medicine, 207*, 1293–1305.

Sonnenberg, G. F., Fouser, L. A., & Artis, D. (2011). Border patrol: regulation of immunity, inflammation and tissue homeostasis at barrier surfaces by IL-22. *Nature Immunology, 12*, 383–390.

Van Belle, A. B., de Heusch, M., Lemaire, M. M., Hendrickx, E., Warnier, G., Dunussi-Joannopoulos, K., et al. (2012). IL-22 is required for imiquimod-induced psoriasiform skin inflammation in mice. *Journal of Immunology, 188*, 462–469.

Wolk, K., & Sabat, R. (2006). Interleukin-22: a novel T- and NK-cell derived cytokine that regulates the biology of tissue cells. *Cytokine & Growth Factor Reviews, 17*, 367–380.

Wolk, K., Kunz, S., Witte, E., Friedrich, M., Asadullah, K., & Sabat, R. (2004). IL-22 increases the innate immunity of tissues. *Immunity, 21*, 241–254.

Xie, M. H., Aggarwal, S., Ho, W. H., Foster, J., Zhang, Z., Stinson, J., et al. (2000). Interleukin (IL)-22, a novel human cytokine that signals through the interferon receptor-related proteins CRF2-4 and IL-22R. *Journal of Biological Chemistry, 275*, 31335–31339.

Zindl, C. L., Lai, J. F., Lee, Y. K., Maynard, C. L., Harbour, S. N., Ouyang, W., et al. (2013). IL-22-producing neutrophils contribute to antimicrobial defense and restitution of colonic epithelial integrity during colitis. *Proceedings of the National Academy of Sciences of the United States of America, 110*, 12768–12773.

Interleukin 23

Ahmet Eken[1,2], Akhilesh K. Singh[1] and Mohamed Oukka[1,3,4]
[1]Center for Immunity and Immunotherapies, Seattle Children's Research Institute, Seattle, WA, USA
[2]Faculty of Medicine, Erciyes University, Medical Biology, Kayseri, Turkey
[3]Department of Immunology, University of Washington, Seattle, WA, USA
[4]Department of Pediatrics, University of Washington, Seattle, WA, USA

Synonyms

IL-23; IL-23A; IL-23p19; p19; SGRF

Definition

IL-23 is a heterodimeric cytokine which belongs to the IL-12 cytokine family. It is composed of a specific alpha chain, p19 (IL-23A), and a shared beta chain, p40 (IL12β) subunit. The p40 subunit can also dimerize with p35 (IL-12A) and makes up the IL-12 cytokine. Thus, the specificity of the IL-23 heterodimer is conferred by p19.

Biosynthesis and Release

Human IL-23p19 gene is located on chromosome 12q13.2 and composed of four exons and three introns. Like alpha subunits of other IL-12 cytokine family members, p19 is a four α-helix protein and has 70 % similarity to its mouse ortholog. It contains five cysteine residues. p19 also has several O-glycosylation sites but no N-glycosylation site. p40 subunit of IL-23 is encoded by its gene located on a different chromosome in humans, 11q1.3. It is composed of eight exons and seven introns. p40, like other beta chain subunits of IL-12 family members, has homology with soluble class I cytokine receptor chains such as IL-6Rα and is composed of three domains (D1-3). p40 is N-glycosylated and can make homodimers. The secretion of biologically active IL-23 requires that both subunits are produced within the same cell, and thus the expression of either subunit alone in a 293 cell line results in poorly secreted proteins, and does not generate biologically active IL-23 (Oppmann et al. 2000). p19 protein by itself also does not appear to have a biological role (except potentially inhibiting the formation of bioactive IL-23 heterodimer formation). The interaction between the p19 and p40 subunits is stabilized by a disulfide bond at the top edge of the interface between p19 residue Cys54 and p40 Cys177 (Lupardus and Garcia 2008).

Functional IL-23 heterodimer is produced mainly by professional antigen-presenting cells (APCs) such as dendritic cells, macrophages, and monocytes. Pattern recognition receptors (PRR) are the major receptors that link extracellular signals to p19 and p40 production. Stimulation of Toll-like receptors (TLR) 2, 3, 4, 5, 7, and 8, C-type lectin receptors, NOD-like receptors, and CD40 by their corresponding ligands leads to production of IL-23 in dendritic cells and macrophages (Langrish et al. 2004). β-Glucan stimulation of APCs activates p19, p40, and p35 production (both IL-12 and IL-23). TLR2 co-stimulation with NOD ligands or β-glucan on the other hand preferentially stimulates IL-23. Also TLR8 and β-glucan co-stimulation results in preferential IL-23 and IL-1β production. CD40L stimulation of intestinal but not splenic APCs preferentially stimulates IL-23 production.

The p40 gene expression is regulated most notably by binding of NFκ-B, CCAAT/enhancer binding protein, and activator protein 1 (AP-1) to the promoter region of p40. The murine and human p19 promoters also have three NFκB binding sites. Two of these binding sites are involved in TLR-mediated activation of p19 transcription. Additionally, Smad3 and AP-1 transcription factors and activating transcription factor 2 (ATF-2) were shown to bind p19 promoter and positively regulate IL-23p19 expression. In both human and murine p19 promoters, there are two binding sites for two interferon regulatory factor (IRF) genes, IRF3 and IRF7. IRF3 appears to be a positive regulator of p19 expression, and thus, its absence leads to downregulation of p19. Lastly, MAPKs including p38, JNK, and ERK were also reported to mediate LPS induced p19 gene expression (Liu et al. 2009).

Biological Activities

IL-23 exerts its biological effects by engaging its heterodimeric receptor (IL-23R). IL-23R was discovered by Parham et al. through screening of a cDNA library generated from Kit225 cell line, which responds to IL-23. IL-23R is composed of two subunits: IL-12Rβ1, which is shared by IL-12 receptor complex, and IL-23R, which is the unique subunit. Both subunits can bind to heterodimeric IL-23. The p19 subunit of IL-23 binds to IL-23R, whereas the p40 subunit interacts with IL-12Rβ1 chain. IL-23R has been conserved in amniotes. In humans and mice, unique IL-23R subunit protein is made up of 629 and 659 amino acids, respectively, and bears 84 % similarity. IL-23R has sequence similarity to IL-12Rβ2 and gp130. IL-23R is located closely to IL-12Rβ2 both in humans and mice on chromosomes 1 and 6, respectively, and is thought to evolve through a gene duplication process (Parham et al. 2002).

IL-23R is expressed by both innate (group 3 ILCs, dendritic cells, macrophages/monocytes, γδ T cells, and recently neutrophils) and adaptive immune cells (Th17, Th22, and some iNKT cells) (Awasthi et al. 2009). Our understanding of IL-23 signaling is based mostly on the information acquired from Kit225 cell line and Th17 cells. IL-23 receptor signaling progresses through JAK kinases and STAT transcription factors. Binding of IL-23 to IL-23R triggers activation of Jak2 and

Tyk2, which then phosphorylate the receptor and create docking sites for the recruitment of STAT proteins. STAT1, 4, 5 are phosphorylated by Jak2 and Tyk2 but STAT3 is the major transcription factor activated by IL-23 stimulation. Other pathways which are activated upon IL-23 binding to its receptor include the P38 MAPK pathway, PI3K-Akt, and NFκ-B pathway (Lankford and Frucht 2003).

IL-23-Responsive Cells

Th17 Cells

The most appreciated function of IL-23 is the maintenance/maturation and expansion of Th17 cells. IL-23 is not required for their initial differentiation from naïve CD4 T cells; however, Th17 cells need IL-23R signaling to maintain their Th17 identity which is performed by IL-23-mediated induction of *Rorc* and *Il17* expression. IL-23 also stimulates expression of its own receptor. The Th17 cells require IL-23 for them to become fully pathogenic, as data from several different investigators' work suggest (Gaffen et al. 2014). Using EAE as a model of multiple sclerosis (MS), various labs showed that Th17 cells are very weak inducers of EAE unless they are differentiated in the presence of IL-23 or express IL-23R. This transcriptional reprogramming of Th17 cells into pathogenic agents is established by IL-23, and the resulting Th17 cells have a unique effector cytokine profile compared to non-pathogenic Th17 cells which are not exposed to IL-23. IL-23 activated Th17 cells express IFN-γ and GMCSF in addition to IL-17 and the associated transcription factors unlike non-pathogenic Th17 cells, which express only IL-17. In addition to MS, a growing list of literature suggest that Th17 cells, hence IL-23R signaling, play critical roles in development of chronic inflammation in multiple diseases such as Crohn's disease, ulcerative colitis, psoriasis, rheumatoid arthritis, systemic lupus erythematosus (SLE), etc.

Group 3 ILCs

IL-23 also binds to its receptor on innate lymphoid cells. Rorγt + group 3 innate lymphoid cells (ILC3) have recently been shown to depend on IL-23 for their production of various effector cytokines including IL-22, IFN-γ, and IL-17. ILC3s are also a heterogeneous population of cells and have an irreplaceable function in protective immunity against extracellular pathogens in the gastrointestinal mucosa and have been recently implicated in the pathogenesis of inflammatory bowel diseases (IBD) (Buonocore et al. 2010; Singh et al. 2014; Sonnenberg et al. 2011)

γδ T Cells

IL-23R is also expressed by a fraction of γδ T cells constitutively. IL-23 stimulation of these cells, depending on the location and phenotype, results in production of IL-22, IL-21, and IL-17. γδ T cells are enriched in the skin and mucosal surfaces, particularly in the intraepithelial compartment, and are involved in protective immunity against various pathogens. Studies with mouse models of various chronic inflammatory diseases revealed that these cells may play roles in IBD, psoriasis, MS, and rheumatoid arthritis (RA) through their effector cytokines (Paul et al. 2015).

Antigen-Presenting Cells, Neutrophils, NKT Cells, and Other Cells

Dendritic cells and macrophages respond to IL-23 and express IL-23R as well (Awasthi et al. 2009). Although these antigen-presenting cells are crucial for the activation of adaptive immunity, and thus the generation of chronic inflammation, how IL-23R signaling within these cells contributes to phenotype, maturation, and functions of APCs or to the pathology is less clear.

Neutrophils were also shown to express IL-23R and respond to IL-23; however, how these cells contribute to disease through Il-23R signaling requires further study (Zindl et al. 2013).

A population of NK1.1$^-$ invariant natural killer T cells (iNKT) in peripheral lymph nodes and skin were shown to express Rorγt and IL-23R and produce IL-17 and IL-22 in response to IL-23 and IL-1β stimulation. The contribution of these cytokines, of iNKT origin, to protective immunity and autoimmunity however is obscure (Doisne et al. 2009).

IL-23R is also expressed by a group of CD3+ CD4-CD8- Rag-dependent T cells. These cells have been reported to increase during disease, specifically in systemic lupus erythematosus and ankylosing spondylitis murine models, in an IL-23 dependent manner (Sherlock et al. 2012).

Pathophysiological Relevance

Inflammatory Bowel Diseases

Genome-wide association studies revealed many variants of the components of the IL-23 signaling pathway (IL-23R, JAK2, TYK2, and STAT3) as risk or resistance factors in IBD pathogenesis. Furthermore, IL-23, IL-23R proteins as well as IL-23R expressing innate and adaptive cells are enriched in the intestines of IBD patients. More importantly, there is a wealth of data from animal studies that implicate IL-23 and its signaling pathway components in the pathogenesis of inflammatory bowel diseases (Eken et al. 2014). This was shown in the ligand level (IL-23) using both the innate and adaptive means of IBD induction in genetically engineered mice. IL-23 transgenic mice develop systemic inflammation, including enterocolitis. Moreover, IL-23$p19^{-/-}$ or Il-12$p40^{-/-}$ KO mice but not IL-12$p35^{-/-}$ KO become resistant to mouse IBD development. This was also confirmed by using neutralizing antibodies against p19, p35, and p40. Therefore, IL-23 but not IL-12 appears to be more critical for the development of mouse IBD.

Involvement of IL-23 in IBD pathogenesis was shown at the receptor level as well. Although the results of these experiments are somewhat more complex and variable from model to model, IL-23R-deficient mice are protected from innate and chemically induced colitis development. In addition, T cells were shown to require IL-23R signaling to initiate adoptive T cell-driven colitis (Ahern et al. 2010; Izcue et al. 2008).

IL-23 contributes to the pathogenesis of murine IBD by promoting the activation and reprogramming or expansion of the abovementioned IL-23R-expressing innate and adaptive immune cells. The signature effector cytokines released by these Rorγt + cells (IFN-γ, IL-22, IL-17A/F) can act on hematopoietic and non-hematopoietic target cells. In turn, these target cells secrete various pro-inflammatory molecules (TNF-α, IL-6, etc.) or chemokines that further recruit inflammatory leukocytes, monocytes/macrophages, and neutrophils. Additionally, this inflammatory milieu was shown to promote in situ hematopoiesis and skew this process toward an inflammatory granulocyte-monocyte fate. GMCSF was shown to support this process, and thus its neutralization ameliorated murine IBD. Blockade of known Rorγt + cell-associated downstream cytokines individually results in more diverse and conflicting outcomes than blocking IL-23 itself based on the model used. Although IL-22 was shown to be protective by stimulating epithelial repair post-injury, studies showed that it can also cause hyperplasia and recruit neutrophils and thus act as a pro-inflammatory cytokine. Similarly, IL-17 blockade mitigated the intestinal pathology in early models; however, later studies reported a protective role for IL-17. In line with this, secukinumab, an IL-17 neutralizing antibody, exacerbated the conditions of Crohn's patients in a recent trial. Lastly, IFN-γ is also produced by some Th17 and ILC3s in IL-23-dependent murine IBD models and was shown to contribute to the pathogenesis (Eken et al. 2014; Griseri et al. 2012).

Multiple Sclerosis

IL-23 is required for the development of murine models of MS, experimental autoimmune encephalomyelitis (EAE). Similar to the other autoimmune diseases, the initial studies were conducted with the manipulation of p40 subunit. With the discovery of IL-23p19, it was quickly realized that IL-23 (p19-p40 dimer) rather than IL-12 (p40-p35 heterodimer) was necessary for the pathogenesis. So, although p35$^{-/-}$ mice develop EAE, p19$^{-/-}$ or p40$^{-/-}$ mice do not. Similarly, deletion of either subunits of IL-23R confers resistance to EAE, whereas IL-12Rβ2 deletion (specific subunit of IL-12R) exacerbates EAE. The pathogenic role of IL-23 in murine EAE attributed to its role in Th17 cells' biology. The contribution of other IL-23-responsive cells (γδ T cells, ILC3,

APCs, and neutrophils) is not as well characterized. IL-23 is not necessary for Th17 cells' initial generation but is required for the reprogramming of Th17 cells into EAE-causing agents. Additionally, pathogenic Th17 cells could be amplified and maintained via IL-23 stimulation. These pathogenic Th17 cells express IL-17, IFN-γ and GMCSF in response to IL-23 stimulation. Currently, Th17 cells that produce IL-17 and IFN at the same time are suggested to be more pathogenic than single producers. More recently, GMCSF, which is produced by Th17 cells in response to IL-23, was shown to be necessary for the development of MS and other autoimmune diseases in mice (Jadidi-Niaragh and Mirshafiey 2011).

Human MS patient samples also show correlation between disease and IL-23-Th17 axis. MS lesions from the CNS tissues of patients presented with an increase in the levels of IL-23 (both subunits), and the source was macrophages/microglia and dendritic cells. Supporting the mouse models, *Il17* and *Rorc* transcripts are enriched in the CNS lesions of MS patients adding further support to the involvement of Th17 cells in the human MS pathogenesis. Despite this overwhelming evidence about the pathogenicity of IL-23 in MS, initial clinical trials that aimed at blocking IL-23 activity via neutralization of p40 subunit (ustekinumab trial) were not encouraging and did not benefit the patient cohorts. The severity of the disease in enrolled patients and the inability of the antibody to cross the blood-brain barrier or alternatively the reliance to IL-12-/IL-23-independent factors for pathogenesis (during the entire disease process or after the progression/initiation of the disease) in MS in humans are some points that may explain the inefficacy of the trial (Cingoz 2009).

Rheumatoid Arthritis

Both IL-23 and IL-17 are found in the serum, synovial fluid, and synovial tissue of RA patients compared with controls. IL-17 also was shown to stimulate production of IL-1β and TNF-α which are involved in synovial inflammation. Importantly, IL-23 levels of serum and synovia correlates with IL-17, TNF-α, and IL-1β. IL-17 was reported to activate synovial fibroblasts through PI3K/Akt, NFkB, and p38 MAPK pathways and stimulate IL-23p19 production which further contributes to the vicious cycle of chronic inflammation. IL-17 also causes bone desorption through stimulation of osteoclastogenesis. Supporting this pathogenic role for IL-23-Th17 axis, neutralization of IL-17 during collagen-induced arthritis in mice alleviates joint swelling, inflammation, and bone erosion (Tang et al. 2012).

Spondyloarthropathies

Spondyloarthropathies are seronegative rheumatic diseases which include, among others, ankylosing spondylitis (AS) and psoriatic arthritis (PsA). The inflammation in spondyloarthropathy is proposed to primarily affect entheses rather than synovia. GWAS studies indicate a link between ankylosing spondylitis and IL-23R variants (Burton et al. 2007). Both IL-23 level and IL-23R-expressing Th17 cells were shown to increase in patients with spondyloarthropathy (Sherlock et al. 2014). More recently, in mouse models of ankylosing spondylitis, IL-23 was shown to have a causal role in the pathogenesis through downstream IL-22 and IL-17, and thus, the neutralization of IL-23 dampened the pathology (Sherlock et al. 2012). A correlation between IL-23-Th17 axis activity and psoriatic arthritis has also been reported.

Psoriasis

Both studies with small animal models and biopsies from patients indicate a role for IL-23 in the pathogenesis of psoriasis. IL-12p40 transgenic mice develop eczematous skin disease resembling psoriasis. More importantly, IL-23 is elevated in basal keratinocytes of p40 transgenic mice but not IL-12. Also, injection of IL-23 into WT mice creates a skin disease similar to that of p40 transgenics. Similarly, IL-23p19 transgenic mice develop inflammation in the skin, along with other organs. Furthermore, skin biopsies from psoriatic human patients show increased p40 expression. Ustekinumab is an FDA-approved monoclonal antibody against p40 and is currently effectively used to treat moderate to severe plaque psoriasis (Teng et al. 2015).

IL-23 stimulates production of effector cytokines IL-22, IFN-γ, and IL-17 by its responding cells in the skin, which includes most notably Th17 cells and γδ T cells. Studies with animal models of psoriasis implicate IFN-γ, IL-17, and, more recently, IL-22 as a driver of pathology as downstream cytokines of IL-23 produced by γδ T and Th17 cells (Zheng et al. 2007).

SLE

SLE is a systemic autoimmune disease that targets multiple organs. The disease is characterized by production of antinuclear autoantibodies and accumulation of the antibody-antigen complexes in the affected organs which then leads to the complement system activation and further recruitment of inflammatory effector cells. Recent studies suggest that the contribution of IL-23 to SLE pathogenesis may occur through Th17 cells. IL-23 and IL-17 cytokines were elevated in the serum of SLE patients. Th17 cell infiltrates have been detected in the kidneys of SLE patients. Moreover, T cells obtained from SLE patients make more IL-17 (Smith et al. 2012). Furthermore, deletion of IL-23R in mice mitigates several aspects of SLE pathology, such as autoantibody production, pro-inflammatory cytokine production, and glomerulonephritis in a lupus model. In line with this report, IL-23 neutralization in a mouse model of lupus appears to delay the nephritis without affecting the anti-dsDNA antibodies (Kyttaris et al. 2010, 2013).

Sjögren's Syndrome

Sjögren's syndrome is an autoimmune disorder that affects primarily exocrine glands (salivary and lacrimal) but can also take systemic form and is accompanied by other diseases, most commonly by RA. Though mechanistic studies in animal models are absent, samples from affected tissues of patients and murine models show correlation between the levels of IL-23, IL-17, and the disease (Voulgarelis and Tzioufas 2010).

Cancer

Both antitumor and pro-tumor activities have been reported for IL-23 in various cancer models. One hypothesis is that the inflammatory milieu which IL-23 favors suppresses the antitumor immunity. Absence of IL-23 or IL-23R results in increased CD8+ T cell counts in the tumor microenvironment, thus providing better immunosurveillance. The increase in CTL responses in IL-23- or IL-23R-deficient mice supports this notion. Additionally, IL-23 can drive the production of factors that may have tumor-promoting properties. IL-22 cytokine is one such example and is very important for antimicrobial immunity and epithelial regeneration in the intestines but was also shown to promote colon tumorigenesis when its quantity is not well adjusted during post-repair after epithelial damage (Ngiow et al. 2013).

Studies show that in various cancer patient samples, IL-23 is elevated and correlates with poor prognosis. GWAS studies also reveal different variants of IL-23R as risk factors for acute myeloid leukemia and solid tumors.

Though controversial, tumor-suppressive effects of high levels of IL-23 have also been reported. The majority of these studies were performed with IL-23 overexpressing cancer cell lines.

Modulation By Drugs

Owing to its central role in several chronic inflammatory conditions discussed above, there are now more than a dozen different companies developing antagonists of IL-23R. Some of these are currently in use; others are in the development-discovery stage, and some have been discontinued see (review by Tang et al. 2012). Table 1 summarizes these therapies and the inflammatory diseases they are used to treat. Most of these antagonists are monoclonal human or humanized antibodies that target specific or common subunits of IL-23.

Ustekinumab (Stelara®)

Ustekinumab (Stelara®) is the only FDA-approved drug that blocks IL-23/IL-12. It is a neutralizing human monoclonal antibody that targets the common subunit p40. Ustekinumab is currently prescribed for the treatment of plaque psoriasis and psoriatic arthritis. However, there are several clinical trials in

Interleukin 23, Table 1 Identified interleukin-23 receptor (IL-23R) antagonists (Adapted and modified from Tang et al. 2012)

Drug	Target	Company	Status	Disease
Ustekinumab	p40 (IL-12R; IL-23R)	Centocor Ortho Biotech and Janssen Research	Launched	Plaque psoriasis
			Launched	Psoriatic arthritis
			Phase III	Crohn's disease
			Phase I	CVID dependent enteropathy
			Phase II	Ankylosing spondylitis, sarcoidosis, atopic dermatitis, rheumatoid arthritis
			Phase II	
			Phase II	
			Phase II	
			Discontinued	Multiple sclerosis
BI 655066	IL-23 p19 antagonist mAb	Boehringer Ingelheim	Phase II	Crohn's disease, psoriasis
LY3074828	Anti-IL-23 humanized antibody	Eli Lilly and Company	Phase I	Psoriasis
Guselkumab	p19 mAb	Janssen Global Services, LLC	Phase III	Psoriasis
Tildrakizumab	IL-23 mAb	Merck	Phase III	Psoriasis
Anti-IL-23 therapeutic treatment	IL-23R	Schering-Plough	Phase I	Chronic inflammatory conditions
MP-196	IL-23R	Effimune	Clinical	Autoimmune disease
IL-12/IL-23 inhibitors	IL-12R; IL-23R	Synta Pharmaceuticals	Discovery	Crohn's disease; rheumatoid arthritis; multiple sclerosis
FM-202	IL-12R; IL-23R	Femta Pharmaceuticals	Discovery	Psoriasis
FM-303	IL-23R	Femta Pharmaceuticals	Discovery	Inflammatory bowel disease
IL-23 adnectin	IL-23R	Bristol-Myers Squibb	Discovery	Immune disorder
IL-23 receptor antagonist	IL-23R	Oscotec	Discovery	Arthritis
Anti-IL-23 immunotherapy	IL-23R	Peptinov SAS	Discovery	Inflammatory disease
ADC-1012	IL-23R	Alligator Bioscience AB	Discontinued	Inflammatory disease; cancer
Anti-IL-12p40/ IL-23p40 HumAbs	IL-12R; IL-23R	Theraclone Sciences	Discontinued	Autoimmune disease
Anti-IL-23 HumAbs	IL-23R	Theraclone Sciences	Discontinued	Autoimmune disease
Apilimod (STA-5326)	blocks NFKB translocation, IL-12, IL-23 production	Synta Pharmaceuticals	Discontinued	Psoriasis; rheumatoid arthritis; common immunodeficiency
LY-2525623	IL-23R	Eli Lilly	Discontinued	Multiple sclerosis; psoriasis
Briakinumab	IL-12R; IL-23R	Abbott	Discontinued	Psoriasis
			Discontinued	Crohn's disease
			Discontinued	Multiple sclerosis

which its effectiveness against a list of autoimmune conditions is being tested. Ustekinumab phase II trials for Crohn's disease and ankylosing spondylitis gave promising results. Ustekinumab is being tested for atopic dermatitis and rheumatoid arthritis as well. Multiple sclerosis trials did not meet the expectations.

Because ustekinumab blocks IL-12 and IL-23 together, both the Th17 and Th1 arm of helper T cells are affected. Although both Th1 and Th17 cells are operative in many autoimmune conditions discussed here (psoriasis, IBD, MS), targeting of IL-23 via p19 antibodies would spare the Th1 arm, NK, and CD8+ T cell responses, which are crucial for immunity against intracellular pathogens and tumors, and therefore will help reduce the risk of getting infections or developing tumors during long-term use of immunosuppression. Guselkumab and tildrakizumab are two different monoclonal anti-p19 neutralizing antibodies that are now actively being tested for treatment of psoriasis and psoriatic arthritis in various clinical trials, and these tests are very likely to be extended to other conditions.

In addition to the targeting of IL-23 directly, several other downstream effector cytokines of IL-23R signaling pathway are being targeted with antibodies to treat autoimmune diseases. These include IL-17 and IL-22 which are now in clinical trials.

IL-23R signals through JAK2/TYK2 kinases. There are several JAK2 inhibitors being tested in clinical trials for cancer and also autoimmune disease treatment. Ruxolitinib is a JAK1/JAK2 inhibitor approved by FDA for myelofibrosis and is now being tested for RA and psoriasis treatment. Baricitinib is another Jak1/Jak2 inhibitor in phase II clinical trials for the treatment of RA. Lastly, lestaurtinib is a JAK2 inhibitor and is in phase II trials for the treatment of psoriasis.

References

Ahern, P. P., Schiering, C., Buonocore, S., McGeachy, M. J., Cua, D. J., Maloy, K. J., et al. (2010). Interleukin-23 drives intestinal inflammation through direct activity on T cells. *Immunity, 33*, 279–288.

Awasthi, A., Riol-Blanco, L., Jager, A., Korn, T., Pot, C., Galileos, G., et al. (2009). Cutting edge: IL-23 receptor gfp reporter mice reveal distinct populations of IL-17-producing cells. *Journal of Immunology, 182*, 5904–5908.

Buonocore, S., Ahern, P. P., Uhlig, H. H., Ivanov, I. I., Littman, D. R., Maloy, K. J., et al. (2010). Innate lymphoid cells drive interleukin-23-dependent innate intestinal pathology. *Nature, 464*, 1371–1375.

Burton, P. R., Clayton, D. G., Cardon, L. R., Craddock, N., Deloukas, P., Duncanson, A., et al. (2007). Association scan of 14,500 nonsynonymous SNPs in four diseases identifies autoimmunity variants. *Nature Genetics, 39*, 1329–1337.

Cingoz, O. (2009). Ustekinumab. *MAbs, 1*, 216–221.

Doisne, J. M., Becourt, C., Amniai, L., Duarte, N., Le Luduec, J. B., Eberl, G., et al. (2009). Skin and peripheral lymph node invariant NKT cells are mainly retinoic acid receptor-related orphan receptor (gamma)t + and respond preferentially under inflammatory conditions. *Journal of Immunology, 183*, 2142–2149.

Eken, A., Singh, A. K., & Oukka, M. (2014). Interleukin 23 in Crohn's disease. *Inflammatory Bowel Diseases, 20*, 587–595.

Gaffen, S. L., Jain, R., Garg, A. V., & Cua, D. J. (2014). The IL-23-IL-17 immune axis: From mechanisms to therapeutic testing. *Nature Reviews Immunology, 14*, 585–600.

Griseri, T., McKenzie, B. S., Schiering, C., & Powrie, F. (2012). Dysregulated hematopoietic stem and progenitor cell activity promotes interleukin-23-driven chronic intestinal inflammation. *Immunity, 37*, 1116–1129.

Izcue, A., Hue, S., Buonocore, S., Arancibia-Carcamo, C. V., Ahern, P. P., Iwakura, Y., et al. (2008). Interleukin-23 restrains regulatory T cell activity to drive T cell-dependent colitis. *Immunity, 28*, 559–570.

Jadidi-Niaragh, F., & Mirshafiey, A. (2011). Th17 cell, the new player of neuroinflammatory process in multiple sclerosis. *Scandinavian Journal of Immunology, 74*, 1–13.

Kyttaris, V. C., Zhang, Z., Kuchroo, V. K., Oukka, M., & Tsokos, G. C. (2010). Cutting edge: IL-23 receptor deficiency prevents the development of lupus nephritis in C57BL/6-lpr/lpr mice. *Journal of Immunology, 184*, 4605–4609.

Kyttaris, V. C., Kampagianni, O., & Tsokos, G. C. (2013). Treatment with anti-interleukin 23 antibody ameliorates disease in lupus-prone mice. *BioMed Research International, 2013*, 861028.

Langrish, C. L., McKenzie, B. S., Wilson, N. J., de Waal Malefyt, R., Kastelein, R. A., & Cua, D. J. (2004). IL-12 and IL-23: Master regulators of innate and adaptive immunity. *Immunological Reviews, 202*, 96–105.

Lankford, C. S., & Frucht, D. M. (2003). A unique role for IL-23 in promoting cellular immunity. *Journal of Leukocyte Biology, 73*, 49–56.

Liu, W., Ouyang, X., Yang, J., Liu, J., Li, Q., Gu, Y., et al. (2009). AP-1 activated by toll-like receptors regulates expression of IL-23 p19. *The Journal of Biological Chemistry, 284*, 24006–24016.

Lupardus, P. J., & Garcia, K. C. (2008). The structure of interleukin-23 reveals the molecular basis of p40 subunit sharing with interleukin-12. *Journal of Molecular Biology, 382*, 931–941.

Ngiow, S. F., Teng, M. W., & Smyth, M. J. (2013). A balance of interleukin-12 and −23 in cancer. *Trends in Immunology, 34*, 548–555.

Oppmann, B., Lesley, R., Blom, B., Timans, J. C., Xu, Y., Hunte, B., et al. (2000). Novel p19 protein engages IL-12p40 to form a cytokine, IL-23, with biological activities similar as well as distinct from IL-12. *Immunity, 13*, 715–725.

Parham, C., Chirica, M., Timans, J., Vaisberg, E., Travis, M., Cheung, J., et al. (2002). A receptor for the heterodimeric cytokine IL-23 is composed of IL-12Rbeta1 and a novel cytokine receptor subunit, IL-23R. *Journal of Immunology, 168*, 5699–5708.

Paul, S., Shilpi, S., & Lal, G. (2015). Role of gamma-delta (gammadelta) T cells in the autoimmunity. *Journal Leukocyte Biology, 97*(2), 259–271.

Sherlock, J. P., Joyce-Shaikh, B., Turner, S. P., Chao, C. C., Sathe, M., Grein, J., et al. (2012). IL-23 induces spondyloarthropathy by acting on ROR-gammat + CD3 + CD4-CD8- entheseal resident T cells. *Nature Medicine, 18*, 1069–1076.

Sherlock, J. P., Buckley, C. D., & Cua, D. J. (2014). The critical role of interleukin-23 in spondyloarthropathy. *Molecular Immunology, 57*, 38–43.

Singh, A. K., Eken, A., Fry, M., Bettelli, E., & Oukka, M. (2014). DOCK8 regulates protective immunity by controlling the function and survival of RORgammat + ILCs. *Nature Communications, 5*, 4603.

Smith, S., Gabhann, J. N., Higgs, R., Stacey, K., Wahren-Herlinius, M., Espinosa, A., et al. (2012). Enhanced interferon regulatory factor 3 binding to the interleukin-23p19 promoter correlates with enhanced interleukin-23 expression in systemic lupus erythematosus. *Arthritis and Rheumatism, 64*, 1601–1609.

Sonnenberg, G. F., Monticelli, L. A., Elloso, M. M., Fouser, L. A., & Artis, D. (2011). CD4(+) lymphoid tissue-inducer cells promote innate immunity in the gut. *Immunity, 34*, 122–134.

Teng, M. W., Bowman, E. P., McElwee, J. J., Smyth, M. J., Casanova, J. L., Cooper, A. M., et al. (2015). IL-12 and IL-23 cytokines: from discovery to targeted therapies for immune-mediated inflammatory diseases. *Nature Medicine, 21*(7), 719–729.

Tang, C., Chen, S., Qian, H., & Huang, W. (2012). Interleukin-23: As a drug target for autoimmune inflammatory diseases. *Immunology, 135*, 112–124.

Voulgarelis, M., & Tzioufas, A. G. (2010). Pathogenetic mechanisms in the initiation and perpetuation of Sjogren's syndrome. *Nature Reviews. Rheumatology, 6*, 529–537.

Zheng, Y., Danilenko, D. M., Valdez, P., Kasman, I., Eastham-Anderson, J., Wu, J., et al. (2007). Interleukin-22, a T(H)17 cytokine, mediates IL-23-induced dermal inflammation and acanthosis. *Nature, 445*, 648–651.

Zindl, C. L., Lai, J. F., Lee, Y. K., Maynard, C. L., Harbour, S. N., Ouyang, W., et al. (2013). IL-22-producing neutrophils contribute to antimicrobial defense and restitution of colonic epithelial integrity during colitis. *Proceedings of the National Academy of Sciences of the United States of America, 110*, 12768–12773.

Interleukin 27

Marcel Batten and Dipti Vijayan
Immunology Division, Garvan Institute of Medical Research, and St. Vincent's Clinical School, University of New South Wales, Sydney, NSW, Australia

Synonyms

EBI3; IL-27; IL27; IL27A; IL-27A; IL-27p28; IL-30; p28

Definition

"Interleukin-27" (IL-27) is the name given to the heterodimeric, non-covalent partnering of the protein IL-27p28 with the protein Epstein-Barr virus-induced gene 3 (EBI3) (Pflanz et al. 2002). IL-27p28 is a classical 4-helical bundle cytokine protein with homology to IL-11, IL-12p35, and others. EBI3 is a soluble receptor-like protein with homology to IL-12p40. Together, they form the cytokine known as IL-27 and signal through a heteromeric receptor comprising glycoprotein 130 (gp130, IL-6ST, IL-6Rβ, CD130) and interleukin 27 receptor alpha (IL-27Rα, TCCR, NPOR, WSX-1) (Pflanz et al. 2004), collectively defined as "IL-27R." By virtue of its structural and signaling characteristics, IL-27 belongs to a subfamily of cytokines that also includes IL-6, IL-12, IL-23, IL-35, and a cytokine comprising CLF-1 and IL-27p28. EBI3 is also a subunit of IL-35.

Biosynthesis and Release

IL-27 is expressed predominantly by human and mouse antigen presenting cells (macrophages, monocytes, dendritic cells) and is rapidly expressed in response to microbial encounter, through toll-like receptors (TLR), or stimulation with the endogenous ligands CD40, IL-1β, or type I or type II interferons (IFN). The IL-27p28 and EBI3 subunits are both regulated at the transcriptional level, but are independently regulated as outlined below.

Both IL-27 subunits are induced by agonism of TLR3 (polyinosinic:polycytidylic acid), TLR4 (LPS), and TLR7/8 (R848). TLR2 activates EBI3 but not IL-27p28 expression. TLR9 is also reported to induce EBI3. EBI3 induction is MyD88, PU.1, and subsequently NFκB (p50/60) dependent, while TLR-induced IL-27p28 is dependent upon both MyD88 and TRIF pathway activation and partially dependent on NFκB (cRel) (Jankowski et al. 2010).

IL-27p28 expression is induced by type I and II IFN stimulation of antigen-presenting cells. The IL-27p28 promoter contains an IFN-stimulated regulatory element site. Activation of interferon regulatory factor (IRF)-1, downstream of MyD88 or IFNαR itself, and IRF3 downstream of TRIF are important for the initial transcription of IL-27p28. Its expression is also directly induced by type I and type II IFN, through activation of IRF-1 and the formation of STAT1/STAT2/IRF9 complexes collectively known as "ISGF3"(Hall et al. 2012b). Upregulation of IL-27p28 expression in response to TLR ligands is at least partially dependent on type I IFN expression. TLRs that enhance IL-27p28 expression are those that induce IFNα production, and blockade of type I IFN diminished TLR-induced IL-27p28 but not EBI3 production. In summary, key stimuli that induce IL-27p28 expression are TLR agonists and type I and II IFN, while the key downstream transcription factors involved are IRF-1, IRF-3, and IRF-9, along with c-rel. It has also been reported that 4-1BB agonism enhances IL-27p28 production while extracellular ATP suppresses its expression (through purinergic (P2) receptors). Expression control has been reviewed in detail here (Jankowski et al. 2010).

IL-27p28 secretion from human cells appears to require co-expression of EBI3, and although murine IL-27p28 can be secreted alone, its secretion is greatly enhanced by EBI3 co-expression (Pflanz et al. 2002). Poor secretion when independently expressed suggests that the two subunits of IL-27 associate intracellularly before secretion. However, several groups have described biological activities for the individual subunits, leaving open the possibility that the subunits may be able to act independently or associate with each other or other cytokine subunits extracellularly. Monomeric IL-27p28 has been reported to act as an inhibitor of IL-27 signaling.

Biological Activities

IL-27 is a cytokine with both pro- and anti-inflammatory effects. It promotes immune responses in tumor models as well as in certain models of infection and inflammation, such as influenza and proteoglycan-induced arthritis. However, IL-27 signals are predominantly immunosuppressive in many contexts. In multiple models of infection, loss of IL-27 signaling led to over-reactive immune responses, and in some cases, such as *Toxoplasma gondii* and *Mycobacterium tuberculosis* infection, the development of immune-mediated tissue pathology despite effective clearance of the pathogens was observed. In the T cell-driven autoimmune disease model experimental autoimmune encephalomyelitis (EAE), loss of IL-27Rα or IL-27p28 also led to a more severe inflammatory disease and administration of recombinant forms of IL-27 suppressed collagen-induced arthritis (Hall et al. 2012b). As outlined below, understanding the effects of IL-27 on particular immune cell subsets has helped to reveal the mechanisms by which IL-27 has these pleiotropic effects.

Receptor Expression and Basic Signaling Characteristics

IL-27 signals through IL-27R receptor consisting of gp130 and IL-27Rα. gp130 is the signal-transducing chain for at least 8 cytokines, including IL-6, and is expressed in many cell and tissue types. The expression of IL-27Rα is more restricted; thus the complete IL-27R is expressed predominantly on hematopoietic cells, having effects on many immune cells including T cells, B cells, dendritic cells (DC), NK cells, mast cells, and eosinophils. IL-27R is also detected on vascular endothelium and keratinocytes (Jankowski et al. 2010). Several of the cytokine receptor alpha chains that engage with gp130 have short or no intracellular domains and therefore rely on gp130 for signal transduction. However, the IL-27Rα chain possesses a cytoplasmic domain that contains a Box1 binding site for JAK proteins and conserved tyrosine residues that could provide docking sites for SH2 domain proteins such as STATs. JAK1 and STAT1 directly associate with the IL-27Rα receptor chain (Hall et al. 2012b); thus, IL-27Rα also contributes to signaling. Indeed, while IL-6 also signals through gp130, the kinetics of STAT1 and STAT3 activation are different following signaling by the two cytokines.

IL-27 stimulation of cells expressing the IL-27R results in the activation of the JAK/STAT pathway, particularly JAK1, STAT1, and STAT3. Activation of JAK2 and TYK2 as well as STAT4 and STAT5 has also been reported. IL-27 activates p38/MAPK and ERK1/2 signaling and induces c-Myc and Pim-1, leading to activation of cyclins D2, D3, and CDK4 in T cells and cyclins A, D2, and D3 in B cells (Charlot-Rabiega et al. 2011).

An alternatively spliced IL-27Rα variant that gives rise to soluble IL-27Rα has been described to be expressed by neuronal cells and to form a humanin receptor in complex with gp130 and CNTFR. However, it remains to be determined whether this soluble form of IL-27Rα could also facilitate trans-signaling, similar to that which occurs with the soluble IL-6R, thereby allowing IL-27 to signal to cells expressing only the gp130 chain (Hall et al. 2012b).

Effects on CD4+ T Cells

The most extensively studied biological effects of IL-27 are those related to CD4+ T helper cells. Key effects on this cell type include supporting cell proliferation and survival, skewing the polarization of T helper cell subsets, and altering cytokine production.

The first reports of IL-27 indicated that it promotes the expansion of naïve CD4+ T cells, activated through the TCR (Pflanz et al. 2002). It was later found that IL-27 could drive cell cycle entry through c-Myc- and Pim-1-dependent induction of cyclin D2, cyclin D3, and CDK4 (Charlot-Rabiega et al. 2011). IL-27 has a much greater impact on the proliferation of naïve T cells compared to memory T cells. IL-27 also promotes survival of activated CD4+ T cells through downregulation of Fas ligand expression and induction of FLICE inhibitory protein (cFLIP), which is a direct inhibitor of Fas-mediated signaling (Kim et al. 2013). These survival effects on activated T cells are thought to contribute to the observed reduction in activated *Il27ra-/-* T cells *in vivo* in colitis and immunization models (Batten et al. 2010; Kim et al. 2013).

IL-27 is extremely important in directing the differentiation of effector CD4+ T helper cell subsets. IL-27 activates STAT1 and induces expression of T-bet, transcription factors that are strongly associated with type 1 CD4+ T cell (TH1) polarization. IL-27 alone has a mild effect on IFNγ production, but it increases IL-12Rβ2 expression on activated T cells and thereby synergizes with IL-12 to enhance IFNγ production (Hall et al. 2012b). IL-27-mediated STAT1 activation also induces expression of adhesion molecule ICAM/LFA-1, which promotes TH1 activity (Hall et al. 2012b). Despite this apparent TH1-promoting activity in vitro, the requirement for IL-27 signaling in in vivo immune responses is context dependent. During infection with multiple pathogens, including those whose clearance is dependent on a TH1 response, IL-27 signaling was not required for IFNγ production or pathogen clearance, although a delay in the TH1 response was noted in some models such as *Leishmania major*. This may have been secondary to enhancement of the TH2 response. In other models, the TH1 response was even enhanced in the absence of IL-27 signals. These data suggest that in terms of TH1 differentiation, other factors may compensate for lack of IL-27 during infection. However, IL-27 *does* appear to be required for effective antitumor responses, which are reliant on TH1 immunity, and reduced IFNγ levels are associated with defective tumor clearance in *Il27ra-/-* mice, as discussed below. Moreover, in the inflammatory PGIA arthritis model, which is TH1 dependent, pathology is ameliorated in the absence of IL-27Rα along with reduced IFNγ levels (Cao et al. 2008).

IL-27 stimulation directly suppresses the differentiation of TH2 and TH17 cells. In vitro studies of the effect of IL-27 on purified CD4+ T cells revealed that it suppressed the expression of the TH2 master transcriptional regulator GATA3 and the Th17 master transcription factor RORc (Jankowski et al. 2010). This in vitro evidence is supported by a plethora of in vivo data, in which loss of IL-27 signaling in mice results in enhanced TH2 and TH17 responses. IL-27 was found to have a protective effect in TH2-type disease models such as asthma and helminth infection. Multiple reports confirm that it controls TH17-mediated inflammation in models such as EAE and *T. gondii* neuroinflammation (Hall et al. 2012b; Jankowski et al. 2010). In in vitro cultures, IL-27 blocks production of IL-17 from CD4+ T cells in a STAT1 and partially STAT3-dependent manner, independent of its TH1-inducing capabilities, since suppression of IL-17 is T-bet and IFNγ independent. TH17 suppression is also independent of IL-10 and SOCS3 induction, which occurs in response to IL-27. In this way, IL-27 suppresses naïve cell differentiation into TH17 cells but seems to have more limited effects on established TH17 cells. Suppression of IL-17 production by TH17-producing memory or effector cells was observed in some situations, such as polyclonal human CD45RO+ memory cells and IL-17-producing cells in toxoplasmic encephalitis, but not in cells isolated from mice with EAE, and a limited reduction in IL-17 was observed in vitro upon second stimulation of in vitro-differentiated TH17 cells (Hall et al. 2012b; Jankowski et al. 2010). Direct suppression of TH9 differentiation has also been observed (Murugaiyan and Saha 2013).

In many in vivo models of infection and inflammation, including autoimmune disease models, bacterial infection, parasitic infection, colitis, asthma, and inflammatory hepatitis, the loss of IL-27 signals led to increased inflammation, suggesting an important immunosuppressive role for IL-27. There are several factors that account for this suppressive activity, including effects on a range of non-T cells. However, IL-27 has important immunosuppressive effects directly on CD4+ T cells. One key outcome of IL-27 signaling to CD4+ T cells is the potent induction of IL-10, which has widespread anti-inflammatory effects. IL-27 directly promotes the production of IL-10 from human and mouse CD4, CD8, and NK cells. Indeed, IL-27, in combination with TGFβ, promotes the differentiation of CD4+ T cells into Foxp3-negative, IL-10-producing type 1 regulatory (Tr1) cells. IL-27 promotes the differentiation of these cells through STAT3-dependent induction of AhR and activation of the transcription factor c-Maf (Murugaiyan and Saha 2013). EGR-2 and Blimp-1 have also been implicated in IL-27-induced IL-10 expression (Iwasaki et al. 2013). Genetic deletion of *Il27ra* results in reduced numbers of IL-10-producing cells and enhanced immune responses in toxoplasmic encephalitis, EAE, *Leishmania major*, and *Listeria monocytogenes*. The suppressive effect of IL-27 in EAE was shown to be reliant on the production of IL-10 by encephalitogenic T cells (Hall et al. 2012b; Jankowski et al. 2010). This induction of IL-10 production, along with the ability of IL-27 to antagonize IL-2 production, are thought to be major contributors to the broad immunosuppressive effects of IL-27.

IL-27 promotes IL-21 production by human and mouse CD4+ T cells. IL-21 is a cytokine produced at high levels by T follicular helper cells (TFH), which are the CD4+ T helper cell subset that provide "help" to B cells within germinal centers (GC). Without TFH cells, GC responses are short-lived and poor affinity maturation and B cell memory ensues. As well as promoting IL-21 expression, IL-27 enhances the survival of TFH cells and the expression of ICOS, a costimulatory molecule important in TFH activity. Loss of IL-27Rα signaling resulted in a T cell-intrinsic GC defect after immunization with TNP-haptenated ovalbumin (OVA) in CFA or during LCMV infection (Batten et al. 2010; Harker et al. 2013), and ameliorated disease was observed in *Il27ra-/-* mice in the pristane-induced lupus model (Batten et al. 2010). *Il27ra-/-* mice display a specific reduction of high-affinity antibody but not overall antibody levels, suggesting that the requirement for IL-27 signaling in the humoral response is restricted to GCs.

IL-27 inhibits the differentiation of Foxp3+ inducible T_{reg} cells during in vitro stimulation in the presence of TGFβ, and *Il27ra-/-* naïve CD4+ T cells were predisposed to develop into Foxp3+ cells after transfer in the CD45RB^{hi} transfer colitis model (Hall et al. 2012b; Jankowski et al. 2010). *Il27ra*-deficient mice also displayed higher numbers of Foxp3+ cells in tumor models (Natividad et al. 2013). However, in unchallenged animals, IL-27Rα deficiency does not alter the basal Foxp3 + T_{reg} number or function, suggesting that IL-27 does not influence the development of "natural" or thymically generated T_{reg} cells. Nor does IL-27 antagonize the function of fully differentiated T_{reg} cells. It has been suggested, however, that IL-27 promotes survival of T_{reg} and expression of T-bet, CXCR3, IL-10, and IFNγ by differentiated T_{reg} cells (Hall et al. 2012a).

Effects on CD8+ T Cells

As in CD4+ T cells, IL-27 signaling to human and mouse CD8+ T cells activates transcription factors STAT1 and STAT3 (as well as STAT2, STAT4, and STAT5 in mice only), enhances proliferation, and promotes the expression of T-bet, eomesodermin, IFNγ, and IL-12Rβ2. IL-27 promotes cytotoxic T lymphocyte (CTL) activity by upregulating the expression of perforin and granzyme resulting in enhanced lysis of target cells. IL-27-mediated granzyme B induction in human cells is reported to be dependent on the stimulation of IL-21 expression by the CD8+ T cells themselves. IL-27 signals have been found to be important in promoting beneficial CTL activity in a number of model tumors (Murugaiyan and Saha 2013).

IL-27 may have a role in regulating the antiviral CD8+ T cell response. One study tested the activity of *Il27ra-/-* CD8+ T cells during mouse models of influenza and *T. gondii* infection and noted that, although normal numbers of antigen-specific CD8+ T cells were detected, CD8+ T cells lacking IL-27R displayed reduced IFNγ production, as detected using genetically modified IFNγ-reporter "Yeti" mice (Mayer et al. 2008). However, other studies have observed that IFNγ production by ex vivo restimulated CD8+ cells derived from influenza or *T. gondii*-infected *Il27ra-/-* mice is not inhibited (Sun et al. 2011; Villarino et al. 2003).

IL-27 also promotes IL-10 secretion by CD8+ T cells. IL-10 production by CTLs during respiratory influenza infection is critical to prevent excess inflammation that can result in tissue pathology during viral clearance. In influenza infection in mice, *Ebi3*-deficient CD8+ cells had normal IFNγ production but impaired IL-10 expression (Sun et al. 2011). IL-2 and IL-27 synergistically upregulate IL-10 by CD8+ T cells through induction of Blimp-1 in both human and mouse cells (Sun et al. 2011). In line with the CD4+ T cell studies, IL-27 has a more pronounced effect during the activation of naïve human CD8+ T cells compared with memory CD8+ cells.

Effects on B Cells

B cells also express the IL-27R, although the levels of expression and downstream effects vary according to the activation and differentiation state. Flow cytometry of human tonsil B cell populations showed IL-27Rα protein expression on resting plasma cells (Cocco et al. 2011) and naïve and memory B cells, but not GC B cells (Larousserie et al. 2006). Anti-CD40 stimulation increased the IL-27Rα expression on naïve, memory, and GC B cells, indicating that IL-27 can signal to activated GC B cells. gp130 was present at low levels on all subsets and was rapidly upregulated after CD40 stimulation. STAT1 and STAT3 were phosphorylated in response to IL-27 stimulation by B cells, although the level of activation was low in GC B cells compared to other subsets. Although naïve and memory B cells expressed comparable IL-27Rα levels, naïve B cells stimulated with IL-27 experienced stronger STAT1 and STAT3 activation, whereas memory B cells had moderate STAT1 and weak STAT3 activation. IL-27 induces greater proliferation of naive than memory human B cells (Larousserie et al. 2006) with IL-27-induced proliferation being dependent on Pim-1, a kinase regulator of cell proliferation that promotes cell proliferation by increasing G1 to S transition (Charlot-Rabiega et al. 2011; Larousserie et al. 2006).

Again mirroring T cell effects, IL-27 enhanced the expression of T-bet in both naïve and memory B cells during anti-CD40 or anti-Ig activation; however, in human B cells, IL-12Rβ1 and β2 expressions were not strongly induced. IL-27 also enhanced CD54, CD86, and CD95 (Fas) expression by human B cells, particularly naïve B cells. In murine B cells, IL-27 enhanced STAT1 activation and T-bet expression and promoted immunoglobulin (Ig) class switching to the IgG2a isotype while inhibiting IgG1 switching (Yoshimoto et al. 2004). Consistent with this, *Il27ra-/-* mice have reduced basal serum IgG2a and reduced IgG2a antibody production in the PGIA model (Cao et al. 2008). However, interpretation of in vivo data from whole-body *Il27ra-/-* mice is complicated by the fact that altered T helper cell responses could cause indirect effects on B cell activity and class switching. In human B cell cultures, IL-27 stimulation did not have a strong influence on Ig switching (Boumendjel et al. 2006; Larousserie et al. 2006), although one study noted that IL-27 weakly enhanced IgG1 production (Boumendjel et al. 2006).

IL-27 and Induction of Tolerogenic DC

IL-27 induces a regulatory function in dendritic cells (DC). In mice, IL-27R is expressed highly on conventional DC (cDC) with very little observed on plasmacytoid DC (pDC) (Mascanfroni et al. 2013). IL-27 exposure of mouse cDCs prior to activation suppressed their maturation, as illustrated by reduced expression of MHC class I and costimulatory molecules CD40, CD80, and CD86. IL-27 also suppressed pro-inflammatory cytokine production by the DCs (IL-12, IL-6, and IL-23) but upregulated expression of immunosuppressive cytokines IL-10 and TGFβ, as well as IL-27 itself, suggesting a positive feedback loop (Mascanfroni et al. 2013). These alterations led to the decreased ability of DC to activate antigen-specific proliferation or pro-inflammatory cytokine production by CD4+ T cells but boosted the differentiation of Foxp3+ and IL-10+ regulatory T cells. In line with the effect of recombinant IL-27 in enhancing tolerogenic DC activity, *Il27ra-/-* DC are hyper-inflammatory (Wang et al. 2007). Tolerogenic DC-specific effects of IL-27 signaling were found to contribute to the protective effect of IL-27 during EAE (Mascanfroni et al. 2013).

In human DCs, IL-27 promotes the expression of B7-H1 (PD-L1), which provides a suppressive signal to T cells (Karakhanova et al. 2011). However, B7-H1 upregulation did not appear to be critical for promoting the tolerogenic DC phenotype, at least in mice, since IL-27 still exerted its suppressive effect when B7-H1-deficient DC were used (Mascanfroni et al. 2013). Instead, CD39 upregulation was essential for the suppressive effects of IL-27 on cDC function (Mascanfroni et al. 2013). CD39 is an ectonuclease that catalyzes the degradation of ATP to ADP, thereby inhibiting inflammasome activation and downstream differentiation of encephalitogenic TH1 and TH17 cells during EAE.

Effects on Other Innate Leukocytes

Many innate cells express the complete IL-27, including natural killer (NK) cells, mast cells, eosinophils, macrophages, and DCs.

IL-27 inhibits the ability of murine macrophages to produce IL-12 and TNF and promotes IL-10 production. *Il27ra-/-* macrophages in *M. tuberculosis* infection show increased IL-12p40, IL-6, and TNF production, which could contribute to the pathogenically exacerbated response to *M. tuberculosis* observed in *Il27ra-/-* mice. In contrast, IL-27 may have a pro-inflammatory role in human monocytes, where it enhanced TLR-driven pro-inflammatory cytokine production. IL-27 also upregulated MHC classes I and II expression along with costimulatory and adhesion molecules such as CD80, CD86, and CD54 by human monocyte cell line THP-1. Thus, IL-27 may promote antigen presentation by human monocytes/macrophages. Through the activation of STAT1, IL-27 is also reported to mediate nitric oxide production by mouse peritoneal macrophages (Hall et al. 2012b; Jankowski et al. 2010; Shimizu et al. 2013).

IL-27 prolonged human eosinophil survival, modulated the expression of adhesion molecules, and induced the release of pro-inflammatory cytokines IL-6, TNF, IL-1β, and chemokines CCL2, CXCL8, and CXCL1. These effects in eosinophils were found to be due to the activation of extracellular signal-regulated kinase (ERK), c-Jun N-terminal kinase, and p38 mitogen-activated protein kinase, as well as NFκB (Hu et al. 2011).

In human mast cells, IL-27, through STAT3, activated the transcription of pro-inflammatory cytokines such as IL-1α, IL-1β, and TNF but did not change Fc-mediated degranulation (Pflanz et al. 2004). In a model of passive cutaneous anaphylaxis, *Il27ra-/-* mast cells exhibit elevated protease activity (Artis et al. 2004).

Hematopoietic Cell Differentiation

IL-27 can signal to hematopoietic stem cells (HSCs) and, in combination with SCF, promote the expansion of human and mouse hematopoietic stem/progenitor cells. Moreover, IL-27 overexpression in transgenic mice leads to enhanced myelopoiesis and impaired B cell lymphopoiesis in the bone marrow and extramedullary hematopoiesis in the spleen (Seita et al. 2008). It is interesting that the IL-27p28 protein has a polyglutamic acid domain that confers hydroxyapatite (i.e., bone)-binding ability to the protein (Tormo et al. 2013), meaning that high local IL-27 concentrations may be found within the bone marrow niche. Nevertheless, IL-27- or IL-27Rα-deficient mice have no overt defects in hematopoiesis.

IL-27 Effects on Non-Hematopoietic Cells

Although IL-27 receptor is predominantly expressed by cells of the hematopoietic lineage, there is evidence for effects of IL-27 on a number of non-hematopoietic cell types, including epithelial cell fibroblasts and keratinocytes. IL-27Rα is expressed on epithelial tumor cells derived from the breast, colon, and melanomas. It stimulates intestinal epithelium to express scavenger receptor DMBT1 and act as an antimicrobial peptide (Hall et al. 2012b). IL-27 is reported to upregulate MHC class II in a human endothelial cell line (Jankowski et al. 2010).

Elevated IL-27R expression in fibroblasts has been noted in systemic sclerosis patients where IL-27 promotes fibroblast proliferation and the production of collagen and induces CXCL10 via p38 MAPK and PI3K-Akt signaling pathways in lung fibroblasts (Dong et al. 2013; Hall et al. 2012b). In chronic eczema, human keratinocytes express IL-27 receptor and respond to IL-27 by increasing MHC class I expression and the production of CXCL10. In keratinocytes and fibroblasts, IL-27 induced the production of anti-inflammatory IL-18-binding protein, an antagonist of IL-18 (Wittmann et al. 2012). IL-27 signals to hepatocytes through STAT1 and STAT3 (Bender et al. 2009).

Pathophysiological Role for IL-27

The biological functions of IL-27 described above have highlighted several areas where IL-27 is likely to be pathophysiologically relevant. Moreover, IL-27 gene polymorphisms have been associated with increased risk for chronic obstructive pulmonary diseases, asthma, and immune thrombocytopenia (Hall et al. 2012b; Zhao et al. 2013). Summarized below are examples of pathological contexts in which IL-27 has been found to have an important role. These diseases are areas of intense interest with regard to the possibility of developing IL-27-related therapeutics.

IL-27 in Autoimmune Disease

The pleiotropic nature of IL-27 in regulating innate and adaptive responses has led to this cytokine being a candidate in the pathogenesis of several human autoimmune disorders. Diseases such as ankylosing spondylitis (Lin et al. 2014), Sjögren's syndrome (Xia et al. 2014), or rheumatoid arthritis (Baek et al. 2012) are characterized by an increased serum IL-27, suggesting that IL-27 may have a pro-inflammatory role in these diseases. However, the increased inflammation occurring, relative to healthy subjects, may lead to elevated IL-27 levels that are not necessarily causative of the disease. An inflammatory role of IL-27 in such diseases may be consistent with the observed reduction in PGIA disease severity in the absence of IL-27 signaling in mice (Cao et al. 2008). However, IL-27 had a protective effect in mouse collagen-induced arthritis (Jankowski et al. 2010). In other autoimmune diseases, such as multiple sclerosis, the levels of IL-27 are significantly reduced (Babaloo et al. 2013), which is in agreement with IL-27 having a protective role in the murine EAE model (Hall et al. 2012b).

Systemic lupus erythematosus (SLE) is a multi-organ, complex autoimmune disorder characterized by the development of autoantibodies and rheumatoid factors. The effects of IL-27 in several mouse models of SLE have been investigated. Absence of IL-27 signaling in *MRL/lpr* mice impacts murine nephritis, such that deficiency of IL-27R or EBI3 directs a phenotype that strongly resembles membranous nephropathy, as opposed to a diffuse proliferative glomerulonephritis seen among control littermates. This skewing occurs due to the impaired ability of *Il27ra-/-* mice to direct a TH1 response and skewing toward a TH2 response with a predominance of IgG1 and IgE antibodies (Igawa et al. 2009; Shimizu et al. 2005). *Il27ra* deficiency in mice resulted in reduced GC activity, and *Il27ra-/-* mice injected with pristane developed a milder form of SLE-like disease compared to their wild-type counterparts (Batten et al. 2010). Moreover, IL-27 also promoted production of B cell stimulatory cytokine IL-21 by human B cells (Batten et al. 2010). In the clinical setting, findings as to the levels of IL-27 in patient blood have been mixed, with two studies finding low IL-27 expression and one finding high IL-27 expression in patients compared to healthy controls (Duarte et al. 2013; Li et al. 2010; Qiu et al. 2013). The later study also found that treatment can have a significant impact on plasma IL-27 concentration, with prednisone resulting in significantly reduced IL-27 (Qiu et al. 2013). While the role of IL-27 in lupus is clearly multifactorial, there is intense interest in IL-27-based therapeutics for the treatment of this disease.

Antitumor Properties of IL-27

Numerous reports have indicated that IL-27 is protective in the tumor context, through various mechanisms including promoting CD8 and NK cell cytotoxicity, IFNγ production, ADCC, antiangiogenic effects, direct suppression of tumor growth, and inhibition of COX-2. Many of these studies used tumor cell line transplant models in which the tumor cells lines were transduced to express recombinant IL-27. In most cases, the mice effectively destroyed the IL-27 expressing tumor cells and developed long-lived immunity against the parental cell line. Studies testing the importance of physiological IL-27 and using models in which the tumors transform and differentiate in situ have confirmed a protective role for IL-27 (Natividad et al. 2013; Shinozaki et al. 2009). Together these data suggest that high IL-27 levels can lead to tumor clearance and long-lived memory against the tumor cells and have generated excitement with respect to the possible development of IL-27-based therapeutic agents.

In the context of cancer, TH1-type responses are beneficial, while T_{reg} cells suppress the antitumor response and allow tumor growth. As described in the sections above, IL-27 can promote TH1 responses and suppress T_{reg} differentiation and, indeed, appears to do so in mice-bearing tumors (Natividad et al. 2013; Salcedo et al. 2009). Exciting observations by Salcedo and colleagues (2009), using a TBJ neuroblastoma cell line model, indicate potent synergism between IL-27 and IL-2, which is in restricted clinical use for the treatment of metastatic renal carcinoma and melanoma. Coadministration of IL-27 completely suppressed the IL-2-induced T_{reg} cell expansion and resulted in a >30-fold increase in serum IFNγ compared with either cytokine individually. The use of the two cytokines together had a powerful protective influence on disseminated TBJ neuroblastoma in mice. Thus, these cytokines may be a particularly effective combination therapy. In terms of its application in the clinic, IL-27 is worthy of further investigation since it significantly alters murine cancer progression as a sole agent (Murugaiyan and Saha 2013). Moreover, there is some suggestion that IL-27 will have an advantage over other cytokine therapies in terms of reduced toxicity.

In most studies, patients with esophageal cancer, breast cancers, or cutaneous T cell lymphomas were found to have increased serum levels of IL-27p28 (Diakowska et al. 2013; Lu et al. 2014; Miyagaki et al. 2010), although one study in esophageal cancer patients found the opposite (Tao et al. 2012). These data are hard to interpret as IL-27 levels may reflect generalized inflammation occurring in response to the tumors. Nonetheless, IL-27 gene polymorphisms (rs153109 and rs17855750) did not increase the susceptibility of patients to colorectal (Wang et al. 2012), nasopharyngeal (Wei et al. 2009), and hepatocellular carcinomas (Peng et al. 2013).

Modulation by Drugs

No drugs that directly target or agonize IL-27 are currently in the clinic.

Cross-References

- ▶ Basophils
- ▶ Cancer and Inflammation
- ▶ Cytokines
- ▶ Dendritic Cells
- ▶ Foxp3
- ▶ Interferon gamma
- ▶ Inflammatory Bowel Disease
- ▶ Inflammatory Bowel Disease Models in Animals
- ▶ Interleukin 2
- ▶ Interleukin 4 and the Related Cytokines (Interleukin 5 and Interleukin 13)
- ▶ Interleukin 6
- ▶ Interleukin 10
- ▶ Interleukin 12
- ▶ Interleukin 17
- ▶ Janus Kinases (Jaks)/STAT Pathway
- ▶ TH17 Response
- ▶ Toll-Like Receptors
- ▶ Type I Interferons

References

Artis, D., Villarino, A., Silverman, M., He, W., Thornton, E. M., Mu, S., et al. (2004). The IL-27 receptor (WSX-1) is an inhibitor of innate and adaptive elements of type 2 immunity. *Journal of Immunology, 173*, 5626–5634.

Babaloo, Z., Yeganeh, R. K., Farhoodi, M., Baradaran, B., Bonyadi, M., & Aghebati, L. (2013). Increased IL-17A but decreased IL-27 serum levels in patients with multiple sclerosis. *Iranian Journal of Immunology: IJI, 10*, 47–54.

Baek, S. H., Lee, S. G., Park, Y. E., Kim, G. T., Kim, C. D., & Park, S. Y. (2012). Increased synovial expression of IL-27 by IL-17 in rheumatoid arthritis. *Inflammation Research: Official Journal of the European Histamine Research Society, 61*, 1339–1345.

Batten, M., Ramamoorthi, N., Kljavin, N. M., Ma, C. S., Cox, J. H., Dengler, H. S., et al. (2010). IL-27 supports germinal center function by enhancing IL-21 production and the function of T follicular helper cells. *The Journal of Experimental Medicine, 207*, 2895–2906.

Bender, H., Wiesinger, M. Y., Nordhoff, C., Schoenherr, C., Haan, C., Ludwig, S., et al. (2009). Interleukin-27 displays interferon-gamma-like functions in human hepatoma cells and hepatocytes. *Hepatology, 50*, 585–591.

Boumendjel, A., Tawk, L., Malefijt Rde, W., Boulay, V., Yssel, H., & Pene, J. (2006). IL-27 induces the production of IgG1 by human B cells. *European Cytokine Network, 17*, 281–289.

Cao, Y., Doodes, P. D., Glant, T. T., & Finnegan, A. (2008). IL-27 induces a Th1 immune response and susceptibility to experimental arthritis. *Journal of Immunology, 180*, 922–930.

Charlot-Rabiega, P., Bardel, E., Dietrich, C., Kastelein, R., & Devergne, O. (2011). Signaling events involved in interleukin 27 (IL-27)-induced proliferation of human naive CD4+ T cells and B cells. *The Journal of Biological Chemistry, 286*, 27350–27362.

Cocco, C., Morandi, F., & Airoldi, I. (2011). Interleukin-27 and interleukin-23 modulate human plasmacell functions. *Journal of Leukocyte Biology, 89*, 729–734.

Diakowska, D., Lewandowski, A., Markocka-Maczka, K., & Grabowski, K. (2013). Concentration of serum interleukin-27 increase in patients with lymph node metastatic gastroesophageal cancer. *Advances in Clinical and Experimental Medicine: Official Organ Wroclaw Medical University, 22*, 683–691.

Dong, S., Zhang, X., He, Y., Xu, F., Li, D., Xu, W., et al. (2013). Synergy of IL-27 and TNF-alpha in regulating CXCL10 expression in lung fibroblasts. *American Journal of Respiratory Cell and Molecular Biology, 48*, 518–530.

Duarte, A. L., Dantas, A. T., de Ataide Mariz, H., dos Santos, F. A., da Silva, J. C., Da Rocha, L. F., Jr., et al. (2013). Decreased serum interleukin 27 in Brazilian systemic lupus erythematosus patients. *Molecular Biology Reports, 40*, 4889–4892.

Hall, A. O., Beiting, D. P., Tato, C., John, B., Oldenhove, G., Lombana, C. G., et al. (2012a). The cytokines interleukin 27 and interferon-gamma promote distinct Treg cell populations required to limit infection-induced pathology. *Immunity, 37*, 511–523.

Hall, A. O., Silver, J. S., & Hunter, C. A. (2012b). The immunobiology of IL-27. *Advances in Immunology, 115*, 1–44.

Harker, J. A., Dolgoter, A., & Zuniga, E. I. (2013). Cell-intrinsic IL-27 and gp130 cytokine receptor signaling regulates virus-specific CD4(+) T cell responses and viral control during chronic infection. *Immunity, 39*, 548–559.

Hu, S., Wong, C. K., & Lam, C. W. (2011). Activation of eosinophils by IL-12 family cytokine IL-27: Implications of the pleiotropic roles of IL-27 in allergic responses. *Immunobiology, 216*, 54–65.

Igawa, T., Nakashima, H., Sadanaga, A., Masutani, K., Miyake, K., Shimizu, S., et al. (2009). Deficiency in EBV-induced gene 3 (EBI3) in MRL/lpr mice results in pathological alteration of autoimmune glomerulonephritis and sialadenitis. *Modern Rheumatology, 19*, 33–41.

Iwasaki, Y., Fujio, K., Okamura, T., Yanai, A., Sumitomo, S., Shoda, H., et al. (2013). Egr-2 transcription factor is required for Blimp-1-mediated IL-10 production in IL-27-stimulated CD4+ T cells. *European Journal of Immunology, 43*, 1063–1073.

Jankowski, M., Kopinski, P., & Goc, A. (2010). Interleukin-27: Biological properties and clinical application. *Archivum Immunologiae et Therapiae Experimentalis, 58*, 417–425.

Karakhanova, S., Bedke, T., Enk, A. H., & Mahnke, K. (2011). IL-27 renders DC immunosuppressive by induction of B7-H1. *Journal of Leukocyte Biology, 89*, 837–845.

Kim, G., Shinnakasu, R., Saris, C. J., Cheroutre, H., & Kronenberg, M. (2013). A novel role for IL-27 in mediating the survival of activated mouse CD4 T lymphocytes. *Journal of Immunology, 190*, 1510–1518.

Larousserie, F., Charlot, P., Bardel, E., Froger, J., Kastelein, R. A., & Devergne, O. (2006). Differential effects of IL-27 on human B cell subsets. *Journal of Immunology, 176*, 5890–5897.

Li, T. T., Zhang, T., Chen, G. M., Zhu, Q. Q., Tao, J. H., Pan, H. F., et al. (2010). Low level of serum interleukin 27 in patients with systemic lupus erythematosus. *Journal of Investigative Medicine: The Official Publication of the American Federation for Clinical Research, 58*, 737–739.

Lin, T.T., J. Lu, C.Y. Qi, L. Yuan, X.L. Li, L.P. Xia, & H. Shen. (2014). Elevated serum level of IL-27 and VEGF in patients with ankylosing spondylitis and associate with disease activity. *Clinical and Experimental Medicine, 15*, 227–231.

Lu, D., Zhou, X., Yao, L., Liu, C., Jin, F., & Wu, Y. (2014). Clinical implications of the interleukin 27 serum level in breast cancer. *Journal of Investigative Medicine: The Official Publication of the American Federation for Clinical Research, 62*, 627–631.

Mascanfroni, I. D., Yeste, A., Vieira, S. M., Burns, E. J., Patel, B., Sloma, I., et al. (2013). IL-27 acts on DCs to suppress the T cell response and autoimmunity by inducing expression of the immunoregulatory molecule CD39. *Nature Immunology, 14*, 1054–1063.

Mayer, K. D., Mohrs, K., Reiley, W., Wittmer, S., Kohlmeier, J. E., Pearl, J. E., et al. (2008). Cutting edge: T-bet and IL-27R are critical for in vivo IFN-gamma production by CD8 T cells during infection. *Journal of Immunology, 180*, 693–697.

Miyagaki, T., Sugaya, M., Shibata, S., Ohmatsu, H., Fujita, H., & Tamaki, K. (2010). Serum interleukin-27 levels in patients with cutaneous T-cell lymphoma. *Clinical and Experimental Dermatology, 35*, e143–e144.

Murugaiyan, G., & Saha, B. (2013). IL-27 in tumor immunity and immunotherapy. *Trends in Molecular Medicine, 19*, 108–116.

Natividad, K. D., Junankar, S. R., Mohd Redzwan, N., Nair, R., Wirasinha, R. C., King, C., et al. (2013). Interleukin-27 signaling promotes immunity against endogenously arising murine tumors. *PLoS ONE, 8*, e57469.

Peng, Q., Qin, X., He, Y., Chen, Z., Deng, Y., Li, T., et al. (2013). Association of IL27 gene polymorphisms and HBV-related hepatocellular carcinoma risk in a Chinese population. *Infection, Genetics and Evolution: Journal of Molecular Epidemiology and Evolutionary Genetics in Infectious Diseases, 16*, 1–4.

Pflanz, S., Timans, J. C., Cheung, J., Rosales, R., Kanzler, H., Gilbert, J., et al. (2002). IL-27, a heterodimeric cytokine composed of EBI3 and p28 protein, induces proliferation of naive CD4(+) T cells. *Immunity, 16*, 779–790.

Pflanz, S., Hibbert, L., Mattson, J., Rosales, R., Vaisberg, E., Bazan, J. F., et al. (2004). WSX-1 and glycoprotein 130 constitute a signal-transducing receptor for IL-27. *Journal of Immunology, 172*, 2225–2231.

Qiu, F., Song, L., Yang, N., & Li, X. (2013). Glucocorticoid downregulates expression of IL-12 family cytokines in systemic lupus erythematosus patients. *Lupus, 22*, 1011–1016.

Salcedo, R., Hixon, J. A., Stauffer, J. K., Jalah, R., Brooks, A. D., Khan, T., et al. (2009). Immunologic and therapeutic synergy of IL-27 and IL-2: Enhancement of T cell sensitization, tumor-specific CTL reactivity and complete regression of disseminated neuroblastoma metastases in the liver and bone marrow. *Journal of Immunology, 182*, 4328–4338.

Seita, J., Asakawa, M., Ooehara, J., Takayanagi, S., Morita, Y., Watanabe, N., et al. (2008). Interleukin-27 directly induces differentiation in hematopoietic stem cells. *Blood, 111*, 1903–1912.

Shimizu, S., Sugiyama, N., Masutani, K., Sadanaga, A., Miyazaki, Y., Inoue, Y., et al. (2005). Membranous glomerulonephritis development with Th2-type immune deviations in MRL/lpr mice deficient for IL-27 receptor (WSX-1). *Journal of Immunology, 175*, 7185–7192.

Shimizu, M., Ogura, K., Mizoguchi, I., Chiba, Y., Higuchi, K., Ohtsuka, H., et al. (2013). IL-27 promotes nitric oxide production induced by LPS through STAT1, NF-kappaB and MAPKs. *Immunobiology, 218*, 628–634.

Shinozaki, Y., Wang, S., Miyazaki, Y., Miyazaki, K., Yamada, H., Yoshikai, Y., et al. (2009). Tumor-specific cytotoxic T cell generation and dendritic cell function are differentially regulated by interleukin 27 during development of anti-tumor immunity. *International Journal of Cancer, 124*, 1372–1378.

Sun, J., Dodd, H., Moser, E. K., Sharma, R., & Braciale, T. J. (2011). CD4+ T cell help and innate-derived IL-27 induce Blimp-1-dependent IL-10 production by antiviral CTLs. *Nature Immunology, 12*, 327–334.

Tao, Y.P., Wang, W. L., Li, S. Y., Zhang, J., Shi, Q. Z., Zhao, F., et al. (2012). Associations between polymorphisms in IL-12A, IL-12B, IL-12Rbeta1, IL-27 gene and serum levels of IL-12p40, IL-27p28 with esophageal cancer. *Journal of Cancer Research and Clinical Oncology, 138*, 1891–1900

Tormo, A. J., Beaupre, L. A., Elson, G., Crabe, S., & Gauchat, J. F. (2013). A polyglutamic acid motif confers IL-27 hydroxyapatite and bone-binding properties. *Journal of Immunology, 190*, 2931–2937.

Villarino, A., Hibbert, L., Lieberman, L., Wilson, E., Mak, T., Yoshida, H., et al. (2003). The IL-27R (WSX-1) is required to suppress T cell hyperactivity during infection. *Immunity, 19*, 645–655.

Wang, S., Miyazaki, Y., Shinozaki, Y., & Yoshida, H. (2007). Augmentation of antigen-presenting and Th1-promoting functions of dendritic cells by WSX-1(-IL-27R) deficiency. *Journal of Immunology, 179*, 6421–6428.

Wang, T., Huang, C., Lopez-Coral, A., Slentz-Kesler, K. A., Xiao, M., Wherry, E. J., et al. (2012). K12/SECTM1, an interferon-gamma regulated molecule, synergizes with CD28 to costimulate human T cell proliferation. *Journal of Leukocyte Biology, 91*, 449–459.

Wei, Y. S., Lan, Y., Luo, B., Lu, D., & Nong, H. B. (2009). Association of variants in the interleukin-27 and interleukin-12 gene with nasopharyngeal carcinoma. *Molecular Carcinogenesis, 48*, 751–757.

Wittmann, M., Doble, R., Bachmann, M., Pfeilschifter, J., Werfel, T., & Muhl, H. (2012). IL-27 regulates IL-18 binding protein in skin resident cells. *PLoS ONE, 7*, e38751.

Xia, S., J. Wei, J. Wang, H. Sun, W. Zheng, Y. Li, Y., et al. (2014). A requirement of dendritic cell-derived interleukin-27 for the tumor infiltration of regulatory T cells. *Journal of Leukocyte Biology.*

Yoshimoto, T., Okada, K., Morishima, N., Kamiya, S., Owaki, T., Asakawa, M., et al. (2004). Induction of IgG2a class switching in B cells by IL-27. *Journal of Immunology, 173*, 2479–2485.

Zhao, H., Zhang, Y., Xue, F., Xu, J., & Fang, Z. (2013). Interleukin-27 rs153109 polymorphism and the risk for immune thrombocytopenia. *Autoimmunity, 46*, 509–512.

Interleukin 32

Sangmin Jeong[2] and Soohyun Kim[1,2]
[1]Laboratory of Cytokine Immunology, Department of Biomedical Science and Technology, Konkuk University, Seoul, South Korea
[2]College of Veterinary School, Konkuk University, Seoul, South Korea

Definition

Interleukin-32 (IL-32) is a cytokine and able to induce crucial inflammatory cytokines such as TNFα and IL-6. IL-32 expression is elevated in various inflammatory autoimmune diseases, certain cancers, and viral infections. IL-32 gene was first cloned from activated T cells; however, IL-32 transcript and protein were also detected in other immune cells as well as nonimmune cells. IL-32 gene is identified in most mammalian species except rodents and transcribed as multiple-spliced variants in the absence of a specific activity by each isoform. IL-32 has been studied mostly in clinical fields such as infection, inflammatory disease, cancer, vascular disease, and pulmonary diseases. The lack of mouse IL-32 gene restricts in vivo studies and restrains further development of IL-32 research in clinical applications although IL-32 new cytokine getting a spotlight as an immune regulatory molecule processing functions in various diseases. In this review, we discuss the structure and function of IL-32 in inflammatory diseases and viral infections.

IL-32 Gene Structure and Function

Although IL-32 does not share sequence homology with known cytokine families, IL-32 induces inflammatory cytokines such as tumor necrosis factor-alpha (TNFα) and IL-8 (also known as CXCL8) in THP-1 human monocytes as well as mouse TNFα and macrophage inflammatory protein 2 (also known as CXCL2) in mouse RAW

264.7 macrophages (Kim et al. 2005). IL-32 stimulates signal pathways of nuclear factor kappa-B (NF-κB) and p38 mitogen-activated protein kinase (p38MAPK) resulted in driving inflammatory immune responses. Human IL-32 is originally described as four splice variants including other species of IL-32 found in the databank, whereas mouse IL-32 has not been identified yet. Although seven IL-32 transcripts were reported in the databank, five IL-32 isoforms exist as proteins because of splicing in untranslated region (unpublished data). Induced in human peripheral lymphocyte cells after mitogen stimulation, in human epithelial cells by IL-1β, interferon gamma (IFNγ), and TNFα, and in natural killer (NK) cells after exposure to the combination of IL-12 plus IL-18, IL-32 is probably involved in a role in inflammatory/autoimmune diseases (Kim et al. 2005; Nishida et al. 2008, 2009).

IL-32α affinity column was used to isolate a putative IL-32 receptor; however, IL-32α affinity column identified an IL-32 binding protein, neutrophil proteinase 3 (PR3) (Novick et al. 2006). PR3 processes a variety of inflammatory cytokines, including TNFα, IL-1β, IL-8, and IL-32, thereby enhancing their biological activities (Sugawara et al. 2001; Coeshott et al. 1999; Robache-Gallea et al. 1995; Padrines et al. 1994). Therefore designed four PR3-cleaved IL-32 separate domains were constructed according to PR3 cleavage sites in the IL-32α and IL-32γ polypeptides. The separate domains of the IL-32α and γ isoforms were more active than the intrinsic IL-32α and γ isoforms. Interestingly, the N-terminal IL-32γ isoform separate domain evidenced the highest activity among the IL-32 separate domains (Kim et al. 2008).

The secondary protein structure of IL-32 revealed coils and α-helixes, but no β sheets. IL-32 also possesses a tripeptide Arg-Gly-Asp (RGD) motif, and the mutation of the RGD motif did not effect on the IL-32β- or IL-32γ-induced cytotoxicity. Although IL-32α interacted with the extracellular part of αVβ3 and αVβ6 integrins, only the αVβ3 binding is inhibited by small RGD peptides. In addition to the IL-32/integrin interactions, IL-32 is also able to interact with intracellular proteins that are involved in integrin and focal adhesion signaling.

IL-32 structure analysis revealed a distinct α-helix structure that is similar to the focal adhesion targeting region of focal adhesion kinase (FAK) (Heinhuis et al. 2011a). Unlike RGD mutation, inhibition of FAK resulted in modulation of the IL-32β- or IL-32γ-induced cytotoxicity. Interestingly, IL-32α binds to paxillin without the RGD motif being involved. FAK blocked IL-32α/paxillin binding, whereas FAK also could interact with IL-32α, demonstrating that IL-32 is a part of the focal adhesion protein complex (Heinhuis et al. 2011a). Although this study demonstrated that IL-32α binds to the extracellular domain of integrins and to intracellular proteins like paxillin and FAK, further studies are necessary to verify that the effect of IL-32α has a similar potency of IL-32β- or IL-32γ-induced cytotoxicity.

IL-32 in Inflammatory Diseases

Inflammatory Bowel Disease

IL-32 synergizes with nucleotide oligomerization domain (NOD) 1 and NOD2 for inflammatory cytokine in peripheral blood mononuclear cells (PBMCs) (Netea et al. 2005). The activation of mucosal immunity requires nonspecific innate signals by various bacterial products via pattern-recognition receptors. IL-32 activity is enhanced by the intracellular NOD. The synergistic effect of IL-32 and NOD2 ligand synthetic muramyl dipeptide (MDP) on the production of inflammatory cytokine is abolished in PBMCs of Crohn's disease (CD) possessing 3020insC mutation (Netea et al. 2005). This in vitro synergism between IL-32 and MDP, NOD2 ligand, is associated with the high expression of IL-32 in human colon epithelial tissues. Furthermore IL-32 synergizes with synthetic ligand of NOD1 FK-156 in cytokine productions, but this effect is absent in NOD1-deficient macrophages (Netea et al. 2005). These data suggest that IL-32 and NODs pathway has important role in mucosal immunity.

Imaeda et al. have identified a new IL-32 isoform from human colonic subepithelial myofibroblasts (SEMFs). The new IL-32 isoform is named as IL-32ε isoform and lacks exons 3 and

4 of the longest IL-32γ isoform. The transcript of IL-32ε isoform is significantly elevated in the inflamed mucosa of inflammatory bowel disease (IBD) patients. TNFα induces transcript of new IL-32ε in a dose- and time-dependent manner (Imaeda et al. 2011). Interestingly, stable transfection of IL-32ε significantly decreased TNF-α-mediated IL-8 transcript in HT-29 cells, but the expression of IL-32α, the shortest isoform lacking exons 3 and 7, has no effect on TNFα-mediated IL-8 transcript. Another study has shown that the level of IL-32α protein and mRNA transcript is evaluated in inflamed epithelial mucosa of IBD patients compared to colonic epithelial cells of normal individuals (Shioya et al. 2007). With intestinal epithelial cell lines, the expression of IL-32α transcript and protein is increased by IL-1β, IFNγ, and TNFα. TNFα plus IFNγ exerts synergistic effect on IL-32α expression, and also IL-32α is highly expressed particularly in epithelial cells of IBD and CD patients. In the ileal tissues of patients with ankylosing spondylitis (AS) and intestinal chronic inflammation, significant upregulation of IL-32 levels found as compared with non-inflamed AS patients and controls (Ciccia et al. 2012). Further studies suggested that the biological activity of IL-32 plays important roles through interaction with other inflammatory cytokines such as TNFα, IL-1β, and IFNγ in the pathophysiology of IBD and CD (Andoh et al. 2008; Fantini et al. 2007; Felaco et al. 2009).

The function of IL-32 in intestinal inflammation is investigated in in vivo experiment by using a transgenic mouse expressing human IL-32γ (IL-32γ TG). Although IL-32γ TG mice are healthy, constitutive serum and colonic tissue levels of TNFα are increased. Compared with wild-type (WT) mice, IL-32γ TG exhibited a modestly enhanced acute inflammation early following the initiation of dextran sodium sulfate (DSS)-induced colitis (Choi et al. 2010). However, after day 6, there is less colonic inflammation and improved survival rate compared with WT mice. Associated with attenuated tissue damage, the colonic level of inflammatory cytokine is significantly reduced in IL-32γ TG-treated with DSS and also constitutive level of IL-32γ itself in colonic tissue is decreased (Choi et al. 2010). These data suggest that IL-32γ emerges as an example of how innate immune response worsens as well as protects intestinal integrity.

IL-32 stimulates monocytes to produce inflammatory cytokines as well as differentiates monocytes into macrophage or dendritic (DC)-like cells (Netea et al. 2008). Also IL-32 directly stimulates neutrophils to produce IL-6 and IL-8 (Choi et al. 2009, 2010; Kim et al. 2010). The differentiated macrophages and DCs are potent producers of crucial inflammatory cytokines in IBD and CD such as TNFα, IL-1β, and IL-6. These inflammatory cytokines in the inflamed area recruit T cells, which are proliferated by the differentiated DCs to protect a host against the pathogens. On the other hand, increased numbers of various immune cells in the absence of proper immune suppressor molecules induce infiltration of neutrophil population in the inflamed area resulted in releasing a large amount of neutrophil proteinase such as elastase, proteinase 3, and cathepsin G. These serine proteinase family enzymes are strong mediators of mucosal tissue damage exacerbating inflammation in IBD and CD. Although IL-32 expressions are elevated in inflamed mucosa epithelial cells of IBD and CD patients, the biological activity of IL-32 in vitro and in vivo is inconsistent. The discrepancy of in vitro and in vivo data could be because each investigator has studied a distinct IL-32 isoform or on the other hand IL-32 regulation and function are complexity. Further studies are necessary to evaluate the precise functions of IL-32 in IBD and CD.

Rheumatoid Arthritis

The biological effects of IL-32γ isoform on the differentiation of osteoclasts and its expression in rheumatoid arthritis (RA) have been studied. CD14$^+$ monocytes of healthy volunteers or RA patients from PBMCs as well as synovial tissue of RA have been used to investigate the role of IL-32 in RA. The levels of IL-32γ are elevated in RA patients, and IL-32 exacerbates mouse models of experimental inflammatory arthritis (Kim et al. 2010; Moon et al. 2012; Xu et al. 2013). The osteoclastogenic effect and resorbed area are enhanced in the presence of soluble receptor activator of nuclear factor kappa-B ligand (sRANKL), and the effect is more significant in the IL-32γ-stimulated cultures than that of IL-17 (Kim et al. 2010). These results suggested that IL-32γ is a potent mediator of active osteoclast generation in the presence of sRANKL. IL-32 is highly expressed in RA synovial tissue biopsies, whereas IL-32 is not observed in synovial tissues from patients with osteoarthritis (Joosten et al. 2006). The level of IL-32 expression is correlated with erythrocyte sedimentation rate, a marker of systemic inflammation. IL-32 is an inducer of prostaglandin E2 (PGE2) release in mouse macrophages and human blood monocytes. In TNFα-deficient mice, IL-32-driven joint swelling is reduced and cell influx is decreased. These results suggested that IL-32 activity is TNFα dependent in RA (Joosten et al. 2006).

Moon et al. have investigated extensively the signal pathway of TNFα-mediated IL-32 induction, and they have characterized that TNF-α-induced IL-32 is regulated through the spleen tyrosine kinase (Syk)/protein kinase C delta (PKCdelta)/c-Jun N-terminal kinase (JNK) pathways in RA synovial fibroblasts (Moon et al. 2012). IL-32 is highly expressed in synovium and fibroblast-like synoviocytes (FLS) from RA, whereas not in OA. TNFα-mediated IL-32 production is specifically inhibited by inhibitors of Syk, PKCdelta, and JNK as well as by siRNA of these kinases (Moon et al. 2012). The levels of IL-32 and TNFα in the active RA groups are higher than those in the stable RA and control groups, and IL-32 level is positively associated with other inflammatory markers in RA (Gui et al. 2013). IL-32 also increases TSLP production in human monocyte THP-1 cell line and PBMCs. IL-32 induces the differentiation of monocytes via TSLP since the blockade of TSLP prevents the monocyte differentiation into macrophage-like cells (Jeong et al. 2013). Gene expression in cultured FLS from RA (RA-FLS) has been compared with gene expression in cultured FLS from OA (OA-FLS) using microarray analysis, and IL-32 is the most prominently differentially expressed gene with higher expression in RA-FLS than in OA-FLS (Cagnard et al. 2005).

IL-17 induces IL-32 expression in the FLSs from RA patients, and conversely IL-32 in the FLSs from RA patients stimulates IL-17 production from CD4$^+$ T cells. Unlike the previous report (Kim et al. 2010), IL-32 and IL-17 synergistically induce the differentiation of osteoclasts. IL-32 and IL-17 also could induce resorption by osteoclasts in a RANKL-dependent manner. Both IL-32 and IL-17 can reciprocally influence each other's production and amplify the function of osteoclastogenesis in the in RA synovium. IL-32 and IL-17 separately stimulated osteoclastogenesis without RANKL, and IL-32 synergistically amplified the differentiation of osteoclasts in the presence of IL-17, which is independent of RANKL stimulation. These results are similar to the effect of IL-32 on osteoclastogenesis, but the co-stimulatory effect of RANKL is distinct from previous report (Kim et al. 2010).

The serum level of IL-32 was assessed by using a clinical study with anti-TNFα therapy. At 24 weeks of treatment, serum samples of Enbrel (also known as etanercept, TNF-binding protein) plus methotrexate responders had decreased IL-6 whereas increased IL-32 and IL-21. However, there were no differences in cytokine levels in nonresponders (Zivojinovic et al. 2011). Pro-inflammatory cytokines contribute to persistent in chronic inflammation of RA, and Enbrel treatment regulates level of serum cytokines. Interestingly, the serum level of IL-32 and IL-21 is specifically increased in Enbrel responders. In contrary, treatment of RA patients with anti-TNFα significantly decreases IL-32 in synovial tissue (Heinhuis et al. 2010). TNFα potently induces IL-32γ expression in FLS, and the elevated TNFα, IL-1β, IL-6, and CXCL8 (also known as IL-8) productions are detected after IL-32γ overexpression in the presence of LPS in THP-1 cells. TNFα stimulation of FLS after IL-32γ/siRNA decreases IL-6 and CXCL8 production, whereas IL-32γ overexpression enhances IL-6 and CXCL8 (Heinhuis et al. 2010). Additional studies are necessary to resolve the inconsistency of IL-32 expression in RA patients.

The overexpression of splice-resistant IL-32γ mutant in THP1 cells or RA synovial fibroblasts increases IL-1β compared with IL-32β (Heinhuis et al. 2011b). The result suggests that splicing to one less active IL-32β appears to be a salvage mechanism to reduce inflammation. The overexpression of primarily IL-32β in RA synovial fibroblasts showed minimal secretion and reduced cytokine production. IL-32β lacks exon 3 possessing 46 amino acids, which contains a weak signal peptide of IL-32γ isoform, whereas the overexpression of splice-resistant IL-32γ mutant in RA synovial fibroblasts enhances IL-32γ secretion. In addition, the level of TNFα and IL-6 production is associated with IL-32γ level in RA patients. These data demonstrate that naturally occurring IL-32γ, the longest isoform with the greatest activity among five IL-32 isoforms (Choi et al. 2009), can be spliced into IL-32β, which is a less active pro-inflammatory mediator.

TLR2, TLR3, and TLR4 ligands as well as IFNγ and TNFα induce IL-32β, γ, and δ mRNA expression by RA FLSs (Alsaleh et al. 2010). Mature IL-32 is expressed intracellularly and released by cells stimulated with the various activators. The IL-32α isoform is expressed intracellularly in response to TNFα and polyriboinosinic polyribocytidylic acid (poly I:C) and not released in culture supernatants. Stimulation of FLS with TNFα, bacterial lipoprotein (BLP), lipopolysaccharide (LPS), or poly I:C concomitant with IFNγ increases IL-32 expression compared with stimulation with IFNγ alone. IL-32 synthesis by FLSs is tightly regulated by innate immunity in RA. Therefore, TNFα, IFNγ, double-stranded RNA (dsRNA), hyaluronic acid, or other damage-associated molecular pattern (DAMP) secretion in synovial tissues of RA patients may trigger IL-32 expression in RA patients. In inflamed synovial spaces, various infiltrated immune cells producing inflammatory cytokines, TNFα, IL-1β, and IL-6 stimulate FLS to induce IL-32 and also DAMPs from death cell synergies with IL-32 further enhancement of inflammatory cytokine productions.

Role of IL-32 in Infectious Diseases

IL-32 in Viral Hepatitis Infection
IL-32 is elevated by hepatitis B virus (HBV) infection both in vitro and in vivo at the level of both transcription and posttranscription. Furthermore, microRNA-29b was identified as a key factor in HBV-regulated IL-32 expression by directly targeting the mRNA 3′-untranslated region of IL-32 (Li et al. 2013). Antiviral analysis showed that IL-32 is not sufficient to suppress HBV replication in HepG2.2.15 cells. The IL-32-γ-treated supernatant of PBMCs exhibited an antiviral activity against multiplex viruses including HBV. The indirect antiviral activity of IL-32γ in the supernatant of PBMCs was characterized by using the known antiviral cytokine. The antiviral factor was IFN-λ1 (Li et al. 2013), and this data suggested that the elevated IL-32 levels during viral infection mediate antiviral effects by stimulating the expression of IFN-λ1.

In addition, IL-32 level in serum of HBV patients is associated with serum biochemical indices of liver function and HBV-related liver failure. Serum IL-32 level is positively correlated with serum alanine transaminase (ALT), aspartate aminotransferase, and total bilirubin, respectively, but IL-32 level is not correlated with HBV DNA load (Zhuang et al. 2013). Unlike the antiviral activity of IL-32 (Li et al. 2013) via immune cell-produced IFN-λ1, the present study suggests that IL-32 level is increased in patients with HBV-related liver failure and associated with the severity of inflammation in liver. Hepatic IL-32 expression is increased in chronic hepatitis B patients and increased with the severity of liver inflammation/fibrosis (Xu et al. 2012). Moreover, the expression level of IL-32 in liver is correlated with serum ALT level but negatively associated with serum albumin level. This data suggest that IL-32 could be implicated in HBV-related liver inflammation/fibrosis.

HBV replicates noncytopathically in hepatocytes, but HBV or proteins encoded by HBV genome could induce cytokine and chemokine expression by hepatocytes. Moreover, liver damage in patients with HBV infection is immune mediated, and cytokines play important roles in immune-mediated liver damage after HBV infection. Hepatitis B virus protein X (HBx) increases IL-32 expression through the promoter of IL-32 at positions from −746 to +25 and in a dose-dependent manner. NF-κB subunits p65 and p50 in Huh7 cells also augment IL-32 expression, and the NF-κB inhibitor blocks the effect of HBx on IL-32 induction (Pan et al. 2011). These results suggest that NF-κB activation is required for HBx-induced IL-32 expression, which might play an important role in inflammatory response after HBV infection.

IL-32 expression is associated with chronic hepatitis C virus (HCV) infection in histological features of steatosis, inflammation, and fibrosis. The effects of IL-32 overexpression and IL-32 silencing on HCV replication revealed that there are positive correlation between hepatic IL-32 mRNA expression and liver steatosis, inflammation, fibrosis, smooth muscle actin (SMA) area, and serum ALT levels (Moschen et al. 2011). Alone, stimulation with IFNα did not induce IL-32 expression in Huh-7.5. However, IFNα exerted a significant additive effect on TNF-α-induced but not IL-1β-induced IL-32 expression, particularly in $CD14^+$ monocytes. This effect is dependent both on NF-κB and Janus kinase (Jak)/signal transducer and activator of transcription (STAT) signaling.

IL-32 in Human Immunodeficiency Virus Infection

IL-32 is induced by IFNγ suggesting roles for IL-32 in innate and adaptive immune responses in viral infections. Human immunodeficiency virus (HIV) promotes IL-32 production at both transcription and protein level. There is a threefold enhancement of IL-32 promoter activity in the present HIV infection. HIV long terminal repeat (LTR) activity is enhanced when endogenous IL-32 is knocked down with siRNA/IL-32 whereas IL-32 overexpression decreased the activity of HIV LTR (Rasool et al. 2008). Microarray study of HIV infection in lymphatic tissue (LT) identified a gene encoding IL-32, and its expression is increased in gut and LT of HIV-infected individuals. IL-32 expression is increased in $CD4^+$ T cells, B cells, macrophages, dendritic cells, and epithelial cells in vivo (Smith et al. 2011). IL-32 induces the expression of immunosuppressive molecules indoleamine 2, 3-dioxygenase (IDO), and Ig-like transcript 4 in immune cells in vitro. IL-32-associated IDO/Ig-like transcript 4 expression in LT macrophages and gut epithelial cells suppresses immune activation resulting in impair host defenses in vivo (Smith et al. 2011). Unlike the previous report (Rasool et al. 2008), IL-32 during HIV infection moderates chronic immune activation to avert associated immunopathology but at the same time dampens the antiviral immune response and thus paradoxically supports HIV replication and viral persistence. A pathological hallmark of HIV infection is chronic activation of the immune system by increasing pro-inflammatory cytokines. The host attempts to counterbalance this prolonged immune activation through compensatory mediators of immune suppression such as IL-32.

HIV replication in macrophages regulated by cytokines and infection is restricted in macrophages activated by type I interferon (IFNα and IFN β) and polarizing cytokines such as IL-4, IL-10, and IL-32. The expression levels of the cellular factors (tripartite motif protein 5-alpha, apolipoprotein B mRNA-editing enzyme-catalytic polypeptide-like 3G, sterile alpha motif, and HD domain-containing protein 1, tripartite motif-containing 22, tetherin, and three prime repair exonuclease 1), and the anti-HIV miRNAs miR-28, miR-150, miR-223, and miR-382 are upregulated by IFNα and IFNβ in macrophages that may account for the inhibitory effect on viral replication and the antiviral state of these cells. Expression of these factors is also increased by IFNγ in the presence or absence of TNFα. IL-32 polarization did not affect the expression of these cellular factors and miRNAs. This data suggest that IL-32 has only a limited role for these cellular factors in restricting HIV replication in macrophages (Cobos Jimenez et al. 2012) although IFNγ and TNFα are the potent inducer of IL-32.

IL-32 in Influenza Virus Infection

IL-32 level is increased in the serum samples of patients infected by influenza A virus comparing to that of healthy individuals. The expression of IL-32 in influenza A virus-infected A549 human lung epithelial cells is suppressed by either selective cyclooxygenase-2 (COX-2) inhibitor NS398 or aspirin indicating that IL-32 is induced by

COX-2 in the inflammatory cascade (Li et al. 2008). Interestingly, COX-2-mediated PGE2 production by influenza A virus infection is suppressed by overexpression of IL-32 but increased by siRNA/IL-32 transfection suggesting that there is a feedback loop between IL-32 and COX-2.

The role of IL-32 in the host immune responses during influenza A infection has been studied. IL-32 production is stimulated by influenza A virus or dsRNA in human PBMCs from healthy volunteers. NF-κB and CREB transcription factors play major roles in the activation of IL-32 promoter in responding to influenza A virus infection (Li et al. 2010). Aberrant epigenetic modifications in the IL-32 promoter are important in the transcriptional regulation of IL-32. One CpG demethylation within the CREB binding site in IL-32 promoter increases the binding of CREB to the IL-32 promoter followed by IL-32 transcriptional activation in influenza A virus-infected cells (Li et al. 2010). Overexpression of assays combined with RNAi exhibit that DNMT-1 and DNMT-3b DNA methyltransferases are critical for methylation in IL-32 promoter for gene silencing prior to influenza A viral infection. Assays of the six IL-32 isoforms during influenza A viral infection provide all IL-32 isoforms possessing antiviral activities with variation in efficiency.

The IL-32 level in the sera of H1N1 influenza A patients is significantly elevated. The antiviral activity of recombinant IL-32γ (rIL-32γ) with WISH cells infected by vesicular stomatitis virus did not show an antiviral activity unlike previous report (Zepp et al. 2011). Therefore the supernatant of rIL-32γ-treated THP-1 cells was examined since this cell line effectively responded to rIL-32γ in inducing inflammatory cytokine and chemokine. The supernatant of rIL-32γ-treated THP-1 cell possessed an antiviral effect, and in addition, an agonistic monoclonal antibody further enhanced a specific antiviral activity of rIL-32γ (Bae et al. 2012). The fractionation and mass spectrometer analysis of the THP-1 cell supernatant revealed that the antiviral activity of rIL-32γ is via a THP-1 cell-produced factor, transferrin, rather than the direct effects of rIL-32γ on epithelial cells (Bae et al. 2012). These results suggest that IL-32γ expression and its genetic variation in individual could be an important aspect of viral infections.

Conclusion

Initial discovery of cytokine IL-32 is an elaborate work with in vitro experiment of microarray by using A549/IL-18Rβ stable cells (Kim et al. 2005); however, actual regulation of IL-32 in vivo is probably controlled by IFNγ (Kim 2014) in Fig. 1. Various pathogens such as virus, intracellular bacteria, and intracellular parasite (protozoan) sufficiently induce IFNγ, which is a hallmark of Th1 cytokine in elevating cellular immune response. Extracellular parasites stimulate Th2 immune response where IL-4, IL-5, IL-10, and IL-13 play prominent roles in elevating humoral immune responses. Helminth antigen drives Th2 immune response via IL-18 or IL-33 alone whereas virus, *M. Tuberculosis*, and *M. Leprae* infections induce Th1 immune response (Schenk et al. 2012; Bai et al. 2010; Netea et al. 2006) through IL-12 plus IL-1 family cytokines such as IL-1α, IL-1β, IL-18, and IL-33 (Xu et al. 2012; Rasool et al. 2008; Smith et al. 2011; Li et al. 2010; Bae et al. 2012; Schenk et al. 2012; Bai et al. 2010; Oh et al. 2013; Huang et al. 2013; Cao et al. 2013; Soyka et al. 2012; Keswani et al. 2011; Nishimoto et al. 2008). Activated T cells and natural killer cells produce a large amount of IFNγ. IL-12/IL-1 family cytokine-induced IFNγ combined with viral RNA/DNA is the strongest inducer of IL-32 expression in vitro and in vivo. On the other hand, pathogenic intracellular bacterial or viral infection directly stimulates nonspecific immune response and activate neutrophils to produce serine proteinases (PR3, elastase, and cathepsin G), which are master regulators for inflammatory cytokines such as IL-32, TNFα, and IL-1β. The nonspecific and specific immunity-mediated chronic local inflammation may contribute to IL-32-associated local inflammatory disorders. Further investigation is necessary to understand specific roles of IL-32 in infections and immune disorders.

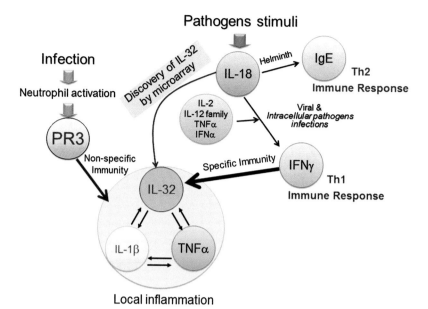

Interleukin 32, Fig. 1 The suggested general regulation of IL-32 in vivo. The experiment of microarray in vitro using A549 stable cells expressing IL-18Rβ (also known as IL-1R7) treated with IL-18 has identified *IL-32 gene* that is indicated by *blue arrow* (Kim et al. 2005). However, the regulation of IL-32 in vivo is the downstream of IFNγ. Th2 immune response is induced by IL-18 after helminth infection whereas intracellular pathogens such as virus, *M. Tuberculosis*, and *M. Leprae* trigger Th1 immune response through IL-2, IL-12 family cytokines, TNFα, and IFNα plus IL-1 family cytokines (IL-1α, IL-1β, IL-18, and IL-33). IFNγ from Th1 T-cells and natural killer cells plus viral RNA are potent inducers of IL-32 through activation of specific immunity; on the other hand, infections may directly activate neutrophil and release a large amount of serine proteinases from neutrophils. The neutrophil-released serine proteinases process IL-32, TNFα, IL-1α, and IL-1β resulting in enhanced activities of these cytokines. The unrestrained nonspecific immunity and specific immunity provoke local inflammation via cross induction of cytokines probably involved in IL-32-related inflammatory disorders

Acknowledgements This work was supported by the National Research Foundation of Korea (NRF) (MEST 2012R1A2A1A01001791).

References

Alsaleh, G., et al. (2010). Innate immunity triggers IL-32 expression by fibroblast-like synoviocytes in rheumatoid arthritis. *Arthritis Research & Therapy, 12*(4), R135.

Andoh, A., et al. (2008). Mucosal cytokine network in inflammatory bowel disease. *World Journal of Gastroenterology, 14*(33), 5154–5161.

Bae, S., et al. (2012). Characterizing antiviral mechanism of interleukin-32 and a circulating soluble isoform in viral infection. *Cytokine, 58*(1), 79–86.

Bai, X., et al. (2010). IL-32 is a host protective cytokine against Mycobacterium tuberculosis in differentiated THP-1 human macrophages. *Journal of Immunology, 184*(7), 3830–3840.

Cagnard, N., et al. (2005). Interleukin-32, CCL2, PF4F1 and GFD10 are the only cytokine/chemokine genes differentially expressed by in vitro cultured rheumatoid and osteoarthritis fibroblast-like synoviocytes. *European Cytokine Network, 16*(4), 289–292.

Cao, H., et al. (2013). Interleukin-32 expression is induced by hepatitis B virus. *Zhonghua Gan Zang Bing Za Zhi, 21*(6), 442–445.

Choi, J. D., et al. (2009). Identification of the most active interleukin-32 isoform. *Immunology, 126*(4), 535–542.

Choi, J., et al. (2010). Paradoxical effects of constitutive human IL-32{gamma} in transgenic mice during experimental colitis. *Proceedings of the National Academy of Sciences of the United States of America, 107*(49), 21082–21086.

Ciccia, F., et al. (2012). Increased expression of interleukin-32 in the inflamed ileum of ankylosing spondylitis patients. *Rheumatology (Oxford), 51*(11), 1966–1972.

Cobos Jimenez, V., et al. (2012). Differential expression of HIV-1 interfering factors in monocyte-derived macrophages stimulated with polarizing cytokines or interferons. *Science Reports, 2*, 763.

Coeshott, C., et al. (1999). Converting enzyme-independent release of tumor necrosis factor alpha and IL-1beta from a stimulated human monocytic cell line in the presence of activated neutrophils or purified proteinase 3. *Proceedings of the National Academy of Sciences of the United States of America, 96*(11), 6261–6266.

Fantini, M. C., Monteleone, G., & Macdonald, T. T. (2007). New players in the cytokine orchestra of inflammatory bowel disease. *Inflammatory Bowel Diseases, 13*(11), 1419–1423.

Felaco, P., et al. (2009). IL-32: A newly-discovered proinflammatory cytokine. *Journal of Biological Regulators and Homeostatic Agents, 23*(3), 141–147.

Gui, M., et al. (2013). Clinical significance of interleukin-32 expression in patients with rheumatoid arthritis. *Asian Pacific Journal of Allergy and Immunology, 31*(1), 73–78.

Heinhuis, B., et al. (2010). Tumour necrosis factor alpha-driven IL-32 expression in rheumatoid arthritis synovial tissue amplifies an inflammatory cascade. *Annals of the Rheumatic Diseases, 70*(4), 660–667.

Heinhuis, B., et al. (2011a). Interleukin 32 (IL-32) contains a typical alpha-helix bundle structure that resembles focal adhesion targeting region of focal adhesion kinase-1. *Journal of Biological Chemistry, 287*(8), 5733–5743.

Heinhuis, B., et al. (2011b). Inflammation-dependent secretion and splicing of IL-32{gamma} in rheumatoid arthritis. *Proceedings of the National Academy of Sciences of the United States of America, 108*(12), 4962–4967.

Huang, Y., et al. (2013). The expression of interleukin-32 is activated by human cytomegalovirus infection and down regulated by hcmv-miR-UL112-1. *Virology Journal, 10*, 51.

Imaeda, H., et al. (2011). A new isoform of interleukin-32 suppresses IL-8 mRNA expression in the intestinal epithelial cell line HT-29. *Molecular Medicine Reports, 4*(3), 483–487.

Jeong, H. J., et al. (2013). Inhibition of IL-32 and TSLP production through the attenuation of caspase-1 activation in an animal model of allergic rhinitis by Naju Jjok (Polygonum tinctorium). *International Journal of Molecular Medicine, 33*(1), 142–150.

Joosten, L. A., et al. (2006). IL-32, a proinflammatory cytokine in rheumatoid arthritis. *Proceedings of the National Academy of Sciences of the United States of America, 103*(9), 3298–3303.

Keswani, A., et al. (2011). Differential expression of interleukin-32 in chronic rhinosinusitis with and without nasal polyps. *Allergy, 67*(1), 25–32.

Kim, S. (2014). Interleukin-32 in inflammatory autoimmune diseases. *Immune Network, 14*(3), 123–127.

Kim, S. H., et al. (2005). Interleukin-32: A cytokine and inducer of TNFalpha. *Immunity, 22*(1), 131–142.

Kim, S., et al. (2008). Proteinase 3-processed form of the recombinant IL-32 separate domain. *BMB Reports, 41*(11), 814–819.

Kim, Y. G., et al. (2010). Effect of interleukin-32gamma on differentiation of osteoclasts from CD14+ monocytes. *Arthritis and Rheumatism, 62*(2), 515–523.

Li, W., et al. (2008). Activation of interleukin-32 pro-inflammatory pathway in response to influenza A virus infection. *PLoS ONE, 3*(4), e1985.

Li, W., et al. (2010). IL-32: A host proinflammatory factor against influenza viral replication is upregulated by aberrant epigenetic modifications during influenza A virus infection. *Journal of Immunology, 185*(9), 5056–5065.

Li, Y., et al. (2013). Inducible interleukin 32 (IL-32) exerts extensive antiviral function via selective stimulation of interferon lambda1 (IFN-lambda1). *Journal of Biological Chemistry, 288*(29), 20927–20941.

Moon, Y. M., et al. (2012). IL-32 and IL-17 interact and have the potential to aggravate osteoclastogenesis in rheumatoid arthritis. *Arthritis Research & Therapy, 14*(6), R246.

Moschen, A. R., et al. (2011). Interleukin-32: A new proinflammatory cytokine involved in hepatitis C virus-related liver inflammation and fibrosis. *Hepatology, 53*(6), 1819–1829.

Netea, M. G., et al. (2005). IL-32 synergizes with nucleotide oligomerization domain (NOD) 1 and NOD2 ligands for IL-1beta and IL-6 production through a caspase 1-dependent mechanism. *Proceedings of the National Academy of Sciences of the United States of America, 102*(45), 16309–16314.

Netea, M. G., et al. (2006). Mycobacterium tuberculosis induces interleukin-32 production through a caspase-1/IL-18/interferon-gamma-dependent mechanism. *PLoS Medicine, 3*(8), e277.

Netea, M. G., et al. (2008). Interleukin-32 induces the differentiation of monocytes into macrophage-like cells. *Proceedings of the National Academy of Sciences of the United States of America, 105*(9), 3515–3520.

Nishida, A., et al. (2008). Phosphatidylinositol 3-kinase/Akt signaling mediates interleukin-32alpha induction in human pancreatic periacinar myofibroblasts. *American Journal of Physiology. Gastrointestinal and Liver Physiology, 294*(3), G831–G838.

Nishida, A., et al. (2009). Interleukin-32 expression in the pancreas. *Journal of Biological Chemistry, 284*(26), 17868–17876.

Nishimoto, K. P., Laust, A. K., & Nelson, E. L. (2008). A human dendritic cell subset receptive to the Venezuelan equine encephalitis virus-derived replicon particle constitutively expresses IL-32. *Journal of Immunology, 181*(6), 4010–4018.

Novick, D., et al. (2006). Proteinase 3 is an IL-32 binding protein. *Proceedings of the National Academy of Sciences of the United States of America, 103*(9), 3316–3321.

Oh, H. A., et al. (2013). Evaluation of the effect of kaempferol in a murine allergic rhinitis model. *European Journal of Pharmacology, 718*(1–3), 48–56.

Padrines, M., et al. (1994). Interleukin-8 processing by neutrophil elastase, cathepsin G and proteinase-3. *FEBS Letters, 352*(2), 231–235.

Pan, X., et al. (2011). Interleukin-32 expression induced by hepatitis B virus protein X is mediated through activation of NF-. *Molecular Immunology, 48*(12–13), 1573–1577.

Rasool, S. T., et al. (2008). Increased level of IL-32 during human immunodeficiency virus infection suppresses HIV replication. *Immunology Letters, 117*(2), 161–167.

Robache-Gallea, S., et al. (1995). In vitro processing of human tumor necrosis factor-alpha. *Journal of Biological Chemistry, 270*(40), 23688–23692.

Schenk, M., et al. (2012). NOD2 triggers an interleukin-32-dependent human dendritic cell program in leprosy. *Nature Medicine, 18*(4), 555–563.

Shioya, M., et al. (2007). Epithelial overexpression of interleukin-32alpha in inflammatory bowel disease. *Clinical and Experimental Immunology, 149*(3), 480–486.

Smith, A. J., et al. (2011). The immunosuppressive role of IL-32 in lymphatic tissue during HIV-1 infection. *Journal of Immunology, 186*(11), 6576–6584.

Soyka, M. B., et al. (2012). Regulation and expression of IL-32 in chronic rhinosinusitis. *Allergy, 67*(6), 790–798.

Sugawara, S., et al. (2001). Neutrophil proteinase 3-mediated induction of bioactive IL-18 secretion by human oral epithelial cells. *Journal of Immunology, 167*(11), 6568–6575.

Xu, Q., et al. (2012). Increased interleukin-32 expression in chronic hepatitis B virus-infected liver. *Journal of Infection, 65*(4), 336–342.

Xu, W. D., et al. (2013). IL-32 with potential insights into rheumatoid arthritis. *Clinical Immunology, 147*(2), 89–94.

Zepp, J. A., et al. (2011). Protection from RNA and DNA viruses by IL-32. *Journal of Immunology, 186*(7), 4110–4118.

Zhuang, G. L., et al. (2013). Interleukin-32 expression in serum of patients with HBV-related liver failure and its significance. *Zhonghua Shi Yan He Lin Chuang Bing Du Xue Za Zhi, 27*(4), 247–249.

Zivojinovic, S. M., et al. (2011). Tumor necrosis factor blockade differentially affects innate inflammatory and Th17 cytokines in rheumatoid arthritis. *Journal of Rheumatology, 39*(1), 18–21.

Interleukin-33

Sangmin Jeong[2] and Soohyun Kim[1,2]
[1]Laboratory of Cytokine Immunology, Department of Biomedical Science and Technology, Konkuk University, Seoul, South Korea
[2]College of Veterinary School, Konkuk University, Seoul, South Korea

Synonyms

DVS27-related protein; IL-1F11; IL-33; Interleukin-1 family member 11; Interleukin-33; Nuclear factor for high endothelial venules; Nuclear factor from high endothelial venules

Definition of IL-33 and Its Receptor

Interleukin (IL)-33 is a member of the IL-1 cytokine family and was originally identified as a nuclear factor in high endothelial venules (Baekkevold et al. 2003). The IL-33 receptor complex is comprised of ST2 (also known as IL-1RL1, DER4, Fit-1, and T1; initially reported as an orphan receptor (Bergers et al. 1994); now renamed IL-1R4), the ligand-binding chain, and IL-1 receptor accessory protein (IL-1RAcP, also known as IL-1R3), the signal-transducing chain (Schmitz et al. 2005). The signal-transducing chain of IL-33 receptor is common to all IL-1 family ligands such as IL-1α, IL-1β, IL-33, IL-36α, IL-36β, and IL-36γ and contributes to downstream intracellular signaling (Fig. 1). The nine members of the IL-1 receptor family, apart from IL-1R8 (also known as SIGIRR), have three immunoglobulin-like domains in the extracellular domain with a single transmembrane domain. The intracellular domain of the IL-1 receptor family contains a Toll/interleukin-1 receptor (TIR) homology domain which interacts with adaptor molecules for downstream inflammatory pathways, whereas IL-1R2 lacks the TIR due to a short intracellular domain (Fig. 1).

Biosynthesis and Release

Following synthesis of the IL-33 precursor, the cytokine is primarily located in the intracellular compartment. The expression of endogenous IL-33 has been described most frequently in endothelial cells, where essentially nuclear, full-length 30 kDa IL-33 precursor was detected (Kuchler et al. 2008). Although the IL-33 precursor does not have a typical hydrophobic signal sequence at the N-terminus for secretion, it is released into the extracellular compartment as a mature form. IL-33 is released from cultured astrocytes following stimulation by LPS and ATP (Hudson et al. 2008) under conditions resembling those required for the processing and release of mature IL-1β from monocytes and macrophages (MacKenzie et al. 2001). A constitutively high level of IL-33 expression is seen in the vascular

Interleukin-33, Fig. 1 IL-33 shares the second chain of its receptor with other members of the IL-1 family cytokines. IL-33 has an independent ligand-binding chain, which was originally named ST2 (also known as IL-1R4) and shares the second chain with IL-1α, IL-1β, IL-36α, IL-36β, and IL-36γ. All eight members of the IL-1R family, except IL-1R2 and IL-18 binding protein, possess a toll/interleukin-1 receptor (TIR) homology domain with three separate compartments

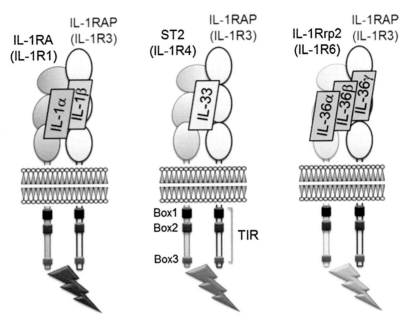

tree of tissue from healthy subjects (Baekkevold et al. 2003), and it is released upon necrosis of endothelial or epithelial cells.

IL-33 Processing Enzyme

Usually IL-1 family cytokines such as IL-1β and IL-18 lack hydrophobic signal peptides for secretion of precursor proteins, which are presented as inactive forms in the cytoplasm (Dinarello 1998, 2009). Following activation of IL-1 converting enzyme (ICE, also known as caspase-1) by inflammasome, precursors of IL-1β and IL-18 are cleaved and biologically matured and then secreted through an unconventional mechanism (Dinarello 1998, 2009). When IL-33 was first reported, it was believed that IL-33 undergoes a similar process to that of the other members of the IL-1 cytokine family in order to be activated and secreted (Schmitz et al. 2005). However, the full-length protein itself is biologically active without any maturation process (Cayrol and Girard 2009; Luthi et al. 2009; Talabot-Ayer et al. 2009). In addition, the caspase-mediated proteolysis occurring during apoptosis associated with activation of inflammasome is not required for activation and/or release of IL-33. Further study of IL-33 processing revealed that apoptotic caspases process pro-IL-33 in apoptotic cells resulting in inactivation of pro-IL-33, whereas pro-IL-33 released from necrotic cell is spontaneously active (Luthi et al. 2009). The IL-33 leaking from necrotic cells without any maturation process is also biologically active (Cayrol and Girard 2009; Ohno et al. 2009). The authors suggested that the cleavage site of caspase-1 does not reside at the site initially proposed (Asp^{110}–Ser^{111}) but rather at the amino acid residues Asp^{178}–Gly^{179}, which is the consensus site for cleavage by caspase-3 (Cayrol and Girard 2009; Luthi et al. 2009; Schroder and Kaufman 2005). Apoptotic caspase-3 and caspase-7 destroy the biological activity of pro-IL-33 by cleavage of amino acid residues Asp^{178}–Gly^{179}. Nevertheless, these studies were performed in vitro; the biological activity of the pro-IL-33 was shown only by NFκ-B luciferase assay (Luthi et al. 2009). Another study suggested calpain-dependent processing of pro-IL-33, as occurs with IL-1α, which was shown by treatment with calcium ionophore. Calpain processes pro-IL-33, which produces mature IL-33 in human epithelial and endothelial cells without proving the biological activity of the calpain-cleaved IL-33 (Hayakawa et al. 2009). In contrast to this result, Ohno et al. (2009) reported that caspase-1/8 and calpain are dispensable for IL-33 release from macrophage and mast cells.

Murine IL-33 was released spontaneously, and its secretion was increased by LPS or phorbol 12-myristate 13-acetate plus ionomycin from the peritoneal macrophages of caspase-1-deficient BALB/c mice (Ohno et al. 2009). The discrepancy in these results may be explained by the fact that the experiment on calpain-dependent IL-33 processing was carried out on human cell lines, whereas the conflicting report by Ohno et al. was performed with the peritoneal macrophages of caspase-1-deficient mice (Ohno et al. 2009). Although there is significant sequence homology between human and mouse IL-33, they share only 55% identity at the amino acid level.

Interestingly, the splice variant of IL-33 that lacks exon 3 was reported to be more active than the full-length IL-33 without any maturation process (Hong et al. 2011). Additionally, the cleaved form of IL-33 generated by neutrophil elastase, cathepsin G, and PR3 shows tenfold higher activity than that of full-length IL-33 (Bae et al. 2012; Lefrancais et al. 2012). PR3 is able to both activate and inactivate pro-IL-33, which was shown by prolonged incubation with IL-33 leading to degradation of the cytokine in vitro (Bae et al. 2012). The PR3-cleaved precursor IL-33 induces inflammatory cytokines, and the biological activity of mature IL-33 at the N-terminus, which was produced by the PR3 cleavage site, was highly active (Bae et al. 2012). In recombinant IL-33, consensus amino acid sequencing of PR3 cleavage sites revealed ABZ-$X_3X_2X_1$-ANB-NH_2, as the active sequence (Bae et al. 2012; Wysocka et al. 2008). The identification of the predicted PR3-cleaved fragments by the enzymatic activity of PR3 was extremely difficult because PR3 cleaves pro-IL-33 transiently and eventually abolishes IL-33 activity by fragmentation (Bae et al. 2012).

Biological Activities

IL-33 has functions similar to those of IL-1α and high-mobility group box 1 (HMGB1) as an endogenous "danger" signal to initiate immune responses following trauma or infection. It has been proposed that both precursor and mature IL-33 possess biological activities, although the precise role of caspases in IL-33 maturation continues to be disputed (Cayrol and Girard 2009; Bae et al. 2012; Lefrancais et al. 2012).

IL-33 functions as an alarmin when it is released from necrotic cells in response to prototypic RNA/DNA virus replication in mice. During viral infection, cells die due to necrosis and release their contents, including IL-33, which affects the polarization of immune competent cells for antiviral activities. For example, IL-33 plays an important role in $CD8^+$ T cell (CTL) responses to RNA/DNA virus infections in mice. IFNγ induction in *IL-33$^{-/-}$* and *IL-1R4$^{-/-}$* mice with the soluble decoy receptor FcIL-1R4 was investigated following viral infection; the results of this study showed a significantly impaired protective antiviral CTL response in *IL-33$^{-/-}$* and *IL-1R4$^{-/-}$* mice (Bonilla et al. 2012) (Fig. 2).

Early studies on IL-33 mainly focused on Th2 cytokine production since ST2 is primarily expressed on eosinophils, basophils, and mast cells (Allakhverdi et al. 2009; Cherry et al. 2008; Moritz et al. 1998). IL-33 receptor ligand-binding chain, IL-1R4, is a selective marker for Th2 cytokine-secreting immune cells. IL-1R4 is selectively expressed during various stages of development of the mast cell lineage. Furthermore, through IL-33, it also directly stimulates primary human mast cells to produce Th2 cytokines (Allakhverdi et al. 2009). IL-33 enhances degranulation in response to IgE cross-linking stimulus and induces IL-4 production by IL-3-prestimulated basophils (Suzukawa et al. 2011). IL-33 induces the type II cytokines, IL-4 and IL-13, in concert with IL-3 and/or FcɛRI activation (Pecaric-Petkovic et al. 2009). In addition, circulating $CD34^+$ hematopoietic progenitor cells express ST2 (IL-1R4) and respond to IL-33 by releasing high levels of Th2 cytokines (Allakhverdi et al. 2009). Th2 effector functions have been demonstrated through administration of an anti-IL-1R4 neutralizing antibody or an Fc-fused recombinant IL-1R4 protein, which suppresses airway inflammation via IL-4 and IL-5 following adoptive transfer of type II cytokine-secreting immune cells (Cherry et al. 2008) (Fig. 2).

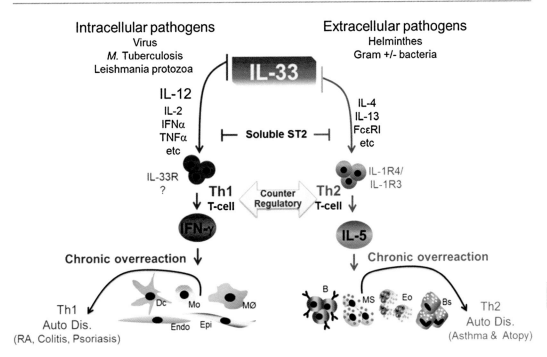

Interleukin-33, Fig. 2 The distinctive biological effects of IL-33 depend on the pathogen which induces its production. IL-33 can act as a Th1 as well as a Th2 cytokine. Intracellular pathogens such as virus, *Mycobacterium tuberculosis*, and *Leishmania protozoa* induce a Th1 immune response. IL-33 plus IL-12 family cytokines, including IL-2, IFNα, and TNFα, stimulate naïve T cells to differentiate into IFNγ-producing Th1 T cells. The chronic overreaction of Th1 cytokines thus unleashed on local immune cells provokes Th1 autoimmune diseases. To the contrary, extracellular pathogens such as helminths and gram-positive and gram-negative bacteria induce a Th2 immune response. IL-33 plus IL-4, IL-13, and FcεR1 agonist stimulate naïve T cells to differentiate into IL-5-producing Th2 T cells. The chronic overreaction of Th2 cytokines thus unleashed on local immune cells provokes Th2 autoimmune diseases

However, studies have demonstrated that IL-33 also has properties of Th1 cytokines such as the induction of IFNγ in the presence of IL-2, IL-12, or the specific iNKT cell ligand, a-GC (Smithgall et al. 2008; Blom and Poulsen 2012; Guo et al. 2009; Yang et al. 2011). Moreover, airway hyperreactivity in vivo induced by asperamide B-activated pulmonary iNKT cells depends on IL-33 activity (Albacker et al. 2013). In the presence of IL-12, IL-33 stimulates IFNγ production in type I cytokine-secreting immune cells and synergizes with IL-12 to promote CD8$^+$ T cell effector function (Yang et al. 2011) (Fig. 2). These studies indicate that the known IL-33 receptors (IL-1R4 and co-receptor IL-1R3) do not mediate the capability of IL-33 to induce the secretion of the Th1 cytokine IFNγ. The induction by IL-12 plus IL-18 of IFNγ occurs through the expression by IL-12 of IL-18 receptor beta chain on the cell surface of IFNγ secreting NK cells and T cells (Ahn et al. 1997). Similar to this role of IL-12 in IL-18-induced IFNγ, the costimulatory function of IL-12 in IL-33-mediated IFNγ production may also be due to the induction of an unknown IL-33 co-receptor on the cell surface of Th1 cytokine-producing immune cells. Further studies to identify an IL-33 co-receptor are necessary to clarify the precise mechanism of IL-33-mediated IFNγ production in Th1 immune cells.

IL-18 acts to stimulate the production of both Th1 and Th2 cytokine production in immune cells (Yoshimoto et al. 1999). Similar to IL-18, recent studies on IL-33 have revealed that it is a dual-function cytokine, enhancing both Th1 and Th2 cytokine production (Blom and Poulsen 2012) (Fig. 2). In the presence of IL-12, IL-18 and IL-33 both induce Th1 cytokines; however, in the absence of IL-12, IL-18 and IL-33 mediate

Th2 cytokine production (Blom and Poulsen 2012; Nakanishi et al. 2001).

IL-33 is expressed in various types of immune cells like mast cells, macrophages, and dendritic cells and also in nonimmune cells such as endothelial cells, epithelial cells, smooth muscle cells, and fibroblasts (Oboki et al. 2010). The molecule is identical to DVS27 and nuclear factor of high endothelial venules (NF-HEV) since it is upregulated in vasospastic cerebral arteries following subarachnoid hemorrhage and the transcript is expressed in endothelial nuclei (Baekkevold et al. 2003; Onda et al. 1999). It has been suggested that IL-33 acts as a transcriptional repressor in vitro since IL-33 is associated with heterochromatin via a helix-turn-helix motif of the N-terminal part (Carriere et al. 2007; Roussel et al. 2008). Furthermore, IL-33 like IL-1α and HMGB1 acts as a nuclear factor (Andersson and Tracey 2011; Werman et al. 2004).

Pathophysiological Relevance

IL-33 in Th2 Immune Response

IL-33 directly stimulates mature and precursor human mast cells in the presence or absence of TSLP (Allakhverdi et al. 2007), and in addition, bone marrow-derived mast cells release inflammatory cytokines in response to IL-33 (Ali et al. 2007; Haenuki et al. 2012). Released from injured necrotic cells, IL-33 activates mast cells and Th2 type immune cells. The mechanism of action of IL-33 involves binding to ST2 (also known as IL-1R4) on mast cells which triggers subsequent activation of NF-κB and transcription of inflammatory cytokines like IL-1β, IL-6, IL-13, TNFα, chemokines, and prostaglandins (Ohno et al. 2012). IL-33 produces Th2 cytokines either in the presence or absence of co-stimulation of mast cells via IgE/antigen-FcεRI signals (Haenuki et al. 2012; Iikura et al. 2007). As a consequence, IL-33 primes mast cells for activation by IgG immune complexes and increases their survival and adhesion (Kaieda et al. 2012).

The IL-33/ST2 axis induces degranulation of mast cells in response to IgE cross-linking stimuli and enhanced basophil migration to eotaxin without an effect on CCR3 expression (Smithgall et al. 2008; Chan et al. 2001). IL-33 also synergizes with IL-3 to induce IL-4 and CD11b production by basophils which enhances basophil adhesiveness (Smithgall et al. 2008). Basophils also produce inflammatory cytokines such as IL-1β, IL-4, IL-5, IL-6, IL-13, and GM-CSF like mast cells (Pecaric-Petkovic et al. 2009; Smithgall et al. 2008; Suzukawa et al. 2008).

IL-33 mediates blood eosinophilia and eosinophilic infiltration in airway inflammation in mice (Stolarski et al. 2010). This likely involves IL-33 stimulation of IL-5-dependent eosinophil differentiation from $CD117^{+}$ progenitors. In human eosinophils, IL-33 mediates survival; it upregulates ICAM-1 on the cell surface and suppresses ICAM-3 and selectin. IL-33 induces IL-6 and chemokines IL-8 and CCL2 and promotes Siglec-8-mediated apoptosis of eosinophils (Chow et al. 2010; Na et al. 2012). In IL-33 null mice, eosinophil infiltration and cytokine production were decreased (Louten et al. 2011), and IL-33 stimulates eosinophils to produce superoxide and degranulation (Cherry et al. 2008). In humans, a correlation between blood and pulmonary eosinophilia on the one hand and elevated IL-33 serum level on the other has been reported (Kim et al. 2010).

It is well known that IL-33 induces Th2 cytokines and chemotaxis of in vitro polarized Th2 cells (Schmitz et al. 2005; Ohno et al. 2012; Komai-Koma et al. 2007). In the mouse respiratory system, antigen-specific Th2 cells stimulated by IL-33 produce IL-5 and IL-13, but not IL-4, and these cells are called atypical Th2 cells (Kurowska-Stolarska et al. 2008). This result was also observed in BAL fluid, since IL-5 and IL-13 levels were increased while IL-4 did not change following intranasal administration of IL-33. This result suggests that IL-33 is involved in IL-4-independent Th2 cell differentiation (Louten et al. 2011). IL-33 induces IL-5 and IL-13, and impressively, IFNγ, a Th1 cytokine, is also enhanced by Th2 in vitro among skewed cells in HDM-specific T-cell culture (Smithgall et al. 2008). IL-33 is involved in Th2 responses to allergens like house dust mites and peanuts in experimental animal models of asthma and food allergy, respectively (Chu et al. 2013). Interestingly, ST2 or IL-33 null mice show normal levels of Th2 differentiation (Hoshino et al. 1999; Townsend et al. 2000). In keeping with this, iNKT cells express membrane ST2 and produce both Th1 and Th2 cytokines under stimulation by IL-33 (Smithgall et al. 2008). This result indicates that IL-33 promotes Th2 cytokines and under certain conditions Th1 cytokines as well. These findings suggest that another component may exist that contains the IL-33 receptor complex and which transduces a Th1 signal following stimulation by IL-33.

IL-33 in Innate Lymphoid Cells

In recent studies, various IL-5- and IL-13-producing Lin$^-$ c-kit$^+$ Sca-1$^{-/+}$ innate lymphoid cells (ILCs), such as natural helper (NH) cells, multipotent progenitor type 2 (MPP^{type2}) cells, nuocytes, and innate helper type 2 (Ih2) cells, have been reported as a distinct subset of cells from lymphoid progenitors, lymphoid tissue inducer cells, and RORγt$^+$ ILCs (Moro et al. 2010; Neill et al. 2010; Price et al. 2010; Koyasu and Moro 2011a; Saenz et al. 2010; Barlow and McKenzie 2011). GATA-3 was identified as a key transcription factor for their development (Hoyler et al. 2012). Lin$^-$ c-kit$^+$ Sca-1$^+$ NH cells were found in fat-associated lymphoid clusters in visceral adipose tissue, which express ST2 and induce IL-5 and IL-6 but not IL-4 more than basophils or mast cells in response to IL-33 with or without IL-2 and IL-25. They are also involved in defense against helminths causing IL-5- and IL-13-dependent eosinophilia and goblet cell hyperplasia (Moro et al. 2010; Koyasu and Moro 2011b). ST2-expressing Lin$^-$ c-kit$^+$ Sca-1$^+$ nuocytes and Lin$^-$ c-kit$^+$ Sca-1$^-$ IH2 cells are present in the mesenteric lymph nodes, spleen, and liver of IL-25- or IL-33-injected or helminth-infected mice. All these NH cells, nuocytes, and Th2 cells are important for host defense against *Nippostrongylus brasiliensis* infection (Neill et al. 2010; Price et al. 2010; Liang et al. 2011) and the onset of allergic airway inflammation (Wilhelm et al. 2011; Barlow et al. 2012). However, expansion of Lin$^-$ c-kit$^+$ Sca-1$^+$ MPP^{type2} cells does not lead to expression of ST2, but they were observed in mesenteric lymph nodes and GALT of IL-25 or helminth-infected mice (Saenz et al. 2010). IL-33 and IL-25 are key factors for expansion of IL-5- and IL-13-producing ILCs, but it is still controversial whether ILCs express IL-5 and/or IL-13 under stimulation by IL-33 or IL-25 except in NH cells.

Meanwhile, IL-33 responsive c-kit$^-$ Sca-1$^+$ CD25$^+$ cells have been shown to be a subset of cells distinct from c-kit$^+$ NH cells (Brickshawana et al. 2011). In the human and mouse lungs, Lin$^-$ c-kit$^+$ Sca-1$^+$ CD90$^+$ CD25$^+$ IL-7Rα$^+$ ST2$^+$ residential cells differentiated by a transcription factor, Id2, express IL-5 and IL-13 in response to IL-33 (Monticelli et al. 2011). Two subsets of ILCs that solely express IL-5 or IL-13 in the presence of IL-33 or IL-25 in the peritoneal cavity, lung, and gut of mice were also reported (Ikutani et al. 2012). A subset of Lin$^-$ c-kit$^+$ Sca-1$^+$ CD25$^+$ ILCs has been identified that expresses IL-5 and IL-13 in response to IL-33 but not to IL-25 from the mouse lung (Bartemes et al. 2012).

IL-33 in Allergic Diseases

In comparison to healthy individuals, IL-33 levels are increased in the serum of allergic rhinitis patients and in the bronchial mucosa of asthma patients increasing with the severity of disease (Prefontaine et al. 2009; Sakashita et al. 2008). Major cellular sources of IL-33 in asthma patients are lung epithelial cells and airway smooth muscle cells (Prefontaine et al. 2009, 2010), and TNFα and IFNγ induce IL-33 in smooth muscle cells (Prefontaine et al. 2009). When IL-33 is given directly to mouse lung, IL-33 provokes eosinophilic inflammation and airway hyperresponsiveness (AHR) (Kurowska-Stolarska et al. 2008; Kondo et al. 2008). In ovalbumin (OVA)-induced airway inflammation models, administered IL-33 induces IL-4-independent inflammation during antigen sensitization (Kurowska-Stolarska et al. 2008). Similarly, administration of neutralizing antibodies against IL-33 or IL-33 during the airway challenge with OVA relieves eosinophilic pulmonary inflammation and AHR (Liu et al. 2009; Lohning et al. 1998; Coyle et al. 1999). When the same experiment was performed in ST2 null mice, AHR was not affected and eosinophilic inflammation was not changed or even became worsen (Hoshino et al. 1999; Mangan et al. 2007). In contrast, inflammation was attenuated in ST2 null mice in a short-term primed model of asthma (Kurowska-Stolarska et al. 2008). These differences in ST2 null mice may be due to the dominant role of mast cells in the short-term model of asthma (Kung et al. 1995). Actually, in a mast cell-dependent asthma model, inflammation was markedly reduced in ST2 null mice (Liew et al. 2010).

Serum sST2 levels were increased in patients with diseases associated with an abnormal Th2-like immune response, such as atopic dermatitis and asthma, and with autoimmune diseases, including SLE, and the sST2 concentration was correlated with the severity of disease (Mok et al. 2010; Sahlander et al. 2010; Oshikawa et al. 2001). Following IgE sensitization, IL-33 activates mast cells and induces degranulation, which may lead to anaphylactic shock (Pushparaj et al. 2009). IL-33 administered to systemically IgE-sensitized animals exacerbates antigen-induced anaphylactic shock and induces the degranulation of IgE-sensitized mast cells in the skin even in the absence of antigen (Pushparaj et al. 2009). A corollary to this finding is that the levels of IL-33 are substantially increased in the blood of atopic patients during anaphylactic shock and in inflamed skin tissue of atopic patients (Pushparaj et al. 2009).

Modulation by Drugs

The role of IL-33 provides novel insights into the mechanisms of allergic diseases and other inflammatory disorders. IL-33 is produced by various types of cells during tissue injury in response to inflammatory stimuli, resulting in the promotion of the severity of allergic diseases. IL-33 has a critical role during the development and worsens certain types of airway inflammation. The baseline IL-33 transcript in epithelial cells from chronic rhinosinusitis patients with nasal polyps was found to be much higher in resistant disease than irresponsive disease (Reh et al. 2010). This finding may be extrapolated to other allergic diseases, especially to less responsive diseases. Results in allergen-immunized mice on treatment with ST2-Fc chimera protein are promising for future therapy (Leung et al. 2004). IL-33 is a potent inflammatory cytokine, activated in an allergen- or non-allergen-dependent manner. Therefore, IL-33/ST2 axis may be a therapeutic target not only for asthma or anaphylaxis but also for other inflammatory diseases.

Possible Application of IL-33 for Drug Development

The precise mechanism of IL-33 induction in vivo is still unclear. It is controversial whether IL-33 is matured by a specific enzyme, but it is possible that this cytokine can be activated by multiple processes. Furthermore, the diverse effects of IL-33 on T-cell immunity suggest the existence of another component of the IL-33 receptor. Unlike the established action of IL-33 through ST2 and the IL-1RAcP receptor complex in Th2 immunity, an additional component of the IL-33 receptor may be involved in Th1 immunity. Future studies on a dual function of IL-33, a mechanism for differential polarization of distinct Th1/Th2 cytokines regulated by IL-33, and in vivo studies in mouse disease models with IL-33 neutralizing antagonists will help clarify the critical role of IL-33 in Th1/Th2 cytokine production by immune cells in association with inflammatory disorders.

Acknowledgments This work was supported by the National Research Foundation of Korea (NRF) (MEST 2012R1A2A1A01001791).

References

Ahn, H. J., et al. (1997). A mechanism underlying synergy between IL-12 and IFN-gamma-inducing factor in enhanced production of IFN-gamma. *Journal of Immunology, 159*(5), 2125–2131.

Albacker, L. A., Chaudhary, V., Chang, Y. J., Kim, H. Y., Chuang, Y. T., Pichavant, M., et al. (2013). Invariant natural killer T cells recognize a fungal glycosphingolipid that can induce airway hyperreactivity. *Nature Medicine, 19*(10), 1297–1304.

Ali, S., et al. (2007). IL-1 receptor accessory protein is essential for IL-33-induced activation of T lymphocytes and mast cells. *Proceedings of the National Academy of Sciences of the United States of America, 104*(47), 18660–18665.

Allakhverdi, Z., et al. (2007). Cutting edge: The ST2 ligand IL-33 potently activates and drives maturation of human mast cells. *Journal of Immunology, 179*(4), 2051–2054.

Allakhverdi, Z., et al. (2009). CD34+ hemopoietic progenitor cells are potent effectors of allergic inflammation. *Journal of Allergy and Clinical Immunology, 123*(2), 472–478.

Andersson, U., & Tracey, K. J. (2011). HMGB1 is a therapeutic target for sterile inflammation and infection. *Annual Review of Immunology, 29*, 139–162.

Bae, S., et al. (2012). Contradictory functions (activation/termination) of neutrophil proteinase 3 enzyme (PR3) in interleukin-33 biological activity. *Journal of Biological Chemistry, 287*(11), 8205–8213.

Baekkevold, E. S., et al. (2003). Molecular characterization of NF-HEV, a nuclear factor preferentially expressed in human high endothelial venules. *American Journal of Pathology, 163*(1), 69–79.

Barlow, J. L., & McKenzie, A. N. (2011). Nuocytes: Expanding the innate cell repertoire in type-2 immunity. *Journal of Leukocyte Biology, 90*(5), 867–874.

Barlow, J. L., et al. (2012). Innate IL-13-producing nuocytes arise during allergic lung inflammation and contribute to airways hyperreactivity. *Journal of Allergy and Clinical Immunology, 129*(1), 191–198 e1-4.

Bartemes, K. R., et al. (2012). IL-33-responsive lineage-CD25+ CD44(hi) lymphoid cells mediate innate type 2 immunity and allergic inflammation in the lungs. *Journal of Immunology, 188*(3), 1503–1513.

Bergers, G., et al. (1994). Alternative promoter usage of the Fos-responsive gene Fit-1 generates mRNA isoforms coding for either secreted or membrane-bound proteins related to the IL-1 receptor. *EMBO Journal, 13*(5), 1176–1188.

Blom, L., & Poulsen, L. K. (2012). IL-1 family members IL-18 and IL-33 upregulate the inflammatory potential of differentiated human Th1 and Th2 cultures. *Journal of Immunology, 189*(9), 4331–4337.

Bonilla, W. V., et al. (2012). The alarmin interleukin-33 drives protective antiviral CD8(+) T cell responses. *Science, 335*(6071), 984–989.

Brickshawana, A., et al. (2011). Lineage(-)Sca1 + c-Kit(-) CD25+ cells are IL-33-responsive type 2 innate cells in the mouse bone marrow. *Journal of Immunology, 187*(11), 5795–5804.

Carriere, V., et al. (2007). IL-33, the IL-1-like cytokine ligand for ST2 receptor, is a chromatin-associated nuclear factor in vivo. *Proceedings of the National Academy of Sciences of the United States of America, 104*(1), 282–287.

Cayrol, C., & Girard, J. P. (2009). The IL-1-like cytokine IL-33 is inactivated after maturation by caspase-1. *Proceedings of the National Academy of Sciences of the United States of America, 106*(22), 9021–9026.

Chan, W. L., et al. (2001). Human IL-18 receptor and ST2L are stable and selective markers for the respective type 1 and type 2 circulating lymphocytes. *Journal of Immunology, 167*(3), 1238–1244.

Cherry, W. B., et al. (2008). A novel IL-1 family cytokine, IL-33, potently activates human eosinophils. *Journal of Allergy and Clinical Immunology, 121*(6), 1484–1490.

Chow, J. Y., et al. (2010). Intracellular signaling mechanisms regulating the activation of human eosinophils by the novel Th2 cytokine IL-33: Implications for allergic inflammation. *Cellular and molecular immunology, 7*(1), 26–34.

Chu, D. K., et al. (2013). IL-33, but not thymic stromal lymphopoietin or IL-25, is central to mite and peanut allergic sensitization. *Journal of Allergy and Clinical Immunology, 131*(1), 187–200 e1-8.

Coyle, A. J., et al. (1999). Crucial role of the interleukin 1 receptor family member T1/ST2 in T helper cell type 2-mediated lung mucosal immune responses. *Journal of Experimental Medicine, 190*(7), 895–902.

Dinarello, C. A. (1998). Interleukin-1 beta, interleukin-18, and the interleukin-1 beta converting enzyme. *Annals of the New York Academy of Sciences, 856*, 1–11.

Dinarello, C. A. (2009). Immunological and inflammatory functions of the interleukin-1 family. *Annual Review of Immunology, 27*, 519–550.

Guo, L., et al. (2009). IL-1 family members and STAT activators induce cytokine production by Th2, Th17, and Th1 cells. *Proceedings of the National Academy of Sciences of the United States of America, 106*(32), 13463–13468.

Haenuki, Y., et al. (2012). A critical role of IL-33 in experimental allergic rhinitis. *Journal of Allergy and Clinical Immunology, 130*(1), 184–194 e11.

Hayakawa, M., et al. (2009). Mature interleukin-33 is produced by calpain-mediated cleavage in vivo. *Biochemical and Biophysical Research Communications, 387*(1), 218–222.

Hong, J., et al. (2011). Identification of constitutively active interleukin 33 (IL-33) splice variant. *Journal of Biological Chemistry, 286*(22), 20078–20086.

Hoshino, K., et al. (1999). The absence of interleukin 1 receptor-related T1/ST2 does not affect T helper cell type 2 development and its effector function. *Journal of Experimental Medicine, 190*(10), 1541–1548.

Hoyler, T., et al. (2012). The transcription factor GATA-3 controls cell fate and maintenance of type 2 innate lymphoid cells. *Immunity, 37*(4), 634–648.

Hudson, C. A., et al. (2008). Induction of IL-33 expression and activity in central nervous system glia. *Journal of Leukocyte Biology, 84*(3), 631–643.

Iikura, M., et al. (2007). IL-33 can promote survival, adhesion and cytokine production in human mast cells. *Laboratory Investigation, 87*(10), 971–978.

Ikutani, M., et al. (2012). Identification of innate IL-5-producing cells and their role in lung eosinophil regulation and antitumor immunity. *Journal of Immunology, 188*(2), 703–713.

Kaieda, S., et al. (2012). Interleukin-33 primes mast cells for activation by IgG immune complexes. *PLoS One, 7*(10), e47252.

Kim, H. R., et al. (2010). Levels of circulating IL-33 and eosinophil cationic protein in patients with hypereosinophilia or pulmonary eosinophilia. *Journal of Allergy and Clinical Immunology, 126*(4), 880–882 e6.

Komai-Koma, M., et al. (2007). IL-33 is a chemoattractant for human Th2 cells. *European Journal of Immunology, 37*(10), 2779–2786.

Kondo, Y., et al. (2008). Administration of IL-33 induces airway hyperresponsiveness and goblet cell hyperplasia in the lungs in the absence of adaptive immune system. *International Immunology, 20*(6), 791–800.

Koyasu, S., & Moro, K. (2011a). Type 2 innate immune responses and the natural helper cell. *Immunology, 132*(4), 475–481.

Koyasu, S., & Moro, K. (2011b). Innate Th2-type immune responses and the natural helper cell, a newly identified lymphocyte population. *Current Opinion in Allergy and Clinical Immunology, 11*(2), 109–114.

Kuchler, A. M., et al. (2008). Nuclear interleukin-33 is generally expressed in resting endothelium but rapidly lost upon angiogenic or proinflammatory activation. *American Journal of Pathology, 173*(4), 1229–1242.

Kung, T. T., et al. (1995). Mast cells modulate allergic pulmonary eosinophilia in mice. *American Journal of Respiratory Cell and Molecular Biology, 12*(4), 404–409.

Kurowska-Stolarska, M., et al. (2008). IL-33 induces antigen-specific IL-5+ T cells and promotes allergic-induced airway inflammation independent of IL-4. *Journal of Immunology, 181*(7), 4780–4790.

Lefrancais, E., et al. (2012). IL-33 is processed into mature bioactive forms by neutrophil elastase and cathepsin G. *Proceedings of the National Academy of Sciences of the United States of America, 109*(5), 1673–1678.

Leung, B. P., et al. (2004). A novel therapy of murine collagen-induced arthritis with soluble T1/ST2. *Journal of Immunology, 173*(1), 145–150.

Liang, H. E., et al. (2011). Divergent expression patterns of IL-4 and IL-13 define unique functions in allergic immunity. *Nature Immunology, 13*(1), 58–66.

Liew, F. Y., Pitman, N. I., & McInnes, I. B. (2010). Disease-associated functions of IL-33: The new kid in the IL-1 family. *Nature Reviews Immunology, 10*(2), 103–110.

Liu, X., et al. (2009). Anti-IL-33 antibody treatment inhibits airway inflammation in a murine model of allergic asthma. *Biochemical and Biophysical Research Communications, 386*(1), 181–185.

Lohning, M., et al. (1998). T1/ST2 is preferentially expressed on murine Th2 cells, independent of interleukin 4, interleukin 5, and interleukin 10, and important for Th2 effector function. *Proceedings of the National Academy of Sciences of the United States of America, 95*(12), 6930–6935.

Louten, J., et al. (2011). Endogenous IL-33 enhances Th2 cytokine production and T-cell responses during allergic airway inflammation. *International Immunology, 23*(5), 307–315.

Luthi, A. U., et al. (2009). Suppression of interleukin-33 bioactivity through proteolysis by apoptotic caspases. *Immunity, 31*(1), 84–98.

MacKenzie, A., et al. (2001). Rapid secretion of interleukin-1beta by microvesicle shedding. *Immunity, 15*(5), 825–835.

Mangan, N. E., et al. (2007). T1/ST2 expression on Th2 cells negatively regulates allergic pulmonary inflammation. *European Journal of Immunology, 37*(5), 1302–1312.

Mok, M. Y., et al. (2010). Serum levels of IL-33 and soluble ST2 and their association with disease activity in systemic lupus erythematosus. *Rheumatology (Oxford), 49*(3), 520–527.

Monticelli, L. A., et al. (2011). Innate lymphoid cells promote lung-tissue homeostasis after infection with influenza virus. *Nature Immunology, 12*(11), 1045–1054.

Moritz, D. R., et al. (1998). The IL-1 receptor-related T1 antigen is expressed on immature and mature mast cells and on fetal blood mast cell progenitors. *Journal of Immunology, 161*(9), 4866–4874.

Moro, K., et al. (2010). Innate production of T(H)2 cytokines by adipose tissue-associated c-Kit(+)Sca-1(+) lymphoid cells. *Nature, 463*(7280), 540–544.

Na, H. J., Hudson, S. A., & Bochner, B. S. (2012). IL-33 enhances Siglec-8 mediated apoptosis of human eosinophils. *Cytokine, 57*(1), 169–174.

Nakanishi, K., et al. (2001). Interleukin-18 regulates both Th1 and Th2 responses. *Annual Review of Immunology, 19*, 423–474.

Neill, D. R., et al. (2010). Nuocytes represent a new innate effector leukocyte that mediates type-2 immunity. *Nature, 464*(7293), 1367–1370.

Oboki, K., et al. (2010). IL-33 and IL-33 receptors in host defense and diseases. *Allergology International, 59*(2), 143–160.

Ohno, T., et al. (2009). Caspase-1, caspase-8, and calpain are dispensable for IL-33 release by macrophages. *Journal of Immunology, 183*(12), 7890–7897.

Ohno, T., et al. (2012). Interleukin-33 in allergy. *Allergy, 67*(10), 1203–1214.

Onda, H., et al. (1999). Identification of genes differentially expressed in canine vasospastic cerebral arteries after subarachnoid hemorrhage. *Journal of Cerebral Blood Flow and Metabolism, 19*(11), 1279–1288.

Oshikawa, K., et al. (2001). Elevated soluble ST2 protein levels in sera of patients with asthma with an acute exacerbation. *American Journal of Respiratory and Critical Care Medicine, 164*(2), 277–281.

Pecaric-Petkovic, T., et al. (2009). Human basophils and eosinophils are the direct target leukocytes of the novel IL-1 family member IL-33. *Blood, 113*(7), 1526–1534.

Prefontaine, D., et al. (2009). Increased expression of IL-33 in severe asthma: Evidence of expression by airway smooth muscle cells. *Journal of Immunology, 183*(8), 5094–5103.

Prefontaine, D., et al. (2010). Increased IL-33 expression by epithelial cells in bronchial asthma. *Journal of Allergy and Clinical Immunology, 125*(3), 752–754.

Price, A. E., et al. (2010). Systemically dispersed innate IL-13-expressing cells in type 2 immunity. *Proceedings of the National Academy of Sciences of the United States of America, 107*(25), 11489–11494.

Pushparaj, P. N., et al. (2009). The cytokine interleukin-33 mediates anaphylactic shock. *Proceedings of the National Academy of Sciences of the United States of America, 106*(24), 9773–9778.

Reh, D. D., et al. (2010). Treatment-recalcitrant chronic rhinosinusitis with polyps is associated with altered epithelial cell expression of interleukin-33. *American Journal of Rhinology & Allergy, 24*(2), 105–109.

Roussel, L., et al. (2008). Molecular mimicry between IL-33 and KSHV for attachment to chromatin through the H2A-H2B acidic pocket. *EMBO Reports, 9*(10), 1006–1012.

Saenz, S. A., Noti, M., & Artis, D. (2010). Innate immune cell populations function as initiators and effectors in Th2 cytokine responses. *Trends in Immunology, 31*(11), 407–413.

Sahlander, K., Larsson, K., & Palmberg, L. (2010). Increased serum levels of soluble ST2 in birch pollen atopics and individuals working in laboratory animal facilities. *Journal of Occupational and Environmental Medicine, 52*(2), 214–218.

Sakashita, M., et al. (2008). Association of serum interleukin-33 level and the interleukin-33 genetic variant with Japanese cedar pollinosis. *Clinical and Experimental Allergy, 38*(12), 1875–1881.

Schmitz, J., et al. (2005). IL-33, an interleukin-1-like cytokine that signals via the IL-1 receptor-related protein ST2 and induces T helper type 2-associated cytokines. *Immunity, 23*(5), 479–490.

Schroder, M., & Kaufman, R. J. (2005). ER stress and the unfolded protein response. *Mutation Research, 569*(1–2), 29–63.

Smithgall, M. D., et al. (2008). IL-33 amplifies both Th1- and Th2-type responses through its activity on human basophils, allergen-reactive Th2 cells, iNKT and NK cells. *International Immunology, 20*(8), 1019–1030.

Stolarski, B., et al. (2010). IL-33 exacerbates eosinophil-mediated airway inflammation. *Journal of Immunology, 185*(6), 3472–3480.

Suzukawa, M., et al. (2008). An IL-1 cytokine member, IL-33, induces human basophil activation via its ST2 receptor. *Journal of Immunology, 181*(9), 5981–5989.

Suzukawa, M., et al. (2011). Leptin enhances survival and induces migration, degranulation, and cytokine synthesis of human basophils. *Journal of Immunology, 186*(9), 5254–5260.

Talabot-Ayer, D., et al. (2009). Interleukin-33 is biologically active independently of caspase-1 cleavage. *Journal of Biological Chemistry, 284*(29), 19420–19426.

Townsend, M. J., et al. (2000). T1/ST2-deficient mice demonstrate the importance of T1/ST2 in developing primary T helper cell type 2 responses. *Journal of Experimental Medicine, 191*(6), 1069–1076.

Werman, A., et al. (2004). The precursor form of IL-1alpha is an intracrine proinflammatory activator of transcription. *Proceedings of the National Academy of Sciences of the United States of America, 101*(8), 2434–2439.

Wilhelm, C., et al. (2011). An IL-9 fate reporter demonstrates the induction of an innate IL-9 response in lung inflammation. *Nature Immunology, 12*(11), 1071–1077.

Wysocka, M., et al. (2008). Design of selective substrates of proteinase 3 using combinatorial chemistry methods. *Analytical Biochemistry, 378*(2), 208–215.

Yang, Q., et al. (2011). IL-33 synergizes with TCR and IL-12 signaling to promote the effector function of CD8 + T cells. *European Journal of Immunology, 41*(11), 3351–3360.

Yoshimoto, T., et al. (1999). IL-18, although antiallergic when administered with IL-12, stimulates IL-4 and histamine release by basophils. *Proceedings of the National Academy of Sciences of the United States of America, 96*(24), 13962–13966.

Interleukin 36 Cytokines

Frank L. van de Veerdonk
Department of Internal Medicine, Radboud University Nijmegen Medical Centre, GA, Nijmegen, The Netherlands
Department of Medicine, University of Colorado Denver, Aurora, CO, USA

Synonyms

FIL1delta; FIL1epsilon; FIL1eta; IL-1F5; IL-1F6; IL-1F8; IL-1F9; IL-1H1; IL-1H2; IL-1H3; IL1HY1; IL-1L1; IL-1RP2; IL-1RP3; IL-36α; IL-36β; IL-36γ

Definition

Several new members of the IL-1 gene family have been discovered over the last decade. Novel cytokines, such as IL-36 cytokines, were renamed in order of their date of publication (Sims et al. 2001). It was proposed in 2010 that these IL-1 family members should each be assigned an individual interleukin designation based on their biological function (Dinarello et al. 2010). The cytokines that were renamed IL-36α, IL-36β, IL-36γ, and IL-36 receptor antagonist (Ra) are collectively called the IL-36 cytokines (Table 1).

Biosynthesis and Release

A specific feature of the IL-1 family cytokines is that they require N-terminal cleavage to process them into full bioactive cytokines. The IL-36 cytokines are no exception and like IL-1 and IL-18 also need to be processed in order to gain full bioactivity (Towne et al. 2011). Processing of IL-36α, IL-36β, and IL-36γ will result in a 1000–10,000-fold increased activity of these cytokines in comparison to their non-truncated forms. IL-36Ra gains full antagonistic activity only when the N-terminal methionine of IL-36Ra is removed, probably explaining why initial studies did not show consistent results regarding the antagonistic property of IL-36Ra (Towne et al. 2004, 2011; Debets et al. 2001). The protease(s) responsible for cleaving IL-36 cytokines is currently unknown. Since the amino acids surrounding the truncation sites do not resemble a caspase-1 cleaving site, proteases other than caspase-1 are responsible (Towne et al. 2011). Moreover, no proteolytic cleavage of IL-36α could be observed in macrophages with confirmed caspase-1 activity (Martin et al. 2009). Notably, most of the early studies have been performed with non-processed IL-36 agonists, such as intraventricular injection of murine IL-36γ, and therefore may not have revealed the genuine role of IL-36 cytokines in inflammatory responses in the brain.

IL-36 cytokines do not have a signal peptide for secretion. They belong to the family of the so-called leaderless secretory proteins. The first study investigating the possible underlying mechanism of secretion of these IL-36 cytokines observed that IL-36α is secreted in macrophages upon stimulation with LPS/ATP. This suggested that IL-36α could be externalized in a stimulus-dependent manner comparable to IL-1β (Martin et al. 2009). However, they also demonstrate that IL-36α has no caspase-1 cleavage consensus and the release of IL-36α is not inhibited in the presence of a caspase-1 inhibitor (Martin et al. 2009). Another study however suggested that the release of IL-36γ by keratinocytes is caspase-1 dependent and that the IL-36γ transcription is dependent on caspase-1 (Lian et al. 2012). It remains to be defined which enzymes are responsible for the processing of the IL-36 cytokines and which mechanism is responsible for secreting the IL-36 cytokines.

Biological Function of the IL-36 Cytokines

IL-36Ra shares about 47–52 % homology with IL-1Ra (Mulero et al. 1999; Barton et al. 2000). Initial reports demonstrated that IL-36Ra does not have similar biological activity as IL-1Ra (Barton et al. 2000). The two other family members that were discovered in that period were IL-36β and IL-36γ, and they were found to share the core 12-fold, β-trefoil structure, and lack a signal peptide, which are both classical features of the IL-1 family of cytokines (Busfield et al. 2000; Kumar et al. 2000). They shared only 27 % and 20 % homology with IL-1Ra, respectively (Dunn et al. 2001; Smith et al. 2000). Shortly thereafter, IL-36α was identified next to IL-36Ra, and the predicted three-dimensional structure of IL-36α was published (Smith et al. 2000). IL-36Ra is constitutively expressed in keratinocytes, whereas IL-36γ expression in keratinocytes is rapidly induced after stimulation with TNF or PMA (Busfield et al. 2000). Additional studies revealed that IL-36γ does not bind to the currently known IL-1 family receptors IL-1R, IL-18R, or ST2 (Kumar et al. 2000; Smith et al. 2000). However, it was reported that IL-36γ could activate NFκB through IL-1Rrp2 and that IL-36Ra was able to block this activation, providing the first evidence that IL-36γ is an agonist and IL-36Ra is an antagonist (Debets et al. 2001).

All new IL-1 family member genes are located in a cluster on chromosome 2, and the gene order from centromere to telomere is *IL-1A-IL-1B-IL-37-IL-36G-IL-36A-IL-36B-IL-36RN-IL1F10-IL-*

Interleukin 36 Cytokines, Table 1 IL-36 cytokines

	New nomenclature	Alternative names	Signaling	Homology IL-1Ra (%)	Homology IL-1β (%)	Processing for optimal bioactivity	Expression	Induction
IL-36 cytokines	IL-36Ra	IL-1F5, IL1HY1, IL-1L1, IL-1RP3, IL-1H3, FIL1delta	Binds IL-1Rrp2 and blocks IL-1racp recruitment Recruitment of SIGIRR	52 47	26	Removal of N-terminal methionine (V2)	Monocytes, B lymphocytes, DCs, keratinocytes, skin, uterus, placenta, heart, brain, kidney	
	IL-36α	IL-1F6, FIL1epsilon	Binds IL-1Rrp2 Recruitment of IL-1racp Activation of NFκb, MAPK, ERK1/2, JNK	24	30	Cleavage at 9 amino acids N-terminal to a conserved A-X-Asp (K6)	Monocytes, T/B lymphocytes, spleen, bone marrow, tonsils, lymph nodes, skin	IL-17, TNFα, and IL-22 EGF
	IL-36β	IL-1F8, IL-1H2, FIL1eta		27	31	Cleavage at 9 amino acids N-terminal to a conserved A-X-Asp (R5)	Monocytes, T/B lymphocytes, bone marrow, tonsils, heart, lung, testis, colon, neuron cells, glial cells	IL-36β IL-17, TNFα, and IL-22 EGF IL-1β LPS
	IL-36γ	IL-1F9, IL-1RP2, IL-1H1		20	31	Cleavage at 9 amino acids N-terminal to a conserved A-X-Asp (S18)	Peripheral blood lymphocytes, keratinocytes, bronchial epithelial cells, THP-1	IL-17, TNFα, and IL-22 IL-1 and TNFα IL-17 TLR3 *Pseudomonas aeruginosa* *E. coli* LPS *P. gingivalis* LPS

IRN. Only *IL-1A, IL-1B*, and *IL-36B* are transcribed toward the centromere (Dunn et al. 2001). These new genes probably arose from a common ancestral gene, which is most likely a primordial IL-1 receptor antagonist gene (Mulero et al. 2000).

IL-36α, IL-36β, and IL-36γ are agonists and can induce proinflammatory cytokine responses. IL-36γ is able to induce NFκB activation in Jurkat cells transfected with the IL-1Rrp2 receptor (IL-36R) (Debets et al. 2001). IL-36 agonists can all activate the MAPK, Erk1/2, and JNK through IL-36R/IL-1RAcP, which targets the IL-8 promoter and results in IL-6 secretion (Towne et al. 2004). Additional studies have confirmed that IL-36α, IL-36β, and IL-36γ all have agonistic characteristics and that they signal through the IL-36R (Towne et al. 2004; Magne et al. 2006; Chustz et al. 2011). IL-36 cytokines are expressed in a more tissue-specific manner. They are mainly expressed in keratinocytes, bronchial epithelium, brain tissue, and monocytes/macrophages (Table 1) (Smith et al. 2000; Barksby et al. 2009). Interestingly, THP-1 cells, a human monocytic cell line, specifically express IL-36γ, but not IL-36α or IL-36β, after stimulation with LPS derived from *E. coli* or *P. gingivalis*. Although monocytes and macrophages can thus be a source of IL-36 cytokines, the significance of this observation with respect to homeostasis and pathogenesis remains to be elucidated. In addition, T lymphocytes can express IL-36α and IL-36β, and peripheral blood lymphocytes are able to express IL-36γ in response to α-particles used in cancer therapy. Furthermore, γ T cells can express IL-36β under specific conditions, meaning that some of the IL-36 cytokines can be expressed in specific lymphocyte subsets. Also CD138-positive plasma cells have been reported to express IL-36a.

Several lines of evidence suggest that there might be an important role for IL-36α and IL-36β in skin homeostasis. Epidermal growth factor signaling regulates the expression of IL-36α and IL-36β in the skin, and mice deficient in fibroblast growth factor receptors (FGFRs) have a defective skin barrier that leads to activation of keratinocytes and γδ T cells and significant expression of IL-36β in the epidermis. Moreover, ADAM17 is a downstream of one of these FGFR receptors, and ADAM17-deficient mice show a similar skin phenotype to FGFR-deficient mice with an increased expression of IL-36α in the skin compared to wild-type mice. Under inflammatory conditions, IL-36 cytokine signaling is also likely to be important, since IL-17 and TNF can induce expression of IL-36α, IL-36β, and IL-36γ in keratinocytes. This can be synergized by the cytokine IL-22, which is an important cytokine for skin repair and the induction of antimicrobial peptides that kill microorganisms.

Stimulation of bronchial epithelial cells with proinflammatory cytokines, *Pseudomonas aeruginosa*, or TLR3 ligands results in a significant expression of IL-36γ. In addition, IL-36γ induces the chemokine IL-8 and the Th17 chemokine CCL20 in human lung fibroblasts (Chustz et al. 2011). These data suggest that IL-36 cytokines play a role in the host defense against viral and bacterial pulmonary infection and most likely play a significant role in the regulation of neutrophilic airway inflammation.

Initial studies have investigated the role of the IL-36 cytokines in inflammatory brain responses because the IL-36R is highly expressed in microglial cells and astrocytes. Recombinant mouse IL-36β or IL-36γ does not induce any of the classical IL-1-like responses in mixed glial cells in vitro, and injection of murine IL-36γ intracerebroventricular does not result in fever or modification of food intake and body weight in mice (Berglöf et al. 2003). The agonist IL-36β is expressed in neuron cells and in glial cells and cannot be upregulated by LPS or IL-1β stimulation in the brain. However, IL-36β can induce expression of itself and thus appears to have an autocrine/paracrine loop similar to IL-1. IL-36Ra is significantly upregulated in an animal model where brain micromotion is simulated. IL-36Ra expression is increased in astrocytes and microglia cells that are subjected to low-magnitude cyclical strain. In the presence of IL-36Ra, primary cortical neurons that are stretched show a significant decrease of TNF superfamily member B11 (TNFRSF11b). Subsequently, pro-apoptotic genes are significantly upregulated suggesting that IL-36Ra can regulate apoptotic pathways via decreasing TNFRSF11b in low-magnitude strain induced by brain tissue inflammation. However, the exact role of IL-36 cytokines and its signaling pathway in brain homeostasis remains to be determined.

Although IL-36Ra shares homology with IL-1Ra and it has been suggested that IL-36Ra acts similar to IL-1Ra, several observations suggest otherwise. In contrast to IL-1Ra, IL-36Ra needs to be processed in order to gain antagonistic properties (Towne et al. 2011). IL-36Ra itself can induce mRNA of IL-4 and protein expression in glial cells in vitro, and this induction is blocked in the presence of anti-SIGIRR antibodies, and in vivo experiments show that the anti-inflammatory action of IL-36Ra in the brain is dependent on IL-4 and SIGIRR. These differences could be explained by the observation that IL-36Ra significantly differs in loop conformations from IL-1Ra and that 147 is an aspartic acid in both IL-1β and IL-36Ra but a lysine in IL-1Ra. Interestingly, changing the lysine of IL-1Ra to an aspartic acid will result in a partial agonistic character of IL-1Ra. Another difference between IL-36Ra and IL-1Ra is that IL-36Ra does not have a classical dose-dependent inhibitory effect on Th17 characteristic cytokines induced by fungi (van de Veerdonk et al. 2012; Gresnigt et al. 2013). These reports suggest that IL-36Ra might be able to activate an anti-inflammatory signaling pathway or recruit alternative anti-inflammatory IL-1 orphan receptors, like SIGIRR, and under specific conditions does not act as the classical receptor antagonist IL-1Ra. However, it must be noted that several reports have shown that IL-36Ra inhibits IL-36γ-induced NFκB activation (Debets et al. 2001; Towne et al. 2004) and that IL-36Ra acts similar to IL-1Ra, by eliciting its antagonistic effects through binding of the IL-36R and blocking the recruitment of the second receptor IL-1RAcP (Towne et al. 2011).

In humans, the expression of the IL-36R within the human monocytic cell line is unique to dendritic cells (DCs) (Mutamba et al. 2012). Monocyte-derived DCs (MDDCs) and plasmacytoid DCs express the IL-36R. IL-36β induces the production of IL-12 and IL-18 in MDDCs resulting in the proliferation of IFNγ-producing T lymphocytes (Mutamba et al. 2012). Furthermore, human MDDCs increase their expression of HLA-DR and CD83 in response to IL-36β and IL-36γ. Dendritic cells stimulated with IL-36α will increase CXCL1 and CXCL2, TNFα, and CD40 expression and have the ability to induce T-cell proliferation. The IL-36 cytokines also have a significant effect on murine dendritic cells (DCs). In murine DCs, IL-36 agonists upregulate CD80, CD86, and MHCII and induce IL-6 and IL-12 production that is completely dependent on IL-36R signaling (Vigne et al. 2011). The IL-36R is highly expressed on naïve T cells, and IL-36 agonists can stimulate proliferation and IL-2 production by naïve T cells. In the presence of mitogenic stimulation (aCD3/aCD28), IL-36 agonists enhance the capability of T-helper cells to produce and secrete IFNγ, IL-4, and IL-17 in a dose-dependent manner. IL-36β can act as an adjuvant that specifically enhances Th1 responses in vivo in a murine immunization model using intradermal BSA injections (Vigne et al. 2011). The adjuvanticity of IL-36β in this model was shown to be completely dependent on IL-36R signaling (Vigne et al. 2011). In addition, IL-36 plays a significant role in modulating human T-helper (Th) responses (Vigne et al. 2012). Human peripheral blood mononuclear cells that are stimulated with fungal pathogens produce less IL-17 and IL-22 in the presence of IL-36Ra (van de Veerdonk et al. 2012; Gresnigt et al. 2013). Therefore, IL-36 cytokines can modulate the immune system through their effects on antigen-presenting cells, such as DCs, and in this way can polarize T-helper responses or via direct effect on T-cell subsets.

Pathophysiological Relevance of IL-36 Cytokines

The importance of the IL-36 cytokines in regulating skin inflammation is underscored by the observation that mutations in the gene encoding IL-36Ra, namely, *IL-36RN*, can cause general pustular psoriasis (GPP), which is a rare life-threatening disease (Marrakchi et al. 2011). The mutations in *IL-36RN* resulted in a misfolded IL-36Ra protein that is less stable and poorly expressed (Marrakchi et al. 2011). Moreover this misfolded IL-36Ra was found to have less affinity with the IL-36R compared to the wild-type IL-36Ra protein and therefore cannot sufficiently dampen IL-36 cytokine-mediated inflammation (Marrakchi et al. 2011). These data indicate that IL-36 signaling plays a significant role in regulating skin inflammation and that IL-36Ra is a crucial regulator of IL-36R signaling. Notably, mutations in the gene encoding IL-36Ra have not been found to date in Chinese patients with GPP, and only 2 out of 14 Japanese patients with GPP carried mutations in *IL-36RN* (Farooq et al. 2012; Li et al. 2012).

In line with the human data, transgenic mice that overexpress the IL-36α gene in basal keratinocytes display characteristics of psoriatic skin lesions (Blumberg et al. 2007). In addition, IL-36Ra deficiency exacerbated skin lesions in IL-36α transgenic mice, suggesting an antagonistic effect of IL-36Ra in vivo for IL-36-mediated inflammation in the skin (Blumberg et al. 2007). Inflammatory conditions with macroscopic and histological similarities to human psoriasis can be chemically induced in mouse skin overexpressing IL-36α (Blumberg et al. 2010). When human psoriatic skin is transplanted into immunodeficient mice, the maintenance of the characteristic inflammation seen in psoriasis is dependent on IL-36R signaling (Blumberg et al. 2010). Several reports have shown that psoriatic skin lesions have increased expression of IL-36Ra and IL-36 agonists (Debets et al. 2001; Blumberg et al. 2007; Johnston et al. 2011). Anti-TNF treatment in patients with psoriasis results in improved clinical outcome, which is associated with a decreased expression of the IL-36 agonists and IL-36Ra (Johnston et al. 2011). It has been observed that there is an interplay between TNFα, IL-22 and IL-17, and IL-36 cytokines. Increased expression of the *IL-36* agonists correlates with Th17 cytokines in human psoriatic skin lesions (Carrier et al. 2011), although the IL-17-induced expression of IL-36Ra keratinocytes derived from patients with psoriasis does not differ from healthy controls. In addition, psoriasiform dermatitis was found to be mediated by DC-keratinocyte cross talk that was dependent on IL-36 cytokines. These data suggest an important role for IL-36 cytokines in the pathogenesis of psoriasis.

Alopecia areata is a skin disease characterized by patchy hair loss with T-cell infiltration in the hair follicles. An early study reports that a polymorphism in IL-36Ra is associated with the development of alopecia areata. Another skin disorder is Kindler syndrome, which is a rare syndrome characterized by skin blistering, increased photosensitivity of the skin, and progressive generalized poikiloderma. Kindler patients can also develop mucocutaneous fibrosis, such as stenosis of the esophagus and urethral stenosis. The IL-36R signaling pathway might play an important role in this disease, since keratinocytes isolated from patients with Kindler syndrome were found to have increased IL-36Ra expression.

IL-36β is specifically expressed in mouse and human joints, and it is constitutively expressed in human articular chondrocytes (Magne et al. 2006). In addition, recombinant IL-36β induces proinflammatory cytokine responses in both synovial fibroblasts and articular chondrocytes (Magne et al. 2006). Polymorphisms in IL-36β are associated with spondylitis ankylopoetica in a Caucasian cohort but not in Asian cohorts. Polymorphisms in a region in the IL-1 locus that includes the gene encoding IL-36β were associated with a higher susceptibility to psoriatic arthritis. Although these data suggest an important role for IL-36R signaling in the development of arthritis, three different experimental arthritis models in IL-36R−/− mice did not show a difference compared to wild-type mice. In addition, serum IL-36β concentrations of healthy volunteers compared to serum concentrations in rheumatoid arthritis also showed no significant differences (Lamacchia et al. 2013).

IL-36 cytokines are involved in pulmonary diseases. Inflammatory mediators in plasma from healthy volunteers, patients with COPD, and patients with COPD that have an acute exacerbation demonstrated that IL-36α and IL-36Ra are significantly lower in patients with COPD with an acute exacerbation. The expression of IL-36γ in the lung is increased in asthma-susceptible mice suggesting that IL-36R signaling might play a role in allergen-induced inflammation in the lung. Moreover, the gene encoding IL-36γ is located in an allergen-induced bronchial hyperresponsiveness-1 locus (*Abhr1*) in mice, and mice that are challenged with house dust mite (HDM) have significant increased IL-36γ expression. The role of IL-36R signaling in the lung is similar to IL-1R signaling in regarding the recruitment of neutrophils. When recombinant IL-36γ is given intratracheally, there is a significant influx of neutrophils in the lung but not an eosinophilic influx in the lungs of mice, suggesting that IL-36γ is involved in the regulation of neutrophilic airway inflammation like IL-1. IL-36α intratracheally can induce neutrophil chemokines CXCL1 and CXCL2 and expression of the IL-36R in the lungs of both wild-type and IL-1αβ −/− mice. This supports the concept that IL-36 signaling mimics IL-1 signaling but can play a role in neutrophil recruitment independent of IL-1α or IL-1β. Furthermore, rhinovirus-infected primary bronchial epithelial cells from patients with asthma show a higher expression of IL-36γ compared to infected cells from healthy controls. These data collectively provide evidence that IL-36 cytokines play a significant role in regulating airway inflammation.

Several other diseases have been associated with IL-36 cytokines. Increased IL-36α expression has been reported in eosinophilic esophagitis, which indicates a role of IL-36 cytokines in Th2 immune responses in the gut. Local overexpression of IL-36α in the kidney has been reported to be associated with tubulointerstitial lesions in mouse models of chronic glomerulonephritis, SLE, nephrotic syndrome, and streptozocin-induced diabetes. However, concentrations of IL-36 cytokines in the serum from patients with SLE are comparable with healthy controls. IL-36Ra is upregulated in a mouse model of biliary atresia, but the significance of this finding remains to be determined. IL-36Ra is also expressed in differentiated pre-adipocytes that can be downregulated by proinflammatory cytokines, such as TNFα. In addition, IL-36α, but not IL-36γ, can be induced by LPS in adipose tissue-resident macrophages, and IL-36α and IL-36β can induce inflammatory gene expression in mature adipocytes. Polymorphisms in IL-36β and IL-36Ra can influence the concentrations of IL-1Ra plasma levels, and IL-1Ra plasma levels in humans are associated with metabolic conditions.

Modulation of IL-36R Signaling by Drugs

Although the IL-36 cytokines and the IL-36R pathway have many similarities with the classical IL-1 and IL-1R pathway, it also has unique non-redundant properties. This is specifically highlighted by diseases such as deficiency of IL-1Ra (DIRA) and deficiency of IL-36Ra (DITRA), which have distinct clinical features (Marrakchi et al. 2011; Aksentijevich et al. 2009). Interestingly, although this distinction is there, DITRA can be treated with IL-1R blocking (Tauber et al. 2014; Rossi-Semerano et al. 2013), which emphasizes the interaction between IL-1 family members and IL-1 receptors. But it must however be noted that mice that overexpress IL-36α in basal keratinocytes have skin abnormalities that are dependent on IL-36R but not IL-1R1 (Blumberg et al. 2007). The IL-36 cytokines are expressed in a more tissue-restricted way than IL-1. Therefore, IL-36 cytokines are involved in IL-1-independent inflammatory responses and have most likely evolved to regulate tissue-specific immune responses; thus, IL-36 cytokines cannot simply be seen as surrogates of IL-1. Therefore, the development of IL-36R-targeted therapy is being developed and will probably be first explored in psoriasis.

However, several questions need to be addressed before novel therapeutic strategies targeting IL-36 cytokines can be fully explored. One of the most important questions that remain to be resolved is why IL-36 cytokines need to be processed to gain full antagonistic activity and which protease is responsible for this? Why does the removal of one amino acid, namely, the methionine, of the N-terminal gain such increased activity? Another important issue is the presence of a unique mutation in the TIR domain of IL-36R, which is only shared by the inhibitory IL-1 orphan receptor SIGIRR. This mutation has no consequence for agonistic activity of IL-36 ligands when IL-1RAcP is recruited to IL-36R (Towne et al. 2011). This is elegantly proven by the fact that the chimera of the extracellular IL-1R1 with the intracellular IL-36R that has the mutated TIR domain still has full agonistic activity of IL-1 (Towne et al. 2011). However, it is notable that this mutation exists in the TIR domain of the IL-36R and it remains to be determined whether it has functional consequences in other specific conditions.

In conclusion, IL-36 cytokines play an important role in tissue homeostasis and diseases. They predominantly are important in regulating skin inflammation, which has opened the opportunity to develop novel treatment strategies in skin diseases such as generalized pustular psoriasis and psoriasis. IL-36 cytokines might also have important modulatory functions in inflammatory diseases affecting the lung, the brain, and the gut, which makes them an attractive target for novel treatment strategies in many inflammatory diseases.

Cross-References

▶ Autoinflammatory Syndromes
▶ Cytokines
▶ Dendritic Cells
▶ Inflammasomes
▶ Pathogen-Associated Molecular Patterns (PAMPs)
▶ Th17 Response

References

Aksentijevich, I., Masters, S. L., Ferguson, P. J., Dancey, P., Frenkel, J., van Royen-Kerkhoff, A., et al. (2009). An autoinflammatory disease with deficiency of the interleukin-1-receptor antagonist. *The New England Journal of Medicine, 360*(23), 2426–2437.

Barksby, H. E., Nile, C. J., Jaedicke, K. M., Taylor, J. J., & Preshaw, P. M. (2009). Differential expression of immunoregulatory genes in monocytes in response to Porphyromonas gingivalis and *Escherichia coli* lipopolysaccharide. *Clinical and Experimental Immunology, 156*, 479–487.

Barton, J. L., Herbst, R., Bosisio, D., Higgins, L., & Nicklin, M. J. (2000). A tissue specific IL-1 receptor antagonist homolog from the IL-1 cluster lacks IL-1, IL-1ra, IL-18 and IL-18 antagonist activities. *European Journal of Immunology, 30*(11), 3299–3308.

Berglöf, E., Andre, R., Renshaw, B. R., Allan, S. M., Lawrence, C. B., Rothwell, N. J., et al. (2003). IL-1Rrp2 expression and IL-1F9 (IL-1H1) actions in brain cells. *Journal of Neuroimmunology, 139*, 36–43.

Blumberg, H., Dinh, H., Trueblood, E. S., Pretorius, J., Kugler, D., Weng, N., et al. (2007). Opposing activities of two novel members of the IL-1 ligand family regulate skin inflammation. *The Journal of Experimental Medicine, 204*(11), 2603–2614.

Blumberg, H., Dinh, H., Dean, C., Trueblood, E. S., Bailey, K., Shows, D., et al. (2010). IL-1RL2 and its ligands contribute to the cytokine network in psoriasis. *Journal of Immunology, 185*, 4354–4362.

Busfield, S. J., Comrack, C. A., Yu, G., Chickering, T. W., Smutko, J. S., Zhou, H., et al. (2000). Identification and gene organization of three novel members of the IL-1 family on human chromosome 2. *Genomics, 66*(2), 213–216.

Carrier, Y., Ma, H. L., Ramon, H. E., Napierata, L., Small, C., O'Toole, M., et al. (2011). Inter-regulation of Th17 cytokines and the IL-36 cytokines in vitro and in vivo: Implications in psoriasis pathogenesis. *The Journal of Investigative Dermatology, 131*(12), 2428–2437.

Chustz, R. T., Nagarkar, D. R., Poposki, J. A., Favoreto, S., Jr., Avila, P. C., Schleimer, R. P., et al. (2011). Regulation and function of the IL-1 family cytokine IL-1F9 in human bronchial epithelial cells. *American Journal of Respiratory Cell and Molecular Biology, 45*(1), 145–153.

Debets, R., Timans, J. C., Homey, B., Zurawski, S., Sana, T. R., Lo, S., et al. (2001). Two novel IL-1 family members, IL-1 delta and IL-1 epsilon, function as an antagonist and agonist of NF-kappa B activation through the orphan IL-1 receptor-related protein 2. *Journal of Immunology, 167*(3), 1440–1446.

Dinarello, C., Arend, W., Sims, J., Smith, D., Blumberg, H., O'Neill, L., et al. (2010). IL-1 family nomenclature. *Nature Immunology, 11*(11), 973.

Dunn, E., Sims, J. E., Nicklin, M. J., & O'Neill, L. A. (2001). Annotating genes with potential roles in the immune system: Six new members of the IL-1 family. *Trends in Immunology, 22*(10), 533–536.

Farooq, M., Nakai, H., Fujimoto, A., Fujikawa, H., Matsuyama, A., Kariya, N., et al. (2012). Mutation analysis of the IL36RN gene in 14 Japanese patients with generalized pustular psoriasis. *Human Mutation, 34*, 176–183.

Gresnigt, M. S., Rosler, B., Jacobs, C. W., Becker, K. L., Joosten, L. A., van der Meer, J. W., et al. (2013). The IL-36 receptor pathway regulates *Aspergillus fumigatus*-induced Th1 and Th17 responses. *European Journal of Immunology, 43*(2), 416–426.

Johnston, A., Xing, X., Guzman, A. M., Riblett, M., Loyd, C. M., Ward, N. L., et al. (2011). IL-1F5, −F6, −F8, and -F9: A novel IL-1 family signaling system that is active in psoriasis and promotes keratinocyte antimicrobial peptide expression. *Journal of Immunology, 186*, 2613–2622.

Kumar, S., McDonnell, P. C., Lehr, R., Tierney, L., Tzimas, M. N., Griswold, D. E., et al. (2000). Identification and initial characterization of four novel members of the interleukin-1 family. *The Journal of Biological Chemistry, 275*(14), 10308–10314.

Lamacchia C, Palmer G, Rodriguez E, Martin P, Vigne S, Seemayer C. A., et al. (2013). The severity of experimental arthritis is independent of IL-36 receptor signaling. *Arthritis Res Ther, 15*(2), R38.

Li, M., Lu, Z., Cheng, R., Li, H., Guo, Y., & Yao, Z. (2012). IL36RN gene mutations are not associated with sporadic generalized pustular psoriasis in Chinese patients. *The British Journal of Dermatology, 168*, 452–455.

Lian, L. H., Milora, K. A., Manupipatpong, K. K., & Jensen, L. E. (2012). The double-stranded RNA analogue polyinosinic-polycytidylic acid induces keratinocyte pyroptosis and release of IL-36gamma. *The Journal of Investigative Dermatology, 132*(5), 1346–1353.

Magne, D., Palmer, G., Barton, J. L., Mézin, F., Talabot-Ayer, D., Bas, S., et al. (2006). The new IL-1 family member IL-1F8 stimulates production of inflammatory mediators by synovial fibroblasts and articular chondrocytes. *Arthritis Research & Therapy, 8*, R80.

Marrakchi, S., Guigue, P., Renshaw, B. R., Puel, A., Pei, X. Y., Fraitag, S., et al. (2011). Interleukin-36-receptor antagonist deficiency and generalized pustular psoriasis. *The New England Journal of Medicine, 365*(7), 620–628.

Martin, U., Scholler, J., Gurgel, J., Renshaw, B., Sims, J. E., & Gabel, C. A. (2009). Externalization of the leaderless cytokine IL-1F6 occurs in response to lipopolysaccharide/ATP activation of transduced bone marrow macrophages. *Journal of Immunology, 183*, 4021–4030.

Mulero, J. J., Pace, A. M., Nelken, S. T., Loeb, D. B., Correa, T. R., Drmanac, R., et al. (1999). IL1HY1: A novel interleukin-1 receptor antagonist gene. *Biochemical and Biophysical Research Communications, 263*(3), 702–706.

Mulero, J. J., Nelken, S. T., & Ford, J. E. (2000). Organization of the human interleukin-1 receptor antagonist gene IL1HY1. *Immunogenetics, 51*(6), 425–428.

Mutamba, S., Allison, A., Mahida, Y., Barrow, P., & Foster, N. (2012). Expression of IL-1Rrp2 by human myelomonocytic cells is unique to DCs and facilitates DC maturation by IL-1F8 and IL-1F9. *European Journal of Immunology, 42*(3), 607–617.

Rossi-Semerano, L., Piram, M., Chiaverini, C., De Ricaud, D., Smahi, A., & Kone-Paut, I. (2013). First clinical description of an infant with interleukin-36-receptor antagonist deficiency successfully treated with anakinra. *Pediatrics, 132*(4), e1043–e1047.

Sims, J. E, Nicklin, M. J, Bazan, J. F, Barton, J. L, Busfield, S. J, Ford, J. E., et al. (2001). A new nomenclature for IL-1-family genes. *Trends Immunol, 22*(10), 536–537.

Smith, D. E., Renshaw, B. R., Ketchem, R. R., Kubin, M., Garka, K. E., & Sims, J. E. (2000). Four new members expand the interleukin-1 superfamily. *The Journal of Biological Chemistry, 275*(2), 1169–1175.

Tauber, M., Viguier, M., Alimova, E., Petit, A., Liote, F., Smahi, A., et al. (2014). Partial clinical response to anakinra in severe palmoplantar pustular psoriasis. *The British Journal of Dermatology, 171*, 646–649.

Towne, J. E., Garka, K. E., Renshaw, B. R., Virca, G. D., & Sims, J. E. (2004). Interleukin (IL)-1F6, IL-1F8, and IL-1F9 signal through IL-1Rrp2 and IL-1RAcP to activate the pathway leading to NF-kappaB and MAPKs. *The Journal of Biological Chemistry, 279*(14), 13677–13688.

Towne, J. E., Renshaw, B. R., Douangpanya, J., Lipsky, B. P., Shen, M., Gabel, C. A., et al. (2011). Interleukin-36 (IL-36) ligands require processing for full agonist (IL-36alpha, IL-36beta, and IL-36gamma) or antagonist (IL-36Ra) activity. *The Journal of Biological Chemistry, 286*(49), 42594–42602.

van de Veerdonk, F. L., Stoeckman, A. K., Wu, G., Boeckermann, A. N., Azam, T., Netea, M. G., et al. (2012). IL-38 binds to the IL-36 receptor and has biological effects on immune cells similar to IL-36 receptor antagonist. *Proceedings of the National Academy of Sciences of the United States of America, 109*, 3001–3005.

Vigne, S., Palmer, G., Lamacchia, C., Martin, P., Talabot-Ayer, D., Rodriguez, E., et al. (2011). IL-36R ligands are potent regulators of dendritic and T cells. *Blood, 118*(22), 5813–5823.

Vigne, S., Palmer, G., Martin, P., Lamacchia, C., Strebel, D., Rodriguez, E., et al. (2012). IL-36 signaling amplifies Th1 responses by enhancing proliferation and Th1 polarization of naive $CD4^+$ T cells. *Blood, 120*(17), 3478–3487.

J

Janus Kinase Inhibitors

Jean-Baptiste Telliez
Inflammation and Immunology, Pfizer Inc.,
Cambridge, MA, USA

Synonyms

Janus kinase inhibitors; JAK inhibitors

Definition

Janus kinase (JAK) inhibitors are small molecule ATP competitive kinase inhibitors that block the kinase activity of one or more isoforms of the four JAK family members.

Introduction

Multiple JAK inhibitors are in various stages of clinical development for the treatment of inflammatory diseases, myeloproliferative disorders, or cancers (Clark et al. 2014; Sonbol et al. 2013). Currently, there are two approved JAK inhibitors in clinical practice. Tofacitinib is an oral JAK inhibitor for the treatment of rheumatoid arthritis, and ruxolitinib is an oral JAK inhibitor for the treatment of certain myeloproliferative disorders.

Chemical Structures and Properties

These two molecules share a common pyrrolopyrimidine core substituted at the same position, the rest of the molecule being different (Fig. 1). In the case of both molecules, this heterocycle represents a pharmacophoric subunit involved in interaction with the hinge region of the kinase domain of JAK proteins (Table 1).

Both ruxolitinib and tofacitinib have chiral centers: two in the case of tofactinib and one in ruxolitinib (Fig. 1). However, they are defined as a specific active isomer in each case. The logP values indicate that they are lipid soluble. Both are weak bases but their pKa values indicate that they are mainly unionized at physiological pH values. These values favor passive diffusion through cell membranes.

Metabolism, Pharmacokinetics, and Dosage

Both ruxolitinib and tofacitinib are rapidly absorbed with a peak concentration around 1 h after oral administration. Tofacitinib is primarily cleared by the liver (70 %) while the remaining 30 % is excreted unchanged in urine (Dowty et al. 2014a). A total of approximately 80 % and 14 % of the radioactivity are recovered in urine and feces, respectively. Similar data were obtained following the administration of a single dose of 25 mg

© Springer International Publishing AG 2016
M.J. Parnham (ed.), *Compendium of Inflammatory Diseases*,
DOI 10.1007/978-3-7643-8550-7

Janus Kinase Inhibitors, Fig. 1 Structures of tofacitinib and ruxolitinib. Chiral centers are shown (*)

Tofacitinib

Ruxolitinib

Janus Kinase Inhibitors, Table 1 Physicochemical properties of ruxolitinib and tofacitinib

Drug	Ruxolitinib	Tofacitinib
Molecular mass (Daltons)	306.4	312.4
logP	2.48–2.94	1.808
Number of chiral centers	1	2
pKa	5.51 (base), 13.89 (acid)	7.13 (base), 8.46 (acid)
Available salt	Phosphate	Citrate
Major species at pH 7.4	Unionized	Unionized

of 14C-labeled ruxolitinib with 74 % and 22 % recovered in urine and feces, respectively (Shilling et al. 2010), with only 1 % in the form of the parent compound.

They are also both primarily metabolized by CYP3A4 with additional contribution by CYP2C9 in the case of ruxolitinib and CYP2C19 in the case of tofacitinib. Most (69.4 %) of the ^{14}C-labeled tofacitinib extracted from the plasma of healthy male volunteers is the parent compound with the remainder being metabolites resulting primarily from oxidation of the pyrrolopyrimidine and piperidine ring as well as oxidation of the piperidine ring side chain, N-demethylation, and glucoronidation (Dowty et al. 2014a). In the case of ruxolitinib, 66 % of ^{14}C-labeled ruxolitinib extracted from the plasma 2 h post-dose was the parent compound indicating a low concentration of metabolites (Shilling et al. 2010).

One notable difference between ruxolitinib and tofacitinib is the amount bound to protein, primarily albumin, in vivo with 40 % bound for tofacitinib and 97 % bound for ruxolitinib. The half-lives of elimination of ruxolitinib and tofacitinib are both about 3 h in healthy subjects (Dowty et al. 2014a; Shilling et al. 2010), and both drugs are commonly prescribed to be taken twice a day. Analysis of preclinical data as well as clinical data in RA patients indicates the importance of the daily average concentration as driver of efficacy rather than maximum and minimum plasma concentrations (Dowty et al. 2014b). For example, dosage of tofacitinib at 20 mg once daily produces a similar change as 10 mg twice daily in the rheumatoid arthritis disease activity score (DAS) measured in an exposure-response (E-R) model yielding an equilibration half-life of 3.2 weeks (Lamba et al. 2015). It is probable that similar exposure-response relationship would apply to ruxolitinib.

Pharmacological Activities

The JAK family of nonreceptor tyrosine kinases is composed of four isoforms; JAK1, JAK2, JAK3, and TYK2 that are required, in pairs, for signaling through type I/II cytokine receptors (Ghoreschi et al. 2009).

Ruxolitinib primarily inhibits JAK1 and JAK2 resulting in inhibition of signaling elicited by cytokines that signal via JAK1 and/or JAK2. The pharmacological activity of ruxolitinib in the treatment of myeloproliferative disorders could be driven by two different but related mechanisms of action. First, the mechanism of action of ruxolitinib, associated with JAK2 inhibition, could result from inhibiting the constitutive activation of the JAK/STAT pathway resulting from the mutation JAK2V617F or other phenocopy mutations (Tefferi 2010), therefore inhibiting the clonal myeloproliferation. Second, by inhibiting JAK1, ruxolitinib downregulates the signaling of many cytokines and also downregulates the expression of cytokines such as TNF and IL-6 that together could lead to a reduction in the inflammatory burden. The inflammatory burden is characterized by an increase in circulating inflammatory cytokines such as IL-6, TNF, IL-8, IL-2, and soluble IL-2 receptor alpha, as well as other cytokines and factors that vary depending on the specific myeloproliferative disorder considered. The role and level of inflammatory cytokines also varies depending of the type of myeloproliferative disorder considered. Hemoglobin values decreased by 10–12 % below baseline between weeks 8 and 12 and subsequently stabilized over time to a new steady state slightly below baseline by week 24 (Verstovsek et al. 2013).

Tofacitinib primarily inhibits JAK1 with activity also against JAK3, resulting in inhibition of signaling of JAK1- or JAK3-dependent cytokines, including type I and II interferons (IFN), IL-6, and the cytokines signaling through the common γ-chain receptor (IL-2, IL-4, IL-7, IL-9, IL-15, and IL-21). It inhibits to a lesser extent JAK2-dependent cytokines such as IL-12, IL-23, and EPO. It is noteworthy that tofacitinib has often been mischaracterized in the literature as a JAK3-selective inhibitor. Indeed tofacitinib was originally described as a JAK3-selective inhibitor (Changelian et al. 2003) in biochemical assays but extensive work since then clearly characterized tofacitinib as a potent JAK1 and JAK3 inhibitor both in biochemical and cellular assays (Clark et al. 2014). In a Phase 2a study, tofacitinib reduced the absolute neutrophil count (ANC) in patients with rheumatoid arthritis where it was shown that the disease induces an increase in baseline ANC (Gupta et al. 2009). Data modeling, as well as data with adalimumab, supports that the reduction in ANC is the result of an indirect effect of tofacitinib reducing the inflammatory burden rather than a direct effect on granulopoiesis (Gupta et al. 2009). Despite having increased cardiovascular morbidity and mortality, rheumatoid arthritis patients often have reduced levels of low-density lipoprotein (LDL) cholesterol and total cholesterol. The high-density lipoprotein (HDL) levels and composition are also often altered in RA patients. A Phase I open-label mechanism of action study has shown that tofacitinib treatment reduced cholesterol ester catabolism, which is increased in rheumatoid arthritis patients, resulting in an increase in cholesterol levels toward those in healthy volunteers (Charles-Schoeman et al. 2015). Importantly, markers of antiatherogenic HDL function were also improved (Charles-Schoeman et al. 2015).

Other changes including changes in lymphocytes count, low neutrophil count, and low red blood cell count as well as serious bacterial, mycobacterial, fungal, and viral infections have also been observed with ruxolitinib and tofacitinib treatments.

Clinical Use

Ruxolitinib is approved for the treatment of polycythemia vera and certain forms of myelofibrosis (MF) including primary MF, post-polycythemia vera MF, and post-essential thrombocythemia MF. In clinical trials of MF, ruxolitinib therapy was associated with significant improvements in splenomegaly, MF-associated symptoms, and quality of life (Kantarjian et al. 2013) as well as survival (Verstovsek et al. 2015). In polycythemia vera, ruxolitinib showed greater rates of response (hematocrit control and spleen volume reduction) and complete hematological remission compared to best standard of care (Incyte Corporation 2014). Ruxolitinib is currently in further clinical trials for multiple oncology indications.

Tofacitinib is approved for the treatment of patients with moderate to severe rheumatoid arthritis who have had an inadequate response or intolerance to methotrexate. A series of six Phase 3 clinical trials has shown that tofacitinib is efficacious compared to placebo and methotrexate, when administered as monotherapy or in combination with methotrexate or other nonbiologic

DMARDs in rheumatoid arthritis patients (Wollenhaupt et al. 2014). Two long-term open-label extension studies have shown tofacitinib to be consistently efficacious (84 months) (Wollenhaupt et al. 2015). Notably, tofacitinib demonstrated that it inhibits the progression of structural damage and is also superior to methotrexate as a monotherapy to reduce signs and symptoms and improve physical function. Four Phase 3 studies in psoriasis have been completed demonstrating significant efficacy in the treatment of moderate-to-severe plaques psoriasis (Papp et al. 2015). Tofacitinib is also currently being explored in additional clinical trials for ulcerative colitis as well as psoriatic arthritis and juvenile idiopathic arthritis.

Safety and Adverse Events

In the case of ruxolitinib, dose-dependent anemia and thrombocytopenia were the most common adverse events. Mean platelet counts decreased during the first 8–12 weeks of treatment and remained relatively stable subsequently (Verstovsek et al. 2013). The most common nonhematologic AEs that occurred more frequently with ruxolitinib than with placebo in the primary analysis were ecchymosis, dizziness, and headache (Tefferi 2010). Urinary tract and herpes zoster infections were also observed in patients treated with ruxolitinib (Verstovsek et al. 2013).

In the case of tofacitinib, its safety profile was consistent across the Phase 3 studies. Common adverse effects included nasopharyngitis, upper respiratory tract infection, headache, and diarrhea. The most common serious reported adverse effects were serious infections (including pneumonia, cellulitis, herpes zoster, and urinary tract infection). Malignancies (excluding nonmelanoma skin cancer), lymphoma, and gastrointestinal perforations have been reported in patients receiving tofacitinib. Changes in laboratory test values have included increases in high-density lipoprotein, low-density lipoprotein, total cholesterol, serum creatinine, and liver transaminases and decreases in absolute neutrophil and lymphocyte counts (Pfizer Inc. 2015).

Drug Interactions

Fluconazole and ketoconazole increase the total exposure (AUC values) of tofacitinib by 79 % and 103 %, respectively (Gupta et al. 2014). Increases in tofacitinib systemic exposure is likely to happen when coadministered with other moderate to potent CYP3A4 inhibitors.

Acknowledgments This article was funded by Pfizer Inc. Editorial support was provided by Complete Medical Communications and funded by Pfizer Inc.

Cross-References

▶ Disease-Modifying Antirheumatic Drugs: Overview
▶ Interleukin-6 Inhibitor: Tocilizumab
▶ Rheumatoid Arthritis
▶ Tumor Necrosis Factor (TNF) Inhibitors

References

Changelian, P. S., Flanagan, M. E., Ball, D. J., Kent, C. R., Magnuson, K. S., Martin, W. H., et al. (2003). Prevention of organ allograft rejection by a specific Janus kinase 3 inhibitor. *Science, 302*, 875–878.

Charles-Schoeman, C., Fleischmann, R., Davignon, J., Schwartz, H., Turner, S., Beysen, C., et al. (2015). Potential mechanisms leading to the abnormal lipid profile in patients with rheumatoid arthritis versus healthy volunteers and reversal by tofacitinib. *Arthritis and Rheumatology, 67*, 616–625.

Clark, J. D., Flanagan, M. E., & Telliez, J. B. (2014). Discovery and development of Janus kinase (JAK) inhibitors for inflammatory diseases. *Journal of Medicinal Chemistry, 57*, 5023–5038.

Dowty, M. E., Lin, J., Ryder, T. F., Wang, W., Walker, G. S., Vaz, A., et al. (2014a). The pharmacokinetics, metabolism, and clearance mechanisms of tofacitinib, a Janus kinase inhibitor, in humans. *Drug Metabolism and Disposition, 42*, 759–773.

Dowty, M. E., Jesson, M. I., Ghosh, M., Lee, J., Meyer, D. M., Krishnaswami, S., et al. (2014b). Preclinical to clinical translation of tofacitinib, a Janus kinase inhibitor, in rheumatoid arthritis. *Journal of Pharmacology and Experimental Therapeutics, 348*, 165–173.

Ghoreschi, K., Laurence, A., & O'Shea, J. J. (2009). Janus kinases in immune cell signaling. *Immunological Reviews, 228*, 273–287.

Gupta, P., Friberg, L. E., Karlsson, M. O., French, J., & Krishnaswami, S. (2009). Semi-mechanistic modeling of the effect of CP-690,550 on circulating neutrophils in patients with rheumatoid arthritis (RA). *Clinical Pharmacology and Therapeutics, 85*, S7.

Gupta, P., Chow, V., Wang, R., Kaplan, I., Chan, G., Alvey, C., et al. (2014). Evaluation of the effect of fluconazole and ketoconazole on the pharmacokinetics of tofacitinib in healthy adult subjects. *Clinical Pharmacology in Drug Development, 3*, 72–77.

Incyte Corporation. Jakafi prescribing information. Available at: http://www.jakafi.com/pdf/prescribing-information.pdf. Last updated December 2014. Accessed 21 Oct 2015.

Kantarjian, H. M., Silver, R. T., Komrokji, R. S., Mesa, R. A., Tacke, R., & Harrison, C. N. (2013). Ruxolitinib for myelofibrosis – An update of its clinical effects. *Clinical Lymphoma Myeloma and Leukemia, 13*, 638–645.

Lamba, M., Furst, D. E., Dikranian, A., Dowty, M., Hutmacher, M. M., Conrado, D., et al. (2015). *Evaluating pharmacokinetic predictors of tofacitinib clinical response in rheumatoid arthritis*. Arthritis and rheumatology conference. American College of Rheumatology/Association of Rheumatology Health Professionals Annual Scientific Meeting.

Papp, K. A., Menter, M. A., Abe, M., Elewski, B., Feldman, S. R., Gottlieb, R., et al. (2015). Tofacitinib, an oral Janus kinase inhibitor, for the treatment of chronic plaque psoriasis: Results from two randomized, placebo-controlled, phase III trials. *British Journal of Dermatology, 173*, 949–961.

Pfizer Inc. Xeljanz prescribing information. Available at: http://labeling.pfizer.com/ShowLabeling.aspx?id=959. Last updated June 2015.

Shilling, A. D., Nedza, F. M., Emm, T., Diamond, S., McKeever, E., Punwani, N., et al. (2010). Metabolism, excretion, and pharmacokinetics of [14C] INCB018424, a selective Janus tyrosine kinase 1/2 inhibitor, in humans. *Drug Metabolism and Disposition, 38*, 2023–2031.

Sonbol, M. B., Firwana, B., Zarzour, A., Morad, M., Rana, V., & Tiu, R. V. (2013). Comprehensive review of JAK inhibitors in myeloproliferative neoplasms. *Therapeutic Advances in Hematology, 4*, 15–35.

Tefferi, A. (2010). Novel mutations and their functional and clinical relevance in myeloproliferative neoplasms: JAK2, MPL, TET2, ASXL1, CBL, IDH and IKZF1. *Leukemia, 24*, 1128–1138.

Verstovsek, S., Mesa, R. A., Gotlib, J., Levy, R. S., Gupta, V., DiPersio, J. F., et al. (2013). Efficacy, safety and survival with ruxolitinib in patients with myelofibrosis: Results of a median 2-year follow-up of COMFORT-I. *Haematologica, 98*, 1865–1871.

Verstovsek, S., Mesa, R. A., Gotlib, J., Levy, R. S., Gupta, V., DiPersio, J. F., et al. (2015). Efficacy, safety, and survival with ruxolitinib in patients with myelofibrosis: Results of a median 3-year follow-up of COMFORT-I. *Haematologica, 100*, 479–488.

Wollenhaupt, J., Silverfield, J., Lee, E. B., Curtis, J. R., Wood, S. P., Soma, K., et al. (2014). Safety and efficacy of tofacitinib, an oral Janus kinase Inhibitor, for the treatment of rheumatoid arthritis in open-label, long-term extension studies. *Journal of Rheumatology, 41*, 837–852.

Wollenhaupt, J., Silverfield, J., Lee, E. B., Wood, S. P., Terry, K., Nakamura, H., et al. (2015). Tofacitinib, an oral Janus kinase inhibitor, for the treatment of rheumatoid arthritis: Safety and efficacy in open-label, long-term extension over 6 years. *Annals of Rheumatic Diseases, 74*, 259.

Janus Kinases (JAKs)/STAT Pathway

Behdad Afzali[1,2] and Susan John[3]
[1]Molecular Immunology and Inflammation Branch, National Institute of Arthritis and Musculoskeletal and Skin Diseases, National Institutes of Health, Bethesda, MD, USA
[2]Division of Transplant Immunology and Mucosal Biology, MRC Centre for Transplantation, King's College London, London, UK
[3]Department of Immunobiology, Kings College London, London, UK

Synonyms

Hopscotch; JAK1; JAK2; JAK3; Janus kinases; Signal transducers and activators of transcription; STAT1; STAT2; STAT3; STAT4; STAT5a; STAT5b; STAT6; STAT92E; Tyk2

Definition and Background

The JAK/STAT pathway is a rapid signal transduction circuit, which is activated upon cytokine or growth factor binding to its cognate transmembrane receptor, resulting in direct translation of an extracellular signal to a transcriptional response in the nucleus. This apparently simple pathway is facilitated by only a few principal proteins that belong to two families: Janus kinases (JAKs), which are cytoplasmic non-receptor tyrosine kinases, and signal transducers and activators of transcription (STATs), which are multifunctional proteins that are cytoplasmic signaling mediators and nuclear transcription factors. This pathway was initially identified in the laboratories of James Darnell, George Stark, and Ian Kerr while studying cellular mediators of the Type I and Type II interferon response (Stark and Darnell 2012). Subsequently, the JAK/STAT signaling pathway was established as a paradigm for signal transduction by virtually all Type I and II cytokine and growth factor receptors. There are four JAKs (JAK1, JAK2, JAK3 and Tyk2) and seven STAT proteins (STAT1, STAT2, STAT3, STAT4, STAT5a, STAT5b, and STAT6) in the mammalian genome (O'Shea et al. 2002) The JAK/STAT pathway is evolutionarily conserved, from slime molds, worms, and flies to mammals (Hou et al. 2002). In *Drosophila*, there is one JAK protein, most similar to mammalian JAK2, called *hopscotch* and one STAT protein, most similar to mammalian STAT3 and STAT5, called STAT92E.

Biosynthesis and Release

JAKs and STATs have common overall structures and are organized into distinct functional domains (Fig. 1). JAK1, JAK2, and Tyk2 are ubiquitously expressed, while JAK3 expression is restricted to hematopoietic cells. JAK protein synthesis is regulated primarily at the post-translational level by proteasomal degradation to maintain their homeostatic levels. STATs are also ubiquitously expressed, although there is selective enrichment in different cell types; in immune cells, STATs 2, 4, 5a, 5b, and 6 are expressed at high levels, while STAT1 and STAT3 are expressed at lower levels. Despite their ubiquitous expression, the levels of some STAT proteins can be increased in an autoregulatory manner following cytokine or mitogenic stimulation; STAT1 expression is increased by interferons (β and γ) and STAT3 by IL-6, while STAT1, STAT3, and STAT5 expression is increased by mitogen treatment of peripheral blood mononuclear cells. The existence of this autoregulatory loop for the biosynthesis of STAT proteins has important implications in the functional competition for receptor interaction between the different STAT proteins that can be activated by the same cytokine receptor, which would produce a very different downstream gene expression

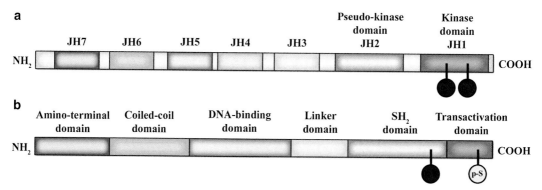

Janus Kinases (JAKs)/STAT Pathway, Fig. 1 Schematic of the structural organization of JAKs and STATs. (**a**) JAKs are organized into seven JAK homology (JH) domains JH1–JH7. The catalytic kinase domain (JH1) is located at the extreme C-terminus of the protein and is immediately preceded by a pseudo-kinase regulatory domain (JH2). The N-terminal JH5–JH7 domains are critical for receptor binding and signal transduction and is called the FERM domain, while an SH2 domain is encompassed within JH3–JH4. JAKs are phosphorylated at conserved tyrosine residues (p-Y) in the JH1 domain. (**b**) STAT proteins are also organized into modular functional domains. The N-domain is involved in unphosphorylated STAT dimerization and phosphorylated tetramer formation; the coiled-coil domain mediates protein-protein interactions and is followed by a DNA-binding domain that is linked to an SH2 domain via a linker domain. The SH2 domain contains the conserved tyrosine residue that is phosphorylated upon activation and mediates dimer formation between STAT monomers. The extreme C-terminus of STAT proteins contains a transactivation domain, containing a serine residue that is phosphorylated (p-S) and enhances transcriptional activation in some STATs

profile depending on which STAT protein is predominantly activated. Recent genome-wide studies of microRNAs (miRs) have revealed that STAT protein biosynthesis is intricately regulated directly and indirectly by miRs, a finding that may have particular relevance in diseases (Kohanbash and Okada 2012). STAT proteins can be posttranslationally modified by phosphorylation, acetylation, methylation, sumoylation, ubiquitination, and ISGylation, though not all modifications have been identified for all STAT proteins (Shuai and Liu 2003).

Activation and Release

The canonical JAK/STAT pathway is activated (Fig. 2) when a ligand (cytokine, growth factor, or hormone) binds to its receptor and induces a conformational change in the cytoplasmic domain of the heteromerized receptor resulting in juxtapositioning and activation of latent cytoplasmic non-receptor tyrosine kinases, the JAK proteins. Receptor-associated JAKs transphosphorylate each other, leading to their activation. The activated JAK then phosphorylates conserved tyrosine residues on the cytoplasmic tail of their associated receptors, creating phosphotyrosine

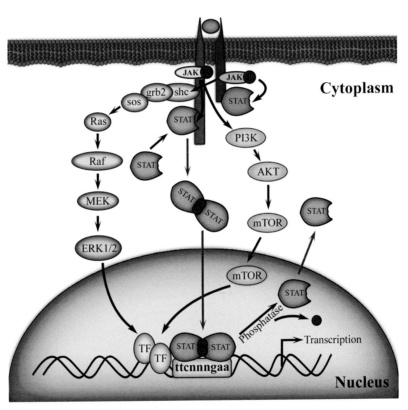

Janus Kinases (JAKs)/STAT Pathway, Fig. 2 The canonical JAK/STAT signal transduction pathway. Ligand-induced heterodimerization of the receptor leads to the activation of latent cytoplasmic JAKs, which get activated by transphosphorylation and subsequently phosphorylate the receptor, creating phosphotyrosine docking sites for latent cytoplasmic STAT proteins. The JAKs phosphorylate the STAT proteins, which then are released from the receptor and form homo- or heterodimers and are actively transported to the nucleus, where they bind to regulatory regions of genes to activate or in some cases, repress their transcription. STATs are inactivated by dephosphorylation by nuclear phosphatases and are actively translocated to the cytoplasm. Unphosphorylated STAT proteins continuously shuttle between the cytoplasm and nucleus in a cytokine-independent manner. The JAK/STAT pathway functionally integrates with the RAS-MAPK and PI3K-Akt pathways, which are also activated in response to ligand stimulation of receptors, leading to the downstream activation of target transcription factors (TF), which function cooperatively with STATs to regulate gene expression

docking domains for the specific association of the latent cytoplasmic transcription factors, the STAT proteins. Elegant structure-function studies have shown that the recruitment and activation of selective STATs to specific cytokine receptors is at least in part dictated by short peptide recognition motifs on the receptor subunits, surrounding the conserved tyrosine residues. Once docked on the receptor, the JAKs phosphorylate the monomeric STAT proteins on a conserved tyrosine residue in their SH2 domain (Fig. 1). The phosphorylated STAT protein immediately dissociates from the receptor and dimerizes with another phosphorylated STAT monomer, via its phosphorylated SH2 domain to form homo- or heterodimers. Phosphorylated STAT dimers are then actively transported, via the RAN-GTP/importin-α cargo system to the nucleus, where they bind to and regulate transcription of genes containing cognate recognition motifs for STAT DNA binding called GAS motifs (gamma-activated sequences). STATs can bind DNA as dimers or as higher-order tetrameric complexes, where pairs of dimers interact with each other via conserved amino acids in the N-domains, thus increasing the stability of the chromatin-bound STAT by protein-DNA and protein-protein interactions (John et al. 1999). The phosphorylated STAT proteins are retained here by DNA binding and are exported back to the cytoplasm following their dephosphorylation, by the action of nuclear phosphatases and/or degradation of the phosphorylated protein via the ubiquitin-mediated proteasomal degradation pathway. Thus, nuclear retention, rather than translocation, regulates the STAT-mediated genomic functions. In addition to the traditional mechanisms of activation of STAT proteins, they can also be activated downstream by some G-protein-coupled receptors such as angiotensin and melanocortin receptor and receptor tyrosine kinases such as EGFR, PDGFR, and VEGFR2 and also by some cytosolic non-receptor tyrosine kinases such as lck, v-abl, and bcr-abl.

Noncanonical U-STAT Pathway

Several deviations from the canonical view of the JAK/STAT pathway have emerged since their first characterization, which reveal a more complex modus operandi for this signaling pathway. Biochemical and crystallography studies have revealed that several STAT proteins (STAT1, STAT2, STAT3, and STAT5) exist as unphosphorylated homo- or heterodimers, rather than monomers in unstimulated cells (Ota et al. 2004). X-ray crystallography studies have shown that the unphosphorylated dimer adopts an antiparallel conformation while the phosphorylated SH2-mediated dimer adopts a parallel conformation on DNA. What then is the relevance of the unphosphorylated dimer? The unphosphorylated N-domain-mediated dimer has been shown to be important for receptor-mediated activation of STAT4 and STAT2, following cytokine stimulation, but not for the activation of the other STATs, which can interact with the receptor as monomers.

Despite the early notion that STAT proteins exclusively reside in the cytoplasm prior to cytokine or growth factor stimulation, subsequent studies revealed the highly dynamic nature of these proteins, wherein the unphosphorylated STAT proteins (specifically STAT1, STAT3, and STAT5) continuously shuttle between the cytoplasm and nucleus in an energy- and carrier-independent manner using the export receptor CRM1 (Vinkemeier 2004). In a further interesting deviation from the mechanism of conventional STAT regulation of gene expression, recent studies have shown that unphosphorylated STAT1, STAT2, and STAT3 (U-STAT1, USTAT2, U-STAT3) can activate transcription of a set of genes that are distinct from the ones activated by the phosphorylated form of these proteins (Yang and Stark 2008). U-STAT can associate with DNA either directly or indirectly, as in the case of U-STAT3 via complex formation with another transcription factor, NF-κB. It remains to be established whether U-STAT forms of the remaining STAT proteins (STAT4 and STAT6) can also behave as transcriptional regulators. It is interesting to note here that some of the downstream target genes for U-STAT3 that have been identified have well-defined roles in oncogenesis, such a *MET*, *MRAS*, and indeed *STAT3* itself, a finding that may have significance in light of the role of STAT3 in oncogenesis.

Nonnuclear Associations of STATs

U-STAT3 has also been shown to have non-genomic functions in the mitochondria, where it regulates cellular respiration, and is involved in RAS-dependent cellular transformation, while U-STAT5 has been identified in the endoplasmic reticulum (ER) and is involved in the ER stress response and in the mitochondria, where it is associated with mitochondrial DNA and may therefore be involved with gene regulation in this organelle (Chueh et al. 2010; Gough et al. 2009; Wegrzyn et al. 2009). U-STAT1 has also been identified in the mitochondria of cardiac fibroblasts, where it plays a role in mitophagy (Bourke et al. 2014). Both U-STAT5 and U-STAT6 are also constitutively associated with the Golgi apparatus in specific cell types and are required for the stability of this organelle, as demonstrated by knockdown studies (Sehgal 2014). Thus, recent studies have radically changed our understanding of these proteins from just simple nucleocytoplasmic signaling proteins to more diverse regulators of cell intrinsic functions.

Biological Activities

The range of biological activities for the JAK/STAT pathway has been gleaned from studies of knockout animals for each of the JAKs and STATs or conditional deletion of these proteins, where they are lethal (O'Shea et al. 2002). For some JAK and STAT proteins, the murine knockout phenotypes have been mirrored by the phenotypes of human patients bearing inactivating mutations or deletions of these JAK and STAT proteins, suggesting that the functions have been broadly conserved through mammalian evolution.

JAK1

JAK1-deficient mice show perinatal lethality due to neurological insufficiencies. They show deficient lymphopoiesis and associated severe combined immunodeficiency (SCID), but normal myelopoiesis. Consistent with in vitro data, JAK1-deficient cells are unable to respond to interferons, gamma-c (γ_c) family of cytokines (IL-2, IL-4, IL-7, IL-9, IL-15, IL-21), and the gp130 family of cytokines (IL-6, IL-11, CNTF, osm, LIF). The latter defect may explain the neurological deficits, while the SCID phenotype can be attributed to the loss of responsiveness to γ_c cytokines.

JAK2

JAK2-deficient mice show embryonic lethality due to defective erythropoiesis, which can be attributed to loss of responsiveness to erythropoietin and thrombopoietin, IL-3, IL-5, and GMCSF. JAK2-deficient cells are also important for interferon-γ (IFNγ) responses. Conditional deletion of JAK2 reveals that this protein plays an important role in postnatal life, and deletion of JAK2 in young adult mice resulted in severely impaired erythropoiesis and thrombopoiesis, modestly affected granulopoiesis, and monocytopoiesis and had no effect on lymphopoiesis.

JAK3

JAK3 is the only JAK with restricted cellular expression and shows exclusive association with the gamma-chain of the γ_c family of cytokine receptors. Consistently, the JAK3-deficient mice exhibit a phenotype that is identical to the γ_c knockout mice, displaying severe immunodeficiency, due to the lack of T, B, and NK cells, establishing a vital role for the γ_c cytokines in lymphoid cell development. Humans bearing deleterious mutations in the JAK3 gene similarly display a SCID phenotype; however, they are characterized as T-B + NK- SCID. The few T cells that survive in both mice and human are functionally defective and paradoxically show both increased survival and an apoptotic phenotype, the former being attributed to loss of Bcl2 and Bax expression while the latter is due to the loss of activation-induced cell death (AICD) associated with FAS activation, which is regulated by IL-2. IFN-γ responses are also reduced in JAK3-deficient cells, as JAK3 activation is critical for permissive epigenetic regulation of the IFNγ locus.

Tyk2

Tyk2-deficient mice are viable and relatively healthy. Tyk2 is activated by interferon α/β

receptor activation and by IL-6, IL-10, IL-12, IL-23, and IL-27. Indeed, these responses are only partially affected in the Tyk2 knockout mice. IL-12-induced NK cell cytotoxicity and IFNγ production was severely affected though not abrogated, and IFNα/β responses were only affected at low doses. Nevertheless, Tyk2-deficient mice show specific defects in response to viral or bacterial challenges, notably resistance to LCMV-induced bone marrow aplasia associated with IL-7-induced AICD, and a defect in nitric oxide production by macrophages during lipopolysaccharide-induced endotoxin shock. Tyk2 is also essential for the maintenance of mitochondrial respiration. The rare human mutations of Tyk2 reveals a more severe phenotype than its murine counterpart, characterized by immunodeficiency and increased susceptibility to infections due to complete loss of IL-12 and IFNα/β responses and impaired but not complete loss of IL-10 activation.

Nuclear JAKs

While the familiar view of JAKs is their strict cytoplasmic location and function as tyrosine kinases that activate STAT proteins, this notion has been amended by several studies that have demonstrated the presence of JAK1 and JAK2 in the nucleus, where they function as epigenetic regulators and phosphorylate non-STAT proteins, including histones (Dawson et al. 2009; Ito et al. 2004; Lobie et al. 1996). The presence of nuclear JAKs is usually noted in cells with high levels of proliferation, such as tumor cells or regenerating cells, and still awaits confirmation in normal hematopoietic cells (Rinaldi et al. 2010; Zouein et al. 2011).

STAT1

STAT1-deficient mice confirm the importance of this protein for mediating IFNα/β and IFNγ responses and are highly susceptible to viral and bacterial infections and also to tumor formation, implicating a tumor suppressor role for STAT1. The increased tumor formation is most likely due to loss of IFNγ responsiveness resulting in decreased apoptosis of tumor cells and reduced expression of MHC class I expression, which allows the tumor cells to more efficiently escape the host tumor surveillance mechanisms. IFNγ, a key cytokine for CD4 TH1 cell differentiation, activates the Th1 master regulatory gene, Tbet, which is only partly compromised in STAT1-deficient mice, indicating a non-STAT1-dependent mode of Tbet gene regulation. STAT1-deficient mice show impaired TH1 differentiation. Human patients, bearing loss of function mutations of STAT1, show enhanced susceptibility to infections with atypical environmental or weakened mycobacteria, salmonella, intramacrophagic bacteria, fungi, and parasites, but interestingly not to viral infections, suggesting that the main biological action of STAT1 is to facilitate the actions of the IFNγ receptor. This notion is also supported by the phenotype of knock-in mice expressing cooperative DNA-binding-deficient STAT1, which shows a specific defect for antibacterial (Type II) but not antiviral (Type I) interferon responses (Begitt et al. 2014).

STAT2

The biological activity of STAT2 is to facilitate IFNα responses, and accordingly, STAT2-deficient mice are viable and fertile but defective for immunity to viral infections (vesicular stomatitis virus, dengue, lymphocytic choriomeningitis virus, Semliki Forest virus, Theiler's Murine encephalomyelitis virus, measles virus). Interestingly, macrophages from STAT2-deficient mice do not show a defect in IFNα responsiveness, highlighting tissue-specific differences in STAT2 activity.

STAT3

STAT3 is activated downstream of several cytokines/growth factors, though it was originally identified as the acute phase response factor (APRF) activated by IL-6. STAT3-deficient mice are embryonically lethal due to developmental defects, and the biological activity of this protein has been deduced by cell-type-specific deletions of STAT3. STAT3-deficient T cells have severely impaired response to IL-6 and partially impaired IL-2 responses. Macrophages and neutrophils lacking STAT3 show a loss of IL-10 responsiveness and consequently exaggerated proinflammatory cytokine production in response to LPS challenge. Keratinocyte-specific STAT3 deletion reveals a role for STAT3 in wound healing, hair regrowth, and migration of epidermal cells. Mammary gland-specific deletion of STAT3 shows defects in involution timing and a decrease in apoptosis, likely associated with impaired IGF signaling. There are two forms of STAT3 generated by alternative splicing, STAT3α (full-length) and STAT3β truncated with an alternative C-terminus, which have overlapping and distinct biological actions, as revealed by mice specifically deficient for either isoform. Thus, STAT3α mediates cellular responses to IL-6 and IL-10 function in macrophages, while STAT3β regulates LPS-induced endotoxin-mediated inflammation. Recent, comprehensive mapping by ChIP-seq and RNA-seq analyses of transcriptional modules regulated by STAT3 and STAT1 downstream of their activating cytokines, IL-6 and IL-27 respectively, has demonstrated that significant co-operation and overlap exists between these STAT proteins. Notably, intact STAT3 signaling is a requisite for the complete transcriptional output of Stat1, which depends on STAT3 to bind to STAT1-regulated sites (Hirahara et al. 2015). Human patients bearing dominant negative STAT3 mutations were identified as the genetic background to autosomal dominant hyper IgE syndrome (AD-HIES, also known as Job's syndrome). Molecular characterization of the wide range of physiological deficiencies associated with AD-HIES established dominant biological roles for STAT3 in IL-6, IL-10, IL-17, IL-21, and IL-11 mediated functions.

STAT4

STAT4 is primarily activated by IL-12 and the related family cytokines, IL-23 and IL-35, which are involved in CD4 Th1, Th17, and iTr35 lineage differentiation. However, STAT4-deficient mice predominantly show impaired Th1 differentiation; IFNγ production by CD4, CD8, and dendritic cells; and enhanced TH17-mediated autoimmune pathologies. STAT4, like STAT3, exists in different isoforms, STAT4α (full-length) and STAT4β (C-terminal truncation). Studies using mice expressing either isoform alone reveal that each isoform differentially regulates inflammatory cytokines; STAT4β promotes proinflammatory cytokines, IFNγ, and IL-17, while STAT4α promotes IL-10 production, resulting in distinct effects on the onset and severity of experimental autoimmune encephalitis. The exact biological relevance of STAT4 for IL-35-mediated actions remains to be established.

STAT5

STAT5a and STAT5b are encoded by two tandemly linked genes on the same chromosome and have >90 % identity at the protein level. Like STAT3, STAT5a and STAT5b are activated by a wide variety of cytokines and growth factors. STAT5a and STAT5b single-knockout mice are viable and reveal overlapping and distinct biological roles for the two proteins. STAT5a-deficient mice show a lactation defect associated with impaired prolactin signaling, while STAT5b mice exhibit impaired sexually dimorphic growth associated with a defect in growth hormone signaling. Both mice show significant impairments in the immune compartment since both IL-2 signaling and associated functions such as defective T-cell cycle progression and activation-induced cell death are clearly impaired. STAT5b-deficient mice have a more profound loss of NK cells, while STAT5a mice have significantly decreased γδ T cells. The STAT5a/b double-knockout mice die within a few weeks of birth and exhibit lymphopenia, thrombocytopenia, neutrophilia, impaired granulopoiesis, mild anemia, hypocellularity of bone marrow compartment due to reduced repopulation by hematopoietic progenitors and reduced numbers of B cells. Thymic development is not perturbed in the double-knockout mice; however, the peripheral T cells display an activated phenotype associated with a loss of AICD, and these mice develop autoimmunity, consistent with a loss of IL-2 signaling and consequently affecting regulatory T cell function, whose maintenance and activity are acutely dependent on IL-2. In recent years, human patients bearing mutations of the STAT5b gene have been identified, initially described as showing insensitivity to growth hormone and being of small stature. The identification of these patients, who have normal STAT5a proteins, confirms that the two proteins, although highly homologous, have distinct functions that cannot be compensated by the other. In the immune compartment, these patients have very few peripheral $CD4^+CD25^+$ $FoxP3^+$ Treg cells, indicating an important, non-redundant biological role for STAT5b in the function and maintenance of these cells and in the prevention of autoimmunity. Murine and/or human studies have also shown that STAT5, activated by IL-2, is important for Th1 and Th2 differentiation and for the inhibition of Th17 differentiation, while constitutive expression of STAT5 can drive self-renewal of human cord blood CD34+ cells and inhibit their myeloid differentiation.

STAT6
STAT6 is activated by IL-4 and IL-13 and is essential for CD4+ Th2 differentiation by these cytokines. STAT6-deficient mice show impaired Th2 responses, including expulsion of helminthic parasites, and reduced asthma-associated pathology in animal models. In addition to its importance in CD4+ TH2 differentiation, innate lymphoid cell type 2 development and function are also dependent on STAT6 and contribute to antiparasite immunity. In contrast, $STAT6^{-/-}$ mice show exaggerated Th1 responses, IFNγ production, and pathology associated with increased inflammation in animal models of EAE, but also increased clearance of bacteria in infection models, highlighting the role of STAT6 in cross-regulating opposing CD4 Th1 differentiation program. STAT6 is important for the alternative activation of macrophages, which regulates tissue repair and anti-inflammatory responses and B-cell isotype switching to IgE. Surprisingly, STAT6 is also important for efficient viral clearance in a non-IL-4-/IL-13-dependent manner. Here, the innate immune sensor STING, which is activated by viral RNA, directly recruits and activates STAT6 in a non-JAK-dependent mechanism, via Tbk1 (Chen et al. 2011). STAT6 plays an inhibitory role in tumor immune surveillance as revealed by STAT6−/− mice, which exhibit enhanced ability to reject or delay primary tumor growth, prevent recurrence of primary tumors, and/or reject established, spontaneous metastatic disease, which is attributed to the IL-13-induced immunosuppressive functions of NKT cells.

The biological activity of the JAK/STAT signaling pathway is significantly influenced by integration with other signaling pathways that are simultaneously elicited by cytokine stimulation. The two main pathways that have extensive cross talk with the JAK/STAT pathway are the RAS-MAP kinase and the PI3-kinase pathways (Fig. 2). JAKs can phosphorylate other signaling/adaptor proteins such as p85 PI3K, RAF1, GAB2, and shc, in addition to STATs, linking JAK signaling directly to the MAP kinases and PI3K activation. Additionally, suppressor of cytokine signaling protein-3 (SOCS3), which is a downstream target gene and negative feedback regulator of the JAK/STAT pathway, can bind to the GTPase RasGAP, inhibiting its activity resulting in increased activation of the RAS-MAPK pathway (Cacalano et al. 2001). Similarly, the negative regulators of MAPK signaling, the dual-specificity phosphatases (DUSP), which are downstream target genes of MAPK activation, can also inhibit STAT activation resulting in attenuation of JAK/STAT signaling (Huang et al. 2012).

Pathophysiological Relevance

Immunodeficiencies
The critical importance of the JAK/STAT signaling pathway for the maintenance of normal immune homeostasis is underscored by the pathological consequences that ensue when components of this pathway are perturbed (Casanova et al. 2012; O'Shea et al. 2013). The study of human mutations of JAKs and STATs, aided greatly by the advent of deep sequencing technologies, has provided a wealth of information on the important physiological roles of these proteins. To date, germline mutations in the genes encoding Tyk2, JAK3, STAT3, STAT1, and STAT5b have been identified. They cover a range of primary, combined, and severe combined immunodeficiencies (CID, SCID), which can be attributed to

the loss of pleiotropic biological actions of the cytokines affected, resulting in the lack of/or reduced numbers of functional T, B, and NK cells. The most well-documented pathology associated with mutations of the JAK/STAT pathway is SCID or CID. Inactivating mutations of JAK3 protein, which exclusively binds to the shared γ_c receptor subunit, is the most common cause of SCID associated with this pathway, along with mutations in γ_c. The clinical phenotype of JAK3-deficient patients ranges from life-threatening SCID to milder immunodeficiencies in individuals with hypomorphic mutations who retain a partially functioning immune system.

Individuals with deleterious STAT1 mutations have recurrent bacterial infections, while Tyk2-deficient patients are susceptible to both viral and bacterial infections. STAT1 gain-of-function mutant patients clearly illustrate the critical physiological importance of cytokine cross-regulation, as these patients present with chronic mucocutaneous candidiasis (CMC), a fungal infection caused by the pathogen *Candida albicans*, which is specifically controlled by CD4+ effector cells of the Th17 lineage (Puel et al. 2012). CMC patients have increased IFNγ responses, which act to suppress Th17 differentiation in these individuals, leading to loss of immunity to *Candida albicans*. Therefore, these studies reveal an important biological role for STAT1 in the negative regulation of IL-17 responses and Th17 differentiation (Liu et al. 2011).

The main pathological consequences of STAT5b mutations in humans are attributed to defective growth hormone signaling or IL-2 signaling. STAT5b-deficient patients are characterized by short stature, facial dysmorphism, severe infections, and lymphoid interstitial pneumonitis, owing to growth hormone insensitivity (GHIS) and lack of functional Treg cells. The GHIS phenotype of STAT5b mutant patients is akin to those of patients bearing mutations of the growth hormone receptor (GHR), indicating the critical importance of STAT5b in GHR signaling. The fact that these patients are immunocompromised, despite being hypomorphic for STAT5 function because they possess intact STAT5a protein, establishes the non-redundant nature of the two closely related STAT5 proteins in humans. The loss of functional $CD4^+CD25^{hi}$ $FoxP3^+$ Tregs in STAT5b-deficient patients is associated with defective IL-2-induced FoxP3 expression, resulting in impaired immune regulation, increased autoimmunity, and T-cell lymphopenia, symptoms that are reminiscent of those of CD25-deficient patients. These lessons from nature clearly reveal the vital role of the IL-2R signaling system in the maintenance and function of normal immune homeostasis. Similarly, the essential physiological role of STAT3 has been revealed by lessons from nature in the identification of patients with hypomorphic mutations of STAT3. As homozygous mutations of STAT3 would be predicted to be embryonically lethal, the STAT3 mutations identified are autosomal dominant. Autosomal dominant mutations in STAT3 give rise to a condition called hyper IgE syndrome (HIES), where the patients have greatly elevated levels of IgE and defective antibody synthesis. Additionally, other abnormalities are observed in STAT3-mutant patients, including characteristic facial appearance, impaired neutrophil chemotaxis, eczema, lung infections, and disposition to infections by *Staphylococcus aureus* and *Candida albicans*. The disease phenotype of AD-HIES confirms the importance of STAT3 in the biological actions of cytokines belonging to the gp130 family and IL-21, IL-23, and IL-10 family (IL-10, IL-22). Impaired development of CD4+ TH1 responses of lymphoma patients after autologous stem cell transplantation has been mapped to loss of STAT4 expression resulting in reduced IFNγ, TNFα, and IL-12Rβ2 expression, demonstrating the physiological importance of STAT4 for Th1 lineage development.

Cancer

In addition to pathologies resulting from germline mutations of JAK/STAT family proteins, another major disease area associated with these proteins is cancer. Consistent with their roles in regulating cytokine-dependent inflammation and immunity, STAT1, STAT3, and STAT5, in particular, are central in determining whether immune responses in the tumor microenvironment promote or inhibit cancer. Constitutive activation of STAT3 and

STAT5 and to a lesser extent STAT1 is a hallmark feature of many hematopoietic and non-hematopoietic malignancies of viral or non-viral origins and confers upon the cancerous cell a cytokine-independent means of achieving proliferation, survival, and eventually invasion, while suppressing antitumor immunity. STAT3 is particularly important in this respect, as it can promote tumor inflammation and escape from host tumor immune surveillance by promoting pro-oncogenic inflammatory pathways, including NF-κB and IL-6 signaling, while opposing STAT1-mediated Th1 antitumor immune responses. Constitutive activation of STAT5 and the formation of N-domain-mediated STAT5 tetramers are essential in human leukemogenesis and provide a promising avenue for therapeutic intervention (Timofeeva and Tarasova 2013). Translocations of the JAK2 gene have been identified in a number of myeloid and lymphoid leukemias (e.g., T-ALL, B-ALL, CML, AML, myelodysplastic syndrome (MDS)). In all cases of JAK2 translocations, the catalytic kinase (JH1) domain of the JAK2 protein is fused to a multimerization domain of a partnering protein, resulting in a fusion protein where there is constitutive activation of the JAK2 kinase domain, leading to abnormal myeloid hematopoiesis.

Inflammation

Consistent with the intimate connection between regulation of inflammation and the JAK/STAT pathway, the activation or inactivation of JAK/STAT proteins has been identified in various inflammatory diseases. Examples of inflammatory diseases and their dysregulated STAT activity include liver disease (STAT1, STAT2, STAT3, STAT4, STAT5, and STAT6), asthma (STAT6), Crohn's disease (STAT3), psoriasis (STAT3), and inflammatory/rheumatoid arthritis (RA), where increased activity of STAT1 and STAT3, reflecting abnormal IFNγ and IL-6 levels in the rheumatoid synovium, has been implicated in the pathogenesis of rheumatoid arthritis. The aberrant activation of STATs can lead to the dysregulated expression of SOCS proteins, which act in a negative feedback loop to attenuate receptor signaling by interfering with JAK activation directly or by blocking its interaction with the receptor. Impaired STAT-mediated induction of SOCS proteins has been shown to contribute to the pathogenicity of JAK/STAT signaling in inflammatory arthritis and in atopic dermatitis and asthma. Thus, immunodeficiencies, inflammation, and cancer are closely linked to the activities of the JAK/STAT signaling pathway.

Modulation by Drugs

The therapeutic targeting of the JAK/STAT pathway is strongly indicated in treatment of inflammatory disorders and cancer. As JAKs are placed hierarchically above the STATs in mediating the actions of the inflammatory cytokines, their inactivation would lead to immunosuppression and resolution of disease. Similarly, the identification of the constitutively active mutant JAK2 in MDS and lymphoid cancers has resulted in the development of small chemical compounds for the treatment of specific cancers. Currently, two small molecule JAK inhibitors have been approved for clinical use; ruxolitinib (Incyte), which is a potent inhibitor of JAK1 and JAK2, is used in the treatment of an MDS disease polycythemia vera and is also being studied for use in the treatment of autoimmune diseases such as RA (Eghtedar et al. 2012). Tofacitinib (Pfizer) a potent inhibitor of JAK3, and to a lesser extent of JAK1, is used as an immunosuppressant in transplantation therapy and also in the treatment of autoimmune diseases (Fox 2012). As these drugs can also affect the wild-type JAK proteins, these inhibitors have specific side effects associated with the treatments, and this is generally a drawback in the safety considerations of small molecule inhibitors. Tofacitinib is also being tested as an immunosuppressant in the treatment of other inflammatory diseases, such as colitis, psoriasis, juvenile idiopathic arthritis, and others. In addition to these, there are several other small molecule inhibitors targeting JAKs in the pipeline, at various stages of clinical trials, which may emerge as being efficacious but with less side effects.

STAT3 and STAT5 inhibitors would be highly desirable in the treatment of several malignancies, as they are involved in the development, progression, and survival of tumors (Timofeeva and Tarasova 2013). A plethora of animal cancer model studies have shown that inhibition of STAT signaling can lead to tumor regression suggesting that STAT activation serves as a common mechanistic convergence point from aberrant upstream activation signals during cellular transformation.

Indeed, cancer cells are more dependent than their normal counterparts on STAT activity, which provides a powerful reason for targeting STATs in malignancies to disrupt their vital source of survival. STAT activity can be inhibited at the level of dimerization via their SH2 domain or by blocking their DNA binding. However, in contrast to JAK inhibitors, disappointingly, there are no STAT inhibitors currently approved for therapeutic purposes. The targeting of transcription factors poses a number of challenges, particularly with respect to drug delivery and stability issues. Nevertheless, STAT targeting is an area of active research, and a number of approaches have been tried in the development of specific STAT inhibitors, including non-peptide small-molecule inhibitors, peptide inhibitors or peptidomimetics, antisense oligonucleotides, decoy oligonucleotides, or SH2-domain inhibitors to block STAT DNA binding (Furqan et al. 2013). However, no STAT inhibitors have reached beyond Phase 1 clinical trials, due to metabolic instability and drug delivery considerations. Future studies directed at developing STAT inhibitors for the treatment of cancers should hopefully yield more potent, druggable compounds.

Cross-References

- Antiviral Responses
- Asthma
- Cancer and Inflammation
- Cytokines
- FoxP3
- Interleukin 4 and the Related Cytokines (Interleukin 5 and Interleukin 13)
- Interleukin 17
- Interleukin 23
- Interleukin 27
- Interleukin 6
- Interleukin 9
- Mammalian Target of Rapamycin (mTOR)
- MAP Kinase Pathways
- Natural Killer Cells
- Rheumatoid Arthritis
- TH17 Response
- Tumor Necrosis Factor Alpha (TNFalpha)
- Type I Interferons

References

Begitt, A., Droescher, M., Meyer, T., Schmid, C. D., Baker, M., Antunes, F., et al. (2014). STAT1-cooperative DNA binding distinguishes type 1 from type 2 interferon signaling. *Nature Immunology, 15*, 168–176.

Bourke, L. T., Knight, R. A., Latchman, D. S., Stephanou, A., & McCormick, J. (2014). Signal transducer and activator of transcription-1 localizes to the mitochondria and modulates mitophagy. *JAKSTAT, 2*, e25666.

Cacalano, N. A., Sanden, D., & Johnston, J. A. (2001). Tyrosine-phosphorylated SOCS-3 inhibits STAT activation but binds to p120 RasGAP and activates Ras. *Nature Cell Biology, 3*, 460–465.

Casanova, J. L., Holland, S. M., & Notarangelo, L. D. (2012). Inborn errors of human JAKs and STATs. *Immunity, 36*, 515–528.

Chen, H., Sun, H., You, F., Sun, W., Zhou, X., Chen, L., et al. (2011). Activation of STAT6 by STING is critical for antiviral innate immunity. *Cell, 147*, 436–446.

Chueh, F. Y., Leong, K. F., & Yu, C. L. (2010). Mitochondrial translocation of signal transducer and activator of transcription 5 (STAT5) in leukemic T cells and cytokine-stimulated cells. *Biochemical and Biophysical Research Communications, 402*, 778–783.

Dawson, M. A., Bannister, A. J., Gottgens, B., Foster, S. D., Bartke, T., Green, A. R., et al. (2009). JAK2 phosphorylates histone H3Y41 and excludes HP1alpha from chromatin. *Nature, 461*, 819–822.

Eghtedar, A., Verstovsek, S., Estrov, Z., Burger, J., Cortes, J., Bivins, C., et al. (2012). Phase 2 study of the JAK kinase inhibitor ruxolitinib in patients with refractory leukemias, including postmyeloproliferative neoplasm acute myeloid leukemia. *Blood, 119*, 4614–4618.

Fox, D. A. (2012). Kinase inhibition–a new approach to the treatment of rheumatoid arthritis. *New England Journal of Medicine, 367*, 565–567.

Furqan, M., Akinleye, A., Mukhi, N., Mittal, V., Chen, Y., & Liu, D. (2013). STAT inhibitors for cancer therapy. *Journal of Hematology & Oncology, 6*, 90.

Gough, D. J., Corlett, A., Schlessinger, K., Wegrzyn, J., Larner, A. C., & Levy, D. E. (2009). Mitochondrial

STAT3 supports Ras-dependent oncogenic transformation. *Science, 324,* 1713–1716.

Hirahara, K., Onodera, A., Villarino, A. V., Bonelli, M., Sciumè, G., Laurence, A., et al. (2015). Asymmetric action of STAT transcription factors drives transcriptional outputs and cytokine specificity. *Immunity, 19,* 877–889.

3Hou, S. X., Zheng, Z., Chen, X., & Perrimon, N. (2002). The Jak/STAT pathway in model organisms: emerging roles in cell movement. *Developmental Cell, 3,* 765–778.

Huang, C. Y., Lin, Y. C., Hsiao, W. Y., Liao, F. H., Huang, P. Y., & Tan, T. H. (2012). DUSP4 deficiency enhances CD25 expression and CD4+ T-cell proliferation without impeding T-cell development. *European Journal of Immunology, 42,* 476–488.

Ito, M., Nakasato, M., Suzuki, T., Sakai, S., Nagata, M., & Aoki, F. (2004). Localization of janus kinase 2 to the nuclei of mature oocytes and early cleavage stage mouse embryos. *Biology of Reproduction, 71,* 89–96.

John, S., Vinkemeier, U., Soldaini, E., Darnell, J. E., Jr., & Leonard, W. J. (1999). The significance of tetramerization in promoter recruitment by Stat5. *Molecular and Cellular Biology, 19,* 1910–1918.

Kohanbash, G., & Okada, H. (2012). MicroRNAs and STAT interplay. *Seminars in Cancer Biology, 22,* 70–75.

Liu, L., Okada, S., Kong, X. F., Kreins, A. Y., Cypowyj, S., Abhyankar, A., et al. (2011). Gain-of-function human STAT1 mutations impair IL-17 immunity and underlie chronic mucocutaneous candidiasis. *The Journal of Experimental Medicine, 208,* 1635–1648.

Lobie, P. E., Ronsin, B., Silvennoinen, O., Haldosen, L. A., Norstedt, G., & Morel, G. (1996). Constitutive nuclear localization of Janus kinases 1 and 2. *Endocrinology, 137,* 4037–4045.

O'Shea, J. J., Gadina, M., & Schreiber, R. D. (2002). Cytokine signaling in 2002: new surprises in the Jak/Stat pathway. *Cell, 109*(Suppl), S121–S131.

O'Shea, J. J., Holland, S. M., & Staudt, L. M. (2013). JAKs and STATs in immunity, immunodeficiency, and cancer. *New England Journal of Medicine, 368,* 161–170.

Ota, N., Brett, T. J., Murphy, T. L., Fremont, D. H., & Murphy, K. M. (2004). N-domain-dependent nonphosphorylated STAT4 dimers required for cytokine-driven activation. *Nature Immunology, 5,* 208–215.

Puel, A., Cypowyj, S., Marodi, L., Abel, L., Picard, C., & Casanova, J. L. (2012). Inborn errors of human IL-17 immunity underlie chronic mucocutaneous candidiasis. *Current Opinion in Allergy and Clinical Immunology, 12,* 616–622.

Rinaldi, C. R., Rinaldi, P., Alagia, A., Gemei, M., Esposito, N., Formiggini, F., et al. (2010). Preferential nuclear accumulation of JAK2V617F in CD34+ but not in granulocytic, megakaryocytic, or erythroid cells of patients with Philadelphia-negative myeloproliferative neoplasia. *Blood, 116,* 6023–6026.

Sehgal, P. B. (2014). Non-genomic STAT5-dependent effects at the endoplasmic reticulum and Golgi apparatus and STAT6-GFP in mitochondria. *JAKSTAT, 2,* e24860.

Shuai, K., & Liu, B. (2003). Regulation of JAK-STAT signalling in the immune system. *Nature Reviews Immunology, 3,* 900–911.

Stark, G. R., & Darnell, J. E., Jr. (2012). The JAK-STAT pathway at twenty. *Immunity, 36,* 503–514.

Timofeeva, O. A., & Tarasova, N. I. (2013). Alternative ways of modulating JAK-STAT pathway: looking beyond phosphorylation. *JAKSTAT, 1,* 274–284.

Vinkemeier, U. (2004). Getting the message across, STAT! Design principles of a molecular signaling circuit. *The Journal of Cell Biology, 167,* 197–201.

Wegrzyn, J., Potla, R., Chwae, Y. J., Sepuri, N. B., Zhang, Q., Koeck, T., et al. (2009). Function of mitochondrial Stat3 in cellular respiration. *Science, 323,* 793–797.

Yang, J., & Stark, G. R. (2008). Roles of unphosphorylated STATs in signaling. *Cell Research, 18,* 443–451.

Zouein, F. A., Duhe, R. J., & Booz, G. W. (2011). JAKs go nuclear: emerging role of nuclear JAK1 and JAK2 in gene expression and cell growth. *Growth Factors, 29,* 245–252.

Kawasaki Disease

Ho-Chang Kuo[1] and Kai-Sheng Hsieh[2]
[1]Kawasaki Disease Center and Department of Pediatrics, Division of Allergy, Immunology and Rheumatology, Kaohsiung Chang Gung Memorial Hospital, Chang Gung University College of Medicine, Kaohsiung city, Taiwan
[2]Department of Pediatrics, Kaohsiung Chang Gung Memorial Hospital, Chang Gung University College of Medicine, Kaohsiung city, Taiwan

Synonyms

Kawasaki disease; Kawasaki syndrome; Lymph node syndrome; Mucocutaneous lymph node syndrome

Definition

Prolonged fever (>5 days), with four of the following five phenotypes: conjunctivitis, diffuse oral mucosal inflammation, polymorphous skin rashes, indurative edema of the hands and feet associated with the peeling of finger tips, and non-suppurative neck lymphadenopathy.

Epidemiology and Genetics

The incidence of KD is increasing globally, especially in Japan (Fig. 1). In Taiwan, Korea, and Japan, the incidence ranges from 67.3, 113, to 239.36 cases per 100,000 children under 5 years of age, respectively (Nakamura et al. 2012). The highest incidence is in Japan, and the lowest is in the United Kingdom (8.39 per 100,000 children under 5 years of age, almost 1/30 of the incidence in Japan). The annual incidence rates increase in Japan but not in Taiwan. Epidemiologic surveys of KD found that 1.5 % of all cases were recurrent (had second episode of KD after first episode) (Huang et al. 2009). In Taiwan, KD occurs most frequently in the summer (April to June) and least frequently in winter; for unknown reasons, its seasonal occurrence varies in other countries. In Japan, monthly number of patients peaked during winter to spring months; lower peaks were noted during summer months. The environmental factor's trigger for KD could be wind-borne in Japan, Hawaii, and San Diego. The association of KD and vaccination is still unclear. A review of RotaTeq (rotavirus vaccine live) clinical trial data revealed higher but not statistically significant in KD rates among RotaTeq vaccines than placebo recipients. But the Vaccine Adverse Event Reporting System (VAERS) for all US-licensed vaccines reported that not suggest an elevated KD risk for RotaTeq or other vaccines. In conclusion, the incidence was increased globally, especially in Japan, and environmental factors may play a role in the epidemiology of KD but not vaccination.

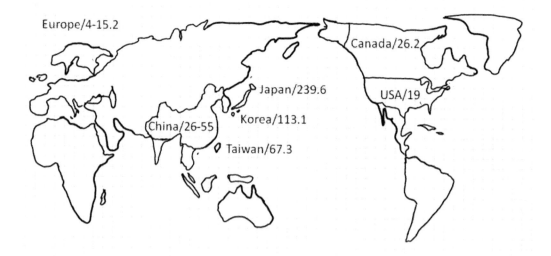

Kawasaki Disease, Fig. 1 Global prevalence of Kawasaki disease (case per 100,000 children younger than 5 years)

Genetic Study

The incidence of KD in Asian countries is higher than American and Europe, male predominant, no certain infectious source found suggests that genetic predisposition might play an important role in the susceptibility to this disease. There is also evidence that the incidence of KD is higher among siblings than in the general population (about ten times higher). A number of reports provided evidence that genetic polymorphisms contribute to the susceptibility to KD and/or coronary artery lesions (CAL) formation (Table 1). For example, single-nucleotide polymorphisms (SNPs) in the monocyte chemoattractant protein 1 (MCP-1), IL-10, CD40L, CD40, IL-4, CASP3, IL-18, IL-1B, HLA-E, C-C chemokine receptor 5 (CCR5), ITPKC, and TGF-β receptors have been reported to be associated with the susceptibility of KD. Although genetic association studies have been widely performed in KD, several studies have produced inconsistent results. The possibilities of these inconsistencies may come from sample size, different genetic backgrounds within populations, environmental factors, and/or infectious agents between countries.

Since 2009, first genome-wide association study (GWAS) on 119 Caucasian KD cases and 135 matched controls was reported. This insightful work identified SNP within the N-acetylated alpha-linked acidic dipeptidase-like two gene (NAALADL2), which was significantly associated with the susceptibility of KD. Although the function of NAALADL2 remains unclear, mutations in the gene may be involved in the development of Cornelia de Lange syndrome. In 2010, GWAS of Korean which reported a locus in the 1p31 region was identified as a susceptibility locus for KD, and PELI1 gene locus in the 2p13.3 region was confirmed to associate with CAL. Three novel loci was found including COPB2 (coatomer protein complex beta-2 subunit), ERAP1 (endoplasmic reticulum amino peptidase 1), and immunoglobulin heavy chain variable region genes in Han Chinese population residing in Taiwan (Tsai et al. 2011). In 2011, 5 independent sample collections from the USA, Asia, and Europe reported GWAS from 2,173 individuals with KD and 9,383 controls and showed a functional polymorphism in the IgG receptor gene FCGR2A (encoding an H131R substitution) with the A allele (coding for histidine) conferring elevated disease risk for KD (Khor et al. 2011). In 2012, two independent research groups published GWAS data from Taiwanese and Japanese populations (Lee et al. 2012; Onouchi et al. 2012) and suggested that BLK

Kawasaki Disease, Table 1 Genes associated with the susceptibility or coronary artery lesions (CAL) formation of KD

Gene	Abbreviation	Locus	Phenotypes
C-reactive protein	CRP	1q21-q23	Susceptibility
Tissue inhibitor of metalloproteinase 4	TIMP4	3p25	CAL
C-C chemokine receptor 5	CCR5	3p21	Susceptibility
Angiotensin-II type-1 receptor	AGTR1	3q21-q25	CAL
Vascular endothelial growth factor receptor 2	VEGFR2	4q12	CAL
Interleukin-4	IL-4	5q31.1	Susceptibility
CD14 antigen	CD14	5q31.1	CAL
Vascular endothelial growth factor	VEGFA	6p12	Susceptibility CAL
Lymphotoxin-alpha	LTA	6p21.3	Susceptibility
Tumor necrosis factor-alpha	TNF-α	6p21.3	CAL
Interleukin-18	IL-18	11q22.3-q22.3	Susceptibility
Matrix metalloproteinase-3	MMP3	11q22.3	CAL
Matrix metalloproteinase-13	MMP13	11q22.3	CAL
Angiotensin-1 converting enzyme	ACE	17q23	Susceptibility CAL
Tissue inhibitor of metalloproteinase 2	TIMP2	17q25	CAL
CD209	CD209	19p13	Susceptibility
Inositol-trisphosphate 3-kinase C	ITPKC	19q13.1	Susceptibility CAL
CD40	CD40	20q12-q13.2	Susceptibility CAL
Macrophage migration inhibitory factor	MIF	22q11.2	CAL
CD40 ligand	CD40L	Xq26	CAL

Modified from Kue et al. (Kuo and Chang 2011)

(encoding B-lymphoid tyrosine kinase) and CD40 are novel susceptibility genes for KD. This is the first time the same genetic results were found from a different area using GWAS. GWAS is very useful to identify novel loci for KD, global study cooperation to diminish bias from sample size and environment factors, meta-analysis and functional studies are needed to confirm these meaningful findings.

Pathophysiology

Kawasaki disease (KD), an acute systemic vasculitis, occurs mainly in infants and children under 5 years of age and was first described by Dr. Tomisaku Kawasaki et al. (Fig. 2) in 1967 in Japanese and later in 1974 in English. Currently, it is the leading cause of acquired heart disease in children in oriental countries; however, its etiology remains unknown (Kuo et al. 2012). The etiology of KD remains unknown and may be attributed to the combined effects of infection, immune response, and genetic susceptibility. Standard treatment with high-dose aspirin (80–100 mg/kg/day, acetylsalicylic acid; ASA) and high-dose intravenous immune globulin (2 g/kg, IVIG, or intravenous gammaglobulin, IVGG) has been shown to decrease the rate of coronary artery aneurysm development from 25–25 % to 3–5 %. In children with acute KD, a single large dose of IVIG is more effective than the conventional regimen of four smaller daily doses (Burns and Glode 2004).

In the acute stage of KD, activation of numerous immunologic factors including T cell activation, cytokine production (IL-1, IL-4, IL-5, IL-6, IL-8, IL-10, IL-13, IL-17, GM-CSF, TNF-α, IFN-γ, TGF-β, MIP-1, MCP-1, CCL17, MMP-9, CD40L, etc.) (Burns and Glode 2004; Hsieh et al. 2011; Kuo et al. 2011, 2012; Lin et al. 2012), nitric oxide (NO) production, autoantibody production, and enhanced adhesion molecule expression is well documented. Pathologic examination of the coronary arteritis in the acute stage of KD indicates that activated T lymphocyte-dependent processes characterized by transmural infiltration of activated T lymphocytes occur with

Kawasaki Disease, Fig. 2 Dr. Tomisaku Kawasaki (*left*) and Dr. Ho-Chang Kuo (*right*) at 2008 International Kawasaki Disease Symposium, Taipei, Taiwan

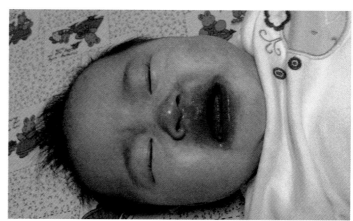

Kawasaki Disease, Fig. 3 Face of children with Kawasaki disease showed redness over the perioral area, fissure lip, and strawberry tongue

accumulation of CD8+ T cells in vascular lesions. Macrophage activation and altered T helper (Th1/Th2) and regulatory cell functions are also implicated in dysregulation of the immune response in patients with KD (Wang et al. 2005).

Clinical Presentation

As shown in Figs. 3, 4, 5, 6, 7, and 8, the clinical characteristics of KD patients include fever lasting >5 days, diffuse mucosal inflammation, bilateral nonpurulent conjunctivitis, dysmorphic skin rashes, indurative angioedema over the hands and feet, and cervical lymphadenopathy. KD should be considered in the differential diagnosis of any young child with unexplained fever more than 5 days. Changes in the extremities are prominent

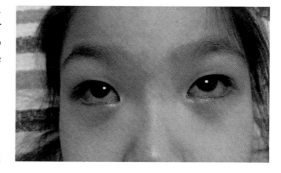

Kawasaki Disease, Fig. 4 Bilateral nonpurulent conjunctivitis

with redness and edematous change. Erythema of the palms and soles occur along with firm, sometimes painful induration of the hands or feet in the early phase of KD. A polymorphous exanthema usually appears within 5 days of the onset of fever.

Kawasaki Disease, Fig. 5 Dysmorphism skin rash

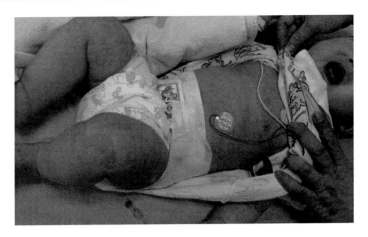

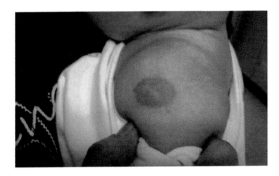

Kawasaki Disease, Fig. 6 BCG injection site induration

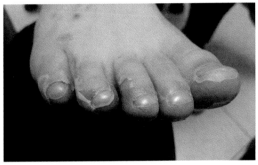

Kawasaki Disease, Fig. 8 Desquamation over the tip of toes

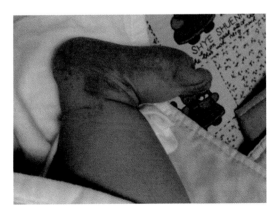

Kawasaki Disease, Fig. 7 Palm erythematic change

The rash may present various forms on the trunk and extremities, including an urticarial exanthema, a maculopapular morbilliform eruption, a scarlatiniform erythroderma, an erythema multiforme-like rash, and a fine micropustular eruption. Any kind of skin rash should be considered KD-related when suspected. Neck lymph node enlargement, usually >1.5 cm in diameter, also occurs in the early phase. Bilateral conjunctival injection usually begins shortly after the onset of fever. It typically involves the bulbar conjunctivae much more than the palpebral or tarsal conjunctivae and is not associated with an exudate. It is usually painless. Desquamation of the fingers and toes usually begins 1–3 weeks after onset of fever in the periungual region. A summary of the clinical features from 350 cases with KD seen in the Kaohsiung Chang Gung Memorial Hospital, Taiwan, is shown in Table 2.

In addition to the diagnostic criteria, there is a broad range of nonspecific clinical features, including irritability, uveitis, aseptic meningitis, cough, vomiting, diarrhea, abdominal pain, gallbladder hydrops, urethritis, arthralgia, arthritis, hypoalbuminemia, liver function impairment,

Kawasaki Disease, Table 2 Clinical symptoms and signs of Kawasaki disease. A summary of the clinical features from 350 cases observed in Kaohsiung, Taiwan

Clinical presentations	Percentage (%)
Conjunctivitis	93
Strawberry tongue	92
Desquamation over finger tip	91
Polymorphism skin rash	91
Indurations of palm	75
BCG scar indurations	42
Neck lymphadenopathy	38

urinary tract infection, hearing impairment, and heart failure. Sensory hearing loss was reported to be under diagnosis in KD and more prevalent than CAL. Hearing loss was reported that 22/40 (55 %) disclosed within the first 30 days. In 12/40 (30 %), the hearing loss persisted after 6 months. Transient sensorineural hearing loss (20–35 dB) is a frequent complication of acute KD and may be related to salicylate toxicity in some patients. Persistent sensorineural hearing loss is uncommon. So, parents, primary care providers, and pediatricians should be more aware of the potential for persistent sensorineural hearing loss following KD, but routine audiologic screening of KD patients does not appear to be warranted.

In some countries where newborn babies receive Calmette-Guérin bacillus (BCG) vaccination, KD can be associated with erythematous induration or even ulceration of BCG scars in one-third of cases. It may be useful for differential diagnosis of suspect KD patient, but the incidence is lower than 50 % of KD patients. The reason why BCG scar indurations in KD patients are still under investigation is because it may provide a hint for further investigations about the immunopathogenesis of KD.

Diagnosis

The diagnostic criteria for Kawasaki disease are listed as Table 3. We have established the "Kuo mnemonic" for rapid memorization of the diagnostic criteria of KD (Table 4).

To date, there is no specific diagnostic laboratory test for KD available. The diagnosis is based on the clinical phenotype, i.e., the presence of fever lasting longer than 5 days and the fulfillment

Kawasaki Disease, Table 3 Diagnostic criteria for Kawasaki disease

Fever more than 5 days
1. Diffuse mucosal inflammation (fissure lip, strawberry tongue)
2. Bilateral nonpurulent conjunctivitis
3. Polymorphous skin rashes (skin rash, wheal formation, urticaria like, any kinds of skin rash should be considered with KD-related when suspect KD)
4. Indurative angioedema of the hands and feet (desquamation in the subacute stage)
5. Cervical lymphadenopathy (unilateral lymphadenopathy, more than 1.5 cm in diameter)
The diagnosis of Kawasaki disease is considered confirmed by the presence of fever, and 4 of the remaining 5 criteria if other known diseases can be excluded

Kawasaki Disease, Table 4 "Kuo mnemonic" for the diagnostic criteria regarding the rapid memorization of Kawasaki Disease

Number	Mnemonic	Clinical signs
1	"One" mouth	Diffuse mucosal inflammation with strawberry tongue and fissure lips
2	"Two" eyes	Bilateral non-purulent conjunctivitis
3	"Three" fingers palpation neck lymph nodes	Unilateral cervical lymphadenopathy
4	"Four" limbs changes	Indurative angioedema over both the hands and feet
5	"Five" = multiple skin rash	Dysmorphic skin rashes

of four of five specific clinical criteria. In Japan, at least five of six criteria (fever and five other clinical criteria) should be fulfilled for the diagnosis of KD. However, a diagnosis of KD can be made when coronary aneurysm or dilatation is identified in patients with four of the principal clinical features. According to the Japanese Circulation Society Joint Working Groups criteria (JCS Joint Working Group 2010), KD can be diagnosed even when fever lasts less than 5 days. However, according to the American Heart Association (AHA) criteria, fever lasting more than 5 days is essential for the diagnosis of KD.

Some patients who do not fulfill the criteria have been diagnosed with "incomplete" or "atypical" KD, a diagnosis often based on echocardiographic identification of CAL. The term "incomplete" may be preferable to "atypical" because these patients have insufficient criteria instead of atypical presentation.

In countries with a bacillus Calmette-Guérin (BCG) vaccine policy (i.e., Taiwan, Turkey, and Japan), KD with erythematous induration or even ulceration of the BCG scar has been observed in near 50 % of KD patients (the incidence of BCG site induration is higher than that of neck lymphadenopathy in these countries). Redness or the formation of a crust at the BCG inoculation site is a useful diagnostic sign for KD in children aged 3–20 months. Even if patients exhibit 4 or fewer signs of the clinical criteria for KD, physicians should consider the redness or crust formation at the BCG inoculation site as a possible indicator of KD.

Incomplete cases of KD are not uncommon (up to 15–20 %). The incidence of CAL in patients exhibiting four principal symptoms of KD is slightly higher than that in patients with five to six principal symptoms. Presentation of a small number (<4) of principal symptoms does not indicate a milder form of the disease. Patients with at least four principal symptoms require the same treatment as patients with complete (typical) presentation of KD, and those with three or fewer principal symptoms should be treated similarly when they meet the supplementary criteria. Herein, common supplementary criteria for the diagnosis of incomplete KD are introduced.

Incomplete KD is more common in young infants than in older children, making an accurate diagnosis and timely treatment especially important in these young patients, who are at high risk of developing coronary abnormalities. The incidence of KD is actually higher than that previously reported worldwide, partly because earlier reports did not take incomplete forms into account. Patients with fever for 5 days or more (with two or three principal clinical features for KD) without other causes should undergo laboratory testing, and if there is evidence of systemic inflammation, an echocardiogram should be obtained even if the patient does not fully meet the clinical criteria for KD. Likewise, infants 6 months or younger with fever for 7 days or more without other causes should undergo laboratory testing, and if evidence of systemic inflammation is found, an echocardiogram should be obtained even if the infant fulfills no clinical criteria for KD. The 2004 AHA supplemental laboratory criteria are described in Table 5. The flow chart reference for the management of refractory or incomplete case of KD is showed in Fig. 9.

If a patient has more than three supplementary criteria, incomplete KD is diagnosed and IVIG should be prescribed before performing echocardiography (Newburger et al. 2004).

Kawasaki Disease, Table 5 2004 AHA supplement criteria for Kawasaki disease

Fever of >5 days associated with 2 or 3 clinical criteria for KD, C-reactive protein >3.0 mg/dL, and/or erythrocyte sedimentation rate >40 mm/h with the following criteria
1. Albumin ≤3.0 g/dL
2. Anemia for age
3. Elevation of alanine aminotransferase (ALT)
4. Platelets after 7 days ≥450,000/mm^3
5. White blood cell count ≥15,000/mm^3
6. Urine ≥10 white blood cells/high-power field

Therapy

Intravenous Immunoglobulin (IVIG or IVGG)

IVIG was used for treatment of KD since 1983 reported by Furusho et al. (1983), more than 10 years after first report of KD by Dr. Kawasaki. A randomized control trail revealed that high-dose IVIG (400 mg/kg/day for 4 days) is safe and effective in reducing the prevalence of CAL from 20 % to 3–5 % when administered in those at the acute stage of KD. A study by Newburger et al. contributed greatly in this regard, and a single high dose of IVIG (2 g/kg) is considered to be the gold standard therapy in the acute stage of KD in 1991 (Newburger et al. 1991). The mechanism of IVIG in reducing inflammation in KD is not clearly understood. Possible

**Kawasaki Disease,
Fig. 9** Flow chart reference for the management of refractory Kawasaki disease (*KD*) (Modified from Newburger et al. (2004))

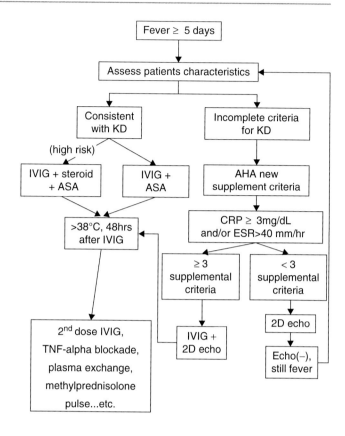

mechanisms include Fc receptor blockade, neutralization of the causative agents or a toxin produced by an infectious agent, an immunomodulating effect, induction of suppressor activity, and modulation of the production of cytokines and cytokine antagonists. IVIG treatment reduced the activity of oxidative stress, which provokes vasculitis in KD. The mechanism of IVIG action is still under investigation. IVIG appears to have a generalized anti-inflammatory effect. Possible mechanisms of action include modulation of cytokine production, neutralization of bacterial superantigens or other etiologic agents, augmentation of regulatory T cell activity (TGF-β), suppression of antibody synthesis and inflammatory markers (CD40-CD40L, nitric oxide, and iNOS expression), provision of anti-idiotypic antibodies, Fc-gamma receptor, and balancing Th1/Th2 immune responses (Kuo et al. 2009, 2011; Wang et al. 2002, 2003).

KD patients should be treated with a single 12-h infusion of IVIG, 2 g/kg in a single infusion, together with aspirin in acute phase with fever. This therapy should be performed within 10 days of illness onset, and, if this is not possible, within 7 days of illness onset. Treatment of KD before day 5 of illness appears no more likely to prevent cardiac sequelae than treatment on days 5–9. It may, however, be associated with an increased need for IVIG re-treatment. In the presence of four of five classic criteria for KD, US and Japanese experts agree that only 4 days of fever are necessary before initiating treatment with IVIG.

The efficacy of treating patients using IVIG after 10 days of illness is unknown; therefore, early diagnosis and treatment is desired. IVIG should be administered to children presenting after day 10 of illness (i.e., children with delayed diagnosis or incomplete KD) if they have either persistent fever without explanation or aneurysms and ongoing systemic inflammation, as manifested by elevated ESR or CRP. Any child with KD who has evidence of persisting inflammation, including fever or high concentrations of inflammatory markers with or without coronary artery abnormalities, should be treated even if the diagnosis is made after 10 days of illness. For IVIG-resistant patients, earlier and more effective anti-inflammatory therapy might not be emphasized enough to reduce the risk of CAL.

Aspirin

Now, aspirin is forbidden to be used in children for antipyretics regiment except in cases of KD. Aspirin has been prescribed in the treatment of KD for many years, even before the usage of IVIG. Although aspirin has important anti-inflammatory (high dose) and antiplatelet (low dose, 3–5 mg/kg/day) effects, it does not appear to reduce the frequency of CAL formation and IVIG resistance rate. Anti-inflammatory doses of aspirin are recommended in conjunction with IVIG, but controversy remains regarding anti-inflammatory doses of aspirin. In North America, high-dose (80–100 mg/kg per day) aspirin is most widely used during the acute phase. In Japan, concern about hepatic toxicity has led to the use of moderate-dose (30–50 mg/kg per day) aspirin as a recommended standard therapy in the acute phase. The results of our previous study indicate that treatment without aspirin in acute-stage KD had no effect on the response rate of IVIG therapy, duration of fever, or CAL incidence (Hsieh et al. 2004). High- or medium-dose aspirin therapy may be unnecessary in acute KD when the available data show no appreciable benefits in preventing the failure of IVIG therapy or CAL formation or in shortening fever duration. Aspirin has been reported to have hemolytic potential in individuals with glucose-6-phosphate dehydrogenase (G6PD) deficiency. Whether or not high-dose aspirin should be utilized in the acute stage of KD needs further multiple center randomized control trials before making a definitive recommendation.

Steroids

Although corticosteroids are the treatment of choice in other forms of vasculitis, the usefulness of steroids in treatment of KD is not well established. Corticosteroids were used as the initial therapy for KD long before the first report of IVIG efficacy. An early study suggested that steroids exerted a negative effect when used as the initial therapy for KD, but some studies have shown possible benefit. Corticosteroid therapy combined with IVIG as the initial treatment was showed rapidly ameliorated symptoms by reducing cytokine levels in children with KD. Methylprednisolone and IVIG were also effective and safe as a primary treatment for high-risk KD patients. However, Newburger et al. (2007) demonstrated that their data do not provide support for the addition of a single-pulsed dose of intravenous methylprednisolone to IVIG for routine primary treatment of children with KD. Meta-analysis of comparison of the incidence of CAL between IVIG plus corticosteroid therapy and IVIG therapy alone for the initial treatment of KD showed that IVIG plus corticosteroid therapy significantly reduced the risk of coronary abnormality (Kobayashi et al. 2012). In conclusion, steroids are beneficial for acute stage of KD especially in cases involving the high-risk group but not necessarily for methylprednisolone pulse as an additional treatment to initial IVIG.

Others

A new class of therapies directed against specific cytokines has expanded treatment for KD. Infliximab is a monoclonal antibody to TNF-α and has been effective in the treatment of patients with refractory KD. Treatment with infliximab might also be an initial therapy for high-risk KD patients. Acute KD can lead to the development of large coronary artery aneurysms that may persist for years. Abciximab, a platelet glycoprotein IIb/IIIa receptor inhibitor, is associated with resolution of thrombi and vascular remodeling in adults with acute coronary syndromes. KD patients who were treated with abciximab and demonstrated greater regression in aneurysm diameter at early follow-up were reported. Abciximab seems to benefit KD patients, especially those who developed aneurysms.

There are still no well-defined treatments for refractory KD (all see Fig. 9). Cyclosporine

A (CyA) treatment is considered safe and well tolerated and may serve as a promising option for patients with refractory KD. Hyperkalemia developed in 9/28 (32 %) patients 3–7 days after commencing CyA treatment. Adverse effects such as arrhythmias should be monitored with CyA.

Specific changes in inflammatory markers (such as white blood cell count, neutrophil count, C-reactive protein, IL-6, soluble IL-2 receptor, T-helper-type 17/regulatory T cell imbalance, and IL-1 pathway) have been reported to disturb immunological functions and result in KD with IVIG resistance and CAL formation. This indicates the possible treatment role of plasma exchange (PE) for KD with IVIG resistance. PE is considered safe and effective in the prevention of CAL in KD that is refractory to IVIG therapy. PE could be performed at an early stage, as soon as fractional increases in inflammatory markers are found after first or second dosage of IVIG therapy.

Outcome

Coronary Artery Lesions (CAL) and Aneurysm

Even after the use of high-dose IVIG treatment, there are still 3–5 % of KD patients that developed CAL. This may indicate that anti-inflammatory regiments other than IVIG are needed for high-risk KD patients. The most serious complication of KD is the development of CAL, including myocardial infarction, coronary artery fistula formation (Liang et al. 2009), coronary artery dilatation, and coronary artery aneurysm. Definition of CAL (also known as coronary artery abnormality) is based on the Japanese Ministry of Health criteria: >3 mm maximum absolute internal diameter of the coronary artery in children younger than 5 years of age or >4 mm in children 5 years and older, or a segmental diameter 1.5 times greater than that of an adjacent segment, or the presence of luminal irregularity.

Coronary arteries should be corrected relative to body surface area (BSA) (if body weight and body height were available) and expressed as standard deviation units from the mean

Kawasaki Disease, Table 6 Percentage of coronary artery lesion in Kawasaki disease patients

Initial (during admission)	35 %
1 month after disease onset	17.2 %
2 month after disease onset	10.2 %
1 year after disease onset	4 %

(Z scores). Several studies have analyzed CAL, including aortic root dimension, and transient CAL (the definition of "transient" varies among studies, from 30 days to 6–8 weeks after diagnosis of disease). As such, KD patients with coronary artery ectasia or dilatation, which disappears within the first 8 weeks after disease onset, are defined as transient ectasia or dilatation (transient CAL). According to our previous report on CAL analysis including 341 KD patients, 35 % of KD patients had dilatation during the acute phase of admission, 17.2 % had dilatation 1 month after disease onset, 10.2 % had dilatation at 2 months, and 4 % had persistent CAL for more than 1 year (Table 6). Echocardiography is most common tool for follow-up cardiovascular complication of KD patients, the time schedule varying according to severity of CAL. Computed tomography (CT) scan also provides good image of coronary artery. Figures 10 and 11 show coronary artery dilation or aneurysm formation with calcification in KD patient from cardioechography and CT scan.

IVIG Resistance or Nonresponsiveness

However, there are still 7.8–38.3 % of children that are unresponsive to initial IVIG treatment. Recent studies have identified demographic and laboratory characteristics as predictors of IVIG resistance, including age, illness day, platelet count, erythrocyte sedimentation rate, hemoglobin concentration, C-reactive protein, eosinophil, lactate dehydrogenase, albumin, alanine aminotransferase (ALT), G-CSF, and sonographic gallbladder abnormalities (Abe et al. 2008; Kuo et al. 2010). Because IVIG-resistant patients are at a higher risk for CAL formation, it is important to identify those who may benefit from a more aggressive therapy.

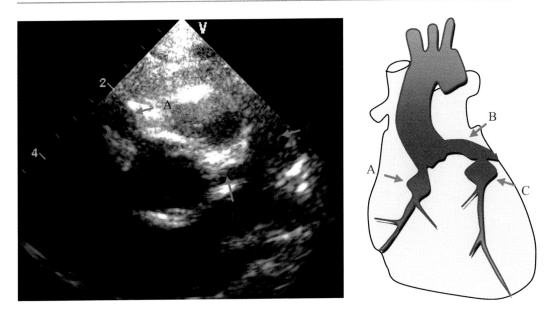

Kawasaki Disease, Fig. 10 Echocardiography from 1-year-old Kawasaki disease patient showed multiple dilatation of coronary artery of 7 mm in RCA (**a**), LCA (**b**), and 10 mm in diameter over LAD (**c**)

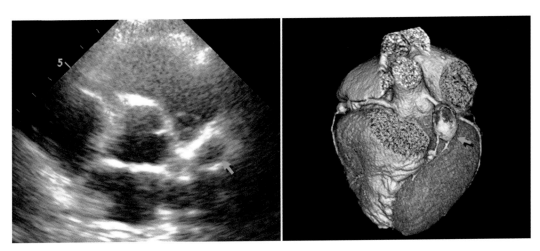

Kawasaki Disease, Fig. 11 Calcification of coronary artery (*green arrow*) of an 18-year-old women with Kawasaki disease history at 2 years old, CT scan (*right*) and echocardiography (*left*), 15 mm aneurysm noted over left circumflex coronary artery

Allergic Diseases Followed Kawasaki Disease

Atopic dermatitis (AD) was shown to increase the incidence among children with KD than that of controls. KD was reported to be associated with AD, allergy, elevated total serum IgE levels, and eosinophilia and that may be related to the effects of IL-4 (Burns et al. 2005; Kuo et al. 2007). KD may be a risk factor for subsequent allergic disease and occurs more frequently in children at risk of immune disequilibrium, with an initial abnormal inflammatory response, and, subsequently, more allergic manifestations as well as admitted with asthma/allergy. T-helper (Th) type 2 immune response was elevated in the acute stage of KD, including eosinophils, IL-4, IL-5, and eotaxin than in age-matched control. There is a fair amount of evidence to support abnormal Th1/Th2 balance in KD patients. Comparison of eosinophils in KD and enterovirus (EV) patients with IVIG treatment demonstrated a more significant eosinophil increase in KD patients than patient with EV after IVIG treatment (Lin et al. 2012). This may indicate an imbalance of the Th1/Th2 immune response, with a skewed Th2 response in KD. The skewed Th2 immune response in patient with KD may lead to allergic diseases late after disease onset.

Cross-References

- Allergic Disorders
- Atopic Dermatitis
- Autoinflammatory Syndromes
- Henoch-Schönlein Purpura
- Idiopathic Thrombocytopenic Purpura
- Rheumatic fever

References

Abe, J., Ebata, R., Jibiki, T., Yasukawa, K., Saito, H., & Terai, M. (2008). Elevated granulocyte colony-stimulating factor levels predict treatment failure in patients with Kawasaki disease. *The Journal of Allergy and Clinical Immunology, 122*(5), 1008–1013 e1008.

Burns, J. C., & Glode, M. P. (2004). Kawasaki syndrome. *Lancet, 364*(9433), 533–544.

Burns, J. C., Shimizu, C., Shike, H., Newburger, J. W., Sundel, R. P., Baker, A. L., et al. (2005). Family-based association analysis implicates IL-4 in susceptibility to Kawasaki disease. *Genes and Immunity, 6*(5), 438–444.

Furusho, K., Sato, K., Soeda, T., Matsumoto, H., Okabe, T., Hirota, T., et al. (1983). High-dose intravenous gammaglobulin for Kawasaki disease. *Lancet, 2*(8363), 1359.

Hsieh, K. S., Weng, K. P., Lin, C. C., Huang, T. C., Lee, C. L., & Huang, S. M. (2004). Treatment of acute Kawasaki disease: Aspirin's role in the febrile stage revisited. *Pediatrics, 114*(6), e689–e693.

Hsieh, K. S., Lai, T. J., Hwang, Y. T., Lin, M. W., Weng, K. P., Chiu, Y. T., et al. (2011). IL-10 promoter genetic polymorphisms and risk of Kawasaki disease in Taiwan. *Disease Markers, 30*(1), 51–59.

Huang, W. C., Huang, L. M., Chang, I. S., Chang, L. Y., Chiang, B. L., Chen, P. J., et al. (2009). Epidemiologic features of Kawasaki disease in Taiwan, 2003–2006. *Pediatrics, 123*(3), e401–e405.

JCS Joint Working Group (2010). Guidelines for diagnosis and management of cardiovascular sequelae in Kawasaki disease (JCS 2008)–digest version. *Circ J, 74*(9), 1989–2020.

Khor, C. C., Davila, S., Breunis, W. B., Lee, Y. C., Shimizu, C., Wright, V. J., et al. (2011). Genome-wide association study identifies FCGR2A as a susceptibility locus for Kawasaki disease. *Nature Genetics, 43*(12), 1241–1246.

Kobayashi, T., Saji, T., Otani, T., Takeuchi, K., Nakamura, T., Arakawa, H., et al. (2012). Efficacy of immunoglobulin plus prednisolone for prevention of coronary artery abnormalities in severe Kawasaki disease (RAISE study): A randomised, open-label, blinded-endpoints trial. *Lancet, 379*(9826), 1613–1620.

Kuo, H. C., & Chang, W. C. (2011). Genetic polymorphisms in Kawasaki disease. *Acta Pharmacologica Sinica, 32*(10), 1193–1198.

Kuo, H. C., Yang, K. D., Liang, C. D., Bong, C. N., Yu, H. R., Wang, L., et al. (2007). The relationship of eosinophilia to intravenous immunoglobulin treatment failure in Kawasaki disease. *Pediatric Allergy and Immunology, 18*(4), 354–359.

Kuo, H. C., Wang, C. L., Liang, C. D., Yu, H. R., Huang, C. F., Wang, L., et al. (2009). Association of lower eosinophil-related T helper 2 (Th2) cytokines with coronary artery lesions in Kawasaki disease. *Pediatric Allergy and Immunology, 20*(3), 266–272.

Kuo, H. C., Liang, C. D., Wang, C. L., Yu, H. R., Hwang, K. P., & Yang, K. D. (2010). Serum albumin level predicts initial intravenous immunoglobulin treatment failure in Kawasaki disease. *Acta Paediatrica, 99*(10), 1578–1583.

Kuo, H. C., Onouchi, Y., Hsu, Y. W., Chen, W. C., Huang, J. D., Huang, Y. H., et al. (2011a). Polymorphisms of transforming growth factor-beta signaling pathway and Kawasaki disease in the Taiwanese population. *Journal of Human Genetics, 56*(12), 840–845.

Kuo, H. C., Yang, K. D., Juo, S. H., Liang, C. D., Chen, W. C., Wang, Y. S., et al. (2011b). ITPKC single nucleotide polymorphism associated with the Kawasaki disease in a Taiwanese population. *PloS One, 6*(4), e17370.

Kuo, H. C., Yang, K. D., Chang, W. C., Ger, L. P., & Hsieh, K. S. (2012a). Kawasaki disease: An update on diagnosis and treatment. *Pediatrics and Neonatology, 53*(1), 4–11.

Kuo, H. C., Yang, Y. L., Chuang, J. H., Tiao, M. M., Yu, H. R., Huang, L. T., et al. (2012b). Inflammation-induced hepcidin is associated with the development of anemia and coronary artery lesions in Kawasaki disease. *Journal of Clinical Immunology, 32*(4), 746–752.

Lee, Y. C., Kuo, H. C., Chang, J. S., Chang, L. Y., Huang, L. M., Chen, M. R., et al. (2012). Two new susceptibility loci for Kawasaki disease identified through genome-wide association analysis. *Nature Genetics, 44*(5), 522–525.

Liang, C. D., Kuo, H. C., Yang, K. D., Wang, C. L., & Ko, S. F. (2009). Coronary artery fistula associated with Kawasaki disease. *American Heart Journal, 157*(3), 584–588.

Lin, I. C., Kuo, H. C., Lin, Y. J., Wang, F. S., Wang, L., Huang, S. C., et al. (2012a). Augmented TLR2 expression on monocytes in both human Kawasaki disease and a mouse model of coronary arteritis. *PloS One, 7*(6), e38635.

Lin, L. Y., Yang, T. H., Lin, Y. J., Yu, H. R., Yang, K. D., Huang, Y. C., et al. (2012b). Comparison of the laboratory data between Kawasaki disease and enterovirus after intravenous immunoglobulin treatment. *Pediatric Cardiology, 33*(8), 1269–1274.

Nakamura, Y., Yashiro, M., Uehara, R., Sadakane, A., Tsuboi, S., Aoyama, Y., et al. (2012). Epidemiologic features of Kawasaki disease in Japan: Results of the 2009–2010 nationwide survey. *Journal of Epidemiology, 22*(3), 216–221.

Newburger, J. W., Takahashi, M., Beiser, A. S., Burns, J. C., Bastian, J., Chung, K. J., et al. (1991). A single intravenous infusion of gamma globulin as compared with four infusions in the treatment of acute Kawasaki syndrome. *The New England Journal of Medicine, 324*(23), 1633–1639.

Newburger, J. W., Takahashi, M., Gerber, M. A., Gewitz, M. H., Tani, L. Y., Burns, J. C., et al. (2004). Diagnosis, treatment, and long-term management of Kawasaki disease: A statement for health professionals from the Committee on Rheumatic Fever, Endocarditis and Kawasaki Disease, Council on Cardiovascular Disease in the Young, American Heart Association. *Circulation, 110*(17), 2747–2771.

Newburger, J. W., Sleeper, L. A., McCrindle, B. W., Minich, L. L., Gersony, W., Vetter, V. L., et al. (2007). Randomized trial of pulsed corticosteroid therapy for primary treatment of Kawasaki disease. *The New England Journal of Medicine, 356*(7), 663–675.

Onouchi, Y., Ozaki, K., Burns, J. C., Shimizu, C., Terai, M., Hamada, H., et al. (2012). A genome-wide association study identifies three new risk loci for Kawasaki disease. *Nature Genetics, 44*(5), 517–521.

Tsai, F. J., Lee, Y. C., Chang, J. S., Huang, L. M., Huang, F. Y., Chiu, N. C., et al. (2011). Identification of novel susceptibility Loci for kawasaki disease in a Han chinese population by a genome-wide association study. *PloS One, 6*(2), e16853.

Wang, C. L., Wu, Y. T., Lee, C. J., Liu, H. C., Huang, L. T., & Yang, K. D. (2002). Decreased nitric oxide production after intravenous immunoglobulin treatment in patients with Kawasaki disease. *The Journal of Pediatrics, 141*(4), 560–565.

Wang, C. L., Wu, Y. T., Liu, C. A., Lin, M. W., Lee, C. J., Huang, L. T., et al. (2003). Expression of CD40 ligand on CD4+ T-cells and platelets correlated to the coronary artery lesion and disease progress in Kawasaki disease. *Pediatrics, 111*(2), E140–E147.

Wang, C. L., Wu, Y. T., Liu, C. A., Kuo, H. C., & Yang, K. D. (2005). Kawasaki disease: Infection, immunity and genetics. *The Pediatric Infectious Disease Journal, 24*(11), 998–1004.

Kinins

Julio Scharfstein
Immunobiology, Institute of Biophysics Carlos Chagas Filho, Federal University of Rio de Janeiro, Rio de Janeiro, Brazil
Center of Health Sciences (CCS), Cidade Universitária, Rio de Janeiro, Brazil

Definitions

Kinins. The term "kinins" is a general designation for the collection of peptides that are structurally related to bradykinin (BK), a hypotensive and proinflammatory nonapeptide (NH2-RPPGFSPFR-COOH) discovered by Rocha e Silva et al. (1949). Generation of BK or kallidin (lysyl-BK; LBK) requires the proteolytic processing of high- or low-molecular-weight kininogen (HK/LK) by specialized serine proteases (kallikreins). While plasma kallikrein (PK) releases BK from circulating HK, tissue kallikrein (KLK1) predominantly releases vasoactive kinins from LK. Once liberated from kininogens, intact kinins and their metabolites exert their paracrine signaling functions by triggering two distinct subtypes of seven-transmembrane G-protein-coupled bradykinin receptors, designated as B1R and B2R. The biological responses resulting from activation of the kallikrein-kinin system (KKS) are regulated by overlapping mechanisms, such as PKa inactivation by the serpin C1 inhibitor and enzymatic degradation/processing of BK/LBK by specialized metallopeptidases. Although the KKS was traditionally viewed as proteolytic network that modulates vascular homeostasis, inflammation, and pain sensations via activation of BRs, recent developments in this field have linked activation of the procoagulative contact system to innate immunity and thrombogenesis. More than a century after the description of kallikrein as the pancreatic activity that generated a hypotensive substance in human urine (Abelous and Bardier 1909), the KKS is now perceived as a proteolytic system implicated in a broad range of inflammatory diseases (reviewed by Bader 2011).

Biosynthesis and Release

The Kininogens

Synthesized by alternative splicing of a gene (KNG) located in human chromosome 3, HK and LK are multidomain glycoproteins synthesized by alternative splicing of RNA. Out of the six domains present in HK, only four of these (D1–D4, N-terminally located), including the BK sequence (D4), are shared by LK (Fig. 1). The biosynthesis of the kininogens by hepatocytes is stimulated by estrogens and involves transactivation of the farnesoid receptor (reviewed by Alhenc-Gelas and Girolami 2012). Unlike humans, mice have two kininogen genes, both of which coding for precursors that release bioactive kinins upon cleavage by kallikreins (Cardoso et al. 2004). Rats have conserved the classical kininogen gene, but additionally, they also produce a LK-like glycoprotein (T-kininogen) that is refractory to cleavage by kallikreins. Noteworthy, the substrate specificity of kallikreins from human, mouse, and rats is not identical, reflecting peculiarities in the amino acid sequences flanking the kinin moiety of kininogens from each species (Fogaça et al. 2004).

After the primary sequence of HK/LK was elucidated (Kellermann et al. 1986), the domain organization of these glycoproteins was systematically dissected with the help of domain-specific antibodies (Kauffman et al. 1993). Starting from the N-terminus, HK and LK share three cystatin-like domains (D1/D2/D3). Two of these N-terminal domains (D2/D3) display cysteine protease inhibitory function, while D3 contains a noncontiguous docking site for the endothelium (Herwald et al. 1995; Herwald et al. 1996a; Joseph et al. 1996) followed by the BK sequence (D4) (Fig. 1). At the C-terminal end, HK has two extra domains: D5H, a negatively charged (histidine-rich) sequence that (i) promotes the binding of HK to endothelial cell surfaces via interaction to heparan sulfate (Renné et al. 2000) and (ii) contains overlapping endothelial binding sites for gC1qR, cytokeratin 1, and urokinase plasminogen activator (reviewed by Kaplan and Joseph 2014). At the C-terminus, HK has a domain (D6H) that contains binding sites for the apple domains of the contact factors prekallikrein and FXI (Fig. 1). Although HK is essential nonenzymatic cofactor of the contact pathway of

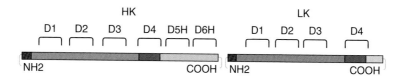

- D1: cystatin - like domain (inactive)
- D2: cystatin - like domain (cysteine protease inhibitor)
- D3: cystatin - like domain (cysteine protease inhibitor; surface binding site)
- D4: bradykinin (hypotensive, inducer of vascula permeability, proangiogenic)
- D5H: histidine-rich domine (surface binding site, anti-angiogenic in Hka)
- D6H: docking sites for contact phase zymogens (prekallikrein and FXI)

Kinins, Fig. 1 Multidomain structure of kininogens. Due to differential splicing of the kininogen transcript, HK has two C-terminal domains that are absent in LK. The specific domains of HK (*green*) are the following: D5H, a segment that includes a negatively charged (histidine-rich) sequence that promotes HK tethering to the surface of mammalian cells via interaction with sulfated glycosaminoglycans. In addition, D5H promotes HK binding to endothelial surfaces through the binding of gC1qR, cytokeratin 1, and urokinase plasminogen activator. Located at the C-terminal end of HK, D6H binds to the apple domains of the contact factors prekallikrein and FXI. At the N-terminal side, HK and LK share three cystatin-like domains (*D1*, *D2*, and *D3*, *blue*) and the nonapeptide bradykinin (*D4*, *pink*), flanked by amino acid sequences that are sensitive to cleavage by kallikreins

coagulation, bleeding is not increased in individuals with kininogen deficiency nor in mice genetically deficient of the circulating forms of kininogens (Merkulov et al. 2008). Interestingly, ischemic brain injury is attenuated in mice deficient in kininogen-1, the protection of the blood barrier being ascribed to reduced thrombosis, inflammation, and neurodegeneration (Langhauser et al. 2012). Prior to this finding, it was reported that experimentally induced thrombosis is attenuated in transgenic mice deficient of FXII and FXI (Renné 2012), as discussed later in this chapter.

KLK1 is constitutively activated in extravascular tissues as well as in vascular cells. The vasodilating function of kinins depends at least in part on KLK1-driven generation of LBK (kallidin) (Regoli 2015). Although KLK1 releases LBK from both LK and HK, the former is thought to be the main substrate of KLK1 because its plasma concentration is 3X fold higher than HK. A large body of studies suggests that tissue kallikrein (KLK1)-mediated generation of LBK promotes vascular remodeling via activation of B2R (Ju et al. 2000; Plendl et al. 2000; reviewed by Madeddu and Kränkel 2011). Conversely, in vitro studies suggest that HKa (i.e., the two-chained cleaved form of HK – devoid of the internal kinin moiety) exerts anti-angiogenic activity by inhibiting endothelial cell migration/proliferation and apoptosis (Colman et al. 2000; Sun and McCrae 2006). Molecular studies linked the anti-angiogenic function of HKa to DH5-mediated inhibition of urokinase plasminogen binding to its receptor (uPAR), a GPI-linked surface protein that otherwise drives the proangiogenic β1-integrin/VEGFR2 pathway (Larusch et al. 2013).

Exploring the role of the KKS in immunity, Yang et al. (2014) have recently reported that HK/HKa opsonizes apoptotic cells by binding to phosphatidyl serine ("eat me signal"). The evidence that HK/HKa triggers efferocytosis (phagocytic uptake of apoptotic cells) via the uPAR/RAC1 pathway suggests that HK may limit inflammation and regulate innate immunity in tissues exposed to apoptotic cells. A contrasting picture emerged from studies of the role of kininogens in microbial immunity. According to Frick et al. (2006), the proteolytic processing of kininogens may liberate an antibacterial peptide of 26 amino acids (NAT-26) from domain D3. Noteworthy in this context, phylogenetic studies suggested that kininogen genes underwent evolutionary adaptation in mammals, with sequence conservation being pronounced in the flanking sequences of NAT26 and of the kinin moiety (Cagliani et al. 2013). Another example of an immune function exerted by kininogen emerged from studies in mice infected systemically with Trypanosoma cruzi, the intracellular parasitic protozoan that causes human Chagas disease. Monteiro et al. (2007) demonstrated that kinins released in peripheral/lymphoid tissues activated endritic cells (DCs) via the B2R pathway, converting these specialized antigen-presenting cells into drivers of IL-12-dependent TH1 polarization (Scharfstein et al. 2007).

Plasma Kallikrein

Codified by a gene (KLB1) localized in chromosome 4, plasma prekallikrein (PK) is the zymogen of the major kinin-forming enzyme circulating in the blood. Synthesized in liver hepatocytes as an inactive polypeptide precursor of 619 amino acids, PK displays structural homology to the Factor XI, a serine protease that initiates fibrin formation via activation of the intrinsic pathway of coagulation (Kaplan and Josephs 2014). As mentioned in the previous section, PK and HK form stable molecular complexes in the bloodstream (Fig. 2). Activation of the contact system in the plasma is triggered by a diverse group of exogenous (microbial) or endogenous substances (reviewed by Renné 2012). Among the latter group, negatively charged polymers stand out as examples of contact activators that are released from intracellular stores of activated circulating cells, such as platelet-derived polyphosphates (Müller et al. 2009), RNA (Kannemeier et al. 2007) and nucleosomes extruded from activated neutrophils (extracellular traps/NETs) (Oehmcke et al. 2009) (Fig. 2). After interacting with contact activators, FXII (Hageman Factor) undergoes autocatalytic cleavage, forming

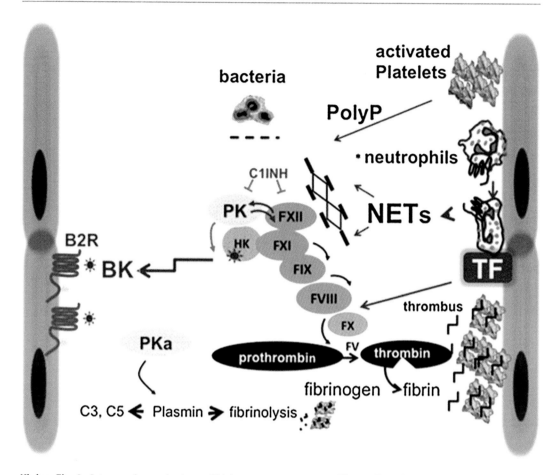

Kinins, Fig. 2 Intravascular mechanisms of kinin generation. The scheme illustrates the cascade type of enzymatic reactions (contact system) that couples fibrin deposition on the nascent thrombi (intrinsic pathway of coagulation) to the intravascular generation of BK. Activation of the contact system is initiated when FXII (Hageman factor) circulating in the bloodstream interacts with "contact" surfaces of endogenous or exogenous origin. Included in the former group are negatively charged polymers such as polyphosphates released from activated platelets or DNA from neutrophil extracellular traps (*NETs*). In the scheme (*top/right side of the panel*), the release of polyphosphates or DNA/NETs by activated platelets/neutrophils is preceded by tissue factor-dependent triggering of the extrinsic pathway of coagulation (*top/right size of panel*). Alternatively, some pathogens display docking sites for the contact phase zymogens (*FXII/PK/FXI*) and the cofactor HK. Formation of trace levels of FXIIa in the bloodstream is sufficient to convert plasma prekallikrein (*PK*) into PKa. This serine protease in turn reciprocally cleaves FXII zymogens. The cyclic activation of FXII/PK leads to a rapid rise in the blood levels of PKa/FXIIa, eventually surpassing the inhibitory effects of the serpin C1INH. Further downstream, the expansion of the proteolytic cascade has different functional outcomes. On one hand, FXIIa generates FXIa, an enzyme that in turn converts FIX into FXIa. This serine protease (*FIXa*) is the effector of the intrinsic pathway of coagulation (*right side of panel*). Simultaneously, PKa promotes the excision of BK from an internal moiety of HK (*left side of panel*). Acting as a paracrine signal, the short-lived BK activates B2R, a GPCR subtype that is constitutively expressed by endothelial cells and by other cell types, such as smooth muscle cells, sensory fibers, and/or immune sentinel cells (not represented in the scheme). Beyond inducing plasma leakage via release of BK, PKa activates the fibrinolytic pathway via generation of plasmin (*bottom of panel*), a serine protease that activates complement C3 and C5, thus linking the KKS network to the complement system

FXIIa. The trace levels of FXIIa are sufficient to convert prekallikrein (complexed to HK) into PKa, which in turn reciprocally activates FXII, generating increased levels of FXIIa (Fig. 2).

The amplification forged by this feedback cycle culminates with the cleavage of FXI by surface-bound FXIIa (reviewed by Kaplan and Josephs 2014). Further downstream, FXIa generates

FIXa, the effector of the intrinsic pathway of coagulation (Fig. 2). During the expansion of this proteolytic cascade, PKa promotes inflammation by liberating the vasoactive BK from circulating forms of HK. In parallel, PKa generates plasmin, an effector of fibrinolysis. Beyond promoting degradation of fibrins, plasmin links the KKS to the complement cascade via cleavage of C3 and C5 (Amara et al. 2010) (Fig. 2). Noteworthy, the serpin C1 inhibitor (C1INH) regulates the KKS (and the classical pathway of the complement system) by inactivating FXIIa and PKa (Fig. 2), in addition to C1 esterases. Individuals with genetic deficiency of C1INH develop hereditary angioedema, a disease characterized by frequent episodes of edema caused by excessive formation of BK by PKa (Cugno et al. 2009).

While it is well accepted that FXIIa is a key enzyme involved in PKa formation under pathophysiological states, there is evidence that prekallikrein displayed at the surface of endothelial cells might be activated by FXII-independent pathways. For example, Shariat-Madar et al. (2002) suggested that this pericellular function might be exerted by prolylcarboxypeptidase, a lysosomal enzyme secreted by endothelial cells. Alternatively, interactions of prekallikrein with the heat shock protein HSP-90 (in the presence of HK) are thought to generate PKa by autocatalysis (reviewed by Kaplan and Joseph 2014). Of further interest, formation of the ternary contact phase enzyme system is not limited to surface of endothelial cells; there is now awareness that kinin-forming proteases are also assembled at the surface of platelets (Gustafson et al. 1986), human neutrophils (Henderson et al. 1994), and monocytes/macrophages (Barbasz and Kozik 2009). Evidence of mast cell-dependent triggering of BK-induced lung inflammation was recently reported in mice (Oschatz et al. 2011). Accordingly, mast cell degranulation releases heparin, a negatively charged polymer that activates FXII/PK, generation of BK extravascularly. In a subsequent work, Moreno-Sanchez et al. (2012) reported that mast cell degranulation also releases polyphosphates, a well-characterized activator of contact phase enzymes.

Tissue Kallikrein
The main kininogenases acting in extravascular tissues is KLK1, a member of a family of 15 homologous serine proteases whose genes (KRP) are clustered in a single locus on human chromosome 19q13.2–13.4. Synthesized as a zymogen (prekallikrein) of 262 amino acids, KLK1 is rendered catalytically active following removal of a 17 amino acid pro-fragment. Active forms of KLK1 are readily detected in exocrine glandular cells, absorptive epithelial cells, vascular cells, and peripheral and central nervous system neuronal cells (reviewed by Bhoola et al. 1992). In epithelial cells, KLK1 is trafficked to the apical side before being secreted in urine, saliva, and pancreatic juice. In renal tissues, KLK1 is secreted on the basolateral region of kidney cells, this being the reason why the protease is found in interstitial fluid and human plasma. Whether secreted in trace levels in the bloodstream or interstitial spaces, the active KLK1 protease is inactivated by kallistatin, a serpin inhibitor (kallistatin) produced by vascular cells. Endowed with an extended active site pocket, KLK1 is so far the only member of the KRP family that accommodates the kininogen substrate (Brillard-Bourdet et al. 1995). Unlike PKa, which exclusively release BK from HK, KLK1 preferentially cleaves LK (present at threefold higher concentrations as HK in the plasma) – generating LBK (kallidin). Apart from modulating vascular tonus and inducing microvascular leakage via release of LBK, KLK1 promotes B2R-dependent neovascularization by upregulating the expression of VEGF by stromal fibroblasts (reviewed by Madeddu and Kränkel 2011). Of further interest, there is evidence that classical kallikreins (KLK1 and PK) secreted in the tumor environment increase angiogenesis and tissue invasiveness of cancer cells through the activation of matrix metalloproteases (MMPs) (Bhoola et al. 2001).

Proteolytic release of kinins in inflammatory exudates. During inflammation, oxidized kininogens are preferentially processed by neutrophil elastase and mast cell tryptase (Kozik et al. 1998). Biochemical studies revealed that oxidation of a methionine residue located upstream of the N-terminal flanking site of the

Kinins, Table 1 Regulation of kinin receptor signaling

Target molecule	Regulator	Functional outcome
FXIIa/PKa	C1 inhibitor	Reduced formation of bradykinin
BK/KD	ACE/kininase II	Down-modulation of B2R signaling
BK/KD	NEP	Down-modulation of B2R signaling
BK/KD	CPN (s-kininase I [a])	Generation of des-Arg-kinins (B1R agonists)
BK/KD	CPM (m-kininase I [a])	Generation of des-Arg-kinins (B1R agonists)
B2R	GRK4α/β-arrestin	Desensitization/endocytosis
B1R	GASP	Constitutive lysosomal degradation
B1R	Des-Arg-kinins	Agonist-induced stabilization of surface B1R expression

[a] s-kininase I: (soluble), m-kininase I: membrane-bound

internal kinin moiety of kininogens inhibits tissue kallikrein-mediated cleavage, without interfering with kinin release by the mast cell tryptase/neutrophil elastase pathway. More recently, studies with patients with vasculitis indicated that neutrophil proteinase 3 (PR3) generates an extended kinin peptide (PR3-kinin) containing two extra amino acids on both sides of the kinin moiety (Met-Lys-BK-Ser-Ser) (Kahn et al. 2009).

Kinin-degrading enzymes. The biological activity of kinins is modulated by the action of a small set of metallopeptidases, generically referred to as "kininases." Clinical and experimental studies indicate that the levels of kinin agonists that trigger each subtype of G-protein-coupled kinin receptors (B1R or B2R) depend on the extent of kinin degradation by ACE/kininase II or levels of [des-Arg]-kinins generated by kininase I (Erdös 1990) (Table 1).

Kininase I

This terminology is used to identify soluble versus cell surface-associated metallopeptidases that generate the B1R agonists, i.e., des-Arg^9-BK (DABK) or des-Arg^{10}-LBK (DALBK) by removing the C-terminal arginine residue from intact kinins (BK or LBK). In the plasma, this task is executed by carboxypeptidase N (CPN), the soluble metallopeptidase that inactivates complement anaphylatoxins C5a and C3a. CPN is a tetramer consisting of two identical catalytic and regulatory subunits, respectively, coded by sequences localized in chromosome 10 and 3. In addition to the processing role of CPN in the fluid phase, B1R agonists are efficiently generated on cell surfaces by CPM, a GPI-anchored carboxypeptidase (Skidgel et al. 1989) (Table 1). In a follow-up study, Zhang et al. (2008) reported that CPM physically associates to the B1R within caveolae-related lipid rafts, hence providing an efficient way to redirect the signaling activity of kinins in inflamed tissues from B2R (constitutive) to the B1R (inducible) pathway.

Kininase II

Traditionally referred as ACE, kininase II is a transmembrane dipeptidyl carboxypeptidase (Erdös 1990) that promotes vasoconstriction by inactivating intact kinins (Table 1) while converting the inactive angiotensin I precursor into the vasopressor angiotensin II (Table 1). Located on human chromosome 17q in human, the ACE gene arose from a duplication of an ancestral gene. Folded as a single polypeptide chain (1,306 amino acids), the homologous domains of ACE code for two active sites, both exposed in the lumen of vascular endothelium (Soubrier et al. 1988). ACE destroys the ligand-binding function of intact kinins (BK or LBK) by sequentially removing two dipeptides from their C-terminus, generating fragments that are no longer able to function as B2R agonists. Differences in the plasma levels of ACE were associated with genetic susceptibility to cardiovascular and other diseases (Cousterousse et al. 1993). In patients that are treated with ACE inhibitors, additional enzymes are involved in kinin metabolism (Table 1): neutral endopeptidase 24.11 (neprilysin, NEP), aminopeptidase P (which inactivates both intact kinins and des-Arg^9-kinins), and kininase I (CPM/CPN). ACE is constitutively expressed by endothelial and epithelial cells, neurons, and lymphocytes, but its expression is upregulated in vascular smooth muscle cells under various pathological conditions, e.g., (i) hypertension (ii) cardiomyocytes during hemodynamic and postischemic remodeling, and (iii) macrophage differentiation (Metzger et al. 2011). Since ACE expression is upregulated in the endothelium of the lung, the bulk of kinins released in the bloodstream are destroyed during circulation in these microvascular beds. In the kidney, NEP has a prominent role in kinin degradation in the kidney, compensating for the downregulation of ACE, required to spare the sensitive kidney tubules from angiotensin II-mediated vasoconstriction (Metzger et al. 2011).

Biological Activities of Kinins

Kinin Receptors

As previously stated, kinins exert their pleiotropic functions by activating two distinct subtypes of seven-transmembrane G-protein-coupled receptors, B2R and B1R (Leeb-Lundberg et al. 2005). Expressed constitutively by a broad range of cell types, including vascular endothelial cells, nociceptive fibers, and innate sentinel cells (macrophages and dendritic cells), B2R is activated by the intact kinin ligands (BK or LBK) excised from kininogens. In contrast, B1R is activated by kinin metabolites generated by the proteolytic removal of the C-terminal arginine residue from intact kinins, such as DAB-K. Mapped to human chromosome 14q32 and rat, the genes coding for B1R (BDKRB1) and B2R (BDKRB2) consist of 353 and 364 amino acids, respectively, showing 36 % of homology in the protein sequence (reviewed by Leeb-Lundberg et al. 2005). Although B1R is hardly detectable in most tissues under steady-state conditions, its transcription is upregulated following NF-κB activation by proinflammatory cytokines (IL-1β, TNF-α, and IFN-γ) (McLean et al. 2000a; Schanstra et al. 1998; Sabourin et al. 2002; Medeiros et al. 2004). A broad range of cells exposed to proinflammatory cues upregulate B1R, including macrophages, neutrophils, mast cells, sensory C-fibers, epithelial cells, fibroblasts, and smooth muscle and cardiac cells (Leeb-Lundberg et al. 2005). Studies with human fibroblasts provided evidences of homologous transactivation of B1R/NF-κB during tissue repair: first, B1R agonists activate NF-κB, which in turn stimulates endogenous synthesis of IL-1β and/or TNF-α. Upon secretion, these proinflammatory cytokines further drive activation of NF-κB, which in turn upregulates B1R transcription in feedback manner (Schanstra et al. 1998). Another distinction between B2R and B1R concerns their sensitivity to agonist-induced desensitization (Table 1). Mammalian cells transfected with individual G-protein-coupled receptor kininases (GRK) revealed that GRK4α drastically increased the level of serine phosphorylation at the C-terminal tail of the constitutively expressed B2R (Table 1). In contrast, the inducible B1R is not efficiently phosphorylated nor desensitized because it lacks the conserved serine and threonines that are present at the C-terminal tail of B2R (Blaukat et al. 2001; Leeb-Lundberg et al. 2001). Although B1R (in the absence of agonist binding) undergoes clathrin-mediated endocytosis and GAP-mediated lysosomal degradation, these events are inhibited upon ligation of DABK, which stabilizes B1R expression at the plasma membrane of activated endothelium (Enquist et al. 2007) (Table 1).

Vascular responses transmitted by B2R are primarily coupled to $G_{\alpha q}$, or alternatively to $G_{\alpha i}$, $G_{\alpha s}$, and $G_{\alpha 12/13}$ (Leeb-Lundberg et al. 2005). BK activates endothelial B2R sequestered into caveolae where it associates with NO synthase (eNOS) (Haasemann et al. 1998). Upon ligand binding, eNOS is released from B2R and reaches the endothelial cytoplasm, from where it produces NO through pathways that depend on calcium mobilization and PI3K/AKT-dependent phosphorylation (Kuhr et al. 2010; reviewed by Leeb-Lundberg et al. 2005). Noteworthy, BK induces the release of tissue plasminogen activator from human vasculature via B2R through signal transduction pathways that are independent of NOS or activation of cyclooxygenase (Brown et al. 2000). As mentioned earlier in this chapter, the proangiogenic kinins (B2R agonists) stimulate cardiac neovascularization by upregulating endothelial NO production via transactivation of the

Kinins, Table 2 KKS roles in inflammation and immunity

Biological response	Key effector molecules	Cellular target and receptor pathway
Thrombus stabilization	FXII/PKa/HK	FXI>>thrombin>fibrin
Microvascular leakage and inflammation	Kinin peptides	Endothelium/B2R/B1R
Nitric oxide-dependent vasodilation	Kinin peptides	Endothelium/B2R/B1R
Induction of angiogenesis	Kinin peptides	Endothelial cells/B2R/VEGF
Inhibition of angiogenesis	HKa (D5H)	Endothelium/uPAR
Promotion of fibrinolysis	Kinin peptides	Endothelial cells/stromal cells/B2R
Nociception	Kinin peptides	C-type sensitive fibers/B2R/B1R
Upregulation of ICAM, E-selectin	Kinin peptides	Endothelium/B2R/B1R
ECM degradation via MMP9	Kinin peptides	Neutrophils/B1R
Phagocytosis and microbicidal activity	Kinin peptides	Macrophages/B2R/B1R
Generation of antimicrobial peptides	HK/NAT-26	Bacteria
Immune regulation (efferocytosis)	HK/HKa peptides	Apoptotic cells/macrophages/uPAR
Link between innate/adaptive immunity	Kinin peptides	Dendritic cells/B2R
Inhibition of T cell migration (BBB)	Kinin peptides	Endothelium/uPAR
Stimulus of TH1 cell infiltration (BBB)	Kinin peptides	Effector CD4 TH1 cells/B1R
Neovascularization/cardioprotection	Kinin peptides	Cardiovascular cells/monocytes/B2R
Reduced formation of atheroma	Kinin peptides	Endothelium/B1R
Increased chronic inflammation/fibrosis	Kinin peptides	Renal tissues/heart/B1R
Parasite infectivity		Cardiovascular cells/B2R/B1R

vascular endothelial growth factor (VEGDF) receptor KDR/Flk-1 pathway (Thuringer et al. 2002) (Table 2). Independent studies performed by other workers (Plendl et al. 2000; Smith et al. 2008) linked the KLK1- and B2R-driven angiogenesis to the caveolae-associated signaling of Janus-activated kininase-signal transducers of transcription activation (JAK/STAT pathway) (Ju et al. 2000). In vivo studies showed that the infusion of both KLK1 and kinin induced neovascularization in models of limb ischemia (Smith et al. 2008). This, notwithstanding, the concept that KL1 promotes angiogenesis by releasing B2R agonist was challenged by studies showing that B2R is directly activated by tissue kallikrein (Hecquet et al. 2000; Chao et al. 2008). Of further interest, it has been recently reported that B2R expression in circulating progenitor cells is critical for recruitment and induction of proangiogenic endothelial function in injured tissues (Kränkel et al. 2008, 2013).

Regarding the role of kinins in the pathogenesis of pain, B2R and B1R are expressed in most nociceptors and in several central structures related to pain transduction, including the spinal cord and the cerebral cortex (reviewed by Calixto et al. 2011). Both subtypes of BRs transmit excitatory responses through indirect interactions with TRVP1 and TRPA1. Acting on small-diameter afferent nociceptive fibers, kinin peptides promote nociceptive sensitization through the release of prostanoids and sympathetic amines via activation of phospholipase-C (PLC)-diacylglycerol (DAG)/protein kinase C (PKC) pathway (Ferreira et al. 2008). Using mice models of mechanical and thermal hyperalgesia, Pesquero et al. (2000) found that B1R-deficient mice (unlike wild-type littermates) do not upregulate the production of proinflammatory mediators that are required for nociceptor activation. Further exploring the differential roles of B2R and B1R in mechanical hyperalgesia, Cunha et al. (2007) verified that BK (B2R agonist) induces pain sensations in naïve mice through cytokine-independent release of prostanoids and sympathetic amines. Conversely, B1R-driven hyperalgesia in LPS-treated mice depends on TNF-α-/IL-1-β-dependent release of prostanoids and sympathetic amines. Extending the breadth of these studies to models of visceral hyperalgesia, Kawabata et al. (2006) showed that HOE-140 (B2R antagonist) inhibited visceral pain induced by a selective PAR 2 agonist and provided evidences that activation of TRPA1 channels of intestinal afferents depends on the functional interplay between PAR2, B2R, and TRPA1.

Whether inflicted by trauma or infection, the induction of tissue injury frequently impairs the integrity of the endothelium barrier, enabling the influx of plasma proteins and blood-borne nutrients into extravascular tissues. Recent studies showed that BK dismantles the junctional complexes formed between VE-cadherin-catenin by inducing Src-dependent phosphorylation and endothelial internalization of cadherin via B2R (Orsenigo et al. 2012) (Table 2), reminiscent of the effects that shear stress cause in the venular endothelium. Campos and Calixto (1995) examined the roles of B2R and B1R in the dynamics of inflammatory edema by injecting BK or B1R agonist (des-Arg ^{9}-BK, heretofore abbreviated as DABK) in naïve or inflamed animals. Although BK induced potent swelling responses in normal mice, DABK failed to evoke detectable edema. However, animals that were repeatedly injected with BK over a period of 24 h became sensitive to DABK, despite the fact that they were desensitized to the B2R agonist. Collectively, these results suggested that B1R is upregulated in inflamed tissues subjected to repetitive stimulation of the kinin/B2R pathway. Using the air pouch as a model to study the proinflammatory responses orchestrated by B1R, Ahluwalia and Perreti (1996) linked leukocyte transmigration to IL-1β-dependent induction of B1R. Whether using B1R blockers (Vianna and Calixto 1998) or B1R knockout mice (Pesquero et al. 2000), studies in mice models of lung inflammation linked neutrophil infiltration to the activation of B1R. Exploring the roles of B1R in the induction of proinflammatory chemokines, Duchene et al. (2007) linked these pathways to upregulated endothelial expression of the chemokine CXCL5. In a study focusing on human neutrophils, Ehrenfeld et al. (2009) demonstrated that B1R (i) upregulated chemotaxis and surface expression of LFA-1 and Mac-1 integrins and (ii) promoted ERK1/2 and p38 MAPK-dependent release of matrix metalloproteinase 9 (MMP9) and myeloperoxidase (MPO) (Table 2).

Studies of dendritic cell (DC) maturation in the TH2-dependent BALB/c model of allergic lung inflammation induced by alum/ovalbumin immunization provided the initial evidences that BK, acting as an endogenous danger signal, induces IL-12-dependent TH1 responses via the kinin/B2R pathway (Aliberti et al. 2003; reviewed by Scharfstein and Svensjö 2011). Extending this investigation to the context of parasitic diseases, Monteiro et al. (2007) demonstrated that type 1 immunity was impaired in B2R$^{-/-}$ mice challenged (systemically) with the parasitic protozoan T. cruzi. Analysis of the susceptible phenotype of B2R$^{-/-}$-infected mice revealed that (i) splenic dendritic cells (DCs) were refractory to activation (IL-12 and costimulatory molecules) by the kinin-releasing parasites and that (ii) the defective B2R$^{-/-}$ DC function had profound impact on adaptive immunity, reducing the frequencies of IFN-γ-producing (immunoprotective) effector CD4$^+$ and CD8$^+$ T cells in the spleen while reciprocally increasing the frequency of antiparasite TH17 responses (Monteiro et al. 2007). A second example of immune dysfunction in parasitic diseases came from the studies in B2R-deficient mice infected by Leishmania chagasi, the etiologic agent of visceral leishmaniosis. Reminiscent of the phenotype observed in T. cruzi infection, B2R-deficient mice showed increased parasite burden (liver) and hepatosplenic pathology associated with decreased type 1 immunity (Nico et al. 2012). In another report, B2R$^{-/-}$ mice were increasingly susceptible to infection by the intracellular bacterium Listeria monocytogenes, albeit the role of effector CD8$^+$ T cells was not explored in this particular work (Kaman et al. 2009). It is still unclear whether B2R agonists are able to reprogram the highly heterogeneous human DCs. In a study involving monocyte-derived human DCs (moDCs), Gulliver et al. (2011) found that the expression of B1R is higher in activated moDCs from asthma patients as compared to asthma-free subjects. Additional studies are required to determine whether endogenously released DABK may recruit moDCs to the inflamed lung tissues of asthma patients.

Although the KKS seems to provide a bridge between innate and adaptive immunity, it is still unclear whether B1R signaling may modulate T cell effector-memory development during the course of infection. The dissection of the role of B1R in immunity is complicated by the fact that this GPCR is upregulated by a broad range of tissue-resident and migratory innate sentinel cells (Table 2). For example, there is evidence that activation of B1R-expressing sensory C-fibers and mast cells leads to the release of substance P and histamine. After diffusing through the subendothelial matrix, substance P and histamine prolong inflammation by, respectively, (i) stimulating leukocyte transmigration via signaling of NK1/NK3 neurokinin receptors and/or (ii) destabilizing the endothelial barrier through the signaling of H1-histamine receptors (McLean et al. 2000b; Leeb-Lundberg et al. 2005). As already mentioned (see section on Plasma Kallikrein), studies in models of allergic inflammation demonstrated that histamine-induced microvascular leakage may enable BK release by heparin/polyphosphates released from activated mast cells (Oschatz et al. 2011; Moreno-Sanchez et al. 2012).

Pathophysiological Relevance

Hereditary Angioedema

Inherited as an autosomal dominant trait, HAE is characterized as recurrent localized edema of the skin or mucosal tissues (gastrointestinal tract or larynx) in individuals heterozygous for C1 inhibitor (C1INH) deficiency. Type I HAE is due to mutations that reduce the plasma levels of C1INH, whereas in 15 % of the patients (type-II HAE), the mutations result in dysfunctional C1INH protein (Cugno et al. 2009). Originally characterized as a selective inhibitor of C1 (C1r and C1s) of the complement system, C1INH also targets the contact factors FXIIa and PKa (see previous section), thrombin, tissue plasminogen activator, and plasmin. Studies in the early 1960s showed that C1INH deficiency resulted in defective control of kallikrein activity in HAE (Cugno et al. 2009). Involvement of BK in HAE was also supported by the evidence that BK and cleaved HK accumulated in the plasma after attacks and by the evidence that B2R blockade averted microvascular leakage in C1INH-knockout mice (Han et al. 2012). Recently, auf dem Keller et al. (2013) shed new light into the mechanism governing the inhibitory activity of C1INH during the progression of inflammatory edema in the mouse skin. Proteomic analysis (N-terminome) of edematous tissues of wild-type versus $MMP2^{-/-}$ mice revealed that C1INH is cleaved/inactivated by MMP2. Additional data presented in this study suggested that the transient drop in the extravascular levels of C1INH may upregulate PKa-dependent liberation of proinflammatory kinins in peripheral tissues (auf dem Keller et al. 2013). It remains to be determined whether MMP2-dependent inactivation of C1INH amplifies BK-induced edema in inflammatory human diseases.

Cardiovascular Diseases

Early studies in the 1960s showed that kinins applied locally or systemically increased coronary flow and improved myocardial metabolism, where intracoronary application of the B2R antagonist icatibant (HOE140) had the opposite effect. Towards the mid-1990s, progress in this field was accelerated by the development of animal models in transgenic mice deficient in B2R (Borkowski et al. 1995) or B1R (reviewed by Sales et al. 2011). Studies in the $B2R^{-/-}$ transgenic line (J129 strain) revealed that cardiac dilatation worsened with aging, the heart dysfunction being attributed to upregulated angiotensin II signaling in the absence of B2R (Emanueli et al. 1999; reviewed by Madeddu and Krankel 2011). Parallel investigations in other models of heart inflammatory diseases supported the concept that KLK1, acting via B2R, counterbalances the adverse effects of profibrotic peptides (Griol-Charhbili et al. 2005), e.g., angiotensin II, aldo sterone, and endothelin. The beneficial effects of KLK1 were observed in studies in mice models of streptozotocin-induced diabetic cardiopathy, showing that KLK1 (human) overexpression in the heart improved LV end-diastolic pressure (reviewed by Spillmann and Tschöpe 2011). Bergaya et al. (2001) showed evidence that KLK1 is partially responsive for the flow-dependent vasodilation orchestrated via the B2E/NO pathway. Part of the protective effects mediated by KLK1 may result to B2R-driven attenuation of endothelial injury caused by oxidative stress, possibly a response associated with inhibition of cytokine secretion and leukocyte transmigration. It is also known that KLK1 degrades collagen and induces angiogenesis by upregulating VEGF in stromal fibroblasts via activation of the B2R pathway (reviewed by Spillmann and Tschöpe 2011). The finding that the B2R antagonist icatibant reverted the cardioprotective effects associated with KLK1 overexpression linked the antifibrotic action of tissue kallikrein to B2R-mediated activation of the nitric oxide/cyclic GMP-dependent pathway (reviewed by Spillmann and Tschöpe 2011). In another step forward, Kränkel et al. (2008) demonstrated that a subset of circulating $CD34^+CXCR4^+$ mononuclear cells expressing B2R promote the repair of injured arterial walls. Adding weight to this concept, clinical studies have recently showed that B2R expression in the corresponding subset of human mononuclear cells is reduced in patients with more severe coronary arterial disease (Kränkel et al. 2013).

Although the deficiency of functional B2R has worsened cardiovascular function in aging mice (J129 strain) (reviewed by Madeddu and Krankel 2011), there are a few reports showing that absence of B1R may compromise vascular remodeling, aggravating heart disease. For example, it was reported that coronary resistance is worsened in B1R-deficient mice (Lauton-Santos et al. 2007), the adverse phenotype being ascribed to the reduced expression of eNOS (NOS-3) in the aorta. Along similar lines, studies of B1R function in myocardial infarction suggested that the therapeutic effects of AT1 blockade are at least in part due to the functional engagement of B1R (reviewed by Spillmann and Tschöpe 2011). Noteworthy in this context, Morand-Contant et al. (2010) showed evidence that vascular smooth muscle cells activated by angiotensin I and endothelin upregulate expression of B1R, raising the possibility that DABK might exert compensatory vascular responses in some pathological conditions. Another example of the beneficial functions brought about by B1R came from studies of atherosclerosis in Apo $E^{-/-}$ mice subjected to cholesterogenic diet (Merino et al. 2009). These authors reported that the incidence of abdominal aneurisms was higher in mice with double deficiency for apoE and B1R as compared to single $ApoE^{-/-}$ littermates. Highlighting the dual function of the KKS, studies of the pathogenic outcome of streptozotocin (SPZ)-induced diabetic cardiomyopathy revealed that B1R-deficient mice were protected from inflammation, fibrosis, and oxidative stress. Consistent with this favorable outcome, the diabetic $B1R^{-/-}$ mice displayed improved left ventricular functions and systolic/diastolic responses (reviewed by Spillmann and Tschöpe 2011).

Role of the Contact Phase in Thrombosis and Inflammation

As mentioned earlier in this chapter (see section on plasma kallikreins), the risk of hemorrhage is not significantly increased in patients with genetic deficiency of contact factors (reviewed by Renné 2012). In the last decade, research on the contact system was revitalized by the findings that thrombosis is attenuated in mice genetically deficient of Factor XII (Kleinschnitz et al. 2006), FXI (Rosen et al. 2002), or kininogen-1 (Merkulov et al. 2008), without significantly increasing bleeding times (reviewed by Renné 2012). Currently, the contact system is regarded as a proinflammatory proteolytic circuit that, acting cooperatively with the TF-driven extrinsic pathway of coagulation, stabilizes thrombus development through increased deposition of fibrin (Stoll et al. 2008). As previously mentioned, activation of the contact system generates PKa, a serine protease that promptly liberates proinflammatory BK from HK. While arterial thrombosis is associated with the formation of atherosclerotic plaques, venous thrombosis is caused by depressed venous blood flow in the absence of overt endothelial injury. Animal models of venous deep vein thrombosis indicate that neutrophils (thrombus resident) are indispensable for the propagation of fibrin formation through the externalization of neutrophil-derived nucleosomes (von Brühl et al. 2012) (Fig. 2).

KKS: Integrating Immunity and Thrombosis in Bacterial Infection

Acting at the endothelium interface, the contact system is an example of a proteolytic network that reciprocally couples thrombosis/inflammation to innate immunity (reviewed by Engelman and Massberg 2013). Intravital microscopy in models of deep vein thrombosis revealed that the expression of tissue factor (TF) is upregulated by activated neutrophils/monocytes crawling at the endothelium lining. Following TF-mediated triggering of the extrinsic pathway of coagulation, platelets are activated, forming the nascent thrombi, which is further stabilized at the expense of FXII-dependent coagulation (Fig. 2). Engelman and Massberg (2013) have recently proposed that neutrophil-induced thrombosis promotes the entrapment of pathogenic bacteria within fibrin scaffolds that are forged by the sequential activation of the extrinsic (tissue factor dependent) and intrinsic pathway of coagulation. Consistent with this view, transgenic mice lacking neutrophil serine proteases (elastase and cathepsin G) were unable to efficiently retain E. coli within liver microvessels. Mechanistic studies revealed that formation of the fibrin scaffold that retains E. coli intravascularly is crucially dependent on inactivation of tissue factor pathway inhibitor by neutrophil elastase/cathepsin G. Further optimized by FXII-driven (contact system) fibrin formation, the thrombogenic response limits the systemic spread of pathogenic bacteria while favoring immune attack by activated leukocytes (Engelman and Massberg 2013).

Microbial cysteine proteases are determinants of virulence/pathogenicity. Although hypotensive bacteria activates the contact system via LPS or teichoic acid (reviewed by Potempa and Herwald 2011), studies in the late 1990s demonstrated that HK binds to M-proteins from Streptococcus pyogenes and showed evidence that this interaction translates into kinin release (Ben-Nasr et al. 1997). Subsequent studies demonstrated that kininogens release kinins upon processing by cysteine proteases (streptopain) from *S. pyogenes* (Herwald et al. 1996b), whereas *Staphylococcus aureus* exert this task via concerted action between two staphopains (A and B) (reviewed by Potempa and Herwald 2011). Although staphopain B by itself does not generate a vasoactive peptide from kininogens, this cysteine protease is required to generate a longer kinin peptide (Leu-Met-Lys-BK) from kininogen fragments released by staphopain A (Imamura et al. 2005). Porphyromonas gingivalis, a Gram-negative anaerobic bacterium commonly associated with periodontitis, is another example of a pathogen that evokes inflammation by generating kinins via the combined activity of cysteine proteases (gingipains R and K) (reviewed by Potempa and Herwald 2011). Intriguingly, Rapala-Kozik et al. (2011) reported that the hemagglutinin/adhesion domains of gingipains might serve as docking sites for contact phase assembly and activation at bacterial surfaces.

Using a mouse model of intramucosal infection by *Porphyromonas gingivalis*, Monteiro et al. (2009) demonstrated that LPS – an atypical TLR2 agonist expressed by *P. gingivalis* – induces a subtle microvascular leakage by activating neutrophils/endothelium via TLR2. By inducing endothelial leakiness, the activated neutrophils enabled rapid diffusion/accumulation of plasma proteins (including kininogens) into bacteria-laden tissues. Further downstream, gingipains liberate proinflammatory kinins from the kininogens, hence coupling the early-phase innate immunity (TLR2/neutrophil-dependent) to the proteolytic phase (KKS) of the inflammatory cascade (Fig. 3). Immunological studies in mice infected intramucosally by *P. gingivalis* WT and gingipain-deficient mutants revealed that generation of kinins upregulated the formation of Fimbria-specific effector TH1 responses (C57BL/6 mice) or TH1/TH17 responses in submandibular lymph nodes (Monteiro et al. 2009). Recall assays performed with experienced T cells linked TH1 effector development in C57BL/6 mice to the early activation of innate immunity (TLR2/neutrophil>kinin/B2R pathway). In

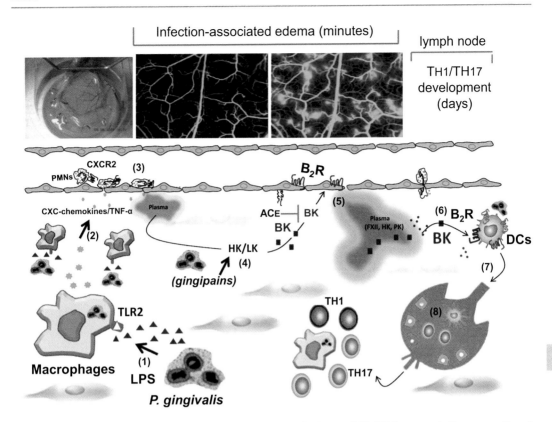

Kinins, Fig. 3 Kinins integrate innate to adaptive immunity: functional interplay between TLR2 agonists and microbial cysteine proteases. The panels illustrate the dynamics of inflammation evoked by the Gram-negative periodontal bacterium *P. gingivalis* (Adapted from Monteiro et al. 2009). *Upper panel*: the intravital microscopy images of the hamster cheek pouch (HCP) (*left panel*) depict the microvascular permeability responses (leakage of dextran-FITC through post capillary venules) induced by the topical application of the periodontal bacterium. *Lower panel*: The scheme depicts the cooperation between LPS (an atypical TLR2 agonist; *left side of panel*) and cysteine proteases (gingipain R and K, center of panel) in *P. gingivalis*-evoked (i) edema (time frame: minutes) and (ii) generation of effector CD4$^+$ T cells in draining lymph nodes (time frame: days). Inflammatory edema is initiated when LPS from *P. gingivalis* activates TLR2-expressing innate sentinel cells (for simplicity, here represented by tissue macrophages). In response to the secretion of chemokines/cytokines (e.g., TNF-α), the endothelium/neutrophils are reciprocally activated. A subtle destabilization of the endothelial barrier allows for the influx of plasma proteins (including the kininogens) into peripheral sites of infection (*top/left side* of panel). Next, kininogens accumulating extravascularly undergo proteolytic processing by the microbial-derived cysteine proteases (gingipain R and K). While not excluding cooperation of host kininogenases (e.g., kallikreins), the short-lived kinins released in peripheral sites of infection fuel the TLR2-/neutrophil-driven inflammatory wave by triggering endothelial B2R. The trans-endothelial leakage is further increased, thus setting in motion iterative cycles of edema and inflammation. Further downstream, immature dendritic cells (DCs) recruited to the inflamed subgengival tissues internalize the periodontal pathogen. Subjected to activation by microbial-derived molecules and/or endogenous danger signals (BK and other proinflammatory mediators), the antigen-bearing DCs undergo functional reprogramming "en route" to the submandibular lymph nodes (*right side* of the panel). Upon entry into T cell-rich areas (*bottom* of panel, *right side*), the DCs present fimbriae-epitopes to virgin T cells. Synaptic contact between activated DCs and antigen-specific T cells promotes development of TH1 effector cells (C57/BL6 mice) or TH1/TH17 effector cells (BALB/c mice). Operationally termed TLR2/B2R "cross talk,'" the transcellular activation pathway forged by LPS and gingipains interconnect the early-phase inflammatory edema (TLR2/B2R pathway) to the cytokine circuitry that links innate to adaptive immunity. Similar mechanistic principles were described in the parasitic infection (Chagas disease) caused by the protozoan T. cruzi

contrast, infection-associated activation of the kinin/B2R pathway led to mixed response (TH1/TH17) in BALB/c mice infected intramucosally (Monteiro et al. 2009).

Similar to the proinflammatory phenotype of *P. gingivalis*, infective forms of *T. cruzi* (trypomastigotes) amplify/propagate inflammation by releasing kinins from plasma-borne kininogens (Monteiro et al. 2006). Intravital microscopy studies in the hamster cheek pouch combined to studies in TLR2-deficient mice revealed that inflammation is initiated by tGPI, a potent lipid anchor (TLR2 ligand) shed by the trypomastigotes (Schmitz et al. 2009). Once activated by KC/MIP-2 (CXC chemokines), neutrophils induce a discrete leak in the endothelial barrier via activation of CXCR2/ETaR/ETbR (Schmitz et al. 2009; Andrade et al. 2012). Further downstream, cruzipain liberates kinins and C5 anaphylatoxin from plasma-borne kininogens and native C5 diffusing through peripheral tissues (Schmitz et al. 2014). Analysis of the TH profile of antiparasite effector T cells generated in the draining lymph nodes revealed that ACE/kininase II limits the functional coupling between innate immunity (TLR2/CXCR2-pathway) and the C5aR/B2R-driven proteolytic cascades (Monteiro et al. 2006; Schmitz et al. 2014). The hypothesis that *T. cruzi* trypomastigotes may reciprocally benefit from the formation of inflammatory edema (Andrade et al. 2012) was motivated by the finding that cruzipain potentiates host cell invasion by liberating kinins from surface-bound kininogens (Del Nery et al. 1997; Lima et al. 2002). Acting as paracrine mediators, the short-lived kinins stimulate $[Ca^{2+}]$-regulated endocytic uptake of the pathogens via cross talk between B2R (Scharfstein et al. 2000) and B1R (Todorov et al. 2003). More recently, Andrade et al. (2012) demonstrated that B2R and endothelin receptors (ETaR/ETbR) assembled in cholesterol-rich lipid domains integrate the multimolecular "gateway" of *T. cruzi* entry in cardiovascular cells. Based on these collective findings, Scharfstein et al. (2013) proposed that cruzipain might fuel intracardiac parasitism by upregulating the proteolytic release of kinins and further downstream in the progression of the inflammatory cascade, kinins and endothelins propel *T. cruzi* invasion via cross talk between B2R, B1R, and ETRs (reviewed by Scharfstein et al. 2013).

KKS in Experimental Autoimmune Encephalitis

The first precedent linking activation of B1R to T cell function came from a study with patients with multiple sclerosis (Prat et al. 1999), showing that exogenous DABK inhibited T cell migration. Extending these studies to animal models of experimental autoimmune encephalitis (EAE), Schulze-Topphoff et al. (2009) showed milder CNS immunopathology in SJL mice treated with a B1R agonist (R838). Although these results suggested that B1R signaling inhibits the transendothelial migration of encephalitogenic TH17 effector cells in the SJL brain, opposite results were observed in EAE models established in the C57BL/6 mouse strain, i.e., CNS pathology and spatial reference memory deficits were attenuated both in B1R-knockout mice and in wild-type mice treated with B1R antagonists (Göbel et al. 2011; Dutra et al. 2011). In a recent study of EAE in C57BL/6 background, Dutra et al. (2013) suggested that B1R contributes to deficits of spatial reference memory that precede motor dysfunction. Additional studies are required to understand how mouse genetic influence the contrasting phenotypes of C57BL/6 mice versus SJL mice in EAE.

Targeting of the KKS

Perspectives for treatment of inflammatory diseases. To this date, HAE remains the best example of an inflammatory disease that can be efficiently treated with drugs that target key components of the KKS (reviewed by Marceau 2011). HAE patients have been successfully treated with C1INH replacement therapy or with two alternative synthetic pharma: the PKa inhibitor ecallantide (DX-88, Kalbitor) or the antagonist of B2R (icatibant). The rationale to treat HAE with inhibitors of PKa was supported by evidences that patients have dysregulated PKa, leading to the release of high levels of BK from circulating HK. Acting as an inhibitor of PKa, ecallantide is a 60-residue peptide engineered by phage display from a different serpin scaffold (human tissue factor pathway inhibitor) to achieve high selectivity and potency for the human plasma kallikrein target. In two clinical trials, subcutaneous injection of ecallantide efficiently aborted attacks in HAE patients (Cicardi et al. 2012). The other drug registered for treatment of HAE in several countries is icatibant (HOE-140, Firazyr), a B2R antagonist that was rendered resistant to proteolytic degradation due to the incorporation of nonnatural amino acid residues (reviewed by Marceau 2011). Administered subcutaneously, icatibant prevented edema attacks in HAE patients (Cicardi et al. 2012) and efficiently prevented acute tissue swelling, resulting from treatment with ACE inhibitors (Stahl et al. 2014). The side effects of icatibant are fairly mild, presumably resulting from low residual agonist activity. These adverse reactions were no longer observed upon treatment with nonpeptide antagonists of B2R, such as Anatibant (L16-0687).

As discussed earlier in this chapter, ischemic stroke destabilizes the blood-brain barrier through the formation of kinins via activation of the contact phase. Owing to the formation of inflammatory edema, secondary damage worsens the CNS pathology. Studies of ischemic stroke in B1R-deficient mice revealed that volume of brain infarct, edema, and inflammation were attenuated in this transgenic line (Austinat et al. 2009). More recently, Raslan et al. (2010) reevaluated the role of the KKS in stroke using a model of brain injury induced by cryogenic cortical trauma and found out that brain lesions were attenuated in the absence of B1R, but not in the absence of B2R function. Efforts to develop therapeutic drugs for B1R can be traced to the 1970s, when Regoli and coworkers introduced a lysine to the N-terminus of the modified B1R agonist, producing a B1R antagonist (Lys-[Leu^8]des-Arg^9-BK) with high affinity for human B1R (Regoli and Barabé 1980). Years later, Gera et al. (2008) incorporated nonnatural amino acids into the peptide backbone of the first generation of B1R antagonists, obtaining a more stable compound (B-9958). Although pharmaceutical companies have met success in developing several B1R antagonists that are suitable for oral administration for treatment of pain, the clinical trials with MK-0686 (brain-penetrant compound) did not meet expectations. Despite the setback in the treatment of pain, the new generations of B1R antagonists (Gobeil et al. 2013) are promising options for therapeutic intervention in myocardial infarction (Cubedo et al. 2013), stroke (Raslan et al. 2010), diabetic myocardiopathy (Westermann et al. 2005), renal fibrosis associated with obstructive nephropathy (Klein et al. 2009), multiple sclerosis (Dutra et al. 2013), inflammatory bowel disease (IBD) (Stadnicki et al. 2005), cancer (da Costa et al. 2014; Dillenburg-Pilla et al. 2013), and chronic degenerative diseases (Bicca et al. 2014; Jong et al. 2002).

References

Abelous, J. E., & Bardier, E. (1909). Les substances hypotensives de l urine humaine normale. *Comptes Rendus Biologies, 66*, 511–520.

Ahluwalia, A., & Perretti, M. (1996). Involvement of bradykinin B1 receptor in the polymorphonuclear leukocyte accumulation induced by IL-1 beta in vivo in the mouse. *Journal of Immunology, 156*, 269–274.

Aliberti, J., Viola, J. P., Vieira-de-Abreu, A., Bozza, P. T., Sher, A., & Scharfstein, J. (2003). Cutting edge: Bradykinin induces IL-12 production by dendritic cells: A danger signal that drives Th1 polarization. *Journal of Immunology, 170*(11), 5349–5353.

Amara, U., Flierl, M. A., Rittirsch, D., Klos, A., Chen, H., Acker, B., et al. (2010). Molecular intercommunication between the complement and coagulation systems. *Journal of Immunology, 185*(9), 5628–5636.

Andrade, D., Serra, R., Svensjö, E., Lima, A. P., Ramos, E. S., Jr., Fortes, F. S., et al. (2012). Trypanosoma cruzi invades host cells through the activation of endothelin and bradykinin receptors: A converging pathway leading to chagasic vasculopathy. *British Journal of Pharmacology, 165*, 1333–1347. http://www.ncbi.nlm.nih.gov/pubmed/21797847?dopt=Abstract.

AufdemKeller, U., Prudova, A., Eckhard, U., Fingleton, B., & Overall, C. M. (2013). Systems-level analysis of proteolytic events in increased vascular permeability and complement activation in skin inflammation. *Science Signaling, 6*(258), rs2. http://www.ncbi.nlm.nih.gov/pubmed/23322905?dopt=Abstract.

Austinat, M., Braeuninger, S., Pesquero, J. B., Brede, M., Bader, M., Stoll, G., et al. (2009). Blockade of bradykinin receptor B1 but not bradykinin receptor B2 provides protection from cerebral infarction and brain edema. *Stroke, 40*(1), 285–293. http://www.ncbi.nlm.nih.gov/pubmed/18988906?dopt=Abstract.

Bader, M. (2011). Kinins: History and outlook. In M. Bader (Ed.), *Kinins* (pp. 1–5). Berlin: Walter de Gruyter GmbH & Co. KG.

Barbasz, A., & Kozik, A. (2009). The assembly and activation of kinin-forming systems on the surface of human U-937 macrophage-like cells. *Biological Chemistry, 390*, 269–275. http://www.ncbi.nlm.nih.gov/pubmed/19090728?dopt=Abstract.

Ben-Nasr, A. B., Olsén, A., Sjobring, U., Muller-Esterl, W., & Bjork, L. (1997). Absorption of kininogen from human plasma by Streptococcus pyogenes is followed by the release of bradykinin. *The Biochemical Journal, 326*, 657–660. http://www.ncbi.nlm.nih.gov/pubmed/9307013?dopt=Abstract.

Bergaya, S., Meneton, P., Bloch-Faure, M., Mathieu, E., Alhenc-Gelas, F., Lévy, B. I., et al. (2001). Decreased flow-dependent dilation in carotid arteries of tissue kallikrein-knockout mice. *Circulation Research, 88*, 593–599. http://www.ncbi.nlm.nih.gov/pubmed/11282893?dopt=Abstract.

Bhoola, K. D., Figueroa, C. D., & Worthy, K. (1992). Bioregulation of kinins: Kallikreins, kininogens, and kininases. *Pharmacological Reviews, 44*(1), 1–80. http://www.ncbi.nlm.nih.gov/pubmed/1313585?dopt=Abstract.

Bhoola, K., Ramsaroop, R., Plendl, J., Cassim, B., Dlamini, Z., & Naicker, S. (2001). Kallikrein and kinin receptor expression in inflammation and cancer. *Biological Chemistry, 382*, 77–89. http://www.ncbi.nlm.nih.gov/pubmed/11258677?dopt=Abstract.

Bicca, M. A., Costa, R., Loch-Neckel, G., Figueiredo, C. P., Medeiros, R., & Calixto, J. B. (2014). B2 receptor blockage prevents Aβ-induced cognitive impairment by neuroinflammation inhibition. *Behavioural Brain Research, 278C*, 482–491.

Blaukat, A., Pizard, A., Breit, A., Wernstedt, C., Alhenc-Gelas, F., Muller-Esterl, W., et al. (2001). Determination of bradykinin B2 receptor in vivo phosphorylation sites and their roles in receptor function. *The Journal of Biological Chemistry, 276*, 40431–40440. http://www.ncbi.nlm.nih.gov/pubmed/11517230?dopt=Abstract.

Borkowski, J. A., Ransom, R. W., Seabrook, G. R., Trumbauer, M., Chen, H., Hill, R. G., et al. (1995). Targeted disruption of a B2 bradykinin receptor gene in mice eliminates bradykinin action in smooth muscle and neurons. *The Journal of Biological Chemistry, 270*(23), 13706–13710. http://www.ncbi.nlm.nih.gov/pubmed/7775424?dopt=Abstract.

Brillard-Bourdet, M., Moreau, T., & Gauthier, F. (1995). Substrate specificity of tissue kallikreins: Importance of an extended interaction site. *Biochimica et Biophysica Acta, 1246*(1), 47–52. http://www.ncbi.nlm.nih.gov/pubmed/7811730?dopt=Abstract.

Brown, N. J., Gainer, J. V., Murphey, L. J., & Vaughan, D. E. (2000). Bradykinin stimulates tissue plasminogen activator release from human forearm vasculature through B(2) receptor-dependent, NO synthase-independent, and cyclooxygenase-independent pathway. *Circulation, 102*(18), 2190–2196. http://www.ncbi.nlm.nih.gov/pubmed/11056091?dopt=Abstract.

Cagliani, R., Forni, D., Riva, S., Pozzoli, U., Colleoni, M., Bresolin, N., et al. (2013). Evolutionary analysis of the contact system indicates that kininogen evolved adaptively in mammals and in human populations. *Molecular Biology and Evolution, 30*(6), 1397–1408. http://www.ncbi.nlm.nih.gov/pubmed/23505046?dopt=Abstract.

Calixto, J. B., Dutra, R. C., Bento, A. F., Marcon, R., & Campos, M. M. (2011). Kallikrein-kinin system in pain. In M. Bader (Ed.), *Kinins* (pp. 247–260). Berlin: Walter de Gruyter GmbH & Co. KG.

Campos, M. M., & Calixto, J. B. (1995). Involvement of B1 and B2 receptors in bradykinin-induced rat paw edema. *British Journal of Pharmacology, 114*, 1005–1013. http://www.ncbi.nlm.nih.gov/pubmed/7780633?dopt=Abstract.

Cardoso, C. C., Garrett, T., Cayla, C., Meneton, P., Pesquero, J. B., & Bader, M. (2004). Structure and expression of two kininogen genes in mice. *Biological Chemistry, 385*(3–4), 295–301. http://www.ncbi.nlm.nih.gov/pubmed/15134344?dopt=Abstract.

Chao, J., Yin, H., Gao, L., Hagiwara, M., Shen, B., Yang, Z. R., et al. (2008). Tissue kallikrein elicits cardioprotection by direct kinin B2 receptor activation independent of kinin formation. *Hypertension, 52*, 715–720. http://www.ncbi.nlm.nih.gov/pubmed/18768400?dopt=Abstract.

Cicardi, M., Bork, K., Caballero, T., Craig, T., Li, H. H., Longhurst, H., et al. (2012). Evidence-based recommendations for the therapeutic management of angioedema owing to hereditary C1 inhibitor deficiency: Consensus report of an International Working Group. *Allergy, 67*(2), 147–157. http://www.ncbi.nlm.nih.gov/pubmed/22126399?dopt=Abstract.

Colman, R. W., Jameson, B. A., Lin, Y., Johnson, D., & Mousa, S. A. (2000). Domain 5 of high molecular weight kininogen (kininostatin) down-regulates endothelial cell proliferation and migration and inhibits angiogenesis. *Blood, 95*(2), 543–550. http://www.ncbi.nlm.nih.gov/pubmed/10627460?dopt=Abstract.

Costerousse, O., Danilov, S., & Alhenc-Gelas, F. (1993). Genetics of angiotensin I-converting enzyme. *Clinical and Experimental Hypertension, 9*(5–6), 659–669.

Cubedo, J., Ramaiola, I., Padro, T., Martin-Yuste, V., Sabate-Tenas, M., & Badimon, L. (2013). High-molecular weight kininogen and the intrinsic coagulation pathway in patients with de novo acute myocardial infarction. *Thrombosis and Haemostasis, 110*(6), 1121–1134. http://www.ncbi.nlm.nih.gov/pubmed/24096571?dopt=Abstract.

Cugno, M., Zanichelli, A., Foieni, F., Acacia, S., & Cicardi, M. (2009). C1-inhibitor deficiency and angioedema: Molecular mechanisms and clinical progress. *Trends in Molecular Medicine, 15*(2), 69–78. http://www.ncbi.nlm.nih.gov/pubmed/19162547?dopt=Abstract.

Cunha, T. M., Verri, W. A., Jr., Fukada, S. Y., Guerrero, A. T., Santodomingo-Garzón, T., Poole, S., et al. (2007). TNF-alpha and IL-1beta mediate inflammatory hypernociception in mice triggered by B1 but not B2 kinin receptor. *European Journal of Pharmacology, 573*(1–3), 221–229. http://www.ncbi.nlm.nih.gov/pubmed/17669394?dopt=Abstract.

da Costa, P. L., Sirois, P., Tannock, I. F., & Chammas, R. (2014). The role of kinin receptors in cancer and therapeutic opportunities. *Cancer Letters, 345*(1), 27–38. http://www.ncbi.nlm.nih.gov/pubmed/24333733?dopt=Abstract.

Del Nery, E., Juliano, M. A., Lima, A. P., Scharfstein, J., & Juliano, L. (1997). Kininogenase activity by the major cysteinyl proteinase (cruzipain) from Trypanosoma cruzi. *The Journal of Biological Chemistry, 272*, 25713–25718. http://www.ncbi.nlm.nih.gov/pubmed/9325296?dopt=Abstract.

Dillenburg-Pilla, P., Maria, A. G., Reis, R. I., Floriano, E. M., Pereira, C. D., De Lucca, F. L., et al. (2013). Activation of the kinin B1 receptor attenuates melanoma tumor growth and metastasis. *PloS One, 8*(5), e64453. http://www.ncbi.nlm.nih.gov/pubmed/23691222?dopt=Abstract.

Duchene, J., Lecomte, F., Ahmed, S., Cayla, C., Pesquero, J., Bader, M., et al. (2007). A novel inflammatory pathway involved in leukocyte recruitment: Role for the kinin B1 receptor and the chemokine CXCL5. *Journal of Immunology, 179*(7), 4849–4856.

Dutra, R. C., Leite, D. F., Bento, A. F., Manjavachi, M. N., Patrício, E. S., Figueiredo, C. P., et al. (2011). The role of kinin receptors in preventing neuroinflammation and its clinical severity during experimental autoimmune encephalomyelitis in mice. *PloS One, 6*(11), e27875. http://www.ncbi.nlm.nih.gov/pubmed/22132157?dopt=Abstract.

Dutra, R. C., Moreira, E. L., Alberti, T. B., Marcon, R., Prediger, R. D., & Calixto, J. B. (2013). Spatial reference memory deficits precede motor dysfunction in an experimental autoimmune encephalomyelitis model: The role of kallikrein-kinin system. *Brain, Behavior, and Immunity, 33*, 90–101. http://www.ncbi.nlm.nih.gov/pubmed/23777652?dopt=Abstract.

Ehrenfeld, P., Matus, C. E., Pavicic, F., Toledo, C., Nualart, F., Gonzalez, C. B., et al. (2009). Kinin B1 receptor activation turns on exocytosis of matrix metalloprotease-9 and myeloperoxidase in human neutrophils: Involvement of mitogen-activated protein kinase family. *Journal of Leukocyte Biology, 86*(5), 1179–1189. http://www.ncbi.nlm.nih.gov/pubmed/19641039?dopt=Abstract.

Emanueli, C., Maestri, R., Corradi, D., Marchione, R., Minasi, A., Tozzi, M. G., et al. (1999). Dilated and failing cardiomyopathy in BK B2 receptor knockout mice. *Circulation, 100*, 2359–2365. http://www.ncbi.nlm.nih.gov/pubmed/10587341?dopt=Abstract.

Engelmann, B., & Massberg, S. (2013). Thrombosis as an intravascular effector of innate immunity. *Nature, 13*, 34–45.

Enquist, J., Skröder, C., Whistler, J. L., & Leeb-Lundberg, L. M. F. (2007). Kinins promote B2R receptor endocytosis and delay constitutive B1 receptor endocytosis. *Molecular Pharmacology, 71*, 494–507. http://www.ncbi.nlm.nih.gov/pubmed/17110500?dopt=Abstract.

Erdös, E. G. (1990). Angiotensin I converting enzyme and the changes in our concepts through the years. Lewis K. Dahl memorial lecture. *Hypertension, 16*(4), 363–370. http://www.ncbi.nlm.nih.gov/pubmed/2170273?dopt=Abstract.

Ferreira, J., Trichês, K. M., Medeiros, R., Cabrini, D. A., Mori, M. A., Pesquero, J. B., et al. (2008). The role of kinin B1 receptors in the nociception produced by peripheral protein kinase C activation in mice. *Neuropharmacology, 54*(3), 597–604. http://www.ncbi.nlm.nih.gov/pubmed/18164734?dopt=Abstract.

Fogaça, S. E., Melo, R. L., Pimenta, D. C., Hosoi, K., Juliano, L., & Juliano, M. A. (2004). Differences in substrate and inhibitor sequence specificity of human, mouse and rat tissue kallikreins. *The Biochemical Journal, 380*(3), 775–781. http://www.ncbi.nlm.nih.gov/pubmed/15040788?dopt=Abstract.

Frick, I. M., Akesson, P., Herwald, H., Mörgelin, M., Malmsten, M., Nägler, D. K., et al. (2006). The contact system-a novel branch of innate immunity generating antibacterial peptides. *The EMBO Journal, 25*(23), 5569–5578. http://www.ncbi.nlm.nih.gov/pubmed/17093496?dopt=Abstract.

Gera, L., Stewart, J. M., Fortin, J. P., Morissette, G., & Marceau, F. (2008). Structural modification of the highly potent peptide bradykinin B1 receptor antagonist B9958. *International Immunopharmacology, 8*(2), 289–292. http://www.ncbi.nlm.nih.gov/pubmed/18182242?dopt=Abstract.

Gobeil, F., Jr., Sirois, P., & Regoli, D. (2013). Preclinical pharmacology, metabolic stability, pharmacokinetics and toxicology of the peptidic kinin B1 receptor antagonist R-954. *Peptides, 52*, 82–89. http://www.ncbi.nlm.nih.gov/pubmed/24361511?dopt=Abstract.

Göbel, K., Pankratz, S., Schneider-Hohendorf, T., Bittner, S., Schuhmann, M. K., Langer, H. F., et al. (2011). Blockade of the kinin receptor B1 protects from autoimmune CNS disease by reducing leukocyte trafficking. *Journal of Autoimmunity, 36*(2), 106–114. http://www.ncbi.nlm.nih.gov/pubmed/21216565?dopt=Abstract.

Griol-Charhbili, V., Messadi-Laribi, E., Bascands, J. L., Heudes, D., Meneton, P., Giudicelli, J. F., et al. (2005). Role of tissue kallikrein in the cardioprotective effects of ischemic and pharmacological preconditioning in myocardial ischemia. *The FASEB Journal, 19*(9), 1172–1174. http://www.ncbi.nlm.nih.gov/pubmed/15860541?dopt=Abstract.

Gulliver, R., Baltic, S., Misso, N. L., Bertram, C. M., Thompson, P. J., & Fogel-Petrovic, M. (2011). Lysdes[Arg9]- bradykinin alters migration and production of interleukin-12 in monocyte-derived dendritic cells. *American Journal of Respiratory Cell and Molecular Biology, 45*(3), 542–549. PubMed.

Gustafson, E. J., Schutsky, D., Knight, L. C., & Schmaier, A. H. (1986). High molecular weight kininogen binds to unstimulated platelets. *The Journal of Clinical Investigation, 78*(1), 310–318. PubMed.

Haasemann, M., Cartaud, J., Müller-Esterl, W., & Dunia, I. (1998). Agonist-induced redistribution of bradykinin B2 receptor in caveolae. *Journal of Cell Science, 111* (Pt 7), 917–928. PubMed.

Han, E. D., MacFarlane, R. C., Mulligan, A. N., Scafidi, J., & Davis, A. E., 3rd. (2012). Increased vascular permeability in C1 inhibitor-deficient mice mediated by the bradykinin type 2 receptor. *The Journal of Clinical Investigation, 109*(8), 1057–1063.

Hecquet, C., Tan, F., Marcic, B. M., & Erdös, E. G. (2000). Human bradykinin B(2) receptor is activated by kallikrein and other serine proteases. *Molecular Pharmacology, 58*(4), 828–836. PubMed.

Henderson, L., Figueroa, C. D., Müller-Esterl, W., & Bhoola, K. D. (1994). Assembly of the contact-phase factors on the surface of the human neutrophil-membrane. *Blood, 84*, 474–482. PubMed.

Herwald, H., Hasan, A. A., Godovac-Zimmermann, J., Schmaier, A. H., & Müller-Esterl, W. (1995). Identification of an endothelial cell binding site on kininogen domain D3. *The Journal of Biological Chemistry, 270*(24), 14634–14642. PubMed.

Herwald, H., Dedio, J., Kellner, R., Loos, M., & Müller-Esterl, W. (1996a). Isolation and characterization of the kininogen-binding protein p33 from endothelial cells. Identity with the gC1q receptor. *The Journal of Biological Chemistry, 271*(22), 13040–13047. PubMed.

Herwald, H., Collin, M., Müller-Esterl, W., & Björk, L. (1996b). Streptococcal cysteine proteinase release kinins: A novel virulence mechanism. *The Journal of Experimental Medicine, 184*, 665–673. PubMed.

Imamura, T., Tanase, S., Szmyd, G., Kozik, A., Travis, J., & Potempa, J. (2005). Induction of vascular leakage through the release of bradykinin and a novel kinin by cysteine proteinases from Staphylococcus aureus. *The Journal of Experimental Medicine, 201*, 1669–1676. PubMed.

Jong, Y. J., Dalemar, L. R., Seehra, K., & Baenziger, N. L. (2002). Bradykinin receptor modulation in cellular models of aging and Alzheimer's disease. *International Immunopharmacology, 2*(13–14), 1833–1840. PubMed.

Joseph, K., Ghebrehiwet, B., Peerschke, E. I., Reid, K. B., & Kaplan, A. P. (1996). Identification of the zinc-dependent endothelial cell binding protein for high molecular weight kininogen and factor XII: Identity with the receptor that binds to the globular "heads" of C1q (gC1q-R). *Proceedings of the National Academy of Sciences of the United States of America, 93*(16), 8552–8557. PubMed.

Ju, H., Venema, V. J., Liang, H., Harris, M. B., Zou, R., & Venema, R. C. (2000). BK activates the Janus-activated kininase/signal transducers and activators of transcription (JAK/STAT) pathway in vascular endothelial cells: Localization of JAK/STAT signalling proteins in plasmalemmal caveolae. *The Biochemical Journal, 351*, 257–264. PubMed.

Kahn, R., Hellmark, T., Leeb-Lundberg, L. M., Akbari, N., Todiras, M., Olofsson, T., et al. (2009). Neutrophil-derived proteinase 3 induces kallikrein-independent release of a novel vasoactive kinin. *Journal of Immunology, 182*(12), 7906–7915.

Kaman, W. E., Wolterink, A. F., Bader, M., Boele, L. C., & van der Kleij, D. (2009). The bradykinin B2 receptor in the early immune response against Listeria infection. *Medical Microbiology and Immunology, 198*(1), 39–46. PubMed.

Kannemeier, C., Shibamiya, A., Nakazawa, F., Trusheim, H., Ruppert, C., Markart, P., et al. (2007). Extracellular RNA constitutes a natural procoagulant cofactor in blood coagulation. *Proceedings of the National Academy of Sciences of the United States of America, 104*(15), 6388–6393. PubMed.

Kaplan, A. P., & Joseph, K. (2014). Pathogenic mechanisms of bradykinin mediated diseases: Dysregulation of an innate inflammatory pathway. *Advances in Immunology, 121*, 41–89. PubMed.

Kaufmann, J., Haasemann, M., Modrow, S., & Müller-Esterl, W. (1993). Structural dissection of the multidomain kininogens. Fine mapping of the target epitopes of antibodies interfering with their functional properties. *The Journal of Biological Chemistry, 268*(12), 9079–9091. PubMed.

Kawabata, A., Kawao, N., Kitano, T., Matsunami, M., Satoh, R., Ishiki, T., et al. (2006). Colonic hyperalgesia triggered by proteinase-activated receptor-2 in mice: Involvement of endogenous bradykinin. *Neuroscience Letters, 402*(1–2), 167–172. PubMed.

Kellermann, J., Lottspeich, F., Henschen, A., & Müller-Esterl, W. (1986). Completion of the primary structure of human high-molecular-mass kininogen. The amino acid sequence of the entire heavy chain and evidence for its evolution by gene triplication. *European Journal of Biochemistry, 154*(2), 471–478. PubMed.

Klein, J., Gonzalez, J., Duchene, J., Esposito, L., Pradère, J. P., Neau, E., et al. (2009). Delayed blockade of the kinin B1 receptor reduces renal inflammation and fibrosis in obstructive nephropathy. *The FASEB Journal, 23*(1), 134–142. PubMed.

Kleinschnitz, C., Stoll, G., Bendszus, M., Schuh, K., Pauer, H. U., Burfeind, P., et al. (2006). Targeting coagulation factor XII provides protection from pathological thrombosis in cerebral ischemia without interfering with hemostasis. *The Journal of Experimental Medicine, 203*(3), 513–518. PubMed.

Kozik, A., Moore, R. B., Potempa, J., Imamura, T., Rapala-Kozik, M., & Travis, J. (1998). A novel mechanism for bradykinin production at inflammatory sites. Diverse effects of a mixture of neutrophil elastase and mast cell tryptase versus tissue and plasma kallikreins on native and oxidized kininogens. *The Journal of Biological Chemistry, 273*(50), 33224–33229. PubMed.

Kränkel, N., Katare, R. G., Siragusa, M., Barcelos, L. S., Campagnolo, P., Mangialardi, G., et al. (2008). Role of kinin B2 receptor signalling in the recruitment of circulating progenitor cells with neovascularization potential. *Circulation Research, 103*(11), 1335–1343. PubMed.

Kränkel, N., Kuschnerus, K., Müller, M., Speer, T., Mocharla, P., Madeddu, P., et al. (2013). Novel insights into the critical role of bradykinin and the kinin B2 receptor for vascular recruitment of circulating endothelial repair-promoting mononuclear cell subsets: Alterations in patients with coronary disease. *Circulation, 127*(5), 594–603. PubMed.

Kuhr, F., Lowry, J., Zhang, Y., Brovkovych, V., & Skidgel, R. A. (2010). Differential regulation of inducible and endothelial nitric oxide synthase by kinin B1 and B2 receptors. *Neuropeptides, 44*(2), 145–154. PubMed.

Langhauser, F., Göb, E., Kraft, P., Geis, C., Schmitt, J., Brede, M., et al. (2012). Kininogen deficiency protects from ischemic neurodegeneration in mice by reducing thrombosis, blood-brain barrier damage, and inflammation. *Blood, 120*(19), 4082–4092. PubMed.

Larusch, G. A., Merkulova, A., Mahdi, F., Shariat-Madar, Z., Sitrin, R. G., Cines, D. B., et al. (2013). Domain 2 of uPAR regulates single-chain urokinase-mediated angiogenesis through B1-integrin and VEGFR2. *American Journal of Physiology. Heart and Circulatory Physiology, 305*(3), H305–H320. PubMed.

Lauton-Santos, S., Guatimosim, S., Castro, C. H., Oliveira, F. A., Almeida, A. P., Dias-Peixoto, M. F., et al. (2007). Kinin B1 receptor participates in the control of cardiac function in mice. *Life Sciences, 81*(10), 814–822. PubMed.

Leeb-Lundberg, L. M., Kang, D. S., Lamb, M. E., & Fathy, D. B. (2001). The human B1 bradykinin receptor exhibits high ligand-independent, constitutive activity. Roles of residues in the fourth intracellular and third transmembrane domains. *The Journal of Biological Chemistry, 276*(12), 8785–8792. PubMed.

Leeb-Lundberg, L. M., Marceau, F., Müller-Esterl, W., Pettibone, D. J., & Zuraw, B. L. (2005). International union of pharmacology. XLV. Classification of the kinin receptor family: From molecular mechanisms to pathophysiological consequences. *Pharmacological Reviews, 57*(1), 27–77. PubMed.

Lima, A. P., Almeida, P. C., Tersariol, I. L., Schmitz, V., Schmaier, A. H., Juliano, L., et al. (2002). Heparan sulfate modulates kinin release by Trypanosoma cruzi through the activity of cruzipain. *The Journal of Biological Chemistry, 277*(8), 5875–5881. PubMed.

Madeddu, P., & Kränkel, N. (2011). Kallikrein-kinin system in the vessel wall. In M. Bader (Ed.), *Kinins* (pp. 137–153). Berlin: Walter de Gruyter GmbH.

Marceau, F. (2011). Drugs in the kallikrein-kinin system. In M. Bader (Ed.), *Kinins* (pp. 69–83). Berlin: Walter de Gruyter GmbH.

McLean, P. G., Perretti, M., & Ahluwalia, A. (2000a). Kinin B(1) receptors and the cardiovascular system: Regulation of expression and function. *Cardiovascular Research, 48*(2), 194–210. PubMed.

McLean, P. G., Ahluwalia, A., & Perretti, M. (2000b). Association between kinin B(1) receptor expression and leukocyte trafficking across mouse mesenteric postcapillary venules. *The Journal of Experimental Medicine, 192*(3), 367–380. http://www.ncbi.nlm.nih.gov/pubmed/10934225?dopt=Abstract.

Medeiros, R., Cabrini, D. A., Ferreira, J., Fernandes, E. S., Mori, M. A., Pesquero, J. B., et al. (2004). Bradykinin B1 receptor expression induced by tissue damage in the rat portal vein: A critical role for mitogen-activated protein kinase and nuclear factor-kappa-B signalling pathways. *Circulation Research, 94*(10), 1375–1382. http://www.ncbi.nlm.nih.gov/pubmed/15087417?dopt=Abstract.

Merino, V. F., Todiras, M., Mori, M. A., Sales, V. M., Fonseca, R. G., Saul, V., et al. (2009). Predisposition to atherosclerosis and aortic aneurysms in mice deficient in kinin B1 receptor and apolipoprotein E. *Journal of Molecular Medicine, 87*(10), 953–963. http://www.ncbi.nlm.nih.gov/pubmed/19618151?dopt=Abstract.

Merkulov, S., Zhang, W. M., Komar, A. A., Schmaier, A. H., Barnes, E., Zhou, Y., et al. (2008). Deletion of murine kininogen gene 1 (mKng1) causes loss of plasma kininogen and delays thrombosis. *Blood, 111*(3), 1274–1281. http://www.ncbi.nlm.nih.gov/pubmed/18000168?dopt=Abstract.

Metzger, R., Franke, F. E., Bohle, R. M., Alhenc-Gelas, F., & Danilov, S. M. (2011). Heterogeneous distribution of angiotensin I-converting enzyme (CD143) in the human and rat vascular systems: Vessel, organ and species specificity. *Microvascular Research, 81*(2), 206–215. http://www.ncbi.nlm.nih.gov/pubmed/21167844?dopt=Abstract.

Monteiro, A. C., Schmitz, V., Svensjö, E., Gazzinelli, R. T., Almeida, I. C., Todorov, A., et al. (2006). Cooperative activation of TLR2 and bradykinin B2 receptor is required for induction of type-1 immunity in a mouse model of subcutaneous infection by Trypanosoma cruzi. *Journal of Immunology, 177*(9), 6325–6335.

Monteiro, A. C., Schmitz, V., Morrot, A., de Arruda, L. B., Nagajyothi, F., Granato, A., et al. (2007). Bradykinin B2 Receptors of dendritic cells, acting as sensors of kinins proteolytically released by Trypanosoma cruzi, are critical for the development of protective type-1 responses. *PLoS Pathogens, 3*(11), e185. http://www.ncbi.nlm.nih.gov/pubmed/18052532?dopt=Abstract.

Monteiro, A. C., Scovino, A., Raposo, S., Gaze, V. M., Cruz, C., Svensjö, E., et al. (2009). Kinin danger signals proteolytically released by gingipain induce Fimbriae-specific IFN-gamma- and IL-17-producing T cells in mice infected intramucosally with Porphyromonas gingivalis. *Journal of Immunology, 183*(6), 3700–3711.

Morand-Contant, M., Anand-Srivastava, M. B., & Couture, R. (2010). Kinin B1 receptor upregulation by angiotensin II and endothelin-1 in rat vascular smooth muscle cells: Receptors and mechanisms. *American Journal of Physiology. Heart and Circulatory*

Physiology, 299(5), H1625–H1632. http://www.ncbi.nlm.nih.gov/pubmed/20833961?dopt=Abstract.

Moreno-Sanchez, D., Hernandes-Ruiz, L., & Docampo, R. (2012). Polyphosphate is a novel pro-inflammatory regulator of mast cells and is located in acidocalcisomes. The Journal of Biological Chemistry, 287(34), 28435–28444. http://www.ncbi.nlm.nih.gov/pubmed/22761438?dopt=Abstract.

Müller, F., Mutch, N. J., Schenk, W. A., Smith, S. A., Esterl, L., Spronk, H. M., et al. (2009). Platelet polyphosphates are proinflammatory and procoagulant mediators in vivo. Cell, 139(6), 1143–1156. http://www.ncbi.nlm.nih.gov/pubmed/20005807?dopt=Abstract.

Nico, D., Feijó, D. F., Maran, N., Morrot, A., Scharfstein, J., Palatnik, M., et al. (2012). Resistance to visceral leishmaniasis is severely compromised in mice deficient of bradykinin B2-receptors. Parasites & Vectors, 5, 261.

Oehmcke, S., Mörgelin, M., & Herwald, H. (2009). Activation of the human contact system on neutrophil extracellular traps. Journal of Innate Immunity, 1(3), 225–230. http://www.ncbi.nlm.nih.gov/pubmed/20375580?dopt=Abstract.

Orsenigo, F., Giampietro, C., Ferrari, A., Corada, M., Galaup, A., Sigismund, S., et al. (2012). Phosphorylation of VE-cadherin is modulated by haemodynamic forces and contributes to the regulation of vascular permeability in vivo. Nature Communications, 3, 1208. http://www.ncbi.nlm.nih.gov/pubmed/23169049?dopt=Abstract.

Oschatz, C., Maas, C., Lecher, B., Jansen, T., Björkqvist, J., Tradler, T., et al. (2011). Mast cells increase vascular permeability by heparin-initiated bradykinin formation in vivo. Immunity, 34(2), 258–268. http://www.ncbi.nlm.nih.gov/pubmed/21349432?dopt=Abstract.

Pesquero, J. B., Araujo, R. C., Heppenstall, P. A., Stucky, C. L., Silva, J. A., Jr., Walther, T., et al. (2000). Hypoalgesia and altered inflammatory responses in mice lacking kinin B1 receptors. Proceedings of the National Academy of Sciences of the United States of America, 97, 8140–8145. http://www.ncbi.nlm.nih.gov/pubmed/10859349?dopt=Abstract.

Plendl, J., Snyman, C., Naidoo, S., Sawant, S., Mahabeer, R., & Bhoola, K. D. (2000). Expression of tissue kallikrein and kinin receptors in angiogenic microvascular cells. Biological Chemistry, 381, 1103–1115. http://www.ncbi.nlm.nih.gov/pubmed/11154068?dopt=Abstract.

Potempa, J., & Herwald, H. (2011). Kinins in bacterial infections. In M. Bader (Ed.), Kinins (pp. 307–320). Berlin: Walter de Gruyter GmbH.

Prat, A., Weinrib, L., Becher, B., Poirier, J., Duquette, P., Couture, R., et al. (1999). Bradykinin B1 receptor expression and function on T lymphocytes in active multiple sclerosis. Neurology, 53(9), 2087–2092. http://www.ncbi.nlm.nih.gov/pubmed/10599786?dopt=Abstract.

Rapala-Kozik, M., Bras, G., Chruscicka, B., Karlowska-Kuleta, J., Sroka, A., Herwald, H., et al. (2011). Adsorption of components of the plasma kinin-forming system on the surface of Porphyromonas gingivalis involves gingipains as the major docking platforms. Infection and Immunity, 79(2), 797–805. http://www.ncbi.nlm.nih.gov/pubmed/21098107?dopt=Abstract.

Raslan, F., Schwarz, T., Meuth, S., Austinat, M., Bader, M., Renné, T., et al. (2010). Inhibition of bradykinin receptor B1 protects mice from focal brain injury by reducing blood-brain barrier and inflammation. Journal of Cerebral Blood Flow & Metabolism, 30, 1477–1486.

Regoli, D. (2015). Six decades with kinins. International Meeting on Kinin System and peptide Receptors, 28 June–1 July, Sao Paulo.

Regoli, D., & Barabé, J. (1980). Pharmacology of bradykinin and related kinins. Pharmacological Reviews, 32(1), 1–46. http://www.ncbi.nlm.nih.gov/pubmed/7015371?dopt=Abstract.

Renné, T. (2012). The procoagulant and proinflammatory plasma contact system. Seminars in Immunopathology, 34, 31–41. http://www.ncbi.nlm.nih.gov/pubmed/21858560?dopt=Abstract.

Renné, T., Dedio, J., David, G., & Müller-Esterl, W. (2000). High molecular weight kininogen utilizes heparan sulfate proteoglycans for accumulation on endothelial cells. The Journal of Biological Chemistry, 275(43), 33688–33696. http://www.ncbi.nlm.nih.gov/pubmed/10843988?dopt=Abstract.

Rocha e Silva, M., Beraldo, W. T., & Rosenfeld, G. (1949). Bradykinin, a hypotensive and smooth muscle stimulating factor released from plasma globulin by snake venoms and by trypsin. American Journal of Physiology, 156(2), 261–273. http://www.ncbi.nlm.nih.gov/pubmed/18127230?dopt=Abstract.

Rosen, E. D., Gallani, D., & Castellino, F. J. (2002). FXI is essential for thrombus formation following FeCl3-induced injury of the carotid artery in the mouse. Thrombosis and Haemostasis, 87(4), 774–776. http://www.ncbi.nlm.nih.gov/pubmed/12008966?dopt=Abstract.

Sabourin, T., Morissette, G., Bouthillier, J., Levesque, L., & Marceau, F. (2002). Expression of kinin B1 receptor in fresh or cultured rabbit aortic smooth muscle: Role of NF-kappa B. The American Journal of Physiology, 283(1), H227–H237.

Sales, V. M. T., TuraÓa, L. T., & Pesquero, J. B. (2011). Animal models in the kinin field. In M. Bader (Ed.), Kinins (pp. 51–68). Berlin: Walter de Gruyter GmbH & Co. KG.

Schanstra, J. P., Bataillé, E., Marin-Castaño, M. E., Barascud, Y., Hirtz, C., Pesquero, J. B., et al. (1998). The B1-agonist [des-Arg10]-kallidin activates transcription factor NF-kappa B and induces homologous upregulation of the bradykinin B1-receptor in cultured human lung fibroblasts. The Journal of Clinical Investigation, 101(10), 2080–2091. http://www.ncbi.nlm.nih.gov/pubmed/9593764?dopt=Abstract.

Scharfstein, J., & Andrade, D. (2011). Infection-associated vasculopathy in experimental Chagas disease pathogenic roles of endothelin and kinin pathways. Advances in *Parasitology, 76*, 101–127. http://www.ncbi.nlm.nih.gov/pubmed/21884889?dopt=Abstract.

Scharfstein, J., & Svensjö, E. (2011). The kallikrein-kinin system in parasitic infections. In M. Bader (Ed.), *Kinins* (pp. 321–336). Berlin: Walter de Gruyter GmbH & Co. KG.

Scharfstein, J., Schmitz, V., Morandi, V., Capella, M. M., Lima, A. P., Morrot, A., et al. (2000). Host cell invasion by Trypanosoma cruzi is potentiated by activation of bradykinin B(2) receptors. *The Journal of Experimental Medicine, 192*(9), 1289–1300. http://www.ncbi.nlm.nih.gov/pubmed/11067878?dopt=Abstract.

Scharfstein, J., Schmitz, V., Svensjö, E., Granato, A., & Monteiro, A. C. (2007). Kininogens coordinate adaptive immunity through the proteolytic release of bradykinin, an endogenous danger signal driving dendritic cell maturation. *Scandinavian Journal of Immunology, 66*(2–3), 128–136. http://www.ncbi.nlm.nih.gov/pubmed/17635790?dopt=Abstract.

Scharfstein, J., Andrade, D., Svensjö, E., Oliveira, A. C., & Nascimento, C. R. (2013). The kallikrein-kinin system in experimental Chagas disease: A paradigm to investigate the impact of inflammatory edema on GPCR-mediated pathways of host cell invasion by Trypanosoma cruzi. *Frontiers in Immunology, 3*, 396. http://www.ncbi.nlm.nih.gov/pubmed/23355836?dopt=Abstract.

Schmitz, V., Svensjö, E., Serra, R. R., Teixeira, M. M., & Scharfstein, J. (2009). Proteolytic generation of kinins in tissues infected by Trypanosoma cruzi depends on CXC-chemokine secretion by macrophages activated via toll-like 2 receptors. *Journal of Leukocyte Biology, 85*(6), 1005–1014. http://www.ncbi.nlm.nih.gov/pubmed/19293401?dopt=Abstract.

Schmitz, V., Almeida, L. N., Svensjö, E., Monteiro, A. C., Köhl, J., & Scharfstein, J. (2014). C5a and bradykinin receptor cross-talk regulates innate and adaptive immunity in Trypanosoma cruzi infection. *Journal of Immunology, 193*(7), 3613–3623.

Schulze-Topphoff, U., Prat, A., Prozorovski, T., Siffrin, V., Paterka, M., Herz, J., et al. (2009). Activation of kinin receptor B1 limits encephalitogenic T lymphocyte recruitment to the central nervous system. *Nature Medicine, 15*(7), 788–793. http://www.ncbi.nlm.nih.gov/pubmed/19561616?dopt=Abstract.

Shariat-Madar, Z., Mahdi, F., & Schmaier, A. (2002). Identification and characterization of prolylcarboxypeptidase as an endothelial cell prekallikrein activator. *The Journal of Biological Chemistry, 277*(20), 17962–17969. http://www.ncbi.nlm.nih.gov/pubmed/11830581?dopt=Abstract.

Skidgel, R. A., Davies, R. M., & Tan, F. (1989). Human carboxypeptidase M. Purification and characetrization of a membrane-bound carboxypeptidase that cleaves peptide hormones. *The Journal of Biological Chemistry, 264*, 2236–2239.

Smith, R. S., Jr., Gao, L., Chao, L., & Chao, J. (2008). Tissue kallikrein and kinin infusion promotes neovascularization in limb ischemia. *Biological Chemistry, 389*, 725–730. http://www.ncbi.nlm.nih.gov/pubmed/18627294?dopt=Abstract.

Soubrier, F., Alhenc-Gelas, F., Hubert, C., Allegrini, J., John, M., Tregear, G., et al. (1988). Two putative active centers in human angiotensin I-converting enzyme revealed by molecular cloning. *Proceedings of the National Academy of Sciences of the United States of America, 85*(24), 9386–9390. http://www.ncbi.nlm.nih.gov/pubmed/2849100?dopt=Abstract.

Spillmann, F., & Tschöpe, C. (2011). Kallikrein-kinin system in the heart. In M. Bader (Ed.), *Kinins* (pp. 117–136). Berlin: Walter de Gruyter GmbH & Co. KG.

Stadnicki, A., Pastucha, E., Nowaczyk, G., Mazurek, U., Plewka, D., Machnik, G., et al. (2005). Immunolocalization and expression of kinin B1R and B2R receptors in human inflammatory bowel disease. *American Journal of Physiology. Gastrointestinal and Liver Physiology, 289*(2), G361–G366. http://www.ncbi.nlm.nih.gov/pubmed/15805101?dopt=Abstract.

Stahl, M. C., Harris, C. K., Matto, S., & Bernstein, J. A. (2014). Idiopathic nonhistaminergic angioedema successfully treated with ecallantide, icatibant, and C1 esterase inhibitor replacement. *The Journal of Allergy and Clinical Immunology: In Practice, 2*(6), 818–819. http://www.ncbi.nlm.nih.gov/pubmed/25439384?dopt=Abstract.

Stoll, G., Kleinschnitz, C., & Nieswandt, B. (2008). Molecular mechanism of thrombus formation in ischemic stroke: Novel insights and targets for treatment. *Blood, 112*(9), 3555–3562. http://www.ncbi.nlm.nih.gov/pubmed/18676880?dopt=Abstract.

Sun, D., & McCrae, K. R. (2006). Endothelial-cell apoptosis induced by cleaved high-molecular-weight kininogen (HKa) is matrix dependent and requires the generation of reactive oxygen species. *Blood, 107*(12), 4714–4720. http://www.ncbi.nlm.nih.gov/pubmed/16418331?dopt=Abstract.

Thuringer, D., Maulon, L., & Frelin, C. (2002). Rapid transactivation of the vascular endothelial growth factor receptor KDR/Flk-1 by the bradykinin B2 receptor contributes to endothelial nitric-oxide synthase activation in cardiac capillary endothelial cells. *The Journal of Biological Chemistry, 277*(3), 2028–2032. http://www.ncbi.nlm.nih.gov/pubmed/11711543?dopt=Abstract.

Todorov, A. G., Andrade, D., Pesquero, J. B., Araujo Rde, C., Stewart, J., Gera, L., et al. (2003). Trypanosoma cruzi induces edematogenic responses in mice and invades cardiomyocytes and endothelial cells in vitro by activating distinct kinin receptor (B1/B2) subtypes. *The FASEB Journal, 17*(1), 73. http://www.ncbi.nlm.nih.gov/pubmed/12424228?dopt=Abstract.

Vianna, R. M., & Calixto, J. B. (1998). Characterization of the receptor and the mechanisms underlying the inflammatory response induced by des-Arg9-BK in mouse pleurisy. *British Journal of Pharmacology, 123*(2), 281–291. http://www.ncbi.nlm.nih.gov/pubmed/9489617?dopt=Abstract.

von Brühl, M. L., Stark, K., Steinhart, A., Chandraratne, S., Konrad, I., Lorenz, M., et al. (2012). Monocytes, neutrophils, and platelets cooperate to initiate and propagate venous thrombosis in mice in vivo. *The Journal of Experimental Medicine, 209*(4), 819–835.

Westermann, D., Walther, T., Savvatis, K., Escher, F., Sobirey, M., Riad, A., et al. (2005). Gene deletion of the kinin receptor B1 attenuates cardiac inflammation and fibrosis during the development of experimental diabetic cardiomyopathy. *Diabetes, 58*(6), 1373–1381.

Yang, A., Dai, J., Xie, Z., Colman, R. W., Wu, Q., Birge, R. B., et al. (2014). High molecular weight kininogen binds phosphatidylserine and opsonizes urokinase plasminogen activator receptor-mediated efferocytosis. *Journal of Immunology, 192*(9), 4398–4408.

Zhang, X., Tan, F., Zhang, Y., & Skidgel, R. A. (2008). Carboxypeptidase M and kinin B1 receptors interact to facilitate efficient b1 signaling from B2 agonists. *The Journal of Biological Chemistry, 283*(12), 7994–8004. http://www.ncbi.nlm.nih.gov/pubmed/18187413?dopt=Abstract.

L

Leflunomide

Kevin D. Pile[1] and Garry G. Graham[2,3]
[1]Campbelltown Hospital, School of Medicine, University of Western Sydney, Campbelltown, NSW, Australia
[2]Department of Pharmacology, School of Medical Sciences, University of New South Wales, Sydney, NSW, Australia
[3]Department of Clinical Pharmacology and Toxicology, St Vincent's Hospital, Sydney, NSW, Australia

Synonyms

Anti-metabolite; Dihydroorotate dehydrogenase inhibitor; Pyrimidine synthesis inhibitor

Definition

Leflunomide is a small-molecular-weight drug which is a member of a group of drugs known as disease-modifying anti-rheumatic drugs (DMARDs) or slow-acting anti-rheumatic drugs (SAARDs).

Chemical Structures and Properties

Chemically, leflunomide and its active metabolite, teriflunomide, are unrelated to any other marketed DMARD. Both leflunomide and teriflunomide are neutral, lipid-soluble compounds with logP values of 2.5 and 2.1, respectively. Their lipid solubilities indicate and promote their passive transport through cell membranes, including their oral absorption. Teriflunomide is formed from leflunomide by ring opening and exists as two interchangeable isomers with the equilibrium favoring the Z isomer.

Metabolism and Pharmacokinetics

Leflunomide is rapidly and almost completely converted to the active open chain metabolite, teriflunomide, by first-pass metabolism in the gut wall and liver (Fig. 1). The bioavailability of leflunomide in man is not known but, in experimental animals, 75–90 % of an oral dose is absorbed as teriflunomide. At the usual maintenance dose of 20 mg daily, the plasma concentrations of the active metabolite, teriflunomide, are approximately 30 mg/l (110 μmol/l). With only about 0.6 % unbound in plasma (Rozman 2002), the unbound therapeutic concentrations in plasma are about 0.7 μmol/l. Teriflunomide is not bound to red blood cells.

Teriflunomide has a long half-life of between 15 and 18 days because of its enterohepatic recirculation (Rozman 2002). About 90 % of a single dose of leflunomide is eliminated, about half in urine primarily as metabolites, while about 50 % is secreted in bile as teriflunomide

© Springer International Publishing AG 2016
M.J. Parnham (ed.), *Compendium of Inflammatory Diseases*,
DOI 10.1007/978-3-7643-8550-7

Leflunomide,
Fig. 1 Structures of leflunomide and its active metabolite teriflunomide (formerly known as A771726)

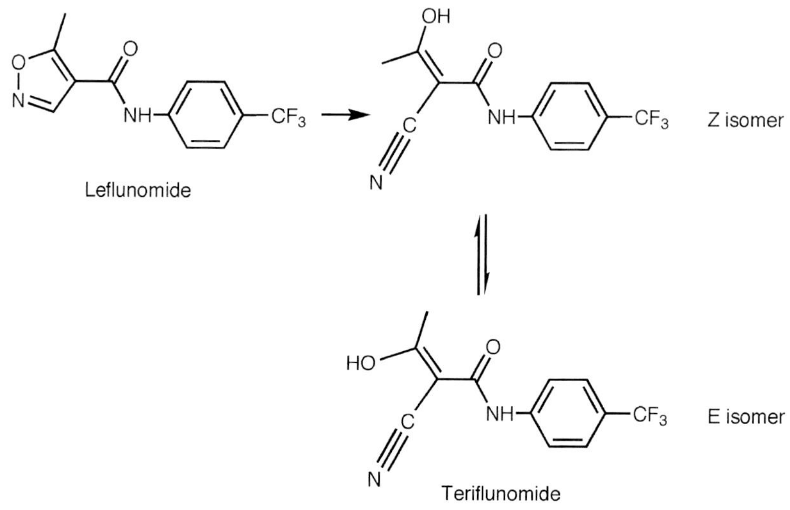

and is ultimately excreted in feces. Because teriflunomide relies heavily on biliary excretion for its clearance, and also given its risk of hepatotoxicity, leflunomide is contraindicated in patients with hepatic impairment.

At a fixed dosage regimen, teriflunomide may take 15–20 weeks to reach steady-state plasma concentrations because of its long half-life. In order to achieve therapeutic concentrations rapidly, it is common to administer loading doses. Thus, the recommended dose of leflunomide is 100 mg once daily for 3 days with a maintenance dose of 20 mg once daily. In practice, the loading dose is often decreased or not used with the expectation of less "nuisance" problems with diarrhea or nausea, both of which may influence early patient compliance.

Teriflunomide binds strongly to cholestyramine within the gastrointestinal tract. The result is that its plasma half-life is reduced to approximately 1 day and cholestyramine is used when rapid elimination of the active metabolite is required (see section on "Adverse Effects" below). Teriflunomide is not well cleared by dialysis because of its high binding to plasma proteins. Consequently, the dosage regimen of leflunomide is unchanged in patients undergoing dialysis.

Pharmacological Activities

The key mode of action of teriflunomide is the reversible inhibition of dihydroorotate dehydrogenase (Davis et al. 1996) (Fig. 2). Leflunomide is inactive and inhibition of dihydroorotate dehydrogenase is due entirely to teriflunomide. The equilibrium dissociation constant (Ki) of the binding of teriflunomide to dihydroorotate dehydrogenase is 179 nmol/l (Davis et al. 1996), well within the therapeutic range of unbound teriflunomide in plasma. Dihydroorotate dehydrogenase is the rate-limiting step in the pyrimidine synthesis that is accelerated in the activated CD4+ T cells that proliferate rapidly during the progression of rheumatoid arthritis. This antiproliferative effect on activated lymphocytes is likely the key effect of leflunomide on the pathophysiology of rheumatoid arthritis. Teriflunomide also directly inhibits several tyrosine kinases, cyclooxygenase 1, and cyclooxygenase 2 although the concentrations in vitro are supratherapeutic and these effects are unlikely to be related to the therapeutic actions of leflunomide (Hamilton et al. 1999).

In experimental animals, leflunomide has many effects consistent with its clinical anti-inflammatory effect. These actions include inhibition of adjuvant disease of rats and a systemic lupus erythematosus – like disease in rats

Leflunomide, Fig. 2 Synthesis of pyrimidine bases showing the step catalyzed by dihydroorotate dehydrogenase which is potently inhibited by teriflunomide, the active metabolite of leflunomide

(Bartlett et al. 1991). Leflunomide also inhibits rejection of transplanted skin and kidney in rats and graft-versus-host disease in mice.

Clinical Uses and Efficacy

Leflunomide has similar efficacy to methotrexate and sulfasalazine in the treatment of rheumatoid arthritis (Kalden et al. 2003). It not only decreases symptoms and increases function and quality of life in rheumatoid arthritis but also retards radiographic joint damage (Weinblatt et al. 1999). Decreased symptoms of rheumatoid arthritis commence within about 4 weeks with continuing improvement for 4–6 months. Clinical improvement has been sustained for up to 5 years. Combination therapy of leflunomide and methotrexate is also effective and well tolerated in patients responding inadequately to methotrexate alone (Weinblatt et al. 1999). Leflunomide is useful in the treatment of psoriasis and psoriatic arthritis (Sehgal and Verma 2013) and has been used in a small number of patients with systemic lupus erythematosus with benefit of various features of the diseases and relief of the associated arthritis (Remer et al. 2001). Leflunomide has also been tested in individual patients with a variety of skin diseases, such as atopic dermatitis and Wegener's granulomatosis, but good proof is lacking (Sehgal and Verma 2013).

The metabolite, teriflunomide, has promising activity in the treatment of multiple sclerosis. Both alone and in combination with interferon beta, teriflunomide shows lesser numbers of brain lesions than placebo. Consequently, it has been approved by FDA for the treatment of this disease although its long-term usefulness is unclear at this stage.

Adverse Effects

A substantial number of adverse effects of leflunomide have been noted. The cytostatic effect of leflunomide may explain some of the side effect profile, such as reversible alopecia and, conversely, the lack of opportunistic infections. Most memory T cells circulate in the G_0 phase and, therefore, do not require dihydroorotate for any de novo pyrimidine synthesis and are therefore not susceptible to the antiproliferative effect of leflunomide. In addition, because of the sparing of the salvage pathway, the replicating cells in the gastrointestinal tract and hemopoietic system are relatively unaffected, thus explaining the lack of mucositis or marrow toxicity.

Any suspected toxicity may be evaluated by the use of a short course (1–2 days) of cholestyramine at lower dose (4 g three times daily). This will often reverse the adverse effect, be it rash or diarrhea or other, quite quickly.

Adverse effects and recommended monitoring of treatment with leflunomide and teriflunomide include:

- Teratogenic (lymphomas). Leflunomide and teriflunomide are absolutely contraindicated in women who are or may become pregnant. All women of childbearing age should use effective forms of contraception and have a negative pregnancy test before beginning the drug. Termination of pregnancy is recommended generally if the patient has been on leflunomide, even though there have been a number of reported cases of delivery of full-term healthy infants (Bermas 2014). If pregnancy is contemplated, teriflunomide should be allowed to fall below 0.02 mg/l. This may take several months because of its long half-life. Alternatively, the elimination of the active metabolite, teriflunomide, can be accelerated by cholestyramine. Teriflunomide diffuses into breast milk although toxicity is unknown. At this stage, however, it is contraindicated in nursing mothers.
- Hepatitis. Reactivation of hepatitis B and C are possible. Check antibodies before treatment. High hepatic transaminase levels have been noted in 5–15 % of patients, but these effects were generally mild (less than twofold elevation) and usually resolved while continuing treatment. Post-marketing surveillance shows that almost all cases of hepatic dysfunction had other confounding factors present. Liver function tests (hepatic transaminases) should be monitored every 4–6 weeks for at least 6 months and longer if patients are taking leflunomide with methotrexate or other hepatotoxic drugs. Thereafter, liver function tests should be repeated at least every 3 months, more frequently if transaminases have increased. Intake of alcohol should be limited because of possible liver impairment.
- Hematopoietic. Decreased numbers of blood cells are uncommon, but regular blood counts (red blood cells, leukocytes, and platelets) are recommended.
- Cardiovascular. Leflunomide may increase blood pressure. Blood pressure should be measured before treatment and regularly during therapy.
- Gastrointestinal. Diarrhea is common but usually lessens with continued treatment or reduction in dosage.
- Cutaneous. Rash and reversible alopecia are common. Severe skin reactions, such as Stevens-Johnson syndrome, are very uncommon, but, if they are suspected, removal by cholestyramine is strongly recommended.
- Neuropathy. Apparent signs of peripheral neuropathy (such as intermittent paraesthesiae of the fingertips) have been reported.
- Lipoproteins. Progressive increases in plasma concentrations of cholesterol and low-density lipoproteins have been noted, but long-term effects of this are unknown.

Drug Interactions

- As discussed above, the anion exchange resin (cholestyramine) binds teriflunomide avidly. This interaction is used to decrease the plasma concentrations of teriflunomide if toxicity is suspected.
- Leflunomide is usefully combined with other SAARDs such as antimalarials, methotrexate, or sulfasalazine.
- Leflunomide may be used in combination with corticosteroids and/or nonsteroidal anti-inflammatory drugs (NSAIDs), particularly when the antirheumatic effects of leflunomide are developing. Ultimately, it may be possible to withdraw treatment with corticosteroids and/or NSAIDs if the patients respond well to leflunomide or its combinations with other SAARDs.
- Leflunomide inhibits cytochrome P4502C9 and potentiates the actions of phenytoin, oral anticoagulants (warfarin), and tolbutamide. Caution should be taken if leflunomide is commenced in patients stabilized on these drugs.

Cross-References

▶ Atopic Dermatitis
▶ Corticosteroids
▶ Granulomatosis with Polyangiitis (GPA)
▶ Methotrexate
▶ Non-steroidal Anti-inflammatory Drugs: Overview
▶ Rheumatoid Arthritis
▶ Sulfasalazine and Related Drugs

References

Bartlett, R. R., Dimitrijevic, M., Mattar, T., Zielinski, T., Germann, T., Rude, E., et al. (1991). Leflunomide (HWA 486), a novel immunomodulating compound for the treatment of autoimmune disorders and reactions leading to transplantation rejection. *Agents and Actions, 32*(1/2), 10–21.

Bermas, B. (2014) Non-steroidal anti-inflammatory drugs, glucocorticoids and disease modifying anti-rheumatic drugs for the management of rheumatoid arthritis before and during pregnancy. *Curr Opin Rheumatol, 26*(3), 334–340

Davis, J. P., Cain, G. A., Pitts, W. J., Magolda, R. L., & Copeland, R. A. (1996). The immunosuppressive metabolite of leflunomide is a potent inhibitor of human dihydroorotate dehydrogenase. *Biochemistry, 35*(4), 1270–1273.

Hamilton, L. C., Vojnovic, I., & Warner, T. D. (1999). A771726, the active metabolite of leflunomide, directly inhibits the activity of cyclo-oxygenase-2 in vitro and in vivo in a substrate-sensitive manner. *British Journal of Pharmacology, 127*(7), 1589–1596.

Kalden, J. R., Schattenkirchner, M., Sorensen, H., Emery, P., Deighton, C., Rozman, B., et al. (2003). The efficacy and safety of leflunomide in patients with active rheumatoid arthritis: A five-year follow up study. *Arthritis & Rheumatism, 48*(6), 1513–1520.

Remer, C. F., Weisman, M. H., & Wallace, D. J. (2001). Benefits of leflunomide in systemic lupus erythematosus: A pilot observational study. *Lupus, 10*(7), 480–483.

Rozman, B. (2002). Clinical pharmacokinetics of leflunomide. *Clinical Pharmacokinetics, 41*(6), 421–430.

Sehgal, V. N., & Verma, P. (2013). Leflunomide: Dermatologic perspective. *Journal of Dermatological Treatment, 24*(2), 89–95.

Weinblatt, M. E., Kremer, J. M., Coblyn, J. S., Maier, A. L., Helfgott, S. M., Morrell, M., et al. (1999). Pharmacokinetics, safety, and efficacy of combination treatment with methotrexate and leflunomide in patients with active rheumatoid arthritis. *Arthritis & Rheumatism, 42*(7), 1322–1328.

Leukocyte Recruitment

Ioannis Kourtzelis and Ioannis Mitroulis
Department of Clinical Pathobiochemistry and Institute for Clinical Chemistry and Laboratory Medicine, Faculty of Medicine, Technische Universität Dresden, Dresden, Germany

Synonyms

Leukocyte infiltration; Leukocyte transmigration

Inflammatory Processes and Cells

Definition

Leukocyte recruitment at the site of pathogen evasion or sterile tissue injury is a critical adaptation for the preservation of tissue integrity. Neutrophils are the cell population that acutely responds to the alterations of inflammatory microenvironment and especially endothelium. Neutrophil infiltration that takes place within 6–8 h from the initiation of the inflammatory process is followed by the recruitment of other cell populations, like monocytes, lymphocytes, and eosinophils, which either promote or drive the resolution of inflammation. Leukocyte infiltration into sites of infection or sterile inflammation is a tightly regulated process that follows a sequence of adhesive events, termed as leukocyte adhesion cascade. From the initial selectin-dependent leukocyte tethering to endothelial cells to the final migration of leukocytes into the subendothelium, this process depends on the interplay between leukocyte receptors and endothelial cell counter-receptors, as well as on the presence of endogenous inhibitors of leukocyte adhesion, enabling the targeted recruitment of leukocytes to inflamed tissues. Leukocyte recruitment can be divided in four steps: starting with capturing and rolling of leukocytes to the vessel wall, adhesion to endothelial cells and post-adhesion strengthening, leukocyte crawling, and transendothelial migration.

Leukocyte Recruitment

Leukocyte recruitment mainly takes place at the level of postcapillary venules. However, migration through the arterial wall is observed in atherosclerosis. To enable the infiltration of leukocytes at the site of inflammation, a series of alterations in endothelial cells and leukocytes takes place:

(a) Regulation of the expression of adhesion molecules in leukocytes
(b) Increased secretion of chemokines by endothelial cells
(c) Increased expression of adhesion molecules in the luminal surface of endothelial cells

Endothelial Cells

Under noninflamed conditions, the interaction between endothelial cells and leukocytes is minimal, enabling the unhindered flow of leukocytes. Resting endothelial cells do not express adhesion molecules on their luminal surface, including selectins and the integrin ligands, vascular cell adhesion molecule-1 (VCAM-1), and intercellular adhesion molecule-1 (ICAM-1). Upon inflammation, a rapid activation of endothelial cells takes place, which can be divided in rapid adaptations that do not depend on newly expressed proteins and in changes that depend on increased expression of proteins, regulated at the transcriptional level. The net result of this process is the increased blood flow at the site of inflammation, the increased endothelial leakage, and the recruitment and activation of leukocytes (Pober and Sessa 2007).

Vasoactive mediators, like histamine, act rapidly in venular endothelial cells via G-protein-coupled receptors (GPCRs), mediating the contraction of endothelial cells, increasing vascular leakiness and formation of inflammatory exudate, and promoting the release to the endothelial surface of proteins already formed and stored in cytosolic vesicles (Weibel-Palade bodies), including P-selectin and chemokines like IL-8 (CXC-chemokine ligand 8) and eotaxin-3, which regulate chemotaxis of granulocytes. Additionally, GPCR-dependent signaling induces the expression of lipid mediators, like platelet-activating factor (PAF), which activates leukocytes and platelets (Pober and Sessa 2007).

The effect of GPCR activation is short term and self-limited, and it is followed by more sustained changes in endothelial cells that depend on the effect of inflammatory cytokines, including IL-1 and TNF. These cytokines induce the expression E-selectin, ICAM-1, and VCAM-1 by endothelial cells. Signaling through IL-1 receptor and TNF receptor 1 also promotes the production and release of the chemokines IL-8 and CC-chemokine ligand 2 (CCL2), which are major neutrophil and monocyte activators, respectively. Additionally, endothelial cells express Toll-like receptors, which enable their direct response to pathogen-derived products or endogenous danger signals and induction of similar responses (Pober and Sessa 2007, 2015).

Leukocyte Recruitment Cascade

The inflammation-dependent alterations in endothelial cells are the initiatory event that enables the recruitment of inflammatory leukocytes in a continuum of events, which are based on adhesive interactions between leukocytes and endothelium (Fig. 1).

Tethering and Initiation of Rolling The initial step in leukocyte adhesion cascade is the capturing and rolling of circulating leukocytes to endothelial cells of postcapillary venules. This initiatory event is mediated by selectins (Herter and Zarbock 2013; Zarbock et al. 2011). Selectins are a group of type I membrane glycoproteins, composed of an aminoterminal lectin domain, an epidermal growth factor (EGF)-like domain, a different number of consensus repeat units, a transmembrane domain, and an intracellular cytoplasmic tail. Three subsets of selectins have been described, playing a critical role in leukocyte recruitment. E- and P-selectins are expressed by endothelial cells, and their expression on endothelial surface is induced under inflammation. P-selectin is stored in cytosolic vesicles (Weibel-Palade bodies) and released to endothelial surface after activation, while the expression of E-selectin is regulated at the transcriptional level. L-selectin is expressed in leukocytes. L-selectin shedding

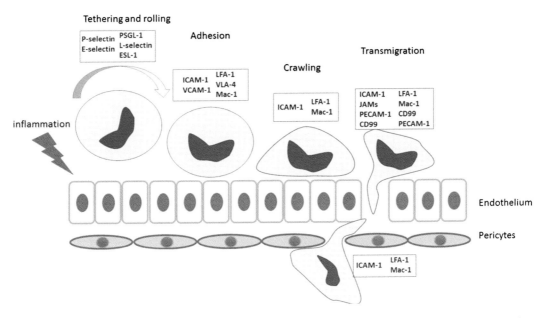

Leukocyte Recruitment, Fig. 1 The leukocyte recruitment cascade. Depiction of the different steps of leukocyte adhesion cascade showing the adhesion molecules involved in the process (*red*, endothelial/pericyte adhesion molecules; *blue*, leukocyte counter-receptors)

from leukocyte surface upon their activation alters leukocyte rolling velocity and modulates leukocyte recruitment. P-selectin glycoprotein ligand-1 (PSGL-1), CD44, and E-selectin ligand-1 (ESL-1) have been identified as receptors for selectins (Herter and Zarbock 2013; Zarbock et al. 2011; Hidalgo et al. 2007). Even though selectin-mediated adhesive interactions are transient and unable to immobilize leukocytes, they act as a break for the reduction of leukocyte velocity. The resultant exposure of leukocytes to chemokines expressed on endothelial surface mediates intracellular signals that enable integrin activation, which in turn mediates the subsequent steps of leukocyte adhesion cascade.

Slow Rolling and Adhesion Leukocyte integrins are the key players in the next steps of leukocyte adhesion cascade. Interaction between leukocyte integrins and their endothelial ligands promotes slow rolling, leukocyte adhesion and crawling, and finally transendothelial migration. Integrins are a family of transmembrane heterodimeric adhesion molecules, consisted of one α- and one β-subunit (Herter and Zarbock 2013; Mitroulis et al. 2015). Among the several members of this family, the α4 integrin, α4β1 (very late antigen-4, VLA-4), expressed in monocytes and T lymphocytes; the α4β7 (LPAM-1), expressed in lymphocytes; the β2 integrin, αLβ2 integrin (lymphocyte function-associated antigen-1, LFA-1), expressed in neutrophils, lymphocytes, and monocytes; and the αMβ2 integrin (macrophage-1 antigen, Mac-1; complement receptor 3, CR3), expressed in neutrophils and monocytes, participate in leukocyte recruitment. Binding of VLA-4 to VCAM-1 and binding of LFA-1 and Mac-1 to ICAM-1 and ICAM-2 are critical events in the induction of slow rolling and adhesion of leukocytes to endothelium (Herter and Zarbock 2013; Schmidt et al. 2013).

Different cell populations depend on different integrins during this step of adhesion cascade. Neutrophils depend on LFA-1 for the slowing of rolling and adhesion. On the other hand, VLA-4 is critical for slow rolling and adhesion in monocytes and lymphocytes, whereas LFA-1 participates in lymphocyte adhesion and β2 integrins in a lesser extent in monocyte adhesion (Herter and Zarbock 2013).

Even though integrin expression, and especially Mac-1, can be regulated by inflammatory stimuli at the transcriptional level (Sumagin et al. 2010), their function depends on the conformational status of their extracellular domain, which regulates the strength of a single integrin bond with the integrin receptors, termed as affinity (Herter and Zarbock 2013; Luo and Springer 2006; Luo et al. 2007). The overall strength of integrin-based adhesion, which depends on the affinity and the number of adhesive bonds (valency), is termed as avidity. Upon adhesion to the endothelial surface, integrins are clustered at the binding surface increasing valency (Herter and Zarbock 2013). Under steady-state conditions, integrins exist in a bent, inactive conformation and need activation signals to change their conformation status toward an extended, active conformation of their extracellular domain (Luo and Springer 2006; Luo et al. 2007). This process increases integrin affinity to its ligands, enabling the exposure of their binding site and promoting the accomplishment of their role in leukocyte recruitment (Luo and Springer 2006; Luo et al. 2007). To swift from the inactive conformation toward the high-affinity conformation, inside-out activation of integrins by chemokines or through PSGL-1 is needed (Alon et al. 2003; Herter and Zarbock 2013; Mitroulis et al. 2015; Zarbock et al. 2012). On the other hand, the initial integrin binding exerts outside-in signaling, further causing conformational changes and increasing integrin affinity (Luo and Springer 2006; Luo et al. 2007).

E-selectin binding to PSGL-1 promotes a cascade of intracellular signaling events that, in conjunction with chemokine-derived signaling, enables inside-out integrin activation. This pathway involves the Src family kinases – Fgr, Hck, and Lyn – as well as the adaptor proteins DAP12 and FcRγ and the spleen tyrosine kinase. The subsequent activation of phospholipase C (PLC)γ2 results in the activation of the small GTPase RAS-related protein 1 (Rap1), in a pathway that involves CalDAG-GEFI and p38 MAP kinase, which promotes integrin affinity (Herter and Zarbock 2013; Mitroulis et al. 2015; Zarbock et al. 2012).

In addition to signals derived from PSGL-1, chemokines are able to cause conformational changes in integrins, through GPCRs. The intracellular signaling that follows GPCR activation by chemokines is complex and involves GEF-dependent activation of Rap1; phospholipase C activation, with subsequent intracellular Ca2+ flux from the endoplasmatic reticulum; generation of inositol 1,4,5-trisphosphate; and interaction of proteins like kindlin-3, talin-1, cytohesin-1, and 14-3-3, with the cytoplasmin domain of the β subunit of integrins, which promotes the separation of the two intracellular domains, resulting in the change of the conformational status of the extracellular domains (Alon et al. 2003; Herter and Zarbock 2013; Mitroulis et al. 2015).

Crawling After their adhesion on endothelial surface, leukocytes actively move, in order to find an optimal site for their migration into the inflamed tissue. During this process, which is termed as intraluminal crawling, it is believed that leukocytes are guided by intraluminal chemokine gradients and move sometimes perpendicularly or against the blood flow toward an endothelial junction (Herter and Zarbock 2013; Nourshargh and Alon 2014). Crawling also depends on integrins of the β2 family (Phillipson et al. 2006; Sumagin et al. 2010); however, there are differences among different leukocyte subsets. Even though LFA-1 is the critical β2 integrin in neutrophil adhesion, Mac-1 is essential for their crawling, as shown by the reduction in the percentage, the distance, and the velocity of crawling neutrophils in the in vivo cremaster muscle inflammatory model, when Mac-1 is blocked or genetically ablated (Phillipson et al. 2006; Sumagin et al. 2010). In monocytes, LFA-1 has a critical role in their crawling onto noninflammed endothelium, known as patrolling. During inflammation however, Mac-1 expression in leukocytes is increased and takes over as the main integrin in monocyte crawling (Sumagin et al. 2010).

After leukocyte adhesion to endothelial cells, the interaction between integrins and their counter-receptors on endothelial cell surface induces an outside-in integrin signaling that is able to cause alterations in cytoskeleton and is important for cell crawling and transmigration. The signaling cascade that follows integrin activation includes activation of Src tyrosine kinases and recruitment and phosphorylation of Syk kinase and the GEF VAV-1.

Transmigration The final step of leukocyte recruitment cascade is transmigration, also known as diapedesis or extravasation. When leukocytes reach an appropriate site, they migrate mainly paracellulary, through the endothelial junction (Nourshargh and Alon 2014). However, transcellular migration has also been described and is thought to play a significant role in the transmigration in the central nervous system, due to the decreased permeability of blood vessels (Engelhardt and Ransohoff 2012). Leukocyte transendothelial migration (TEM) is based on bidirectional interactions between leukocytes and adhesion molecules that are expressed in the endothelial junctions. Homophilic and heterophilic interactions among adhesion molecules expressed on endothelial cells and leukocytes enable the accomplishment of this process (Muller 2011; Nourshargh and Alon 2014).

The β2 integrins, Mac-1 and LFA-1, and their counter-receptors ICAM-1 and ICAM-2 on endothelial cells are critical for this step of the leukocyte recruitment cascade. Additional adhesion molecules that have a role are platelet/endothelial cell adhesion molecule-1 (PECAM, CD31); the junctional adhesion molecules JAM-A, JAM-B, and JAM-C; and CD99 (Nourshargh and Alon 2014). In order to open the endothelial junctions and overcome the endothelial barriers, signaling that follows leukocyte integrin interaction with ICAM-1 is needed. This signaling involves the increase of intracellular Ca2+ in endothelial cells and results in cytoskeletal changes in endothelial cells. ICAM-1 ligation further promotes the formation of clusters, which are enriched with ICAM-1 and VCAM-1. These clusters, known as transmigratory cups, assist leukocyte firm adhesion and TEM. ICAM-1 interaction with integrins further mediates the translocation of ICAM-1 clusters to caveolae- and actin-rich domains, which initiates the recruitment of caveolae and caveolae-like vesicles, which form the vesiculo-vacuolar organelle (VVO) and lead to the formation of transmigration pores (Nourshargh and Alon 2014; Schmidt et al. 2013).

Leukocytes that reach the endothelial junction interact with several adhesion molecules including JAMs, PECAM-1, and CD99. Homophilic interactions between PECAM-1 expressed on endothelial cells and leukocytes and CD99 homophilic and heterophilic interactions with CD99L2 further participate in leukocyte transmigration. VE-cadherin has a critical role in the gatekeeping of the integrity of endothelial barrier. To overcome this barrier, a transient loss of VE-cadherin via a reversible endocytosis takes place in close proximity to transmigrating leukocytes (Muller 2011; Nourshargh and Alon 2014; Schmidt et al. 2013).

Even though a lot is known regarding TEM, fewer studies have focused on the subsequent steps of endothelial migration through venule walls. Except from endothelial cells, leukocytes have to migrate through pericytes and the vascular basement membrane. It is known that neutrophils that have passed through the endothelial cell layer crawl along pericytes in order to find gaps for transmigration, in a process that depends on neutrophil LFA-1 and Mac-1 and pericyte ICAM-1. Inflammatory cytokines have been shown to cause changes in pericyte shape, which results in the enlargement of these gaps, enabling neutrophil migration (Proebstl et al. 2012).

Inhibitors of Leukocyte Recruitment

The fine tuning of leukocyte recruitment is achieved not only through the regulation of the expression of adhesion molecules on endothelial cells and regulation of counter-receptor activity in leukocytes but also through the presence of endogenous inhibitors.

Developmental endothelial locus-1 (Del-1) is the first characterized endogenous inhibitor of leukocyte recruitment. It is a glycoprotein produced and released by endothelial cells and associated with endothelial cell membrane and extracellular matrix. Del-1 has been identified as a negative regulator of leukocyte recruitment by binding to LFA-1 and therefore antagonizing its interaction with ICAM-1 (Mitroulis et al. 2015; Choi et al. 2008).

Pentraxin-3 (PTX-3), a member of the long pentraxin family of soluble pattern-recognition molecules, is another inhibitor of leukocyte recruitment, expressed by neutrophils and released during inflammation. PTX-3 acts as the initial stages of the leukocyte recruitment cascade. By binding to P-selectin, it antagonizes its interaction with PSGL-1, inhibiting leukocyte rolling (Deban et al. 2010; Herter and Zarbock 2013).

Growth differentiation factor (GDF)-15, a member of the transforming growth factor (TGF)-β superfamily, is an endogenous inhibitor, whose expression is restricted to coronary arteries. By activating Cdc42 GTPase and inhibiting of Rap1 GTPase, GDF-15 attenuates chemokine-mediated activation of β2 integrins in the context of ischemic injury (Kempf et al. 2011; Mitroulis et al. 2015).

Leukocyte Recruitment in Human Disease

The importance of the leukocyte recruitment cascade in the shaping of inflammatory responses has been clearly confirmed by the discovery of a group of clinical entities named leukocyte adhesion deficiency (LAD) syndromes. These rare pathologies give rise to primary immunodeficiency, associated with defects in leukocyte migration, and each one is inherited as an autosomal recessive trait (Schmidt et al. 2013).

Until now, three types of LAD have been identified, each one demonstrating defects in different steps of the cascade. The most common type, reported in several hundreds of patients, is LAD-I. In this type, leukocytes fail to perform firm adhesion to endothelium. Specifically, mutations in the ITGB2 gene, which encodes the β2 integrin subunit CD18, have been linked to LAD-I. Therefore, the β2 subunit is absent or produced in very low levels, and its expression levels have been inversely correlated with the severity of the observed symptoms (Kishimoto et al. 1987; Schmidt et al. 2013).

Impaired leukocyte rolling, early in the leukocyte recruitment cascade, has been described in LAD-II (also called congenital disorder of glycosylation, type IIc). The disease is extremely rare with only a few reported cases. The development of this phenotype has been attributed to defective fucose metabolism. This is caused due to mutations in the gene SLC35C1 that is responsible for the expression of the GDP-fucose transporter. The mutated version of this enzyme lacks the ability to convert GDP mannose to fucose. The fucosylation of the selectin ligand sialyl Lewis x that is present on leukocyte surface is impaired, thus affecting dramatically leukocyte rolling process. Another feature of LAD-II is the Bombay blood type as a result of the absence of the H antigen that also incorporates fucose (Hidalgo et al. 2003; Luhn et al. 2001; Schmidt et al. 2013).

LAD-III (also known as LAD-I variant) is also an extremely rare disease, with only a few reported cases. Mutations in the gene kindlin-3 (fermitin family homolog 3, FERMT3) have been linked to the pathophysiology of LAD-III. This gene encodes the cytoplasmic protein kindlin-3, expressed in hematopoietic cells including platelets and erythrocytes. Given the essential role of kindlin-3 on the activation of the integrins beta 1, beta 2, and beta 3, mutations in this protein result in disrupted inside-out signaling and subsequent deregulation of leukocyte and platelet adhesion (Schmidt et al. 2013; Svensson et al. 2009).

The severity of the symptoms in patients diagnosed with LAD varies depending on the components of the cascade that have been affected. Leukocytosis, associated with extreme neutrophilia, and increased susceptibility to recurrent bacterial infections and periodontitis characterize all types of LAD. LAD-I is associated with life-threatening bacterial infections, which develops during infancy or childhood. An early finding is the delay in umbilical cord separation. In severe forms, the disease is fatal within the first 2 years of life, whereas patients with a less severe form can survive in adulthood. Impaired wound healing has been associated with LAD-I. Moreover, the absence of signs of inflammation and pus formation at the site of infection is observed in patients with LAD-I and LAD-III. Growth defects and mental retardation are characteristic findings among the patients with LAD-II. Growth and mental retardation dominate the phenotype of this syndrome, while infections are not as severe and life-threatening as in other forms of LAD. LAD-III has a similarity to LAD-I phenotype, with symptoms starting during infancy or early childhood. Except from the impaired leukocyte adhesion, bleeding disorders have also been reported in LAD-III. This is due to defects in the platelet integrin αIIbβ3, thus demonstrating symptoms similar to those observed in Glanzmann thrombasthenia. Moreover, impaired osteoclast function can result in development of osteopetrosis.

A lot of effort focusing on the development of efficient therapies for the treatment of LAD has been undertaken. Antibiotics are often administered in an effort to treat the repeated infections occurring in patients with LAD. In some cases, LAD-II patients responded to oral administration of fucose supplementation. Patients diagnosed with LAD-III are required to receive blood transfusions. However, hematopoietic stem cell transplantation is considered as the only curative therapeutic approach in LAD-I and LAD-III (van de Vijver et al. 2013).

Taken together, the investigation of the underlying pathophysiology of LAD syndromes provides novel insights into the orchestration of immune and inflammatory responses.

Inhibitors of Leukocyte Recruitment in Clinical Use

The involvement of leukocyte recruitment in the pathogenesis of a broad spectrum of human disorders, ranging from infectious disorders to autoimmune and autoinflammatory syndromes, rendered several molecules that participate in this process as potential therapeutic targets. Monoclonal antibodies targeting integrins have been effectively used in clinical practice for the treatment of autoimmune and inflammatory diseases. However, despite their efficiency, their use is restricted due to the appearance of severe side effects.

Efalizumab, a humanized monoclonal IgG1 antibody that blocks LFA-1 interaction with ICAM-1, has been used for the treatment of patients with psoriasis. Even though it showed effectiveness against the dermatologic features of the disease, it has been withdrawn in 2009 due to the high risk of John Cunningham (JC) polyomavirus reactivation and development of progressive multifocal leukoencephalopathy (PML) (Dedrick et al. 2002; Gordon et al. 2003; Mitroulis et al. 2015).

Natalizumab, a humanized monoclonal IgG4 antibody against the α4-integrin subunit, blocks both α4β1 and α4β7 integrins. Natalizumab has been shown to be effective in the treatment of multiple sclerosis by inhibiting leukocyte migration into the central nervous system. Natalizumab is also associated with an increased risk for PML development, restricting its use (Bloomgren et al. 2012; Mitroulis et al. 2015).

Vedolizumab (Millennium Pharmaceuticals) and etrolizumab are humanized monoclonal antibodies targeting α4β7 integrin, inhibiting lymphocyte interaction with endothelial MAdCAM, used for the treatment of severe forms of inflammatory bowel disease. By sparing α4β1, the use of these antibodies is not associated with increased risk for PML (Bamias et al. 2013; Mitroulis et al. 2015).

Conclusion

Leukocyte recruitment at the sites of inflammation is a complex process that involves endothelial cells in one side and leukocytes on the other. Even though it is separated in different steps, it is a tightly regulated continuum of events that depend on each other, enabling the appropriate infiltration of leukocytes according to the needs. The presence of regulatory molecules in the cascade ensures its proper function avoiding phenomena of aberrant and uncontrolled activation. The discovery of LAD syndromes, where defective operation of the leukocyte recruitment takes place, addresses the importance of the smooth function of the cascade in the shaping of proper immune responses. The further elucidation of mechanisms that govern the regulation of leukocyte recruitment may lead to the development of targeted effective therapeutic approaches for the treatment of inflammation-associated pathologies.

References

Alon, R., Grabovsky, V., & Feigelson, S. (2003). Chemokine induction of integrin adhesiveness on rolling and arrested leukocytes local signaling events or global stepwise activation? *Microcirculation, 10*, 297–311.

Bamias, G., Clark, D. J., & Rivera-Nieves, J. (2013). Leukocyte traffic blockade as a therapeutic strategy in inflammatory bowel disease. *Current Drug Targets, 14*, 1490–1500.

Bloomgren, G., Richman, S., Hotermans, C., Subramanyam, M., Goelz, S., Natarajan, A., et al. (2012). Risk of natalizumab-associated progressive multifocal leukoencephalopathy. *The New England Journal of Medicine, 366*, 1870–1880.

Choi, E. Y., Chavakis, E., Czabanka, M. A., Langer, H. F., Fraemohs, L., Economopoulou, M., et al. (2008). Del-1, an endogenous leukocyte-endothelial adhesion inhibitor, limits inflammatory cell recruitment. *Science, 322*, 1101–1104.

Deban, L., Russo, R. C., Sironi, M., Moalli, F., Scanziani, M., Zambelli, V., et al. (2010). Regulation of leukocyte recruitment by the long pentraxin PTX3. *Nature Immunology, 11*, 328–334.

Dedrick, R. L., Walicke, P., & Garovoy, M. (2002). Anti-adhesion antibodies efalizumab, a humanized anti-CD11a monoclonal antibody. *Transplant Immunology, 9*, 181–186.

Engelhardt, B., & Ransohoff, R. M. (2012). Capture, crawl, cross: The T cell code to breach the blood–brain barriers. *Trends in Immunology, 33*, 579–589.

Gordon, K. B., Papp, K. A., Hamilton, T. K., Walicke, P. A., Dummer, W., Li, N., et al. (2003). Efalizumab for patients with moderate to severe plaque psoriasis: A randomized controlled trial. *JAMA, 290*, 3073–3080.

Herter, J., & Zarbock, A. (2013). Integrin regulation during leukocyte recruitment. *The Journal of Immunology, 190*, 4451–4457.

Hidalgo, A., Ma, S., Peired, A. J., Weiss, L. A., Cunningham-Rundles, C., & Frenette, P. S. (2003). Insights into leukocyte adhesion deficiency type 2 from a novel mutation in the GDP-fucose transporter gene. *Blood, 101*, 1705–1712.

Hidalgo, A., Peired, A. J., Wild, M. K., Vestweber, D., & Frenette, P. S. (2007). Complete identification of E-selectin ligands on neutrophils reveals distinct functions of PSGL-1, ESL-1, and CD44. *Immunity, 26*, 477–489.

Kempf, T., Zarbock, A., Widera, C., Butz, S., Stadtmann, A., Rossaint, J., et al. (2011). GDF-15 is an inhibitor of leukocyte integrin activation required for survival after myocardial infarction in mice. *Nature Medicine, 17*, 581–588.

Kishimoto, T. K., Hollander, N., Roberts, T. M., Anderson, D. C., & Springer, T. A. (1987). Heterogeneous mutations in the beta subunit common to the LFA-1, Mac-1, and p150,95 glycoproteins cause leukocyte adhesion deficiency. *Cell, 50*, 193–202.

Luhn, K., Wild, M. K., Eckhardt, M., Gerardy-Schahn, R., & Vestweber, D. (2001). The gene defective in leukocyte adhesion deficiency II encodes a putative GDP-fucose transporter. *Nature Genetics, 28*, 69–72.

Luo, B. H., & Springer, T. A. (2006). Integrin structures and conformational signaling. *Current Opinion in Cell Biology, 18*, 579–586.

Luo, B. H., Carman, C. V., & Springer, T. A. (2007). Structural basis of integrin regulation and signaling. *Annual Review of Immunology, 25*, 619–647.

Mitroulis, I., Alexaki, V. I., Kourtzelis, I., Ziogas, A., Hajishengallis, G., & Chavakis, T. (2015). Leukocyte integrins: Role in leukocyte recruitment and as therapeutic targets in inflammatory disease. *Pharmacology and Therapeutics, 147*, 123–135.

Muller, W. A. (2011). Mechanisms of leukocyte transendothelial migration. *Annual Review of Pathology, 6*, 323–344.

Nourshargh, S., & Alon, R. (2014). Leukocyte migration into inflamed tissues. *Immunity, 41*, 694–707.

Phillipson, M., Heit, B., Colarusso, P., Liu, L., Ballantyne, C. M., & Kubes, P. (2006). Intraluminal crawling of neutrophils to emigration sites: A molecularly distinct process from adhesion in the recruitment cascade. *The Journal of Experimental Medicine, 203*, 2569–2575.

Pober, J. S., & Sessa, W. C. (2007). Evolving functions of endothelial cells in inflammation. *Nature Reviews Immunology, 7*, 803–815.

Pober, J. S., & Sessa, W. C. (2015). Inflammation and the blood microvascular system. *Cold Spring Harbor Perspectives in Biology, 7*, a016345.

Proebstl, D., Voisin, M. B., Woodfin, A., Whiteford, J., D'Acquisto, F., Jones, G. E., et al. (2012). Pericytes support neutrophil subendothelial cell crawling and breaching of venular walls in vivo. *The Journal of Experimental Medicine, 209*, 1219–1234.

Schmidt, S., Moser, M., & Sperandio, M. (2013). The molecular basis of leukocyte recruitment and its deficiencies. *Molecular Immunology, 55*, 49–58.

Sumagin, R., Prizant, H., Lomakina, E., Waugh, R. E., & Sarelius, I. H. (2010). LFA-1 and Mac-1 define characteristically different intralumenal crawling and emigration patterns for monocytes and neutrophils in situ. *The Journal of Immunology, 185*, 7057–7066.

Svensson, L., Howarth, K., McDowall, A., Patzak, I., Evans, R., Ussar, S., et al. (2009). Leukocyte adhesion deficiency-III is caused by mutations in KINDLIN3 affecting integrin activation. *Nature Medicine, 15*, 306–312.

van de Vijver, E., van den Berg, T. K., & Kuijpers, T. W. (2013). Leukocyte adhesion deficiencies. *Hematology/Oncology Clinics of North America, 27*, 101–116, viii.

Zarbock, A., Ley, K., McEver, R. P., & Hidalgo, A. (2011). Leukocyte ligands for endothelial selectins: Specialized glycoconjugates that mediate rolling and signaling under flow. *Blood, 118*, 6743–6751.

Zarbock, A., Kempf, T., Wollert, K. C., & Vestweber, D. (2012). Leukocyte integrin activation and deactivation: Novel mechanisms of balancing inflammation. *Journal of Molecular Medicine (Berlin), 90*, 353–359.

Leukotrienes

Magnus Bäck
Center for Molecular Medicine, Department of Medicine, Karolinska Institutet, Stockholm, Sweden

Definition

Leukotrienes, i.e., leukotrienes (LT) B_4, C_4, D_4, and E_4, are lipid mediators of inflammation, derived from the 5-lipoxygenase pathway of arachidonic acid metabolism (Samuelsson 1983; Fig. 1). The name originates from their leukocyte origin (leuko-) and their three conjugated double bonds (triene). Leukotrienes C_4, D_4, and E_4 all contain cysteine (Fig. 1) and are therefore referred to as the cysteinyl-leukotrienes in order to distinguish them from LTB_4, which is a non-cysteine-containing dihydroxy-leukotriene. Before the structures of the cysteinyl-leukotrienes were known, a slow-reacting substance of anaphylaxis (SRS-A) was shown to be released from animal lungs after challenge with a snake venom (for original references, see Bäck 2002). This substance induced slow contractile responses of guinea pig intestinal preparations and was later related to anaphylactic reactions and bronchoconstriction.

Leukotrienes exert their actions via 7-transmembrane G-protein-coupled receptors divided into two subclasses: BLT receptors, activated by LTB_4, and CysLT receptors, activated by the cysteinyl-leukotrienes. There are four cloned leukotriene receptors leading to further subdivision of the BLT and CysLT receptors into BLT_1 and BLT_2 and $CysLT_1$ and $CysLT_2$, respectively (Bäck et al. 2011). In addition, leukotrienes may transduce their biological effects also through additional receptors (Bäck et al. 2014), which will be further discussed below. A detailed description of leukotriene receptors is available at http://www.iuphar-db.org/DATABASE/FamilyMenuForward?familyId=35.

Classically, LTB_4 is described as a mediator of neutrophil chemotaxis, whereas the cysteinyl-leukotrienes are mainly associated with bronchoconstriction and asthma. However, as will be discussed in this entry, leukotrienes induce pro-inflammatory signaling in several cell types, with pathophysiological significance in a wide spectrum of inflammatory diseases (Fig. 2; Peters-Golden and Henderson 2007).

Biosynthesis and Release

Biochemistry

The release of SRS-A described above illustrates the induction of LT synthesis by phospholipase A_2 (PLA_2), which is the major enzymatic component of snake venom. Phospholipases hydrolyze phospholipids to release fatty acids, and the PLA_2 family consists of 15 different members, which include secreted ($sPLA_2$), lipoprotein-associated ($Lp-PLA_2$) and cytosolic ($cPLA_2$) forms (Burke and Dennis 2009). The intracellular $cPLA_2$ (Group IVA) displays calcium-dependent activation to metabolize phospholipids containing arachidonic acid bound at the sn-2 position. The resulting liberation of arachidonic acid provides the substrate for the formation of leukotrienes (Fig. 1; Burke and Dennis 2009).

In the following enzymatic step, 5-lipoxygenase (5-LO) oxygenates carbon number 5 of arachidonic acid to form 5-HpETE (5-hydroperoxyeicosatetraenoic acid; Fig. 1). Subsequently, removal of hydrogen at carbon 10 from 5-HpETE leads to formation of the epoxide intermediate LTA_4, which serves as precursor for leukotriene synthesis (Shimizu et al. 1984). While it is detected in the cytosolic or nucleosolic fraction of resting cells, cellular activation leads to 5-lipoxygenase translocation to the nuclear envelope in a calcium-dependent

Leukotrienes, Fig. 1 Leukotriene biosynthesis through the 5-lipoxygenase (5-LO) pathway of arachidonic acid metabolism. *Abbreviations*: 5-HpETE 5-hydroperoxyeicosatetraenoic acid, *FLAP* 5-LO-activating protein, *γ-GT* γ-glutamyltranspeptidase, *LT* leukotriene, LTA_4H LTA_4 hydrolase, LTC_4S LTC_4 synthase

manner. This reaction requires a 5-lipoxygenase-activating protein (FLAP), which, in addition to the actual enzyme, is a target for inhibitors of leukotriene formation (Fig. 1, Evans et al. 2008).

The enzyme LTA_4 hydrolase is located in the cytosol and stereospecifically adds water to carbon 12 of LTA_4, leading to formation of LTB_4 (Haeggstrom 2004). As an integral component of inactivating pathways, ω-oxidation of LTB_4 leads to the formation of 20-hydroxy-LTB_4 and 20-carboxy-LTB_4, which also exhibit biological activities at BLT receptors (Fig. 1; Yokomizo et al. 2001).

On the other hand, the enzyme LTC_4 synthase (LTC_4S) is a microsomal glutathione S-transferase, which conjugates LTA_4 with glutathione (GSH) to form LTC_4 (Fig. 1; Lam and Austen 2002). Subsequently, LTC_4 is transported to the extracellular space by an ATP-dependent transporter (MRP) that recognizes its glutathione moiety. LTC_4, being a glutathionyl eicosatetraenoic acid, shares its subsequent metabolism with GSH as a result of the activity of the membrane-bound enzyme γ-glutamyltranspeptidase (γ-GT), which cleaves the γ-glutamyl group of the GSH side chain of LTC_4, thus yielding LTD_4 (Fig. 1). The

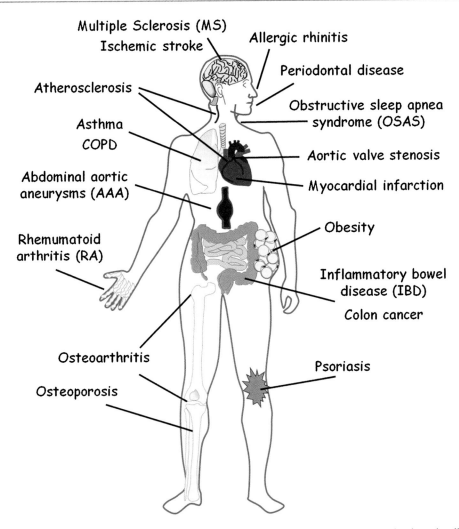

Leukotrienes, Fig. 2 Examples of inflammatory diseases, which have been associated with leukotriene signaling

subsequent metabolism of LTD_4 involves a dipeptidase that cleaves the peptide bond between the cysteinyl and glycyl residues in the LTD_4 side chain. The product, a leukotriene with a cysteinyl group at carbon number 6, is referred to as LTE_4 (Fig. 1).

Cellular Sources

The formation of LTA_4 via the 5-lipoxygenase pathway takes place in myeloid cells, e.g., leukocytes, macrophages, and mast cells (Samuelsson et al. 1987). However, 5-lipoxygenase expression is regulated by promoter methylation (Katryniok et al. 2010) and may be induced also in nonmyeloid cells through epigenetic mechanisms (Nagy and Bäck 2012). The next step in cysteinyl-leukotriene synthesis, conjugation of LTA_4 with GSH, can take place in endothelial cells, vascular smooth muscle cells, platelets, mast cells, and macrophages as well as in eosinophilic granulocytes (Claesson and Haeggström 1988). These differential activities mean that in some cells (e.g., eosinophils, macrophages, and mast cells), LTC_4 can be formed from arachidonic acid, whereas in cells that have LTC_4 synthase but lack 5-lipoxygenase, LTA_4 is needed as a substrate. The LTA_4 formed in one cell type can be donated to another cell type in order to form LTC_4, which is referred to as transcellular biosynthesis (Sala et al. 2010).

Biological Activities

LTB_4 and BLT Receptors

Chemotaxis, one of the principal biological activities of LTB_4, occurs via activation of the BLT_1 receptor subtype, which is the high-affinity LTB_4 receptor (Yokomizo et al. 2001). Binding studies, in addition, identified low-affinity sites in human granulocytes mediating lysosomal degranulation. Subsequent studies confirmed the presence of a second subtype of BLT receptor, which was identified during the analysis of the BLT_1 promoter. This receptor was named BLT_2 and exhibits low affinity for LTB_4 but high affinity for the by-product of thromboxane A_2 biosynthesis, 12-HHT (12(S)-Hydroxyheptadeca-5Z, 8E, 10E-trienoic acid; (Nakamura and Shimizu 2011).

In addition to neutrophil chemotaxis and secretion, LTB_4 may also stimulate monocytes/macrophages in terms of migration, upregulation of integrin expression, cyto- and chemokine production, and increased phagocytosis (Bäck et al. 2011; Nakamura and Shimizu 2011). The immunomodulatory effects of LTB_4 are not only mediated through effects on phagocytic leukocytes but also through adaptive immune circuits. For example, LTB_4 stimulation of T lymphocytes in vitro induces migration, enhanced production of interleukins and IFN-γ, as well as inhibition of differentiation toward immunosuppressive T regulatory cells. Studies in murine in vivo models have identified a key role of LTB_4 signaling in T lymphocyte recruitment (Bäck et al. 2011). Finally, mast cells migrate in response to LTB_4, which may have implications for allergic diseases (Bäck et al. 2011).

In addition to the effects described above on inflammatory cells, receptors for LTB_4 are expressed on nonmyeloid cells, such as smooth muscle and endothelial cells (Bäck et al. 2005), and LTB_4 for example, induces vascular smooth muscle cell migration and proliferation.

Cysteinyl-Leukotrienes and CysLT Receptors

The potent bronchoconstrictive effects of cysteinyl-leukotrienes in asthma (Capra et al. 2013; Drazen 2003) are mediated through the $CysLT_1$ receptor, which is the target for the clinically introduced antiasthmatic leukotriene receptor antagonists. The second CysLT receptor, $CysLT_2$, exhibits a wider expression pattern, extending also to cardiovascular and cerebral tissues.

In addition to mediating slow smooth muscle contractions and increase vascular permeability, which are the classical responses to cysteinyl-leukotrienes, CysLT receptors are also expressed on several leukocyte populations, such as monocytes, macrophages, eosinophil granulocytes, and dentritic cells, participating in leukocyte recruitment and activation at different sites of inflammation. For example, cysteinyl-leukotriene stimulation of these cells is coupled to cell adhesion and migration, as well as release of cytokines and other inflammatory mediators.

A recent observation is the nuclear/perinuclear expression pattern of CysLT receptors, which may be coupled to nuclear calcium signaling, cell survival, mitogenesis, and alterations of gene expression in different cell populations (Nagy et al. 2011; Nielsen et al. 2005).

Other Receptors Mediating the Biological Effects of Leukotrienes

In addition to activation of BLT receptors, LTB_4 is an endogenous ligand for peroxisome proliferator-activated receptor (PPAR) α, a ligand-activated transcription factor that belongs to the nuclear hormone receptor superfamily (Devchand et al. 1996). In addition, LTB_4 induces activation of the transient receptor potential cation channel subfamily V member 1 (TRPV1, the vanilloid receptor), albeit with tenfold lower affinity compared with capsaicin (Bäck et al. 2011).

The presence of additional CysLT receptors has received support from functional in vitro studies, radioligand binding and mice lacking both $CysLT_1$ and $CysLT_2$ receptors (Capra et al. 2013). Cysteinyl-LT signaling through

P2Y12 receptors and the orphan receptors GPR17 and GPR99 (Fig. 1) has been described, although not consequently replicated in all settings (Bäck et al. 2014; Capra et al. 2013).

Pathophysiological Relevance

Leukotriene signaling has been associated with several inflammatory diseases, some of which are depicted in Fig. 2. This paragraph provides an overview of some of the pathophysiological roles of leukotrienes. Further details on this subject can be found in recent reviews (Bäck et al. 2011, 2014; Capra et al. 2013; Peters-Golden and Henderson 2007).

Cardiovascular Disease

The atherosclerotic plaque is a site of chronic inflammation, at which leukotriene synthesizing enzymes and leukotriene receptors have been shown to be expressed (Bäck et al. 2005; Spanbroek et al. 2003). Leukotriene signaling has been implicated during several stages of atherosclerosis development (Fig. 2). For example, leukocyte recruitment and activation is a key feature of the early phase consisting of lipid retention and foam cell accumulation within the vascular wall, as well as in the development of advanced atherosclerotic lesions. The direct stimulating effects on vascular smooth muscle cell proliferation induced by leukotrienes may in addition contribute to a thickening of the vascular wall, referred to as intimal hyperplasia. Furthermore, leukotrienes activate matrix metalloproteinases (MMP), and degradation of extracellular matrix plays a major role in atherosclerotic plaque rupture. Finally, studies of either myocardial or cerebral ischemia and reperfusion indicate that leukotriene signaling participates in several pathophysiological aspects of both chronic and acute coronary and cerebrovascular disease (Bäck 2009). Either genetic or pharmacological targeting of 5-lipoxygenase and FLAP have, however, generated contradictory results in terms of atherosclerosis development in hyperlipidemic mouse models, whereas targeting LT receptors in general is associated with beneficial effects in different animal models of atherosclerosis and vascular injury (Bäck 2009).

In addition to atherosclerosis, the pro-inflammatory effects of leukotrienes, as well as their direct effects on vascular smooth muscle cells and MMPs, have been implicated in the weakening of the vascular wall associated with aneurysm formation. In abdominal aortic aneurysms (AAA; Fig. 2), leukotriene biosynthesis takes place both in neutrophil granulocytes within the intraluminal thrombus, which is formed at the aneurysmal lesion and in macrophage-rich areas within the adventitia (Houard et al. 2009). Studies of human AAA specimens have revealed neutrophil-derived LTB_4 as a major chemotactic factor released from the intraluminal thrombus, and mice lacking the BLT_1 receptor are protected from AAA development.

Inflammation is also a key feature in the calcification and thickening of the aortic valve, leading to aortic stenosis (Fig. 2). The leukotriene pathway was recently identified in human stenotic aortic valves, in which expression levels correlated with clinical data on stenosis severity (Nagy et al. 2011). Furthermore, LTC_4 induces activation of the interstitial cells of heart valves, associated with alterations of gene expression and in vitro calcification.

Respiratory Diseases

Asthma is a chronic inflammatory disease of the airways causing wheezing, breathlessness, and cough associated with bronchoconstriction, increased mucus formation, edema of the airway mucosa, increase in airway responsiveness to a variety of stimuli, and, in the long term, also structural changes in the airway wall (Fig. 2). The clinically introduced leukotriene synthesis inhibitors and receptor antagonists significantly reduce bronchoconstriction in asthmatics (Capra et al. 2013; Drazen 2003). In addition to bronchoconstriction, cysteinyl-leukotrienes also increase bronchial mucus secretion in human airways in vitro and in animal models in vivo, promote bronchial mucosal edema through increased airway plasma exudation, and increase airway responsiveness to methacholine in experimental models (Capra et al. 2013).

Although most studies on leukotrienes in asthma have focused on cysteinyl-leukotrienes, it should be mentioned that LTB$_4$ may also participate in chronic airway inflammation, as suggested by increased concentrations of LTB$_4$ in bronchial alveolar lavage (BAL) and exhaled breath condensates from patients with asthma. In addition, LTB$_4$ has been associated with increased bronchial responsiveness in asthmatics, and BLT$_1$ receptor targeting induces beneficial effects on airway lymphocyte- and eosinophil-mediated immunological responses in a murine model of ovalbumin (OVA) sensitization and challenge. In contrast, BLT$_2$-deficient mice exhibit more severe eosinophilic inflammation, induced by sensitization and elicitation by ovalbumin, accompanied by the reduced accumulation of IL-13 in the allergic airways, suggesting a dual role of LTB$_4$ signaling in airway inflammation depending on which receptor subtype is activated (Nakamura and Shimizu 2011). Finally, given the importance of airway neutrophil recruitment in chronic obstructive pulmonary disease (COPD), LTB$_4$ has also been suggested as a therapeutic target in this context (Bäck et al. 2011; Peters-Golden and Henderson 2007).

Obstructive sleep apnea (OSA) is characterized by recurrent episodes of partial or complete upper airway obstruction occurring during sleep, leading to chronic intermittent hypoxia. Leukotriene pathway activation by intermittent hypoxia has been implicated as a molecular link between OSA and the increased atherosclerosis observed in this patient group (Stanke-Labesque et al. 2014). Furthermore, leukotrienes have been associated with hypertrophy of adenotonsillar tissues, which is the predominant etiological factor involved in pediatric OSA.

Neurological Diseases

Multiple sclerosis (MS; Fig. 2) is an inflammatory disorder of the central nervous system associated with blood-brain barrier breakdown, inflammatory cell accumulation, and myelin degradation. Cerebrospinal fluid from patients with MS contains increased levels of LTB$_4$, and the leukotriene synthesizing enzyme, 5-lipoxygenase, was identified among the most strongly upregulated genes by microarray analysis of both human MS lesions and in brains from mice after induction of experimental allergic encephalomyelitis (EAE), which is a murine model of MS. In further support of LTB$_4$ as a potential therapeutic target in MS, either genetic or pharmacological targeting of the BLT$_1$ receptor induces beneficial effects in EAE in terms of blocking recruitment of eosinophils into the spinal cord and complete inhibition of paralysis. In addition, CysLT$_1$-receptor antagonism also reduces central nervous system inflammation and demyelination, associated with decreased EAE disease scores (Bäck et al. 2011). In contrast, 5-lipoxygenase-deficient mice exhibit an exacerbated EAE compared with WT mice, which suggests that endogenous anti-inflammatory mediators derived from 5-lipoxygenase metabolism, such as lipoxins, may also be involved in regulating this disease model and that leukotriene receptor antagonism may be a potential therapeutic option to explore in MS (see Bäck et al. 2011 and references therein).

Rheumatoid Arthritis

Rheumatoid arthritis is an autoimmune disease, which induces a chronic inflammatory response of the joints (Fig. 2). Leukotrienes are detected in synovial fluid from patients with rheumatoid arthritis and psoriasis arthritis. Several experimental studies have shown that mice lacking the BLT$_1$ and/or the BLT$_2$ receptor for LTB$_4$ are protected from the development of arthritis using different models, in association with decreased articular neutrophil recruitment. In addition, the CysLT$_1$ receptor antagonist, montelukast, reduces the number of mast cells in the inflamed paws of mice with collagen-induced arthritis (Bäck et al. 2011).

Gastrointestinal Diseases

Patients with Crohn's disease and ulcerative colitis present with chronic inflammation of the small intestines and/or colon and are referred to collectively as inflammatory bowel disease (Fig. 2). Leukotrienes are produced by infiltrating leukocytes in the colonic mucosa, and several findings support leukotrienes as mediators of inflammatory bowel disease. In contrast, mice deficient in the

BLT$_2$ subtype of LTB$_4$ receptor exhibit a more severe colitis induced by dextran sulfate, associated with a loss of intestinal barrier function (Nakamura and Shimizu 2011), suggesting a protective role of the BLT$_2$ receptor in intestinal inflammation. It should, however, be considered that other ligands than LTB$_4$ may be involved in the exacerbated colitis observed after BLT$_2$R knockout since the BLT$_2$ receptor agonist 12-HHT is highly increased in inflamed colonic biopsies.

Inflammatory bowel disease may predispose to the development of colon cancer, and in this context, CysLT receptors may transduce colon adenocarcinoma cell proliferation and survival, suggesting direct tumorigenic effects of leukotriene signaling (Nielsen et al. 2005).

Cancer

In addition to the findings mentioned above in colon cancer, leukotriene signaling has also been associated with cancer of the prostate, breast, bladder, brain, etc. Several signaling pathways are activated by leukotrienes in tumor cell lines, in association with tumor cell proliferation, differentiation, and reduced apoptosis. In addition to cancer growth, leukotrienes may also stimulate cancer invasiveness and metastasis, for example, through stimulation of MMPs.

Modulation by Drugs

Leukotriene Synthesis Inhibitors

Reducing the iron in the ferrous form at the active site of 5-lipoxygenase will inactivate leukotriene biosynthetic capacity of the enzyme, and drugs that either alter redox state or chelate iron have been developed as 5-lipoxygenase inhibitors. An example of an iron ligand inhibitor is zileuton, which was the first antileukotriene to be approved for clinical use in asthma (Drazen 2003). The relatively short half-life of 2–3 h, however, determines the requirement for quadruple daily doses of zileuton. The follow-up compound, atreleuton (ABT-761, VIA2291), with a longer half-life, was recently reported to reduce leukotriene synthesis in patients with acute coronary syndrome and in a subgroup of 34 patients examined by coronary CT and to significantly reduce atherosclerotic plaque volume after 24 weeks of atreleuton treatment compared with placebo (Bäck 2009).

FLAP was first identified on the basis of its ability to bind the indole LT synthesis inhibitor MK886 (Evans et al. 2008), which has been shown to inhibit leukotriene biosynthesis after allergen challenge in asthmatic subjects. Also another FLAP antagonist, the quinoline veliflapon (BAYx1005, DG031) has been evaluated in clinical trials of asthma, COPD, and cardiovascular biomarkers (References in Bäck 2009).

An alternative for blockade of leukotriene biosynthesis is inhibition of LTA$_4$ hydrolase, and such compounds have been developed either from LTA$_4$ analogs or from inhibitors of the dipeptidase activity of the enzyme (Haeggström 2004). For example, the angiotensin-converting enzyme (ACE) inhibitor captopril and the aminopeptidase inhibitor bestatin have been reported to inhibit LTB$_4$ formation through interaction with LTA$_4$ hydrolase. Studies on the carboxylic acid JNJ-26993135 suggested that LTA$_4$ hydrolase inhibition blocks LTB$_4$ production without affecting cysteinyl-leukotriene production, suggesting that this class of drugs may not shunt LTA$_4$ metabolism toward cysteinyl-leukotriene biosynthesis but rather toward anti-inflammatory lipoxins.

BLT Receptor Antagonists

BLT receptor antagonists have been evaluated in trials in rheumatoid arthritis, COPD, solid tumors, psoriasis, and cystic fibrosis (Peters-Golden and Henderson 2007). An example of this class of antileukotrienes is the biphenylyl substituted chroman carboxylic acid CP105696, which is a selective BLT$_1$ receptor antagonist used in several experimental studies and which inhibits LTB$_4$-induced CD11b/CD18 expression after oral administration. Amelubant (BIIL284) is a prodrug which inhibits Mac-1 expression on neutrophils and whose active metabolites exhibit antagonistic activities against both the BLT$_1$ and BLT$_2$ receptors. The hydroxyacetophenone, LY293111 (VML 295), reduced the number of neutrophils in BAL derived from treated asthmatic patients compared with placebo. Finally, the trisubstituted benzene, ONO-4057, is a dual antagonist of both BLT$_1$ and BLT$_2$ receptors developed for ulcerative colitis, psoriasis, and Behcet's disease. It should, however, be pointed out that none of these antagonists have been introduced to the market (references for each compound and study are provided in Bäck 2009).

CysLT Receptor Antagonists

A number of selective and potent CysLT$_1$ receptor antagonists have been developed, and montelukast (MK0476), zafirlukast (ICI204219), and pranlukast (ONO1078) have been clinically introduced for the treatment of asthma and allergic rhinitis (Capra et al. 2013). Recently, a pharmacoepidemiological study revealed, in addition, a decreased cardiovascular risk associated with leukotriene receptor antagonist exposure. In the latter study, the use of the CysLT$_1$ receptor antagonist montelukast conferred significant protective effects against recurrent stroke in subjects not taking angiotensin-modifying drugs (ACE inhibitors and/or angiotensin receptor blockers) and a decreased risk of recurrent myocardial infarction in males but not in females. Taken together, these observations suggested a generalizability of the anti-inflammatory effects of clinically used leukotriene modifiers beyond pulmonary disease (Ingelsson et al. 2012).

References

Bäck, M. (2002). Studies of receptors and modulatory mechanisms in functional responses to cysteinyl-leukotrienes in smooth muscle. *Acta Physiologica Scandinavica Supplementum, 648*, 1–55.

Bäck, M. (2009). Inhibitors of the 5-lipoxygenase pathway in atherosclerosis. *Current Pharmaceutical Design, 15*(27), 3116–3132.

Bäck, M., Bu, D. X., Branstrom, R., Sheikine, Y., Yan, Z. Q., & Hansson, G. K. (2005). Leukotriene B4 signaling through NF-kappaB-dependent BLT1 receptors on vascular smooth muscle cells in atherosclerosis and intimal hyperplasia. *Proceedings of the National Academy of Sciences of the United States of America, 102*(48), 17501–17506. doi:10.1073/pnas.0505845102.

Bäck, M., Dahlen, S. E., Drazen, J. M., Evans, J. F., Serhan, C. N., Shimizu, T., et al. (2011). International union of basic and clinical pharmacology. LXXXIV: Leukotriene receptor nomenclature, distribution, and pathophysiological functions. *Pharmacological Reviews, 63*(3), 539–584. doi:10.1124/pr.110.004184.

Bäck, M., Powell, W. S., Dahlen, S. E., Drazen, J. M., Evans, J. F., Serhan, C. N., et al. (2014). Update on leukotriene, lipoxin and oxoeicosanoid receptors: IUPHAR review 7. *British Journal of Pharmacology, 171*(15), 3551–3574. doi:10.1111/bph.12665.

Burke, J. E., & Dennis, E. A. (2009). Phospholipase A2 structure/function, mechanism, and signaling. *Journal of Lipid Research, 50*(Suppl), S237–S242. doi:10.1194/jlr.R800033-JLR200.

Capra, V., Bäck, M., Barbieri, S. S., Camera, M., Tremoli, E., & Rovati, G. E. (2013). Eicosanoids and their drugs in cardiovascular diseases: Focus on atherosclerosis and stroke. *Medicinal Research Reviews, 33*(2), 364–438. doi:10.1002/med.21251.

Claesson, H. E., & Haeggström, J. (1988). Human endothelial cells stimulate leukotriene synthesis and convert granulocyte released leukotriene A4 into leukotrienes B4, C4, D4 and E4. *European Journal of Biochemistry, 173*(1), 93–100.

Devchand, P. R., Keller, H., Peters, J. M., Vazquez, M., Gonzalez, F. J., & Wahli, W. (1996). The PPARalpha-leukotriene B4 pathway to inflammation control. *Nature, 384*(6604), 39–43. doi:10.1038/384039a0.

Drazen, J. M. (2003). Leukotrienes in asthma. *Advances in Experimental Medicine and Biology, 525*, 1–5.

Evans, J. F., Ferguson, A. D., Mosley, R. T., & Hutchinson, J. H. (2008). What's all the FLAP about?: 5-lipoxygenase-activating protein inhibitors for inflammatory diseases. *Trends in Pharmacological Sciences, 29*(2), 72–78. doi:10.1016/j.tips.2007.11.006.

Haeggstrom, J. Z. (2004). Leukotriene A4 hydrolase/aminopeptidase, the gatekeeper of chemotactic leukotriene B4 biosynthesis. *Journal of Biological Chemistry, 279*(49), 50639–50642. doi:10.1074/jbc.R400027200.

Houard, X., Ollivier, V., Louedec, L., Michel, J. B., & Bäck, M. (2009). Differential inflammatory activity across human abdominal aortic aneurysms reveals neutrophil-derived leukotriene B4 as a major chemotactic factor released from the intraluminal thrombus. *FASEB Journal, 23*(5), 1376–1383. doi:10.1096/fj.08-116202.

Ingelsson, E., Yin, L., & Bäck, M. (2012). Nationwide cohort study of the leukotriene receptor antagonist montelukast and incident or recurrent cardiovascular disease. *Journal of Allergy and Clinical Immunology, 129*(3), 702–707. doi:10.1016/j.jaci.2011.11.052.e702.

Katryniok, C., Schnur, N., Gillis, A., von Knethen, A., Sorg, B. L., Looijenga, L., et al. (2010). Role of DNA methylation and methyl-DNA binding proteins in the repression of 5-lipoxygenase promoter activity. *Biochimica et Biophysica Acta, 1801*(1), 49–57. doi:10.1016/j.bbalip.2009.09.003.

Lam, B. K., & Austen, K. F. (2002). Leukotriene C4 synthase: A pivotal enzyme in cellular biosynthesis of the cysteinyl leukotrienes. *Prostaglandins & Other Lipid Mediators, 68–69*, 511–520.

Nagy, E., & Bäck, M. (2012). Epigenetic regulation of 5-lipoxygenase in the phenotypic plasticity of valvular interstitial cells associated with aortic valve stenosis. *FEBS Letters, 586*(9), 1325–1329. doi:10.1016/j.febslet.2012.03.039.

Nagy, E., Andersson, D. C., Caidahl, K., Eriksson, M. J., Eriksson, P., Franco-Cereceda, A., et al. (2011). Upregulation of the 5-lipoxygenase pathway in human aortic valves correlates with severity of stenosis and leads to leukotriene-induced effects on valvular myofibroblasts. *Circulation, 123*(12), 1316–1325. doi:10.1161/CIRCULATIONAHA.110.966846.

Nakamura, M., & Shimizu, T. (2011). Leukotriene receptors. *Chemical Reviews, 111*(10), 6231–6298. doi:10.1021/cr100392s.

Nielsen, C. K., Campbell, J. I., Ohd, J. F., Morgelin, M., Riesbeck, K., Landberg, G., et al. (2005). A novel localization of the G-protein-coupled CysLT1 receptor in the nucleus of colorectal adenocarcinoma cells. *Cancer Research, 65*(3), 732–742.

Peters-Golden, M., & Henderson, W. R., Jr. (2007). Leukotrienes. *New England Journal of Medicine, 357*(18), 1841–1854.

Sala, A., Folco, G., & Murphy, R. C. (2010). Transcellular biosynthesis of eicosanoids. *Pharmacological Reports, 62*(3), 503–510.

Samuelsson, B. (1983). Leukotrienes: Mediators of immediate hypersensitivity reactions and inflammation. *Science, 220*, 568–575.

Samuelsson, B., Dahlén, S. E., Lindgren, J., Rouzer, C. A., & Serhan, C. N. (1987). Leukotrienes and lipoxins: Structures, biosynthesis, and biological effects. *Science, 237*(4819), 1171–1176.

Shimizu, T., Radmark, O., & Samuelsson, B. (1984). Enzyme with dual lipoxygenase activities catalyzes leukotriene A4 synthesis from arachidonic acid. *Proceedings of the National Academy of Sciences of the United States of America, 81*(3), 689–693.

Spanbroek, R., Grabner, R., Lotzer, K., Hildner, M., Urbach, A., Ruhling, K., et al. (2003). Expanding expression of the 5-lipoxygenase pathway within the arterial wall during human atherogenesis. *Proceedings of the National Academy of Sciences of the United States of America, 100*(3), 1238–1243. doi:10.1073/pnas.242716099.

Stanke-Labesque, F., Pepin, J. L., Gautier-Veyret, E., Levy, P., & Bäck, M. (2014). Leukotrienes as a molecular link between obstructive sleep apnoea and atherosclerosis. *Cardiovascular Research, 101*(2), 187–193. doi:10.1093/cvr/cvt247.

Yokomizo, T., Izumi, T., & Shimizu, T. (2001). Leukotriene B4: Metabolism and signal transduction. *Archives of Biochemistry and Biophysics, 385*(2), 231–241. doi:10.1006/abbi.2000.2168.

Lymphocyte Homing and Trafficking

Luc de Chaisemartin
INSERM UMRS 996, Universté Paris Sud, Chatenay-Malabry, France

Synonyms

Lymphocyte migration

Definition

Lymphocytes are central effector cells of adaptive immune responses. Their ability to migrate from one place to another is a major feature of their action since the initiation of immune responses is centralized in secondary lymphoid organs while the effector phase takes place where needed in peripheral tissues. Here the mechanisms responsible for this ability and major trafficking pathways of lymphocyte in normal and pathological conditions will be discussed.

Structure and Functions

Molecular Basis of Lymphocyte Trafficking

Lymphocyte homing and trafficking are essential functions of adaptive immune responses. The basic molecular processes involved in trafficking are chemotaxis and cellular adhesion. Chemotaxis is mediated mainly by chemokines and inflammatory response by-products (anaphylatoxins and prostaglandins), while cellular adhesion is mediated by selectins interacting with sugars (weak interactions) and integrins (strong interactions).

Chemotaxis

The Chemokine/Receptor System Chemokines (from "chemotactic cytokines," CKs) are small chemoattractant proteins of 6–12 kDa that play a central role in lymphocyte homeostasis and immune response organization. Through their 7-transmembrane domains (G protein-coupled receptors), CKs are capable of inducing mobility, extracellular matrix interaction, activation, and survival signals (Balkwill 2004). They are involved in T-cell intrathymic development, secondary lymphoid organ organization, and T-cell migration to specific places under normal or pathological conditions (Kim 2005; Viola and Luster 2008). The interaction of chemokines with their receptors has also been linked with neoplasia either directly (tumor growth factors, metastatic cells migration) or indirectly (immune cells recruitment).

Currently, 42 chemokines have been described in humans, belonging to 4 different groups depending on the number and position of conserved N-terminal cysteines (C, CC, CXC, and CX_3C chemokines; Table 1). However, only 19 chemokine receptors (CKRs) have been described so far, implying an importantly redundant system. Indeed, most chemokines can bind several receptors and vice versa.

However, CKRs are specific of a given CK group (except some decoy receptors), and 6 of them only bind one chemokine (CCR6, CCR9, CXCR4, CXCR5, CXCR6, and CX3CR1). Chemokine decoy receptors (mostly Duffy, D6, and CXCR7) have been described. These receptors bind to the chemokines without inducing intracellular signaling, and they are involved in negative regulation of CK effects. Duffy receptors and D6 can bind numerous chemokines, mostly those produced during inflammation, and participate in local inflammatory response control (Mantovani et al. 2006). In contrast, CXCR7 can only bind CXCL11 and CXCL12 (Naumann et al. 2010).

CKR are expressed on numerous cell types and notably on lymphocytes, and the CKR expression profile varies considerably according to cell type, differentiation, and activation status. This has been extensively studied in lymphocytes. For example, naïve T cells recirculate permanently between blood and secondary lymphoid organs and express CCR7 that is necessary to enter lymph nodes (Cyster 1999). Once activated, T cells downregulate CCR7 expression and upregulate CKR binding inflammatory chemokines. The exact expression profile depends partly on T-cell polarization. The expression of CCR5, CXCR3, and CXCR6 has been associated with CD4 + Th1 T cells and cytotoxic CD8+ T cells, while CCR4 and CCR8 have been associated with CD4+ Th2 T cells (Sallusto and Lanzavecchia 2000; Bromley et al. 2008). These associations have been confirmed at molecular level since Th1 polarizing transcription factor T-bet upregulates CXCR3 while Th2-polarizing GATA-3 upregulates CCR4 (Bromley et al. 2008). However, while this has been well characterized on in vitro cytokine-activated T cells (Sallusto and Lanzavecchia 2000), CKR expression profile in vivo is more complex and less well known. For example, many peripheral blood lymphocytes co-express CCR4 and CXCR3, and CCR4 is not only expressed by Th2 T cells but also by skin T cells that are mostly Th1 (Bromley et al. 2008). Moreover, if CCR4 and CCR7 have been associated to regulatory T-cell migration by several authors, the exact set of CKRs expressed by these cells is controversial and varies depending on the study model (Wei et al. 2006).

As far as Th17 T cells are concerned, they have been reported to express many different CKRs, the most frequently found receptor being CCR6 (Acosta-Rodriguez et al. 2007). Finally, follicular helper T-cell subpopulation is characterized by CXCR5 expression, the same receptor that is responsible for B-cell follicle formation.

Molecular Basis of CK Effects CKs bind on their receptors mostly as a monomer, though it has been shown that oligomerization of both CK and CKR is possible. However the possible physiological relevance of such oligomers is not demonstrated yet (Thelen and Stein 2008). G proteins coupled to CKR are made of three subunits

Lymphocyte Homing and Trafficking, Table 1 Chemokines and their receptors described in humans

CC Family			CXC Family		
Chemokines		Receptors	Chemokines		Receptors
CCL1	CCL17	CCR1	CXCL1	CXCL9	CXCR1
CCL2	CCL18	CCR2	CXCL2	CXCL10	CXCR2
CCL3	CCL19	CCR3	CXCL3	CXCL11	CXCR3
CCL4	CCL20	CCR4	CXCL4	CXCL12	CXCR4
CCL5	CCL21	CCR5	CXCL5	CXCL13	CXCR5
CCL7	CCL22	CCR6	CXCL6	CXCL14	CXCR6
CCL8	CCL23	CCR7	CXCL7	CXCL16	CXCR7
CCL11	CCL24	CCR8	CXCL8		
CCL13	CCL25	CCR9			
CCL14	CCL26	CCR10			
CCL15	CCL27				
CCL16	CCL28				
XC family			CX3C family		
Chemokines		Receptors	Chemokines		Receptors
XCL1		XCR1	CX3CL1		CX3CR1
XCL2					

(α, β, γ) and are pertussis toxin-sensitive. After CK binding, β- and γ-subunits of the G protein activate phospholipase C (PLC) which synthetize inositol triphosphate (InsP3) and diacyl glycerol, leading to calcic signaling and activation of effectors like Rap-1-GTP. This protein is responsible for a quick activation of integrins and participates in cytoskeleton modification necessary for cell migration (Ley et al. 2007; Thelen and Stein 2008).

All chemokines are excreted in extracellular space except CX3CL1 and CXCL16 which are produced membrane-bound. These membrane CKs may work as adhesion molecules between cells expressing them and cells expressing their receptors (Shimaoka et al. 2004). However, they also have a cleavage site for metalloproteinases ADAM10 and ADAM17 and can be released in extracellular space. Some CK can bind to glycan residues on endocyte luminal surface, which allow them not to be diluted in blood flow and to bind to circulating lymphocytes CKR (Mantovani et al. 2006). The most studied property of chemokines is their capacity to induce cell migration toward a concentration gradient. This capacity is central to most current leukocyte migration models. Repulsion induction has also been observed with high CK concentration of CXCL12 (Poznansky et al. 2000). Moreover, CKs bound to extracellular matrix allow lymphocytes to crawl on a surface (haptotactic migration) (Kim 2005). Finally, even without concentration gradient, CKs induce an increase of random motility of lymphocytes, called chemokinesis.

Other Chemoattractant Molecules CKs are not the only molecules capable of attracting lymphocytes. Some by-products of inflammatory response like C3a and C5a anaphylatoxins, leukotriene B4, prostaglandin D2, or β-defensin. All these molecules have specific G protein-coupled receptors, except β-defensin which binds CCR6 (Lai and Gallo 2009). They participate in linking inflammatory response and adaptive immune response. Some bacterial components like fMLP (formyl-methionyl-leucyl-phenylalanine) or some cytokines (IL-8, IL-16) can also attract lymphocytes. Moreover, phospholipids like sphingosine-1-phosphate constitutively present in peripheral blood and lymph have an important chemoattractant activity. Most of these molecules attract lymphocyte to inflamed tissues, except sphingosine-1-phosphate which is mostly responsible of lymphocyte exit from the lymph node and thymus to peripheral blood (Cyster 2005).

Selectins

Selectins are calcium-dependent transmembrane glycoproteins of C-type lectin family that can make low-affinity binding to glucidic residues. Three selectins have been described so far, E, L, and P or CD62E, CD62L, and CD62P. CD62E is constitutively expressed on the luminal side of skin endothelium and can be induced by inflammation on the other endothelia. CD62P is expressed on alpha granules of platelets and on intracytoplasmic structures of endothelial cells called Weibel-Palade bodies and is transported to the membrane under inflammatory conditions. CD62L, on the other hand, is not expressed on endothelium but on leukocytes (myeloid cells, B cells, naïve and central-memory T cells) (Ley and Kansas 2004; Sperandio 2006).

The main function of selectins is to help slow down transendothelial migration of circulating leukocytes by weak adhesion to glycoprotein from leukocytes or endothelium. Many selectin ligands have been identified in vitro, but few have demonstrated in vivo activity. The main in vivo ligands are PSG-L1 (P-selectin glycoprotein ligand 1) for CD62P, CD44 for CD62E, and MAdCAM (mucosal vascular addressin cell adhesion molecule) and PNAd (peripheral lymph node addressin) for CD62L. PSG-L1 and CD44 are expressed by all leukocytes, PNAd is expressed on HEVs (high endothelial venules), and MadCAM is expressed in intestinal endothelium. These ligands can be upregulated on inflammatory endothelium.

Integrin and Cell Adhesion Molecules

Integrins are transmembrane αβ-heterodimers involved in cell adhesion to extracellular matrix. They can also be involved in cell-cell interactions, including immune cells (Hynes 2002). There are 18 α-subunits that can associate with 8 β-subunits, but only 24 couples have been described. The subunits αL, αM, αV, αX, β2, and β7 are expressed only on leukocytes. Lymphocytes express mainly αLβ2 (also called LFA-1), α4β1, and α4β7. In their resting conformation, integrins are inactive. They must be activated by a cellular signal like CK binding and BCR or TCR activation (inside-out signaling) (Evans et al. 2009). Active integrins may bind extracellular matrix proteins (collagen, laminin, fibronectin, vitronectin) (Borland and Cushley 2004), coagulation proteins (fibrinogen, thrombospondin, factor X), but also immunoglobulin superfamily proteins as CAMs (cell adhesion molecules).

These transmembrane molecules feature an immunoglobulin-like domain and can engage in homodimeric or heterodimeric binding with integrins. The main integrin-binding CAMs are ICAM-1,2,3,5 (intercellular adhesion molecule), V-CAM-1 (vascular cell adhesion molecule), and MadCAM-1 (mucosal vascular addressin cell adhesion molecule). They are mostly expressed on endothelia and their expression is regulated by vessel location and inflammatory environment. In the context of immune response, integrins play a major role in lymphocyte recruitment for peripheral blood, phagocytosis, and immunological synapse formation. Apart from their binding functions, integrins can also transduce signals (outside-in signaling) involved in immune cell costimulation and polarization (Luo et al. 2007).

Major Functions of Lymphocyte Migration

Chemokines, integrins, and selectins are all involved in circulating lymphocyte recruitment to lymph node through HEVs. Furthermore, CKR expression regulates T-cell/DC and T-cell/B-cell cooperation in lymph nodes, making it an essential component of the priming of immune adaptive responses.

In inflammatory conditions, the trafficking of T cells to the affected tissue is affected by multiple molecules and depends on the type of lymphocyte and its polarization and the kind of pathological process involved. However, some lymphocytes have tissue-specific homing properties (mostly for the skin and gut) that are imprinted by their activating DC. Since all these features are related to immune responses, they will be discussed in the next chapter.

Pathological Relevance

T-Cell Progenitors Recruitment and Trafficking in the Thymus

The first migration in T-cell early life is the migration of pluripotent progenitor cells from bone marrow to the thymus, where irreversible engagement in T lineage occurs. While the exact nature of the progenitors is still matter of debate, it is established that cells need to express CCR9 and CCR7 to enter the thymus. The ligands of these receptors are expressed on thymic endothelium and inside the thymus. The P-selectin is also expressed on thymic endothelium and it seems that its ligand PSGL-1 also contributes to entry into the thymus.

Inside the thymus, thymocytes migrate as they mature, from cortex-medulla junction to subcapsular cortex with CCR9. Then they migrate back into the inner cortex in response to chemotactic factors still to be determined (CXCR4 being suspected) where they undergo positive selection. This selection upregulates the expression of CCR7, which allows thymocytes to leave the cortex to the medulla where they undergo the negative selection. When lymphocytes are mature, they egress the thymus by upregulating a receptor to sphingosin-1-phosphate (S1P), S1P1. This chemotactic phospholipid is highly concentrated in blood and lymph and thus allows newly formed lymphocyte to enter blood circulation.

Lymphocyte Recruitment to Secondary Lymphoid Organs and Initiation of Adaptive Immune Responses

Lymphocytes are continuously recruited to lymph nodes at steady state by a multistep interplay between lymphocytes and endothelium. The first step is the low-affinity binding of CD62L on T cells and sulfated glycan residues of a glycoprotein group called PNAd, expressed only on lymphoid organs and specialized blood vessels, the HEVs (high endothelial venules). This weak interaction leads to lymphocyte slowing down and rolling along the endothelium. The CK CCL19 and CCL21, expressed on luminal surface of HEV, can then signal through CCR7, their common receptor, and induce activation of LFA-1 integrin (Miyasaka and Tanaka 2004). CXCR4 and its ligand CXCL12 have been shown to participate to a lesser degree to this step (Bai et al. 2009). After activation, LFA-1 binds to ICAM-1 and ICAM-2, constitutively expressed on HEV, and leads to lymphocyte arrest and crawling on endothelium. Lymphocytes are then going to cross endothelium (diapedesis) via paracellular and transcellular transmigration (Denucci et al. 2009).

Inside the lymph node, T and B cells migrate along a network of stromal cells and separate into two distinct zones. B cells are guided to B-cell follicles by CXCL13/CXCR5 interaction, while T cells enter T-cell-rich zone due to CCL21/CCR7 molecules. These chemokines are mostly expressed by stromal cells and are responsible for cell migration to and motility inside each zone (Evans et al. 2009). CCR7 and CD62L are thus considered the main lymph node addressing molecules and are used to characterize naïve and central-memory T cells.

After activation by an antigen in peripheral tissues, antigen-presenting dendritic cells (DCs) express CCR7 and are then able to migrate through lymphatic vessels expressing CCL19 and CCL21 toward lymph nodes. Once in the lymph node, mature DCs migrate along the stromal cell network, and since they express the same CKR than T cells, they settle in the T-cell zone and present their antigen to lymphocytes, thus starting adaptive immune responses. During the humoral response priming, T-B-cell cooperation in the lymph node also depends on chemokines. After first activation, T cells downregulate CCR7 and transiently express CXCR5, while B cells upregulate CCR7. These combined migration pattern changes allow B and T cells to migrate toward each other at the frontier between T- and B-cell zones and to exchange the signals necessary for B-cell terminal differentiation (Stein and Nombela-Arrieta 2005). Furthermore, while lymphocyte continuously enters lymph node at steady state, this migration is greatly increased upon inflammatory conditions. Blood and lymphatic flows are increased, and it has been shown that pro-inflammatory chemokines can be expressed on HEV surface and promote CCR7-independent lymphocyte recruitment (especially with CXCR3 and its ligand CCL9) (Guarda et al. 2007).

Finally, the egress of lymphocytes from the lymph node is regulated in the same way than thymic egress by S1P and its receptor S1P1.

Lymphocyte Migration Toward Effector Sites

During immune responses, activated lymphocytes migrate from secondary lymphoid organs to inflammatory sites to exercise their effector functions. Pro-inflammatory cytokines IL-6 and TNF-α are able to induce expression of ICAM-1, V-CAM-1, and MAdCAM-1 on endothelium, facilitating lymphocyte recruitment not only in lymph nodes but also in peripheral tissues. Numerous CKs have been associated to inflammation, including CCR5, CCR6, CXCR1, and cphCXCR3 ligands, which participate in different lymphocyte subpopulation recruitments (Bromley et al. 2008). For example, the expression of CXCL9, CXCL10, CXCL11, and CCL5 is induced by interferons and then plays a major part in recruiting cytotoxic CD8+ and Th1 CD4+ T cells in infectious settings. The expression of theses chemokines has been shown to be mandatory for bacterial clearance in some models (Olive et al. 2010). Similarly, these pro-inflammatory CKs are expressed in many autoimmune conditions in humans and contribute to the physiopathology by recruiting effector T cells (Lee et al. 2009). CCR5 and CXCR3 have also been linked to alloreactive T-cell recruitment in kidney and lung allografts, and migration inhibitors are also tested in this application (Panzer et al. 2004).

Th2-cytokines as IL-4 and IL-13 can induce CK expression, particularly from eotaxin family (CCL1, CCL24, and CCL26) which binds CCR3. In some pathological contexts as allergy or antiparasitic responses, CCR3+ Th2 T cells are recruited with mast cells and eosinophils and contribute to tissular lesions (Mackay 2008). However, Th2 T-cell recruitment is less documented than Th1, probably because of the preponderant role of myeloid cells in allergic reaction physiopathology.

Concerning Tregs and Th17, no specific CKR expression pattern has been described as they seem to be recruited in many pathological contexts, probably by several types of CKR. For example, in most autoimmune disease models, Th17 T cells are recruited by CCR6, while CCR2-, CCR6-, and CCR4-dependent recruitments have been documented for Tregs (Wei et al. 2006; Lim et al. 2008; Hirota et al. 2007).

While migration pattern of lymphocytes depends on subtype, polarization, and pathophysiological mechanisms, some T lymphocyte may also display tissue-specific homing properties. For example, T cells primed in Peyer's patches (intestinal mucosa-associated secondary lymphoid tissue) receive a special addressing signal by the DC activating them. This signal leads to upregulation of α4β7 integrin and CCR9 receptor. The ligand of CCR9, CCL25, is expressed by intestinal epithelial cells, and when binding CCR9 on circulating lymphocytes, it activates α4β7 integrin. This integrin can bind MAdCAM-1 which is strongly expressed on Peyer's patches HEV compared to PNAd and thus allows T cell primed in the gut to locate in the *lamina propria* (Sigmundsdottir and Butcher 2008). At the skin level, lymphocyte recruitment requires E selectin, constitutively expressed by skin vessels and its ligand CLA (cutaneous lymphocyte antigen) expressed on activated and memory T cells with cutaneous tropism. The CKR CCR10 is also involved, its ligand CCL27 being constitutively expressed by keratinocytes (Kim 2005; Sigmundsdottir and Butcher 2008). Finally, although not totally elucidated, migration of lymphocytes into central nervous system through brain-blood barrier has been shown to be largely dependent on α4 integrins. This is one of the first successful translations of lymphocyte migration data into clinical applications since a humanized anti-α4 integrin monoclonal antibody (natalizumab) is now used to treat multiple sclerosis.

Interactions with Other Processes and Drugs

Since they participate in many pathological processes, lymphocyte recruitment mechanisms are major therapeutic targets, and many migration inhibitors are currently in development. It has been shown in particular that disruption of chemokine-driven recruitment of inflammatory cells could improve diseases in animal models, and in CKR-deficient patients, consequently many chemokine inhibitors are currently tested in clinical trials (Allegretti et al. 2012). Most of them are aiming autoimmune diseases and some cancer. However, the only commercialized products so far are a CCR5 antagonist (maraviroc) blocking HIV entry into target cells and a CXCR4 antagonist (plerixafor) used for stem-cell mobilization along with G-SCF.

The only successful approach targeting lymphocyte migration was blocking a4 integrin with a monoclonal antibody (natalizumab) that despite concerns about viral side effects is now licensed for treating relapsing multiple sclerosis.

With these three effective drugs and many others on the way, there is no doubt that controlling lymphocyte migration is about to become a major tool in many inflammatory conditions.

Cross-References

▶ Leukocyte Recruitment
▶ Mechanisms of Macrophage Migration in 3-Dimensional Environments
▶ Platelets, Endothelium, and Inflammation

References

Acosta-Rodriguez, E., Rivino, L., et al. (2007). Surface phenotype and antigenic specificity of human interleukin 17-producing T helper memory cells. *Nature Immunology, 8*(6), 639–646.
Allegretti, M., Cesta, M. C., et al. (2012). Current status of chemokine receptor inhibitors in development. *Immunology Letters, 145*(1–2), 68–78.
Bai, Z., Hayasaka, H., et al. (2009). CXC chemokine ligand 12 promotes CCR7-dependent naive T cell trafficking to lymph nodes and Peyer's patches. *The Journal of Immunology, 182*(3), 1287–1295.
Balkwill, F. (2004). Cancer and the chemokine network. *Nature Reviews Cancer, 4*(7), 540–550.
Borland, G., & Cushley, W. (2004). Positioning the immune system: Unexpected roles for alpha6-integrins. *Immunology, 111*(4), 381–383.
Bromley, S. K., Mempel, T. R., et al. (2008). Orchestrating the orchestrators: Chemokines in control of T cell traffic. *Nature Immunology, 9*(9), 970–980.
Cyster, J. G. (1999). Chemokines and cell migration in secondary lymphoid organs. *Science, 286*(5447), 2098–2102.
Cyster, J. (2005). Chemokines, sphingosine-1-phosphate, and cell migration in secondary lymphoid organs. *Annual Review of Immunology, 23*, 127–159.
Denucci, C. C., Mitchell, J. S., et al. (2009). Integrin function in T-cell homing to lymphoid and nonlymphoid sites: Getting there and staying there. *Critical Reviews in Immunology, 29*(2), 87–109.
Evans, R., Patzak, I., et al. (2009). Integrins in immunity. *Journal of Cell Science, 122*(Pt 2), 215–225.
Guarda, G., Hons, M., et al. (2007). L-selectin-negative CCR7- effector and memory CD8+ T cells enter reactive lymph nodes and kill dendritic cells. *Nature Immunology, 8*(7), 743–752.
Hirota, K., Yoshitomi, H., et al. (2007). Preferential recruitment of CCR6-expressing Th17 cells to inflamed joints via CCL20 in rheumatoid arthritis and its animal model. *Journal of Experimental Medicine, 204*(12), 2803–2812.
Hynes, R. (2002). Integrins: Bidirectional, allosteric signaling machines. *Cell, 110*(6), 673–687.
Kim, C. (2005). The greater chemotactic network for lymphocyte trafficking: Chemokines and beyond. *Current Opinion in Hematology, 12*(4), 298–304.
Lai, Y., & Gallo, R. L. (2009). AMPed up immunity: How antimicrobial peptides have multiple roles in immune defense. *Trends in Immunology, 30*(3), 131–141.
Lee, E., Lee, Z., et al. (2009). CXCL10 and autoimmune diseases. *Autoimmunity Reviews, 8*(5), 379–383.
Ley, K., & Kansas, G. (2004). Selectins in T-cell recruitment to non-lymphoid tissues and sites of inflammation. *Nature Reviews Immunology, 4*(5), 325–335.
Ley, K., Laudanna, C., et al. (2007). Getting to the site of inflammation: The leukocyte adhesion cascade updated. *Nature Reviews Immunology, 7*(9), 678–689.
Lim, H., Lee, J., et al. (2008). Human Th17 cells share major trafficking receptors with both polarized effector T cells and FOXP3+ regulatory T cells. *The Journal of Immunology, 180*(1), 122–129.
Luo, B. H., Carman, C. V., et al. (2007). Structural basis of integrin regulation and signaling. *Annual Review of Immunology, 25*, 619–647.
Mackay, C. R. (2008). Moving targets: Cell migration inhibitors as new anti-inflammatory therapies. *Nature Immunology, 9*(9), 988–998.
Mantovani, A., Bonecchi, R., et al. (2006). Tuning inflammation and immunity by chemokine sequestration: Decoys and more. *Nature Reviews Immunology, 6*(12), 907–918.
Miyasaka, M., & Tanaka, T. (2004). Lymphocyte trafficking across high endothelial venules: Dogmas and enigmas. *Nature Reviews Immunology, 4*(5), 360–370.
Naumann, U., Cameroni, E., et al. (2010). CXCR7 functions as a scavenger for CXCL12 and CXCL11. *PloS One, 5*(2), e9175.
Olive, A., Gondek, D., et al. (2010). CXCR3 and CCR5 are both required for T cell-mediated protection against *C. trachomatis* infection in the murine genital mucosa. *Mucosal Immunology, 4*, 208–216.
Panzer, U., Reinking, R., et al. (2004). CXCR3 and CCR5 positive T-cell recruitment in acute human renal allograft rejection. *Transplantation, 78*(9), 1341–1350.

Poznansky, M., Olszak, I., et al. (2000). Active movement of T cells away from a chemokine. *Nature Medicine, 6*(5), 543–548.

Sallusto, F., & Lanzavecchia, A. (2000). Understanding dendritic cell and T-lymphocyte traffic through the analysis of chemokine receptor expression. *Immunology Reviews, 177*, 134–140.

Shimaoka, T., Nakayama, T., et al. (2004). Cell surface-anchored SR-PSOX/CXC chemokine ligand 16 mediates firm adhesion of CXC chemokine receptor 6-expressing cells. *Journal of Leukocyte Biology, 75*(2), 267–274.

Sigmundsdottir, H., & Butcher, E. (2008). Environmental cues, dendritic cells and the programming of tissue-selective lymphocyte trafficking. *Nature Immunology, 9*(9), 981–987.

Sperandio, M. (2006). Selectins and glycosyltransferases in leukocyte rolling in vivo. *FEBS Journal, 273*(19), 4377–4389.

Stein, J., & Nombela-Arrieta, C. (2005). Chemokine control of lymphocyte trafficking: A general overview. *Immunology, 116*(1), 1–12.

Thelen, M., & Stein, J. (2008). How chemokines invite leukocytes to dance. *Nature Immunology, 9*(9), 953–959.

Viola, A., & Luster, A. (2008). Chemokines and their receptors: Drug targets in immunity and inflammation. *Annual Review of Pharmacology and Toxicology, 48*, 171–197.

Wei, S., Kryczek, I., et al. (2006). Regulatory T-cell compartmentalization and trafficking. *Blood, 108*(2), 426–431.

Macrophage Heterogeneity During Inflammation

Nathalie Dehne, Michaela Jung,
Christina Mertens, Javier Mora and
Andreas Weigert
Faculty of Medicine, Institute of Biochemistry I,
Goethe University Frankfurt, Frankfurt, Germany

Synonyms

Macrophage activation; Macrophage ontogeny; Macrophage phenotypes; Macrophage polarization; Macrophage subsets

Definition

Inflammation is a protective response against infection or injury that is characterized by a set of dynamic and progressive events that ultimately serve to restore tissue homeostasis. The first step in the inflammatory cascade is elimination of the stimulus or agent provoking the infection/injury by inducing acute pro-inflammatory signaling events. This is followed by clearance of primary immune effectors once the inflammatory stimulus/agent is removed and the shift to an anti-inflammatory environment. The final step is the active resolution of inflammation by clearing secondary immune effectors and reestablishing the original tissue integrity and architecture. This includes activation of stem and progenitor cells to regenerate the epithelium, reprogramming of stromal cells to prevent excessive fibrosis, and revascularization of the affected tissue (Gilroy and De Maeyer 2015; Headland and Norling 2015; Ortega-Gomez et al. 2013).

Macrophages, professional guardians of homeostasis, are intimately involved in the regulation of all these, somewhat antithetical, processes. To accomplish this, macrophages need to express a broad sensory arsenal to detect disturbances in tissue integrity and exhibit a remarkable functional plasticity, enabling them to adapt to the needs of an ever-changing microenvironment during the course of an inflammatory response. The remarkable functional spectrum of macrophages encompasses the killing of pathogens and malignant or transformed cells, rearranging extracellular matrix, taking up and recycling cellular and molecular debris, initiating cellular growth cascades, and favoring directed migration of cells. As it is true for most of such tightly regulated processes, disturbances in macrophage-dependent tissue homeostasis may result in disease. Pathologies may develop due to overshooting macrophage activity that, under physiological circumstances, would be an integral, albeit temporally limited part of the program to restore homeostasis (Lavin et al. 2015; Wynn et al. 2013).

Structure and Function

Macrophages populate each tissue of the adult organism, where they fulfill homeostatic functions that are specific to the environmental niche characterizing the tissue. The resulting functional and morphological heterogeneity had the result that macrophages were "discovered" multiple times predominantly during early histological studies and have thus been assigned different names. Among tissue macrophage populations are microglia (brain macrophages), Kupffer cells (liver macrophages), Langerhans cells and dermal macrophages (skin macrophages), marginal-zone macrophages, red-pulp macrophages, subcapsular sinus, medullary macrophages and metallophilic macrophages (spleen macrophages), alveolar macrophages (lung macrophages), and osteoclasts (macrophages in the bone). The morphological heterogeneity is also present on the molecular level, making it so far impossible to identify all tissue macrophages by a common molecular marker (Gautier et al. 2012). Rather, a combination of markers is required to separate macrophages from other cell populations in tissues. In the mouse, these include CD11b, CD64, MerTK, and CSF-1R as well as rather macrophage subset-specific markers such as CD11c, CD169, F4/80, and MHC-II. In humans, accepted macrophage markers are CD64, CD163, MerTK, and CSF1-R (Lavin and Merad 2013; Lavin et al. 2015).

Functional plasticity of macrophages during steady state and during an inflammatory insult is shaped on (at least) three levels (Fig. 1). First, macrophages of distinct developmental origin populate tissues under homeostatic conditions or following an inflammatory insult, which may respond differently to an inflammatory stimulus. Second, the unique microenvironmental factors characterizing a specific tissue affect the phenotype of such tissue-resident or recruited macrophages. Third, external factors or changes in the tissue microenvironment activate the sensory repertoire of resident or recruited macrophages to adapt to changing requirements. The impact on functional macrophage heterogeneity is discussed in the following paragraphs.

Origin of Tissue Macrophages

Tissue macrophage ontogeny has been a matter of intense debate following their discovery by Elie Metchnikoff in 1887, for which he received the Nobel Prize in 1908 (Kaufmann 2008; Naito et al. 1996). Following the introduction of the mononuclear phagocyte system by van Furth, Cohn, and others (1972), the prevalent view was that tissue macrophages were terminally differentiated, nonproliferating, short-lived cells originating from bone marrow-derived monocytes. However, in parallel observations were reported that challenged this view by suggesting that tissue macrophages might stem from early fetal hematopoiesis in the yolk sac or the fetal liver, without the requirement of monocytes, and that tissue-resident macrophages can be long-lived cells that self-renew/proliferate in situ. This discrepancy of views is getting resolved by a number of recent studies utilizing, among others, fate-mapping and parabiosis techniques. In fact, a unifying concept of tissue macrophage provenance and maintenance that gives merit to both theories mentioned above is approaching (Ginhoux and Jung 2014; Lavin et al. 2015). According to this concept, the body is populated with macrophages in at least three different, probably overlapping waves during ontogeny.

The earliest macrophages develop from progenitors in the extraembryonic yolk sac from where they directly populate the embryo as soon as a functional vasculature is established (Ginhoux and Jung 2014; Lavin et al. 2015). This primitive system of hematopoiesis is required since macrophages play important roles in determining tissue architecture already in the early embryo, where they are involved, among others, in the generation of bone and adipose tissue and in branching morphogenesis (Wynn et al. 2013). The microglia in the brain are direct descendants of these primitive macrophages, self-renewing throughout the life-span of the organism with no contribution from other hematopoietic sources. In other tissues, yolk sac macrophages appear to be consecutively replaced by a second wave of macrophages that again arise from early yolk sac progenitors. However, these progenitors migrate to the fetal liver, where they differentiate

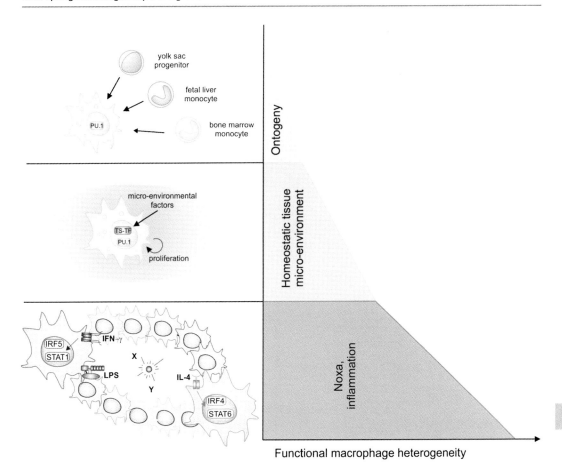

Macrophage Heterogeneity During Inflammation, Fig. 1 Tissue macrophage heterogeneity. Macrophage phenotype diversity is regulated on three levels. First, distinct macrophage precursors arise during ontogeny to generate mixed populations of tissue macrophages. Second, tissue-enriched microenvironmental cues activate tissue-specific transcription factors (*TS-TF*) alongside the lineage-determining transcription factor PU.1 to equip macrophages for tissue-specific functional requirements. Third, disturbances of tissue homeostasis trigger a multitude of signals (indicated by *X*, *Y*) that create a large spectrum of polarized macrophages. Two prominent phenotypes in this spectrum, the stimuli and downstream transcription factor shaping them, are depicted. Details are described in the main text. *IFN* interferon, *IL* interleukin, *IRF* interferon regulatory factor, *LPS* lipopolysaccharide, *STAT* signal transducer and activator of transcription

to monocytes and enter embryonic tissues to generate macrophages (Ginhoux and Jung 2014; Lavin et al. 2015). Fetal monocyte-derived macrophages self-renew locally and maintain their presence in the adult organism in most tissues, for instance, in the lung (alveolar macrophages) or the epidermis (Langerhans cells). After birth finally a third wave of macrophage progenitors is generated, bone marrow-derived monocytes. In some organs they replace macrophages of embryonic sources, such as in the intestines or the dermis. Adult monocyte-derived macrophages appear to show poor proliferative potential and are continuously replaced by newly recruited monocyte-derived macrophages (Bain and Mowat 2014). The question arises whether distinct functional properties of fetal versus adult bone marrow-derived macrophages exist due to their distinct origin, as targeting a particular population might be of benefit for therapeutic approaches. Although there is a limited amount of evidence that this might be true, e.g., adult monocyte-derived macrophages in the peritoneal cavity are superior in terms of some inflammatory properties (antigen presentation and nitric oxide (NO) production) compared to their embryonic cohabitants (Brune et al. 2013), recent reports investigating macrophage turnover when the steady state is challenged by an inflammatory insult suggest otherwise. In such a situation, the profound numbers of inflammatory monocytes that are recruited to the site of inflammation from the circulation and differentiate to macrophages in situ may take over the function of embryonic macrophages that succumbed to the inflammatory insult. An exception is the brain, where monocytes do not permanently contribute to the macrophage pool (Epelman et al. 2014; Ginhoux and Jung 2014). Thus, the local microenvironment might be superior in determining macrophage function as compared to ontogeny.

The Tissue Microenvironment in Determining Macrophage Function

The functionality is determined ultimately at the genetic level, i.e., gene expression. This depends on the chromatin structure, allowing or restricting gene transcription, the expression of cell type-specific (lineage-determining) transcription factors (Smale et al. 2014). Thus, the chromatin structure is defined by transcription factors that are upregulated during cell differentiation and whose persistent expression is required to maintain cell identity (i.e., chromatin structure). The lineage-determining transcription factor of macrophages is PU.1 (Jenkins and Hume 2014). Its expression is induced downstream of colony-stimulating factor-1 (CSF1) receptor by its ligands CSF1 or interleukin-34 (IL-34). PU.1 provides the basis of transcriptional availability, i.e., regions of open chromatin, on top of which other transcription factors can be installed to generate the vast spectrum of different tissue macrophages. Currently, only a small number of transcription factors that are involved in the stimulus-specific differentiation of tissue-resident macrophages have been identified. Examples are myocyte-specific enhancer factor 2c (MEF2c) and SMAD transcription factors in microglia, which are induced by transforming growth factor-ß (TGF-β), and peroxisome proliferator-activated receptor γ (PPARγ) in alveolar macrophages that is induced by granulocyte macrophage colony-stimulating factor (GM-CSF). Furthermore, heme uptake promotes the expression of PU.1-related factor (SPI-C) to generate iron-recycling macrophages in the spleen and bone marrow, and GATA binding protein 6 (GATA6) expression is induced by retinoic acid in peritoneal macrophages (Lavin et al. 2014). These examples illustrate that the tissue microenvironment dictates the genetic signature of its resident macrophages by inducing the expression of specific transcription factors. The notion is supported by transplantation experiments using fully differentiated macrophages derived from bone marrow precursors and transferring them to several organs. Transplanted macrophages rapidly acquired a chromatin structure resembling that of tissue-resident embryonic macrophages and consequently functionally integrated into the recipient tissue macrophage pool (Lavin et al. 2014). Moreover, stimulation of peritoneal macrophages with TGF-β in vitro resulted in a reshaped chromatin landscape, closely resembling that of microglia (Gosselin et al. 2014). In conclusion, these data support the view that macrophages from different developmental origins can be influenced by the local environment to fulfill tissue-specific demands (Epelman et al. 2014; Ginhoux and Jung 2014). However, in some tissues, distinct macrophage subpopulations with different functions are found virtually side-by-side (e.g., in the skin, spleen, peritoneum). Although the transcriptional profile of these distinct macrophage subpopulations is similar enough to suggest exposure to common tissue-specific cues, the remaining differences might allow for an impact of ontogeny (Gosselin et al. 2014).

The plasticity of macrophages that is illustrated by the transplantation experiments mentioned above is far excelled when the microenvironment is drastically altered during inflammation, leading, for instance, to alterations in oxygen pressure (pO_2) or production of cytokines that activate stimulus-specific transcription factors. These transcription factors bind to regions of open chromatin that are shaped by PU.1 and the tissue macrophage-specific transcription factors, creating a nearly infinite number of possible macrophage phenotypes, a process that is known as macrophage polarization.

Macrophage Polarization
Independent of genetic imprinting due to ontogeny or differentiation in a specific microenvironment, macrophages retain a high functional plasticity, enabling them to respond to varying stimuli during an inflammatory insult (Brune et al. 2013; Murray and Wynn 2011). Hereby, the identity of a stimulus as well as its concentration and the duration of exposure determine the resulting macrophage phenotype. Historically, two discreet activation states of macrophages were identified. Macrophage activation by activated T helper 1 (TH1) cell-derived interferon-γ (IFN-γ), in combination with tumor necrosis factor (TNF)-α or the activation of toll-like receptors (TLR) by bacterial cell wall components such as lipopolysaccharides (LPS), creates cells with a strong pro-inflammatory profile. IFN–γ stimulated macrophages show a transcriptional signature defined by activation of signal transducer and activator of transcription 1 (STAT1), as well as interferon regulatory factor 5 (IRF 5) (Krausgruber et al. 2011; Xue et al. 2014). These transcription factors enable classically activated macrophages to generate a multitude of pro-inflammatory mediators such as TNF-α, interleukin (IL)-1β, IL-6, IL-12, IL-23, reactive oxygen (ROS), and nitrogen species and to present antigens to T cells via induction of major histocompatibility complex (MHC) class II molecules. Classically activated macrophages show a high potency in host defense against microbes as well as tumors. In contrast, macrophage activation by activated T helper 2 (TH2) cell-derived IL-4 or IL-13 produces an alternative set of cytokines and chemokines opposing the repertoire of classically activated macrophages. In addition to that, they express specific phagocytic receptors such as the mannose receptor (CD206) and produce extracellular matrix (ECM) and growth factors to promote tissue remodeling and to combat extracellular parasites (Brune et al. 2013). Their transcription factor profile is dominated by STAT6 and IRF4.

The two antithetic macrophage phenotypes were designated as M1 and M2 macrophages, respectively, to reflect the main cellular triggers of their activity, namely, TH1 and TH2 cells. Although these terms are helpful, they are also somewhat misleading as they suggest a relatively low remaining plasticity in their functional repertoire as observed in terminally differentiated TH cells. In reality, macrophage activation states are rather transient than stable, and activated macrophages maintain functional flexibility. This is exemplified by observations that functional patterns of activated macrophages change over time and revert back to the original state after a certain amount of time (3–7 days) and that M2 macrophages readily acquire M1-associated functions when being stimulated with TLR ligands or IFN-γ (Cassetta et al. 2011; Stout et al. 2005). The ability to switch phenotypes enables macrophages to perform different tasks including the killing of pathogens, engulf and digest dead cells and debris, stimulate different types of adaptive immune cells, and promote trophic functions such as epithelial and endothelial regeneration, according to the requirements during the initiation, progression, and resolution of inflammation. In addition to the retained plasticity of macrophages, the M1/M2 terminology tends to obscure the view to the fact that a variety of environmental changes that simultaneously occur during inflammation can be sensed by macrophages, rarely resulting in "pure" M1/M2 polarization but a plethora of intermediate, mixed, or completely unrelated phenotypes as confirmed by transcriptome studies (Xue et al. 2014). This heterogeneity has to be considered when looking at macrophage function during disease, as differently polarized macrophages are likely found in distinct environmental niches in a particular disease state.

It has to be noted that a line of literature suggests that different macrophage phenotypes during inflammation, inflammatory versus resolving macrophages, are a result of different immediate progenitors, which are blood monocytes. At least two monocyte subpopulations are distinguished in mouse and human blood based on the expression of different cell surface receptors. In the murine system, Ly6C+ CCR2+ and CX3CR1- monocytes (CD14+ monocytes in humans) are rapidly recruited to the site of inflammation in response to the CCR2 ligand CCL2, where they differentiate to inflammatory macrophages or monocyte-derived DCs. Ly6C- CCR2- and CX3CR1+ monocytes (CD14low CD16+ monocytes in humans) do not initially migrate to sites of infection, rather being recruited at later stages, upon production of the CX3CR1 ligand fractalkine, where they differentiate preferentially to macrophages with an anti-inflammatory phenotype. Nevertheless, this different differentiation pattern probably reflects the impact of the current microenvironment at the time of recruitment to the inflammatory site, as recent studies show that both monocyte subsets can be activated to support inflammation as well as its resolution and repair (Ginhoux and Jung 2014). In conclusion, the microenvironment under steady state and most prominently under inflammatory conditions appears as the main generator of macrophage heterogeneity. The plethora of data illustrating macrophage plasticity in specific inflammatory diseases is discussed exemplarily below.

Pathological Relevance

Prolonged or uncontrolled macrophage responses are associated with a number of pathologies, which is not surprising given the prominent role of macrophages during inflammation. The following examples serve to illustrate how macrophage heterogeneity might be harmful to the host if not properly controlled.

Sepsis

Classically activated macrophages are promoting resistance against intracellular pathogens. However, uncontrolled M1 macrophage-dependent inflammation in response to acute (systemic) infection, e.g., with *E. coli*, may cause sepsis (Biswas et al. 2012; Sica and Mantovani 2012). Sepsis is one of the leading causes of patient death in intensive care units. In the initial phase, hyper-inflammatory macrophages produce overshooting cytokine levels, the so-called cytokine storm which in the end accounts for multi-organ dysfunction and patient death. Once patients survive this initial hyper-inflammatory phase, immune paralysis develops, which again poses a high risk for patients, due to the inability to fight secondary opportunistic infection. Immune paralysis is also observed in macrophages of late-phase septic patients, being a compensatory mechanism to limit the duration of inflammatory responses. This illustrates the transient nature of macrophage polarization as outlined above.

Obesity

Obesity-associated chronic inflammation is a major cause of the metabolic syndrome, which is accompanied by insulin resistance. Obesity induces enhanced macrophages recruitment into adipose tissue by adipocyte-released chemoattractants. These obesity-associated adipose tissue macrophages display a pro-inflammatory M1-like phenotype, releasing cytokines such as TNF-α, IL-1, and IL-6, which promote insulin resistance (Biswas et al. 2012; Sica and Mantovani 2012). Classical macrophage polarization is apparently a direct result of the present noxa, including free fatty acids, in combination with a disturbed microenvironment (e.g., tissue hypoxia). In contrast, macrophages infiltrating nonobese adipose tissue show an anti-inflammatory phenotype, and fasting is able to reverse obesity-associated classical macrophage activation to a protective M2-like phenotype. Again, these findings illustrate the transient nature of macrophage polarization that is reverted once a specific stimulus has been withdrawn.

Asthma/Allergy

Allergy and asthma are considered prototypical TH2-driven pathologies. Since the inflammatory insult can persist over time without a chance of final clearance (such as continuous exposure to dust mites), allergy and asthma are associated with prolonged IL-4/IL-13-dependent macrophage activation. Prolonged M2 macrophage polarization results in abnormal tissue remodeling leading, e.g., to increased collagen disposition, a characteristic of asthma (Biswas et al. 2012). However, some reports also suggest a suppressive/protective role of M2 macrophages in some models or types of allergy and asthma, which may be explained by studies suggesting that chronic TH17-driven inflammatory responses may also be involved in asthma (Biswas et al. 2012; Murray and Wynn 2011). These may be suppressed by anti-inflammatory IL-4/IL-13-activated macrophages.

Cancer

Tumors are highly complex pathological entities that are populated by a number of different cell types besides the tumor cells themselves, including innate immune cells. Among these, tumor-associated macrophages (TAMs) take the center stage at least when looking at abundance. Their presence in various malignancies, including mammary carcinoma, prostate cancer, bladder cancer, glioma, and lymphoma, is associated with poor patient prognosis (Galdiero et al. 2013). Just like resident and recruited macrophages during inflammation, TAMs contribute to diverse phases of cancer progression, including initiation and promotion by supporting angiogenesis, promoting steps to metastasis, and maintaining low-grade inflammation that blocks tumoricidal immunity. It is not surprising that TAMs, fulfilling these diverse functions, cannot be classified according to the M1/M2 paradigm. Rather, TAMs are educated by the tumor to display the features of both macrophage subsets (Van Ginderachter et al. 2006), at least when looking at the whole population. M1 mediators promote angiogenesis and maintain low-grade inflammation to promote tumor cell proliferation, whereas M2 mediators block antitumor immunity and shape the extracellular matrix in tumors. Macrophages in tumors face IL-4 as well as damage-associated molecular patterns (DAMPs) derived from necrotic or stressed cells inducing classical activation via activating TLRs (Kuraishy et al. 2011; Noy and Pollard 2014). Consequently, TAMs may, as a whole population, simultaneously be subjected to transcriptional programs operating in M1 and M2 macrophages. However, recent studies suggest that these different functional patterns can be attributed to discreet macrophage subpopulations in experimental and human tumors (Noy and Pollard 2014). In breast cancer, two TAM populations were identified according to their surface marker expression profile. One population showed a surface phenotype of resident mammary tissue macrophages, whereas the other resembled monocyte-derived macrophages. The first population showed high expression of alternative (IL-4-driven) macrophage markers such as CD206 and the second a more pronounced pro-inflammatory profile (Olesch et al. 2015). Since both macrophage subsets are replenished by adult monocytes, which might be expected when considering tumors as "new" organs that develop in the adult (Franklin et al. 2014; Noy and Pollard 2014), the differences in their activation profile might be explained by their functional polarization in distinct microenvironmental niches characterized by different compositions of cytokines, growth factors, oxygen, and nutrient levels (Laoui et al. 2011). Thus, tumors are a great example to illustrate the heterogeneity of macrophage responses in an inflammatory setting. Understanding their diversity will serve to identify TAM subsets to target and to spare in cancer therapy. This may be an improvement above existing therapy approaches aiming at TAM depletion (MacDonald et al. 2010).

Interaction with Other Processes and Drugs

Interfering with macrophages in disease so far is convincingly performed at the level of recruitment or survival, i.e., macrophage depletion can be achieved by substances such as CSF1-neutralizing antibodies (MacDonald et al. 2010). Drugs that achieve targeted conversion of distinct macrophage activation programs, such as shifting macrophage polarization from M1 to M2, especially when the highly plastic macrophages are exposed to the strong influences of a particular microenvironment, are so far elusive. Therefore, the last part of this essay will briefly introduce major microenvironmental factors that modulate macrophage activation during inflammation.

The Microenvironment: Hypoxia

During acute inflammation, the shift from pro-inflammatory cytokines, such as TNF-α or IL-1β, to anti-inflammatory cytokines, such as TGFβ, IL-10, or IL-13, induces transcription factors activated during these different phases of inflammation, which alters macrophage phenotypes through changes in their transcriptome, as outlined above. This versatility of macrophages is not restricted to responses to cytokines and other mediators of inflammation but extends to a vast variety of stimuli in the microenvironment, including parameters such as pH, pO_2, and NO levels as well as nutrient availability. The respective sensory repertoire enables macrophages to detect possible harmful alterations in the tissues they are located in (Chovatiya and Medzhitov 2014). The inflammatory microenvironment, for example, after wounding or in tumors, is characterized by reduced pO_2, increased NO and ROS production, reduced nutrient availability, and increased lactate production, resulting in a decreased pH (Brune et al. 2013). One consequence is activation of hypoxia-inducible factor (HIF). HIF is a dimeric transcription factor in which the α-subunit is tightly regulated by pO_2 levels in the microenvironment. A low pO_2 level leads to the stabilization and nuclear translocation of the α-subunit and consequently to the formation of the active HIF dimer. In addition to pO_2, HIF is activated by increased NO and ROS levels or low pH. Activation of HIF in macrophages without any additional stimulus does not lead to an inflammatory response but rather to the activation of several pathways to restore oxygen homeostasis such as promotion of angiogenesis, production of anaerobic energy, and reorganization of the extracellular matrix. When HIF is activated in an inflammatory setting, HIF target genes function to guarantee adequate ATP production in the hypoxic microenvironment, which is necessary to ensure production of NO and cytokines such as IL-6 and TNF-α and to increase angiogenesis in the affected tissues. In this manner, macrophages are able to adapt to the hypoxic microenvironment in order to fulfill their tasks during the pro-inflammatory phase of inflammation.

The Microenvironment: Dying Cells

Inflammatory processes often result in a dramatic increase in the number of dying cells in spatially defined areas. For instance, reduced nutrient and/or oxygen availability induces cellular necrosis (Kuraishy et al. 2011), and clearance of inflammatory neutrophils is preceded by activation-induced apoptosis (Gilroy and De Maeyer 2015; Headland and Norling 2015). The mode of cell death and different factors released and/or expressed on the surface of dying cells define the type of immune response elicited by macrophages.

Necrotic cells lose their plasma membrane integrity, releasing the cellular contents into the extracellular space. The endogenous necrosis-derived DAMPs or alarmins initiate a pro-inflammatory response that resembles the recognition of microbes regarding the macrophage surface receptors (e.g., TLRs) as well as the downstream mediators involved (cytokines, chemokines) (Kuraishy et al. 2011). In contrast, apoptotic cells maintain plasma membrane integrity, although its composition is altered to discriminate apoptotic from living cells. As a consequence the apoptotic bodies are recognized through specific macrophage surface receptors and removed by phagocytosis. For this reason this type of cell death is immunologically silent and does not provoke inflammation under steady-state conditions (Duprez et al. 2009). However, during inflammation, recognition of apoptotic cells by macrophages blocks inflammatory responses in macrophages, in turn triggering a wound healing phenotype through factors such as vascular endothelial growth factor (VEGF), TGF-β, prostaglandin E_2, IL-10, and sphingosine-1-phosphate (Sangiuliano et al. 2014; Weigert et al. 2009). Thus, the phagocytosis of dying pro-inflammatory effectors does not only serve to remove these cells (neutrophil death can be activately induced by macrophages) but also promotes transition to resolution by switching macrophage phenotypes.

Cross-References

▶ Allergic Disorders
▶ Cytokines
▶ Dendritic Cells
▶ Janus Kinases (JAKs)/STAT Pathway
▶ Leukocyte Recruitment
▶ Obesity and Inflammation
▶ Sepsis
▶ Toll-Like Receptors

References

Bain, C. C., & Mowat, A. M. (2014). Macrophages in intestinal homeostasis and inflammation. *Immunological Reviews, 260*(1), 102–117. doi:10.1111/imr.12192.

Biswas, S. K., Chittezhath, M., Shalova, I. N., & Lim, J. Y. (2012). Macrophage polarization and plasticity in health and disease. *Immunologic Research, 53*(1–3), 11–24. doi:10.1007/s12026-012-8291-9.

Brune, B., Dehne, N., Grossmann, N., Jung, M., Namgaladze, D., Schmid, T., et al. (2013). Redox control of inflammation in macrophages. *Antioxid Redox Signal 19*(6), 595–637.

Cassetta, L., Cassol, E., & Poli, G. (2011). Macrophage polarization in health and disease. *ScientificWorldJournal, 11*, 2391–2402. doi:10.1100/2011/213962.

Chovatiya, R., & Medzhitov, R. (2014). Stress, inflammation, and defense of homeostasis. *Molecular Cell, 54*(2), 281–288. doi:10.1016/j.molcel.2014.03.030.

Duprez, L., Wirawan, E., Vanden Berghe, T., & Vandenabeele, P. (2009). Major cell death pathways at a glance. *Microbes and Infection, 11*(13), 1050–1062. doi:10.1016/j.micinf.2009.08.013. S1286-4579(09)00204-4 [pii].

Epelman, S., Lavine, K. J., & Randolph, G. J. (2014). Origin and functions of tissue macrophages. *Immunity, 41*(1), 21–35. doi:10.1016/j.immuni.2014.06.013.

Franklin, R. A., Liao, W., Sarkar, A., Kim, M. V., Bivona, M. R., Liu, K., et al. (2014). The cellular and molecular origin of tumor-associated macrophages. *Science 344* (6186):921–925.

Galdiero, M. R., Garlanda, C., Jaillon, S., Marone, G., & Mantovani, A. (2013). Tumor associated macrophages and neutrophils in tumor progression. *Journal of Cellular Physiology, 228*(7), 1404–1412. doi:10.1002/jcp.24260.

Gautier, E. L., Shay, T., Miller, J., Greter, M., Jakubzick, C., Ivanov, S., et al. (2012). Gene-expression profiles and transcriptional regulatory pathways that underlie the identity and diversity of mouse tissue macrophages. *Nature Immunology, 13*(11), 1118–1128. doi:10.1038/ni.2419.

Gilroy, D., & De Maeyer, R. (2015). New insights into the resolution of inflammation. *Seminars in Immunology, 27*(3), 161–168. doi:10.1016/j.smim.2015.05.003.

Ginhoux, F., & Jung, S. (2014). Monocytes and macrophages: Developmental pathways and tissue homeostasis. *Nature Reviews Immunology, 14*(6), 392–404. doi:10.1038/nri3671.

Gosselin, D., Link, V. M., Romanoski, C. E., Fonseca, G. J., Eichenfield, D. Z., Spann, N. J., et al. (2014). Environment drives selection and function of enhancers controlling tissue-specific macrophage identities. *Cell, 159*(6), 1327–1340. doi:10.1016/j.cell.2014.11.023 S0092-8674(14)01500-1 [pii].

Headland, S. E., & Norling, L. V. (2015). The resolution of inflammation: Principles and challenges. *Seminars in Immunology, 27*(3), 149–160. doi:10.1016/j.smim.2015.03.014.

Jenkins, S. J., & Hume, D. A. (2014). Homeostasis in the mononuclear phagocyte system. *Trends in Immunology, 35*(8), 358–367. doi:10.1016/j.it.2014.06.006. S1471-4906(14)00111-2 [pii].

Kaufmann, S. H. (2008). Immunology's foundation: The 100-year anniversary of the Nobel Prize to Paul Ehrlich and Elie Metchnikoff. *Nature Immunology, 9*(7), 705–712.

Krausgruber, T., Blazek, K., Smallie, T., Alzabin, S., Lockstone, H., Sahgal, N., et al. (2011). IRF5 promotes inflammatory macrophage polarization and TH1-TH17 responses. *National Immunology, 12*(3), 231–238. doi:10.1038/ni.1990 ni.1990 [pii].

Kuraishy, A., Karin, M., & Grivennikov, S. I. (2011). Tumor promotion via injury- and death-induced inflammation. *Immunity, 35*(4), 467–477.

Laoui, D., Movahedi, K., Van Overmeire, E., Van den Bossche, J., Schouppe, E., Mommer, C., et al. (2011). Tumor-associated macrophages in breast cancer: Distinct subsets, distinct functions. *International Journal Developmental Biology, 55*(7–9), 861–867.

Lavin, Y., & Merad, M. (2013). Macrophages: Gatekeepers of tissue integrity. *Cancer Immunology Research, 1*(4), 201–209. doi:10.1158/2326-6066.CIR-13-0117. 1/4/201 [pii].

Lavin, Y., Winter, D., Blecher-Gonen, R., David, E., Keren-Shaul, H., Merad, M., et al. (2014). Tissue-resident macrophage enhancer landscapes are shaped by the local microenvironment. *Cell, 159*(6), 1312–1326. doi:10.1016/j.cell.2014.11.018 S0092-8674(14)01449-4 [pii].

Lavin, Y., Mortha, A., Rahman, A., & Merad, M. (2015). Regulation of macrophage development and function in peripheral tissues. *Nature Reviews Immunology, 15*(12), 731–744. doi:10.1038/nri3920.

MacDonald, K. P., Palmer, J. S., Cronau, S., Seppanen, E., Olver, S., Raffelt, N. C., et al. (2010). An antibody against the colony-stimulating factor 1 receptor depletes the resident subset of monocytes and tissue- and tumor-associated macrophages but does not inhibit inflammation. *Blood, 116*(19), 3955–3963. doi:10.1182/blood-2010-02-266296.

Murray, P. J., & Wynn, T. A. (2011). Protective and pathogenic functions of macrophage subsets. *Nature Reviews Immunology, 11*(11), 723–737.

Naito, M., Umeda, S., Yamamoto, T., Moriyama, H., Umezu, H., Hasegawa, G., et al. (1996). Development, differentiation, and phenotypic heterogeneity of murine tissue macrophages. *Journal Leukocyte Biology, 59*(2), 133–138.

Noy, R., & Pollard, J. W. (2014). Tumor-associated macrophages: From mechanisms to therapy. *Immunity, 41*(1), 49–61. doi:10.1016/j.immuni.2014.06.010.

Olesch, C., Sha, W., Angioni, C., Sha, L. K., Acaf, E., Patrignani, P., et al. (2015). MPGES-1-derived PGE2 suppresses CD80 expression on tumor-associated phagocytes to inhibit anti-tumor immune responses in breast cancer. *Oncotarget, 6*(12), 10284–10296.

Ortega-Gomez, A., Perretti, M., & Soehnlein, O. (2013). Resolution of inflammation: An integrated view.

EMBO Molecular Medicine, 5(5), 661–674. doi:10.1002/emmm.201202382.

Sangiuliano, B., Perez, N. M., Moreira, D. F., & Belizario, J. E. (2014). Cell death-associated molecular-pattern molecules: Inflammatory signaling and control. *Mediators of Inflammation, 2014*, 821043. doi:10.1155/2014/821043.

Sica, A., & Mantovani, A. (2012). Macrophage plasticity and polarization: In vivo veritas. *Journal of Clinical Investigation, 122*(3), 787–795. doi:10.1172/JCI59643

Smale, S. T., Tarakhovsky, A., & Natoli, G. (2014). Chromatin contributions to the regulation of innate immunity. *Annual Review of Immunology, 32*, 489–511. doi:10.1146/annurev-immunol-031210-101303.

Stout, R. D., Jiang, C., Matta, B., Tietzel, I., Watkins, S. K., & Suttles, J. (2005). Macrophages sequentially change their functional phenotype in response to changes in microenvironmental influences. *Journal of Immunology, 175*(1), 342–349.

van Furth, R., Cohn, Z. A., Hirsch, J. G., Humphrey, J. H., Spector, W. G., & Langevoort, H. L. (1972). The mononuclear phagocyte system: A new classification of macrophages, monocytes, and their precursor cells. *Bulletin of the World Health Organization, 46*(6), 845–852.

Van Ginderachter, J. A., Movahedi, K., Hassanzadeh Ghassabeh, G., Meerschaut, S., Beschin, A., Raes, G., et al. (2006). Classical and alternative activation of mononuclear phagocytes: Picking the best of both worlds for tumor promotion. *Immunobiology, 211*(6–8), 487–501.

Weigert, A., Jennewein, C., & Brune, B. (2009). The liaison between apoptotic cells and macrophages – The end programs the beginning. *Biological Chemistry, 390*(5–6), 379–390.

Wynn, T. A., Chawla, A., & Pollard, J. W. (2013). Macrophage biology in development, homeostasis and disease. *Nature, 496*(7446), 445–455. doi:10.1038/nature12034. nature12034 [pii].

Xue, J., Schmidt, S. V., Sander, J., Draffehn, A., Krebs, W., Quester, I., et al. (2014). Transcriptome-based network analysis reveals a spectrum model of human macrophage activation. *Immunity, 40*(2), 274–288. doi:10.1016/j.immuni.2014.01.006.

Mammalian Target of Rapamycin (mTOR)

Estela Jacinto and Guy Werlen
Department of Biochemistry and Molecular Biology, Rutgers-Robert Wood Johnson Medical School, Piscataway, NJ, USA

Synonyms

Dominant rapamycin-resistant mutation (DRR); FKBP12-rapamycin-associated protein (FRAP); Mammalian target of rapamycin; Mechanistic target of rapamycin; Rapamycin and FKBP12 target (RAFT); Rapamycin target (RAPT); TOR

Definition

The mechanistic target of rapamycin (mTOR) was discovered in yeast as a protein that is inhibited by the natural compound, rapamycin. Rapamycin, isolated from the bacteria *Streptomyces hygroscopicus*, forms a complex with the cyclophilin FKBP12 and together allosterically inhibits the serine/threonine kinase TOR or mTOR (Heitman et al. 1991). The inhibition of TOR by rapamycin arrests cells in G1 phase of the cell cycle and elicits a starvation-like phenotype (Barbet et al. 1996; Heitman et al. 1991). Moreover, rapamycin induces autophagy, a starvation response that degrades and recycles cellular components (Blommaart et al. 1995). Further genetic studies in yeast revealed that TOR promotes protein synthesis when nutrient conditions are favorable (Barbet et al. 1996). In mammalian cells including T lymphocytes, rapamycin blocks the phosphorylation and activation of the translation regulators, p70-S6K (S6K) and 4E-BP. The phosphorylation of S6K and its substrate, the ribosomal protein S6, is also triggered by the presence of amino acids and growth factors. Since rapamycin and amino acids have opposite effects on autophagy and protein synthesis, these findings connect mTOR to the control of protein synthesis via nutrient and growth factor signals.

In multicellular organisms, growth is coordinated between cells and the expansion of organs or tissues. Thus, in addition to nutrients, other extracellular inputs, such as growth factors, cytokines, and hormones, control the growth processes. The phosphatidylinositol 3-kinase (PI3K)/Akt signaling pathway couples signals from growth factors to the control of gene expression and cellular growth responses (Manning and Cantley 2007). Engagement of growth factor receptors activates PI3K, which phosphorylates the lipid second messenger phosphatidylinositol 4,5-bisphosphate (PIP2) to produce phosphatidylinositol (3,4,5)-trisphosphate (PIP3) (Fig. 1).

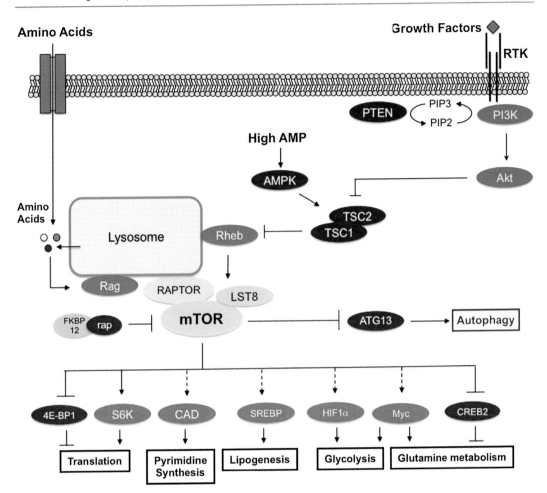

Mammalian Target of Rapamycin (mTOR), Fig. 1 mTORC1 responds to growth signals to control cellular metabolism. mTORC1, composed of mTOR, raptor, and LST8, is activated in response to growth signals such as amino acids and growth factors/cytokines. This active complex is found to be associated with the Rag GTPase in the lysosomal membrane compartment. Signals from growth factors and cytokines activate the PI3K/Akt pathway, which dissociates the TSC from the lysosome and thus promotes Rheb-mediated activation of mTORC1. Active mTORC1 promotes anabolic metabolism that includes increased translation or protein synthesis along with lipid and nucleic acid synthesis. It enhances aerobic glycolysis and glutamine metabolism while inhibiting autophagy, to promote cell growth and proliferation

PIP3 recruits a number of signaling molecules into the membrane including Akt. Active Akt phosphorylates the tuberous sclerosis complex 2 (TSC2) and downregulates its activity. The tuberous sclerosis protein complexes 1 and 2 (TSC1/TSC2) connect PI3K/Akt signals to mTOR (Huang and Manning 2008). Numerous signals including energy, nutrients, and growth factors that converge on TSC1/TSC2 altogether modulate mTOR signaling. TSC1/TSC2, by acting as a GTPase-activating protein (GAP) for the small G protein, Rheb, downregulates mTOR (as part of mTOR complex 1 (mTORC1); see below). When TSC1/TSC2 signals are low, Rheb is active in its GTP-bound form and directly interacts and stimulates mTORC1 (Huang and Manning 2008). Another family of small G proteins, the Rag GTPases, modulates the subcellular localization of mTORC1 in response to amino acid signals (Sancak et al. 2010). Amino acids promote the GTP loading of RagA/B and the heterodimer's interaction with the mTORC1 component, raptor, resulting in enhanced translocation of mTORC1 to the lysosomal surface. In this compartment, mTORC1 becomes activated possibly via Rheb, which is found throughout the endomembrane system. On the other hand, the absence of amino acids inactivates RagA/B and instead recruits the TSC complex to the lysosome, while the presence of insulin signals acutely dissociates TSC from this compartment (Menon et al. 2014). Thus, nutrient and growth factor signals are integrated to regulate the activation and subcellular localization of mTORC1.

Structure and Function

mTOR is an atypical protein kinase that belongs to the phosphatidylinositol 3-kinase-related kinase (PIKK) family. The PIKKs share homology in the catalytic domain with lipid kinases including PI3K but they possess serine/threonine kinase activity (Wu et al. 2012). mTOR activity and target specificity are regulated via association with distinct protein partners. The rapamycin-sensitive mTOR complex 1 (mTORC1) is generated by the association of mTOR with raptor and mLST8 (Fig. 1). The rapamycin-insensitive mTORC2 consists of mTOR, rictor (mAVO3), SIN1, and mLST8 (Fig. 2). In addition to these conserved partners, the mTORCs also associate with other less well-conserved proteins that could regulate mTOR activity and function (Wu et al. 2012). The well-characterized function of mTOR in mRNA translation is mediated by the mTORC1-triggered phosphorylation of translation regulators including S6K and 4E-BP1 (Ma and Blenis 2009). S6K is part of the AGC (protein kinase A/PKG/PKC) family of Ser/Thr protein kinases and is phosphorylated by mTOR at conserved motifs known as the hydrophobic (HM) and turn (TM) motifs. On the other hand, mTORC2 phosphorylates Akt, another AGC kinase family member, at its hydrophobic motif site (HM; Ser473), leading to optimal activation of this kinase. Although Akt phosphorylation at its conserved HM site is not acutely sensitive to rapamycin and is thus used as a hallmark of mTORC2 activity, other direct substrates of mTORC2 are emerging from recent studies (Wu et al. 2012). mTORC2 has also been linked to other cellular functions, such as actin cytoskeleton reorganization, translation, and protein maturation or processing (Oh and Jacinto 2011). It phosphorylates Akt at its TM and is required for phosphorylation of PKC at both TM and HM sites (Oh and Jacinto 2011). The TM phosphorylation occurs cotranslationally and is required for proper folding and stabilization of the kinase domain. Consistent with these latter functions, mTORC2 was also found to associate with ribosomes (Oh et al. 2010; Zinzalla et al. 2011). Thus, both mTORCs function during translation, albeit by distinct mechanisms. Whereas mTORC1 promotes translation in response to amino acids, it remains unclear what activates mTORC2. It is however considered to be part of the PI3K pathway because growth factor/PI3K signals induce the mTORC2-mediated allosteric activation of Akt (by phosphorylation of Akt hydrophobic motif). Moreover, augmented mTORC2 activity towards Akt in cancer cells correlates with enhanced ribosome association of the complex as well as upregulated PI3K activity, supporting mTORC2's activation via the PI3K pathway (Zinzalla et al. 2011). Nevertheless, recent phosphoproteomic studies identified numerous direct and indirect targets of both mTORCs (Hsu et al. 2011; Yu et al. 2011). However, most of what is currently known about mTORC1 and mTORC2 is based on analysis of the canonical substrates S6K (mTORC1 substrate) and Akt (mTORC2 substrate).

mTOR is not only involved in protein synthesis but is also a central controller of anabolic metabolism (Shimobayashi and Hall 2014). It integrates signals from nutrients and other growth factor signals to positively regulate biosynthetic processes that produce macromolecules required for cell growth and proliferation. These processes include the synthesis of lipid moieties, nucleic acids/nucleotides, and amino acids. At the same time, mTOR negatively regulates catabolic processes such as autophagy, lipid oxidation, and lipolysis. mTOR controls a number of these metabolic processes via its role in translation and gene transcription. For example, mTOR's involvement

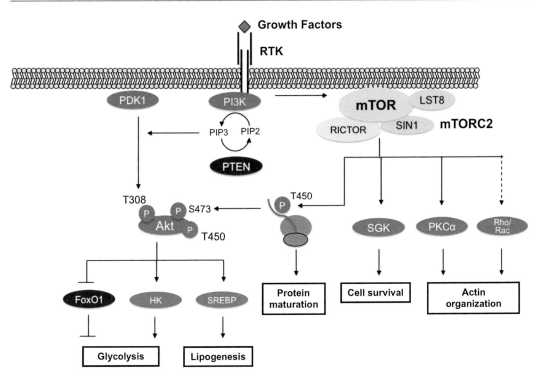

Mammalian Target of Rapamycin (mTOR), Fig. 2 Growth factor/PI3K signals are linked to active mTORC2 signaling. mTORC2 associates with membrane compartments and actively translating ribosomes. It phosphorylates conserved motifs in AGC (protein kinase A, PKG, PKC families) kinases to allosterically activate (such as Akt hydrophobic motif (HM) phosphorylation at Ser473) or stabilize (such as Akt turn motif (TM) T450 phosphorylation, PKC HM, and TM phosphorylation) the protein. mTORC2 signals have been linked to regulation of the actin cytoskeleton reorganization, cell survival, protein maturation, and anabolic metabolism

in glucose metabolism, which is enhanced in highly proliferating cells, primarily occurs via the control of HIF1 and Myc, two critical transcription factors that play a role in the metabolic switch (Duvel et al. 2010). In turn, these factors regulate the expression of glycolytic genes and nutrient transporters. Highly proliferating cells additionally upregulate the metabolism of glutamine, which is a nonessential amino acid that can serve as a carbon source for energy production and a nitrogen source for biosynthetic reactions. In a feedback loop manner, mTORC1 controls glutaminolysis and the increased glutamine levels from this metabolic process in turn enhance mTORC1 activity (Shimobayashi and Hall 2014). mTORC1 and mTORC2 also control enzymes that regulate the pentose phosphate pathway to generate nucleic acids. Furthermore, mTORC1 specifically controls de novo synthesis of pyrimidines. mTOR modulates lipogenesis via the control of the sterol regulatory element-binding protein (SREBP), a master regulator of genes involved in lipid synthesis. Moreover, ongoing work in cancer and immune cells continues to uncover mTOR's relevance to cellular metabolism and how it controls metabolic reprogramming.

Pathological Relevance

mTOR Complexes in Early T-Cell Development

The development of T lymphocytes from hematopoietic stem cells to a mature functional T cell has served as a paradigm to understand how distinct environmental stimuli can influence cell differentiation via specific signal transduction pathways and metabolic processes. While the use of rapamycin has laid the groundwork to unravel the role of mTOR in T-cell ontogeny, the availability of conditional knockout models that temporally and/or tissue-specifically delete mTORC component genes (via promoter-*Cre* recombinase driver system) has allowed to more precisely examine mTORC relevance at distinct stages of T-cell development. We will discuss how the strategies of conditionally deleting mTOR complexes affect T-cell differentiation in the thymus and formulate an emerging picture of how the mTORCs are specifically involved in T-cell ontogeny and the control of immune responses.

Hematopoietic stem cells migrate from the bone marrow into the thymus where as immature thymocytes that do not yet express the coreceptors CD4 or CD8 (CD4$^-$CD8$^-$ double-negative (DN)), they become committed to the T-lymphocyte lineage that will express a polymorphic T-cell receptor (TCR) as a hallmark (Fig. 3). The mTOR complex components raptor, rictor, and SIN1 have been conditionally deleted in the early stages of thymocyte development (Fig. 4). Similar to rapamycin treatment, inducible raptor deletion leads to thymic atrophy (Hoshii et al. 2014; Luo et al. 1994). However, whereas rapamycin treatment blocks differentiation from the DN3 to DP stage, raptor deficiency constrains DN1 to DN2 transition (Hoshii et al. 2014). There are pronounced decreases in the S/G2/M phases of *raptor*-deficient DN1 cells with concomitant increases in G0/G1, indicating that mTORC1 can promote proliferation in early T-cell progenitors.

Distinct promoters driving the *Cre* recombinase have been used to delete *rictor* in thymocytes. Whereas *Mx1-Cre* promoter is effective at abolishing *rictor* at the earliest stage of thymocyte maturation (DN1), *Lck-Cre* ablates rictor at the DN2 and later stages. Although thymic cellularity is highly decreased, mature T cells still exist in the periphery in these studies (Chou et al. 2014; Tang et al. 2012; Lee et al. 2012), suggesting that rictor is not absolutely required for T-cell maturation. Decreased proliferation rather than increased apoptosis accounts in part for the reduced thymocyte cellularity in the absence of rictor, while a consistent defect in early thymocyte differentiation also massively reduces production of mature T cells. On the other hand, BM cellularity appears normal following *Mx1-Cre* ablation of *rictor*$^{fl/fl}$ (Tang et al. 2012), implying that mTORC2 is not essential for general hematopoiesis.

The ablation of *rictor*$^{fl/fl}$ by Lck-cre starting at the DN2 dramatically affects the total number of thymocytes (Chou et al. 2014). The proportion of DN3 (CD25$^+$CD44$^-$) subset is pronouncedly increased, accompanied by an attenuation of the DN4 (CD25$^-$CD44$^-$) population (Chou et al. 2014). A similar phenotype is recapitulated in rictor-deleted bone marrow chimera mice as well as when *Mx1-Cre* or *vav-cre* is used to delete rictor instead of *Lck-cre* (Lee et al. 2012; Tang et al. 2012). In addition, there is a partial developmental block at the CD8$^+$-immature SP (CD8$^+$-ISP) stage that leads to a dramatically decreased proportion and number of DP thymocytes (Chou et al. 2014). The DN3 and CD8-ISP developmental blocks in the absence of rictor correlates with diminished expression of receptors that are required to drive thymocyte maturation (Chou et al. 2014). Indeed, the expression of Notch, the pre-TCR/TCR, as well as CD147, CD8, and CD4, is significantly attenuated on the surface of rictor-deficient thymocytes, while the level of CD127 is elevated, suggesting that specific receptors are affected differently by the disruption of mTORC2. How mTORC2 can impact receptor expression remains to be unraveled.

When thymocytes from rictor-deficient mice (*Lck-Cre* x *rictor*$^{fl/fl}$) are cultured on OP9 stromal cells expressing the Notch ligand DL1 (OP9-DL1 cells), proliferation decreases and DN to DP transition is impaired (Lee et al. 2012). Compared to wild-type controls, there is no difference in size and glycolytic rate of cultured rictor-deficient thymocytes. In the absence of rictor, phosphorylation of FoxO1/3a as well as Akt at Ser473 is attenuated in DN subsets, while the levels of ICN (intracellular domain of Notch) and the phosphorylation of the ribosomal protein S6 are comparable to purified wild-type cells (Lee et al. 2012). Interestingly, overexpressing a constitutively active Akt construct (Myr-Akt) in rictor-depleted thymocytes triggers proliferation and promotes development into DP cells, even in the absence of Notch ligand. Phosphomimetic constructs of

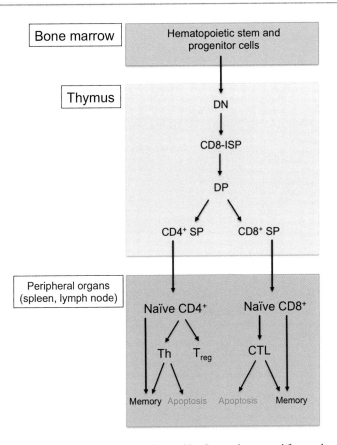

Mammalian Target of Rapamycin (mTOR), Fig. 3 Schematic diagram of T-cell development. Hematopoietic stem and progenitor cells migrate into the thymus where they differentiate and commit to a T-lineage phenotype. $CD4^+$ single-positive (SP) or $CD8^+$ SP cells migrate into peripheral organs where they encounter antigens presented by antigen-presenting cells (APCs), which triggers their asymmetric division and differentiation into effector and memory cells. The extent of stimulation along with other environmental factors determines their cell fate. Activated T cells undergo clonal expansion and differentiation into effector cells. Effector cells eventually undergo apoptosis after pathogen clearance, but a fraction of these cells differentiate into memory cells. Activated $CD4^+$ SP cells differentiate into various T helper subsets (effector cells) while $C8^+$ SP cells differentiate into cytotoxic T lymphocytes. Cells that are suboptimally stimulated differentiate into suppressor T-regulatory cells

Akt that harbor mutations at the activation and HM sites (Akt-DD) rescue both differentiation and proliferation of rictor-deficient cells in the presence of Notch ligand. In contrast, mutating solely Akt's activation site (Akt-DA) only rescues proliferation but not differentiation into DP cells, supporting a crucial role for the mTORC2-mediated HM phosphorylation in thymocyte maturation (Lee et al. 2012). Akt-DD, but not Akt-DA, also rescues NF-κB nuclear localization in Notch-stimulated rictor-deficient thymocytes. Supporting the defective Nf-κB nuclear induction in the absence of rictor, mRNA levels of several genes regulated by Notch, FoxO, or NF-κB are decreased. Thus, in addition to modulating receptor processing during translation (Chou et al. 2014), mTORC2 also plays a role in transcriptional regulation of genes required for early T-cell development (Lee et al. 2012).

mTOR Complexes in Peripheral T-Cell Differentiation

Circulating mature but naïve T cells encounter foreign antigens in the secondary lymphoid organs (spleen, lymph nodes, or Peyer's patches), which triggers their asymmetric division and differentiation into effector and memory (central memory) cells with similar antigenic recognition

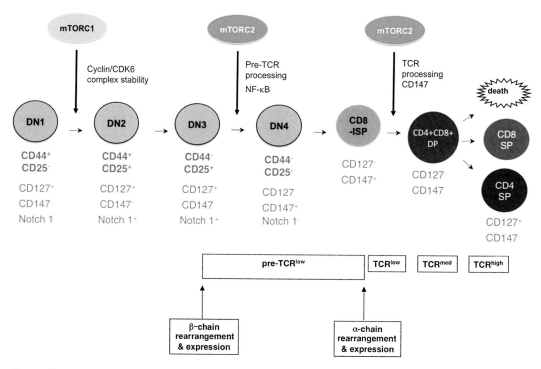

Mammalian Target of Rapamycin (mTOR), Fig. 4 Early T-cell development in the thymus. Hematopoietic stem cells from the bone marrow that have migrated into the thymus undergo differentiation towards a T-lymphocyte lineage. The differentiating subsets are distinguished by characteristic cell surface markers. The $CD4^-CD8^-$ double-negative subset is divided into four stages (DN1-DN4). Commitment to the T-lineage occurs during DN3 to DN4 stages wherein TCR β-chain gene rearrangement and expression commences. The functionality of the pre-TCR is assessed at the pre-TCR checkpoint between the DN3 and DN4 stages. Prior to expressing both CD4 and CD8, DN4 cells transition into CD8-immature CD8-expressing single-positive cells (CD8-ISP). The TCR α-chain gene rearrangement commences at this stage. Functionality of the αβTCR is assessed at the TCR checkpoint during positive selection of double-positive (DP) thymocytes and their differentiation to either the CD4 or the CD8 single-positive (SP) lineage. When mTORC1 is conditionally deleted in HSCs, development from DN1 to DN2 stage is impaired. mTORC2 is involved in DN3 to DN4 and CD8-ISP to DP differentiation. Processing and expression of pre-TCR/TCR receptors, along with other critical receptors during developmental stages including CD147, are defective upon mTORC2 disruption

potential (Chang et al. 2007). Although the TCR must recognize its cognate antigenic peptide/ MHC on the surface of an antigen-presenting cell (APC), co-engagement of CD28 by B7.1 or B7.2 (CD80 and CD86, respectively) expressed on the same APC is also required to trigger clonal expansion. Together, these activation signals along with nutrients and signals from engaged cytokine or chemokine receptors reprogram the metabolism of activated naïve T cells towards enhanced aerobic glycolysis and glutamine oxidation (MacIver et al. 2013). Whereas stimulated $CD8^+$ T cells readily gain cytotoxic effector activity, $CD4^+$ T cells differentiate further into one of many distinct types of T helper effector cells (Th1, Th2, Th17, T follicular helper (Tfh), or T regulatory (T_{reg})) based on the presence of specific cytokines and other extracellular cues. Particular transcription factors have been linked to the differentiation of naïve CD4 T cells to a distinct effector lineage. Furthermore, while most effector cells die at the end of an immune response, the surviving T lymphocytes also differentiate into memory cells (peripheral memory) that can be reactivated upon secondary infection of the same pathogen. The latter response is more rapid and efficient, and this process has been exploited for the prevention of diseases by vaccination. Distinct metabolic programs maintain regulatory and memory cells as compared to the differentiation of naïve into effector cells (Michalek et al. 2011).

CD4+ T Cells

Deletion of mTOR (*Frap1*) at the DP (CD4+CD8+) stage of thymocyte maturation and onward using the *CD4-Cre* driver generates normal ratios of CD4:CD8 cells in the periphery (Delgoffe et al. 2009). The specific disruption of mTORC1 by *CD4-Cre*-driven abrogation of *raptor*$^{f/f}$ generates a more striking phenotype as compared to the direct deletion of mTOR. Both in vitro and in vivo stimulations dramatically decrease proliferation of *raptor*-deficient CD4 T cells (Yang et al. 2013), in part due to a substantial reduction in IL-2 production. While raptor is essential for both antigen-specific and lymphopenia-induced proliferation, its deficiency abolishes in vivo antibacterial immune responses as well. In naïve T cells, raptor is required to exit quiescence and enter the S phase, but it is dispensable once the cells have entered the cell cycle. Corroborating the importance of raptor in cell cycle entry, global expression profiles revealed that genes related to cell cycle entry such as cyclin D2, cyclin E, CDK2, and CDK6 become downregulated in the absence of raptor (Yang et al. 2013). Furthermore, the expression of nutrient receptors such as CD98 and the transferrin receptor CD71 is reduced as well as the levels of metabolic programs including glycolysis, lipogenesis, and oxidative phosphorylation. However, the kinetic difference of metabolic and cell cycle gene expression suggests that raptor plays a role in the transition from metabolic reprogramming to cell cycle entry.

On the other hand, when rictor is deleted at the DP stage using a distal *Lck-Cre* (*dLck-Cre*) driver, CD4 T cells develop normally while CD8 SP cells decrease modestly (Lee et al. 2010). Costimulation of rictor-deficient CD4 T cells diminishes phosphorylation of Akt and its downstream substrate GSK3. Although their proliferation is reduced, the effect is less profound as compared to raptor deficiency (Lee et al. 2010; Yang et al. 2013). Thus, proliferation of CD4 T cells is more highly dependent on mTORC1 than mTORC2, while neither complex seems to affect cell death even in the presence of cytokines.

Regulatory T Cells (T_{reg})

Regulatory T cells are generated naturally in the thymus (natural T_{reg} (nT_{reg})), as well as in the periphery (induced T_{reg} (iT_{reg})) as a result of TGFβ or IL-35 (Tr35) exposure following TCR engagement of naïve CD4+ T cells (Fig. 5). They are important for the induction of peripheral tolerance. Indeed, they curb the proliferation of potentially self-reactive T cells by competing with them for the costimulatory signals provided by APCs or via secreting inhibitory factors such as IL-10 and TGFβ. Due to this ability to modulate T-cell homeostasis, T_{reg} cells serve as promising treatment for autoimmune disorders and acute graft-versus-host disease. Besides CD4, T_{reg} constitutively express CD25 (IL-2Rα) on their surface, while the presence of the transcription factor FoxP3 prevents upregulation of IL-2 expression following TCR engagement (Huynh et al. 2014; Sawant and Vignali 2014). Thus, FoxP3 is absolutely essential for the ontogeny and function of regulatory T cells. Notably, transducing naïve peripheral CD4 T cells with the *foxp3* gene converts them into T_{reg}, while transient rather than sustained TCR signals contribute to FoxP3 induction in developing thymocytes and the generation of nT_{reg}. The duration of TCR signaling is thus also a crucial parameter that modulates regulatory T-cell fate determination.

T_{regs} exhibit low levels of mTOR activity, consistent with the transient TCR signals that favor their development. Their differentiation is facilitated by culture conditions that block or reduce mTOR signaling, such as suboptimal antigen stimulation or depletion of essential amino acids (Cobbold et al. 2009; Sauer et al. 2008). Inhibition of mTOR by rapamycin upregulates the FoxP3 and CD25 genes (Sauer et al. 2008) and promotes the generation of T_{reg} under specific conditions (Battaglia et al. 2005; Kim et al. 2010) and despite the absence of TGFβ (Kang et al. 2008; Valmori et al. 2006). Although the total number of T cells decreases in mice upon rapamycin treatment, the proportion of T_{reg} cells that display

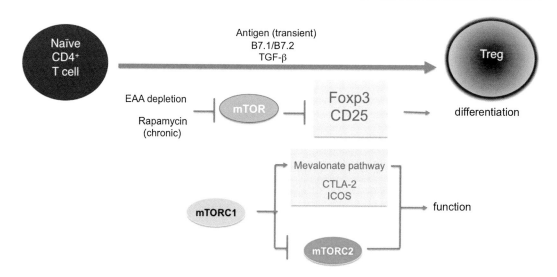

Mammalian Target of Rapamycin (mTOR), Fig. 5 mTORC1 is required for T_{reg} function but downregulation of both mTORC complexes promotes T_{reg} differentiation. Downregulating mTORC1 and mTORC2 signals via diminished nutrients or chronic rapamycin treatment promotes T_{reg} differentiation via mechanisms that lead to induction of Foxp3 and enhanced CD25 expression. However, mTORC1 signals are required for T_{reg} suppressive function and maintenance via regulation of lipid metabolism (particularly mevalonate pathway), expression of the suppressive mediators CTLA4 and ICOS, and negative regulation of mTORC2

immunosuppressive ability is enhanced (Qu et al. 2007), owing in part to the preferential expansion of Foxp3$^+$ T cells under those conditions. Similarly, IL-2 and TCR/CD28 costimulation triggers the expansion of purified human CD4$^+$CD25^{high} T cells in the presence of rapamycin, whereas CD4$^+$CD25^{low} or CD4$^+$CD25$^-$ cells do not proliferate (Strauss et al. 2007). The expanded population expresses markers for T_{reg} and displays suppressive ability. Together, these findings suggest that downregulating mTOR activity, including via rapamycin treatment, can be effective in specifically triggering T_{reg} differentiation and proliferation.

More recent studies using conditional knockout of raptor, or the upstream regulator Rheb1, shed further light on the intricate mechanisms by which mTORC1 contributes to T_{reg} differentiation and function. Crossing *raptor$^{fl/fl}$* with *FoxP3-Cre* mice specifically generates raptor deficiency in T_{reg} cells, but not in other T cells (Zeng et al. 2013). These animals develop severe autoimmune diseases due to profound loss of T_{reg} suppressive activity despite the presence of an increased number of Foxp3$^+$ T_{reg} cells (Zeng et al. 2013). In this same study, deletion of raptor using *CD4-Cre* system also generates T_{regs} with an intrinsic defect in suppressive activity (Zeng et al. 2013). However, in the latter scenario, there is a small reduction in the number of T_{reg} cells residing in peripheral lymphoid organs.

While mTORC1 controls the suppressive activity of T_{reg}, it remains unclear how downregulating its signals precisely contributes to T_{reg} differentiation. The deletion of Rheb1, which downregulates mTORC1 signaling, blocks development of FoxP3$^+$ T cells (Delgoffe et al. 2009), suggesting that mTORC1 plays a role in T_{reg} differentiation. However, when TGFβ is present in this model, FoxP3 expression increases and differentiation into inducible T_{reg} cells ensues. Thus, T_{reg} differentiation can occur in the absence of mTORC1 if other signals that bypass the complex, such as TGFβ, promote FoxP3 expression.

Blocking mTORC2 signals seem to promote T_{reg} generation. The proportion of nT_{reg} cells is enhanced in the thymus of lethally irradiated mice that were reconstituted with *Sin1*-deficient fetal liver HSCs (Chang et al. 2012). In contrast, enhanced mTORC2 signaling abrogates the generation and maintenance of T_{reg} (Zeng et al. 2013). Similarly, increased signaling of mTORC2's effector, Akt, strongly represses the entry into T_{reg} phenotype (Haxhinasto et al. 2008). Constitutively active Akt reduces TGFβ-induced Foxp3 expression in a rapamycin-sensitive manner (Haxhinasto et al. 2008). Furthermore, the transcription factors Foxo1/3a control T_{reg} differentiation and function by possibly altering the expression of FoxP3 as well as other T_{reg}-associated proteins including CTLA4 (Kerdiles et al. 2010; Ouyang et al. 2010). While the Foxo1/3a transcription factors are negatively regulated by mTORC2-Akt, it remains to be elucidated whether other downstream effectors of mTORC2 are critically involved in T_{reg} differentiation or maintenance.

Despite a requirement for mTORC1 in maintaining T_{reg} homeostasis and functionality (Zeng et al. 2013), it seems crucial that very low total mTOR signals (mTORC1 and mTORC2 combined) are transduced to preserve these processes. Indeed, unless combined with genetic inactivation of rictor, low levels of rapamycin generate only weak T_{reg} induction. On the other hand, pharmacologic inhibitors that affect both mTORC1 and mTORC2 efficiently promote the generation of $Foxp3^+$ T_{reg} (Delgoffe et al. 2011). Thus, signaling from both mTORC1 and mTORC2 need to be blocked simultaneously to induce maximal T_{reg} differentiation (Delgoffe et al. 2011). Inhibiting both complexes could thus become a very significant therapeutic tool for immunosuppression.

Effector T Helper 17 (Th17) Cells

The Th17 effector lineage plays a critical role in the clearance of bacterial infections as well as the induction of tissue inflammation, while their uncontrolled activation is involved in the development of autoimmune disorders such as rheumatoid arthritis and multiple sclerosis (Nagai et al. 2013). Similar to T_{reg} cells, Th17 come in two flavors; the induced Th17 (iTh17) cells differentiate in the peripheral lymphoid organs from naïve CD4+ T cells in response to antigens and cytokines (Fig. 6). In contrast, natural nTh17 cells acquire the capacity to produce IL-17 during their commitment to the $CD4^+$ T-cell lineage in the thymus. Common features of Th17 cells include

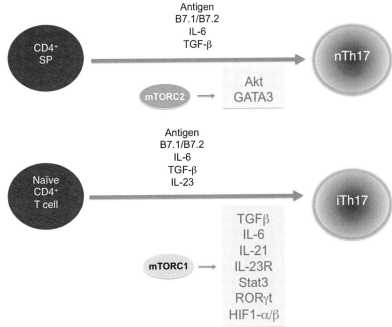

Mammalian Target of Rapamycin (mTOR), Fig. 6 mTORC involvement in Th17 differentiation. mTORC1 promotes the differentiation of activated $CD4^+$ T cells into iTh17 in response to antigens and cytokines in peripheral lymphoid organs. Disrupting mTORC1 signals affect expression of cytokines, receptors, and transcription factors that are crucial for iTh17 differentiation. mTORC2 appears to be specifically required for differentiation of nTh17 cells in the thymus

the expression of the transcription factor retinoic acid-related orphan receptor γ (RORγT) that drives the expression of their characteristic cytokines such as IL-17, IL-22, and IL-21. IL-21 is an autocrine hormone that activates the transcription factor signal transducer and activator of transcription 3 (STAT3) during the initial differentiation phase of iTh17, while IL-17 and IL-22 are secreted by effector Th17 to activate stromal cells at the site of infection. iTh17 cells arise when TGFβ and IL-6 are present in the peripheral organs while IL-4 and IL-12 are absent. Furthermore, IL-23 secreted by APCs is also important for the expansion and maintenance of the Th17 cell population.

The development of IL-17-secreting $CD4^+$ T cells is severely diminished by the absence of mTORC1 or its defective signaling (Delgoffe et al. 2009, 2011; Kurebayashi et al. 2012). In the absence of mTOR, IL-17$^+$ cells isolated from Peyer's patches are reduced in number as compared to their high frequency in wild-type mice. Furthermore, cultured mTOR-deficient $CD4^+$ T cells fail to differentiate into Th17 effectors despite the presence of TGFβ and IL-6 (Delgoffe et al. 2009). Furthermore, deleting the mTORC1 upstream regulator Rheb1 in T cells also leads to a profound reduction in iTh17 differentiation (Delgoffe et al. 2011; Kim et al. 2013). So far, mTORC1 has been linked to the generation of iTh17 via several mechanisms. First, mTOR signaling triggers HIF1-α activity in Th17 cells (Shi et al. 2011), which mediates enhanced glycolytic metabolism as well as augmented transcriptional activation of RORγT (Dang et al. 2011). Furthermore, the mild hypoxic conditions that stabilize HIF1-α during Th17 differentiation also activate mTORC1 (Ikejiri et al. 2012), while the absence of ARNT (aryl hydrocarbon receptor nuclear translocator, aka HIF1-β), the HIF1-α dimerizing partner, led to defective generation of iTh17 cells (Kim et al. 2013). Hence, mTORC1 controls Th17 differentiation via its crucial role in the promotion of glycolytic processes. Second by partly mediating the expression of S6K2, mTORC1 also indirectly regulates the nuclear translocation of RORγT during the maturation of Th17 cells (Delgoffe et al. 2009; Kurebayashi et al. 2012).

Indeed, TCR signaling and mTORC1 signaling promote the association of RORγT to the nuclear localization signal-bearing S6K2 and thus facilitate its relocation to the nucleus (Kurebayashi et al. 2012). In addition, mTORC1 prevents the expression of the Th17 negative regulator, Gfi1, via the activation of S6K1 (Kurebayashi et al. 2012).

mTOR, via either mTORC1 or mTORC2, regulates the STAT transcription factors relevant for Th17 differentiation. In the absence of mTOR, IL-6 is unable to induce the phosphorylation of STAT3, and consequently, IL-21, IL-23R, and RORγT fail to be upregulated, which results in defective Th17 differentiation (Delgoffe et al. 2009). Similarly, deleting Rheb also reduces the tyrosine phosphorylation of STAT3 and STAT4 (Delgoffe et al. 2011), whereas rapamycin treatment or raptor deletion has no effect (Kurebayashi et al. 2012; Lee et al. 2010). These results would point towards mTORC2 as the complex by which mTOR regulates STAT activity during Th17 differentiation. However, rictor ablation does not significantly affect the generation of iTh17 cells (Delgoffe et al. 2011; Lee et al. 2010). mTORC2's contribution to STAT3/4 phosphorylation remains to be examined.

In contrast, mTORC2 appears to have a specific role in promoting the differentiation of nTh17 cells. Indeed, ablating *rictor* via the CD4-Cre system results in defective development of nTh17 cells in the thymus (Lee et al. 2010), whereas the number of iTh17 cells isolated from the small intestinal lamina propria is unaffected. In contrast, losing Rheb has no effect on nTh17 differentiation (Lee et al. 2010), suggesting that the development of this subset is more dependent on mTORC2 rather than on mTORC1. This is supported by the requirement of Akt activation and Foxo inhibition for maximal nTh17 differentiation (Lee et al. 2010). On the other hand, rictor is not required for the generation of iTh17 cells (Delgoffe et al. 2011; Lee et al. 2010).

Effector T Helper 1 (Th1) Cells

T helper 1 (Th1) cells are effective against intracellular microbes. Their differentiation from naïve $CD4^+$ T cells relies on the engagement of the TCR

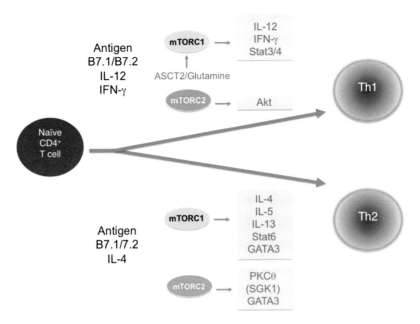

Mammalian Target of Rapamycin (mTOR), Fig. 7 mTORC involvement in T helper 1 (Th1) and Th2 differentiation. mTORC1 promotes differentiation of activated CD4⁺ SP cells into Th1 or Th2 cells via production of cytokines and transcription factors that define each of these lineages. The amino acid transporter ASCT2, which is required for glutamine uptake, is essential for Th1 differentiation and TCR-stimulated activation of mTORC1. mTORC2 has been linked to Th differentiation via regulation of AGC kinases and the transcription factor GATA3 (for Th2). The mTORC2 target SGK1 has been shown to be required for Th2 differentiation but whether this is via mTORC2 remains unclear

and the CD28 coreceptor as well as the presence of IL-12 and IFN-γ, two cytokines that are produced by innate APCs during the early stages of an intracellular pathogen infection (Fig. 7). IFN-γ activates the transcription factors STAT1 and T-bet that commit CD4⁺ T cells to the Th1 lineage by switching on the expression of the IL-12 receptor (IL-12R). In concert, IL-12 promotes further expansion and differentiation of committed Th1 cells via the recruitment and activation of STAT4 downstream of IL-12R.

Several lines of evidence support an essential role of mTORC1 in Th1 differentiation. Rapamycin treatment of activated naïve CD4 T cells leads to aberrant STAT3 and STAT4 recruitment, as well as decreased IFN-γ production upon IL-12 treatment (Kusaba et al. 2005). Cytokine stimulation to promote Th1 differentiation leads to defective T-bet expression in the absence of mTOR (Delgoffe et al. 2009). IL-12-induced phosphorylation of STAT4 is also reduced in mTOR-deficient T cells despite normal expression of IL-12R (Delgoffe et al. 2009). Similarly, CD4⁺ T cells lacking Rheb fail to differentiate towards a Th1 phenotype and show reduced STAT4 phosphorylation in response to IL-12 (Delgoffe et al. 2011). Furthermore, mTOR-deficient CD4⁺ T cells fail to differentiate into Th1 effector cells when challenged with *Vaccinia virus*, a pathogen that normally induces a strong Th1 response (Delgoffe et al. 2009). Likewise, the induction of Th1 (along with Th17) cells is impaired in the different models of murine immunity and autoimmunity, when the amino acid transporter ASCT2 is deficient (Nakaya et al. 2014). This transporter is required for increased glutamine uptake, which is necessary for TCR-stimulated activation of mTORC1. The enhanced nutrient uptake upon activation of naïve T cells is used to reprogram the metabolism as well as to promote clonal cell expansion.

In contrast to mTORC1, mTORC2's role in Th1 differentiation is less clear. While *Lck-Cre-induced rictor* deficiency leads to fewer Th1 and

Th2 cells (Lee et al. 2010), deleting rictor with the *CD4-Cre* driver has no significant effect on Th1 differentiation (Delgoffe et al. 2011). It is however noteworthy that a constitutive active allele of Akt but not the mutant allele that has Ala mutation in the HM site rescues the lack of Th1 differentiation in *Lck-Cre*-driven *rictor deficiency*, suggesting that mTORC2 is involved in Th1 differentiation (Lee et al. 2010).

Effector T Helper 2 (Th2) Cells

Th2 cells play a role in immunity against extracellular pathogenic infections. Their differentiation is promoted by IL-4-induced STAT6 activation along with TCR and CD28 engagement (Fig. 7). Th2 cells express high levels of the transcription factor GATA3 and secrete cytokines including IL-4, IL-5, and IL-13. mTOR has been linked to Th2 differentiation as part of both mTORCs (Delgoffe et al. 2009, 2011; Heikamp et al. 2014; Lee et al. 2010; Yang et al. 2013). In mTOR-deficient T cells, IL-4-induced STAT6 phosphorylation is decreased, leading to reduced GATA3 expression and diminished Th2 differentiation, despite proper expression of the IL-4 receptor (Delgoffe et al. 2009). In contrast, mTOR deficiency still allows differentiation into Th2 cells in a Rheb-abrogated background, even under conditions that normally promote differentiation into Th1 and/or Th17 cells (Delgoffe et al. 2011). The deletion of raptor impairs proper differentiation into Th2 effector cells owing to abnormal cytokine receptor expression, Stat activation, as well as decreased IL-4 production under Th2-promoting conditions (Yang et al. 2013). Glucose levels and metabolic rates directly influence the amount of cytokine receptors expressed in these cells. Interestingly, raptor deficiency diminishes lung inflammation and leukocyte infiltration in a model of Th2 cell-dependent allergic airway disease (Yang et al. 2013), suggesting that mTORC1 plays a role in both Th2 cell differentiation and response. Moreover, deleting rictor in a raptor-deficient background does not further reduce IL-4 as compared to raptor deficiency alone (Yang et al. 2013), suggesting a dominant role of mTORC1 over mTORC2 in Th2 cell differentiation. However, deleting *rictor*$^{fl/fl}$ with the *dLck-Cre* driver compromises the generation of Th2 cells as well as Th1 and T$_{reg}$, while Th17 cell development is not significantly affected (Lee et al. 2010). A constitutively active allele of PKC-θ increases the expression of GATA3 in rictor-deficient CD4$^+$ T cells and rescues their differentiation towards Th2 cells, while expressing a constitutively active Akt mutant does not (Lee et al. 2010). Thus, Th2 cell differentiation involves mTORC1 and mTORC2 activation wherein the upstream inputs are independent of Rheb or Akt. The nature of these inputs remains to be determined.

CD8$^+$ T Cells

Cytotoxic CD8$^+$ T cells eradicate virally infected cells and selectively target and destroy tumor cells via the release of cytokines and cytolytic mediators. A few of the effector cells can differentiate into memory CD8 T cells that are capable of reactivation as effector cells upon reencounter of the same pathogen. The state of mTOR activation and consequently the expression of critical cytokines and transcription factors as well as metabolic reprogramming have been strongly linked to the two distinct CD8 cell fates.

While both IL-7 and IL-15 cytokines control homeostatic survival of naïve and memory CD8$^+$ T cells, only IL-15 is critical for the long-term survival and proliferation of the latter subset. In contrast, IL-2 is required for naïve, but not memory, cell clonal expansion following antigenic stimulation. In addition to cytokines, distinct transcription factors also control the differentiation of naïve CD8$^+$ T cells into effector and long-lived memory subsets (Fig. 8) (Banerjee et al. 2010; Pipkin et al. 2010). T-bet (*Tbx21*) is the master transcription factor regulating effector differentiation, while another T-box-containing transcription factor, Eomesodermin (Eomes), has been suggested to promote memory cell development (McLane et al. 2013). Expression of T-bet is high in short-lived effector cells, but low in the long-lived memory niche, suggesting that its expression levels influence development into these distinct cell fates. T-bet's expression is also critical for the production of effector cytokines such as IFN-γ. IL-12 further enhances and sustains the

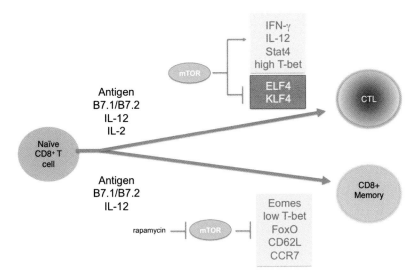

Mammalian Target of Rapamycin (mTOR), Fig. 8 mTOR signals promote CD8$^+$ differentiation into effector cells, whereas diminished mTOR signals favor differentiation into memory cells. While the precise contributions of each mTOR complexes in CD8$^+$ differentiation remain to be elucidated, downregulating mTORC1 signals promotes memory cell differentiation. The role of mTORC2 is yet to be examined, but sustained Akt signals favor differentiation into effector cells while FoxO1 deficiency impairs memory cell generation

levels of T-bet. In contrast, the expression of Eomes gradually increases during an immune response from effector to memory cells (Banerjee et al. 2010), which indicates that it preferentially promotes memory formation over effector differentiation.

mTOR has been linked to effector CD8$^+$ T-cell differentiation via up- or downregulation of critical transcription factors. Indeed, IL-12-induced STAT4 activity maintains mTOR phosphorylation, which in turn triggers T-bet expression (Rao et al. 2010). Sustained mTOR activity promotes naïve CD8$^+$ T cells to differentiate into effector cells in the presence of IL-12. Rapamycin reverses the effect of IL-12 and promotes the expression of Eomes instead of T-bet, resulting in memory cell maturation (Rao et al. 2010). Thus, mTOR could control CD8$^+$ differentiation by positively regulating the expression of T-bet while negatively regulating Eomes. Indeed, a balance of expression between these two transcription factors has previously been shown to dictate CD8$^+$ T-cell differentiation into effector or memory cells (Intlekofer et al. 2005). Besides upregulating the expression of "differentiation" transcription factors, mTOR is also involved in downregulating factors that maintain naïve T-cell quiescence. mTOR inhibits the expression of ELF4 (E74-like factor), a transcription factor that blocks the proliferation of naïve CD8$^+$ T cells by promoting the expression of the cell cycle inhibitor Krüppel-like factor 4 (KLF4) (Yamada et al. 2010). When the ERK pathway is activated upon TCR stimulation, ELF4 is downregulated in an mTOR-dependent manner, thereby inhibiting the expression of KLF4 and promoting CD8$^+$ T-cell proliferation.

The kinetics and strength of mTOR activity also trigger distinct T-cell differentiation fates. Whereas a strong activation of the kinase promotes the maturation of CD8 effector cells, suboptimal signaling favors the generation of memory cells. Effector cells develop in the presence of IL-2 in part due to sustained mTOR activation. On the other hand, IL-2 also downregulates the expression of receptors critical for the homing of naïve and self-renewing memory T cells such as the adhesion receptor L-selectin (CD62 ligand (CD62L)) and the chemokine receptor CCR7. Rapamycin blocks this receptor downregulation (Sinclair et al. 2008), implying that mTORC1 is involved in this process as well as the generation of CD8$^+$ memory T cells. The generation of CD8$^+$ memory T cells is augmented when low doses of rapamycin are used to dampen the activation of the mTORC1 pathway (Araki et al. 2009). Furthermore, enhanced memory T-cell responses occur in rapamycin-treated mice following virus infection as well as rapamycin-treated primates following vaccination (Araki et al. 2009). Thus, rapamycin is contributing to both increasing the number of memory precursors during naïve T-cell expansion and accelerating the reprograming of effector T cells during their transition to the memory niche.

The role of distinct mTORC components such as raptor and rictor in the generation, differentiation, and function of CD8$^+$ T cells remains poorly understood. So far in the absence of raptor, there is a modest reduction of peripheral CD8$^+$ T cells under steady state, while CD3- and CD28-induced proliferation is markedly diminished (Yang et al. 2013). Ablating raptor as well as mTORC1 effectors such as S6K1 increases memory differentiation of antigen-specific CD8 T cells (Yang et al. 2013), supporting earlier reports that document enhanced memory subsets when mTORC1 signaling is blocked. The ablation of rictor during the early stages of T-cell ontogeny leads to a significant decrease in peripheral CD8 cells due to partial developmental blocks at the DN3 and CD8-ISP stages (Chou et al. 2014). How rictor is involved in CD8 effector or memory cell differentiation and function remains to be examined. Nevertheless, the downstream mTORC2 effectors, Akt and FoxO1, were shown to control the transition from effector to memory cells as well as to maintain the function of the latter cells. Whereas sustained Akt signals favor differentiation into effector cells, low activation of this kinase promotes the generation of memory cells (Macintyre et al. 2011; Tejera et al. 2013). In contrast, FoxO1 deficiency impairs the latter (Tejera et al. 2013). These findings suggest that downregulating mTORC2 signals could also contribute to memory differentiation of CD8$^+$ T cells.

Interaction with Other Processes and Drugs

Due to their positive effects on immune tolerance (inhibition of effector T-cell differentiation while promoting T$_{reg}$ generation), rapamycin and its analogs (rapalogs) remain in clinical use to prevent organ graft rejection. Their immunosuppressive properties also make them promising for the treatment of autoimmune disorders including type I diabetes, arthritis, multiple sclerosis, and lupus (Kato and Perl 2014; Monti et al. 2008). However, adverse side effects such as immune dysfunction, cardiovascular complications, and metabolic disorders oftentimes occur while using rapamycin to prevent allograft rejection as well as to treat autoimmunity. Nevertheless, these adverse effects are found to be mostly moderate in impact and manageable through palliative therapy (Kaplan et al. 2014). Furthermore, they can be minimized by tailoring the doses, the timing of treatment, and the combination with other immunosuppressants based on the patients' health profiles and risk factors (Kaplan et al. 2014). Intriguingly and despite its immunosuppressive activity, rapamycin had the surprising effect of limiting cancer incidence in organ-transplanted patients (Blagosklonny 2013). While this could be accounted for by blocked mTOR hyperactivation that oftentimes occurs in malignant cells, it could also be due to a mechanism by which rapamycin slows down the aging process. The latter, which has been linked to changes in the metabolism, is a major risk factor in cancer development. Thus, rapamycin could contribute to preventing malignancy by significantly impacting the metabolism and growth of specific T-cell subsets. Currently, we can maximize the therapeutic benefits associated with blocking the mTOR pathway by gaining better understanding of the immunosuppressive versus immunostimulatory properties of dampening this signaling pathway.

Rapalogs have undergone clinical trials for hematologic malignancies. Phase I/II study of everolimus in patients with relapsed or refractory hematologic malignancies revealed that it is well tolerated and could be promising for patients with myelodysplastic syndrome (Yee et al. 2006). Another rapalog, deforolimus, also is well tolerated in patients with relapsed or refractory hematologic malignancies and shown antitumor activity (Rizzieri et al. 2008). However, inhibition of mTORC1 by rapalogs/rapamycin has limited efficacy in cancer treatment possibly due to induction of feedback loops that consequently activate the PI3K pathway. Thus, current strategies have focused on inhibiting mTOR kinase activity either as monotherapy or in combination with other inhibitors or chemotherapeutic agents.

While the monotherapeutic use of mTOR inhibitors (rapalogs in particular) to treat leukemias had mainly cytostatic effects, mTOR kinase inhibitors (MKIs) and combined chemotherapy were more successful in inducing cytotoxicity. Unlike rapamycin, which allosterically inhibits mTOR and blocks some of mTORC1 functions, MKIs are ATP-competitive inhibitors that bind to the kinase domain of mTOR and thus have the potential to simultaneously impede both mTOR complex activities. Preclinical studies using MKIs effectively decrease cell viability of T-ALL in cell lines and patients by inducing cell cycle arrest, apoptosis, and autophagy (Evangelisti et al. 2011). In patient samples, MKIs target a specific subpopulation of leukemic initiating cells. Whether MKIs will be efficacious in treating T-cell malignancies in the clinic awaits further investigation.

Targeting metabolic pathways has been an effective strategy for the treatment of hematologic malignancies. For example, the enzyme L-asparaginase is used in the clinic to treat ALL and related lymphomas. While mTOR plays a pivotal role in protein synthesis, its inhibition by rapamycin could thus decrease the expression of enzymes regulating amino acid biosynthesis including asparagine synthetase (Peng et al. 2002). Other chemotherapeutic agents that target metabolic pathways in leukemias are nucleoside analogs and antifolates, which block DNA synthesis. By decreasing the expression of dihydrofolate reductase (DHFR), mTOR inhibitors have been shown to effectively synergize with the antifolate methotrexate (Teachey et al. 2008). Thus, due to the extensive involvement of mTOR in the control of metabolic pathways, mTOR inhibitors serve as effective synergistic therapy with antimetabolites against hematologic malignancies.

Cross-References

▶ Antiviral Responses
▶ Cancer and Inflammation
▶ Costimulatory Receptors
▶ Cytokines
▶ Foxp3
▶ Interleukin 17
▶ Interleukin 2
▶ Janus Kinases (Jaks)/STAT Pathway
▶ MAP Kinase Pathways
▶ Methotrexate
▶ NFkappaB
▶ TH17 Response

References

Araki, K., Turner, A. P., Shaffer, V. O., Gangappa, S., Keller, S. A., Bachmann, M. F., et al. (2009). mTOR regulates memory CD8 T-cell differentiation. *Nature, 460*, 108–112.

Banerjee, A., Gordon, S. M., Intlekofer, A. M., Paley, M. A., Mooney, E. C., Lindsten, T., et al. (2010). Cutting edge: The transcription factor eomesodermin enables $CD8^+$ T cells to compete for the memory cell niche. *The Journal of Immunology, 185*, 4988–4992.

Barbet, N. C., Schneider, U., Helliwell, S. B., Stansfield, I., Tuite, M. F., & Hall, M. N. (1996). TOR controls translation initiation and early G1 progression in yeast. *Molecular Biology of the Cell, 7*, 25–42.

Battaglia, M., Stabilini, A., & Roncarolo, M. G. (2005). Rapamycin selectively expands $CD4^+CD25^+FoxP3^+$ regulatory T cells. *Blood, 105*, 4743–4748.

Blagosklonny, M. V. (2013). Immunosuppressants in cancer prevention and therapy. *Oncoimmunology, 2*, e26961.

Blommaart, E. F., Luiken, J. J., Blommaart, P. J., van Woerkom, G. M., & Meijer, A. J. (1995). Phosphorylation of ribosomal protein S6 is inhibitory for autophagy in isolated rat hepatocytes. *The Journal of Biological Chemistry, 270*, 2320–2326.

Chang, J. T., Palanivel, V. R., Kinjyo, I., Schambach, F., Intlekofer, A. M., Banerjee, A., et al. (2007). Asymmetric T lymphocyte division in the initiation of adaptive immune responses. *Science, 315*, 1687–1691.

Chang, X., Lazorchak, A. S., Liu, D., & Su, B. (2012). Sin1 regulates Treg-cell development but is not required for T-cell growth and proliferation. *European Journal of Immunology, 42*, 1639–1647.

Chou, P. C., Oh, W. J., Wu, C. C., Moloughney, J., Ruegg, M. A., Hall, M. N., et al. (2014). Mammalian target of rapamycin complex 2 modulates alphabetaTCR processing and surface expression during thymocyte development. *The Journal of Immunology, 193*, 1162–1170.

Cobbold, S. P., Adams, E., Farquhar, C. A., Nolan, K. F., Howie, D., Lui, K. O., et al. (2009). Infectious tolerance via the consumption of essential amino acids and mTOR signaling. *Proceedings of the National Academy of Sciences of the United States of America, 106*, 12055–12060.

Dang, E. V., Barbi, J., Yang, H. Y., Jinasena, D., Yu, H., Zheng, Y., et al. (2011). Control of T(H)17/T(reg) balance by hypoxia-inducible factor 1. *Cell, 146*, 772–784.

Delgoffe, G. M., Kole, T. P., Zheng, Y., Zarek, P. E., Matthews, K. L., Xiao, B., et al. (2009). The mTOR kinase differentially regulates effector and regulatory T cell lineage commitment. *Immunity, 30*, 832–844.

Delgoffe, G. M., Pollizzi, K. N., Waickman, A. T., Heikamp, E., Meyers, D. J., Horton, M. R., et al. (2011). The kinase mTOR regulates the differentiation of helper T cells through the selective activation of signaling by mTORC1 and mTORC2. *Nature Immunology, 12*, 295–303.

Duvel, K., Yecies, J. L., Menon, S., Raman, P., Lipovsky, A. I., Souza, A. L., et al. (2010). Activation of a metabolic gene regulatory network downstream of mTOR complex 1. *Molecular Cell, 39*, 171–183.

Evangelisti, C., Ricci, F., Tazzari, P., Tabellini, G., Battistelli, M., Falcieri, E., et al. (2011). Targeted inhibition of mTORC1 and mTORC2 by active-site mTOR inhibitors has cytotoxic effects in T-cell acute lymphoblastic leukemia. *Leukemia, 25*, 781–791.

Haxhinasto, S., Mathis, D., & Benoist, C. (2008). The AKT-mTOR axis regulates de novo differentiation of CD4$^+$Foxp3$^+$ cells. *The Journal of Experimental Medicine, 205*, 565–574.

Heikamp, E. B., Patel, C. H., Collins, S., Waickman, A., Oh, M. H., Sun, I. H., et al. (2014). The AGC kinase SGK1 regulates TH1 and TH2 differentiation downstream of the mTORC2 complex. *Nature Immunology, 15*, 457–464.

Heitman, J., Movva, N. R., & Hall, M. N. (1991). Targets for cell cycle arrest by the immunosuppressant rapamycin in yeast. *Science, 253*, 905–909.

Hoshii, T., Kasada, A., Hatakeyama, T., Ohtani, M., Tadokoro, Y., Naka, K., et al. (2014). Loss of mTOR complex 1 induces developmental blockage in early T-lymphopoiesis and eradicates T-cell acute lymphoblastic leukemia cells. *Proceedings of the National Academy of Sciences of the United States of America, 111*, 3805–3810.

Hsu, P. P., Kang, S. A., Rameseder, J., Zhang, Y., Ottina, K. A., Lim, D., et al. (2011). The mTOR-regulated phosphoproteome reveals a mechanism of mTORC1-mediated inhibition of growth factor signaling. *Science, 332*, 1317–1322.

Huang, J., & Manning, B. D. (2008). The TSC1-TSC2 complex: A molecular switchboard controlling cell growth. *The Biochemical Journal, 412*, 179–190.

Huynh, A., Zhang, R., & Turka, L. A. (2014). Signals and pathways controlling regulatory T cells. *Immunology Reviews, 258*, 117–131.

Ikejiri, A., Nagai, S., Goda, N., Kurebayashi, Y., Osada-Oka, M., Takubo, K., et al. (2012). Dynamic regulation of Th17 differentiation by oxygen concentrations. *International Immunology, 24*, 137–146.

Intlekofer, A. M., Takemoto, N., Wherry, E. J., Longworth, S. A., Northrup, J. T., Palanivel, V. R., et al. (2005). Effector and memory CD8$^+$ T cell fate coupled by T-bet and eomesodermin. *Nature Immunology, 6*, 1236–1244.

Kang, J., Huddleston, S. J., Fraser, J. M., & Khoruts, A. (2008). De novo induction of antigen-specific CD4$^+$CD25$^+$Foxp3$^+$ regulatory T cells in vivo following systemic antigen administration accompanied by blockade of mTOR. *Journal of Leukocyte Biology, 83*, 1230–1239.

Kaplan, B., Qazi, Y., & Wellen, J. R. (2014). Strategies for the management of adverse events associated with mTOR inhibitors. *Transplantation Reviews, 28*, 126–133.

Kato, H., & Perl, A. (2014). Mechanistic target of rapamycin complex 1 expands Th17 and IL-4+ CD4-CD8- double-negative T cells and contracts regulatory T cells in systemic lupus erythematosus. *The Journal of Immunology, 192*, 4134–4144.

Kerdiles, Y. M., Stone, E. L., Beisner, D. R., McGargill, M. A., Ch'en, I. L., Stockmann, C., et al. (2010). Foxo transcription factors control regulatory T cell development and function. *Immunity, 33*, 890–904.

Kim, B. S., Kim, I. K., Park, Y. J., Kim, Y. S., Kim, Y. J., Chang, W. S., et al. (2010). Conversion of Th2 memory cells into Foxp3$^+$ regulatory T cells suppressing Th2-mediated allergic asthma. *Proceedings of the National Academy of Sciences of the United States of America, 107*, 8742–8747.

Kim, J. S., Sklarz, T., Banks, L. B., Gohil, M., Waickman, A. T., Skuli, N., et al. (2013). Natural and inducible TH17 cells are regulated differently by Akt and mTOR pathways. *Nature Immunology, 14*, 611–618.

Kurebayashi, Y., Nagai, S., Ikejiri, A., Ohtani, M., Ichiyama, K., Baba, Y., et al. (2012). PI3K-Akt-mTORC1-S6K1/2 axis controls Th17 differentiation by regulating Gfi1 expression and nuclear translocation of RORgamma. *Cell Reports, 1*, 360–373.

Kusaba, H., Ghosh, P., Derin, R., Buchholz, M., Sasaki, C., Madara, K., et al. (2005). Interleukin-12-induced interferon-gamma production by human peripheral blood T cells is regulated by mammalian target of rapamycin (mTOR). *The Journal of Biological Chemistry, 280*, 1037–1043.

Lee, K., Gudapati, P., Dragovic, S., Spencer, C., Joyce, S., Killeen, N., et al. (2010). Mammalian target of rapamycin protein complex 2 regulates differentiation of Th1 and Th2 cell subsets via distinct signaling pathways. *Immunity, 32*, 743–753.

Lee, K., Nam, K. T., Cho, S. H., Gudapati, P., Hwang, Y., Park, D. S., et al. (2012). Vital roles of mTOR complex 2 in Notch-driven thymocyte differentiation and leukemia. *The Journal of Experimental Medicine, 209*, 713–728.

Luo, H., Duguid, W., Chen, H., Maheu, M., & Wu, J. (1994). The effect of rapamycin on T cell development in mice. *European Journal of Immunology, 24*, 692–701.

Ma, X. M., & Blenis, J. (2009). Molecular mechanisms of mTOR-mediated translational control. *Nature Reviews Molecular Cell Biology, 10*, 307–318.

Macintyre, A. N., Finlay, D., Preston, G., Sinclair, L. V., Waugh, C. M., Tamas, P., et al. (2011). Protein kinase B controls transcriptional programs that direct cytotoxic T cell fate but is dispensable for T cell metabolism. *Immunity, 34*, 224–236.

MacIver, N. J., Michalek, R. D., & Rathmell, J. C. (2013). Metabolic regulation of T lymphocytes. *Annual Review of Immunology, 31*, 259–283.

Manning, B. D., & Cantley, L. C. (2007). AKT/PKB signaling: Navigating downstream. *Cell, 129*, 1261–1274.

McLane, L. M., Banerjee, P. P., Cosma, G. L., Makedonas, G., Wherry, E. J., Orange, J. S., et al. (2013). Differential localization of T-bet and Eomes in CD8 T cell memory populations. *The Journal of Immunology, 190*, 3207–3215.

Menon, S., Dibble, C. C., Talbott, G., Hoxhaj, G., Valvezan, A. J., Takahashi, H., et al. (2014). Spatial control of the TSC complex integrates insulin and nutrient regulation of mTORC1 at the lysosome. *Cell, 156*, 771–785.

Michalek, R. D., Gerriets, V. A., Jacobs, S. R., Macintyre, A. N., MacIver, N. J., Mason, E. F., et al. (2011). Cutting edge: Distinct glycolytic and lipid oxidative metabolic programs are essential for effector and regulatory $CD4^+$ T cell subsets. *The Journal of Immunology, 186*, 3299–3303.

Monti, P., Scirpoli, M., Maffi, P., Piemonti, L., Secchi, A., Bonifacio, E., et al. (2008). Rapamycin monotherapy in patients with type 1 diabetes modifies $CD4^+CD25^+FOXP3^+$ regulatory T-cells. *Diabetes, 57*, 2341–2347.

Nagai, S., Kurebayashi, Y., & Koyasu, S. (2013). Role of PI3K/Akt and mTOR complexes in Th17 cell differentiation. *The Annals of the New York Academy of Sciences, 1280*, 30–34.

Nakaya, M., Xiao, Y., Zhou, X., Chang, J. H., Chang, M., Cheng, X., et al. (2014). Inflammatory T cell responses rely on amino acid transporter ASCT2 facilitation of glutamine uptake and mTORC1 kinase activation. *Immunity, 40*, 692–705.

Oh, W. J., & Jacinto, E. (2011). mTOR complex 2 signaling and functions. *Cell Cycle, 10*, 2305–2316.

Oh, W. J., Wu, C. C., Kim, S. J., Facchinetti, V., Julien, L. A., Finlan, M., et al. (2010). mTORC2 can associate with ribosomes to promote cotranslational phosphorylation and stability of nascent Akt polypeptide. *EMBO Journal, 29*, 3939–3951.

Ouyang, W., Beckett, O., Ma, Q., Paik, J. H., DePinho, R. A., & Li, M. O. (2010). Foxo proteins cooperatively control the differentiation of $Foxp3^+$ regulatory T cells. *Nature Immunology, 11*, 618–627.

Peng, T., Golub, T. R., & Sabatini, D. M. (2002). The immunosuppressant rapamycin mimics a starvation-like signal distinct from amino acid and glucose deprivation. *Molecular and Cellular Biology, 22*, 5575–5584.

Pipkin, M. E., Sacks, J. A., Cruz-Guilloty, F., Lichtenheld, M. G., Bevan, M. J., & Rao, A. (2010). Interleukin-2 and inflammation induce distinct transcriptional programs that promote the differentiation of effector cytolytic T cells. *Immunity, 32*, 79–90.

Qu, Y., Zhang, B., Zhao, L., Liu, G., Ma, H., Rao, E., et al. (2007). The effect of immunosuppressive drug rapamycin on regulatory $CD4^+CD25^+Foxp3^+$ T cells in mice. *Transplant Immunology, 17*, 153–161.

Rao, R. R., Li, Q., Odunsi, K., & Shrikant, P. A. (2010). The mTOR kinase determines effector versus memory $CD8^+$ T cell fate by regulating the expression of transcription factors T-bet and Eomesodermin. *Immunity, 32*, 67–78.

Rizzieri, D. A., Feldman, E., Dipersio, J. F., Gabrail, N., Stock, W., Strair, R., et al. (2008). A phase 2 clinical trial of deforolimus (AP23573, MK-8669), a novel mammalian target of rapamycin inhibitor, in patients with relapsed or refractory hematologic malignancies. *Clinical Cancer Research, 14*, 2756–2762.

Sancak, Y., Bar-Peled, L., Zoncu, R., Markhard, A. L., Nada, S., & Sabatini, D. M. (2010). Ragulator-Rag complex targets mTORC1 to the lysosomal surface and is necessary for its activation by amino acids. *Cell, 141*, 290–303.

Sauer, S., Bruno, L., Hertweck, A., Finlay, D., Leleu, M., Spivakov, M., et al. (2008). T cell receptor signaling controls Foxp3 expression via PI3K, Akt, and mTOR. *Proceedings of the National Academy of Sciences of the United States of America, 105*, 7797–7802.

Sawant, D. V., & Vignali, D. A. (2014). Once a Treg, always a Treg? *Immunology Reviews, 259*, 173–191.

Shi, L. Z., Wang, R., Huang, G., Vogel, P., Neale, G., Green, D. R., et al. (2011). HIF1alpha-dependent glycolytic pathway orchestrates a metabolic checkpoint for the differentiation of TH17 and Treg cells. *The Journal of Experimental Medicine, 208*, 1367–1376.

Shimobayashi, M., & Hall, M. N. (2014). Making new contacts: The mTOR network in metabolism and signalling crosstalk. *Nature Reviews Molecular Cell Biology, 15*, 155–162.

Sinclair, L. V., Finlay, D., Feijoo, C., Cornish, G. H., Gray, A., Ager, A., et al. (2008). Phosphatidylinositol-3-OH kinase and nutrient-sensing mTOR pathways control T lymphocyte trafficking. *Nature Immunology, 9*, 513–521.

Strauss, L., Whiteside, T. L., Knights, A., Bergmann, C., Knuth, A., & Zippelius, A. (2007). Selective survival of naturally occurring human CD4$^+$CD25$^+$Foxp3$^+$ regulatory T cells cultured with rapamycin. *The Journal of Immunology, 178*, 320–329.

Tang, F., Wu, Q., Ikenoue, T., Guan, K. L., Liu, Y., & Zheng, P. (2012). A critical role for Rictor in T lymphopoiesis. *The Journal of Immunology, 189*, 1850–1857.

Teachey, D. T., Sheen, C., Hall, J., Ryan, T., Brown, V. I., Fish, J., et al. (2008). mTOR inhibitors are synergistic with methotrexate: an effective combination to treat acute lymphoblastic leukemia. *Blood, 112*, 2020–2023.

Tejera, M. M., Kim, E. H., Sullivan, J. A., Plisch, E. H., & Suresh, M. (2013). FoxO1 controls effector-to-memory transition and maintenance of functional CD8 T cell memory. *The Journal of Immunology, 191*, 187–199.

Valmori, D., Tosello, V., Souleimanian, N. E., Godefroy, E., Scotto, L., Wang, Y., et al. (2006). Rapamycin-mediated enrichment of T cells with regulatory activity in stimulated CD4$^+$ T cell cultures is not due to the selective expansion of naturally occurring regulatory T cells but to the induction of regulatory functions in conventional CD4$^+$ T cells. *The Journal of Immunology, 177*, 944–949.

Wu, C.C., Chou, P., & Jacinto, E. (2012). The target of rapamycin: Structure and functions. In Da Silva Xavier, G. (Eds.), *Protein kinases* (pp. 1–40). Intechopen.com.

Yamada, T., Gierach, K., Lee, P. H., Wang, X., & Lacorazza, H. D. (2010). Cutting edge: Expression of the transcription factor E74-like factor 4 is regulated by the mammalian target of rapamycin pathway in CD8$^+$ T cells. *The Journal of Immunology, 185*, 3824–3828.

Yang, K., Shrestha, S., Zeng, H., Karmaus, P. W., Neale, G., Vogel, P., et al. (2013). T cell exit from quiescence and differentiation into Th2 cells depend on Raptor-mTORC1-mediated metabolic reprogramming. *Immunity, 39*, 1043–1056.

Yee, K. W., Zeng, Z., Konopleva, M., Verstovsek, S., Ravandi, F., Ferrajoli, A., et al. (2006). Phase I/II study of the mammalian target of rapamycin inhibitor everolimus (RAD001) in patients with relapsed or refractory hematologic malignancies. *Clinical Cancer Research, 12*, 5165–5173.

Yu, Y., Yoon, S. O., Poulogiannis, G., Yang, Q., Ma, X. M., Villen, J., et al. (2011). Phosphoproteomic analysis identifies Grb10 as an mTORC1 substrate that negatively regulates insulin signaling. *Science, 332*, 1322–1326.

Zeng, H., Yang, K., Cloer, C., Neale, G., Vogel, P., & Chi, H. (2013). mTORC1 couples immune signals and metabolic programming to establish T(reg)-cell function. *Nature, 499*, 485–490.

Zinzalla, V., Stracka, D., Oppliger, W., & Hall, M. N. (2011). Activation of mTORC2 by association with the ribosome. *Cell, 144*, 757–768.

MAP Kinase Pathways

John M. Kyriakis[1] and Joseph Avruch[2,3]
[1]Mercury Therapeutics, Inc., Woburn, MA, USA
[2]Diabetes Unit, Medical Services and Department of Molecular Biology, Massachusetts General Hospital, Boston, MA, USA
[3]The Department of Medicine, Harvard Medical School, Simches Research Center, Massachusetts General Hospital, Boston, MA, USA

Synonyms

Extracellular signal-regulated kinase (ERK); Jun-N-terminal kinase (JNK); p38 MAPK

Definition

Mitogen-activated protein kinases (MAPKs) are a family of highly conserved eukaryotic protein Ser/Thr kinases, most of which are activated by concomitant Tyr and Thr phosphorylation. For the majority of MAPKs, this phosphorylation occurs within a closely spaced motif Thr-X-Tyr (X is any amino acid, but for mammalian cells, these are generally Glu, Pro, or Gly) in the T-loop of subdomain VIII of the catalytic domain. MAPKs can shuttle between the cytosol and the nucleus and phosphorylate numerous substrates in each compartment (Kyriakis and Avruch 2012).

MAPK activation is catalyzed by a cohort of equally conserved, dual-specificity MAPK kinases (Map2Ks) that can phosphorylate both the Tyr and Thr residues. Map2Ks, in turn, are regulated by Ser/Thr phosphorylation catalyzed by several families of protein kinases broadly referred to as Map2K kinases (Map3Ks). The so-called MAPK core pathway is defined as Map3K → Map2K → MAPK. Extracellular stimuli can typically recruit several diverse MAPK core pathways to exert profound effects on cellular physiology (Kyriakis and Avruch 2012).

The presence in eukaryotic cells of multiple MAPK pathways enables coordinated and integrated responses to divergent extracellular inputs including hormones and growth factors that signal through receptor Tyr kinases, cytokines, agents acting through G-protein-coupled receptors, environmental stresses, TGF-β-related agents that recruit Ser-Thr kinase receptors, and pathogen-associated molecular patterns (PAMPs) and danger-associated molecular patterns (DAMPs) that signal via pattern recognition receptors (PRRs) (Kyriakis and Avruch 2012; Arthur and Ley 2013).

MAPK Groups

ERKs

The extracellular signal-regulated kinases (ERK1 and ERK2; also called MAPK-3 and MAPK-1, respectively – Fig. 1, Table 1) are recruited by agonists that engage the Ras proto-oncoprotein. Ras, in turn, recruits the Raf family of Map3Ks. The Rafs activate two ERK-specific Map2Ks: MEK1 and MEK2 (Map2K-1 and Map2K-2, respectively – Table 2) (McKay and Morrison 2007). ERKs can also be activated, in a Ras-independent way, by proinflammatory stimuli including cytokines of the TNF family, PAMPs (e.g., bacterial lipopolysaccharide), and by danger-associated molecular patterns (DAMPs – endogenous, "sterile" inflammatory agonists, i.e., oxidized LDL in dyslipidemia and atherosclerosis or crystalline uric acid in gout). PAMPs/DAMPs engage pattern recognition receptors (PRRs) which utilize Tpl-2 rather than the Rafs as the Map3K.

The ERKs are encoded by two genes: *mapk3* encodes ERK1, while *mapk1* encodes ERK2 (Table 1). The ERKs require dual activating Tyr and Thr phosphorylation at two adjacent resides separated by a Glu residue: Thr185-Glu-Tyr187 (ERK2) and Thr203-Glu-Tyr205 (ERK1) (McCay and Morrison 2007).

JNKs

JNKs (Fig. 1, Table 1) are activated by mitogens, environmental stresses (heat shock, ionizing radiation, oxidants), genotoxins (topoisomerase inhibitors and alkylating agents), ischemic reperfusion injury, mechanical shear stress, vasoactive peptides, proinflammatory cytokines, PAMPs/DAMPs, translational inhibitors (cycloheximide and anisomycin) (Dérijard et al. 1994; Kyriakis et al. 1994), and agents that cause endoplasmic reticulum (ER) stress (tunicamycin) (Kyriakis et al. 1994).

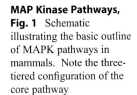

MAP Kinase Pathways, Fig. 1 Schematic illustrating the basic outline of MAPK pathways in mammals. Note the three-tiered configuration of the core pathway

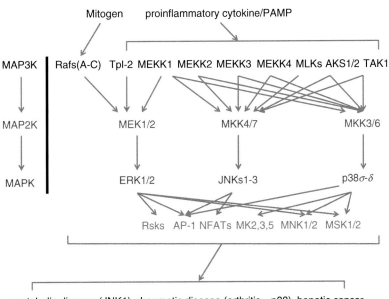

MAP Kinase Pathways, Table 1 Mammalian MAPKs

Name	Synonyms	Substrates	Human genome accession	Mouse genome accession
ERK1	*Mapk-3*, p44-MAPK	RSK, MNKs, MSKs, Elk1	NP_002737.2	NP_036082.1
ERK2	*Mapk-1*, p42-MAPK	RSK, MNKs, MSKs, Elk1	NP_620407.1	NP_036079.1
JNK1 *isoforms:*	*Mapk-8*, SAPK-γ, SAPK1c	c-Jun, Jun D, ATF2, Elk1	NP_620637.1 (JNK1α2)	NP_057909.1 (JNK1α2)
	Isoforms:			
JNK1β2	SAPK-p54γ1			
JNK1α2	SAPK-p54γ2			
JNK1β1	SAPK-p46γ1			
JNK1α1	SAPK-p46γ2			
JNK2	*Mapk-9*, SAPK-α, SAPK1a	c-Jun, JunD, ATF2, Elk1	NP_002743.3 (JNK2α2)	NP_997575.2 (JNK2α2)
Isoforms:	*Isoforms:*			
JNK2β2	SAPK-p54α1			
JNK2α2	SAPK-p54α2			
JNK2β1	SAPK-p46α1			
JNK2α1	SAPK-p46α2			
JNK3	*Mapk-10*, SAPK-β, SAPK1b	c-Jun, JunD, ATF2, Elk1	NP_620448.1 (JNK3α2)	NP_001075036.1 (JNK3α2)
Isoforms:	*Isoforms:*			
JNK3β2	SAPK-p54β1			
JNK3α2	SAPK-p54β2			
JNK3β1	SAPK-p46β1			
JNK3α1	SAPK-p46β2			
p38α	*Mapk-14*, SAPK2a, CSBP1	MK-2/MK-3, MSKs, ATF2, Elk1, MEF2C	NP_001306.1	NP_001161980.1
p38β	*Mapk-11*, SAPK2b, p38-2	MK-2/MK-3, MSKs, ATF2	NP_002742.3	NP_034991.4
p38γ	*Mapk-12*, SAPK3 ERK6	ATF2	NP_002960.2	NP_038899.1
p38δ	*Mapk-13*, SAPK4	ATF2	NP_002745.1	NP_036080.2

As with the ERKs, JNKs are activated by dual phosphorylation at the characteristic MAPK Thr-X-Tyr motif phosphoacceptor loop: Thr183-Pro-Tyr185.

There are three genes: JNK1–3 – also called *Mapk-8, Mapk-9* and *Mapk-10*, respectively (Dérijard et al. 1994; Kyriakis et al. 1994) (Table 1). Subsequent to transcription, each of the JNK hnRNAs undergoes differential splicing, both within a region encoding catalytic domain subdomains IX and X (resulting in α and β JNKs, respectively, Table 1) and at a segment encoding the extreme carboxyl terminus (producing 46- and 54-kDa polypeptides – respectively, type 1 and 2 JNKs, Table 1). This gives rise to a total of 12 JNK polypeptides. The functional differences between the type 1 and 2 isoforms are unclear (Dérijard et al. 1994; Gupta et al. 1996; Kyriakis et al. 1994).

p38 MAPKs

The p38 MAPKs, a third group of MAPKs, are activated by the same suite of proinflammatory stimuli and environmental stresses that recruit JNKs. p38s are generally not activated by Ras-dependent mitogens that engage receptor Tyr kinases. The canonical MAPK activating phosphoacceptor sites on p38s are separated by a Gly residue (e.g., Thr180/Tyr182 for p38α).

MAP Kinase Pathways, Table 2 Mammalian Map2Ks

Name	Synonyms	Substrate(s)	Human genome accession	Mouse genome accession
MEK-1	*Map2k-1*, MKK-1	ERK1, ERK2	NP_002746.1	NP_032953.1
MEK-2	Map2k-2, MKK-2	ERK1, ERK2	NP_109587.1	NP_075627.2
MKK-4	*Map2k-4*, SEK1, JNK kinase (JNKK)-1, MEK-4, SAPK-kinase (SKK)-1	JNKs	NP_003001.1	NP_033183.1
MKK-7	*Map2k-7*, JNKK2, MEK-7, SKK4	JNKs	NP_660186.1	NP_001157644.1
MKK-3	*Map2k-3*, MEK-3, SKK2	p38s	NP_659731.1	NP_032954.1
MKK-6	*Map2k-6*, MEK-6, SKK3	p38s	NP_002749.2	NP_036073.1

There are four p38 genes (Table 1): p38α–δ, encoded by *Mapk-14* (p38α), *Mapk-11* (p38β), *Mapk-12* (p38γ), and *Mapk-13* (p38δ) (Freshney et al. 1994; Goedert et al. 1997; Han et al. 1994; Jiang et al. 1996, 1997; Rouse et al. 1994).

Map2K Groups

All conventional mammalian MAPK groups are activated by concomitant Tyr/Thr phosphorylation catalyzed by a family of dual-specificity Map2Ks. Each of the above MAPK groups has a relatively selective group of Map2Ks (Kyriakis and Avruch 2012).

MEK-1/MEK-2

MAPK kinase-1 and MAPK kinase-2 (MAPKK-1/MAPKK-2) – also called MAPK/ERK kinase-1 or MAPK/ERK kinase-2 (MEK-1/MEK-2 or MKK-1/MKK-2, encoded by *Map2k-1* and *Map2k-2*, respectively) (Fig. 1, Table 2) are Map2Ks that phosphorylate and activate ERK1 and ERK2 (Kyriakis and Avruch 2012; McKay and Morrison 2007).

MKK-4, MKK-7

JNKs are activated by phosphorylation at Thr183 and Tyr185 catalyzed by two Map2Ks, MAPK-kinase-4 (MKK-4, encoded by *Map2k-4* and also called Map2K-4, SAPK/ERK kinase-1, SEK1; MEK-4; JNK kinase-1, JNKK1; and SAPK-kinase-1, SKK1, Table 2) and MKK-7 (encoded by *Map2k-7* and also called Map2K-7, MEK-7, JNKK2, and SKK4, Table 2). MKK-4 can also phosphorylate and activate p38 (Dérijard et al. 1995; Sánchez et al. 1994) (below).

MKK-7 does not activate p38 at all (Tournier et al. 1997; Wu et al. 1997).

MKK-3, MKK-6

The p38 MAPKs are activated by three Map2Ks: MKK-4 (above) and two p38-specific Map3Ks, MKK-3 (encoded by *Map2k-3* and also called Map2K-3, MEK-3, and SKK2) and MKK-6 (encoded by *Map2k-6* and also called Map2K-6, MEK-6, and SKK3, Fig. 1, Table 2) (Cuenda et al. 1996; Dérijard et al. 1995; Raingeaud et al. 1996).

Map3K Groups

Map3Ks are the most proximal components of MAPK core signaling pathways. In contrast to the high degree of structural conservation among MAPKs and among Map2Ks, Map3Ks are highly diverse and are divided among several families.

TAK1 and Tpl-2, Map3Ks Recruited by Proinflammatory Stimuli

TGF-β-activated kinase-1 (TAK1, ~60 kDa, encoded by *Map3k7* Table 3) and tumor progression locus-2 (Tpl-2), also called Cot (encoded by *Map3k8*, Table 3) are likely the predominant Map3Ks recruited by proinflammatory stimuli (Arthur and Ley 2013; Kyriakis and Avruch 2012).

TAK1 is regulated by a family of TAK1-binding proteins (TAB-1–TAB-3) that are essential for coupling TAK1 to upstream signals (Ishitani et al. 2003; Kanayama et al. 2004; Shibuya et al. 1996). Thus, in response to IL-1, TNF, and PAMPs, TAK1 is activated upon the

MAP Kinase Pathways, Table 3 Mammalian Map3Ks

Name	Synonyms	Substrates/effectors	Human genome accession	Mouse genome accession
Raf-1		MEK-1, MEK-2	NP_002871.1	NP_084056.1
A-Raf		MEK-1, MEK-2	NP_001645.1	NP_033833.1
B-Raf		MEK-1, MEK-2	NP_004324.2	NP_647455.2
MEKK1	*Map3k1*	MKK-4, MKK-7	NP_005912.1	NP_036075.2
MEKK2	*Map3k2*	MKK-4, MEK-1	NP_006600.3	NP_036076.2
MEKK3	*Map3k3*	MKK-4, MEK-1, MKK-3, MKK-6	NP_976226.1	NP_036077.1
MEKK4	*Map3k4*, MKT (MAP three kinase 1)	MKK-4, MKK-3, MKK-6	NP_005913.2	NP_036078.2
ASK1	*Map3k5*	MKK-4, MKK-3, MKK-6	NP_005914.1	NP_032606.4
ASK2	*Map3k6*	MKK-4, MKK-3, MKK-6	NP_004663.3	NP_057902.3
TAK1	*Map3k7*	MKK-4, MKK-3, MKK-6	NP_033342.1	NP_663304.1
Tpl-2	*Map3k8*, Cot	MEK-1, SEK1	NP_005195.2	NP_031772.1
MLK1	*Map3k9*	MKK-4, MKK-7	NP_149132.2	NP_001167578.1
MLK2	*Map3k10*, MST	MKK-4, MKK-7	NP_002437.2	NP_001074761.1
MLK3	*Map3k11*, SPRK, PTK1	MKK-4, MKK-7	NP_002410.1	NP_071295.2
DLK	*Map3k12*, MUK, ZPK	MKK-4, MKK-7	NP_001180440.1	NP_033608.3

binding of free or anchored polyubiquitin chains (Lys63-linked, exclusively) which bind to TAB-2 and TAB-3. These polyubiquitin chains are synthesized, in a stimulus-dependent manner, by E3 ubiquitin ligases of the TNF receptor-associated factor (TRAF) family (notably, TRAF-2 and TRAF-6) (Kanayama et al. 2004; Takaesu et al. 2000; Xia et al. 2009).

TAK1 can phosphorylate and activate MKK-3, MKK-4, and MKK-6 as well as IKKs and may do so in lymphocytes (Ishitani et al. 2003; Kanayama et al. 2004; Ninomiya-Tsuji et al. 1999; Wang et al. 2001; Xia et al. 2009; Yamaguchi et al. 1995). Despite these findings, genetic studies indicate that TAK1 is, surprisingly, a negative physiological regulator of myeloid cell p38, JNK, and IKK activation by IL-1, TNF, and PAMPs (Ajibade et al. 2012) (Fig. 1, Table 3).

Tpl-2 is a 58-kDa protein Ser/Thr kinase (Patriotis et al. 1993). Genetic studies indicate that Tpl-2 is recruited by proinflammatory cytokines (TNF) and that, in most instances, MEK-1 and the ERKs are relevant Tpl-2 targets (Dumitru et al. 2000; Kyriakis and Avruch 2012) (Fig. 1, Table 3).

In resting cells, the Tpl-2 exists as a heteromer in complex with the p105 NF-κB1 subunit and a second protein, A20-binding inhibitor of NF-κB2 (ABIN2). In response to extracellular stimuli (TNF, LPS), p105 is phosphorylated by the IκB-kinase-β (IKKβ). p105 phosphorylation triggers the ubiquitination and proteasomal degradation of p105 and ABIN2 dissociation, which is followed by the consequent autophosphorylation-dependent activation of Tpl-2 (Beinke et al. 2004; Papoustopoulou et al. 2006; Stafford et al. 2006).

MAPK-ERK Kinase Kinases (MEKKs)

MEKKs (MEKK1–MEKK4) share a structurally similar catalytic domain strikingly homologous to that of the *Saccharomyces cerevisiae* Map3K Ste11p. Outside of the catalytic domains, however, the structures are completely different (Kyriakis and Avruch 2012).

MEKK1 (encoded by *Map3k1*), in addition to a carboxyl terminal kinase domain, contains a plant homeodomain (PHD)/really interesting new gene (RING) domain with intrinsic E3 ubiquitin (Ub) ligase activity (AAs 438–486). This activity assembles Lys48-linked polyubiquitin chains exclusively and can promote autoubiquitination as well as the ubiquitination of ERK2, a reaction which may attenuate Ras-dependent ERK activation (Kyriakis and Avruch 2012; Lu et al. 2002). MEKK1, possibly through its PHD domain, can also directly ubiquitinate and degrade the transcription factor c-Jun (Xia et al. 2007). It is important to note, however, that the MEKK1 PHD/RING domain also binds and can activate the E3 Ub ligase Itch (Venuprasad et al. 2006). Accordingly, it is somewhat unclear if MEKK1 directly or indirectly ubiquitinates c-Jun.

MEKK1 can activate MKK-4 and MKK-7 and, through these, JNKs. Accordingly, MEKK1 is a relatively selective activator of JNKs. However, in some circumstances, MEKK1 can also activate the ERKs and p38 (Kyriakis and Avruch 2012). MEKK1 can be activated by TNF (Baud et al. 1999) and the B lymphocyte receptor protein CD40 (Matsuzawa et al. 2008) (Fig. 1, Table 3).

MEKK2 and MEKK3 (encoded by *Map3k2* and *Map3k3*, respectively) are closely related ~70 and 71-kDa polypeptides (Blank et al. 1996). Genetic studies suggest that MEKK3 is a physiological MKK-6 kinase (Huang et al. 2004). Biochemical and some genetic studies have also shown that MEKK3 can, in response to TNF, directly activate IκB kinases (IKKs) and, through these, NF-κB pathway (Huang et al. 2004) (Fig. 1, Table 3).

The ~150-kDa MEKK4 (encoded by *Map3k4*, Table 3) binds proteins of the growth arrest and DNA damage (GADD)45 family; and the GADD45s (notably GADD45γ) may function to activate MEKK4. There is some uncertainty as to the precise substrate spectrum for MEKK4 with results suggesting either that MEKK4 is selective for MKK-4 and JNKs, MKK-3, MKK-4, and MKK-6 and, through them, JNKs and p38s or p38s exclusively (Chi et al. 2004; Gerwins et al. 1997; Takekawa and Saito 1998).

Mixed Lineage Kinases (MLKs)

Mixed lineage kinases (MLKs) consist of an amino terminal kinase domain, one to two leucine zippers, a Cdc42/Rac interaction and binding (CRIB) domain, and a carboxyl terminal domain with several consensus SH3 binding motifs. Although entirely Ser-/Thr-specific, the kinase domains of MLKs contain structural features found in both Ser/Thr and Tyr kinases. Four MLKs, MLK-1–MLK-3 and dual leucine zipper kinase (DLK), are encoded by *Map3k9*, *Map3k10*, *Map3k11*, and *Map3k12*, respectively (Fig. 1 and Table 3) (Gallo and Johnson 2002). MLKs may function in innate immunity and metabolic regulation (Gallo and Johnson 2002; Kyriakis and Avruch 2012).

ASKs, Map3Ks Recruited by Oxidant Stress

Apoptosis signal-regulating kinase 1 and apoptosis signal-regulating kinase 2 (ASK1, encoded by *Map3k5*, and ASK2, encoded by *Map3k6*) are a pair of very closely related ~150-kDa polypeptides. ASK1 competitively binds polypeptides of the TNF receptor-associated factor (TRAF) family and redox sensing enzyme thioredoxin such that reactive oxygen species (exogenous or generated through cytokine actions) trigger thioredoxin dissociation, TRAF binding, and consequent ASK1 activation. ASKs can activate MKK-4, MKK-3, and MKK-6 and, through these, JNKs and p38s (Ichijo et al. 1997; Liu et al. 2000; Nishitoh et al. 1998; Saitoh et al. 1998).

ASK2 is expressed primarily in the skin and gut epithelium, whereas ASK1 is ubiquitously expressed. ASK1 and ASK2 function as heteromers, and genetic disruption of *ask1* leads to the constitutive degradation of ASK2; thus, ASK2 cannot function biochemically in the absence of ASK1, whereas, biochemically, ASK1 can function alone (Iriyama et al. 2009; Takeda et al. 2007). Despite the dependence of ASK2 on ASK1, ASK1 and ASK2 perform specific, nonredundant functions in vivo where they regulate cell survival and inflammation in tumorigenesis (discussed below) (Iriyama et al. 2009).

Key MAPK Effectors

Protein Kinases

Rsks The ribosomal S6 kinases (Rsks, not the dominant somatic cell S6 kinases, despite the name) are a family of Ser/Thr kinases that contain two tandem protein kinase domains that fall broadly into the AGC subfamily. Rsks are activated in a complex process either through the combined activity of ERKs and 3-phosphoinositide-dependent protein kinase 1 (PDK1) or selectively in dendritic cells and, possibly, macrophages treated with agents that recruit Toll-like receptors, via p38 which, in turn, stimulates the activities of MK-2 and MK-3 (see section immediately following). In this instance, MK-2/MK-3 phosphorylate and activate Rsks independently of ERK and PDK1 (Dalby et al. 1998; Richards et al. 1999; Zaru et al. 2007). Rsks, in turn, phosphorylate a number of key substrates including c-Fos, a component of the activator protein-1 (AP-1) transcription factor (see below) (Kyriakis and Avruch 2012). The p38-dependent mechanism of Rsk activation apparently functions selectively in TLR-stimulated proinflammatory dendritic cell phagocytosis and pinocytosis of antigens, as well as for the subsequent induction of IL-10 during the resolution of inflammation (Fig. 1) (Zaru et al. 2007).

Mitogen-Activated Protein Kinase-Activated Protein Kinase (MK)-2, MK-3 and MK-5 MK-2, MK-3, and MK-5 are a small family of Ser/Thr kinases strongly activated by stresses, proinflammatory cytokines, and PAMPs/DAMPs. MK-2, MK-3, and MK-5 are phosphorylated and activated by p38α and p38β (but not by p38γ or p38δ). In addition to the p38-dependent regulation of Rsks mentioned above, MK-2 can also bind and phosphorylate polypeptides that bind to mRNAs and modulate mRNA stability. These include tristetraprolin (TTP), which destabilizes the mRNA for TNF. MK-2 phosphorylation of TTP suppresses this destabilization in response to TLR engagement, thus enhancing TNF production (Dean et al. 2001; Gaestel 2006; New et al. 1998; Ronkina et al. 2007; Rousseau et al. 2002).

MAPK Interacting Kinases (MNKs) MNKs are Ser/Thr kinases that are phosphorylated and activated both by ERK1 and ERK2 (in response to insulin and mitogens) and by p38s (in response to cytokines and stress) (Fukunaga and Hunter 1997; Waskiewicz et al. 1997, 1999). Insulin, mitogens, and environmental stresses also stimulate the MNK-catalyzed, regulatory phosphorylation of the mammalian elongation factor eIF-4E at Ser209. The role of this phosphorylation is somewhat unclear but may contribute to eIF-4E's oncogenic potential (Buxade et al. 2008; Furic et al. 2010; Ueda et al. 2010).

Mitogen- and Stress-Activated Protein Kinase 1 and MSK2 (MSK1/MSK2) MSK1/MSK2, like Rsks, contain two tandem protein kinase domains. MSK1 is activated by both the ERKs and p38s (α and β). MSKs can directly phosphorylate the bZIP transcription factor cAMP response element binding protein (CREB, also a substrate of Rsks and MK-2). During inflammation, MSKs, through CREB, may act redundantly to suppress innate immunity (Anaieva et al. 2008; Deak et al. 1998).

Transcription Factors

Activator Protein-1 (AP-1) AP-1 is a catchall term that describes a group of heterodimeric transcription factors that consist of bZIP transcription factors of the Jun group paired either with members of the *fos* (usually c-Fos) or activating transcription factor (ATF, usually ATF2) families. AP-1 can bind, via Jun family components, to the tetradecanoyl phorbol acetate (TPA) response element (TRE) or, via the ATFs, to the cAMP response element (the CRE, also targeted by CREB). AP-1 is critical to the expression of numerous genes recruited during inflammation, including those encoding interleukin-1 and interleukin-2, CD40, CD30, TNF, and c-Jun itself. AP-1 also mediates the transcriptional induction of proteases and cell adhesion proteins, such as E-selectin, which are important to inflammation (Karin and Gallagher 2005; Shaulian and Karin 2002).

MAPKs can regulate AP-1 at multiple levels. JNKs phosphorylate the c-Jun and JunD *trans* activating domains in reactions that correlate with increased transcriptional activation. Similarly, JNKs and p38s phosphorylate the ATF2 *trans* activation domain coincident with the activation of *trans* activating activity (Gupta et al 1995; Kallunki et al. 1996; Kyriakis and Avruch 2012; Ventura et al. 2003).

JNKs also regulate c-Jun stability. JNK binding and phosphorylation of c-Jun target c-Jun for

Lys48-linked polyubiquitination and degradation. JNKs also can phosphorylate and activate the HECT domain E3 Ub ligase Itch, which then promotes c-Jun ubiquitination and degradation. MEKK1, independent of JNK, may also activate Itch-dependent c-Jun degradation (Musti et al. 1997; Venuprasad et al. 2006).

Lastly, MAPKs can activate transcription factors that participate in the induction of AP-1 components. The serum response factor (SRF), as part of a heterodimer with members of the ternary complex family (Elk-1 and Sap-1a), is critical to the induction of c-*fos*. JNKs and ERKs can phosphorylate Elk-1, while p38s phosphorylate Sap-1a. In both cases, phosphorylation enhances heterodimerization with SRF, enhanced *trans* acting activity, and c-*fos* induction (Gille et al. 1992; Janknecht and Hunter 1997; Marais et al. 1993; Whitmarsh et al. 1995). Likewise, p38α can phosphorylate transcription factors of the myocyte enhancer factor 2 (MEF2) group (MEF2-A and MEF2-C). This correlates with enhanced MEF2A/C *trans* activating activity. Inasmuch as a MEF2C *cis* element resides in the promoter for c-*jun*, this permits p38-dependent induction of c-*jun* (Han et al. 1997).

Nuclear Factor of Activated T Cells (NFATs)
NFATs are activated, in part, when the calcium/calmodulin-dependent phosphatase calcineurin dephosphorylates phosphoacceptor sites phosphorylated in resting cells by glycogen synthase kinase-3 and casein kinase-2. Often acting in concert with AP-1, NFATs then promote the induction of numerous proinflammatory and immunogenic genes (IL-2, IL-4, IL-5, and CD40L are examples).

This calcium-dependent activation of NFATs is suppressed by serum factors that impair nuclear translocation of NFATs. In the immune system, this suppression is seen most predominantly in T cells and is mediated by JNK phosphorylation (of NFAT-2 and NFAT-4), a process that blunts the binding of calcineurin. NFAT2 function is essential for the polarization of T_H cells to the T_H2 phenotype. Accordingly, JNK-dependent NFAT2 suppression is consistent with the idea that JNK activation favors the T_H1 trajectory at least in part through NFAT2 inhibition. Consistent with this, knockout of either *jnk1* or both *jnk1* and *jnk2* favors T_H2 polarization (Chow et al. 2000; Dong et al. 1998, 2000).

Biological Activities

Innate Immunity and Pattern Recognition Receptor (PRR) Responses

Tpl-2 in Innate Immune Signaling
Activation of the ERK pathway by PAMPs and TNF requires the Tpl-2 Map3K. Treatment of *tpl-2−/−* mice with LPS results in considerably reduced TNF production and, consequently, significantly reduced LPS-induced systemic toxicity. This has been attributed to Tpl-2-mediated ERK-dependent nucleocytoplasmic transport of TNF mRNA (Dumitru et al. 2000).

Consistent with this, Tpl-2 is critical to host defense. Thus, *tpl-2−/−* mice show impaired interferon-gamma (IFN-γ) production and consequent host defense against *Toxoplasma gondii*. This is a T cell-mediated process inasmuch as reconstitution of *rag2−/−* mice with *tpl-2−/−* T cells recapitulates this defect in host defense (Watford et al. 2008).

Tpl-2 also plays a role in T cell polarization – enhancing the T_H1 trajectory. T_H1 cells naturally express more Tpl-2 polypeptide. Accordingly, *tpl-2−/−* T cells, in a manner arising from weakened ERK activation, are biased toward an exaggerated T_H2 response (Watford et al. 2010).

TAK1 as a Negative Regulator of Innate Immune Signaling
In T cells, TAK1 is required for T cell receptor, IL-2, IL-7, and IL-15 activation of JNK and in fibroblasts and B-lymphocytes for JNK activation by PAMPs and cytokines (Sato et al. 2005; Schuman et al. 2009).

Although biochemical studies have dissected a mechanism by which PRRs can recruit TAK1, and knockout studies indicate a role for TAK1 in lymphocyte recruitment of JNK and p38, genetic knockout studies have revealed an unexpected negative signaling role for myeloid cell TAK1. Thus, disruption of neutrophil *map3k7* enhances JNK and p38 activation by LPS and renders mice more susceptible to LPS-induced cytokine (TNF, IL-1) production and systemic inflammation. The basis for this is unclear (Ajibade et al. 2012).

MSK1/MSK2 as Negative Regulators of Inflammation

The ERK and p38 target kinases MSK1 and MSK2 also negatively regulate inflammation by promoting the induction of IL-10. The combined disruption of *msk1/msk2* renders mice hypersensitive to LPS-induced systemic inflammation and makes mice more prone to contact eczema (Anaieva et al. 2008).

JNK Pathways and T Cell Maturation and Apoptosis

JNK1/JNK2 are largely dispensable for the activation of naïve T cells to the T_H state. However, JNKs function to promote differentiation to the T_H1 phenotype. Thus, the disruption of either *jnk1* or *jnk2* biases T cells toward the T_H2 trajectory. The mechanisms by which different JNK isoforms do this differ. *jnk2−/−* T cells fail to differentiate into T_H1. JNK1, by contrast, actively suppresses T_H2 differentiation by inhibiting the production of IL-4, IL-5, IL-10, and IL-13 – cytokines required for the T_H2 trajectory. This inhibition of T_H2 differentiation by JNK1 may be due to JNK1's negative regulation of NFAT recruitment of the E3 ligase Itch. Itch can destabilize both c-Jun and JunB, which are critical to T_H2 polarization (Chow et al. 2000; Dong et al. 1998, 2000; Sabapathy et al. 1999; Venuprasad et al. 2006).

JNKs and p38s, through MKK-4, may protect cells from Fas-induced apoptosis. Thus, when *rag2*-deficient blastocysts were microinjected with *mkk4−/−* ES cells to produce chimeric mice with a wild-type background and an MKK-4-deficient immune system, the chimeric mouse T cells were remarkably hypersensitive to FasL-induced apoptosis (Nishina et al. 1997).

p38 Pathways and T Cell Differentiation

GADD45 – especially GADD45β and γ – may function to interact with and activate MEKK4, (Takekawa and Saito 1998). Disruption of *mekk4/Map3k4* indicates that this Map3K is required for efficient T cell receptor activation of CD4$^+$ T cell p38. In vivo MEKK4 appears to be required for T_H1 differentiation inasmuch as disruption of *mekk4* impairs T cell IFN-γ production. Ectopic expression of GADD45β or γ promotes T_H1 function (IFN-γ production) – but only in *mekk4+/+* T cells – implicating the MEKK4-GADD45 axis in T_H1 polarization (Chi et al. 2004).

T_H17 and Treg Differentiation and MEKK2/MEKK3

T cell-specific deletion of both *Map3k2* and *Map3k3* produces mice with excess levels of both proinflammatory T_H17 and anti-inflammatory Treg cells. The double knockout cells are hyperresponsive to TGF-β, which is required for the differentiation of both T_H17 and Treg cells. Disruption of *Map3k2* and *Map3k3* in T cells suppresses the inhibitory ERK-catalyzed phosphorylation of the TGF-β targets similar to mothers against decapentaplegic (SMAD)-2 and SMAD-3 (Massagué et al. 2005), thus enhancing the intensity of TGF-β signaling. Although the disruption of *Map3k2* and *Map3k3* enhances the levels of both T_H17 and Treg cells, it is likely that the T_H17 effect predominates inasmuch as the double knockout mice experience a more severe inflammatory autoimmune response when subjected to treatments that elicit the multiple sclerosis-like condition experimental autoimmune encephalitis (Chang et al. 2011).

p38s and the Promotion of Cytokine Release via Posttranscriptional Regulation of mRNAs Encoding Proinflammatory Proteins

PAMP/TLR-dependent stabilization of mRNAs for cytokines, including those encoding TNF and IL-1, is a well-established mechanism for promoting inflammation. This posttranscriptional mRNA stabilization is mediated largely by p38α and p38β and their target kinase MK-2. MK-2, in turn, phosphorylates and inhibits the mRNA destabilizing protein tristetraprolin (TTP).

Indeed, genetic disruption of *mk2* severely impairs LPS induction of TNF, and mice are resistant to LPS-induced systemic inflammation and show a reduced ability to raise a strong host defense response against invading pathogens (Gaestel 2006; Ronkina et al. 2007).

Pathophysiological Relevance

JNKs and p38 Pathways and Inflammatory Oncogenesis

JNK1 from the nonhepatocyte cell population provides to hepatocytes signals from the pro-inflammatory/pro-tumorigenic cytokines IL-6 and TNF in response to the tumorigenic compound diethyl nitrosamine (DEN) – thus inducing hepatocellular carcinoma (HCC). DEN-induced HCC is strikingly reduced (but not eliminated) upon whole body disruption of *jnk1*, but not *jnk2*. Neither JNK1 nor JNK2 has any intrinsic role in hepatocyte proliferation.

Thus, a global knockout of *jnk1* and a hepatocyte-specific knockout of *jnk2*, despite modestly attenuating hepatocyte proliferation, did not affect DEN-induced HCC. By contrast, a global knockout of *jnk2* accompanied by a disruption of *jnk1* in *all* liver cells (hepatocytes and nonparenchymal cells including stellate and Kupffer cells) did reduce DEN-induced HCC (Das et al. 2011; Sakurai et al. 2006, 2008).

In contrast to JNKs, p38α appears to antagonize DEN-induced HCC. Thus, hepatocyte disruption of *p38α/mapk14* increases reactive oxygen species (ROS) production which, in turn, leads to the increased production of IL-1α. The increased production of IL-1α promotes hepatocarcinogenesis (Hui et al. 2007; Sakurai et al. 2008).

The Map3Ks ASK1 and ASK2 play differential roles in inflammatory oncogenesis. ASK2 is necessary – along with ASK1 – for suppressing gut epithelial cell tumorigenesis. By contrast, ASK1 itself is required for the expression of pro-tumorigenic cytokines needed for inflammatory oncogenesis in the gut (Iriyama et al. 2009; Takeda et al. 2007). Similarly, ASK2, functioning with ASK1, is necessary for suppressing 7,12-dimethylbenz[a]anthracene (DMBA)/phorbol ester (PMA) skin tumorigenesis by promoting DMBA-induced keratinocyte apoptosis. Disruption of *ask1* destabilizes ASK2. Consistent with this, keratinocyte p38 activation by DMBA is also reduced. ASK1 is required for the induction, in the skin, in response to DMBA/TPA, of proinflammatory/pro-oncogenic cytokines – notably TNF and IL-6 (Iriyama et al. 2009).

MAPKs and Inflammatory Signaling in Rheumatoid Arthritis (RA)

While the role of MAPKs in cytokine signaling suggests a function in rheumatic diseases such as RA, and while (below) MAPK inhibitors have been analyzed for efficacy in RA treatment, the functions of MAPKs in RA pathogenesis are still nebulous. IL-1 induction of synovial matrix metalloprotease (MMP)-1 is a well-established feature of RA. This process is reduced upon pharmacological inhibition of ERK. Synovial fibroblasts from JNK1- or JNK2-knockout mice express reduced levels of MMP-3 and MMP-13. However, disruption of either *mapk8* or *mapk9* has only a small effect on joint inflammation (but does reduce joint destruction), suggesting that both isoforms may have to be targeted for effective, comprehensive therapy (Thalhamer et al. 2008).

p38 inhibition is more promising. Pharmacological inhibition of p38α/β reduces synovial TNF and IL-1 production, along with consequent inflammation, as does inhibition of the p38 effector MK-2 (consistent with MK-2's role in TNF mRNA stabilization). Inhibition of p38 also blunts bone resorption, suggesting that joint destruction might also be targeted by inhibiting p38 (Thalhamer et al. 2008).

MAPKs and Inflammatory Signaling in Type 2 Diabetes and Metabolic Disease

Obesity and insulin resistance progressing to type 2 diabetes mellitus are associated with a chronic inflammation consequent to elevated levels of proinflammatory cytokines (notably, TNF, IL-1, and IL-6). This inflammation arises in adipose tissue which, when stressed by excess lipids, produces abundant TNF. A common result of this is the recruitment by adipose cells of circulating myeloid cells. The myeloid cells surround stressed or dying adipocytes forming so-called crown-like structures that, in turn, produce cytokines and other inflammatory mediators which can influence metabolic control in ways that are still incompletely understood. It has been proposed that the suppression of this inflammation could be salutary in the treatment of metabolic disease. In this context, signaling by JNKs, notably JNK1, contributes substantially to insulin resistance.

Feeding mice a high-fat diet (HFD) results in JNK activation in the muscle, fat, and liver. Global knockout of *jnk1* protects mice against HFD-induced obesity and strongly attenuates the consequent insulin resistance through improved insulin signaling (Hirosumi et al. 2002). Disruption of *jnk1* reduces HFD-induced adipocyte hypertrophy, hyperglycemia, hyperinsulinemia, glucose tolerance, and overall adiposity (Hirosumi et al. 2002; Özcan et al. 2004).

Several tissue-specific JNK1 knockouts have provided some insight into how JNK1 contributes to whole body insulin resistance. Disruption of adipose cell *jnk1* protects HFD-fed mice from the development of whole body insulin resistance. This improvement is entirely attributable to improved hepatic insulin action. Muscle glucose uptake showed little change. The *jnk1*-deficient adipose tissue exhibited a marked reduction in HFD-induced IL-6 mRNA and IL-6 blood levels. This, in turn, may reduce adipose inflammation and prevent consequent fat accumulation and insulin resistance in the liver. Thus, the beneficial effect of *jnk1* disruption in adipose tissue is essentially a hormonal process (Sabio et al. 2008).

Not surprisingly, disruption of murine skeletal muscle *jnk1* results in improved insulin sensitivity upon HFD. Somewhat unexpectedly, this improvement is not accompanied by an amelioration of glucose intolerance. This is likely because skeletal muscle *jnk1* disruption does not reduce overall adiposity, adipose tissue inflammation, or hepatic steatosis (Sabio et al. 2010a).

Despite the relevance of hepatic insulin resistance to obesity and metabolic disease, hepatocyte-specific ablation of *jnk1* results in liver steatosis, insulin resistance, and glucose intolerance (Sabio et al. 2009). Whether this suggests a cryptic role for JNK2 in hepatic metabolic control or a compensatory effect manifested by nonparenchymal cells remains to be determined.

The fact that crown-like structures consist of myeloid cell infiltration into the adipose tissues suggested that myeloid cell JNK1 was important to the insulin resistance of obesity. However, both myeloid-specific *jnk1* disruption and adoptive transfer experiments indicate that, on balance and in contrast to the effect of global *jnk1* disruption, it is unlikely that myeloid (as opposed to parenchymal) JNK1 contributes significantly and directly to HFD-induced adipogenesis and obesity; however, this conclusion is somewhat controversial (Sabio et al. 2008; Solinas et al. 2007; Vallerie et al. 2008).

By contrast, myeloid JNK1 appears to play an important role in inflammation-induced, systemic insulin resistance. Again, however, some studies have not completely ruled out a role for parenchymal JNK1 in this process (Sabio et al. 2008; Solinas et al. 2007; Vallerie et al. 2008).

As noted above, while global *jnk1* disruption leads to both an improvement in insulin sensitivity and resistance to weight gain upon HFD challenge, adipose-specific *jnk1* deletion improves insulin sensitivity without an accompanying resistance to weight gain. This difference may be due to the impact of CNS JNK1 in the global knockout setting. In support of this, *jnk1* disruption in the CNS produces mice which, when fed an HFD, exhibit considerably less weight gain, improved insulin sensitivity in most organs, a somewhat lower fasting glucose, but a similar glucose tolerance when compared to HFD-fed *jnk1+/+* controls (Sabio et al. 2010b). The protection from HFD-induced adiposity is largely attributable to a markedly increased level of energy expenditure, arising from the activation of the hypothalamic-pituitary-thyroid axis.

Based on these data, JNK1 is seen as a potential target for antidiabetic treatments.

Upstream of JNKs, Tpl-2 has also been shown to play a key role in metabolic disease. Thus, disruption of *tpl-2*, likely through the regulation of both JNK and ERK, reduces HFD-induced adipose crown-like structures and hepatic steatosis and improves insulin sensitivity (Perfield et al. 2011).

Modulation by Drugs

Small molecule inhibitors, most of them ATP competitors, of several inflammation-induced MAPK pathway protein kinases have been developed, and a number of them are in clinical trial as potential treatments (Arthur and Ley 2013). The primary indication is rheumatoid arthritis (RA), a major unmet medical need in which blunting cytokine-induced proinflammatory signaling will likely prove beneficial. These compounds are described in Table 4.

What is notable is that assuming reasonable bioavailability in humans, essentially none of these compounds have had a profound, sustained

MAP Kinase Pathways, Table 4 Small molecule drugs targeting proinflammatory MAPK components

Company	Inhibitor	Main target	Disease indication	Clinical trial phase	Comments	References
Pfizer	PH797804	p38α	Rheumatoid arthritis (RA), COPD, pain	II	No information available	Goldstein et al. 2010
Bristol-Myers Squibb	BMS-582949	p38α	RA, psoriasis, atherosclerosis	II	No information available	Goldstein et al. 2010
Array BioPharma	Arry-797	p38α	RA, pain	Ib/II	Unsuccessful for RA, effective for analgesia	Goldstein et al. 2010
GlaxoSmithKline	SB681323 (dilmapimod)	p38α	RA, COPD, pain	II	Reduced cytokine (TNF) is pain and COPD	Anand et al. 2011
GlaxoSmithKline	GW856553 (iosmapimod)	p38α	RA, COPD, cardiovascular disease, pain	II	Reduction in plasma fibrinogen (COPD)	Lomas et al. 2012
Roche	Pamapimod	p38α	RA	II	Less effective than methotrexate, modest drop in CRP	Cohen et al. 2009
Vertex	VX-702	p38α	RA	II	Very limited improvement, modest drop in CRP	Damjanov et al. 2009
Boehringer Ingelheim	BIRB 796 (doramapimod)	All p38s	Crohn's disease	II	No sustained improvement, modest drop in CRP	Goldstein et al. 2010
Pfizer	PF-364022	MK-2, MK-3	Not determined	Preclinical	Reduced clinical scores in a rat model of RA	Mourey et al. 2010
Merck Serono	MSC2032964A	ASK1	Not determined	Preclinical	Reduced clinical scores in a mouse model of multiple sclerosis	Guo et al. 2010

effect on RA (or other designated indications), and none are past the stage of phase II trials (Arthur and Ley 2013). One take-home message from these findings is that the MAPK inhibitors alone may be insufficient for efficacy against chronic inflammatory diseases. Such compounds may have to be used in conjunction with existing or other novel therapies.

In addition, it is noteworthy that all of the compounds target the p38s, their effectors (MK-2/MK-3), or key activators (ASK1 – which can also activate JNK). No direct JNK inhibitors are currently in clinical trial – despite the potential attractiveness, noted above, of targeting JNK (notably JNK1) to treat metabolic diseases and, potentially, hepatocellular carcinoma.

References

Ajibade, A. A., Wang, Q., Cui, J., Zou, J., Xia, X., Wang, M., et al. (2012). TAK1 negatively regulates NF-κB and p38 MAP kinase activation in Gr-1$^+$CD11b$^+$ neutrophils. *Immunity, 36*, 43–54.

Anaieva, O., Darragh, J., Johansen, C., Carr, J. M., McIlrath, J., Park, J. M., et al. (2008). The kinases MSK1 and MSK2 act as negative regulators of Toll-like receptor signaling. *Nature Immunology, 9*, 1028–1036.

Anand, P., Shenoy, R., Palmer, J. E., Baines, A. J., Lai, R. Y., Robertson, J., et al. (2011). Clinical trial of the p38 MAP kinase inhibitor dilmapimod in neuropathic pain following nerve injury. *European Journal of Pain, 15*, 1040–1048.

Arthur, J. S. C., & Ley, S. C. (2013). Mitogen-activated protein kinases in innate immunity. *Nature Reviews Immunology, 13*, 679–692.

Baud, V., Liu, Z.-g., Bennett, B., Suzuki, N., Xia, Y., & Karin, M. (1999). Signaling by proinflammatory cytokines: Oligomerization of TRAF2 and TRAF6 is sufficient for JNK and IKK activation and target gene induction via an amino terminal effector domain. *Genes and Development, 13*, 1297–1308.

Beinke, S., Robinson, M. J., Hugunin, M., & Ley, S. C. (2004). Lipopolysaccharide activation of the TPL-2/MAP2K/extracellular signal-regulated kinase mitogen-activated protein kinase cascade is regulated by IκB kinase-induced proteolysis of NF-κB1 p105. *Molecular and Cellular Biology, 24*, 9658–9667.

Blank, J. L., Gerwins, P., Elliot, E. M., Sather, S., & Johnson, G. L. (1996). Molecular cloning of mitogen activated protein/ERK kinase kinases (MEKK) 2 and 3. *Journal of Biological Chemistry, 271*, 5361–5368.

Buxade, M., Parra-Palau, J. L., & Proud, C. G. (2008). The Mnks: MAP kinase-interacting kinases (MAP kinase signal-integrating kinases). *Frontiers in Bioscience, 13*, 5359–5373.

Chang, X., Liu, F., Wang, X., Lin, A., Zhao, H., & Su, B. (2011). The kinases MEKK2 and MEKK3 regulate transforming growth factor- β-mediated helper T cell differentiation. *Immunity, 34*, 201–212.

Chi, H., Lu, B., Takekawa, M., Davis, R. J., & Flavell, R. A. (2004). GADD45β/GADD45γ and MEKK4 comprise a genetic pathway mediating STAT4-independent IFN-γ production in T cells. *EMBO Journal, 23*, 1576–1586.

Chow, C.-W., Dong, C., Flavell, R. A., & Davis, R. J. (2000). c-Jun NH2-terminal kinase inhibits targeting of the protein phosphatase calcineurin to NFATc1. *Molecular and Cellular Biology, 20*, 5227–5234.

Cohen, S. B., Chen, T. T., Chindalore, V., Damjanov, N., Burgos-Vargas, R., Delora, P., et al. (2009). Evaluation of the efficacy and safety of pamapimod, a p38 MAP kinase inhibitor, in a double-blind, methotrexate-controlled study of patients with active rheumatoid arthritis. *Arthritis and Rheumatology, 60*, 335–344.

Cuenda, A., Alonso, G., Morrice, N., Jones, M., Meier, R., Cohen, P., et al. (1996). Purification and cDNA cloning of SAPKK3, the major activator of RK/p38 in stress- and cytokine-stimulated monocytes and epithelial cells. *EMBO Journal, 15*, 4156–4164.

Dalby, K. N., Morrice, N., Caudwell, F. B., Avruch, J., & Cohen, P. (1998). Identification of regulatory phosphorylation sites in mitogen-activated protein kinase (MAPK)-activated protein kinase-1a/p90rsk that are inducible by MAPK. *Journal of Biological Chemistry, 273*, 1496–1505.

Damjanov, N., Kauffmann, R. S., & Spencer-Green, G. T. (2009). Efficacy, pharmacodynamics, and safety of VX-702, a novel p38 MAPK inhibitor, in rheumatoid arthritis: Results of two randomized, double-blind, placebo-controlled clinical studies. *Arthritis and Rheumatology, 60*, 1232–1241.

Das, M., Garlick, D. S., Greiner, D. L., & Davis, R. J. (2011). The role of JNK in the development of hepatocellular carcinoma. *Genes and Development, 25*, 634–645.

Deak, M., Clifton, A. D., Lucocq, J., & Alessi, D. R. (1998). Mitogen- and stress-activated protein kinase-1 (MSK1) is directly activated by MAPK and SAPK2/p38, and may mediate activation of CREB. *EMBO Journal, 17*, 4426–4441.

Dean, J. L., Wait, R., Mahtani, K. R., Sully, G., Clark, A. R., & Saklatvala, J. (2001). The 3′ untranslated region of tumor necrosis factor alpha mRNA is a target of the mRNA-stabilizing factor HuR. *Molecular and Cellular Biology, 21*, 721–730.

Dérijard, B., Hibi, M., Wu, L.-H., Barrett, T., Su, B., Deng, T., et al. (1994). JNK1: A protein kinase stimulated by UV light and Ha-Ras that binds and phosphorylates the c-Jun transactivation domain. *Cell, 76*, 1025–1037.

Dérijard, B., Raingeaud, J., Barrett, T., Wu, L.-H., Han, J., Ulevitch, R. J., et al. (1995). Independent human MAP kinase signal transduction pathways defined by MEK and MKK isoforms. *Science, 267*, 682–685.

Dong, C., Yang, D. D., Wysk, M., Whitmarsh, A. J., Davis, R. J., & Flavell, R. A. (1998). Defective T cell differentiation in the absence of JNK1. *Science, 282*, 2092–2095.

Dong, C., Yang, D. D., Tournier, C., Whitmarsh, A. J., Xu, J., Davis, R. J., et al. (2000). JNK is required for effector T-cell function but not for T cell activation. *Nature, 405*, 91–94.

Dumitru, C. D., Ceci, J. D., Tsatsanis, C., Kontoyiannis, D., Stamatakis, K., Lin, J. H., et al. (2000). TNF-α induction by LPS is regulated posttranscriptionally via a Tpl2/ERK-dependent pathway. *Cell, 103*, 1071–1083.

Freshney, N. W., Rawlinson, L., Guesdon, F., Jones, E., Cowley, S., Hsuan, J., et al. (1994). Interleukin-1 activates a novel protein kinase cascade that results in the phosphorylation of Hsp27. *Cell, 78*, 1039–1049.

Fukunaga, R., & Hunter, T. (1997). MNK1, a new MAP kinase-activated protein kinase, isolated by a novel expression screening method for identifying protein kinase substrates. *EMBO Journal, 16*, 1921–1933.

Furic, L., Rong, L., Larsson, O., Koumakpayi, I. H., Yoshida, K., Brueschke, A., et al. (2010). eIF4E phosphorylation promotes tumorigenesis and is associated with prostate cancer progression. *Proceedings of the National Academy of Sciences of the United States of America, 107*, 14134–14139.

Gaestel, M. (2006). MAPKAP kinases—MKs—two's company, three's a crowd. *Nature Reviews Molecular Cell Biology, 7*, 120–130.

Gallo, K. A., & Johnson, G. L. (2002). Mixed-lineage kinase control of JNK and p38 pathways. *Nature Reviews Molecular Cell Biology, 3*, 663–672.

Gerwins, P., Blank, J. L., & Johnson, G. L. (1997). Cloning of a novel mitogen-activated protein kinase-kinase-kinase, MEKK4, that selectively regulates the c-Jun amino terminal kinase pathway. *Journal of Biological Chemistry, 272*, 8288–8295.

Gille, H., Sharrocks, A. D., & Shaw, P. E. (1992). Phosphorylation of transcription factor p62TCF by MAP kinase stimulates ternary complex formation at c-fos promoter. *Nature, 358*, 414–417.

Goedert, M., Cuenda, A., Craxton, M., Jakes, R., & Cohen, P. (1997). Activation of the novel stress-activated protein kinase SAPK4 by cytokines and cellular stresses is mediated by SKK3 (MKK6); comparison of its substrate specificity with that of other SAP kinases. *EMBO Journal, 16*, 3563–3571.

Goldstein, D. M., Kuglstatter, A., Lou, Y., & Soth, M. J. (2010). Selective p38α inhibitors clinically evaluated for the treatment of chronic inflammatory disorders. *Journal of Medicinal Chemistry, 53*, 2345–2353.

Guo, X., Harada, C., Namekata, K., Matsuzawa, A., Camps, M., Ji, H., et al. (2010). Regulation of the severity of neuroinflammation and demyelination by TLR-ASK1-p38 pathway. *EMBO Molecular Medicine, 2*, 504–515.

Gupta, S., Campbell, D., Dérijard, B., & Davis, R. J. (1995). Transcription factor ATF2 regulation by the JNK signal transduction pathway. *Science, 267*, 389–393.

Gupta, S., Barrett, T., Whitmarsh, A. J., Cavanagh, J., Sluss, H. A., Dérijard, B., et al. (1996). Selective interaction of JNK protein kinase isoforms with transcription factors. *EMBO Journal, 15*, 2760–2770.

Han, J., Lee, J.-D., Bibbs, L., & Ulevitch, R. J. (1994). A MAP kinase targeted by endotoxin and hyperosmolarity in mammalian cells. *Science, 265*, 808–811.

Han, J., Jiang, Y., Li, Z., Kravchenko, V. V., & Ulevitch, R. J. (1997). MEF2C participates in inflammatory responses via p38-mediated activation. *Nature, 386*, 563–566.

Hirosumi, J., Tuncman, G., Chang, L., Görgün, C. Z., Uysal, K. T., Maeda, K., et al. (2002). A central role for JNK in obesity and insulin resistance. *Nature, 420*, 334–337.

Huang, Q., Yang, J., Lin, Y., Walker, C., Cheng, J., Liu, Z.-g., et al. (2004). Differential regulation of interleukin 1 receptor and Toll-like receptor signaling by MEKK3. *Nature Immunology, 5*, 98–103.

Hui, L., Bakiri, L., Mairhorfer, A., Schweifer, N., Haslinger, C., Kenner, L., et al. (2007). p38α suppresses normal and cancer cell proliferation by antagonizing the JNK-c-Jun pathway. *Nature Genetics, 39*, 741–749.

Ichijo, H., Nishida, E., Irie, K., ten Dijke, P., Saitoh, M., Moriguchi, T., et al. (1997). Induction of apoptosis by ASK1, a mammalian MAPKKK that activates SAPK/JNK and p38 signaling pathways. *Science, 275*, 90–94.

Iriyama, T., Takeda, K., Nakamura, H., Morimoto, Y., Kuroiwa, T., Mizukami, J., et al. (2009). ASK1 and ASK2 differentially regulate the counteracting roles of apoptosis and inflammation in tumorigenesis. *EMBO Journal, 28*, 843–853.

Ishitani, T., Takaesu, G., Ninomiya-Tsuji, J., Shibuya, H., Gaynor, R. B., & Matsumoto, K. (2003). Role of the TAB2-related protein TAB3 in IL-1 and TNF signaling. *EMBO Journal, 22*, 6277–6288.

Janknecht, R., & Hunter, T. (1997). Convergence of MAP kinase pathways on the ternary complex factor Sap-1a. *EMBO Journal, 16*, 1620–1627.

Jiang, Y., Chen, C., Li, Z., Guo, W., Gegner, J. A., Lin, S., et al. (1996). Characterization of the structure and function of a new mitogen-activated protein kinase (p38β). *Journal of Biological Chemistry, 271*, 17920–17926.

Jiang, Y., Gram, H., Zhao, M., New, L., Gu, J., Feng, L., et al. (1997). Characterization of the structure and function of the fourth member of the p38 group mitogen-activated protein kinases, p38δ. *Journal of Biological Chemistry, 272*, 30122–30128.

Kallunki, T., Deng, T., Hibi, M., & Karin, M. (1996). c-Jun can recruit JNK to phosphorylate dimerization partners via specific docking interactions. *Cell, 87*, 929–939.

Kanayama, A., Seth, R. B., Sun, L., Ea, C. K., Hong, M., Shaito, A., et al. (2004). TAB2 and TAB3 activate the

NF-κB pathway through binding to polyubiquitin chains. *Molecular Cell, 15*, 535–548.

Karin, M., & Gallagher, E. (2005). From JNK to pay dirt: jun kinases, their biochemistry, physiology and clinical importance. *IUBMB Life, 57*, 283–295.

Kyriakis, J. M., & Avruch, J. (2012). Mammalian MAPK signal transduction pathways activated by stress and inflammation: A 10 year update. *Physiological Reviews, 92*, 689–737.

Kyriakis, J. M., Banerjee, P., Nikolakaki, E., Dai, T., Rubie, E. A., Ahmad, M. F., et al. (1994). The stress-activated protein kinase subfamily of c-Jun kinases. *Nature, 369*, 156–160.

Liu, H., Nishitoh, H., Ichijo, H., & Kyriakis, J. M. (2000). Activation of apoptosis signal-regulating kinase-1 (ASK1) by TNF receptor-associated factor-2 requires prior dissociation of the ASK1 inhibitor thioredoxin. *Molecular and Cellular Biology, 20*, 2198–2208.

Lomas, D. A., Lipson, D. A., Miller, B. E., Willits, L., Keene, O., Barnacle, H., et al. (2012). An oral inhibitor of p38 MAP kinase reduces plasma fibrinogen in patients with chronic obstructive pulmonary disease. *Journal of Clinical Pharmacology, 52*, 416–424.

Lu, Z., Xu, S., Joazerio, C., Cobb, M. H., & Hunter, T. (2002). The PHD domain of MEKK1 acts as an E3 ubiquitin ligase and mediates ubiquitination and degradation of ERK1/2. *Molecular Cell, 9*, 945–956.

Marais, R., Wynne, J., & Treisman, R. (1993). The SRF accessory protein Elk-1 contains a growth factor-regulated transcriptional activation domain. *Cell, 73*, 381–393.

Massagué, J., Seoane, J., & Wottonm, D. (2005). Smad transcription factors. *Genes and Development, 19*, 2783–2810.

Matsuzawa, A., Tseng, P.-H., Vallabhapurapu, S., Luo, J.-L., Zhang, W., Wang, H., et al. (2008). Essential cytoplasmic translocation of a cytokine receptor-assembled signaling complex. *Science, 321*, 663–668.

McKay, M. M., & Morrison, D. K. (2007). Integrating signals from RTKs to ERK/MAPK. *Oncogene, 26*, 3113–3121.

Mourey, R. J., Burnette, B. L., Brustkern, S. J., Daniels, J. S., Hirsch, J. L., Hood, W. F., et al. (2010). A benzothiophene inhibitor of mitogen-activated protein kinase-activated protein kinase 2 inhibits tumor necrosis factor-α production and has oral anti-inflammatory efficacy in acute and chronic models of inflammation. *Journal of Pharmacology and Experimental Therapeutics, 333*, 797–807.

Musti, A. M., Treier, M., & Bohmann, D. (1997). Reduced ubiquitin-dependent degradation of c-Jun after phosphorylation by MAP kinases. *Science, 275*, 400–402.

New, L., Jiang, Y., Zhao, M., Liu, K., Zhu, W., Flood, L. J., et al. (1998). PRAK, a novel protein kinase regulated by the p38 MAP kinase. *EMBO Journal, 17*, 3372–3384.

Ninomiya-Tsuji, J., Kishioto, K., Hiyama, A., Inoue, J. I., Cao, Z., & Matsumoto, K. (1999). The kinase TAK1 can activate the NIK-IκB as well as the MAP kinase cascade in the IL-1 signalling pathway. *Nature, 398*, 252–256.

Nishina, H., Fischer, K. D., Radvanyi, L., Shahinian, A., Razqallah, H., Rubie, E. A., et al. (1997). Stress-signalling kinase Sek1 protects thymocytes from apoptosis mediated by CD95 and CD3. *Nature, 385*, 350–353.

Nishitoh, H., Saitoh, M., Mochida, Y., Takeda, K., Nakano, H., Rothe, M., et al. (1998). ASK1 is essential for JNK/SAPK activation by TRAF2. *Molecular Cell, 2*, 389–395.

Özcan, U., Cao, Q., Yilmaz, E., Lee, A.-H., Iwakoshi, N. N., Özdelen, E., et al. (2004). Endoplasmic reticulum stress links obesity, insulin action and type 2 diabetes. *Science, 306*, 457–461.

Papoutsopoulou, S., Symons, A., Tharmalingham, T., Belich, M. P., Kaiser, F., Kioussis, D., et al. (2006). ABIN-2 is required for optimal activation of Erk MAP kinase in innate immune responses. *Nature Immunology, 7*, 606–615.

Patriotis, C., Makris, A., Bear, S. E., & Tsichlis, P. N. (1993). Tumor progression locus 2 (Tpl-2) encodes a protein kinase involved in the progression of rodent T-cell lymphomas and in T-cell activation. *Proceedings of the National Academy of Sciences of the United States of America, 90*, 2251–2255.

Perfield, J. W., Lee, Y., Shulman, G. I., Samuel, V. T., Jurczak, M. H., Ghang, E., et al. (2011). Tumor progression locus 2 (TPL2) regulates obesity-associated inflammation and insulin resistance. *Diabetes, 60*, 1168–1176.

Raingeaud, J., Whitmarsh, A. J., Barett, T., Dérijard, B., & Davis, R. J. (1996). MKK3- and MKK6-regulated gene expression is mediated by the p38 mitogen-activated protein kinase signal transduction pathway. *Molecular and Cellular Biology, 16*, 1247–1255.

Richards, S. A., Fu, J., Romanelli, A., Shimamura, A., & Blenis, J. (1999). Ribosomal S6 kinase 1 (RSK1) activation requires signals dependent on and independent of the MAP kinase ERK. *Current Biology, 9*, 810–820.

Ronkina, N., Kotlyarov, A., Dittrich-Breiholz, O., Kracht, M., Hitti, E., Milarski, K., et al. (2007). The mitogen-activated protein kinase (MAPK)-activated protein kinases MK2 and MK3 cooperate in stimulation of tumor necrosis factor biosynthesis and stabilization of p38 MAPK. *Molecular and Cellular Biology, 27*, 170–181.

Rouse, J., Cohen, P., Trigon, S., Morange, M., Alonso-Llamazares, A., Zamanillo, D., et al. (1994). A novel kinase cascade triggered by stress and heat shock that stimulates MAPKAP kinase-2 and phosphorylation of the small heat shock proteins. *Cell, 78*, 1027–1037.

Rousseau, S., Morrice, N., Peggie, M., Campbell, D. G., Gaestel, M., & Cohen, P. (2002). Inhibition of SAPK2a/p38 prevents hnRNP A0 phosphorylation by MAPKAP-K2 and its interaction with cytokine mRNAs. *EMBO Journal, 21*, 6505–6514.

Sabapathy, K., Hu, Y., Kallunki, T., Schreiber, M., David, J.-P., Jochum, W., et al. (1999). JNK2 is required for efficient T-cell activation and apoptosis but not for normal lymphocyte development. *Current Biology, 9*, 116–125.

Sabio, G., Das, M., Mora, A., Zhang, Z., Jun, J. Y., Ko, H. J., et al. (2008). A stress signaling pathway in adipose tissue regulates hepatic insulin resistance. *Science, 322*, 1539–1543.

Sabio, G., Cavanagh-Kyros, J., Ko, H. J., Jung, D. Y., Gray, S., Jun, J. Y., et al. (2009). Prevention of steatosis by hepatic JNK1. *Cell Metabolism, 10*, 491–498.

Sabio, G., Kennedy, N. J., Cavanagh-Kyros, J., Jung, D. Y., Ko, H. J., Ong, H., et al. (2010a). Role of muscle c-Jun-NH_2-terminal kinase 1 in obesity-induced insulin resistance. *Molecular and Cellular Biology, 30*, 106–115.

Sabio, G., Cavanagh-Kyros, J., Barrett, T., Jung, D. Y., Ko, H. J., Ong, H., et al. (2010b). Role of the hypothalamic-pituitary-thyroid axis in metabolic regulation By JNK1. *Genes and Development, 24*, 256–264.

Saitoh, M., Nishitoh, H., Fujii, M., Takeda, K., Tobiume, K., Sawada, Y., et al. (1998). Mammalian thioredoxin is a direct inhibitor of apoptosis signal-regulating kinase (ASK) 1. *EMBO Journal, 17*, 2596–2606.

Sakurai, T., Maeda, S., Chang, L., & Karin, M. (2006). Loss of hepatic NF-κB activity enhances chemical hepatocarcinogenesis through sustained c-Jun-N-terminal kinase activation. *Proceedings of the National Academy of Sciences, 103*, 10544–10551.

Sakurai, T., He, G., Matsuzawa, A., Yu, G. Y., Maeda, S., Hardiman, G., et al. (2008). Hepatocyte necrosis induced by oxidative stress and IL-1α release mediate carcinogen-induced compensatory proliferation and liver tumorigenesis. *Cancer Cell, 14*, 156–165.

Sánchez, I., Hughes, R. T., Mayer, B. J., Yee, K., Woodgett, J. R., Avruch, J., et al. (1994). Role of SAPK/ERK kinase-1 in the stress-activated pathway regulating transcription factor c-Jun. *Nature, 372*, 794–798.

Sato, S., Sanjo, H., Takeda, K., Ninomiya-Tsuji, J., Yamamoto, M., Kawai, T., et al. (2005). Essential function for the kinase TAK1 in innate and adaptive immune responses. *Nature Immunology, 6*, 1087–1095.

Schuman, J., Chen, Y., Podd, A., Yu, M., Liu, H. H., Wen, R., et al. (2009). A critical role of TAK1 in B-cell receptor-mediated nuclear factor-κB activation. *Blood, 113*, 4566–4574.

Shaulian, E., & Karin, M. (2002). AP-1 as a regulator of cell life and death. *Nature Cell Biology, 4*, E131–E136.

Shibuya, H., Yamaguchi, K., Shirakabe, K., Tonegawa, A., Gotoh, Y., Ueno, N., et al. (1995). TAB1: An Activator of the TAK1 MAPKKK in TGF-β Signal Transduction. *Science, 272*, 1179–1182.

Solinas, G., Vilcu, C., Neels, J. G., Bandyopadhyay, G. K., Luo, J.-L., Naugler, W., et al. (2007). JNK1 in hematopoietically derived cells contributes to diet-induced inflammation and insulin resistance without affecting obesity. *Cell Metabolism, 6*, 386–397.

Stafford, M. J., Morrice, N. A., Peggie, M. W., & Cohen, P. (2006). Interleukin-1 stimulated activation of the COT catalytic subunit through the phosphorylation of Thr290 and Ser62. *FEBS Letters, 580*, 4010–4014.

Takaesu, G., Kishida, S., Hiyama, A., Yamaguchi, K., Shibuya, H., Irie, K., et al. (2000). TAB2, a novel adaptor protein, mediates activation of TAK1 MAPKKK by linking TAK1 to TRAF6 in the IL-1 signal transduction pathway. *Molecular Cell, 5*, 649–650.

Takeda, K., Shimozono, R., Noguchi, T., Umeda, T., Morimoto, Y., Naguro, I., et al. (2007). Apoptosis signal-regulating kinase (ASK) 2 functions as a mitogen-activated protein kinase kinase kinase in a heteromeric complex with ASK1. *Journal of Biological Chemistry, 282*, 7522–7531.

Takekawa, M., & Saito, H. (1998). A family of stress-inducible GADD45-like proteins mediate activation of the stress-responsive MKT/MEKK4 MAPKKK. *Cell, 95*, 521–530.

Thalhamer, T., McGrath, M. A., & Harnett, M. M. (2008). MAPKs and their relevance to arthritis and inflammation. *Rheumatology, 47*, 409–414.

Tournier, C., Whitmarsh, A. J., Cavanagh, J., Barrett, T., & Davis, R. J. (1997). Mitogen-activated protein kinase kinase 7 is an activator of the c-Jun NH2-terminal kinase. *Proceedings of the National Academy of Sciences of the United States of America, 94*, 7337–7342.

Ueda, T., Sasaki, M., Elia, A. J., Chio, I. I., Hamada, K., Fukunaga, R., et al. (2010). Combined deficiency for MAP kinase interacting kinase 1 and 2 (Mnk1 and Mnk2) delays tumor development. *Proceedings of the National Academy of Sciences of the United States of America, 107*, 13984–13990.

Vallerie, S. N., Furuhashi, M., Fucho, R., & Hotamisligil, G. (2008). A predominant role for parenchymal c-Jun amino terminal kinase (JNK) in the regulation of systemic insulin sensitivity. *PLoS One, 3*, e3151.

Ventura, J. J., Kennedy, N. J., Lamb, J. A., Flavell, R. A., & Davis, R. J. (2003). c-Jun NH_2-terminal kinase is essential for the regulation of AP-1 by tumor necrosis factor. *Molecular and Cellular Biology, 23*, 2871–2882.

Venuprasad, K., Elly, C., Gao, M., Salek-Ardakani, S., Harada, Y., Luo, J. L., et al. (2006). Convergence of itch-induced ubiquitination with MEKK1-JNK signaling in Th2 tolerance and airway inflammation. *Journal of Clinical Investigation, 116*, 1117–1126.

Wang, C., Deng, L., Hong, M., Akkaraju, G. R., Inoue, J., & Chen, Z. J. (2001). TAK1 is a ubiquitin-dependent kinase of MKK and IKK. *Nature, 412*, 346–351.

Waskiewicz, A. J., Flynn, A., Proud, C. G., & Cooper, J. A. (1997). Mitogen-activated protein kinases activate the serine/threonine kinases Mnk1 and Mnk2. *EMBO Journal, 16*, 1909–1920.

Waskiewicz, A. J., Johnson, J. C., Penn, B., Mahalingam, M., Kimball, S. R., & Cooper, J. A. (1999). Phosphorylation of the cap-binding protein eukaryotic translation initiation factor 4E by protein kinase Mnk1 in vivo. *Molecular and Cellular Biology, 19*, 1871–1880.

Watford, W. T., Hissong, B. D., Durant, L. O., Yamane, H., Muul, L. M., Kanno, Y., et al. (2008). Tpl2 kinase regulates T cell interferon-gamma production and host resistance to Toxoplasma gondii. *Journal of Experimental Medicine, 205*, 2803–2812.

Watford, W. T., Wang, C. C., Tsatsanis, C., Mielke, L. A., Eliopoulos, A. G., Daskalakis, C., et al. (2010). Ablation of tumor progression locus 2 promotes a type 2 Th cell response in ovalbumin-immunized mice. *Journal of Immunology, 184*, 105–113.

Whitmarsh, A. J., Shore, P., Sharrocks, A. D., & Davis, R. J. (1995). Integration of MAP kinase signal transduction pathways at the serum response element. *Science, 269*, 403–407.

Wu, Z., Wu, J., Jacinto, E., & Karin, M. (1997). Molecular cloning and characterization of human JNKK2, a novel Jun NH$_2$-terminal kinase-specific kinase. *Molecular and Cellular Biology, 17*, 7407–7416.

Xia, Y., Wang, J., Xu, S., Johnson, G. L., Hunter, T., & Lu, Z. (2007). MEKK1 mediates the ubiquitination and degradation of c-Jun in response to osmotic stress. *Molecular and Cellular Biology, 27*, 510–517.

Xia, Z. P., Sun, L., Chen, X., Pineda, G., Jiang, X., Adhikari, A., et al. (2009). Direct activation of protein kinases by unanchored polyubiquitin chains. *Nature, 461*, 114–119.

Yamaguchi, K., Shirakabi, K., Shibuya, H., Irie, K., Oishi, I., Ueno, N., et al. (1995). Identification of a member of the MAPKKK family as a potential mediator of TGF-b signal transduction. *Science, 270*, 2008–2011.

Zaru, R., Ronkina, N., Gaestel, M., Arthur, J. S., & Watts, C. (2007). The MAPK-activated kinase Rsk controls an acute toll-like receptor signaling response in dendritic cells and is activated through two distinct pathways. *Nature Immunology, 8*, 1227–1235.

Mast Cells

Bernhard F. Gibbs[1] and Madeleine Ennis[2]
[1]Medway School of Pharmacy, University of Kent, The Universities of Greenwich and Kent at Medway, Chatham, Kent, UK
[2]Centre for Infection and Immunity, School of Medicine, Dentistry and Biomedical Sciences, The Queen's University of Belfast, Belfast, UK

Definition

Mast cells are usually found in connective tissue and are filled with granules that stain metachromatically. Paul Ehrlich called them Mastzellen because he thought that they derived from connective tissue cells which had been "fattened up" deriving the name from the German verb mästen. Mast cells from different species and from different sites even in the same species respond differently to both activators and inhibitors.

Structure and Functions

Mast cells are found throughout the animal kingdom; however, this review will only discuss human mast cells, mainly from primary cells, though some data has been derived from studies with human mast cell lines. Mast cells are characterized by their granularity, although usually round, they can also take an elongated shape. They are found throughout the body, often in close contact with blood vessels or nerves, in particularly high numbers in those areas that come into contact with the outside world, e.g., skin, gastrointestinal tract, and respiratory tract. Mast cells are characterized based on their protease content: those that contain tryptase and chymase are designated MC-TC, whereas those with only tryptase are called MC-T. There are also some reports of a third type which only contains chymase.

On activation mast cells release a myriad of mediators some of which are stored, e.g., histamine and heparin, others are rapidly generated, e.g., prostaglandins and leukotrienes, and others which require protein synthesis, e.g., cytokines (Fig. 1). The mediators released exert many different functions, and depending on the stimulus, not all are released; thus, mast cells have the capacity to display differential biological functions. A detailed description of the actions of each mediator is beyond the scope of this article. Histamine and heparin, which are located together in the granules, exert profound effects increasing vascular permeability, causing smooth muscle contraction, and acting as an anticoagulant. The prostaglandins and leukotrienes contract smooth muscle and cause bronchoconstriction, increase vascular permeability, stimulate mucus secretion, and can also be chemotaxins. Platelet-activating factor (PAF) attracts and activates leukocytes. Through the chemotactic factors CXCL8 and CCL5, mast cells recruit neutrophils and

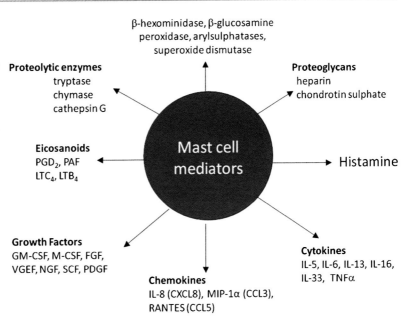

Mast Cells, Fig. 1 Summary of major mediators released by human mast cells. Most data have been derived from human mast cells, though some has been taken from human mast cell lines (Further information can be found in Metcalfe (2008) and Sismanopoulos et al. (2012))

eosinophils to inflamed tissue sites. Enzymes such as tryptase and chymase cause tissue damage, and tryptase can activate PAR-2. Chymase can act in a pro-inflammatory manner as it activates pro-IL-1β and pro-IL-18 to their active counterparts. It also can be anti-inflammatory as it can degrade IL-6 and IL-13, eotaxin, and substance P. Mast cell-derived cytokines promote the growth of the different cell types and cause vasodilation, neovascularization, and angiogenesis as well as orchestrating inflammatory responses, including recruitment and proliferation of leukocytes in inflamed tissues.

Classically, mast cells are associated with allergic disorders, but they also play important roles in wound healing, angiogenesis, as well as both the innate and adaptive immunity. They have several roles in pathologies other than allergy, some of which will be described in the section below (Table 1).

Pathological Relevance

Allergy

Mast cells are involved in type 1 hypersensitivity reactions. These occur when high-affinity IgE receptors (FcεRI) are first sensitized with allergen-specific IgE antibodies which then become cross-linked with subsequent allergen exposure. Cross-linking adjacent IgE molecules bound to FcεRI causes activation leading to degranulation and release of histamine as well as several protease enzymes. This is then followed by the production of various eicosanoids and

Mast Cells, Table 1 Some diseases where mast cell involvement is postulated

Allergic disease	Asthma, rhinitis, atopic dermatitis, eczema
Autoimmune disorders	Rheumatoid arthritis, bullous pemphigoid, chronic idiopathic urticaria
Cancer	Systemic mastocytosis, lung, breast, pancreas, skin, and gastrointestinal tract tumors
Cardiovascular disease	Atherosclerosis, abdominal aortic aneurysm
Skin disease (nonallergic)	Psoriasis
Gastrointestinal tract	Crohn's disease, ulcerative colitis
Neurological disease	Alzheimer's disease, multiple sclerosis

cytokines which vary according to the mast cell phenotype. Mast cell numbers have been shown to increase in number and reactivity in atopic diseases such as atopic asthma and atopic dermatitis. For example, mast cells retrieved by bronchoalveolar lavage from asthmatic patients release more histamine than those obtained from control subjects.

Cancer

Mast cells are involved in many cancers. However, they have been shown to exert both positive and negative effects. This is perhaps not surprising as mediators released by mast cells can help tumor progression (e.g., by stimulating angiogenesis, degrading the extracellular matrix) and also inhibit tumor progression (e.g., by cytotoxic actions of, e.g., TNF-α, recruiting and activating other inflammatory cells). Increased mast cell number in tumors has been found to be associated with poorer outcome in diverse cancers such as colorectal, prostate, lung, and pancreatic cancer (Khazaie et al. 2011). In pancreatic adenocarcinoma, the number of mast cells in the tumor correlates with tumor grade, and high mast cell numbers are associated with shorter recurrence-free survival and disease-specific survival. In contrast, high mast cell and eosinophil numbers predict better survival in colorectal cancer. In breast cancer, both positive and negative effects of mast cells have been reported (Marichal et al. 2013).

Cardiovascular Disease

Mast cell numbers are increased in atherosclerotic lesions. They are found in fatty streaks, and the number further increases in advanced lesions (Xu and Shi 2012). In infarct-related coronary arteries, the number of mast cells is lowest in the normal intima, is higher in the non-ruptured plaque and highest in the ruptured plaques. This increase in number is mirrored by the increase in degranulated mast cells (Laine et al. 1999). Studies in the human mast cell line HMC-1 have demonstrated that oxidized LDL elicits the release of pro-inflammatory cytokines (TNF-α and IL-6), suggesting one mechanism for the activation of coronary mast cells (Meng et al. 2013). Mast cells have also been associated with abdominal aortic aneurysm (AAA), where increased numbers are found in the media and adventitia (Xu and Shi 2012).

Joint Disease

Increased mast cell numbers are found not only in synovial tissue in rheumatoid arthritis but also in osteoarthritis (Nigrovic and Lee 2007). Mast cells isolated from these tissues can be activated in vitro, and studies have reported the release of histamine, TNF-α, and IL-8 (He et al. 2001; Kashiwakura et al. 2013). Indeed, synovial fluid from patients with both osteo- and rheumatoid arthritis contains histamine and tryptase (Eklund 2007).

Mastocytosis

Mastocytosis is the name given to a heterogeneous clonal disorder where mast cell populations have increased abnormally in tissues such as the skin, liver, spleen, bone marrow, lymph nodes, and gastrointestinal tract. The two main classifications are cutaneous and systematic (da Silva et al. 2014). However, the World Health Organization has detailed criteria describing the different classes of the disease (Brockow and Metcalfe 2010). The symptoms reported are those which are associated with the action of mast cell-derived mediators, for example, itch and reddening of the skin and gastrointestinal problems such as nausea, vomiting, and diarrhea. Although serum tryptase levels are raised in some patients, they are not a universal diagnostic tool.

Neurological Disease

Mast cell numbers are increased in the white matter, and demyelinated lesions in the brains of people with multiple sclerosis and tryptase levels are raised in the cerebrospinal fluid (da Silva et al. 2014). However, given the difficulty of human studies, most work has concentrated on animal models.

Skin Disease

Increased skin mast cell numbers are observed not only in atopic disorders such as atopic dermatitis but also in psoriasis, bullous pemphigoid and chronic idiopathic urticaria, as well as in mastocytosis (Harvima et al. 2008). They also play a major role in wound healing (Wulff and Wilgus 2013). Their increased presence in chronic wounds such as chronic venous leg ulcers suggests that they play an important pathological role. In vitro studies of human skin mast cells have shown that they respond not only to stimulation by anti-IgE but also to neuropeptides such as substance P and drugs such as morphine (Church and Clough 1999).

Interactions with Other Processes and Drugs

Mast Cell Activators

Mast cells have multiple biological roles owing to both their varied locations in a number of different organs and tissues and their potential to respond to a wide variety of stimuli, many of these caused by external agents such as certain pathogens, drugs, and various environmental allergens. Of these, the ability of mast cells to respond to allergen provocation in allergic individuals is best known. Certain drugs are known to activate mast cells because of an existing allergic sensitization to the drug, the most notable being penicillin and analogs which share the ß-lactam ring. A summary of possible activators of human mast cells is shown in Fig. 2.

Activation of FcɛRI on mast cells is not, however, restricted to allergen provocation but can also occur in an IgE-specific manner with various autoantigens and parasitic antigens. Some parasitic glycoproteins can even cause IgE-dependent mast cell activation by binding to nonspecific regions of IgE, which may also occur with various lectins (e.g., concanavalin A) or viruses (e.g., gp120 glycoproteins present on HIV). Additionally, in certain forms of urticaria (hives), mast cells can be activated through FcɛRI stimulation caused by auto-anti-IgE or even auto-anti-FcɛRI antibodies (reviewed in Fiebiger et al. 1996). Although IgE-/FcɛRI-dependent activation of mast cells is probably the most important stimulatory route caused by immunoglobulins in these cells, some human mast cells can also be stimulated by IgG antibody-mediated cross-linking owing to the presence of several stimulatory IgG (Fcγ) receptor subtypes (especially FcγRI). Their role, however, has not yet been fully elucidated, especially with respect to human mast cell function.

Other immunological triggers of mast cells include complement factors, especially C5a, and they are further thought to play an important role in innate immunity and inflammation by responding to various pathogens via toll-like receptors (TLR). However, although human mast cells express a number of different TLRs (especially, TLR2, TLR3, TLR4, TLR5, and TLR9), stimulation through these receptors rarely results in degranulation (associated with histamine release), whereas eicosanoid and cytokine release has been more widely observed, especially following TLR2-dependent stimulation.

One of the many important locations of mast cells is around nerve endings. These can release substance P which is an efficacious activator of connective tissue (MC-TC) mast cells but not mucosal-like (MC-T) mast cells. The release of substance P from C-fiber nerve endings is thought to contribute importantly to nonallergic (intrinsic) responses in asthma which particularly affect individuals with chronic allergic inflammation due to increased exposure of C-fibers resulting from tissue remodeling and mast cell hyperplasia. The release of histamine from mast cells, caused by substance P stimulation, results in the reciprocal activation of these nerve endings since they express H1 receptors. Structurally, substance P is a polybasic amine, and several natural as well as synthetic basic amine analogs (e.g., polylysine, compound 48/80) cause substantial and rapid degranulation of connective tissue (MC-TC) mast cell phenotypes. Environmental polyamine mast cell activators include bee and wasp venoms (peptide 401 and mastoparan, respectively) which directly stimulate skin mast cells independent of whether an individual is allergic to these insects or not. Snake venoms have also been shown to activate mast cells, and it is thought that the resulting release of mast cell-associated proteases may facilitate venom destruction and thus support survival to snake bites (Metz et al. 2006).

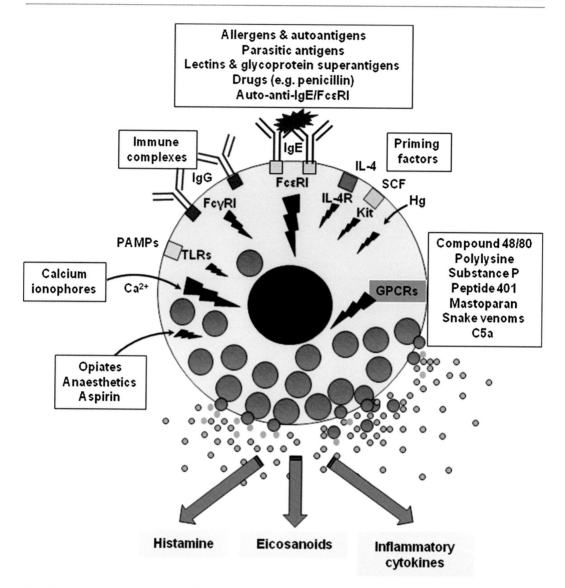

Mast Cells, Fig. 2 Summary of possible activators of human mast cells. IgE-/FcεRI-mediated activators may depend on the presence of specific IgE antibodies in sensitized individuals to certain allergens, autoantigens, and drugs. Nonspecific mechanisms include the effects of certain lectins, parasitic antigens, and superantigens or antibodies against either IgE or FcεRI itself. Other immunological activators include IgG complexes, priming factors (which mainly require an additional stimulus), and pathogen-associated molecular patterns (PAMPs) which bind to various TLRs on human mast cells. There are many different GPCR-mediated triggers which cause substantial degranulation in connective tissue (MC-TC) mast cell phenotypes. Activation of all mast cell phenotypes occurs with calcium ionophores, whereas drugs such as aspirin, opiates, and anesthetics only affect mast cells in certain individuals

Mast cells can be activated by a number of drugs and other non-immunological stimuli. The most effective stimuli are calcium ionophores, such as A23187, originally designed as antibiotics, which cause the production and release of all major inflammatory mediator types from these cells by increasing intracellular calcium concentrations. Opiates and certain anesthetics can also cause degranulation of certain mast cell subtypes in some individuals which is often not dependent on IgE-induced mechanisms. Aspirin can stimulate the production and release of cysteinyl leukotrienes from mast cells in aspirin-exacerbated respiratory disease, although the mechanisms are not fully understood. In addition, heavy metals, such as mercury, can lead to toxic release of histamine from mast cells and, at lower concentrations, potentiate inflammatory mediator production and release caused by IgE-dependent stimuli.

Certain growth factors and cytokines also potentiate human mast cell histamine release and the production of newly synthesized mediators, particularly following IgE-induced stimulation. Stem cell factor (SCF) is a prominent priming factor in this regard, in addition to its crucial role in mast cell growth and differentiation. Compared to rodent mast cells, however, the number of priming cytokines that can enhance human mast cell responses seems more restricted, IL-4 being a notable exception which has differential priming effects and can stimulate human mast cells, in conjunction with SCF, to produce IL-13 following long-term culture (Lorentz et al. 2005).

Inhibitors of Mast Cell Function

The ability of mast cells to respond to allergens can be abrogated by preventing the binding of circulating IgE to FcεRI on the cell surface using omalizumab. This monoclonal antibody approach neutralizes unbound IgE and, over time, reduces the amounts of antibody bound to the cells. As a result, sensitization of mast cells with allergen-specific IgE diminishes to such an extent that mast cells no longer respond to allergen provocation. FcεRI expressions on the surface of mast cells also decrease with reduced levels of IgE sensitization thus limiting further potential sensitization. However, this therapy is comparatively expensive, and the process of reducing IgE sensitization of mast cells takes several weeks before a reduction in mast cell-specific responses is observed (e.g., wheal and flare responses in the skin) (Beck et al. 2004).

Pharmacological agents based on chromones (e.g., cromoglycate) are often portrayed as "mast cell stabilizers." There is, however, currently no agent known to specifically inhibit mast cell activation per se, although chromones have been shown to be effective in mild allergic disorders and can reduce activation of certain mast cell types. In humans, cromoglycate, nedocromil, and similar analogs display inhibitory effects on mucosal-like (MC-T) mast cells but are relatively ineffective at blocking the responses of connective tissue mast cells present in the skin. This is in stark contrast to certain rodent species (where, e.g., connective tissue mast cells of the rat respond well to the inhibitory actions of nedocromil) which highlights considerable pharmacological heterogeneity among mast cell populations from different tissues and species (Pearce et al. 1989). The mode of action of chromone-like drugs, however, is still not fully understood, which is exemplified by the problem that, clinically, these agents are used prophylactically, while in vitro, they display tachyphylaxis.

Possibly the most common agents currently used to prevent mast cell responses in allergy increase intracellular levels of cAMP. Elevating this second messenger leads to the prevention of increased calcium influx which is an essential step for mast cell degranulation and mediator synthesis. To some extent, mast cells limit their own activities due to the expression of H2 receptors, which are GPCRs that increase cAMP upon stimulation with histamine, although therapeutic use of H2 agonists is precluded to due side effects. Beta agonists, particularly long-acting ones such as salmeterol, can affect mast cell function by elevating cAMP through similar GPCR mechanisms. Methylxanthines (e.g., theophylline, caffeine, and theobromine) are highly effective inhibitors of mast cell activation by blocking phosphodiesterase (PDE), an enzyme which degrades cAMP. More selective PDE4 have recently been developed for use in allergic diseases, particularly in asthma. However, although these drugs effectively reduce mast cell function, they display a relatively narrow therapeutic window for clinical – especially systemic – use other than for severe allergic diseases.

There are several intracellular signal transduction pathways which can be targeted by drugs that inhibit mast cell responses, particularly those resulting from FcεRI-mediated activation. These include Syk inhibitors, calcineurin inhibitors, and PI 3-kinase inhibitors. Syk is one of the earliest stimulatory signals activated by FcεRI triggering; thus, Syk inhibitors potentially are limited to blocking IgE-dependent mast cell responses, and several trials have recently reported their clinical potential. Calcineurin inhibitors, such as ascomycin and cyclosporin, are highly potent inhibitors of mast cell degranulation and are more commonly used topically for the treatment of atopic dermatitis. There are concerns about their systemic use for mast cell-related diseases, and, as a result, they often reserved for only managing severe steroid-resistant asthma (NB steroids are generally not effective inhibitors of mast cell function). Signal transduction inhibitors that were primarily designed to inhibit Kit signaling can also downregulate IgE-dependent responses, such as dasatinib, owing to its actions on other mast cell tyrosine kinases in addition to Kit. Kit, the receptor for stem cell factor, plays a crucial role in mast cell development, and agents targeting its activation have been used to treat mastocytosis and related myeloid leukemias. The combined effects of inhibiting Kit signaling with other kinases, due to the nonspecific actions of some of these drugs, have shown therapeutic potential for treating steroid-resistant asthma (Humbert et al. 2009).

In addition to agents which block stimulatory signaling in mast cells, recent studies have also highlighted the potential of triggering inhibitory signaling pathways in these cells. The phosphodiesterase SHIP-1, for example, which is involved in downregulating PI 3-kinase-mediated responses in mast cells, has been shown to be increased by SHIP-1 activators (Stenton et al. 2013). Other approaches are currently being investigated which target inhibitory receptors on these cells such as CD300a, CD200R, and various members of the Siglec family (reviewed in Harvima et al. 2014). These inhibitory mechanisms have not yet been fully elucidated but may play a vital role in causing mast cell anergy during certain types of allergen immunotherapy.

There are many other drugs which have been reported to inhibit human mast cell function but are currently not widely used clinically for the treatment of allergies and other mast cell-related disorders. Statins, which are mainly employed to prevent cardiovascular disease, also inhibit mast cell growth and function (Krauth et al. 2006). The diuretic agent furosemide (frusemide) has been known for decades to moderately inhibit IgE-dependent histamine release human lung mast cells and other allergic effector cells. Similar inhibitory effects have also been reported for the

Mast Cells, Table 2 Agents that inhibit human mast cell function

Drug/agent	Mechanism
Omalizumab	Prevents IgE from binding to FcεRI
Chromones (e.g., cromoglycate, nedocromil)	Unclear, but drugs mainly only affect MC-T (mucosal mast cells) in humans
Kit/Src family tyrosine kinase inhibitors (e.g., dasatinib)	Blockade of Kit-mediated signaling and other tyrosine kinases
Syk inhibitors	Blockade of early IgE-dependent stimulatory signaling
PI 3-kinase inhibitors	Blockade of PI-3 kinase (activated by several different stimuli)
Calcineurin inhibitors (e.g., ascomycin, cyclosporine)	Blockade of calcineurin but unknown additional mechanisms involved in their effects on degranulation
Inhibitory signal activators	Stimulation of inhibitory phosphatases (e.g., SHIP-1)
β-agonists (e.g., salbutamol, salmeterol)	Increase cAMP levels
Phosphodiesterase inhibitors (e.g., theophylline)	Increase cAMP levels by preventing the action of phosphodiesterase
Loop diuretics (e.g., furosemide/frusemide)	Unclear
Ambroxol	Unclear; may involve blockade of p38 MAPK and ROS-mediated signaling
Curcumin	Unclear; may involve blockade of p38 MAPK
Statins (e.g., cerivastatin, atorvastatin)	Unclear; may affect receptor-dependent signaling by disrupting lipid rafts

mucolytic agent ambroxol. Many natural products are also known to affect mast cell mediator release, such as curcumin. These, and other agents that prevent mast cell activation, are summarized in Table 2.

Cross-References

▶ Allergic Disorders
▶ Anaphylaxis (Immediate Hypersensitivity): From Old to New Mechanisms
▶ Anti-asthma Drugs, Overview
▶ Asthma
▶ Atopic Dermatitis
▶ Basophils
▶ Beta2 Receptor Agonists
▶ Complement C5a Receptors
▶ Corticosteroids
▶ Cytokines
▶ Interleukin 4 and the Related Cytokines (Interleukin 5 and Interleukin 13)
▶ Immunoglobulin Receptors and Inflammation
▶ Leukotrienes
▶ Phosphodiesterase 4 Inhibitors: Apremilast and Roflumilast
▶ Prostanoids
▶ Rheumatoid Arthritis
▶ Substance P in Inflammation
▶ Theophylline
▶ Toll-Like Receptors
▶ Tumor Necrosis Factor Alpha (TNFalpha)

References

Beck, L. A., Marcotte, G. V., MacGlashan, D., Togias, A., & Saini, S. (2004). Omalizumab-induced reductions in mast cell Fce psilon RI expression and function. *Journal of Allergy and Clinical Immunology, 114*, 527–530.

Brockow, K., & Metcalfe, D. D. (2010). Mastocytosis. *Chemical Immunology and Allergy, 95*, 110–124.

Church, M. K., & Clough, G. F. (1999). Human skin mast cells: In vitro and in vivo studies. *Annals of Allergy, Asthma, and Immunology, 83*, 471–475.

da Silva, E. Z., Jamur, M. C., & Oliver, C. (2014). Mast cell function: A new vision of an old cell. *Journal of Histochemistry and Cytochemistry, 62*, 698–738.

Eklund, K. K. (2007). Mast cells in the pathogenesis of rheumatic diseases and as potential targets for antirheumatic therapy. *Immunological Reviews, 217*, 38–52.

Fiebiger, E., Stingl, G., & Maurer, D. (1996). Anti-IgE and anti-Fc epsilon RI autoantibodies in clinical allergy. *Current Opinion in Immunology, 8*, 784–789.

Harvima, I. T., Nilsson, G., Suttle, M. M., & Naukkarinen, A. (2008). Is there a role for mast cells in psoriasis? *Archives of Dermatological Research, 300*, 461–478.

Harvima, I. T., Levi-Schaffer, F., Draber, P., Friedman, S., Polakovicova, I., Gibbs, B. F., et al. (2014). Molecular targets on mast cells and basophils for novel therapies. *Journal of Allergy and Clinical Immunology, 134*, 530–544.

He, S., Gaça, M. D., & Walls, A. F. (2001). The activation of synovial mast cells: Modulation of histamine release by tryptase and chymase and their inhibitors. *European Journal of Pharmacology, 412*, 223–229.

Humbert, M., de Blay, G., Garcia, G., Prud'homme, A., Leroyer, C., Magnan, A., et al. (2009). Masitinib, a c-kit/PDGF receptor tyrosine kinase inhibitor, improves disease control in severe corticosteroid-dependent asthmatics. *Allergy, 64*, 1194–1201.

Kashiwakura, J., Yanagisawa, M., Lee, H., Okumura, Y., Sasaki-Sakamoto, T., Saito, S., et al. (2013). Interleukin-33 synergistically enhances immune complex-induced tumor necrosis factor alpha and interleukin-8 production in cultured human synovium-derived mast cells. *International Archives of Allergy and Immunology, 161*(Suppl 2), 32–36.

Khazaie, K., Blatner, N. R., Khan, M. W., Gounari, F., Gounaris, E., Dennis, K., et al. (2011). The significant role of mast cells in cancer. *Cancer and Metastasis Reviews, 30*, 45–60.

Krauth, M. T., Majlesi, Y., Sonneck, K., Samorapoompichit, P., Ghannadan, M., Hauswirth, A. W., et al. (2006). Effects of various statins on cytokine-dependent growth and IgE-dependent release of histamine in human mast cells. *Allergy, 61*, 281–288.

Laine, P., Kaartinen, M., Penttilä, A., Panula, P., Paavonen, T., & Kovanen, P. T. (1999). Association between myocardial infarction and the mast cells in the adventitia of the infarct-related coronary artery. *Circulation, 99*, 361–369.

Lorentz, A., Wilke, M., Sellge, G., Worthmann, H., Klempnauer, J., Manns, M. P., et al. (2005). IL-4-induced priming of human intestinal mast cells for enhanced survival and Th2 cytokine generation is reversible and associated with increased activity of ERK1/2 and c-Fos. *Journal of Immunology, 174*, 6751–6756.

Marichal, T., Tsai, M., & Galli, S. J. (2013). Mast cells: Potential positive and negative roles in tumor biology. *Cancer Immunology Research, 1*, 269–279.

Meng, Z., Yan, C., Deng, Q., Dong, X., Duan, Z. M., Gao, D. F., et al. (2013). Oxidized low-density lipoprotein induces inflammatory responses in cultured human mast cells via Toll-like receptor 4. *Cellular Physiology and Biochemistry, 31*, 842–853.

Metcalfe, D. D. (2008). Mast cells and mastocytosis. *Blood, 112*, 946–956.

Metz, M., Piliponsky, A. M., Chen, C. C., Lammel, V., Abrink, M., Pejler, G., et al. (2006). Mast cells can enhance resistance to snake and honeybee venoms. *Science, 313*, 526–530.

Nigrovic, P. A., & Lee, D. M. (2007). Synovial mast cells: Role in acute and chronic arthritis. *Immunological Reviews, 217*, 19–37.

Pearce, F. L., Al-Laith, M., Bosman, L., Brostoff, J., Cunniffe, T. M., Flint, K. C., et al. (1989). Effects of sodium cromoglycate and nedocromil sodium on histamine secretion from mast cells from various locations. *Drugs, 37*(Suppl 1), 37–43.

Sismanopoulos, N., Delivanis, D. A., Alysandratos, K. D., Angelidou, A., Therianou, A., Kalogeromitros, D., et al. (2012). Mast cells in allergic and inflammatory diseases. *Current Pharmaceutical Design, 18*, 2261–2277.

Stenton, G. R., Mackenzie, L. F., Tam, P., Cross, J. L., Harwig, C., Raymond, J., et al. (2013). Characterization of AQX-1125, a small-molecule SHIP1 activator: Part 1. Effects on inflammatory cell activation and chemotaxis in vitro and pharmacokinetic characterization in vivo. *British Journal of Pharmacology, 168*, 1506–1518.

Wulff, B. C., & Wilgus, T. A. (2013). Mast cell activity in the healing wound: More than meets the eye? *Experimental Dermatology, 22*, 507–510.

Xu, J., & Shi, G. (2012). Emerging role of mast cells and macrophages in cardiovascular and metabolic diseases. *Endocrine Reviews, 33*, 71–108.

Mechanisms of Macrophage Migration in 3-Dimensional Environments

Isabelle Maridonneau-Parini[1] and
Celine Cougoule[2]
[1]Institut de Pharmacologie et de Biologie Structurale - CNRS, UMR 5089, Toulouse, France
[2]Département Mécanismes Moléculaires des Infections Mycobactériennes, CNRS, Institute of Pharmacology and Structural Biology, Toulouse, France

Synonyms

Cell adhesion; Extracellular matrix; Macrophage tissue infiltration in pathologies; Phagocyte; Podosomes; Proteases; Tissue migration

Definition: Tissue Infiltration of Macrophages

Macrophages play a vital role in both homeostasis maintenance by clearing the interstitial environment of extraneous cellular materials and as key players in the immune response (Mosser and Edwards 2008). Macrophages are present in virtually all tissues. They differentiate from circulating peripheral-blood mononuclear cells (PBMCs) which migrate to tissues both in steady state and in response to inflammation. During steady state, monocytes migrate from blood to tissues to replenish the stocks of long-lived tissue-specific macrophages of the bone (osteoclasts), lung alveoli, central nervous system (microglial cells), connective tissue (histiocytes), gastrointestinal tract, liver (Kupffer cells), spleen, and peritoneum. Following injury or infection, monocyte migration is induced by stimuli that are rapidly generated by the wounded tissue. Tissue infiltration of macrophages is essential to host defense, tissue repair, and immune regulation. Moreover, macrophages have a remarkable phenotypic plasticity that allows them to respond efficiently to environmental signals (Gordon and Mantovani 2011), for example, in tissues, they display specific functions required for tissue homeostasis. Macrophage tissue infiltration is also a hallmark of several pathological situations including cancer, neurodegenerative disorders, and chronic inflammation. Hence, deciphering the mechanisms of macrophage migration across a variety of tissues holds great potential for novel therapies (Mackay 2008; Qualls and Murray 2010; Ruhrberg and De Palma 2010).

Structure and Functions

Macrophage Migration, Introduction
The trafficking of phagocytes from the bloodstream to inflammatory or infectious sites takes place in two-dimensional and three-dimensional environments.

Diapedesis consists in crossing the endothelial wall and the basal membrane of capillaries. The thickness of these two barriers is 2 μm and

100–300 nm, respectively, while the diameter of leukocytes ranges from 10 to 20 μm. This implies that these cells are not constrained in 3D environments and diapedesis is thus considered as a 2D migration process (Verollet et al. 2011). The next step takes place in interstitial tissues, a 3D environment which exerts constraints onto migrating cells. Cell interactions with the extracellular matrix (ECM) provide chemical and mechanical signals which impact their differentiation, adhesion, and migration (Mierke 2011). Despite the fact that a large part of cell migration occurs in 3D substrates, most of the studies to date have been performed in two dimensions, on rigid (plastic or glass) surfaces coated or not with a thin layer of extracellular matrix proteins. The migration in 2D and 3D environments has, however, been shown to require distinct mechanisms (Harunaga and Yamada 2011). Recently, with the improvement of live cell imaging, it has become possible to examine the in vivo behavior of phagocytes in 3D environments by intravital microscopy (Auffray et al. 2007; Lämmermann et al. 2008; McDonald et al. 2010; Pflicke and Sixt 2009; Woodfin et al. 2011) and in "artificial" in vitro matrices such as Matrigel or collagen I (Cougoule et al. 2010, 2012; Jevnikar et al. 2012; Sabeh et al. 2009; Steadman et al. 1997; Van Goethem et al. 2010, 2011).

In Vitro Migration of Macrophages in Extracellular Matrices and Tissue-Like Environments

With the aim of mimicking the diversity of the environments encountered by macrophages in vivo, several matrices have been designed in which several parameters have been modified: the biochemical composition, the viscoelasticity, and the architecture. Although these matrices still poorly mimic the complexity and diversity of in vivo tissues, they are suitable to investigate the effect of a constraining 3D environment on the migration capacity of macrophages and the consequences of variable mechanical, structural, and biochemical properties on cell migration.

Using these matrices polymerized as a thick layer (approximately 1 mm) in the upper chamber of transwells with pores of 8 μm diameter (Fig. 1), it has been observed that both human monocyte-derived macrophages (MDMs) and mouse bone marrow-derived macrophages (BMDMs) can use two distinct migration modes: amoeboid and mesenchymal (Cougoule et al. 2010; Guiet et al. 2012; Van Goethem et al. 2010). Interestingly, the architecture of the matrix rather than the composition appears to dictate the choice between these two migration modes (Van Goethem et al. 2010). These migration modes were originally described for tumor cells (Friedl et al. 1998; Sabeh et al. 2009; Sahai and Marshall 2003; Wolf et al. 2003) and appear to apply also for macrophages with only a few discrepancies. The amoeboid mode takes place in porous matrices such as denatured collagen I polymerized as fibers (so-called fibrillar collagen I at concentrations between 2 and 4 mg/ml). This mode is characterized by a spherical cell shape, a rapid migration (40 μm/h), and the involvement of Rho/ROCK signaling pathway (Van Goethem et al. 2010). Integrins are dispensable in tumor cells using the amoeboid mode (Harunaga and Yamada 2011). The mesenchymal mode takes place in dense, poorly porous matrices (Matrigel® polymerized at concentrations from 8 to 12 mg/ml or collagen I polymerized as a gel at 5 mg/ml) in which the estimated diameter of the pores is inferior to 1–2 μm. It is characterized by an elongated cell shape, a low velocity (10 μm/h), and the requirement of proteases to degrade and create paths in the ECM (Van Goethem et al. 2010). Integrins are required in tumor cells using the mesenchymal mode, but their role has not yet been studied in macrophages (Friedl et al. 1998; Sahai and Marshall 2003; Wolf et al. 2003). The protease-dependent migration of macrophages also involves ingestion of the degraded matrix and compaction of the matrix (Van Goethem et al. 2011). The result is the formation of tunnels which might close up more or less rapidly after the transit of macrophages, depending on the viscoelastic properties of the matrix.

The mesenchymal migration is not shared with other leukocytes (neutrophils, monocytes, T lymphocytes) which are unable to proteolyze the ECM (Cougoule et al. 2012; Friedl and Weigelin 2008; Rowe and Weiss 2009) and are

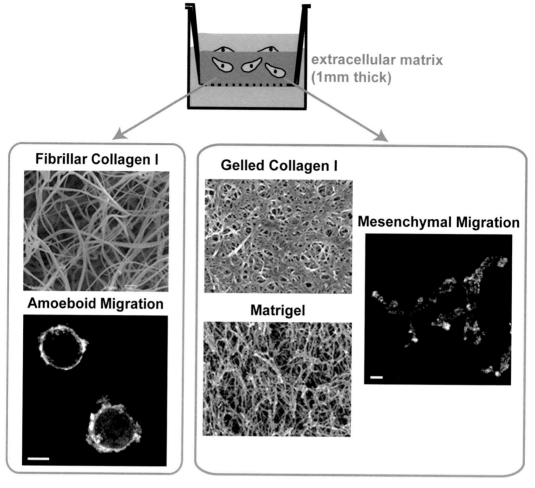

Mechanisms of Macrophage Migration in 3-Dimensional Environments, Fig. 1 Amoeboid and mesenchymal 3D migration of macrophages is dictated by the matrix architecture. The indicated matrices were polymerized as thick layers (1 mm) in transwells. The lower chamber is filled with a culture medium containing a chemoattractant. Scanning electron microscopy images show the matrix architectures. Macrophages are seeded on top of the matrices. Confocal microscopy images show that macrophages exhibit a rounded cell shape in fibrillar collagen I, characteristic of the amoeboid migration mode, while in gelled collagen I or Matrigel, macrophages have an elongated cell shape characteristic of the mesenchymal migration mode. Scale bar = 10 μm

thus unable to infiltrate dense matrices in vitro (Cougoule et al. 2012). A recent report also shows that, in vivo and ex vivo, T lymphocytes are unable to infiltrate dense collagen areas surrounding human tumors (Salmon et al. 2012). This observation supports the idea that these cells cannot use the mesenchymal mode. However, further studies are clearly required to examine whether leukocytes, including macrophages, can use the mesenchymal mode in vivo.

Since macrophages can use two distinct modes of migration, a possible hypothesis is that two subpopulations of macrophages might coexist, one performing the amoeboid migration and the other one the mesenchymal migration. However, it appears that a single macrophage can switch from one mode to another when it is progressing from a matrix with porous architecture to a dense matrix and vice versa (IMP, manuscript in preparation). Thus, a single macrophage is able to use

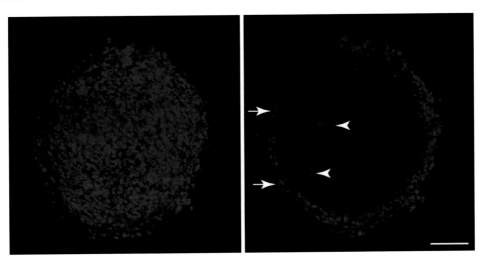

Mechanisms of Macrophage Migration in 3-Dimensional Environments, Fig. 2 Human macrophages infiltrate a tumor cell spheroid. Human macrophages stained with CellTracker (*red*) are co-cultured for 3 days with spheroids of human breast carcinoma SUM159PT cells. Cells are fixed and stained with DAPI and examined by two-photon microscopy. *Left panel*: a two-photon Z-stack projection (1.2 μm step) of a spheroid infiltrated by human macrophages (*red*). *Right panel*: a section illustrates human macrophages outside the spheroid (*arrows*) and macrophages infiltrated into the spheroid (*arrowheads*). Scale bar = 100 μm

the two migration modes and has the ability to adapt its migration mode as a function of the environment.

Infiltration of macrophages into tumor cell spheroids has also been investigated. Tumor cells grown in these three-dimensional structures will provide cohesive spheres which contain several ECM proteins secreted by the tumor cells thus mimicking tissues (Fig. 2, Guiet et al. 2011). Macrophages co-incubated with spheroids of 0.5 mm diameter have been shown to infiltrate these tissue-like structures using both migration modes (Guiet et al. 2011).

Interestingly, when macrophages infiltrate Matrigel or gelled collagen I, they use lysosomal proteases and matrix metalloproteases (MMPs) are dispensable (Van Goethem et al. 2010). In contrast, when macrophages infiltrate tumor cell spheroids, MMPs are critical (Guiet et al. 2011). It is likely that macrophages have the possibility to commit receptors and/or distinct signaling pathways depending on the biochemical, biophysical, and architectural properties of the extracellular environments, which may lead to the release of different sets of protease. The involvement of particular proteases in certain tissues is a new area of research which may lead to the identification of pharmacological targets for the control of macrophage migration.

Podosomes Are Involved in the Mesenchymal but Not the Amoeboid Migration

Podosomes are F-actin-rich structures which are constitutively formed in only a few cell types of the body, all derived from monocytes: macrophages, immature dendritic cells, and osteoclasts (Linder et al. 2011). Podosomes have been well studied in cells cultured in two dimensions (2D). They are defined as contacts between the cell and the matrix with the unique property of degrading the matrix (Fig. 3). A recent proteomic analysis of podosomes identified more than 200 proteins belonging to these structures, some of these being shared with focal adhesions and invadopodia (Cervero et al. 2012). Podosomes are very dynamic structures, which perform fusion with other podosomes and fission, and with a life span which fluctuates between 2 and 10 min. Podosomes are adhesion structures which contain integrins and other matrix-binding proteins such as CD44. Several podosome-related structures have been shown to display

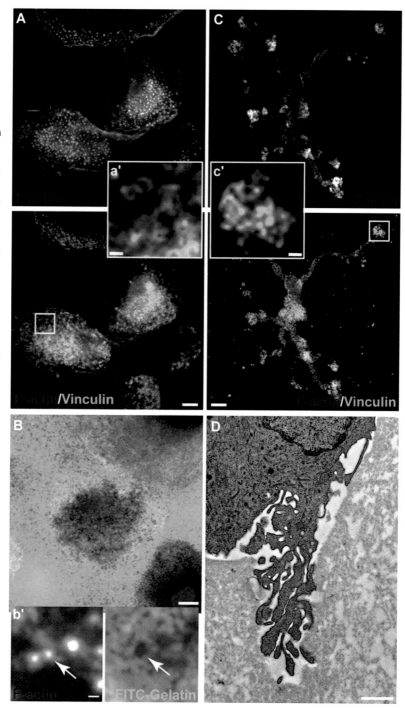

Mechanisms of Macrophage Migration in 3-Dimensional Environments, Fig. 3 Macrophages form proteolytic cell structures called podosomes. (**a**) Human macrophages seeded on fibrinogen-coated glass coverslips form individual podosomes characterized by a core of F-actin (*red*) surrounded by a ring of vinculin (*green*) (*inset* **a'**, scale bar = 1 μm). (**b**) When human macrophages are seeded on FITC-coupled gelatin, dark holes of gelatin degradation are visible underneath podosomes (*inset* **b'** *arrow*, scale bar = 1 μm). (**c**) Human macrophages infiltrating a dense matrix such as gelled collagen I form cell protrusions with F-actin (*red*) and podosome markers, here vinculin (*green*), enriched at the tip (*inset* b') scale bar = 10 μm. (**d**) Electron microscopy picture of a human macrophage in a Matrigel matrix showing matrix degradation as a hole surrounding the tip of a cell protrusion (scale bar = 2 μm)

mechanosensing properties (Collin et al. 2006, 2008) mediated by actomyosin contraction and growth of actin filaments and are subjected to periodic fluctuations of their stiffness (Labernadie et al. 2010). The height and mean stiffness of podosomes are not influenced by the nature of the matrix protein. Podosomes can organize as superstructures, called sealing zone in osteoclasts (Duplat et al. 2007; Jurdic et al. 2006; Teitelbaum and Ross 2003), and podosome rosettes (rings of podosomes) in activated macrophages (Cougoule et al. 2010; Poincloux et al. 2006), which both exert matrix proteolytic activity on large surfaces.

In 3D environments, when macrophages use the mesenchymal mode, they form cell protrusions which accumulate F-actin at their tip, as well as proteins described in podosomes such as paxillin, vinculin, talin, gelsolin, beta-integrins, etc. (Fig. 3, Van Goethem et al. 2010, 2011). In addition, a proteolytic activity visualized by DQ-collagen fluorescence signal was detected at these protrusions (Van Goethem et al. 2010). Since macrophages also ingest matrix in the endocytic compartment, it is difficult to conclude definitively whether the fluorescence signal is the result of matrix degradation outside or inside the cells. Electron microscopy examination of macrophages inside the matrix shows that extracellular matrix degradation occurs (Fig. 3), but again whether this takes place at cell protrusions or at other areas of the cell body is difficult to distinguish with this technique. Taken together, however, these observations led to these cell protrusions formed by macrophages using the mesenchymal mode being called 3D podosomes (Van Goethem et al. 2011). In contrast, macrophages using the amoeboid mode neither exhibited cell protrusions nor any particular co-accumulation of podosome constituents with F-actin (Van Goethem et al. 2010). Interestingly, macrophages harvested from the peritoneal cavity of healthy mice do not form podosomes, do not degrade the ECM, and do not use the mesenchymal migration mode (Cougoule et al. 2012). In contrast, peritoneal macrophages elicited with thioglycollate form podosomes with proteolytic activity towards the ECM and use the two migration modes (Cougoule et al. 2012). Thus, it appears that depending on the physiopathological conditions, the 3D migration ability of macrophages is controlled along their phenotypic changes.

Additional observations indicate that podosomes play a critical role in the mesenchymal migration mode: only leukocytes which form podosomes are able to use the mesenchymal mode (Cougoule et al. 2012), and deletion of proteins involved in the formation/stability of podosomes strongly affects the mesenchymal migration without impacting the amoeboid mode (see paragraph 5 and Cougoule et al. 2010; Guiet et al. 2012).

In conclusion, podosomes exert a proteolytic activity on the extracellular environment; they are critical for the mesenchymal migration, while they disappear when macrophages use the amoeboid mode.

Hck and Filamin A (FlnA) Are Involved Both in Podosome Formation/Stability and in Mesenchymal Migration

Hck (hematopoietic cell kinase) is a non-receptor tyrosine kinase belonging to the Src family with a phagocyte-restricted expression (Guiet et al. 2008). Hck is expressed as two isoforms, p59Hck and p61Hck, respectively, associated with the plasma membrane and the membrane of lysosomes (Carréno et al. 2000). Ectopic expression of constitutively active p59Hck in fibroblasts triggered the formation of membrane protrusions, while expression of active p61Hck induces the formation of podosome rosettes which degrade the ECM (Carreno et al. 2002; Cougoule et al. 2005). Co-expression of the two isoforms provides fibroblasts with the ability to migrate in a dense Matrigel matrix with a correlation between the percentage of the cells forming podosome rosettes and their migration ability (Cougoule et al. 2010). Hck activation has been shown to stimulate the exocytosis of lysosomes in phagocytes (Guiet et al. 2008). In macrophages, Hck and lysosomal markers are sporadically present at podosome cores (Guiet et al. 2008). By inducing lysosomal secretion with a calcium ionophore, the formation of podosome rosettes can be triggered in fibroblasts suggesting that the formation of podosome and podosome rosettes requires lysosome exocytosis.

In macrophages, Hck is present at podosome cores (Cougoule et al. 2005). When Hck is knocked down, the stability of podosomes is markedly reduced with a 2.5-fold decrease of the podosome life span (Guiet et al. 2012). In macrophages derived from Hck-/- mice, the mesenchymal migration is markedly inhibited, the formation and organization of podosomes as rosettes are reduced, and the ECM degradation is impaired, while the amoeboid migration and 2D migration are not affected (Cougoule et al. 2010; Guiet et al. 2012). During a sterile inflammation induced by an injection of thioglycollate, macrophages hardly infiltrate the peritoneal cavity in Hck-/- mice. Instead, they accumulate in interstitial tissues, suggesting that Hck deficiency leads to defective migration in these particular tissues or to the inability to cross the anatomic barrier between interstitial tissues and the peritoneal cavity (Cougoule et al. 2010). Thus, by controlling the stability and organization of podosomes as rosettes, which are structures necessary for the protease-dependent mesenchymal migration, Hck specifically controls that particular mode of migration without affecting the amoeboid mode. Interestingly, Hck, as a component of the IL4 receptor signaling pathway, has also been linked with macrophage M2 polarization (Bhattacharjee et al. 2011). In M2 macrophages, the formation of podosome rosettes and the mesenchymal migration of macrophages are enhanced compared to their nonpolarized counterpart (Cougoule et al. 2012).

FlnA is a cross-linker of actin filaments, which forms orthogonal branches with actin filaments. It also binds multiple partners including membrane receptors, enzymes, and signaling intermediates (Nakamura et al. 2011). In humans, null mutations in the FlnA gene result in a wide range of anomalies with strong consequences on embryonic development, which could be the result of a defect in cell adhesion and migration (Nakamura et al. 2011). In macrophages, FlnA localizes at the podosome ring and at podosome rosettes; it controls the life span of podosomes and their organization as rosettes and consequently the degradation capacity of macrophages towards the extracellular matrix. FlnA-deficient macrophages (FlnA-/-) present a defect in mesenchymal migration, but the amoeboid migration is normal. Finally, FlnA depletion inhibits Hck-induced podosome formation, suggesting that FlnA either takes part of Hck signaling pathway or is an essential component of podosomes (Guiet et al. 2012). In conclusion, Hck and FlnA are two regulators of the mesenchymal migration, thanks to their role in podosome organization and stability.

Pathological Relevance

Although tissue infiltration of macrophage is associated with essential host functions, it must be tightly controlled since excessive tissue macrophage infiltration and activation can lead to a score of undesirable effects such as host-tissue damage which predisposes surrounding tissue to neoplastic transformation, alteration of the glucose metabolism by promoting obesity and insulin resistance, promotion of tissue fibrosis, and progression of tumors (Osborn and Olefsky 2012; Qian and Pollard 2010; Sica and Mantovani 2012).

For example, the adipose tissue of obese mice and humans is infiltrated with large numbers of macrophages that can comprise up to 40 % of the cells. These adipose-tissue macrophages play a key role in obesity-associated tissue inflammation leading to insulin resistance and type 2 diabetes (Osborn and Olefsky 2012). Moreover, both primary lesions and secondary solid tumors are infiltrated by tumor-associated macrophages (TAMs), and their presence is associated with a bad prognosis as they stimulate tumor growth and formation of metastases (Qian and Pollard 2010; Allavena and Mantovani 2012; Wyckoff et al. 2007). In addition to the well-established paracrine loop of cytokines between tumor cells and macrophages which is critical to enhance tumor cell migration (Joyce and Pollard 2009; Mantovani and Sica 2010), the matrix remodeling activity of macrophages has also been involved in the process of tumor cell invasion (Qian and Pollard 2010). Invasive tumor cells, which migrate by themselves using a MMP-dependent mesenchymal mode, can follow macrophages which have opened paths in the matrix. When they migrate in the presence of macrophages, tumor cells switch to the amoeboid mode as their migration becomes resistant to protease inhibitors (Guiet et al. 2011).

Diversity and plasticity are hallmarks of macrophages. In response to cytokines, macrophages undergo M1 (classical) or M2 (alternative) activation, which represent extremes of a continuum of activation states. M1 or classically activated macrophages are stimulated by bacterial products and T-helper type 1 (Th1) cytokines such as interferon gamma; they produce inflammatory and immunostimulating cytokines to elicit the adaptive immune response, and they secrete reactive oxygen species and may have cytotoxic activity to transform cells. M2 or alternatively activated macrophages differentiate in microenvironments rich in Th2 cytokines such as IL-4 and IL-13; they have high scavenging activity, produce several growth factors that activate the process of tissue repair, and suppress adaptive immune responses (Gordon and Martinez 2010; Sica and Mantovani 2012). In tumors, tumor-associated-macrophages (TAMs) resemble M2-like macrophages, and their M2-related functions help disease progression (Schmieder et al. 2012). Interestingly, the migratory capacity of M1 and M2 macrophages is very different: while M1 macrophages are motionless, M2 macrophages are able to use both the mesenchymal and amoeboid modes (Cougoule et al. 2012). Thus, in the context of M2 TAMs, the mechanic remodeling of the matrix by migrating macrophages is likely to favor tumor cell invasion. Intravital microscopy has actually revealed that TAMs are located at strategic positions inside tumors where they help tumor cells to migrate. Macrophages localize to areas of tumor invasion and are often found close to vessels where cancer cell intravasation into the blood or lymphatic circulation occurs preferentially (Gocheva et al. 2010; Guiet et al. 2011; Wyckoff et al. 2004, 2007). Whether TAMs use the mesenchymal migration mode to help the invasive capacity of tumor cells remains to be established in vivo.

Conclusion and Perspective on the Mesenchymal Migration In Vivo

The migration modes used by macrophages (amoeboid or mesenchymal) have mostly been studied in vitro. The amoeboid mode has been described in vivo for few leukocytes (Lämmermann et al. 2008); it remains to determine whether the mesenchymal migration mode is also observed in vivo. For the moment, despite the development of new probes to label macrophages for intravital imaging approaches (Beckmann et al. 2009; Choe et al. 2010; Foley et al. 2009; Thurlings et al. 2009), the migration mode used by macrophages and the contribution of proteases for opening paths in dense environments have not been investigated. When monocytes/macrophages infiltrate tissues, the presence of MMPs is detected, but what precise role(s) these proteases play remains to be studied: are they involved in remodeling the matrix to allow macrophage infiltration or do they play a role in releasing macrophage chemoattractants from proteins of the matrix? As cell morphology is indicative of the migration mode (amoeboid or mesenchymal) used by the migrating cells, this point has been examined in the few published in vivo imaging experiments with a level of resolution compatible with the determination of cell morphology (Egen et al. 2008; Leimgruber et al. 2009) and noticed that macrophages mostly exhibit an elongated cell shape as observed in the mesenchymal movement.

Thus, the mesenchymal migration mode seems operative in vivo but further work, specifically addressing this point, will be required to definitively conclude on this aspect.

In vivo, both MMPs and cathepsins are involved in macrophage tissue infiltration, and their expression is upregulated when macrophages infiltrate tissues (Kessenbrock et al. 2010; Mason and Joyce 2011; Verollet et al. 2011). In several pathological models, MMPs have been involved in macrophage tissue recruitment either by favoring the process or, on the contrary, by inhibiting it (Corry et al. 2002; Deguchi et al. 2005; Hamada et al. 2008; Manicone et al. 2009; Mukherjee et al. 2005; Schneider et al. 2008; Shipley et al. 1996; Xiong et al. 2009). Similarly, in several studies mostly using cathepsin knockout mice, the role of these proteases has been correlated with macrophage tissue infiltration. In addition, macrophages located at the invasive tumor front adjacent to blood vessels are found to be the major source of active cathepsins (Gocheva et al. 2010; Vasiljeva et al. 2006).

Thus, taken together, these results suggest that macrophages could be using the mesenchymal mode in vivo. In particular, pathologies like cancer and chronic inflammation are characterized by increased density and stiffness of tissues (Egeblad et al. 2010). It is tempting to propose that to infiltrate these pathological tissues, the mesenchymal migration might be preferentially used by macrophages. If this hypothesis turns out to be experimentally verified, proteins that specifically regulate this particular migration mode will be potential pharmacological targets, and thus Hck and Filamin A could be good candidates. Actually, Hck presents the advantage to be specifically expressed in phagocytes; its inhibition would not affect the migration of the other leukocyte populations such as lymphocytes which often play beneficial roles in diseases.

Cross-References

▶ Leukocyte Recruitment

References

Allavena, P., & Mantovani, A. (2012). Immunology in the clinic review series; focus on cancer: Tumour-associated macrophages: Undisputed stars of the inflammatory tumour microenvironment. *Clinical and Experimental Immunology, 167*, 195–205.

Auffray, C., Fogg, D., Garfa, M., Elain, G., Join-Lambert, O., Kayal, S., et al. (2007). Monitoring of blood vessels and tissues by a population of monocytes with patrolling behavior. *Science (New York, N.Y.), 317*, 666–670.

Beckmann, N., Cannet, C., Babin, A., Blé, F.-X., Zurbruegg, S., Kneuer, R., et al. (2009). In vivo visualization of macrophage infiltration and activity in inflammation using magnetic resonance imaging. Wiley interdisciplinary reviews. *Nanomedicine and Nanobiotechnology, 1*, 272–298.

Bhattacharjee, A., Pal, S., Feldman, G. M., & Cathcart, M. K. (2011). Hck is a key regulator of gene expression in alternatively activated human monocytes. *The Journal of Biological Chemistry, 286*, 36709–36723.

Carréno, S., Gouze, M., Schaak, S., Emorine, L., & Maridonneau-Parini, I. (2000). Lack of palmitoylation redirects p59Hck from the plasma membrane to p61Hck-positive lysosomes. *The Journal of Biological Chemistry, 275*, 36223–36229.

Carreno, S., Caron, E., Cougoule, C., Emorine, L. J., & Maridonneau-Parini, I. (2002). p59Hck isoform induces F-actin reorganization to form protrusions of the plasma membrane in a Cdc42- and Rac-dependent manner. *The Journal of Biological Chemistry, 277*, 21007–21016.

Cervero, P., Himmel, M., Krüger, M., & Linder, S. (2012). Proteomic analysis of podosome fractions from macrophages reveals similarities to spreading initiation centres. *European Journal of Cell Biology, 91*, 908–922.

Choe, S.-W., Acharya, A., Keselowsky, B., & Sorg, B. (2010). Intravital microscopy imaging of macrophage localization to immunogenic particles and co-localized tissue oxygen saturation. *Acta Biomaterialia, 6*, 3491–3498.

Collin, O., Tracqui, P., Stephanou, A., Usson, Y., Clement-Lacroix, J., & Planus, E. (2006). Spatiotemporal dynamics of actin-rich adhesion microdomains: Influence of substrate flexibility. *Journal of Cell Science, 119*, 1914–1925.

Collin, O., Na, S., Chowdhury, F., Hong, M., Shin, M. E., Wang, F., et al. (2008). Self-organized podosomes are dynamic mechanosensors. *Current Biology, 18*, 1288–1294.

Corry, D., Rishi, K., Kanellis, J., Kiss, A., Song, L.-Z., Xu, J., et al. (2002). Decreased allergic lung inflammatory cell egression and increased susceptibility to asphyxiation in MMP2-deficiency. *Nature Immunology, 3*, 347–353.

Cougoule, C., Carreno, S., Castandet, J., Labrousse, A., Astarie-Dequeker, C., Poincloux, R., et al. (2005). Activation of the lysosomal-associated p61Hck isoform triggers the biogenesis of podosomes. *Traffic, 6*, 682–694.

Cougoule, C., Le Cabec, V., Poincloux, R., Al Saati, T., Mege, J. L., Tabouret, G., et al. (2010). Three-dimensional migration of macrophages requires Hck for podosome organization and extracellular matrix proteolysis. *Blood, 115*, 1444–1452.

Cougoule, C., Van Goethem, E., Le Cabec, V., Lafouresse, F., Dupré, L., Mehraj, V., et al. (2012). Blood leukocytes and macrophages of various phenotypes have distinct abilities to form podosomes and to migrate in 3D environments. *European Journal of Cell Biology, 91*, 938–949.

Deguchi, J.-O., Aikawa, E., Libby, P., Vachon, J., Inada, M., Krane, S., et al. (2005). Matrix metalloproteinase-13/collagenase-3 deletion promotes collagen accumulation and organization in mouse atherosclerotic plaques. *Circulation, 112*, 2708–2715.

Duplat, D., Chabadel, A., Gallet, M., Berland, S., Bedouet, L., Rousseau, M., et al. (2007). The in vitro osteoclastic degradation of nacre. *Biomaterials, 28*, 2155–2162.

Egeblad, M., Rasch, M. G., & Weaver, V. M. (2010). Dynamic interplay between the collagen scaffold and tumor evolution. *Current Opinion in Cell Biology, 22*, 697–706.

Egen, J., Rothfuchs, A., Feng, C., Winter, N., Sher, A., & Germain, R. (2008). Macrophage and T cell dynamics during the development and disintegration of mycobacterial granulomas. *Immunity, 28*, 271–284.

Foley, L., Hitchens, T., Ho, C., Janesko-Feldman, K., Melick, J., Bayir, H., et al. (2009). Magnetic resonance imaging assessment of macrophage accumulation in mouse brain after experimental traumatic brain injury. *Journal of Neurotrauma, 26*, 1509–1519.

Friedl, P., & Weigelin, B. (2008). Interstitial leukocyte migration and immune function. *Nature Immunology, 9*, 960–969.

Friedl, P., Entschladen, F., Conrad, C., Niggemann, B., & Zänker, K. (1998). CD4+ T lymphocytes migrating in three-dimensional collagen lattices lack focal adhesions and utilize beta1 integrin-independent strategies for polarization, interaction with collagen fibers and locomotion. *European Journal of Immunology, 28*, 2331–2343.

Gocheva, V., Wang, H.-W., Gadea, B., Shree, T., Hunter, K., Garfall, A., et al. (2010). IL-4 induces cathepsin protease activity in tumor-associated macrophages to promote cancer growth and invasion. *Genes & Development, 24*, 241–255.

Gordon, S., & Mantovani, A. (2011). Diversity and plasticity of mononuclear phagocytes. *European Journal of Immunology, 41*, 2470–2472.

Gordon, S., & Martinez, F. O. (2010). Alternative activation of macrophages: Mechanism and functions. *Immunity, 32*, 593–604.

Guiet, R., Poincloux, R., Castandet, J., Marois, L., Labrousse, A., Le Cabec, V., et al. (2008). Hematopoietic cell kinase (Hck) isoforms and phagocyte duties – from signaling and actin reorganization to migration and phagocytosis. *European Journal of Cell Biology, 87*, 527–542.

Guiet, R., Van Goethem, E., Cougoule, C., Balor, S., Valette, A., Al Saati, T., et al. (2011). The process of macrophage migration promotes matrix metalloproteinase-independent invasion by tumor cells. *Journal of Immunology, 187*, 3806–3814.

Guiet, R., Verollet, C., Lamsoul, I., Cougoule, C., Poincloux, R., Labrousse, A., et al. (2012). Macrophage mesenchymal migration requires podosome stabilization by filamin A. *The Journal of Biological Chemistry, 287*, 13051–13062.

Hamada, T., Fondevila, C., Busuttil, R., & Coito, A. (2008). Metalloproteinase-9 deficiency protects against hepatic ischemia/reperfusion injury. *Hepatology (Baltimore, Md.), 47*, 186–198.

Harunaga, J., & Yamada, K. (2011). Cell-matrix adhesions in 3D. *Matrix Biology, 30*, 363–368.

Jevnikar, Z., Mirković, B., Fonović, U., Zidar, N., Svajger, U., & Kos, J. (2012). Three-dimensional invasion of macrophages is mediated by cysteine cathepsins in protrusive podosomes. *European Journal of Immunology, 42*, 3429–3441.

Joyce, J., & Pollard, J. (2009). Microenvironmental regulation of metastasis. *Nature Reviews Cancer, 9*, 239–252.

Jurdic, P., Saltel, F., Chabadel, A., & Destaing, O. (2006). Podosome and sealing zone: Specificity of the osteoclast model. *European Journal of Cell Biology, 85*, 195–202.

Kessenbrock, K., Plaks, V., & Werb, Z. (2010). Matrix metalloproteinases: Regulators of the tumor microenvironment. *Cell, 141*, 52–67.

Labernadie, A., Thibault, C., Vieu, C., Maridonneau-Parini, I., & Charrière, G. (2010). Dynamics of podosome stiffness revealed by atomic force microscopy. *Proceedings of the National Academy of Sciences of the United States of America, 107*, 21016–21021.

Lämmermann, T., Bader, B., Monkley, S., Worbs, T., Wedlich-Söldner, R., Hirsch, K., et al. (2008). Rapid leukocyte migration by integrin-independent flowing and squeezing. *Nature, 453*, 51–55.

Leimgruber, A., Berger, C., Cortez-Retamozo, V., Etzrodt, M., Newton, A., Waterman, P., et al. (2009). Behavior of endogenous tumor-associated macrophages assessed in vivo using a functionalized nanoparticle. *Neoplasia (New York, N.Y.), 11*, 459.

Linder, S., Wiesner, C., & Himmel, M. (2011). Degrading devices: Invadosomes in proteolytic cell invasion. *Annual Review of Cell and Developmental Biology, 27*, 185–211.

Mackay, C. R. (2008). Moving targets: Cell migration inhibitors as new anti-inflammatory therapies. *Nature Immunology, 9*, 988–998.

Manicone, A., Birkland, T., Lin, M., Betsuyaku, T., van Rooijen, N., Lohi, J., Keski-Oja, J., et al. (2009). Epilysin (MMP-28) restrains early macrophage recruitment in Pseudomonas aeruginosa pneumonia. *Journal of Immunology (Baltimore, Md.: 1950), 182*, 3866–3876.

Mantovani, A., & Sica, A. (2010). Macrophages, innate immunity and cancer: Balance, tolerance, and diversity. *Current Opinion in Immunology, 22*, 231–237.

Mason, S., & Joyce, J. (2011). Proteolytic networks in cancer. *Trends in Cell Biology, 21*, 228–237.

McDonald, B., Pittman, K., Menezes, G., Hirota, S., Slaba, I., Waterhouse, C., et al. (2010). Intravascular danger signals guide neutrophils to sites of sterile inflammation. *Science (New York, N.Y.), 330*, 362–366.

Mierke, C. (2011). The biomechanical properties of 3d extracellular matrices and embedded cells regulate the invasiveness of cancer cells. *Cell Biochemistry and Biophysics, 61*, 217–236.

Mosser, D., & Edwards, J. (2008). Exploring the full spectrum of macrophage activation. *Nature Reviews Immunology, 8*, 958–969.

Mukherjee, R., Bruce, J., McClister, D., Allen, C., Sweterlitsch, S., & Saul, J. (2005). Time-dependent changes in myocardial structure following discrete injury in mice deficient of matrix metalloproteinase-3. *Journal of Molecular and Cellular Cardiology, 39*, 259–268.

Nakamura, F., Stossel, T., & Hartwig, J. (2011). The filamins: Organizers of cell structure and function. *Cell Adhesion & Migration, 5*, 160–169.

Osborn, O., & Olefsky, J. (2012). The cellular and signaling networks linking the immune system and metabolism in disease. *Nature Medicine, 18*, 363–374.

Pflicke, H., & Sixt, M. (2009). Preformed portals facilitate dendritic cell entry into afferent lymphatic vessels. *The Journal of Experimental Medicine, 206*, 2925–2935.

Poincloux, R., Vincent, C., Labrousse, A., Castandet, J., Rigo, M., Cougoule, C., et al. (2006). Re-arrangements of podosome structures are observed when Hck is activated in myeloid cells. *European Journal of Cell Biology, 85*, 327–332.

Qian, B. Z., & Pollard, J. W. (2010). Macrophage diversity enhances tumor progression and metastasis. *Cell, 141*, 39–51.

Qualls, J., & Murray, P. (2010). A double agent in cancer: Stopping macrophages wounds tumors. *Nature Medicine, 16*, 863–864.

Rowe, R. G., & Weiss, S. J. (2009). Navigating ECM barriers at the invasive front: The cancer cell-stroma interface. *Annual Review of Cell and Developmental Biology, 25*, 567–595.

Ruhrberg, C., & De Palma, M. (2010). A double agent in cancer: Deciphering macrophage roles in human tumors. *Nature Medicine, 16*, 861–862.

Sabeh, F., Shimizu-Hirota, R., & Weiss, S. J. (2009). Protease-dependent versus -independent cancer cell invasion programs: Three-dimensional amoeboid movement revisited. *The Journal of Cell Biology, 185*, 11–19.

Sahai, E., & Marshall, C. (2003). Differing modes of tumour cell invasion have distinct requirements for Rho/ROCK signalling and extracellular proteolysis. *Nature Cell Biology, 5*, 711–719.

Salmon, H., Franciszkiewicz, K., Damotte, D., Dieu-Nosjean, M. C., Validire, P., Trautmann, A., et al. (2012). Matrix architecture defines the preferential localization and migration of T cells into the stroma of human lung tumors. *Journal of Clinical Investigation, 122*, 899–910.

Schmieder, A., Michel, J., Schönhaar, K., Goerdt, S., & Schledzewski, K. (2012). Differentiation and gene expression profile of tumor-associated macrophages. *Seminars in Cancer Biology, 22*, 289–297.

Schneider, F., Sukhova, G., Aikawa, M., Canner, J., Gerdes, N., Tang, S.-M. T., et al. (2008). Matrix-metalloproteinase-14 deficiency in bone-marrow-derived cells promotes collagen accumulation in mouse atherosclerotic plaques. *Circulation, 117*, 931–939.

Shipley, J., Wesselschmidt, R., Kobayashi, D., Ley, T., & Shapiro, S. (1996). Metalloelastase is required for macrophage-mediated proteolysis and matrix invasion in mice. *Proceedings of the National Academy of Sciences of the United States of America, 93*, 3942–3946.

Sica, A., & Mantovani, A. (2012). Macrophage plasticity and polarization: In vivo veritas. *Journal of Clinical Investigation, 122*, 787–795.

Steadman, R., St John, P. L., Evans, R. A., Thomas, G. J., Davies, M., Heck, L. W., et al. (1997). Human neutrophils do not degrade major basement membrane components during chemotactic migration. *International Journal of Biochemistry & Cell Biology, 29*, 993–1004.

Teitelbaum, S. L., & Ross, F. P. (2003). Genetic regulation of osteoclast development and function. *Nature Reviews Genetics, 4*, 638–649.

Thurlings, R., Wijbrandts, C., Bennink, R., Dohmen, S., Voermans, C., Wouters, D., et al. (2009). Monocyte scintigraphy in rheumatoid arthritis: The dynamics of monocyte migration in immune-mediated inflammatory disease. *PLoS ONE, 4*, e7865.

Van Goethem, E., Poincloux, R., Gauffre, F., Maridonneau-Parini, I., & Le Cabec, V. (2010). Matrix architecture dictates three-dimensional migration modes of human macrophages: Differential involvement of proteases and podosome-like structures. *Journal of Immunology, 184*, 1049–1061.

Van Goethem, E., Guiet, R., Balor, S., Charriere, G. M., Poincloux, R., Labrousse, A., et al. (2011). Macrophage podosomes go 3D. *European Journal of Cell Biology, 90*, 224–236.

Vasiljeva, O., Papazoglou, A., Krüger, A., Brodoefel, H., Korovin, M., Deussing, J., et al. (2006). Tumor cell-derived and macrophage-derived cathepsin B promotes progression and lung metastasis of mammary cancer. *Cancer Research, 66*, 5242–5250.

Verollet, C., Charriere, G. M., Labrousse, A., Cougoule, C., Le Cabec, V., & Maridonneau-Parini, I. (2011). Extracellular proteolysis in macrophage migration: Losing grip for a breakthrough. *European Journal of Immunology, 41*, 2805–2813.

Wolf, K., Mazo, I., Leung, H., Engelke, K., von Andrian, U. H., Deryugina, E. I., et al. (2003). Compensation mechanism in tumor cell migration: Mesenchymal-amoeboid transition after blocking of pericellular proteolysis. *The Journal of Cell Biology, 160*, 267–277.

Woodfin, A., Voisin, M.-B., Beyrau, M., Colom, B., Caille, D., Diapouli, F.-M., et al. (2011). The junctional adhesion molecule JAM-C regulates polarized transendothelial migration of neutrophils in vivo. *Nature Immunology, 12*, 761–769.

Wyckoff, J., Wang, W., Lin, E. Y., Wang, Y., Pixley, F., Stanley, E. R., et al. (2004). A paracrine loop between tumor cells and macrophages is required for tumor cell migration in mammary tumors. *Cancer Research, 64*, 7022–7029.

Wyckoff, J., Wang, Y., Lin, E., Li, J.-F., Goswami, S., Stanley, E., et al. (2007). Direct visualization of macrophage-assisted tumor cell intravasation in mammary tumors. *Cancer Research, 67*, 2649–2656.

Xiong, W., Knispel, R., MacTaggart, J., Greiner, T., Weiss, S., & Baxter, B. (2009). Membrane-type 1 matrix metalloproteinase regulates macrophage-dependent elastolytic activity and aneurysm formation in vivo. *The Journal of Biological Chemistry, 284*, 1765–1771.

Medicinal Fatty Acids

Lisa Stamp[1] and Leslie Cleland[2]
[1]University of Otago, Christchurch, New Zealand
[2]Rheumatology Unit, Royal Adelaide Hospital, Adelaide, SA, Australia

Synonyms

Eicosanoids; Inflammation; Omega-3 fatty acids

Definition

Nutrition plays an important role in the management of many chronic diseases. Dietary advice to patients with diabetes, heart disease, gout, celiac disease, and obesity is routinely given as part of clinical care. However, the role of diet in inflammatory diseases and in particular inflammatory rheumatic diseases such as rheumatoid arthritis is less well recognized. The omega-three fatty acids can be considered medicinal fatty acids due to their beneficial effects on the inflammatory response and thus some inflammatory diseases. This review will discuss the biochemistry of omega-three fatty acids, the effects of omega-three fatty acids on the inflammatory process, and their effects in a prototypic inflammatory disease, namely, rheumatoid arthritis.

Omega-3 Fatty Acid Biochemistry

There are three groups of fatty acids based on the number of double bonds they contain: (i) saturated fatty acids (no double bond), (ii) monounsaturated fatty acids (one double bond), and (iii) polyunsaturated fatty acids (PUFA) (≥ 2 double bonds). PUFA are further subdivided according to the site of the first double bond from the methyl (omega) terminus, with omega-6 (n-6) and omega-3 (n-3) being the principal subgroups. Vertebrates do not have the enzymes required to introduce double bonds in the n-3 and n-6 positions, and these fatty acids are therefore known as essential fatty acids as they must be obtained from the diet.

The Western diet characteristically contains more n-6 fats than n-3 fats due to the dominance in processed foods and visible fats of soybean, safflower, sunflower, and corn oils, which contain the n-6 fat linoleic acid (LA; 18:2n-6). The n-3 homologue of LA, α-linolenic acid (ALA; 18:3n-3), is present in flaxseed oil which is generally a minor dietary component. LA and ALA may be used in energy metabolism or be converted to the C20 fatty acids arachidonic acid (AA; 20:4n-6) and eicosapentaenoic acid (EPA; 20:5n-3) or docosahexaenoic acid (DHA; 22:6n-3), respectively (Table 1). EPA and DHA can also be obtained directly through consumption of fish and fish oils. AA and EPA are incorporated into cell membranes and tissues where they can be metabolized to eicosanoids, which are oxy-lipid autocrine and paracrine messengers derived from C20 PUFA.

Omega-Three Fatty Acids and the Inflammatory Process

The metabolism of AA and EPA to eicosanoids and resolvins provides the link between fatty acids and inflammation. The eicosanoids, which include prostaglandins, thromboxanes, and leukotrienes, and the resolvins (E-series and D-series), protectins, and maresins act as inflammatory mediators.

Eicosanoids

AA is metabolized via cyclooxygenase (COX) to the n-6 eicosanoids, prostaglandin (PG) E_2, and thromboxane (TX) A_2 or via 5-lipoxygenase (5-LOX) to n-6 leukotrienes (LTs). In comparison, EPA is metabolized via COX and 5-LOX to n-3 prostaglandins and n-3 leukotrienes, respectively. While the n-6 eicosanoids are readily produced from AA, EPA is a poor COX substrate such that n-3 PGs are not as readily produced (Table 1). In addition, EPA competitively inhibits production of most n-6 eicosanoids, except prostacyclin (PGI_2).

Medicinal Fatty Acids, Table 1 Metabolism of arachidonic acid (*AA*) and eicosapentaenoic acid (*EPA*)

Fatty acid family	n-6	n-3	
18 carbon fatty acid	Linoleic acid (LA; 18:2n-6)	α-Linolenic acid (ALA; 18:3n-3)	
Dietary sources	Sunflower, corn, and safflower oil	Flaxseed, canola, and rapeseed oil	
Dietary intake	Large intake (7-8% dietary energy)[a]	Minor intake (0.3-1.0% dietary energy)[a]	
Metabolism	↓	↓	
20 carbon fatty acids	Arachidonic acid (AA; 20:4n-6)	Eicosapentaenoic acid (EPA; 20:5n-3)	Docosahexaenoic acid (DHA; 22:6n-3)
Sources	Mainly synthesized from ingested linoleic acid	Mainly from ingested EPA (fish, fish oil)	
Metabolism	COX and LOX ↓	COX and LOX ↓	
Metabolites of C20 fatty acid relevant to inflammation	Proinflammatory n-6 prostaglandins and leukotrienes (TXA$_2$, PGE$_2$, LTB$_4$)	Less inflammatory n-3 prostaglandins and leukotrienes (TXA$_3$, PGE$_3$, LTB$_5$)	D and E-Resolvins anti-inflammatory, protectins and maresins

In general, the n-6 eicosanoids are pro-inflammatory and the n-3 eicosanoids are anti-inflammatory. For example, TXA$_2$ promotes interleukin (IL)-1β and tumor necrosis factor (TNF)-α production by mononuclear cells (Caughey et al. 1997), and PGE$_2$ results in vasodilatation, increased vascular permeability, and hyperalgesia. In comparison, the n-3 eicosanoids are either less potent in their effects or less abundant. For example, PGE$_3$ is edemogenic but little is produced, and LTB$_5$ is 10–30 times less potent than LTB$_4$ as a neutrophil chemotaxin.

Dietary n-3 and n-6 fatty acid consumption alters the availability of AA or EPA for incorporation into the cell membranes and thus the balance of n-3/n-6 eicosanoid production. For example, increased ingestion of EPA and DHA or, to a lesser extent, increased dietary consumption of ALA, which increases endogenous EPA production, increases incorporation of EPA into cell membranes and tissues partly at the expense of AA incorporation. The net result is increased production of n-3 eicosanoids relative to n-6 eicosanoids (Table 1).

Resolvins, Protectins, and Maresins

The resolvins (resolution phase interaction products), protectins, and maresins are relatively recently recognized lipid mediators derived from EPA and DHA. Resolvins derived from EPA are known as E-resolvins, while those derived from DHA are known as D-resolvins (Table 1). They have wide ranging actions on immune effector cells (e.g., neutrophils and macrophages), with effects including inhibition of TNF-induced transcription of IL-1β and inhibition of human polymorphonuclear leukocyte transendothelial migration (for review, see Serhan and Petasis 2011). The inflammation-resolving actions of these mediators are likely contributors to the observed benefits of n-3 fatty acids in inflammatory diseases.

Effect of n-3 Fatty Acids on Pro-inflammatory Cytokine Production

In vitro cell studies, as well as dietary intervention studies in animals and humans, have shown that IL-1β, IL-6, and TNF-α production may be reduced as a consequence of dietary n-3 fatty acid supplementation. This appears to be mediated, at least in part, by alterations in the ratio of n-3/n-6 eicosanoids and effects of resolvins and via direct effects of n-3 fatty acids on intracellular signaling mechanisms including NF-κB and PPAR-γ (Jump and Clarke 1999).

Effects of n-3 Fatty Acids on T Cells

Studies of the effects of n-3 fatty acids on T cells in humans are inconsistent. Some studies have shown that n-3 fatty acids inhibit T-cell proliferation and IL-2 production while other studies have not confirmed these effects.

Effects of n-3 Fatty Acids on MHC Expression and Antigen Presentation

The number of MHC molecules expressed on antigen-presenting cells (APCs) is an important determinant of T-cell responses to antigen. In vitro studies show that monocyte exposed to EPA and/or DHA can lead to reduced expression of HLA-DR and HLA-DP molecules and reduced antigen presentation (Hughes and Pinder 2000). Thus, n-3 fatty acids may have an anti-inflammatory effect via suppression of pathogenic T-cell activation.

Effect of n-3 Fatty Acids on Adhesion Molecule Expression

Engagement of adhesion molecules on leukocytes with their cognate receptors on endothelial cells is an important precursor to the migration of leukocytes from the circulation into tissues. ICAM-1 and its cognate receptor, leukocyte function-associated antigen (LFA)-1, have been shown to be important for migration of leukocytes into inflamed synovium in animal models (Liao and Haynes 1995). n-3 fatty acids have been shown to decrease human monocyte ICAM-1 and LFA-1 expression in in vitro studies (Hughes and Pinder 2000). Dietary n-3 fatty acid supplementation also reduces soluble plasma ICAM-1 and VCAM-1 concentrations (Lopez-Garcia et al. 2004).

In summary, n-3 fatty acids have wide ranging effects on inflammation (Table 2). Given these effects, it is not surprising that consumption of n-3 fatty acids can have beneficial effects in patients with inflammatory diseases such as rheumatoid arthritis.

Dietary Sources of n-3 Fatty Acids

Fish oil is a rich source of the anti-inflammatory long-chain n-3 fatty acids EPA and DHA. For non-fish eaters, plant sources of n-3 fatty acids in the form of ALA include flaxseed and flaxseed oil, walnuts and walnut oil, and canola oil. Studies suggest that incorporation of dietary fatty acids into cell membranes is dose dependent and begins within days and peaks at 1–2 weeks after

Medicinal Fatty Acids, Table 2 Effects of omega-three fatty acids on the inflammatory process

Decreased production of n-6-derived eicosanoids (PGE$_2$, TXA$_2$, LTB$_4$) which have pro-inflammatory effects
Increased production of n-3-derived eicosanoids (PGE$_3$, TXA$_3$, LTB$_5$) which in general are less pro-inflammatory
Decreased IL-1β and TNF-α production
Increased production of resolvins
Decreased MHC II expression by antigen-presenting cells
Decreased adhesion molecule expression – ICAM, VCAM, LFA
Decreased expression of MMPs

increasing dietary intake (Miles and Calder 2012). In general, the dose of EPA and DHA required to obtain an anti-inflammatory effect is 2.7 g or more per day. Tissue levels of EPA and DHA from fish and fish oils are increased when dietary n-6 fatty acid intake is reduced concomitantly through substitution of n-6-rich visible fats (e.g., with a base of corn oil, soy oil, sunflower oil) with unsaturated products with less n-6 fat (with a base of olive oil, rapeseed/canola oil, or flaxseed oil) (Adam et al. 2003). While flaxseed oil is rich in ALA, conversion of this C18 n-3 fatty acid to the long-chain (C20 and C22) n-3 fatty acids EPA and DHA is relatively inefficient, and supplementation with ALA is a suboptimal substitute for dietary supplementation with EPA- and DHA-rich fish oils. Flaxseed oil supplementation can be useful for strict vegetarians and those with a true allergy to scale fish (as distinct from the more common allergy to crustaceans, which do not contribute to fish oils).

Omega-3 Fatty Acids and Rheumatoid Arthritis

The effects of medicinal doses of fish oils rich in EPA and DHA in rheumatoid arthritis (RA) are well established. Anti-inflammatory doses of fish oil have been shown to reduce symptoms and to improve disease control in RA and to reduce cardiovascular risk factors. The latter effects are important because RA is associated with increased mortality from cardiovascular disease. Despite these beneficial effects, dietary n-3 fatty acid supplementation is not always recommended to patients with RA as a component of their management.

Omega-3 Fatty Acids Reduce the Risk of Developing RA

The Seattle Women's Health Study reported a reduced risk of developing RA in those who consumed two or more fish meals per week compared to subjects consuming less than one fish meal per week (adjusted OR = 0.57 (95 % CI 0.35–0.93)) (Shapiro et al. 1996). Another population-based case-controlled study reported a modest decrease in the risk of RA with consumption of oily fish one to seven times per week compared to rare or no fish consumption (OR 0.8; 95 % CI 0.6–1.0) (Rossell et al. 2009). This increased risk persisted after allowing for the presence of rheumatoid factor and anti-cyclic citrullinated peptide antibodies.

Omega-3 Fatty Acids Reduce Inflammatory Disease Activity in RA and NSAID Consumption

The goal of treatment in RA is suppression of the inflammatory process in order to preserve joint structure and function. Disease-modifying anti-rheumatic drugs (DMARDs) such as methotrexate, alone or in combination, are the mainstay of therapy, and treatment is directed toward achieving disease remission.

When combined with conventional DMARDs, anti-inflammatory doses of fish oil can result in improved disease control. In a longitudinal cohort study of patients with RA of less than 12 months' duration, response-driven, intensive combination DMARD therapy was combined with either supplemental fish oil or placebo. At 3 years, those patients compliant with fish oil therapy had improved self-reported function in activities of daily living, lower tender joint counts, lower ESR, higher remission rates (72 % vs. 31 %), and less NSAID use than those who did not consume fish oil (Cleland et al. 2006). The better outcomes for patients taking fish oil are consistent with other randomized controlled trial data showing reduced symptoms with fish oil compared to placebo treatment (Table 3).

Nonsteroidal anti-inflammatory drugs (NSAIDs) are commonly used in patients with joint pain associated with RA. However, despite their pain-relieving properties, NSAIDs have not been shown to mitigate disease progression. Furthermore, NSAIDs are associated with substantially increased risk for a number of serious adverse events, notably gastrointestinal ulceration and bleeding and thrombotic cardiovascular events (Table 4). Anti-inflammatory doses of fish oil have been shown to reduce NSAID requirements in patients with RA (Galarraga et al. 2008; Kjeldsen-Kragh et al. 1992). A recent meta-analysis of ten randomized controlled trials of

Medicinal Fatty Acids, Table 3 Medical conditions and evidence for benefit of omega-3 fatty acid supplementation

Condition	Evidence for n-3 fatty acid supplementation
Rheumatoid arthritis	Evidence for improved inflammatory disease activity and reduction in NSAID requirement with EPA plus DHA >2.7 g/day
Cardiovascular disorders	Reduction in cardiovascular events and cardiac death, small but significant reduction in blood pressure
Dementia and cognitive decline in elderly	No evidence for benefit
Cancers	Experimental models show modulation of carcinogenesis by n-3 fatty acids
	Limited evidence suggests EPA may be beneficial as adjuvant therapy in non-small-cell lung cancer
Preterm birth	Improved neural development
Inflammatory bowel disease	EPA plus DHA reduces relapse rates in patients at high risk
	Insufficient data to make any recommendation about use in active disease
Osteoporosis	Conflicting evidence in small randomized controlled trials
Type II diabetes	ALA may be associated with a lower risk of diabetes
	EPA + DHA no beneficial effect or harm observed

Medicinal Fatty Acids, Table 4 Comparison between NSAIDs and anti-inflammatory doses of fish oil

	NSAIDs	Fish oil
COX inhibition	COX-1/COX-2 selectivity varies depending between agents	Nonselective
NSAID sparing	No	Yes
Serious cardiovascular events	Increased (except naproxen)	Reduced
Blood pressure	Increased	Reduced
Cardiac failure	Increased	Reduced
Renal function	Compromised	Progression to failure reduced
TNF-α and IL-1β	Increased	Reduced
Upper GI bleeding	Increased	Not reported
Mortality	Increased	Reduced (especially sudden cardiac death)
Time to effect	Prompt	Delayed (up to 3 months)

supplementation with >2.7 g/day of omega-3 fatty acids for >3 months in patients with RA reported a significant reduction in NSAID consumption (Lee et al. 2012). Thus, fish oil can deliver health benefits directly and indirectly through reduction in discretionary NSAID use (Table 4).

Role of Omega-3 Supplementation in Rheumatic Diseases with Increased CVD Risk

The association between increased cardiovascular mortality and inflammatory rheumatic diseases including RA, SLE, gout, and psoriatic arthritis is well recognized. In the general population, n-3 fatty acid supplementation has been shown to be beneficial in the primary and secondary prevention of ischemic heart disease (Lavie et al. 2009). There are a number of potential mechanisms for this observation including stabilization of the myocardium leading to reduction in cardiac arrhythmias, reduced blood pressure, stabilization of atheromatous plaques, reduced triglycerides and increased HDL, decreased platelet thromboxane release, increased vascular prostacyclin release, and anti-inflammatory effects.

There are no specific studies of the cardiovascular benefits of n-3 fatty acid supplementation in patients with RA or other inflammatory rheumatic diseases. However, a study of n-3 fatty acid supplementation in patients with early RA has shown a reduction in triglycerides, increased "good" HDL cholesterol, less NSAID use, greater disease suppression, and reduced platelet synthesis of TXA_2, all of which have the potential to reduce cardiovascular risk (Cleland et al. 2006).

Omega-3 Fatty Acid Supplementation in Other Medical Conditions

There is growing interest in the role of omega-3 fatty acids in other medical conditions. A detailed review of the evidence in all conditions is beyond the scope of this paper. However, a brief outline is presented below and in Table 4.

A recent Cochrane review of the evidence of n-3 fatty acids in preventing cognitive decline and dementia in patients >60 years of age concluded that, while well tolerated, there was no benefit (Sydenham et al. 2012). While some epidemiological studies suggest that omega-3 fatty acids may be beneficial in osteoporosis, the evidence from small randomized controlled trials is less convincing. These studies are limited by small sample size, failure to use fracture as an outcome measure, and concomitant administration of calcium. Given the anti-inflammatory effects of n-3 fatty acids, one would anticipate that they would be beneficial in inflammatory bowel disease (IBD, which includes ulcerative colitis and Crohn's disease). Benefit has been shown with regard to maintenance of remission in Crohn's disease (Turner et al. 2009). With regard to ulcerative colitis and effects on active IBD, data are inconclusive, and the use of fish oil is not generally recommended. Omega-3 fatty acids have been shown to be effective in the prevention of cardiovascular events and cardiac death (Delgado-Lista et al. 2012) as well as in leading to a small but significant reduction in blood pressure (Cabo et al. 2012). While experimental models suggest n-3 fatty acids can modulate carcinogenesis, there is little evidence to support their use in patients with cancer except in non-small-cell lung cancer where ALA may be beneficial as an adjuvant to standard chemotherapy (Gerber 2012).

Practical Considerations

There are a number of practical considerations that should be explained to patients and physicians when recommending medicinal doses of fish oil as means of supplementing the diet with long-chain n-3 fatty acids.

Dosing

Firstly, the dose of EPA plus DHA of >2.7 g/day is generally more than that with which patients will self-prescribe. In general, this dose means ten or more standard fish oil capsules daily. Patient preferences vary, but bottled fish oil layered on juice is the most efficient way of taking an anti-inflammatory dose. Fish oil is best tolerated with food and not on an empty stomach. The term "fish oil" defines oil prepared from fish bodies which is different from cod liver oil, which is prepared from fish livers and is rich in the fat-soluble vitamins A and D. Standard fish oil contains more EPA+DHA (30 % w/w) than cod liver oil (EPA + DHA ~20 %) and is the preferred material.

Latency of Beneficial Effects

Secondly, like most standard DMARDs, there is a latent period of up to 15 weeks before the symptomatic benefits of anti-inflammatory doses of fish oil are experienced. Intravenous administration of n-3 fats can reduce this latent period but is costly, inconvenient, and not generally available.

Unwanted Effects

Like any medication, there are potential adverse effects associated with use of n-3 fatty acid supplements. The most common adverse effects associated with fish oil are a fishy aftertaste, gastrointestinal upset, and nausea, which can be dose limiting. No serious toxicity has been associated with anti-inflammatory doses of n-3 fatty acids. Although prolonged bleeding time and hence tendency toward bleeding have been observed in Greenland Eskimos who consume very high amounts of EPA and DHA (~twice the anti-inflammatory dose), an increased bleeding tendency has not emerged as a problem with long-term use of anti-inflammatory doses of fish oil.

Avoidance of Contaminants

Environmental contaminants including methylmercury, polychlorinated biphenyls (PCBs), and dioxins, which are concentrated in large carnivorous fish, are excluded from fish oil prepared for therapeutic use. The FDA has accorded "generally regarded as safe" status to intakes of up to 3 g/day of long-chain n-3 fatty acids (EPA plus DHA) from marine sources.

Summary

Omega-3 fatty acids are well recognized as medicinal by virtue of their anti-inflammatory properties. There is convincing evidence of the benefit in patients with rheumatoid arthritis and cardiovascular disease. There is less convincing evidence of their benefit in some other conditions, but further larger randomized controlled trials are required.

Cross-References

▶ Inflammatory Bowel Disease
▶ Non-steroidal Anti-inflammatory Drugs: Overview

References

Adam, O., Beringer, C., Kless, T., Lemmen, C., Adam, A., Wiseman, M., et al. (2003). Anti-inflammatory effects of a low arachidonic acid diet and fish oil in patients with rheumatoid arthritis. *Rheumatology International, 23,* 27–36.

Cabo, J., Alonso, R., & Mata, P. (2012). Omega-3 fatty acids and blood pressure. *The British Journal of Nutrition, 107*(Suppl2), S195–S200.

Caughey, G. E., Pouliot, M., Cleland, L. G., & James, M. J. (1997). Regulation of tumor necrosis factor-a and IL-1b synthesis by thromboxane A_2 in nonadherent human monocytes. *The Journal of Immunology, 158,* 351–358.

Cleland, L., Caughey, G., James, M., & Proudman, S. (2006). Reduction of cardiovascular risk factors with longterm fish oil treatment in early rheumatoid arthritis. *The Journal of Rheumatology, 33*(10), 1973–1979.

Delgado-Lista, J., Perez-Martinez, P., Lopez-Miranda, J., & Perez-Jimenez, F. (2012). Long chain omega-3 fatty acids and cardiovascular disease: A systematic review. *The British Journal of Nutrition, 107*(Suppl 2), S201–S213.

Galarraga, B., Ho, M., Youssef, H., Hill, A., McMahon, H., Hall, C., et al. (2008). Cod liver oil (*n*-3 fatty acids) as a non-steroidal anti-inflammatory drug sparing agent in rheumatoid arthritis. *Rheumatology, 47,* 665–669.

Gerber, M. (2012). Omega-3 fatty acids and cancers: A systematic update review of epidemiological studies. *The British Journal of Nutrition, 107*(Suppl 2), S228–S239.

Hughes, D. A., & Pinder, A. C. (2000). n-3 Polyunsaturated fatty acids inhibit the antigen-presenting function of human monocytes. *The American Journal of Clinical Nutrition, 71*(Suppl), 357S–360S.

Jump, D., & Clarke, S. (1999). Regulation of gene expression by dietary fat. *Annual Review of Nutrition, 19,* 63–90.

Kjeldsen-Kragh, J., Lund, J. A., Riise, T., Finnanger, B., Haaland, K., Finstad, R., et al. (1992). Dietary omega-3 fatty acid supplementation and naproxen treatment in patients with rheumatoid arthritis. *The Journal of Rheumatology, 19*(10), 1531–1536.

Lavie, C., Milani, R., Mehra, M., & Ventura, H. (2009). Omega-3 polyunstaturated fatty acids and cardiovascular disease. *Journal of the American College of Cardiology, 54*(7), 585–594.

Lee, Y.-H., Bae, S.-C., & Song, G.-G. (2012). Omega-3 polyunsaturated fatty acids and the treatment of rheumatoid arthritis: A meta-analysis. *Archives of Medical Research, 43,* 356–362.

Liao, H.-X., & Haynes, B. F. (1995). Role of adhesion molecules in the pathogenesis of rheumatoid arthritis. *Rheumatic Disease Clinics of North America, 21*(3), 715–740.

Lopez-Garcia, E., Schulze, M., Manson, J., Meigs, J., Albert, C., Rifai, N., et al. (2004). Consumption of (n-3) fatty acids is related to plasma biomarkers of inflammation and endothelial activation in women. *The Journal of Nutrition, 134,* 1806–1811.

Miles, E., & Calder, P. C. (2012). Influence of marine n-3 polyunsaturated fatty acids on immune function and a systematic review of their effects on clinical outcomes in rheumatoid arthritis. *The British Journal of Nutrition, 107*(S2), S171–S184.

Rossell, M., Wesley, A., Rydin, K., Klareskog, L., Alfredsson, L., & group, a. t. E. s. (2009). Dietary fish and fish oil and the risk of rheumatoid arthritis. *Epidemiology, 20*(6), 896–901.

Serhan, C., & Petasis, N. (2011). Resolvins and protectins in inflammation-resolution. *Chemical Reviews, 111*(10), 5922–5943.

Shapiro, J. A., Koepsell, T. D., Voigt, L. F., Dugowson, C. E., Kestin, M., & Nelson, J. L. (1996). Diet and rheumatoid arthritis in women: A possible protective effect of fish consumption. *Epidemiology, 7*(3), 256–263.

Sydenham, E., Dangour, A., & Lim, W. (2012). Omega 3 fatty acid for the prevention of cognitive decline and dementia. *Cochrane Database of Systematic Reviews, 13*(6), CD005379.

Turner, D., Zlotkin, S., Shah, P., & Griffiths, A. (2009). Omega 3 fatty acids (fish oil) for maintenance of remission in Crohn's disease. *Cochrane Database of Systematic Reviews, 1,* CD006320.

Methotrexate

Kevin D. Pile[1] and Garry G. Graham[2,3]
[1]Campbelltown Hospital, School of Medicine, University of Western Sydney, Campbelltown, NSW, Australia
[2]Department of Pharmacology, School of Medical Sciences, University of New South Wales, Sydney, NSW, Australia
[3]Department of Clinical Pharmacology and Toxicology, St Vincent's Hospital, Sydney, NSW, Australia

Synonyms

Amethopterin; MTX

Definition

Methotrexate is an analogue of folates (Fig. 1) and is a cytotoxic drug used in the treatment of several malignancies. At low doses, methotrexate is a member of the drug group slow-acting antirheumatic drugs (SAARDs). Methotrexate is also a member of a group of drugs known as conventional synthetic disease-modifying antirheumatic drugs (csDMARDs).

Chemical Structures and Properties

The molecular mass of methotrexate is 454. Like the folates, methotrexate is a dicarboxylic acid and has pKa values of 4.8 and 5.5. It is therefore highly ionized with two negative charges at physiological pH values (Figs. 1 and 2). Methotrexate therefore has low lipid solubility in the body. The logP of the unionized form is low -1.85 indicating low lipid solubility even of unionized form. These physicochemical properties indicate that it should not diffuse passively through cell membranes. Transporters will be required.

Pharmacokinetics, Metabolism, and Dosage

Features of the pharmacokinetics and metabolism of methotrexate include:

- Methotrexate is approximately 25 % bound to plasma proteins.
- The oral bioavailability of methotrexate is about 85 % compared to its subcutaneous absorption (Wilson et al. 2013). Interpatient variability is relatively small.
- Following oral, subcutaneous, or intramuscular dosage, the initial half-life of methotrexate is about 7 h. This is followed by a very slow phase with a half-life of about 5 days (Seideman et al. 1993).

Methotrexate, Fig. 1 Comparative structures of methotrexate and folate

Methotrexate, Fig. 2 Metabolism of methotrexate. A range of polyglutamates with up to four additional glutamate residues in addition to one glutamate in methotrexate. Methotrexate and its glutamate metabolites can be hydroxylated to the 7-hydroxymethotrexate by aldehyde oxidase, the enzyme which converts the anti-gout drug, allopurinol, to its active metabolite, oxypurinol. The methotrexate glutamates are, however, hydroxylated at decreased rates with increased glutamate conjugation. The deglutamated metabolite, 2,4-diamino-N^{10}–methylpteroate, is formed in the gastrointestinal tract by removal of the single glutamate residue of methotrexate

- Methotrexate contains a single glutamate moiety (Fig. 2), and after entering cells, up to four additional glutamates are added by the action of folylpolyglutamyl synthase, allowing the polyglutamate form of methotrexate to have up to five glutamates (Fig. 2) (Stamp and Roberts 2011). The polyglutamation of methotrexate maintains low cellular levels of methotrexate because only methotrexate is transported out of cells. Further, glutamation enhances its pharmacological actions because the methotrexate polyglutamates are inhibitors of dihydrofolate reductase (see below).
- There is gradual cellular buildup of methotrexate polyglutamates over about 6 months in red blood cells (see below) with considerable interpatient variability (Stamp and Roberts 2011).
- Methotrexate is hydroxylated to 7-hydroxymethotrexate by an aldehyde oxidase (Bannwarth et al. 1996; Rosowsky et al. 1990) (Fig. 2). Less than 10 % of an oral dose of methotrexate is hydroxylated, but its total plasma concentrations often exceed those of the parent methotrexate although the unbound concentrations of the 7-hydroxymethotrexate are lower as the unbound metabolite is approximately 75 % bound to plasma proteins.
- The polyglutamate metabolites of methotrexate can be hydroxylated to the polyglutamate derivatives of 7-hydroxymethotrexate but at decreasing rates with increasing numbers of glutamate residues (Rosowsky et al. 1990) (Fig. 2).
- 7-hydroxymethotrexate is polyglutamated in a similar fashion to the parent, methotrexate (Fig. 2).
- A deglutamated metabolite, 2,4-diamino-N^{10}–methylpteroate, is formed by intestinal flora by removal of the single glutamate residue in methotrexate (Fig. 2). It is then absorbed but appears inactive.

These pharmacokinetic parameters explain much of the clinical pharmacology of methotrexate, including its once-weekly dosage, the slow onset, and variable clinical effect. However, no consistent relationship has been observed between the clinical effect of methotrexate and any of its pharmacokinetic parameters in rheumatoid arthritis (Bannwarth et al. 1996). Genetic variants of several transporters of methotrexate have been found but have not been associated consistently with the clinical response to the drug (Badagnani et al. 2006; Plant et al. 2014; Stamp and Roberts 2011).

In rats, the metabolite, 7-hydroxymethotrexate, has considerably lesser activity than methotrexate in adjuvant arthritis (Baggott et al. 1998), although at very high doses the metabolite has similar hepatic and renal toxicity to methotrexate.

Methotrexate is transported into cells by the reduced folate carrier 1 (RFC1, SLC19A1), while the free non-glutamated methotrexate is removed from the cell by transporters belonging to the ATP-binding cassette (ABC) family (Stamp and Roberts 2011). In addition, methotrexate is a substrate of organic anion-transporting polypeptide 1A2, a transporter which is present in several tissues. This transporter is present in the distal kidney tubule indicating that it mediates the renal resorption of methotrexate and therefore limits its renal clearance (Badagnani et al. 2006).

Methotrexate is primarily eliminated unchanged in urine. It follows therefore that:

- Serum creatinine should be measured as a measure of renal function.
- A lower dose of methotrexate should be used in patients with chronic renal impairment. It is difficult to adjust the dose of methotrexate in dialyzed patients, and it has been recommended that these patients should receive other DMARDs, particularly a tumor necrosis factor inhibitor (TNF inhibitor) (Al-Hasani and Roussou 2011).
- A temporary cessation of methotrexate treatment may be required at times of volume depletion (such as perioperatively).
- Dosage should generally be higher in younger patients than in elderly patients because renal function decreases with increasing age (Ranganath and Furst 2007).
- Co-prescription of agents which impair renal function, such as aminoglycosides and cyclosporin, should be undertaken with caution.
- The prolonged use of methotrexate itself may reduce renal function and hence its own clearance (Kremer et al. 1995), a possible mechanism being increased plasma adenosine concentrations as a consequence of methotrexate activating A_1 receptors in the renal parenchyma (Cronstein 1996), thereby diminishing renal blood flow and salt and water excretion.
- Probenecid decreases the renal excretion of methotrexate. The combination should, in general, be avoided although, with considerable care, a reduced dose of methotrexate could be administered.

Pharmacological Activities

The mechanism of action of methotrexate is still not completely understood although it is generally accepted that methotrexate inhibits some folate-dependent enzymes, particularly because of its structural similarity to folate. Some aspects include (Stamp and Roberts 2011):

- In the treatment of tumors, the major action of methotrexate is inhibition of dihydrofolate reductase, the result being the blockade of the intracellular production of reduced tetrahydrofolate which is important in the transfer of one-carbon units. These are necessary for the synthesis of some amino acids and thymine, a nucleic acid base. An action on dihydrofolate reductase is indicated at the low dose in rheumatoid arthritis because the trough concentration of unbound methotrexate (Seideman et al. 1993) exceeds the approximate dissociation constant of methotrexate from dihydrofolate reductase (2 ng/l) (Schweitzer et al. 1989).
- An important effect of methotrexate may be inhibition of thymidylate synthase, an enzyme which catalyzes the formation of

Methotrexate, Fig. 3 Proposed mechanism of action of methotrexate (MTX). Inhibition of the conversion of 5-aminoimidazole-4-carboxamide ribonucleotide (AICAR) to 5-formyl-aminoimidazole-4-carboxamide ribonucleotide (FAICAR) by methotrexate leads to increased levels of AICAR which inhibits the enzymic deamination of adenosine and adenosine monophosphate. The result is a buildup of adenosine which mediates the effects of methotrexate on inflammatory cells. *IMP* inosine monophosphate, *AMP* adenosine monophosphate

deoxythymidine monophosphate (dTMP) from deoxyuridine monophosphate (dUMP):

5,10-methylenetetrahydrofolate + dUMP
→ dihydrofolate + dTMP

dTMP is then phosphorylated further and incorporated into DNA. This effect could be a mechanism of the cytotoxic anticancer effect of methotrexate but may also be involved in the effect of methotrexate on immature and inflammatory monocytes in inflamed synovium (see below).

- The major anti-inflammatory effect of low-dose methotrexate may be the increased extracellular levels of adenosine which has anti-inflammatory and immunosuppressant actions (Cronstein 1996). The mechanism is proposed to be inhibition of the enzyme 5-aminoimidazole-4-carboxamide ribonucleotide (AICAR) transformylase which is a folate-dependent enzyme (Fig. 3). The resulting elevated levels of AICAR are proposed to block the deamination of adenosine and adenosine monophosphate (AMP), resulting in increased conversion to adenosine in the extracellular space (Fig. 3) (Cronstein 1996). It is suggested that the higher levels of extracellular adenosine then bind to the transmembrane G-protein-coupled adenosine cell surface receptors (A_1, A_{2a}, A_{2b}, A_3) resulting in decreased inflammation and immune suppression.

According to this hypothesis, methotrexate acts indirectly via ligation of the A_{2a} receptors that are present on neutrophils, macrophage-monocytes, lymphocytes, and basophils. Binding increases intracellular cAMP leading to immunosuppression by inhibition of phagocytosis; inhibition of secretion of TNF, IFNγ, IL-2, IL-6, IL-8, and HLA expression; and increased secretion of IL-10, an anti-inflammatory cytokine. Binding of adenosine to A_3 receptors on macrophage-monocytes leads to inhibition of secretion of TNF, IL-12, IFNγ, and IL-1ra. Results on A_{2A} and A_3 knockout mice are consistent with adenosine mediating the anti-inflammatory effects of methotrexate because methotrexate does not have anti-inflammatory activity in mice lacking either receptor. Further, methotrexate increases adenosine concentrations in air pouch exudates, a model of inflammation (Montesinos et al. 2003). By contrast, methotrexate does not increase the blood concentration of adenosine in patients, although changes at peripheral sites cannot be excluded (Smolenska et al. 1999).

- AICAR is an intermediate in the synthesis of adenosine, but there is evidence that inhibition of AICAR transformylase is less significant than any subsequent AICAR-dependent inhibition of the deamination of adenosine (Cronstein 1996) (Fig. 3).
- As methotrexate inhibits intracellular folate systems, interpatient variations in the levels of folate derivatives and their polyglutamated forms may contribute to the markedly variable differences in the clinical response to methotrexate (Becker et al. 2012). However, no genetic polymorphisms have been associated consistently with the interpatient differences in response to anti-inflammatory and immunosuppressant effects of methotrexate (Plant et al. 2014; Stamp and Roberts 2011).
- There are several other suggested mechanisms of action of methotrexate. One proposal is inhibition of methotrexate inhibited proliferation of activated lymphocytes. There is, however, no convincing evidence that lymphocyte proliferation is inhibited directly by methotrexate in rheumatoid arthritis. Methotrexate may, however, decrease the involvement of lymphocytes in rheumatoid arthritis by inhibiting recruitment of immature and inflammatory monocytes into inflammatory sites and reduce their survival in the inflamed synovium, but with little or no effect on tissue-infiltrating monocytes and resident macrophages (Cutolo et al. 2000). This action may be secondary to increased adenosine tone (see above).

Clinical Use and Efficacy

Methotrexate was originally developed in the 1940s as a cytotoxic drug for the treatment of various tumors. An older folate analogue, aminopterin, was shown to be useful in treatment of rheumatoid arthritis in 1951 (Gubner et al. 1951). Aminopterin is more toxic than methotrexate and its use was stopped in favor of methotrexate. Double-blind clinical trials of methotrexate in the 1980s established the value of methotrexate in the treatment of rheumatoid arthritis (Braun and Rau 2009; Ranganath and Furst 2007; Stamp and Roberts 2011). Features of its clinical use and efficacy include:

- Methotrexate is used at lower doses in inflammatory joint diseases than when used as a cytotoxic drug in the treatment of malignancies.
- A variety of dose strategies of methotrexate have been used in inflammatory diseases, but a review of dosing strategies found a starting dose of 15 mg/week orally, escalating at 5 mg/month to 25 to 30 mg/week or the highest tolerable dose, yields the best results in rheumatoid arthritis (Visser and van der Heijde 2009). Overall, the aim is to increase the dosage up to a level which produces satisfactory suppression of the activity of the disease with limited adverse effects.
- The clinical response to methotrexate develops slowly with considerable interpatient variability which may be due to interpatient differences in the clearance of methotrexate or its polyglutamate metabolites although, as discussed above, there is no consistent correlation between response in rheumatoid arthritis and the concentrations of methotrexate or its metabolites in plasma or red blood.

- As outlined above, associations between genetic polymorphisms in transporters and biochemical aspects of methotrexate action have been reported, but at this stage, no genetic factor is used in the selection of the methotrexate dosage in arthritic diseases (Plant et al. 2014).
- Methotrexate is usually administered orally but may also be administered by subcutaneous or intramuscular injection if excessive nausea occurs when the drug is taken orally and the nausea is not controlled by folic acid (see adverse effects below).
- The utility of methotrexate in rheumatoid arthritis is seen from the high maintenance on treatment with over 50 % of patients still taking methotrexate for 5 years or more after initiation of the treatment. This retention rate is generally greater than seen with the older DMARD.
- Methotrexate retards, but does not entirely block, joint damage of rheumatoid arthritis. Its efficacy in limiting joint damage is slightly less than that of the TNF inhibitors.
- Methotrexate shows no significant activity in the treatment of spinal inflammation of ankylosing spondylitis.
- Methotrexate is useful in the treatment of psoriasis and psoriatic arthritis.

The present view on methotrexate is that it should be considered for all patients at the time of diagnosis of rheumatoid arthritis. Individual factors such as pregnancy and excessive alcohol intake may impact on that decision but methotrexate needs to be considered. Methotrexate is frequently used with other csDMARDs depending upon the progress of the treatment of rheumatoid arthritis. The combination of methotrexate with biological DMARDs is considered to be the most efficacious presently used in the treatment of rheumatoid arthritis.

Adverse Effects

Low-dose methotrexate produces a large number of adverse reactions as listed below. Withdrawals due to adverse effects are common. Ten to 35 % of patients treated with methotrexate cease therapy due to toxicity, which is less than for sulfasalazine and gold complexes but higher than for the antimalarials and the biological DMARDs.

Common Adverse Effects

Oral ulceration, nausea, and fatigue occur very frequently and are probably related to intracellular depletion of folates, resulting in increased adenosine and homocysteine. These adverse effects are reduced by supplementary administration of folic acid. Various doses of folic acid have been recommended, but a common advice is that a single dose of 5 mg folic acid should be administered to all patients on the morning following the dose of methotrexate (Whittle and Hughes 2004). Supplementation at this level does not reduce the antirheumatic efficacy of low-dose methotrexate.

Treatment of Hematotoxicity and Overdose

The metabolic effects of folinic acid, the fully reduced form of folic acid, are not blocked by methotrexate and are used to treat methotrexate-induced hematotoxicity and overdose with the drug. The complete blood count and tests of liver and renal function should be monitored every month for 6 months and then every 1–2 months subsequently.

Hepatitis

This is a significant adverse effect of methotrexate and it should not be used in patients with significant liver disease. At the onset of therapy, liver function tests including measurements of aspartate aminotransferase, alanine aminotransferase, albumin, and alkaline phosphatase should be undertaken. The prevalence of raised liver enzymes to above twice the upper limit of normal is around 13–15 %. Treatment with methotrexate should be stopped in patients with transaminase concentrations persistently at twice the upper limit of normal or at three times the upper level of normal at any time. Tests for hepatitis B and C should be conducted in patients who are at risk of these diseases. Liver fibrosis and cirrhosis have been reported but the reported incidence is conflicting. Moderate intake of alcohol does not appear to increase the risk of hepatitis (Braun and

Rau 2009). At this stage, plasma concentrations do not correlate with the incidence of methotrexate-induced hepatitis. Liver biopsy is required only for those patients who need to continue methotrexate and who have sustained enzyme abnormalities.

Pneumonitis

Interstitial pneumonitis is a serious side of methotrexate and occurs in 2–7 % of patients and is potentially fatal. Treatment consists of cessation of methotrexate treatment, general supportive measures, and high doses of corticosteroids. Although most patients with methotrexate-induced lung disease have a complete recovery, some have permanent lung damage. The strongest predictors for lung injury are age above 60 years, diabetes mellitus, rheumatoid pulmonary involvement, previous use of a csDMARD, and hypoalbuminemia (Alarcon et al. 1997). Methotrexate should not be used in patients with pulmonary diseases and a chest X-ray should be taken at the onset of treatment with metformin. Methotrexate should not be reintroduced after recovery from pneumonitis. Although pneumonitis is clearly an adverse effect of methotrexate, many reported cases of pneumonitis were the result of pulmonary infections which were not differentiated from methotrexate-induced pneumonitis.

Infections

Unlike the TNF inhibitors and other biological slow-acting antirheumatic drugs, methotrexate does not appear to increase the risk of serious infections, including herpes zoster, and the drug could provide reduced cardiovascular mortality.

Malignancy

There is no strong evidence of increased risk of malignancy but the data is insufficient to draw strong conclusions.

Rheumatoid Nodules

An unexpected side effect of methotrexate is the accelerated formation of rheumatoid nodules, particularly around the fingers. This may be due to activation of adenosine A1 receptors leading to the development of multinucleated giant cells and the nodules (Merrill et al. 1997). Colchicine may prevent their formation.

Teratogenic Effects

As is the case during treatment with most DMARDs, pregnancy should be avoided during treatment with low-dose methotrexate because of the high risk of teratogenic effects. Consequently, treatment with methotrexate, as well as with other DMARD, should be stopped 3 months before conception and not restarted until after delivery. Fortunately, the disease activity generally decreases during pregnancy but exacerbations can be treated with low-dose corticosteroids.

Drug Interactions

- Methotrexate is commonly administered with other DMARDs. As noted above, the combination of methotrexate and a biological agent, such as a TNF inhibitor, is widely considered to be the most efficacious treatment of rheumatoid arthritis. Methotrexate also appears useful to reduce immunological reactions to the biological agent.
- Bone marrow suppression has occasionally been seen with the combination of co-trimoxazole and methotrexate, probably because co-trimoxazole has weak antifolate activity in humans (Bannwarth et al. 1996).
- NSAIDs commonly reduce renal function, but there appears to be no interaction with NSAIDs (Colebatch et al. 2012) with the exception of anti-inflammatory doses of aspirin (>2 g daily) which increase the unbound fraction of methotrexate in plasma and decrease the unbound renal clearance of methotrexate (Stewart et al. 1991). With care, however, the combination can be used.
- There are recommendations that alcohol should never be taken with methotrexate. However, it appears that alcohol can be taken in moderation with methotrexate (Braun and Rau 2009).
- As noted above, care should be taken if methotrexate is taken with cyclosporin or aminoglycoside antibiotics which may reduce

renal function. Cyclosporin also blocks the hydroxylation of methotrexate to 7-hydroxymethotrexate producing a small increase (about 18 % in the plasma concentrations of unchanged methotrexate).

Cross-References

▶ Antimalarial Drugs
▶ Disease-Modifying Antirheumatic Drugs: Overview
▶ Gold Complexes
▶ Rheumatoid Arthritis
▶ Sulfasalazine and Related Drugs
▶ Tumor Necrosis Factor (TNF) Inhibitors

References

Alarcon, G. S., Kremer, J. M., Macaluso, M., Weinblatt, M. E., Cannon, G. W., Palmer, W. R., et al. (1997). Risk factors for methotrexate-induced lung injury in patients with rheumatoid arthritis. A multicenter, case-control study. Methotrexate-Lung Study Group. *Annals of Internal Medicine, 127*(5), 356–364.

Al-Hasani, H., & Roussou, E. (2011). Methotrexate for rheumatoid arthritis patients who are on hemodialysis. *Rheumatology International, 31*(12), 1545–1547.

Badagnani, I., Castro, R. A., Taylor, T. R., Brett, C. M., Huang, C. C., Stryke, D., et al. (2006). Interaction of methotrexate with organic-anion transporting polypeptide 1A2 and its genetic variants. *Journal of Pharmacology and Experimental Therapeutics, 318*(2), 521–529.

Baggott, J. E., Morgan, S. L., & Koopman, W. J. (1998). The effect of methotrexate and 7-hydroxymethotrexate on rat adjuvant arthritis and on urinary aminoimidazole carboxamide excretion. *Arthritis and Rheumatism, 41*(8), 1407–1410.

Bannwarth, B., Pehourcq, F., Schaeverbeke, T., & Dehais, J. (1996). Clinical pharmacokinetics of low-dose pulse methotrexate in rheumatoid arthritis. *Clinical Pharmacokinetics, 30*(3), 194–210.

Becker, M. L., van Haandel, L., Gaedigk, R., Thomas, B., Hoeltzel, M. F., Lasky, A., et al. (2012). Red blood cell folate concentrations and polyglutamate distribution in juvenile arthritis: predictors of folate variability. *Pharmacogenetics and Genomics, 22*(4), 236–246.

Braun, J., & Rau, R. (2009). An update on methotrexate. *Current Opinion in Rheumatology, 21*, 216–223.

Colebatch, A. N., Marks, J. L., van der Heijde, D. M., & Edwards, C. J. (2012). Safety of nonsteroidal antiinflammatory drugs and/or paracetamol in people receiving methotrexate for inflammatory arthritis: a Cochrane systematic review. *The Journal of Rheumatology, 39*(Supplement 90), 62–73.

Cronstein, B. N. (1996). Molecular therapeutics Methotrexate and its mechanism of action. *Arthritis and Rheumatism, 39*, 1951–1960.

Cutolo, M., Bisso, A., Sulli, A., Felli, L., Briata, M., Pizzorni, C., et al. (2000). Antiproliferative and antiinflammatory effects of methotrexate on cultured differentiating myeloid monocytic cells (THP-1) but not on synovial macrophages from patients with rheumatoid arthritis. *The Journal of Rheumatology, 27*, 2551–2557.

Gubner, R., August, S., & Ginsberg, V. (1951). Therapeutic suppression of tissue reactivity. II. Effect of aminopterin in rheumatoid arthritis and psoriasis. *Americal Journal of Medical Sciences, 221*, 176–182.

Kremer, J. M., Petrillo, G. F., & Hamilton, R. A. (1995). Pharmacokinetics and renal function in patients with rheumatoid arthritis receiving a standard dose of oral weekly methotrexate: association with significant decreases in creatinine clearance and renal clearance of the drug after 6 months of therapy. *Journal of Rheumatology, 22*(1), 38–42.

Merrill, J. T., Shen, C., & Schreibman, D. (1997). Adenosine A1 receptor promotion of multinucleated giant cell formation by human monocytes: a mechanism for methotrexate-induced nodulosis in rheumatoid arthritis. *Arthritis and Rheumatism, 40*, 1308–1315.

Montesinos, M. C., Desai, A., Delano, D., Chen, J. F., Fink, J. S., Jacobson, M. A., et al. (2003). Adenosine A2A or A3 receptors are required for inhibition of inflammation by methotrexate and its analog MX-68. *Arthritis and Rheumatism, 48*(1), 240–247.

Plant, D., Wilson, A. G., & Barton, A. (2014). Genetic and epigenetic predictors of responsiveness to treatment in RA. *Nature Reviews. Rheumatology, 10*, 329–337.

Ranganath, V. K., & Furst, D. E. (2007). Disease-modifying antirheumatic drug use in the elderly rheumatoid arthritis patient. *Rheumatic Diseases Clinics of North America, 33*, 197–217.

Rosowsky, A., Wright, J. E., Holden, S. A., & Waxman, D. J. (1990). Influence of lipophilicity and carboxyl group content on the rate of hydroxylation of methotrexate derivatives by aldehyde oxidase. *Biochemical Pharmacology, 40*(4), 851–957.

Schweitzer, B. I., Srimatkandada, S., Gritsman, H., Sheridan, R., Venkataraghavan, R., & Bertino, J. R. (1989). Probing the role of two hydrophobic active site residues in the human dihydrofolate reductase by site-directed mutagenesis. *Journal of Biological Chemistry, 264*(34), 20786–20795.

Seideman, P., Beck, O., Eksborg, S., & Wennberg, M. (1993). The pharmacokinetics of methotrexate and its 7-hydroxy metabolite in patients with rheumatoid arthritis. *British Journal of Clinical Pharmacology, 35*, 409–412.

Smolenska, Z., Kaznowska, Z., Zarowny, D., Simmonds, H. A., & Smolenski, R. T. (1999). Effect of methotrexate on blood purine and pyrimidine levels in patients with rheumatoid arthritis. *Rheumatology (Oxford), 38*, 997–1002.

Stamp, L. K., & Roberts, R. L. (2011). Effect of genetic polymorphisms in the folate pathway on methotrexate therapy in rheumatic diseases. *Pharmacogenomics, 12*(10), 1449–1463.

Stewart, C. F., Fleming, R. A., Germain, B. F., Seleznick, M. J., & Evans, W. E. (1991). Aspirin alters methotrexate disposition in rheumatoid arthritis patients. *Arthritis and Rheumatism, 34*, 1514–1520.

Visser, K., & van der Heijde, D. (2009). Optimal dosage and route of administration of methotrexate in rheumatoid arthritis: a systematic review of the literature. *Annals of the Rheumaic Diseases, 68*, 1094–1099.

Whittle, S., & Hughes, R. A. (2004). Folate supplementation and methotrexate treatment in rheumatoid arthritis: a review. *Rheumatology (Oxford), 43*, 267–271.

Wilson, A., Patel, V., Chande, N., Ponich, T., Urquhart, B., Asher, L., et al. (2013). Pharmacokinetic profiles for oral and subcutaneous methotrexate in patients with Crohn's disease. *Alimentary Pharmacology and Therapeutics, 37*(3), 340–345.

Microvascular Responses to Inflammation

Daniel Neil Granger[1] and Stephen F. Rodrigues[2]
[1]Department of Molecular and Cellular Physiology, LSU Health Sciences Center, Shreveport, LA, USA
[2]Department of Clinical and Toxicological Analyses, School of Pharmaceutical Sciences, University of Sao Paulo, Sao Paulo, Brazil

Synonyms

Microvasculature-driven inflammatory process

Definition

Inflammation is a microcirculation-dependent tissue response to an endogenous or exogenous aggressor. Several cardinal signs characterize an inflammatory process, and all of them are generated by alterations in the microcirculatory function. These include redness, heat, pain, swelling, and loss of tissue function.

Structure and Function

Microvessels are small fluid conduits that transport blood or lymph within tissues. These structures are important for controlling of several vital tissue functions such as absorption of nutrients, nourishment of cells, elimination of cell waste products, transport of gases, control of blood pressure, communication with distant organs and tissues, and host defense. As a component of the host defense mechanism, inflammation is largely dependent on the microcirculation. Both lymphatic and blood vessels play an important role in an inflammatory response. While lymphatic microvessels differ in size from collecting to conducting vessels, with a gradual thickening of the smooth muscle coat, the physiological function of these vessels remain unchanged over the size range. Blood vessels, on the other hand, exhibit more heterogeneity in both structure and function between arterioles, capillaries, and venules. Arterioles extend and branch out from an artery to form capillaries. A few layers of smooth muscle (one or two) in arterioles account for the role of these vessels in the maintenance of vascular tone, control of blood flow, and regulation of blood pressure (Martinez-Lemus 2012). Arterioles carry oxygen- and nutrient-rich blood at pressures of 40–70 mmHg (with normal mean systemic blood pressure by 100 mmHg) (Gore 1974). During inflammation, arterioles dilate to increase tissue blood flow, which accounts for the redness and local heating of inflamed tissue.

Capillaries are the smallest blood vessels in the body. They measure 5–10 μm in diameter and connect arterioles to venules. Capillary pressure differs between tissues, and there is a gradient of pressure between the arterial and venous ends of these vessels. Capillaries allow for the ready exchange of gases such as oxygen and carbon dioxide, water, nutrients, and cellular waste products between the blood and interstitial compartments. Capillaries can be divided into two types (continuous or fenestrated) depending on the tightness of their inter-endothelial junctions. Continuous capillaries are formed by a continuous layer of endothelial cells that offer some restriction to the exchange of water and solutes between blood and perivascular cell compartments (Salmon and Satchell 2012). Endothelial cells in capillaries are held together at the intercellular junctions by transmembrane or cytoplasmic proteins that include claudins, occludins, and junctional adhesion molecules (JAMs) (Sawada 2013). Continuous-type capillaries are found in muscle (skeletal and heart), lung, and brain tissue. Fenestrated-type capillaries, which are found in the gastrointestinal and endocrine tissues and kidney, are characterized by circular openings (pores) within the endothelial cells that can be subtended by a diaphragm (layer of fine fibrillar material). Open fenestrae (not covered by a diaphragm) are highly permeable to water, small solutes (e.g., glucose), and macromolecules (albumin). The presence of diaphragms increases the resistance to albumin permeation. A gradient in capillary permeability, increasing from arterial to venous ends, is evidenced in capillaries due to an increased number and width of pores along the vessel length. During inflammation, capillary permeability increases, allowing an increased extravasation of protein and water and ultimately interstitial edema. The inflammation-induced change in capillary permeability results from the action of different mediators (e.g., histamine, bradykinin) on endothelial cells, which open the intercellular junctions by eliciting cell contraction and the mobilization of junctional proteins into the cell. Chronic inflammation is associated with the proliferation of capillaries, i.e., angiogenesis.

Venules, situated downstream from capillaries, receive oxygen-poor blood and exhibit a lower intravascular pressure (Gore 1974). While the postcapillary venule is devoid of a smooth muscle coat near the capillaries, smooth muscle appears on the media of larger venules (muscular venules) that drain the postcapillary venules. The passive, distensible nature of the postcapillary and muscular venules accounts for the ability of these microvessels to store and mobilize significant quantities of blood in certain organs. The postcapillary venules represent the segment of the microvasculature that is most reactive to inflammation. When venular endothelial cells assume an inflammatory phenotype, leukocytes adhere to the venular wall and emigrate into the adjacent interstitial compartment via dilated inter-endothelial junctions. The trafficking of leukocytes through inter-endothelial cell junctions is associated with an accelerated leakage of plasma proteins via the same pathway. The sequential and highly coordinated events associated with ▶ leukocyte recruitment are mediated by adhesion molecules expressed on the endothelial cells, which bind to counter-receptors expressed by circulating leukocytes.

Lymphatic vessels can be divided into capillaries and collecting lymphatic vessels. Lymph capillaries (terminal lymphatics) are comprised of a single layer of endothelial cells, while the collecting lymphatics are larger and have smooth muscle coat as well as valves that prevent the reflux of lymph. The terminal lymphatics drain the interstitial fluid that results from capillary fluid and protein filtration, while the collecting ducts transport the duly formed lymph to the lymph nodes and eventually back to the blood circulation. The lymph vessels also clear the interstitial compartment of immune cells. During inflammation, the lymph vessels serve several functions, including the clearance of the excessive amounts of fluid and proteins that enter the interstitium from the blood capillaries, the delivery of leukocytes and antigen-presenting cells to the lymph nodes, and the prevention of inflammatory mediator accumulation in the affected tissue (Alitalo et al. 2005). Chronic inflammation is associated with an enhanced production of lymph vessels, i.e., lymphangiogenesis.

Pathophysiological Relevance

Microcirculatory dysfunction is a common feature of different disease states. Impaired blood flow regulation, increased capillary permeability, inflammatory cell accumulation, oxidative stress, enhanced thrombus development, and excess proliferation of blood and lymph capillaries represent both a cause and consequence of the tissue injury response that accompanies different inflammatory diseases.

Increased vascular permeability and/or lymphatic failure can result in a profound accumulation of fluid and protein in the interstitial compartment, resulting in tissue swelling (edema) with consequent pain, organ dysfunction, and tissue damage and death (Alitalo et al. 2005). Specific failure of the barrier function of blood (but not lymphatic) microvessels is evident in certain pathological conditions such as hemorrhagic or ischemic stroke. Hemorrhagic stroke is a neurological emergency that kills more than one million people worldwide each year (Sacco et al. 2009). It is characterized by a permanent or transitory reduction of the blood flow in part of the brain caused by a ruptured blood vessel. Hemorrhagic stroke accounts for approximately 10–15 % of all strokes. As a consequence of the barrier failure, there is intensive cell death in the core of the hemorrhage zone with damage of surrounding tissue resulting from a coagulum-induced pressure effect, edema, neurotoxicity mediated by products released after cell lysis, and inflammation. In ischemic stroke, which results from partial or total obstruction of blood flow in a region of the brain, microvessel barrier failure is more of a consequence than cause of the ensuing tissue injury and inflammatory response. The microvascular dysfunction that accompanies ischemic stroke is associated with excess generation of reactive oxygen species, ionic imbalances, glutamate release, and impaired glycolysis; all of which lead to cell apoptosis and necrosis. Ischemic stroke accounts for approximately 85–90 % of all strokes (Bakhai 2004) and is associated with a mortality rate of approximately 20 % and an even larger number of survivors that are permanently disabled (Roger et al. 2012).

Another inflammatory disease characterized by profound microvascular dysfunction is septic shock. Septic shock is the main cause of death in intensive care units in the United States, and it is associated with nearly a 30 % mortality rate (Angus et al. 2001). All segments of the microcirculation, i.e., arterioles, capillaries, and venules, exhibit functional abnormalities during the developmental and maintenance phases of sepsis (Lundy and Trzeciak 2011). Capillary leak (an increased permeability), reduced vasomotor tone with hypotension, reduced tissue perfusion, and acidosis are all evidence in the early stages of septic shock. These responses can ultimately result in dysfunction and failure of a variety of organs including the lungs, kidneys, liver, and heart and the central nervous system. The multiple organ failure then leads to death (Parker et al. 1987). The pronounced alterations in microvascular function that occur in sepsis are linked to microorganism-derived cell components (e.g., lipopolysaccharide) that stimulate the release of a myriad of inflammatory mediators, including ▶ cytokines, chemokines, complement components, and proteolytic enzymes.

Enhanced leukocyte transport mediated by an altered endothelial cell gene expression pattern and consequent overexpression of cell adhesion molecules has been implicated in the initiation and perpetuation of inflammatory conditions such as ulcerative colitis and Crohn's disease (Rivera-Nieves et al. 2005), the two major forms of ▶ inflammatory bowel disease (IBD). While the etiology of IBD remains unclear, microvascular dysfunction is a common finding in the diseased tissue, both in human disease and in animal models of IBD (Fiocchi 2002). The local intense tissue inflammation that accompanies IBD produces the expected changes in microvascular function (Fig. 1) that inevitably exacerbate the tissue injury and inflammatory responses. These changes include impaired vasomotor function (an early vasodilation and late vasoconstriction), enhanced thrombus development (due to increased platelet reactivity and activation of the coagulation cascade), leukocyte and platelet recruitment (resulting from increased adhesion molecule expression), interstitial edema (resulting from increases in capillary pressure and vascular permeability), and enhanced angiogenesis and lymphangiogenesis (initiated by the release of different isoforms of VEGF). These changes have been linked to the activation of cells that normally circulate in the blood (leukocytes, platelets) or reside in the vessel wall (endothelial cells, vascular smooth muscle) or in the perivascular space (macrophages, mast cells).

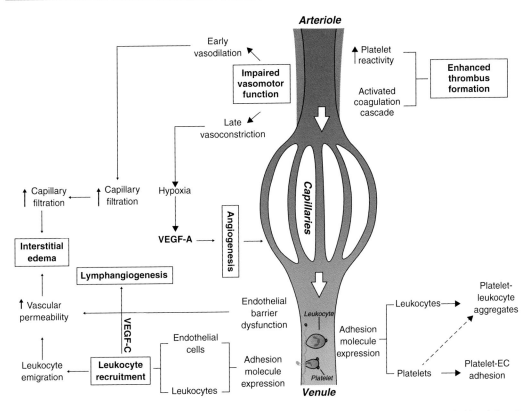

Microvascular Responses to Inflammation, Fig. 1 Summary of the microvascular responses to inflammation. The changes in microvascular function that accompany acute and chronic inflammation include impaired vasomotor function (an early vasodilation and late vasoconstriction), enhanced thrombus development (due to increased platelet reactivity and activation of the coagulation cascade), leukocyte and platelet recruitment (resulting from increased adhesion molecule expression), interstitial edema (resulting from increases in capillary pressure and vascular permeability), and enhanced angiogenesis and lymphangiogenesis (initiated by the release of different isoforms of VEGF). These changes have been linked to the activation of cells that normally circulate in the blood (leukocytes, platelets) or reside in the vessel wall (endothelial cells, vascular smooth muscle) or in the perivascular space (macrophages, mast cells). The chemical messengers elaborated from these activated cells stimulate different signaling pathways that ultimately elicit the phenotypic microvascular responses that accompany inflammation

The chemical messengers elaborated from these activated cells stimulate different signaling pathways that ultimately elicit the phenotypic microvascular responses that accompany inflammation. Pharmacological interventions directed toward preventing some of the microvascular alterations in inflammation (e.g., angiogenesis) provide novel and potentially effective targets for treatment of IBD.

In contrast to the microvascular dysfunction observed in conditions associated with the activation of the inflammatory pathways, such changes are not observed in disease states that are characterized by defects in leukocyte trafficking, such as leukocyte adhesion deficiency (LAD), Wiskott-Aldrich syndrome (WAS), and WHIM syndrome (Anderson and Springer 1987). LAD is an autosomal recessive trait that leads to recurrent bacterial infections, persistent granulocytosis, and impaired wound healing. It is caused by defects in one of the following genes: (i) CD18, being classified as LAD-1; (ii) GDP-fucose transporter, which participates in the synthesis of selectin ligands, grouped LAD-2; and (iii) kindlin-3, a protein which mediates integrin protein activation, the LAD-3 (Badolato 2013).

Several hundred people have been diagnosed with LAD-1 (Crowley et al. 1980), while LAD-2 (Marquardt et al. 1999) and LAD-3 (Meller et al. 2012) are much less common. WAS syndrome results from a mutation in the WAS protein (WASp) that regulates the actin polymerization. The WAS syndrome is an X-linked primary immunodeficiency characterized by microthrombocytopenia, eczema, and frequent bacterial infection. The recurrent bacterial infections occur because of the impaired ability of leukocytes to migrate across the vessel wall into an inflammatory locus. WHIM syndrome is a rare autosomal dominant immune disease that is characterized by warts, hypogammaglobulinemia, infections, and myelokathexis, which is the increased retention of mature neutrophils in the bone marrow. These clinical features form the acronym for the syndrome. The WHIM syndrome is caused by a genetic defect in a chemokine receptor, more precisely, CXCR4, that increases the rate of binding of CXCR4 and its ligand, CXCL12 (also named SDF-1). The higher rate of chemokine-receptor binding occurs as consequence of a lower internalization of the receptor, which leads to a longer permanency of the leukocytes, mainly neutrophils, in the marrow, with consequent apoptosis of the mature cells at this site (Dotta et al. 2011). This results in a reduced response to an inflammatory stimulus. HPV infections are common in these patients, which may explain the warts that can progress to carcinoma.

An impaired function of microscopic lymphatics can lead to an acceleration of interstitial fluid accumulation during acute and chronic inflammation. Chronic lymphedema, which affects over 100 million people worldwide, can result in fibrosis, immune deficiency, and degeneration of the connective tissue. There is also evidence indicating that chronic obstruction of lymph flow can elicit an inflammatory response, which likely results from the diminished clearance of immune cells and their activation products from the interstitial compartment by microlymphatics. While there are some genetic causes of lymphedema (e.g., Milroy disease), this condition is more commonly caused by filarial (nematode) infections in tropical countries and by surgical resection of lymph nodes following mastectomy and/or radiation therapy as part of cancer treatment in industrialized countries. It is estimated that one in five women develop lymphedema after mastectomy and these women are more susceptible to inflammation in the affected arm (DiSipio et al. 2013). Lymphangiogenesis is stimulated under conditions of chronic lymphatic obstruction by VEGF-C and VEGF-D released from infiltrating monocytes; however, the rate of lymph microvessel formation often appears insufficient to restore the fluid, protein, and immune cell clearing capacity needed to effectively prevent interstitial edema.

The Microvasculature: A Target for Therapeutic Intervention in Inflammation

It is now well recognized that inflammation is a critical factor in the pathogenesis of diseases as diverse as atherosclerosis, diabetes, stroke, cancer, and multiple sclerosis. This recognition has led massive effort to better understand the mechanisms that initiate and regulate the inflammatory response. An outgrowth of this effort is the growing appreciation that all three segments of the microcirculation (arterioles, capillaries, venules) are affected by, and contribute to, both the initiation and perpetuation phases of inflammation. Drugs that target one or more responses of the microvasculature to inflammation are receiving more attention as potential therapeutic agents for acute and chronic inflammatory diseases.

Of the microvascular changes that contribute to inflammation, tissue injury, and mortality in different disease states, leukocyte-endothelial cell adhesion, vascular permeability, and angiogenesis have received the most attention as potential therapeutic targets. Interfering with leukocyte-endothelial cell adhesion has shown promise in a number of preclinical and clinical studies of inflammatory disease. For example, the humanized monoclonal antibody natalizumab, which immunoneutralizes the cell adhesion molecule α-4 integrin and prevents the adhesion and subsequent transendothelial migration of leukocytes in the microcirculation, is used to treat multiple sclerosis and Crohn's disease (Ulbrich et al. 2003). Preclinical studies of asthma, stroke, myocardial infarction, psoriasis, and other inflammatory diseases have demonstrated some benefit to anti-adhesion therapy; however, clinical trials have revealed little or no benefit. A major side effect of humanized monoclonal antibodies against endothelial CAMs is an increased risk of opportunistic infections. Small-molecule antagonists, which are less likely to trigger adverse immune reactions, are now receiving more attention as potential therapeutic agents that target this microvascular component of the inflammatory response.

Microvascular hyperpermeability has been implicated in the pathogenesis of a variety of pathological conditions characterized by an inflammatory component. These include cancer, sepsis, allergies, and diabetes. In sepsis, for example, pulmonary capillaries can become hyperpermeable, leading to excess fluid and protein filtration and the eventual accumulation of interstitial fluid in alveoli, which reduces local pulmonary blood flow and impedes oxygen exchange. Multiple inflammatory agents, including cytokines, VEGF, thrombin, and reactive oxygen species, have been implicated in the alveolar capillary hyperpermeability that accompanies acute respiratory distress syndrome (ARDS) that co-occurs with sepsis. However, clinical trials employing anti-cytokine, antioxidant, and anti-thrombin agents have not led to reduced mortality from ARDS (Matthay et al. 2012). Similar efforts to block the vascular hyperpermeability and ARDS associated with third-degree burns have shown little promise in the clinical setting.

Anti-angiogenic therapy has received much attention for the treatment of cancer and for blindness associated with age-related macular degeneration. In both conditions, VEGF-blocking antibodies and small-molecule inhibitors have shown benefit. There is a growing body of evidence that angiogenesis and inflammation are closely related processes. This contention is supported by the appearance of newly formed blood vessels in granulation tissue and the dual functionality of angiogenic factors, i.e., they exhibit both pro-inflammatory and proangiogenic effects. While inflammation and angiogenesis are capable of potentiating each other, these processes are distinct and separable. In recent years, preclinical studies have revealed that the angiogenesis that accompanies chronic inflammation tends to prolong and intensify the inflammatory response. This view is supported by reports describing a worsening of disease activity, tissue injury, and colonic inflammation in experimental IBD by administration or genetic overexpression of VEGF-A, while treatment of colitic mice with anti-angiogenic agents or genetic overexpression of soluble VEGFR-1 had the opposite effect. These findings have led to the proposed use of anti-angiogenesis drugs in the treatment of IBD (Granger and Senchenkova 2010).

Cross-References

▶ Cytokines
▶ Inflammatory Bowel Disease
▶ Leukocyte Recruitment
▶ Reactive Oxygen Species

References

Alitalo, K., Tammela, T., & Petrova, T. V. (2005). Lymphangiogenesis in development and human disease. *Nature, 438*, 946–953.

Anderson, D. C., & Springer, T. A. (1987). Leukocyte adhesion deficiency: An inherited defect in the Mac-1, LFA-1, and p150,95 glycoproteins. *The Annual Review of Medicine, 38*, 175–194.

Angus, D. C., Linde-Zwirble, W. T., Lidicker, J., Clermont, G., Carcillo, J., & Pinsky, M. R. (2001). Epidemiology of severe sepsis in the United States: Analysis of incidence, outcome, and associated costs of care. *Critical Care Medicine, 29*, 1303–1310.

Badolato, R. (2013). Defects of leukocyte migration in primary immunodeficiencies. *European Journal of Immunology, 43*, 1436–1440.

Bakhai, A. (2004). The burden of coronary, cerebrovascular and peripheral arterial disease. *PharmacoEconomics, 22*, 11–18.

Crowley, C. A., Curnutte, J. T., Rosin, R. E., André-Schwartz, J., Gallin, J. I., Klempner, M., et al. (1980). An inherited abnormality of neutrophil adhesion. Its genetic transmission and its association with a missing protein. *The New England Journal of Medicine, 302*, 1163–1168.

DiSipio, T., Rye, S., Newman, B., & Hayes, S. (2013). Incidence of unilateral arm lymphoedema after breast cancer: A systematic review and meta-analysis. *The Lancet Oncology, 14*, 500–515.

Dotta, L., Tassone, L., & Badolato, R. (2011). Clinical and genetic features of Warts, Hypogammaglobulinemia, Infections and Myelokathexis (WHIM) syndrome. *Current Molecular Medicine, 11*, 317–325.

Fiocchi, C. (2002). Inflammatory bowel disease: Dogmas and heresies. *Digestive and Liver Disease, 34*, 306–311.

Gore, R. W. (1974). Pressures in cat mesenteric arterioles and capillaries during changes in systemic arterial blood pressure. *Circulation Research, 34*, 581–591.

Granger, D. N., & Senchenkova, E. (2010). *Inflammation and the microcirculation*. San Rafael: Morgan & Claypool Life Sciences.

Lundy, D. J., & Trzeciak, S. (2011). Microcirculatory dysfunction in sepsis. *Critical Care Nursing Clinics of North America, 23*, 67–77.

Marquardt, T., Brune, T., Lühn, K., Zimmer, K. P., Körner, C., Fabritz, L., et al. (1999). Leukocyte adhesion deficiency II syndrome, a generalized defect in fucose metabolism. *Journal of Pediatrics, 134*, 681–688.

Martinez-Lemus, L. A. (2012). The dynamic structure of arterioles. *Basic & Clinical Pharmacology & Toxicology, 110*, 5–11.

Matthay, M. A., Ware, L. B., & Zimmerman, G. A. (2012). The acute respiratory distress syndrome. *The Journal of Clinical Investigation, 122*, 2731–2740.

Meller, J., Malinin, N. L., Panigrahi, S., Kerr, B. A., Patil, A., Ma, Y., et al. (2012). Novel aspects of Kindlin-3 function in humans based on a new case of leukocyte adhesion deficiency III. *Journal of Thrombosis and Haemostasis, 10*, 1397–1408.

Parker, M. M., Shelhamer, J. H., Natanson, C., Alling, D. W., & Parrillo, J. E. (1987). Serial cardiovascular variables in survivors and nonsurvivors of human septic shock: Heart rate as an early predictor of prognosis. *Critical Care Medicine, 15*, 923–929.

Rivera-Nieves, J., Olson, T., Bamias, G., Bruce, A., Solga, M., Knight, R. F., et al. (2005). L-selectin, alpha 4 beta 1, and alpha 4 beta 7 integrins participate in $CD4^+$ T cell recruitment to chronically inflamed small intestine. *Journal of Immunology, 174*, 2343–2352.

Roger, V. L., Go, A. S., Lloyd-Jones, D. M., Benjamin, E. J., Berry, J. D., Borden, W. B., et al. (2012). Executive summary: Heart disease and stroke statistics–2012 update: A report from the American Heart Association. *Circulation, 125*, 188–197.

Sacco, S., Marini, C., Toni, D., Olivieri, L., & Carolei, A. (2009). Incidence and 10-year survival of intracerebral hemorrhage in a population-based registry. *Stroke, 40*, 394–399.

Salmon, A. H., & Satchell, S. C. (2012). Endothelial glycocalyx dysfunction in disease: Albuminuria and increased microvascular permeability. *The Journal of Pathology, 226*, 562–574.

Sawada, N. (2013). Tight junction-related human diseases. *Pathology International, 63*, 1–12.

Ulbrich, H., Eriksson, E. E., & Lindbom, L. (2003). Leukocyte and endothelial cell adhesion molecules as targets for therapeutic interventions in inflammatory disease. *Trends in Pharmacological Sciences, 24*, 640–647.

Modulation of Inflammation by Key Nutrients

Luc Cynober[1,2] and Jean-Pascal De Bandt[1,2]
[1]Department of Experimental, Metabolic and Clinical Biology, Paris Descartes University, Paris, France
[2]Service de Biochimie, Hôpital Cochin, Paris, France

Synonyms

Amino acids; Food; Free fatty acids; Trace elements; Vitamins

Definition

A number of nutrients play a regulatory role in inflammation, being anti- or pro-inflammatory. Hence, the dietary balance of nutrients warrants consideration and additional lever for controlling the inflammation process, prompting an intensive research effort to dampen inflammation using diets that are either rich or poor in a number of specific nutrients (Fig. 1). Inflammation and oxidative stress are tightly connected (Wei et al. 2008) and, if excessive or too long lasting, can lead to systemic inflammatory response syndrome (SIRS), which involves septic shock with multi-organ failure. Moreover, they interact closely with wound healing and insulin sensitivity. This article focuses on the literature addressing inflammation or, in some cases, outcome (i.e., morbi–mortality) when the effect is likely related to inflammation.

Most diseases are marked by some degree of inflammation. Due to space limitations, this entry concentrates mainly on acute inflammation.

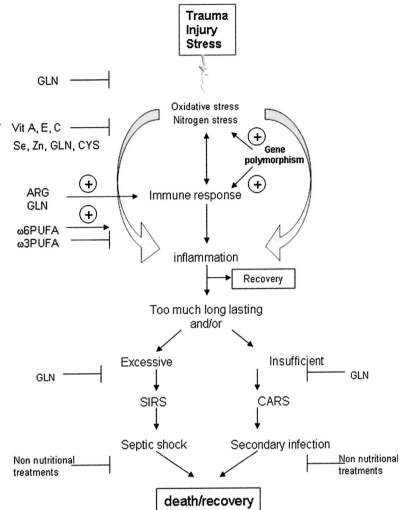

Modulation of Inflammation by Key Nutrients, Fig. 1 From trauma to recovery or death: targets for nutrients. Most nutrients act upstream in this cascade. Nutrients are not magic bullets: only therapeutic intervention (e.g., control of blood pressure, antibiotics) is able to influence the final outcome in patients with overly long-lasting and/or inadequate immune response. *GLN* glutamine, *Vit* vitamins, *Se* selenium, *Zn* zinc, *CYS* cysteine, *ARG* arginine, *PUFAs* polyunsaturated fatty acids, *SIRS* systemic inflammatory response syndrome, *CARS* compensatory anti-inflammatory response syndrome

Readers interested in the effects of nutrients in chronic inflammation states (e.g., obesity, aging, heart or pulmonary failure) are invited to refer to some excellent reviews (Field et al. 2002; Cevenini et al. 2013; Wu 2009).

Structures and Functions

Amino Acids

Glutamine (GLN)
GLN is the amide of a dicarboxylic amino acid, glutamate.

Glutamine displays anti-inflammatory properties through a number of mechanisms. First, while resting cells primarily use glucose, GLN is a major energy fuel for activated immune cells, as it is totally oxidized via a pathway called "glutaminolysis."

Second, GLN is a nitrogen (N) donor for the synthesis of purines and pyrimidines that are required for RNA transcription and DNA replication and therefore mandatory for cell proliferation (Roth 2007). This property, combined with its role as an energy source, makes GLN indispensable for an adequate immune response to stress and for wound healing through fibroblast proliferation (Stechmiller 2010).

Third, GLN is a key precursor for the glutamate pool used for the synthesis of glutathione (Roth 2007), a major antioxidant.

Fourth, GLN may directly, or indirectly through glutathione, repress NFkB activation (Roth 2007) or activate the MAP kinase cascade, thus facilitating tissue repair.

Fifth, GLN upregulates the expression of heat shock proteins (Hsps), notably Hsp70 which helps protect cells from various stresses (Roth 2007).

Arginine (ARG)

ARG is a dibasic amino acid.

ARG pools are often depleted in inflammatory situations, especially when acute (Roth 2007). This results from activation of ureagenesis and upregulation of the inducible form of °NO synthase (iNOS) at whole-body level. In immune cells, °NO production depends on the iNOS expression–arginase expression balance, where arginase overexpression diverts ARG from °NO synthesis (Morris 2007). Now, ARG is the only endogenous direct precursor of °NO to exert potent anti-inflammatory action. Note that in the absence of ARG, NOS works in a so-called uncoupled mode generating a very potent pro-inflammatory free radical called peroxynitrite (Roth 2007).

In addition, ARG is a precursor of aliphatic polyamines and of creatine and an intermediary in the urea cycle.

Cysteine (CYS)

CYS is a sulfur-containing amino acid.

CYS is easily oxidized into its dipeptide cystine (CYS-CYS).

CYS is the second limiting amino acid for glutathione synthesis (Wu 2009).

Fatty Acids (FAs)

FAs play a complex role in the control of inflammation. As ω3 and ω6 polyunsaturated fatty acids (PUFAs) use the same pathway for elongation and desaturation (Carpentier 1993), they compete for the same enzyme. Therefore, providing large amounts of one type of PUFA decreases the synthesis of the derivatives of the other.

ω3 Polyunsaturated Fatty Acids (ω3 PUFAs)

The main ω3 PUFAs are α-linolenic acid, eicosapentaenoic acid (EPA), and docosahexaenoic acid (DHA). All three are anti-inflammatory and modulate immune function by acting on physical properties of the cell membrane, signal transduction, and gene expression (Calder and Grimble 2011; Field et al. 2002). Besides their role as precursors of anti-inflammatory eicosanoids, ω3 PUFAs – in contrast with ω6 PUFA derivatives – lead to the synthesis of resolvins and protectins involved in the resolution of inflammation (Ott et al. 2011).

ω6 Polyunsaturated Fatty Acids (ω6 PUFAs)

The main ω6 PUFA is linoleic acid. ω6 PUFAs are pro-inflammatory and immunosuppressive (Calder and Grimble 2011).

Trace Elements and Vitamins

Severe inflammation such as that seen after burn injury is characterized by heavy depletion in trace elements and vitamins as a result of oxidative stress, tissue sequestration, and cutaneous and/or urinary losses (Cynober et al. 2005; Hardy et al. 2012; Shenkin 2011).

Zinc (Zn)

Zn is an important antioxidant due to its role in superoxide dismutase and glutathione activities. Zn also modulates immune function and glucose homeostasis and promotes wound healing (Taylor and Krenitsky 2010).

Selenium (Se)

Se has antioxidant, immunological, and anti-inflammatory properties likely related to its presence in selenoproteins (Hardy et al. 2012). The anti-inflammatory action of Se results from inhibition of the activation of the NFkB cascade (Hardy et al. 2012). The immune properties of Se include stimulation of T-cell proliferation and macrophage cytotoxicity and the expression of IL-2R (Field et al. 2002).

Vitamin C (Vit C)

Vit C modulates immune function through an array of mechanisms (see Field et al. 2002 for details).

Vit C is a potent antioxidant by itself due to its redox properties, as well as through the regeneration of Vit E into its reduced form.

Vitamin A (Vit A)

Vit A affects immune functions by eliciting lymphocyte proliferation by activating the retinoic acid receptor RAR-α and negatively regulating IFN-γ secretion, thus favoring Th2 versus Th1 response (Field et al. 2002).

Vit A promotes wound healing through its action on the inflammation phase (by increasing the number of macrophages) and on the epithelialization phase (by increasing collagen deposition by fibroblasts) (Stechmiller 2010).

Vitamin E (Vit E)

Vit E is a potent antioxidant, or more precisely an oxidant scavenger, that regulates both specific and nonspecific immune response (Field et al. 2002). Vit E functions as a nonspecific chain-breaking antioxidant, thus preventing propagation of free radical reactions, in particular, in PUFA.

Pathological Relevance

Amino Acids

Glutamine (GLN)

Depletion of GLN pools has been associated to poor prognosis in critically ill patients, probably due to its pleiotropic properties (Roth 2007). Conversely, a vast number of experimental studies and some clinical trials have shown that dietary supplementation with GLN decreases inflammation; in particular, GLN promotes the expression of Th2 lymphocytes (Cynober 2000).

There is broad consensus that GLN decreases SIRS (systemic inflammatory response syndrome) in critically ill patients, with any discrepancies likely related to inadequate study design. The importance of study design in situations such as inflammation was highlighted in a recent clinical trial by Cavalcante et al. (2012) showing that high-dose (30 g/day) GLN supplementation may have only marginal effects on inflammation and oxidative stress in ICU patients with SIRS if administered for a too short period, i.e., 2 days (Cavalcante et al. 2012). Moreover, crossover study designs (Cavalcante et al. 2012) are frequently ill-adapted to ICU patients in whom clinical picture can abruptly change.

Of particular interest in the context of this entry are burn patients (Cynober et al. 2005), since severe burn injury combines both local and systemic inflammation. In this setting, quality and speed of wound healing and prevention of infectious complications are critical factors. GLN-supplemented burn patients exhibit more complete wound healing, lower systemic inflammation (judged on plasma CRP levels), and lower bacterial complications than controls. Results are fuzzier (better with GLN or no effect) in terms of ICU length of stay and mortality (Cynober et al. 2005).

Ornithine α-ketoglutarate (OKG) is a potent GLN precursor (Le Boucher et al. 1999). Several clinical trials have been performed in burn patients, including a dose-ranging study (Cynober et al. 2005). OKG-supplemented enteral nutrition has a huge positive effect on wound healing. In addition, several experimental studies performed in various models where immunological status is compromised (e.g., burn injury, corticosteroid administration, endotoxemia) indicate that OKG has a potent action on immune suppression or overactivation (Cynober 2000). Based on these results, French guidelines (Cynober et al. 2005) advocate a 30 g/day GLN or OKG supplementation of the diet in burn patients.

Arginine (ARG)

A number of studies (Stechmiller 2010) have shown that ARG supplementation enhances wound healing, largely by enhancing wound tensile strength. Experiments in iNOS knockout mice indicate that this effect is related to °NO production within the wound (Shi et al. 2000). The secretagogue effect of ARG on growth hormone secretion may also be involved (Cynober 2000).

Note that OKG is not only a GLN precursor (see above) but also a potent ARG precursor (Cynober 2004).

Cysteine (CYS)
Plasma cysteine concentration is decreased in conditions of severe infection (Roth 2007). Experimental models of sepsis demonstrate an increased requirement for sulfur-containing AAs, probably due to increased glutathione synthesis to limit oxidative stress (Malmezat et al. 1998, 2000). Provided in its acetylated form N-acetyl-CYS, CYS appears to improve outcome in ICU patients (Spapen et al. 1998; Ortolani et al. 2000).

Leucine (LEU)
Inflammation decreases muscle protein synthesis through inhibition of the mTOR pathway. LEU is a strong activator of this pathway and has been shown to increase protein synthesis in a large array of pathological situations, but its effect is blunted in sepsis conditions (Lang et al. 2007). In addition, there is no data to show that LEU modulates inflammation per se (De Bandt and Cynober 2006).

Fatty Acids (FA)

ω3 Polyunsaturated Fatty Acids (ω3 PUFAs)
Clinical trials performed in ICU subjects (Taylor and Krenitsky 2010) show that patients receiving ω3 PUFA-containing parenteral nutrition (PN) have lower plasma CRP than those receiving standard PN. In septic patients, fish oil-based PN decreases inflammation, infection rate, and hospital length of stay (Calder and Grimble 2011).

ω6 Polyunsaturated Fatty Acids (ω6 PUFAs)
Studies indicate that it is important to cut ω6 PUFA intake in patients with acute inflammation (Calder and Grimble 2011). There is a general consensus that conventional soybean oil-based lipid emulsions contain excessive amounts of ω6 PUFA, which can be decreased in absolute amount or by adding ω3 PUFA to increase ω3/ω6 ratio (Sobotka et al. 2011).

Trace Elements and Vitamins

Zinc (Zn)
While the importance of Zn has been demonstrated in several chronic conditions (Prasad 2009), studies in severe acute situations post disappointing results. The fact that most such studies associate Zn with several other antioxidants may be a matter of concern (see below).

Selenium (Se)
The few limited clinical trials on Se supplementation in ICU patients have posted fairly disappointing results (Taylor and Krenitsky 2010), especially for intention-to-treat analysis (Hardy et al. 2012). A recent review by Hardy et al. (2012) reached a more optimistic conclusion but stressed that more studies are required to conclude on the efficacy of Se supplementation in the critically ill and underlined the importance of type of salt used (i.e., organic versus inorganic), daily dosage, and administration protocol (a loading dose followed by continuous infusion seems better). They suggested providing 500–1600 mg Se/day to critically ill patients, but this dosage may prove higher than the lowest observed adverse effect level (LOAEL) of Se, which is 850 mg/day.

Vitamin C (Vit C)
Vit C supplementation has been shown to promote wound healing (Stechmiller 2010) but mainly promotes the noninflammatory phases of the process.

Vitamin A (Vit A)
To enhance wound healing in injured patients, an oral intake of 10,000–50,000 IU vit A/day is recommended (Stechmiller 2010).

Vitamin E (Vit E)
The recommended dose for adult patients is 9.1 mg/day (Dupertuis et al. 2011).

The So-Called Immune-Enhancing Diets (IEDs) or Immune-Modulating Nutrients

The discovery made in the 1980s that some key nutrients are able to modulate inflammation, oxidative stress, and related clinical events has led investigators to test (and then companies to formulate) a loosely defined group of enteral nutrition products enriched with ARG, GLN, ω3 PUFA, nucleotides, vitamins, and/or trace elements, called IEDs. Several meta-analyses, such as from Marimuthu et al. (2012), conclude that IEDs decrease postoperative infectious and noninfectious complications and hospital length of stay in surgical patients. In ICU patients, debate continues over whether or not IEDs increase mortality and thus whether or not ARG, through its purported excessive °NO production, is to blame. In-depth discussion of this point is beyond the scope of this entry, but the interested reader is invited to look at some of the reviews and editorials available (Cynober 2003; Cynober et al. 2012; Heyland et al. 2001; Luiking et al. 2005; Manzanares & Heyland 2012). What is certain is that the key nutrients described above may have positive effects when given alone but may interact negatively when given in combination (Hamani et al. 2010).

Interaction with Other Processes

Hardy et al. (2012) rightly reminded us that the activity of a number of key enzymes involved in the action of nutrients is influenced by gene polymorphism. Therefore, providing a nutrition enriched in a given nutrient may prove beneficial, neutral, or harmful depending on the patient involved. A good example is patients with innate high iNOS expression who may be more prone to circulatory shock during sepsis if given excessive ARG supply. There is no doubt that, in the future, critically ill patients will be able to benefit from gene screening at entrance in the ICU, which will provide a sound basis for feeding (or not) patients with pharmacological dosages of anti-inflammatory nutrients.

Another key aspect is the time schedule of administration of certain nutrients, given that the metabolic and immune status of patients with acute inflammation can change very fast. Thus, nutrition should be personalized not only to patient phenotype but to stage of the inflammatory response. This type of nutrition product (Modulis™, Nestlé Clinical Nutrition) has already been proposed and includes two units. The first contains antioxidant and should be given in the early phase of inflammation to prevent SIRS; the second contains five key AAs (including GLN and CYS) and should be given later on to prevent compensatory anti-inflammatory response syndrome (Schiffrin 2007).

Summary

A number of amino acids (glutamine, arginine, cysteine), lipids (ω3 polyunsaturated fatty acids), and trace elements (zinc, selenium, vitamins C, A, E), delivered separately or in cocktails, exert a regulatory role on inflammation, with all but one (ω6 polyunsaturated fatty acids) blunting this process. From a nutritional standpoint, inflammation and oxidative stress are tightly connected, making it difficult to determine which is the *prime mover*. Tackling inappropriate (excessive and/or overly long-lasting) inflammation is a major therapeutic goal, especially in ICU patients where it can be responsible for septic shock or fatal secondary infection.

Cross-References

▶ Autophagy and Inflammation
▶ Cancer and Inflammation
▶ Non-steroidal Anti-inflammatory Drugs: Overview
▶ Sepsis Models in Animals
▶ Skin Inflammation Models in Animals

References

Calder, P. C., & Grimble, R. F. (2011). Nutrients that influence inflammation and immunity: ω3 fatty acids. In L. Sobotka (Ed.), *Basics in clinical nutrition* (4th ed., pp. 292–299). Galen: Prague.

Carpentier, Y. A. (1993). Lipid emulsions. In P. Fürst (Ed.), *New strategies in clinical nutrition* (pp. 52–63). München: W. Zuckschwerdt Verlag.

Cavalcante, A. A. M., Campelo, M. W. S., de Vasconcelos, M. P. P., et al. (2012). Enteral nutrition supplemented with L-glutamine in patients with systemic inflammatory response syndrome due to pulmonary infection. *Nutrition, 28*, 397–402.

Cevenini, E., Monti, D., & Franceschi, C. (2013). Inflamm-aging. *Current Opinion in Clinical Nutrition and Metabolic Care, 16*, 14–20.

Cynober, L. (2000). Control of inflammatory mediators by nitrogenous nutrients. *Nutrition Clinique et Métabolisme, 14*, 194–200.

Cynober, L. (2003). Immune-enhancing diets for stressed patients with a special emphasis on arginine content: Analysis of the analysis. *Current Opinion in Clinical Nutrition and Metabolic Care, 6*, 189–193.

Cynober, L. (2004). Ornithine α-ketoglutarate as a potent precursor of arginine and nitric oxide: A new job for an old friend. *Journal of Nutrition, 134*(Suppl), 2858–2862.

Cynober, L., Bargues, L., Berger, M. M., et al. (2005). Nutritional recommendations for severe burn victims. *Nutrition Clinique et Métabolisme, 19*, 166–194.

Cynober, L., Moinard, C., & Charrueau, C. (2012). If the soup tastes bad, it doesn't mean the potatoes are the culprit. *Critical Care Medicine, 40*, 2540–2541.

De Bandt, J. P., & Cynober, L. (2006). Therapeutic use of branched-chain amino acids in burn, trauma and sepsis. *Journal of Nutrition, 136*(suppl), 308–313.

Dupertuis, Y. M., Cai, F., & Pichard, C. (2011). Nutrients that influence immunity: Experimental and clinical data. In L. Sobotka (Ed.), *Basics in clinical nutrition* (4th ed., pp. 299–307). Gaten: Prague.

Field, C. J., Johnson, I. R., & Schley, P. D. (2002). Nutrients and their role in host resistance to infection. *Journal of Leukocyte Biology, 71*, 16–32.

Hamani, D., Kuhn, M., Charrueau, C., et al. (2010). Interactions between ω3 polyunsaturated fatty acids and arginine on nutritional and immunological aspects in severe inflammation. *Clinical Nutrition, 29*, 654–662.

Hardy, G., Hardy, I., & Manzanares, W. (2012). Selenium supplementation in the critically ill. *Nutrition in Clinical Practice, 27*, 21–33.

Heyland, D. K., Novak, F., Drover, J. W., et al. (2001). Should immunonutrition become routine in critically ill patients? *JAMA, 286*, 22–29.

Lang, C. L., Frost, R. A., & Vary, T. C. (2007). Regulation of muscle protein synthesis during sepsis and inflammation. *American Journal of Physiology. Endocrinology and Metabolism, 203*, E453–E459.

Le Boucher, J., Farges, M. C., Minet, R., et al. (1999). Modulation of immune response with ornithine a-ketoglutarate in burn injury: An arginine or glutamine dependency? *Nutrition, 15*, 773–777.

Luiking, Y. C., Poeze, M., Ramsay, G., et al. (2005). The role of arginine in infection and sepsis. *JPEN Journal of Parenteral and Enteral Nutrition, 29*(Suppl), 70–74.

Malmezat, T., Breuillé, D., Pouyet, C., et al. (1998). Metabolism of cysteine is modified during the acute phase of sepsis in rats. *Journal of Nutrition, 128*, 97–105.

Malmezat, T., Breuillé, D., Pouyet, C., et al. (2000). Methionine transsulfuration is increased during sepsis in rats. *American Journal of Physiology. Endocrinology and Metabolism, 279*, E1391–E1397.

Manzanares, W., & Heyland, D. K. (2012). Pharmaconutrition with arginine decreases bacterial translocation in an animal model of severe trauma. Is a clinical study justified? The time is now! *Critical Care Medicine, 40*, 350–352.

Marimuthu, K., Varadhan, K. K., Ljungqvist, O., et al. (2012). A meta-analysis of the effect of combinations of immune modulating nutrients on outcome in patients undergoing major open gastrointestinal surgery. *Annals of Surgery, 255*, 1060–1068.

Morris, S. M., Jr. (2007). Arginine metabolism: Boundaries of our knowledge. *Journal of Nutrition, 137* (Suppl), 1602–1609.

Ortolani, O., Conti, A., De Gaudio, A. R., et al. (2000). The effect of glutathione and N-acetylcysteine on lipoperoxidative damage in patients with early septic shock. *American Journal of Respiratory and Critical Care Medicine, 161*, 1907–1911.

Ott, J., Hiesgen, C., & Mayer, K. (2011). Lipids in critical care medicine. *Prostaglandins, Leukotrienes, and Essential Fatty Acids, 85*, 267–273.

Prasad, A. S. (2009). Zinc: Role in immunity, oxidative stress and chronic inflammation. *Current Opinion in Clinical Nutrition and Metabolic Care, 12*, 646–652.

Roth, E. (2007). Immune and cell modulation by amino acids. *Clinical Nutrition, 26*, 535–544.

Schiffrin, E. (2007). [Enteral nutrition in the ICU. The Nestlé Modulisä innovation]. Physiopathology of the traumatized patient in the ICU. *Nutr Clin Métabol, 21*, S6–S10.

Shenkin, A. (2011). Physiological function and deficiency states of vitamins. In L. Sobotka (Ed.), *Basics in clinical nutrition* (4th ed., pp. 145–153). Galen: Prague.

Shi, H. P., Efron, D. T., Most, D., et al. (2000). Supplemental dietary arginine enhances wound healing in normal but not inducible nitric oxide synthase knockout mice. *Surgery, 128*, 374–378.

Sobotka, L., Soeters, P. B., Jolliet, P., et al. (2011). Nutritional support in critically ill and septic patients. In L. Sobotka (Ed.), *Basics in clinical nutrition* (4th ed., pp. 444–451). Galen: Prague.

Spapen, H., Zhang, H., Demanet, C., et al. (1998). Does N-acetyl-L-cysteine influence cytokine response during early human septic shock? *Chest, 113*, 1616–1624.

Stechmiller, J. K. (2010). Understanding the role of nutrition and wound healing. *Nutrition in Clinical Practice, 25*, 61–68.

Taylor, B., & Krenitsky, J. (2010). Nutrition in the intensive care unit: Year in review 2008–2009. *JPEN Journal of Parenteral and Enteral Nutrition, 34*, 21–31.

Wei, Y., Chen, K., Whaley-Connell, A. T., et al. (2008). Skeletal muscle insulin resistance: Role of inflammatory cytokines and reactive oxygen species. *American Journal of Physiology. Regulatory, Integrative and Comparative Physiology, 294*, R673–R680.

Wu, G. (2009). Amino acids: Metabolism, functions, and nutrition. *Amino Acids, 37*, 1–17.

Natural Killer Cells

Paul Rouzaire[1], Sébastien Viel[2], Jacques Bienvenu[2] and Thierry Walzer[2]
[1]Service d'Immunologie Biologique, CHU de Clermont-Ferrand, Université d'Auvergne, ERTICa EA4677, Clermont-Ferrand, France
[2]Université de Lyon, INSERM U1111, Lyon, France

Synonyms

Innate cytotoxic lymphocytes

Definition

Initially considered as "background noise" in specific T cell cytolytic assays (Vivier 2006), natural killer (NK) cells were first characterized in 1975 by their ability to kill tumor cells without any prior sensitization (Kiessling et al. 1975a, b). Since then, these cells have been the focus of lots of scientific publications, and we know today that NK cells represent a key component of the innate immune system, widespread through lymphoid and nonlymphoid tissues, displaying two major functions: cytotoxicity against tumor, infected or stressed cells, and cytokine production (Vivier et al. 2008). Morphologically, NK cells are very similar to cytotoxic T cells: large granular lymphocytes containing numerous vesicles rich in cytolytic proteins such as perforin and granzymes (Trinchieri 1989). However, unlike T cells, NK cells do not display somatically rearranged receptors but are regulated by a fixed repertoire of germ-line-encoded activating and inhibitory receptors (Stewart et al. 2006).

Structure and Functions

NK Cell Development and Localization

As a key component of the innate immune system, NK cells are widely distributed in the body. They are mainly produced in the bone marrow, and during their development, they reach peripheral tissues, through blood circulation. Indeed they represent 3–10 % of lymphocytes in the blood, the spleen, the liver, and the lung. They are also present, but at lower frequency, in the lymph nodes, thymus, dermis, or bowel lamina propria (Gregoire et al. 2007). Furthermore, it is interesting to note the massive recruitment of NK cells to uterine tissue during pregnancy (Moffett-King 2002).

Human NK cell development program is classically divided in five steps, from $CD34^+$ hematopoietic stem cells to the fully mature NK cells (Freud and Caligiuri 2006). During the first three steps, progenitors progressively engage into the NK lineage. The following steps correspond to the functional maturation of NK cells. These last steps are distinguished by the expression level of the CD56 surface molecule. $CD56^{bright}$ NK cells are mainly localized in lymphoid organs and further differentiate into CD56^{dim} NK cells that circulate in the blood. Various phenotypic and functional differences exist between these two subtypes of NK cells. First, although all NK cells express the IL-2 receptor, only CD56^{bright} NK cells express the high affinity heterotrimeric receptor, allowing them to proliferate in response to weaker doses of IL-2 than CD56^{dim} NK cells (Caligiuri et al. 1990). Second, CD56^{dim} and CD56^{bright} NK cells express different sets of chemokine receptors and adhesion molecules, which probably account for their different localizations in the body. Indeed, CD56^{bright} NK cells express CCR7, CXCR3, and the selectin CD62-L, explaining their preferential localization in the secondary lymphoid tissues, whereas CD56^{dim} NK cells, mainly distributed in the peripheral blood, do not express CCR7 and lose the expression of CD62-L during their differentiation (Walzer and Vivier 2011). Third, CD56^{dim} and CD56^{bright} NK cells differ in their expression of activating and inhibitory receptors. Notably, CD56^{bright} NK cells do not express CD16 (FcγRIII), the low affinity receptor of gamma immunoglobulin, whereas 95 % of CD56^{dim} NK cells express it, making them capable of antibody-dependent cell cytotoxicity (ADCC) (Cooper et al. 2001). Fourth, CD56^{bright} and CD56^{dim} display different effector functions as outlined below.

NK Cell Receptors

Unlike T cells, NK cells do not express a unique antigen receptor but an array of activating and inhibitory receptors, called natural killer cell receptors (NKRs), which form the recognition repertoire of NK cells. In human, NKRs are divided into two structural groups: the immunoglobulin-like receptors (KIRs) and the C-type lectin receptors (as CD94 coupled to NKG2 receptors) (Stewart et al. 2006).

Many inhibitory NK cell receptors recognize MHC-I molecules, either classical MHC-I molecules, recognized by KIRs, or nonconventional MHC molecules, recognized, for instance, by the heterodimeric receptor CD94-NKG2A (a lectin-C-type receptor). All these receptors transduce inhibitory signals via immunotyrosine inhibitory motifs (ITIM) in their cytoplasmic tail. Other inhibitory receptors do not recognize MHC molecules. For example, KLRG1 recognizes E/N/R cadherins.

The group of activating NKRs is more diverse, both in terms of structure and ligands. Among them, CD16 (FcγRIII) is responsible for ADCC, enabling the recognition of IgG-opsonized cells. The group of natural cytotoxicity receptors (NCRs) comprises several members (notably NKp30, NKp44, and NKp46 in humans) that recognize viral proteins. They also bind to other ligands that remain unknown. NKG2D is a C-type lectin-like activating immunoreceptor which ligands are MHC-related molecules (such as MIC-A, MIC-B, and Rae-1 in humans). The expression of NKG2D ligands is weak on cells at basal state and induced by different cellular stresses (such as infection or tumoral transformation). Lastly, other structural groups (KIRs and lectin-type receptors) also comprise activating members. All these activating receptors have to associate with adaptor molecules comprising immunotyrosine activating motifs (ITAM) or other activating motifs for the transduction of activating signals to NK cells.

NK Cell Recognition of Danger

The emblematic property of NK cells is their capacity to recognize "missing self," i.e., the absence of expression of MHC-I molecules by cells, as discovered by the group of K. Karre in 1990s (Karre 2008). A reduced expression of MHC-I levels is often observed on tumoral or virus-infected cells. As MHC-I molecules are the ligands of numerous inhibitory NKRs, their downregulation leads to the lack of inhibitory signals and thus to the activation of NK cells.

More recent works have shown that NK cells could also recognize the "stress-induced self" and the "infectious nonself" (Luci and Tomasello 2008). The archetypic receptor involved in stress-induced self-recognition is NKG2D which ligands are induced by cellular stress caused by infections or tumors. Several studies have illustrated the strong interaction between activating NKRs and viral proteins.

Figure 1 represents these different situations of tolerance/activation of NK cells.

Natural Killer Cells, Fig. 1 NK cell activation

It appears today that these three situations are closely intertwined, and that the activation of NK cells is dictated by the integration of all inhibitory and activating signals received through the different NKRs.

NK Cell Education

As the downregulation or the absence of MHC-I molecules can induce NK cell activation, the mirror situation, i.e., the absence of expression of any inhibitory NKRs for self-MHC-I molecules by NK cells, might result in similar consequences. The existence of NK cells that do not express any inhibitory NKRs for self-MHC-I has been demonstrated both in mice and humans. According to the missing self theory, these cells should thus be autoreactive. Actually, such cells seem rather hyporesponsive and are not able to lyse target cells, expressing or not MHC-I molecules. These findings led to the NK cell "education" or "licensing" theory that proposes that during their development, NK cells, to become fully mature and functional, have to be "educated" via the engagement of at least one of their inhibitory NKRs on a self-ligand (Elliott and Yokoyama 2011).

Effector Functions of NK Cells

NK cells are characterized by two major effector functions: cytotoxicity and chemokine/cytokine production.

NK cell cytotoxicity is mediated by degranulation, i.e., the exocytosis of lytic granules containing perforin, granzymes, and granulysin (in humans) (Luci and Tomasello 2008). The destruction of target cells by NK cells requires the establishment of an immunological synapse between the two cells, comparable in many aspects to the cytotoxic T cell synapse (Orange 2008). This phenomenon allows the recognition of the target cells, the polarization of the NK cells, and the secretion of lytic granules responsible for the target cell death. Moreover, NK cells can also induce cell death by the activation of "death receptors," notably the pathways Fas/Fas-L and TRAIL-R/TRAIL (Arase et al. 1995; Kayagaki et al. 1999).

The second major function of NK cells is the production of several chemokines and cytokines. Among them, IFN-γ is particularly important; indeed, NK cells provide a massive and early source of this cytokine (Martin-Fontecha et al. 2004), necessary for Th1 polarization and antiviral and antitumoral immunity.

Innate or Adaptative Cells?

It is classically admitted in immunology that only T and B lymphocytes can adapt to the encountered antigen and mount an "adaptive" immune response, characterized by a clonal expansion of antigen-specific lymphocytes, followed by a contraction phase with the persistence of a memory component able to build a robust and quick secondary response against the same antigen. However, recent findings suggest the existence of NK cells endowed with "adaptive-like" properties, i.e., antigenic specificity, expansion following antigen encounter, and memory. Indeed, independent studies demonstrated that in mouse models of delayed-type hypersensitivity (DTH), in the absence of adaptive effectors (T and B cells), NK cells could mount antigen-specific responses directed against different haptens or viral particles (O'Leary et al. 2006; Paust et al. 2010; Rouzaire et al. 2012). These antigen-specific NK cells are furthermore characterized by their capacity to persist for long periods of time and to mount a memory response, even several weeks after the first encounter with the hapten or the viral particles. The cellular and molecular mechanisms of these memory properties are still unclear today. It is also still unclear whether "memory" NK cells participate to secondary responses in physiological contexts.

Pathological Relevance

NK Cells and Viral Infections

The role of NK cells in antiviral defenses is largely demonstrated. Alteration of NK cell function by genetic or experimental means results in a major increase in susceptibility to infections by viruses such as *Ectromelia*, *Ebola*, *Influenza*, or *Cytomegalovirus* in mouse models of infection (Lee et al. 2007). NK cells have also been reported to play an important role in the defense against hepatitis C virus (Cheent and Khakoo 2011), human immunodeficiency virus (Biron et al. 1999), or yellow fever virus (Andoniou et al. 2006).

This antiviral role of NK cells is mediated by their two main effector functions: cytokine production (notably IFN-γ) and cytotoxicity (either via the exocytosis of lytic granules or via their death ligands) (Lee et al. 2007). Interestingly, viruses of the Herpesviridae group have developed strategies to escape NK cell surveillance, for instance, by blocking the expression of NKG2D ligands by the infected cell (Jonjic et al. 2008).

NK Cells and Cancer

NK cells were identified in 1975 on the basis of their capacity to lyse leukemic cells without prior sensitization (Kiessling et al. 1975a, b). Since then, numerous in vitro studies (using human or other mammalian cells) and in vivo studies (in mouse or rat models) have demonstrated that NK cells can recognize and lyse tumor cells. NK cells participate thus, along with other immune cells, to the tumor "immunosurveillance" process. Different mouse models have thereby demonstrated the role of NK cells in the rejection of different transplant tumors, in a manner dependant of the presence (or absence) or certain NKR ligand on tumor cells, such as the absence of

Natural Killer Cells, Fig. 2 NK cell functions

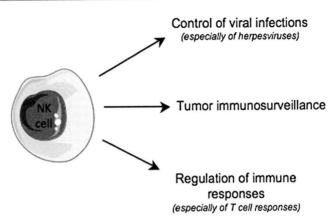

MHC-I molecules (Karre et al. 1986), or the overexpression of NKG2D ligands (Diefenbach et al. 2001). Furthermore, NKG2D-deficient mouse models have shown an important role of the NKR in the tumor immunosurveillance (Guerra et al. 2008).

In humans, very few selective NK cell deficiencies have been described: the characterization of NK cell involvement in tumor immunosurveillance is thus not possible (Orange 2006). However, it has been reported an inverse relation between NK cell cytotoxic activity in peripheral blood and the cancer risk (Imai et al. 2000). The function of NK cells infiltrating solid tumors is often altered, suggesting that many tumors secrete factors inhibiting NK cell function as a means to escape their tumoricidal activity. Several strategies have been used in clinics to boost NK cell function in cancer patients as detailed below.

Regulation of T Cell Responses by NK Cells

In addition to their undisputed role in innate immunity, NK cells have been more recently shown to display important functions in shaping adaptive immunity, especially T cell responses. Indeed, it has been established that NK cells, upon stimulation, are recruited to lymph nodes and that this recruitment correlates with $CD4^+$ T cell lineage priming toward a Th1 response, via an early and important production of IFN-γ (Martin-Fontecha et al. 2004). Furthermore, it has also been shown that NK cells can control $CD8^+$ T cell priming (Soderquest et al. 2011). Indeed, NK cells regulate the amplitude of $CD8^+$ T cell differentiation, limiting their clonal expansion. They also switch this differentiation toward an effector memory (T_{EM}) phenotype. As a result, the number of central memory T cells (T_{CM}) is higher when naïve $CD8^+$ T cells are primed in a NK cell-free environment, paving the way for improving new vaccine strategies against tumors and infections.

In vivo functions of NK cells are schemed in Fig. 2.

Interactions with Other Processes and Drugs

Manipulation of NK Cells in Clinics: Applications in Tumor Therapy

Since their description in the 1970s, the manipulation of antitumoral activity of NK cells has been considered as a promising strategy for cancer immunotherapy.

Allogenic Hematopoietic Stem Cell Transplantation (HSCT)

HSCT has revolutionized the treatment of numerous hematological malignancies for years (Thomas et al. 1975). It has been established that donor T cells can fight against malignant cells (the so-called *graft-versus-tumor* (GVT) effect) and reconstitute the immunity of the recipient. However, some complications have limited the widespread use of this therapy, especially the fact that allogenic immune activation leads to *graft-versus-host disease* (GVHD). Today, many efforts are made to develop strategies to manipulate donor NK cells in the graft, for at least two main reasons: NK cells are among the first cells to arise in the recipient following HSCT (Moretta et al. 2009), and it has been shown that NK cells, by contrast to T cells, can exert potent GVT effect without inducing GVHD (Ruggeri et al. 2002).

Anti-KIR

NK cell activity is regulated by a complex balance between activating and inhibitory signals, perceived through the array of, respectively, activating and inhibitory NKRs bound to their ligands on stressed or transformed target cells. As a result, the blockade of inhibitory signals can shift the balance to activation of NK cells. To achieve this aim, fully humanized KIR-specific monoclonal antibodies have been generated to block, in vivo, the interaction between inhibitory KIRs and HLA-C molecules (Alici 2010). Using this strategy of blocking inhibitory signals, anti-KIR antibodies lead to activation of NK cells toward tumoral cells (which express stress-induced ligands for activating receptors) and spare healthy cells (which do not express sufficient amount of activating ligands). First clinical trials seem to demonstrate promising results that give hope of innovative NK cell-based cancer therapies.

IMiDs

Thalidomide was originally used to treat morning sickness, but it was withdrawn in the 1960s following numerous reports of congenital birth defects. Many years later, thalidomide and its derivatives – IMiDs (immunomodulatory drugs) – constitute a "novel" class of drugs used for their antitumoral and anti-inflammatory properties in the treatment of erythema nodosum leprosum (ENL) or diverse cancers, such as relapsed/refractory multiple myeloma (Teo 2005). Among their numerous properties, these "new old" products have been shown to enhance NK cell activity and numbers and to enhance ligands for activating NKRs. These effects justify the development of clinical trials combining several NK cell-based therapies, such as anti-KIR and IMiDs (Benson et al. 2011).

Altogether, these different NK cell-based anticancer therapies have not yet firmly demonstrated efficacy. However, the number of clinical trials involving strategies to enhance NK cell activity against tumors gives hope for the improvement of anticancer treatments.

Cross-References

▶ Antiviral Responses
▶ Cancer and Inflammation
▶ Immunoglobulin Receptors and Inflammation
▶ Lymphocyte Homing and Trafficking

References

Alici, E. (2010). IPH-2101, a fully human anti-NK-cell inhibitory receptor mAb for the potential treatment of hematological cancers. *Current Opinion in Molecular Therapeutics, 12*, 724–733.

Andoniou, C. E., Andrews, D. M., & Degli-Esposti, M. A. (2006). Natural killer cells in viral infection: More than just killers. *Immunological Reviews, 214*, 239–250.

Arase, H., Arase, N., & Saito, T. (1995). Fas-mediated cytotoxicity by freshly isolated natural killer cells. *The Journal of Experimental Medicine, 181*, 1235–1238.

Benson, D. M., Jr., Bakan, C. E., Zhang, S., Collins, S. M., Liang, J., Srivastava, S., et al. (2011). IPH2101, a novel anti-inhibitory KIR antibody, and lenalidomide combine to enhance the natural killer cell versus multiple myeloma effect. *Blood, 118*, 6387–6391.

Biron, C. A., Nguyen, K. B., Pien, G. C., Cousens, L. P., & Salazar-Mather, T. P. (1999). Natural killer cells in antiviral defense: Function and regulation by innate cytokines. *Annual Review of Immunology, 17*, 189–220.

Caligiuri, M. A., Zmuidzinas, A., Manley, T. J., Levine, H., Smith, K. A., & Ritz, J. (1990). Functional consequences of interleukin 2 receptor expression on resting human lymphocytes. Identification of a novel natural killer cell subset with high affinity receptors. *The Journal of Experimental Medicine, 171*, 1509–1526.

Cheent, K., & Khakoo, S. I. (2011). Natural killer cells and hepatitis C: Action and reaction. *Gut, 60*, 268–278.

Cooper, M. A., Fehniger, T. A., & Caligiuri, M. A. (2001). The biology of human natural killer-cell subsets. *Trends in Immunology, 22*, 633–640.

Diefenbach, A., Jensen, E. R., Jamieson, A. M., & Raulet, D. H. (2001). Rae1 and H60 ligands of the NKG2D receptor stimulate tumour immunity. *Nature, 413*, 165–171.

Elliott, J. M., & Yokoyama, W. M. (2011). Unifying concepts of MHC-dependent natural killer cell education. *Trends in Immunology, 32*, 364–372.

Freud, A. G., & Caligiuri, M. A. (2006). Human natural killer cell development. *Immunological Reviews, 214*, 56–72.

Gregoire, C., Chasson, L., Luci, C., Tomasello, E., Geissmann, F., Vivier, E., et al. (2007). The trafficking of natural killer cells. *Immunological Reviews, 220*, 169–182.

Guerra, N., Tan, Y. X., Joncker, N. T., Choy, A., Gallardo, F., Xiong, N., et al. (2008). NKG2D-deficient mice are defective in tumor surveillance in models of spontaneous malignancy. *Immunity, 28*, 571–580.

Imai, K., Matsuyama, S., Miyake, S., Suga, K., & Nakachi, K. (2000). Natural cytotoxic activity of peripheral-blood lymphocytes and cancer incidence: An 11-year follow-up study of a general population. *Lancet, 356*, 1795–1799.

Jonjic, S., Polic, B., & Krmpotic, A. (2008). Viral inhibitors of NKG2D ligands: Friends or foes of immune surveillance? *European Journal of Immunology, 38*, 2952–2956.

Karre, K. (2008). Natural killer cell recognition of missing self. *Nature Immunology, 9*(5), 477–480.

Karre, K., Ljunggren, H. G., Piontek, G., & Kiessling, R. (1986). Selective rejection of H-2-deficient lymphoma variants suggests alternative immune defence strategy. *Nature, 319*, 675–678.

Kayagaki, N., Yamaguchi, N., Nakayama, M., Takeda, K., Akiba, H., Tsutsui, H., et al. (1999). Expression and function of TNF-related apoptosis-inducing ligand on murine activated NK cells. *Journal of Immunology, 163*, 1906–1913.

Kiessling, R., Klein, E., Pross, H., & Wigzell, H. (1975a). "Natural" killer cells in the mouse. II. Cytotoxic cells with specificity for mouse Moloney leukemia cells. Characteristics of the killer cell. *European Journal of Immunology, 5*, 117–121.

Kiessling, R., Klein, E., & Wigzell, H. (1975b). "Natural" killer cells in the mouse. I. Cytotoxic cells with specificity for mouse Moloney leukemia cells. Specificity and distribution according to genotype. *European Journal of Immunology, 5*, 112–117.

Lee, S. H., Miyagi, T., & Biron, C. A. (2007). Keeping NK cells in highly regulated antiviral warfare. *Trends in Immunology, 28*, 252–259.

Luci, C., & Tomasello, E. (2008). Natural killer cells: Detectors of stress. *International Journal of Biochemistry & Cell Biology, 40*, 2335–2340.

Martin-Fontecha, A., Thomsen, L. L., Brett, S., Gerard, C., Lipp, M., Lanzavecchia, A., et al. (2004). Induced recruitment of NK cells to lymph nodes provides IFN-gamma for T(H)1 priming. *Nature Immunology, 5*, 1260–1265.

Moffett-King, A. (2002). Natural killer cells and pregnancy. *Nature Reviews. Immunology, 2*, 656–663.

Moretta, L., Locatelli, F., Pende, D., Mingari, M. C., & Moretta, A. (2009). Natural killer alloeffector responses in haploidentical hemopoietic stem cell transplantation to treat high-risk leukemias. *Tissue Antigens, 75*, 103–109.

O'Leary, J. G., Goodarzi, M., Drayton, D. L., & von Andrian, U. H. (2006). T cell- and B cell-independent adaptive immunity mediated by natural killer cells. *Nature Immunology, 7*, 507–516.

Orange, J. S. (2006). Human natural killer cell deficiencies. *Current Opinion in Allergy and Clinical Immunology, 6*, 399–409.

Orange, J. S. (2008). Formation and function of the lytic NK-cell immunological synapse. *Nature Reviews. Immunology, 8*, 713–725.

Paust, S., Gill, H. S., Wang, B. Z., Flynn, M. P., Moseman, E. A., Senman, B., et al. (2010). Critical role for the chemokine receptor CXCR6 in NK cell-mediated antigen-specific memory of haptens and viruses. *Nature Immunology, 11*, 1127–1135.

Rouzaire, P., Luci, C., Blasco, E., Bienvenu, J., Walzer, T., Nicolas, J. F., et al. (2012). Natural killer cells and T cells induce different types of skin reactions during recall responses to haptens. *European Journal of Immunology, 42*, 80–88.

Ruggeri, L., Capanni, M., Urbani, E., Perruccio, K., Shlomchik, W. D., Tosti, A., et al. (2002). Effectiveness of donor natural killer cell alloreactivity in mismatched hematopoietic transplants. *Science, 295*, 2097–2100.

Soderquest, K., Walzer, T., Zafirova, B., Klavinskis, L. S., Polic, B., Vivier, E., et al. (2011). Cutting edge: CD8+ T cell priming in the absence of NK cells leads to enhanced memory responses. *Journal of Immunology, 186*, 3304–3308.

Stewart, C. A., Vivier, E., & Colonna, M. (2006). Strategies of natural killer cell recognition and signaling. *Current Topics in Microbiology and Immunology, 298*, 1–21.

Teo, S. K. (2005). Properties of thalidomide and its analogues: Implications for anticancer therapy. *The AAPS Journal, 7*, E14–E19.

Thomas, E., Storb, R., Clift, R. A., Fefer, A., Johnson, F. L., Neiman, P. E., et al. (1975). Bone-marrow transplantation (first of two parts). *New England Journal of Medicine, 292*, 832–843.

Trinchieri, G. (1989). Biology of natural killer cells. *Advances in Immunology, 47*, 187–376.

Vivier, E. (2006). What is natural in natural killer cells? *Immunology Letters, 107*, 1–7.

Vivier, E., Tomasello, E., Baratin, M., Walzer, T., & Ugolini, S. (2008). Functions of natural killer cells. *Nature Immunology, 9*, 503–510.

Walzer, T., & Vivier, E. (2011). G-protein-coupled receptors in control of natural killer cell migration. *Trends in Immunology, 32*, 486–492.

Neutrophil Extracellular Traps

Viviana Marin-Esteban[1], Lorena Barrientos[2] and Sylvie Chollet-Martin[1]
[1]Faculté de Pharmacie, Université Paris Sud, INSERM UMR 996, Châtenay-Malabry, France
[2]Faculté de Pharmacie, Université Paris Sud, INSERM UMR 996, Paris, France

Synonyms

Netosis; NETs

Definition

In response to various dangers, cells respond by releasing soluble mediators that initiate the inflammatory response. Neutrophils represent the first wave of immune cells which are abundantly recruited to the site of inflammation with the objective of eliminating the danger. In particular, neutrophils are crucial for the efficient elimination of microbes through mechanisms involving either intracellular and/or extracellular killing. In both cases, microbes are killed by the combined action of different antimicrobial molecules stored in the cytoplasmic granules or generated *de novo*.

Among the extracellular killing mechanisms, the release of neutrophil extracellular traps (NETs) is of main importance. NETs were originally described in 2004 (Brinkmann et al. 2004) and correspond to tangled webs of neutrophil DNA harboring a large diversity of proteins expelled out of the cell (Fig. 1). NETs fulfill important antimicrobial functions but are also involved in several inflammatory and autoimmune diseases.

NET release by activated leukocytes is a conserved innate antimicrobial strategy, described in fish, chicken, and different mammals including mouse and human (Guimaraes-Costa et al. 2012). The release of extracellular DNA structures decorated with proteins from the different cell compartments is not restricted to neutrophils. This phenomenon has also been documented in eosinophils, basophils, monocytes, and mast cells (Schorn et al. 2012).

Structure and Functions

Formation of NETs

A large diversity of stimuli prompts neutrophil activation leading to NET release. The formation of NETs in response to pathogens and their microbicidal effect are probably the main physiological functions of NETs and will be discussed below. During inflammation, NET release can be induced by proinflammatory mediators, such as TNFα, IL-8, platelet activating factor, the combination of granulocyte/macrophage colony-stimulating factor (GM-CSF)/complement factor 5a (C5a), or GM-CSF/LPS. Moreover, NET release can also be triggered by activated platelets or endothelial cells, by circulating or extravascular antibodies and immune complexes during autoimmune diseases, by amyloid fibrils, and by monosodium urate crystals from synovial fluid of patients with gout (Azevedo et al. 2012; Chen et al. 2012; Schorn et al. 2012). Importantly, NETs can be efficiently generated in vitro by pharmacological agents. Phorbol 12-myristate 13-acetate (PMA), a prominent activator of protein kinase C, is the most extensively studied stimulus to induce NET release by human neutrophils; interestingly, the response of murine neutrophils to PMA is accompanied by a scanty and delayed release of NETs. Ionomycin, calcium ionophore A23187, ROS, and reactive nitrogen species donors also lead to NET release (Fuchs et al. 2007; Kaplan and Radic 2012; Patel et al. 2010).

To date, *three different cell mechanisms* have been identified giving origin to NETs (Fig. 2).

In the first mechanism, NETs are formed as the end product of a distinct cell death program called netosis, activated upon neutrophil stimulation by PMA. During netosis, chromatin, the supporting structure of these NETs, gets unwound while nuclear envelope disintegrates allowing the mixing of chromatin with granular and cytoplasmic proteins. Later on, 3 h after neutrophil activation, the integrity of extracellular membrane is

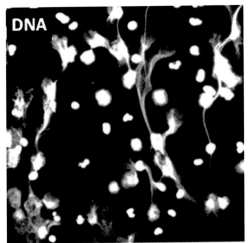
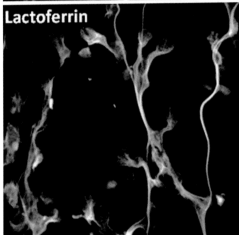
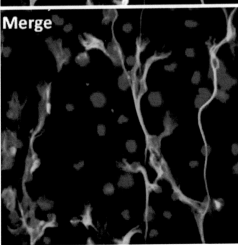

Neutrophil Extracellular Traps, Fig. 1 **Visualization of NETs after in vitro neutrophil stimulation.** Various stimuli activate neutrophils to release their chromatin decorated with proteins from the different cell compartments.

lost, and the chromatin, harboring a large diversity of neutrophil proteins, is released into the extracellular medium giving origin to the NETs (Fuchs et al. 2007). Different cellular processes and molecular mechanisms participate in the netosis process. The neutrophil NADPH-oxidase (NOX2), which generates large amounts of reactive oxygen species (ROS), is necessary for netosis. Consistently, netosis is inhibited in the presence of pharmacological ROS scavengers and is absent in patients with chronic granulomatous disease (CGD) with a genetic deficiency of NOX2. A second mandatory enzyme in netosis is the peptidylarginine deiminase 4 (PAD4), activated during the inflammatory response and catalyzing the citrullination of arginine residues. As a result, the density on positively charged amino acid residues is reduced, weakening the histone-DNA electrostatic interactions. This is a crucial event for chromatin decondensation during netosis. Other molecules involved in the netosis cascade include the Raf–MEK–ERK kinase pathway and the Rac2, a GTPase associated to NOX2. Elastase and myeloperoxidase (MPO) need to translocate to the nucleus where they participate in histone proteolysis and chromatin decondensation (Brinkmann and Zychlinsky 2012). In addition, an extensive formation of double-membrane vacuoles is observed in the cytoplasm of PMA-activated neutrophils. During the first minutes, these vesicles have the characteristics of autophagic vesicles, engulfing granules and ribosomes that are eventually degraded (Remijsen et al. 2011). At later times, a different type of double-membrane vesicles appears that are not typical autophagic and might be remnants of the nuclear envelope (Fuchs et al. 2007). In

Neutrophil Extracellular Traps, Fig. 1 (continued) Upon human neutrophil activation with PMA, NETs are released and are visualized by fluorescence microscopy after DNA staining with DAPI (*upper panel*) and lactoferrin staining with specific antibodies (*middle panel*). The merging of both images allows the identification of neutrophils which retain their membrane integrity, preventing intracellular lactoferrin staining. More interestingly, some nuclei have a more diffused staining and large extracellular structures are formed where DNA and lactoferrin colocalize. These structures correspond to NETs

a PMA activation

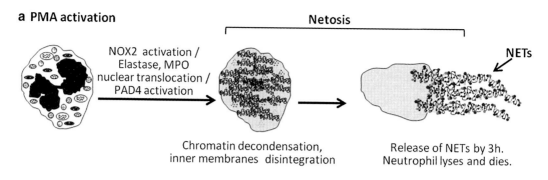

b GM-CSF priming / C5a or LPS activation

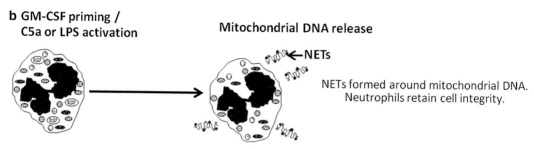

c In vivo activation: LPS, *E. coli* or *S. aureus*

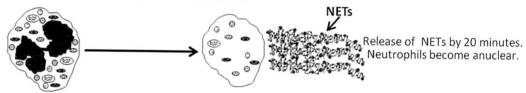

Neutrophil Extracellular Traps, Fig. 2 Models of NET formation. The formation of NETs seems to be related to different mechanisms according to the activation stimulus. (**a**) In vitro activation of human neutrophils with PMA induces netosis, a cell death program that ends with the release of NETs. NETs correspond to the decondensed chromatin decorated with proteins form granular and cytoplasmic compartments. (**b**) Priming with GM-CSF followed by activation with C5a or LPS leads to a rapid release (15 min) of mitochondrial DNA decorated with proteins from cytoplasmic granules. Neutrophils retain cell integrity and show increased survival due to activation. (**c**) In some models of bacterial infection and sepsis, after 20 min of infection, neutrophils release their chromatin decorated with proteins from granular compartments. Neutrophils become enucleated but retain granules to accomplish pathogen phagocytosis and killing

fact, autophagy is activated along netosis, and its blockade, by pharmacological inhibitors, redirects neutrophil death toward apoptosis (Remijsen et al. 2011).

In the second mechanism identified as giving origin to NETs, neutrophils primed with GM-CSF and subsequently activated with LPS or C5a release mitochondrial DNA decorated with granule-derived proteins (Fig. 2b). This process is rapid, with efficient mitochondrial release at 15 min, depending on ROS production. However, it is not directly accompanied by neutrophil cell death; rather, neutrophils show an increased survival due to the activation stimulus (Remijsen et al. 2011).

The third mechanism at the origin of NETs encompasses the rapid release of nuclear DNA and maintains functional enucleated neutrophils able to accomplish phagocytosis and classic bacterial killing (Fig. 2c). This process was originally characterized in vitro upon neutrophil coculture with the gram-positive bacteria *Staphylococcus aureus* and was recently proved in vivo in models of systemic or skin bacterial infections

(McDonald et al. 2012; Pilsczek et al. 2010; Yipp et al. 2012). Along this process, short stretches of nuclear DNA can be rapidly exported to the extracellular medium in vesicles formed from the outer nuclear membrane. Concomitantly, the whole chromatin is decondensed, and the nucleus loses its morphology, expands, and can be released out of the cell forming the NETs. This process is not dependent on ROS and occurs through a very rapid mechanism with DNA-loaded vesicles detected at 5 min after bacterial contact in vitro (Pilsczek et al. 2010). In vivo, accumulation of extracellular chromatin is detectable by intravital microscopy at 20 min after infection (Yipp et al. 2012). The signaling pathways controlling the release of mitochondrial DNA and the release of nuclei, leaving enucleated neutrophils, are completely uncharacterized. It is important to stress that ROS, elastase, and myeoloperoxidase are not required in different in vivo and in vitro models of NET release (Kaplan and Radic 2012; Yipp et al. 2012). For all these processes, further explorations are required to clearly dissect the mechanisms leading to the release of NETs.

Structure of NETs

The best characterized NETs are those produced during netosis, upon neutrophil activation with PMA. The central structure of NETs is the chromatin, i.e., DNA wrapped around histones. The smaller NET filaments have a diameter of 15–17 nm, probably corresponding to modified nucleosomes, whose reference diameter is 10 nm. Along these filaments, multiple globular domains of around 50 nm diameter are present, where granule-derived and cytoplasmic-derived proteins concentrate. Other regions of NETs, with different diameters, can be identified according to the level of chromatin decondensation. In vitro, NETs occupy a volume 10–15 times larger than the volume of the initial neutrophil (Brinkmann et al. 2004; Brinkmann and Zychlinsky 2012; Urban et al. 2009). This might also be the case of NETs produced in vivo in not confined spaces like blood, lung or peritoneum.

At least 30 different proteins have been identified associated to NETs. The core histones H2A, H2B, H3 and H4 are the most abundant proteins and account for 70 % of the total proteins (Urban et al. 2009). Along netosis, histones undergo post-transcriptional modifications, such as proteolysis and citrullination, which translate into an apparent weight decrease of 2–5 kDa (Urban et al. 2009). Apart histones, NETs comprise other nuclear proteins as well as proteins from granules, cytoplasm (Urban et al. 2009) and membrane proteins such as NOX2 and CD11b (Marin-Esteban et al. 2012; Munafo et al. 2009). Most of the granule proteins associated with NETs come from azurophilic granules: elastase, the most abundant granular protein corresponding to around 5 % of the total NET proteins, cathepsin G, proteinase-3, defensins, myeoloperoxidase, azurocidin and lysozyme (Urban et al. 2009). Calprotectin, or S100-calcium binding protein, is the most abundant cytoplasmic protein detected in NETs, also corresponding to around 5 % of the total NET proteins. Proteins from cytoskeleton and from different cell organelles also relate to NETs and represent a small fraction of the total proteins. Some of the proteins associated to NETs, including histones, are highly charged cationic proteins that can tightly interact with DNA (Fuchs et al. 2007).

Protein composition of NETs can be modified according to the activation or the priming state of the cells. For instance, human neutrophils primed with rhIFNa or with IFNa-rich serum from patients before PMA in vitro stimulation, generate NETs carrying a significantly larger amount of LL37 and HMGB1, as compared with non primed neutrophils (Garcia-Romo et al. 2011).

Main Role of NETs: Response to Infection

NETs are formed in response to infection and to other inflammatory conditions. The main physiologic role attributed to NETs is to contribute to host defense. In vitro, bacteria, fungi and protozoa parasites can induce the release of NETs. The net-like structure of NETs enables them to snare microbes. As a consequence, microbes get in tight contact with the various enzymes, microbicidal peptides, histones and other molecules immobilized in NETs.

The participation of NETs in controlling infections relays on three strategies (Brinkmann and Zychlinsky 2012). First, trapping microbes limits their dissemination. The in vivo relevance of this effect has been demonstrated in mouse models of *Escherichia coli* sepsis (McDonald et al. 2012) and polymicrobial sepsis (Meng et al. 2012). In these models, NETs are generated by activated neutrophils that adhere to the vascular endothelium while interacting with platelets (McDonald et al. 2012). Dismantling NETs, by treatment with deoxyribonucleases (DNases), leads to increased bacterial spreading to different organs. NETs might be important in limiting *Streptococcus pneumoniae*, Group *AStreptococci* (GAS) and *Mycobacterium tuberculosis* spreading, as NETs have no direct microbicidal effect on these pathogens (Papayannopoulos and Zychlinsky 2009). Second, NETs weaken virulence of microbes by degrading toxins and virulence factors. *Shigella flexneri* and *Staphylococcus aureus* entangled into NETs lose part of their virulence factors: the invasion plasmid antigen (IpaB) and the a-toxin, respectively, can be degraded by elastase or by other proteases associated to NETs (Brinkmann et al. 2004). Finally, microbicidal proteins associated with NETs display an enhanced and synergic effect, leading to a more efficient microbe killing by mechanisms involving disruption of cell walls and membranes by cationic proteins and peptides, growth inhibition by chelating iron and zinc ions, cytotoxic effect by histones and the generation of ROS. This synergic effect is proved by the significant reduction of NET microbicidal effect after DNase treatment.

Amongst other bacteria, NETs display a bactericidal action against *S. aureus*, gram-negative bacteria, by a mechanism dependent on the peroxidase activity of MPO. Gram-positive bacteria, including *Escherichia coli*, *S. flexnery* and *Salmonella typhimurium* can also be killed by NETs (Brinkmann et al. 2004; Brinkmann and Zychlinsky 2012). Different fungi induce NET formation: *Aspergillus fumigatus*, *Aspergillus nidalus*, *Candida albicans* and the yeast *C. gattii*. The antifungal activity of NETs has been attributed to calprotectins, which chelate zinc and, less efficiently, manganese ions necessary for fungi survival. A gene therapy addressed to recover NOX2 activity in a child with a CGD allowed to recover the capacity of his neutrophils to form NETs and, concomitantly, the capacity to eradicate an extensive aspergillosis by *A nidalus* he was suffering from. This result points to the prominent role of neutrophils on antifungal defense and the contribution of NETs as part of the engaged mechanisms to fulfill this function (Brinkmann and Zychlinsky 2012). NETs also participate in anti-viral defense. They accumulate in lungs of mice infected by influenza A virus, though their presence is not necessary for virus clearance (Guimaraes-Costa et al. 2012). Conversely, neutrophils recognize Human Immunodeficiency Virus (HIV) through Toll-like Receptor 7 (TLR-7) and TLR-8 and respond by producing ROS and releasing NET. These NETs exhibit an anti-HIV activity dependent on MPO and a-defensins (Saitoh et al. 2012). Lastly, NETs are also formed during protozoan infections. They are observed in blood of *Plasmodium falciparum*-infected patients, snaring free trophozoites and infected red blood cells. NETs are also present in biopsies from patients with cutaneous leishmaniasis. Whether NETs have an anti-protozoan activity remains to be clearly established as contradictory results exist concerning a potential cytotoxic effect of histones against amastigotes of different *Leishmania* species (Guimaraes-Costa et al. 2012).

The identification of various pathogen strategies to evade NETs witnesses a long co-existence of neutrophils and pathogens during evolution. Extracellular DNases and other extracellular nucleases constitute virulent factors for *Staphylococcus*, *S. pneumonia* and GAS and they can disassemble the DNA backbone of NETs. In addition, certain serotypes of *S. pneumonia* reduce NET interaction by forming a capsule and elude the action of cationic bactericidal components of NETs by electrostatic repulsion as a result of introducing positively charged groups on the membrane-associated lipoteichoic acid. *Haemophilus influenzae* forms biofilms that incorporate structural elements of NETs, thereby eluding NET action and phagocytosis by neutrophils (Brinkmann and Zychlinsky 2012; Guimaraes-Costa et al. 2012; Kaplan and Radic 2012). Pathogens can also promote a cytokine environment opposed to NET formation, as has been described for HIV that activates IL-10 production by dendritic cells dampening neutrophil activation (Saitoh et al. 2012).

Pathological Relevance

Tissue Cytotoxicity and Pro-Thrombotic Activity

The structure of NETs restrains the diffusion of neutrophil-derived mediators, limiting the extension of bystander lesions. However, the high local concentration of cytotoxic molecules increases the magnitude of local lesions. This harmful local effect has been particularly proved in an in vivo sepsis model, where integrity of NETs in the liver vasculature correlated with hepatocellular damage, tissue necrosis and sinusoidal congestion (McDonald et al. 2012). Tissue injury and deficient oxygenation have been associated to NET accumulation in other inflammatory pathologies such as liver injury, preeclampsia and different neutrophilic lung inflammatory conditions, including transfusion-related acute lung injury, cystic fibrosis and allergic asthma (Brinkmann and Zychlinsky 2012; Kaplan and Radic 2012; Saffarzadeh and Preissner 2013). Two major deleterious effects of NETs are at the origin of these lesions: cytotoxicity and a pro-thrombotic activity.

Cytotoxicity of NETs against endothelial and epithelial cell is mainly mediated by histones and elastase (Brinkmann and Zychlinsky 2012; Marin-Esteban et al. 2012; Saffarzadeh and Preissner 2013). The pro-thrombotic activity of NETs relies on multiple factors. Thrombi are formed by cross-linked fibrin polymers trapping activated platelets and red blood cells. These fibrin polymers can be formed by two different coagulation pathways; one activated by foreign solid bodies in the blood, the other activated by proteases released by the injured endothelium.

NETs activate both coagulation pathways (Brinkmann and Zychlinsky 2012; Saffarzadeh and Preissner 2013). On one hand, NETs provide a foreign surface that triggers the contact activation pathway. On the other hand, endothelium injured by NETs releases the tissue factor that activates the extrinsic coagulation pathway. Elastase and cathepsin-G present in NETs boost this latter pathway by degrading some of the endogenous inhibitors. In consequence, NETs have a significant prothrombotic potential that has been evidenced in different in vivo models of thrombosis, deep vein thrombosis and infection-related abdominal aneurysms (Brinkmann and Zychlinsky 2012; Saffarzadeh and Preissner 2013).

The release of NETs is acknowledged as a physiological part of the neutrophil response during inflammation. Correspondingly, regulatory mechanisms exist to dismantle NETs and to regulate their production to counterbalance the extent of the inflammatory response. NETs are mainly dismantled by DNases constantly present in plasma and NETs production can be downregulated by anti-inflammatory cytokines such as IL-10 (Brinkmann and Zychlinsky 2012; Saitoh et al. 2012).

Autoimmunity

The identification of NETs as a normal response to infection, exposing a large variety of neutrophil proteins to the extracellular space, was early identified as a risky maneuver with possible consequences in braking of tolerance towards self-antigens (Brinkmann et al. 2004). Autoantibodies against NET components have already been implicated in several autoimmune diseases.

NETs have been implicated in the physiopathology of small-vessel vasculitis (SVV) positive for anti-neutrophil cytoplasmic antibodies (ANCA). ANCA-SVV is characterized by the presence of autoantibodies, mostly directed against proteinase-3 or MPO, molecules stocked in neutrophil granules but also associated with NETs and with the cell surface on primed neutrophils. ANCA isolated from patients' sera are able to induce NET release by primed neutrophils. By exposing the autoantigens targeted by ANCA,

NETs increase the production of these pathogenic antibodies that, in turn, will target new neutrophils to induce NET release, leading thus to an amplification loop of the autoimmune response. Concomitantly, during SVV, NETs can be trapped in small-vessels inducing endothelial injury. Fluorescent microscopy analysis of kidney biopsies from SVV patients have provided evidence of NETs tightly associated to sites of glomeruli endothelial and interstitial lesions (Kessenbrock et al. 2009). Importantly, a link between NETs and the specific immune response in SVV has been identified (Sangaletti et al. 2012). Myeloid dendritic cells loaded with NETs become fully activated and able to induce, in mice, ANCA production and autoimmune vasculitis accompanied by parenchyma lesions in kidney and lungs. Immunogenicity of NETs in this model is dependent on the DNA integrity (Sangaletti et al. 2012). NETs themselves have also been identified as able to directly prime T lymphocytes, lowering their activation threshold (Kaplan and Radic 2012).

Systemic lupus erythematosus (SLE) is an autoimmune disease characterized by the presence of circulating immune complexes. In SLE, autoantibodies are mainly directed against nuclear antigens, including double-stranded DNA and different nuclear proteins. Recently, antibodies directed against neutrophil proteins have also been identified; in particular against LL-37, an antimicrobial peptide of the cathelicidin family, abundantly associated to NETs formed by neutrophils from SLE patients (Dorner 2012). During the last 3 years, various mechanisms have been described demonstrating the prolonged exposition and modified immunogenicity of self antigens associated to NETs in SLE. The half life of NETs in SLE patients can be increased by multiple factors including both the protection of DNA against the action of nucleases by the different proteins and autoantibodies associated with NETs, and the direct inhibition of serum DNase-1 by the complement fragment C1q deposited on NETs. The immunogenicity of self-antigens associated with NETs is susceptible of being modified by the presence of several danger signals in the same structure. In SLE, the presence of LL-37 and HMGB-1 in NETs renders self-DNA immunogenic and is suspected of initiating an insidious pathogenic loop. These immunogenic NETs engage the TLR-9 receptors on plasmacytoid dendritic cells (pDC) and elicit the production of IFNα. In turn, IFNα primes neutrophils enhancing their efficacy to produce NETs upon activation by circulating immune complexes. Concomitantly, by exposing self-antigens such as self DNA and ribonucleoproteins, NETs might boost the production of auto-antibodies, increasing the rate of circulation immune complexes that activate neutrophils to produce NETs, thus closing this pathogenic loop (Brinkmann and Zychlinsky 2012; Dorner 2012; Saffarzadeh and Preissner 2013). This mechanism first described for human SLE, has been confirmed in mice. The complex DNA / Cramp (murin orthologue of LL37) activates the production of INFaby pDC and leads to the production of antinuclear autoantibodies and to pathological manifestations of SLE (Kaplan and Radic 2012; Saffarzadeh and Preissner 2013).

The immunogenicity of the self-antigens associated to NETs can also be altered by the post-transcriptional modifications undergone during NET formation (e.g., oxidation, citrullination). This is the case of citrullinated histones that become the target of auto-antibodies during Felty's syndrome, a form of arthritis accompanied by neutropenia. Auto-antibodies present in sera from patients with Felty's syndrome bind to NETs, in particular to citrullinated histones (Brinkmann and Zychlinsky 2012; Kaplan and Radic 2012).

Atherosclerosis

Atherosclerosis is a chronic inflammatory disease affecting arterial endothelium. Atherosclerotic lesions start as lipid streaks in the intima, the innermost layer of endothelial walls. These streaks are characterized by the presence of macrophages with a lipoprotein-rich cytoplasm: the foam cells. The evolution of atherosclerotic lesions is progressive, forming prominent atheromatous plaques with a lipid rich core surrounded by a fibrous layer and the intima, infiltrated by various leukocyte types, mainly macrophages, lymphocytes and pDCs. (Drechsler et al. 2011).

Neutrophils are not abundantly found in atherosclerosis lesions. However, recent studies demonstrate interesting correlations between atherosclerosis lesions and the presence of NETs (Doring et al. 2012; Drechsler et al. 2011; Megens et al. 2012).

NETs are present in atherosclerosis lesions in mice and human (Doring et al. 2012; Megens et al. 2012). On early atheromas in mice, extracellular structures with colocalized signals of DNA and Cramp (murin orthologue of LL37) are present in necrotic core areas (Doring et al. 2012). NETS are also detected in the luminal area of these lesions (Megens et al. 2012). This DNA / Cramp complex has a proven pathogenic role in SLE that has also been demonstrated in a murine model of atherosclerosis. By activating local pDC through TLR9, DNA / Cramp complex induce the production of IFNα (Doring et al. 2012). This cytokine primes neutrophils, enhancing their capacity to produce NETs (Dorner 2012). Moreover, human and mice suffering from atherosclerosis have circulating anti-DNA autoantibodies (Doring et al. 2012), possibly boosted by NETs themselves, as has been proposed in the case of SLE (Brinkmann and Zychlinsky 2012; Dorner 2012; Saffarzadeh and Preissner 2013). Immune complexes formed with these antibodies might activate NET formation (Dorner 2012). Hence, NETs contribute to atherosclerosis enhancing endothelial lesions by their cytotoxicity, providing immunogenic stimulus to innate immune cells and adding an autoimmune component to this disease.

Interactions With Other Processes and Drugs

The release of NETs conveys beneficial but also adverse consequences. NETs accomplish a non disputable role in controlling infections. However, their cytotoxic effects impact the integrity of host tissues and they are also implicated in the breaking of immune tolerance. Therapeutic strategies to control NET formation and / or their clearance can be oriented either to increase their production to better control infections, or to reduce their presence in order to avoid their adverse effects.

The antimicrobial function of neutrophils cannot be entirely and permanently substituted by pharmacologic antimicrobial treatments, as drug resistance hampers their use. Neutrophils and, in particular, NETs are necessary to control infections. This has been clearly demonstrated through the aforementioned case of a CGD patient in whom a gene therapy recovering NOX2 function allowed the recovery of neutrophils' capacity to perform oxidative burst and to produce NETs. Both features correlated to the capacity of patient to control a drug refractory pulmonary aspergillosis (Brinkmann and Zychlinsky 2012). Developing strategies to enhance NET formation could be inspired by the case with statins; therapeutic molecules used to reduce serum cholesterol levels. Patients following statins treatment show a reduced risk to bacterial infections, correlated with the capacity of these molecules to boost the production of NETs (Chow et al. 2010).

On the contrary, inhibiting NET production might be possible with various pharmacological approaches. In particular, ROS scavengers, PAD4 inhibitors and MEK-ERK kinase pathway inhibitors reduce NET production in vitro (Kaplan and Radic 2012). However, none of these inhibitors is specific for NET formation as they have a broad effect spectrum. Recently, IL-10 has been described as a physiological inhibitor of NET formation (Saitoh et al. 2012). Alternatively, dismantling NETs by DNases or neutralizing the action of some their components, mainly histones, decreases tissue injury in both in vitro and in vivo pathology models (Saffarzadeh and Preissner 2013). Neutralizing histones can be achieved by several mechanisms: the use of activated protein C or plasmin that degrade histones, the use of polysialic acids that outcompetes histone binding to DNA or to negative charged surface, or by using specific antibodies (Saffarzadeh and Preissner 2013).

In conclusion, a better knowledge of the mechanisms controlling the formation of NETs and their effects is required to develop therapeutic tools that could specifically enhance or inhibit the release of NETs or that allow blocking individual components of NETs involved in their adverse effects.

Cross-References

▶ Neutrophil Oxidative Burst
▶ Platelets, Endothelium, and Inflammation
▶ Sepsis Models in Animals

References

Azevedo, E. P., Guimaraes-Costa, A. B., Torezani, G. S., Braga, C. A., Palhano, F. L., Kelly, J. W., et al. (2012). Amyloid fibrils trigger the release of Neutrophil Extracellular Traps (NETs), causing fibril fragmentation by NET-associated elastase. *The Journal of Biological Chemistry, 287*, 37206–37218.

Brinkmann, V., & Zychlinsky, A. (2012). Neutrophil extracellular traps: Is immunity the second function of chromatin? *The Journal of Cell Biology, 198*, 773–783.

Brinkmann, V., Reichard, U., Goosmann, C., Fauler, B., Uhlemann, Y., Weiss, D. S., et al. (2004). Neutrophil extracellular traps kill bacteria. *Science, 303*, 1532–1535.

Chen, K., Nishi, H., Travers, R., Tsuboi, N., Martinod, K., Wagner, D. D., et al. (2012). Endocytosis of soluble immune complexes leads to their clearance by FcγRIIIB but induces neutrophil extracellular traps via FcγRIIA in vivo. *Blood, 120*, 4421–4431.

Chow, O. A., von Kockritz-Blickwede, M., Bright, A. T., Hensler, M. E., Zinkernagel, A. S., Cogen, A. L., et al. (2010). Statins enhance formation of phagocyte extracellular traps. *Cell Host & Microbe, 8*, 445–454.

Doring, Y., Drechsler, M., Wantha, S., Kemmerich, K., Lievens, D., Vijayan, S., et al. (2012). Lack of neutrophil-derived CRAMP reduces atherosclerosis in mice. *Circulation Research, 110*, 1052–1056.

Dorner, T. (2012). SLE in 2011: Deciphering the role of NETs and networks in SLE. *Nature Reviews. Rheumatology, 8*, 68–70.

Drechsler, M., Doring, Y., Megens, R. T., & Soehnlein, O. (2011). Neutrophilic granulocytes: Promiscuous accelerators of atherosclerosis. *Thrombosis and Haemostasis, 106*, 839–848.

Fuchs, T. A., Abed, U., Goosmann, C., Hurwitz, R., Schulze, I., Wahn, V., et al. (2007). Novel cell death program leads to neutrophil extracellular traps. *The Journal of Cell Biology, 176*, 231–241.

Garcia-Romo, G. S., Caielli, S., Vega, B., Connolly, J., Allantaz, F., Xu, Z., et al. (2011). Netting neutrophils are major inducers of type I IFN production in pediatric systemic lupus erythematosus. *Science Translational Medicine, 3*, 73ra20.

Guimaraes-Costa, A. B., Nascimento, M. T., Wardini, A. B., Pinto-da-Silva, L. H., & Saraiva, E. M. (2012). ETosis: A microbicidal mechanism beyond cell death. *Journal of Parasitology Research, 2012*, 929743.

Kaplan, M. J., & Radic, M. (2012). Neutrophil extracellular traps: Double-edged swords of innate immunity. *Journal of Immunology, 189*, 2689–2695.

Kessenbrock, K., Krumbholz, M., Schonermarck, U., Back, W., Gross, W. L., Werb, Z., et al. (2009). Netting neutrophils in autoimmune small-vessel vasculitis. *Nature Medicine, 15*, 623–625.

Marin-Esteban, V., Turbica, I., Dufour, G., Semiramoth, N., Gleizes, A., Gorges, R., et al. (2012). Afa/Dr diffusely adhering *Escherichia coli* strain C1845 induces neutrophil extracellular traps that kill bacteria and damage human enterocyte-like cells. *Infection and Immunity, 80*, 1891–1899.

McDonald, B., Urrutia, R., Yipp, B. G., Jenne, C. N., & Kubes, P. (2012). Intravascular neutrophil extracellular traps capture bacteria from the bloodstream during sepsis. *Cell Host & Microbe, 12*, 324–333.

Megens, R. T., Vijayan, S., Lievens, D., Doring, Y., van Zandvoort, M. A., Grommes, J., et al. (2012). Presence of luminal neutrophil extracellular traps in atherosclerosis. *Thrombosis and Haemostasis, 107*, 597–598.

Meng, W., Paunel-Gorgulu, A., Flohe, S., Hoffmann, A., Witte, I., Mackenzie, C., et al. (2012). Depletion of neutrophil extracellular traps in vivo results in hypersusceptibility to polymicrobial sepsis in mice. *Critical Care, 16*, R137.

Munafo, D. B., Johnson, J. L., Brzezinska, A. A., Ellis, B. A., Wood, M. R., & Catz, S. D. (2009). DNase I inhibits a late phase of reactive oxygen species production in neutrophils. *Journal of Innate Immunity, 1*, 527–542.

Papayannopoulos, V., & Zychlinsky, A. (2009). NETs: A new strategy for using old weapons. *Trends in Immunology, 30*, 513–521.

Patel, S., Kumar, S., Jyoti, A., Srinag, B. S., Keshari, R. S., Saluja, R., et al. (2010). Nitric oxide donors release extracellular traps from human neutrophils by augmenting free radical generation. *Nitric Oxide, 22*, 226–234.

Pilsczek, F. H., Salina, D., Poon, K. K., Fahey, C., Yipp, B. G., Sibley, C. D., et al. (2010). A novel mechanism of rapid nuclear neutrophil extracellular trap formation in response to *Staphylococcus aureus*. *Journal of Immunology, 185*, 7413–7425.

Remijsen, Q., Vanden Berghe, T., Wirawan, E., Asselbergh, B., Parthoens, E., De Rycke, R., et al. (2011). Neutrophil extracellular trap cell death requires both autophagy and superoxide generation. *Cell Research, 21*, 290–304.

Saffarzadeh, M., & Preissner, K. T. (2013). Fighting against the dark side of neutrophil extracellular traps in disease: Manoeuvres for host protection. *Current Opinion in Hematology, 20*, 3–9.

Saitoh, T., Komano, J., Saitoh, Y., Misawa, T., Takahama, M., Kozaki, T., et al. (2012). Neutrophil extracellular traps mediate a host defense response to human immunodeficiency virus-1. *Cell Host & Microbe, 12*, 109–116.

Sangaletti, S., Tripodo, C., Chiodoni, C., Guarnotta, C., Cappetti, B., Casalini, P., et al. (2012). Neutrophil extracellular traps mediate transfer of cytoplasmic neutrophil antigens to myeloid dendritic cells toward ANCA induction and associated autoimmunity. *Blood, 120*, 3007–3018.

Schorn, C., Janko, C., Latzko, M., Chaurio, R., Schett, G., & Herrmann, M. (2012). Monosodium urate crystals induce extracellular DNA traps in neutrophils, eosinophils, and basophils but not in mononuclear cells. *Frontiers in Immunology, 3*, 277.

Urban, C. F., Ermert, D., Schmid, M., Abu-Abed, U., Goosmann, C., Nacken, W., et al. (2009). Neutrophil extracellular traps contain calprotectin, a cytosolic protein complex involved in host defense against *Candida albicans*. *PLoS Pathogens, 5*, e1000639.

Yipp, B. G., Petri, B., Salina, D., Jenne, C. N., Scott, B. N., Zbytnuik, L. D., et al. (2012). Infection-induced NETosis is a dynamic process involving neutrophil multitasking in vivo. *Nature Medicine, 18*, 1386–1393.

Neutrophil Oxidative Burst

Jamel El-Benna[1], Pham My-Chan Dang[1] and Margarita Hurtado-Nedelec[2]
[1]INSERM U1149, CNRS ERL8252, Centre de Recherche sur l'Inflammation, Paris, France
[2]AP-HP, Centre Hospitalier Universitaire Xavier Bichat, UF Dysfonctionnements Immunitaires, Paris, France

Synonyms

ROS production by the phagocyte NADPH oxidase

Definition

Polymorphonuclear neutrophils comprise more than 50 % of the circulating white blood cells in humans. They constitute one of the most powerful means of host defense against bacteria and fungi. Circulating neutrophils are freely flowing at resting state. At the infection site, phagocytosis of the pathogen agent in a closed vacuole called the phagosome initiates the activation process of neutrophils that induces the release into the vacuole of antibacterial peptides, proteases, other enzymes, and reactive oxygen species (ROS), all of which contribute to the death and destruction of the pathogen. These toxic molecules could also inflict harm to nearby tissues, thus participating to tissue injury during the inflammatory process. Neutrophil-derived ROS are believed to be involved in several inflammatory diseases such as rheumatoid arthritis, lung inflammation, and inflammatory bowel diseases. In this chapter, the way in which neutrophils make ROS and their role in host defense and inflammation will be described.

Structure and Function of the Phagocyte NADPH Oxidase

The discovery of ROS production by the phagocyte NADPH oxidase started in the 1880s, when Elie Metchnikoff observed that certain cells were capable of ingesting bacteria and thus discovered phagocytosis as a defense mechanism in eukaryotes (Metchnikoff 1891). In 1933, it was shown that when neutrophils are incubated with bacteria, oxygen uptake dramatically increased (Baldridge and Gerard 1933). It was believed that this increase in oxygen uptake was due to an increase in mitochondrial oxidative phosphorylation, and the process was therefore called "oxidative burst" or "respiratory burst." In 1959, it was shown that mitochondria were not responsible for the neutrophil oxidative burst because it was not affected by cyanide, a mitochondrial respiratory chain inhibitor (Sbarra and Karnovsky 1959). In 1961, Iyer and colleagues showed that neutrophils were able to produce hydrogen peroxide (H_2O_2) (Iyer et al. 1961). In 1967, Klebanoff's group showed that myeloperoxidase (MPO) uses H_2O_2 and Cl^- in order to produce hypochlorous acid (HOCl), a highly microbicidal and toxic agent (Klebanoff 1967). In 1969, McCord and Fridovich discovered the superoxide dismutase (SOD) activity of erythrocuprein, an abundant protein in many organisms (1969). In 1973, Babior and colleagues used SOD to show that neutrophils produce superoxide anion (O_2^-), which dismutates into H_2O_2, making O_2^- the source of other ROS (Babior et al. 1973). The enzyme responsible for ROS generation in neutrophils was called NADPH oxidase because it uses cytocolic NADPH to reduce oxygen. In 1978, the core enzyme of NADPH oxidase was identified (Segal et al. 1998), and recently homologues of this enzyme were identified in cells other than neutrophils and in tissues (Bedard and Krause 2007).

The phagocyte NADPH oxidase is the enzyme that produces O_2^-, the precursor of the other ROS. It is an enzyme complex, composed of several proteins that are localized in membranes and the cytosol of resting cells (Babior et al. 1997). The central membrane-associated component of the NADPH oxidase is called the flavocytochrome b_{558}, which is composed of a glycosylated 91-kDa protein subunit (gp91phox; phox: phagocyte oxidase), non-covalently bound in a 1:1 stabilized complex to a 22-kDa subunit (p22phox). The cytosolic proteins are p47phox, p67phox, and p40phox and the small G-proteins, Rac1 (in monocytes) or Rac2 (in neutrophils). In response to stimulation, the cytosolic components migrate to the membrane where they assemble with the flavocytochrome b_{558} to form the active enzyme, a process that is tightly regulated by protein-protein interactions and by phosphorylation (El-Benna et al. 2009).

The gp91phox subunit (also called NOX2) comprises binding sites for FAD, NADPH, and two hemes, making it the electron transfer chain of the active NADPH oxidase. Long thought to be specific to phagocytes, it is now known to belong to a large family of proteins expressed in many different cell types and called NOX for NADPH oxidase, of which NOX2 is the phagocyte protein (Bedard and Krause 2007). In resting neutrophils, 60–70 % of the flavocytochrome b_{558} is located in the membranes of the specific granules and 20–25 % in the membranes of the gelatinase granules, and the remainder is found in the plasma membranes and in the membranes of secretory vesicles.

The structure of the cytosolic protein p47phox has been well-studied and contains several protein domains, including two src-homology 3 (SH3) domains, one phox homology (PX) domain, an autoinhibitory domain, and a proline-rich region (El-Benna et al. 2009). Two SH3 domains are also found in the sequence of p67phox, which also contains four tetratricopeptide-rich regions (TPR), an activation domain that regulates the catalytic activity of the cytochrome b_{558} in the active enzyme, a PB1/PC domain, and one proline-rich region. The third component of the cytosolic complex, p40phox, is a 339-amino-acid protein that was initially identified through its binding to p67phox. It contains one SH3 domain, one PB1/PC domain, and one PX domain. The specified protein domains found in the components of the NADPH oxidase regulate their interaction with each other, with membrane proteins and with membrane phospholipids.

The structure and the stoichiometry of the cytosolic complex of resting cells is not entirely defined; however, p67phox has been shown to associate tightly with p40phox, while p47phox has the ability to interact with both p40phox and p67phox. In resting cells, Rac2, which is more abundant in human neutrophils than Rac1 (92 % homologous to Rac2), is also present in the cytosol but is not part of the cytosolic complex. In its resting GDP-form, cytosolic Rac2 is bound to its inhibitor, rho-GDI.

When neutrophils are activated and exocytosis of granules occurs, fusion of granule membranes with the plasma membrane increases the cell surface expression of the cytochrome b_{558}, which seems to be the central docking site for the 10–20 % of the cytosolic components that translocate to the plasma membrane during activation (El-Benna et al. 2008). The activation is accompanied with extensive phosphorylation of p47phoxon several serines located in the polybasic region of the carboxy-terminal portion of the protein, designated as the autoinhibitory region (El-Benna et al. 2009). Resulting in the unmasking of the cryptic SH3 domains, which can then bind the proline-rich region of p22phox. Other interactions with gp91phox/NOX2 via less-defined domains and with phosphatidyl inositol 3,4 phosphate and phosphatidic acid via its PX domain tether p47phox at the membrane. The p47phox subunit is thought to be responsible for transporting the cytosolic complex to the membrane during oxidase activation and is considered as the organizer of the NADPH oxidase active complex. At the membrane, p67phox binds to the cytochrome b_{558} and regulates its activity via the activation domain. It also binds to Rac2 (or Rac1), which translocates to the plasma membrane independently and interacts with the p67phox/cytochrome b_{558} complex at the plasma membrane.

The function of the phagocyte NADPH oxidase (or NOX2) is to produce $O_2^{-\cdot}$, which is the source of the other ROS generated in the phagosome (Hampton et al. 1998; Winterbourn 2008). This process is accompanied by an increase in glucose oxidation via the hexose monophosphate shunt to provide cytosolic NADPH. Superoxide anion is produced by monovalent reduction of oxygen:

$$2\ O_2 + NADPH \longrightarrow 2\ O_2^{-\cdot} + NADP^+ + H^+$$

Being a free radical, $O_2^{-\cdot}$ is unstable and reacts quickly by reduction or oxidation. In addition, it has limited plasma membrane permeability.

$O_2^{-\cdot}$ is transformed into hydrogen peroxide (H_2O_2) by spontaneous dismutation in the presence of protons at acidic pH in the phagosome:

$$O_2^{-\cdot} + O_2^{-\cdot} + 2H^+ \longrightarrow H_2O_2 + O_2$$

Hydrogen peroxide is very stable in the absence of metabolizing enzymes. It is an oxidizing agent that reacts with thiols, metals, and $O_2^{-\cdot}$. H_2O_2 is membrane permeable and diffuses readily in cells and tissues.

Through the Haber-Weiss reaction in the presence of a transition metal (or the Fenton reaction in the presence of iron), H_2O_2 and $O_2^{-\cdot}$ can interact, giving rise to the hydroxyl radical ($OH°$):

$$O_2^{-\cdot} + H_2O_2 \longrightarrow OH° + OH^- + O_2$$

The hydroxyl radical is not stable and reacts with most biological molecules such as DNA, proteins, lipids, and carbohydrates.

Myeloperoxidase (MPO), released from azurophilic granules, catalyzes the transformation of H_2O_2 in the presence of a halogen (Cl^-, Br^-, I^-) into highly toxic molecules:

$$H_2O_2 + H^+Cl^- \longrightarrow HOCl^- + HO$$

HOCl is a membrane-permeable weak acid and a strong oxidizing agent. Most of the hypochlorous acid (OCl^-) reacts with amines to generate toxic chloramines:

$$H^+ + OCl^- + R - NH_2 \longrightarrow R - NHCl + HO$$

Pathological Relevance

The requirement of the neutrophil oxidative burst for host defense was demonstrated by a rare inherited immunodeficiency disease called **chronic granulomatous disease (CGD)**, in which neutrophils are unable to efficiently produce ROS (Heyworth et al. 2003). CGD is the most clinically significant genetic disorder of neutrophil function that usually starts during childhood and results in recurrent and often life-threatening bacterial and fungal infections as well as inflammatory granulomas, a hallmark of this neutrophil disorder. CGD results from gene mutations of the NADPH oxidase components, namely, the CYBB gene (Xp21) that encodes the gp91phox subunit and the CYBA (16q24), NCF1 (7q11), and NCF2 genes (1q25), which encode p22phox, p47phox, and p67phox, respectively. Neutrophils and monocytes from homozygous CGD patients are reported as having null or very decreased ROS production. The most frequent form of CGD (approximately 70 % of all cases) is the X-linked gp91phox-deficient form, followed by the autosomal recessive form deficient in p47phox (approximately 25 %). Less frequent forms are the autosomal recessive CGD deficient in p67phox (<5 %) or p22phox (<5 %). Only one case of rac2 gene mutation and one case of p40phox gene mutation have been described so far. Common infectious syndromes include pneumonia and lung abscesses, skin and soft tissue infections, lymphadenopathy, suppurative lymphadenitis, osteomyelitis, and hepatic abscesses.

The most common pathogens encountered in CGD patients are gram-positive bacteria (*Staphylococcus aureus*), gram-negative bacteria (*Salmonella*, *Pseudomonas cepacia*, *Serratia marcescens*, etc.), and fungi (*Aspergillus*, *Candida albicans*). *Aspergillus* species can cause intractable pneumonia and sometimes septicemia in CGD patients and are a frequent cause of death. Histologically, CGD is characterized by the formation of large granulomas resulting from the fusion of macrophages that have phagocytosed bacteria but are unable to destroy them because NADPH oxidase activity is lacking. The granulomas can obstruct vital structures such as the gastrointestinal and genitourinary tracts.

CGD illustrated the need of NADPH oxidase-derived ROS in innate immunity. In addition to neutrophils, eosinophils, monocytes, and macrophages, the phagocyte NADPH oxidase, NOX2, is also expressed in **dendritic cells**, EBV-transformed lymphocytes and T lymphocytes. Recently, it was shown that the phagocyte NADPH oxidase is essential for antigen cross-presentation in dendritic cells (Savina et al. 2006) and for T- and B-lymphocyte functions such as cytokines and antibodies production. Thus, NOX2 seems to play a larger role in the immune system than initially believed.

Excessive ROS production by the neutrophil oxidative burst induces **tissue injury** at the origin of many inflammatory diseases (Babior 2000; Lambeth 2007). ROS react with biological molecules causing DNA oxidation and strand breaks, enzyme activation/inactivation, and lipid peroxidation. A large body of evidence has now indicated that ROS regulate biological pathways, acting as an intracellular second messenger. Relevant to inflammation, H_2O_2 is known to activate the nuclear factor NF-kB (Janssen-Heininger et al. 2000) and to induce pro-inflammatory cytokine production. In p47phox-deficient mice in which phagocytes do not produce ROS, LPS-induced NF-kB activation was decreased (Koay et al. 2001). ROS can also regulate the activation of the transcription factor AP1/Jun/Fos complex (Karin and Shaulian 2001). Hydrogen peroxide induces the activation of the heterotrimeric G-proteins at the membrane level and of the small GTPase Ras (Nishida et al. 2000). Several protein kinases and phosphatases are also regulated by ROS in macrophages (Forman and Torres 2002) and in neutrophils (Fialkow et al. 2007). Recently, it was shown that the protein tyrosine kinase Lyn is a redox sensor that mediates leukocyte wound attraction in vivo (Yoo et al. 2011). Because of these effects, excessive ROS production by the neutrophil oxidative burst participates in tissue injury at the origin of many inflammatory diseases such as acute respiratory distress syndrome (ARDS), rheumatoid arthritis (RA), and other diseases (Babior 2000; Lambeth 2007).

Although the pathogenesis of **acute respiratory distress syndrome (ARDS)** is complex, several observations point to an important role of ROS derived from neutrophils in this disease. A subpopulation of neutrophils with high capacity to generate ROS after ex vivo stimulation was found in ARDS patients (Chollet-Martin et al. 1992). Furthermore, the degree of H_2O_2 hyperproduction by neutrophils correlated with elevated plasma levels of TNFα in ARDS patients. TNFα is known to induce upregulation or priming of neutrophil ROS production, suggesting that TNFα-primed neutrophils may play a major role in the pathogenesis of ARDS-associated lung injury. The antioxidant enzymes such as SOD and catalase have been shown to prevent lung injury in animal models (Koyama et al. 1992).

Rheumatoid arthritis (RA) is a systemic inflammatory disorder most commonly targeting the joints. The pathophysiology of RA involves dysregulated cytokine production and neutrophil accumulation in synovial fluids. Both excessive production of ROS and release of degradative enzymes by neutrophils have been implicated in rheumatoid tissue damage (Firestein and Zvaifler 1992). Massive neutrophil accumulation in the inflamed joints and massive ROS release are believed to contribute to tissue injury in rheumatoid arthritis. Neutrophils isolated from synovial fluids of RA patients are primed for ROS production and exhibit phosphorylation of p47phox on Ser345 as well as activation of both ERK1/2 and p38MAPK, the kinases responsible for p47phox phosphorylation (Dang et al. 2006). The presence of carbonyl groups in proteins present in synovial liquid from arthritis patients reflects ROS-induced oxidation of proteins. In mouse models of arthritis, SOD has been shown to have a beneficial effect (Parizada et al. 1991).

Interactions with Other Processes and Drugs

The phagocyte NADPH oxidase plays a key role in host defense against microbial and fungal pathogens by generating superoxide anion and other ROS. However, excessive ROS release by neutrophils and their reactions with lipids, proteins, and DNA can also damage surrounding host tissues, thereby amplifying the inflammatory reaction. Thus, ROS are believed to be involved in several inflammatory diseases. Developing new therapeutic agents that will limit extracellular ROS production by neutrophils will be a novel approach for new anti-inflammatory drugs. Future research is necessary to understand how ROS production by neutrophils is regulated by inflammatory mediators and to identify new biological targets affected by ROS.

Indeed a number of NADPH oxidase inhibitors have been described and have been shown to inhibit inflammation. Some biological agents such as NO, steroids, adrenaline, IL-10, and IL-4 inhibit NADPH oxidase activity and expression. Numerous pharmacological NADPH oxidase inhibitors are known; the most used is diphenylene iodonium (DPI) which inhibits electrons transport by gp91phox. Apocynin, a methoxy-substituted catechol, is a natural molecule which inhibits NADPH oxidase. Other molecules such as phenylarsine oxide (PAO), 4-(2-aminoethyl)-benzenesulfonyl fluoride (AEBSF), and N-alpha-tosyl phenylalanine chloromethyl cétone (TPCK) inhibit NADPH oxidase by inhibiting complex assembly. These molecules exert an anti-inflammatory effect in vivo in animal models of inflammation such as arthritis, intestinal, and lung inflammation.

Cross References

▶ Acute Exacerbations of Airway Inflammation
▶ Leukocyte Recruitment
▶ Spondyloarthritis

References

Babior, B. M. (2000). Phagocytes and oxidative stress. *American Journal of Medicine, 109*, 33–44.

Babior, B. M., Kipnes, R. S., & Curnutte, J. T. (1973). Biological defense mechanisms. The production by leukocytes of superoxide, a potential bactericidal agent. *Journal of Clinical Investigation, 52*, 741–744.

Babior, B. M., El Benna, J., Chanock, S. J., & Smith, R. M. (1997). The NADPH oxidase of leukocytes: The respiratory burst oxidase. In J. G. Scandalios (Ed.), *Oxidative stress and the molecular biology of antioxidant defenses* (pp. 737–783). Cold Spring Harbor: Cold Spring Harbor Laboratory Press.

Baldridge, C. W., & Gerard, R. W. (1933). The extra respiration of phagocytosis. *American Journal of Physiology, 103*, 235.

Bedard, K., & Krause, K. H. (2007). The NOX family of ROS-generating NADPH oxidases: Physiology and pathophysiology. *Physiological Reviews, 87*, 245–313.

Chollet-Martin, S., Montravers, P., Gibert, C., Elbim, C., Desmonts, J. M., Fagon, J. Y., et al. (1992). Subpopulation of hyperresponsive polymorphonuclear neutrophils in patients with adult respiratory distress syndrome. Role of cytokine production. *American Review of Respiratory Disease, 146*, 990–996.

Dang, P. M., Stensballe, A., Boussetta, T., Raad, H., Dewas, C., Kroviarski, Y., et al. (2006). A specific p47phox -serine phosphorylated by convergent MAPKs mediates neutrophil NADPH oxidase priming at inflammatory sites. *Journal of Clinical Investigation, 116*, 2033–2043.

El-Benna, J., Dang, P. M., & Gougerot-Pocidalo, M. A. (2008). Priming of the neutrophil NADPH oxidase activation: Role of p47phox phosphorylation and NOX2 mobilization to the plasma membrane. *Seminars in Immunopathology, 30*, 279–289.

El-Benna, J., Dang, P. M., Gougerot-Pocidalo, M. A., Marie, J. C., & Braut-Boucher, F. (2009). p47phox, the phagocyte NADPH oxidase/NOX2 organizer: Structure, phosphorylation and implication in diseases. *Experimental and Molecular Medicine, 41*, 217–225.

Fialkow, L., Wang, Y., & Downey, G. P. (2007). Reactive oxygen and nitrogen species as signaling molecules regulating neutrophil function. *Free Radical Biology and Medicine, 42*, 153–164.

Firestein, G. S., & Zvaifler, N. J. (1992). Rheumatoid Arthritis: A disease of disordered immunity. In Inflammation. In J. I. Gallin, I. M. Goldstein, & R. Snyderman (Eds.), *Basic principles and clinical correlates* (2nd ed., pp. 959–977). New York: Raven Press.

Forman, H. J., & Torres, M. (2002). Reactive oxygen species and cell signaling: Respiratory burst in macrophage signaling. *American Journal of Respiratory and Critical Care Medicine, 166*, S4–S8.

Hampton, M. B., Kettle, A. J., & Winterbourn, C. C. (1998). Inside the neutrophil phagosome: Oxidants, myeloperoxidase, and bacterial killing. *Blood, 92*, 3007–3017.

Heyworth, P. G., Cross, A. R., & Curnutte, J. T. (2003). Chronic granulomatous disease. *Current Opinion in Immunology, 15*, 578–584.

Iyer, G. Y. N., Islam, M. F., & Quastel, J. H. (1961). Biochemical aspects of phagocytosis. *Nature, 192*, 535–541.

Janssen-Heininger, Y. M., Poynter, M. E., & Baeuerle, P. A. (2000). Recent advances towards understanding redox mechanisms in the activation of nuclear factor kappaB. *Free Radical Biology and Medicine, 28*, 1317–1327.

Karin, M., & Shaulian, E. (2001). AP-1: Linking hydrogen peroxide and oxidative stress to the control of cell proliferation and death. *IUBMB Life, 52*, 17–24.

Klebanoff, S. J. (1967). A peroxidase-mediated antimicrobial system in leukocytes. *Journal of Clinical Investigation, 46*, 1478.

Koay, M. A., Christman, J. W., Segal, B. H., Venkatakrishnan, A., Blackwell, T. R., Holland, S. M., et al. (2001). Impaired pulmonary NF-kappaB activation in response to lipopolysaccharide in NADPH oxidase-deficient mice. *Infection and Immunity, 69*, 5991–5996.

Koyama, S., Kobayashi, T., Kubo, K., Sekiguchi, M., & Ueda, G. (1992). Recombinant-human superoxide dismutase attenuates endotoxin-induced lung inhuur in awake sheep. *American Review of Respiratory Disease, 145*, 1404–1409.

Lambeth, J. D. (2007). Nox enzymes, ROS, and chronic disease: An example of antagonistic pleiotropy. *Free Radical Biology and Medicine, 43*, 332–347.

McCord, J. M., & Fridovich, I. (1969). Superoxide dismutase. An enzymic function for erythrocuprein (hemocuprein). *Journal of Biological Chemistry, 244*, 6049–6055.

Metchnikoff, E. (1891). Lecture on phagocytosis and immunity. *British Medical Journal, 1*, 213–217.

Nishida, M., Maruyama, Y., Tanaka, R., Kontani, K., Nagao, T., & Kurose, H. (2000). G alpha(i) and G alpha(o) are target proteins of reactive oxygen species. *Nature, 408*, 492–495.

Parizada, B., Werber, M. M., & Nimrod, A. (1991). Protective effects of human recombinant MnSOD in adjuvant arthritis and bleomycin-induced lung fibrosis. *Free Radical Research Communications, 15*, 297–301.

Savina, A., Jancic, C., Hugues, S., Guermonprez, P., Vargas, P., Moura, I. C., et al. (2006). NOX2 controls phagosomal pH to regulate antigen processing during crosspresentation by dendritic cells. *Cell, 126*, 205–218.

Sbarra, A. J., & Karnovsky, M. L. (1959). The biochemical basis of phagocytosis. I. Metabolic changes during the ingestion of particles by polymorphonuclear leukocytes. *Journal of Biological Chemistry, 234*, 1355–1362.

Segal, A. W., Jones, O. T., Webster, D., & Allison, A. C. (1998). Absence of a newly described cytochrome b from neutrophils of patients with chronic granulomatous disease. *Lancet, 2*, 446–449.

Winterbourn, C. C. (2008). Reconciling the chemistry and biology of reactive oxygen species. *Nature Chemical Biology, 4*, 278–286.

Yoo, S. K., Starnes, T. W., Deng, Q., & Huttenlocher, A. (2011). Lyn is a redox sensor that mediates leukocyte wound attraction in vivo. *Nature, 480*, 109–112.

NFkappaB

Christine V. Möser[1] and Ellen Niederberger[1,2]
[1]Pharmazentrum Frankfurt/ZAFES, Klinikum der Johann Wolfgang Goethe-Universität Frankfurt, Frankfurt am Main, Germany
[2]JW Goethe University Frankfurt, Frankfurt, Germany

Synonyms

Avian reticuloendotheliosis; **c-rel**; C-Rel proto-oncogene protein; DNA binding factor KBF1; DNA-binding factor KBF1; DNA-binding factor KBF2; EBP-1; IREL; I-Rel; KBF1; Lymphocyte translocation chromosome 10 protein; LYT10; LYT-10; NF-kappaB; NF-kappa-B; NF-kappa-B p65delta3; NF-kappabeta; **NFKB1**; **NFKB2**; **NFKB3**; NFKB-p105; NFKB-p50; Nuclear factor kappa-B DNA-binding subunit; Nuclear factor NF-kappa-B; Nuclear factor NF-kappa-B p105 subunit; Nuclear factor NF-kappa-B p50 subunit; Nuclear factor NF-kappa-B p65 subunit; Nuclear factor of kappa light chain gene enhancer in B-cells 2; Nuclear factor of kappa light polypeptide gene enhancer in B-cells 3; Nuclear factor of kappa light polypeptide gene enhancer in B-cells 3; Oncogene Lyt-10; Oncogene REL; **p100/p52**; p105; **p105/p50**; p50; p52; p52 H2TF1; p65; Proto-oncogene c-Rel; **RelB**; RELB; REL-B; Subunit; Transcription factor p65; Transcription factor RelB; v-rel avian reticuloendotheliosis viral oncogene homolog; v-rel avian reticuloendotheliosis viral oncogene homolog B (nuclear factor of kappa light polypeptide gene enhancer in B-cells 3); v-rel reticuloendotheliosis viral oncogene homolog B; v-rel reticuloendotheliosis viral oncogene homolog B

Definition

The eukaryotic transcription factor, nuclear factor kappa B (NF-κB), was identified in 1986 by Sen and Baltimore as a protein that binds to a specific DNA sequence (5′-GGGACTTTCC-3′) within the intronic enhancer of the immunoglobulin kappa light chain gene in mature B and plasma cells (Sen and Baltimore 1986). In the decades following its discovery, NF-κB has been shown to exist in nearly all cell types, and specific NF-κB binding sites, termed κB, have been identified in the promoters/enhancers of a very large number of inducible genes. Similarly, the range of biological factors and environmental conditions known to induce NF-κB activity is remarkably large and diverse (May and Ghosh 1998). One of the key NF-κB-functions is its role in several phases of inflammatory responses. It is involved in the primary pro-inflammatory phase, in the regulation of the adaptive phase, and finally also in the resolution of inflammation, when anti-inflammatory genes are expressed and apoptosis is induced (Lawrence et al. 2001). Dysregulations of NF-κB have been observed in many painful inflammatory diseases, such as rheumatoid arthritis, atherosclerosis, inflammatory bowel diseases, and multiple sclerosis in which NF-κB has an impact on the production of inflammatory molecules (Tak and Firestein 2001; Gasparini and Feldmann 2012). For this reason, NF-κB has been considered as a possible target in the treatment of inflammatory diseases and has been extensively studied. The recent understanding that NF-κB can regulate both pro-inflammatory and anti-inflammatory mechanisms makes it challenging to define which mechanisms could be the most relevant to be targeted for selective blockage of specific functions (Gasparini and Feldmann 2012). Additionally, activation of NF-κB is important in innate and adaptive immunity, in which NF-κB promotes the differentiation and development of immune cells to clear microbial invasion and dissemination. Besides its function in inflammation, NF-κB plays a role in the life cycle of a number of viruses, for example, HIV1, CMV, and SV40, that contain κB sites within their promoters and as such make use of their hosts transcriptional apparatus to direct their own expression and infectivity (Ghosh 2007). Finally, NF-κB has been shown to be involved in many aspects of cell growth, differentiation, and proliferation via the induction of certain growth and transcription factors, as well as by acting as a protective factor against programmed cell death (apoptosis), and, thus, also plays a role during initiation and progression of cancer (Bassères and Baldwin 2006; Dutta et al. 2006).

Biosynthesis and Release

In the majority of cell types, NF-κB is composed of homo- or heterodimers of a family of structurally related proteins in the cytoplasm. Mammalian cells express five members of the NF-κB family: p65 (RelA), RelB, c-Rel, p52, and p50 (Fig. 1). All five proteins exhibit a conserved N-terminal region called Rel homology domain (RHD), which mediates subunit dimerization, interaction with IκB proteins, and DNA binding. At its C-terminus, the RHD contains a nuclear translocation domain (NLS) (Karin et al. 2004). Members of the Rel subfamily contain an additional transactivation domain (TAD), which is crucial for target gene activation (Yamamoto and Gaynor 2004). In contrast to the other family members, p50 and p52 are initially produced as the large precursors, p105 and p100, and are proteolytically processed to p50 and p52, respectively (Karin and Ben-Neriah 2000). The five NF-κB proteins have been shown to form almost every combination as homo- or heterodimers with different functions being ascribed to different dimers (Yamamoto and Gaynor 2004; Häcker and Karin 2006). For example, the classical and ubiquitously expressed p50/p65 heterodimer is found in the cytoplasm of

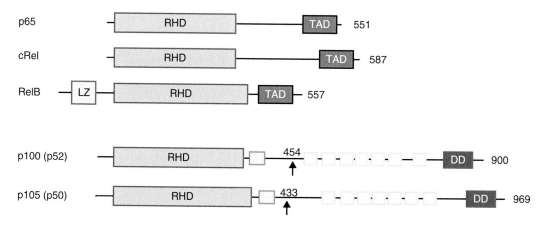

NFkappaB, Fig. 1 The five members of the NF-κB family: p65 (RelA), cRel, RelB, and precursor p100 processed to p52 (cut marked by *arrow*) and precursor p105 processed to p50 (cut marked by *arrow*). All members of the NF-κB family harbor an N-terminal Rel homology domain (*RHD*). RelB contains a leucine zipper (*LZ*) at its N-terminal region additionally. Three family members (p65, cRel, and RelB) contain a C-terminal transactivation domain (*TAD*). *Green squares*, glycine-rich repeat; *orange squares*, ankyrin repeats; *DD* death domain

most cells and is the important dimer for the production of primary inflammatory mediators. On the other hand, cRel, which is triggered in the hematopoietic compartment, is the active nuclear NF-κB in mature B cells as cRel/p50 dimer. Dimers of p52/c-Rel, p65/c-Rel, p65/p65, p50/p50, p52/p52, and p50/p52 have also been identified. RelB is a notable exception in that it only forms dimers with p50 or p52, possibly due to the characteristic leucine zipper domain (LZ) positioned in its N-terminal domain (Huang et al. 2005). The biological functions of the homo- and heterodimers are dependent on their differences in DNA-binding affinity and specificity and on their interactions with other factors that may be soluble or bound to neighboring DNA sites (Gasparini and Feldmann 2012).

Biological Activities

In most quiescent cells, NF-κB dimers are bound to inhibitory molecules of the IκB family of proteins (inhibitors of NF-κB). These inhibitors are characterized by ankyrin repeats, which interact with the RHD domains of NF-κB and, at least partly, mask the NLS of the transcription factors thereby making them transcriptionally inactive. The best studied inhibitors are IκBα, IκBβ, and IκBε, which have distinct and overlapping specificities with differences in tissue distribution. Since different IκB molecules might control the regulation of distinct genes in various tissues by inhibiting specific NF-κB subsets, IκB proteins could be attractive targets for specific therapies. Interestingly, p105 and p100, the precursors of p50 and p52, also contain ankyrin repeats, which are cleaved upon maturation and can therefore function both as reservoirs for the mature p50 and p52 subunits and as IκBs. In contrast to the other members of the NF-κB family, these two proteins do not contain a transactivation domain (Ghosh and Karin 2002). As a consequence, dimers of p50 and p52, which bind to NF-κB elements of gene promoters, act as transcriptional repressors (Solan et al. 2002). However, when p50 or p52 are bound to a member containing a transactivation domain, such as p65 or RelB, they constitute a transcriptional activator. The complexity of this transcriptional regulation system is also augmented by the fact that different NF-κB dimers have differential preferences for variations of the DNA-binding sequence. Thus, various target genes are differentially induced by distinct NF-κB dimers (Wong et al. 2011). Furthermore, NF-κB subunits also contain sites for phosphorylation and other posttranslational modifications, e.g., acetylation and methylation, which are important for activation and cross talk with other signaling pathways and transcription factors (Oeckinghaus and Ghosh 2009). Binding of NF-κB dimers to IκB molecules does not only prevent binding to DNA but also shift the steady-state localization of the complex to the cytosol. Nevertheless, shuttling between cytosol and nucleus does occur (Birbach et al. 2002).

In general, activation of most forms of NF-κB, especially the most common p50/p65 dimer, depends on phosphorylation-induced ubiquitination of IκB proteins mediated by IκB kinases (IKK). After activation of cells by a variety of stimuli including cytokines and bacterial lipopolysaccharides, IκB is phosphorylated at specific N-terminal serine residues by different IKKs, which form variable complexes dependent on the activating stimulus. The phosphorylation subsequently induces IκB-ubiquitinylation through the $E3^{I\kappa B}$ ubiquitin ligase complex and finally the degradation of IκB by the 26S proteasome. Thereby, NF-κB dimers are released from the cytoplasmic trapping complex and can translocate to the nucleus. NF-κB then binds to κB enhancer elements in the promoter region of a great number of target genes, inducing transcription of pro-inflammatory genes (Pahl 1999; Karin et al. 2004; Yamamoto and Gaynor 2004). Phosphorylation of IκB by different IκB kinases (IKK) is a crucial step in the NF-κB activation cascade. The activation of these kinases can be induced by various pathways. In addition, they provide a basis for manifold cross talk with other signaling pathways, as well as complex feedback loops allowing for a fine-tuning of the response.

A cascade essential for the activation of innate immunity and inflammation as well as inhibition of apoptosis is the **so-called classical, canonical pathway** (Fig. 2). It is stimulated by tumor necrosis factor (TNF), interleukin-1 β (IL-1β), toll-like, or T-cell receptor ligands and leads to activation of an IKK complex consisting of the regulatory subunit IKKγ (also known as NF-κB essential modulator (NEMO)) and the catalytical subunits IKKα and IKKβ. Activation of the IKKα/β/γ complex is mainly associated with nuclear translocation of p50/p65 NF-κB complexes (Yamamoto and Gaynor 2004). Another pathway, which is linked to B-cell activation, lymphoid organogenesis, and humoral immunity, is well known as the **alternative noncanonical pathway**. This pathway is stimulated by other activators such as B-cell activating factor (BAFF), lymphotoxin (LT) receptor α/β, and receptor activator for nuclear factor kappa B (RANK) or CD40L, which activate the NF-κB-inducing kinase (NIK) which phosphorylates and thereby activates IKKα homodimers. This leads to phosphorylation and degradation of the RelB-inhibitor protein p100 resulting in p52/RelB activation (Viatour et al. 2005; Häcker and Karin 2006). Another IKK complex which is activated by phorbol esters (PMA), lipopolysaccharide (LPS), cytokines, and viral components is composed of IKK epsilon (IKKε, also known as IKKi) and tank-binding kinase 1 (TBK1) which are structurally related to the classical IKKs, IKKα, and IKKβ. It has been shown that IKKε and TBK1 play major roles in the response to viral infections since both are involved in phosphorylation of interferon regulatory factors (IRF) 3 and 7 and the subsequent activation of interferon-I (Sharma et al. 2003; McWhirter et al. 2004). Furthermore, they also appear to play a role in the NF-κB activation pathway by phosphorylation of several proteins including IκBα, IKKβ, p65, or c-Rel (Peters and Maniatis 2001; Buss et al. 2004; Adli and Baldwin 2006; Harris et al. 2006; Mattioli et al. 2006).

After the liberation of NF-κB dimers, they are normally **translocated to the nucleus**, and their RHD can bind cognate DNA sequences in the enhancer elements of target gene promoters. Depending on the accessibility of the genome regulated by epigenetic mechanisms and the cell type, thousands of different target genes can be transcriptionally activated (Hoesel and Schmid 2013). This activation is further controlled by interaction of NF-κB with additional transcription factors, kinases, or inhibitors, which can either enhance or reduce the effect of NF-κB. Besides the interactions with its inhibitors, e.g., IκBα, IκBβ, or IκBε, NF-κB has been shown to interact with upstream kinases, with chromatin modifiers such as histone deacetylases (HDACs) or CREB-binding protein (CBP), and also with other transcription factors like STATs, IRFs, p53, CREB,

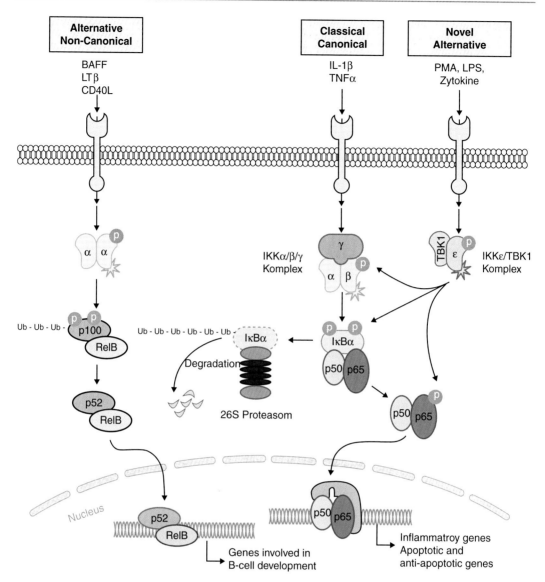

NFkappaB, Fig. 2 Schematic overview of different NF-κB activation pathways. The noncanonical pathway is activated by binding of the CD40 ligand, B-cell activating factor (*BAFF*), or lymphotoxin β (*LTβ*) to their receptors, leading to an activation of IKKα, which then induces the processing of the inhibitory NF-κB protein p100. p100 proteolysis releases p52, which can then translocate as a heterodimer with RelB into the nucleus and bind to the promoter of genes that are frequently involved in B-cell development. This pathway is independent of IKKγ but relies on IKKα homodimers. The classical canonical NF-κB-activating pathway can be induced by inflammatory stimuli such as chemokines and cytokines. This pathway is crucially dependent on activation of the classical IKKα, IKKβ, and IKKγ complex, which leads to the phosphorylation of I-κBα at serines 32 and 36 with subsequent ubiquitination (Ub, ubiquitin) and degradation of the inhibitory protein via the proteasome pathway. Then, the heterodimer p50/p65 is released and migrates to the nucleus, where it binds to specific κB sites and activates a variety of NF-κB target genes. A novel, alternative pathway is represented by an IKK complex consisting of IKKε and, most likely, TANK-binding kinase 1 (*TBK1*). This complex is activated by different stimuli, such as phorbol esters (PMA, phorbol 12-myristate 13-acetate), by LPS, but also, like the classical pathway, by cytokines and results in phosphorylation of several targets in the NF-κB activation pathway leading to NF-κB activation

NFkappaB, Table 1 NF-κB activation in inflammatory diseases (Tak and Firestein 2001)

Atherosclerosis
Rheumatoid arthritis (RA)
Chronic inflammatory demyelinating polyradiculoneuritis
Inflammatory bowel disease
Helicobacter pylori-associated gastritis
Asthma
Multiple sclerosis (MS)
Systemic inflammatory response syndrome (SIRS)

ATFs, SMAD3, and AP-1 (Gerritsen et al. 1997; Mengshol et al. 2000; Ashburner et al. 2001; Frank et al. 2011). Termination of the transcriptional activity of NF-κB is mainly achieved by the fact that NF-κB upregulates its own inhibitors of the IκB family (Pahl 1999). In addition, negative regulators of the NF-κB signaling pathway such as A20 and CYLD are upregulated by NF-κB (Trompouki et al. 2003; Wertz et al. 2004).

In acute inflammation, these negative feedback loops usually result in deactivation of NF-κB. However, in chronic inflammatory conditions and in the pathogenesis of several inflammatory diseases (Table 1), the permanent presence of NF-κB-activating factors seems to exceed the inhibitory feedback loops and leads to an elevated constitutive activity of NF-κB, which induces transcription of pro-inflammatory cytokines (i.e., TNF-α, IL-1β, IL-6, IL-8, IL-10), chemokines (i.e., MCP-1, IP-10, RANTES), adhesion molecules (i.e., E-selectin, VCAM-1, ICAM-1), matrix metalloproteinases (MMPs), cyclooxygenase (Cox)-2, and inducible nitric oxide synthase (iNOS) (Karin and Ben-Neriah 2000; Li and Verma 2002; Hayden and Ghosh 2004; Karin et al. 2004; Ghosh and Hayden 2008).

Pathophysiological Activities

Due to space limitations, this section will only cover the possible role of NF-κB during inflammation-linked diseases. One should keep in mind that abnormal regulation of NF-κB also plays a role in neurological disorders and many cancers.

Atherosclerosis is triggered by a large number of growth factors, chemokines, and cytokines released from damaged endothelial and smooth muscle cells of the vessel wall. Activation of NF-κB by circulating cytokines has been linked to atherosclerosis and thrombosis and NF-κB. Furthermore, NF-κB regulation of genes involved in the inflammatory response and the control of cellular proliferation are relevant for initiation and progression of atherosclerosis (Monaco and Paleolog 2004). Similar observations have been made in the context of *rheumatoid arthritis* (RA), where NF-κB is overexpressed in the inflamed synovium. The accompanying enriched NF-κB activity enhances the production of pro-inflammatory mediators like IL-1, IL-6, IL-8, and TNF-α and the recruitment of immune cells. In addition, NF-κB binding to DNA and subsequent pro-inflammatory cytokine expression are much higher in RA compared with osteoarthritis controls (Han et al. 1998). Interestingly, mice with a deletion of the p50 subunit of NF-κB do not develop acute and chronic arthritis, respectively (Campbell et al. 2000). Further animal models showed that increasing synovial NF-κB activity during the progression of disease precedes the development of clinical joint inflammation in murine collagen-induced arthritis (CIA) (Tsao et al. 1997). In addition to RA, NF-κB has been implicated in several other tissue-specific models of inflammation. NF-κB activation induces production of pro-inflammatory cytokines in nephritic glomeruli, resulting in experimental *glomerulonephritis* in rats (Sakurai et al. 1996). IL-2 knockout mice develop *inflammatory colitis* that is marked by enhanced NF-κB activation and increased IL-1 expression in colonic epithelium (Yang et al. 1999). *Helicobacter pylori*-associated *gastritis* in humans is also marked by increased NF-κB activity in gastric epithelial cells, and the number of NF-κB positive cells correlates with the degree of gastritis. Similarly, there is evidence of NF-κB activation in lamina propria macrophages in *inflammatory bowel diseases*, including Crohn's disease and ulcerative colitis (van Den Brink et al. 2000). NF-κB activation is seen in mucosal biopsy specimens from patients with active Crohn's disease and ulcerative colitis. *Inflammatory airway disease* (e.g., asthma) in humans has also been associated with increased cytokine and chemokine expression, which correlate with activation of NF-κB in bronchial biopsies from asthma patients (Hart et al. 1998). The increased NF-κB activity is especially apparent in airway epithelial cells, where there is abundant expression of pro-inflammatory cytokines, chemokines, iNOS, and Cox-2. In allergen-induced asthma, loss of the p50 locus in knockout mice ablates the eosinophilic airway response (Yang et al. 1998). Similarly, c-Rel knockout mice show decreased airway hyperresponsiveness and eosinophil infiltration as well as lower levels of serum IgE (Donovan et al. 1999). Activation of NF-κB has been found in brain tissue of *multiple sclerosis* (MS) patients in or near central nervous system (CNS) lesions (Gveric et al. 1998). In addition, microarray analysis of MS brain tissue identified upregulated expression of genes related to NF-κB (Mycko et al. 2003). Similar to what is seen in the brain of MS patients, activation of NF-κB could be detected in the spinal cord of experimental autoimmune encephalomyelitis (EAE)-diseased mice (Pahan and Schmid 2000).

However, one should keep in mind that NF-κB not only plays a role in the onset of inflammation by induction of cytokine production or leukocyte recruitment but also contributes to the *resolution of inflammation*, for example, by promoting leukocyte apoptosis. It has been also shown that NF-κB is involved in the feedback control of inflammation by various mechanisms to affect the magnitude and duration of the inflammatory response (Lawrence et al. 2001; Hoffmann and Baltimore 2006; Lawrence 2009). Thus, NF-κB inhibition as a therapeutic approach in the treatment of chronic diseases should be considered carefully.

Modulation by Drugs

The crucial role of NF-κB in several diseases and inflammatory processes indicates that pharmacological modulation of the NF-κB signaling cascade might provide beneficial effects. The NF-κB-activating pathways offer many targets for various drugs which might interfere either with receptors such as toll-like or TNF receptors, with IκB kinases, I-κB, ubiquitin ligases, the 26S proteasome, and decoy of κB sites at the DNA or NF-κB itself. Also, well-known drugs like NSAIDS (nonsteroidal anti-inflammatory drugs) and glucocorticoids are known to interfere with NF-κB activation pathways as has been described for several immunosuppressive agents. Due to space limitations, this section will focus only on direct NF-κB modifying drugs, which modulate NF-κB nuclear translocation and DNA binding.

Inhibition of Active NF-κB Dimers

One possible approach to inhibit NF-κB activation is to overexpress its inhibitor protein, IκB, and prevent its phosphorylation. For IkBα this could be achieved by generation of a protein mutated at N-terminal serine residues 32 and 36, which are normally phosphorylated (IκB super-repressor) (DiDonato et al. 1996). The IκB super-repressor showed beneficial effects in a rat model of rheumatoid arthritis (Blackwell et al. 2004). Furthermore, transgenic mice with overexpressed IκBα showed a decreased expression of pro-inflammatory proteins such as CCL2 and TGF-β2, less symptoms, and a strong improvement of the functional recovery after spinal cord injury (Fu et al. 2007).

Inhibition of Nuclear Transport

SN50 constitutes a small cell-permeable peptide carrying the nuclear localization sequence of the p50 NF-κB subunit fused to a linker sequence that allows the passage through cell membranes. This mechanism inhibits NF-κB nuclear translocation and activation in response to various stimuli such as TNF-α and LPS and is able to reduce inflammatory responses in vivo (D'Acquisto et al. 2000; D'Acquisto and Ianaro 2006). Unfortunately, this drug has been reported to lack specificity since it also inhibits the nuclear import of AP-1, NFAT, and STAT1 too (Torgerson et al. 1998).

Small Nucleic Acid Molecules: Gene Silencing Strategies

Antisense oligonucleotides (ODNs) target the mRNA sequence of NF-κB and thereby effectively inhibit its expression. These structures have already been assessed in in vitro and in vivo inflammatory models and successfully inhibited the expression of several NF-κB-dependent genes (Lawrance et al. 2003; Wu and Chakravarti 2007).

Another possibility is NF-κB decoy ODNs, synthetic double-strand oligonucleotides with the NF-κB-specific consensus sequence. The ODNs bind to activated NF-κB and inhibit DNA binding of the transcription factor at the promoter of NF-κB-regulated genes (Morishita et al. 2004). NF-κB decoy ODNs were used to inhibit p65/p50 activity in mouse models of inflammatory bowel disease or asthma, by reducing the NF-κB-dependent production of inflammatory cytokines (Desmet et al. 2004; De Vry et al. 2007). In patients with rheumatoid arthritis, NF-κB decoy inhibited the expression of inflammatory cytokines and MMPs in synovial cells (Tomita et al. 2000).

RNA interference (RNAi) is based on short interfering RNA (siRNA) that leads to specific cleavage and degradation of target mRNA and is being considered as a treatment strategy for inflammatory diseases by targeting NF-κB. p65 siRNA resulted in decreased transcription of NF-κB-regulated proteins such as COX-2 or MMPs in in vitro studies (Lianxu et al. 2006). In vivo, an adenovirus-delivered p65 siRNA has been found to inhibit the activation of NF-κB and the expression of p65 in cartilage and synovium of the knee and thus reduce the degradation of cartilage in the early phase of experimental osteoarthritis (Chen et al. 2008). Some of the major concerns arising with the use of siRNA in vivo, as with all small-molecule nucleic acid agents, are tissue specificity and short half-life through degradation by nucleases.

Inhibition of DNA Acetylation

In addition to phosphorylation, NF-κB p65 is also regulated by diverse acetylations, which can either increase or decrease its activation. Several inhibitors have been developed to target the increased NF-κB activation by acetylation. Anacardic acid, a well-known histone acetyltransferase inhibitor, has been shown to inhibit NF-κB activation by inhibiting p65 acetylation as well as regulation of other components of the canonical NF-κB pathway (Sung et al. 2008). Furthermore, the natural product gallic acid which is an ingredient, for instance, of oak bark and green tea was able to decrease NF-κB activation by inhibiting histone acetyltransferase activity (Choi et al. 2009). The modulation of NF-κB acetylation could thus be linked to anti-inflammatory effects.

Although direct NF-κB inhibitors have shown beneficial effects in the described models, one has to be aware that complete inhibition of NF-κB activation also bears risk. Genetically modified mice with deletions of NF-κB subunits, particularly p50 knockout mice, are lethal indicating that NF-κBs are essential for development and physiological mechanisms. Inhibition of disease-specific NF-κB pathways might, therefore, be a useful strategy to reduce only pathological NF-κB activity.

Cross-References

▸ Corticosteroids

References

Adli, M., & Baldwin, A. S. (2006). IKK-i/IKKepsilon controls constitutive, cancer cell-associated NF-kappaB activity via regulation of Ser-536 p65/RelA phosphorylation. *The Journal of Biological Chemistry, 281*, 26976–26984.

Ashburner, B. P., Westerheide, S. D., & Baldwin, A. S. (2001). The p65 (RelA) subunit of NF-kappaB interacts with the histone deacetylase (HDAC) corepressors HDAC1 and HDAC2 to negatively regulate gene expression. *Molecular and Cellular Biology, 21*, 7065–7077.

Bassères, D. S., & Baldwin, A. S. (2006). Nuclear factor-kappaB and inhibitor of kappaB kinase pathways in oncogenic initiation and progression. *Oncogene, 25*, 6817–6830.

Birbach, A., Gold, P., Binder, B. R., Hofer, E., de Martin, R., & Schmid, J. A. (2002). Signaling molecules of the NF-kappa B pathway shuttle constitutively between cytoplasm and nucleus. *The Journal of Biological Chemistry, 277*, 10842–10851.

Blackwell, N. M., Sembi, P., Newson, J. S., Lawrence, T., Gilroy, D. W., & Kabouridis, P. S. (2004). Reduced infiltration and increased apoptosis of leukocytes at sites of inflammation by systemic administration of a membrane-permeable IkappaBalpha repressor. *Arthritis and Rheumatism, 50*, 2675–2684.

Buss, H., Dörrie, A., Schmitz, M. L., Hoffmann, E., Resch, K., & Kracht, M. (2004). Constitutive and interleukin-1-inducible phosphorylation of p65 NF-{kappa}B at serine 536 is mediated by multiple protein kinases including I{kappa}B kinase (IKK)-{alpha}, IKK{beta}, IKK{epsilon}, TRAF family member-associated (TANK)-binding kinase 1 (TBK1), and an unknown kinase and couples p65 to TATA-binding protein-associated factor II31-mediated interleukin-8-transcription. *The Journal of Biological Chemistry, 279*, 55633–55643.

Campbell, I. K., Gerondakis, S., O'Donnell, K., & Wicks, I. P. (2000). Distinct roles for the NF-kappaB1 (p50) and c-Rel transcription factors in inflammatory arthritis. *Journal of Clinical Investigation, 105*, 1799–1806.

Chen, L. X., Lin, L., Wang, H. J., Wei, X. L., Fu, X., Zhang, J. Y., et al. (2008). Suppression of early experimental osteoarthritis by in vivo delivery of the adenoviral vector-mediated NF-kappaBp65-specific siRNA. *Osteoarthritis and Cartilage/OARS, Osteoarthritis Research Society, 16*, 174–184.

Choi, K.-C., Lee, Y.-H., Jung, M. G., Kwon, S. H., Kim, M.-J., Jun, W. J., et al. (2009). Gallic acid suppresses lipopolysaccharide-induced nuclear factor-kappaB signaling by preventing RelA acetylation in A549 lung cancer cells. *Molecular Cancer Research, 7*, 2011–2021.

D'Acquisto, F., & Ianaro, A. (2006). From willow bark to peptides: The ever widening spectrum of NF-kappaB inhibitors. *Current Opinion in Pharmacology, 6*, 387–392.

D'Acquisto, F., Ialenti, A., Ianaro, A., Di Vaio, R., & Carnuccio, R. (2000). Local administration of transcription factor decoy oligonucleotides to nuclear factor-kappaB prevents carrageenin-induced inflammation in rat hind paw. *Gene Therapy, 7*, 1731–1737.

De Vry, C. G., Prasad, S., Komuves, L., Lorenzana, C., Parham, C., Le, T., et al. (2007). Non-viral delivery of nuclear factor-kappaB decoy ameliorates murine inflammatory bowel disease and restores tissue homeostasis. *Gut, 56*, 524–533.

Desmet, C., Gosset, P., Pajak, B., Cataldo, D., Bentires-Alj, M., Lekeux, P., et al. (2004). Selective blockade of NF-kappa B activity in airway immune cells inhibits the effector phase of experimental asthma. *Journal of Immunology (Baltimore, Md. : 1950), 173*, 5766–5775.

DiDonato, J., Mercurio, F., Rosette, C., Wu-Li, J., Suyang, H., Ghosh, S., et al. (1996). Mapping of the inducible IkappaB phosphorylation sites that signal its ubiquitination and degradation. *Molecular and Cellular Biology, 16*, 1295–1304.

Donovan, C. E., Mark, D. A., He, H. Z., Liou, H. C., Kobzik, L., Wang, Y., et al. (1999). NF-kappa B/Rel transcription factors: c-Rel promotes airway hyperresponsiveness and allergic pulmonary inflammation. *Journal of Immunology (Baltimore, Md. : 1950), 163*, 6827–6833.

Dutta, J., Fan, Y., Gupta, N., Fan, G., & Gélinas, C. (2006). Current insights into the regulation of programmed cell death by NF-kappaB. *Oncogene, 25*, 6800–6816.

Frank, A. K., Leu, J. I.-J., Zhou, Y., Devarajan, K., Nedelko, T., Klein-Szanto, A., et al. (2011). The codon 72 polymorphism of p53 regulates interaction with NF-{kappa}B and transactivation of genes involved in immunity and inflammation. *Molecular and Cellular Biology, 31*, 1201–1213.

Fu, E. S., Zhang, Y. P., Sagen, J., Yang, Z. Q., & Bethea, J. R. (2007). Transgenic glial nuclear factor-kappa B inhibition decreases formalin pain in mice. *Neuroreport, 18*, 713–717.

Gasparini, C., & Feldmann, M. (2012). NF-kB as a target for modulating inflammatory responses. *Current Pharmaceutical Design, 18*, 5735–5745.

Gerritsen, M. E., Williams, A. J., Neish, A. S., Moore, S., Shi, Y., & Collins, T. (1997). CREB-binding protein/p300 are transcriptional coactivators of p65. *Proceedings of the National Academy of Sciences of the United States of America, 94*, 2927–2932.

Ghosh, S. (2007). *Handbook of transcription factor NF-kappaB* (pp. 1–246). Boca Raton: CRC Press.

Ghosh, S., & Hayden, M. S. (2008). New regulators of NF-kappaB in inflammation. *Nature Reviews Immunology, 8*, 837–848.

Ghosh, S., & Karin, M. (2002). Missing pieces in the NF-kappaB puzzle. *Cell, 109*(Suppl), S81–S96.

Gveric, D., Kaltschmidt, C., Cuzner, M. L., & Newcombe, J. (1998). Transcription factor NF-kappaB and inhibitor I kappaBalpha are localized in macrophages in active multiple sclerosis lesions. *Journal of Neuropathology and Experimental Neurology, 57*, 168–178.

Häcker, H., & Karin, M. (2006). Regulation and function of IKK and IKK-related kinases. *Science's STKE: Signal Transduction Knowledge Environment, 2006*, re13.

Han, Z., Boyle, D. L., Manning, A. M., & Firestein, G. S. (1998). AP-1 and NF-kappaB regulation in rheumatoid arthritis and murine collagen-induced arthritis. *Autoimmunity, 28*, 197–208.

Harris, J., Olière, S., Sharma, S., Sun, Q., Lin, R., Hiscott, J., et al. (2006). Nuclear accumulation of cRel following C-terminal phosphorylation by TBK1/IKK epsilon. *Journal of Immunology (Baltimore, Md. : 1950), 177*, 2527–2535.

Hart, L. A., Krishnan, V. L., Adcock, I. M., Barnes, P. J., & Chung, K. F. (1998). Activation and localization of transcription factor, nuclear factor-kappaB, in asthma.

American Journal of Respiratory and Critical Care Medicine, 158, 1585–1592.

Hayden, M. S., & Ghosh, S. (2004). Signaling to NF-kappaB. *Genes & Development, 18*, 2195–2224.

Hoesel, B., & Schmid, J. A. (2013). The complexity of NF-κB signaling in inflammation and cancer. *Molecular Cancer, 12*, 1–15.

Hoffmann, A., & Baltimore, D. (2006). Circuitry of nuclear factor kappaB signaling. *Immunological Reviews, 210*, 171–186.

Huang, D.-B., Vu, D., & Ghosh, G. (2005). NF-kappaB RelB forms an intertwined homodimer. *Structure (London, England: 1993), 13*, 1365–1373.

Karin, M., & Ben-Neriah, Y. (2000). Phosphorylation meets ubiquitination: The control of NF-[kappa]B activity. *Annual Review of Immunology, 18*, 621–663.

Karin, M., Yamamoto, Y., & Wang, Q. M. (2004). The IKK NF-kappa B system: A treasure trove for drug development. *Nature Reviews Drug Discovery, 3*, 17–26.

Lawrance, I. C., Wu, F., Leite, A. Z. A., Willis, J., West, G. A., Fiocchi, C., et al. (2003). A murine model of chronic inflammation-induced intestinal fibrosis down-regulated by antisense NF-kappa B. *Gastroenterology, 125*, 1750–1761.

Lawrence, T. (2009). The Nuclear Factor NF- B pathway in inflammation. *Cold Spring Harbor Perspectives in Biology, 1*, a001651–a001651.

Lawrence, T., Gilroy, D. W., Colville-Nash, P. R., & Willoughby, D. A. (2001). Possible new role for NF-kappaB in the resolution of inflammation. *Nature Medicine, 7*, 1291–1297.

Li, Q., & Verma, I. M. (2002). NF-kappaB regulation in the immune system. *Nature Reviews Immunology, 2*, 725–734.

Lianxu, C., Hongti, J., & Changlong, Y. (2006). NF-kappaBp65-specific siRNA inhibits expression of genes of COX-2, NOS-2 and MMP-9 in rat IL-1beta-induced and TNF-alpha-induced chondrocytes. *Osteoarthritis and Cartilage/OARS, Osteoarthritis Research Society, 14*, 367–376.

Mattioli, I., Geng, H., Sebald, A., Hodel, M., Bucher, C., Kracht, M., et al. (2006). Inducible phosphorylation of NF-kappa B p65 at serine 468 by T cell costimulation is mediated by IKK epsilon. *The Journal of Biological Chemistry, 281*, 6175–6183.

May, M. J., & Ghosh, S. (1998). Signal transduction through NF-kappa B. *Immunology Today, 19*, 80–88.

McWhirter, S. M., Fitzgerald, K. A., Rosains, J., Rowe, D. C., Golenbock, D. T., & Maniatis, T. (2004). IFN-regulatory factor 3-dependent gene expression is defective in Tbk1-deficient mouse embryonic fibroblasts. *Proceedings of the National Academy of Sciences of the United States of America, 101*, 233–238.

Mengshol, J. A., Vincenti, M. P., Coon, C. I., Barchowsky, A., & Brinckerhoff, C. E. (2000). Interleukin-1 induction of collagenase 3 (matrix metalloproteinase 13) gene expression in chondrocytes requires p38, c-Jun N-terminal kinase, and nuclear factor kappaB: differential regulation of collagenase 1 and collagenase 3. *Arthritis and Rheumatism, 43*, 801–811.

Monaco, C., & Paleolog, E. (2004). Nuclear factor kappaB: A potential therapeutic target in atherosclerosis and thrombosis. *Cardiovascular Research, 61*, 671–682.

Morishita, R., Tomita, N., Kaneda, Y., & Ogihara, T. (2004). Molecular therapy to inhibit NFkappaB activation by transcription factor decoy oligonucleotides. *Current Opinion in Pharmacology, 4*, 139–146.

Mycko, M. P., Papoian, R., Boschert, U., Raine, C. S., & Selmaj, K. W. (2003). cDNA microarray analysis in multiple sclerosis lesions: detection of genes associated with disease activity. *Brain, 126*, 1048–1057.

Oeckinghaus, A., & Ghosh, S. (2009). The NF-kappaB family of transcription factors and its regulation. *Cold Spring Harbor Perspectives in Biology, 1*, a000034.

Pahan, K., & Schmid, M. (2000). Activation of nuclear factor-kB in the spinal cord of experimental allergic encephalomyelitis. *Neuroscience Letters, 287*, 17–20.

Pahl, H. L. (1999). Activators and target genes of Rel/NF-kappaB transcription factors. *Oncogene, 18*, 6853–6866.

Peters, R. T., & Maniatis, T. (2001). A new family of IKK-related kinases may function as I kappa B kinase kinases. *Biochimica et Biophysica Acta, 1471*, M57–M62.

Sakurai, H., Hisada, Y., Ueno, M., Sugiura, M., Kawashima, K., & Sugita, T. (1996). Activation of transcription factor NF-kappa B in experimental glomerulonephritis in rats. *Biochimica et Biophysica Acta, 1316*, 132–138.

Sen, R., & Baltimore, D. (1986). Multiple nuclear factors interact with the immunoglobulin enhancer sequences. *Cell, 46*, 705–716.

Sharma, S., tenOever, B. R., Grandvaux, N., Zhou, G.-P., Lin, R., & Hiscott, J. (2003). Triggering the interferon antiviral response through an IKK-related pathway. *Science, 300*, 1148–1151.

Solan, N. J., Miyoshi, H., Carmona, E. M., Bren, G. D., & Paya, C. V. (2002). RelB cellular regulation and transcriptional activity are regulated by p100. *The Journal of Biological Chemistry, 277*, 1405–1418.

Sung, B., Pandey, M. K., Ahn, K. S., Yi, T., Chaturvedi, M. M., Liu, M., et al. (2008). Anacardic acid (6-nonadecyl salicylic acid), an inhibitor of histone acetyltransferase, suppresses expression of nuclear factor-kappaB-regulated gene products involved in cell survival, proliferation, invasion, and inflammation through inhibition of the inhibitory subunit of nuclear factor-kappaBalpha kinase, leading to potentiation of apoptosis. *Blood, 111*, 4880–4891.

Tak, P. P., & Firestein, G. S. (2001). NF-κB: A key role in inflammatory diseases. *Journal of Clinical Investigation, 107*, 7–11.

Tomita, T., Takano, H., Tomita, N., Morishita, R., Kaneko, M., Shi, K., et al. (2000). Transcription factor decoy for

NFkappaB inhibits cytokine and adhesion molecule expressions in synovial cells derived from rheumatoid arthritis. *Rheumatology (Oxford, England), 39,* 749–757.
Torgerson, T. R., Colosia, A. D., Donahue, J. P., Lin, Y. Z., & Hawiger, J. (1998). Regulation of NF-kappa B, AP-1, NFAT, and STAT1 nuclear import in T lymphocytes by noninvasive delivery of peptide carrying the nuclear localization sequence of NF-kappa B p50. *Journal of Immunology (Baltimore, Md.: 1950), 161,* 6084–6092.
Trompouki, E., Hatzivassiliou, E., Tsichritzis, T., Farmer, H., Ashworth, A., & Mosialos, G. (2003). CYLD is a deubiquitinating enzyme that negatively regulates NF-kappaB activation by TNFR family members. *Nature, 424,* 793–796.
Tsao, P. W., Suzuki, T., Totsuka, R., Murata, T., Takagi, T., Ohmachi, Y., et al. (1997). The effect of dexamethasone on the expression of activated NF-kappa B in adjuvant arthritis. *Clinical Immunology and Immunopathology, 83,* 173–178.
van Den Brink, G. R., ten Kate, F. J., Ponsioen, C. Y., Rive, M. M., Tytgat, G. N., van Deventer, S. J., et al. (2000). Expression and activation of NF-kappa B in the antrum of the human stomach. *Journal of Immunology (Baltimore, Md. : 1950), 164,* 3353–3359.
Viatour, P., Merville, M.-P., Bours, V., & Chariot, A. (2005). Phosphorylation of NF-kappaB and IkappaB proteins: implications in cancer and inflammation. *Trends in Biochemical Sciences, 30,* 43–52.
Wertz, I. E., O'Rourke, K. M., Zhou, H., Eby, M., Aravind, L., Seshagiri, S., et al. (2004). De-ubiquitination and ubiquitin ligase domains of A20 downregulate NF-kappaB signalling. *Nature, 430,* 694–699.
Wong, D., Teixeira, A., Oikonomopoulos, S., Humburg, P., Lone, I. N., Saliba, D., et al. (2011). Extensive characterization of NF-κB binding uncovers non-canonical motifs and advances the interpretation of genetic functional traits. *Genome Biology, 12,* R70.
Wu, F., & Chakravarti, S. (2007). Differential expression of inflammatory and fibrogenic genes and their regulation by NF-kappaB inhibition in a mouse model of chronic colitis. *Journal of Immunology (Baltimore, Md. : 1950), 179,* 6988–7000.
Yamamoto, Y., & Gaynor, R. B. (2004). IkappaB kinases: key regulators of the NF-kappaB pathway. *Trends in Biochemical Sciences, 29,* 72–79.
Yang, L., Cohn, L., Zhang, D. H., Homer, R., Ray, A., & Ray, P. (1998). Essential role of nuclear factor kappaB in the induction of eosinophilia in allergic airway inflammation. *The Journal of Experimental Medicine, 188,* 1739–1750.
Yang, F., de Villiers, W. J., Lee, E. Y., McClain, C. J., & Varilek, G. W. (1999). Increased nuclear factor-kappaB activation in colitis of interleukin-2-deficient mice. *The Journal of Laboratory and Clinical Medicine, 134,* 378–385.

Non-steroidal Anti-inflammatory Drugs: Overview

Richard O. Day and Garry G. Graham
Department of Pharmacology, School of Medical Sciences, University of New South Wales, Sydney, NSW, Australia
Department of Clinical Pharmacology and Toxicology, St Vincent's Hospital, Sydney, NSW, Australia

Synonyms

Antipyretic analgesics; Aspirin-like drugs; Cyclo-oxygenase inhibitors; Nonselective NSAIDs; Selective COX-2 inhibitors; Simple analgesics

Definitions

The term nonsteroidal anti-inflammatory drugs (NSAIDs) refers to a group of drugs whose major therapeutic activities are the suppression of pain (analgesia), reduction of body temperature in fever (antipyresis), and decreasing signs of inflammation (anti-inflammatory activity).

Chemical Structures

A variety of chemical structures lead to NSAID activity (Fig. 1). Most nonselective NSAIDs are lipid-soluble acids with pKa values in the range 4–5. They are therefore highly ionized at physiological pH values. The selective COX-2 inhibitors are also lipid soluble but have variable acidic properties. Celecoxib and valdecoxib are weak acids (pKa $\approx$ 11), while lumiracoxib is a carboxylic acid (pKa $\approx$ 4.7) and rofecoxib is a neutral substance.

Metabolism and Excretion

Being lipid soluble in the unionized forms and highly bound to plasma proteins, only low

Non-steroidal Anti-inflammatory Drugs: Overview, Fig. 1 The chemical structures of various groups of NSAIDs

Salicylates
- Salicylic acid R = R' = R" = H
- Aspirin R = R' = H, R" = acetyl
- Salicylsalicylic acid R = R' = H, R" = Salicylate
- Diflunisal R = difluorophenyl, R' = R" = H

Propionic acid derivatives (Profens)
R-profen → S-profen
R = Group containing aromatic ring system

Acetic acid derivatives
R—CH$_2$—COOH
Indomethacin, diclofenac, tolmetin, etodolac, lumiracoxib
R = Group containing aromatic ring systems

Sulfonamides
- Oxicams R' = CH$_3$, R and R" within common ring system containing N and S
- Celecoxib, valdecoxib R = Group containing aromatic ring systems, R' = R" = H
- Nimesulide R = CH$_3$, R' = H, R" = Group containing aromatic ring systems

Sulfones
Etoricoxib, Rofecoxib
R = Group containing aromatic ring systems

Aminobenzoic acid derivatives
Fenamates
R = Group containing aromatic ring system

proportions of NSAIDs are excreted unchanged in urine. The only exception is salicylate which undergoes considerable urinary excretion, particularly when urinary pH is above 7. Metabolites of NSAIDs are formed mainly by oxidative processes in the liver.

Pharmacokinetics

The NSAIDs are well absorbed orally, but, depending on their chemical structure, their half lives of elimination are very variable. The pharmacokinetics of NSAIDs are discussed in the sections on individual groups of NSAIDs (propionic acid derivatives (profens); salicylates).

The NSAIDs are strongly bound to proteins in plasma, often over 99 %. The total concentrations (total and unbound) are generally recorded, but these are very much higher than unbound or free concentrations. It is a principle of pharmacology that the effects of drugs are related to the unbound concentrations. Consequently, when conducting pharmacological experiments in vitro, the actions of NSAIDs should be correlated with the unbound plasma concentrations, not the total concentrations.

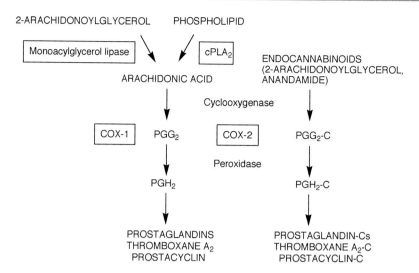

Non-steroidal Anti-inflammatory Drugs: Overview, Fig. 2 Synthesis of prostaglandins and related compounds. Cyclooxygenase-1 (*COX-1*) catalyzes the synthesis of prostaglandin (*PG*) intermediates, PGG_2 and PGH_2. COX-2 catalyzes not only the synthesis of PG intermediates from arachidonic acid but also PG cannabinoid intermediates (PGG_2-C and PGH_2-C). COX-1 and COX-2 are bifunctional enzymes with cyclooxygenase and peroxidase functions. The nonselective NSAIDs inhibit the formation of PG intermediates by both COX-1 and COX-2, whereas the selective COX-2 inhibitors block the synthesis of PG and PG cannabinoid intermediates by COX-2. R-enantiomers of the profens selectively inhibit the synthesis of PGH_2-C by COX-2. The final production of PGs and PG-Cs is synthesized by specific enzymes

Pharmacological Activities

The major proven mechanism of action of the NSAIDs is inhibition of the cyclooxygenase enzymes (COX-1 and COX-2) which are central enzymes in the synthesis of prostaglandins (Fig. 2). In this regard, there are three major groups, as is outlined below.

Classes of NSAIDs

The NSAIDs can be separated into three major groups by their effects on the synthesis of prostaglandins and related compounds:

- Nonselective NSAIDs, such as aspirin, diclofenac, and indomethacin. These drugs have anti-inflammatory, analgesic, and antipyretic activities. They are also well known to produce gastrointestinal damage, inhibit platelet aggregation, decrease kidney function, and precipitate aspirin-induced asthma. The activity of these drugs is due to inhibition of two central enzymes (cyclooxygenase-1 (COX-1) and COX-2) involved in the synthesis of *prostaglandins* and related compounds (Fig. 2). The prostaglandins are mediators of pain, fever and inflammation, and inhibition of platelet aggregation as well as gastroprotection. Consequently, inhibition of their synthesis leads to the characteristic therapeutic effects and adverse actions of the NSAIDs.

- COX-2-selective inhibitors (coxibs or COX-1 sparing agents, CSIs) have similar therapeutic activities to the nonselective NSAIDs. Celecoxib and valdecoxib inhibit COX-2 (Fig. 2) with very little effect on COX-1. These selective COX-2 inhibitors have better gastrointestinal tolerance than the nonselective NSAIDs and no significant antiplatelet effect and do not produce asthma in aspirin-sensitive patients. However, the distinction between the nonselective NSAIDs and the selective COX-2 inhibitors is not absolutely clear, and some NSAIDs, such as meloxicam and etodolac, are moderately selective.

- Drugs which are selective inhibitors of the oxidation of endocannabinoids by COX-2 but do not inhibit the oxidation of arachidonic acid by COX-2 (Fig. 2). This group includes the R-enantiomers of the propionic acid derivatives (profens) such as ibuprofen, naproxen, ketoprofen, and flurbiprofen. The S-enantiomers of these drugs are typical nonselective NSAIDs, but the R-enantiomers have contrasting pharmacological and clinical effects. Inhibition of the oxidation of the endocannabinoids is a recent discovery about the R profens and is discussed in another chapter of this encyclopedia (Graham and Day, Propionic Acid Derivatives (Profens)).

Aspirin is an unusual NSAID. It is a nonselective NSAID which irreversibly inhibits COX-1 and COX-2 due to covalent acetylation of the enzymes (Fig. 1). Currently, it is used mainly at low doses to block the synthesis of thromboxane A_2 by platelets and subsequently to inhibit platelet aggregation. As an analgesic, antipyretic, and anti-inflammatory agent, the metabolism of aspirin to salicylate provides a second mechanism of action. Salicylate has different actions to those of aspirin, and the basic and clinical pharmacology of aspirin and salicylate is compared in a separate chapter of this encyclopedia (Graham and Day, Salicylates). Overall, the actions of salicylate resemble those of the selective COX-2 inhibitors.

Paracetamol (acetaminophen) is not described as a NSAID. It is an effective antipyretic. It also has analgesic properties which are generally weaker than those of the NSAIDs and has weak anti-inflammatory activity. Paracetamol is often used in preference to both classes of NSAIDs because of its lesser adverse effects (Day and Graham 2013). Paracetamol is usually stated to be a pure analgesic. However, it is as active as NSAIDs in relieving the inflammation, as well as the pain, following oral surgery (Graham et al. 2013). Paracetamol's mechanism of action has been unclear for many years, but recent analysis indicates that it inhibits the synthesis of prostaglandins although by a different mechanism to those of the nonselective NSAIDs and the selective COX-2 inhibitors (Graham et al. 2013).

Cyclooxygenase-Independent Actions

The NSAIDs have many cyclooxygenase-independent actions in vitro. These actions include a variety of immunological and biochemical changes which have largely been observed in cellular incubations (Tegeder et al. 2001). The problem is that these pharmacological investigations have often been conducted in media containing either no plasma proteins or low concentrations, typically 10 % fetal calf serum or 0.5 % bovine serum albumin. The concentrations of NSAIDs are often similar to the total concentrations in plasma during therapy. However, the unbound concentrations in these low protein media (or protein free media) are much higher than in whole plasma. The overall conclusion is that many actions of NSAIDs, other than those on COX-1 COX-2, appear clinically insignificant. Examples of clinically irrelevant actions of NSAIDs are discussed in the chapter on profens and salicylates.

Clinical Use

Anti-inflammatory Effects

The anti-inflammatory and analgesic effects of the nonselective NSAIDs and the selective COX-2 inhibitors are clinically significant. In summary:

- Both classes are useful in decreasing the severe inflammation of diseases such as rheumatoid arthritis and psoriatic arthritis.
- The NSAIDs not only reduce the inflammation of these inflammatory diseases but also relieve the associated pain. The decreased pain may be associated with the reduced inflammation but may also be due, at least in part, to a direct analgesic activity of the NSAIDs.
- While the symptoms of rheumatoid arthritis are relieved by both classes of NSAIDs, neither class has any clear effect on the progression of the disease (Barraclough 2002; McCormack 2011). Joint damage appears to continue. Consequently, the NSAIDs are very commonly used with the disease-modifying antirheumatic drugs which may slow the degeneration of joints.

- Both the nonselective NSAIDs and the selective COX-2 inhibitors are valuable in the treatment of acute gout (Rubin et al. 2004; Schumacher et al. 2002). The pain and inflammation are relieved rapidly. Other drugs may also be used for acute gout. They are firstly the old drug colchicine, which is not a general anti-inflammatory drug. Corticosteroids may also be used.
- The NSAIDs are useful in the treatment of acute painful conditions in which there may be mild inflammation. For example, the NSAIDs have good activity in the relief of pain and inflammation after oral surgery to remove third molar (wisdom) teeth, dysmenorrhea, and acute soft tissue injury, such as sprains (Day and Graham 2013).
- Both major classes of NSAIDs (nonselective and selective COX-2 inhibitors) are widely used for the treatment of the pain of osteoarthritis but are only modestly effective. Other measures, such as weight loss and muscle strengthening are also very important in the treatment of osteoarthritis (Day and Graham 2013).

Antidiabetic Actions

There is evidence for the involvement of inflammatory processes in type 2 diabetes mellitus (T2DM) and obesity, which is often associated with insulin resistance. An inflammatory state is indicated in T2DM and obesity by increased levels of acute phase reactants, such as C-reactive protein (Donath and Sheldon 2011) whose concentrations are increased in inflammation, infection, tumors, tissue injury, and necrosis. Inflammation of the Islets of Langerhans has also been detected in patients and animals with diabetes (Donath and Sheldon 2011). Consequently, it is not surprising that NSAIDs and paracetamol have antidiabetic actions because of their anti-inflammatory actions (Graham et al. 2013). The largest clinical trial has been conducted with salsalate, a prodrug of salicylate (Goldfine et al. 2013), although further clinical trials are required to evaluate fully the clinical utility of the NSAIDs and paracetamol.

Anticancer Actions

The NSAIDs, particularly aspirin and celecoxib, are being widely examined for their prevention of colorectal cancers. Such prevention has been shown in carcinogen-induced colon cancers in experimental animals and in clinical studies in man. The use of NSAIDs is consistent with the high expression of COX-2 in approximately 80 % of patients with colorectal cancer. Even low dose of aspirin (80–100 mg daily) may be preventative (Thun et al. 2012). NSAIDs may also protect against some ovarian cancers and may decrease cachexia (tissue wasting associated with severe chronic diseases, particularly severe cancers).

The routine use of NSAIDs to prevent cancers is limited by their considerable toxic effects which are outlined below.

Adverse Effects

The adverse effects of both the nonselective NSAIDs and the selective COX-2 inhibitors are of major clinical significance because the adverse effects are severe in some patients and because of the widespread use of the NSAIDs. Their adverse effects, risk factors, and monitoring of patients are summarized below. The adverse effects of these drugs have been reviewed widely in the past 4–5 years, and some conclusions are discussed. However, it should be noted that not all published conclusions are consistent.

Gastrointestinal

Toxicity to the upper gastrointestinal tract (stomach and duodenum) is a major adverse effect of the nonselective NSAIDs. Toxicity is greatest in older patients or patients with a history of gastrointestinal blood loss, ulcer, or perforation of the stomach or duodenum. Features of the causes, epidemiology, and measures to decrease the toxicity include:

- Prostaglandins are cytoprotective in the stomach and small intestine.
- Inhibition of prostaglandin synthesis in the upper gastrointestinal tract is associated with

gastrointestinal damage to the upper gastrointestinal tract (stomach and duodenum).
- Serious cases of gastrointestinal damage affect nearly 1 % of chronic users of the nonselective NSAIDs per year.
- Dyspepsia and heartburn are often associated with treatment with nonselective NSAIDs but do not correlate with gastrointestinal damage.
- Gastrointestinal tolerance of the nonselective NSAIDs is improved with enteric coating, coprescription of antacids, ingestion with food, or rectal or parental routes of administration, but there is still a considerable risk of serious upper gastrointestinal bleeding.
- The toxicity of the nonselective NSAIDs is reduced by administration with misoprostol (a prostaglandin analogue) which is cytoprotective. However, it produces diarrhea and is contraindicated in pregnant patients because it may produce miscarriage. A better alternative is to administer NSAIDs with proton pump inhibitors which decrease acid secretion by the stomach (Day and Graham 2013).
- Paracetamol is well-tolerated by the gastrointestinal tract.

The COX-2-selective inhibitors were developed in order to decrease the gastrointestinal toxicity of the NSAIDs. This was successful to a degree. The selective COX-2 inhibitors have a lower rate of dyspepsia and serious gastrointestinal damage than is seen with the nonselective drugs (Bombardier et al. 2000; Latimer et al. 2009).

- The lower rate of gastrointestinal toxicity of the selective COX-2 inhibitors is reduced further by concomitant dosage with proton pump inhibitors (Latimer et al. 2009).
- Selective COX-2 inhibitors still may cause dyspepsia.
- Low doses of aspirin decrease the gastrointestinal sparing effects of the COX-2-selective inhibitors. These patients are now often prescribed a nonselective NSAID together with a proton pump inhibitor, as well as aspirin.

Obstruction of the lower gastrointestinal tract may occur with both classes of NSAIDs even when taken with a proton pump inhibitor (Maiden 2009). Patients should see a physician immediately if there are signs of gastrointestinal bleeding (vomiting blood or "coffee ground" feces or severe pain (Maiden 2009)).

Thrombosis

The selective COX-2 inhibitors have been associated to a varying degree with stroke and myocardial infarction, but these effects are, in general, increased by escalating dose and the individual patient's risk. The mechanism of the tendency to stroke and myocardial infarction with the selective COX-2 inhibitors is considered through the following mechanisms:

- Thromboxane A_2 is synthesized by a COX-1-dependent pathway in platelets. Platelet aggregation is therefore not affected by the COX-2-selective inhibitors.
- The selective COX-2 inhibitors block the synthesis of prostacyclin, a vasodilator and antithrombotic factor that is largely synthesized through COX-2.
- The balance of effects of selective COX-2 inhibitors on thrombosis therefore favors thrombosis.

A COX-2-selective inhibitor, rofecoxib, has been withdrawn because of the increased occurrence of stroke and myocardial infarction. The major currently used COX-2-selective inhibitor, celecoxib, has no significant tendency to cause myocardial infarction at low doses but an increased risk at higher doses.

Despite the well-developed theory for the association between selective COX-2 inhibitors and thrombosis, some of the nonselective NSAIDs are still associated with thrombosis. The best known are the following:

- The nonselective NSAID, diclofenac, is associated with an increased incidence of fatal myocardial infarction (CNT 2013; Latimer et al. 2009; Strand 2007; Trelle et al. 2011).
- Ibuprofen is not associated with cardiovascular adverse effects at doses up to 1,200 mg daily

(Latimer et al. 2009; McGettigan and Henry 2011), but recent meta analysis of clinical trials shows that doses of about 2,400 mg are associated with an increased risk of myocardial infarction (CNT 2013).

Of all the NSAIDs (nonselective and selective COX-2 inhibitors), naproxen appears to produce the least risk of myocardial infarction although safety cannot be guaranteed in all patients (CNT 2013; Latimer et al. 2009; McGettigan and Henry 2011).

An increased risk of stroke has often been shown with NSAIDs (Trelle et al. 2011), but the most recent and largest meta analysis of randomized, controlled trials shows no increased risk with either COX-2-selective drugs or high-dose nonselective NSAIDs, ibuprofen, diclofenac, or celecoxib (CNT 2013).

Renal Impairment and Cardiac Failure

Both the nonselective NSAIDs and the selective COX-2 inhibitors may precipitate renal failure and worsen cardiac failure.

Risk factors include:

- Age over 60
- Preexisting renal impairment
- Dehydration
- Cirrhosis
- Preexisting congestive cardiac failure
- Salt restricted diets
- Concomitant treatment with diuretics or inhibitors of angiotensin formation or action
- After general anesthesia

The renal function of patients in these situations is considered to be more dependent on the function of prostaglandins than normal subjects, and consequently, inhibition of prostaglandin synthesis by NSAIDs may markedly reduce renal function which may worsen cardiac failure. As far as possible, NSAIDs should be avoided in patients with substantial risk factors. If NSAIDs are used, greater care is required with increasing risk factors for renal impairment.

Hypertension

Blood pressure may rise, in some cases, quite substantially during treatment with either the nonselective NSAIDs or the COX-2-selective agents (Barraclough 2002; Day and Graham 2013; Whelton et al. 2002).

- Blood pressure should be monitored if dosage with the nonselective NSAIDs or the COX-2-selective drugs is commenced in patients taking antihypertensives.
- The dosage of antihypertensives should be adjusted, if required, during long-term treatment with nonselective NSAIDs and selective COX-2 inhibitors.
- NSAIDs should be avoided in patients with unstable blood pressure.

Asthma

Asthma is a concern for many patients taking nonselective NSAIDs.

- Asthma is precipitated in up to 20 % of asthmatics by aspirin and other nonselective NSAIDs (Jenkins et al. 2004).
- Almost total cross-sensitivity between aspirin and other nonselective NSAIDs.
- This reaction appears to be produced by inhibition of COX-1.
- Asthma is not induced by the selective COX-2 inhibitors (West and Fernandez 2003).
- Paracetamol is also safer in aspirin-sensitive asthmatics but does produce mild asthma in occasional patients (Jenkins et al. 2004).

Summary

The NSAIDs are widely used drugs. They decrease inflammation of rheumatoid arthritis and gout and are useful in the treatment of fever and mild to moderate pain. However, their adverse effects are becoming increasingly clear, and consequently, their use should be restricted as far as possible. Aspirin is a nonselective NSAID but is largely used in present-day medicine for its antiplatelet effects.

Cross-References

▶ Asthma
▶ Coxibs
▶ Fenamates
▶ Gout
▶ Non-steroidal Anti-inflammatory Drugs: Overview
▶ Osteoarthritis
▶ Propionic Acid Derivative Drugs (Profens)
▶ Prostanoids
▶ Rheumatoid Arthritis
▶ Salicylates

References

Barraclough, D. R. (2002). Considerations for the safe prescribing and use of COX-2-specific inhibitors. *Medical Journal of Australia, 176*(7), 328–331.

Bombardier, C., Laine, L., Reicin, A., Shapiro, D., Burgos-Vargas, R., Davis, B., et al. (2000). Comparison of upper gastrointestinal toxicity of rofecoxib and naproxen in patients with rheumatoid arthritis. VIGOR Study Group. *New England Journal of Medicine, 343*(21), 1520–1530.

CNT. (2013). Coxib and traditional NSAID Trialists' (CNT) Collaboration. Vascular and upper gastrointestinal effects of non-steroidal anti-inflammatory drugs: Meta-analyses of individual participant data from randomised trials. *Lancet, 382*(9894), 769–779. doi:10.1016/S0140-6736(1013)60900-60909.

Day, R. O., & Graham, G. G. (2013). Non-steroidal anti-inflammatory drugs (NSAIDs). *British Medical Journal* (Clinical research ed.). 346, 2013. Article Number: f3195.

Donath, M. Y., & Sheldon, S. E. (2011). Type 2 diabetes as an inflammatory disease. *Nature Reviews Immunology, 11*(2), 98–107.

Goldfine, A. B., Fonseca, V., Jablonski, K. A., Chen, Y. D. A., Tipton, L., Staten, M. A., et al. (2013). Salicylate (salsalate) in patients with type 2 diabetes. *Annals of Internal Medicine, 159*(1), 1–12.

Graham, G. G., Davies, M. J., Day, R. O., Mohamudally, A., & Scott, K. F. (2013). The modern pharmacology of paracetamol: Therapeutic actions, mechanism of action, metabolism, toxicity and recent pharmacological findings. *Inflammopharmacology, 21*(3), 201–232.

Jenkins, C., Costello, J., & Hodge, L. (2004). Systematic review of prevalence of aspirin induced asthma and its implications for clinical practice. *British Medical Journal, 328*(7437), 434.

Latimer, N., Lord, J., Grant, R., O'Mahony, R., Dickson, J., & Conaghan, P. (2009). Cost effectiveness of COX 2 selective inhibitors and traditional NSAIDs alone or in combination with a proton pump inhibitor for people with osteoarthritis. *British Medical Journal, 339*, b2538.

Maiden, L. (2009). Capsule endoscopic diagnosis of non-steroidal antiinflammatory drug-induced enteropathy. *Journal of Gastroenterology, 44*(Supplement XIX), 64–71.

McCormack, P. L. (2011). Celecoxib: A review of its use for symptomatic relief in the treatment of osteoarthritis, rheumatoid arthritis and ankylosing spondylitis. *Drugs, 71*(18), 2457–2489.

McGettigan, P., & Henry, D. (2011). Cardiovascular risk with non-steroidal anti-inflammatory drugs: Systematic review of population-based controlled observational studies. *PLoS Medicine, 8*(9), e1001098.

Rubin, B. R., Burton, R., Navarra, S., Antigua, J., Londono, J., Pryhuber, K. G., et al. (2004). Efficacy and safety profile of treatment with etoricoxib 120 mg once daily compared with indomethacin 50 mg three times daily in acute gout: A randomized controlled trial. *Arthritis and Rheumatism, 50*(2), 598–606.

Schumacher, H. R., Boice, J. A., Daikh, D. I., Mukhopadhyay, S., Malmstrom, K., Ng, J., et al. (2002). Randomised double blind trial of etoricoxib and indomethacin in treatment of acute gouty arthritis. *British Medical Journal, 324*(7352), 1488–1492.

Strand, V. (2007). Are COX-2 inhibitors preferable to non-selective non-steroidal anti-inflammatory drugs in patients with risk of cardiovascular events taking low-dose aspirin? *Lancet, 370*(9605), 2138–2151.

Tegeder, I., Pfeilschifter, J., & Geisslinger, G. (2001). Cyclooxygenase-independent actions of cyclooxygenase inhibitors. *FASEB Journal, 15*(12), 2057–2072.

Thun, M. J., Jacobs, E. J., & Patrono, C. (2012). The role of aspirin in cancer prevention. *Nature Reviews. Clinical Oncology, 9*(5), 259–267.

Trelle, S., Reichenbach, S., Wandel, S., Hildebrand, P., Tschannen, B., Villiger, P. M., et al. (2011). Cardiovascular safety of non-steroidal anti-inflammatory drugs: Network meta-analysis. *British Medical Journal, 342*, c7086.

West, P. M., & Fernandez, C. (2003). Safety of COX-2 inhibitors in asthma patients with aspirin hypersensitivity. *Annals of Pharmacotherapy, 37*(10), 1497–1501.

Whelton, A., White, W. B., Bello, A. E., Puma, J. A., & Fort, J. G. (2002). Effects of celecoxib and rofecoxib on blood pressure and edema in patients > or =65 years of age with systemic hypertension and osteoarthritis. *American Journal of Cardiology, 90*(9), 959–963.

Nuclear Receptor Signaling in the Control of Inflammation

Nicolas Venteclef[1], Tomas Jakobsson[2] and Eckardt Treuter[3]
[1]Centre de recherche des Cordeliers, INSERM UMRS 1138, Université Pierre et Marie Curie, Paris, France
[2]Department of Laboratory Medicine, Division of Clinical Chemistry, Karolinska Institutet C1:62, Karolinska University Hospital Huddinge, Stockholm, Sweden
[3]Department of Biosciences and Nutrition, Karolinska Institutet Huddinge, Stockholm, Sweden

Synonyms

Anti-inflammatory transrepression; Coregulators; Inflammation; Nuclear receptors; Transcription

Definition

Nuclear receptors (NRs) are prototypic ligand-regulated transcription factors that coordinate selective gene expression programs governing nearly every aspect of human physiology and disease. In humans, 48 distinct receptors are differentially expressed in metabolic tissues and communicating immune cells. By acting as sensors for hormones, lipid metabolites, xenobiotics, and pharmaceuticals, nuclear receptors modulate closely linked metabolic and inflammatory – referred to as "metaflammatory" – pathways in response to dietary challenges, environmental cues, disease-associated alterations, and therapeutic interventions. Nuclear receptors regulate inflammatory gene expression by a variety of distinct mechanisms, with anti-inflammatory transrepression being the most common and best characterized to date.

Biosynthesis and Release

Unlike classic mediators of inflammation, NRs are intracellular regulatory proteins that are expressed in the same cell types where they function as transcription factors. Their expression is mainly regulated at the level of mRNA synthesis (transcription), and each NR is expressed in a highly selective spatiotemporal fashion. Upon protein biosynthesis (translation) in the cytoplasm, NRs usually become posttranslationally modified and enter the nucleus, their main site of action. Nuclear import or export can be modulated by ligands, modifications, and cofactors.

Molecular and Biological Activities of NRs in the Immune System

Nuclear Receptor Structure and Function

Humans express 7 NR subfamilies with 48 members (Table 1) encoded by separate genes, many of which are alternatively spliced giving rise to multiple NR isoforms (Gronemeyer et al. 2004; www.nursa.org). Most NRs function as receptors for small lipophilic ligands that directly modulate their activities, thereby regulating gene expression programs linked to physiological processes. Ligands remain to be identified for a substantial number of orphan receptors, many of which may require ligand-independent regulation via posttranslational modifications (PTMs). Nonetheless, many NRs are already established as and all NRs are considered to be pharmaceutical targets for the development of synthetic compounds (agonists, antagonists, selective modulators) in key therapeutic areas such as cancer, metabolic disease, and inflammation (Gronemeyer et al. 2004; Saijo et al. 2010).

NRs are modular transcription factors that execute three key functions necessary for genomic signaling: (1) binding to DNA response elements in the regulatory regions (promoters, enhancers) of target genes, (2) binding to small molecule ligands, and (3) recruitment of coregulators (coactivators, corepressors, PTM modifiers, etc.) (Fig. 1). Additionally, NRs interact and "cross talk" with other primary transcription factors and signaling pathways by means of many different mechanisms, most significantly by anti-inflammatory "transrepression" (as outlined below). Further, some NRs possess so-called

Nuclear Receptor Signaling in the Control of Inflammation, Table 1 Members of the human nuclear receptor family and key functions

NR group	Name	Symbol[a]	Gene[b]	Ligands	Functions in metabolism and development	Functions in inflammation
Thyroid hormone receptors	TRα TRβ	NR1A1 NR1A2	THRA THRB	Thyroid hormone	Crucial developmental and metabolic regulators	?
Retinoic acid receptors	RARα RARβ RARγ	NR1B1 NR1B2 NR1B3	RARA RARB RARC	Retinoic acid	Crucial developmental regulators	Pleiotropic modulation of immune functions, T-cell differentiation, mucosal immunity
Peroxisome proliferator-activated receptors	PPARα PPARβ/δ PPARγ	NR1C1 NR1C2 NR1C3	PPARA PPARD PPARC	Fatty acids, fibrates Fatty acids Fatty acids, PGJ2, TZDs	Lipid, glucose metabolism Lipid, glucose metabolism Adipogenesis	Liver (acute phase response) Vascular endothelium Macrophages, intestine
Rev-ErbA orphan receptors	Rev-ErbAα Rev-ErbAβ	NR1D1 NR1D2	NR1D1 NR1D2	Heme, synthetic agonists	Circadian metabolism Adipogenesis	?
RAR-related orphan receptors	RORα RORβ RORγ	NR1F1 NR1F2 NR1F3	RORA RORB RORG	Synthetic antagonists	Circadian metabolism Brain, bone development Th-17 differentiation	Th-17 differentiation (anti-inflammatory via inhibitory drugs)
Liver X receptors	LXRα LXRβ	NR1H3 NR1H2	NR1H3 NR1H2	Oxysterols, GW3965	Cholesterol homeostasis	Macrophages, T-cells, B-cells, liver (acute phase response), intestinal epithelium
Farnesoid X receptor	FXR	NR1H4	NR1H4	Bile acids, GW6064	Bile acid synthesis	Intestinal epithelium
Vitamin D receptor	VDR	NR1I1	VDR	Vitamin D, bile acids	Bone homeostasis, wound healing, detoxification	Broad anti-inflammatory effects, anti-fibrotic in the liver
Pregnane X receptor	PXR/SXR	NR1I2	NR1I2	Xenobiotics	Hepatic detoxification	Chronic liver inflammation
Androstane receptor	CAR	NR1I3	NR1I3	Xenobiotics	Hepatic detoxification	?
Hepatocyte nuclear factor 4 receptors	HNF4α HNF4γ	NR2A1 NR2A2	HNF4A HNF4G	Fatty acids (linoleic acid) ?	Lipid homeostasis	?
Retinoid X receptors	RXRα RXRβ RXRγ	NR2B1 NR2B2 NR2B3	RXRA RXRB RXRC	9-cis retinoic acid	Lipid and glucose metabolism	Macrophages (pro- and anti-inflammatory) Intestinal epithelium
Testicular receptors	TR2 TR4	NR2C1 NR2C2	NR2C1 NR2C2	?	Gonad development, lipid homeostasis	Possibly pro-inflammatory in adipose tissue and the liver
Tailless-like receptor	TLL	NR2E2	TLL1	?	?	?
Photoreceptor-specific receptor	PNR	NR2E3	NR2E3	?	Retina-specific pathways	?
COUP-transcription factor receptors	COUP-TF1 COUP-TF2	NR2F1 NR2F2	NR2F1 NR2F2	?	Pleiotropic metabolic regulation, adipogenesis	?
V-ErbA-related receptor	EAR2	NR2F6	NR2F6		T-lymphocyte metabolism	Th-1/Th-17 suppression

(*continued*)

Nuclear Receptor Signaling in the Control of Inflammation, Table 1 (continued)

NR group	Name	Symbol[a]	Gene[b]	Ligands	Functions in metabolism and development	Functions in inflammation
Estrogen receptors	ERα ERβ	NR3A1 NR3A2	ESR1 ESR2	17β-estradiol, tamoxifen, raloxifene, genistein, 27-hydroxycholesterol	Reproduction, breast cancer, pleiotropic metabolic regulation (in females)	Vascular endothelium Neuro-inflammation (astrocytes, glia cells)
Estrogen receptor-related receptors	ERRα ERRβ ERRγ	NR3B1 NR3B2 NR3B3	ESRRA ESRRB ESRRG	Synthetic antagonists	Mitochondrial biogenesis, Glucose and lipid metabolism Stem cell proliferation	Arthritis Pulmonary inflammation (COPD)
Glucocorticoid receptor	GR	NR3C1	NR3C1	Cortisol, NSAIDs	Metabolic homeostasis adipogenesis	Pleiotropic anti-inflammatory action
Mineralocorticoid receptor	MR	NR3C2	NR3C2	Aldosterone, cortisol	Regulation of salt and water transport, blood pressure	Vascular and pulmonary inflammation ?
Progesterone receptor	PR	NR3C3	PGR	Progesterone	Female reproduction Embryogenesis	Vascular inflammation through leukocyte trafficking
Androgen receptor	AR	NR3C4	AR	Testosterone	Reproduction, prostate cancer, pleiotropic metabolic regulation (in males)	Inflammation-dependent prostate cancer, wound healing (skin, hair)
Nur-related factor receptors	NGFIB NURR1 NURR77	NR4A1 NR4A2 NR4A3	NR4A1 NR4A2 NR4A3	*Excluded (lack a ligand-binding pocket)*	Lipid, glucose metabolism	Macrophages, vascular endothelium, neuro-inflammation
Steroidogenic factor 1 Liver receptor homologue 1	SF-1 LRH-1	NR5A1 NR5A2	NR5A1 NR5A2	Phospholipids Synthetic agonists	Steroidogenesis, reproduction Lipid, glucose metabolism Bile acid synthesis	? Liver (acute phase response) Intestinal glucocorticoid synthesis
Germ cell nuclear factor	GCNF	NR6A1	NR6A1	?	Embryogenesis, germ cell proliferation	?
NR0B orphan receptors[c]	DAX-1 SHP	NR0B1 NR0B2	NR0B1 NR0B2	*Excluded (lack a ligand-binding pocket)*	Steroidogenesis Liver metabolism Bile acid biosynthesis	? Cytoplasmic inhibitor of TLR activation in macrophages

[a]NRNC symbol according to the Nuclear Receptors Nomenclature Committee (Nuclear Receptors Nomenclature 1999); see also www.nursa.org
[b]NCBI Gene Symbol
[c]Atypical orphan receptors that lack a DNA-binding domain, function as NR corepressors

"non-genomic" activities, for example, by communicating with rapid membrane/cytoplasmic signaling cascades, some of which may contribute to the modulation of inflammation.

NRs have typically two conserved "signature domains" named DBD (DNA-binding domain) and LBD (ligand-binding domain) and less-conserved adjacent domains (Fig. 1). The DBD is responsible for recognizing specific NR response elements in the regulatory regions (i.e., promoters, enhancers) of target genes. Most NRs function as dimers, either as homodimers (i.e., class I steroid receptors) or as heterodimers with 9-cis retinoic acid receptor RXR (i.e., class II receptors). Thus, DNA-binding specificity is determined by both the sequence specificity of

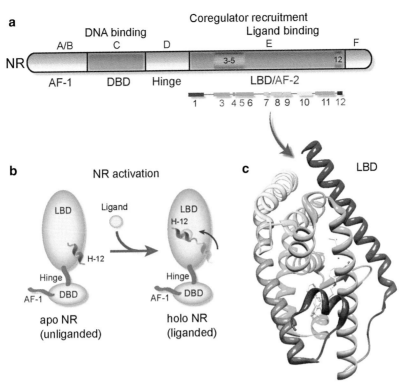

Nuclear Receptor Signaling in the Control of Inflammation, Fig. 1 NR domain structure and function. Highlighted are the common NR domain nomenclature and the structure–function relationship with emphasis on the multifunctional LBD. (**a**) NR domains A–F and associated key functions that control the transcriptional activity of NRs. The conserved LBD consists of 12 α-helices with AF-2 helices 3 to 5 (static part) and helix 12 (dynamic part) decorated in *red*. (**b**) Conserved mechanism of ligand-dependent NR activation. Major conformational changes occur within the LBD (H-12 relocation) and affect the entire NR structure and function. Usually, many apo-NRs bind corepressors, while holo-NRs bind either coactivators (when bound to agonists) or corepressors (when bound to antagonists). Some orphan receptors may not undergo the ligand-dependent switch and adopt only one conformation to function constitutively active or repressive. (**c**) Structural view of the conserved LBD, with helices decorated in colors as in (**a**). The example shows the LXRβ LBD in complex with the synthetic agonist GW3965 (Farnegardh et al. 2003). *Abbreviations: NR* nuclear receptor; *DBD* DNA-binding domain; *LBD* ligand-binding domain; *AF-1* N-terminal activation function 1 (ligand independent); *AF-2* C-terminal activation function 2 (ligand dependent, coregulator LXXLL-binding surface); *H-12* LBD/AF-2 helix 12

the individual DBDs and the type of dimer. Recent progress in applying ChIP-sequencing allowed determining genome-wide NR binding sites (referred to as cistromes) in vivo. These studies revealed additional specificity determinates such as the chromatin landscape (epigenomes) and the dependence on pioneer factors (e.g., GATA, FOXA1, PU.1).

The LBD is a multifunctional domain responsible for ligand binding, NR dimerization, and recruitment of diverse coregulators and PTM modifiers (Fig. 1). Structure determination revealed a conserved LBD fold consisting of 12 α-helices. Helices 3 to 5 and 12 form the common coactivator-interaction surface called AF-2, which usually is exposed upon binding of activating ligands (agonists) but blocked by inhibitory ligands (antagonists) (Gronemeyer et al. 2004). Amino acid residues from different helices shape the NR-specific ligand-binding pocket accounting for ligand selectivity. Helices 10 and 11 form the common dimerization surface. While structure information was until recently limited to the isolated DBDs and LBDs, the first structures of full-length NR dimers bound to DNA suggest a high level of conformational flexibility

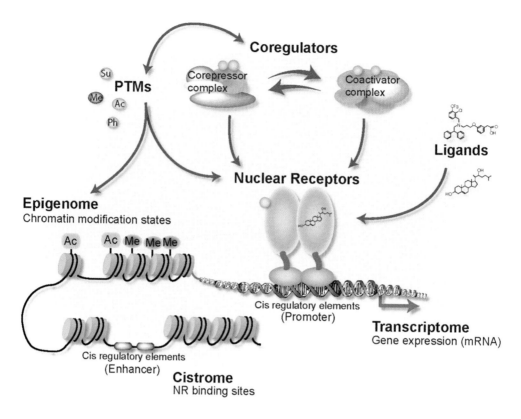

Nuclear Receptor Signaling in the Control of Inflammation, Fig. 2 Key components of genomic NR signaling. NR signaling has to be understood as the complex interplay of various key components which determine genomic responses linked to gene expression and physiological outcomes in a highly cell-type selective manner. Ligands, PTMs, and coregulators are major regulatory components that directly control NR activity, while cistromes and epigenomes specify the genome-wide target gene range (transcriptomes). Coregulators often function as "writers," "erasers," or "readers" of specific PTMs toward NRs and chromatin, and some of these activities are uniquely responsible to trigger anti-inflammatory NR pathways (see Fig. 3). *Abbreviations: PTMs* posttranslational modifications, such as Su (SUMOylation), Ac (acetylation), Me (methylation), Ph (phosphorylation)

and massive changes to occur upon ligand binding.

NR signaling is complex and various key components have to be taken into account when trying to understand the role of an individual NR and the respective ligands (Fig. 2). NRs, ligands (hormones, metabolites, synthetic compounds, endocrine disrupters), posttranslational modifications (PTMs, e.g., phosphorylation, SUMOylation), coregulators/chromatin modifiers (coactivators, corepressors), cistromes (genomic NR binding sites), transcriptomes (NR-regulated genes), and epigenomes (chromatin/DNA and histone modifications). Due to the presence or absence of individual key components, many NR signaling pathways are highly cell-type specific, and some differ significantly between species (such as rodents and humans).

Despite substantial progress in characterizing each of these components, in particular at the genome-wide level, it is the control of NR activity by ligands and coregulators that is probably best understood. Structural and functional data have demonstrated that ligand binding induces conformational changes within NRs, in particular within the LBD/AF-2 region. Coregulators recognize distinct ligand-dependent NR conformations and establish interactions with the transcriptional machinery and chromatin (Rosenfeld et al. 2006). Thereby, they function either as coactivators, by recognizing the active NR state and promoting transcription, or as corepressors, by recognizing the inactive NR state and repressing NR-dependent transcription. Coregulators often function in larger multiprotein complexes to establish cell-type- and ligand-dependent epigenomes by "writing" or "reading" reversible epigenetic chromatin modifications linked to transcription (e.g., acetylation or methylation of histone tails). Evidence is emerging that alterations in function or expression of coregulators cause disease by propagating disease-specific epigenomes linked to dysregulation of transcription, but most of the underlying mechanisms remain to be elucidated.

To appreciate the impact of NR signaling on inflammatory gene expression, it must be noted that many NR coregulators participate in the control of pro-inflammatory transcription factors and thereby act as shared components to trigger transcriptional cross talk between NR and inflammatory signaling (Fig. 3). In particular, it has become clear that NR corepressors interact with inflammatory transcription factors to prevent their activation, thereby putting a molecular break on inflammation (Glass and Saijo 2010; Perissi et al. 2010; Toubal et al. 2013; Treuter and Venteclef 2011; Venteclef et al. 2011). Upon pro-inflammatory signaling, corepressors are normally cleared from promoters and replaced by coactivators to induce gene transcription linked to acute and chronic inflammation. Studies within the last decade suggest that many NRs and their ligands are powerful inhibitors of pro-inflammatory activation pathways, raising a substantial interest in understanding the underlying anti-inflammatory mechanisms and the biological and therapeutic implications, all of which should be discussed in this review.

Transrepression as Key Molecular Mechanism of Anti-inflammatory NR Action in Immune Cells

The study of the glucocorticoid receptor (GR) in the immune system, along with the development of potent synthetic immunosuppressive GR ligands, has paved the way for our principal molecular understanding of anti-inflammatory NR action in the immune system (Huang and Glass 2010; Ratman et al. 2013). Natural and synthetic glucocorticoids activate the GR and are the probably best-characterized and clinically most significant anti-inflammatory NR ligands to date. Subsequently, lipid-sensing NRs such as peroxisome proliferator-activated receptors (PPARs, receptors for fatty acids) and liver X receptors (LXRs, receptors for oxysterols) have emerged as potent inhibitors of inflammatory processes (Glass and Saijo 2010; Hong and Tontonoz 2008; Zelcer and Tontonoz 2006). These metabolic NRs appear particularly suited in sensing metabolic states, allowing them to control multiple pathways involved in cholesterol, fatty acid, and glucose homeostasis. Furthermore, clinical observations and mechanistic studies suggest anti-inflammatory action of additional NRs and their ligands, such as the receptors for estrogens (ERs), vitamin D (VDR), vitamin A derivatives (retinoids) (RARs, RXRs), and bile acids (FXR). The anti-inflammatory aspect of NR signaling is intriguing as it provides clues for how endogenous hormones and metabolites communicate with inflammatory processes (and vice versa), and as it provides opportunities for selective therapeutic intervention with intrinsically inflammatory components and pathways of common metabolic diseases (Glass and Saijo 2010).

The initial characterization of anti-inflammatory cross talk has established molecular mechanisms by which these NRs repress inflammatory gene expression. A key feature is that repression occurs "in trans," i.e., without direct binding of NRs to DNA response elements, therefore termed "transrepression" (Glass and Saijo 2010; Medzhitov and Horng 2009; Saijo et al. 2010; Venteclef et al. 2011). Common to the diverse transrepression mechanisms is that these NRs, in response to ligand activation, interfere with the inflammatory signal-dependent activation of pro-inflammatory transcription factors such as nuclear factor kappa B (NFκB), activator protein 1 (AP-1) family members (FOS, JUN),

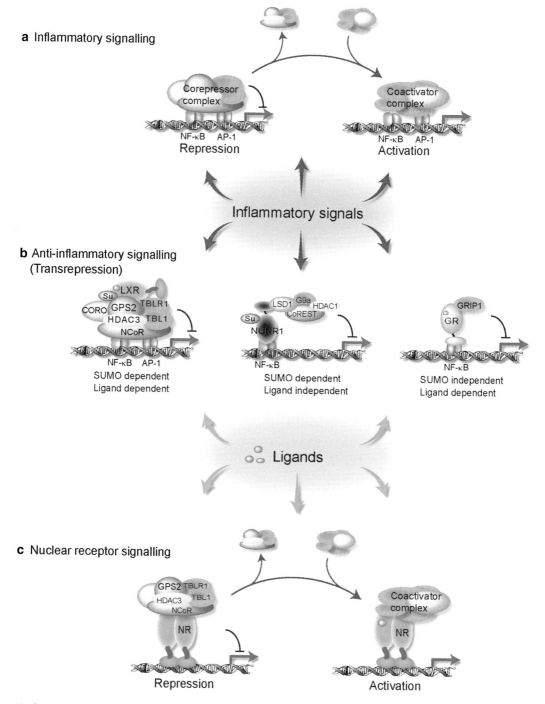

Nuclear Receptor Signaling in the Control of Inflammation, Fig. 3 Transcriptional cross talk between inflammatory and NR signaling. Illustrated are some of the principal transcriptional components and mechanisms of inflammatory signaling (**a**) and NR signaling (**c**), as well as the cross talk between the two pathways during anti-inflammatory transrepression (**b**). (**a**) In the pre-inflammatory state, inflammatory transcription factors such as NFκB and AP-1 (as an example, there are many more factors involved) recruit corepressors, which upon inflammatory signaling become exchanged by coactivators. (**c**) NR signaling follows related principles,

interferon response factors (IRFs), CCAAT-box enhancer binding proteins (C/EBPs), ETS-domain transcription factors (ETS), nuclear factor of activated T-cells (NFAT), and signal transducers and activators of transcription (STATs). According to current models (Chinenov et al. 2012; Ratman et al. 2013), direct interactions of GR with pro-inflammatory transcription factors largely account for transrepression, and coregulators such as GRIP1 (also SRC2, TIF2, NCOA2) may further specify the anti-inflammatory responses (Fig. 3).

A number of pioneering studies, studying the role of PPARγ and LXRs in inhibiting toll-like receptor (TLR)-dependent gene expression in mouse macrophages exposed to bacterial lipopolysaccharide (LPS), have identified novel molecular components of anti-inflammatory transrepression (Saijo et al. 2010; Treuter and Venteclef 2011; Venteclef et al. 2011). First, unlike GR, PPAR and LXRs need to be modified by members of the small ubiquitin-related modifier (SUMO) family to enter the transrepression pathway. SUMOylation can be modulated by ligand binding, in particular for some of the SUMO acceptor lysine residues located within the LBD. Specifically, ligand activation has been demonstrated to promote the conjugation of SUMO-1 to PPARγ, involving PIAS E3 ligases, or of SUMO-2/3 to LXRs, involving HDAC4 as putative E3. Second, it appears that SUMOylated NRs dock to a conserved NR corepressor complex (Oberoi et al. 2011), containing the core subunits NR corepressor (NCOR/SMRT), transducin β-like 1-related proteins (TBLR1/TBL1), G-protein pathway suppressor 2 (GPS2), coronin 2 A (CORO2A), and histone deacetylase 3 (HDAC3), at the promoters of inflammatory genes. Thus, the dual involvement of the corepressor complex in both NR and inflammatory transcription forms the basis for this specific type of coregulator-mediated transrepression and anti-inflammatory NR cross talk. Third, docking of SUMOylated PPARγ or LXRs inhibits the release of the corepressor complex upon inflammatory signaling, resulting in the maintenance of the repressed state. Notably, some endogenous cholesterol metabolites and synthetic ligands can differently modulate LXR transrepression, and TLR2 signaling can abolish LXR transrepression of lipopolysaccharide (LPS)-dependent TLR4 signaling. Thus, the anti-inflammatory capacity of LXRs appears to be dynamically regulated by ligands and intracellular signaling.

It is currently unclear which NRs are capable of utilizing the SUMO-corepressor-dependent pathway, but it has been noted that many NR family members that are expressed in immune cells also can be SUMOylated (Treuter and Venteclef 2011). However, many details of these probably multiple

Nuclear Receptor Signaling in the Control of Inflammation, Fig. 3 (continued) with the principal difference that ligands act directly on NRs without triggering intracellular signaling cascades. Intriguingly, the major NR corepressor complex (containing core subunits NCOR, SMRT, GPS2, and HDAC3) acts as a shared component of both inflammatory and NR signaling and appears therefore well positioned to mediate cross talk of the two pathways. (b) Anti-inflammatory NRs control mechanistically distinct transrepression pathways, some of which require NR-specific modification by SUMO. Common to the pathways appears the functional interaction of NRs with corepressors at inflammatory genes, resulting in stabilization of the repressed state despite pro-inflammatory signaling. GPS2 and CORO2, subunits of the NCOR/SMRT/HDAC3 complex, have been implicated in recognizing SUMOylated NRs in different cellular contexts and with different SUMO subtype specificity. The examples refer to transrepression pathways governed by the oxysterol receptors LXRs in macrophages and hepatocytes, by the orphan receptor NURR1 in microglia and astrocytes, and by the glucocorticoid receptor GR in macrophages. *Abbreviations: NFκB* nuclear factor kappa B (p50/p65 dimers); *AP-1* activator protein 1 (FOS/JUN dimers); *NCOR* NR corepressor (also NCOR1); *SMRT* silencing mediator of retinoid and thyroid receptors (also NCOR2); *HDAC1/3* histone deacetylase 1/3; *GPS2* G-protein pathways suppressor 2; *TBL1/TBLR1* transducing β-like 1 proteins; *CORO2A* coronin 2 A; *GRIP1* GR-interacting protein 1 (also SRC2, TIF2, NCOA2); *CoREST* corepressor of REST, subunit of the NRSF complex; *LSD1* lysine-specific demethylase 1 (also KDM1A); *G9a* lysine methyltransferase (also EHMT2); *LXR* liver X receptor (also NR1H2/3); *NURR1* NR-related protein 1 (also NR4A2); *GR* glucocorticoid receptor (also NR3C1); *Su* Small ubiquitin-related modifier (SUMO)

transrepression mechanisms remain to be clarified, such as the specific requirement of receptor SUMOylation, the role of the common heterodimerization partner RXR, or the docking mechanism of NRs to the complex. Also, the determination of anti-inflammatory NR, SUMO-NR, and corepressor cistromes (i.e., genome-wide binding sites) remains a challenge, which may be related to the indirect tethering mode of chromatin binding. Evidence is emerging that related transrepression pathways may occur in a variety of metabolic tissues cell types that communicate with immune cells, such as hepatocytes, adipocytes, or intestinal epithelial cells, as outlined in the next chapters. Therefore, it is likely that NR activation by ligands may have much broader effects on inflammation by triggering transrepression pathways in distinct sets of "metaflammatory" cell types.

Biological Role of NRs in Macrophages (Innate Immunity)

Macrophages, including dendritic cells, are the sentinels of innate immunity and take residence in nearly every tissue and display marked heterogeneity in their cell surface markers, location, and function (Dalmas et al. 2011; Davies et al. 2013; Olefsky and Glass 2010; www.macrophages.com). Metabolic tissues, particularly under disease conditions, are enriched in subsets of specialized resident macrophages including brain microglia, liver Kupffer cells, intestinal macrophages, peritoneal macrophages, alveolar macrophages, adipose tissue macrophages, and atherosclerotic plaque foam cells. Differentiation of myeloid precursors into monocytes and macrophages is an important determinant of macrophage function. Efforts are ongoing to classify macrophage subtypes according to specific gene expression signatures, referred to as transcriptomes. These data revealed that a majority of NR family members is expressed in different macrophage subtypes (Barish et al. 2005; www.biogps.org; www.macrophages.com; www.nursa.org).

Macrophages can rapidly adjust their gene expression programs and immune phenotypes in response to injury, infection, and the tissue microenvironments. Two major immune phenotypes are distinguished, M1 and M2, and some NRs play major roles in specifying these. Classically activated M1 macrophages are characterized by a pro-inflammatory expression profile of cytokines, chemokines, and effector molecules to facilitate pathogen-killing activity. Alternatively activated M2 macrophages are low cytokine producers with major functions in tissue repair, parasite clearance, and immune modulation. Alteration in the ratio of M1/M2 is associated with tissue dysfunction leading to autoimmunity, insulin resistance, and neuro-inflammation.

A variety of NRs have already emerged as being particularly important regulators of gene expression programs linked to differentiation, polarization, metabolism, and inflammatory responses in macrophages. Selective synthetic ligands and genetically modified mouse models have been, and remain to be, critical tools to elucidate the multifaceted functions of NRs in vivo. Mutual interactions exist between these NRs and inflammation induced either by pathogens (acute, hot inflammation) or by metabolic disease conditions (chronic, cold inflammation, metaflammation). Also, monocytes and macrophages produce endogenous lipids, for example, diverse oxysterols, prostaglandin D2 metabolites, and oxidized LDL-derived fatty acids, that serve as endogenous ligands for lipid-sensing NRs such as LXRs and PPARs.

Here a few selected NRs that already play significant roles in macrophage biology are highlighted. Clearly, based on the NR expression profiles and on the possibility for novel endogenous ligands, additional NRs have to be considered as modulators of macrophage biology.

LXRs: LXR ligands potently inhibit expression of pro-inflammatory mediators such as IL6, TNFα, iNOS/NOS2, and MCP1/CCL2 and promote macrophage M2 polarization (Bensinger et al. 2008; Calkin and Tontonoz 2012; Pascual-Garcia and Valledor 2012; Steffensen et al. 2013). However, longtime treatment of mice with LXR agonists also upregulates the expression of TLR4, the major pathogenic LPS receptor, thus counteracting anti-inflammatory action and causing drug resistance. The critical role of LXR in macrophages was proven by bone marrow transplantation from LXR-deficient into wild-type mice, which resulted in elevated inflammation, macrophage apoptosis, and decreased bacterial clearance.

In addition to counteracting pro-inflammatory gene expression by transrepression, LXRs also directly activate target genes linked to key processes in macrophages and thereby further modulate inflammatory responses (Calkin and Tontonoz 2012). For example, LXRs control cholesterol efflux from macrophages and reverse cholesterol transport via regulating the expression of transporters (ABCA1, G1) and lipoprotein acceptors (ApoE, ApoC) (Chawla et al. 2001). Intracellular cholesterol and oxysterol pathways may affect also the proliferative capacity and other functions of immune cells (Bensinger et al. 2008). Furthermore, LXRs regulate the expression of AIM1, an inhibitor of apoptosis in macrophages, and of ARG2, an anti-inflammatory arginase, and of MERTK, a receptor tyrosine kinase critical for phagocytosis. Thereby, LXRs are also critically involved in the clearance of apoptotic cells by macrophages (N-Gonzalez et al. 2009). Very recently, LXRs have been implicated in directly controlling the expression of enzymes involved in the synthesis of anti-inflammatory ω-3 fatty acids (Li et al. 2013) and to control the functional specialization of splenic macrophages (N-Gonzalez et al. 2013).

Overall, the precise contribution of the two distinct LXR α and β forms to distinct aspects of macrophage biology remains to be scrutinized, as it might offer possibilities for more selective drug targeting. Since there are no genetic, clinical, or pharmacological data (e.g., using approved LXR agonists), the translation of LXR biology into human macrophage physiology and disease remains to be established.

PPARs: All three PPAR subtypes are potent anti-inflammatory NRs in many different immune cell types including macrophages (Chawla 2010). Most evidence exists for the PPARγ subtype, which is viewed as the dominant PPAR subtype in macrophages and for which synthetic TZD ligands have been clinically applied (Ahmadian et al. 2013; Cariou et al. 2012). It is likely that the proven antidiabetic effect of these compounds is in part due to the combined metabolic and anti-inflammatory action in immune cells. Notably, macrophages are a major source of endogenous PPARγ ligands such as derivatives of prostaglandin D2 and oxidized LDL (Ricote et al. 1998). Myeloid-specific knockout mice revealed that PPARγ is not essential for macrophage differentiation but indispensible for M2 macrophage polarization via function cooperation with the IL-4/STAT6 pathway (Odegaard et al. 2007). Unlike LXRs, PPARγ expression is significantly downregulated by LPS/TLR4 activation, leading to some controversy as to its anti-inflammatory potency, particularly in human macrophages (Hall and McDonnell 2007).

PPARβ/δ is also involved in M2 macrophage phenotypic switching, as demonstrated by myeloid-specific depletion, but regulates in part separate aspects than PPARγ (Barish et al. 2008; Odegaard et al. 2008). Mechanistically, PPARβ/δ exerts its anti-inflammatory actions by sequestering the transcriptional repressor BCL-6 away from the promoters of inflammatory genes such as CCL2 and IL1β, and it is unclear whether PPARβ/δ also utilizes the SUMO/compressor-type mechanism.

Evidence for a potential role of PPARα in inflammation came from the discovery that PPARα knockout mice displayed a prolonged inflammatory response (Rigamonti et al. 2008). In M1 macrophages, PPARα activation inhibits the production of various pro-inflammatory molecules such as MMP9 and TNFα, although the mechanistic details of transrepression remain to be elucidated, in particular the role of SUMOylation. Myeloid-specific knockout mice have not yet been reported so its role in M2 polarization is so far uncertain. In conjunction with its function in cholesterol, fatty acid, and glucose homeostasis in macrophages, PPARα is a main player in the repression of vascular inflammation in atherosclerosis.

RXRs and RARs. Active metabolites of vitamin A, named retinoids, are potent NR ligands that play major roles in innate immunity. 9-cis retinoic acid (9-cis RA) functions via RXRs and RARs, while the isomer all-trans retinoid acid

(ATRA) only binds to RARs. The receptors exist in three subtypes each and affect multiple aspects of differentiation, metabolism, and inflammation in myeloid cells (Kiss et al. 2013; Roszer et al. 2013).

RXRs are the common heterodimer partners of most macrophage NRs (except GR) and thereby modulate the activities of those receptors, particularly in pathways that require NR dimer-dependent DNA binding to target genes. Currently less clear is whether RXRs are also required for the transrepression pathway by LXRs or PPARs, which does not involve direct DNA binding and thus may not require hetero-dimerization. First insights from macrophage-specific knockout mice revealed a key function of RXRα in innate inflammation. RXR is required for the clearance of apoptotic cells, the propagation of anti-inflammatory M2 macrophages, and the recruitment of leukocytes to sites of inflammation. Interestingly, RXRα deficiency led to downregulation of chemokines such as CCL6 and CCL9, and RXR probably functions as homodimer in these pathways (Nunez et al. 2010). Open is the question whether 9-cis retinoic acid is the most significant endogenous RXR ligand in macrophages.

RARs are key regulators of many differentiation programs and appear critical also for myeloid differentiation, mostly via directly regulating crucial differentiation such as macrophage colony-stimulating factor (M-CSF), CD18, and CD38. RARs act also anti-inflammatory but the underlying mechanisms are so far poorly characterized.

GR: Glucocorticoids (GCs) are considered the "gold standard" for immune suppression and remain the most powerful anti-inflammatory NR compounds (Beck et al. 2009; Ratman et al. 2013). GCs generally suppress cytokine expression in macrophages and T-lymphocytes. However, the exact function of the GR in macrophages and dendritic cells has remained controversial. Without doubt, GR is a powerful inhibitor of inflammatory gene expression by directly interfering with NFκB, JNK, and IRFs pathways, but the classic model that GR transrepression is independent of DNA binding, dimerization, and corepressor action has recently been questioned. GCs regulate the expression of NLRP3, a component of the inflammasome complex, to augment pro-inflammatory response by governing IL1β and IL18 secretion. Likely, many aspects of GR action in macrophage may involve the direct regulation of target genes (Uhlenhaut et al. 2013).

Biological Role of NRs in T-Helper Cells (Adaptive Immunity)

GR, PPARs, LXRs, and lately the orphan receptors RORs have been implicated in CD4+ T-helper cell differentiation (Glass and Saijo 2010). Upon antigenic stimulation, these cells initiate a developmental program marked by an expansion and differentiation into five distinct subsets, characterized by the production of specific cytokines. Briefly, Th1 cells play a critical role in protecting hosts from intracellular pathogens, Th2 cells are required for protective immunity from extracellular infections, and Tregs and Th17 cells play major roles in regulating autoimmunity and inflammatory neutrophilic responses to pathogens. While these T-helper cell subsets are generally thought to be distinct, considerable plasticity in effector function has been observed in a context-specific manner. Given that metabolites can influence cellular fate and function, it is well accepted that metabolic alterations, some of which are directly linked to NR activation, emerge as powerful regulators of T-helper cell differentiation.

GR: In pro-inflammatory T-cells (Th1 and Th17), GCs induce apoptosis, whereas in regulatory T-cells (Treg) they exert pro-survival effects. Indeed, mice lacking GR in T-cells displayed resistance to glucocorticoid-induced apoptosis. GCs inhibit type-1 cytokines that cause a shift toward the anti-inflammatory Th2 profile, and they inhibit adhesion molecules on antigen-presenting cells (APCs).

PPARs: PPAR signaling influences T-helper cell differentiation and mediate T-cell function in inflammatory diseases. Pharmacologic activation of PPARγ reduces interferon γ production in T-cells. PPARγ activation further impairs IL2-dependent T-cell proliferation through a NFAT-dependent mechanism. PPARγ regulates Th17 differentiation by interfering with RORγ in response to TGF and IL6 via a SMRT-dependent transrepression pathway. However, PPARγ activation promotes Treg cells, which counteract the progression of inflammation and contribute to the improvement of insulin resistance mediated by TZD treatment. PPARα plays different roles as deficient T-cells in mice were predisposed to a Th1 response at the expense of Th2 function. Repressive action of PPARα is due to blocking NF-kB and c-Jun function, thereby modulating the production of Th1-related pro-inflammatory cytokines, critical players in autoimmune disease. Finally, PPARβ/δ activation also appears to influence T-cell differentiation and protects against autoimmune diseases, in part by altering the proliferative capacity of Th17 cells and inhibition of IFNγ production through repression of the transcription factor T-bet.

LXRs. Recent studies suggest a role for LXRs in the regulation of Th17 cells and neutrophilic immune responses (N-Gonzalez et al. 2013; Solt et al. 2012). The molecular mechanism by which LXRs modulate Th17 response appears to be multifactorial. In addition to its transrepressive activity, LXRs may indirectly interfere with IL17 transcription via the LXR target SREBP1 that antagonizes AHR action at the IL17 promoter.

RORs. The retinoid acid receptor-related orphan receptors RORγt (one isoform of RORγ) and RORα are considered master regulators of Th17 proliferation and indispensible for function that emerge as significant targets for the treatment of Th17-dependent autoimmune diseases (Solt and Burris 2012). Overexpression of RORγt in CD4+ T-cells drives Th17 development, while genetic depletion impairs it. RORα/γ double knockout mice completely lack Th17 cells and appear resistant to several autoimmune diseases. RORs probably directly regulate the transcription of IL17 cytokines, but how exactly they cooperate with additional transcription factors, including NRs, remains to be clarified. Intriguingly, RORs cross talk with LXR signaling in Th17 cells (Solt et al. 2012), and they bind to structurally related ligands, such as endogenous oxysterols and synthetic ligands (Fig. 4).

Pathophysiological Relevance: NRs as Targets to Combat Inflammatory Diseases

NRs and Obesity-Associated Inflammation of Adipose Tissue (Metaflammation)

Obesity is a major health problem characterized by chronic low-grade inflammation (referred to as "metaflammation") and insulin resistance. The interplay between metabolic and inflammatory pathways has been particularly well studied in the context of adipose tissue, where the recruitment of immune cells (such as macrophages and regulatory T- and B-cells) to adipocytes is viewed as a hallmark of obesity (Gregor and Hotamisligil 2011; Olefsky and Glass 2010; Osborn and Olefsky 2012; Toubal et al. 2013b). In response to overnutrition, adipocyte reprogramming triggers the secretion of hormones (called adipokines) and inflammatory mediators such as MCP1/CCL2 and IL6 that attract monocytes and induce the switch of resident M2 macrophages into pro-inflammatory M1 macrophages. On top, dietary fatty acids can directly trigger inflammatory responses by activating the TLR4 pathway.

Lipid-sensing NRs, in particular the PPARs, offer exciting opportunities to pharmacologically modulate macrophage phenotypes and obesity-associated adipose tissue inflammation. All PPARs are expressed in adipocytes and macrophages where they act as sensors for dietary lipids and derivatives with specific signaling functions in macrophages (as outlined above). A substantial number of in vivo studies, both in mice and humans, suggest that treatment with synthetic PPAR agonists generally improves adipose tissue inflammation and insulin sensitivity.

PPARγ is commonly considered the most significant biological regulator and drug target among the three PPARs in adipose tissue (Ahmadian et al. 2013). (1) PPARγ is a key metabolic regulator of adipogenesis that directly responds to dietary fatty acids for the following reasons. (2) PPARγ activation improves insulin sensitivity, partially through promoting fatty acid storage as triglycerides in adipocytes, and it reduces lipotoxicity, linked to pro-inflammatory programs in obese adipocytes. (3) PPARγ has

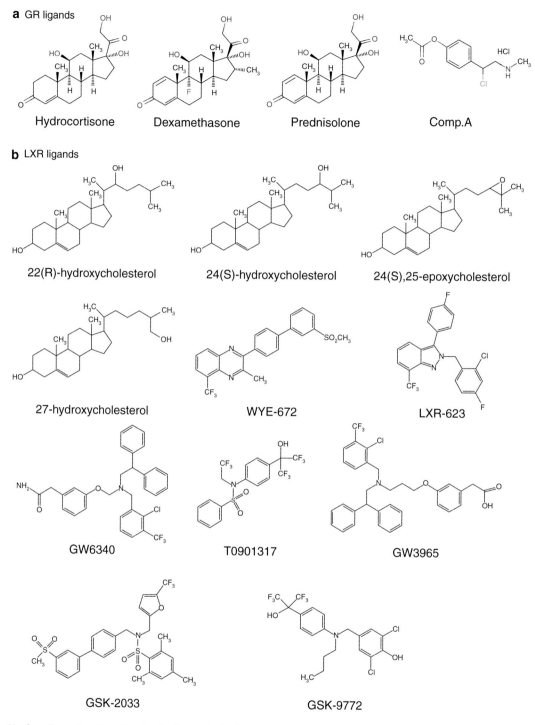

Nuclear Receptor Signaling in the Control of Inflammation, Fig. 4 (continued)

Nuclear Receptor Signaling in the Control of Inflammation, Fig. 4 Examples of natural and synthetic NR ligands with immune-modulatory and anti-inflammatory properties.

(a) *GR ligands.* **Hydrocortisone (cortisol)** – endogenous steroid hormone/glucocorticoid produced by the adrenal gland; **DEX – dexamethasone** (brand name Oradexo) and **prednisolone** are clinically approved synthetic GR agonists, so far the most successful anti-inflammatory NR drug class. **Comp. A (compound A/CpdA)** – a plant-derived nonsteroidal GR modulator (phenyl aziridine precursor). "Dissociating" compound with enhanced anti-inflammatory transrepression properties by interfering with GR dimerization, thereby preventing activation of direct (metabolic) GR target genes. Induces anti-inflammatory macrophage M2 switch and beneficially anti-inflammatory properties in adipose tissue-associated immune cells including macrophages (ATMs), T-cells, and B-cells, which may contribute to the potency of antidiabetic PPARγ ligands in the context of inflamed adipose tissue. Notably, PPARγ was recently shown to modulate the accumulation and function of regulatory T-cells (Treg) in visceral adipose tissue, and TZD treatment increased the Treg population in wild-type but not knockout mice fed with a high-fat diet.

(4) Recent findings also suggest that PPARγ may modulate inflammation by regulating the expression of anti-inflammatory transcription factors (such as TWIST1) and coregulators (such as the GPS2-/SMRT-containing NR corepressor complex) directly in the adipocytes (Toubal et al. 2013a). Interestingly, the expression of TWIST1, GPS2, and SMRT appears reduced in adipocytes of human obese subjects, an event which presumably causes increased inflammation

Nuclear Receptor Signaling in the Control of Inflammation, Fig. 4 (continued) modulates regulatory T-cells in autoimmune diseases.

(b) *LXR ligands.* Hydroxycholesterols (oxysterols) are candidates for being major endogenous LXR ligands but their role in regulating cholesterol homeostasis is debated and may differ between rodents and humans. Oxysterols can pass the blood–brain barrier and may contribute to the protective effects of LXRs in neuro-inflammation and neurodegenerative disorders. **24(S)-Hydroxycholesterol** is considered the most potent endogenous LXR agonist and acts anti-inflammatory, along with 22(R)-hydroxycholesterol and 24(S),25-epoxycholesterol, by inducing LXR SUMOylation. **27-Hydroxycholesterol** is comparatively weak and unable to trigger the SUMO-dependent transrepression pathway. It is the most abundant oxysterol in circulation and also functions as a selective estrogen receptor modulator (SERM) in the vascular system. **GW3965**, most common synthetic agonist in today's preclinical LXR research with potent anti-inflammatory properties, induces LXRβ SUMOylation. **T0901317** (also T1317, T09) – potent synthetic LXR agonist with properties distinct from GW3965 and also a high-affinity inverse agonist for ROR orphan receptors, which promote differentiation of inflammatory T_h17 T-helper cells. Thus, T0901317 may act anti-inflammatory through the combined modulation of LXRs and RORs, resulting in the inhibition of T_h17 differentiation and function in autoimmune diseases. **LXR-623** (WAY-252623) – anti-atherogenic LXR modulator that lowered LDL cholesterol without increasing hepatic lipogenesis. **GW6340**, intestinal-specific LXR agonist, promotes macrophage RCT and is potentially anti-atherogenic, without promoting undesired LXR activation in the liver. **WYE-672**, LXRβ-selective agonist with a tissue-selective activity profile, discriminates between macrophage and hepatic LXRs to promote atherosclerotic lesion regression without triggering deleterious hepatic lipid effects. **GSK2033** – tertiary sulfonamide, first described synthetic LXR antagonist; **GSK9772**, N-phenyl tertiary amine, first synthetic LXR agonist with enhanced anti-inflammatory properties, induces LXRβ SUMOylation.

(c) *PPAR ligands.* *PPARα ligands:* **Fenofibrate** (trade name Tricor, Triglide, Antara), cholesterol- and TG-lowering drug; **LTB4** – leukotriene 4 – a chemoattractant produced in leukocytes and neutrophils; *PPARγ ligands:* **PGJ2** – **15-deoxy-delta12,14-prostaglandin J2**, first identified endogenous PPARγ ligand, signals in macrophages. **Rosiglitazone** (trade name Avandia, Avandamet in combination with metformin) – insulin sensitizing, antidiabetic agonists of the thiazolidinedione (TZD) class. Until 2006 it is a top-selling NR drug, and then drastic sales decline due to controversy about safety concerns such as increased cardiovascular events. In 2010, the FDA recommended restrictions in use, while the EMA suggested withdrawal of the drug.

(d) *FXR ligands.* **CDCA** – **chenodeoxycholic acid**, primary bile acid in the liver and intestine, major product of hepatic cholesterol metabolism. **GW4064** – most common synthetic FXR agonist that also activates ERRs, thus activities of the compound in the immune system may not necessarily be linked to FXR activation. **Z-guggulsterone** – a plant (guggul tree)-derived steroidal FXR antagonist.

(e) *ER ligands.* **E2, 17β-estradiol**, most abundant endogenous estrogen (in females, levels drop after menopause). Clinical and experimental evidence supports the anti-inflammatory and protective properties of estrogens, particularly in the area of cardiovascular and neuro-inflammatory disorders. However, in breast cancer cells, estrogen and inflammatory signaling act in part synergistic. **ADIOL** – **5-androsten-3β,17β-diol**, ERβ-selective agonist with enhanced anti-inflammatory properties in the brain, androgenic (AR ligand). **Tamoxifen** (trade name Nolvadex), **raloxifene** (trade name Evista) – selective ER modulators (SERM), as antiestrogens used in breast cancer treatment and prevention; raloxifene has less side effects than tamoxifen in the uterus and is a potent estrogen in the bone, thus beneficial in breast cancer patients with osteoporosis. **Genistein** – a plant (soy)-derived "phytoestrogen," isoflavone with partial agonist/antagonist (SERM) activity and enhanced ERβ selectivity. Overall, the precise role of SERMs in inflammation is unclear and may depend on the cellular context, endogenous estrogen levels, and the relative abundance of the two receptors ERα and ERβ in adipose tissue. More recently, another subunit of the NCOR/SMRT complex called TBL1-related protein 1 (TBLR1) was also found to be involved in the regulation in metaflammatory events in adipose tissue. Thus, the NR corepressor complex appears a key regulator of inflammation in adipocytes, just like in macrophages, and could be a mediator of the anti-inflammatory PPARγ action that specifically affects signals originating from the inflamed adipocytes.

The role of PPARα and PPARβ/δ in adipose tissue inflammation remains to be clarified. Although both subtypes are not essential for adipogenesis and their expression in adipose tissue is considerably lower than PPARγ, they may modulate adipogenesis indirectly and they may act anti-inflammatory in macrophages. PPARα knockout mice display increased inflammation of adipose tissue under high-fat diet conditions, but it is not clear whether the effect is due to specific PPARα depletion in the adipose tissue or rather systemic. Intriguingly, PPARα regulates the hepatic expression of FGF21, which as circulating growth factor may modulate PPARγ activity in adipose tissue. Also PPARβ/δ has been suggested to modulate adipose tissue mass perhaps indirectly by functioning in the liver and skeletal muscle (a major tissue of action).

NRs and Inflammatory Liver Diseases

Both acute inflammation/infections and obesity-associated chronic inflammation are known to affect liver (patho)physiology and can cause severe alterations leading to nonalcoholic fatty liver disease (NAFLD) with its subtypes steatosis and nonalcoholic steatohepatitis (NASH), with increased fibrosis and cirrhosis, and eventually leading to hepatocellular carcinoma (HCC). According to current models, overnutrition results in hepatic lipid accumulation. When the liver's capacity to accumulate TG is overloaded (lipotoxicity), inflammatory pathways become activated resulting to insulin resistance. Liver macrophages, including Kupffer cells, and stellate cells accumulate at the sites of inflammation to promote tissue repair and fibrosis, but they also secret inflammatory mediators (such as IL6 and TNFα) that accelerate the progression of NAFLD into cirrhosis and HCC.

Current evidence suggests that pathological liver inflammation linked to disease is accompanied, and in part caused, by alterations in NR expression and function (Wagner et al. 2011). It has also been demonstrated that the pharmacological activation of NRs in the liver may be a promising strategy to inhibit the inflammatory side of liver diseases. Among the better-characterized anti-inflammatory NRs in the liver are PPARα, LXRs, FXR, and ERα, where knockout mice and the availability of selective ligands allowed defining specific roles in controlling inflammatory pathways in the liver.

While most of the anti-inflammatory NR pathways may occur in the different immune cell types found in the diseased liver, there is also evidence that NRs function anti-inflammatory within the hepatocytes. Lipid-sensing NRs such as PPARα and LXRβ and the orphan receptor LRH1 have recently been demonstrated to be powerful inhibitors of the hepatic "acute phase response" (Venteclef et al. 2011). Liver inflammation, especially IL6 and IL1β, is accompanied by a production of "acute phase proteins" (e.g., C-reactive protein/CRP, haptoglobin, serum amyloid A/SAA, and fibrinogen), which play diverse roles in acute defense and repair but also modulate lipoprotein metabolism and chronic inflammation under conditions of obesity. As demonstrated for LXRβ and LRH1, anti-inflammatory transrepression in hepatocytes, as in macrophages, requires receptor SUMOylation and interactions with the GPS2-containing corepressor complex at the promoters/enhancers of acute phase genes (Venteclef et al. 2010). LRH1 plays additionally roles by regulating the expression of the anti-inflammatory cytokine IL1RA in hepatocytes.

Liver cancer (HCC) is one of the human cancers where the connection between obesity-associated chronic low-grade systemic inflammation acute inflammation/infections and tumor development is well known (Sun and Karin 2008). HCC develops in hepatocytes and inflammatory transcription factors such as NFκB and STAT3 are pivotal players. A first example of endogenously occurring anti-inflammatory NR cross talk and its importance for HCC have been discovered by showing that estrogens are protective against HCC development in woman because they are potent inhibitors of IL6 expression in liver macrophages (Naugler et al. 2007). Conceivably, additional NRs have to be considered as important players in HCC development and protection. The bile acid receptor FXR emerges as a recent prominent candidate, but the FXR-regulated interplay between proliferative and anti-inflammatory pathways in hepatocytes remains to be clarified (Calkin and Tontonoz 2012).

NRs in Inflammatory Bowel Disease (IBD)

Ulcerative colitis and Crohn's disease are the two major forms of idiopathic IBD, characterized by an unrelenting of the gut mucosa. IBD is a complex inflammatory disease that usually develops in the second or third period of life. Several immune cell types are involved in the onset of the disease including intestinal cells (epithelial, paneth, and goblet cells), innate cells (DCs and macrophage), and adaptive (lymphocytes) cells. IBD is associated with a Th1- and Th2-type immune response, with excessive production of TNFα, IFNγ, IL13, and IL12. TNFα and TLR signaling pathways toward NFκB appear key drivers of IBD and central to therapeutic strategies, which include the application of anti-inflammatory NR ligands (Taylor and Irving 2011).

A variety of NRs seem to be involved in the protection against the development of IBD. Using mouse models of chemical-induced IBD, in combination with selective ligands, it was shown that PPARs, FXR, and LRH1 play anti-inflammatory and protective roles in the gut, albeit utilizing distinct mechanisms. In the case of PPARγ, the combined anti-inflammatory activities in epithelial mucosa cells and immune cells may ameliorate colitis. PPARγ also controls the expression of β-defensin 1 (DEBF1) expression, involved in the killing of microbes. Thus, maintaining colonic PPARγ activity through therapeutic or nutritional approaches may contribute to the prevention of colonic inflammation by restoring antimicrobial immunity in IBD. In contrast to PPARs, FXR appears expressed only in the intestinal epithelial cells, where it plays an essential role in preserving the integrity of the epithelial barrier and in coordinating mucosal immune response (Gadaleta et al. 2011; Modica et al. 2010). FXR ligands have anti-inflammatory properties and protected against colitis in wild-type mice but not in knockout mice. FXR-deficient mice develop intestinal inflammation-driven fibrosis in the colon. As a last example, LRH1 was demonstrated to limit gut inflammation by an alternative pathway that controls the synthesis of anti-inflammatory GCs. Both LRH1 haplo-insufficiency and somatic deficiency in the intestinal epithelium rendered mice more susceptible to colitis, probably due to reducing anti-inflammatory GR action in the gut.

NRs and Atherosclerosis

The control of cholesterol homeostasis is of critical importance in the pathogenesis of atherosclerosis, as imbalance of cholesterol influx and efflux will lead to excessive accumulation of cholesterol in macrophages and their transformation into foam cells. Additionally, atherogenesis is associated with pro-inflammatory macrophage activation as well as the inflammation of the vascular wall cells including smooth muscle cells and endothelial cells.

Clearly, both aspects are controlled by the lipid-sensing PPARs and LXRs that emerge as main players and promising therapeutic targets (Calkin and Tontonoz 2010). Intriguingly, foam cells enrich endogenous ligands for these receptors, namely, oxidized LDL (oxLDL) and its derivatives (PPAR ligands) or desmosterol (LXR ligand) (Spann et al. 2012). PPARγ and LXRs are robustly expressed in macrophages taking residence in atherosclerotic plaque. Pharmacological and genetic studies in murine models of atherosclerosis (LDLR-/- and APOE -/-) demonstrated the crucial function of PPARγ and both LXRs in inhibiting plaque formation and inflammation in the vessel wall. Even if LXRs have also key functions in the liver, bone marrow transplantation experiments suggest that the beneficial action of LXR in atherosclerosis models was mainly driven by LXR in bone marrow-derived cells. While PPARβ/δ also attenuates atherogenic inflammation by acting in vascular cells and macrophages, the anti-atherogenic action of PPARα is suggested to be mainly driven by its role in the liver.

NRs and Neurodegenerative Diseases

Neuro-inflammation is a key component of diseases of the central nervous system, characterized by a progression of microglial (brain macrophages) and astrocyte inflammation, in conjunction with activation of auto-reactive T-cells. Not surprisingly, neurodegenerative diseases such as Alzheimer's disease, Parkinson's disease, and multiple sclerosis emerge as key therapeutic areas of anti-inflammatory intervention strategies, which involve the consideration of various NRs and their ligands (Saijo et al. 2010). Notably, oxysterols are important signaling molecules in the brain suspected for their involvement and protective functions in neurodegenerative disorders (Spann and Glass 2013). Specifically, 24-hydroxycholesterol is synthesized exclusively in the brain and signals in part via activating LXRs in glia cells and astrocytes.

Multiple sclerosis (MS) is characterized by activation of immune cells (glia cells, astrocytes, Th17 cells, dendritic cells) in the perivascular region of the brain and in the spinal cord. MS is also classified as an autoimmune disease, suggesting that NRs such as GR and RORs with major function in T-cell biology may be useful for therapeutic targeting. Indeed, anti-inflammatory GCs such as methylprednisolone are used in MS since the 1980s to manage acute exacerbations in short-time treatments, while longtime treatments are not recommended due to severe side effects. GCs may mainly act in T-cells to suppress autoimmune-related pathways. Pharmacological and genetic approaches have validated the beneficial contribution of PPARγ and of LXRs in MS regression. In particular, LXR ligands alone or in conjunction with RXR ligands have been demonstrated to reduce inflammation in microglia and decrease the severity of MS. A surprising but significant finding was that the ERβ-specific ligand 5-androsten-3β,17β-diol (ADIOL; see Fig. 4) and some synthetic agonists, but not 17-β-estradiol, suppress inflammatory responses of microglia and astrocytes (Saijo et al. 2011). The study demonstrated that ERβ activation prevented autoimmune encephalomyelitis (EAE, an experimental mouse model for MS). Thus, the selective targeting of ERβ may need to be evaluated as possible MS therapy.

Parkinson's disease (PD) is associated with the loss of dopaminergic neurons. In addition to monogenic mutations (e.g., α-synuclein), environmental factors, inflammation, and aging contribute to PD onset and progression. Loss of dopaminergic neurons is often associated with, or caused by, an accumulation of inflammatory effectors driven by the activation of microglia, astrocytes, and lymphocytes. Conceivably, a number of anti-inflammatory NRs including PPARs and LXR may protect PD progression, but direct experimental evidence has yet to be provided. However, the orphan receptor NURR1 and more recently LXRs have been independently implicated in controlling the development of dopaminergic neurons. Notably, 24(S),25-epoxycholesterol (24,25-EC) was found to be the most abundant LXR ligand in the developing midbrain and shown to promote dopaminergic differentiation of embryonic stem cells, suggesting that NR ligands can be employed for future cell replacement regenerative therapies envisaged for PD (Theofilopoulos et al. 2013). While the role of anti-inflammatory LXR action in PD is unknown, NURR1 contributes to the pathology of PD by limiting inflammatory response in microglia and astrocytes, thus protecting dopaminergic neurons from neurotoxic death (Saijo et al. 2009). NURR1 trans-represses NFκB by recruiting a CoREST corepressor complex (Fig. 3), thereby expanding the complexity of anti-inflammatory mechanisms to an orphan receptor.

Alzheimer's disease (AD) is characterized by the appearance of senile plaques as hallmark, composed of protein aggregates containing amyloid-β (A-b), neurofibrillary tangles, and activated immune cells such as microglia involved in antigen presentation and secretion of pro-inflammatory mediators. The two main NRs known for their involvement in AD are LXRs and PPARs (Kummer and Heneka 2008; Terwel et al. 2011). Indeed, in mouse models for AD, LXR agonists trigger the clearance of A-β through

ApoE-mediated proteolysis and repress the production of pro-inflammatory mediators. Consistently, deletion of LXRα or LXRβ aggravates the symptoms of AD. The effect of PPARγ and β/δ in AD seems to be dependent on their anti-inflammatory actions in microglia and astrocytes, but improvement of glucose homeostasis may influence AD progression. Finally, a modestly regression of AD symptoms was reported in a phase 2 clinical trial with a synthetic PPARγ agonist.

NRs and Autoimmune Diseases

The accumulation of pro-inflammatory immune cells, mainly Th17 cells and dendritic cells, is a hallmark of autoimmune diseases such rheumatoid arthritis (RA), systemic lupus erythematosus (SLE), and multiple sclerosis (MS, discussed above). GCs interfere with T-cell and DC differentiation and polarization and are effective in treating the abovementioned autoimmune diseases. As pointed out in the previous section, the targeting of RORγt promises a main strategy to prevent Th17 differentiation but has yet to be shown to function in the clinics. PPARs and LXRs also play an important role in repressing pathological expansion of innate and adaptive immune cells (see above). Although LXR agonist prevented the evolution of collagen-induced arthritis in mice, the benefit of LXRs targeting in RA or SLE is controversial. However, mice lacking LXRs manifested a breakdown in self-tolerance and developed autoantibodies and autoimmune glomerulonephritis, and treatment with an LXR agonist ameliorated disease progression in a mouse model of lupus-like autoimmunity (N-Gonzalez et al. 2009). Concerning PPARs, several studies reported the suppression of RA and SLE in mouse models upon administration of PPARγ ligands (endogenous or synthetic), supporting the proof of concept. This effect was mainly driven by a decrease of plasma levels of IL6, TNFα, and IL1β as well as repression Th17 cells and favoring Treg expansion. In a recent clinical trial, supplementation of the usual treatment with a TZD (Pioglitazone) resulted in a marked reduction in disease activity scores and serum levels of the inflammatory marker CRP compared with the usual treatment alone. However, side effects on bone fracture were observed, thus limiting TZD treatment in autoimmune diseases.

Modulation of Metabolic and Inflammatory NR Signaling by Drugs

NRs have been the focus of drug discovery efforts since the early years of the twentieth century when a series of findings that certain natural steroid and thyroid hormones could be used to treat diseases. Clinical and epidemiological observations have early linked the deficiency in hormones and vitamins, many of which are now known as powerful NR ligands and endocrine regulators, to disturbances in innate and adopted immune responses. Today, NRs are significant drug targets because of their fundamental role in a wide variety of physiological and pathophysiological processes, including the immune response, toxin clearance, vascular and cardiac function, carbohydrate and lipid metabolism, and cancer. Virtually all NRs for which ligands have been identified are successful therapeutic targets, with either natural or synthetic forms of the ligands having been converted to marketed drugs (see Marc Via: Nuclear Receptors: The pipeline outlook, Insight Pharma Report.com, 2010). It is estimated that NR drugs account for approximately 13 % of all drugs in global pharmaceutical sales, next to GPCRs (Overington et al. 2006). Notably, anti-inflammatory GR ligands are the most frequently prescribed NR drugs to date.

A substantial number of NRs are expressed in the different cell types of the immune system (see datasets in www.nursa.org; www.biogps.org). Due to their specific SUMOylation and interaction with anti-inflammatory corepressors, many of these NRs have to be considered as "transrepression-competent" (Treuter and Venteclef 2011), making them attractive pharmacological targets. Some of these NRs have additional, seemingly unrelated, functions in the immune system by affecting differentiation, specification and polarization, and lipid metabolism, all of which may contribute to the net anti-inflammatory action of drugs targeting these NRs.

Synthetic NR modulators have been identified by compound screens or designed by structure-guided approaches based on the structure of the natural cognate ligands. This has remained challenging for many orphan receptors, with the notable exception of the orphan receptors ROR (a key driver of Th17 immune cell differentiation) and LRH1 (a metabolic and anti-inflammatory NR expressed throughout the enterohepatic axis). It can be assumed that all NRs having a ligand-binding pocket are also amendable to synthetic drug design. Perhaps, some orphan receptors, such as those lacking metabolic functions and/or endogenous ligands, may be particularly promising targets to achieve anti-inflammatory pathway selectivity of synthetic drugs.

A common feature of today's NR drugs – irrespective of functioning as pure agonists, partial agonists/antagonists, inverse agonists, or pure antagonists – is that all of them have to some extent undesired side effects. This is due to expression in multiple cell types and tissues, to ligand-selective interactions with cell-type-specific coregulators, to ligand-selective modulation of the PTM status, and to the fact that NRs carry out a variety of distinct physiological functions. Moreover, it has been realized that the ligand-binding pockets of some NRs are flexible and can accommodate structurally unrelated compounds. Therefore, different NRs can bind to the same ligand, as demonstrated for both natural and synthetic ligands. For example, bile acids such as CDCA activate the bile acid receptor FXR but also with the vitamin D receptor VDR, now considered the second bile acid receptor. Furthermore, synthetic LXR agonists can also bind to RORs and modulate their immune functions in T(h)17 cells. Another example is the synthetic FXR agonist GW4064, which is commonly used to dissect FXR biology. Using FXR knockout mice, it was shown that GW4064 activates ERRs, which then induce the powerful coactivator PGC-1, thereby affecting energy metabolism. Therefore, experimental data obtained using some synthetic NR compounds have to be interpreted with caution.

Lastly, despite being the most valuable tools to dissect NR biology in preclinical research, most of the currently available anti-inflammatory NR compounds (except GR ligands) have yet to make it into clinical applications. Thus, a major goal for future drug development must be to obtain compounds, referred to as selective NR modulators (SNRMs), with improved NR, cell type, and pathway selectivity. This is of major concern in the case of anti-inflammatory NR drugs, such as those targeting established anti-inflammatory receptors including the GR, the PPARs, and the LXRs as well as many candidate NRs that need to be further scrutinized for their anti-inflammatory action and drugability.

The chemical structures and key properties of some significant NR ligands with immune-modulatory and anti-inflammatory properties are summarized in Fig. 4.

Cross-References

▶ Corticosteroids
▶ Obesity and Inflammation

References

Ahmadian, M., Suh, J. M., Hah, N., Liddle, C., Atkins, A. R., Downes, M., et al. (2013). Ppargamma signaling and metabolism: The good, the bad and the future. *Nature Medicine, 19*(5), 557–566. doi:10.1038/nm.3159.

Barish, G. D., Downes, M., Alaynick, W. A., Yu, R. T., Ocampo, C. B., Bookout, A. L., et al. (2005). A nuclear receptor atlas: Macrophage activation. *Molecular Endocrinology, 19*(10), 2466–2477. doi:10.1210/me.2004-0529.

Barish, G. D., Atkins, A. R., Downes, M., Olson, P., Chong, L. W., Nelson, M., et al. (2008). Ppardelta regulates multiple proinflammatory pathways to suppress atherosclerosis. *Proceedings of the National Academy of Sciences of the United States of America, 105*(11), 4271–4276. doi:10.1073/pnas.0711875105.

Beck, I. M., Vanden Berghe, W., Vermeulen, L., Yamamoto, K. R., Haegeman, G., & De Bosscher, K. (2009). Crosstalk in inflammation: The interplay of glucocorticoid receptor-based mechanisms and kinases and phosphatases. *Endocrine Reviews, 30*(7), 830–882. doi:10.1210/er.2009-0013.

Bensinger, S. J., Bradley, M. N., Joseph, S. B., Zelcer, N., Janssen, E. M., Hausner, M. A., et al. (2008). Lxr signaling couples sterol metabolism to proliferation in the acquired immune response. *Cell, 134*(1), 97–111. doi:10.1016/j.cell.2008.04.052.

Calkin, A. C., & Tontonoz, P. (2010). Liver x receptor signaling pathways and atherosclerosis. *Arteriosclerosis, Thrombosis, and Vascular Biology, 30*(8), 1513–1518. doi:10.1161/ATVBAHA.109.191197.

Calkin, A. C., & Tontonoz, P. (2012). Transcriptional integration of metabolism by the nuclear sterol-activated receptors lxr and fxr. *Nature Reviews. Molecular Cell Biology, 13*(4), 213–224. doi:10.1038/nrm3312.

Cariou, B., Charbonnel, B., & Staels, B. (2012). Thiazolidinediones and ppargamma agonists: Time for a reassessment. *Trends in Endocrinology and Metabolism, 23*(5), 205–215. doi:10.1016/j.tem.2012.03.001.

Chawla, A. (2010). Control of macrophage activation and function by ppars. *Circulation Research, 106*(10), 1559–1569. doi:10.1161/CIRCRESAHA.110.216523.

Chawla, A., Boisvert, W. A., Lee, C. H., Laffitte, B. A., Barak, Y., Joseph, S. B., et al. (2001). A ppar gamma-lxr-abca1 pathway in macrophages is involved in cholesterol efflux and atherogenesis. *Molecular Cell, 7*(1), 161–171.

Chinenov, Y., Gupte, R., Dobrovolna, J., Flammer, J. R., Liu, B., Michelassi, F. E., et al. (2012). Role of transcriptional coregulator grip1 in the anti-inflammatory actions of glucocorticoids. *Proceedings of the National Academy of Sciences of the United States of America, 109*(29), 11776–11781. doi:10.1073/pnas.1206059109.

Dalmas, E., Clement, K., & Guerre-Millo, M. (2011). Defining macrophage phenotype and function in adipose tissue. *Trends in Immunology, 32*(7), 307–314. doi:10.1016/j.it.2011.04.008. S1471-4906(11)00076-7 [pii].

Davies, L. C., Jenkins, S. J., Allen, J. E., & Taylor, P. R. (2013). Tissue-resident macrophages. *Nature Immunology, 14*(10), 986–995. doi:10.1038/ni.2705.

Farnegardh, M., Bonn, T., Sun, S., Ljunggren, J., Ahola, H., Wilhelmsson, A., et al. (2003). The three-dimensional structure of the liver x receptor beta reveals a flexible ligand-binding pocket that can accommodate fundamentally different ligands. *Journal of Biological Chemistry, 278*(40), 38821–38828. doi:10.1074/jbc.M304842200.

Gadaleta, R. M., van Erpecum, K. J., Oldenburg, B., Willemsen, E. C., Renooij, W., Murzilli, S., et al. (2011). Farnesoid x receptor activation inhibits inflammation and preserves the intestinal barrier in inflammatory bowel disease. *Gut, 60*(4), 463–472. doi:10.1136/gut.2010.212159.

Glass, C. K., & Saijo, K. (2010). Nuclear receptor transrepression pathways that regulate inflammation in macrophages and t cells. *Nature Reviews. Immunology, 10*(5), 365–376. doi:10.1038/nri2748.

Gregor, M. F., & Hotamisligil, G. S. (2011). Inflammatory mechanisms in obesity. *Annual Review of Immunology, 29*, 415–445. doi:10.1146/annurev-immunol-031210-101322.

Gronemeyer, H., Gustafsson, J. A., & Laudet, V. (2004). Principles for modulation of the nuclear receptor superfamily. *Nature Reviews. Drug Discovery, 3*(11), 950–964. doi:10.1038/nrd1551.

Hall, J. M., & McDonnell, D. P. (2007). The molecular mechanisms underlying the proinflammatory actions of thiazolidinediones in human macrophages. *Molecular Endocrinology, 21*(8), 1756–1768. doi:10.1210/me.2007-0060.

Hong, C., & Tontonoz, P. (2008). Coordination of inflammation and metabolism by ppar and lxr nuclear receptors. *Current Opinion in Genetics and Development, 18*(5), 461–467. doi:10.1016/j.gde.2008.07.016.

Huang, W., & Glass, C. K. (2010). Nuclear receptors and inflammation control: Molecular mechanisms and pathophysiological relevance. *Arteriosclerosis, Thrombosis, and Vascular Biology, 30*(8), 1542–1549. doi:10.1161/ATVBAHA.109.191189.

Kiss, M., Czimmerer, Z., & Nagy, L. (2013). The role of lipid-activated nuclear receptors in shaping macrophage and dendritic cell function: From physiology to pathology. *The Journal of Allergy and Clinical Immunology, 132*(2), 264–286. doi:10.1016/j.jaci.2013.05.044.

Kummer, M. P., & Heneka, M. T. (2008). Ppars in alzheimer's disease. *PPAR Research, 2008*, 403896. doi:10.1155/2008/403896.

Li, P., Spann, N. J., Kaikkonen, M. U., Lu, M., da Oh, Y., Fox, J. N., et al. (2013). Ncor repression of lxrs restricts macrophage biosynthesis of insulin-sensitizing omega 3 fatty acids. *Cell, 155*(1), 200–214. doi:10.1016/j.cell.2013.08.054.

Medzhitov, R., & Horng, T. (2009). Transcriptional control of the inflammatory response. *Nature Reviews. Immunology, 9*(10), 692–703. doi:10.1038/nri2634.

Modica, S., Gadaleta, R. M., & Moschetta, A. (2010). Deciphering the nuclear bile acid receptor fxr paradigm. *Nuclear Receptor Signaling, 8*, e005. doi:10.1621/nrs.08005.

Naugler, W. E., Sakurai, T., Kim, S., Maeda, S., Kim, K., Elsharkawy, A. M., et al. (2007). Gender disparity in liver cancer due to sex differences in myd88-dependent il-6 production. *Science, 317*(5834), 121–124. doi:10.1126/science.1140485.

N-Gonzalez, A., Bensinger, S. J., Hong, C., Beceiro, S., Bradley, M. N., Zelcer, N., et al. (2009). Apoptotic cells promote their own clearance and immune tolerance through activation of the nuclear receptor lxr. *Immunity, 31*(2), 245–258. doi:10.1016/j.immuni.2009.06.018.

N-Gonzalez, A., Guillen, J. A., Gallardo, G., Diaz, M., de la Rosa, J. V., Hernandez, I. H., et al. (2013). The nuclear receptor lxralpha controls the functional specialization of splenic macrophages. *Nature Immunology, 14*(8), 831–839. doi:10.1038/ni.2622.

Nuclear Receptors Nomenclature Committee (1999). A unified nomenclature system for the nuclear receptor superfamily. *Cell, 97*(2):161–3. PubMed PMID:10219237.

Nunez, V., Alameda, D., Rico, D., Mota, R., Gonzalo, P., Cedenilla, M., et al. (2010). Retinoid x receptor alpha controls innate inflammatory responses through the up-regulation of chemokine expression. *Proceedings of the National Academy of Sciences of the United States of America, 107*(23), 10626–10631. doi:10.1073/pnas.0913545107.

Oberoi, J., Fairall, L., Watson, P. J., Yang, J. C., Czimmerer, Z., Kampmann, T., et al. (2011). Structural basis for the assembly of the smrt/ncor core transcriptional repression machinery. *Nature Structural & Molecular Biology, 18*(2), 177–184. doi:10.1038/nsmb.1983

Odegaard, J. I., Ricardo-Gonzalez, R. R., Goforth, M. H., Morel, C. R., Subramanian, V., Mukundan, L., et al. (2007). Macrophage-specific ppargamma controls alternative activation and improves insulin resistance. *Nature, 447*(7148), 1116–1120. doi:10.1038/nature05894.

Odegaard, J. I., Ricardo-Gonzalez, R. R., Red Eagle, A., Vats, D., Morel, C. R., Goforth, M. H., et al. (2008). Alternative m2 activation of kupffer cells by ppardelta ameliorates obesity-induced insulin resistance. *Cell Metabolism, 7*(6), 496–507. doi:10.1016/j.cmet.2008.04.003.

Olefsky, J. M., & Glass, C. K. (2010). Macrophages, inflammation, and insulin resistance. *Annual Review of Physiology, 72*, 219–246. doi:10.1146/annurev-physiol-021909-135846.

Osborn, O., & Olefsky, J. M. (2012). The cellular and signaling networks linking the immune system and metabolism in disease. *Nature Medicine, 18*(3), 363–374. doi:10.1038/nm.2627. nm.2627 [pii].

Overington, J. P., Al-Lazikani, B., & Hopkins, A. L. (2006). How many drug targets are there? *Nature Reviews. Drug Discovery, 5*(12), 993–996. doi:10.1038/nrd2199.

Pascual-Garcia, M., & Valledor, A. F. (2012). Biological roles of liver x receptors in immune cells. *Archivum Immunologiae et Therapiae Experimentalis, 60*(4), 235–249. doi:10.1007/s00005-012-0179-9.

Perissi, V., Jepsen, K., Glass, C. K., & Rosenfeld, M. G. (2010). Deconstructing repression: Evolving models of co-repressor action. *Nature Reviews. Genetics, 11*(2), 109–123. doi:10.1038/nrg2736.

Ratman, D., Vanden Berghe, W., Dejager, L., Libert, C., Tavernier, J., Beck, I. M., et al. (2013). How glucocorticoid receptors modulate the activity of other transcription factors: A scope beyond tethering. *Molecular and Cellular Endocrinology, 380*(1–2), 41–54. doi:10.1016/j.mce.2012.12.014.

Ricote, M., Li, A. C., Willson, T. M., Kelly, C. J., & Glass, C. K. (1998). The peroxisome proliferator-activated receptor-gamma is a negative regulator of macrophage activation. *Nature, 391*(6662), 79–82. doi:10.1038/34178.

Rigamonti, E., Chinetti-Gbaguidi, G., & Staels, B. (2008). Regulation of macrophage functions by ppar-alpha, ppar-gamma, and lxrs in mice and men. *Arteriosclerosis, Thrombosis, and Vascular Biology, 28*(6), 1050–1059. doi:10.1161/ATVBAHA.107.158998.

Rosenfeld, M. G., Lunyak, V. V., & Glass, C. K. (2006). Sensors and signals: A coactivator/corepressor/epigenetic code for integrating signal-dependent programs of transcriptional response. *Genes & Development, 20*(11), 1405–1428. doi:10.1101/gad.1424806.

Roszer, T., Menendez-Gutierrez, M. P., Cedenilla, M., & Ricote, M. (2013). Retinoid x receptors in macrophage biology. *Trends in Endocrinology and Metabolism, 24*(9), 460–468. doi:10.1016/j.tem.2013.04.004.

Saijo, K., Winner, B., Carson, C. T., Collier, J. G., Boyer, L., Rosenfeld, M. G., et al. (2009). A nurr1/corest pathway in microglia and astrocytes protects dopaminergic neurons from inflammation-induced death. *Cell, 137*(1), 47–59. doi:10.1016/j.cell.2009.01.038.

Saijo, K., Crotti, A., & Glass, C. K. (2010). Nuclear receptors, inflammation, and neurodegenerative diseases. *Advances in Immunology, 106*, 21–59. doi:10.1016/S0065-2776(10)06002-5.

Saijo, K., Collier, J. G., Li, A. C., Katzenellenbogen, J. A., & Glass, C. K. (2011). An adiol-erbeta-ctbp transrepression pathway negatively regulates microglia-mediated inflammation. *Cell, 145*(4), 584–595. doi:10.1016/j.cell.2011.03.050.

Solt, L. A., & Burris, T. P. (2012). Action of rors and their ligands in (patho)physiology. *Trends in Endocrinology and Metabolism, 23*(12), 619–627. doi:10.1016/j.tem.2012.05.012.

Solt, L. A., Kamenecka, T. M., & Burris, T. P. (2012). Lxr-mediated inhibition of cd4+ t helper cells. *PloS One, 7*(9), e46615. doi:10.1371/journal.pone.0046615.

Spann, N. J., & Glass, C. K. (2013). Sterols and oxysterols in immune cell function. *Nature Immunology, 14*(9), 893–900. doi:10.1038/ni.2681.

Spann, N. J., Garmire, L. X., McDonald, J. G., Myers, D. S., Milne, S. B., Shibata, N., et al. (2012). Regulated accumulation of desmosterol integrates macrophage lipid metabolism and inflammatory responses. *Cell, 151*(1), 138–152. doi:10.1016/j.cell.2012.06.054.

Steffensen, K. R., Jakobsson, T., & Gustafsson, J. A. (2013). Targeting liver x receptors in inflammation. *Expert Opinion on Therapeutic Targets, 17*(8), 977–990. doi:10.1517/14728222.2013.806490.

Sun, B., & Karin, M. (2008). Nf-kappab signaling, liver disease and hepatoprotective agents. *Oncogene, 27*(48), 6228–6244. doi:10.1038/onc.2008.300.

Taylor, K. M., & Irving, P. M. (2011). Optimization of conventional therapy in patients with ibd. *Nature Reviews. Gastroenterology & Hepatology, 8*(11), 646–656. doi:10.1038/nrgastro.2011.172.

Terwel, D., Steffensen, K. R., Verghese, P. B., Kummer, M. P., Gustafsson, J. A., Holtzman, D. M., et al. (2011). Critical role of astroglial apolipoprotein e and liver x receptor-alpha expression for microglial abeta phagocytosis. *Journal of Neuroscience, 31*(19), 7049–7059. doi:10.1523/JNEUROSCI.6546-10.2011.

Theofilopoulos, S., Wang, Y., Kitambi, S. S., Sacchetti, P., Sousa, K. M., Bodin, K., et al. (2013). Brain endogenous liver x receptor ligands selectively promote midbrain neurogenesis. *Nature Chemical Biology, 9*(2), 126–133. doi:10.1038/nchembio.1156.

Toubal, A., Clement, K., Fan, R., Ancel, P., Pelloux, V., Rouault, C., et al. (2013a). Smrt-gps2 corepressor pathway dysregulation coincides with obesity-linked adipocyte inflammation. *Journal of Clinical Investigation, 123*(1), 362–379. doi:10.1172/JCI64052.

Toubal, A., Treuter, E., Clement, K., & Venteclef, N. (2013b). Genomic and epigenomic regulation of adipose tissue inflammation in obesity. *Trends in Endocrinology and Metabolism.* doi:10.1016/j.tem.2013.09.006.

Treuter, E., & Venteclef, N. (2011). Transcriptional control of metabolic and inflammatory pathways by nuclear receptor sumoylation. *Biochimica et Biophysica Acta, 1812*(8), 909–918. doi:10.1016/j.bbadis.2010.12.008.

Uhlenhaut, N. H., Barish, G. D., Yu, R. T., Downes, M., Karunasiri, M., Liddle, C., et al. (2013). Insights into negative regulation by the glucocorticoid receptor from genome-wide profiling of inflammatory cistromes. *Molecular Cell, 49*(1), 158–171. doi:10.1016/j.molcel.2012.10.013.

Venteclef, N., Jakobsson, T., Ehrlund, A., Damdimopoulos, A., Mikkonen, L., Ellis, E., et al. (2010). Gps2-dependent corepressor/sumo pathways govern anti-inflammatory actions of lrh-1 and lxrbeta in the hepatic acute phase response. *Genes & Development, 24*(4), 381–395. doi:10.1101/gad.545110.

Venteclef, N., Jakobsson, T., Steffensen, K. R., & Treuter, E. (2011). Metabolic nuclear receptor signaling and the inflammatory acute phase response. *Trends in Endocrinology and Metabolism, 22*(8), 333–343. doi:10.1016/j.tem.2011.04.004.

Wagner, M., Zollner, G., & Trauner, M. (2011). Nuclear receptors in liver disease. *Hepatology, 53*(3), 1023–1034. doi:10.1002/hep.24148.

Zelcer, N., & Tontonoz, P. (2006). Liver x receptors as integrators of metabolic and inflammatory signaling. *Journal of Clinical Investigation, 116*(3), 607–614. doi:10.1172/JCI27883.

Obesity and Inflammation

Giuseppe Matarese[1,2], Claudio Procaccini[3] and Veronica De Rosa[3,4]
[1]Dipartimento di Medicina e Chirurgia, Università di Salerno, Salerno, Italy
[2]IRCCS Multimedica, Milan, Italy
[3]Laboratorio di Immunologia, Istituto di Endocrinologia e Oncologia Sperimentale, Consiglio Nazionale delle Ricerche (IEOS-CNR), Università di Napoli, Napoli, Italy
[4]Unità di NeuroImmunologia, IRCCS Fondazione Santa Lucia, Rome, Italy

Synonyms

Adipose tissue; Inflammation; Metabolism; Obesity; Regulatory T cells

Definition

Obesity has been considered for decades to be the result of the complex intersection between genes and environment and its pathogenesis is still unresolved. Obesity is strongly associated with insulin resistance, hypertension, and dyslipidemia, and accumulating evidence indicates that a state of chronic inflammation has a crucial role in the pathogenesis of obesity-related metabolic dysfunction. Excess adipose mass is associated with increased levels of the pro-inflammatory marker, and interventions aimed at causing weight loss lead to a reduction in the levels of pro-inflammatory proteins. In this context, the discovery of the adipose tissue-derived hormone, leptin, and other adipocytokines has shed fundamental insights on the basic mechanisms governing immune tolerance in the context of metabolic disease susceptibility. Here the rapidly expanding body of animal and clinical data that support a potential role for inflammation in the pathogenesis of obesity and its associated complications is reviewed.

Structure and Function

Pathogenesis of Obesity and Neural Control of Food Intake

Over the past 20 years, obesity has become a world epidemic pathological condition, particularly in the Western world. Obesity could be defined as a state in which the total amount of triglyceride stored in adipose tissue is abnormally increased, and it is associated with a wide variety of adverse health outcomes, such as type 2 diabetes (T2D), insulin resistance, inflammation, cardiovascular disease, and tumors, which reduce life expectancy and together have huge economic and social consequences. This pathological condition results from a chronic, positive imbalance between energy intake and energy expenditure, and usually it derives from the interaction of genetic factors with abundance of caloric intake and the decline of physical activities (Friedman 2009; Hotamisligil 2006).

In this context, there are clear lessons from monogenic disorders: indeed, it had long been known that human obesity could result from a disorder in a single gene. There are now at least 20 single-gene disorders that clearly result in an autosomal form of human obesity. In spite of this, the vast majority of "common" obesity is the result of a complex interaction between genes and environment; the prevalence of pathogenic single-gene mutations in subjects with hyperphagia and early onset obesity is estimated to be around 10–15 % of all cases of obesity, suggesting that the pathogenic story of obesity can only be clarified studying interaction among genes, environment, and epigenetic factors (O'Rahilly 2009).

Generally the main two peripheral signals which control food intake and energy expenditure are represented by leptin, which activates the hypothalamic POMC neurons inducing an anorexigenic signal (inhibition of food intake) and by ghrelin, which activates the hypothalamic AgRP/NPY neurons that induce an orexigenic response (induction of food intake) (Dietrich and Horvath 2009; Gao and Horvath 2008; Horvath and Bruning 2006). Leptin is a cytokine-like hormone mainly produced by adipose tissue in proportion to body fat mass. As a hormone, leptin regulates food intake and basal metabolism, while as a cytokine, it can affect thymic homeostasis and the secretion of acute-phase reactants, promoting T helper 1 (T_H1)-cell differentiation and the onset and progression of autoimmune responses (La Cava and Matarese 2004). Genetic deficiency in the leptin/LepR system cause early onset obesity in mice and humans, and all these data suggest that the pathogenic story of obesity can only be clarified studying interaction among genes, environment, and epigenetic factors.

The Link Between Obesity and Chronic Inflammation

Obesity has been associated with a series of consequences such as increased risk of cardiovascular disorders including atherosclerosis, diabetes, fatty liver disease, inflammation, and cancer. All these pathological conditions are closely associated with chronic inflammation, characterized by abnormal cytokine production, increased acute-phase reactants, and activation of a network of inflammatory signaling pathways. They seem to be consequent to the long-term "low-degree" chronic inflammation typical of obesity (Symonds et al. 2009). Conversely, a reduction in body weight is accompanied by a decrease or even a normalization of these biological parameters (van Dielen et al. 2004).

A new field of study that investigates the interface among immune response, nutrition, and metabolism has recently developed, and it has been found that certain genetic alterations (i.e., mutation, loss of function, among others) of leptin (Lep), leptin receptor (LepR), pro-opiomelanocortin (POMC), pro-protein convertase 1 (PCSK1), and melanocortin-4 receptor (MC4-R), can cause obesity and also significantly affect immune responses (Montague et al. 1997; Clément et al. 1998; Krude et al. 1998; Vaisse et al. 1998; Yeo et al. 1998). Therefore, the immune function in obesity has become a factor of particular interest and relevance to better understand and possibly modulate the inflammatory condition associated with this disorder.

Recent studies have documented the unusual properties of adipocytes and centrally placed adipose tissue as a crucial site in the generation of inflammatory responses and mediators. The finding that tumor necrosis factor-a(TNF-α and interleukin-6 (IL-6) are overexpressed in the adipose tissue of obese mice and humans and when administered exogenously leads to insulin resistance provided the first clear link between obesity, diabetes, and chronic inflammation (Hotamisligil and Spiegelman 1994). Moreover, adipocytes share with a diverse set of immune cells (including T cells, macrophages, and dendritic cells) several features, such as complement activation, production of inflammatory mediators to pathogen sensing, and phagocytic properties (Dixit 2008). Interestingly, numerous genes that code for transcription factors, cytokines, inflammatory signaling molecules, and fatty acid transporters are essential for adipocyte biology and are also expressed and functional in macrophages (Totonoz et al. 1998).

Adipose tissue is a mix of adipocytes, stromal preadipocytes, immune cells, and endothelium, and it can respond rapidly and dynamically to alterations in nutrient excess through adipocyte hypertrophy and hyperplasia (Halberg et al. 2008). With obesity and progressive adipocyte enlargement, the blood supply to adipocytes may be reduced with consequent hypoxia (Cinti et al. 2005). Hypoxia has been proposed to be an inciting etiology of necrosis and macrophage infiltration into adipose tissue, leading to an overproduction of pro-inflammatory factors like inflammatory chemokines. This results in a localized inflammation in adipose tissue which propagates an overall systemic inflammation associated with the development of obesity-related comorbidities (Cinti et al. 2005).

Immune Cells Infiltration During Obesity

Obesity can lead to changes in the cellular composition of the fat pad as well as to the modulation of individual cell phenotypes. Adipose tissue during obesity is infiltrated by a large number of macrophages, and this recruitment is linked to systemic inflammation and insulin resistance (Weisberg et al. 2003); conversely weight loss results in a reduction in the number of adipose tissue macrophages with consequent decrease of the pro-inflammatory profiles of obese individuals (Cancello et al. 2005).

Large-scale studies of gene expression using microarray approaches have already highlighted that in WAT from rodent genetic models of obesity, the expression of genes coding for proteins involved in inflammatory processes was markedly altered (Soukas et al. 2000). It was observed that these variations in gene expression in WAT were essentially related to a macrophage infiltration in WAT of these obese mice (Weisberg et al. 2003). These locally present macrophages are responsible for the major part of the locally produced TNF-α, IL-6, and inducible nitric oxide synthase (iNOS). It is noteworthy that a reduction in body weight is accompanied, not only by an improvement in the inflammatory process and the comorbidities but also by a decrease in the expression of genes coding for inflammation-related proteins (Clement et al. 2004).

The recruitment of macrophages into adipose tissue is the initial event in obesity-induced inflammation and insulin resistance. As a general model, overnutrition causes adipocytes to secrete chemokines such as monocyte chemotactic protein-1 (MCP-1), leukotriene B4 (LTB4), and others, providing a chemotactic gradient that attracts monocytes into the adipose tissue, where they become activated. Once pro-inflammatory macrophages migrate into adipose tissue, they also secrete their own chemokines, attracting additional macrophages and setting up a feed-forward inflammatory process (Fig. 1).

It has been recently reported that macrophages accumulate in adipose tissues during the early phase of weight loss, presumably as a result of adipose tissue lipolysis (Kosteli et al. 2010). Adipose tissue also contains fibroblasts, which produce extracellular matrix components. Recently, it has been shown that metabolically dysfunctional adipose tissue produces excess matrix components that may interfere with adipose mass expansion and contribute to metabolic dysregulation (46). In a paper by Cinti et al., the classification of metabolically dysfunctional obese individuals correlated with the presence of crown-like structures, which are histological features that represent an accumulation of macrophages around dead adipocytes in inflamed adipose tissue (Cinti et al. 2005), and since a key function of macrophages is to remove apoptotic cells in an immunologically silent manner to prevent the release of dangerous substances, it is reasonable to speculate that the presence of crown-like structures in adipose tissue reflects a pro-inflammatory state that is due, in part, to an impairment of the macrophage-mediated phagocytic process.

Different subsets of macrophages are involved in obesity-induced adipose tissue inflammation. Macrophages that accumulate in the adipose tissues of obese mice mainly express genes associated with an M1 or "classically activated" macrophage phenotype, whereas adipose tissue macrophages from lean mice express genes associated with an M2 or "alternatively activated" macrophage phenotype (Lumeng et al. 2007). Both M1-like and M2-like populations express F4/80 and CD11b, and the M1-like macrophages

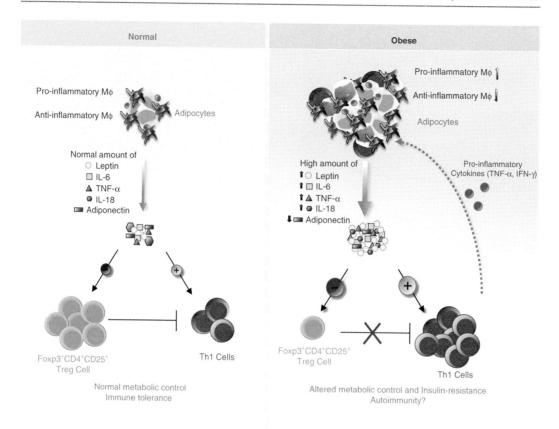

Obesity and Inflammation, Fig. 1 Model of immune effects of leptin and other adipocytokines in normal and obese subjects. In normal subjects, secretion of adipocyte-derived leptin and other cytokines (such as IL-6, IL-18, TNF-α, and adiponectin) associates with a normal control of metabolic functions and with a balance between the number of Th1 cells and Treg cells, which are functionally able to suppress immune and autoimmune responses. Conversely, during obesity, the high amount of leptin secreted by adipocytes accounts for an altered control of metabolic functions, often associated with insulin resistance; a high frequency and expansion of Th1 cells, on one side; and a low proportion and proliferation of Treg cells infiltrating adipose tissue, on the other. Obesity-induced perturbations in the balance between Th1- and Th2-type signals may influence the recruitment and activation of macrophages in adipose tissues, thereby generating a pathogenic and inflammatory environment. Tissue macrophages respond to changes in the local environment by changing their polarization status, and along with the increased numbers of Th1 cells and macrophages in adipose tissue, a higher number of CD8$^+$ T cells and mast cells have been reported. Whether these phenomena are reflection of an ongoing adipose tissue autoimmunity needs to be further investigated

also expresses CD11c (Lumeng et al. 2007). In the obese state, M1-like macrophages can accumulate lipids, taking on a foamy appearance in the adipose tissue (Gordon 2003). Stimulation with T helper 1 (Th1)-type cytokines, including interferon-γ (IFNγ), or with bacterial products leads to the generation of M1 macrophages, which produce pro-inflammatory cytokines (including TNF and IL-6), express inducible nitric oxide synthase (iNOs), and produce reactive oxygen species (ROs) and nitrogen intermediates (Gordon 2003). Conversely, Th2-type cytokines such as IL-4 and IL-13 polarize macrophages to the M2 phenotype, which upregulate IL-10 production and downregulate synthesis of pro-inflammatory cytokines. In addition, the transcription of several genes, including those encoding arginase 1, macrophage mannose receptor, 1 and IL-1 receptor antagonist, is upregulated in M2 macrophages (Gordon 2003). Thus, obesity-induced perturbations in the balance between Th1- and Th2-type signals may influence the recruitment and activation of macrophages in adipose tissues, thereby generating either a pathogenic and inflammatory environment or a noninflammatory and protective environment (Fig. 1). Tissue macrophages respond to changes in the local environment by changing their polarization status, and, thus, the M1 and M2 classifications are oversimplifications of the more dynamic and varied polarization states of macrophages that can be observed in vivo.

The Novel View of Obesity as an Autoimmune Disorder

The presence of an abundant immune cell infiltrate in adipose tissue of obese subjects is considered one of the classical pathologic lesions present in obesity. The real significance of these infiltrates is still unknown and has been, until now, considered directly or indirectly the result of a massive attraction exerted by adipocytes toward immune cells, particularly of the natural immunity compartment (i.e., macrophages, neutrophils, natural killer cells, dendritic cells) through the secretion of adipocytokines and chemokines. Strikingly, a series of recent studies have shown in mice that T cells in the adipose tissue show specific T-cell receptor (TCR) rearrangements, suggesting that there are clonal T-cell populations infiltrating adipose tissue. These data along with extensive macrophage infiltration and Th1 cytokine secretion account for the consequent insulin resistance in adipocytes and chronic inflammation typical of obesity (Lumeng et al. 2009). Taken together these data can lead to the hypothesis to consider obesity as an autoimmune disorder. Typically, criteria to consider a pathological condition as "autoimmune" include: (1) infiltration by immune cells of self-target organ and its consequent tissue damage; (2) the presence of circulating autoantibodies that react against self-antigens and subsequent complement system activation; (3) the clonality of TCRs from infiltrating T cells; (4) secretion of pro-inflammatory Th1 cytokines; (5) quantitative or qualitative alterations of Treg cells; and (6) association with other autoimmune disease. In the case of obesity, most of the abovementioned points have been detected (Feuerer et al. 2009; Winer et al. 2009). However, the self-antigen present in the adipose tissue is still unknown. Identifying these antigens and the corresponding antigen-presenting cells in fat is clearly the next challenge for the field. It is also important to note that obesity and type 2 diabetes are associated with high leptin levels and higher frequency of psoriasis (Sterry et al. 2007), autoimmune thyroiditis (Radetti et al. 2008), and higher risk of developing multiple sclerosis (Munger et al. 2009). All these conditions are characterized by lower Treg cells, higher adipose tissue with consequent higher leptin secretion. Moreover, also Treg cells have been found preferentially localized in the adipose tissue of normal individuals. Their role in this context is still object of extensive investigations, but interestingly they appear massively reduced in obesity, further suggesting their possible role to control an autoimmune attack against adipose tissue. Furthermore, adipose tissue Treg cells express (and are thought to secrete) an unusually high amount of the anti-inflammatory cytokine IL-10, which in lean mice could help to suppress adipose tissue inflammation. Winer et al. have also shown that Treg cells are able to protect against insulin resistance and hyperglycemia (Winer et al. 2009).

Another open question is represented by the notion that it is not known what comes first in the temporal events leading to obesity: the "autoimmune attack" of adipose tissue that leads to an unbalanced metabolic control of food intake or an imbalanced hypothalamic/neural control of feeding leading to increased adipose tissue deposition able to stimulate the break of immune self-tolerance. The possibility that metabolic mediators could induce hypothalamic dysregulation and inflammation leading to uncontrolled food intake is at the moment a very accredited hypothesis. Indeed excess of nutrients such as saturated fatty acids (i.e., palmitate) can induce hypothalamic inflammation via cell-autonomous mechanisms through stimulation of the toll-like receptor 4 (TLR4) (Fessler et al. 2009), reactive oxygen species (ROS) (Horvath et al. 2009), and endoplasmic reticulum stress (Hotamisligil 2008). Energy homeostasis is achieved by hypothalamic sensing and integration of afferent negative feedback from hormones such as leptin and insulin, which act in the brain to limit weight gain. During high-fat feeding, neuronal inflammation causes resistance to these hormones, thereby favoring weight gain.

Adipocytokines

It is now well established that adipose tissue is not only involved in energy storage but also functions as an endocrine organ that secretes various bioactive substances, which are collectively referred to as adipokines. The dysregulated expression of these factors, caused by excess adiposity and adipocyte dysfunction, has been linked to the pathogenesis of various disease processes through altered immune responses.

Leptin

Leptin is the product of the ob gene. It is involved in the regulation of energy homeostasis (Zhang et al. 1994) and is almost exclusively expressed and produced by WAT and more particularly by differentiated mature adipocytes. Circulating levels and adipose tissue mRNA expression of leptin are strongly associated with BMI and fat mass in obesity, while reduction of leptin secretion occurs after fasting (Friedman and Halaas 1998).

Although leptin acts mainly at the level of the central nervous system to regulate food intake and energy expenditure, there is a relationship between leptin and the low-grade inflammatory state in obesity, suggesting that leptin could exert peripheral biological effects as a function of its cytokine-like structure (Procaccini et al. 2012a). Indeed, leptin receptors belong to the cytokine class I receptor family and recent evidence has shown that leptin has a role as pro-inflammatory cytokine. In particular, leptin stimulates, in innate immune responses, the production of several pro-inflammatory mediators such as IL-1, IL-6, IL-12, and TNF; moreover, it is able to activate neutrophils chemotaxis and stimulate production of reactive oxygen species (ROS). Leptin promotes activation and phagocytosis by monocytes/macrophages and their secretion of leukotriene B_4 (LTB_4), cyclooxygenase 2 (COX2), and nitric oxide, while on NK cells, this hormone sustains their cytotoxicity through the activation of signal transducer and activator of transcription 3 (STAT3) and IL-2 (La Cava and Matarese 2004), as testified by the evidence that db/db mice have a deficit in NK cell development.

Leptin has different effects on proliferation and cytokine production by human naive ($CD45RA^+$) and memory ($CD45RO^+$) $CD4^+$ T cells (both of which express LepRb). Leptin promotes proliferation and IL-2 secretion by naive T cells, whereas it minimally affects the proliferation of memory cells (on which it promotes a bias toward T_H1-cell responses). Another important role of leptin in adaptive immunity is highlighted by the observation that leptin deficiency in ob/ob mice is associated with immunosuppression and thymic atrophy (Lord et al. 1998). On the other hand, recently, it has been reported that leptin can act as a negative signal for the expansion of human naturally occurring $Foxp3^+CD4^+CD25^{high}$ regulatory T cells (Treg) (De Rosa et al. 2007), a cellular subset involved in the prevention of autoimmune diseases. De Rosa et al. showed that freshly isolated human Treg cells produce leptin and express high levels of leptin receptor (LepR). In vitro neutralization with anti-leptin monoclonal antibody (mAb) following anti-CD3/CD28 stimulation resulted in Treg cells proliferation (De Rosa et al. 2007). Leptin has been shown to inhibit rapamycin-induced proliferation of Tregs, by increasing the activation of the mTOR, a 289-kDa serine/threonine protein kinase that is inhibited by rapamycin. In addition, under normal conditions, Tregs secreted leptin, which activated mTOR in an autocrine manner to maintain their state of hyporesponsiveness (Fig. 1). Finally, it has been shown that Treg cells from db/db mice exhibited a decreased mTOR activity and increased proliferation compared with that of wild-type cells (Procaccini et al. 2010, 2012b). More recently, it has been shown that leptin-induced mTOR pathway activation defines a specific molecular and transcriptional signature, which controls $CD4^+$ T effector (Teff) responses (Procaccini et al. 2012c). Together, these data suggest that the leptin–mTOR axis sets the threshold for the responsiveness of Tregs and that this pathway might integrate cellular energy status with metabolic-related signaling in both Treg cells and Teff that use this information to control immune tolerance.

Moreover, leptin-deficient mice are resistant to the induction of active and adoptively transferred experimental autoimmune encephalomyelitis (EAE) (La Cava and Matarese 2004). Importantly, a surge of serum leptin anticipates the onset of clinical manifestations of EAE, and this event correlates with inflammatory anorexia, weight loss, and the development of pathogenic T-cell responses against myelin. Interestingly, in relapsing–remitting multiple sclerosis (RRMS) patients, an inverse correlation between serum leptin and percentage of circulating Treg cells was also observed. Moreover, treatment of WT mice with soluble LepR fusion protein (LepR:Fc) increased the percentage of Tregs and ameliorated the clinical course and progression of disease in relapsing-experimental autoimmune encephalomyelitis (R-EAE), an animal model of RRMS (De Rosa et al. 2006). Two recent reports have shown that adipose tissue in normal individuals is a preferential site of accumulation of Treg cells. Their precise role in this tissue is still object of extensive investigation but what is clear is that in obese, insulin-resistant mice, their number is reduced, and their supplementation with adoptive transfer experiments was able to dampen the inflammatory state and insulin resistance associated with obesity, by increasing IL-10 and Th2/regulatory-type cytokines (Feuerer et al. 2009; Winer et al. 2009). All these data indicate that leptin could be the molecular link between obesity and reduced number and probably impaired function of Treg cell observed in this condition, and these alterations could determine a further amplification of the inflammatory process through the recruitment of other cell types of adaptive and innate immunity such as $CD8^+$, Th1, mast cells, and macrophages (Nishimura et al. 2009; Liu et al. 2009).

TNF-α

TNF-α is a pro-inflammatory cytokine produced by a variety of cell types but mainly by macrophages and lymphocytes and partly by adipose tissue. This cytokine plays a central role in inflammatory and autoimmune diseases and in the pathophysiology of insulin resistance in rodents (Hotamisligil et al. 1993) through the phosphorylation of the insulin receptor substrate-1 (IRS-1) protein on serine residues. This could prevent its interaction with the insulin receptor beta subunit and stop the insulin signaling pathway. TNF expression is increased in the adipose tissue of experimental models of obesity and T2D and reduction of body weight associates with a decreased TNF expression. Moreover, neutralization of TNF in obese mice leads to an improvement in insulin sensitivity associated with an enhanced insulin signaling in muscle and adipose tissue (Hotamisligil et al. 1993). Recent evidence has suggested that adipose tissue is not directly implicated in the increased circulating TNF-α levels observed in obesity in human (Fig. 1). It can be hypothesized that other mechanisms involving a systemic effect of leptin or of other adipokines may induce TNF-α secretion by other cell types such as macrophages. Nevertheless, the precise role of TNF-α in human obesity requires further investigation.

Interleukin-6

IL-6 is produced by many cell types (fibroblasts, endothelial cells, monocytes) and many tissues including adipose tissue. It is now well known that IL-6 production by adipose tissue is enhanced in obesity (Fried et al. 1998) (Fig. 1), and weight loss leads to a reduction in IL-6 levels. It is thought that 15–30 % of circulating IL-6 levels derives from adipose tissue production in the absence of an acute inflammation. Interleukin-6 is a multifunctional cytokine acting on many cells and tissues. One of the main effects of IL-6 is the induction of hepatic C reactive protein (CRP) production, which is now known to be an independent, major risk marker of cardiovascular complications (Fried et al. 1998).

Moreover T2D patients display higher levels of this cytokine, and increased IL-6 levels are predictive of the development of T2D. Senn et al. have recently shown that IL-6 suppresses insulin-stimulated metabolic actions in hepatocytes through a mechanism that is mediated by the induction of SOCS3 expression (Senn et al. 2003). On the contrary, reduction of IL-6 in adipose tissue (by ablation of JNK) protects against the development of insulin resistance.

Interleukin-18

IL-18 is a pro-inflammatory cytokine, produced by adipose tissues (Fig. 1) (Wood et al. 2005). Its levels are increased in obese individuals and in atherosclerotic lesions, and they decline following weight loss. Overexpression of IL-18 in rats results in increased expression of endothelial cell adhesion molecules, macrophage infiltration of the blood vessel wall, and vascular abnormalities, whereas IL-18 deficiency led to smaller lesions in a mouse model of atherosclerosis (Mallat et al. 2001; Elhage et al. 2003).

Adiponectin

Adiponectin is mainly produced in white adipose tissue (WAT) by mature adipocytes, with increasing expression and secretion during adipocyte differentiation and by nonfat cells, but it also can be found in skeletal muscle cells, cardiac myocytes, and endothelial cells.

Adiponectin levels inversely correlate with visceral obesity and insulin resistance, and weight loss is a potent inducer of adiponectin synthesis, thus suggesting that adiponectin may play a protective role against atherosclerosis and insulin resistance (Fig. 1). TNF suppress adiponectin secretion in adipocyte, and its production is regulated also by other pro-inflammatory cytokines such as IL-6 (Ouchi and Walsh 2007). Moreover, adiponectin may modulate the TNFα-induced inflammatory response, since it has been shown that adiponectin reduces TNF-α secretion of macrophages (Ouchi and Walsh 2007).

In contrast to leptin, early studies have indicated that adiponectin has anti-inflammatory effects on endothelial cells inhibiting NF-kB activation and TNF-induced adhesion molecule expression (vascular cell adhesion molecule-1 (VCAM-1), endothelial-leukocyte adhesion molecule-1 (E-selectin), intracellular adhesion molecule-1 (ICAM-1). Adiponectin induces the secretion of some anti-inflammatory cytokines, such as IL-10 and IL-1RA (receptor antagonist), by human monocytes, macrophages, and dendritic cells and suppress the production of INF-γ (Wolf et al. 2004).

Interestingly, it was observed that the addition of adiponectin results in a significant diminution of antigen-specific T-cell proliferation and cytokines production. A paper by Tsang et al. suggests that the immunomodulatory effect of adiponectin on immune response could be at least partially be mediated by its ability to alter dendritic cell functions (Tsang et al. 2011). Indeed, adiponectin-treated dendritic cells show a lower production of IL-12p40 and a lower expression of CD80, CD86, and histocompatibility complex class II (MHCII). Moreover, in coculture experiments of T cells and adiponectin-treated dendritic cells, a reduction in T-cell proliferation and IL-2 production and a higher percentage of $CD4^+CD25^+Foxp3^+$ Treg cells were observed suggesting that adiponectin could also control regulatory T-cell homeostasis.

Pathological Relevance

Caloric Restriction, Immune Tolerance in Metabolic Dysregulation, and Autoimmunity

It is well established from literature that in more affluent countries, where increased metabolic overload and obesity are more frequent, chronic inflammation, obesity, cancer, and autoimmunity are more common (Symonds et al. 2009). A primary link has been provided by the notion that obesity and type 2 diabetes are now being considered closed to autoimmune/chronic inflammatory disorders rather than classically metabolic/endocrine dysregulation. This because of the presence of abundant immune cell infiltrates in the adipose tissue of obese individuals is considered a classical pathologic lesion in obesity.

Several studies have shown that calorie restriction (CR) without malnutrition prevents many age-associated, chronic diseases and prolongs the lifespan of mammals. Indeed dietary restriction causes metabolic and physiologic changes that have beneficial effects against obesity, insulin resistance, inflammation, oxidative stress, and cardiovascular diseases (Duan et al. 2003).

A recent study by Galgani et al. shows that nutritional status, through leptin, directly affects survival and proliferation of autoreactive T cells, modulating the activity of the survival protein Bcl-2, the Th1/Th17 cytokines secretion, and the nutrient/energy-sensing AKT-mTOR pathway (Galgani et al. 2010). This is in line with the epidemiological evidence that susceptibility to autoimmune diseases, in some circumstances, correlates with increased body fat mass and higher body weight at birth (Munger et al. 2009). Moreover, other studies published by Piccio et al. and our group have shown that either nutritional deprivation (Piccio et al. 2008) and the survival of chronically food-restricted mice is higher than ad libitum-fed mice, suggesting that nutritional and metabolic state influence the break of self-immune tolerance. Similar results have also been obtained in mice, where chronic rapamycin treatment increased significantly their overall survival (Harrison et al. 2009). The precise mechanisms for these results are still not fully elucidated, but it is well known that rapamycin treatment is able to induce pharmacologically a "frugal phenotype" similar to that observed in CR animals. Indeed, rapamycin, through mTOR inhibition, is able to dampen the metabolic overload through reduction of absorption of amino acids and glucose and also to dampen the level of a series of pro-inflammatory adipocytokines produced by adipocytes, including leptin. Recent reports have also confirmed this evidence in mouse models of autoimmunity in which rapamycin has been shown to improve disease curse and progression, particularly EAE and type 1 diabetes, by dampening Th1/Th17 responses and increasing regulatory T-cell responses (Donia et al. 2009; Esposito et al. 2010). The fact that chronic leptin deficiency (ob/ob mice) or rapamycin-induced leptin deficiency can reduce the survival of autoreactive $CD4^+$ T cells indicates that the nutritional status can control survival of potentially autoreactive $CD4^+$ T cells – through leptin/mTOR. Drugs that target nutrient-sensing pathways to obtain the health benefits of dietary restriction are realistic, and notably, considering that nutritional deprivation or CR reduces EAE, it could be suggested that manipulation of the leptin axis and in general of the nutritional status could represent a new means to modulate T-cell tolerance in autoimmunity.

Interactions with Other Processes and Drugs

Drugs That Affect Metabolism Have an Impact on Self-Immune Tolerance

Several pharmacologic compounds that affect glucose and cholesterol metabolism and adipocyte development also have immunomodulatory activities. These effects can be seen for autoreactive T cells and pro-inflammatory immune responses in obesity, insulin resistance, atherosclerosis, and autoimmunity. Current strategies to treat obesity-related metabolic disturbances (such as insulin resistance and glucose tolerance) utilize drugs that were considered to only act on metabolism but are now known to also downregulate immune responses. Examples include metformin, thiazolidinediones, and statins, which are, at different levels, capable of reducing pro-inflammatory cytokines and leptin secretion. For example, metformin (the most widely used drug for T2D that mediates its action through the activation of the AMPK) has anti-inflammatory properties and inhibits T-cell-mediated immune responses and the production of Th1 or Th17 cytokines, while inducing generation of IL-10-secreting Treg cells (Nath et al. 2009). Another example is represented by thiazolidinediones, which are potent activators of the transcription factor peroxisome–proliferator-activating receptor (PPAR)-γ and are key players in the development and differentiation of the adipose tissue. PPAR-agonists downmodulate the transcription of pro-inflammatory cytokines such as leptin, TNF-α, and IL-6. This class of drugs is utilized in the treatment of insulin resistance in obese and T2D patients for its capacity to exert insulin-sensitizing actions (Diab et al. 2002). In EAE, these drugs are able to stop disease progression and neural inflammation. Finally, statins – inhibitors of the 3-hydroxy-3-methyl-glutaryl-CoA (HMG-CoA)-reductase – inhibit cholesterol synthesis and thus lower cholesterol and also dampen autoreactive immune responses by promoting the release of Th2 cytokines and by reducing pro-inflammatory cytokines such as leptin and TNF-α (Zhang and Markovic-Plese 2008). Statins are widely utilized in T2D and insulin resistance and in EAE have been found highly effective as therapeutic agents. Overall, drugs that have long been considered to affect metabolism and insulin actions are also immune modulators that reduce pro-inflammatory cytokines and increase the number and function of Treg and Th2/regulatory-type cytokine release, all important in the control of immune tolerance and autoimmunity.

Concluding Remarks

It is clear that there is a strong relationship between obesity and chronic inflammation. The precise temporal/sequential development of obesity and inflammation are unclear. The recognition of some of the obesity-related alterations as an immune-based dysregulation leading to autoimmunity against adipose tissue represents a novel and intriguing vision of the pathogenic processes resulting in the most common form of obesity. They also indicate that type 2 diabetes is not just directly linked to single-gene mutations leading to obesity. A question is still open: how can adipose tissue infiltration lead to a precise behavioral change in food intake and basal metabolism so that obese individuals are totally or partially unable to control food intake? Based on the hypothesis that an autoimmune process is present in obesity, pharmacological manipulation leading to immune-downmodulation before starting any nutritional/pharmacological intervention could represent a completely novel approach to treat obesity and its metabolic consequences.

In this context, many compelling questions remain to be addressed in the next few years: (1) Which hypothalamic cell type triggers inflammation in response to high-fat feeding? (2) Are neuronal microglia present in the hypothalamus participating these processes? (3) Do genetic factors that affect obesity susceptibility act by modifying hypothalamic inflammation? (4) What is the specific link between hypothalamic inflammation, leptin resistance, and obesity? Answers to these questions may help to inform new approaches to the prevention and treatment of obesity.

Acknowledgments G.M. is supported by grants from the EU Ideas Programme, ERC-Starting Independent Grant "menTORingTregs" n. 310496, the Fondazione Italiana Sclerosi Multipla (FISM) n. 2012/R/11 and Medicina Personalizzata CNR Grant.

V.D.R. is supported by the Ministero della Salute grant n. GR-2010-2315414 and the Fondo per gli Investimenti della Ricerca di Base (FIRB) grant n. RBFR12I3UB_004.

The authors wish to thank Dr. Fortunata Carbone for the artwork and critically reading of the manuscript. This work is dedicated to the memory of Eugenia Papa and Serafino Zappacosta.

Cross-References

▶ Leukocyte Recruitment
▶ Modulation of Inflammation by Key Nutrients

References

Cancello, R., Henegar, C., Viguerie, N., Taleb, S., Poitou, C., Rouault, C., et al. (2005). Reduction of macrophage infiltration and chemoattractant gene expression changes in white adipose tissue of morbidly obese subjects after surgery-induced weight loss. *Diabetes, 54*, 2277–2286.

Cinti, S., Mitchell, G., Barbatelli, G., Murano, I., Ceresi, E., Faloia, E., et al. (2005). Adipocyte death defines macrophage localization and function in adipose tissue of obese mice and humans. *Journal of Lipid Research, 46*, 2347–2355.

Clément, K., Vaisse, C., Lahlou, N., Cabrol, S., Pelloux, V., Cassuto, D., et al. (1998). A mutation in the human leptin receptor gene causes obesity and pituitary dysfunction. *Nature, 392*, 398–401.

Clement, K., Viguerie, N., Poitou, C., Carette, C., Pelloux, V., Curat, C. A., et al. (2004). Weight loss regulates inflammation-related genes in white adipose tissue of obese subjects. *FASEB Journal, 18*, 1657.

De Rosa, V., Procaccini, C., La Cava, A., Chieffi, P., Nicoletti, G. F., Fontana, S., et al. (2006). Leptin neutralization interferes with pathogenic T cell autoreactivity in autoimmune encephalomyelitis. *Journal of Clinical Investigation, 116*, 447–455.

De Rosa, V., Procaccini, C., Calì, G., Pirozzi, G., Fontana, S., Zappacosta, S., et al. (2007). A key role of leptin in the control of regulatory T cell proliferation. *Immunity, 26*, 241–255.

Diab, A., Deng, C., Smith, J. D., Hussain, R. Z., Phanavanh, B., Lovett-Racke, A. E., et al. (2002). Peroxisome proliferator-activated receptor-γ agonist 15-deoxy-Δ12,14-prostaglandin J2 ameliorates experimental autoimmune encephalomyelitis. *Journal of Immunology, 168*, 2508–2515.

Dietrich, M. O., & Horvath, T. L. (2009). Feeding signals and brain circuitry. *European Journal of Neuroscience, 30*, 1688–1696.

Dixit, V. D. (2008). Adipose-immune interactions during obesity and caloric restriction: Reciprocal mechanisms regulating immunity and health span. *Journal of Leukocyte Biology, 84*, 882–892.

Donia, M., Mangano, K., Amoroso, A., Mazzarino, M. C., Imbesi, R., Castrogiovanni, P., et al. (2009). Treatment with rapamycin ameliorates clinical and histological signs of protracted relapsing experimental allergic encephalomyelitis in Dark Agouti rats and induces expansion of peripheral CD4$^+$CD25$^+$Foxp3$^+$ regulatory T cells. *Journal of Autoimmunity, 33*, 135–140.

Duan, W., Guo, Z., Jiang, H., Ware, M., & Mattson, M. P. (2003). Reversal of behavioral and metabolic abnormalities, and insulin resistance syndrome, by dietary restriction in mice deficient in brain-derived neurotrophic factor. *Endocrinology, 144*, 2446–2453.

Elhage, R., Jawien, J., Rudling, M., Ljunggren, H. G., Takeda, K., Akira, S., et al. (2003). Reduced atherosclerosis in interleukin-18 deficient apolipoprotein E-knockout mice. *Cardiovascular Research, 59*, 234–240.

Esposito, M., Ruffini, F., Bellone, M., Gagliani, N., Battaglia, M., Martino, G., et al. (2010). Rapamycin inhibits relapsing experimental autoimmune encephalomyelitis by both effector and regulatory T cells modulation. *Journal of Neuroimmunology, 220*, 52–63.

Fessler, M. B., Rudel, L. L., & Brown, J. M. (2009). Toll-like receptor signaling links dietary fatty acids to the metabolic syndrome. *Current Opinion in Lipidology, 20*, 379–385.

Feuerer, M., Herrero, L., Cipolletta, D., Naaz, A., Wong, J., Nayer, A., et al. (2009). Lean, but not obese, fat is enriched for a unique population of regulatory T cells that affect metabolic parameters. *Nature Medicine, 15*, 930–399.

Fried, S. K., Bunkin, D. A., & Greenberg, A. S. (1998). Omental and subcutaneous adipose tissues of obese subjects release interleukin-6: Depot difference and regulation by glucocorticoid. *Journal of Clinical Endocrinology and Metabolism, 83*, 847–850.

Friedman, J. M. (2009). Obesity: Causes and control of excess body fat. *Nature, 21*, 340–342.

Friedman, J. M., & Halaas, J. L. (1998). Leptin and the regulation of body weight in mammals. *Nature, 395*, 763–770.

Galgani, M., Procaccini, C., De Rosa, V., Carbone, F., Chieffi, P., La Cava, A., et al. (2010). Leptin modulates the survival of autoreactive CD4+ T cells through the nutrient/energy-sensing mammalian target of rapamycin signaling pathway. *Journal of Immunology, 185*, 7474–7479.

Gao, Q., & Horvath, T. L. (2008). Neuronal control of energy homeostasis. *FEBS Letters, 9*, 132–141.

Gordon, S. (2003). Alternative activation of macrophages. *Nature Reviews Immunology, 3*, 23–35.

Halberg, N., Wernstedt-Asterholm, I., & Scherer, P. E. (2008). The adipocyte as an endocrine cell. *Endocrinology Metabolism Clinics of North America, 37*(3), 753–768.

Harrison, D. E., Strong, R., Sharp, Z. D., Nelson, J. F., Astle, C. M., Flurkey, K., et al. (2009). Rapamycin fed late in life extends lifespan in genetically heterogeneous mice. *Nature, 460*, 392–395.

Horvath, T. L., & Bruning, J. C. (2006). Developmental programming of the hypothalamus: A matter of fat. *Nature Medicine, 12*, 52–53.

Horvath, T. L., Andrews, Z. B., & Diano, S. (2009). Fuel utilization by hypothalamic neurons: Roles for ROS. *Trends in Endocrinology and Metabolism, 20*, 78–87.

Hotamisligil, G. S. (2006). Inflammation and metabolic disorders. *Nature, 14*, 860–867.

Hotamisligil, G. S. (2008). Inflammation and endoplasmic reticulum stress in obesity and diabetes. *International Journal of Obesity, 32*, 52–54.

Hotamisligil, G. S., & Spiegelman, B. M. (1994). Tumor necrosis factor alpha: A key component of the obesity-diabetes link. *Diabetes, 43*, 1271–1278.

Hotamisligil, G. S., Shargill, N. S., & Spiegelman, B. M. (1993). Adipose expression of tumor necrosis factor-a: Direct role in obesity-linked insulin resistance. *Science, 259*, 87–91.

Kosteli, A., Sugaru, E., Haemmerle, G., Martin, J. F., Lei, J., Zechner, R., et al. (2010). Weight loss and lipolysis promote a dynamic immune response in murine adipose tissue. *Journal of Clinical Investigation, 120*, 3466–3479.

Krude, H., Biebermann, H., Luck, W., Horn, R., Brabant, G., & Grüters, A. (1998). Severe early-onset obesity, adrenal insufficiency and red hair pigmentation caused by POMC mutations in humans. *Nature Genetics, 19*, 155–157.

La Cava, A., & Matarese, G. (2004). The weight of leptin in immunity. *Nature Reviews Immunology, 4*, 371–379.

Liu, J., Divoux, A., Sun, J., Zhang, J., Clément, K., Glickman, J. N., et al. (2009). Genetic deficiency and pharmacological stabilization of mast cells reduce diet-induced obesity and diabetes in mice. *Nature Medicine, 15*, 940–945.

Lord, G. M., Matarese, G., Howard, J. K., Baker, R. J., Bloom, S. R., & Lechler, R. I. (1998). Leptin modulates the T-cell immune response and reverses starvation-induced immunosuppression. *Nature, 394*, 897–901.

Lumeng, C. N., Bodzin, J. L., & Saltiel, A. R. (2007). Obesity induces a phenotypic switch in adipose tissue macrophage polarization. *Journal of Clinical Investigation, 117*, 175–184.

Lumeng, C. N., Maillard, I., & Saltiel, A. R. (2009). T-ing up inflammation in fat. *Nature Medicine, 15*, 846–847.

Mallat, Z., Corbaz, A., Scoazec, A., Besnard, S., Lesèche, G., Chvatchko, Y., et al. (2001). Expression of interleukin-18 in human atherosclerotic plaques and relation to plaque instability. *Circulation, 104*, 1598–1603.

Montague, C. T., Farooqi, I. S., Whitehead, J. P., Soos, M. A., Rau, H., Wareham, N. J., et al. (1997). Congenital leptin deficiency is associated with severe early-onset obesity in humans. *Nature, 387*, 903–908.

Munger, K. L., Chitnis, T., & Ascherio, A. (2009). Body size and risk of MS in two cohorts of US women. *Neurology, 73*, 1543–1550.

Nath, N., Khan, M., Paintlia, M. K., Singh, I., Hoda, M. N., & Giri, S. (2009). Metformin attenuated the autoimmune disease of the central nervous system in animal models of multiple sclerosis. *Journal of Immunology, 182*, 8005–8014.

Nishimura, S., Manabe, I., Nagasaki, M., Eto, K., Yamashita, H., Ohsugi, M., et al. (2009). CD8+ effector T cells contribute to macrophage recruitment and adipose tissue inflammation in obesity. *Nature Medicine, 15*, 914–920.

O'Rahilly, S. (2009). Human genetics illuminates the paths to metabolic disease. *Nature, 19*, 307–314.

Ouchi, N., & Walsh, K. (2007). Adiponectin as an anti-inflammatory factor. *Clinica Chimica Acta, 380*, 24–30.

Piccio, L., Stark, J. L., & Cross, A. H. (2008). Chronic calorie restriction attenuates experimental autoimmune encephalomyelitis. *Journal of Leukocyte Biology, 8*, 940–948.

Procaccini, C., De Rosa, V., Galgani, M., Abanni, L., Calì, G., Porcellini, A., et al. (2010). An oscillatory switch in mTOR kinase activity sets regulatory T cell responsiveness. *Immunity, 33*, 929–941.

Procaccini, C., Jirillo, E., & Matarese, G. (2012a). Leptin as an immunomodulator. *Molecular Aspects of Medicine, 33*, 35–45.

Procaccini, C., Galgani, M., De Rosa, V., & Matarese, G. (2012b). Intracellular metabolic pathways control immune tolerance Trends. *Immunology, 33*, 1–7.

Procaccini, C., De Rosa, V., Galgani, M., Carbone, F., Cassano, S., Greco, D., et al. (2012c). Leptin-induced mTOR activation defines a specific molecular and transcriptional signature controlling CD4$^+$ effector T cell responses. *Journal of Immunology, 189*, 2941–2953.

Radetti, G., Kleon, W., Buzi, F., Crivellaro, C., Pappalardo, L., Di Lorgi, N., et al. (2008). Thyroid function and structure are affected in childhood obesity. *Journal of Clinical Endocrinology and Metabolism, 93*, 4749–4754.

Senn, J. J., Klover, P. J., Nowak, I. A., Zimmers, T. A., Koniaris, L. G., Furlanetto, R. W., et al. (2003). Suppressor of cytokine signaling-3 (SOCS-3), a potential mediator of interleukin-6-N dependent insulin resistance in hepatocytes. *Journal of Biological Chemistry, 278*, 13740–13746.

Soukas, A., Cohen, P., Socci, N. D., & Friedman, J. M. (2000). Leptin-specific patterns of gene expression in white adipose tissue. *Genes and Development, 14*, 963.

Sterry, W., Strober, B. E., & Menter, A. (2007). Obesity in psoriasis: The metabolic, clinical and therapeutic implications. *British Journal of Dermatology, 157*, 649–655.

Symonds, M. E., Sebert, S. P., Hyatt, M. A., & Budge, H. (2009). Nutritional programming of the metabolic syndrome. *Nature Reviews Endocrinology, 5*, 604–610.

Totonoz, P., Nagy, L., Alvarez, J. G., Thomazy, V. A., & Evans, R. M. (1998). PPARgamma promotes monocyte/macrophage differentiation and uptake of oxidized LDL. *Cell, 93*, 241.

Tsang, J. Y., Li, D., Ho, D., Peng, J., Xu, A., Lamb, J., et al. (2011). Novel immunomodulatory effects of adiponectin on dendritic cell functions. *International Immunopharmacology, 11*, 604–609.

Vaisse, C., Clement, K., Guy-Grand, B., & Froguel, P. (1998). A frameshift mutation in human MC4R is associated with a dominant form of obesity. *Nature Genetics, 20*, 113–114.

Van Dielen, F. M., Buurman, W. A., Hadfoune, M., Nijhuis, J., & Greven, J. W. (2004). Macrophage inhibitory factor, plasminogen activator inhibitor-1, other acute phase proteins, and inflammatory mediators normalize as a result of weight loss in morbidly obese subjects treated with gastric restrictive surgery. *Journal of Clinical Endocrinology and Metabolism, 89*, 4062.

Weisberg, S. P., McCann, D., Desai, M., Rosenbaum, M., Leibel, R. L., & Ferrante, A. W., Jr. (2003). Obesity is associated with macrophage accumulation in adipose tissue. *Journal of Clinical Investigation, 112*, 1796–1808.

Winer, S., Chan, Y., Paltser, G., Truong, D., Tsui, H., Bahrami, J., et al. (2009). Normalization of obesity-associated insulin resistance through immunotherapy. *Nature Medicine, 15*, 921–929.

Wolf, A. M., Wolf, A. D., Rumpold, H., Enrich, B., & Tilg, H. (2004). Adiponectin induces the anti-inflammatory cytokines IL-10 and IL-1RA in human leukocytes. *Biochemical and Biophysical Research Communications, 323*, 630–635.

Wood, I. S., Wang, B., Jenkins, J. R., & Trayhurn, P. (2005). The pro-inflammatory cytokine IL-18 is expressed in human adipose tissue and strongly upregulated by TNFα in human adipocytes. *Biochemical and Biophysical Research Communications, 337*, 422–429.

Yeo, G. S., Farooqi, I. S., Aminian, S., Halsall, D. J., Stanhope, R. G., & O'Rahilly, S. (1998). A frameshift mutation in MC4R associated with dominantly inherited human obesity. *Nature Genetics, 20*, 111–112.

Zhang, X., & Markovic-Plese, S. (2008). Statins' immunomodulatory potential against Th17 cell-mediated autoimmune response. *Immunologic Research, 41*, 165–174.

Zhang, Y., Proenca, R., Maffei, M., Barone, M., Leopold, L., & Friedman, J. M. (1994). Positional cloning of the mouse obese gene and its human homologue. *Nature, 372*, 425.

Osteoarthritis

Anne-Marie Malfait and Joel A. Block
Division of Rheumatology, Department of Internal Medicine, Rush University Medical Center, Chicago, IL, USA

Synonyms

Degenerative joint disease; Osteoarthrosis

Definition

Unlike most forms of arthritis, osteoarthritis (OA) is not primarily an inflammatory disease. While local inflammation of involved joints is common, the pathophysiology of OA is primarily degenerative, and there is minimal or no systemic inflammation. OA is a highly heterogeneous disease process that is characterized clinically by pain and functional loss.

OA often remains undertreated because many physicians still assume that it is a "normal" part of aging. Nonetheless, it results in vast direct societal costs and significant loss of work. It is the leading indication for total joint replacement and is a major cause of work disability. Throughout much of the twentieth century, OA was considered to be primarily a degenerative disease of articular cartilage. As our understanding of pathophysiology has progressed, however, it has become clear that the OA disease process involves the entire joint. A formal definition of OA is that it is a *painful* degenerative process involving progressive deterioration of all joint structures and remodeling of subchondral bone (SCB) and which is not primarily inflammatory (Block and Scanzello 2015). It is important to distinguish true OA from asymptomatic structural degeneration of joints, which is universal during normal aging.

Epidemiology and Genetics

OA is overwhelmingly the most prevalent form of arthritis and one of the most common diagnoses in clinical practice (Neogi and Zhang 2013). Since it is primarily a disease of aging, its prevalence will rise dramatically in the coming decades as global populations age. In large epidemiologic studies, OA is often defined based on radiographic assessment. Most people have radiographic evidence of OA in at least one joint, and cartilage lesions are detectable in all knee joints at time of autopsy by the eighth decade (Loeser 2000). The presence of both radiographic OA and symptoms (pain, stiffness) in the same joint indicates symptomatic OA. The lifetime risk of developing symptomatic knee OA is estimated to be ~45 % (40 % in men and 47 % in women) based upon Johnston County Osteoarthritis Project data, with risks increasing to 60.5 % among obese persons, which is approximately double the risk of normal-weight subjects (Neogi 2013).

Risk factors for OA include age, sex, and obesity (Neogi and Zhang 2013). *Female sex* is associated with higher incidence and severity of OA, particularly after menopause. Females develop more hand, foot, and knee OA than men. *Age* is one of the strongest predictors of OA development, and after the age of 50, radiographic changes in the joint become more common while joints often start to hurt (Loeser 2013). Worldwide, OA is a leading cause of disability among older adults. *Obesity*, rapidly becoming a global public health concern, significantly increases the risk of incident hip and knee OA, particularly in women, and constitutes a major risk for its radiographic progression. Meta-analyses suggest that a dose–response relationship exists between obesity and the risk for knee OA. For every 5-unit increase in body mass index (BMI), there is an associated 35 % increased risk of knee OA (Jiang et al. 2012), and this relationship is stronger in women.

Multiple *joint-level risk factors* may predispose to OA. *Repetitive joint (over) use* is associated with increased incidence of OA in that joint; a classical example is OA of the wrist and elbow in pneumatic drill operators. *Joint injury* presents a major risk for developing OA. This has been best documented in the knee, where meniscal tears and anterior cruciate ligament injuries lead to OA (Riordan et al. 2014). *Knee malalignment* is one of the strongest predictors of OA progression. Lastly, *joint shape/anatomy* may contribute to OA risk, as has been described for hip OA, where mild acetabular dysplasia predisposes to OA.

Genetics

Congenital skeletal dysplasias result from mutations in genes encoding major cartilage molecules, including collagens, aggrecan, and matrilin 3, and often lead to early-onset and severe secondary OA. Risk for primary OA is complex and multifactorial and includes a strong heritable component, with genetic factors estimated to account for about 50 % of the risk for developing some forms of OA, such as generalized OA. In recent years, a wealth of candidate gene studies and genome-wide association scans in large populations have identified specific genes associated with OA risk. It has become clear that no individual loci contribute substantial risk, but rather, OA susceptibility is determined by alleles with small effects (with odds ratios typically below 1.2). Currently available data show that genetic factors are often joint specific and ethnicity dependent. They manifest through multiple biological pathways, particularly pathways involved in endochondral ossification and chondrogenesis. Susceptibility genes do not typically encode structural proteins, but rather factors involved in transcriptional regulation (Reynard and Loughlin 2013).

The intensity and quality of OA pain vary widely between subjects (Neogi 2013). Variations in pain may be genetically determined, and several association studies in patients with painful OA have been reported (Malfait and Schnitzer 2013), though polymorphisms in *TRPV1* and *PCSK6* are the only associations that have been reproduced in multiple cohorts to date.

Pathophysiology

The Normal Joint

The synovial joint operates as an organ, its proper function enabled by extensive mechanical and biochemical cross talk between different joint tissues (Fig. 1). Healthy articular cartilage transmits applied load across joint surfaces and allows smooth articulation with low friction. This low-friction load-bearing function requires the cartilage extracellular matrix (ECM) to have exceptional tensile strength, elasticity, and resistance to compressive as well as shear forces. These properties are endowed by its unique macromolecular organization: A network of type II collagen fibers entraps very large aggregates of the proteoglycan, aggrecan, a negatively charged macromolecule that avidly retains water. Cartilage ECM is synthesized and maintained by the resident cells, chondrocytes; in general, turnover of ECM is very slow. Underlying the articular cartilage is the SCB, which, being much softer than the cortical bone, is highly susceptible to deformation and serves as a major shock absorber, thereby protecting the articular cartilage from damage. The synovial lining consists of a thin layer of cells of fibroblast or macrophage lineage overlying a vascularized stroma. The synovium is the source for synovial fluid components such as hyaluronan and provides blood supply for the avascular cartilage. Finally, ligaments, tendons, and periarticular muscles stabilize the joint, and the latter play a crucial role in handling mechanical load (for instance, the quadriceps muscle in the knee). Except for cartilage, which is aneural, all joint tissues are abundantly innervated. The sensory innervation of the joint consists mostly of proprioceptors and nociceptors, indicating the vital importance of joint position sense and awareness of injurious movement to joint health.

The Osteoarthritic Joint

OA represents the failure of the synovial joint as an organ (Fig. 1). Radiographically, this is manifested as joint space narrowing due to cartilage loss, SCB sclerosis, sometimes bone cysts, and osteophytes (bony spurs at the joint margin) (Fig. 2). Pathophysiologically, radiographic OA constitutes the late

Osteoarthritis, Fig. 1 Schematic drawing of a normal (*left*) and OA knee joint (*right*). *Insets* show histology of the different joint tissues, healthy and osteoarthritic. The schematic representation depicts cartilage degradation in the weight-bearing area of the joint, bone remodeling, osteochondral channels (*red*), accumulation of cartilage matrix degradation products, and mononuclear cells in the synovial fluid. Histology shows (*a*) normal synovium vs. (*b*) OA synovium showing hyperplasia of the lining, inflammatory cell infiltration and fibrosis; (*c*) normal meniscus vs. (*d*) OA meniscal changes with tears and loss of proteoglycan (*blue staining*); (*e*) normal joint margin vs. (*f*) chondrophytes/osteophytes (*arrows*); (*g*) toluidine *blue* staining of normal cartilage (stains the proteoglycans *dark blue*) vs. (*h*) loss of proteoglycans, fibrillation of the cartilage surface (due to collagen degradation), chondrocyte cloning and cell death, and advancement of the tidemark (the zone between the articular and the calcified cartilage); (*i*) normal subchondral bone vs. (*l*) subchondral bone sclerosis; and (*j*) normal deeper trabecular bone and (*k*) remodeling (Adapted from Little, C. B. & Hunter, D. J. Nature Reviews Rheumatology, 2013)

stage of a joint remodeling process that has often been going on for many years. This process affects all synovial joint tissues, including articular cartilage, SCB, synovium, ligaments, periarticular muscles, peripheral nerves, and in the knee also the menisci. Imaging modalities that can visualize soft tissues reveal that subjects may have early osteochondral damage and meniscal tears that are asymptomatic but represent risk factors for progressive OA.

Cartilage – Progressive loss of articular cartilage through enzymatic degradation of the ECM is a hallmark of OA pathology. Early on, chondrocytes become very active and assume a catabolic and proinflammatory phenotype. They produce an abundance of proteases (Miller et al. 2013), most notably the metalloproteinases, ADAMTS-4, and ADAMTS-5, which degrade aggrecan and MMP13, the major type II collagen-degrading enzyme. Proinflammatory

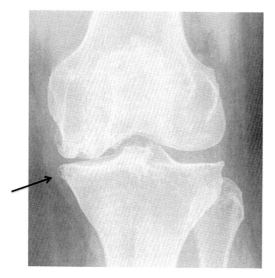

Osteoarthritis, Fig. 2 Radiograph of a knee with osteoarthritis in the medial compartment (*left*). The radiograph shows the characteristics of radiographic OA: joint space narrowing, subchondral bone sclerosis, and osteophyte formation (*arrow*)

cytokines, including IL-1 and TNFα, further boost the catabolic phenotype of chondrocytes, while simultaneously suppressing synthesis of matrix molecules, thus promoting net cartilage loss. Early loss of proteoglycans and damage to the collagen network result in cartilage swelling. Cartilage becomes softer and less resistant to damage, and this perpetuates a vicious cycle. Eventually, chondrocytes undergo apoptosis, and cartilage damage becomes irreversible.

Bone – The OA joint is characterized by marked changes in periarticular bone (Goldring 2012). Increased SCB turnover results in sclerosis, which diminishes the capacity of the bone to absorb energy upon loading. Repetitive loading can cause microdamage in the SCB, with foci of high bone remodeling that likely manifest on magnetic resonance imaging (MRI) as bone marrow lesions. The zone of calcified cartilage that forms the boundary between articular cartilage and SCB displays alterations, including the formation of osteochondral channels extending from the SCB and bone marrow into the cartilage. These channels carry blood vessels and sensory nerves, which may contribute to joint pain (Suri and Walsh 2012).

Osteophytes are pathognomonic for structural OA and may serve to stabilize the joint. They result from an endochondral ossification process that recapitulates the cellular mechanisms of skeletal development.

Synovium – OA is frequently associated with changes in the synovium that are indicative of low-grade synovitis, including hyperplasia, immune cell infiltration, increased vascularization, and fibrosis (Scanzello and Goldring 2012). Synovitis can be detectable by MRI and, in knee OA, its presence is associated with pain and may be predictive of faster rates of cartilage loss.

Meniscus – Reportedly, 82 % of subjects with radiographic knee OA have meniscal damage, mostly degenerative lesions, and 91 % of subjects with symptomatic knee OA have a meniscal tear. While meniscal injuries are clearly a predisposing factor for OA, knee OA itself may also damage the meniscus (Johnson and Hunter 2014).

Periarticular muscles – Muscle weakness, atrophy, and proprioception deficits are common in patients with knee OA. These are considered a consequence of disuse in order to avoid pain, but quadriceps weakness can also exist in subjects with radiographic OA and no pain. Muscle weakness leads to gait changes and altered joint loading, and this may contribute to progressive joint damage (Brandt et al. 2009).

The Role of Biomechanics

Altered biomechanics represent a major drive for OA onset and progression (Felson 2013). Abnormal loading is a major risk factor for incident OA, as is evident in the high incidence of OA in malaligned knees. For instance, knee OA is most common in the medial compartment, but subjects with valgus knees develop OA in the lateral compartment. As OA progresses, worsening cartilage damage, bone remodeling and muscle weakness magnify pathological loading, thus fuelling a vicious cycle of disease progression.

The Role of Inflammation

Local inflammation is increasingly recognized as a driver of OA pathology. Proinflammatory cytokines and chemokines are detectable in OA synovial fluid, and all joint cells can produce and/or respond to these mediators (Goldring and Otero 2011). Low-grade synovitis produces mediators that promote cartilage catabolism. Chondrocytes themselves are also a major source of proinflammatory mediators, including IL-1, TNFα, IL-6, a range of chemokines, nitric oxide, and reactive oxygen species, all of which further trigger release and activation of catabolic enzymes. Biomechanical stress can directly promote an inflammatory and catabolic chondrocyte phenotype, largely regulated by the NF-κB signaling pathway (Goldring and Otero 2011). Matrix degradation and cellular stress give rise to disease-associated molecular patterns (DAMPs) that cause an innate immune response within the joint: S100 alarmins and matrix fragments (e.g., fibronectin fragments, tenascin C, and hyaluronan) signal through pattern recognition receptors, such as toll-like receptors, on chondrocytes to further promote a proinflammatory and catabolic phenotype. Thus, a vicious cycle ensues (Liu-Bryan 2013).

The Role of Other Pathogenic Factors

Age – Aging causes well-documented changes in bone and muscle, but it also causes marked alterations in articular cartilage. These include chondrocyte senescence with abnormal secretory profiles, alterations in the cartilage matrix such as accumulation of advanced glycation end products, reduced thickness of cartilage, and calcium deposition (Loeser 2013). DNA methylation, histone acetylation and methylation, and microRNAs are increasingly studied as means of epigenetic regulation of changes in gene expression associated with aging joint tissues. Age-related changes also occur in meniscus and ligaments. These result in a more vulnerable joint with a diminished capacity to adapt to biomechanical challenges.

Obesity – The contribution of obesity to OA pathogenesis is multifactorial (Thijssen et al. 2015). First, in weight-bearing joints such as the knee and hip, excess body weight increases harmful load on the joint. Secondly, metabolic and inflammatory effects associated with obesity may contribute to OA pathogenesis, which would explain why obesity is also a risk factor for hand OA. Obesity-associated raises in levels of cytokines, chemokines, and adipose-tissue derived mediators like adiponectin and leptin (adipokines) may affect joint tissues. In the knee, the fat pad may also act as a local source of inflammatory mediators.

Female sex – The observation that women often develop symptomatic and more severe radiographic OA after menopause has led to the hypothesis that estrogen plays a pathogenic role, but this remains poorly understood. Other contributing factors are lower strength of bone and muscle and different joint alignment in women.

Mechanisms of OA Pain

Insight into the cellular and molecular mechanisms of OA pain remains limited. Current evidence supports the notion that pain is maintained through continuous nociceptive peripheral input from the OA joint (Malfait and Schnitzer 2013). Mechanical stimuli as well as locally generated inflammatory mediators, including cytokines and chemokines, may contribute to sensitization of joint nociceptors (Miller et al. 2014). Pain may originate from many articular tissues, and a growing literature exploring the relationship between structural joint changes and pain suggests a relation of bone marrow lesions (detected by MRI) and synovitis to pain (Hunter et al. 2013). Nerve growth factor (NGF) has been found paramount to OA pain, and ongoing clinical trials with agents that block NGF show great promise (Schnitzer and Marks 2015).

Clinical investigators have reported signs of central sensitization in some OA patients, and systematic quantitative sensory testing reveals that OA patients display reduced pressure pain thresholds in both affected and unaffected sites (Suokas et al. 2012).

The Heterogeneity of OA

Symptomatic OA is the result of a complex interplay between pathogenic factors in different tissues that contribute to progressive cellular, biochemical, and mechanical changes in the joint. Rather than being a single disorder, OA can be stratified into several overlapping phenotypes depending on the cause (e.g., injury, age, obesity), joint, tissue involvement, and symptomatology (Driban et al. 2010). This complexity in pathogenesis is clinically relevant, since it may affect response to treatment.

Clinical Presentation of Osteoarthritis

As is the case with most musculoskeletal diseases, pain is the most common presenting feature of OA. Some patients may initially have cosmetic complaints, especially bony changes of the small joints in the hands such as prominences at the proximal interphalangeal (Bouchard's nodes) or distal interphalangeal (Heberden nodes) joints (Fig. 3) or of functional loss; however, the vast majority of early complaints are of pain in the affected joints, especially during loading. Importantly, OA is generally a very slowly progressive disease; hence, joint pain may precede radiographic changes by several years.

In contrast to other diseases, the diagnosis of OA depends largely on the clinical presentation. As OA is considered to be a primary disease of the joints, rather than a systemic condition, laboratory and imaging modalities are predominantly useful to exclude competing diagnoses and are generally not greatly helpful in establishing an OA diagnosis.

Symptomatology

Pain is the feature that characterizes clinical OA. While degenerative joint changes are almost universal during normal aging, these changes of so-called "structural OA" are asymptomatic in most people. It is the onset of pain that distinguishes the clinical disease. Although it is often limited to the affected joints, pain can become widespread over time (Neogi 2013). The quality and severity of OA pain is variable, ranging from "aching" joint pain to less localized periarticular or referred pain. In early OA, pain is felt principally during activity or with usage of the involved joints; as the disease advances, however, pain at rest may become severe and can interrupt sleep. In addition to pain, functional loss is common, which may be manifest by limitation of joint motion, locking, a grinding feeling with joint motion, or a sensation of joint instability. Finally, patients may notice esthetic changes, typically in the fingers and feet, but osteophytes at the knees may also be visible. In contrast to inflammatory arthritis, there is minimal morning stiffness, and if present it lasts less than 30 min.

OA Patterns

The pattern of involved joints may offer clues to distinguish OA from other forms of arthritis.

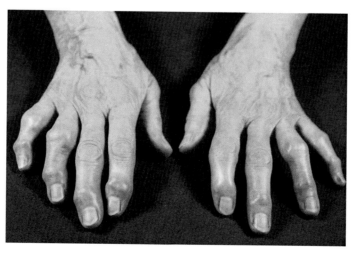

Osteoarthritis, Fig. 3 Bony enlargement can be seen in distal and proximal interphalangeal joints. The changes in proximal interphalangeal joints (Bouchard's nodes) and distal interphalangeal joints (Heberden's nodes) are common findings in degenerative joint disease of the hands. These changes are more frequently found in women after menopause and often show a genetic predisposition (© American College of Rheumatology)

While OA may be monoarticular, it is more typically bilateral and affects multiple joints. Classically, some joints that are commonly affected by the inflammatory arthritides are only rarely involved in primary (not posttraumatic) OA, such as the elbows and ankles, whereas the distal interphalangeal joints of the fingers are commonly affected in OA but rarely in rheumatoid arthritis or lupus.

Lower Extremities

Involvement of the large weight-bearing joints is common; knee OA is substantially more prevalent than hip OA and represents the most common indication for joint replacement surgery. Knee OA may involve any or all of the joint compartments. The medial knee compartment is involved in the majority of knee OA patients and may lead to varus (bow-legged) deformity, whereas OA of the lateral compartment may lead to valgus (knock-kneed) deformity. Patellofemoral compartment OA has been underappreciated but may cause severe pain exacerbated by walking stairs. OA of the weight-bearing joints is frequently accompanied by noninflammatory synovial effusions. In knee OA, there may be synovial cysts and anserine or prepatellar bursitis. In the case of hip OA, pain is typically felt in the groin and may radiate to the anterior thigh or to the knee. Thigh pain may be confused with nonarticular pain caused by trochanteric bursitis or radicular pain from the lumbar spine, and referred pain to the knee may be confused with primary knee pain. Early hip OA is characterized by restricted internal rotation and may progress to limited motion in general and to limb length discrepancy. In the foot, OA typically involves the first MTP, resulting in a bunion deformity, though mid-foot pain is increasingly recognized.

Upper Extremities

The most frequently affected joints of the upper extremities are the small joints of the hands, specifically the distal and proximal interphalangeal joints (DIPs and PIPs, respectively) and first carpometacarpal (CMC) joints. Palpable osteophytes of the DIPs (Heberden's nodes) and of the PIPs (Bouchard's nodes) are often the initial sign of hand OA (Fig. 3) and may be otherwise asymptomatic though these joints may have periods of painful flares. This pattern of hand OA is most common in Caucasian women and is termed "primary generalized OA" or "nodal OA." Involvement of the first CMC joint is usually symptomatic, resulting in difficulty in grasping, opening jars, buttoning clothes, and turning doorknobs. *Erosive* or *inflammatory* OA is a less common but distinct subset in which erosions develop in the DIPs and PIPs and the patient experiences repeated episodes of acute inflammatory symptoms (Anandarajah 2010).

Spine

Spinal OA typically involves the lumbar and cervical regions, affecting the apophyseal (facet) and uncovertebral joints. Low-grade inflammation and bone remodeling can lead to local pain. In addition, osteophytes may cause nerve root compression and result in radicular, radiating pain. Degenerative disk disease often coexists with spinal OA, and together they contribute to spinal stenosis causing muscle weakness, paresthesias, and numbness. In severe cases, spinal cord impingement with myelopathy may result.

Diagnosis

OA is diagnosed primarily through a medical history and physical examination and typically presents with articular pain that is exacerbated by activity. Morning stiffness, a characteristic of the inflammatory arthritides, is absent or limited to less than 30 min. In addition, there should not be complaints of multiple hot, swollen joints; individual joints affected by OA may commonly develop joint effusions, which may result in mild joint warmth but not overtly inflamed joints. The *physical examination* should reveal evidence of the degenerative processes in involved joints. As articular cartilage becomes fibrillated, the smooth articulation characterized by normal joints becomes rough, and crepitance (palpable crackling during passive joint motion) may be appreciated in the more superficial joints, though not in the hips or the spine. Osteophytes can be palpated as bony projections and may result in deformities, especially noticeable in the DIPs and PIPs. Squaring of the first CMC joint and palpable knee osteophytes are also common. These deformities may eventually restrict joint range of motion.

Laboratory Evaluation

As OA is not a systemic disease, tests that assess critical organ function are generally normal, as are conventional markers of inflammation. These include blood count and the comprehensive metabolic panel, as well as the erythrocyte sedimentation rate, conventional CRP, and other acute-phase reactants. It is important to realize that OA is most prevalent among the middle-aged and elderly, a demographic that has a high prevalence of comorbidities; thus, coexisting conditions may confound the interpretation of laboratory testing. In addition, whereas OA is not associated with the production of autoantibodies, low-titer detection of nonspecific rheumatoid factors and ANA may be seen in the normal aging population. When joint effusions are present, examination of the synovial fluid should yield a noninflammatory sample. Total leukocyte counts are typically less than 1500–2000 cells/mm^3 with a predominance of lymphocytes rather than neutrophils. Incidental findings of crystals, such as calcium pyrophosphate dihydrate or of cartilage fragments may sometimes be observed.

As laboratory evaluation yields nonspecific results in OA, it is not surprising that there are no clinically useful *biomarkers* in OA, notwithstanding an aggressive decades-long search for macromolecules measurable in blood, synovial fluid, or urine that might provide prognostic or diagnostic value. A variety of tissue components, typically breakdown or cleavage products of cartilage or bone matrix, have been identified that have a statistical correlation with OA progression or pain, and the National Institutes of Health, through the OA Initiative, is conducting a large longitudinal study of several thousand individuals in an attempt to identify novel body fluid and imaging biomarkers that may provide such information about OA onset and progression. As such, clinically useful biomarkers may be identified in the coming years.

Imaging

Conventional *radiography* remains the primary modality for imaging OA and provides the only structural outcomes recognized by regulatory agencies for determining OA progression. Characteristic features of OA include narrowed joint space, osteophytes, and subchondral bone sclerosis (Fig. 2). Not all patients have all three features, and there is an imperfect relationship between radiographic appearance and clinical symptoms. Disease progression can be monitored by longitudinal imaging, both by qualitative grading and by quantitative assessment of joint space narrowing. In contrast to radiography, in which cartilage is radiolucent, *MRI* provides detailed images of each of the joint structures, detecting even subtle defects of articular cartilage. MRI also permits identification of subchondral bone marrow lesions that are associated with symptomatic disease. Nonetheless, MRI remains predominantly a research tool in OA. Clinically, it has little role in routine OA evaluation and management, though it is often used to exclude other potential sources of pain, such as degenerative menisci, ligamentous tears, and other intra-articular pathology. *Ultrasonography* is increasingly used in office-based imaging because it can provide detailed information of joint structure without exposing the patient to ionizing radiation. Although operator dependent, it has been shown to sensitively identify features such as occult osteophytes undetectable by standard radiography, subtle articular cartilage lesions, and mild local synovitis; it remains unclear whether it can play a significant role in diagnosis.

Therapy

The ultimate goals of OA therapy are similar to the treatment of any disease: to maintain or restore function, relieve symptoms, and prevent disease progression. While at present there are no therapeutic approaches that have been demonstrated to retard or reverse disease progression, palliative modalities are effective, and therapy is focused on pain relief and maintenance of joint function. When contemplating OA therapy, it is important to distinguish structural degeneration, which is a normal consequence of aging and is asymptomatic, from symptomatic OA, which requires treatment. Many physicians persist in the fallacy that symptomatic OA is merely part of "getting old" for which nothing can be done. However, a careful multipronged approach can provide relief and maintain function even among those with advanced OA. Patient education and support are critical; self-help programs alone have been shown to improve outcomes in OA. Patients should be provided thorough information about the natural course of OA, their role in disease management, and appropriate expectations.

Symptom Palliation

A variety of strategies can provide effective pain palliation in OA, either individually or in combination. These include physical measures, medication, and surgery. Evidence-based guidelines have been formulated by multiple organizations internationally, though there is widespread consensus among these recommendations (Block 2014).

Physical Measures

During painful flares and immediately following exertion, application of heat or ice to affected joints can be useful. Exercise has been consistently shown to improve function and to provide significant pain palliation (Golightly et al. 2012). This is especially true for exercises aimed at strengthening muscles surrounding affected joints. Hence, patients should be encouraged to exercise regularly and may benefit from physical therapy for instruction in appropriate strength training and to improve and maintain range of motion. Finally, OA pain may be mediated by aberrant biomechanical loading. Therefore, ambulatory assistive devices, such as canes and walkers, which significantly reduce loads across the knee during gait, can reduce pain and improve stability. Unloading knee braces, which are more cumbersome and are often not tolerated, may also provide substantial pain relief in knee OA.

Medication

As OA progresses, physical measures alone are unlikely to suffice, and most patients will require medication to control pain. OA is particularly prevalent among the elderly, a demographic for which comorbidities are common; hence, the choice of medication is frequently influenced by coexisting medical conditions. Topical agents may reduce risks of systemic adverse effects and are appropriate when only a few joints are symptomatic. Topical diclofenac (Barthel et al. 2009) and salicylates are widely available and effective. Capsaicin is approved for knee OA, but it is only marginally effective and is accompanied by a high adverse effect profile, so it is not often used. If topical agents do not provide sufficient relief, oral analgesics are often the next step. Most specialty organizations, including the American College of Rheumatology (Hochberg et al. 2012), still suggest that acetaminophen may be beneficial, especially among patients with contraindications to nonsteroidal anti-inflammatory drugs (NSAIDs) such as renal dysfunction or cardiac disease. However, acetaminophen, while effective for acute pain, may not be effective for long-term analgesia (Case et al. 2003; Towheed et al. 2003), and chronic use may place OA patients at risk for liver damage and hypertension. As such, acetaminophen may be most properly used for short-term flares of OA pain, typically lasting no more than a few weeks. Other analgesics that have been shown to be effective include tramadol and opiates, but these substantially increase morbidity among the elderly, especially their risk for traumatic falls. Finally, agents used to treat neuropathic pain are widely used for OA. One such medication, duloxetine hydrochloride, a serotonin and norepinephrine reuptake inhibitor (SNRI), was studied systematically in OA (Chappell et al. 2009) and approved by the US FDA in 2010 for the treatment of musculoskeletal pain including OA.

NSAIDs have been demonstrated to be effective for OA pain and may maintain efficacy for years (Bannuru et al. 2015). Although concerns have been raised regarding potential toxicities, these medications remain the mainstay of OA therapy. They may not be a viable option for patients with significant renal, cardiac, or gastrointestinal conditions, though proton pump inhibitors or misoprostol can provide gastric protection in middle-aged and elderly patients and those at risk for gastrointestinal bleeding. Cyclooxygenase-2 inhibitors are an additional option; in the USA, celecoxib is the only agent available in this class.

Intra-articular Therapy

Intra-articular glucocorticoids have been shown to be effective for short-term relief of OA pain (Bellamy et al. 2006a). By convention, each joint is not injected more than four times per year because of theoretical concerns of toxicity to articular cartilage. Hyaluronan derivatives are also available for injection to relieve OA pain, but the magnitude of relief provided beyond placebo is controversial (Bellamy et al. 2006b). This modality is referred to as *viscosupplementation* because extrinsic hyaluronans were initially intended to supplement the viscosity of synovial fluid and thereby improve articular lubrication. However, their intra-articular residence time is too brief to have this effect, and their mechanism of action *in vivo* is poorly understood.

Complementary Approaches

As is the case with other chronic pain conditions, the majority of OA patients try complementary approaches. Among the more popular are glucosamine, chondroitin, and acupuncture. Although none have been demonstrated to substantially retard structural progression and independently funded trials have been consistently negative (Block et al. 2010), many patients derive pain relief. These results may be confounded by the substantial placebo response that is observed in any blinded study of OA pain (Doherty and Dieppe 2009) and there is controversy regarding whether these approaches provide incremental pain relief over that obtained with placebo; nonetheless, many of these approaches can be safely utilized by individuals who feel that they benefit from them.

Surgery

Joint replacement surgery in well-selected patients significantly relieves pain and restores function and represents the most effective therapeutic advance in OA treatment to date. It is generally reserved for those in whom pain or joint dysfunction significantly limits normal activities despite optimal medical and physical management. The presence of advanced structural degeneration of the joints, without severe symptoms, should not be an indication for arthroplasty. The most experience is with knees and hips, but other joints are routinely replaced with good results as well. Even with dramatic advances in material sciences, joint prostheses have limited durability, so joint replacement surgery should be delayed in younger patients when possible. In addition to total joint replacement, a variety of temporizing strategies are also in standard practice and are useful for joints that have less severe structural degeneration, including realignment osteotomy in the knee and hemiarthroplasty.

Maintenance of Function

In addition to providing pain palliation, it is critical to promote full functioning among OA patients. Strategies to retain function and independence include ambulatory assistive devices, such as canes and walkers, which provide stability in addition to reducing loading across arthritic joints. Motorized carts can assist individuals with severe knee or hip OA to retain independence in the community. Physical therapy can help to retain strength and range of motion, and occupational therapy can provide customized assistive devices and braces. Whenever possible, counseling on weight loss for overweight or obese patients should be provided as weight loss can improve both pain and function (Bliddal et al. 2011).

Outcome

The natural history of OA has been well described. Once structural joint degeneration is established, it is likely to progress; in most individuals, the disease progresses slowly over decades, though rapidly progressing subpopulations have been described in hip OA and knee OA and presumably exist for OA of other joints. The causes of this variability remain unclear, but several factors may contribute to an individual's prognosis, as delineated above in the discussion of OA epidemiology.

At present, there are no therapies that have been demonstrated to alter the natural history of OA or delay disease progression, but some strategies may ameliorate symptomatic progression. Obesity may be the most modifiable of the strong OA risk factors. Weight loss in adulthood reduces the risk of incident radiographic and symptomatic OA and reduces pain severity in patients who already have OA (Messier et al. 2013). Exercise is an important component of weight strategies and ameliorates pain, but specific types of exercise have not yet shown consistent preventive effects. Strategies to decrease joint injuries in young athletes are critical to reducing posttraumatic OA, and proper conditioning has been shown to reduce knee injuries among female soccer players. Finally, OA progression is mediated by aberrant loading of joints, suggesting that improved loading patterns should affect structural progression (Block and Shakoor 2010), though no biomechanically active approaches have yet been shown to impact OA progression.

Despite extensive efforts over the past several decades to identify "disease-modifying OA drugs" (DMOADs) that could actively delay OA disease progression, to date none have been developed. Current investigations targeting novel mediators of cartilage and joint tissue metabolism and inflammation hold promise, and tissue engineering approaches and mesenchymal stem cell technology may yield functional joint tissue replacement in the future. Current cartilage replacement techniques are not indicated for OA treatment, but are restricted to patients with isolated chondral defects.

References

Anandarajah, A. (2010). Erosive osteoarthritis. *Discovery Medicine, 9*(48), 468–477.

Bannuru, R. R., Schmid, C. H., Kent, D. M., Vaysbrot, E. E., Wong, J. B., & McAlindon, T. E. (2015). Comparative effectiveness of pharmacologic interventions for knee osteoarthritis: A systematic review and network meta-analysis. *Annals of Internal Medicine, 162*(1), 46–54. doi:10.7326/M14-1231.

Barthel, H. R., Haselwood, D., Longley, S., 3rd, Gold, M. S., & Altman, R. D. (2009). Randomized controlled trial of diclofenac sodium gel in knee osteoarthritis. *Seminars in Arthritis and Rheumatism, 39*(3), 203–212. doi:10.1016/j.semarthrit.2009.09.002.

Bellamy, N., Campbell, J., Robinson, V., Gee, T., Bourne, R., & Wells, G. (2006a). Intraarticular corticosteroid for treatment of osteoarthritis of the knee. *Cochrane Database of Systematic Reviews, 2*, CD005328. doi:10.1002/14651858.CD005328.pub2.

Bellamy, N., Campbell, J., Robinson, V., Gee, T., Bourne, R., & Wells, G. (2006b). Viscosupplementation for the treatment of osteoarthritis of the knee. *Cochrane Database of Systematic Reviews, 2*, CD005321. doi:10.1002/14651858.CD005321.pub2.

Bliddal, H., Leeds, A. R., Stigsgaard, L., Astrup, A., & Christensen, R. (2011). Weight loss as treatment for knee osteoarthritis symptoms in obese patients: 1-year results from a randomised controlled trial. *Annals of the Rheumatic Diseases, 70*(10), 1798–1803. doi:10.1136/ard.2010.142018.

Block, J. A. (2014). Osteoarthritis: OA guidelines: Improving care or merely codifying practice? *Nature Reviews. Rheumatology, 10*(6), 324–326. doi:10.1038/nrrheum.2014.61.

Block, J. A., & Scanzello, C. (2015). Osteoarthritis. In L. Goldman & A.I. Schafer (Eds), *Goldman-Cecil medicine*. Philadelphia: Elsevier.

Block, J. A., & Shakoor, N. (2010). Lower limb osteoarthritis: Biomechanical alterations and implications for therapy. *Current Opinion in Rheumatology, 22*(5), 544–550. doi:10.1097/BOR.0b013e32833bd81f.

Block, J. A., Oegema, T. R., Sandy, J. D., & Plaas, A. (2010). The effects of oral glucosamine on joint health: Is a change in research approach needed? *Osteoarthritis and Cartilage/OARS, Osteoarthritis Research Society, 18*(1), 5–11. doi:10.1016/j.joca.2009.07.005.

Brandt, K. D., Dieppe, P., & Radin, E. (2009). Etiopathogenesis of osteoarthritis. *The Medical Clinics of North America, 93*(1), 1–24, xv. doi:10.1016/j.mcna.2008.08.009.

Case, J. P., Baliunas, A. J., & Block, J. A. (2003). Lack of efficacy of acetaminophen in treating symptomatic knee osteoarthritis: A randomized, double-blind, placebo-controlled comparison trial with diclofenac sodium. *Archives of Internal Medicine, 163*(2), 169–178.

Chappell, A. S., Ossanna, M. J., Liu-Seifert, H., Iyengar, S., Skljarevski, V., Li, L. C., et al. (2009). Duloxetine, a centrally acting analgesic, in the treatment of patients with osteoarthritis knee pain: A 13-week, randomized, placebo-controlled trial. *Pain, 146*(3), 253–260. doi:10.1016/j.pain.2009.06.024.

Doherty, M., & Dieppe, P. (2009). The "placebo" response in osteoarthritis and its implications for clinical practice. *Osteoarthritis and Cartilage/OARS, Osteoarthritis Research Society, 17*(10), 1255–1262. doi:10.1016/j.joca.2009.03.023.

Driban, J. B., Sitler, M. R., Barbe, M. F., & Balasubramanian, E. (2010). Is osteoarthritis a heterogeneous disease that can be stratified into subsets? *Clinical Rheumatology, 29*(2), 123–131. doi:10.1007/s10067-009-1301-1.

Felson, D. T. (2013). Osteoarthritis as a disease of mechanics. *Osteoarthritis and Cartilage/OARS, Osteoarthritis Research Society, 21*(1), 10–15. doi:10.1016/j.joca.2012.09.012.

Goldring, S. R. (2012). Alterations in periarticular bone and cross talk between subchondral bone and articular cartilage in osteoarthritis. *Therapeutic Advances in Musculoskeletal Disease, 4*(4), 249–258. doi:10.1177/1759720X12437353.

Goldring, M. B., & Otero, M. (2011). Inflammation in osteoarthritis. *Current Opinion in Rheumatology, 23*(5), 471–478. doi:10.1097/BOR.0b013e328349c2b1.

Golightly, Y. M., Allen, K. D., & Caine, D. J. (2012). A comprehensive review of the effectiveness of different exercise programs for patients with osteoarthritis. *The Physician and Sportsmedicine, 40*(4), 52–65. doi:10.3810/psm.2012.11.1988.

Hochberg, M. C., Altman, R. D., April, K. T., Benkhalti, M., Guyatt, G., McGowan, J., et al. (2012). American College of Rheumatology 2012 recommendations for the use of nonpharmacologic and pharmacologic therapies in osteoarthritis of the hand, hip, and knee. *Arthritis Care and Research, 64*(4), 465–474.

Hunter, D. J., Guermazi, A., Roemer, F., & Neogi, T. (2013). Structural correlates of pain in joints with osteoarthritis. *Osteoarthritis and Cartilage, 21*(9), 1170–1178.

Jiang, L., Tian, W., Wang, Y., Rong, J., Bao, C., Liu, Y., et al. (2012). Body mass index and susceptibility to knee osteoarthritis: A systematic review and meta-analysis. *Joint, Bone, Spine: Revue du Rhumatisme, 79*(3), 291–297. doi:10.1016/j.jbspin.2011.05.015.

Johnson, V. L., & Hunter, D. J. (2014). The epidemiology of osteoarthritis. *Best Practice and Research Clinical Rheumatology, 28*(1), 5–15. doi:10.1016/j.berh.2014.01.004.

Liu-Bryan, R. (2013). Synovium and the innate inflammatory network in osteoarthritis progression. *Current Rheumatology Reports, 15*(5), 323. doi:10.1007/s11926-013-0323-5.

Loeser, R. F., Jr. (2000). Aging and the etiopathogenesis and treatment of osteoarthritis. *Rheumatic Diseases Clinics of North America, 26*(3), 547–567.

Loeser, R. F. (2013). Aging processes and the development of osteoarthritis. *Current Opinion in Rheumatology, 25*(1), 108–113. doi:10.1097/BOR.0b013e32835a9428.

Malfait, A. M., & Schnitzer, T. (2013). Toward a mechanism-based approach of pain management in osteoarthritis. *Nature Reviews. Rheumatology, 9*, 654–664.

Messier, S. P., Mihalko, S. L., Legault, C., Miller, G. D., Nicklas, B. J., DeVita, P., et al. (2013). Effects of intensive diet and exercise on knee joint loads, inflammation, and clinical outcomes among overweight and obese adults with knee osteoarthritis: The IDEA randomized clinical trial. *JAMA: The Journal of the American Medical Association, 310*(12), 1263–1273. doi:10.1001/jama.2013.277669.

Miller, R. E., Lu, Y., Tortorella, M. D., & Malfait, A. M. (2013). Genetically engineered mouse models reveal the importance of proteases as osteoarthritis drug targets. *Current Rheumatology Reports, 15*(8), 350.

Miller, R. E., Miller, R. J., & Malfait, A. M. (2014). Osteoarthritis joint pain: The cytokine connection. *Cytokine, 70*(2), 185–193. doi:10.1016/j.cyto.2014.06.019.

Neogi, T. (2013). The epidemiology and impact of pain in osteoarthritis. *Osteoarthritis Cartilage, 21*(9), 1145–1153.

Neogi, T., & Zhang, Y. (2013). Epidemiology of osteoarthritis. *Rheumatic Diseases Clinics of North America, 39*(1), 1–19. doi:10.1016/j.rdc.2012.10.004.

Reynard, L. N., & Loughlin, J. (2013). Insights from human genetic studies into the pathways involved in osteoarthritis. *Nature Reviews. Rheumatology, 9*(10), 573–583. doi:10.1038/nrrheum.2013.121.

Riordan, E. A., Little, C., & Hunter, D. (2014). Pathogenesis of post-traumatic OA with a view to intervention. *Best Practice and Research Clinical Rheumatology, 28*(1), 17–30. doi:10.1016/j.berh.2014.02.001.

Scanzello, C. R., & Goldring, S. R. (2012). The role of synovitis in osteoarthritis pathogenesis. *Bone, 51*(2), 249–257. doi:10.1016/j.bone.2012.02.012.

Schnitzer, T. J., & Marks, J. A. (2015). A systematic review of the efficacy and general safety of antibodies to NGF in the treatment of OA of the hip or knee. *Osteoarthritis and Cartilage/OARS, Osteoarthritis Research Society, 23*(Suppl 1), S8–S17. doi:10.1016/j.joca.2014.10.003.

Suokas, A. K., Walsh, D. A., McWilliams, D. F., Condon, L., Moreton, B., Wylde, V., et al. (2012). Quantitative sensory testing in painful osteoarthritis: A systematic review and meta-analysis. *Osteoarthritis and Cartilage/OARS, Osteoarthritis Research Society, 20*(10), 1075–1085. doi:10.1016/j.joca.2012.06.009.

Suri, S., & Walsh, D. A. (2012). Osteochondral alterations in osteoarthritis. *Bone, 51*(2), 204–211. doi:10.1016/j.bone.2011.10.010.

Thijssen, E., van Caam, A., & van der Kraan, P. M. (2015). Obesity and osteoarthritis, more than just wear and tear: Pivotal roles for inflamed adipose tissue and dyslipidaemia in obesity-induced osteoarthritis. *Rheumatology, 54*(4), 588–600.

Towheed, T. E., Judd, M. J., Hochberg, M. C., & Wells, G. (2003). Acetaminophen for osteoarthritis. *Cochrane Database of Systematic Reviews, 2*,

CD004257. [Update in Cochrane Database of Systematic Reviews. 2006;(1):CD004257; PMID: 16437479]. [61 refs].

Osteoarthritis Genetics

Ana M. Valdes[1] and Gwen S. Fernandes[1,2]
[1]Academic Rheumatology, School of Medicine, University of Nottingham, Nottingham, UK
[2]Arthritis Research UK Centre for Sport, Exercise and Osteoarthritis, University of Nottingham, Nottingham, UK

Synonyms

Cartilage damage; Degenerative arthritis; Low-grade inflammation; Osteoarthritis; Pain and disability; Synovitis; Total joint replacement

Definition

Osteoarthritis (OA) is the most common global chronic joint disease. The disease may affect single or multiple joints and even be generalized. OA is a chronic arthropathy affecting the entire joint involving the cartilage, joint lining, ligaments, and underlying bone. In OA cartilage loss, osteophyte formation (bone spurs) and subchondral bone sclerosis lead to pain, disability, and a reduction in quality of life (Dieppe and Lohmander 2005). Structural changes visible on radiography include narrowing of the joint space, osteophyte, and bone remodeling around the joints. OA can arise in any synovial joint in the body but is most common in the large joints (knees and hips), hands, and spine (Dieppe and Lohmander 2005).

OA is believed to result from both biomechanical and molecular changes in the joint brought about by injury, joint malalignment, obesity, aging, and inflammation. OA is classified as idiopathic or secondary to anatomic abnormalities, trauma, or inflammatory arthritis. The American College of Rheumatology (ACR) criteria developed for hand, hip, and knee OA are intended to distinguish OA from other causes of symptoms and are best suited to clinical settings in which a high prevalence of other forms of arthritis and of joint pain is expected (see entries on "▶ Gout," and "▶ Rheumatoid Arthritis"). This is different from the definition used for epidemiological studies given the poor correlation between radiographic disease severity (joint damage) and the joint pain and functional impairment presented by a patient. Thus, OA can be defined pathologically, radiographically, or clinically, but most epidemiological studies have relied upon radiographic features to characterize the disease (Sharma and Kapoor 2007).

Epidemiology and Genetics

OA is one of the most disabling diseases in developed countries. Global estimates are that 9.6 % of men and 18.0 % of women over 60 have symptomatic (painful) OA (Woolf and Pfleger 2003). Eighty percent of patients with OA have limitations in movement and 25 % cannot perform their major daily activities of life. WHO data also showed that OA moved from 12th to 6th leading cause of years lost to disability or morbidity between 2002 and 2007. Increases in life expectancy and aging populations are expected to make OA the fourth leading cause of disability by the year 2020 (Woolf and Pfleger 2003). OA of the large weight-bearing joints (hips and knees) is a major contributor to over 500,000 hip and knee arthroplasties per year in the USA (Losina et al. 2009).

Diagnosis: OA is classified as idiopathic or secondary to anatomic abnormalities, trauma, or inflammatory arthritis. The American College of Rheumatology (ACR) criteria developed for hand, hip, and knee OA are intended to distinguish OA from other causes of symptoms and are best suited to clinical settings in which a high prevalence of other arthritides and of joint pain is expected. This is different from the definition used for epidemiological studies given the poor correlation between radiographic disease severity (joint damage) and the joint pain and functional impairment presented by a patient. Thus, OA can be defined pathologically, radiographically, or clinically, but most epidemiological studies have relied upon radiographic features to characterize the disease (Sharma and Kapoor 2007).

The most widely used system to grade radiographic osteoarthritis is the Kellgren and Lawrence (K/L) system (Sharma and Kapoor 2007) which assigns a value from 0 to 4 with the aid of atlas reproductions (0 = normal; 1 = possible osteophytes; 2 = definite osteophytes and possible joint space narrowing (JSN); 3 = definite or multiple osteophytes and definite joint space narrowing, some sclerosis; 4 = large osteophytes, marked joint space narrowing, severe sclerosis). Importantly this system is osteophytes driven and hence unreliable when there is definite cartilage loss in the absence of osteophytes, as is often the case for hip OA. In this case a measure of minimum joint space width can be used although this may also differ between ethnic groups and age groups. Although MRI is increasingly common in epidemiologic studies, a universally agreed MRI definition has not yet been established.

Ethnicity and OA: The prevalence of OA and patterns of joints affected by OA vary among racial and ethnic groups. Both hip and hand OA are less frequent among Chinese in the Beijing Osteoarthritis Study than in whites in the Framingham Study, but they had significantly higher prevalence of both radiographic and symptomatic knee OA than white women in Framingham Study. The prevalence of individual radiographic features of hip OA also varies between African Americans compared with whites (Zhang and Jordan 2010).

Mortality of OA: Individuals with osteoarthritis (defined both symptomatically and radiographically) at the knee or the hip show a 55 % excess in all-cause mortality. Diabetes (95 % increased risk), cancer (128 % increased risk), cardiovascular disease (38 % increased risk), and the presence of walking disability at baseline (48 % increased risk) are independently associated with the excess of all-cause mortality (Nüesch et al. 2011).

Importantly, deaths from cardiovascular causes are higher in patients with walking disability due to OA (72 % higher) even after adjustment for baseline covariates, indicating that there is an interplay between the underlying OA and the additional comorbid conditions which results in a higher risk of mortality. Thus, although the main clinical symptoms of OA are pain and disability, the consequences of the disease are far reaching.

Risk Factors: Several risk factors have been recognized to affect hip and knee OA, and these are body weight, age, female gender, occupational activity and injury (Zhang and Jordan 2010; Zhang et al. 2011), congenital abnormalities and joint shape, meniscal tears, presence of OA at other joints (Heberden's and Bouchard's nodes), and foot and knee alignment in addition to genetic predisposition. Furthermore, specific inter- and intra-articular patterns of OA may represent subsets that have different risk factor profiles and disease course. The reader is referred to Sharma and Kapoor 2007 for an in-depth review of the epidemiology of OA.

Genetics of OA

Osteoarthritis has an important genetic contribution which has been explored in humans both in monogenic syndromes and in its complex common form. Monogenic syndromes refer to modifications in a single gene that result in an OA-like phenotype. Examples of this are many mutations affecting cartilage extracellular assembly or stability, such as Stickler's syndrome type I, multiple epiphyseal dysplasia, and osteochondritis dissecans (Valdes and Spector 2015).

The complex form of OA also runs in families. This is measured by using the risk ratio for a relative, such as a brother or sister, of an affected individual to that of the general population prevalence. When the occurrence of OA which is tested in siblings of subjects with OA severe enough to lead to surgery is compared to that in the general population, it is found that sibs have a much higher prevalence of OA. The sibling recurrence risk for knee OA/total knee replacement has been estimated to be 2.08–4.80 and for hip OA/total hip replacement 1.78–8.53 (Valdes and Spector 2011). Twin studies show that the influence of genetic factors in radiographic OA of the hand, hip, and knee is between 39 % and 65 %, independent of known environmental or demographic confounding factors (Valdes and Spector 2011).

In the past few years a number of genetic risk factors have been identified that predispose to OA of the hip, the knee, and the hand using genome-wide association studies (Valdes and Spector 2011, 2015). Genome-wide association scans and candidate gene studies have revealed the complexity of OA with many genes contributing modestly to overall disease risk.

Some of the main genes to come out of these studies are as expected involved in cartilage metabolism and chondrogenesis (e.g., *PTHLH, CHST11, IGFBP3 , DOT1L, GDF5)*, and some of them are involved in immune response *(HLA-DQB1, BTNL2)*. For most of the genes identified so far, however, the specific role in the pathogenesis of OA remains to be elucidated (Valdes and Spector 2015).

Clinical Presentation

According to the American College of Rheumatology criteria, diagnosis of OA of the knee, the hip, or the hand can be made based purely on clinical symptoms or be assisted by X-rays (see Bellamy et al. 1999) – hand pain, aching, or stiffness and at least three of the following features: hard tissue enlargement of two or more of hand joints and fewer than three swollen metacarpophalangeal joints and deformity of at least one of the hand joints.

For knee OA, in addition to knee pain, the guidelines require that the patient have at least three of the following six: age > 50 years, stiffness in the morning lasting less than 30 min, crepitus, bony tenderness, bony enlargement, and no palpable warmth. For hip OA, a diagnosis can be made if the following are present: hip pain plus hip internal rotation $<15°$ and ESR $\leq$ 45 mm/h or hip flexion $\leq 115°$ or hip pain, with hip internal rotation $\geq$ 15°, pain in internal rotation, morning stiffness of the hip ≤ 60 min, and age > 50 years.

In general the main symptom of OA is joint pain which is exacerbated by exercise and relieved by rest. Although pain at rest or during the night is not uncommon, it is common in advanced disease. Knee pain due to OA is usually bilateral and felt in and around the knee. Hip pain due to OA is felt in the groin and anterior or lateral thigh. Hip OA pain can also be referred to the knee. Signs of OA include reduced range of joint movement, joint swelling/synovitis (warmth, effusion, synovial thickening), crepitus, periarticular tenderness, bony swelling, and deformity due to osteophytes – in the fingers this presents as swelling at the distal interphalangeal joints (Heberden's nodes) or swelling at the proximal interphalangeal joints (Bouchard's nodes) (Bellamy et al. 1999).

Pathophysiology

At the level of articular cartilage, the pathological progression of OA follows a typical pattern. The earliest indication of pathological change is chondrocyte clustering as a result of increased cell proliferation and a general upregulation of synthetic activity. Increased expression of cartilage-degrading proteinases and matrix proteins suggests an attempt at repair. Gradual loss of proteoglycans appears in the surface region of articular cartilage and this is followed by type II collagen degradation. Cracks develop along the articular surface, producing the histological image termed fibrillation. At later stages of the disease, fibrocartilage forms, probably as a consequence of unsuccessful attempts by chondrocytes to fill in the cracks. Finally, the development of osteophytes is observed (osteophytes are bony structures at the periphery of the joint surface) (Li et al. 2007). In addition to articular cartilage, synovium, tendons, and bones are also involved in disease progression and manifestation. While synovitis (inflammation of the synovium) is a critical characteristic of OA and often considered the driver of the OA process, OA is really a complex, "whole joint" disease. This is because inflammatory processes are initiated via mediators that are released not just by the synovium but by the bone and cartilage too (Goldring and Otero 2011).

OA is no longer considered a noninflammatory arthritis as there is evidence of subclinical low-grade inflammation. The drivers of this inflammation are varied and advances in molecular biology have postulated several theories. The *synovial inflammation* theory is one which affects the synovium, cartilage, and bone in OA. Following a traumatic injury or a repetitive microtrauma to a joint, fragmented cartilage within a joint space provokes a reaction from synovial cells. As the fragments are considered foreign bodies, the synovial cells release inflammatory mediators. These mediators activate chondrocytes and produce metalloproteinases (MMPs) which promote cartilage destruction. These mediators also induce inflammatory cytokines and MMPs by the synovium itself which perpetuates further destruction of the synovium. As part of the synovial inflammatory processes, the formation of osteophytes also decreases (Berenbaum 2013). Altered rates of osteophyte remodeling in OA are due to increased or decreased osteoclastic bone resorption. Findlay and Atkins (Findlay and Atkins 2014) present evidence of chemical communication between chondrocytes and osteophytes. While small molecules can transit between these tissues in vivo, larger molecules such as inflammatory mediators could do the same and this provides evidence OA interactions implicating the whole joint. The *mechanical* theory of inflammation is attributed to biomechanical overloading of a joint that is detected by mechanoreceptors that are located at the joint surface. An abnormal increased joint loading increases the expression and release of inflammatory cytokines, chemokines, and prostaglandins (Sowers and Karvonen-Gutierrez 2010). At an intracellular level, the mechanical forces are translated into chemical signals also known as mechanotransduction that trigger inflammation and gradual onset of OA. The *protein-driven* theory of inflammatory implicates proteins in maintaining synovial inflammation and cartilage degradation in OA. This theory has been verified by proteomic analysis which showed that proteins from synovial fluid of early knee OA patients, such as the interleukin-15 (IL-15), result in the production of inflammatory cytokines. Complement is another group of proteins which has been expressed and activated in abnormally high levels in OA joints in vivo (Wang et al. 2011). Complement proteins are part of the innate immune system and, when expressed, cause the dilation of blood vessels and contribute to the classic signs of inflammation: redness, warmth, swelling, pain, and loss of function as seen in OA. The *arthrosclerosis inflammatory* theory and the *adipokine* theory offer an alternative explanation for systemic inflammatory processes rather than joint localized ones in OA. For instance, the risk of hand OA is increased by twofold in obese patients, suggesting a systemic explanation rather than a biomechanical or abnormal joint loading etiology. Obesity induces an inflammatory environment as adipose tissues express cytokines and adipocytokines which have been identified in the plasma and synovial fluid of OA patients. It destabilizes cartilage homeostasis and contributes to cartilage degradation (Fernandes and Valdes 2015). Obesity can also accelerate the OA process by ischemic diseases at the subchondral level. The direct ischemic effects on bones are known to reduce cartilage nutrition and inflict multiple bone infarcts that are characteristic of advanced OA. Bony and cartilage remnants can be identified in the synovial and capsule space with marked eburnation of the bone surrounding the infarct. In fact, once OA develops, the underlying ischemia or avascular necrosis can no longer be identified which means that it is often missed. Adipokines also induces insulin resistance, endothelial dysfunction, and a systemic inflammation, which are implicated in atherosclerosis and could explain why OA patients have an almost 40 % increased risk of CVD (Fernandes and Valdes 2015). The *aging* theory of inflammation in OA is also implicated when an increase of inflammatory mediators means that oxidated proteins increase in concentration within the joint. These reactive oxygen species cause oxidative damage which can trigger inflammation and promote cell senescence or aging and in particular, chondrocyte aging. In aging, there is a loss in the ability of cells and tissues to maintain homeostasis, particularly when placed under abnormal stresses such as oxidative stress or biomechanical overloading of a joint. This promotes stress-induced senescence of chondrocytes. Another facet of the aging theory involves the formation of advanced glycation end products (AGEs) which is produced in aging tissues. These end products alter the mechanical properties of cartilage, lead to more brittle tissue with an increased fatigue failure, and stimulate the overproduction of proinflammatory cytokines and MMPs (Lotz and Loeser 2012). Overall, the integrity of the joint structure is compromised internally by inflammatory processes and externally by the systemic effects of aging and mechanical loading. The role of gender and, in particular, *hormonal regulation* of inflammatory mediators is another theory of inflammation that could explain the increase in OA incidence in postmenopausal women. Chondrocytes, the synovium, and the subchondral bone all have estrogen receptors (Tanko et al. 2008). Postmenopause, when estrogen levels decrease, there is an increase in the secretion of inflammatory cytokines in the joint space, while the simultaneous decrease in ovarian function increases inflammatory cytokines within plasma. This is thought to contribute to inflammation at both a joint and whole-body level (Goldring and Otero 2011). Therefore, while there seems to be a complex interplay between mediators and inflammatory processes affecting different joint structures, the overall effect over time is one of gradual degradation of tissues, loss of optimal joint integrity and function, and the onset of pain.

Treatment

While many disease-modifying therapies exist for the more aggressive and inflammatory forms of arthritis, such as rheumatoid arthritis, the options available for treating OA patients are limited (Hunter and Lo 2009). As of today, no disease-modifying drugs for OA are approved by the Food and Drug Administration (FDA) or the European Medicines Agency (EMEA). In the absence of disease-modifying drugs, clinical recommendations for treatment available for each level of severity from mildest to most severe are (a) mildest = non-pharmacological conservative interventions such as exercise, weight loss, and footwear; (b) mild = further non-pharmacologic management such as physiotherapy and braces in addition to simple analgesics such as paracetamol; (c) severe = pharmacologic management: non-steroid anti-inflammatory drugs (NSAIDs), opioids, intra-articular corticosteroids, and intra-articular hyaluronates; and (d) end-stage OA = osteotomy, unicompartmental joint replacement, and total joint replacement (Hunter and Lo 2009).

It should be noted that the available options for pharmacotherapy which aim to reduce the symptoms of OA do so only with limited efficacy and leave the patient with a considerable burden of pain (Hunter and Lo 2009).

Some studies have suggested that glucosamine sulfate, chondroitin sulfate, sodium hyaluronan, doxycycline, matrix metalloproteinase (MMP) inhibitors, bisphosphonates, calcitonin, diacerein, and avocado-soybean unsaponifiables may modify disease progression (Abramson et al. 2006). Nevertheless, structure-modifying efficacy has not been convincingly demonstrated for any existing pharmacologic agents and the difference between these proposed drugs and placebo is very small and extremely difficult to detect (Hunter and Lo 2009).

Surgical Options

Total joint replacement (TJR) is a frequently performed and effective procedure that relieves pain and improves functional status in patients with end-stage knee or hip OA.

There are no generally accepted criteria for joint replacement surgery in patients with OA (Katz 2006). Typically, the diagnosis of OA is made on the basis of clinical examination and plain radiographs. However, as it is well recognized that radiographic severity is only loosely correlated with symptom severity and functional limitation, the critical questions that drive recommendations for surgery are whether the patient has contraindications and whether the patient's functional limitations are severe enough to warrant surgery, although the level of pain needed to indicate when surgery is appropriate is unclear given the lack of standardized guidelines. In terms of outcomes, results from population-based studies indicate that approximately 80–90 % of patients having a TJR will have improvement in function and near-complete relief of pain and will be satisfied with the results of these surgeries (Losina et al. 2009).

The major concern in the long-term outcome of TJR is survival of the prosthesis. The only population-based data available suggest over a follow-up of up to 4 years that the rate of failure of THR leading to revision is approximately 4 or 1 % per year (Losina et al. 2009). Wear debris, primarily generated from the prosthetic joint articular surface, remains the major factor limiting the survival of joint implants, contributing to osteolysis and aseptic loosening after THR.

Outcome

Musculoskeletal aging represents the major cause of disability in the over-65 population in Europe. Disabling OA is a very common problem in the elderly which is associated with increased rates of comorbid conditions, increased mortality, and decreased quality of life (Hunter et al. 2011). By 2020 the number of people with OA will have doubled relative to 2000, in large part, because of the exploding prevalence of obesity and the aging of the baby-boomer generation.

OA is a slowly progressing and debilitating process and is characterized clinically by pain, disability, and loss of function. Patients report with a diminished ability to perform the basic activities of daily living such as climbing stairs or changing from the sitting to a standing position (Tanner et al. 2007). In the UK, a recent survey called "OA Nation" found that 81 % of people with OA experience constant pain and face limitations in performing certain tasks. As OA affects a large percentage of the population, the OA Nation survey reported some worrying statistics regarding access to help from the health service, diagnosis of OA, and the use of prescribed as well as nonprescribed medications. For example, an OA diagnosis was usually provided about 18 months after symptoms first appeared with many patients visiting their doctor three to four times before the diagnosis was made (OA Nation 2012). This not only represents a significant cost from a health economic point of view but also from a patient-centered perspective. Combined with the fact that often patients with OA also suffer with comorbidities such as hypertension, heart disease, and diabetes, this group of patients is complex and their management plan needs to be individualized to provide the right type of care at the right time for one particular patient.

Cross-References

▶ Gout
▶ Rheumatoid Arthritis

References

Abramson, S. B., Attur, M., & Yazici, Y. (2006). Prospects for disease modification in osteoarthritis. *Nature Clinical Practice. Rheumatology, 2*, 304–312.

Bellamy, N., Klestov, A., Muirden, K., Kuhnert, P., Do, K. A., O'Gorman, L., et al. (1999). Perceptual variation in categorizing individuals according to American College of Rheumatology classification criteria for hand, knee, and hip osteoarthritis (OA): Observations based on an Australian Twin Registry study of OA. *The Journal of Rheumatology, 26*(12), 2654–2658.

Berenbaum, F. (2013). Osteoarthritis as an inflammatory disease (osteoarthritis is not osteoarthrosis!). *Osteoarthritis and Cartilage, 21*(1), 16–21.

Dieppe, P. A., & Lohmander, L. S. (2005). Pathogenesis and management of pain in osteoarthritis. *Lancet, 365*, 965–973.

Fernandes, G. S., & Valdes, A. M. (2015). Cardiovascular disease and osteoarthritis: Common pathways and patient outcomes. *European Journal of Clinical Investigation.* doi:10.1111/eci.12413.

Findlay, D. M., & Atkins, G. J. (2014). Osteoblast-chondrocyte interactions in osteoarthritis. *Current Osteoporosis Reports, 12*(1), 127–134.

Goldring, M. B., & Otero, M. (2011). Inflammation in osteoarthritis. *Current Opinion in Rheumatology, 23*(5), 471–478.

Hunter, D. J., & Lo, G. H. (2009). The management of osteoarthritis: An overview and call to appropriate conservative treatment. *The Medical Clinics of North America, 93*(1), 127–143.

Hunter, D. J., Guermazi, A., Lo, G. H., Grainger, A. J., Conaghan, P. G., Boundreau, R. M., et al. (2011). Evolution of semi-quantitative whole joint assessment of knee OA: MOAKS (MRI Osteoarthritis Knee Score). *Osteoarthritis and Cartilage, 19*(8), 990–1002.

Katz, J. N. (2006). Total joint replacement in osteoarthritis. *Best Practice & Research. Clinical Rheumatology, 20*(1), 145–153.

Li, Y., Xu, L., & Olsen, B. R. (2007). Lessons from genetic forms of osteoarthritis for the pathogenesis of the disease. *Osteoarthritis and Cartilage, 15*(10), 1101–1105.

Losina, E., Walensky, R. P., Kessler, C. L., Emrani, P. S., Reichmann, W. M., Wright, E. A., et al. (2009). Cost-effectiveness of total knee arthroplasty in the United States: Patient risk and hospital volume. *Archives of Internal Medicine, 169*(12), 1113–1121.

Lotz, M., & Loeser, R. F. (2012). Effects of aging on articular cartilage homeostasis. *Bone, 51*(2), 241–248.

Nüesch, E., Dieppe, P., Reichenbach, S., Williams, S., Iff, S., & Jüni, P. (2011). All cause and disease specific mortality in patients with knee or hip osteoarthritis: Population based cohort study. *BMJ, 342*, d1165.

OA Nation Survey. (2012). Arthritis care. Available online www.arthritiscare.org.uk. Accessed 01 Nov 2014.

Sharma, L., & Kapoor, D. (2007). Epidemiology of osteoarthritis. In R. W. Moskowitz, R. D. Altman, M. C. Hochberg, J. A. Buckwalter, & V. M. Goldberg (Eds.), *Osteoarthritis: Diagnosis and medical/surgical management* (4th ed., pp. 3–26). Philadelphia: Lippincott Williams & Wilkins.

Sowers, M. R., & Karvonen-Gutierrez, C. A. (2010). The evolving role of obesity in knee osteoarthritis. *Current Opinion in Rheumatology, 22*(5), 533–537.

Tanko, L. B., Sondergaard, B. C., Oestergaard, S., Karsdal, M. A., & Christiansen, C. (2008). An update review of cellular mechanisms conferring the indirect and direct effects of estrogen on articular cartilage. *Climacteric, 11*(1), 4–16.

Tanner, S. M., Dainty, K. N., Marx, R. G., & Kirkley, A. (2007). Knee-specific quality of life instruments: Which ones measure symptoms and disabilities most important to patients? *American Journal of Sports Medicine, 35*(9), 1450–1458.

Valdes, A. M., & Spector, T. D. (2011). Genetic epidemiology of hip and knee osteoarthritis. *Nature Reviews. Rheumatology, 7*(1), 23–32.

Valdes, A. M., & Spector, T. D. (2015). Genetics of osteoarthritis. In M. C. Hochberg, A. J. Silman, J. S. Smole, M. E. Weinblatt, & M. H. Weisman (Eds.), *Rheumatology* (6th ed., pp. 1761–1769). Philadelphia: Elsevier.

Wang, Q., Rozelle, A. L., Lepus, C. M., Scanzello, C. R., Song, J. J., Larsen, D. M., et al. (2011). Identification of a central role for complement in osteoarthritis. *Nature Medicine, 17*(12), 1674–1679.

Woolf, A. D., & Pfleger, B. (2003). Burden of major musculoskeletal conditions. *Bulletin of the World Health Organization, 81*(9), 646–656.

Zhang, Y., & Jordan, J. M. (2010). Epidemiology of osteoarthritis. *Clinics in Geriatric Medicine, 26*(3), 355–369.

Zhang, W., McWilliams, D. F., Ingham, S. L., Zhang, W., McWilliams, D. F., Ingham, S. L., et al. (2011). Nottingham knee osteoarthritis risk prediction models. *Annals of the Rheumatic Diseases, 70*(9), 1599–1604.

Osteoclasts in Inflammation

Dávid Győri and Attila Mócsai
Department of Physiology, Semmelweis University, Budapest, Hungary

Synonyms

Bone-resorbing cells in inflammation

Definitions

Osteoclasts are multinucleated giant cells of the myeloid lineage that resorb bone. Mature osteoclasts are polarized cells that undergo structural changes to form tight connection with the bone surface, where they secrete hydrochloric acid and lytic enzymes into a resorption pit to erode the underlying bone.

Osteoimmunology is an interdisciplinary research area which investigates the interplay between the skeletal and immune systems. Numerous immune cell-derived mediators influence osteoclast differentiation and function, while bone cells can regulate the activity of the cells outside the bone.

Structure and Functions

Osteoclasts are derived from monocyte/macrophage precursors, which express various cytokine receptors. A major osteoclastogenic cytokine that stimulates osteoclastogenesis is the macrophage colony-stimulating factor (M-CSF). M-CSF is essential for the development, differentiation, and survival of osteoclasts (Ross and Teitelbaum 2005). M-CSF acts through its receptor on the osteoclasts, colony-stimulating factor 1 receptor (c-fms), a transmembrane tyrosine kinase receptor. Another essential cytokine that regulates osteoclastogenesis is the receptor activator of nuclear factor (NF)-κB (RANK) ligand (RANKL) (Boyle et al. 2003). RANKL is expressed by the bone-forming mesenchymal cells, osteoblasts, while its receptor, RANK, is expressed on the osteoclast precursor cells. Ligand binding of RANK results in trimerization of the receptor and recruitment of the adapter molecule tumor necrosis factor (TNF) receptor-associated factor 6 (TRAF6) which further activates the transcription factors NF-κB and mitogen-activated protein kinases (MAPKs) (Takayanagi 2010). RANK also activates the transcription factor activator protein 1 (AP-1) and nuclear factor of activated T-cell cytoplasmic 1 (NFATc1), the master regulator of osteoclast differentiation. NFATc1 (together with AP-1) regulates the expression of a number of osteoclast-specific proteins, including *tartrate-resistant acid phosphatase, calcitonin receptor, cathepsin K*, and the *β3 integrin* (Takayanagi 2007). Signaling interplay between M-CSF and RANKL is crucially important for osteoclast differentiation. Osteoclast differentiation is inhibited by osteoprotegerin (OPG), a soluble homolog of RANK, which is produced by osteoblasts and binds to RANKL thereby preventing its interaction with RANK. The differentiation of osteoclasts from monocyte/macrophage precursors is seen in Fig. 1.

Tight binding of the osteoclasts to the bone matrix is essential for the resorptive process. This binding is mediated by integrins, transmembrane heterodimers of α/β subunits, which mediate cell-matrix and cell-cell interactions. αvβ3 integrins are highly expressed on the surface of osteoclasts. The integrins assume their high-affinity conformation in the presence of M-CSF and activate a canonical signaling pathway required for osteoclast cytoskeletal rearrangement and polarization for efficient resorption. A critical component of this signaling pathway is the c-Src tyrosine kinase.

Osteoclast development and function also require immunoreceptor-like costimulatory molecules such as osteoclast-associated receptor (OSCAR) and triggering receptor expressed in myeloid cells-2 (TREM2). Those receptors are coupled to immunoreceptor tyrosine-based activation motif (ITAM)-bearing transmembrane adaptors: DNAX-activating protein 12 (DAP12) and the Fc receptor γ-chain (FcRγ). DAP12 and FcRγ recruit the spleen tyrosine kinase (Syk) which further activates phospholipase Cγ2 (PLCγ2). This immunoreceptor-like signaling pathway is essential for osteoclast differentiation and function in mice, and its importance is also underlined by human mutations in the DAP12 and TREM2 genes which cause the Nasu-Hakola disease, a complex human disease with skeletal abnormalities (Humphrey et al. 2005).

Once activated, mature osteoclasts move to the site of bone resorption where they attach to the bone through the sealing zone. The sealing zone is bounded by belts of adhesion structures called podosomes. The sealing zone allows the osteoclast to create a resorption pit that is isolated from the surrounding interstitium. Attachment to the bone matrix is facilitated by the binding of integrins (primarily αvβ3) to specific amino acid motifs (Arg-Gly-Asp) in bone matrix proteins, such as osteopontin. To promote the resorption process, osteoclasts polarize their plasma membrane and form a tubulovesicular membrane network, called ruffled border, on their side juxtaposed to the bone surface. Osteoclasts then acidify the resorption pit to solubilize the mineral content of the bone. This is achieved by intracellular generation of protons (H^+ ions) through the action of carbonic anhydrase, which are then pumped against their concentration gradient into the resorption pit by a vacuolar H^+ ATPase. Charge compensation is provided by chloride ions transported through ClC7 chloride channels. In addition, osteoclasts secrete lysosomal enzymes, including cathepsin K and matrix metalloproteinases (MMPs) to remove the organic components of the bone. Partially digested bone

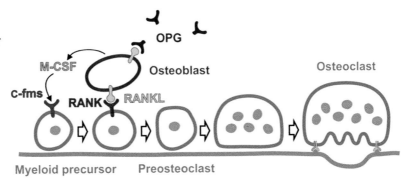

Osteoclasts in Inflammation, Fig. 1 The development of osteoclasts from myeloid precursors

fragments eventually traverse the osteoclasts by transcytosis and leave the cells through the basolateral membrane. Osteoclasts die by apoptotic processes that are regulated by paracrine-acting mediators and cytokines deposited in the bone matrix (Teitelbaum 2000). The bone resorption process by mature osteoclast is depicted in Fig. 2.

Pathological Relevance

Various overlapping and interacting mechanisms have been described between the bone and the immune system and have led to the rapid evolution of the field osteoimmunology. The term osteoimmunology was first introduced following the landmark observation describing that T lymphocytes triggered bone loss by inducing the differentiation of bone-resorbing osteoclasts. This emerging field has provided perspective and framework for studying the mechanisms underlying various bone diseases (Takayanagi 2007).

Bone is a complex tissue with several functions: it supports and protects the internal organs of the body, and it provides site for hematopoiesis and stores minerals such as calcium and phosphorus. Those functions ensure that abnormal bone biology also has significant pathological consequences. The balance between bone resorption and bone formation maintains bone homeostasis. This balance is altered under pathological conditions. Excessive osteoclast activity leads to pathological bone resorption as seen in osteoporosis (see later) and inflammatory bone diseases. Inflammatory bone diseases include rheumatoid

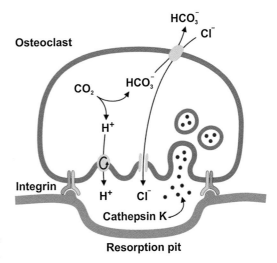

Osteoclasts in Inflammation, Fig. 2 The function of the mature osteoclast

arthritis, periodontitis, osteolytic bone metastases, and several less prevalent disorders.

Chronic inflammation can turn into destructive, self-perpetuating autoimmune disease leading to progressive structural damage. Inflammatory arthritis is a manifestation of chronic inflammation affecting the synovial membrane of the joints and the periarticular bone. Bone destruction in inflammatory arthritis is best described in rheumatoid arthritis. **Rheumatoid arthritis** is a chronic, systemic autoimmune disease, which primarily affects the small synovial joints of the hands and feet. It affects about 1 % of the Western society (Firestein 2003). In human rheumatoid arthritis, inflammation of the synovial joints is accompanied by cartilage and bone destructions. A hallmark of rheumatoid arthritis is erosions of the periarticular bone. Osteoclasts are found to be present at the sites of erosions, and they are responsible for the periarticular bone loss (Schett 2009).

In normal joints, bone resorption and formation are maintained by the balanced function of osteoclasts and osteoblasts. Bone resorption is followed by bone formation, allowing the replacement of old skeletal tissue with new one. In rheumatoid arthritis, the balance between osteoclast and osteoblast activity is severely disrupted. Osteoclastogenesis in the inflamed synovial membrane is enhanced. This pronounced osteoclast formation is on one hand due to the robust influx of monocyte/macrophage osteoclast precursors into the joint space and on the other hand to the maturation of these precursors to bone-resorbing cells by increased expression of osteoclastogenic cytokines. Monocyte/macrophage influx to the intra-articular space is regulated by the increased expression of chemokines, such as CXCL-12, macrophage inflammatory protein 1α (MIP-1α), and CCL20. These chemokines promote the attachment of the cells to synovial vessels and their invasion into the synovial membrane (Schett and Teitelbaum 2009). Under the inflammatory conditions of rheumatoid arthritis, joint infiltration by activated $CD4^+$ T lymphocytes and macrophages results in appearance of proinflammatory cytokines in the synovium. Differentiation of osteoclasts from their monocyte/macrophage precursors requires M-CSF and RANKL. Both cytokines are present in the inflamed synovial membrane, providing the optimal conditions for osteoclastogenesis. Synovial fibroblasts, osteoblasts, and bone marrow stromal cells are the major source of these essential cytokines. Activated T cells, which are present in the inflamed synovium, also express RANKL (Schett 2009). The proinflammatory cytokines are TNF-α, interleukin (IL)-1β, IL-6, and IL-17. TNF-α exerts its osteoclastogenic effects by directly targeting osteoclast precursors and by promoting mesenchymal cell expression of M-CSF and RANKL. IL-1β upregulates the expression of RANK on osteoclast progenitors. Increased RANK signaling results in recruitment of TRAF6 and subsequent activation of the transcription factor NFATc1, which is the master regulator of osteoclastogenesis. IL-17-producing T helper 17 (T_{H17}) cells are an important osteoclastogenic T-cell subset (Takayanagi 2010). IL-17 promotes the expression of RANKL on osteoblasts and other mesenchymal cells. IL-17 also induces local inflammation and increased production of inflammatory cytokines, which further promote RANKL activity. Thus, the interplay between osteoclasts and T lymphocytes plays a crucial role in pathological bone loss induced by arthritis. A new link between the bone and the adaptive immune system was recently provided by showing that autoantibodies against citrullinated vimentin (ACPA) directly led to osteoclast differentiation and bone resorption (Harre et al. 2012).

Proinflammatory cytokines promote the expression of surface molecules on monocytes/macrophages participating in osteoclastogenesis. OSCAR is inducible in peripheral monocytes of patients with rheumatoid arthritis, suggesting that the precursor cells entering the rheumatic joints are predisposed to commitment to the osteoclast lineage. In contrast, T-cell cytokines IFN-γ, IL-4, and IL-10 all inhibit osteoclastogenesis, and IL-12 and IL-18 also negatively affect osteoclast differentiation (Takayanagi 2007).

Disturbances of bone formation have also been implicated in the pathogenesis of rheumatoid arthritis. TNF-α downregulates the expression of key osteoblastogenic transcription factors and increases the expression of Wnt signaling antagonists in inflamed joints. Because Wnt signaling promotes OPG expression, which limits osteoclastogenesis, this regulatory mechanism further enhances bone loss (Goldring and Goldring 2007).

Periodontitis comprises a set of inflammatory diseases affecting the tissues that surround and support the teeth. Induced by infection with various subgingival bacteria, periodontitis is a major cause of tooth loss and is associated with increased risk for cardiovascular thrombotic events. Periodontitis, caused by bacterial infection, is an aggressive local inflammatory reaction that activates the immune system. Amplification of this initial localized response results in the propagation of inflammation through the gingival tissues. Periodontitis induces progressive loss of the alveolar bone around the teeth. Inflammation and bone loss are hallmarks of periodontal disease. Osteoclasts are found to be present at the sites of inflammation and contribute to the periodontal bone loss (Arron and Choi 2000).

Bacteria-derived factors and antigens evoke a local inflammatory reaction and activation of the innate immune system. Similar to rheumatoid arthritis, proinflammatory molecules and cytokine networks play essential roles in this process. TNF-α and IL-1 are the most important cytokines that influence the activity of the cells involved in the lesion. The same cascade of events as in inflammatory arthritis leads to increased osteoclastogenesis via the RANK-RANKL-OPG axis. Increased RANKL/OPG ratio stimulates the differentiation of monocyte/macrophage precursor cells into osteoclasts. Enhanced expression of RANKL on osteoblast and other osteoclastogenic mesenchymal cells also stimulates the maturation and survival of the osteoclasts, leading to periodontal bone loss.

Bone metastases result from the invasion of the primary tumor to bone. Cancers originating from the skeletal system, such as osteosarcoma, chondrosarcoma, and Ewing's sarcoma, are rare. In general, the epithelial tumors give bone metastases and form a solid mass inside the bone. The skeleton is a common location for metastasis. Although any type of cancer is capable of forming bone metastasis, there are particular types of tumors which tend to favor the microenvironment of the bone marrow, such as prostate, breast, lung, renal, and thyroid carcinomas. Either condition induces high morbidity and risk of pathological fractures. Osteotropic cancer cells acquire bone cell-like characteristics, which promote extravasation, survival, and proliferation in the bone microenvironment. The acquisition of bone cell phenotype relies on expression of key osteoclast and osteoblast genes.

Bone metastases are classified as osteolytic and osteosclerotic bone metastases. Osteolytic metastases are characterized by the proliferation of osteoclasts. They are caused by tumor-derived factors that stimulate the activity of osteoclasts. Several factors, including IL-1, IL-6, RANKL, parathyroid hormone-related protein (PTHrP), and MIP-1α, have been implicated as factors that enhance osteoclast formation and bone destruction in patients with osteolytic neoplasia. PTHrP is the major factor produced by cancer cells that induces osteoclastogenesis via upregulation of RANKL. The enhanced osteoclastic bone resorption in turn then releases growth factors from the bone matrix which further stimulate tumor growth. This reciprocal interaction between the cancer cells and the osteoclasts results in a "vicious cycle," in which cancer cells stimulate osteoclast formation and bone resorption, and the increased bone resorption releases factors that enhance tumor growth. Recently, the tumor/bone "vicious cycle" model has been expanded to include T cells as additional regulators of bone tumor growth (Faccio 2011).

Finally, bone loss has also been recognized as complication of some other inflammatory disorders. Excessive bone resorption by osteoclasts may be an extraintestinal manifestation of Crohn's disease and ulcerative colitis, as well as a complication of Paget's disease and multiple myeloma (Lorenzo et al. 2007). Understanding the molecular mechanisms responsible for osteoclast activation in inflammatory diseases may lead to the development of novel therapeutic approaches.

Interactions with Other Processes and Drugs

The bone and the immune system share important regulatory mechanisms. The better understanding of the control of osteoclast formation by the immune cells together with the novel functions of bone cells in the regulation of cells outside the bone can provide key insights into the molecular mechanisms of bone destruction under inflammatory conditions.

In addition to inflammatory bone disorders, altered immune responses may cause other osteolytic diseases. In **postmenopausal osteoporosis**, reduced estrogen levels after menopause are associated with rapid and sustained increase in the rate of bone loss. This phenomenon is the result of the increased bone resorption which is not followed by an equivalent increase in bone formation. Activated T cells may cause rapid bone loss under conditions of estrogen deficiency by enhancing TNF-α production, therefore promoting osteoclastogenesis (Teitelbaum 2004).

The detailed osteoimmunological understanding of the pathogenesis of bone loss by osteoclasts may lead to novel strategies for the treatment of inflammatory diseases, including rheumatoid arthritis, periodontitis, and osteolytic bone metastases. Current therapies applied in the clinical practice for the treatment of rheumatoid arthritis target various aspects of the abovementioned inflammatory pathways. Pharmacological treatment of rheumatoid arthritis includes disease-modifying antirheumatic drugs (DMARDs), anti-inflammatory agents, and analgesics. Novel biological agents such as TNF blockers, IL-1β and IL-6 pathway blockade, monoclonal antibodies against B cells, and T-cell costimulation blockers have been found to produce durable symptomatic remissions in patients with rheumatoid arthritis (Choi et al. 2009). Effective treatments for bone metastases are not yet available. Pharmacological interventions provide palliative treatment, but they do not offer a lifelong benefit to patients with advanced cancer. The reduction of inflammation and attenuation of the host's immune reaction to the microbial plaque are the actual therapies for periodontitis.

The previously mentioned therapies have been shown to slow down progressive tissue damage, and not all patients respond to these therapies with respect to bone destruction. In addition to targeting inflammation, it is highly desirable to find mechanisms that stop and reverse bone destruction in inflammatory diseases.

Bisphosphonates are pyrophosphate analogs that target osteoclasts. Bisphosphonates, which are considered the standard care for reducing bone loss in postmenopausal osteoporosis, are used to prevent bone destruction in rheumatoid arthritis. High potency intravenous bisphosphonates have also been shown to modify progression of skeletal metastasis in several forms of cancer (Choi et al. 2009).

Denosumab, a monoclonal antibody raised against RANKL, has recently been developed and tested in the clinic. The development of this antibody was focused on postmenopausal osteoporosis and RANKL blockade resulted in sustained suppression of the systemic markers of bone turnover. The efficiency of RANKL-specific antibody for rheumatoid arthritis has also been reported (Choi et al. 2009), and its preventive effect on inflammatory bone loss is promising.

Protein kinases are essential components of diverse signaling pathways in immune cells. Their fundamental functions have made them effective therapeutical targets in inflammatory diseases. **Tofacitinib**, a Jak kinase inhibitor which blocks signal transduction by various cytokine receptors, has recently been approved for the treatment of human rheumatoid arthritis, and it provides significant protection from arthritis-induced joint damage. Another tyrosine kinase inhibitor, the Syk inhibitor **fostamatinib**, has also shown promising efficacy in human rheumatoid arthritis. The role of the Syk in various immune processes and the development and function of osteoclasts (Mócsai et al. 2010) raise the possibility that fostamatinib may also provide significant bone protection in an inflammatory environment.

Taken together, the development of new therapeutics is required, and to this end, more profound understanding of the molecular mechanism underlying inflammation-related bone resorption should be achieved. Blockade of osteoclastic intracellular signaling pathways downstream of RANK prevented bone destruction in ovariectomy-induced bone loss in mouse models (Choi et al. 2009). Therefore, inhibition of RANK signaling may be beneficial for protection against bone damage. Protein kinase inhibitors also will likely serve as a major novel group in anti-inflammatory therapy for bone protection during inflammatory diseases.

Cross-References

▶ Rheumatoid Arthritis

References

Arron, J. R., & Choi, Y. (2000). Bone versus immune system. *Nature, 408*, 535–536.

Boyle, W. J., Simonet, W. S., & Lacey, D. L. (2003). Osteoclast differentiation and activation. *Nature, 423*(6937), 337–342.

Choi, Y., Arron, J. R., & Townsend, M. J. (2009). Promising bone-related therapeutic targets for rheumatoid arthritis. *Nature Reviews Rheumatology, 5*, 543–548.

Faccio, R. (2011). Immune regulation of the tumor/bone vicious cycle. *Annals of the New York Academy of Sciences, 1237*, 71–78.

Firestein, G. S. (2003). Evolving concepts of rheumatoid arthritis. *Nature, 423*(6937), 356–361.

Goldring, S. R., & Goldring, M. B. (2007). Eating bone or adding it: The Wnt pathway decides. *Nature Medicine, 13*, 133–134.

Harre, U., Georgess, D., Bang, H., Bozec, A., Axmann, R., Ossipova, E., et al. (2012). Induction of osteoclastogenesis and bone loss by human autoantibodies against citrullinated vimentin. *Journal of Clinical Investigation, 122*(5), 1791–1802.

Humphrey, M. B., Lanier, L. L., & Nakamura, M. C. (2005). Role of ITAM-containing adapter proteins and their receptors in the immune system and bone. *Immunology Reviews, 208*, 50–65.

Lorenzo, J., Horowitz, M., & Choi, Y. (2007). Osteoimmunology: Interaction of the bone and immune systems. *Endocrine Reviews, 29*, 403–440.

Mócsai, A., Ruland, J., & Tybulewicz, V. L. (2010). The SYK tyrosine kinase: A crucial player in diverse biological functions. *Nature Reviews Immunology, 10*(6), 387–402.

Ross, F. P., & Teitelbaum, S. L. (2005). αvβ3 and macrophage colony-stimulating factor: Partners in osteoclast biology. *Immunology Reviews, 208*, 88–105.

Schett, G. (2009). Osteoimmunology in rheumatic diseases. *Arthritis Research & Therapy, 11*, 210.

Schett, G., & Teitelbaum, S. L. (2009). Osteoclasts and arthritis. *Journal of Bone and Mineral Research, 24*(7), 1142–1146.

Takayanagi, H. (2007). Osteoimmunology: Shared mechanisms and crosstalk between the immune and bone systems. *Nature Reviews Immunology, 7*(4), 292–304.

Takayanagi, H. (2010). New immune connections in osteoclast formation. *Annals of the New York Academy of Sciences, 1192*, 117–123.

Teitelbaum, S. L. (2000). Bone resorption by osteoclasts. *Science, 289*(5484), 1504–1508.

Teitelbaum, S. L. (2004). Postmenopausal osteoporosis, T cells, and immune dysfunction. *Proceedings of the National Academy of Sciences, 101*, 16711–16714.

Pathogen-Associated Molecular Patterns (PAMPs)

Sandro Silva-Gomes[1,2], Alexiane Decout[1,2] and Jérôme Nigou[1,2]
[1]Département "Mécanismes moléculaires des infections mycobactériennes", Institut de Pharmacologie et de Biologie Structurale, UMR 5089 CNRS; IPBS/Université Paul Sabatier, Toulouse, France
[2]Université de Toulouse; UPS; IPBS, Toulouse, France

Synonyms

Innate immune targets; Ligands of pattern recognition receptors; Microbe-associated molecular patterns (MAMPs)

Definition

PAMPs are conserved molecular structures produced by microorganisms and recognized as foreign by the receptors of the innate immune system.

Introduction

The term PAMPs was first introduced in 1989 by Janeway in his visionary article proposing the pattern recognition theory (Janeway 1989), to describe microbial components that are not found in multicellular hosts and whose recognition by a limited number of germline-encoded innate immune receptors (referred to as pattern recognition receptors: PRRs) allows detection of nonself, i.e., infection. The definition of this concept, which provided a framework for innate immune recognition, was later validated experimentally and has now become a paradigm (Medzhitov 2009, 2013). Innate immunity is the first line of host defense against invading microorganisms. It is mediated by a range of specialized cells such as phagocytes and, unlike adaptive immunity, is immediately active and does not require a previous encounter with the infective agent. To be suited to innate immune recognition, PAMPs follow three criteria: (i) they have an invariant core structure among a given class of microorganisms, (ii) they are products of pathways that are unique to microorganisms allowing discrimination of self from nonself, and (iii) they are essential for the survival of the microorganism and are therefore difficult for it to alter (Janeway 1989; Medzhitov 2007). These conserved molecular structures are not unique to pathogens and are therefore sometimes referred to as microbe-associated molecular patterns (MAMPs), which would be a more accurate term. However, the historical term PAMPs remains the most widely used. In addition, the appropriateness of the appellation "molecular patterns" has been questioned since innate immunity simply involves the recognition of a molecule of microbial origin by a molecule, i.e., a receptor, of the host (Beutler 2003). Nevertheless, PRRs usually do not recognize the entire molecule of microbial origin (often a macromolecule), but only a small and conserved part of it, that can be referred to as the pattern recognized by the receptor. The innate immune system has evolved to recognize PAMPs with a chemical nature as diverse as lipids, proteins, carbohydrates, or nucleic acids (Table 1).

PRRs are membrane-associated (either surface-exposed or endosomal) or cytosolic evolutionary conserved receptors, which include toll-like receptors (TLRs), NOD-like receptors (NLRs), C-type lectin receptors (CLRs), RIG-I-like receptors (RLRs), and DNA sensors (Table 1). PAMP recognition by PRRs elicits intracellular signaling cascades in immune cells that rapidly lead to the activation of transcription factors, such as NF-κB or interferon-regulated factors (IRF), and to transcriptional expression of inflammatory mediators that ultimately shape the adaptive immune response and coordinate the elimination of pathogens and infected cells (Akira et al. 2006; Iwasaki and Medzhitov 2010; Takeuchi and Akira 2010). However, aberrant activation of this system may lead to pathology, including chronic or acute inflammation, septic shock, immunodeficiency, or induction of autoimmunity (Takeuchi and Akira 2010).

The main PAMPs found in microorganisms and the molecular bases underlying their recognition by the innate immune system will be described in the following sections. Recent reports of the structures of several PRR-ligand complexes determined by X-ray crystallography have been a major breakthrough by allowing definition of these interactions at the atomic level (Kang and Lee 2011) and by providing a rationale to design synthetic molecules that target PRRs and their associated signaling pathways. How this fundamental knowledge might be translated into new therapeutic strategies/tools to be explored will be illustrated by some examples.

Diversity, Structure, and Receptors

Table 1 tentatively lists all the PAMPs known so far. Their structural diversity will be highlighted in the following paragraphs, with a special focus on PAMPs for which the molecular bases underlying their recognition by the innate immune system is clearly defined. For greater clarity, microbial molecules bearing PAMPs will be divided into lipids, proteins, carbohydrates, and nucleic acids, according to the usual biochemical classification of biological molecules (Table 1). Whenever possible and sufficiently chemically defined, PAMPs, i.e., the structural motifs recognized by the innate immune receptors, will be distinguished from the microbial molecule they belong to.

Lipids

Membrane- or cell wall-associated lipids constitute an important class of microbial molecules detected by the innate immune system. They are structurally very diverse with polar head groups of variable chemical nature and size (mono-, oligo-, polysaccharide; peptide, protein, others) linked to aliphatic hydrocarbon chains that can vary in number and length. They are generally recognized via their lipid moiety by TLR4 or TLR2, although the hydrophilic moiety can have a modulating effect or may be recognized by another PRR (Table 1).

Lipopolysaccharide

Lipopolysaccharide (LPS), a highly abundant glycolipid of the outer layer of the Gram-negative bacteria outer membrane, is probably the most widely studied and characterized microbial molecule that is recognized by the innate immune system (Raetz and Whitfield 2002). It is composed of a lipid anchor, namely lipid A, substituted by a polysaccharidic chain made of a relatively conserved core and a highly variable O-antigen chain (Fig. 1a). The LPS structure is very diverse and varies according to the bacterial genus and species, as well as to the environmental conditions.

LPS has been known for a long time as a strong inducer of the pro-inflammatory cytokine response. It is sometimes referred to as Gram-negative bacterial endotoxin as its recognition by the innate immune system can result in an overwhelming and uncontrolled response leading to fatal septic shock (Raetz and Whitfield 2002). LPS is recognized by a stable heterodimer receptor consisting of TLR4 and MD-2 (Poltorak et al. 1998). TLR4-MD-2-mediated LPS recognition plays a critical role in the detection of Gram-negative bacteria. LPS is extracted from the bacterial outer membrane and transferred to TLR4-MD-2 by two accessory proteins,

Pathogen-Associated Molecular Patterns (PAMPs), Table 1 PAMPs diversity and receptors involved in their recognition

Chemical family	Microbial molecule	PAMP	Microorganism	PRR Membrane	Cytosolic	Acessory molecules/ co-receptors
Lipids	Lipopolysaccharide (LPS)	Lipid A	Gram-negative bacteria	TLR4[a]-MD-2		LPB, CD14
	Lipopolysaccharide (LPS) (atypical)	Atypical lipid A	*Leptospira*	TLR2[a]		CD14
	Lipopolysaccharide (LPS) (atypical)	Lipid A	*Helicobacter pylori*	TLR4[a]-MD-2		
		Lewis x antigen		DC-SIGN[d]		
	Lipopolysaccharide (LPS) (atypical)	Lipid A	*Klebsiella pneumoniae*	TLR4[a]		
		Oligomannosides		DC-SIGN[d]		
	Lipoproteins, lipopeptides	Diacylated lipopeptides	*Mycoplasma*, Gram-positive bacteria	TLR2-TLR6[a]		CD14, CD36
	Lipoproteins, lipopeptides	Triacylated lipopeptides	Gram-negative bacteria, mycobacteria	TLR2-TLR1[a]		CD14
	Lipoteichoic acid (LTA)	Glycosylated diacylglycerol anchor	Gram-positive bacteria	TLR2-TLR6[a]		CD14, CD36
	Phosphatidyl-*myo*-inositol mannosides (PIM), Lipomannan (LM), Lipoarabinomannan (LAM)	Mannosyl-phosphatidyl-*myo*-inositol anchor	Mycobacteria	TLR2-TLR1[a]		CD14
		Oligomannosides		MR[d], DC-SIGN[d]		
	Lipophosphoglycan (LPG)	Lipid anchor (?)	*Leishmania* (parasites)	TLR2[a]		
		Oligomannosides		DC-SIGN[d]		
	tGPI-mucin	Lipid anchor (?)	*Trypanosoma* (parasites)	TLR2[a]		CD14
	Glycoinositolphospholipids	ND	*Trypanosoma* (parasites)	TLR4[a]		
	Phospholipomannan	Lipid anchor (?)	*Candida albicans* (fungi)	TLR2[a]		Galectin-3
	Trehalose-6′6′-dimycolate (TDM)	Diacylated trehalose	Mycobacteria	Mincle[d], MCL[d]		
		ND		TLR2[a]		Marco, CD14
	Malassezia Glycolipids	ND	*Malassezia* (fungi)	Mincle[d]		
Proteins	Porins	ND	Bacteria	TLR2[a]		
	Flagellin	Constant domain D1	Flagellated bacteria	TLR5[a]		
	Hemagglutinin	ND	Virus	TLR2[a]		
	Profilin-like molecule	ND	*Toxoplasma gondii* (parasites)	TLR11[a]		

(continued)

Pathogen-Associated Molecular Patterns (PAMPs), Table 1 (continued)

Chemical family	Microbial molecule	PAMP	Microorganism	PRR Membrane	PRR Cytosolic	Acessory molecules/ co-receptors
Carbohydrates	Mycobacterial glycoproteins	Oligomannosides	Mycobacteria	DC-SIGN[d], MR[d]		
	Viral glycoproteins	Oligomannosides	Virus	DC-SIGN[d], MR[d]		
	Parasite glycoproteins	Lewis x antigen	Schistosoma mansoni (parasite)	DC-SIGN[d]		
	Malassezia glycoprotein	Oligomannosides	*Malassezia* (fungi)	Dectin-2[d]		
	($\beta 1 \rightarrow 3$)-glucans	($\beta 1 \rightarrow 3$)-oligoglucosides	Fungi	Dectin-1[d]		
	Cyclic ($\beta 1 \rightarrow 2$)-glucans	ND	*Brucella* (bacteria)	TLR4[a]		
	Fungi α-mannan	ND	Fungi	Dectin-2[d]		
	Glucuronoxylomannan	ND	*Cryptococcus neoformans* (fungi)	TLR2[a], TLR4[a]		CD14
	Peptidoglycan	Diaminopimelic acid motifs	Bacteria		NOD1[b]	
		Muramyl dipeptide			NOD2-NALP1[b]	
Nucleic acids	RNA	ssRNA	Viruses	TLR7[a,f], TLR8[a,f]	NOD2[b]	
		dsRNA	Viruses	TLR3[a,f]	RIG-I[c], MDA5[c]	
		5'-Triphosphate RNA (3pRNA)	Viruses		RIG-I[c]	
		23S rRNA	Bacteria	TLR13[a]		
		mRNA	Bacteria			
	DNA	dsDNA (CpG motifs)	Bacteria, viruses	TLR9[a, f]		
		dsDNA	Bacteria, viruses		DAI[e], AIM2[e], IFI16[e], DDX41[e]-STING, Pol III- RIG-I[c]	
	Cyclic dinucleotides	Cyclic di-GMP	Bacteria		DDX41[e]-STING	
		Cyclic di-AMP	Bacteria		DDX41[e]-STING	
Others	Hemozoin		*Plasmodium* (parasites)	TLR9[a, f]	NALP3[b]	

ND Not determined
[a] Toll-like receptors
[b] NOD-like receptors
[c] RIG1-like receptors
[d] C-type lectins receptors
[e] DNA sensors
[f] Endosomal

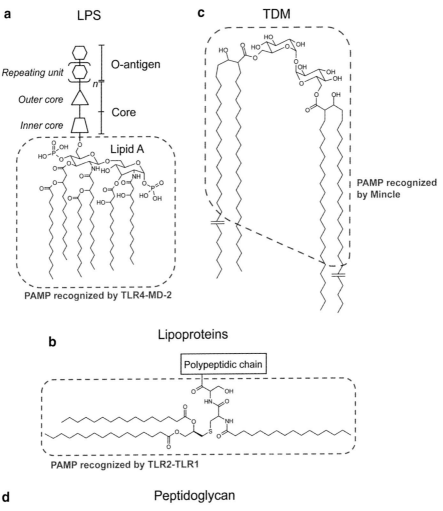
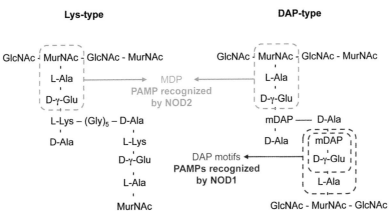

Pathogen-Associated Molecular Patterns (PAMPs), Fig. 1 Generic structure of LPS (**a**), lipoproteins (**b**), TDM (**c**) and peptidoglycan, (**d**) and PAMPs recognized by PRRs. *DAP* diaminopimelic acid, *GlcNAc* N-acetylglucosamine, *LPS* lipopolysaccharide, *MDP* muramyl dipeptide, *MurNAc* N-acetylmuramic acid, *TDM* trehalose-6,6′-dimycolate

LPS-binding protein (LBP) and CD14. LBP transfers LPS to CD14, which in turn transfers LPS to the TLR4-MD2 receptor. The crystal structure of CD14 revealed the presence of a large hydrophobic pocket able to accommodate LPS but also a multiplicity of grooves that probably contribute to the broad ligand specificity of CD14. Indeed, CD14 can bind other microbial compounds such as lipoproteins, lipoteichoic acid (LTA), or lipoglycans (see below).

The lipid A portion, composed of a diglucosamine backbone containing multiple acyl chains (typically four to seven) linked by ester or amide bonds (Fig. 1a) (Raetz and Whitfield 2002), is the conserved structural motif (PAMP) recognized by the TLR4-MD-2 heterodimer (Table 1), although the core can have a modulating effect. A crystal structure of LPS (Ra form of *Escherichia coli*) in complex with a soluble form of TLR4-MD-2 has revealed that LPS binding induces the formation of a multimer composed of two copies of the TLR4-MD-2 LPS complex arranged symmetrically (Kang and Lee 2011; Park et al. 2009). Five of the six lipid chains of LPS are buried deep inside the hydrophobic pocket of MD-2, whereas the remaining one is partially exposed to the surface, forming hydrophobic interactions with the other TLR4 member of the dimer. The phosphate groups of LPS contribute to receptor multimerization by forming ionic interactions with positively charged residues of MD-2 and both TLR4 members of the dimer. The degree of acylation and the fatty acyl chain length of lipid A play a critical role in its pro-inflammatory activity, hexa-acylated forms, with C12 or C14 acyl chains, being the most active. Deletion or addition of one acyl chain dramatically decreases activity. Tetra- or triacylated lipid A can act as TLR4 antagonists, while possibly becoming TLR2 agonists. Some pathogens have evolved immune escape strategies by modifying the structure of their LPS. Two main strategies are found: (i) prevention of innate immune recognition of LPS by TLR4-MD-2, as a result of steric masking of Lipid A or by modulating its acylation degree, as in *Francisella tularensis*, *Coxiella burnetii*, or *Brucella abortus*, among others and (ii) directing LPS toward an immunomodulatory receptor, most particularly by adding carbohydrate motifs recognized by the C-type lectin DC-SIGN, such as Lewis antigens by *Helicobacter pylori* or mannose residues by *Klebsiella pneumoniae* (van Kooyk and Geijtenbeek 2003).

Lipoproteins

Lipoproteins are a class of ubiquitous bacterial proteins that are anchored to the cell envelopes via a conserved N-terminal cysteine modified by lipid chains, mostly palmitoyl groups. Lipoproteins from Gram-negative bacteria and mycobacteria typically have three lipid chains, two of them are attached by ester bonds to a glyceryl backbone that is in turn connected to the sulfur atom of the N-terminal cysteine (S-diacylglyceryl moiety) (Fig. 1b). The third lipid chain is connected to the amino terminus via an amide bond. Lipoproteins from Gram-positive bacteria and mycoplasmas often lack the amide-linked lipid chain and are therefore diacylated only. The polypeptidic chain is highly variable. In Gram-positive bacteria, cell-associated lipoproteins are found in the plasma membrane. In Gram-negative bacteria, a relatively small number of lipoproteins remain in the plasma membrane, the vast majority being located in the outer membrane. In mycobacteria, lipoproteins are located both in the plasma membrane and in the cell wall. Lipoproteins induce strong TLR2-dependent pro-inflammatory responses and can be detected at picomolar levels (Brightbill et al. 1999; Zahringer et al. 2008). Scavenger receptors CD36 and CD14 contribute to lipoprotein recognition and TLR2 signal transduction by increasing bioavailability of the ligands (Table 1). TLR2 generally functions as a heterodimer with either TLR1 or TLR6, which is involved in discrimination of the acylation state of lipoproteins. The acylated cysteinyl moiety is the conserved structure (PAMP) recognized by the receptors (Fig. 1b). Triacylated lipoproteins are preferentially recognized by the TLR2/TLR1 complex, whereas diacylated lipoproteins are recognized by the TLR2/TLR6 complex. However, in addition to the acylation pattern, the nature of the amino acids of the peptidic chain can also modulate the specificity of the recognition by the heterodimers.

Crystal structures with model lipopeptides have established that the triacylated lipopeptide appears to form a bridge between TLR2 and TLR1 with the two ester-bound fatty acyl chains inserted deep into a pocket in the hydrophobic core of TLR2, the third amide-linked acyl chain occupying a hydrophobic channel at the surface of TLR1 and the conserved polar head located at the region of contact between the two receptors (Jin et al. 2007; Kang and Lee 2011). In the case of diacylated lipopeptides, the hydrophobic dimerization interface between TLR2 and TLR6 is increased, which compensates for the lack of the amide-linked lipid chain interaction with the receptors (Kang et al. 2009). In contrast, the amino acids of the polypeptidic chain have only limited interactions with TLR2, TLR1, or TLR6, which is consistent with the fact that the highly variable polypeptide chain of lipoproteins is not included in the pattern that is recognized by the receptors (Jin et al. 2007; Kang and Lee 2011; Kang et al. 2009).

Lipoteichoic Acid

Lipoteichoic acid (LTA) is a term covering macroamphiphiles encountered in the plasma membrane of the majority of low G+C Gram-positive bacteria (Firmicutes) (Sutcliffe 1994). They are composed of a lipid anchor, made of a diacylglycerol unit substituted by a di- or tri-glycoside, that is attached to a chain of polyglycerol or polyribitol units separated by a phosphate group. Chain length can vary from five to double-figure repeated units. Four types of LTA have been described. The most frequent are the simple polyglycerolphosphate (PGP)-LTAs such as in *Staphylococcus aureus*, *Bacillus subtilis*, or *Listeria monocytogenes*. In PGP-LTA, the repeated unit is simply a glycerol phosphate, position 2 of the glycerol being substituted by D-alanine in the majority of the cases, but also by glycosidic units, such as galactosyl, diglucosyl, or N-acetylglucosaminyl residues. However, LTA can have a more elaborate structure, as in *Streptococcus pneumoniae* (Ray et al. 2013).

LTAs have weaker pro-inflammatory activity when compared to lipopeptides (Zahringer et al. 2008). LTAs are recognized not only by TLR2 in cooperation with TLR6, CD36, and CD14 (Table 1) but also by another receptor of the innate immune system, the lectin pathway of the complement. The PAMP recognized by TLR2 is not completely defined although it clearly involves the lipid anchor. A crystal structure of TLR2 in complex with *Streptococcus pneumoniae* LTA showed that the two fatty acyl chains of the diacylglycerol unit are inserted in the hydrophobic pocket of TLR2 (Kang et al. 2009).

Lipoglycans

High G+C Gram-positive bacteria of the suborders Corynebacterineae (including mycobacteria) and Pseudonocardineae do not produce LTA but rather lipoglycans, with a structure based on a mannosyl-phosphatidyl-*myo*-inositol (MPI) anchor (Ray et al. 2013; Sutcliffe 1994). Mycobacteria, which include three major human pathogens, *Mycobacterium tuberculosis*, *Mycobacterium leprae*, and *Mycobacterium ulcerans*, produce a family of MPI-based lipoglycans, namely, phosphatidyl-*myo*-inositol mannosides (PIM), lipomannans (LM), and lipoarabinomannans (LAM). A portion of these molecules is anchored to the plasma membrane, where they are biosynthesized, but they are also found in the outermost layers of the bacterial envelope. Lipoglycans share a conserved MPI anchor, based on a *sn*-glycero-3-phospho-(1-D-*myo*-inositol) unit with one α-D-mannosyl unit linked at the *O*-2 of the *myo*-inositol. This MPI anchor contains four potential sites of acylation distributed along the glycerol unit, the mannosyl unit, and the *myo*-inositol. The *O*-6 of *myo*-inositol can be glycosylated by one or five mannosyl units, yielding PIM. LM correspond to polymannosylated PIM and are built from a conserved (α1→6)-mannopyranosyl chain. LAM correspond to LM with an attached D-arabinan domain. Tri- or tetra-acylated PIM, LM, and LAM are recognized by TLR2 in cooperation with TLR1 and CD14 (Table 1) (Gilleron et al. 2008). The PAMP recognized by the TLR2-TLR1 heterodimer is the MPI anchor

(Ray et al. 2013). However, the carbohydrate moiety modulates the magnitude of TLR2 activation, and LM is a much stronger TLR2 agonist than PIM or LAM (Ray et al. 2013). Indeed, TLR2 activation increases with the length of the (α1→6)-mannopyranosyl chain through a yet unknown mechanism; however, it decreases in LAM, as a result of the masking of the (α1→6)-mannopyranosyl chain by a bulky arabinan domain (Ray et al. 2013).

PIM, LM, and mannose-capped LAM (ManLAM) are also recognized by the C-type lectins DC-SIGN and mannose receptor (MR), through binding of repeated terminal (α1→2)-oligomannosides (Gilleron et al. 2008). DC-SIGN and MR recognize, via conserved carbohydrate recognition domains (CRDs), mannose-containing glycoconjugates in a Ca^{2+}-dependent manner (van Kooyk and Geijtenbeek 2003). High-avidity binding is achieved through multivalent interaction between multiple terminal oligomannosides (which can be repeated on the same microbial molecule and/or whose density is increased by clustering of microbial molecules in solution or at the surface of microorganisms) with multiple CRDs (that are associated, either on the same polypeptidic chain such as for MR or by oligomerization such as for DC-SIGN) (Fig. 2) (Gilleron et al. 2008; Nigou et al. 2001; van Kooyk and Geijtenbeek 2003). Binding to MR or DC-SIGN mediates entry of mycobacteria into phagocytic cells but might also be an immune escape strategy developed by the pathogen, via triggering of immunosuppressive pathways associated with these receptors (Gilleron et al. 2008; Nigou et al. 2001; van Kooyk and Geijtenbeek 2003).

Trehalose Dimycolate

Trehalose-6,6′-dimycolate (TDM; also called cord factor) is a major immunostimulant of the mycobacterial cell wall. It plays a key role in granuloma formation, a characteristic immune process of mycobacterial infection, and contributes to the effect of complete Freund's adjuvant, an emulsion of mycobacterial cell wall components in paraffin oil. It is composed of a trehalose core acylated at the 6- and 6′-OH positions with mycolic acids, which are α-alkylated β-hydroxylated long-chain fatty acids (from 60 to 80 carbon atoms) (Fig. 1c). The strong immunogenicity of TDM is mediated by the C-type lectins Mincle (macrophage-inducible C-type lectin) (Ishikawa et al. 2009) and MCL (macrophage C-type lectin) (Miyake et al. 2013) (Table 1). MCL is a dominant low-affinity TDM receptor in resting cells. Upon stimulation, it triggers the expression of inducible Mincle, which is a high-affinity receptor for TDM. Binding of TDM to optimally expressed Mincle induces innate immune responses (Miyake et al. 2013).

The crystal structure of the extracellular domain of Mincle with trehalose shows a primary Ca^{2+}-dependent binding site for one glucose residue of the trehalose while the second glucose residue interacts with the receptor through a secondary binding site of lower affinity. Situated adjacent to the primary binding site is a hydrophobic groove providing a docking site for one of the acyl chains (Feinberg et al. 2013). These hydrophobic regions are found only in Mincle and MCL and not in other C-type lectins, suggesting a role for the fatty acids in the binding of TDM (Furukawa et al. 2013). Accordingly, structure/function data indicate that the PAMP recognized by the receptors involves the trehalose core as well as a portion of the fatty acids (Fig. 1c). Besides Mincle and MCL, a class A scavenger receptor, MARCO, has been shown to cooperate with TLR2 and CD14 to detect TDM (Table 1) (Bowdish et al. 2009).

Proteins

Very few proteins per se have been described as PAMPs. The best characterized is flagellin from β- and γ-proteobacteria that constitutes the whiplike flagellar filament responsible for locomotion (Akira et al. 2006; Takeuchi and Akira 2010). Flagellin is recognized by TLR5, via its constant domain D1, which is relatively conserved among species (Table 1). A crystal structure of a D1/D2/D3 fragment of *Salmonella* flagellin in complex with a soluble form of TLR5 has revealed that TLR5 primarily interacts with three helices of the D1 domain (Yoon et al. 2012). Flagellin binding induces the formation of a multimer composed of

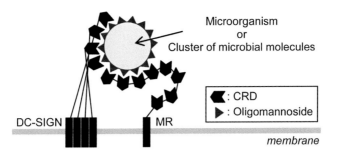

Pathogen-Associated Molecular Patterns (PAMPs), Fig. 2 Principle of high-avidity recognition by C-type lectins through multivalent interaction. *CRD* carbohydrate recognition domain, *DC-SIGN* dendritic cell-specific intercellular adhesion molecule-3-grabbing non-integrin, *MR* mannose receptor

two copies of the TLR5-flagellin complex arranged symmetrically (Yoon et al. 2012).

Carbohydrates

Carbohydrates exhibit a high structural diversity because of branching and complexing of monomers or polymers or through the formation of glycoconjugates, such as glycoproteins or glycolipids. A large variety of microbial glycoconjugates is sensed by the innate immune system. Glycoconjugates can be recognized not only via their aglycone moiety, as described above for glycolipids (Table 1), but also via their carbohydrate moiety. In the latter case, it usually involves lectins that are membrane protein receptors specialized in the recognition of carbohydrates. This is how many glycolipids are also recognized by C-type lectins (see the section "Lipids" and Table 1). Among glycoconjugates, glycoproteins constitute an important family of microbial molecules recognized by C-type lectins, whereas bacterial cell wall peptidoglycan is sensed by cytosolic NOD receptors. Polysaccharides are also widely distributed PAMPs.

Glycoproteins

Glycoproteins bearing *O*-mannoside motifs (mannoproteins) are expressed by a large variety of microorganisms, including mycobacteria, viruses, parasites, and fungi (Table 1). These mannoproteins, such as *M. tuberculosis* Apa or LpqH (also known as the 19 kDa lipoprotein), or HIV gp120, are recognized by the C-type lectins MR or DC-SIGN via their oligomannosides as described above (see section "Lipoglycans"; Fig. 2). A mannoprotein from the fungus *Malassezia* is sensed by Dectin-2 by a yet unknown mechanism (Table 1) (Ishikawa et al. 2013).

Peptidoglycan

Peptidoglycan is an essential component of the bacterial cell wall. It is a polymer of alternating ($\beta 1 \rightarrow 4$)-linked N-acetylglucosamine (GlcNAc) and N-acetylmuramic acid (MurNAc) carbohydrate residues, in which all lactyl groups of MurNAc are substituted with subunit peptides typically containing four alternating L- and D-amino acids (Fig. 1d). Peptide subunits from adjacent strands are often cross-linked, either directly or through short peptides and serve to cross-link the carbohydrate polymers. The structure of the latter is conserved in all bacteria whereas the composition of the peptide subunit varies. Most of the Gram-positive bacteria contain a lysine (Lys) residue at the second position of the chain (Lys type) while Gram-negative bacteria contain an m-diaminopimelic acid (mDAP) residue in this position (DAP type). The peptide subunits not only contain unusual amino acids like mDAP but also D isomers of amino acids that are not common in eukaryotes. Peptidoglycan is therefore ideally suited for innate immune recognition (Sorbara and Philpott 2011).

Muramyl dipeptide (MDP), a motif that is common to all bacteria, is recognized by nucleotide-binding oligomerization domain protein 2 (NOD2) (Table 1; Fig. 1d). It is worth noting that an unusual glycolylated form of MDP produced by mycobacteria is more potent and efficacious at inducing NOD2-mediated host responses (Coulombe et al. 2009). This is consistent with the historical observation that the mycobacterial cell wall exerts exceptional immunogenic activity. DAP-containing motifs, found in Gram-negative bacteria and some Gram-positive bacteria, are recognized by NOD1, D-γ-Glu-mDAP being the minimum structure required (Fig. 1d) (Sorbara and Philpott 2011). It remains to be determined at the atomic level how these peptidoglycan fragments are recognized by NOD proteins.

β-Glucans

β-glucans are polysaccharides found in fungi, such as *Aspergillus*, *Candida*, *Coccidioides*, *Penicillium*, *Pneumocystis*, and *Saccharomyces*, and in some bacteria, such as *Alcaligenes faecalis*. They are an extensive family of glucose polymers showing a variety of chemical structures. A common feature is a (β1→3)-glucan backbone that can bear (β1→6) ramifications (Plato et al. 2013). They can range from small soluble polymers to large insoluble particles. β-glucans are recognized by Dectin-1 (Brown and Gordon 2001), which is a phagocytic receptor belonging to the C-type lectin family, even though calcium ions are not required for ligand binding. Dectin-1 distinguishes between particulate and soluble β-glucans. Particulate β-glucans trigger activation of the signaling pathways leading to phagocytosis of the pathogens and production of reactive oxygen species and pro-inflammatory cytokines. Soluble β-glucans, on the other hand, bind to Dectin-1 without triggering the signaling pathways and can prevent activation by the particulate ligands. Particulate β-glucans cluster Dectin-1 into "phagocytic synapses" allowing downstream signaling. This process may allow the immune cells to discriminate between cell surface-associated molecules requiring phagocytosis to eliminate pathogens from soluble ligands released by remote organisms (Goodridge et al. 2011). How Dectin-1 interacts with β-glucans remains to be characterized. The putative binding site is a shallow hydrophobic groove on the surface of the protein that differs from the carbohydrate binding site found in other C-type lectins. Recognition by the receptor requires the (β1→3)-linked backbone and increases with the length of the polymer and the presence of (β1→6) side chains. The minimal motif required for binding to a soluble form of the receptor is a branched (β1→3)-heptaose with a single (β1→6) branched glucosyl unit.

Besides the widely distributed (β1→3)-glucans, cyclic (β1→2)-glucans are found in *Brucella* species. The latter require TLR4 for activation of innate immune cells but by a yet uncharacterized mechanism (Martirosyan et al. 2012).

Nucleic Acids

PAMPs can be broadly divided in molecular structures associated with microbial envelopes (e.g. lipids, proteins, and carbohydrates) and microbial nucleic acids. These two groups also differ in the cellular location of their cognate receptors. While the former is mainly recognized by receptors at the cellular membrane that probe the extracellular environment, the latter is detected by intracellular (endosomal and cytoplasmic) receptors (Table 1). Microbial nucleic acids are an important group of PAMPs, particularly for the recognition of pathogens that present few conserved molecular patterns, such as viruses. Indeed, nucleic acids are the main virus-derived PAMPs recognized by the innate immune system, and deficiency in receptors involved in their recognition has been linked to susceptibility to viral infections (Zhang et al. 2007). Unlike most other PAMPs, nucleic acids are highly abundant in the host, raising the question of how the immune system is able to discriminate between self and microbial nucleic acids. Although the complete mechanisms are not yet fully understood, discrimination is based on differences in sequence, structure, molecular modifications, and cellular localization. It should be noted that some PRRs involved in nucleic acid sensing make no differentiation between self and foreign molecules, since mislocalized self nucleic acids, such as extranuclear DNA or extracellular RNA, are detected as danger-associated molecular patterns (DAMPs) by the same mechanisms as nucleic acid PAMPs.

RNA

Double-stranded RNA (dsRNA) can be present in the viral genome of RNA viruses such as reoviruses, and is generated by most viruses at some point during their replication. Although short stretches of dsRNA can be found in eukaryotic cells, long dsRNA are exclusively produced during viral infections and therefore are de facto PAMPs. The most characterized receptor for dsRNA is TLR3, which is located in the endosome. Crystallographic studies have shown TLR3 has a defined binding site specific for dsRNA, but it does not distinguish between base sequences of dsRNA ligands. A TLR3 dimer binds to dsRNA oligonucleotides made of at least 45 bp, the minimal length required for signal transduction (Liu et al. 2008). In addition to TLR3, dsRNA can be recognized by the cytoplasmic PRRs retinoic acid-inducible protein I (RIG-I) and melanoma differentiation-associated gene 5 (MDA5). The crystal structure of MDA5 bound to dsRNA shows that the receptor recognizes the internal duplex structure of long dsRNA, whereas RIG-I recognizes the terminus of short dsRNA (Wu et al. 2013). In addition, it provides the molecular basis for the differential recognition of viruses by RIG-I and MDA5: RIG-I is essential for immunity to RNA viruses, including paramyxoviruses, influenza virus, and Japanese encephalitis virus, whereas MDA5 is critical for picornavirus detection (Kato et al. 2006). In addition to dsRNA, RIG-I recognizes short single-stranded RNA (ssRNA) with an uncapped 5′-triphosphate end, which is present in some RNA viruses but is very distinct from the endogenous RNA transcripts that present a 7-methylguanosine group at the 5′-end.

ssRNA is recognized by TLR7 and TLR8. Except for a possible preference of guanosine/uridine (GU)-rich ssRNA by the receptors, there is no specificity for sequence motifs. The distinction between pathogen and self-nucleic acids seems rather to rely on the compartmentalization of TLR7 and TLR8. These receptors are located in the endosome and sense viruses that enter the cell through endocytosis, but not cytoplasmic self ssRNA. Cytoplasmic viral ssRNA was shown to be recognized by NOD2 (Sabbah et al. 2009), which is also a receptor for the bacterial envelope component muramyl dipeptide, a motif of peptidoglycan (see above in paragraph 3). However, the mechanisms by which NOD2 recognizes ssRNA and how it distinguishes viral from self RNA are still unknown. It is likely that differences in methylation as well as other modifications may contribute to the distinction.

RNA PAMPs are also present in bacteria. A conserved 10 bp sequence from bacterial 23S ribosomal RNA (rRNA) is recognized by the mouse TLR13 (Oldenburg et al. 2012), a receptor absent in humans. The innate immune system can also detect bacterial mRNA (Sander et al. 2011), providing a toll to discriminate between live and dead pathogens and to respond accordingly. For this reason, bacterial mRNA was termed "vita-PAMP", i.e., viability-associated PAMP. The molecular recognition of this PAMP was proposed to depend on the absence of 3′-polyadenylation in bacterial mRNA, allowing the host to distinguish between self and foreign mRNA. The receptor for this PAMP remains to be identified. Interestingly, the existence of vita-PAMPs such as mRNA provides one explanation for the well-known fact that live vaccines induce higher immune responses than their killed counterparts.

DNA

Unmethylated cytosine-guanosine (CpG) motifs, present in the DNA of bacteria and viruses, but rare in mammalian genomes, are recognized by TLR9. The endosomal localization of the receptor also prevents the recognition of endogenous DNA. However, DNA is not sensed in the endosome only. Several cytosolic DNA sensors have been identified, although the exact mechanisms of recognition used by most of them are not known. In addition, some of them may be involved in signal translation rather than in direct recognition of the PAMPs. DAI (DNA-dependent activator of IRFs), AIM2 (absent in melanoma 2), IFI16 (IFNγ-inducible protein 16), and DDX41 (DEAD (aspartate-glutamate-alanine-aspartate)-box polypeptide 41) bind cytoplasmic dsDNA (Paludan and Bowie 2013). Both AIM2 and IFI16 contain a C-terminal DNA-binding HIN (hematopoietic interferon-inducible nuclear) domain, and the crystal structures of their HIN domains in complex with dsDNA revealed that non-sequence-specific DNA recognition is accomplished through electrostatic interaction between the positively charged HIN domain residues and the DNA phosphate backbone and not with individual nucleotides (Jin et al. 2012). Finally, AT-rich DNA, characteristic of some pathogen genomes, is recognized by RNA polymerase III, which transcribes it into dsRNA, which in turn is recognized by RIG-I.

Cyclic Dinucleotides

Many bacterial pathogens use cyclic diguanosine monophosphate (c-di-GMP) or cyclic diadenosine monophosphate (c-di-AMP) as secondary messengers. During infection, these dinucleotides act as PAMPs, being recognized by the helicase DDX41 and STING (Parvatiyar et al. 2012). These compounds represent another type of vita-PAMP, which, as discussed above, enable the detection of metabolically active pathogens.

Translation into Public Health Benefits

The discovery of PAMPs and how these shape the immune response opened the possibility of modulating immunity based on PAMPs or PAMP-like molecules. One major field of research that has harvested this knowledge is vaccination. Vaccines composed of attenuated or killed whole-cell microorganisms usually present sufficient immunostimulatory activity by themselves, as they carry PAMPs capable of eliciting a sustained immune response. In contrast, subunit vaccines, i.e., composed of purified or recombinant antigens, tend to be poor immunogens. This is overcome by the addition of adjuvants, which boost the immunogenicity of the vaccine. Until very recently, the development of adjuvants was an empirical process. Aluminum salts (alum), the first adjuvant licensed for human vaccines in the 1920s, nowadays still remains the most widely used adjuvant for vaccination. Although alum is effective at boosting antibody responses, it induces weak Th1 immunity, which is crucial for the development of vaccines against intracellular pathogens. Understanding how PAMPs are recognized at the molecular level and what kind of response they induce in vivo can help in the rational development of safe and more efficient adjuvants. Indeed, adjuvants based on PAMPs are a prospering field of research (Connolly and O'Neill 2012; Savva and Roger 2013).

The first one to be licensed for human use was 3-O-desacyl-4'-monophosphoryl lipid A (MPL), derived from the lipid A portion of LPS isolated from *Salmonella minnesota*, which retains the TLR4 agonist activity but not the toxicity of the parental molecule. MPL is a component in vaccines against the hepatitis B virus (Fendrix®) and the human papillomavirus (Cervarix®), both from GlaxoSmithKline (GSK) Biologicals. Other promising PAMP-based adjuvants under study include flagellin, dsRNA, and CpG oligodinucleotides (Plotkin 2009).

In addition to their use in vaccination, PAMPs and PAMP-based molecules are being investigated as agonists of PRRs, notably TLRs, for the treatment of infectious diseases and cancer. The list is quite extensive and includes agonists of TLR2, TLR4, TLR5, TLR7, TLR8, or TLR9 (Connolly and O'Neill 2012; Savva and Roger 2013). One of the most successful drugs is imiquimod (Aldara®, 3M Pharmaceuticals), an imidazoquinoline analogue of the ssRNA. It acts via a TLR7-dependent pathway to potently induce pro-inflammatory cytokines and is approved for treatment of external genital warts, actinic keratosis, and basal cell carcinoma.

The knowledge acquired from the structure of PAMPs and their mode of action is not limited to the design of new pathways to activate the immune system, but it can be used in the opposite direction, i.e., to reduce immune activation when it is deleterious for the patient. Eritoran (E5564; Eisai Co., Ltd.) is a synthetic lipopolysaccharide, whose structure is based on the nontoxic lipid A portion of *Rhodobacter sphaeroides* LPS, designed to antagonize the toxic effects of LPS. It competes with the lipid A component of LPS for binding to the hydrophobic pocket of MD-2, preventing its dimerization with TLR4 and subsequent downstream intracellular signaling. Therefore, Eritoran is a potentially effective therapeutic agent for treatment of pathologies characterized by deleterious hyperactivation of TLR4.

One such pathology is sepsis, a common cause of death among hospitalized patients. It is characterized by a systemic inflammation that generally arises in response to an infection. PAMP recognition by PRR, namely TLR4, is a central event in the pathogenesis of sepsis and hence targeting PAMP recognition may be a potential therapeutic option. In healthy volunteers, Eritoran blocked the inflammation induced by LPS administration and had no TLR4 agonist activity. However, in a phase III trial of patients with severe sepsis, Eritoran administration failed to demonstrate a significant inhibitory effect on mortality. This study suggests that LPS may not be the only trigger for the sepsis and that other PAMPs may be relevant for this pathology. Currently, Eritoran is being investigated for the treatment of TLR4-mediated pathologies, e.g. hemorrhagic shock and influenza infection (Shirey et al. 2013). Similar strategies have been used to develop antagonists of TLR7 and TLR9, whose activation is deleterious in inflammatory conditions such as systemic lupus erythematosus, rheumatoid arthritis, psoriasis, and multiple sclerosis. Compounds based on DNA, the natural PAMP recognized by TLR7 and TLR9, have been developed by Idera Pharmaceuticals and Dynavax Technologies and are currently in clinical phases of development.

The study of PAMPs, allied with the advances in our understanding of the precise roles of PRRs in the immune response, will continue to contribute to the development of human therapies, either by the use of PAMPs themselves or by the design of synthetic compounds able to modulate the immune response.

Summary

PAMPs are conserved molecular structures produced by microorganisms and recognized as foreign by the receptors of the innate immune system, referred to as pattern recognition receptors (PRRs). Innate immunity is the first line of host defense against invading microorganisms. It is immediately active, does not require a previous encounter with the infective agent, and has evolved to recognize PAMPs with a chemical nature as diverse as lipids, proteins, carbohydrates, or nucleic acids. The main PAMPs found in microorganisms and the molecular bases underlying their recognition by the innate immune system are described. How this fundamental knowledge might be translated into new therapeutic strategies/tools to be explored is illustrated.

Cross-References

▶ Bacterial Lipopolysaccharide
▶ Toll-Like Receptors

References

Akira, S., Uematsu, S., & Takeuchi, O. (2006). Pathogen recognition and innate immunity. *Cell, 124*(4), 783–801.

Beutler, B. (2003). Not "molecular patterns" but molecules. *Immunity, 19*(2), 155–156.

Bowdish, D. M., Sakamoto, K., Kim, M. J., Kroos, M., Mukhopadhyay, S., Leifer, C. A., et al. (2009). MARCO, TLR2, and CD14 are required for macrophage cytokine responses to mycobacterial trehalose dimycolate and Mycobacterium tuberculosis. *PLoS Pathogen, 5*(6), e1000474.

Brightbill, H. D., Libraty, D. H., Krutzik, S. R., Yang, R. B., Belisle, J. T., Bleharski, J. R., et al. (1999). Host defense mechanisms triggered by microbial lipoproteins through toll-like receptors. *Science, 285*(5428), 732–736.

Brown, G. D., & Gordon, S. (2001). Immune recognition. A new receptor for beta-glucans. *Nature, 413*(6851), 36–37.

Connolly, D. J., & O'Neill, L. A. (2012). New developments in toll-like receptor targeted therapeutics. *Current Opinion in Pharmacology, 12*(4), 510–518.

Coulombe, F., Divangahi, M., Veyrier, F., de Leseleuc, L., Gleason, J. L., Yang, Y., et al. (2009). Increased NOD2-mediated recognition of N-glycolyl muramyl dipeptide. *Journal of Experimental Medicine, 206*(8), 1709–1716.

Feinberg, H., Jegouzo, S. A., Rowntree, T. J., Guan, Y., Brash, M. A., Taylor, M. E., et al. (2013). Mechanism for recognition of an unusual mycobacterial glycolipid by the macrophage receptor mincle. *Journal of Biological Chemistry, 288*(40), 28457–28465.

Furukawa, A., Kamishikiryo, J., Mori, D., Toyonaga, K., Okabe, Y., Toji, A., et al. (2013). Structural analysis for glycolipid recognition by the C-type lectins Mincle and

MCL. *Proceedings of the National Academy of Sciences of the United States of America, 110*(43), 17438–17443.

Gilleron, M., Jackson, M., Nigou, J., & Puzo, G. (2008). Structure, biosynthesis, and activities of the phosphatidyl-myo-inositol-based lipoglycans. In M. Daffe & J. Reyrat (Eds.), *The mycobacterial cell envelope* (pp. 75–105). Washington, DC: ASM Press.

Goodridge, H. S., Reyes, C. N., Becker, C. A., Katsumoto, T. R., Ma, J., Wolf, A. J., et al. (2011). Activation of the innate immune receptor Dectin-1 upon formation of a 'phagocytic synapse'. *Nature, 472*(7344), 471–475.

Ishikawa, E., Ishikawa, T., Morita, Y. S., Toyonaga, K., Yamada, H., Takeuchi, O., et al. (2009). Direct recognition of the mycobacterial glycolipid, trehalose dimycolate, by C-type lectin Mincle. *Journal of Experimental Medicine, 206*(13), 2879–2888.

Ishikawa, T., Itoh, F., Yoshida, S., Saijo, S., Matsuzawa, T., Gonoi, T., et al. (2013). Identification of distinct ligands for the C-type lectin receptors Mincle and Dectin-2 in the pathogenic fungus Malassezia. *Cell Host & Microbe, 13*(4), 477–488.

Iwasaki, A., & Medzhitov, R. (2010). Regulation of adaptive immunity by the innate immune system. *Science, 327*(5963), 291–295.

Janeway, C. A., Jr. (1989). Approaching the asymptote? Evolution and revolution in immunology. *Cold Spring Harbor Symposia on Quantitative Biology, 54*(Pt 1), 1–13.

Jin, M. S., Kim, S. E., Heo, J. Y., Lee, M. E., Kim, H. M., Paik, S. G., et al. (2007). Crystal structure of the TLR1-TLR2 heterodimer induced by binding of a tri-acylated lipopeptide. *Cell, 130*(6), 1071–1082.

Jin, T., Perry, A., Jiang, J., Smith, P., Curry, J. A., Unterholzner, L., et al. (2012). Structures of the HIN domain: DNA complexes reveal ligand binding and activation mechanisms of the AIM2 inflammasome and IFI16 receptor. *Immunity, 36*(4), 561–571.

Kang, J. Y., & Lee, J. O. (2011). Structural biology of the Toll-like receptor family. *Annual Review of Biochemistry, 80*, 917–941.

Kang, J. Y., Nan, X., Jin, M. S., Youn, S. J., Ryu, Y. H., Mah, S., et al. (2009). Recognition of lipopeptide patterns by Toll-like receptor 2-Toll-like receptor 6 heterodimer. *Immunity, 31*(6), 873–884.

Kato, H., Takeuchi, O., Sato, S., Yoneyama, M., Yamamoto, M., Matsui, K., et al. (2006). Differential roles of MDA5 and RIG-I helicases in the recognition of RNA viruses. *Nature, 441*(7089), 101–105.

Liu, L., Botos, I., Wang, Y., Leonard, J. N., Shiloach, J., Segal, D. M., et al. (2008). Structural basis of toll-like receptor 3 signaling with double-stranded RNA. *Science, 320*(5874), 379–381.

Martirosyan, A., Perez-Gutierrez, C., Banchereau, R., Dutartre, H., Lecine, P., Dullaers, M., et al. (2012). Brucella beta 1,2 cyclic glucan is an activator of human and mouse dendritic cells. *PLoS Pathogen, 8*(11), e1002983.

Medzhitov, R. (2007). Recognition of microorganisms and activation of the immune response. *Nature, 449*(7164), 819–826.

Medzhitov, R. (2009). Approaching the asymptote: 20 years later. *Immunity, 30*(6), 766–775.

Medzhitov, R. (2013). Pattern recognition theory and the launch of modern innate immunity. *Journal of Immunology, 191*(9), 4473–4474.

Miyake, Y., Toyonaga, K., Mori, D., Kakuta, S., Hoshino, Y., Oyamada, A., et al. (2013). C-type lectin MCL is an FcRgamma-coupled receptor that mediates the adjuvanticity of mycobacterial cord factor. *Immunity, 38*(5), 1050–1062.

Nigou, J., Zelle-Rieser, C., Gilleron, M., Thurnher, M., & Puzo, G. (2001). Mannosylated lipoarabinomannans inhibit IL-12 production by human dendritic cells: evidence for a negative signal delivered through the mannose receptor. *Journal of Immunology, 166*(12), 7477–7485.

Oldenburg, M., Kruger, A., Ferstl, R., Kaufmann, A., Nees, G., Sigmund, A., et al. (2012). TLR13 recognizes bacterial 23S rRNA devoid of erythromycin resistance-forming modification. *Science, 337*(6098), 1111–1115.

Paludan, S. R., & Bowie, A. G. (2013). Immune sensing of DNA. *Immunity, 38*(5), 870–880.

Park, B. S., Song, D. H., Kim, H. M., Choi, B. S., Lee, H., & Lee, J. O. (2009). The structural basis of lipopolysaccharide recognition by the TLR4-MD-2 complex. *Nature, 458*(7242), 1191–1195.

Parvatiyar, K., Zhang, Z., Teles, R. M., Ouyang, S., Jiang, Y., Iyer, S. S., et al. (2012). The helicase DDX41 recognizes the bacterial secondary messengers cyclic di-GMP and cyclic di-AMP to activate a type I interferon immune response. *Nature Immunology, 13*(12), 1155–1161.

Plato, A., Willment, J. A., & Brown, G. D. (2013). C-type lectin-like receptors of the dectin-1 cluster: Ligands and signaling pathways. *International Reviews of Immunology, 32*(2), 134–156.

Plotkin, S. A. (2009). Vaccines: The fourth century. *Clinical and Vaccine Immunology, 16*(12), 1709–1719.

Poltorak, A., He, X., Smirnova, I., Liu, M. Y., Van Huffel, C., Du, X., et al. (1998). Defective LPS signaling in C3H/HeJ and C57BL/10ScCr mice: Mutations in Tlr4 gene. *Science, 282*(5396), 2085–2088.

Raetz, C. R., & Whitfield, C. (2002). Lipopolysaccharide endotoxins. *Annual Review of Biochemistry, 71*, 635–700.

Ray, A., Cot, M., Puzo, G., Gilleron, M., & Nigou, J. (2013). Bacterial cell wall macroamphiphiles: Pathogen-/microbe-associated molecular patterns detected by mammalian innate immune system. *Biochimie, 95*(1), 33–42.

Sabbah, A., Chang, T. H., Harnack, R., Frohlich, V., Tominaga, K., Dube, P. H., et al. (2009). Activation of innate immune antiviral responses by Nod2. *Nature Immunology, 10*(10), 1073–1080.

Sander, L. E., Davis, M. J., Boekschoten, M. V., Amsen, D., Dascher, C. C., Ryffel, B., et al. (2011). Detection of prokaryotic mRNA signifies microbial viability and promotes immunity. *Nature, 474*(7351), 385–389.

Savva, A., & Roger, T. (2013). Targeting Toll-like receptors: Promising therapeutic strategies for the management of sepsis-associated pathology and infectious diseases. *Frontiers in Immunology, 4*, 387.

Shirey, K. A., Lai, W., Scott, A. J., Lipsky, M., Mistry, P., Pletneva, L. M., et al. (2013). The TLR4 antagonist Eritoran protects mice from lethal influenza infection. *Nature, 497*(7450), 498–502.

Sorbara, M. T., & Philpott, D. J. (2011). Peptidoglycan: A critical activator of the mammalian immune system during infection and homeostasis. *Immunology Reviews, 243*(1), 40–60.

Sutcliffe, I. C. (1994). The lipoteichoic acids and lipoglycans of gram-positive bacteria: A chemotaxonomic perspective. *Systematic and Applied Microbiology, 17*, 467–480.

Takeuchi, O., & Akira, S. (2010). Pattern recognition receptors and inflammation. *Cell, 140*(6), 805–820.

van Kooyk, Y., & Geijtenbeek, T. B. (2003). DC-SIGN: Escape mechanism for pathogens. *Nature Reviews Immunology, 3*(9), 697–709.

Wu, B., Peisley, A., Richards, C., Yao, H., Zeng, X., Lin, C., et al. (2013). Structural basis for dsRNA recognition, filament formation, and antiviral signal activation by MDA5. *Cell, 152*(1-2), 276–289.

Yoon, S. I., Kurnasov, O., Natarajan, V., Hong, M., Gudkov, A. V., Osterman, A. L., et al. (2012). Structural basis of TLR5-flagellin recognition and signaling. *Science, 335*(6070), 859–864.

Zahringer, U., Lindner, B., Inamura, S., Heine, H., & Alexander, C. (2008). TLR2 – promiscuous or specific? A critical re-evaluation of a receptor expressing apparent broad specificity. *Immunobiology, 213*(3–4), 205–224.

Zhang, S. Y., Jouanguy, E., Ugolini, S., Smahi, A., Elain, G., Romero, P., et al. (2007). TLR3 deficiency in patients with herpes simplex encephalitis. *Science, 317*(5844), 1522–1527.

Pentraxins

Sébastien Jaillon, Antonio Inforzato, Barbara Bottazzi and Cecilia Garlanda
Humanitas Clinical Research Center, Rozzano (Milano), Italy

Synonyms

C-reactive protein (CRP) or PTX1; Pattern recognition molecule; PTX3; Serum amyloid protein (SAP) or PTX2

Definition

Pentraxins are a superfamily of evolutionary conserved proteins with multifunctional properties in innate immunity, including complement activation and opsonization (Fig. 1). Based on the primary structure, pentraxins are divided into two subfamilies: short and long pentraxins. Prototypes of the short pentraxin family are C-reactive protein (CRP, named for its ability to bind the C polysaccharide of *Streptococcus pneumoniae*) and serum amyloid P (SAP) component, both identified at the beginning of the last century as acute-phase protein in men and mouse, respectively. CRP and SAP are also known as PTX1 and PTX2, respectively. More recently, the long pentraxin PTX3 has been identified as early-induced gene by primary proinflammatory signals (Bottazzi et al. 2010; Garlanda et al. 2005). PTX3 is characterized by a C-terminal pentraxin-like domain homologous to CRP and a long N-terminal domain unrelated to other known proteins. PTX3 and CRP share fundamental functions as fluid-phase pattern recognition molecules (PRMs); however, they differ in structural features and pattern of expression. Unlike CRP, the sequence and regulation of PTX3 are conserved from mouse to man, this favoring a genetic approach to understand the in vivo roles of the protein. Gene-modified animals have been essential to reveal the multifunctional properties of PTX3 at the crossroad between innate immunity, inflammation, matrix deposition, and female fertility.

Biosynthesis and Release

CRP and *SAP* genes are located on chromosome 1 in close physical and genetic linkage. CRP and SAP are expressed by hepatocytes and constitute the main acute-phase proteins in human and mouse, respectively (Bottazzi et al. 2010). In human, the baseline level of circulating CRP is low (≤ 5 μg/ml) and increases rapidly and dramatically (1000-fold) in several pathological inflammatory conditions, with a sharp rise within 6 h of induction and a maximum at approximately 48 h.

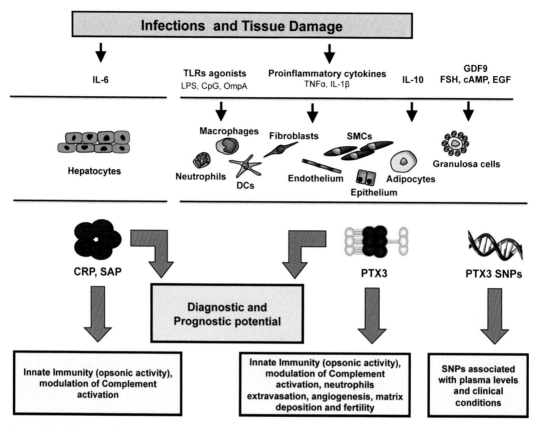

Pentraxins, Fig. 1 Cellular sources and main functions of the classical short pentraxin CRP and long pentraxin PTX3 in infections and tissue damage. CRP and SAP are mainly produced by hepatocytes in response to IL-6, whereas PTX3 is produced by different cells in response to TLR agonists and cytokines. The released pentraxins participate to innate responses and inflammation. PTX3 is also involved in angiogenesis and tissue remodeling. PTX3 polymorphisms have been associated with PTX3 plasma levels and different clinical conditions. Both pentraxins represent useful diagnostic and prognostic biomarkers in inflammatory and cardiovascular conditions

In contrast, SAP is the major acute-phase protein in the mouse, and its concentration level in plasma of healthy human (30–50 μg/ml) is not modified in inflammatory conditions (Mantovani et al. 2013).

The *PTX3* gene is remarkably conserved across evolutionary distant species in terms of sequence, structural organization, and regulation. For example, both human and murine *PTX3* genes localize on chromosome 3 and comprise three exons coding for leader peptide and N- and C-terminal domains of the protein, respectively (see below). Also, the gene's proximal promoters in both species share a number of transcriptional regulatory elements including Pu1, AP-1, NF-κB, SP1, and NF-IL-6 sites (Garlanda et al. 2005). It has been shown that the NF-κB binding site is essential for the transcriptional response to proinflammatory cytokines, where AP-1 controls the basal transcription of the gene. Several pathways of signal transduction have been implicated in the production of PTX3. For instance, high-density lipoproteins trigger PTX3 expression in endothelial cells via the PI3K/Akt axis, whereas the TNF-α-dependent induction of PTX3 in alveolar epithelial cells requires JNK instead of NF-κB pathways (Bottazzi et al. 2010).

PTX3 was originally identified as a cytokine-inducible gene in endothelial cells and fibroblasts. Following studies have reported PTX3 expression in a number of additional cell types (i.e., dendritic cells, monocytes, macrophages, epithelial cells, chondrocytes, and adipocytes) upon stimulation with inflammatory signals (e.g., IL-1β, TNF-α), toll-like receptor (TLR) agonists, microbial moieties (e.g., LPS, OmpA), or intact microorganisms (Garlanda et al. 2005) (Fig. 1). IFN-γ inhibits PTX3 expression in dendritic cells, monocytes, and macrophages, reducing both gene transcription and transcript stability. In contrast, induction of PTX3 by LPS is amplified by IL-10, pointing to a potential role of PTX3 in the IL-10-mediated resolution of inflammation. In this respect, PTX3 deficiency has been associated with increased inflammatory response and tissue damage in both sterile and infectious conditions (Jaillon et al. 2014; Salio et al. 2008). More recently, PTX3 expression has been reported in uroepithelial cells. In these cells, PTX3 is rapidly induced upon infection with uropathogenic *Escherichia coli* (UPEC) in a TLR4- and MyD88-dependent fashion (Jaillon et al. 2014). Also, mammary epithelial and CD11b$^+$ milk cells are sources of PTX3 that is found in human and mouse colostrum and milk and protects neonates from infections (Jaillon et al. 2013).

As opposed to the aforementioned cell types, which synthesize PTX3 de novo, neutrophils contain in their specific granules a constitutive stock of "ready-to-use" protein, which is promptly released in response to microorganisms or TLR agonists (Jaillon et al. 2007). Synthesis of the PTX3 mRNA is temporally confined to immature myeloid cells. Interestingly, when released, PTX3 can partially localize in neutrophil extracellular traps, known to promote the generation of an antimicrobial microenvironment. Given their abundance in the blood, neutrophils are a major reservoir of PTX3 in vivo and provide "premade" protein at earlier times than de novo synthesis (Jaillon et al. 2007). PTX3 gene is also constitutively expressed in both human and murine lymphatic endothelial cells. Resting T and B lymphocytes and natural killer cells do not express PTX3 mRNA (Bottazzi et al. 2010; Garlanda et al. 2002).

The human PTX3 is a multimeric glycoprotein, whose composing subunits are made of 381 amino acids, including a 17-residue signal peptide, and have a molecule mass of 43 kDa (Bottazzi et al. 2010). PTX3 primary sequence is highly conserved among animal species, which suggests a strong evolutionary pressure to maintain the structure/function relationships of this protein. PTX3 is composed of a unique N-terminal region coupled to a 203-amino acid C-terminal domain homologous to the short pentraxins CRP and SAP (57 % similarity) (Bottazzi et al. 2010). The N-terminal region (residues 18–178 of the preprotein) is unrelated to any known protein, and its structural details are yet to be unraveled. Instead, three-dimensional models of the C-terminal domain (residues 179–381 of the preprotein) are available, based on the crystallographic structures of the short pentraxins CRP and SAP. These models show the pentraxin domain of PTX3 to adopt a β-jelly roll topology, which resembles that found in legume lectins. Interestingly, the amino acids residues that form the calcium-binding pocket in CRP and SAP are missing in the pentraxin domain of PTX3, which might explain some binding properties of this long pentraxin (Garlanda et al. 2005). Studies performed with recombinant preparations of the N- and C-terminal domains of PTX3 have indicated that among PTX3 ligands, fibroblast growth factor 2 (FGF2), inter-α-inhibitor (IαI), and conidia of *Aspergillus fumigatus* each bind to the N-terminal region (Bottazzi et al. 2010); C1q and P-selectin interact with the pentraxin-like domain (Bottazzi et al. 2010; Deban et al. 2010), whereas both domains have been implicated in the interaction with complement factor H (Bottazzi et al. 2010).

A single N-glycosylation site has been identified in the C-terminal domain of PTX3 at Asn220 (Bottazzi et al. 2010; Garlanda et al. 2005). This site is fully occupied by complex-type oligosaccharides, mainly fucosylated and sialylated biantennary sugars. Most importantly, PTX3 glycosylation modulates the protein binding to a number of ligands, thus suggesting that changes in the glycosylation status might represent a strategy to tune the biological activity of this long pentraxin (Garlanda et al. 2005; Mantovani et al. 2008).

In addition to the multidomain organization, the human PTX3 protein shows a complex quaternary structure with eight protomer subunits held together by both covalent and non-covalent interactions (i.e., with a molecule mass of 340 kDa) (Inforzato et al. 2008). Mass spectrometry and site-directed mutagenesis have defined the pattern of disulfide bonds that stabilize the protein. In particular, the N-terminal region contains three cysteine residues at positions 47, 49, and 103 that form three interchain disulfide bonds holding four protein subunits in a tetrameric arrangement. The functional octamer originates from additional interchain bridges that link two tetramers via cysteine residues at positions 317 and 318 in the C-terminal domain (Inforzato et al. 2008). Electron microscopy and small-angle X-ray scattering have allowed making a low-resolution model of the intact PTX3 molecule that shows the eight subunits to fold into an elongated structure with a large and a small domain interconnected by a stalk region (Inforzato et al. 2010). This original oligomeric state and the asymmetric shape of the molecule make PTX3 unique among pentraxins, where only SAP from *Limulus polyphemus* forms octamers, however with planar and radial symmetry (Inforzato et al. 2010).

As discussed above, the structural determinants of the PTX3 quaternary organization mainly localize in the N-terminal domain, where this region mediates the association of protomers into tetramers via both covalent (i.e., disulfide bonds) and non-covalent (i.e., interchain coiled coils) interactions (Inforzato et al. 2010). Besides their structural role as building blocks of the PTX3 molecule, the N-terminal domain tetramers act as functional units in the recognition of several PTX3 ligands (Bottazzi et al. 2010). Therefore, the PTX3 quaternary structure plays a key role in dictating the protein-binding properties and, ultimately, its biological functions.

Biological Activities

Complement Regulation and Opsonic Activity

The short pentraxins participate in the activation and regulation of the three complement pathways (i.e., classical, lectin, and alternative). Despite the fact that the recruitment of C1q by ligand-bound CRP leads to complement activation, surface-bound CRP can also recruit factor H, the main soluble regulator of the alternative pathway, and C4BP, a soluble inhibitor of the complement activation via its cofactor activity in the cleavage and inactivation of C4b, suggesting that short pentraxins play a dual role in complement activation (Bottazzi et al. 2010).

Short pentraxins CRP and SAP interact with cell-surface Fcγ receptors (FcγR) and activate leukocyte-mediated phagocytosis and cytokine secretion, suggesting antibody-like functions for pentraxins (Lu et al. 2008). The relationship between interaction with pathogens and activity of CRP and SAP is still a matter of debate. Indeed, despite the fact that CRP interacts with bacteria, fungi, and yeast and promotes their phagocytosis and resistance to infections (e.g., *S. pneumoniae*), CRP can play a protective role in defense against *Salmonella typhimurium* without interacting with this microorganism. SAP binds to LPS, preventing the LPS-mediated complement activation and protecting the host from LPS toxicity. Similarly to CRP, SAP also exhibits a host defense function against pathogens that it does not recognize. In addition, SAP may enhance virulence of microorganisms (e.g., *S. typhimurium*) by protecting the bacteria against phagocytosis (Bottazzi et al. 2010).

PTX3 recognizes self-, nonself-, and modified self-ligands (Bottazzi et al. 2010). C1q, the main activator of the classical complement pathway, was the first ligand identified for PTX3 (Bottazzi et al. 2010). The interaction between plastic-immobilized PTX3 and C1q activates the classical complement cascade; however, PTX3 in the fluid phase inhibits the activation of the classical complement cascade via competitive blocking of relevant interaction sites. This suggests that the interaction between PTX3 and C1q leads either to the activation or inhibition of the complement cascade, depending on the way it is presented (Bottazzi et al. 2010). In parallel, PTX3 interacts with ficolin-1, ficolin-2, and mannose-binding lectin (MBL), three members of the lectin pathway. The deposition of MBL and ficolin-2 on pathogens is enhanced in presence of PTX3, promoting complement-mediated innate immune response. Similarly, the heterocomplex ficolin-1-PTX3 found on the surface of apoptotic cells facilitates the clearance of apoptotic cells by macrophages (Mantovani et al. 2013). PTX3 interacts also with negative complement regulators, e.g., factor H and C4BP, favoring the deposition of these soluble inhibitors and limiting complement cascade activation that would otherwise lead to inflammation and tissue damage (Bottazzi et al. 2010; Mantovani et al. 2013).

PTX3 has the capacity to bind a broad spectrum of microorganisms, including selected fungi (e.g., *Paracoccidioides brasiliensis*), virus (e.g., cytomegalovirus, selected strains of influenza virus), and bacteria (e.g., *Klebsiella pneumoniae*, uropathogenic *Escherichia coli* (UPEC), *Salmonella typhimurium*) (Mantovani et al. 2013). Genetically modified mice have allowed investigating the in vivo functions of PTX3. PTX3-deficient mice showed increased susceptibility to invasive pulmonary aspergillosis, *Pseudomonas aeruginosa* lung infection, and urinary tract infection induced by UPEC (Garlanda et al. 2002; Jaillon et al. 2014). PTX3-deficient phagocytes defectively recognize and clear *A. fumigatus*, *P. aeruginosa*, and UPEC, and treatment with PTX3 reverses this phenotype, demonstrating that PTX3 has an opsonic activity (Garlanda et al. 2002; Jaillon et al. 2014, 2007; Moalli et al. 2010, 2011) (Fig. 1). PTX3 enhances the recognition and phagocytosis of *A. fumigatus* conidia by neutrophils in a FcγRII-, CD11b-, and complement-dependent mechanism. Indeed, PTX3-opsonized conidia interact with FcγRs, which have been proposed as pentraxin receptors, inducing the activation of CR3 (CD11b/CD18) and thereby promoting the phagocytosis of C3-opsonized pathogens (Moalli et al. 2010). Accordingly, CD11b recruitment in the phagocytic cup is defective in PTX3- and FcγR-deficient neutrophils, and recombinant PTX3 reverses this phenotype in PTX3-deficient neutrophils but not in FcγR-deficient neutrophils (Moalli et al. 2010). As observed for *A. fumigatus*, the positive effect of PTX3 on the phagocytosis of *P. aeruginosa* is maintained in C1q-deficient mice but not in C3- and FcγR-deficient mice, suggesting that increased phagocytosis of PTX3-opsonized *P. aeruginosa* occurs through the interplay between complement and FcγRs (Moalli et al. 2011). In agreement with this opsonic activity, PTX3 has therapeutic activity in mouse models of chronic *P. aeruginosa* lung infection (adult and neonate mice), a major cause of morbidity and mortality in cystic fibrosis patients, and invasive pulmonary aspergillosis. In these models, PTX3 reduced pathogen colonization and induced a protective immune response (Garlanda et al. 2002; Moalli et al. 2011).

The interaction between PTX3 and zymosan increases the recognition and elimination of zymosan-containing pathogens (i.e., *P. brasiliensis*) by macrophages. In accordance, PTX3-overexpressing macrophages have an increased phagocytic activity toward *P. brasiliensis* compared to wild-type cells. This mechanism is dependent on the presence of dectin-1, the major cellular receptor involved in the recognition and response to beta-glucans. In fact, in response to zymosan, expression of PTX3 is induced and PTX3-overexpressing macrophages have increased expression of dectin-1, promoting a positive feedback for zymosan phagocytosis (Mantovani et al. 2013).

In a model of urinary tract infection, defective recognition and clearance of UPEC in PTX3-deficient mice was associated with exacerbated inflammation and tissue damage (Jaillon et al. 2014). In addition, to increase the phagocytosis of pathogens, opsonization of microorganisms by PTX3 can enhance and accelerate phagosome maturation in neutrophils, as observed for UPEC (Jaillon et al. 2014). The relevance in human of these data derived from animal infection models has been demonstrated by genetic studies (see below).

Different studies have also highlighted the role of PTX3 in viral infections, such as in defense against cytomegalovirus or against specific strain of influenza virus (Bottazzi et al. 2010). Both human and murine cytomegalovirus are recognized by PTX3, and this interaction reduces viral entry and infectivity in dendritic cells in vitro. Similarly, PTX3 binds to influenza virus (H3N2), inducing a range of antiviral activities (e.g., inhibition of hemagglutination and viral neuraminidase activity and neutralization of virus infectivity) (Mantovani et al. 2013). Consistently, PTX3-deficient mice present increased susceptibility to both cytomegalovirus and H3N2 infection, and PTX3 has therapeutic activities in these infections. In contrast, PTX3 is inefficient in the defense against both seasonal and pandemic H1N1 influenza, due to the absence of interaction between both viral hemagglutinin and neuraminidase from these viruses and the sialic acid of PTX3 (Job et al. 2014).

Inflammation and Neutrophil Extravasation

The inflammatory response can be regulated by PTX3. For instance, the interaction between PTX3 and the outer membrane protein A of *K. pneumoniae* (KpOmpA) does not interfere with the recognition of KpOmpA or cell activation induced by KpOmpA; however, PTX3 enhances the local inflammation induced by KpOmpA in vivo, in terms of production of proinflammatory cytokines and cell recruitment (Jeannin et al. 2005). This mechanism is complement-dependent and abrogated after treatment with complement inhibitors (Cotena et al. 2007). Consistently, PTX3-overexpressing mice show increased production of proinflammatory mediators, including NO and TNF-α after infection with *K. pneumoniae*, leading to protection or faster lethality, depending on the dose of the pathogen (Mantovani et al. 2013). This mechanism cannot be extended to all microbial moieties since the inflammatory response induced by LPS, which is not recognized by PTX3, is not modulated by PTX3 (Cotena et al. 2007).

PTX3 is rapidly induced during acute myocardial ischemia, and PTX3-deficient mice have increased myocardial damage during acute myocardial ischemia. In agreement with the complement regulatory role of PTX3, deposition of C3 is increased in PTX3-deficient mice, a process likely due to the missing interaction between PTX3 and factor H. In addition, deficiency of PTX3 was also associated with enhanced inflammatory response and aortic lesions in a mouse model of atherosclerosis induced by a null mutation in the gene encoding for apolipoprotein E (Mantovani et al. 2013; Salio et al. 2008).

PTX3 selectively binds P-selectin via its N-linked glycosidic moiety, creating a negative feedback loop that prevents excessive P-selectin-dependent recruitment of neutrophils in vivo (Deban et al. 2010). In agreement with this mechanism, PTX3 derived from hematopoietic cells functions locally to reduce neutrophil recruitment and regulate inflammation (Deban et al. 2010). In vivo, PTX3 has a protective role in acute lung injury (ALI), pleurisy, mesenteric inflammation, and acute kidney ischemia and reperfusion injury. In contrast to this protective effect, PTX3 deficiency has also been associated with decreased tissue inflammation, as in a mouse model of intestinal ischemia and reperfusion, suggesting that PTX3 may play different roles in mediating reperfusion injury in the heart and kidney (protective) versus the intestine (deleterious) (Mantovani et al. 2013).

Matrix Remodeling

CRP and SAP bind to extracellular matrix (ECM) proteins, such as fibronectin, collagen IV, laminin, and proteoglycans (Bottazzi et al. 2010). However, specific roles for these interactions were not clearly identified.

An increasing body of evidence points to PTX3 as an important component of the ECM with novel functional implications in tissue remodeling. In this regard, PTX3-deficient mice display severe deficiency in female fertility, where this is due to defective assembly of the viscoelastic hyaluronan (HA)-rich matrix that forms around the oocyte in the preovulatory follicle, namely, the cumulus oophorus complex, and is required for fertilization in vivo (Mantovani et al. 2013). The major component of the cumulus matrix is HA; however, other molecules than PTX3 are required for effective incorporation of HA into the matrix, including the HA-binding protein TSG-6 and the serum proteoglycan IαI. Current literature supports the hypothesis that heavy chains from IαI become covalently attached to HA through reactions involving TSG-6, which acts as both a cofactor and catalyst. The resulting complexes are believed to be cross-linked by PTX3 that through its multimeric structure provides structural integrity to the cumulus matrix (Bottazzi et al. 2010).

Consistent with this, using mutants of PTX3 with varying degrees of oligomerization, the protein tetramer has been identified as the minimal functional unit of PTX3 in matrix organization (Inforzato et al. 2008).

Angiogenesis

The process of angiogenesis, which occurs under normal physiological conditions in growth and development and for successful wound healing, is regulated by the balance between pro- and anti-angiogenic factors. PTX3 binds FGF2, a major angiogenic inducer, with high affinity, and the binding site has been identified in the N-terminal extension of the protein spanning aa97-110 of the preprotein (Mantovani et al. 2013). Both PTX3 and FGF2 are produced by elements of the vessel wall, such as endothelial cells and smooth muscle cells, during inflammation. When in a complex with PTX3, FGF2 is no longer able to interact with its receptors on endothelial cells. Thus PTX3 causes inhibition of the FGF2-dependent endothelial cell proliferation in vitro and angiogenesis in vivo. These inhibitory effects are reversed by TSG-6, through competition with the interaction of FGF2/PTX3, suggesting a novel mechanism of modulation of the angiogenic process whereby the relative levels of TSG-6 and PTX3 dictate the biological activity of FGF2 (Leali et al. 2012).

The PTX3–FGF2 axis might also have important implications in restenosis, the process of blood vessel narrowing that frequently occurs after percutaneous transluminal coronary angioplasty of atherosclerotic arteries. This process initiates with disruption of the endothelial layer followed by smooth muscle cell migration and proliferation. PTX3 attenuates intimal thickening following balloon injury in rat carotid arteries by counteracting the proliferation and chemotaxis of smooth muscle cells that is triggered and sustained by FGF2 (Mantovani et al. 2013).

Recent evidence has suggested that PTX3 might also affect FGF2-driven proliferation and downstream FGF receptor signaling in murine melanoma cells (Ronca et al. 2013b). PTX3 overexpression in these cells blocks transition from epithelial to mesenchymal phenotypes, inhibits proliferation, and reduces motility and invasive capacity in vitro and in vivo (Ronca et al. 2013b). Furthermore, recent studies on multiple myeloma patients have shown that PTX3 impairs the functional effects of FGF2 on endothelial cells and fibroblasts (i.e., viability, chemotaxis, angiogenesis, cytokine secretion) as well as adhesion of these cells to bone marrow plasma cells driving an apoptotic effect (Basile et al. 2013).

PTX3 has been reported to antagonize FGF8b and to inhibit the FGF8b-dependent vascularization and growth of steroid hormone-regulated tumors (Leali et al. 2011). As observed for FGF2, PTX3 prevents recognition of FGF8b by its receptors (Leali et al. 2011). In accordance, PTX3 impairs both proliferation and angiogenic potential of FGF2- or FGF8b-stimulated murine prostate cancer cells inhibiting their angiogenic and tumorigenic potential (Ronca et al. 2013b).

Pathophysiological Relevance

PTX3 as a Marker of Human Pathology

The homology with CRP, a classic diagnostic in humans, suggests that PTX3 may be a new marker of infectious or inflammatory diseases (Fig. 1). Unlike CRP, which is made primarily in the liver in response to IL-6, PTX3 is a rapid marker for primary local activation of innate immunity and inflammation. In healthy subjects, PTX3 basal levels in the blood are below 2 ng/ml; however, a rapid (peak at 6–8 h) and dramatic (up to 800 ng/ml) increase has been observed in different inflammatory or infectious diseases, including cardiovascular diseases, infections, and sepsis.

An increasing number of studies have associated PTX3 with **cardiovascular diseases (CVD)**. A rapid increase in PTX3 plasma levels has been observed in patients with acute myocardial infarction (AMI), with a peak within 8 h from the onset of symptoms (Mantovani et al. 2008, 2013). In the same patients, CRP peaks around 50 h. The rapid increase in PTX3 plasma levels is likely due to the release of the protein stored in neutrophil granules in response to tissue damage induced by ischemia. In a series of 748 patients with AMI and ST elevation, PTX3 emerged as the only independent predictor of mortality within 3 months among other consolidated markers of cardiac damage, i.e., troponin T (TnT) and N-terminal pro-brain natriuretic peptide (NT-proBNP) (Mantovani et al. 2008, 2013). The association of PTX3 with CVD and all-cause mortality has been confirmed in more than 1500 subjects enrolled in the Cardiovascular Health Study or with stable coronary heart disease (CHD) enrolled in the Heart and Soul Study. PTX3 was also associated with CVD risk factors, subclinical CVD, and clinical events in subjects from the Multi-Ethnic Study of Atherosclerosis. Baseline PTX3 blood levels and their changes over time were evaluated in 1457 patients enrolled in the Controlled Rosuvastatin Multinational Trial in Heart Failure (HF) (CORONA) and 1233 patients enrolled in the GISSI-Heart Failure trial (GISSI-HF) (Latini et al. 2012; Mantovani et al. 2013). In these two independent trials, baseline PTX3 was consistently associated with a higher risk of all-cause mortality, cardiovascular mortality, or hospitalization for worsening HF. Circulating levels of PTX3 were also found elevated in patients with unstable angina pectoris and nonrheumatic aortic valve stenosis and associated with inflammation and neointimal thickening after vascular injury (Mantovani et al. 2013). In patients after heart surgery, PTX3 blood levels showed predictive values of short-term functional recovery, and baseline concentration levels of PTX3 were the only variable associated with the 1-year incidence of major adverse cardiovascular events. Despite the association of PTX3 with CVD risk factors, a correlation with metabolic syndrome or obesity is still controversial (Mantovani et al. 2013).

Several clinical studies have demonstrated an association of PTX3 with **kidney dysfunction**. PTX3 plasma levels are increased in patients with chronic kidney disease and are associated with other inflammatory markers such as TNF-α and IL-6 (Speeckaert et al. 2013). PTX3 concentrations are higher in patients undergoing hemodialysis or peritoneal dialysis, possibly as a result of the continuous low-grade systemic inflammation present in those patients (Speeckaert et al. 2013).

Increased PTX3 plasma levels were observed in patients with different **infections**. In particular, in pulmonary aspergillosis, tuberculosis, leptospirosis, dengue, and meningitides, PTX3 levels correlate with severity of disease and mortality (Mantovani et al. 2013). In addition, PTX3 can be measured in the urine of patients with acute pyelonephritis but not in healthy subjects. In these patients, PTX3 urinary levels correlate with hematuria and decrease in response to therapy (Jaillon et al. 2014).

Preclinical evidence has demonstrated the therapeutic potential of PTX3 against pathogens, in particular *A. fumigatus* (Moalli et al. 2011). In agreement with this, ongoing studies are evaluating the possible use of PTX3 as an antifungal agent in immunocompromised patients. Interestingly, a proteolytic cleavage of PTX3 by *A. fumigatus* proteases occurs in the airways of cystic fibrosis patients and may explain, at least in part, the recurrent pulmonary aspergillosis observed in these patients (Hamon et al. 2013). In the same context, the redox-sensitive oligomerization state of PTX3 can predict outcome in septic patients at risk of acute organ failure, better then cardiac damage markers. In these subjects, monomeric PTX3 is inversely correlated with complement activation and cardiac tissue damage (Cuello et al. 2014).

Dramatically high levels of circulating PTX3 were also associated with **sepsis**, severe sepsis, and unfavorable outcome in patients with systemic inflammatory response syndrome and in emergency room patients with suspected infection (Muller et al. 2001). In a limited group of intubated critically ill patients, PTX3 levels in bronchoalveolar lavage (BAL) fluids can predict the presence of lung infection (Mauri et al. 2014). This important observation, if confirmed in larger cohorts of patients, is crucial for the rapid (few hours compared to 2 days for the microbiological results) identification of pneumonia cases requiring antibiotic treatment.

Increased levels of PTX3 were also observed in a set of **autoimmune disorders**, such as in the blood of patients with small vessel vasculitis and in the synovial fluid of patients with rheumatoid arthritis, but not in systemic lupus erythematosus

(SLE). The concentration of PTX3, but not of CRP, was significantly higher in Takayasu arteritis (TA) patients where it reflects progression of the disease (Ramirez et al. 2014). The discrepancy between PTX3 levels in SLE and in other autoimmune diseases may be explained, at least in part, by the presence of anti-PTX3 autoantibodies in SLE patients but not in patients with other autoimmune rheumatic diseases (i.e., polydermatomyositis, systemic sclerosis, Sjogren's syndrome, and psoriatic arthritis). The role of these anti-PTX3 autoantibodies and whether their titers correlate with the activity of the disease remain matter of discussion (Ramirez et al. 2014).

Finally, increasing evidence links PTX3 and cancer. PTX3 expression has been observed in soft tissue **sarcomas, prostate cancer, and lung cancer** (Mantovani et al. 2013). Immunohistochemical analysis of human gliomas with different grade of malignancy demonstrates that PTX3 is more expressed in high-grade tumors, representing a possible new marker of inflammation associated to glioma malignancy (Locatelli et al. 2013). PTX3 levels are also increased in blood of patients with myeloproliferative neoplasms (Mantovani et al. 2013). In pancreatic carcinoma, patients' plasma PTX3 levels are correlated with a more advanced stage of the tumor (Kondo et al. 2013). Besides its role as a biomarker, PTX3 could also play a role in malignant transformation. In fact, in esophageal squamous cell carcinoma, PTX3 acts as oncosuppressor gene. PTX3 promoter hypermethylation that has been observed in the early tumor stages (I and II) is responsible for down modulation of PTX3 expression (Mantovani et al. 2013). In keeping with these data, PTX3 expression is lost in high-grade prostatic intraepithelial neoplasia and in invasive tumor areas (Ronca et al. 2013a). Given these results, a deeper investigation into PTX3 in cancer transformation is mandatory to better define the role of this molecule in specific tumors.

About 40 **single nucleotide polymorphisms** (SNPs) forming as many as 29 different haplotypes have been identified in the *CRP* gene. These different haplotypes can influence circulating CRP levels over the adult lifespan. Despite that massive data showed an association between elevated CRP levels and coronary heart disease, SNPs in *CRP* were not associated with this pathology, suggesting that elevated CRP must be seen as a bystander and not a causal factor in disease progression (Strang and Schunkert 2014).

Despite its strong evolutionary conservation, recent studies have reported a number of SNPs in the *PTX3* gene. In line with the structural and functional conservation of the gene, most of these SNP variations were found in the noncoding regions, and only one of the two described exon substitutions results in amino acid variation of the protein. Three of these polymorphisms combined in specific haplotypes have been associated with PTX3 plasma levels and clinical conditions including infection and female fertility. In particular, a protective effect was attributed to a specific haplotype (for sites rs2305619, rs3816527, and rs1840680) in pulmonary tuberculosis, *Pseudomonas aeruginosa* lung infection in cystic fibrosis Caucasian patients, *A. fumigatus* infection in patients undergoing bone marrow transplantation, and severe urinary tract infections, thus suggesting that variation within PTX3 affects the disease outcome (Jaillon et al. 2014; Mantovani et al. 2013). Since these polymorphisms were associated with protein levels, it has been suggested that they may alter mRNA stability, leading to decreased expression of PTX3, impaired phagocytosis, and clearance of pathogens.

Modulation by Drugs

Data on modulation of PTX3 plasma levels by pharmacological treatment are still sketchy. In different pathological conditions, PTX3 levels correlated with the response to therapy, indicating that they were indirectly influenced by the efficacy of treatment. However, two classes of molecules were shown to directly regulate *PTX3* gene expression: glucocorticoid hormones and statins. Contrasting results were reported on the modulation of PTX3 expression by statins: Both increased or decreased levels of PTX3 were reported in statin-treated patients, while *PTX3* gene expression is down modulated in endothelial cells exposed to statins (Latini et al. 2012; Mantovani et al. 2013). PTX3 levels are increased in patients affected by Cushing syndrome treated with dexamethasone and are significantly decreased in patients affected by hypocortisolism (Garlanda et al. 2005).

Cross-References

▶ Angiogenesis Inhibitors
▶ Antiviral Responses
▶ Bacterial Lipopolysaccharide
▶ Cancer and Inflammation
▶ Complement System
▶ Corticosteroids
▶ Cytokines
▶ Dendritic Cells
▶ Immunoglobulin Receptors and Inflammation
▶ Leukocyte Recruitment
▶ Neutrophil Extracellular Traps
▶ Pathogen-Associated Molecular Patterns (PAMPs)
▶ Toll-Like Receptors
▶ Tumor Necrosis Factor Alpha (TNFalpha)

References

Basile, A., Moschetta, M., Ditonno, P., Ria, R., Marech, I., De Luisi, A., et al. (2013). Pentraxin 3 (PTX3) inhibits plasma cell/stromal cell cross-talk in the bone marrow of multiple myeloma patients. *The Journal of Pathology, 229*(1), 87–98. doi:10.1002/path.4081.

Bottazzi, B., Doni, A., Garlanda, C., & Mantovani, A. (2010). An integrated view of humoral innate immunity: Pentraxins as a paradigm. *Annual Review of Immunology, 28*, 157–183.

Cotena, A., Maina, V., Sironi, M., Bottazzi, B., Jeannin, P., Vecchi, A., et al. (2007). Complement dependent amplification of the innate response to a cognate microbial ligand by the long pentraxin PTX3. *The Journal of Immunology, 179*(9), 6311–6317.

Cuello, F., Shankar-Hari, M., Mayr, U., Yin, X., Marshall, M., Suna, G., et al. (2014). Redox state of pentraxin 3 as a novel biomarker for resolution of inflammation and survival in sepsis. *Molecular and Cellular Proteomics, 13*(10), 2545–2557. doi:10.1074/mcp.M114.039446, M114.039446 [pii].

Deban, L., Russo, R. C., Sironi, M., Moalli, F., Scanziani, M., Zambelli, V., et al. (2010). Regulation of leukocyte recruitment by the long pentraxin PTX3. *Nature Immunology, 11*(4), 328–334.

Garlanda, C., Hirsch, E., Bozza, S., Salustri, A., De Acetis, M., Nota, R., et al. (2002). Non-redundant role of the long pentraxin PTX3 in anti-fungal innate immune response. *Nature, 420*(6912), 182–186.

Garlanda, C., Bottazzi, B., Bastone, A., & Mantovani, A. (2005). Pentraxins at the crossroads between innate immunity, inflammation, matrix deposition, and female fertility. *Annual Review of Immunology, 23*, 337–366.

Hamon, Y., Jaillon, S., Person, C., Ginies, J. L., Garo, E., Bottazzi, B., et al. (2013). Proteolytic cleavage of the long pentraxin PTX3 in the airways of cystic fibrosis patients. *Innate Immunity.* doi:10.1177/1753425913476741, [pii] 1753425913476741.

Inforzato, A., Rivieccio, V., Morreale, A. P., Bastone, A., Salustri, A., Scarchilli, L., et al. (2008). Structural characterization of PTX3 disulfide bond network and its multimeric status in cumulus matrix organization. *The Journal of Biological Chemistry, 283*(15), 10147–10161.

Inforzato, A., Baldock, C., Jowitt, T. A., Holmes, D. F., Lindstedt, R., Marcellini, M., et al. (2010). The angiogenic inhibitor long pentraxin PTX3 forms an asymmetric octamer with two binding sites for FGF2. *The Journal of Biological Chemistry, 285*(23), 17681–17692. doi:10.1074/jbc.M109.085639, [pii] M109.085639.

Jaillon, S., Peri, G., Delneste, Y., Fremaux, I., Doni, A., Moalli, F., et al. (2007). The humoral pattern recognition receptor PTX3 is stored in neutrophil granules and localizes in extracellular traps. *The Journal of Experimental Medicine, 204*(4), 793–804.

Jaillon, S., Mancuso, G., Hamon, Y., Beauvillain, C., Cotici, V., Midiri, A., et al. (2013). Prototypic long pentraxin PTX3 is present in breast milk, spreads in tissues, and protects neonate mice from pseudomonas aeruginosa lung infection. *The Journal of Immunology, 191*(4), 1873–1882. doi:10.4049/jimmunol.1201642, [pii] jimmunol.1201642.

Jaillon, S., Moalli, F., Ragnarsdottir, B., Bonavita, E., Puthia, M., Riva, F., et al. (2014). The humoral pattern recognition molecule PTX3 is a key component of innate immunity against urinary tract infection. *Immunity, 40*(4), 621–632. doi:10.1016/j.immuni.2014.02.015.

Jeannin, P., Bottazzi, B., Sironi, M., Doni, A., Rusnati, M., Presta, M., et al. (2005). Complexity and complementarity of outer membrane protein a recognition by cellular and humoral innate immunity receptors. *Immunity, 22*(5), 551–560.

Job, E. R., Bottazzi, B., Short, K. R., Deng, Y. M., Mantovani, A., Brooks, A. G., et al. (2014). A single amino acid substitution in the hemagglutinin of H3N2 subtype influenza a viruses is associated with resistance to the long pentraxin PTX3 and enhanced virulence in mice. *The Journal of Immunology, 192*(1), 271–281. doi:10.4049/jimmunol.1301814. [pii] jimmunol.1301814.

Kondo, S., Ueno, H., Hosoi, H., Hashimoto, J., Morizane, C., Koizumi, F., et al. (2013). Clinical impact of pentraxin family expression on prognosis of pancreatic carcinoma. *British Journal of Cancer, 109*(3), 739–746. doi:10.1038/bjc.2013.348, [pii] bjc2013348.

Latini, R., Gullestad, L., Masson, S., Nymo, S. H., Ueland, T., Cuccovillo, I., et al. (2012). Pentraxin-3 in chronic heart failure: The CORONA and GISSI-HF trials. *European Journal of Heart Failure, 14*(9), 992–999. doi:10.1093/eurjhf/hfs092, [pii] hfs092.

Leali, D., Alessi, P., Coltrini, D., Ronca, R., Corsini, M., Nardo, G., et al. (2011). Long pentraxin-3 inhibits FGF8b-dependent angiogenesis and growth of steroid hormone-regulated tumors. *Molecular cancer therapeutics, 10*(9), 1600–1610. doi:10.1158/1535-7163. MCT-11-0286, [pii] 1535-7163.MCT-11-0286.

Leali, D., Inforzato, A., Ronca, R., Bianchi, R., Belleri, M., Coltrini, D., et al. (2012). Long pentraxin 3/tumor necrosis factor-stimulated gene-6 interaction: A biological rheostat for fibroblast growth factor 2-mediated angiogenesis. *Arteriosclerosis, Thrombosis, and Vascular Biology, 32*(3), 696–703. doi:10.1161/ATVBAHA.111.243998, [pii] ATVBAHA.111.243998.

Locatelli, M., Ferrero, S., Martinelli Boneschi, F., Boiocchi, L., Zavanone, M., Maria Gaini, S., et al. (2013). The long pentraxin PTX3 as a correlate of cancer-related inflammation and prognosis of malignancy in gliomas. *Journal of Neuroimmunology, 260* (1–2), 99–106. doi:10.1016/j.jneuroim.2013.04.009, [pii] S0165-5728(13)00093-3.

Lu, J., Marnell, L. L., Marjon, K. D., Mold, C., Du Clos, T. W., & Sun, P. D. (2008). Structural recognition and functional activation of FcgammaR by innate pentraxins. *Nature, 456*(7224), 989–992.

Mantovani, A., Garlanda, C., Doni, A., & Bottazzi, B. (2008). Pentraxins in innate immunity: From C-reactive protein to the long pentraxin PTX3. *Journal of Clinical Immunology, 28*(1), 1–13.

Mantovani, A., Valentino, S., Gentile, S., Inforzato, A., Bottazzi, B., & Garlanda, C. (2013). The long pentraxin PTX3: A paradigm for humoral pattern recognition molecules. *Annals of the New York Academy of Sciences, 1285*, 1–14. doi:10.1111/nyas.12043.

Mauri, T., Coppadoro, A., Bombino, M., Bellani, G., Zambelli, V., Fornari, C., et al. (2014). Alveolar pentraxin 3 as an early marker of microbiologically confirmed pneumonia: A threshold-finding prospective observational study. *Critical Care, 18*(5), 562. doi:10.1186/s13054-014-0562-5, [pii] s13054-014-0562-5.

Moalli, F., Doni, A., Deban, L., Zelante, T., Zagarella, S., Bottazzi, B., et al. (2010). Role of complement and Fc {gamma} receptors in the protective activity of the long pentraxin PTX3 against Aspergillus fumigatus. *Blood, 116*(24), 5170–5180.

Moalli, F., Paroni, M., Veliz Rodriguez, T., Riva, F., Polentarutti, N., Bottazzi, B., et al. (2011). The therapeutic potential of the humoral pattern recognition molecule PTX3 in chronic lung infection caused by Pseudomonas aeruginosa. *The Journal of Immunology, 186*(9), 5425–5434. doi:10.4049/jimmunol.1002035, [pii] jimmunol.1002035.

Muller, B., Peri, G., Doni, A., Torri, V., Landmann, R., Bottazzi, B., et al. (2001). Circulating levels of the long pentraxin PTX3 correlate with severity of infection in critically ill patients. *Critical Care Medicine, 29*(7), 1404–1407.

Ramirez, G. A., Maugeri, N., Sabbadini, M. G., Rovere-Querini, P., & Manfredi, A. A. (2014). Intravascular immunity as a key to systemic vasculitis: A work in progress, gaining momentum. *Clinical and Experimental Immunology, 175*(2), 150–166. doi:10.1111/cei.12223.

Ronca, R., Alessi, P., Coltrini, D., Di Salle, E., Giacomini, A., Leali, D., et al. (2013a). Long pentraxin-3 as an epithelial-stromal fibroblast growth factor-targeting inhibitor in prostate cancer. *The Journal of Pathology, 230*(2), 228–238. doi:10.1002/path.4181.

Ronca, R., Di Salle, E., Giacomini, A., Leali, D., Alessi, P., Coltrini, D., et al. (2013b). Long pentraxin-3 inhibits epithelial-mesenchymal transition in melanoma cells. *Molecular cancer therapeutics, 12*(12), 2760–2771. doi:10.1158/1535-7163.MCT-13-0487, [pii] 1535-7163.MCT-13-0487.

Salio, M., Chimenti, S., De Angelis, N., Molla, F., Maina, V., Nebuloni, M., et al. (2008). Cardioprotective function of the long pentraxin PTX3 in acute myocardial infarction. *Circulation, 117*(8), 1055–1064.

Speeckaert, M. M., Speeckaert, R., Carrero, J. J., Vanholder, R., & Delanghe, J. R. (2013). Biology of human pentraxin 3 (PTX3) in acute and chronic kidney disease. *Journal of Clinical Immunology, 33*(5), 881–890. doi:10.1007/s10875-013-9879-0.

Strang, F., & Schunkert, H. (2014). C-reactive protein and coronary heart disease: All said – Is not it? *Mediators of Inflammation, 2014*, 757123. doi:10.1155/2014/757123.

Phosphodiesterase 4 Inhibitors: Apremilast and Roflumilast

Garry G. Graham[1,2] and Kevin D. Pile[3]
[1]Department of Pharmacology, School of Medical Sciences, University of NSW, Sydney, NSW, Australia
[2]Department of Clinical Pharmacology and Toxicology, St Vincent's Hospital, Darlinghurst, Sydney, NSW, Australia
[3]Campbelltown Hospital, School of Medicine, University of Western Sydney, Campbelltown, NSW, Australia

Synonyms

Phosphodiesterase 4 (PDE4) antagonists; Phosphodiesterase 4 (PDE4) inhibitors

Definition

Inhibitors of phosphodiesterase 4 (PDE4) are anti-inflammatory drugs that have been evaluated for the treatment of several inflammatory diseases. The major clinically useful members of this group are apremilast and roflumilast.

Introduction

Cyclic adenosine monophosphate (cAMP) is a major intracellular messenger whose increased intracellular levels mediate the actions of the beta-sympathomimetic agonists (Giembycz and Maurice 2014). Eleven types of PDEs are recorded but PDE4 is the enzyme of major pharmacologic interest as it hydrolyses cAMP. In turn, PDE4 has four isoenzymes, PDE4A to PDE4D.

Inhibition of the breakdown of cAMP has long been considered as a potential mechanism for the development of new drugs. Additionally, the old drug, theophylline, may be an inhibitor of PDE4, although the concentrations at which theophylline shows this activity are considered to be supratherapeutic.

PDE4 is a cAMP-specific PDE and is the dominant PDE in inflammatory cells, vascular endothelial cells, smooth muscle cells, keratinocytes, chondrocytes, and the central nervous system (Abdulrahim et al. 2015). This wide distribution and activity of cAMP indicates that inhibition of PDE4 has many potential pharmacological and clinical actions. Several drugs with PDE4 inhibitory activity have been tested over many years but have failed in clinical trials due to adverse effects or inadequate clinical effects. Consequently, the continuing search for inhibitors of PDE4 was questioned. However, this class of drugs is still considered to have high potential, particularly in combination with other drugs, such as inhaled beta-sympathomimetics and corticosteroids (Giembycz and Maurice 2014). Furthermore, the recent success of apremilast and roflumilast has led to the continuation of the search for novel PDE4 inhibitors, including compounds that may be suitable for inhalation (Spina 2008). Inhalation of these drugs should decrease the systemic adverse effects of the PDE4 inhibitors but no inhaled forms are approved at present.

Apremilast is used for psoriasis and psoriatic arthritis and roflumilast for chronic obstructive pulmonary disease. Both have been approved in several countries and their basic and clinical pharmacology are described in this entry.

Several older PDE4 inhibitors have been discarded or have minor clinical or experimental uses. These PDE4 inhibitors and their pharmacological effects include (Tenor et al. 2011) those discussed below.

Rolipram shows highly selective inhibition of PDE4. Rolipram shows a variety of anti-inflammatory effects within and outside the central nervous system. These effects correlate, at least in part, with downregulation of TNF. A particular effect is the decreased symptoms of experimental autoimmune encephalomyelitis in rats. Consequently, rolipram was tested in the treatment in multiple sclerosis. However, it produced blood-brain barrier disruption due to increased apparent inflammation of the central nervous system and therefore failed in the treatment of multiple sclerosis. Rolipram also showed weak antidepressant activity but its adverse effects, nausea and vomiting, stopped further clinical testing. It is presently used as an experimental agent for inhibition of PDE4 in research in basic pharmacology.

Ibudilast is a weak inhibitor of PDE3, PDE4, and PDE10. A variety of other effects have been observed from in vitro studies including decreased release or synthesis of leukotrienes, histamine, tumor necrosis factor, and interferon-γ. Ibudilast is also an inhibitor of the actions of migration inhibitory factor (MIF). MIF is considered to be an important cytokine in the development of rheumatoid arthritis, atherosclerosis, and possibly also in diabetes and cancer. However, it is not known whether sufficient concentrations are achieved in vivo to produce these various pharmacological effects. Ibudilast is still being evaluated in mice and rats for a variety of possible favorable effects on the central nervous system. These include reductions in paclitaxel-induced neuropathy and pain (Mo et al. 2012). Inhibition of glial cell activation by ibudilast may also lead to decreased neuronal death and lower the self-administration of methylamphetamine (Mizuno et al. 2004; Snider et al. 2013). Overall, ibudilast may preserve neural function due to inhibition of the activation of glial cells.

Ibudilast is registered in Japan for the oral treatment of asthma and poststroke dizziness and, in eye drops, for the treatment of ocular allergies. The most common adverse effects of the oral preparations are nausea and abdominal discomfort.

Tetomilast is a weak inhibitor of PDE4 and a stronger inhibitor of PDE3. So far, tetomilast has shown poor activity in the treatment of ulcerative colitis and is being tested for chronic obstructive pulmonary disease. Nausea and vomiting are the most common adverse effects, particularly at the higher doses.

Cilomilast is an inhibitor of PDE4D and a weaker inhibitor of PDE4B. It has been tested in asthma and chronic obstructive pulmonary disease. However, improvement was slight and inconsistent. Its further clinical testing has been stopped.

Oglemilast potently inhibits all PDE4 subtypes. In addition, other in vitro actions result from inhibition of PDE4, such as the release of tumor necrosis factor in the blood and the oxidative burst of neutrophils. However, oglemilast failed in clinical trials in asthma and chronic obstructive pulmonary disease and its clinical testing has also been stopped.

Up to the present, most attention has been concentrated on compounds which are inhibitors of PDE4 as adverse effects may be caused by inhibition of other PDEs. On the other hand, it is suggested that decreased function of a variety of PDEs may lead to the successful treatment of several inflammatory diseases because of the greater number of potential targets (Spina 2008).

Chemical Structures and Properties

Apremilast is quite lipid soluble with a predicted logP value of 1.8. It has a chiral center and is used as its active S-enantiomer (Fig. 1). The structure of apremilast is complex and its molecular mass (460.5 D) is greater than most conventional drugs. However, its lipid solubility and molecular mass still favor passive diffusion through lipid cellular membranes and, consequently, oral absorption.

Roflumilast is very lipid soluble with a predicted logP value of 4.6. Unlike apremilast, it is a symmetrical molecule and therefore does not exist as isomers. While administered as the neutral species, it is an acid with a pKa of 8.7 but the major form in plasma is the unionized species. It is also a complex molecule (Fig. 1) with a molecular mass of 403.2 but is still a sufficiently small lipid-soluble molecule to undergo passive diffusion through cell membranes.

Phosphodiesterase 4 Inhibitors: Apremilast and Roflumilast, Fig. 1 Structures of apremilast and roflumilast. The chiral center in apremilast is shown (*)

Metabolism, Pharmacokinetics, and Dosage

Apremilast is absorbed orally with a bioavailability of approximately 70 %. The plasma clearance of apremilast is on average about 10 L/hr in healthy subjects, with an elimination half-life of approximately 6–8 h. There may, however, be a slower terminal elimination phase (Hoffman et al. 2011; Shutty et al. 2012). The dosage interval of 12 h is slightly longer than its stated initial half-life of 6–8 h. Apremilast undergoes many metabolic steps with the major pathway being oxidative demethylation followed by glucuronidation. The metabolite is excreted in urine and accounts for about 35 % of the total elimination of the drug (Hoffman et al. 2011). The metabolite is inactive.

Roflumilast is rapidly absorbed with a bioavailability of about 80 %. Its major metabolite is roflumilast N-oxide which is produced mainly by cytochrome P4503A4 and P45031A2 (Pinner et al. 2012; Tenor et al. 2011). The N-oxide is then metabolized by cytochrome P4503A4 and cytochrome P4502C19. The metabolism of roflumilast and its N-oxide by cytochrome P450 isoenzymes indicates that a variety of other drugs may either inhibit or induce the metabolism of the two drugs (Pinner et al. 2012) (see "Drug Interactions" below). The half-lives of both roflumilast and its N-oxide are about 24 h and are therefore consistent with once-daily dosage of roflumilast (Tenor et al. 2011).

The N-oxide is a less potent inhibitor of PDE4 but its unbound plasma concentrations are higher, and complex modeling indicates that approximately 90 % of PDE4 inhibitory activity of oral roflumilast is due to the N-oxide (Pinner et al. 2012).

Pharmacological Activities

Apremilast is a highly selective inhibitor of phosphodiesterase 4 (PDE4) with minimal actions on other PDEs (Abdulrahim et al. 2015; Schafer et al. 2014; Shutty et al. 2012). Consequently, PDE4 inhibition increases intracellular cAMP levels, which in turn downregulates the inflammatory response by downmodulating the expression of several inflammatory cytokines, including TNF, IL-12/IL-23, IL-17, and interferon-α and interferon-γ. Inhibition of the hydrolysis of cAMP also upmodulates levels of anti-inflammatory cytokines such as IL-10 (Abdulrahim et al. 2015). These pro- and anti-inflammatory mediators are involved in psoriasis and psoriatic arthritis. The apremilast-induced inhibition of PDE4 is therefore considered to be the mechanism of its action in psoriasis and psoriatic arthritis.

Roflumilast is also a highly selective inhibitor of PDE4 with very little effect on other PDE enzymes. As noted above, the N-oxide of roflumilast is an inhibitor of PDE4 (Pinner et al. 2012). Compared with placebo, roflumilast decreases the numbers of eosinophils, neutrophils, and lymphocytes in the blood. As outlined in the "Introduction" and in "Clinical Uses and Efficacy," roflumilast may be useful in combination with corticosteroids and beta$_2$-sympathomimetics in the treatment of COPD and asthma. It is suggested that roflumilast may increase the effect of beta-sympathomimetics on T lymphocytes where beta$_2$-adrenoceptor is weak or coupling to a cellular effect is weak (Giembycz and Maurice 2014).

Clinical Uses and Efficacy

The two major inhibitors have quite different clinical applications.

Apremilast is active in the treatment of psoriasis and psoriatic arthritis and is approved for the treatment of these diseases in the USA and several other countries. However, the response to apremilast is generally modest. For example, approximately one third of patients achieve improvement by an ACR20 (Abdulrahim et al. 2015; Shutty et al. 2012). Apart from its approved use in psoriasis, apremilast has been tested in rosacea and Behcet's syndrome with some benefit although larger clinical trials are required. Early studies in rheumatoid arthritis and ankylosing spondylitis have not shown clinically adequate responses (Abdulrahim et al. 2015).

Treatment with apremilast is associated with reduced plasma concentrations of IL-8, TNF, and IL-6 and these effects are consistent with its pharmacological actions (Abdulrahim et al. 2015). Its oral administration dosage is an advantage over the biological agents, such as the inhibitors of tumor necrosis factor and IL-12/23, which have to be injected. However, these biological agents appear to have greater activity than apremilast in the treatment of psoriasis and psoriatic arthritis, although not all patients treated with the biologicals respond adequately (Palfreeman et al. 2013).

Alternative oral systemic drugs, methotrexate, leflunomide, sulfasalazine, and cyclosporin, are also used for psoriasis and psoriatic arthritis, but their use, particularly methotrexate and cyclosporin, requires close monitoring because of their toxicities (Shutty et al. 2012). Methotrexate and cyclosporin may, however, be used with apremilast.

Topical therapy with corticosteroids and vitamin D and phototherapy are widely used in patients with limited areas of plaque psoriasis, although adherence to topical therapy is often poor (Shutty et al. 2012). A novel PDE4 inhibitor (code name AN2728) is in clinical trials as a topical treatment of plaque psoriasis and may also be useful for atopic dermatitis (Moustafa and Feldman 2014). It is a novel boron-containing molecule.

Roflumilast is approved in several countries for the treatment of severe chronic obstructive pulmonary disease. It is approved as an add-on treatment with other bronchodilators, such as beta-sympathomimetics, in patients whose chronic obstructive chronic pulmonary disease is severe with frequent exacerbations (Tenor et al. 2011). With other drugs, roflumilast further improves lung function but clinical results on the number of exacerbations are conflicting (Pinner et al. 2012). Present guidelines are that roflumilast be given to patients already taking the inhaled combination of a long-acting beta-agonist and a corticosteroid for their chronic pulmonary obstructive disease (Giembycz and Maurice 2014).

It has been suggested that the extrapulmonary aspects of chronic obstructive pulmonary disease, including exacerbation of cachexia, muscle wasting, coronary heart disease, and osteoporosis, may be reduced by roflumilast or other PDE4 (Tenor et al. 2011). However, clinical trials are required.

Roflumilast is approved only for chronic obstructive pulmonary disease, although several studies have shown its activity in the treatment of various grades of asthma (Page 2014).

Adverse Effects

Apremilast is generally well tolerated, but headache, diarrhea, and nausea are its most commonly reported adverse effects. Furthermore, about 10 % of patients lose 5–10 % of body weight (Abdulrahim et al. 2015). A risk of depression is noted in the USA prescribing information although this risk is not supported by data from clinical trials on apremilast (Abdulrahim et al. 2015). In order to maximize tolerance to apremilast, it is recommended that its dosage should start at 10 mg on the first day, increasing to 10 mg twice daily with a subsequent a rapid increase to 30 mg twice daily (US label).

Roflumilast is also better tolerated than other PDE4 inhibitors. Vomiting is rare while nausea occurs more commonly than during placebo (5 % vs. 1 %). Diarrhea occurs in approximately 10 % of patients treated with roflumilast and in 3 % of placebo-treated patients. As is the case with apremilast, roflumilast causes loss of body weight (Pinner et al. 2012). The weight loss is about 2 kg in the patients with obstructive pulmonary disease. The weight loss is reversible and body weight is regained on cessation of treatment.

Drug Interactions

As outlined above, apremilast and roflumilast are metabolized by cytochrome P450 enzymes with the consequent likelihood of metabolic drug interactions. There are potential interactions with several other drugs particularly those which inhibit or induce cytochrome P450 isoenzymes. A full range of metabolic interaction studies has not been conducted but known interactions include the following:

Apremilast Ketoconazole reduces the clearance of apremilast by about one third (Hoffman et al. 2011).

Roflumilast Substantial inhibition of the oxidative metabolism of roflumilast has been shown with several drugs. However, less inhibition of the metabolism of roflumilast N-oxide occurs, and, among the recorded studies, only the metabolism of the parent drug and the N-oxide by fluvoxamine and cimetidine was inhibited sufficiently to increase the total PDE4 activity by about 50 % (Pinner et al. 2012). Concerning induction of oxidative metabolism, rifampicin decreases the total PDE4 inhibitory activity after oral administration of roflumilast (Pinner et al. 2012). The administration of roflumilast with other inducers of cytochrome P450 should be avoided until the potential interaction has been checked.

Cross-References

▶ Anti-asthma Drugs, Overview
▶ Asthma
▶ Behçet's Disease
▶ Beta2 Receptor Agonists
▶ Corticosteroids
▶ Disease-Modifying Antirheumatic Drugs: Overview
▶ Glial Cells
▶ Inflammatory Bowel Disease
▶ Methotrexate
▶ NFkappaB
▶ Rheumatoid Arthritis
▶ Theophylline
▶ Tumor Necrosis Factor (TNF) Inhibitors

References

Abdulrahim, H., Thistleton, S., Adebajo, A. O., Shaw, T., Edwards, C., & Wells, A. (2015). Apremilast: A PDE4 inhibitor for the treatment of psoriatic arthritis. *Expert Opinion on Pharmacotherapy, 16*(7), 1099–1108.

Giembycz, M. A., & Maurice, D. H. (2014). Cyclic nucleotide-based therapeutics for chronic obstructive pulmonary disease. *Current Opinion in Pharmacology, 16*, 89–107.

Hoffman, M., Kumar, G., Schafer, P., Cedzik, D., Capone, L., Fong, K. L., et al. (2011). Disposition, metabolism and mass balance of [14C]apremilast following oral administration. *Xenobiotica, 41*(12), 1063–1075.

Mizuno, T., Kurotani, T., Komatsu, Y., Kawanokuchi, J., Kato, H., Mitsuma, N., et al. (2004). Neuroprotective role of phosphodiesterase inhibitor ibudilast on neuronal cell death induced by activated microglia. *Neuropharmacology, 46*(3), 404–411.

Mo, M., Erdely, I., Szigeti-Buck, K., Benbow, J. H., & Ehrlich, B. E. (2012). Prevention of paclitaxel-induced peripheral neuropathy by lithium pretreatment. *FASEB Journal, 26*(22), 4696–4709.

Moustafa, F., & Feldman, S. R. (2014). A review of phosphodiesterase-inhibition and the potential role for phosphodiesterase 4 inhibitors in clinical dermatology. *Dermatology Online Journal, 20*(5), 22608.

Page, C. P. (2014). Phosphodiesterase inhibitors for the treatment of asthma and chronic obstructive pulmonary disease. *International Archives of Allergy and Immunology, 165*(3), 152–164.

Palfreeman, A. C., McNamee, K. E., & McCann, F. E. (2013). New developments in the management of psoriasis and psoriatic arthritis: A focus on apremilast. *Drug Design, Development and Therapy, 7*, 201–210.

Pinner, N. A., Hamilton, L. A., & Hughes, A. (2012). Roflumilast: A phosphodiesterase-4 inhibitor for the treatment of severe chronic obstructive pulmonary disease. *Clinical Therapeutics, 34*(1), 56–66.

Schafer, P. H., Parton, A., Capone, L., Cedzik, D., Brady, H., Evans, J. F., et al. (2014). Apremilast is a selective PDE4 inhibitor with regulatory effects on innate immunity. *Cellular Signalling, 26*(9), 2016–2029.

Shutty, B., West, C., Pellerin, M., & Feldman, S. (2012). Apremilast as a treatment for psoriasis. *Expert Opinion on Pharmacotherapy, 13*(12), 1761–1770.

Snider, S. E., Hendrick, E. S., & Beardsley, P. M. (2013). Glial cell modulators attenuate methamphetamine self-administration in the rat. *European Journal of Pharmacology, 701*(1–3), 124–130.

Spina, D. (2008). Spina PDE4 inhibitors: Current status. *British Journal of Pharmacology, 155*(3), 308–315.

Tenor, H., Hatzelmann, A., Beume, R., Lahu, G., Zech, K., & Bethke, T. D. (2011). Pharmacology, clinical efficacy, and tolerability of phosphodiesterase-4 inhibitors: Impact of human pharmacokinetics. *Handbook of Experimental Pharmacology, 204*, 85–119.

Platelets, Endothelium, and Inflammation

Nadine Ajzenberg and Marie-Geneviève Huisse
Department of Haematology AP-HP, Bichat-Claude Bernard Hospital, University Paris Diderot, Paris, France

Synonyms

Inflammation; Innate immune response; Platelet activation; Platelet-endothelial interaction

Definition

The interactions of platelets and leukocytes with the endothelium proceed under physiological and pathological conditions through a dynamic and time-dependent process depending on the local conditions of shear stress and the phenotype of the endothelium, which is specific to the segment of the vascular tree and the organ considered. Experimental models in animals and in vitro experiments using intravital fluorescence technology have considerably improved our knowledge of this subject and of the consequences for the inflammation process.

This review will focus on platelet-endothelial cell interactions in vivo under physiological conditions and conditions of endothelial activation. Many of the molecules associated with platelet-endothelial interactions initiate or regulate signal transmission. Concentration of these molecules allows new local interactions and induces specific cellular responses, with crucial effects on the physiology and pathology of the endothelium and more generally of the vascular system. Such interactions could be envisaged as molecular targets for different types of therapeutic intervention.

Structure and Function

Properties of the Intact Endothelium as a "Gatekeeper"

The endothelium covers the vascular tree with an uninterrupted layer of cells held together by tight junctions, separating the underlying tissues from the flowing blood. Endothelial cells are characterized by a specific phenotype (arterial, venous, or arteriolar), genotype, and organ specialization (Aird 2012). They constitute zones of exchange between the tissues and the blood, across which molecules and cells transit in response to a wide range of signals. The vascular permeability is strictly regulated by molecules such as sphingosine-1-phosphate and $\alpha v \beta 3$ (Rosen et al. 2007; Su et al. 2012).

The endothelium is also a powerful barrier against platelet activation owing to three enzymatic systems regulating platelet adhesion. The first is mediated by ecto-ADPase, which is an integral component of the endothelial cell membrane (CD39/nucleoside triphosphate diphosphohydrolase). It exerts its antiplatelet effect through the metabolism of ADP, a powerful platelet agonist. The second, prostacyclin or PGI2 (Moncada et al. 1976), is the main product of arachidonic metabolism in endothelial cells. PGI2 is an endothelial-derived relaxing factor which is synthesized in response to many stimuli including shear stress. It is a potent inhibitor of platelet aggregation as it increases cAMP levels.

The nitric oxide (NO) pathway (Radomski et al. 1987) constitutes the third system. NO is derived from L-arginine and oxygen under the control of an NO synthase pathway. Its biological effects including vasodilatation are mediated by activation of soluble guanylate cyclase, which results in increased levels of cGMP. NO is a potent inhibitor of platelet adhesion and activation (Radomski et al. 1987). Furthermore, NO can also inhibit leukocyte adhesion and migration to the vascular endothelium (Kubes et al. 1991) and smooth muscle cell proliferation.

In addition, between the flowing blood and the endothelial cell membrane, the endothelial surface layer or glycocalyx, composed mainly of glycoproteins and proteoglycans, is physiologically active in regulating endothelial permeability, the response to shear stress, and leukocyte-endothelial interactions. It also represents an additional mechanism preventing platelet adhesion as evidenced by loss of this function following its disruption (Noble et al. 2008).

Platelets: Partners of the Endothelium in Maintaining Its Integrity

Platelets are anucleated circulating cells derived from megakaryocytes in the perivascular microenvironment. The migration of platelets into the blood stream is facilitated by the gradient of sphingosine-1-phosphate, which is maximal in the blood (Zhang et al. 2012). Once in the blood stream, platelets are involved in protection and repair of the endothelium.

Platelets are equipped with highly specific surface/membrane receptors linked to signal transduction systems/activation pathways and with different types of storage granules (alpha- and dense granules, lysosomes, canalicular system) containing substances involved in the processes of hemostasis (adhesins, selectins, integrins, nucleotides, factor V, factor XI, plasminogen activator inhibitor-1, plasminogen, protein S), angiogenesis and cell proliferation (angiopoietin, vascular endothelial growth factor), chemotaxis and inflammation (platelet-derived growth factor (PDGF), transforming growth factor beta, CD40L, RANTES, CXCL4, etc.), and proteolysis (matrix metalloproteinase (MMP)2). All these functions are subject to precise mechanisms of regulation in specific situations (Gawaz et al. 2005). An alteration of one of these mechanisms can accelerate the inflammatory process.

Under normal conditions, platelets circulate and roll over the intact endothelium without adhering firmly, although they circulate in close proximity to the vascular wall as compared to leukocytes. This is due the concerted action of the regulatory enzymatic systems cited above.

However, when the vascular endothelium is injured, platelets adhere to the subendothelial matrix of collagen through GP1bα/von Willebrand factor interactions and directly to collagen mainly through GPVI. Subsequent stimulation of the immunoreceptor tyrosine-based activation motif (ITAM)-coupled receptor leads to activation of the cells and their aggregation with other platelets recruited through α2bβ3 integrin/fibrinogen bridges, thus forming a haemostatic plug.

The "Inflamed" Endothelium

It has recently been demonstrated that an endothelial lesion is not necessarily required to generate contacts between platelets and the vascular wall. The adhesion step is indeed facilitated when any one of the regulation processes fails, which may be triggered by different stimuli including variations in wall shear rate, reduction of blood flow, infection, oxidized lipoproteins, or platelet-bound-oxidized low-density lipoprotein in acute coronary syndromes (ACS) (Stellos et al. 2012). In these situations, platelets adhere to the activated endothelium in a multistep process and release from their granules a wide array of mediators and inflammatory cytokines, thereby promoting inflammation.

Adhesive interactions between platelets and endothelial cells under inflamed conditions have been extensively studied by Massberg et al. in a mouse model of ischemia-reperfusion (von Bruhl et al. 2012). The initial loose contact is through P-selectin which is expressed on the endothelium. As soon as P-selectin is exposed on the endothelium, rolling platelets tether and adhere directly to inflamed or activated endothelial cells through P-selectin/P-selectin glycoprotein ligand-1 (PSGL1) or P-selectin/GPIbα binding. Firm adhesion is mediated by α2bβ3 integrin in its active conformation (Gawaz et al. 2005). In response to this process, the platelets are activated and express on their surface much more important levels of P-selectin than are expressed on the endothelium, thereby recruiting leukocytes more easily. These platelets also release CD40L and interleukin (IL) 1β (Gawaz et al. 2005).

IL1β is constitutively synthesized in packaged form by platelet mRNA and the synthesis is dependent on platelet activation and α2bβ3 engagement (Denis et al. 2005). IL1β induces cytokine production in endothelial cells (IL6 and IL8) and increases their expression of intercellular adhesion molecule (ICAM)-1, vascular cell adhesion molecule-1, monocyte chemotactic protein (MCP)-1, and αvβ3. ICAM-1 and MCP-1 upregulation in endothelial cells in response to IL1β is dependent on activation of NF kappa B (Gawaz et al. 1998). These chemokines constitute powerful chemoattractants for leukocyte recruitment (Henn et al. 1998).

Adherent/activated platelets then recruit and activate monocytes through ternary interactions which involve P-selectin/PSGL1 and monocyte Mac-1 (CD11b-CD18) together with integrin b3 and fibrinogen. Thus, platelets initiate the secretion of chemokines, cytokines (IL8, tumor necrosis factor α), and procoagulant tissue factor (TF) and the upregulation of proteases (MMP9, urokinase-type plasminogen activator/urokinase-type plasminogen activator receptor) in monocytes and induce monocyte differentiation into macrophages (Gawaz et al. 2005) and the production of microparticles expressing tissue factor (Wagner and Frenette 2008) (Fig. 1). Alternatively, leukocytes have the capacity to interact directly with inflamed endothelium in a manner very similar to platelets and can themselves initiate thrombus formation (Darbousset et al. 2012).

Although arterial and venous thrombosis represent two separate pathological entities with their own conditions of initiation and development, both involve the platelet adhesion step and commitment of the cells into an activated state.

Pathophysiological Relevance

Platelet Adhesion to Inflamed Endothelium in Atherosclerosis: Experimental Models and Human Pathology (ACS and Stroke)

The importance of the primary role of platelet-endothelial cell adhesion in triggering atherothrombotic events has been clearly established in an experimental model of Apo $E^{-/-}$ mice, in which platelet adhesion develops on hypercholesterolemic zones before the formation of an atherosclerotic plaque (Massberg et al. 2002). This process is inhibited by anti-GPIbα and anti-GPIIb-IIIa antibodies. More recently, it has been shown that GPVI is critically involved in adhesion of platelets (through avb3) to the diseased endothelium of carotid arteries in Apo $E^{-/-}$ mice. This can be prevented by

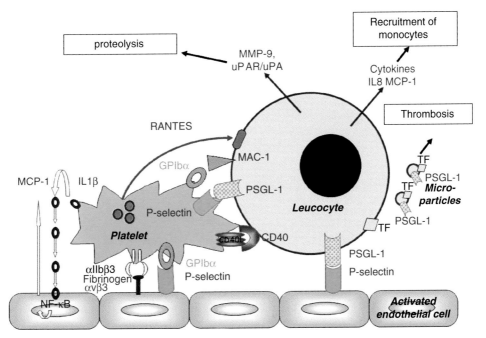

Platelets, Endothelium, and Inflammation, Fig. 1 Schematic representation of the main interactions between activated platelets, activated endothelial cells, and monocytes. *PSGL1* P-selectin glycoprotein ligand-1, *MCP-1* monocyte chemotactic protein-1, *uPAR/uPA* urokinase-type plasminogen activator receptor/urokinase-type plasminogen activator, *TF* tissue factor, *MMP-9* matrix metalloproteinase-9

administration of a dimeric soluble fusion protein, GPVI-Fc (Schönberger et al. 2012), which also improves reperfusion in the model.

All these data provide evidence that platelets are in the front line of the inflammatory process, assisted in the propagation phase by their interactions with other circulating cells, including immune cells.

Whether these observations can be transferred to human pathology is questionable, although there is some evidence that platelets are involved in the early steps of atherosclerosis including in coronary and carotid artery disease. Thus, the surface expression of GPVI is enhanced in patients with ACS, and this heralds an imminent coronary event before irreversible myocardial necrosis is present (Bigalke et al. 2008). Oxidized low-density lipoprotein is also increased on circulating platelets in patients with ACS (Stellos et al. 2012).

Systemic platelet activation and platelet degranulation are associated with enhanced vessel wall thickness in humans with cardiovascular risk factors (Fateh-Moghadam et al. 2005; Koyama et al. 2003). Finally, platelet chemokines (platelet factor 4) and growth factors have been identified in atherosclerotic lesions in humans (Pitsilos et al. 2003).

Platelet Adhesion to Inflamed Endothelium in Deep Vein Thrombosis (DVT): Experimental Models and Human Pathology

Platelets therefore play a key role in both inflammation and thrombosis. Whereas this role has been extensively studied in arteriosclerosis and arterial thrombosis, our knowledge of their contribution to DVT models is more recent. Moreover, there is increasing evidence that inflammatory processes and DVT are closely linked. Venous thrombi are rich in fibrin and red blood cells and they are infiltrated with large numbers of leukocytes. The precise contribution of leukocytes to DVT induction nevertheless remains unclear due to the lack of an appropriate animal model leading to DVT in a large vein. The models to date have used endothelial disruption as a thrombogenic stimulus, which is a rare cause of DVT in humans.

Recently, cooperation between platelets and leukocytes in the pathogenesis of venous thrombosis was studied using a novel mouse model of DVT induced by flow restriction in combination with intravital imaging (von Bruhl et al. 2012). In this new flow perturbation model (in the inferior vena cava), the interruption reached 80 %, and it took 6–12 h to detect small thrombi and 24–48 h to observe occlusive thrombi. These were mostly red thrombi similar to those observed in DVT patients. Using the same DVT model, Massberg et al. demonstrated that adherence of predominantly neutrophils (monocytes represented the remaining 30 %) to the venous endothelium provided the initiating stimulus for DVT development. The cells were already recruited 6 h after flow restriction, whereas endothelial disruption was not detected. Using P-selectin deficient mice, these authors further demonstrated that the leukocyte accumulation depended on endothelial P-selectin.

Since tissue factor is the major initiator of coagulation and some leukocytes express tissue factor, the role of blood cell-derived tissue factor in DVT development has also been investigated. Hematopoietic tissue factor expressed by myeloid cells and not vessel wall tissue factor was found to be responsible for the initiation of DVT in response to flow restriction without endothelial damage. The exact origin of this tissue factor (monocytes or neutrophils) nevertheless remains unclear (Pawlinski et al. 2010). Moreover, platelets can either adhere directly to the intact endothelium or attach to adherent leukocytes. Platelets also contribute to DVT progression by promoting leukocyte recruitment and release of neutrophil extracellular traps (NETs) (Fuchs et al. 2012). Altogether, these results identify immune cells as the primary initiators of the clot formation leading to DVT, through their interaction with platelets, delivery of tissue factor, and release of procoagulant NETs.

Other aspects of the proinflammatory properties of platelets have been studied in another common inflammatory disease, rheumatoid arthritis.

Whereas some inflammatory diseases are known to be associated with an increased risk of thrombosis, in others conflicting data have been reported, as in rheumatoid arthritis, a disorder characterized by chronic inflammation of the synovial lining of the joint. In a case control study conducted in Taiwan in 5,193 patients with DVT and 20,772 controls matched for gender, age, and year of indexing, a significant association between DVT and prior rheumatoid arthritis was observed (OR 1.88; 95 % CI: 1.42–2.58; $p < 0.001$) (Kang et al. 2012).

Furthermore, the implication of platelets in rheumatoid arthritis has recently been demonstrated (Boilard et al. 2010). Platelets are activated through the collagen receptor GPVI and release platelet microparticles (MPs), submicron vesicles shed from the platelet surface. These proinflammatory MPs are enriched in the cytokine IL1 and can exacerbate inflammation by stimulating resident fibroblasts and potentially other cells. Another potential role of platelets – and not microparticles – which could contribute to synovitis through the production of proinflammatory prostacyclin by a transcellular mechanism has also been reported (Boilard et al. 2011).

An enhanced permeability of the synovial microvasculature has been observed in human inflammatory arthritis in association with tissue edema. Murine studies suggested that such vascular leaks facilitate the entry of autoantibodies and could thereby promote joint inflammation. As platelets typically help to promote microvascular integrity, their role in synovial vascular permeability was studied in murine experimental arthritis. Using an in vivo model of autoimmune arthritis, the presence of endothelial gaps in the inflamed synovium was confirmed. Surprisingly, the vascular permeability of the inflamed joints was abrogated if platelets were absent. This effect was mediated by platelet serotonin accumulated through the serotonin transporter and could be antagonized using serotonin-specific reuptake inhibitors. Moreover, the dimensions of the gaps were compatible with those of platelet MPs and with their translocation into the synovial space (Cloutier et al. 2012).

Overall, these observations show that the contributions of platelets to joint inflammation are many and varied and that further studies will be required to elucidate their exact participation in autoimmune and inflammatory pathophysiology.

There is indeed suspicion of a potential role of yet another inflammatory platelet molecule, CD40L (Duffau et al. 2010).

These findings might have therapeutic consequences as pharmacological inhibitors of platelet secretion or antagonists of GPVI could be proposed for the treatment of inflammatory arthritis.

Clinical Studies on the interaction between inflammation and DVT

In order to explore the clinical association between inflammation and DVT, some studies have focused on looking for increased levels of inflammatory parameters in DVT patients. Preliminary data obtained in the JUPITER study suggest a link between inflammatory C-reactive protein and DVT (Glynn et al. 2009). This study was conducted to determine the impact of statin treatment in reducing the risk of DVT, which was found to be 43 % and was associated with a reduction of 37 % in CRP levels (Glynn et al. 2009).

To examine the robustness of this association, a larger study was conducted over more than 16 years in 10,388 volunteers, among whom 903 previously had DVT. A level of CRP of >3 mg/L vs. <1 mg/L was associated with a 2.3-fold increase in DVT, but this was not a causal relation because genetically elevated CRP did not correlate with the risk of DVT (Zacho et al. 2010). Nevertheless, conflicting data concerning this association have been reported (Folsom et al. 2009; Vormittag et al. 2005).

Conversely, a link between platelets and DVT could be indirectly deduced from the clinical efficiency of low-dose aspirin for prevention of the recurrence of unprovoked DVT. In two large trials (ASPIRE, Brighton et al. 2012; WARFASA Becattini et al. 2012), low-dose aspirin significantly reduced the recurrence of unprovoked DVT, either alone or in association with cardiovascular events, during 2 years of follow-up.

Platelets as a Safeguard of the Microvasculature

More recently, platelets were shown to safeguard the microvasculature at sites of leukocyte infiltration. Interestingly, from a mechanistic point of view, the supporting role of platelets in these different vessels does not seem to necessarily involve the well-understood process of platelet plug formation but rather might rely on secretion of various active components from the platelet granules. It is important to note here that the vascular protective action of platelets in inflammation has been demonstrated experimentally in mouse models of severe thrombocytopenia (<5 % of the normal platelet count) and that it can be restored by transfusing platelets to replenish as little as 10 % of the circulating cells (Ho-Tin-Noe et al. 2011).

In conclusion, this review attempts to describe some of the newly updated functions of platelets. Starting from the well-known function of these cells in arresting bleeding when an endothelial lesion occurs, important advances have been made in our comprehension of the role of platelets as guardians of the vascular integrity and major actors in controlling inflammatory processes, in close interaction with cells of the innate immune response. The new concept of cooperation between platelets and leukocytes in the physiopathology of DVT should pave the way for the development of new therapeutic approaches, specifically targeting the cellular factors which initiate DVT such as platelets and/or leukocytes.

Acknowledgement We thank Benoit Ho-Tin-Noe for critical reading of the manuscript

Cross-References

▶ Neutrophil Extracellular Traps
▶ Platelets, Endothelium, and Inflammation

References

Aird, W. C. (2012). Endothelial cell heterogeneity. *Cold Spring Harbor Perspectives in Medicine, 2*(1), a006429.

Becattini, C., Agnelli, G., Schenone, A., Eichinger, S., Bucherini, E., Silingardi, M., et al. (2012). Aspirin for preventing the recurrence of venous thromboembolism. *The New England Journal of Medicine, 366*(21), 1959–1967.

Bigalke, B., Geisler, T., Stellos, K., Langer, H., Daub, K., Kremmer, E., et al. (2008). Platelet collagen receptor glycoprotein VI as a possible novel indicator for the acute coronary syndrome. *American Heart Journal, 156*(1), 193–200.

Boilard, E., Nigrovic, P. A., Larabee, K., Watts, G. F., Coblyn, J. S., Weinblatt, M. E., et al. (2010). Platelets amplify inflammation in arthritis via collagen-dependent microparticle production. *Science, 327*(5965), 580–583.

Boilard, E., Larabee, K., Shnayder, R., Jacobs, K., Farndale, R. W., Ware, J., et al. (2011). Platelets participate in synovitis via Cox-1-dependent synthesis of prostacyclin independently of microparticle generation. *Journal of Immunology, 186*(7), 4361–4366.

Brighton, T. A., et al. (2012). *The New England Journal of Medicine, 367*(21), 1979–1987.

Cloutier, N., Pare, A., Farndale, R. W., Schumacher, H. R., Nigrovic, P. A., Lacroix, S., et al. (2012). Platelets can enhance vascular permeability. *Blood, 120*(6), 1334–1343.

Darbousset, R., Thomas, G. M., Mezouar, S., Frere, C., Bonier, R., Mackman, N., et al. (2012). Tissue factor-positive neutrophils bind to injured endothelial wall and initiate thrombus formation. *Blood, 120*(10), 2133–2143.

Denis, M. M., Tolley, N. D., Bunting, M., Schwertz, H., Jiang, H., Lindemann, S., et al. (2005). Escaping the nuclear confines: Signal-dependent pre-mRNA splicing in anucleate platelets. *Cell, 122*(3), 379–391.

Duffau, P., Seneschal, J., Nicco, C., Richez, C., Lazaro, E., Douchet, I., et al. (2010). Platelet CD154 potentiates interferon-alpha secretion by plasmacytoid dendritic cells in systemic lupus erythematosus. *Science Translational Medicine, 2*(47), 47ra63.

Fateh-Moghadam, S., et al. (2005). *Arteriosclerosis, Thrombosis, and Vascular Biology, 25*(6), 1299–1303.

Folsom, A. R., Lutsey, P. L., Astor, B. C., & Cushman, M. (2009). C-reactive protein and venous thromboembolism. A prospective investigation in the ARIC cohort. *Thrombosis and Haemostasis, 102*(4), 615–619.

Fuchs, T. A., Brill, A., & Wagner, D. D. (2012). Neutrophil extracellular trap (NET) impact on deep vein thrombosis. *Arteriosclerosis, Thrombosis, and Vascular Biology, 32*(8), 1777–1783.

Gawaz, M., Neumann, F. J., Dickfeld, T., Koch, W., Laugwitz, K. L., Adelsberger, H., et al. (1998). Activated platelets induce monocyte chemotactic protein-1 secretion and surface expression of intercellular adhesion molecule-1 on endothelial cells. *Circulation, 98*(12), 1164–1171.

Gawaz, M., Langer, H., & May, A. E. (2005). Platelets in inflammation and atherogenesis. *The Journal of Clinical Investigation, 115*(12), 3378–3384.

Glynn, R. J., Danielson, E., Fonseca, F. A., Genest, J., Gotto, A. M., Jr., Kastelein, J. J., et al. (2009). A randomized trial of rosuvastatin in the prevention

of venous thromboembolism. *The New England Journal of Medicine, 360*(18), 1851–1861.
Henn, V., Slupsky, J. R., Grafe, M., Anagnostopoulos, I., Forster, R., Muller-Berghaus, G., et al. (1998). CD40 ligand on activated platelets triggers an inflammatory reaction of endothelial cells. *Nature, 391*(6667), 591–594.
Ho-Tin-Noe, B., Demers, M., & Wagner, D. D. (2011). How platelets safeguard vascular integrity. *Journal of Thrombosis and Haemostasis, 9*(Suppl 1), 56–65.
Kang, J. H., Keller, J. J., Lin, Y. K., & Lin, H. C. (2012). A population-based case-control study on the association between rheumatoid arthritis and deep vein thrombosis. *Journal of Vascular Surgery, 56*(6), 1642–1648.
Koyama, H., Maeno, T., Fukumoto, S., Shoji, T., Yamane, T., Yokoyama, H., et al. (2003). Platelet P-selectin expression is associated with atherosclerotic wall thickness in carotid artery in humans. *Circulation, 108*(5), 524–529.
Kubes, P., Suzuki, M., & Granger, D. N. (1991). Nitric oxide: An endogenous modulator of leukocyte adhesion. *Proceedings of the National Academy of Sciences of the United States of America, 88*(11), 4651–4655.
Massberg, S., Brand, K., Gruner, S., Page, S., Muller, E., Muller, I., et al. (2002). A critical role of platelet adhesion in the initiation of atherosclerotic lesion formation. *The Journal of Experimental Medicine, 196*(7), 887–896.
Moncada, S., Gryglewski, R., Bunting, S., & Vane, J. R. (1976). An enzyme isolated from arteries transforms prostaglandin endoperoxides to an unstable substance that inhibits platelet aggregation. *Nature, 263*(5579), 663–665.
Noble, M. I., Drake-Holland, A. J., & Vink, H. (2008). Hypothesis: Arterial glycocalyx dysfunction is the first step in the atherothrombotic process. *QJM, 101*(7), 513–518.
Pawlinski, R., Wang, J. G., Owens, A. P., 3rd, Williams, J., Antoniak, S., Tencati, M., et al. (2010). Hematopoietic and nonhematopoietic cell tissue factor activates the coagulation cascade in endotoxemic mice. *Blood, 116*(5), 806–814.
Pitsilos, S., Hunt, J., Mohler, E. R., Prabhakar, A. M., Poncz, M., Dawicki, J., et al. (2003). Platelet factor 4 localization in carotid atherosclerotic plaques: Correlation with clinical parameters. *Thrombosis and Haemostasis, 90*(6), 1112–1120.
Radomski, M. W., Palmer, R. M., & Moncada, S. (1987). Endogenous nitric oxide inhibits human platelet adhesion to vascular endothelium. *Lancet, 2*(8567), 1057–1058.
Rosen, H., Sanna, M. G., Cahalan, S. M., & Gonzalez-Cabrera, P. J. (2007). Tipping the gatekeeper: S1P regulation of endothelial barrier function. *Trends in Immunology, 28*(3), 102–107.
Schönberger, T., et al. (2012). *The American Journal of Physiology – Cell Physiology, 303*(7), C757–C766.
Stellos, K., Sauter, R., Fahrleitner, M., Grimm, J., Stakos, D., Emschermann, F., et al. (2012). Binding of oxidized low-density lipoprotein on circulating platelets is increased in patients with acute coronary syndromes and induces platelet adhesion to vascular wall in vivo–brief report. *Arteriosclerosis, Thrombosis, and Vascular Biology, 32*(8), 2017–2020.
Su, G., Atakilit, A., Li, J. T., Wu, N., Bhattacharya, M., Zhu, J., et al. (2012). Absence of integrin alphavbeta3 enhances vascular leak in mice by inhibiting endothelial cortical actin formation. *American Journal of Respiratory and Critical Care Medicine, 185*(1), 58–66.
von Bruhl, M. L., Stark, K., Steinhart, A., Chandraratne, S., Konrad, I., Lorenz, M., et al. (2012). Monocytes, neutrophils, and platelets cooperate to initiate and propagate venous thrombosis in mice in vivo. *The Journal of Experimental Medicine, 209*(4), 819–835.
Vormittag, R., Vukovich, T., Schonauer, V., Lehr, S., Minar, E., Bialonczyk, C., et al. (2005). Basal high-sensitivity-C-reactive protein levels in patients with spontaneous venous thromboembolism. *Thrombosis and Haemostasis, 93*(3), 488–493.
Wagner, D. D., & Frenette, P. S. (2008). The vessel wall and its interactions. *Blood, 111*(11), 5271–5281.
Zacho, J., Tybjaerg-Hansen, A., & Nordestgaard, B. G. (2010). C-reactive protein and risk of venous thromboembolism in the general population. *Arteriosclerosis, Thrombosis, and Vascular Biology, 30*(8), 1672–1678.
Zhang, L., Orban, M., Lorenz, M., Barocke, V., Braun, D., Urtz, N., et al. (2012). A novel role of sphingosine 1-phosphate receptor S1pr1 in mouse thrombopoiesis. *The Journal of Experimental Medicine, 209*(12), 2165–2181.

Polymyositis and Dermatomyositis

Shiro Matsubara
Department of Neurology, Tokyo Metropolitan Neurological Hospital, Fuchu, Tokyo, Japan

List of Abbreviations

APC	Antigen presenting cell
CADM	Clinically amyopathic dermatomyositis
CAM	Cancer-associated myositis
CK	Creatine kinase
DC	Dendritic cell
DM	Dermatomyositis
IBM	Inclusion body myositis
ILD	Interstitial lung disease

IMNM	Immune-mediated necrotizing myopathy
IVIg	Intravenous immunoglobulin
JDM	Juvenile (childhood) dermatomyositis
MDA5	Melanoma differentiation-associated gene 5
MHC	Major histocompatibility complex
MSA	Myositis specific antibody
OS	Overlap syndrome
PM	Polymyositis
RV	Rimmed vacuole
SRP	Signal recognition particle
SUMO-1	Small ubiquitine-like modifier 1

Synonyms

Idiopathic inflammatory myopathies

Definition

Polymyositis (PM) is an inflammatory myopathy that causes weakness of muscles. Histologically, it shows degeneration of muscle fibers in association with inflammation, but there is no evidence of any causative infectious organism.

Dermatomyositis (DM) is an inflammatory myopathy that causes weakness of muscles in association with distinct skin rash. Histologically, it shows degeneration of muscle fibers and intramuscular blood vessels, mainly in the interstitial tissue of the muscles in association with inflammation. A small proportion of patients have skin manifestations without apparent muscle symptoms.

Inclusion body myositis (IBM) is an inflammatory myopathy in adults older than 30 years, causing muscle weakness, usually of distinct distribution preferentially affecting the anterior thigh and forearm muscles. Histologically, it shows degeneration of muscle fibers with inflammation often associated with formation of rimmed vacuoles (RVs).

Epidemiology and Genetics

Epidemiology
PM and DM are rare diseases with annual incidences ranging from 5.8 to 7.9 per 100,000 person-years and annual prevalence rates of 14.0 to 17.4 per 100,000 person-years in the United States . PM and DM are more common among women than among men, with a female to male ratio of 2.2:1 in the United States (Oddis et al. 1990). This ratio is higher during childbearing ages (15–44 years). Moreover, PM and DM are more frequent among black than white populations. However, it should be noted that many of these studies failed to thoroughly exclude patients with IBM.

About 20 % of patients with DM are thought to have clinically amyopathic DM (CADM), a condition characterized by DM skin manifestations but lacking weakness and other muscle symptoms. The median age of onset of DM ranges from 40 to 60 years, although DM associated with malignancy has a higher age of onset of 62 years (Koh et al. 1993).

The frequency of PM depends on the diagnostic criteria used; PM accounted for 45 % of patients with idiopathic inflammatory myopathy (IIM) when it is diagnosed according to the criteria of Bohan and Peter (1975) (Troyanov et al. 2005). When the criteria of Dalakas and Hohlfeld (2003) were applied, however, its frequency decreased markedly.

Diagnosis of myositis as overlap syndrome (OS) depends on how it is defined. In one study, OS accounted for 24 % of patients with IIM. Systemic sclerosis was relatively common among these patients, being present in 42 % (Troyanov et al. 2005).

Several larger scale studies about association of DM with malignant neoplasm since the 1970s have shown that 11–27 % of patients with DM have malignant neoplasms (DeVere and Bradley 1975). This rate is higher in males than in females and in patients aged ≥ 65 than < 65 years (Marie et al. 1999). Patients with CADM also had a higher rate of malignant neoplasms than the general population, although this rate was not as high as observed in patients with classical DM (Bendewald et al. 2010). DM has been associated with many types of malignant neoplasms. It occurs at about the same frequency in males and females. Roughly equal numbers of patients are found to have neoplasms before, at the same time, or after being diagnosed with DM.

It is unclear whether PM has a higher rate of association with neoplasms than controls, but the rate is much lower in PM than in DM. It is noteworthy that patients with suspected DM but lacking evident skin rash (Dalakas and Hohlfeld 2003) were not excluded from statistical determinations of PM.

Juvenile or childhood dermatomyositis (JDM) has an incidence of 0.2 to 0.3 per 100,000 person-years (Ramanan and Feldman 2002). Among children, DM far exceeds PM in frequency by 10- to 20-fold. The female to male ratio of JDM is about 2:1. The mortality rate of these patients has been found to range from 8.3 % to 70 %, being higher in earlier series (Winkelmann 1982). In only a small proportion of cases are JDM associated with interstitial lung disease (ILD) or myositis-specific antibody (MSA). In a small number of patients, JDM has been associated with malignant conditions including lymphoma and leukemia.

IBM has a prevalence rate of 0.79 per 100,000 person-years in the United States (Wilson et al. 2008). IBM usually affects people older than 30 years, with most affected after the age of 50 years. Males are more susceptible to IBM than females, with male to female ratios of 2.2:1 to 3:1 (Badrising et al. 2005). A small number of patients with genetic background, including the valosin-containing protein gene mutation (Guyant-Marechal et al. 2006), have been reported as having hereditary IBM or inclusion body myopathy.

Immunogenetic Studies

The importance of the human leukocyte antigen (HLA) to the genetic background of myositis has been recognized increasingly in recent years. Genetic overlap with other autoimmune diseases has been observed. However, the frequencies of HLA-DRB1*0301(DR3) and DQB1*0201(DQ2) alleles were significantly increased in white patients with myositis, especially PM, but most strikingly in those with MSAs (Arnett et al. 1996). In Caucasians, DQA1*0501 was increased in patients with childhood or juvenile DM (Reed et al. 1998).

The genetic background to myositis has been studied for each autoantibody. Caucasian patients of DM with antibody against the small ubiquitin-like modifier activation enzyme (SAE) had increased frequency of DQB1*03 allele (Betteridge et al. 2009). Patients of myositis with anti-P155 (TIF1γ) antibody have increased frequency of malignancy and haplotype HLA DQA1*0301(Targoff et al. 2006). African-American patients with anti-signal recognition particle (SRP) antibody were detected to have increased frequency of DQA1*0102 allele (O'Hanlon et al. 2006). Among Japanese, susceptibility to anti-MDA5 antibody-positive DM was associated with DRB1*0101 or DRB1*0405 alleles (Gono et al. 2012).

Apart from the HLA region, polymorphisms of no-HLA genes were associated with myositis. Single-nucleotide polymorphisms (SNPs) of the protein tyrosine phosphatase N22 gene (PTPN22) were associated with juvenile and adult inflammatory myopathy in British Caucasian population (Chinoy et al. 2008). In the Japanese population, a polymorphism in C8orf1-BLK was associated with PM/DM (Sugiura et al. 2014).

Interaction between genetic susceptibility and nongenetic factors was also reported. Susceptibility to anti-histidyl tRNA synthetase (Jo-1) antibody-positive myositis was found to be increased in Europeans with the DRB1*3 allele who were smoking (Chinoy et al. 2012). An increased risk of statin-induced inflammatory myopathy with anti-hydroxy-methyl-glutaryl-coenzyme A reductase antibody was found to be associated with DRB1*11:01 in white and black populations in North America (Mammen et al. 2012).

According to Australian (Needham et al. 2008) and other studies, carriage of DRB1*0303 (DR3) allele was associated with susceptibility to IBM. Polymorphisms, haplotypic for the sIBM-associated 8.1 ancestral haplotype, have been identified in the NOTCH4 gene in Caucasians (Scott et al. 2012).

Pathophysiology

Dermatomyositis (DM)

Muscle tissue shows degeneration, necrosis, and regeneration of the muscle fibers. Perifascicular atrophy, namely, tendency to have atrophic muscle fibers at the periphery of the fasciculus, is thought to be characteristic of DM (Fig. 1a, c) (Dalakas and Hohlfeld 2003). Many of the small fibers at the periphery are regenerating fibers.

Immunological analyses of the muscle usually demonstrate infiltration of inflammatory cells, particularly CD4+ T cells and macrophages, mainly in the perimysium, the interstitial tissue between muscle fascicules, particularly around the blood vessels (Fig. 1b). The B cells, plasma cells and dendritic cells (DCs), predominantly plasmacytoid DCs (Greenberg et al. 2007), and a limited number of CD8+ T cells are also observed. Deposition of the immunoglobulin around the blood vessels is commonly seen. Moderate expression of major histocompatibility complex (MHC) class I antigen is usually observed at the perifascicular zone.

Electron microscopy of the intramuscular blood vessels sometimes shows degenerative changes in endothelial cells (Fig. 1e), occasionally associated with the granulotubular inclusions (Fig. 1d). These findings are consistent with the hypothesis that the main target of autoimmune reactions is the intramuscular blood vessels.

Patients with anti-histidyl-transfer RNA (tRNA) synthetase antibody (Jo-1 antibody) or antibodies against other anti-tRNA synthetases have been diagnosed with anti-synthetase syndrome (ASS) and classified into either DM or PM (see "Serological Tests"). However, recent pathological observations of necrotizing perifascicular myositis in patients with anti-Jo-1 antibody (Mescam-Mancini et al. 2015) and actin aggregation in the myonuclei of patients with ASS (Stenzel et al. 2015) prompted an argument that they should be grouped as ASS-associated myositis separately from other inflammatory myopathies.

Juvenile (Childhood) Dermatomyositis (JDM)

Muscle biopsies from patients with JDN are characterized by prominent vasculitic changes and massive collections of necrotic fibers often accompanying calcification in muscle. Perifascicular atrophy of muscle fibers and infiltration of lymphocytes in the perimysium, particularly in the perivascular area, are features common to JDM and adult DM.

Polymyositis (PM)

In PM muscle, degeneration, necrosis, and regeneration of muscle fibers are seen with infiltration of inflammatory cells in close proximity to individual muscle fibers in the endomysium (Fig. 2a, b). Strong expression of MHC class I antigen is observed on the cell surface of non-necrotic muscle fibers (Fig. 2c). The infiltrating cells mainly consist of CD8+ T cells (Fig. 2d), macrophages, and CD4+ T cells (Fig. 2e). These findings suggest that muscle fibers are undergoing degeneration through the cytotoxic action of CD8+ T cells (Engel and Arahata 1986). Destruction of the muscle fibers is carried out through action of perforin, granzyme, and granulysin.

The antigen presenting cells (APCs) are needed for initiation and maintenance of inflammation by CD8+ T cells. In addition to professional APCs, such as myeloid DCs (Greenberg et al. 2007) and macrophages, the myocytes may act as APCs. Additionally, the presence of B cells and plasma cells (Greenberg et al. 2005) among the infiltrating cells implies a local antigen-driven humoral response (Bradshaw et al. 2007). Furthermore, varieties of cytokines, chemokines, and co-stimulatory factors play important roles. Among the cytokines, the interferons α and β, IL-17 (Kondo et al. 2009), and IL-18 (Tucci et al. 2006) are considered important. Chemokine receptor CCR7 and its ligands CCR7 and CCL21 are found in muscle of PM (Tateyama et al. 2006). The inducible co-stimulatory factors (ICOS) and its ligand are detected in the muscles of PM and IBM (Schmidt et al. 2004). Despite these findings, the main target of the immune reaction remains unidentified.

The CD28(null) cells in the CD4+ and CD8+ T cell phenotype are effector/memory cells which have cytotoxic effects and resistance to apoptosis. Increase in CD28(null) cells in the tissue and peripheral blood has been reported in wide varieties of collagen vascular diseases and infections. In PM and DM, an increment of CD28(null) cells

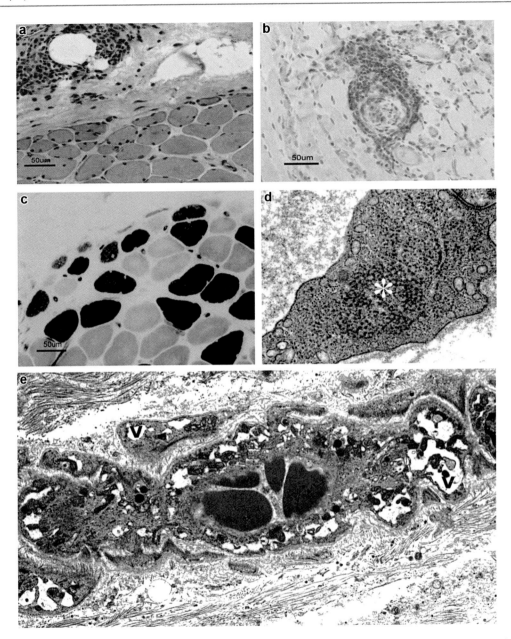

Polymyositis and Dermatomyositis, Fig. 1 Histological features of dermatomyositis. (**a**) Muscle fibers at the perimeter of the fascicles tend to be atrophied (perifascicular atrophy). Inflammatory cell infiltration is observed mainly in connective tissue around the fascicles (perimysium), often around the small blood vessels (HE). (**b**) Many infiltrating cells are CD4+ cells (*brown*). (**c**) Some of the atrophic muscle fibers are type 2C, suggesting they may be undergoing regeneration (routine myosin ATPase activity at pH 10.3). (**d**) Granulotubular inclusions (*asterisk*) in the endothelial cell of a small blood vessel in a patient with juvenile dermatomyositis. (**e**) Highly degenerated blood vessel with vacuoles (V) in a patient with dermatomyositis

among CD4+ and CD8+ T cells in the muscle and peripheral blood was observed (Fasth et al. 2009; Espinosa-Ortega et al. 2015). This may have relevance for the maintenance of inflammation.

Myositis Associated with Other Connective Tissue Diseases (Overlap Syndrome, OS)

Connective tissue diseases which commonly accompany myositis include systemic sclerosis,

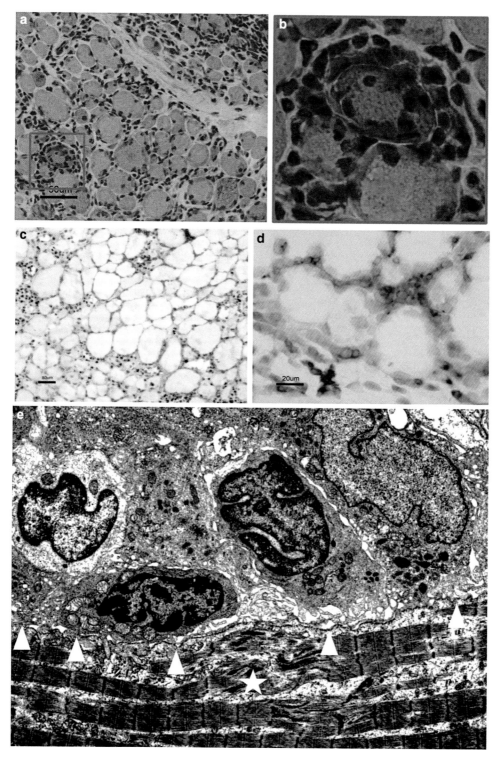

Polymyositis and Dermatomyositis, Fig. 2 Histological features of polymyositis. (**a**) Cell infiltration is distributed mainly in the fascicules around individual muscle fibers. Many fibers are undergoing degeneration and/or regeneration. (**b**) Enlarged image of part of **a**. (**c**) Aberrant expression of MHC class I antigen (*brown*) on the

mixed connective tissue disease, systemic lupus erythematosus, and rheumatoid arthritis. Muscle pathology in these patients varies considerably but is generally characterized by relatively moderate degrees of interstitial cell infiltration, with scattered muscle fibers degeneration. Severe changes comparable to those in PM and DM are rare, and patients with such changes may be considered to have PM or DM accompanying features of other connective tissue diseases.

Myositis Associated with Malignant Neoplasm (Cancer-Associated Myositis, CAM)

The histological features of DM associated with malignant neoplasms do not differ from those of other types of DM. The granulotubular inclusions in vascular endothelial cells have also been observed in patients with DM associated with malignancy. Some cases show only a limited degree of inflammation.

Immune-Mediated Necrotizing Myopathy (IMNM)

Muscle biopsies from patients with IMNM and positivity for anti-signal recognition particle (SRP) antibody show scattered necrotic muscle fibers with no or minimum inflammatory infiltrates in the interstitial tissue, usually around the blood vessels (Fig. 3). Aberrant expression of MHC class I antigen is usually absent or minimum, except for a moderate expression on the degenerated fibers. The infiltrating cells are predominantly macrophages.

Necrotizing myopathy may also be observed in patients with toxic myopathy. Statin drugs have been found to cause myopathy and hyperCKemia. Although this myopathy usually ceases soon after discontinuation of statin treatment, it may persist. Autoantibodies reacting against hydroxy-methylglutaryl Co-A reductase, the target of the statins, were reported in a proportion of patients with myositis and a history of statin intake (Mammen et al. 2011).

Inclusion Body Myositis (IBM)

Histologically, IBM has features characteristic of both inflammatory and degenerative conditions. The inflammatory changes resemble those of PM, namely, infiltration of lymphocytes including CD8+ T cells and macrophages, along with aberrant expression of MHC class I antigen on the surface of non-necrotic muscle fibers. DCs, predominantly myeloid DCs (Greenberg et al. 2007), are present in IBM as in PM. An increment of CD4+CD28(null) cells was reported in IBM (Pandya et al. 2010) as in PM/DM.

Enhanced expression of immunoproteasome subunits and local MHC class I presentation was reported in IBM muscle (Ferrer et al. 2004). The immunoproteasome is upregulated by interferonγ and other proinflammatory cytokines and plays important role in degrading proteins for antigen presentation to the CD8+ T cells. The immunoproteasome may also contribute in regulation of stress response and protection against oxidative damage. The link between accumulation of abnormal protein in IBM muscle and enhanced immunoproteasome function is elusive as it is not specific for IBM but reported also in PM and DM (Ghannam et al. 2014).

In IBM muscle, many muscle fibers show nonspecific degenerative processes, while some others harbor the rimmed vacuoles (RVs) (Fig. 4a). They are autophagosomes which accumulate lysosomal proteins along with numerous other substances including amyloid beta, phosphorylated tau, and TAR DNA-binding protein 43. Many of these proteins are ubiquitinated (Fig. 4b, c). Accumulation of these proteins is similar to that observed in other degenerative conditions such as Alzheimer disease. RVs are usually associated with cytoplasmic tubulofilamentous inclusions, of diameter 18–22 nm

Polymyositis and Dermatomyositis, Fig. 2 (continued) surface of almost all muscle fibers. (**d**) Infiltrating cells around muscle fibers include CD8+ cells (*brown*). (**e**) Numerous infiltrating cells, including lymphocytes and macrophages, in direct contact with the plasma membrane (*arrowheads*) of a muscle fiber, with contractile material undergoing degeneration (*asterisk*)

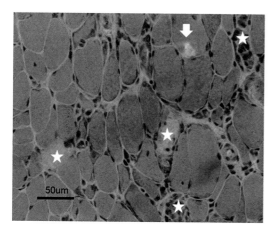

Polymyositis and Dermatomyositis, Fig. 3 Anti-SRP antibody-positive necrotizing myopathy. Scattered necrotic (*asterisks*) muscle fibers, some invaded by macrophages, along with a fiber with a vacuole (*arrow*) (modified Gomori-trichrome stain). Except for macrophages, little inflammatory cell infiltration is observed

(Fig. 4d), which are sometimes observed in the nearby myonuclei. RV is not a histological change specific to IBM but can be seen in many other conditions including distal myopathy with RVs, hereditary inclusion body myopathy associated with Paget disease of bone and frontotemporal dementia, and oculopharyngeal muscular dystrophy.

The significance of these histological features remains unclear. In IBM and cultured human muscle fibers under endoplasmic reticulum (ER) stress, the activity of cathepsin D and B in the lysosomes is decreased significantly. This suggests the possibility that ER stress may contribute to decrease lysosomal proteolytic activity. Furthermore, in normal cell, damaged or misfolded proteins are degraded by autophagosomes which subsequently fuse to lysosomes in which damaged and misfolded proteins are degraded by lysosomal enzymes. As one of the hypotheses for pathogenesis of IBM, it is conceivable that decreased lysosomal proteolytic activity due to ER stress may enhance accumulation of misfolded proteins (Nogalska et al. 2010).

Clinical Presentation

Dermatomyositis (DM)

The muscle symptoms in patients with DM typically start with insidious onset of weakness in the proximal muscles, mostly in the hip girdle. Patients notice difficulties standing up from a low chair and climbing up and down stairs. Some patients find it difficult to raise their arms and lift objects.

Skin manifestations vary in both type and severity. Heliotrope rash is a violaceous discoloration of the eyebrows (heliotrope rash; Fig. 5a), while V-neck and shawl signs are red discolorations of the anterior upper chest and shoulders, respectively. Reddish discoloration with scaling at the back joints, such as the knees, elbows, and fingers, is called Gottron's sign (Fig. 5b). These rashes become increasingly darkened in parts and later have the appearance of poikiloderma. Mechanic's hand is characterized by the thickening and cracking of skin, along with hyperkeratosis at the palmar and radial surfaces of the fingers, particularly the thumb and index finger, mimicking changes associated with heavy manual labor. The association of interstitial lung disease (ILD) occurs in 20–80 % of patients with DM/PM (Fathi et al. 2008). Association of malignant neoplasms will be described in the section on "Myositis Associated with Malignant Neoplasm."

Juvenile (Childhood) Dermatomyositis (JDM)

JDM can occur throughout childhood and is often preceded by skin manifestations. Muscle symptoms appear later. The skin manifestations of JDM are similar to those in adult DM. Erythema with edema on the face is common. The muscles of the hip and shoulder girdles, as well as the neck flexors, become weakened, often accompanied by muscle pain. Severely affected patients may experience difficulties in swallowing and speaking, and the respiratory muscles may be affected.

JDM may be accompanied by systemic symptoms, including fever, fatigue, and joint pain. Vasculitic lesions may cause skin ulcers, particularly when calcified tissue is present underneath the skin. Other infrequent complications include gastrointestinal ulcer and bleeding, renal dysfunction, and hematuria.

Polymyositis (PM)

PM usually begins with weakness of the proximal muscles, particularly of the hip girdle, sometimes

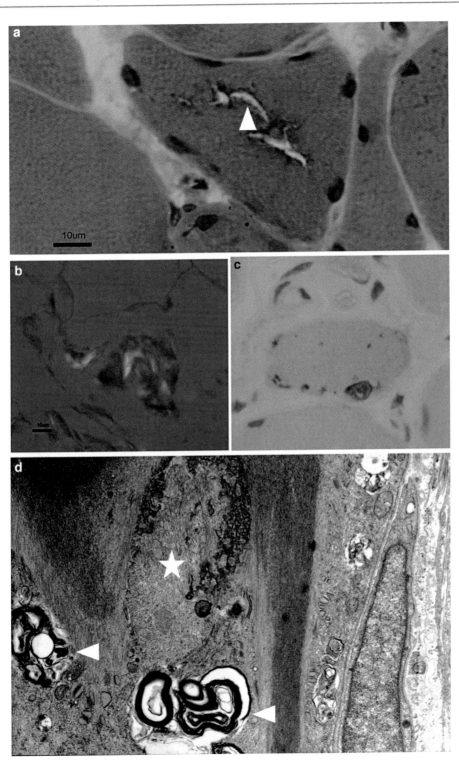

Polymyositis and Dermatomyositis, Fig. 4 Histological features of inclusion body myositis. (**a**) Some muscle fibers harbor vacuoles with violaceous rims (rimmed vacuoles: *arrowheads*). (**b**) Amyloid stain appearing apple green in color under polarized light in a muscle fiber with rimmed vacuoles. (**c**) Ubiquitinated material

accompanied by muscle and joint pain. Systemic symptoms such as fever and malaise occur infrequently, but they are alarming signs of a more severe condition. Muscle weakness progresses at various speeds, later causing muscle atrophy. ILD is also a common complication of PM. Respiratory failure and bulbar involvement are infrequent but possible.

Myositis Associated with Other Connective Tissue Diseases (Overlap Syndrome, OS)

In most patients, OS presents insidiously as moderate weakness and/or atrophy in the proximal muscles of the limbs during the course of other connective tissue diseases. Serum creatine kinase (CK) activity may be moderately elevated.

Myositis Associated with Malignant Neoplasm (CAM)

The clinical features of DM associated with malignant neoplasm do not differ significantly from those of DM without neoplasm. However, CAM tends to show more severe skin manifestations than DM, such as increased frequencies of eruption in the V-neck region, skin ulceration, and necrosis. CAM is associated with interstitial lung disease or collagen vascular disease less frequently than DM.

Immune-Mediated Necrotizing Myopathy (IMNM)

The clinical features observed in patients with anti-SRP antibody-positive necrotizing myopathy resemble those of PM. These symptoms include weakness of the proximal muscles and increased activity of serum CK. Association of other symptoms such as skin rash, Raynaud's phenomenon, arthritis, ILD, or malignant neoplasm, is rare. A small number of patients were found to have long history of muscle weakness, in a few cases since childhood (Suzuki et al. 2012).

Inclusion Body Myositis (IBM)

The initial symptom of IBM is typically weakness of muscles distributed in the knee extensors and forearm muscles, particularly the wrist and finger flexors. However, atypical distribution of weakness is not uncommon. In a small percentage of patients, dysphagia is an early symptom.

Needle electromyography shows changes compatible with inflammatory myopathies. Skeletal muscle MRI demonstrates irregularly distributed high intensity areas on fat suppressed T2 weighted images, accompanied by atrophy of the frontal thigh and forearm muscles. A small number of patients have positive myositis specific antibodies (MSAs), association with ILD, overlap with collagen vascular disease, malignant neoplasm, or infection with a virus such as hepatitis C, HIV or HTLV-1.

Laboratory Examination and Other Tests

Blood Chemistry

Increased serum CK and aldolase (ALD) activities are important indicators of classical DM and PM reflecting disease activity, particularly the destruction of muscle tissue. These enzymes are within normal ranges in patients with CADM and may be moderately increased or remain within normal ranges in patients with overlap syndrome (OS). In patients with IBM, they are moderately increased, being 1.5- to 15-fold higher than the normal upper limit. However, in patients with advanced forms of myositis and marked muscle wasting, CK and ALD can be within normal ranges or even lower.

Serological Tests

Myositis specific antibodies (MSAs) are autoantibodies detected in groups of patients with idiopathic inflammatory myopathies (IIM). They have been detected in ≥60 % of patients with

---

Polymyositis and Dermatomyositis, Fig. 4 (continued) (*brown*) is frequently observed in the cytoplasm and vacuoles of degenerating muscle fibers (anti-ubiquitin). (**d**) Electron micrographs, showing that the vacuoles (*arrowheads*) are filled with the debris of degenerating material (myelin figures). Collections of tubulofilamentous inclusions (*asterisk*) are observed adjacent to the vacuoles

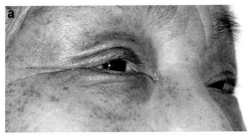
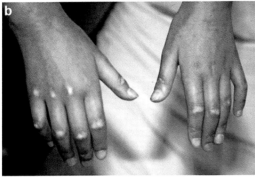

Polymyositis and Dermatomyositis, Fig. 5 Skin manifestations of dermatomyositis. (**a**) Violaceous pigmentation in the upper and lower eyelids (heliotrope rash). (**b**) White and reddish discoloration with scaling at the back of the metacarpophalangeal and interphalangeal joints (Gottron's papules or sign) in a patient with juvenile dermatomyositis

autoimmune myopathies (Gunawardena et al. 2009). Antibodies to aminoacyl-transfer RNA synthetases form a major group among MSAs. One of them, anti-Jo-1 antibody is an antibody to histidyl-transfer RNA synthetase (tRNAS) and has been detected in 25–30 % of patients with PM/DM. Patients with myositis and anti Jo-1 antibody are characterized by high rates of ILD, arthritis, Raynaud's phenomenon, and mechanic's hand. Their condition is often exacerbated by fever. Since patients with antibodies to other tRNAS also show similar manifestations, these characteristics are called anti-synthetase syndrome.

Other antibodies against tRNA synthetases observed in patients with IIM include PL-7 (anti-threonyl-tRNAS), observed in 3–4 % of patients with DM and PM; PL-12 (anti-alanyl-tRNAS), observed in 3–4 %; OJ (anti-isoleucyl-tRNAS), observed in <2 %; EJ (anti-glycyl-tRNAS), observed in <2 %; KS (anti-asparaginyl-tRNAS), observed in <2 %; anti-tyrosine-tRNAS; and Zo (anti-phenylalanyl-tRNAS). Although the pathophysiological role of these anti-tRNAS antibodies remains unclear, some tRNAs have been reported to have proinflammatory properties (Howard et al. 2002).

Unlike the anti-tRNAS antibodies, which occur in patients with both DM and PM, some antibodies are detected almost exclusively in patients with DM. These include anti-Mi-2, anti-P155/140, anti-MJ, anti-MDA5, and anti-SUMO-1 antibodies. Anti-Mi2 is an antibody directed against a component of nucleosome-remodeling deacetylase. Patients with DM positive for anti-Mi-2 tend to have severe skin rashes but respond relatively well to steroid therapy.

An antibody detected in some JDM patients, particularly those with calcinosis, is anti-MJ antibody, which reacts with nuclear matrix protein-2 (NXP-2) (Gunawardena et al. 2009).

Anti-MDA5 antibody, which recognizes the protein encoded by melanoma differentiation-associated gene 5, was found in patients with CADM often with severe ILD (Sato et al. 2009). The antibody to small ubiquitin-like modifier 1 (SUMO-1) was detected in 8 % of DM patients in the United Kingdom and is often associated with dysphagia (Betteridge et al. 2009).

Adult DM patients positive for anti-P155 antibody (Targoff et al. 2006) or anti-P155/140 antibody (Kaji et al. 2007) have a higher risk of malignancy. Both anti-P155 antibody (Targoff et al. 2007) and anti-P155/140 antibody (Fujimoto et al. 2012) recognize a protein transcriptional intermediary factor-1 gamma (TIF1γ) showing high sensitivity and specificity for CAM.

Antigens of the myositis specific autoantibodies (MSA) were found to be expressed in regenerating muscle fibers. They were also expressed strongly in cultured myoblasts and tumor cells but not in normal differentiated muscle fibers and non-tumorous tissues (Casciola-Rosen et al. 2005). It was hypothesized that regenerating muscle fibers may be the main target of the immune reaction in myositis. Tumor cells can be one of the causes of autoantigen production. Furthermore, it is conceivable that in the presence of MSA, muscle fiber regeneration, following muscle damage due to varieties of causes, can induce myositis.

The signal recognition particle (SRP) is a ribonucleoprotein that ubiquitously exists in the cytoplasm of nucleated cells. SRP recognizes the signal sequence of nascent proteins synthesized on the ribosome, binds to these proteins, and transfers them to the ER through a channel on the ER membrane, a process assisted by the docking of SRP with the SRP receptor. Anti-SRP antibody, first reported in a patient with PM (Reeves et al. 1986), has been detected in about 5 % of patients with IIM, particularly those with necrotizing myopathy and no or minimal inflammation (Targoff et al. 1990).

The antibody to 43 kDa protein, cytosolic 5′-nucleotidase 1A (cN1A), was detected in the sera of IBM patients with high sensitivity (70 %) and specificity (92 %) (Larman et al. 2013; Pluk et al. 2013).

Electromyography

Needle electromyography shows myopathic changes often associated with spontaneous activities, including fibrillations and positive sharp waves, sometimes in association with myotonic discharges (Gutierrez-Gutierrez et al. 2012).

Imaging

Skeletal muscle MRI, especially fat-suppressed T2 weighted MRI, is useful in detecting edema and inflammation in muscle tissue (Curriel et al. 2009). This technique can help locate appropriate sites for muscle biopsy and assess the efficacy of therapy, as well as differentiating recurrence of myositis from steroid myopathy. In typical patients, MRI can differentiate IBM from DM and PM by their distinct distribution of muscle atrophy and signal change.

Recent studies have also evaluated the usefulness of ^{18}F fluoro-desoxy-glucose positron emission tomography/computed tomography in patients with inflammatory myopathy and associated conditions including malignant neoplasm and ILD (Al-Nahhas and Jawad 2011).

Differential Diagnosis

Childhood and juvenile DM must be differentiated from Duchenne and Becker types of muscular dystrophy, limb-girdle muscular dystrophy, various congenital myopathies, and spinal muscular atrophy.

Adult DM and PM should be differentiated from muscular dystrophies and congenital myopathies of adult onset, as well as from other inflammatory myopathies, such as IBM, focal myositis, and granulomatous myositis. Among them, facioscapulohumeral muscular dystrophy and dysferlinopathy (limb-girdle muscular dystrophy type 2B and distal muscular dystrophy of Miyoshi) simulate PM, tending to show inflammation in muscle biopsy. Myasthenia gravis may not show typical fluctuations in weakness and may show muscle inflammation. Myalgia, muscle weakness accompanied by elevated serum CK, can be seen in association with viral infection. Drug induced myopathy and rhabdomyolysis sometimes mimic PM. PM and all diseases that may be confused with PM should be differentiated from IBM. Histological confirmation of diagnoses by muscle biopsy is generally advisable.

Therapy and Outcome

Treatment for myositis should be based on a correct diagnosis and the assessment of severity and associated conditions. Treatment of ILD, when present, is of particular importance. The main objective of early stage treatment is to induce remission, defined as halting the progression of muscle weakness and normalizing serum CK activity. The objective of later stage treatment is to maintain remission, regain muscle power, and to taper the intensity of immunotherapy.

Dermatomyositis and Polymyositis

Standard initial therapy for patients with DM and PM consists of oral corticosteroids, usually 1 mg/kg/day prednisolone. The fluoro-substituted corticosteroids such as dexamethasone should be avoided because of their stronger tendency to cause muscle atrophy. Corticosteroid therapy is administered for 1–2 months, until serum CK activity normalizes and muscle strength starts to recover, after which these steroids are tapered gradually over the following months. It usually takes years to discontinue the steroids. However, most patients require a maintenance dose owing to exacerbation of weakness and/or increase of the serum CK during the course of steroid tapering.

Although the appropriate treatment of ILD is the most important factor affecting patient survival, this issue will not be discussed here in detail. Patients may require additional immunotherapy with calcineurin inhibitors, intravenous immunoglobulin (IVIg), and probably rituximab (Mimori et al. 2012).

While some patients respond well to standard therapy, others with more severe symptoms require stronger initial therapy, including the addition of pulsed intravenous methylprednisolone and/or immunosuppressants, such as methotrexate, cyclophosphamide, azathioprine, and the calcineurin inhibitors cyclosporine or tacrolimus (Matsubara et al. 2012). Intravenous immunoglobulin has been reported to be effective in patients with DM. Treatment with rituximab (Rios et al. 2009) and other biological agents may be effective in patients intractable to conventional therapies.

Along with immunotherapy, various methods of physiotherapy are beneficial at all stages of disease to avoid contracture of joints, aggravation of skin manifestations, and bed sores and to assist improving muscle strength (Munters et al. 2013). Despite laborious treatment at one institution, however, about 80 % of patients with DM and PM still had chronic progressive disease (Bronner et al. 2006).

Juvenile (Childhood) Dermatomyositis (JDM)

Corticosteroids are the first line of medication in patients with JDM, but methotrexate and other immunosuppressing agents are started earlier and more frequently in children than in adults to avoid the side effects of steroids (Stringer and Feldman 2006). Patients can also be treated with intravenous cyclophosphamide, and rituximab has shown benefit for patients refractory to corticosteroids. Although the mortality rate of JDM has dropped significantly to 2–3 %, the disease remains chronic in a large number of children (Ramanan and Feldman 2002).

Myositis Associated with Malignant Neoplasm (CAM)

A large study on treatment of CAM has not yet been reported. According to Neri et al. (2014), a limited proportion, less than one third, of CAM patients responded to successful treatment for malignant neoplasm. As for the rest of the patients, about half of them responded to immunotherapy with corticosteroids, some combined with immunosuppresssants or IVIg.

Immune-Mediated Necrotizing Myopathy (IMNM)

Most patients with IMNM and anti-SRP antibody respond poorly to conventional steroid therapy, although a calcineurin inhibitor, IVIg, and rituximab have been reported to be effective (Valiyil et al. 2010). Although these SRP-positive patients tend to have more severe muscle weakness and atrophy than patients with classical PM, their survival rates may not differ (Kao et al. 2004).

Inclusion Body Myositis (IBM)

To date, no form of immunotherapy has shown verified clinical benefits in patients with IBM (Engel and Askanas 2006). In contrast, marginal clinical effects have been observed during early stages of this disease, justifying a short course of immunotherapy for newly diagnosed IBM patients (Ernste and Reed 2013).

Although relatively little is known about the outcome of IBM, the disease is slowly progressive, with a 3–5 % per year decline in muscle strength, leading to major disability due to loss of ambulation and swallowing. Death is frequently caused by respiratory disorders, although life expectancy did not differ from that of a normal population (Cox et al. 2011).

Cross-References

▶ Corticosteroids
▶ Disease-Modifying Antirheumatic Drugs: Overview

References

Al-Nahhas, A., & Jawad, A. S. (2011). PET/CT imaging in inflammatory myopathies. *Annals of the New York Academy of Sciences, 1228*, 39–45.

Arnett, F. C., et al. (1996). Interrelationship of major histocompatibility complex class II alleles and autoantibodies in four ethnic groups with various forms of myositis. *Arthritis and Rheumatism, 39*, 1507–1518.

Badrising, U. A., et al. (2005). Inclusion body myositis. Clinical features and clinical course of the disease in 64 patients. *Journal of Neurology, 252*, 1448–1454.

Bendewald, M. J., et al. (2010). Incidence of dermatomyositis and clinically amyopathic dermatomyositis: A - population-based study in Olmsted County, Minnesota. *Archives of Dermatology, 146*, 26–30.

Betteridge, Z. E., et al. (2009). Clinical and human leucocyte antigen class II haplotype associations of autoantibodies to small ubiquitin-like modifier enzyme, a dermatomyositis-specific autoantigen target, in UK Caucasian adult-onset myositis. *Annals of the Rheumatic Diseases, 68*, 1621–1625.

Bohan, A., & Peter, J. B. (1975). Polymyositis and dermatomyositis (second of two parts). *The New England Journal of Medicine, 292*, 403–407.

Bradshaw, E. M., et al. (2007). A local antigen-driven humoral response is present in the inflammatory myopathies. *Journal of Immunology, 178*, 547–556.

Bronner, I. M., et al. (2006). Long-term outcome in polymyositis and dermatomyositis. *Annals of the Rheumatic Diseases, 65*, 1456–1461.

Casciola-Rosen, L., et al. (2005). Enhanced autoantigen expression in regenerating muscle cells in idiopathic inflammatory myopathy. *The Journal of Experimental Medicine, 201*, 591–601.

Chinoy, H., et al. (2008). The protein tyrosine phosphatase N22 gene is associated with juvenile and adult idiopathic inflammatory myopathy independent of the HLA 8.1 haplotype in British Caucasian patients. *Arthritis and Rheumatism, 58*, 3247–3254.

Chinoy, H., et al. (2012). Interaction of HLA-DRB1*03 and smoking for the development of anti-Jo-1 antibodies in adult idiopathic inflammatory myopathies: A European-wide case study. *Annals of the Rheumatic Diseases, 71*, 961–965.

Cox, F. M., et al. (2011). A 12-year follow-up in sporadic inclusion body myositis: An end stage with major disabilities. *Brain, 134*, 3167–3175.

Curriel, R. V., et al. (2009). Magnetic resonance imaging of the idiopathic inflammatory myopathies: Structural and clinical aspects. *Annals of the New York Academy of Sciences, 1154*, 101–114.

Dalakas, M. C., & Hohlfeld, R. (2003). Polymyositis and dermatomyositis. *Lancet, 362*, 971–982.

DeVere, R., & Bradley, W. G. (1975). Polymyositis: Its presentation, morbidity and mortality. *Brain, 98*, 637–666.

Engel, A. G., & Arahata, K. (1986). Mononuclear cells in myopathies: Quantitation of functionally distinct subsets, recognition of antigen-specific cell-mediated cytotoxicity in some diseases, and implications for the pathogenesis of the different inflammatory myopathies. *Human Pathology, 17*, 704–721.

Engel, W. K., & Askanas, V. (2006). Inclusion-body myositis: Clinical, diagnostic, and pathologic aspects. *Neurology, 66*(2 Suppl 1), S20–S29.

Ernste, F. C., & Reed, A. M. (2013). Idiopathic inflammatory myopathies: Current trends in pathogenesis, clinical features, and up-to-date treatment recommendations. *Mayo Clinic Proceedings, 88*, 83–105.

Espinosa-Ortega, F., et al. (2015). Quantitative T cell subsets profile in peripheral blood from patients with idiopathic inflammatory myopathies: Tilting the balance towards proinflammatory and pro-apoptotic subsets. *Clinical and Experimental Immunology, 179*, 520–528.

Fasth, A. E., et al. (2009). T cell infiltrates in the muscles of patients with dermatomyositis and polymyositis are dominated by CD28null T cells. *The Journal of Immunology, 183*, 4792–4799.

Fathi, M., et al. (2008). Interstitial lung disease in polymyositis and dermatomyositis. Longitudinal evaluation by pulmonary function and radiology. *Arthritis and Rheumatism, 59*, 677–685.

Ferrer, I., et al. (2004). Proteasomal expression, induction of immunoproteasome subunits, and local MHC class I presentation in myofibrillar myopathy and inclusion body myositis. *Journal of Neuropathology and Experimental Neurology, 63*, 484–498.

Fujimoto, M., et al. (2012). Myositis-specific anti-155/140 autoantibodies target transcription intermediary factor 1 family proteins. *Arthritis Rheum, 64*, 513–522.

Ghannam, K., et al. (2014). Upregulation of immunoproteasome subunits in myositis indicates active inflammation with involvement of antigen presenting cells, CD8 T-cells and IFNGamma. *PLoS One, 9*, e104048.

Gono, T., et al. (2012). Association of HLA-DRB1*0101/*0405 with susceptibility to anti-melanoma differentiation-associated gene 5 antibody-positive dermatomyositis in the Japanese population. *Arthritis and Rheumatism, 64*, 3736–3740.

Greenberg, S. A., et al. (2005). Plasma cells in muscle in inclusion body myositis and polymyositis. *Neurology, 65*, 1782–1787.

Greenberg, S. A., et al. (2007). Myeloid dendritic cells in inclusion-body myositis and polymyositis. *Muscle and Nerve, 35*, 17–23.

Gunawardena, H., et al. (2009). Autoantibodies to a 140-kd protein in juvenile dermatomyositis are associated with calcinosis. *Arthritis and Rheumatism, 60*, 1807–1814.

Gutierrez-Gutierrez, G., et al. (2012). Use of electromyography in the diagnosis of inflammatory myopathies. *Reumatología Clínica, 8*, 195–200.

Guyant-Marechal, L., et al. (2006). Valosin-containing protein gene mutations: Clinical and neuropathologic features. *Neurology, 67*, 644–651.

Howard, O. M., et al. (2002). Histidyl-tRNA synthetase and asparaginyl-tRNA synthetase, autoantigens in myositis, activate chemokine receptors on T lymphocytes and immature dendritic cells. *The Journal of Experimental Medicine, 196*, 781–791.

Kaji, K., et al. (2007). Identification of a novel autoantibody reactive with 155 and 140 kDa nuclear proteins in patients with dermatomyositis: An association with malignancy. *Rheumatology (Oxford), 46*, 25–28.

Kao, A. H., et al. (2004). Anti-signal recognition particle antibody in patients with and patients without idiopathic inflammatory myopathy. *Arthritis and Rheumatism, 50*, 209–215.

Koh, E. T., et al. (1993). Adult onset polymyositis/dermatomyositis: Clinical and laboratory features and treatment response in 75 patients. *Annals of the Rheumatic Diseases, 52*, 857–861.

Kondo, M., et al. (2009). Roles of proinflammatory cytokines and the Fas/Fas ligand interaction in the pathogenesis of inflammatory myopathies. *Immunology, 128*, e589–e599.

Larman, H. B., et al. (2013). Cytosolic 5′-nucleotidase 1A autoimmunity in sporadic inclusion body myositis. *Annals of Neurology, 73*, 408–418.

Mammen, A. L., et al. (2011). Autoantibodies against 3-hydroxy-3-methylglutaryl-coenzyme A reductase in patients with statin-associated autoimmune myopathy. *Arthritis and Rheumatism, 63*, 713–721.

Mammen, A. L., et al. (2012). Increased frequency of DRB1*11:01 in anti-hydroxymethylglutaryl-coenzyme A reductase-associated autoimmune myopathy. *Arthritis Care & Research (Hoboken), 64*, 1233–1237.

Marie, I., et al. (1999). Influence of age on characteristics of polymyositis and dermatomyositis in adults. *Medicine (Baltimore), 78*, 139–147.

Matsubara, S., et al. (2012). Effects of tacrolimus on dermatomyositis and polymyositis: A prospective, open, non-randomized study of nine patients and a review of the literature. *Clinical Rheumatology, 31*, 1493–1498.

Mescam-Mancini, L., et al. (2015). Anti-Jo-1 antibody-positive patients show a characteristic necrotizing perifascicular myositis. *Brain, 138*, 2485–2492.

Mimori, T., et al. (2012). Interstitial lung disease in myositis: Clinical subsets, biomarkers, and treatment. *Current Rheumatology Reports, 14*, 264–274.

Munters, L. A., et al. (2013). Improvement in health and possible reduction in disease activity using endurance exercise in patients with established polymyositis and dermatomyositis. A multicentric randomized controlled trial with a 1-year followup. *Arthritis Care and Research, 65*, 1959–1968.

Needham, M., et al. (2008). Sporadic inclusion body myositis: Phenotypic variability and influence of HLA-DR3 in a cohort of 57 Australian cases. *Journal of Neurology, Neurosurgery, and Psychiatry, 79*, 1056–1060.

Neri, R., et al. (2014). Cancer associated myositis: A 35-year retrospective study of a monocentric cohort. *Rheumatology International, 34*, 565–569.

Nogalska, A., et al. (2010). Impaired autophagy in sporadic inclusion-body myositis and in endoplasmic reticulum stress-provoked cultured human muscle fibers. *The American Journal of Pathology, 177*, 1377–1387.

Oddis, C. V., et al. (1990). Incidence of polymyositis-dermatomyositis: A 20-year study of hospital diagnosed cases in Allegheny County, PA 1963–1982. *Journal of Rheumatology, 17*, 1329–1334.

O'Hanlon, T. P., et al. (2006). HLA polymorphisms in African Americans with idiopathic inflammatory myopathy: Allelic profiles distinguish patients with different clinical phenotypes and myositis autoantibodies. *Arthritis and Rheumatism, 54*, 3670–3681.

Pandya, J. M., et al. (2010). Expanded T cell receptor Vbeta-restricted T cells from patients with sporadic inclusion body myositis are proinflammatory and cytotoxic CD28null T cells. *Arthritis and Rheumatism, 62*, 3457–3466.

Pluk, H., et al. (2013). Autoantibodies to cytosolic 5′-nucleotidase 1A in inclusion body myositis. *Annals of Neurology, 73*, 397–407.

Ramanan, A. V., & Feldman, B. M. (2002). Clinical outcomes in juvenile dermatomyositis. *Current Opinion in Rheumatology, 14*, 658–662.

Reed, A. M., et al. (1998). Immunogenetic studies in families of children with juvenile dermatomyositis. *Journal of Rheumatology, 25*, 1000–1002.

Reeves, W. H., et al. (1986). Human autoantibodies reactive with the signal-recognition particle. *Proceedings of the National Academy of Sciences of the United States of America, 83*, 9507–9511.

Rios, F. R., et al. (2009). Rituximab in the treatment of dermatomyositis and other inflammatory myopathies. A report of 4 cases and review of the literature. *Clinical and Experimental Rheumatology, 27*, 1009–1016.

Sato, S., et al. (2009). RNA helicase encoded by melanoma differentiation-associated gene 5 is a major autoantigen in patients with clinically amyopathic dermatomyositis: Association with rapidly progressive interstitial lung disease. *Arthritis and Rheumatism, 60*, 2193–2200.

Schmidt, J., et al. (2004). Upregulated inducible co-stimulator (ICOS) and ICOS-ligand in inclusion body myositis muscle: Significance for CD8+ T cell cytotoxicity. *Brain, 127*, 1182–1190.

Scott, A. P., et al. (2012). Investigation of NOTCH4 coding region polymorphisms in sporadic inclusion body myositis. *Journal of Neuroimmunology, 250*, 66–70.

Stenzel, W., et al. (2015). Nuclear actin aggregation is a hallmark of anti-synthetase syndrome-induced dysimmune myopathy. *Neurology, 84*, 1346–1354.

Stringer, E., & Feldman, B. M. (2006). Advances in the treatment of juvenile dermatomyositis. *Current Opinion in Rheumatology, 18*, 503–506.

Sugiura, T., et al. (2014). Association between a C8orf13-BLK polymorphism and polymyositis/dermatomyositis in the Japanese population: An additive effect with STAT4 on disease susceptibility. *PLoS One, 9*, e90019.

Suzuki, S., et al. (2012). Myopathy associated with antibodies to signal recognition particle: Disease progression and neurological outcome. *Archives of Neurology, 69*, 728–732.

Targoff, I. N., et al. (1990). Antibody to signal recognition particle in polymyositis. *Arthritis and Rheumatism, 33*, 1361–1370.

Targoff, I. N., et al. (2006). A novel autoantibody to a 155-kd protein is associated with dermatomyositis. *Arthritis and Rheumatism, 54*, 3682–3689.

Targoff, I. N., et al. (2007). Autoantibodies to transcriptional intermediary factor-1 gamma (TIF-1γ) in dermatomyositis. *Arthritis and Rheumatism, 54*, S518, [abstract].

Tateyama, M., et al. (2006). Expression of CCR7 and its ligands CCL19/CCL21 in muscles of polymyositis. *Journal of Neurological Sciences, 249*, 158–165.

Troyanov, Y., et al. (2005). Novel classification of idiopathic inflammatory myopathies based on overlap syndrome features and autoantibodies: Analysis of 100 French Canadian patients. *Medicine (Baltimore), 84*, 231–249.

Tucci, M., et al. (2006). Interleukin-18 overexpression as a hallmark of the activity of autoimmune inflammatory myopathies. *Clinical and Experimental Immunology, 146*, 21–31.

Valiyil, R., et al. (2010). Rituximab therapy for myopathy associated with anti-signal recognition particle antibodies: A case series. *Arthritis Care & Research (Hoboken), 62*, 1328–1334.

Wilson, F. C., et al. (2008). Epidemiology of sporadic inclusion body myositis and polymyositis in Olmsted County, Minnesota. *Journal of Rheumatology, 35*(3), 445–447.

Winkelmann, R. K. (1982). Dermatomyositis in childhood. *Clinics in Rheumatic Diseases, 8*(2), 353–368.

Propionic Acid Derivative Drugs (Profens)

Richard O. Day[1,2], Garry G. Graham[1,2] and Kenneth Williams[2]
[1]Department of Pharmacology, School of Medical Sciences, University of New South Wales, Sydney, NSW, Australia
[2]Department of Clinical Pharmacology and Toxicology, St Vincent's Hospital, Sydney, NSW, Australia

Synonyms

Phenylpropanoic acids; Phenylpropionic acids; Phenylpropanoates; Phenylpropionates; 2-methylphenylacetates; 2-methylphenylacetic acids. There are a large number of profens available commercially including: Carprofen; Naproxen; Fenoprofen; Flurbiprofen; Ibuprofen; Ketoprofen; Tiaprofenic acid

Definition

The profens are a category of nonselective, nonsteroidal anti-inflammatory drugs (NSAIDs). They reduce pain (analgesia), body temperature in fever (antipyresis), signs of inflammation (anti-inflammatory activity), and, in mice, slow the development of cancers.

Chemical Structures and Properties

The profens are derivatives of 2-phenylpropanoic acid. All contain a chiral center resulting in the formation of two enantiomers (R and S) of each profen (Fig. 1). The profens are available mostly as their racemates, i.e., equal mixtures of the R and S stereoisomers. The major exception is naproxen which is available as its pure S-enantiomer, but ibuprofen and ketoprofen are also available as the pure S-enantiomers which are termed dexibuprofen and dexketoprofen, respectively.

Propionic Acid Derivative Drugs (Profens), Fig. 1 General structures of R- and S-profens. The chiral centers are shown*. The R-enantiomers of some profens are inverted metabolically to the S-enantiomers (See Fig. 2; Table 1)

Propionic Acid Derivative Drugs (Profens), Table 1 Stereochemical inversion of the profens

Profen	Percentage inversion of R-enantiomer to S-enantiomer
Ibuprofen	>60
Fenoprofen	73
Ketoprofen	10
Flurbiprofen	0
Tiaprofenic acid	0
Carprofen	0.3 (rat)
Naproxen[a]	2 (rat)

[a]R-enantiomer not available for human use

Pharmacokinetics and Metabolism

The profens generally have moderately short initial half-lives of 2–5 h although they have long terminal half-lives. The major exception is naproxen which has a half-life of about 15 h.

Clinically, the most notable feature of the metabolism of some of the profens is their in vivo stereochemical inversion (Table 1; Figs. 1 and 2). This inversion occurs through the intermediate coenzyme A (CoA) conjugates with the consequent opportunity for incorporation of both enantiomers into hybrid, abnormal triglycerides. A well-known example of this process, namely, the inversion of ibuprofen and its incorporation into hybrid triglycerides, is shown in Fig. 2. The profens which are inverted are well tolerated. It follows that the formation of the hybrid triglycerides appears to have little pharmacological significance. However, this area has not been investigated thoroughly with respect to adverse effects or therapeutic benefit.

The profens are highly bound to plasma proteins. As discussed below, this has a major influence on the interpretation of their pharmacological effects as drug action correlates generally with the unbound concentration in plasma rather than the total (unbound plus bound) concentration.

Pharmacological Activities

The S-profens are classical, nonselective NSAIDs which inhibit the synthesis of prostaglandins and thromboxane A_2 (Fig. 3). As a result, the S-profens have analgesic, antipyretic, anti-inflammatory actions and inhibit platelet aggregation. The therapeutic (total) plasma concentration of S-ibuprofen is about 25 μmol/L. By comparison, the IC_{50} value of S-ibuprofen in a cellular COX-2 system in human blood is 1.6 μmol/L (Rainsford 1999) and, therefore, is clearly sufficient to provide a very high degree of inhibition of prostaglandin synthesis in vivo.

For many years, the R-profens were considered to be pharmacologically inactive because they are weak inhibitors of the synthesis of prostaglandins from arachidonic acid in vitro. An example is the very low activity of R-ibuprofen in a cellular system in whole blood where the IC_{50} is >250 μmol/L. This total concentration is very much higher than the IC_{50} of S-ibuprofen in the same system and also well above the plasma concentrations of R-ibuprofen achieved by dosage with racemic ibuprofen (about 25 μmol/L).

Despite the lack of activity of the R-profens on prostaglandin synthesis in vitro, they still have anti-inflammatory and analgesic activity (Tegeder et al. 2001). The efficacy of R-ibuprofen and R-fenoprofen can be explained because these profens are metabolized to their S-isomers (Table 1) which inhibit prostaglandin synthesis by COX-1 and COX-2. R-flurbiprofen also has analgesic activity, both in humans and experimental animals (Bishay et al. 2010; Lötsch et al. 1995). This observation was particularly puzzling as flurbiprofen is not converted to the S-isomer which is an inhibitor of prostaglandin

Propionic Acid Derivative Drugs (Profens), Fig. 2 The metabolic inversion of R-ibuprofen to S-ibuprofen. R-ibuprofen is first converted to the Coenzyme A (CoA) thioester derivative which is then inverted through a planar intermediate to the S-ibuprofen-CoA which is hydrolyzed to S-ibuprofen. Direct metabolism of S-ibuprofen to the CoA derivative does not occur. Consequently, S-ibuprofen is not inverted to R-ibuprofen. The CoA conjugates of R- and S-ibuprofen can be utilized in the synthesis of lipids (triglycerides) in the place of fatty acids. The results are hybrid triglycerides where some of the fatty acid residues are replaced by ibuprofen. Several other R-profens are inverted by the same mechanism (Table 1)

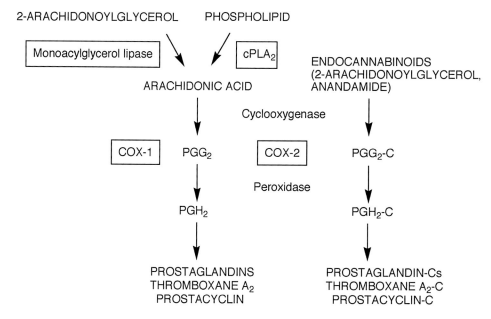

Propionic Acid Derivative Drugs (Profens), Fig. 3 Synthesis of prostaglandins and related compounds. Cyclooxygenase (COX-1) and COX-2 are bifunctional enzymes with cyclooxygenase and peroxide functions. COX-1 catalyzes the synthesis of prostaglandin (PG) intermediates, PGG2 and PGH2. COX-2 not only catalyzes the synthesis of PG intermediates from arachidonic acid but also PG cannabinoid intermediates (PGG2-C and PGH2-C). The S-profens, like other nonselective NSAIDs, inhibit the formation of PG intermediates by both COX-1 and COX-2. The R-profens do not inhibit the synthesis of prostaglandin intermediates but selectively inhibit the synthesis of PGH_2-C by COX-2. The final production of PGs and PG-Cs is synthesized by specific enzymes

synthesis (Graham and Williams 2004). However, a recent finding explains why the R-enantiomers have anti-inflammatory and analgesic activities. The discovery is that the R-profens inhibit the oxidation of endocannabinoids (Fig. 3) (Duggan et al. 2011). The major two endocannabinoids are arachidonoylglycerol and anandamide (N-arachidonoylethanolamine) and, as the names indicate, are conjugates of arachidonic acid. The endocannabinoids are involved in dampening pain pathways within the central nervous system, and maintenance of endocannabinoid levels is now considered essential for the analgesic and anti-inflammatory activity of the R-profens. Like other NSAIDs, R-flurbiprofen also reduces the development of model colon and prostate cancers in mice (Tegeder et al. 2001).

All NSAIDs, including the profens, have a variety of in vitro biochemical and immunological actions which appear unrelated to their inhibition of synthesis of endocannabinoids (nonsteroidal anti-inflammatory drugs (NSAIDs)). For example, high concentrations of ibuprofen variably modify immunological parameters in cells incubated in the absence of plasma albumin. Thus, racemic ibuprofen (100 μmol/L) increases the production of tumor necrosis factor (TNF) by mouse macrophages, while a similar concentration inhibits the production of TNF by human monocytes (Rainsford 1999). However, these concentrations are far higher than the unbound concentrations achieved in plasma after recommended dosage (approximately 0.17 μmol/L and 0.1 μmol/L for S- and R-ibuprofen, respectively). R-flurbiprofen has been studied, in particular because, as discussed above, of its lack of conversion to S-flurbiprofen (Table 1). However, any immunological effects of R-flurbiprofen are also produced only at grossly supratherapeutic concentrations (Tegeder et al. 2001) making its in vitro immunological actions most unlikely to be related to its analgesic and anti-inflammatory effects.

In summary, racemic profens and individual S-profens are nonselective NSAIDs. The R-profens also have analgesic activity, even if their metabolic inversion is low or negligible.

Apart from inhibition of production of prostaglandin, prostaglandin cannabinoids and related mediators, other biochemical and immunological effects of the profens in vitro probably have no clinical significance because they have been reported only at supratherapeutic concentrations.

Clinical Uses

The racemic profens, dexibuprofen, dexketoprofen, and S-naproxen, are all active in the treatment of pain, fever, inflammation, dysmenorrhea, acute gout, and migraine. R-flurbiprofen and R-ketoprofen are the most interesting in clinical research as they show little or no inversion to their S-enantiomers in humans (nonsteroidal anti-inflammatory drugs (NSAIDs)) and should be considered as distinctly different from the S-enantiomers with potentially contrasting clinical actions. While these R-profens have not been studied widely, R-flurbiprofen does inhibit at least one type of painful stimulus in man (Lötsch et al. 1995).

Adverse Effects

All the racemic profens and the individual S-profens (naproxen, dexketoprofen and dexibuprofen) produce the classical adverse gastrointestinal effects of the nonselective NSAIDs (nonsteroidal anti-inflammatory drugs (NSAIDs)). At total daily doses of 1,600 mg or less, racemic ibuprofen is associated with a low risk of adverse gastrointestinal adverse effects, lower than other nonselective NSAIDs. However, the lesser toxicity of racemic ibuprofen is not seen at higher doses (Henry et al. 1999).

As outlined in the general chapter on NSAIDs, the various profens have differing adverse effects on the cardiovascular system:

- Ibuprofen is not associated with cardiovascular adverse effects at doses up to 1,200 mg daily (Latimer et al. 2009; McGettigan and Henry 2011), but a recent meta-analysis of clinical

trials shows that doses of about 2,400 mg are associated with increased myocardial infarction (CNT 2013).
- Of all the NSAIDs (nonselective and selective COX-2 inhibitors), naproxen appears to produce the least risk of myocardial infarction (CNT 2013) although safety cannot be guaranteed in all patients.

In a short-term clinical study, R-flurbiprofen was better tolerated than racemic ketoprofen (Jerussi et al. 1998), while in rats, R-flurbiprofen does not produce gastric ulceration (Wechter et al. 1998).

Drug Interactions

The interactions of the racemic profens and the individual S-profens (naproxen, dexketoprofen, and dexibuprofen) are the same as those of the nonselective NSAIDs (nonsteroidal anti-inflammatory drugs (NSAIDs)). An important interaction is that all the racemic profens and naproxen block the antiplatelet action of aspirin (Catella-Lawson et al. 2001). There is no data on the influence of R-flurbiprofen.

Cross-References

▶ Asthma
▶ Coxibs
▶ Gout
▶ Non-steroidal Anti-inflammatory Drugs: Overview
▶ Osteoarthritis
▶ Prostanoids
▶ Rheumatoid Arthritis
▶ Salicylates

References

Bishay, P., Schmidt, H., Marian, C., Haussler, A., Wijnvoord, N., Ziebell, S., et al. (2010). R-flurbiprofen reduces neuropathic pain in rodents by restoring endogenous cannabinoids. *PLoS ONE, 5*(5), e10628.

Catella-Lawson, F., Reilly, M. P., Kapoor, S. C., Cucchiara, A. J., DeMarco, S., Tournier, B., et al. (2001). Cyclooxygenase inhibitors and the antiplatelet effects of aspirin. *New England Journal of Medicine, 345*(25), 1809–1817.

CNT, & Coxib and traditional NSAID Trialists' (CNT) Collaboration. (2013). Vascular and upper gastrointestinal effects of non-steroidal anti-inflammatory drugs: Meta-analyses of individual participant data from randomised trials. *Lancet.* doi:10.1016/S0140-6736(1013)60900-60909.

Duggan, K. C., Hermanson, D. J., Musee, J., Prusakiewicz, J. J., Scheib, J. L., Carter, B. D., et al. (2011). (R)-Profens are substrate-selective inhibitors of endocannabinoid oxygenation by COX-2. *Nature Chemical Biology, 7*(11), 803–809.

Graham, G. G., & Williams, K. M. (2004). Metabolism and pharmacokinetics of ibuprofen. In K. D. Rainsford (Ed.), *Aspirin and related drugs* (pp. 157–180). London: Taylor & Francis.

Henry, D., Drew, A., & Beuzeville, S. (1999). Gastrointestinal adverse drug reactions attributed to ibuprofen. In K. D. Rainsford (Ed.), *Ibuprofen* (pp. 433–458). London: Taylor & Francis.

Jerussi, T. P., Caubet, J. F., McCray, J. E., & Handley, D. A. (1998). Clinical endoscopic evaluation of the gastroduodenal tolerance to (R)- ketoprofen, (R)-flurbiprofen, racemic ketoprofen, and paracetamol: A randomized, single-blind, placebo-controlled trial. *Journal of Clinical Pharmacology, 38*(2 Suppl), 19S–24S.

Latimer, N., Lord, J., Grant, R., O'Mahony, R., Dickson, J., & Conaghan, P. (2009). Cost effectiveness of COX 2 selective inhibitors and traditional NSAIDs alone or in combination with a proton pump inhibitor for people with osteoarthritis. *British Medical Journal, 339*, b2538.

Lötsch, J., Geisslinger, G., Mohammadian, P., Brune, K., & Kobal, G. (1995). Effects of flurbiprofen enantiomers on pain-related chemo-somatosensory evoked potentials in human subjects. *British Journal of Clinical Pharmacology, 40*(4), 339–346.

McGettigan, P., & Henry, D. (2011). Cardiovascular risk with non-steroidal anti-inflammatory drugs: Systematic review of population-based controlled observational studies. *PLoS Medicine, 8*(9), e1001098.

Rainsford, K. D. (1999). Pharmacology and toxicology of ibuprofen. In K. D. Rainsford (Ed.), *Ibuprofen*, Taylor & Francis, London (pp. 145–275).

Tegeder, I., Pfeilschifter, J., & Geisslinger, G. (2001). Cyclooxygenase-independent actions of cyclooxygenase inhibitors. *FASEB Journal, 15*(12), 2057–2072.

Wechter, W. J., McCracken, J. D., Kantoci, D., Murray, E. D., Quiggle, D., Leipold, D., et al. (1998). Mechanism of enhancement of intestinal ulcerogenicity of S-aryl propionic acids by their R-enantiomers in the rat. *Digestive Diseases & Sciences, 43*(6), 1264–1274.

Prostanoids

Rolf M. Nüsing
Department of Clinical Pharmacology, Johann Wolfgang Goethe-University, Frankfurt, Germany

Synonyms

Prostaglandins

Definition

Prostanoids are a lipid subclass of bioactive eicosanoids formed from 20-carbon fatty acids by the cyclooxygenase pathway and consist of prostaglandins, thromboxanes, and prostacyclins. Prostanoid-like compounds (isoprostanes, phytoprostanes, isofuranes) are formed by free radical mechanisms independent of the cyclooxygenase reaction. The structure of the prostanoids is based on the hypothetical C-20 parent saturated acid called prostanoic acid containing a cyclopentane ring. All prostanoids have an OH group at C-15 position (except thromboxanes) and a trans double bond at the C-13 position. Hundreds of these compounds have been described, and their nomenclature is based on functional groups (hydroxyl or keto groups and their position) in the cyclopentane ring, indicated by a letter (A, B, C, D, E, F, G, H, I, and J). Most interest has been focused on the prostaglandin (PG) types D, E, F, I, and J and thromboxane (TX) A. Thromboxanes are similar but have a six-member ring instead of the cyclopentane ring (Fig. 1). Based on the respective unsaturated C-20 fatty acid substrates, eicosapentaenoic acid with five double bonds, arachidonic acid with four double bonds, or dihomo-γ-linolenic acid with three double bonds, prostanoids of the 3-series (e.g., PGE_3), 2-series (e.g., PGE_2), or 1-series (e.g., PGE_1), respectively, are generated, and the numerical subscript indicates the number of double bonds in the carbon chains. The most important roles are attributed to the 2-series prostanoids derived from arachidonic acid (Fig. 1), not only with regard to inflammatory processes but also in relation to absolute amounts formed in various tissues. Less information is available in respect of series 1 and series 3 prostanoids, mostly due to the low amounts of these metabolites, often at the limit of detection.

Prostanoids are referred to as autacoids or local hormones because they are short-lived (with half-lives of seconds to minutes) and only alter activities of adjoining cells. Upon synthesis of large amounts, prostaglandins reach the circulation and may cause additional distant systemic effects.

Prostanoids are ubiquitous throughout the body. They are not stored in cell granules but are synthesized quickly, as needed, from C-20 fatty acids, liberated from membrane phospholipids, mainly ethanolamine glycerophospholipids, via the hydrolysis of the sn-2 bond by phospholipase A_2 (PLA_2) enzymes. The amount of free arachidonic acid in the cell is very low because of rapid reacylation and transfer into various phospholipid pools by acyltransferases and transacylases. Therefore, release of arachidonic acid from cell membrane phospholipids determines the amounts of prostanoids produced.

Prostanoids play a paramount role in inflammation, mediating various manifestations of the acute response, including fever, hyperalgesia, increased vascular permeability, and edema, and participate in chronic manifestations of inflammation. They are also involved in resolution of inflammation.

Biosynthesis and Release

Cyclooxygenases: COX-1 and COX-2

Following its liberation, arachidonic acid is metabolized by the enzyme cyclooxygenase (COX), also named prostaglandin endoperoxide synthase (PTGS), to the main 2-series prostaglandins PGE_2, PGD_2, and $PGF_{2\alpha}$, to thromboxane TXA_2, and to prostacyclin PGI_2. Inhibition of COX by nonsteroidal anti-inflammatory drugs (NSAIDs) such as ibuprofen or diclofenac and by analgesics/antipyretics such as acetaminophen or metamizole suppresses prostanoid synthesis,

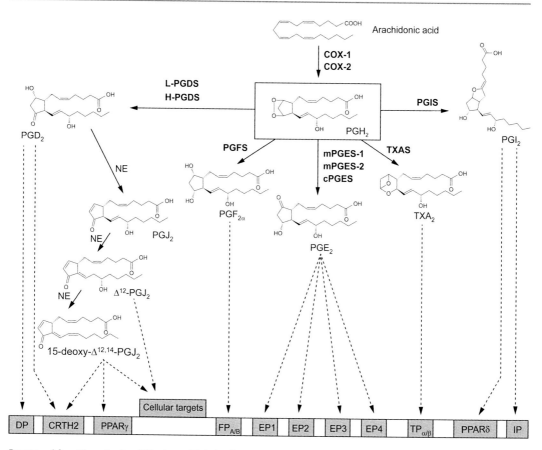

Prostanoids, Fig. 1 Arachidonic acid-derived prostanoids and their relevant receptors and targets (*NE* nonenzymatically formed)

which explains the clinical usefulness of these drugs (Day and Graham 2013).

COX is a bifunctional heme-containing glycoprotein, converting arachidonic acid to the prostaglandin endoperoxide PGH_2 in a two-step process involving a cyclooxygenase and a hydroperoxidase (Smith et al. 2000). Although three different isoforms of COX have been described, COX-1, COX-2, and COX-3 (or COX-1b, a splice variant of COX-1, which in humans is a nonfunctional protein), COX-1 and COX-2 are of clinical interest. COX-1 and COX-2 are membrane bound as a dimer to the endoplasmic reticulum and nuclear envelope. They are encoded by two distinct genes, PTGS-1 and PTGS-2, which exhibit about 65 % amino acid sequence homology and present near-identical catalytic sites. The active site is created by a long hydrophobic channel with the arachidonate-binding site located in the upper half. The channel is also the site of nonsteroidal anti-inflammatory binding. All NSAIDs except aspirin act non-covalently in a competitive manner. Aspirin inhibits prostanoid formation by both COX isoforms via covalent modification of serine, located at the active sites of the enzymes. The most important difference between the COX isoforms is the substitution of an isoleucine in the substrate channel in COX-1 with a valine in COX-2. The smaller valine residue in COX-2 allows access to a hydrophobic side pocket in the enzyme. Presence of this unique structure allowed the development of specific inhibitors for COX-2, the so-called coxibs, e.g., celecoxib or etoricoxib used as anti-inflammatory and analgesic drugs (Marnett 2009).

The profile of prostanoid production is determined by the differential expression of the PGH_2 metabolizing enzymes within cells present at sites of inflammation. Although most cell types are able to produce several types of prostanoids, they often are characterized by the production of single prostanoids. For example, mast cells predominantly generate PGD_2 and endothelial cells PGI_2, whereas platelets produce TXA_2. Especially, macrophages and dendritic cells vary their prostanoid product depending on the state of activation. In the resting state, they produce more TXA_2 than PGE_2, favoring PGE_2 upon activation, e.g., by LPS. Prostanoids are mainly formed intracellularly, primarily within the endoplasmic reticulum. Upon release, they interact with specific seven transmembrane domain receptors of the heterotrimeric G-protein-coupled, rhodopsin-type superfamily. Prostanoid receptor nomenclature is based on the ligand bound by the receptor, EP for PGE_2, DP for PGD2, FP for $PGF_{2\alpha}$, IP for PGI_2, and TP for TXA_2. Some receptor subtypes exist. Other non-prostanoid-type receptors can be stimulated by prostanoids, including the chemoattractant receptor-homologous molecule expressed on Th2 cells (CRTH2) and the peroxisome proliferator-activated receptors (PPAR).

PGE_2 Synthesis

PGE_2 is synthesized from COX-derived PGH_2 by three forms of PGE synthase (PGES) (Hara et al. 2010): membrane-associated PGES type 1 (mPGES-1), membrane-associated PGES type 2 (mPGES-2), and cytosolic PGES (cPGES).

mPGES-1 (molecular weight, MW, 17 kD) belongs to the superfamily of membrane-associated proteins involved in eicosanoid and glutathione metabolism (MAPEG) and requires glutathione as an essential cofactor for activity. It is functionally coupled to COX-2 in preference to COX-1. Constitutive levels of mPGES-1 are normally low. The induction of mPGES-1 both in vitro and in vivo following proinflammatory stimuli (IL-1β, TNFα, LPS) strongly suggests that mPGES-1, often regulated in concert with COX-2, is an essential component of PGE_2 production during inflammatory reactions. Typically, it can be downregulated by glucocorticoids.

mPGES-2 is a 41 kD protein structurally distinct from mPGES-1 and does not require glutathione for catalysis. It forms a dimer, attaches to the lipid membrane, and couples with both COX-2 and COX-1. Though constitutively expressed in various cells and tissues, in contrast to mPGES-1, its expression is not induced markedly during inflammation or tissue damage. mPGES-2-deficient mice show no specific phenotype, and it is generally concluded that mPGES-2 is not involved in PGE_2 synthesis under inflammatory conditions.

cPGES is a 23 kD cytosolic protein and requires GSH as an essential cofactor for its activity. It is identical to the heat shock protein 90 (HSP-90)-associated protein p23 and is associated with protein kinase CK2. Following cell activation, cPGES, CK2, and HSP-90 form a stoichiometric complex of 1:1:1 resulting in phosphorylation of cPGES. Most likely, this complex is translocated to the nucleus membrane to form an assembly with COX-1, but not with COX-2, for PGE_2 synthesis. Expression of cPGES is constitutive and is unaffected by proinflammatory stimuli in most cases. Whether cPGES is involved in inflammatory processes is unclear. There is evidence that cPGES plays a role in mediating early responses during spinal nociceptive processing.

PGD_2 Synthesis

PGD synthase (PGDS) catalyzes the isomerization of PGH_2 to produce PGD_2 with 9-hydroxy and 11-keto groups in the presence of sulfhydryl compounds. Two distinct types of PGDS have been identified: (1) the lipocalin-type enzyme (L-PGDS) and (2) the hematopoietic-type enzyme (H-PGDS). The two different PGDS enzymes have no significant homology at the amino acid level. In addition, they have different tertiary structures for catalysis and different evolutional origins and therefore are thought to result from functional convergence (Urade and Eguchi 2002).

L-PGDS was shown to be identical to the long known protein β-trace, a major protein present in human cerebrospinal fluid. It is a monomeric protein with a MW of 21 kD and belongs to the lipocalin gene family (lipophilic ligand-carrier protein family) presenting it as a dual functional protein. In specific cells, L-PGDS is localized in the rough endoplasmic reticulum and the nuclear membrane in a functional relationship to COX for PGD_2 formation. L-PGDS contributes to the production of PGD_2 in the central nervous system, ocular tissues, cardiovascular systems, and male genital organs of various mammals.

H-PGDS is a cytosolic homodimer with a MW of 23 kD and is a member of the glutathione S-transferase gene family which absolutely requires glutathione for enzymatic activity. Stimulation-dependent translocation of H-PGDS from the cytosol to perinuclear membranes, where COX is localized, occurs. H-PGDS is expressed in peripheral tissues by dendritic cells, mast cells, megakaryocytes, and Th2 cells but not Th1 lymphocytes and is responsible for the bulk of PGD_2 production during allergic responses. PGD_2 is formed from liberated arachidonic acid either by COX-1 or by COX-2 in association with H-PGDS. Studies in mouse mast cells indicate that the early phase of PGD_2 production is COX-1 dependent, whereas COX-2 is responsible for more prolonged PGD_2 formation and also for PGD_2 production in chronically inflamed tissues. In human Th2 cells, PGD_2 is produced in response to antibody challenge as a consequence of coordinated induction of COX-2 and H-PGDS.

PGD_2 is an extremely short-lived prostanoid ($t_{1/2}$ in the blood is about 6 s) and is rapidly metabolized to 13,14-dihydro-15-keto-PGD_2 by 15-PG dehydrogenase or to 11β-$PGF_{2α}$ by PGD 11-ketoreductase. Alternatively, PGD_2 is dehydrated nonenzymatically to produce the J-series of prostaglandins with a cyclopentenone structure, such as PGJ_2, Δ^{12}-PGJ_2, and 15-deoxy-$\Delta^{12,14}$-PGJ_2 (Fig. 1). Given this instability, PGD_2 must act discretely in close proximity to the source of its synthesis.

PGI_2 Synthesis

PGI_2 is formed from PGH_2 by prostacyclin synthase (PGIS), which is a membrane-bound heme-protein with a MW of 52 kD and belongs to the cytochrome P450 family (classified as CYP8). PGIS is widely expressed in human tissues and is found constitutively particularly in endothelial cells and smooth muscle cells but also in nonvascular cells including neurons. Its expression can be induced by inflammatory cytokines such as IL-1β, IL-6, and TNFα. In resting cells, PGIS is colocalized with COX-1, and the level of PGI_2 formed is mainly dependent on the activity of COX. Following inflammatory stimulation, PGIS is also colocalized with COX-2 to the nuclear envelope and the endoplasmic reticulum. Formed PGI_2 is very unstable and dehydrates in aqueous solution to 6-keto-$PGF_{1α}$ with a $t_{1/2}$ of ~60 s. 6-keto-$PGF_{1α}$ is much more stable and is used as a surrogate in order to determine PGI_2 concentration in vitro and in vivo.

TXA_2 Synthesis

TXA_2 is produced by thromboxane synthase (TXS), a heme-protein with a MW of 59 kD which belongs to the cytochrome P450 family (classified as CYP5). TXS is widely expressed in the lung, spleen, and thymus, with particular abundance in platelets, monocytes/macrophages, and dendritic cells. It catalyzes the conversion of PGH_2 either to TXA_2 by an isomerase reaction or to 12-L-hydroxy-5, 8,10-heptadecatrienoic acid (12-HHT) and malondialdehyde (MDA) by a fragmentation reaction in a 1:1:1 ratio (TXA_2/HHT/MDA). TXA_2 is relatively labile with a half-life of ~30 s in aqueous solution and hydrolyses nonenzymatically to the biologically inactive TXB_2 which is determined as a marker of TXA_2 formation in vitro and in vivo. Whether MDA and 12-HHT are merely by-products of TXA_2 synthesis or exert biological functions is unclear.

$PGF_{2α}$ Synthesis

Biosynthesis of $PGF_{2α}$ is proposed to occur via three distinct pathways (Watanabe 2011): (1) the 9,11-endoperoxide group of PGH_2 is reduced to $PGF_{2α}$ by PGF synthase (PGFS, also named PGD 11-ketoreductase), belonging to the aldo-keto reductase (AKR) family; (2) the 11-keto group of PGD_2 is reduced by PGD_2 11-ketoreductase to yield 9α,11β-PGF_2 which is a stereoisomer of $PGF_{2α}$ but has biological functions like $PGF_{2α}$; and (3) the 9-keto group of PGE_2 is reduced by PGE 9-ketoreductase to yield $PGF_{2α}$. In peripheral circulation, $PGF_{2\alpha}$ exhibits a half-life of less than 1 min as it is rapidly degraded enzymatically to the more stable but biologically inactive metabolite 15-keto-dihydro-$PGF_{2\alpha}$. PGFS and PGE 9-ketoreductase are expressed primarily in myometrial tissue. PGFS is also found in the lung, in peripheral blood lymphocytes, and in neurons.

PGJ_2 Synthesis

PGD_2, synthesized by PGDS, undergoes spontaneous dehydration in aqueous solutions to yield PGJ_2. In the presence of serum albumin, PGJ_2 isomerizes to Δ^{12}-PGJ_2 by intramolecular rearrangement of the 13,14-double bond, followed by dehydration to form 15-deoxy-$\Delta^{12,14}$-PGJ_2. These PGJ metabolites are characterized by a cyclopentenone ring structure with an electrophilic α,β-unsaturated carbonyl group. Unlike other prostaglandins, which interact with membrane-bound receptors, cyclopentenone prostaglandins are taken up by cells via an active transport process and accumulate intracellularly and are further transported to the nucleus. Cyclopentenone prostaglandins are rapidly metabolized in cells via glutathione transferase-mediated conjugation to glutathione.

Biological Activities

COX

Different, partially opposing roles have been attributed to COX-1 and COX-2, based on differences in regulation and expression. COX-1 is ubiquitous and is constitutively expressed in the gastrointestinal tract, the kidney, vascular smooth muscle, and platelets. COX-1 is presumably, but not exclusively, involved in housekeeping functions of prostanoids, such as gastroprotection, platelet function and maintenance of renal perfusion, and electrolyte homeostasis. The use of traditional NSAIDs is associated with renal and gastrointestinal toxicity and retardation of coagulation mechanisms arising from the inhibition of COX-1.

Predominantly, but also not exclusively, COX-2-dependent prostanoid synthesis is observed under pathological conditions, such as pain, tumor, and inflammation. COX-2 is inducible by a variety of mitogens, tumor promoters, and proinflammatory stimuli and delivers under these conditions large amounts of prostanoids, PGE_2 in particular. The COX-2 gene is activated by multiple signaling pathways, and in contrast to COX-1, expression of COX-2 is inhibited by glucocorticoids such as cortisol and dexamethasone. Apart from the pathological roles of induced COX-2, constitutive COX-2 at a low level is present in fetal and adult kidney and in the brain. COX-2 is involved in cytoprotection of gastrointestinal mucosa when the latter is inflamed or ulcerated, and induced COX-2 at the site of injury produces large amounts of prostanoids, mainly PGE_2, involved in the healing process. Therefore, both COX-1 and COX-2 need to be inhibited to induce NSAID-induced gastric injury. The cardiotoxicity associated with the prolonged use of COX-2 selective inhibitors confirms a homeostatic role for COX-2 in the cardiovascular system (Grosser et al. 2006). Hence, these functions indicate a physiological role for COX-2 in adaptation of the body to different "stress situations."

PGE_2

PGE_2 is one of the most abundant prostanoids in the body. Under physiological conditions, it is an important mediator of many biological functions, such as regulation of blood pressure, kidney function, gastrointestinal integrity, fertility, and immune responses (Table 1). Under inflammatory conditions, PGE_2 is highly detectable and represents the key mediator of inflammation and pain. A typical setting is the inflamed joint of rheumatoid arthritis patients where proinflammatory cytokine-activated cells, including synovial cells, chondrocytes, macrophages, and monocytes, are the primary sources of PGE_2.

PGE_2 exerts its actions on target cells by interacting with one or more of its four G-protein-coupled receptor subtypes EP1, EP2, EP3, and EP4 (Sugimoto and Narumiya 2007). These receptors are linked to different signaling pathways that involve phosphoinositide hydrolysis and Ca^{2+} increase (EP1), activation of adenylate cyclase (EP2, EP4), and inhibition of adenylate

Prostanoids, Table 1 Prostanoids and their main functions

Prostanoid	Main physiological functions	Main pathophysiological functions
PGE_2	Blood pressure regulation	Mediator of cardinal signs of acute inflammation
	Renal salt and water homeostasis	Redness, swelling, edema, pain
	Renin release and renal perfusion	Hyperalgesia and allodynia
	Fertility and patency of ductus arteriosus	Modulation of immune cells in chronic inflammation
	Cervical ripening and myometrial contraction	Progression of tumorigenesis
	Modulation of immune responses	Neuropathic pain
	Gastrointestinal integrity	
	Controlling stress-induced impulse behavior	
	Modulation of osteogenesis	
PGD_2	Inhibition of platelet aggregation	Mediator of allergic diseases
	Relaxation of vascular smooth muscle cells	Allergic bronchoconstriction
	Sleep modulation	Skin flushing
	Increase in intraocular pressure	Immune cell chemotaxis and infiltration
		Induction of proinflammatory cytokines
		Suppression of cell migration and degranulation
PGI_2	Inhibition of platelet aggregation	Edema, hyperalgesia, and pain in acute inflammation
	Relaxation of vascular smooth muscle cells	Induction of chemokines in chronic inflammation
	Antimitogenesis and antimetastasis	Modulation of lymphocyte function
	Cardioprotection	
TXA_2	Shape change and platelet aggregation	Contributor to inflammatory processes
	Vaso- and bronchoconstriction	Mediator of inflammatory tachycardia
		Progression of atherogenesis
$PGF_{2\alpha}$	Ovulation and luteolysis	Mediator of inflammatory tachycardia
	Myometrial contraction	
	Endometrial proliferation	
	Increase in uveoscleral outflow	
15-deoxy-$\Delta^{12,14}$-PGJ_2	Anti-inflammation	
	Resolution of inflammation	

cyclase (EP3). The different PGE_2 receptor types are expressed in various cells and organs showing a specific pattern of expression, and the heterogeneity of the PGE_2 receptors explains the diversity of PGE_2 actions in various organs.

PGD_2

PGD_2 is released from platelets and causes inhibition of platelet aggregation and relaxation of vascular smooth muscle cells (Table 1). It is also formed in the brain and is involved in the regulation of sleep and central pain perception. Local allergen challenge in patients with allergic rhinitis, bronchial asthma, allergic conjunctivitis, and atopic dermatitis results in an instant increase in local PGD_2 levels. PGD_2 is the major arachidonic acid product in mast cells and modulates inflammatory and immune reactions and elicits characteristic symptoms of allergic asthma, such as bronchoconstriction and airway eosinophil infiltration.

The effects of PGD_2 in the inflammatory setting are mediated by interaction with two types of receptors, the prostanoid-type receptor DP and the chemoattractant receptor-homologous molecule expressed on Th2 cells, CRTH2. Although both receptor types bind the same ligand, there is very little homology between the two receptors. The DP receptor, also referred to as DP1, binds PGD_2 with high affinity, and its coupling to $G_{\alpha s}$ leads to cAMP elevation. DP receptor is widely expressed, in eosinophils, basophils, dendritic cells, Th1 and Th2 cells, central nervous system, retina, nasal mucosa, lungs, and intestine. It is also expressed by vascular smooth muscle cells and platelets, mediating vasodilation and inhibition of platelet aggregation (Woodward et al. 2011). In contrast to DP, CRTH2 activation results in cAMP decrease via $G_{\alpha i}$ and elevation of intracellular Ca^{2+} by $G_{\beta\gamma}$-subunit-mediated formation of IP3. In Th2 cells, phosphatidylinositol 3-kinase and Ca^{2+}/calcineurin/nuclear factor of activated T-cell pathways are activated by the PGD_2-CRTH2 axis triggering chemotaxis and production of proinflammatory cytokines (Xue et al. 2007). In striking contrast to the DP receptor, the CRTH2 receptor is not strictly selective for PGD_2. Metabolites of PGD_2, such as 13,14-dihydro-15-keto-PGD_2, and prostanoids of the J-series, such as PGJ_2, Δ^{12}-PGJ_2, and 15-deoxy-$\Delta^{12,14}$-PGJ_2, are endogenous ligands. The isoprostane 15R-PGD_2, which may be generated by oxidative stress independently of the COX pathway of arachidonic acid metabolism, is also a potent CRTH2 agonist.

PGI_2

PGI_2 is an important prostanoid that regulates cardiovascular homeostasis. It acts as a potent vasodilator and as an inhibitor of platelet aggregation, leukocyte adhesion, and vascular smooth muscle cell proliferation (Moncada et al. 1976) (Table 1). Vascular endothelial cells and vascular smooth muscle cells are the major source of PGI_2, and PGI_2 appears to play an important role in protecting against elevated blood pressure. PGI_2 analogues such as iloprost or epoprostenol are widely used in the clinical management of pulmonary arterial hypertension (Benedict et al. 2007). Moreover, PGI_2 is suggested to be an antimitogenic and antimetastatic prostanoid.

PGI_2 effects are mediated by the prostanoid-type receptor IP. Following binding to the IP receptor, PGI_2 activates a G_s-protein leading to activation of adenylate cyclase and subsequently of protein kinase A. IP receptors are widely distributed throughout the body, with predominant cardiovascular expression in vascular smooth muscle cells, endothelial cells, and blood platelets. Vasodilation and inhibition of platelet aggregation are achieved by activation of the IP receptor. PGI_2 is also able to signal via the peroxisome proliferator-activated receptor δ (PPARδ) within the nucleus.

TXA_2

One of the most important biological actions of TXA_2 is the activation of platelets, leading to platelet shape change, aggregation, and secretion, thereby promoting thrombus formation and thrombosis (Table 1). TXA_2 is also a potent vasoconstrictor and bronchoconstrictor. Ozagrel, an inhibitor of TXA_2 synthesis, and seratrodast, a TXA_2 receptor antagonist, are used for the treatment of asthma. TXA_2 is predominantly derived from platelets, but it can also be produced by other cell types, especially monocytes and macrophages. The balance between circulating PGI_2 and TXA_2 levels is considered to be the most important regulator of vascular hemostasis. Imbalance in favor of TXA_2 is suggested as an important cause for enhanced cardiovascular risk in patients taking NSAIDs or coxibs (Grosser et al. 2006).

TXA_2 signals via the prostanoid-type receptor TP, which is expressed in a number of tissues including the thymus, lung, kidney, spleen, and placenta. TP couples mainly to G_q family G-proteins resulting in phospholipase C-catalyzed phosphoinositide hydrolysis, which in turn mobilizes intracellular Ca^{2+} and protein kinase C. In addition to G_q, the TP receptor is able to communicate with G_{12} family G-proteins which activate RhoGEF following the activation of Rho-mediated signaling. In platelets, activation of $G_{12/13\alpha}$ triggers platelet shape change via ROCK activation, while activation of $G_{q\alpha}$/PLCβ mediates integrin activation, aggregation, and secretion. In humans, TXA_2 signals through two TP receptor isoforms, termed TPα and TPβ, that arise by differential splicing and differ in their carboxyl-terminal cytoplasmic tail region, whereas TPα is the predominant isoform in platelets. In mouse and rat, only TPα is present. 8-iso-PGF$_{2\alpha}$, an isoprostane synthesized by a radical-based mechanism under conditions of oxidative stress such as ischemic heart disease, also activates the TP receptor and causes vasoconstriction, albeit with lower affinity compared to TXA$_2$.

PGF$_{2\alpha}$

PGF$_{2\alpha}$ is a vasopressor prostanoid, found in large quantities in peripheral plasma and in urine. PGF$_{2\alpha}$ is produced in the female reproductive system, where it plays an important role in ovulation and luteolysis, and the amount of PGF$_{2\alpha}$ increases at the time of parturition to induce myometrial contraction (Table 1). Further, PGF$_{2\alpha}$ is produced in ocular tissue regulating intraocular pressure, the basis for the clinical use of PGF$_{2\alpha}$ analogues, such as latanoprost or travoprost to treat open-angle glaucoma.

PGF$_{2\alpha}$ acts via FP receptor which is coupled to G$_q$-protein leading to IP3/DAG generation and mobilization of intracellular Ca^{2+}. Beyond G$_q$, the FP receptor also couples to G$_{12/13}$ with activation of Rho causing changes in cell morphology and in the cell cytoskeleton. Two splice forms of FP receptor (FP$_A$ and FP$_B$) exist in humans, but the physiological significance of FP isoforms has not been completely elucidated. The FP receptor is highly expressed in ovarian tissue, myometrium, endometrium, deciduae, kidney, and eye.

PGJ$_2$

PGJ$_2$ is most likely biologically inactive, but rapidly converted to 15-deoxy-Δ^{12,14}-PGJ$_2$. Although 15-deoxy-Δ^{12,14}-PGJ$_2$ is able to serve as a ligand for CRTH2 receptor and may exert some proinflammatory activity, it is mainly responsible for anti-inflammatory actions during the resolution phase of inflammation (Table 1). Neither for PGJ$_2$ nor for 15-deoxy-Δ^{12,14}-PGJ$_2$ has a specific prostanoid-type receptor been identified. At least two molecular mechanisms underlie the anti-inflammatory effects of 15-deoxy-Δ^{12,14}-PGJ$_2$: (1) 15-deoxy-Δ^{12,14}-PGJ$_2$ acts as a ligand for PPARγ (Kliewer et al. 1995) leading to alterations in gene expression, and (2) the reactive carbonyl moiety of the cyclopentenone rapidly forms Michael adducts with cellular thiols, covalently modifying critical proteins in multiple pathways, altering their conformation, and, subsequently, influencing their function.

Pathophysiological Relevance and Modulation by Drugs

COX

COX-1 and COX-2 are expressed in inflammatory cells. Constitutive COX-1 is thought to contribute to the initial phase of acute inflammation, though COX-2 appears as the dominant source of prostanoids, upregulated within 30 min to hours and declining within hours to days. Chronic inflammation is dominated by ongoing high COX-2 expression, its activity being responsible for the increased prostanoid production seen, e.g., in inflamed joint tissues of arthritic patients. Induced COX-2 is also present in human osteoarthritis-affected cartilage. In various animal models of arthritis, extinguishing COX-2 enzyme markedly suppresses not only PGE$_2$ levels but also synovial inflammation. In humans, selective COX-2 inhibitors such as celecoxib or etoricoxib show high efficacy in reducing the signs and symptoms of rheumatoid arthritis and osteoarthritis, as indicated by a large number of randomized, double-blind, multicenter studies (Chen et al. 2008).

The role of COX-2 in other chronic inflammatory diseases is less established. In patients suffering inflammatory bowel disease, despite a clear relationship between endoscopic activity and relative levels of COX-2 mRNA, contrasting effects in human clinical studies question the benefit of COX-2 inhibitor (coxib) treatment for inflammatory bowel disease. Similarly, correlation of levels of COX-2 protein in the brain with the severity of amyloidosis and clinical dementia and also epidemiologic data indicate that inhibition of COX by NSAIDs may prevent the development of Alzheimer's disease. However, clinical trials using COX unselective NSAIDs and COX-2 selective drugs failed to demonstrate slowing of cognitive decline in patients with mild cognitive impairment or symptomatic Alzheimer's disease (Jaturapatporn et al. 2012).

PGE_2 System

mPGES-1 plays a dominant role in inflammatory, especially arthritic diseases. It is particularly abundant in the intima of synovial biopsy specimens from rheumatoid arthritis patients and more pronounced in active disease. mPGES-1 is also expressed in superficial layers of osteoarthritic cartilage, where inflammatory damage first appears. Data from animal models of arthritis support the importance of the COX-2/mPGES-1/PGE_2 axis in inflammatory joint disease, e.g. mice deficient in mPGES-1 showed reduced joint damage, and diminished articular proteoglycan loss. Induction of mPGES-1 together with COX-2 also exercises a decisive role in central nervous system symptoms, i.e., hyperalgesia, fever, malaise, fatigue, and accompanying inflammatory arthritis. Induced expression occurs in brain tissue, in endothelial cells along the blood-brain barrier, and also in the parenchyma of the paraventricular hypothalamic nucleus with enhanced PGE_2 levels in cerebrospinal fluid, spinal cord, and brain extracts. Moreover, mPGES-1 has emerged as a powerful pathogenic component in idiopathic inflammatory myopathies, periodontitis, atherosclerosis, and multiple sclerosis. The relevance of mPGES-1 as a promising target for anti-inflammatory drugs has been demonstrated by the use of mPGES-1 inhibitors in animal models. The specific mPGES-1 inhibitor MF63 strongly blocked guinea pig mPGES-1 activity and suppressed lipopolysaccharide-induced pyresis, hyperalgesia, and iodoacetate-induced osteoarthritic pain (Xu et al. 2008). In adjuvant-induced arthritis in rats, the mPGES-1 inhibitor compound II ameliorated acute and delayed inflammatory paw swelling (Leclerc et al. 2013). mPGES-1 inhibitors are still in preclinical development, but, most likely, will enter the clinic in the near future.

PGE_2 constitutes a key factor in the development of the four cardinal signs of acute inflammation: redness, swelling, pain, and fever. These clinical signs are mediated by different types of the EP receptors. Locally produced PGE_2 through EP2 and EP4 receptor signaling causes dilatation of vascular smooth muscle cells and thereby enhances blood flow, resulting in the characteristic erythema (redness) of inflammation. PGE_2 triggers mast cell activation via an EP3 receptor-mediated mechanism eliciting histamine release and thereby increases vascular permeability, contributing to fluid extravasation and the appearance of edema (swelling). PGE_2 sensitizes afferent nerve fibers acting both at peripheral sensory neurons at the site of inflammation and at central sites within the spinal cord and brain to evoke hyperalgesia (pain). All four EP receptor types are expressed in the periphery and in dorsal root ganglions, and they all contribute to mediation of pain sensation in a context-dependent manner. Inflammatory pain sensation through the TRPV1 cation channel, which detects several pain stimuli, is augmented by activation of EP1 receptor, and spinal EP1 receptor is also involved in mediation of allodynia. In peripheral inflammation induced by injection of Freund's complete adjuvant, hyperalgesia is mediated by the EP4 receptor, whereas combined treatment with LPS and acetic acid causes inflammatory nociception involving EP3 receptor. PGE_2 also augments the processing of inflammatory pain information in the spinal cord. In an EP2-dependent manner, PGE_2 reduces the extent of neurotransmission mediated by the inhibitory transmitter glycine, facilitating the transmission of nociceptive input to higher brain areas. Moreover, PGE_2 is a potent pyretic mediator (fever). Exogenous (bacterial lipopolysaccharide) and endogenous (IL-1, IL-6, TNFα) pyrogens induce expression of both COX-2 and mPGES-1 in brain microvessels of the preoptic area of the organosum vasculatum lamina terminalis causing enhanced PGE_2 release. Upon activation by PGE_2, inhibitory input from EP3-containing neurons in the preoptic area on hypothalamic mediated thermoregulatory responses is blocked causing an increase in body temperature (fever) (Nakamura et al. 2005).

Although the outstanding role of PGE_2 in inflammatory diseases is well established, less information is available regarding the types and roles of PGE_2 receptors involved in chronic inflammation. Several experimental studies point toward an important role of EP2 and/or EP4 receptors. In different experimental models of arthritis, antagonism or deletion of EP2 and/or EP4 receptor led to the suppression of inflammatory signs, such as swelling or erythema. Recent work suggests that the EP2/EP4 system acts as an amplifier of Th1 and Th17 responses and their cytokine generation. As a corollary, antagonists suppress Th1/Th17 responses in experimental models of chronic inflammation, and PTGER4, the gene for EP4 receptor, is a locus associated with Crohn's disease and multiple sclerosis (Libioulle et al. 2007). In the setting of allergic asthma, EP3 subtype receptor acts as an important negative modulator of allergic reactions and suppresses mast cell degranulation, chemokine synthesis, and eosinophil infiltration.

PGD$_2$ System

The role of L-PGDS in inflammation is not clearly resolved, and pro- as well as anti-inflammatory roles are claimed. Overexpression of L-PGDS increases inflammatory reactions in allergic models, and pathological increases of L-PGDS expression together with COX-2 occur in the mucosa of colitis patients. Inhibition of L-PGDS activity caused significant attenuation of the inflammatory response in a model of chronic allergic dermatitis (Satoh et al. 2006) and in experimental colitis. In contrast, in some infectious models, overexpression of L-PGDS is protective and exhibits anti-inflammatory effects. The L-PGDS-PGD$_2$ system also plays a decelerating role in the evolution of atherosclerosis as deficiency of L-PGDS aggravates atherosclerosis along with enhanced expression of MCP-1 and IL-1β.

H-PGDS is reported to be involved in disease-dependent pro- as well as anti-inflammatory settings. In mouse models of asthma and allergic disease, H-PGDS acts proinflammatorily, and inhibition of H-PGDS reduces eosinophilia, airway hyperreactivity, mucus production, and Th2 cytokine levels. In contrast, H-PGDS acts protectively in the delayed-type hypersensitivity responses, where H-PGDS negatively regulates severity and duration. In collagen-induced arthritis in DBA/1 mice, both L-PGDS and H-PGDS were shown to be upregulated, and thereby formed PGD$_2$ was associated with the dampening of inflammatory response and joint damage (Maicas et al. 2012).

PGD$_2$ is mainly involved in inflammatory diseases like allergy and asthma. Allergen leads to rapid production of PGD$_2$ in the airways of asthmatics, the nasal mucosa of allergic rhinitis, and in the skin of patients with atopic dermatitis. PGD$_2$ is the predominant prostanoid produced by activated mast cells, which initiate IgE-mediated type I acute allergic responses, but is also produced by other immune cells, such as antigen-presenting dendritic cells, basophils, and Th2 cells. The DP receptor has been proposed to mediate production of chemokines and cytokines that recruit inflammatory cells. Regarding allergic rhinitis, activation of DP receptor contributes to the pathological blood flow changes during allergic responses leading to swelling of the nasal mucosa and causing congestion and enhanced leakage of plasma proteins, whereas DP antagonists block rhinitis, conjunctivitis, and eosinophil infiltration. The DP receptor also plays a key role in mediating the effects of PGD$_2$ released by mast cells during an asthmatic response, and DP$^{-/-}$ mice exhibit reduced airway hypersensitivity and pulmonary inflammation in response to ovalbumin challenge. However, on the other hand, experimental evidence indicates an anti-inflammatory role of DP receptor in inflammation. DP activation is able to inhibit the function of neutrophils, basophils, dendritic cells, Langerhans cells, Th1 cells, and natural killer cells. In T cells and other immune cells, PGD$_2$ suppresses cell migration and cell activation via the DP receptor, and local injection of the DP agonist BW245C reduces allergic responses in the skin. PGD$_2$ inhibits eosinophil degranulation and acts anti-apoptotically in eosinophils, also being solely mediated by DP receptors. In the mouse, inhalation of a DP agonist suppresses asthma symptoms by downregulation of lung dendritic cell function and induction of regulatory T cells. In dendritic cells, activation of DP receptor by BW245C suppresses the production of cytokines including IL-12 and, therefore, is relevant in controlling the polarization of T cells to either the Th1 or Th2 phenotype.

CRTH2 accounts for many proinflammatory effects of PGD_2. These include chemotaxis of immune cells such as eosinophils, Th2 cells, basophils, monocytes, and macrophages, respiratory burst of eosinophils, histamine release from basophils, and cytokine release and enhanced survival of Th2 cells. Activation of the CRTH2 receptor is associated with the initiation and perpetuation of allergic inflammation. PGD_2 and several of its metabolites are potent chemoattractants for eosinophils and basophils into the inflammatory site and are also capable of stimulating the release of mature eosinophils from the bone marrow, and these effects rely on CRTH2 activation. CRTH2 is also expressed on monocytes and macrophages and mediates their migration induced by PGD_2. Activation of the CRTH2 receptor on Th2 cells induces the production of proinflammatory cytokines (IL-4, IL-5, IL-10, IL-13) and upregulates the expression of CD40L and delays their apoptosis. In some cells, PGD_2 acts on CRTH2 and on DP as an antagonist. For example, in basophils, CRTH2 activation enhances expression of the adhesion molecule CD11b and causes histamine release, while activation of the DP receptor has opposing effects.

In animal models, exogenously administered CRTH2 agonists induce eosinophil infiltration into the lungs and skin and aggravate the pathology of allergic responses. In contrast, CRTH2 antagonists ameliorate allergen-induced cutaneous, pulmonary, and upper respiratory inflammation. In man, the severity of asthma has been linked with single nucleotide polymorphisms in the CRTH2 gene (Huang et al. 2004). Therefore, CRTH2 antagonists are currently considered a highly promising approach to the treatment of allergic diseases and asthma.

PGI_2 System

PGI_2 acts as an important mediator of the edema and pain accompanying acute inflammation. Bradykinin induces PGI_2 formation leading to enhancement of microvascular permeability and edema. PGI_2 augments peripheral nociceptive sensitization to inflammatory stimuli and also facilitates pain transmission. Deletion of the IP gene causes complete suppression of induced inflammatory swelling and pain. In addition to peripheral inflammatory pain transmission, spinal transmission is at least in part mediated by PGI_2 during early inflammation.

The importance of the PGI_2-IP system in the pathogenesis of chronic inflammation is demonstrated by high concentrations of PGI_2 in synovial fluids from knee joints of arthritic patients and by the reduced clinical and histological arthritic scores in $IP^{-/-}$ mice. In combination with IL-1β, PGI_2-IP signaling induces a variety of arthritis-related genes, such as IL-11, VEGF, FGF-2, and RANKL, supporting a significant role in the effector mechanisms of inflammation (Honda et al. 2006). Furthermore, PGI_2 contributes to the induction of chemokines and resultant infiltration of inflammatory cells by synergizing with IL-1β to augment expression of CXCL7, a chemokine for neutrophils, fibroblasts, and endothelial cells (Honda et al. 2006).

The capacity of blood vessels to generate PGI_2 is essential to the integrity of the endothelium. In human atherosclerotic diseases, endothelial dysfunction and deletion caused by oxidized low-density lipoproteins result in reduced expression of PGIS. Moreover, tyrosine nitration of PGIS mediated by reactive nitrogen species formed during inflammation reduces PGI_2 formation. In models of atherogenesis, deletion of the IP gene causes partial endothelial disruption and enhanced expression of ICAM-1 and thereby platelet activation and leukocyte-endothelial cell interaction.

TXA_2 System

The function of TXA_2 under inflammatory conditions is not definitely clarified, but TXA_2 at least contributes to allergic processes in rhinitis, atopic dermatitis, renal and bowel inflammation and in atherosclerosis. Elevated TXB_2 concentrations are present in the nasal lavage fluid of allergic rhinitis patients after allergen provocation and urinary excretion of TXA_2 metabolites is enhanced in patients with active lupus nephritis. Application of TP receptor antagonist Bay-u3405 attenuates antigen-induced increase in nasal airway resistance and local eosinophil infiltration in sensitized guinea pigs, and TXS inhibitor DP-1904 improves renal function in experimental nephritis. In patients with atopic dermatitis, TP receptor polymorphisms are strongly related to high serum IgE concentrations. TXA_2 is also produced in excess in inflammatory bowel disease, and inhibition of TXS is effective in suppressing colonic inflammation in a murine model (Howes et al. 2007). BM-573, a combined TXS inhibitor and TP antagonist, slows atherogenesis in the LDL receptor-deficient mice and inhibits the development of atherosclerotic lesions in Apo $E^{-/-}$ mice by downregulation of ICAM-1 and VCAM-1 expression. Compound apo$E^{-/-}TP^{-/-}$ mice exhibit a significant delay in atherogenesis. Most likely, TXA_2 promotes initiation and progression of atherogenesis through control of platelet activation and leukocyte-endothelial cell interaction. In pulmonary inflammation, PGE_2 elicits bronchoprotection by restraining TP signaling (Liu et al. 2012). Systemic infection induces various adaptive responses including tachycardia. TXA_2 and also $PGF_{2\alpha}$ directly act on the atrium to increase beating rate, and the early phase in inflammation-associated tachycardia is dominated by COX-1/TXA_2/TP, the late phase by COX-2/$PGF_{2\alpha}$/FP.

$PGF_{2\alpha}$ System

A role of $PGF_{2\alpha}$ in inflammation is not unambiguously clarified. In animal models, induction of acute inflammation by endotoxemic challenge is associated with increased $PGF_{2\alpha}$ formation, and induced tachycardia is greatly attenuated in mice deficient in FP receptor. Although $PGF_{2\alpha}$ is intimately associated with various types of chronic inflammatory diseases, the precise role of $PGF_{2\alpha}$ is not yet recognized. Elevated biosynthesis of $PGF_{2\alpha}$ does occur in patients suffering different forms of arthritis, such as rheumatic, psoriatic, or osteoarthritis, and the levels of $PGF_{2\alpha}$ can be effectively reduced by NSAID treatment. However, it remains to be clarified whether $PGF_{2\alpha}$ contributes to arthritic degeneration or appears as a by-product of the boosted COX system.

Evoked synthesis of $PGF_{2\alpha}$ may also coincide with free radical-catalyzed generation of F-ring isoprostanes, indices of lipid peroxidation. Therefore, the level of $PGF_{2\alpha}$ may not solely depend on enzymatic activity of COX- or $PGF_{2\alpha}$-synthesizing enzymes. A clearer role for $PGF_{2\alpha}$ is seen in bleomycin-induced pulmonary fibrosis, where $PGF_{2\alpha}$ is abundant in bronchoalveolar lavage fluid and stimulates proliferation and collagen production of lung fibroblasts via FP receptor and where deletion of the FP receptor attenuates pulmonary fibrosis (Oga et al. 2009).

PGJ_2 System

Several findings support an anti-inflammatory role for 15-deoxy-$\Delta^{12,14}$-PGJ_2 as a key negative feedback regulator of the arachidonic acid cascade in the COX pathway. Intracerebroventricular infusion of 15-deoxy-$\Delta^{12,14}$-PGJ_2 reduces LPS-induced fever in rats in association with inhibition of LPS-induced COX-2 expression in the hypothalamus (Tsubouchi et al. 2001). In adjuvant-induced arthritis, 15-deoxy-$\Delta^{12,14}$-PGJ_2 suppresses IL-1β-induced PGE_2 synthesis in rheumatoid synovial fibroblasts by inhibition of COX-2 and cPLA2 expression. In experimental multiple sclerosis, 15-deoxy-$\Delta^{12,14}$-PGJ_2 decreases clinical severity, IL-12 production, and differentiation of Th1 cells, and in experimental enterocolitis, 15-deoxy-$\Delta^{12,14}$-PGJ_2 attenuated intestinal NF-κB activation and inflammatory response, both by activation of PPARγ.

PPARγ expression is induced to high levels in activated macrophages but is also present in a variety of immune cell types, including lymphocytes, neutrophils, mast cells, and bone marrow precursors. PPARγ activation represents a major anti-inflammatory pathway through which 15-deoxy-$\Delta^{12,14}$-PGJ_2 suppresses immune reactivity, and abundant evidence implicates NF-κB as a major target. 15-deoxy-$\Delta^{12,14}$-PGJ_2 represses NF-κB activity by PPARγ-mediated upregulation of NF-κB inhibitor IκB. The PPARγ-independent anti-inflammatory mechanism of 15-deoxy-$\Delta^{12,14}$-PGJ_2 includes alkylation of specific cysteine residues in IκK kinase and thereby inhibition of the activator of NF-κB. Negative regulation of the activity of NF-κB or other transcription factors, such as AP-1, results in the downregulation of a number of inflammatory response genes, including genes for IL-1β, TNFα, gelatinase B, MMPs, iNOS, and COX-2. Based on these observations, 15-deoxy-$\Delta^{12,14}$-PGJ$_2$ is considered to be a potential endogenous counterbalance to proinflammatory responses.

Cross-References

- ▶ Alzheimer's Disease
- ▶ Asthma
- ▶ Corticosteroids
- ▶ Coxibs
- ▶ Cytokines
- ▶ Dendritic Cells
- ▶ IkappaB
- ▶ Inflammatory Bowel Disease
- ▶ Interleukin 6
- ▶ Mast Cells
- ▶ Natural Killer Cells
- ▶ NFkappaB
- ▶ Non-steroidal Anti-inflammatory Drugs: Overview
- ▶ Reactive Oxygen Species
- ▶ Rheumatoid Arthritis
- ▶ Tumor Necrosis Factor Alpha (TNFalpha)

References

Benedict, N., Seybert, A., & Mathier, M. A. (2007). Evidence-based pharmacologic management of pulmonary arterial hypertension. *Clinical Therapeutics, 29*, 2134–2153.

Chen, Y. F., Jobanputra, P., Barton, P., Bryan, S., Fry-Smith, A., Harris, G., et al. (2008). Cyclooxygenase-2 selective non-steroidal anti-inflammatory drugs (etodolac, meloxicam, celecoxib, rofecoxib, etoricoxib, valdecoxib and lumiracoxib) for osteoarthritis and rheumatoid arthritis: a systematic review and economic evaluation. *Health Technology Assessment, 12*, 1–278, iii.

Day, R. O., & Graham, G. G. (2013). Non-steroidal anti-inflammatory drugs (NSAIDs). *BMJ, 346*, f3195.

Grosser, T., Fries, S., & FitzGerald, G. A. (2006). Biological basis for the cardiovascular consequences of COX-2 inhibition: Therapeutic challenges and opportunities. *Journal of Clinical Investigation, 116*, 4–15.

Hara, S., Kamei, D., Sasaki, Y., Tanemoto, A., Nakatani, Y., & Murakami, M. (2010). Prostaglandin E synthases: Understanding their pathophysiological roles through mouse genetic models. *Biochimie, 92*, 651–659.

Honda, T., Segi-Nishida, E., Miyachi, Y., & Narumiya, S. (2006). Prostacyclin-IP signaling and prostaglandin E2-EP2/EP4 signaling both mediate joint inflammation in mouse collagen-induced arthritis. *Journal of Experimental Medicine, 203*, 325–335.

Howes, L. G., James, M. J., Florin, T., & Walker, C. (2007). Nv-52: A novel thromboxane synthase inhibitor for the treatment of inflammatory bowel disease. *Expert Opinion on Investigational Drugs, 16*, 1255–1266.

Huang, J. L., Gao, P. S., Mathias, R. A., Yao, T. C., Chen, L. C., Kuo, M. L., et al. (2004). Sequence variants of the gene encoding chemoattractant receptor expressed on Th2 cells (CRTH2) are associated with asthma and differentially influence mRNA stability. *Human Molecular Genetics, 13*, 2691–2697.

Jaturapatporn, D., Isaac, M. G., McCleery, J., & Tabet, N. (2012). Aspirin, steroidal and non-steroidal anti-inflammatory drugs for the treatment of Alzheimer's disease. *Cochrane Database of Systematic Reviews, 2*, CD006378.

Kliewer, S. A., Lenhard, J. M., Willson, T. M., Patel, I., Morris, D. C., & Lehmann, J. M. (1995). A prostaglandin J2 metabolite binds peroxisome proliferator-activated receptor gamma and promotes adipocyte differentiation. *Cell, 83*, 813–819.

Leclerc, P., Pawelzik, S. C., Idborg, H., Spahiu, L., Larsson, C., Stenberg, P., et al. (2013). Characterization of a new mPGES-1 inhibitor in rat models of inflammation. *Prostaglandins & Other Lipid Mediators, 102–103*, 1–12.

Libioulle, C., Louis, E., Hansoul, S., Sandor, C., Farnir, F., Franchimont, D., et al. (2007). Novel Crohn disease locus identified by genome-wide association maps to a gene desert on 5p13.1 and modulates expression of PTGER4. *PLoS Genetics, 3*, e58.

Liu, T., Laidlaw, T. M., Feng, C., Xing, W., Shen, S., Milne, G. L., et al. (2012). Prostaglandin E2 deficiency uncovers a dominant role for thromboxane A2 in house dust mite-induced allergic pulmonary inflammation. *Proceedings of the National Academy of Sciences of the United States of America, 109*, 12692–12697.

Maicas, N., Ibanez, L., Alcaraz, M. J., Ubeda, A., & Ferrandiz, M. L. (2012). Prostaglandin D2 regulates joint inflammation and destruction in murine collagen-induced arthritis. *Arthritis and Rheumatism, 64*, 130–140.

Marnett, L. J. (2009). The COXIB experience: A look in the rearview mirror. *Annual Review of Pharmacology and Toxicology, 49*, 265–290.

Moncada, S., Gryglewski, R. J., Bunting, S., & Vane, J. R. (1976). An enzyme isolated from arteries transforms prostaglandin endoperoxides to an unstable substance that inhibits platelet aggregation. *Nature, 263*, 663–665.

Nakamura, Y., Nakamura, K., Matsumura, K., Kobayashi, S., Kaneko, T., & Morrison, S. F. (2005). Direct pyrogenic input from prostaglandin EP3 receptor-expressing preoptic neurons to the dorsomedial hypothalamus. *The European Journal of Neuroscience, 22*, 3137–3146.

Oga, T., Matsuoka, T., Yao, C., Nonomura, K., Kitaoka, S., Sakata, D., et al. (2009). Prostaglandin F(2alpha) receptor signaling facilitates bleomycin-induced pulmonary fibrosis independently of transforming growth factor-beta. *Nature Medicine, 15*, 1426–1430.

Satoh, T., Moroi, R., Aritake, K., Urade, Y., Kanai, Y., Sumi, K., et al. (2006). Prostaglandin D2 plays an essential role in chronic allergic inflammation of the skin via CRTH2 receptor. *Journal of Immunology, 177*, 2621–2629.

Smith, W. L., DeWitt, D. L., & Garavito, R. M. (2000). Cyclooxygenases: Structural, cellular, and molecular biology. *Annual Review of Biochemistry, 69*, 145–182.

Sugimoto, Y., & Narumiya, S. (2007). Prostaglandin E receptors. *Journal of Biological Chemistry, 282*, 11613–11617.

Tsubouchi, Y., Kawahito, Y., Kohno, M., Inoue, K., Hla, T., & Sano, H. (2001). Feedback control of the arachidonate cascade in rheumatoid synoviocytes by 15-deoxy-Delta(12,14)-prostaglandin J2. *Biochemical and Biophysical Research Communications, 283*, 750–755.

Urade, Y., & Eguchi, N. (2002). Lipocalin-type and hematopoietic prostaglandin D synthases as a novel example of functional convergence. *Prostaglandins & Other Lipid Mediators, 68–69*, 375–382.

Watanabe, K. (2011). Recent reports about enzymes related to the synthesis of prostaglandin (PG) F(2) (PGF (2alpha) and 9alpha, 11beta-PGF(2)). *Journal of Biochemistry, 150*, 593–596.

Woodward, D. F., Jones, R. L., & Narumiya, S. (2011). International Union of Basic and Clinical Pharmacology. LXXXIII: Classification of prostanoid receptors, updating 15 years of progress. *Pharmacological Reviews, 63*, 471–538.

Xu, D., Rowland, S. E., Clark, P., Giroux, A., Cote, B., Guiral, S., et al. (2008). MF63 [2-(6-chloro-1H-phenanthro[9,10-d]imidazol-2-yl)-isophthalonitrile], a selective microsomal prostaglandin E synthase-1 inhibitor, relieves pyresis and pain in preclinical models of inflammation. *Journal of Pharmacology and Experimental Therapeutics, 326*, 754–763.

Xue, L., Gyles, S. L., Barrow, A., & Pettipher, R. (2007). Inhibition of PI3K and calcineurin suppresses chemoattractant receptor-homologous molecule expressed on Th2 cells (CRTH2)-dependent responses of Th2 lymphocytes to prostaglandin D(2). *Biochemical Pharmacology, 73*, 843–853.

Protease-Activated Receptors

Morley D. Hollenberg
Inflammation Research Network-Snyder Institute for Chronic Disease, Department of Physiology and Pharmacology and Department of Medicine, Cumming School of Medicine, University of Calgary, Calgary, AB, Canada

Abbreviations

Amino Acids	Are designated by their on-letter codes, e.g., A = alanine, R = arginine, etc.
CNS	Central nervous system
PAR	Proteinase-activated receptor (PAR1, PAR2, PAR3, PAR4)
TL	Tethered ligand that activates PARs upon proteolytic unmasking

Definitions: Proteinase-Mediated Signalling

Proteinase-Mediated Signalling: a "First Step" in the Inflammatory Response

It is now known that more than 2 % of the human genome codes for either proteinases (colloquially termed: proteases) or their enzyme-targeted inhibitors (Puente et al. 2005). It is thus no surprise that proteinases can serve multiple biological functions. As a consequence, in addition to their "classical roles" as digestive enzymes for food metabolism, proteinases can now be seen to be "hormone-like" mediators that regulate tissue function by both receptor and non-receptor mechanisms (Fig. 1a). This signalling role is particularly important in the setting of an inflammatory response generated either by tissue injury and tissue pathology or by infectious agents. All individuals have an experience with this "hormonal"

Conflict of Interest Disclosure The author asserts that there is no conflict of interest related to the information included in this article.

role of proteinases when they cut their finger. Thus, shortly after the immediate "pain response" to the cut, the extrinsic and intrinsic coagulation pathways are triggered, resulting in the amplified proteolytic cascade that generates thrombin and the counterregulatory proteinase, activated protein-C (APC). The responses that follow are: (1) the blood clots, due to the proteolytic conversion of fibrinogen to fibrin, and (2) all of the hallmark signs of inflammation are triggered: more pain (dolor), increased blood flow (rubor), resulting in increased warmth at the tissue surface (calor), increased fluid to the injury site with swelling (tumor), and subsequently, scarring and possibly reduced function (functio laesa). Shortly after these responses are generated by enzymes of the coagulation cascade, the cellular "shock troops" of inflammation (neutrophils) arrive at the site of injury to initiate the slightly delayed reactions involving further tissue responses and the generation of cytokines. All of these inflammatory responses represent a protective process to remove the injurious stimuli and to initiate the healing process; and all of these responses are regulated by proteinase-mediated signalling, involving proteinases both of the coagulation cascade and from enzymes secreted by the invading neutrophils and other cells in the injury microenvironment. These signalling processes, as illustrated in Fig. 1a, will be summarized in the sections that follow. Although, as illustrated in the figure, the signalling mechanisms triggered by proteinases are complex, involving multiple substrate targets, this overview will focus in large part on signalling by the unique G-protein-coupled receptor family members that are activated by proteolysis (the PARs), rather than signalling by conventional circulating receptor agonists like adrenaline or cytokines. The overview will, with apologies to the reader, use illustrations drawn primarily from our own work related to inflammatory processes and will not be an exhaustive review of this field for which readers are referred elsewhere (Adams et al. 2011; Ramachandran and Hollenberg 2008; Ramachandran et al. 2012; Soh et al. 2010; Zhao et al. 2014). The aim of this chapter is to provide an overview of (1) the hormone-like actions of inflammation-related proteinases, (2) the molecular pharmacology of proteinase-triggered PAR signalling, and (3) the impact of receptor-mediated signalling by proteinases in a number of settings of inflammation.

Hormone-Like Signalling by Proteinases: Insulin-Like Anabolic and Mitogenic Actions

In the mid-1960s, it was discovered that peptide hormones are generated from precursor "prohormones" to stimulate tissue function and are then metabolized by proteolytic processing to downregulate signalling (Chretien 2012; Steiner et al. 1967; Steiner 2011). However, it was also realized at that time that enzymes like trypsin and chymotrypsin have hormone-like metabolic-anabolic actions on tissues and cells akin to the effects of insulin (Rieser and Rieser 1964; Rieser 1967; Kono and Barham 1971). This "insulin-like" action of trypsin on fat cells and rat diaphragm tissue can now be seen to result from the cleavage of the extracellular alpha-subunit of the insulin receptor. That cleavage removes a domain of the extracellular insulin receptor alpha-subunit to release its inhibitory control of receptor function (Shoelson et al.1988). At higher concentrations, trypsin cleavage removes the insulin binding site, thus "disarming" the receptor and preventing insulin-triggered signal transduction (Cuatrecasas 1969, 1971). In principle, inflammatory tissue proteinases can thus have a "bidirectional" impact on signalling by the insulin and insulin-like growth factor (IGF) receptors, both to activate and silence signalling. To date, this impact of tissue proteinases on insulin receptor signalling has been largely ignored and has not yet been explored in any depth.

In addition to proteinase-triggered insulin-like responses in adipocytes and striated muscle, it was also clear by the 1970s that thrombin and trypsin could also stimulate cell division in cultured cells by a receptor-like mechanism. These actions mimic the mitogenic receptor-stimulated actions of polypeptide growth factors such as insulin and epidermal growth factor (EGF: Hollenberg and Cuatrecasas 1973, 1975; Burger 1970; Sefton and Rubin 1970; Chen and Buchanan 1975; Carney and Cunningham 1977, 1978). However, the

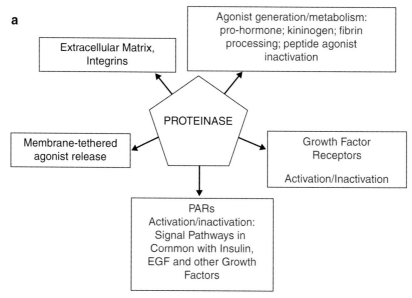

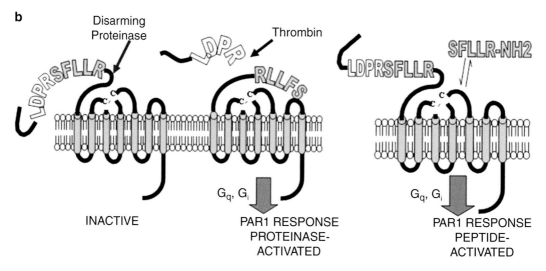

Protease-Activated Receptors, Fig. 1 Mechanisms of proteinase-mediated signaling (**a**) and distinct mechanisms of activation of proteinase-activated receptors (PARs) (**b**). (**a**) (*Upper*): The cartoon shows five different mechanisms by which proteinases can trigger cell signaling, including the generation or degradation of agonist peptides (*top*) and the activation of proteinase-activated receptors (PARs: *bottom*). (**b**) (*Lower*): The cartoon shows the proteolytic triggering of PAR1 (non-activated receptor on *left*) either by revealing a receptor-tethered activating ligand (TL) (*middle*: SFLLR – for PAR1) or by the action of a synthetic peptide with a sequence representing the revealed "tethered ligand" (*right*: SFLLR-NH$_2$)

mechanisms whereby the proteolytic enzymes stimulated mitogenesis were a mystery at the time. As outlined in the sections that follow, it was not until about 10 years later that a "receptor mechanism" for these "hormone-like" actions of proteinases was discovered. That discovery came from a search for the "thrombin receptor," responsible for (1) stimulating human platelet aggregation and (2) triggering mitogenesis in cultured hamster cells. Very unexpectedly, it turned out that thrombin causes these effects by triggering a proteolytically activated G-protein-coupled, receptor (GPCR: PAR) (Rasmussen et al. 1991; Vu et al. 1991; Coughlin 2005). Of the multiple mechanisms whereby proteinases can signal to cells, as outlined in Fig. 1, the main focus of this overview will be on these so-called proteinase-activated receptors or "PARs", for which the Coughlin lab discovered a unique mechanism of proteolytic activation (Fig. 1b) (Vu et al. 1991). The PAR and non-PAR mechanisms of proteinase signalling (Fig. 1) can be seen as complementary mechanisms whereby tissue-derived proteinases can play a role in inflammation.

Thrombin Signalling and Discovery of the Four-Member GPCR Family of PARs

In the search for the "thrombin" receptor, two laboratories independently cloned a G-protein-coupled receptor responsible for the ability of thrombin to stimulate cell replication in cultured hamster cells (Rasmussen et al. 1991) and to activate human platelets (Vu et al. 1991). A key signalling mechanism of the cloned G-protein-coupled receptor observed in both laboratories was its ability to mobilize intracellular calcium when activated by thrombin, via triggering Gq (Rasmussen et al. 1991; Vu et al. 1991). Thrombin was already also known to inhibit adenylyl cyclase, presumably by activating Gi (Vouret-Craviari et al. 1992). Thus, it was realized that thrombin, via its G-protein-coupled receptor (now termed PAR1, with its gene designated, F2R), can signal by coupling to multiple G-proteins. The unusual mechanism discovered for activating this enzyme-activated G-protein-coupled "thrombin" receptor (Vu et al. 1991) involves the proteolytic unmasking of a cryptic N-terminal cell-attached receptor sequence that then acts as a "tethered ligand" (TL), to stimulate signalling (Fig. 1b, middle panel). Of particular note, synthetic peptides with sequences corresponding to the proteolytically exposed N-terminal "TL" amino acid sequence were found to stimulate receptor signalling without the need for receptor proteolysis (Vu et al. 1991) (Fig. 1b, right-hand panel). Thus on its own, the PAR1-derived peptide sequence, SFLLRNPNDKYEPF, can mimic the action of thrombin on human platelets (Vu et al. 1991). However, it was soon realized that the actions of this "thrombin-receptor-activating peptide" (so-called: TRAP) could not stimulate rabbit and rat platelets (Kinlough Rathbone et al. 1993). These data indicated that there had to be a "thrombin receptor" on rodent platelets that is distinct from the one activated by thrombin on human platelets. That observation led us to show pharmacologically, using the principles developed by Ahlquist (1948) for distinguishing adrenoceptors according to their relative agonist potencies, that there is a receptor for the PAR-activating peptides in the vascular endothelium that differs from the one in gastric smooth muscle and human platelets (Hollenberg et al. 1993). The two receptors differed in their responses to either thrombin or trypsin. Unfortunately, we were not able at that time to clone the "second" receptor that was activated by the peptides and trypsin in vascular tissues and that differs from the "thrombin receptor" in human platelets. That vascular endothelial receptor turned out to be proteinase-activated receptor-2, or PAR2/F2RL1, as described in the following section.

The serendipitous cloning of murine PAR2 resulted from querying a genomic library for a substance K receptor (Nystedt et al. 1994). That search found a "new" receptor (now termed, PAR2/F2RL1) that had sequence homology with PAR1, known to be present on human platelets. But, this "second" new PAR was activated by trypsin and not thrombin. Human PAR2/F2RL1 was also cloned shortly after the cloning of murine PAR2 (Nystedt et al. 1995; Bohm et al. 1996). Based on the tethered ligand sequence of PAR2, we were able to deduce that it is the endothelial receptor responsible for the vasorelaxant action caused by trypsin (Al-Ani et al. 1995). The cloning of PARs 1 and 2 heralded the cloning of the other two PAR family members, namely, PAR3/F2RL2 (Connolly et al. 1996; Ishihara et al. 1997) and PAR4/F2RL2 (Kahn et al. 1998; Xu et al. 1998). In their N-terminal extracellular sequences, all of the PARs have a principal serine proteinase-targeted arginine at which cleavage exposes a distinct "tethered ligand" for each

Protease-Activated Receptors, Table 1 The proteinase-activated receptor (PAR) family, their proteolytically unmasked tethered ligands and their non-biased and biased peptide agonists

Canonical and noncanonical PAR cleavage sites and tethered ligand sequences				
Receptor/gene designation	Canonical tethered ligand (human)	Receptor-selective PAR-activating peptides	Noncanonical tethered ligands	Biased PAR-activating peptides
PAR1/F2R	–//SFLLRN–	TFLLR-NH$_2$	MMP1: –// PRSFLLRN– APC: –// NPNDKYEPF–	PRSFLLRN; NPNDKYEPF; YFLLRN
PAR2/F2RL1	–//SLIGKV–	SLIGRL-NH$_2$; 2-furoyl-LIGRL-NH$_2$	NE: –// VLTGKL–	SLAAAA- NH$_2$
PAR3/F2RL2	–//TFRGAP–	TL-derived peptides activate PARs 1 and 2	Not known	Not known
PAR4/F2RL3	–//GYPGQV–	AYPGQV-NH$_2$; AYPGKV-NH$_2$	Not known	Not known

The "canonical" PAR "tethered ligands" resulting from cleavage-activation of the PARs (cleavage site shown as: //) correspond to the sequences of the receptor-selective PAR-activating peptides. The sequences of synthetic receptor-selective "canonical" PAR-activating peptides based on the thrombin (PARs 1 and 4) or trypsin (PAR2)-unmasked tethered ligands are shown along with "noncanonical" tethered ligands unmasked (//) by the inflammation-associated enzymes, neutrophil elastase (NE: PAR 2), matrix metalloproteinase-1 (MMP1: PAR1), or activated protein-C (APC: PAR1). Further, biased synthetic PAR-activating peptides derived from the "noncanonical" enzyme-unmasked tethered ligands are shown in the last column on the right

receptor, as summarized in Table 1. Thrombin is able to activate PARs 1 and 4, both of which are present on human platelets, whereas trypsin activates PAR4 and PAR2, now known to be present in vascular endothelial cells and inflammatory cells, among other sites including intestinal epithelial cells (Nystedt et al. 1994, 1995; Bohm et al. 1996). Activation of PARs 1 and 4 by thrombin and PARs 2 and 4 by trypsin stimulates cell signalling that involves a number of G-proteins (Gq, Gi, G12/13). Moreover, the synthetic peptides based on the revealed tethered ligand sequences can also stimulate comparable signalling via the multiple G-proteins with which each PAR can couple. As shown in Table 1, we have developed PAR-selective-activating peptides to evaluate the impact of signalling by PARs 1, 2, and 4 in a variety of cultured cell and in vivo contexts, without the requirement of proteinases to activate the receptors. PAR3 appears not to signal on its own but to function primarily as a "cofactor" for activation of PAR1 (Nakanishi-Matsui et al. 2000), and peptides derived from its tethered ligand sequence are able to activate both PARs 1 and 2 (Hansen et al. 2004). In certain circumstances, PAR3 is reported to generate a cellular signal (Ostrowska and Reiser 2008), but its general role in regulating tissue function remains to be fully elucidated.

In summary, serine proteinases are now known to regulate tissues by activation of PARs via a "tethered ligand mechanism," in addition to stimulating signalling by the other processes outlined in Fig. 1a. The unique features of the PARs are now well appreciated (Fig. 1b) in terms of: (1) their unusual mechanism of proteolytic activation that involves the unmasking of a tethered ligand (middle panel, Fig. 1b); (2) their activation by receptor-selective synthetic-activating peptides, based on their proteolytically revealed tethered ligand sequences (right panel, Fig. 1b and Table 1); and (3) their susceptibility to activation or inactivation/disarming (left-hand panel, Fig. 1b, left panel, red arrow) by a variety of serine and/or other proteinases that can be found at sites of inflammation. For example, thrombin activates PARs 1 and 4, but not PAR2; trypsin at low concentrations activates PAR2 and PAR4, but at higher concentrations, it disarms PAR1 by removing its tethered ligand (Kawabata et al. 1999). Similarly *Pseudomonas*-derived and neutrophil-derived elastase can disarm trypsin-mediated activation of PAR2 (Dulon et al. 2005; Ramachandran et al. 2011). Thus, a proteinase-like elastase which cleaves C-terminal to the tethered ligand domain sequence shown by the red arrow in the left panel of Fig. 1b (**LDPRSFLLR**) will "disarm" the PAR to prevent its activation by an enzyme (e.g., thrombin or trypsin) that acts by exposing "tethered ligand." The PARs can therefore be seen to have both their circulating proteinase "agonists" that unmask the tethered ligand as well as circulating "antagonist" proteinases that silence the receptors by removing the tethered ligand from the receptor. This "dual action" of proteinases is to be expected of enzymes released at the sites of inflammation.

A number of questions that as yet remain unanswered in detail are: (1) Do the synthetic receptor-selective PAR-activating peptides stimulate signals that are *identical* to those triggered by the proteinase-revealed tethered receptor sequences and (2) which endogenous proteinases regulate PAR function in vivo, e.g., at the sites of inflammation? As will be seen in the following paragraphs, answers to these questions have revealed the ability of the PARs to signal in a "biased" way, as shown for G-protein-coupled receptors in Fig. 2. As summarized in the sections that follow, this "functional selectivity" of PAR signalling is now known to be stimulated either by "biased agonist receptor-activating peptides" (Table 1, column 4) or by inflammation-associated proteinases that cleave the N-terminal PAR sequences at sites differing from the "classical" tethered ligand domain.

Structural Dynamics: The Mobile Receptor Paradigm and "Functional Selectivity" of PAR Signalling

Multiplicity of Receptor Signalling

By the mid-1970s, it was appreciated that receptors like the one for insulin and epidermal growth factor (EGF) as well as G-protein-coupled receptors like the those for ACTH or glucagon could signal via common biochemical pathways (e.g., adenylyl cyclase activation by ACTH and glucagon or common anabolic enzyme pathways stimulated by insulin and EGF). Further, it was realized that a single receptor could simultaneously signal via multiple signal pathways (e.g., the coincident activation of calcium signalling as well as inhibition of adenylyl cyclase by the angiotensin AT1 receptor). This flexibility of signalling by an individual receptor was rationalized in terms of a "floating" or "mobile" receptor model (de Haën 1976; Jacobs and Cuatrecasas 1976). This model proposes that an individual receptor can potentially interact with multiple membrane-localized "effectors." Further, it was proposed that multiple receptors in the same membrane environment could simultaneously regulate the same effector. This "modular" model for "growth factor" receptors (e.g., for insulin, EGF, and platelet-derived growth factor) encompasses the multiple common signalling pathways driven by interactions between the tyrosine-phosphorylated receptors and their multiple SH2-SH3-domain-containing effectors (Jin and Pawson 2012). For G-protein-coupled receptors, this model predicts that different agonists can drive "selective" signalling or "functional selectivity" by "forcing" interactions with a subset of the G-proteins with which an individual receptor can couple. This process is outlined in Fig. 2 and is summarized in depth elsewhere (Kenakin 2012,

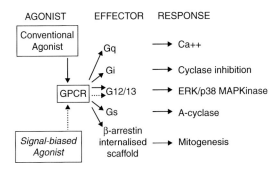

Protease-Activated Receptors, Fig. 2 Functionally selective or biased signaling. In accord with the "floating" or "mobile receptor" model outlined in the text, the diagram shows that a "conventional" agonist can stimulate a G-protein-coupled receptor to affect multiple membrane-localized "G-protein effectors" (e.g., Gq, Gi, G12/13). However, a "functionally selective or biased agonist" stimulates a selective interaction of the receptor with only one (e.g., G12/13) of several available G-protein effectors (Adapted from Hollenberg et al. (2014), with permission)

2013). PARs are included in this paradigm of "biased" signalling or "functional selectivity" (Hollenberg et al. 2014; Zhao et al. 2014). As will be seen, this kind of selective signalling via the PARs is indeed relevant to signalling by inflammation-related proteinases like neutrophil elastase and proteinase-3.

Early on in work with the PAR-activating peptides (PAR-APs) it was appreciated that proteolytic signalling (e.g., by thrombin for PAR1) was not completely duplicated by the receptor-activating peptides. For instance, the PAR-activating peptide agonist for the "thrombin receptor" did not reproduce the mitogenic-MAPKinase-stimulating action of thrombin in hamster fibroblasts (Vouret-Craviari et al. 1992, 1993). Further, studies of mutated PAR1 with changes in its extracellular domains revealed that the receptor sites for binding the PAR1-activating peptides differed significantly from the sites involved in binding thrombin-revealed tethered ligand PAR1 tethered ligand to activate cells. In brief, mutations that affected the binding and action of synthetic receptor-activating peptides did not compromise cell activation caused by binding of the comparable tethered ligand sequence unmasked by thrombin (Blackhart et al. 2000). Similarly, in our own work with PAR2, we found that as a tethered ligand an alanine-substituted sequence (R//SLAAAA—), when unmasked by trypsin cleavage (//) was able to stimulate receptor calcium signalling. However the corresponding synthetic peptide (SLAAAA-amide versus SLIGRL-amide) failed to activate PAR2 calcium signalling. Nonetheless, the same "calcium-inactive" peptide was able to activate PAR2 MAPKinase signalling (Table 2 in Al-Ani et al. 2004; Ramachandran et al. 2009). Thus, for both PARs 1 and 2, signalling by the proteolytically unmasked tethered ligand sequences differs from signalling triggered by the synthetic receptor-activating peptides. Further, the peptide SLAAAA-amide proved to be a "biased PAR2 agonist" that triggers MAPKinase selectively without affecting calcium signalling. Previous work with a PAR1 receptor-activating peptide, YFLLRNP, showed that it acts as a "biased" partial agonist for PAR1, stimulating human platelet shape change, but not aggregation (Rasmussen et al. 1993). Thus, like other G-protein-coupled receptors, the PARs are capable of "functional selectivity" or "biased signalling," as shown schematically in Fig. 2 (Kenakin 2011, 2012, 2013; Kenakin and Miller 2010). What was not determined was whether activation of the PARs by endogenous proteinases, like those released at sites of inflammation, might also be able to stimulate "biased way" PAR signalling. Further, the mechanisms that could account for biased PAR signalling had not been determined. Those issues are dealt with in the following sections:

Proteinase-Triggered Biased PAR Signalling

Based on our data obtained with PAR2 tethered ligand mutants (Al-Ani et al. 2004), we hypothesized that proteolytic activation of PAR2 by enzymes that do not target the cleavage-activation sequence might also signal in a "biased way" like the PAR2-activating peptide, SLAAAA-amide, that can stimulate MAPKinase but not calcium signalling. Since tissue inflammation, in which we knew PAR2 participates, involves the influx of neutrophils, we decided to evaluate the ability of neutrophil proteinases (elastase, proteinase-3, cathepsin-G) to regulate PAR2 signalling. Indeed, as for our observations with *Pseudomonas* elastase (Dulon et al. 2005), the three neutrophil enzymes we studied were all able to "disarm" PAR2, thus preventing trypsin from stimulating PAR2 calcium signalling (see Fig. 1 in Ramachandran et al. 2011). However, while "disarming" PAR2 calcium signalling by trypsin, neutrophil elastase caused an activation of MAPKinase. In contrast, neither cathepsin-G nor proteinase-3 activated PAR2 MAPKinase signalling (Ramachandran et al. 2011). This "biased PAR2 signalling" stimulated by neutrophil elastase results from the proteolytic unmasking of a receptor-activating "tethered ligand" that is downstream from the "canonical" tryptic arginine-/-serine tethered ligand cleavage site (Table 1, column 4) (Ramachandran et al. 2011).

Similar to their impact on PAR2, neutrophil elastase and proteinase-3 also "disarm" thrombin-stimulated PAR1 signalling; but, both enzymes can activate PAR1 MAPKinase signalling (Mihara et al. 2013). The two proteinases do so by cleaving and unmasking "tethered ligand sequences" that are distinct from the one generated by thrombin and also distinct for each of the two neutrophil-derived proteinases (Table 1, column 4). In a similar way, both activated protein-C and matrix metalloproteinase-1 are also able to cleave the N-terminus of PAR1 to stimulate "biased signalling" (Boire et al. 2005; Trivedi et al. 2009; Mosnier et al. 2012; Schuepbach et al. 2012). Yet, none of these inflammation-associated proteinases trigger PAR1-mediated calcium signalling. Thus, depending on the activating proteinases, different cryptic sequences can be revealed in the N-terminal domains of PAR1 so as to drive signalling in distinct biased ways (Table 1). Further, as mentioned, the inflammation-related proteinases which to not of themselves trigger calcium signalling disarm PAR1 for activation by thrombin, preventing calcium signalling altogether. Thus, in the environment of inflammation-related proteinases released by invading neutrophils, PAR1 signalling will be qualitatively different than its signalling stimulated in the setting of coagulation. Whether PAR4 also possesses different cryptic tethered ligand sequences that are differentially unmasked by different proteinases remains to be evaluated. Thus, in an inflammatory environment in which different proteinases are active, this versatility of signalling by the PARs could in principle result in quite different effects, depending on the nature of the inflammation-related proteinases that are released.

PAR Mobility, Internalization, and Signalling: Seeing is Believing

As shown in Fig. 2, the ability of an individual receptor to activate multiple signalling pathways, as described by the "mobile" or "floating" receptor model summarized above (Jacobs and Cuatrecasas 1976; de Haën 1976), depends on its potential interactions with multiple "effectors" in the plane of the membrane (e.g., different G-proteins) and in the case of GPCRs like PAR2, to interact with beta-arrestin, forming an internalized G-protein-independent signalling scaffold (DeFea 2008). We therefore set the goal of visualizing the PARs in the course of enzyme and peptide-stimulated signalling.

To this end, we have generated receptors tagged C-terminally with YFP that can be used in cell expression systems to follow the dynamics of the activation of the PARs. This C-terminally tagged receptor can be used to monitor interactions with other effectors like luciferase-tagged beta-arrestins by measuring bioluminescence energy transfer (BRET) between the interacting partners (Ramachandran et al. 2011). The internalization of the trypsin-activated YFP-tagged receptor can also be monitored (see Fig. 9 in Ramachandran et al. 2011) as was done with PAR2-targeted antibody (see Fig. 1 in Al-Ani et al. 1999). Further, to visualize the activation process of receptor cleavage, we have now generated "dually tagged" PARs as shown for PAR1 in Fig. 3. When intact, the "green" fluorescence of the YFP fluorogen on the C-terminus is present along with the "red" mCherry fluorogen attached to the receptor N-terminus, so that the intact receptor appears "yellow" (lower, left-hand panel, Fig. 3). Activation of PAR1 by thrombin removes the N-terminal mCherry tag, so that the "thrombin-activated" receptor appears "green" due to the remaining C-terminal YFP tag (lower, middle panel, Fig. 3). Thrombin activation causes both the "green"-activated receptor and the released "red" N-terminal domain to internalize (green and red "dots," lower, middle panel, Fig. 3). Activated "green" PAR1 is also clustered at the cell membrane (dots at plasma membrane: lower, middle panel, Fig. 3). In contrast with the activation of PAR1 by thrombin, the dynamics of the peptide-activated receptor differ, in that "yellow" intact receptor becomes internalized (lower, right-hand panel, Fig. 3). Thus, activation of PAR1 by either thrombin or the PAR-activating peptide results in what appears to be comparable levels of receptor internalization; yet, the mitogenic signalling response of cells to thrombin is not duplicated by the action of the PAR1-activating peptide (Vouret-Craviari et al. 1993). Thus, even though internalized, the peptide-activated receptor must interact with different "effectors" than the thrombin-activated receptor.

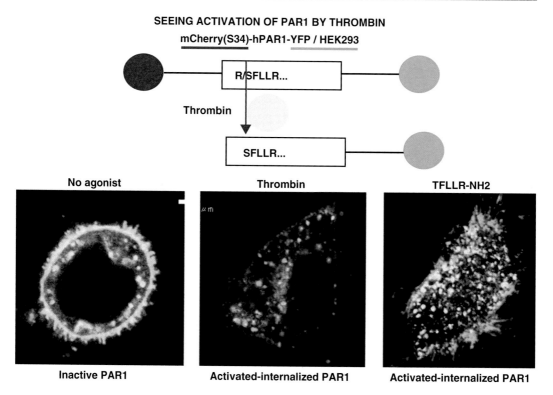

Protease-Activated Receptors, Fig. 3 Imaging activation of PAR1 with distinct labels on its N- and C-terminus. *As shown in the upper scheme*, PAR1 labeled with an N-terminal, mCherry, and a C-terminal YFP appears "yellow" when expressed in HEKs in its non-activated state. Thrombin cleavage activation of the receptor releases the mCherry tag so that the remaining C-terminally YFP-tagged activated receptor appears "green." *The lower left panel* shows the intact receptor expressed in human embryonic kidney (*HEK*) cells that appears largely "yellow" at the cell surface. When activated by thrombin, the mCherry tag is released from the receptor and the activated YFP-retaining receptor appears as "green" internalized *dots*. The released mCherry tag is also internalized (*red dots*). When activated by the PAR1-activating peptide, TFLLR-NH$_2$, both tags are retained on the activated receptor that internalizes as "yellow" *dots* (Dually labeled PAR1: Adapted from Mihara et al. (2013), with permission)

Those differential interactions remain to be determined.

Visualizing the Distinct Dynamics of Proteinase-Stimulated Biased PAR Signalling Caused by Inflammation-Related Proteinases

Using the imaging approach to monitor PAR cleavage and activation of the dually tagged receptor, we next compared the dynamics of PAR1 cleavage either by its "traditional" agonist, thrombin, which stimulates both MAPKinase and calcium signalling or by the "biased" agonist, neutrophil elastase (NE), which triggers PAR1-mediated MAPKinase activation, but not calcium signalling (discussed above for PAR2 and shown also for PAR1: Mihara et al. 2013). As illustrated in Fig. 4, dually tagged inactive PAR1 (panel a, Fig. 4), which appears predominantly "yellow" at the cell surface (panel b, Fig. 4) is rapidly internalized when activated by thrombin (green internal dots, panel c, Fig. 4). However, when exposed to neutrophil elastase, a large proportion of the cleaved "green" receptor remains at the cell surface in the course of activating MAPKinase signalling (panel d, Fig. 4). Thus, it has become possible to follow the unique cleavage-activation of PARs in an inflammatory milieu by

Protease-Activated Receptors, Fig. 4 Visualizing biased signaling by neutrophil elastase (NE). (**a**) The scheme on the upper left illustrates the "colors" of the intact receptor ("yellow"), the N-terminus-cleaved receptor ("green"), and the cleaved/released N-terminus ("red"). The untreated, non-activated dually tagged intact receptor expressed in HEK cells, as per Fig. 6, appears mainly "yellow" (panel **b**). The thrombin-activated N-terminally cleaved receptor appears "green" (panel **c**), and the proteolytically released N-terminal fragment appears "red" (panels **c**, **d**). Thrombin activation triggers the internalization of "green" mCherry-free PAR1 (panel **c**), as per Fig. 3. However, cleavage activation of PAR1 by neutrophil elastase (*NE*) while generating an mCherry-free "green" receptor does not trigger internalization of the receptor that remains "green" at the cell surface (panel **d**) (Adapted from Mihara et al. (2013), with permission)

tracking receptor N-terminal cleavage and location. This tracking is also possible with the use of different receptor-targeted antisera that visualize the intact versus cleaved receptor (Wang et al. 2003).

Taken together, our data show that like other G-protein-coupled receptors, PARs can exhibit biased signalling when activated either by receptor-selective "biased" activating peptides or by proteinases that cleave at "noncanonical" sites to unmask "biased" tethered ligand sequences (Table 1). Further, the results indicate that topographically, biased signalling appears to originate from distinct membrane locations (e.g., plasma membrane vs. internalized signalling complexes), where the PARs can, as per the mobile receptor paradigm outlined in Fig. 2, interact with distinct signalling "effectors." In the setting of an inflammatory response, the differential cleavage of PARs by released microenvironment proteinases can thus trigger complex responses: (1) activation of multiple G-protein-coupled responses, in step with receptor internalization; (2) biased signalling via cleavage of the PARs at "noncanonical" activation sites without receptor internalization; or (3) silencing of PAR signalling by disarming the receptors, leaving the receptor at the cell surface unable to respond to other proteinases.

Pathophysiology: PARs as Sensors for Autocrine-Paracrine Proteinase-Mediated Signalling in Cancer and Inflammation

Visualizing PARs in Receptor-Expressing Prostate Cancer-Derived Epithelial Cells

When PAR2 was first cloned, its expression was observed in a variety of novel locales, including the eye and prostate gland (Nystedt et al. 1994, 1995; Bohm et al. 1996). Further, we subsequently became aware of a rat model of noninfectious prostatitis, wherein the exposure of the prostate epithelium to an ethanol-dinitrobenzene sulfonate stimulus generates many of the hallmarks of inflammation (Lang et al. 2000). Because the prostate is well recognized as a source of the kallikrein-related peptidases (KLK) family of serine proteinases, we proposed that the KLKs might themselves be upregulated in the rat prostatitis model and might be able to regulate PAR activity. We thus suspected that the prostate might potentially host an autocrine-paracrine system for proteinase-regulated PAR activation that might relate to inflammation of the prostate and to prostate cancer. Indeed, we found that the KLKs can regulate PAR activation and that the KLK family members can be considered as important modulators of cell signalling in the setting of dermatitis, prostatitis, prostate cancer, and other inflammatory pathologies (Oikonomopoulou et al. 2006, 2010; Stanton et al. 2013; Hollenberg 2014). With this kind of role in mind, we evaluated the activation status of PAR1 when expressed as a dually tagged receptor in prostate cancer-derived PC3 cells, which originated from a metastatic prostate cancer (Fig. 5a, lower right micrography). To our surprise, the PC3-expressed dually tagged PAR1 was localized as a "green" receptor at the cell surface (Fig. 5a, lower right panel) instead of its "yellow" appearance as an intact receptor when expressed in an embryonic human kidney HEK cell (Fig. 5a, lower left panel). Our new data thus suggest that PC3 cells secrete PAR-regulating proteinases that can in an autocrine-paracrine way regulate PC3 PAR signalling. This kind of "autocrine" signalling may play a role in the settings of cancer and inflammatory diseases. A comparable process very likely also plays a part in infection-driven inflammation, as described for a *Pseudomonas* model of prostatitis (Nelson et al. 2009) or in a model of noninfectious prostatitis (Stanton et al. 2013).

Detecting PAR-Regulating Proteinases Produced by Prostate Cancer Cells

To quantify the presence of PAR-cleaving proteinases secreted by both normal and tumor-derived cells in culture, we have designed a new approach in which PARs 1 and 2 are expressed with an N-terminal chromogenic tag in a background PAR-deficient indicator cell (Fig. 5b). When exposed to supernatants from either normal or tumor-derived cells (e.g., PC3 or DU145 prostate cancer cells), a release of the PAR-attached chromogen into the supernatant serves as an index of the production of PAR-cleaving proteinases. With this assay, we have been able to detect multiple PAR-cleaving proteinases produced constitutively and released into the growth medium by prostate cancer-derived cells (Liu et al. 2014). These preliminary data underline the relevance of proteinases and their PAR targets as important localized regulators of tissue function in diseases where the PARs can serve as "sentinels" of tissue inflammation as well as regulators of tumor cell growth and invasion. As summarized in the following sections, this autocrine-paracrine proteinase-mediated process most likely plays a key role in inflammatory processes in vivo.

PARs and Models of Inflammatory Disease

In view of the hypothesis outlined above, proteinases released in an inflammatory microenvironment can act in an autocrine-paracrine way to regulate local tissue function. We have therefore explored a number of rodent inflammation models to evaluate the role(s) that PARs and their activating proteinases might play. As summarized in Table 2 and elaborated in more detail elsewhere (Ramachandran and Hollenberg 2008; Ramachandran et al. 2012), PAR activation is involved in a range of inflammatory processes ranging from arthritis to colitis and central nervous system neurodegeneration (Figs. 6, 7, and 8).

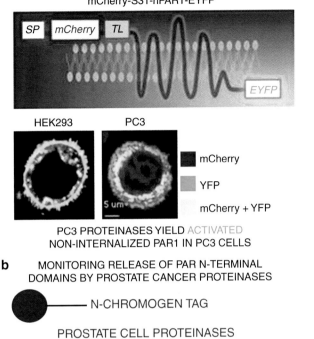

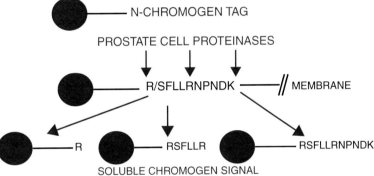

Protease-Activated Receptors, Fig. 5 Autocrine cleavage of PAR1 in prostate PC3 cancer cells and monitoring PAR cleavage by microenvironment inflammatory proteinases. (**a**) Autocrine cleavage. Dually tagged PAR1 (N-terminal, mCherry; C-terminal eYFP: upper cartoon, panel **a**) was expressed in HEK293 (*lower left*, panel **a**) and PC3 cells (*lower right*, panel **a**). Upon expression in HEK293 cells, the intact receptor appears "yellow" (*lower left*, panel **a**). However, when expressed in the PC3 cells, proteinases in the PC3 cell supernatant constitutively remove the N-terminus from PAR1 resulting in its "green" appearance at the cell surface, contrasting with its intact "yellow" appearance in the submembrane region prior to externalization. (**b**) Monitoring PAR-cleaving proteinases released into the supernatant by PC3 cells. PAR1 and PAR2, each N-terminally tagged with a fluorescent chromogen, as shown for PAR1 in the upper cartoon, were expressed separately in a background PAR-free "indicator cell." When the chromogen-tagged PAR expressed in the indicator cell is exposed to supernatant medium harvested from cultured PC3 cells, the N-terminal chromogen is released into the supernatant as a sensitive index of released PAR-cleaving proteinases. The scale for the micrographs (5 mm) is shown below the PC3 cell (Adapted from Liu et al. 2014)

Of particular note, work with PAR2-null mice, complemented by PAR2 antagonists and proteinase inhibitors, indicates that this receptor is a fruitful therapeutic target for arthritis and colitis (Ferrell et al. 2003; Kelso et al. 2006; Lohman et al. 2012a, b; Cenac et al. 2002; Nguyen et al. 2003; Hansen et al. 2005; Hyun et al. 2008). Along with PAR2, PAR4 also mediates the pain component of the inflammatory response of the inflamed joint (Russell

Protease-Activated Receptors, Table 2 PARs and models of inflammatory disease

Disease model	PAR involvement	References
Paw edema/peripheral inflammation and pain	PAR1, PAR2	Vergnolle et al. 1999, 2009; Steinhoff et al. 2000; Sevigny et al. 2011
Colitis	PAR2	Kong et al. 1997; Cenac et al. 2002; Hansen et al. 2005; Motta et al. 2011, 2012; Lohman et al. 2012b
Arthritis and joint pain	PAR2, PAR4	Ferrell et al. 2003; Kelso et al. 2006; Russell et al. 2010, 2012; Lohman et al. 2012a;
Central nervous system inflammation	PAR1, PAR2	Boven et al. 2003; Noorbakhsh et al. 2003, 2006
Pain: nociception and analgesia	PAR1; PAR2;	Asfaha et al. 2002; Martin et al. 2009; Amadesi et al. 2006; Poole et al. 2013; Cattaruzza et al. 2014;
Multiple sclerosis	PAR2	Noorbakhsh et al. 2006
Asthma	PAR2	Ebeling et al. 2005, 2007; Adam et al. 2006; Arizmendi et al. 2011; Polley et al. 2013; Davidson et al. 2013

et al. 2010, 2012), possibly via a neurogenic mechanism (Steinhoff et al. 2000). In the setting of intestinal inflammation, the overexpression of the elastase-selective enzyme inhibitor, elafin, is able to attenuate colitis in a murine model (Motta et al. 2011). Thus, the production of elafin by engineered bacteria is being considered as a therapeutic approach for intestinal inflammation (Motta et al. 2012). Additional data obtained by us and others indicates that blocking activation of the PARs may also be a therapeutic strategy for inflammation both in the periphery (Vergnolle et al. 1999; Sevigny et al. 2011) as well as in the central nervous system (CNS), where PARs are present on both neuronal and non-neuronal cells (Noorbaksh et al. 2003; Luo et al. 2007). In the CNS, PAR1 activation generates inflammation (Boven et al. 2003; Noorbakhsh et al. 2006). In this context, a mouse model of multiple sclerosis points to a role of PAR2 in the demyelination that leads to the behavioral pathology (Noorbakhsh et al. 2003, 2006).

PARs as Neurogenic Inflammatory Mediators of Pain in the Periphery

When evaluating the inflammatory effects of PAR2 activation in the rat paw, we found that sensory nerve-derived agonists play a key role (Steinhoff et al. 2000). This neurogenic component of the inflammatory response, due to the activation of sensory nerves by the PARs, singled out the role that PARs can play in neurons. This link between PAR2 activation and sensory nerve activation is now known to involve cross-talk between PAR2 and the transient receptor potential vanilloid family of receptors, TRPV1 (the capsaicin receptor, sensing heat) and TRPV4 (senses blood flow and pressure). Thus, PAR2 can (1) enhance nociception by sensitizing TRPV1 via a kinase-C-epsilon phosphorylation mechanism (Amadesi et al. 2006) and (2) sustain inflammatory signalling by increasing cellular calcium influx via TRPV4. This influx results from PAR2-stimulated phosphorylation of a tyrosine residue on TRPV4 (Poole et al. 2013). In contrast, PAR1 activation appears to play an antinociceptive role (Asfaha et al. 2002). Thus, the pro-nociceptive actions of PAR2 in peripheral tissues, that can be activated by proteinases in an inflammatory milieu, are mediated via cross-talk between PAR2 and the TRPV channels, TRPV1, TRPV4, and TRPA1 (summarized in Poole et al. 2013). As a consequence, PAR2 antagonists or proteinase inhibitors that target PAR-activating enzymes may prove of therapeutic utility to treat inflammatory pain and inflammation (Cattaruzza et al. 2014). On the other hand, activating PAR1 may prove to be analgesic (Asfaha et al. 2002; Martin et al. 2009).

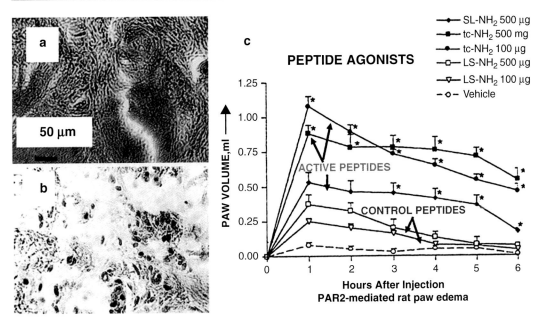

Protease-Activated Receptors, Fig. 6 PAR2-induced inflammation of the rat paw. The inflammatory effects of activating PAR2 in an intact tissue was evaluated by the direct administration into the rat paw of either the PAR2-activating peptide, SLIGRL-NH$_2$, or its reverse sequence PAR2-inactive analogue, LRGILS-NH$_2$. (**a, b**) Tissue histology: at 6 h after the injection of the control PAR-inactive peptide (LRGILS-NH2) (**a**) or the PAR2-activating peptide SLIGRL-NH$_2$ (**b**), animals were sacrificed and the PAR2-treated paws were fixed and stained to evaluate tissue morphology. The scale bar in panel **a**, for both micrographs = 50 µm. (**c**) Paw oedema: the time course of swelling after administration of the different doses of PAR2-activating peptides into the paw shown in the *inset* was measured using a hydroplethismometer. Paw volume was measured for up to 6 h at which time animals were sacrificed for obtaining the hindpaw tissue for fixation and conducting the histochemical analysis. Swelling was maximal at about 1 h for the active peptides (*red font-designated arrows*) relative to the PAR2-inactive peptides or peptide-free buffer (*blue font-designated arrows*) (Adapted from Vergnolle et al. (1999), with permission)

Which CNS Proteinases Might Trigger PAR-Induced Neuroinflammation?

Since both PARs 1 and 2 are implicated in peripheral and central nervous system neuroinflammation, a key question is: Which enzymes might regulate the PARs in the CNS? Proteinases of the coagulation cascade (thrombin, activated protein-C) are likely candidates, originating both from the circulation as a result of stroke or head trauma and from the CNS tissues themselves, in which (pro)thrombin mRNA has been detected. Another interesting PAR-targeting candidate was identified some time ago as a trypsin-related serine proteinase derived from the rat spinal cord (Scarisbrick et al. 1997). This enzyme, designated at that time as rat "myelencephalon-specific protease (MSP)," has turned out to belong to the kallikrein-related peptidase family of serine proteinases, namely, KLK6 (Bernett et al. 2002). This enzyme has been found to exacerbate glutamate neurotoxicity via activation of both PAR1 and PAR2 (Yoon et al. 2013). Of direct relevance to this neurotoxic action of KLK6 involving the PARs and pertinent to our finding of a role for PAR2 in multiple sclerosis, our studies with KLK6 show that this enzyme, found in MS lesions both in humans and rodents (Scarisbrick et al. 2002), can target the cleavage-activation site of both PAR1 and PAR2 (Oikonomopoulou et al. 2006). Thus, the kallikrein-related peptidases which can target the PARs in the CNS can be seen as important inflammatory mediators of neurotoxicity and axonopathies in the CNS (Burda et al. 2013; Radulovic et al. 2013. Therefore, as proposed for the prostate cancer cells described above, we suggest that in the CNS, a "paracrine-autocrine" loop can regulate tissue function involving the

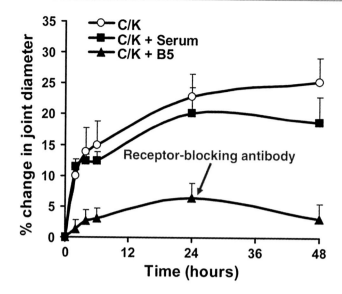

Protease-Activated Receptors, Fig. 7 PAR2-blocking antibody attenuates blocking adjuvant-induced arthritis with a PAR2-blocking antibody that prevents proteolytic activation. Adjuvant-induced arthritis was caused by the intra-articular injection of a carrageenan/kaolin (*C/K*) suspension with or without prior intra-articular administration of a PAR2-targeted antibody (B5) that occludes the receptor cleavage activation site, thereby blocking proteinase activation or PAR2. The adjuvant-induced increase in joint diameter over a 2-day time period was markedly diminished by pretreatment with the cleavage-blocking B5 antiserum (*C/K + B5*), but not by nonimmune rabbit serum (*CK+ serum*) (Adapted from Kelso et al. (2006), with permission)

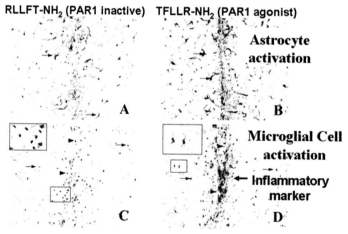

Protease-Activated Receptors, Fig. 8 CNS inflammation caused by PAR1 activation. A PAR1-selective activating peptide (TFLLR-NH$_2$: panels **b** and **d**) or its reverse sequence PAR1-inactive peptide (RLLFT-NH2: panels **a** and **c**) were administered stereotactically into the striatum of anesthetized CD-1 male mice. After 7 days, the animals were sacrificed and fixed brain sections were prepared for immunohistochemical analysis to detect activation of astrocyte and microglial cells (brown staining). Staining reactivity indicative of neuroinflammation was observed in the PAR1-activating peptide-treated tissue (**c** and **d**) but was minimal in the reverse peptide-treated tissue (**a** and **b**) (Adapted from Boven et al. (2003), with permission)

secretion-activation of KLK6 or other proteinases from glial and/or neuronal cells, which in turn can drive inflammation and neurodegeneration in part by activating PARs 1 and 2 in the microenvironment.

Allergenic Pathogen Proteinases, PARs, and Asthma

For some time, it has been known that allergens like the ones from dust mite, *Dermatophagoides pteronyssinus* (Der-P), cockroaches, or *Alternaria alternata* mold contain enzymes that are capable of activating PARs (Adam et al. 2006; Arizmendi et al. 2011; Davidson et al. 2013; Boitano et al. 2011). This principle, whereby pathogen proteinases use PAR signalling to affect mammalian hosts, is widespread, ranging from asthma allergens to trypanosomes (Grab et al. 2009). Biochemically, we have been able to identify three cockroach allergen serine proteinases and one *Alternaria* serine proteinase, each of which has distinct enzymatic properties and each of which is capable of activating PAR2 (Polley et al. 2012). Collectively, the mixture of enzymes in the cockroach allergen plays a role in mucosal allergic sensitization in a mouse model of asthma via activation of PAR2 (Arizmendi et al. 2011). Thus, by neutralizing the trypsin-like proteinase activity in the cockroach allergen with soya trypsin inhibitor, allergic sensitization (increased neutrophil and eosinophil infiltration) is markedly decreased, and likewise, administering a PAR2-blocking antibody also attenuates inflammatory cell migration into the bronchoalveolar lavage fluid (Arizmendi et al. 2011). Of note, sensitization by the cockroach allergen is reduced in PAR2-null mice. A comparable result is observed by blocking PAR2 activation with an activation-cleavage site-targeted antibody in a murine sensitization protocol using dust mite Der-P allergen (Davidson et al. 2013). Since the three proteinases present in the cockroach allergen do not stimulate an immune response directly, it appears that the proteinase-mediated activation of PAR2 caused by the allergen enzymes works in a synergistic way with other antigens in the allergen to cause sensitization. The mechanism whereby PAR2 activation causes its synergistic sensitizing action has yet to be determined. Nonetheless, agents targeting either allergen-triggered PAR2 signalling or the allergen proteinases(s) that activate PAR2 would appear to be of value in the treatment of allergenic sensitization (Vliagoftis and Forsythe 2008).

Taken together, the information summarized in the above sections indicates that PARs and their activating proteinases can be involved in a wide variety of inflammatory pathologies. The key will be to move these observations from the successes observed in the animal models to their implementation in humans (Ramachandran et al. 2012).

PARs as Therapeutic Targets

Given the roles that the PARs can play in cardiovascular and inflammatory diseases, PARs have become attractive targets for the development of therapeutically useful antagonists. That said, the unique mechanism of receptor activation involving a proteinase-unmasked tethered ligand has proved a challenge in terms of developing clinically effective antagonists (Ramachandran et al. 2012). Although thrombin inhibitors have proved to be very effective in controlling thrombosis, their risk of bleeding complications spurred the development of PAR1 antagonists for the platelet-targeted control of thrombosis. This strategy would not interfere with the coagulation pathway. To this end, a number of PAR1 antagonists have been developed, as reviewed elsewhere (see Fig. 3 in Ramachandran et al. 2012). However, to date only vorapaxar, despite its association with increased bleeding in individuals with a previous history of transient ischemic attack or stroke, has been approved for the secondary prevention of atherothrombotic events in individuals after a heart attack or in patients with a history of peripheral artery disease, who are also being treated with standard antiplatelet therapy (French et al. 2015; reviewed by Baker et al. 2014). PAR4 remains an attractive target for antiplatelet therapy, especially in a subpopulation of black individuals (Edelstein et al. 2014). However, clinically successful PAR4 antagonists have not yet been developed. Similarly, as outlined above, preclinical data indicate that PAR2 is a fruitful target for inflammatory diseases like arthritis and colitis, but no clinically successful antagonists are yet available. Given the key pathophysiological roles that the PARs 2 and 4 can play, one can look forward optimistically to the development of PAR2 and PAR4 antagonists with clinical utility in the future.

Acknowledgements Work described in this overview was supported in large part by operating grants from the Canadian Institutes of Health Research, as well as by funds from Prostate Cancer Canada and from the Calgary Ride for Dad. I am most grateful for the contributions of my co-authors/collaborators listed along with my name in the reference section. These are the ones included in the collective "we" used in the text to denote those who made many of the observations described. Those individuals, to a person, have been essential contributors for the discoveries that have been made collectively over the years to understand the molecular pharmacology and inflammatory pathophysiology of the PARs and their activating proteinases. Whatever progress has been made is a tribute to the cooperative Inflammation Research Network atmosphere in which my colleagues and I are able to work.

Cross-References

▶ Asthma

References

Adam, E., Hansen, K. K., Astudillo Fernandez, O., Coulon, L., Bex, F., Duhant, X., et al. (2006). The house dust mite allergen Der p 1, unlike Der p 3, stimulates the expression of interleukin-8 in human airway epithelial cells via a proteinase-activated receptor-2-independent mechanism. *Journal of Biological Chemistry, 281*(11), 6910–6923.

Adams, M. N., Ramachandran, R., Yau, M. K., Suen, J. Y., Fairlie, D. P., Hollenberg, M. D., et al. (2011). Structure, function and pathophysiology of protease activated receptors. *Pharmacology and Therapeutics, 130*(3), 248–282.

Ahlquist, R. P. (1948). A study of the adrenotropic receptors. *American Journal of Physiology, 153*(3), 586–600.

Al-Ani, B., Saifeddine, M., & Hollenberg, M. D. (1995). Detection of functional receptors for the proteinase-activated-receptor-2-activating polypeptide, SLIGRL-NH2, in rat vascular and gastric smooth muscle. *Canadian Journal of Physiology and Pharmacology, 73*(8), 1203–1207.

Al-Ani, B., Saifeddine, M., Kawabata, A., Renaux, B., Mokashi, S., & Hollenberg, M. D. (1999). Proteinase-activated receptor 2 (PAR2): Development of a ligand-binding assay correlating with activation of PAR2 by PAR1- and PAR2-derived peptide ligands. *Journal of Pharmacology and Experimental Therapeutics, 290*(2), 753–760.

Al-Ani, B., Hansen, K. K., & Hollenberg, M. D. (2004). Proteinase-activated receptor-2: Key role of amino-terminal dipeptide residues of the tethered ligand for receptor activation. *Molecular Pharmacology, 65*(1), 149–156.

Amadesi, S., Cottrell, G. S., Divino, L., Chapman, K., Grady, E. F., Bautista, F., et al. (2006). Protease-activated receptor 2 sensitizes TRPV1 by protein kinase Cepsilon- and A-dependent mechanisms in rats and mice. *Journal of Physiology, 575*(Pt 2), 555–571.

Arizmendi, N. G., Abel, M., Mihara, K., Davidson, C., Polley, D., Nadeem, A., et al. (2011). Mucosal allergic sensitization to cockroach allergens is dependent on proteinase activity and proteinase-activated receptor-2 activation. *Journal of Immunology, 186*(5), 3164–3172.

Asfaha, S., Brussee, V., Chapman, K., Zochodne, D. W., & Vergnolle, N. (2002). Proteinase-activated receptor-1 agonists attenuate nociception in response to noxious stimuli. *British Journal of Pharmacology, 135*(5), 1101–1106.

Baker, N. C., Lipinski, M. J., Lhermusier, T., & Waksman, R. (2014). Overview of the 2014 Food and Drug Administration Cardiovascular and Renal Drugs Advisory Committee meeting about vorapaxar. *Circulation, 130*(15), 1287–1294.

Bernett, M. J., Blaber, S. I., Scarisbrick, I. A., Dhanarajan, P., Thompson, S. M., & Blaber, M. (2002). Crystal structure and biochemical characterization of human kallikrein 6 reveals that a trypsin-like kallikrein is expressed in the central nervous system. *Journal of Biological Chemistry, 277*(27), 24562–24570.

Blackhart, B. D., Ruslim-Litrus, L., Lu, C. C., Alves, V. L., Teng, W., Scarborough, R. M., et al. (2000). Extracellular mutations of protease-activated receptor-1 result in differential activation by thrombin and thrombin receptor agonist peptide. *Molecular Pharmacology, 58*(6), 1178–1187.

Bohm, S. K., Kong, W., Bromme, D., Smeekens, S. P., Anderson, D. C., Connolly, A., et al. (1996). Molecular cloning, expression and potential functions of the human proteinase-activated receptor-2. *Biochemical Journal, 314*(Pt 3), 1009–1016.

Boire, A., Covic, L., Agarwal, A., Jacques, S., Sherifi, S., & Kuliopulos, A. (2005). PAR1 is a matrix metalloprotease-1 receptor that promotes invasion and tumorigenesis of breast cancer cells. *Cell, 120*(3), 303–313.

Boitano, S., Flynn, A. N., Sherwood, C. L., Schulz, S. M., Hoffman, J., Gruzinova, I., et al. (2011). Alternaria alternata serine proteases induce lung inflammation and airway epithelial cell activation via PAR2. *American Journal of Physiology - Lung Cellular and Molecular Physiology, 300*(4), L605–L614.

Boven, L. A., Vergnolle, N., Henry, S. D., Silva, C., Imai, Y., Holden, J., et al. (2003). Up-regulation of proteinase-activated receptor 1 expression in astrocytes during HIV encephalitis. *Journal of Immunology, 170*(5), 2638–2646.

Burda, J. E., Radulovic, M., Yoon, H., & Scarisbrick, I. A. (2013). Critical role for PAR1 in kallikrein 6-mediated oligodendrogliopathy. *Glia, 61*(9), 1456–1470.

Burger, M. M. (1970). Proteolytic enzymes initiating cell division and escape from contact inhibition of growth. *Nature, 227*(5254), 170–171.

Carney, D. H., & Cunningham, D. D. (1977). Initiation of check cell division by trypsin action at the cell surface. *Nature, 268*(5621), 602–606.

Carney, D. H., & Cunningham, D. D. (1978). Transmembrane action of thrombin initiates chick cell division. *Journal of Supramolecular Structure, 9*(3), 337–350.

Cattaruzza, F., Amadesi, S., Carlsson, J. F., Murphy, J. E., Lyo, V., Kirkwood, K., et al. (2014). Serine proteases and protease-activated receptor 2 mediate the proinflammatory and algesic actions of diverse stimulants. *British Journal of Pharmacology, 171*(16), 3814–3826.

Cenac, N., Coelho, A. M., Nguyen, C., Compton, S., Andrade-Gordon, P., MacNaughton, W. K., et al. (2002). Induction of intestinal inflammation in mouse by activation of proteinase-activated receptor-2. *American Journal of Pathology, 161*(5), 1903–1915.

Chen, L. B., & Buchanan, J. M. (1975). Mitogenic activity of blood components. I. Thrombin and prothrombin. *Proceedings of the National Academy of Sciences of the United States, 72*(1), 131–135.

Chretien, M. (2012). My road to Damascus: How I converted to the prohormone theory and the proprotein convertases. *Biochemistry and Cell Biology = Biochimie et Biologie Cellulaire, 90*(6), 750–768.

Connolly, A. J., Ishihara, H., Kahn, M. L., Farese, R. V., Jr., & Coughlin, S. R. (1996). Role of the thrombin receptor in development and evidence for a second receptor. *Nature, 381*(6582), 516–519.

Coughlin, S. R. (2005). Protease-activated receptors in hemostasis, thrombosis and vascular biology. *Journal of Thrombosis and Haemostasis, 3*(8), 1800–1814.

Cuatrecasas, P. (1969). Interaction of insulin with the cell membrane: The primary action of insulin. *Proceedings of the National Academy of Sciences of the United States of America, 63*(2), 450–457.

Cuatrecasas, P. (1971). Properties of the insulin receptor of isolated fat cell membranes. *Journal of Biological Chemistry, 246*(23), 7265–7274.

Davidson, C. E., Asaduzzaman, M., Arizmendi, N. G., Polley, D., Wu, Y., Gordon, J. R., et al. (2013). Proteinase-activated receptor-2 activation participates in allergic sensitization to house dust mite allergens in a murine model. *Clinical and Experimental Allergy, 43*(11), 1274–1285.

de Haën, C. (1976). The non-stoichiometric floating receptor model for hormone sensitive adenylyl cyclase. *Journal of Theoretical Biology, 58*(2), 383–400.

Defea, K. (2008). Beta-arrestins and heterotrimeric G-proteins: Collaborators and competitors in signal transduction. *British Journal of Pharmacology, 153* (Suppl 1), S298–S309.

Dulon, S., Leduc, D., Cottrell, G. S., D'Alayer, J., Hansen, K. K., Bunnett, N. W., et al. (2005). Pseudomonas aeruginosa elastase disables proteinase-activated receptor 2 in respiratory epithelial cells. *American Journal of Respiratory Cell and Molecular Biology, 32*(5), 411–419.

Ebeling, C., Forsythe, P., Ng, J., Gordon, J. R., Hollenberg, M., & Vliagoftis, H. (2005). Proteinase-activated receptor 2 activation in the airways enhances antigen-mediated airway inflammation and airway hyperresponsiveness through different pathways. *Journal of Allergy and Clinical Immunology, 115*(3), 623–630.

Ebeling, C., Lam, T., Gordon, J. R., Hollenberg, M. D., & Vliagoftis, H. (2007). Proteinase-activated receptor-2 promotes allergic sensitization to an inhaled antigen through a TNF-mediated pathway. *Journal of Immunology, 179*(5), 2910–2917.

Edelstein, L. C., Simon, L. M., Lindsay, C. R., Kong, X., Teruel-Montoya, R., Tourdot, B. E., et al. (2014). Common variants in the human platelet PAR4 thrombin receptor alter platelet function and differ by race. *Blood, 124*(23), 3450–3458.

Ferrell, W. R., Lockhart, J. C., Kelso, E. B., Dunning, L., Plevin, R., Meek, S. E., et al. (2003). Essential role for proteinase-activated receptor-2 in arthritis. *Journal of Clinical Investigation, 111*(1), 35–41.

French, S. L., Arthur, J. F., Tran, H. A., & Hamilton, J. R. (2015). Approval of the first protease-activated receptor antagonist: Rationale, development, significance, and considerations of a novel anti-platelet agent. *Blood Reviews, 29*(3), 179–189.

Grab, D. J., Garcia-Garcia, J. C., Nikolskaia, O. V., Kim, Y. V., Brown, A., Pardo, C. A., et al. (2009). Protease activated receptor signaling is required for African trypanosome traversal of human brain microvascular endothelial cells. *PLoS Neglected Tropical Diseases, 3*(7), e479.

Hansen, K. K., Saifeddine, M., & Hollenberg, M. D. (2004). Tethered ligand-derived peptides of proteinase-activated receptor 3 (PAR3) activate PAR1 and PAR2 in Jurkat T cells. *Immunology, 112*(2), 183–190.

Hansen, K. K., Sherman, P. M., Cellars, L., Andrade-Gordon, P., Pan, Z., Baruch, A., et al. (2005). A major role for proteolytic activity and proteinase-activated receptor-2 in the pathogenesis of infectious colitis. *Proceedings of the National Academy of Sciences of the United States of America, 102*(23), 8363–8368.

Hollenberg, M. D. (2014). KLKs and their hormone-like signaling actions: A new life for the PSA-KLK family. *Biological Chemistry, 395*(9), 915–929.

Hollenberg, M. D., & Cuatrecasas, P. (1973). Epidermal growth factor: Receptors in human fibroblasts and modulation of action by cholera toxin. *Proceedings of the National Academy of Sciences of the United States of America, 70*(10), 2964–2968.

Hollenberg, M. D., & Cuatrecasas, P. (1975). Insulin and epidermal growth factor. Human fibroblast receptors related to deoxyribonucleic acid synthesis and amino acid uptake. *Journal of Biological Chemistry, 250*(10), 3845–3853.

Hollenberg, M. D., Laniyonu, A. A., Saifeddine, M., & Moore, G. J. (1993). Role of the amino- and carboxyl-terminal domains of thrombin receptor-derived polypeptides in biological activity in vascular endothelium and gastric smooth muscle: Evidence for receptor subtypes. *Molecular Pharmacology, 43*(6), 921–930.

Hollenberg, M. D., Mihara, K., Polley, D., Suen, J. Y., Han, A., Fairlie, D. P., et al. (2014). Biased signalling and proteinase-activated receptors (PARs): Targeting inflammatory disease. *British Journal of Pharmacology, 171*(5), 1180–1194.

Hyun, E., Andrade-Gordon, P., Steinhoff, M., & Vergnolle, N. (2008). Protease-activated receptor-2 activation: A major actor in intestinal inflammation. *Gut, 57*(9), 1222–1229.

Ishihara, H., Connolly, A. J., Zeng, D., Kahn, M. L., Zheng, Y. W., Timmons, C., et al. (1997). Protease-activated receptor 3 is a second thrombin receptor in humans. *Nature, 386*(6624), 502–506.

Jacobs, S., & Cuatrecasas, P. (1976). The mobile receptor hypothesis and "cooperativity" of hormone binding. Application to insulin. *Biochimica et Biophysica Acta, 433*(3), 482–495.

Jin, J., & Pawson, T. (2012). Modular evolution of phosphorylation-based signalling systems. *Philosophical Transactions of the Royal Society of London. Series B: Biological Sciences, 367*(1602), 2540–2555.

Kahn, M. L., Zheng, Y. W., Huang, W., Bigornia, V., Zeng, D., Moff, S., et al. (1998). A dual thrombin receptor system for platelet activation. *Nature, 394*(6694), 690–694.

Kawabata, A., Saifeddine, M., Al-Ani, B., Leblond, L., & Hollenberg, M. D. (1999). Evaluation of proteinase-activated receptor-1 (PAR1) agonists and antagonists using a cultured cell receptor desensitization assay: Activation of PAR2 by PAR1-targeted ligands. *Journal of Pharmacology and Experimental Therapeutics, 288*(1), 358–370.

Kelso, E. B., Lockhart, J. C., Hembrough, T., Dunning, L., Plevin, R., Hollenberg, M. D., et al. (2006). Therapeutic promise of proteinase-activated receptor-2 antagonism in joint inflammation. *Journal of Pharmacology and Experimental Therapeutics, 316*(3), 1017–1024.

Kenakin, T. (2011). Functional selectivity and biased receptor signaling. *Journal of Pharmacology and Experimental Therapeutics, 336*(2), 296–302.

Kenakin, T. (2012). Biased signalling and allosteric machines: New vistas and challenges for drug discovery. *British Journal of Pharmacology, 165*(6), 1659–1669.

Kenakin, T. (2013). New concepts in pharmacological efficacy at 7TM receptors: IUPHAR review 2. *British Journal of Pharmacology, 168*(3), 554–575.

Kenakin, T., & Miller, L. J. (2010). Seven transmembrane receptors as shapeshifting proteins: The impact of allosteric modulation and functional selectivity on new drug discovery. *Pharmacological Reviews, 62*(2), 265–304.

Kinlough-Rathbone, R. L., Rand, M. L., & Packham, M. A. (1993). Rabbit and rat platelets do not respond to thrombin receptor peptides that activate human platelets. *Blood, 82*(1), 103–106.

Kong, W., McConalogue, K., Khitin, L. M., Hollenberg, M. D., Payan, D. G., Böhm, S. K., et al. (1997). Luminal trypsin may regulate enterocytes through proteinase-activated receptor 2. *Proceedings of the National Academy of Sciences of the United States of America, 94*(16), 8884–8889.

Kono, T., & Barham, F. W. (1971). Insulin-like effects of trypsin on fat cells. Localization of the metabolic steps and the cellular site affected by the enzyme. *Journal of Biological Chemistry, 246*(20), 6204–6209.

Lang, M. D., Nickel, J. C., Olson, M. E., Howard, S. R., & Ceri, H. (2000). Rat model of experimentally induced abacterial prostatitis. *Prostate, 45*, 201–206.

Liu, Y. U., Mihara, K., Ramachandran, R., Saifeddine, M., Hansen, K. K., Hyndman, M. E., et al. (2014). Constitutive secretion of enzymes that regulate proteinase-activated receptors (PARs) by cultured prostate cancer cells: Implications for affecting the tumour microenvironment. Abstract presentation, 10th conference on signalling in normal and cancer cells, Banff, 23–27 Mar 2014.

Lohman, R. J., Cotterell, A. J., Barry, G. D., Liu, L., Suen, J. Y., Vesey, D. A., et al. (2012a). An antagonist of human protease activated receptor-2 attenuates PAR2 signaling, macrophage activation, mast cell degranulation, and collagen-induced arthritis in rats. *FASEB Journal, 26*(7), 2877–2887.

Lohman, R. J., Cotterell, A. J., Suen, J., Liu, L., Do, A. T., Vesey, D. A., et al. (2012b). Antagonism of protease-activated receptor 2 protects against experimental colitis. *Journal of Pharmacology and Experimental Therapeutics, 340*(2), 256–265.

Luo, W., Wang, Y., & Reiser, G. (2007). Protease-activated receptors in the brain: Receptor expression, activation, and functions in neurodegeneration and neuroprotection. *Brain Research, 56*(2), 331–345.

Martin, L., Augé, C., Boué, J., Buresi, M. C., Chapman, K., Asfaha, S., et al. (2009). Thrombin receptor: An endogenous inhibitor of inflammatory pain, activating opioid pathways. *Pain, 146*(1-2), 121–129.

Mihara, K., Ramachandran, R., Renaux, B., Saifeddine, M., & Hollenberg, M. D. (2013). Neutrophil elastase and proteinase-3 trigger G protein-biased signaling through proteinase-activated receptor-1 (PAR1). *Journal of Biological Chemistry, 288*(46), 32979–32990.

Mosnier, L. O., Sinha, R. K., Burnier, L., Bouwens, E. A., & Griffin, J. H. (2012). Biased agonism of protease-activated receptor 1 by activated protein C caused by noncanonical cleavage at Arg46. *Blood, 120*(26), 5237–5246.

Motta, J. P., Magne, L., Descamps, D., Rolland, C., Squarzoni-Dale, C., Rousset, P., et al. (2011). Modifying the protease, antiprotease pattern by elafin overexpression protects mice from colitis. *Gastroenterology, 140*(4), 1272–1282.

Motta, J. P., Bermúdez-Humarán, L. G., Deraison, C., Martin, L., Rolland, C., Rousset, P., et al. (2012). Food-grade bacteria expressing elafin protect against inflammation and restore colon homeostasis. *Science Translational Medicine, 4*(158), 158ra144.

Nakanishi-Matsui, M., Zheng, Y. W., Sulciner, D. J., Weiss, E. J., Ludeman, M. J., & Coughlin, S. R. (2000). PAR3 is a cofactor for PAR4 activation by thrombin. *Nature, 404*(6778), 609–613.

Nelson, L. K., D'Amours, G. H., Sproule-Willoughby, K. M., Morck, D. W., & Ceri, H. (2009). Pseudomonas aeruginosa las and rhl quorum-sensing systems are important for infection and inflammation in a rat prostatitis model. *Microbiology, 155*(Pt 8), 2612–2619.

Nguyen, C., Coelho, A. M., Grady, E., Compton, S. J., Wallace, J. L., Hollenberg, M. D., et al. (2003). Colitis induced by proteinase-activated receptor-2 agonists is mediated by a neurogenic mechanism. *Canadian Journal of Physiology and Pharmacology, 81*(9), 920–927.

Noorbakhsh, F., Vergnolle, N., Hollenberg, M. D., & Power, C. (2003). Proteinase-activated receptors in the nervous system. *Nature Review Neuroscience, 4*(12), 981–990.

Noorbakhsh, F., Tsutsui, S., Vergnolle, N., Boven, L. A., Shariat, N., Vodjgani, M., et al. (2006). Proteinase-activated receptor 2 modulates neuroinflammation in experimental autoimmune encephalomyelitis and multiple sclerosis. *Journal of Experimental Medicine, 203*(2), 425–435.

Nystedt, S., Emilsson, K., Wahlestedt, C., & Sundelin, J. (1994). Molecular cloning of a potential proteinase activated receptor. *Proceedings of the National Academy of Sciences of the United States of America, 91*(20), 9208–9212.

Nystedt, S., Emilsson, K., Larsson, A. K., Strömbeck, B., & Sundelin, J. (1995). Molecular cloning and functional expression of the gene encoding the human proteinase-activated receptor 2. *European Journal of Biochemistry, 232*(1), 84–89.

Oikonomopoulou, K., Hansen, K. K., Saifeddine, M., Tea, I., Blaber, M., Blaber, S. I., et al. (2006). Proteinase-activated receptors, targets for kallikrein signaling. *Journal of Biological Chemistry, 281*(43), 32095–32112.

Oikonomopoulou, K., Diamandis, E. P., & Hollenberg, M. D. (2010). Kallikrein-related peptidases: Proteolysis and signaling in cancer, the new frontier. *Biological Chemistry, 391*(4), 299–310.

Ostrowska, E., & Reiser, G. (2008). The protease-activated receptor-3 (PAR-3) can signal autonomously to induce interleukin-8 release. *Cellular and Molecular Life Sciences, 65*(6), 970–981.

Polley, D., Mihara, K., Saifeddine, M., Renaux, B., Vliagoftis, H., Boitano, S., et al. (2012). Allergen-derived proteinases: Isolation, characterization and signaling via proteinase-activated receptors (PARs). *FASEB Journal, 26*, 664.10.

Poole, D. P., Amadesi, S., Veldhuis, N. A., Abogadie, F. C., Lieu, T., Darby, W., et al. (2013). Protease-activated receptor 2 (PAR2) protein and transient receptor potential vanilloid 4 (TRPV4) protein coupling is required for sustained inflammatory signaling. *Journal of Biological Chemistry, 288*(8), 5790–5802.

Puente, X. S., Sanchez, L. M., Gutierrez-Fernandez, A., Velasco, G., & Lopez-Otin, C. (2005). A genomic view of the complexity of mammalian proteolytic systems. *Biochemical Society Transactions, 33*(Pt 2), 331–334.

Radulovic, M., Yoon, H., Larson, N., Wu, J., Linbo, R., Burda, J. E., et al. (2013). Kallikrein cascades in traumatic spinal cord injury: In vitro evidence for roles in axonopathy and neuron degeneration. *Journal of Neuropathology and Experimental Neurology, 72*(11), 1072–1089.

Ramachandran, R., & Hollenberg, M. D. (2008). Proteinases and signalling: Pathophysiological and therapeutic implications via PARs and more. *British Journal of Pharmacology, 153*(Suppl 1), S263–S282.

Ramachandran, R., Mihara, K., Mathur, M., Rochdi, M. D., Bouvier, M., Defea, K., et al. (2009). Agonist-biased signaling via proteinase activated receptor-2: Differential activation of calcium and mitogen-activated protein kinase pathways. *Molecular Pharmacology, 76*(4), 791–801.

Ramachandran, R., Mihara, K., Chung, H., Renaux, B., Lau, C. S., Defea, K. A., et al. (2011). Neutrophil elastase acts as a biased agonist for proteinase activated receptor-2 (PAR2). *Journal of Biological Chemistry, 286*, 24638–24648.

Ramachandran, R., Noorbakhsh, F., Defea, K., & Hollenberg, M. D. (2012). Targeting proteinase-activated receptors: Therapeutic potential and challenges. *Nature Reviews Drug Discovery, 11*(1), 69–86.

Rasmussen, U. B., Vouret-Craviari, V., Jallat, S., Schlesinger, Y., Pages, G., Pavirani, A., et al. (1991). cDNA cloning and expression of a hamster alpha-thrombin receptor coupled to Ca2+ mobilization. *FEBS Letters, 288*(1-2), 123–128.

Rasmussen, U. B., Gachet, C., Schlesinger, Y., Hanau, D., Ohlmann, P., Van Obberghen-Schilling, E., et al. (1993). A peptide ligand of the human thrombin receptor antagonizes alpha-thrombin and partially activates platelets. *Journal of Biological Chemistry, 268*(19), 14322–14328.

Rieser, P. (1967). The insulin-like action of pepsin and pepsinogen. *Acta Endocrinologica, 54*(2), 375–379.

Rieser, P., & Rieser, C. H. (1964). Anabolic responses of diaphragm muscle to insulin and to other pancreatic proteins. *Proceedings of the Society for Experimental Biology and Medicine, 116*, 669–671.

Russell, F. A., Veldhoen, V. E., Tchitchkan, D., & McDougall, J. J. (2010). Proteinase-activated receptor-4 (PAR4) activation leads to sensitization of rat joint primary afferents via a bradykinin B2 receptor-dependent mechanism. *Journal of Neurophysiology, 103*(1), 155–163.

Russell, F. A., Schuelert, N., Veldhoen, V. E., Hollenberg, M. D., & McDougall, J. J. (2012). Activation of PAR (2) receptors sensitizes primary afferents and causes leukocyte rolling and adherence in the rat knee joint. *British Journal of Pharmacology, 167*(8), 1665–1678.

Scarisbrick, I. A., Towner, M. D., & Isackson, P. J. (1997). Nervous system-specific expression of a novel serine protease: Regulation in the adult rat spinal cord by excitotoxic injury. *Journal of Neuroscience, 17*(21), 8156–8168.

Scarisbrick, I. A., Blaber, S. I., Lucchinetti, C. F., Genain, C. P., Blaber, M., & Rodriguez, M. (2002). Activity of a newly identified serine protease in CNS demyelination. *Brain, 125*(Pt 6), 1283–1296.

Schuepbach, R. A., Madon, J., Ender, M., Galli, P., & Riewald, M. (2012). Protease activated receptor-1 cleaved at R46 mediates cytoprotective effects. *Journal of Thrombosis and Haemostasis, 10*, 1675–1684.

Sefton, B. M., & Rubin, H. (1970). Release from density dependent growth inhibition by proteolytic enzymes. *Nature, 227*(5260), 843–845.

Sevigny, L. M., Zhang, P., Bohm, A., Lazarides, K., Perides, G., Covic, L., et al. (2011). Interdicting protease-activated receptor-2-driven inflammation with cell-penetrating pepducins. *Proceedings of the National Academy of Sciences of the United States of America, 108*(20), 8491–8496.

Shoelson, S. E., White, M. F., & Kahn, C. R. (1988). Tryptic activation of the insulin receptor. Proteolytic truncation of the alpha-subunit releases the beta-subunit from inhibitory control. *Journal of Biological Chemistry, 263*(10), 4852–4860.

Soh, U. J., Dores, M. R., Chen, B., & Trejo, J. (2010). Signal transduction by protease-activated receptors. *British Journal of Pharmacology, 160*(2), 191–203.

Stanton, M. M., Nelson, L. K., Benediktsson, H., Hollenberg, M. D., Buret, A. G., & Ceri, H. (2013). Proteinase-activated receptor-1 and immunomodulatory effects of a PAR1-activating peptide in a mouse model of prostatitis. *Mediators of Inflammation, 2013*, 748395.

Steiner, D. F. (2011). On the discovery of precursor processing. *Methods in Molecular Biology, 768*, 3–11.

Steiner, D. F., Cunningham, D., Spigelman, L., & Aten, B. (1967). Insulin biosynthesis: Evidence for a precursor. *Science, 157*(789), 697–700.

Steinhoff, M., Vergnolle, N., Young, S. H., Tognetto, M., Amadesi, S., Ennes, H. S., et al. (2000). Agonists of proteinase-activated receptor 2 induce inflammation by a neurogenic mechanism. *Nature Medicine, 6*(2), 151–158.

Trivedi, V., Boire, A., Tchernychev, B., Kaneider, N. C., Leger, A. J., O'Callaghan, K., et al. (2009). Platelet matrix metalloprotease-1 mediates thrombogenesis by activating PAR1 at a cryptic ligand site. *Cell, 137*(2), 332–343.

Vergnolle, N. (2009). Protease-activated receptors as drug targets in inflammation and pain. *Pharmacology and Therapeutics, 123*(3), 292–309.

Vergnolle, N., Hollenberg, M. D., Sharkey, K. A., & Wallace, J. L. (1999). Characterization of the inflammatory response to proteinase-activated receptor-2 (PAR2)-activating peptides in the rat paw. *British Journal of Pharmacology, 127*(5), 1083–1090.

Vliagoftis, H., & Forsythe, P. (2008). Should we target allergen protease activity to decrease the burden of allergic airway inflammation? *Inflammation & Allergy-Drug Targets, 7*(4), 288–295.

Vouret-Craviari, V., Van Obberghen-Schilling, E., Rasmussen, U. B., Pavirani, A., Lecocq, J. P., & Pouyssegur, J. (1992). Synthetic alpha-thrombin receptor peptides activate G protein-coupled signaling pathways but are unable to induce mitogenesis. *Molecular Biology of the Cell, 3*(1), 95–102.

Vouret-Craviari, V., Van Obberghen-Schilling, E., Scimeca, J. C., Van Obberghen, E., & Pouyssegur, J. (1993). Differential activation of p44mapk (ERK1) by alpha-thrombin and thrombin-receptor peptide agonist. *Biochemical Journal, 289*(Pt 1), 209–214.

Vu, T. K., Hung, D. T., Wheaton, V. I., & Coughlin, S. R. (1991). Molecular cloning of a functional thrombin receptor reveals a novel proteolytic mechanism of receptor activation. *Cell, 64*(6), 1057–1068.

Wang, J., Zheng, H., Hollenberg, M. D., Wijesuriya, S. J., Ou, X., & Hauer-Jensen, M. (2003). Up-regulation and activation of proteinase-activated receptor 2 in early and delayed radiation injury in the rat intestine: Influence of biological activators of proteinase-activated receptor 2. *Radiation Research, 160*(5), 524–535.

Xu, W. F., Andersen, H., Whitmore, T. E., Presnell, S. R., Yee, D. P., Ching, A., et al. (1998). Cloning and characterization of human protease-activated receptor 4. *Proceedings of the National Academy of Sciences of the United States of America, 95*(12), 6642–6646.

Yoon, H., Radulovic, M., Wu, J., Blaber, S. I., Blaber, M., Fehlings, M. G., et al. (2013). Kallikrein 6 signals through PAR1 and PAR2 to promote neuron injury and exacerbate glutamate neurotoxicity. *Journal of Neurochemistry, 127*(2), 283–298.

Zhao, P., Metcalf, M., & Bunnett, N. W. (2014). Biased signaling of protease-activated receptors. *Frontiers in Endocrinology (Lausanne), 5*, 67.

Reactive Oxygen Species

Brent J. Ryan[1], M. Letizia Lo Faro[3], Matthew Whiteman[2] and Paul G. Winyard[2]
[1]Department of Physiology, Anatomy and Genetics, University of Oxford, Oxford, UK
[2]Inflammation Research Group, University of Exeter Medical School, Exeter, UK
[3]Oxford Transplant Centre, University of Oxford, Nuffield Department of Surgical Sciences, Churchill Hospital, Oxford, UK

Synonyms

Free radicals; Nitrative stress; Oxidative stress; Post-translational modifications

Definition

Reactive oxygen species (ROS) is a collective term given to a group of oxygen-containing intermediates, many of which react with biomolecules such as DNA, lipids, or proteins. ROS include (but are not limited to) hydrogen peroxide (H_2O_2), the superoxide radical anion ($O_2^{\cdot-}$), the hydroxyl radical ($^{\cdot}OH$), and singlet oxygen ($^1O_2$). ROS sometimes have one or more unpaired electrons – as denoted by a superscript dot "$^{\cdot}$".

In addition, the term ROS is sometimes used as an umbrella term for other reactive oxygen-containing species such as nitric oxide ($^{\cdot}NO$), peroxynitrite ($ONOO^-$) (also referred to as reactive nitrogen species, RNS), or hypochlorite (^-OCl) (also referred to as reactive chlorine species, RCS). ROS may react directly with proteins, lipids, or DNA or through secondary products of these reactions to mediate numerous physiological processes including the innate immune response and cell signaling. Oxidative stress is a term given to a situation in which the production of oxidants and ROS outweighs the antioxidant capacity of a cell or tissue and highlights the constant dynamic between oxidant production and antioxidant "buffering," particularly during inflammation. The delicate balance between the physiological and pathophysiological roles of ROS appears to be a key factor in the pathophysiology of many diseases including cardiovascular disease, diabetes, autoimmune diseases, and neurodegeneration.

Biosynthesis and Release

A number of cellular processes and enzymes are capable of generating ROS *in vivo*, in response to a wide variety of stimuli. The enzymatic generation of ROS is largely governed by enzyme expression/activity, subcellular localization, and the presence of cofactors, whereas nonenzymatic ROS generation is usually stochastic in nature. ROS are produced during normal homeostasis with mitochondrial oxidative phosphorylation as a major source. Electrons may be lost at numerous steps in the electron transport chain, chiefly at complexes I and III, generating $O_2^{\cdot-}$, through the one-electron reduction of O_2 and/or H_2O_2 (Quinlan et al. 2012). The major product of this "electron leak" from mitochondrial complexes is thought to be $O_2^{\cdot-}$. However, $O_2^{\cdot-}$ is often rapidly converted enzymatically by manganese superoxide dismutase (MnSOD) to membrane permeable H_2O_2. Superoxide is not membrane permeable, whereas H_2O_2 is able to diffuse out through the mitochondrial membrane. Therefore MnSOD activity influences the amount of ROS-induced mitochondrial damage in cells.

Nicotinamide adenine dinucleotide phosphate-oxidases (NADPH oxidases; NOXs) are another major source of enzymatically derived ROS. These enzymes use NADPH as an electron source to reduce O_2 and generate $O_2^{\cdot-}$. The prototypical NOX is the one present in phagocytic cells, which is able to generate large amounts of $O_2^{\cdot-}$ ("respiratory burst") during phagocytosis. NOXs are membrane-bound enzyme complexes that are activated in response to inflammatory stimuli, such as TNF-α (see ▶ Tumor Necrosis Factor Alpha (TNFalpha)). Stimulation results in the translocation of the cytosolic subunits to the membrane-bound subunits and enzyme activation.

Myeloperoxidase (MPO) is another enzyme associated with the respiratory burst that generates ROS. MPO is present in neutrophils and monocytes, in particular macrophages, and is capable of oxidizing a number of substrates to form ROS. Principally, MPO acts on H_2O_2 to form HOCl which can act as an oxidant, either directly or via secondary products (e.g., chloramines), to exert antimicrobial activity by damaging pathogen lipids, proteins, or DNA.

Nitric oxide synthase (NOS) catalyzes the production of ˙NO from O_2 and L-arginine, in the presence of several cofactors, including NADH, calmodulin, and tetrahydrobiopterin. ˙NO produced by inducible nitric oxide synthase (iNOS) in macrophages or glial cells is the primary source of ˙NO in the inflammatory response. Regulation of iNOS, unlike constitutive NOS isoforms (endothelial NOS and neuronal NOS), primarily occurs at the transcriptional level by induction through the ERK/JNK and NF-κB pathways in response to inflammatory stimuli such as TNF-α, IFN-γ, or LPS (Chan and Riches 2001). The levels of ˙NO produced by NOS are influenced by the availability of the cofactor tetrahydrobiopterin. In the absence of sufficient tetrahydrobiopterin, NOS produces $O_2^{\cdot-}$ instead of ˙NO (NOS uncoupling), serving as a source of $O_2^{\cdot-}$ and $ONOO^-$ (produced upon interaction of ˙NO and $O_2^{\cdot-}$) and contributing to oxidative stress (Crabtree and Channon 2011). Levels of ˙NO and it's metabolites, such as nitrite and nitrate, are known to correlate with a number of inflammatory diseases including rheumatoid arthritis (Tarr et al. 2010).

A number of molecules in cells are redox active and able to generate ROS themselves upon oxidation. One such class of molecules is neurotransmitters such as noradrenaline and dopamine. These molecules are capable of being oxidized producing $O_2^{\cdot-}$ or H_2O_2 and generating quinones/semiquinones which can themselves damage proteins. Redox-active transition metals, such as Fe^{2+} (ferrous ions) and Cu^+ (cuprous ions), are able to promote the formation of a variety of ROS through Fenton reactions. The ability of these metal ions to participate in redox reactions is usually limited *in vivo*, due to the presence of ligands coordinating to these metals, e.g., transferrin binds ferrous ions. However some enzymes, such as MPO and NOS, use Fe^{2+}-containing heme groups to catalyze redox reactions and are capable of generating ROS.

Biological Activities

One important activity of physiologically produced ROS is the regulation of the immune response. In response to pathogen-associated molecular patterns (PAMPs), such as lipopolysaccharide (LPS), neutrophils, monocytes, and macrophages (among others) are capable of engulfing (phagocytosing) pathogens. These phagocytes then undergo a respiratory burst and rapidly produce large amounts of ROS to kill pathogens. As previously stated, NADPH oxidases, MPO, and iNOS are important for generating a wide range of ROS (either directly or indirectly), including

$O_2^{·-}$, ·OH, ONOO⁻, and HOCl, which kill the pathogen by damaging lipids, DNA, and proteins. Human subjects with a decreased ability to produce ROS during the respiratory burst usually have an increased susceptibility to bacterial infections.

Besides damaging macromolecules in the activation of the immune response, ROS can also alter enzymatic activities with disparate effects on cells and tissues. H_2O_2 has been shown to have a central role in cell redox signaling in numerous cell types. H_2O_2 appears to signal through the specific and reversible oxidation of cysteine residues in proteins (Sundaresan et al. 1995; Junn et al. 2000; Szabó-Taylor et al. 2013). In accordance with its role as a signaling molecule, H_2O_2 levels are finely regulated. Indeed, H_2O_2 can be enzymatically converted to oxygen and water in the presence of catalase and can be reduced to water in reactions catalyzed by peroxidase enzymes, such as glutathione peroxidases and peroxiredoxins. Peroxiredoxins, in particular, play a major role in H_2O_2-mediated signal transduction (Wood et al. 2003; Szabó-Taylor et al. 2013). The reaction of peroxiredoxin with H_2O_2 proceeds by initial oxidation of the peroxiredoxin cysteine to the sulfenic (-SOH) acid form, followed by reduction through interaction with thioredoxin or sulfiredoxin. Cysteine residues can also be further oxidized by ROS to form sulfinic (-SOOH) acid form. Cysteine oxidation is the main mechanism by which H_2O_2 is thought to interact with proteins, activating many transcription factors (e.g., AP-1, NF-κB) or redox-sensitive proteins and enzymes (e.g., bacterial OxyR, Trx, HO-1, etc.).

Besides H_2O_2, other reactive species have important roles in cellular signaling. ·NO is a key signaling molecule with an important role in the regulation of the vasculature. To exert this function, ·NO binds to the heme group of soluble guanylate cyclase, present in smooth muscle, resulting in activation of the enzyme, increased cGMP production and lower intracellular Ca^{2+}, which causes vasorelaxation (Palmer et al. 1987).

Pathophysiological Relevance

Some ROS (e.g., ·OH) readily react with both DNA and RNA causing modification of both purine and pyrimidine bases and resulting in the formation of products such as 8-oxo-2′deoxyguanosine (Cooke et al. 2003). Two main repair mechanisms exist in cells to deal with such lesions: base excision repair and nucleotide excision repair, although these have the potential to result in deletions or mutations. ROS-induced DNA damage has been shown to contribute to somatic DNA mutations that have been identified in inflammatory diseases, such as rheumatoid arthritis (Bottini and Firestein 2013). These ROS-induced somatic DNA mutations have been demonstrated to increase with age and are associated with a number of diseases. Given the large amounts of ROS production in the mitochondria, mitochondrial DNA (mtDNA) is particularly prone to accumulating somatic mutations. This accumulation of mtDNA mutations is thought to contribute to mitochondrial dysfunction in diseases such as Parkinson's disease (Ryan et al. 2015).

Proteins are a major target of cellular ROS, either during homeostasis or pathology, resulting in posttranslational modifications. Numerous amino acids are susceptible to ROS, but the modifications induced by a given oxidant in a protein depend on several factors including: the concentrations of the oxidant and the given amino acid, the rate constant for the reaction between the oxidant and the amino acid, the steric hindrance around the target amino acid, and the charge of the surrounding residues. As general examples, H_2O_2 will oxidize cysteine and methionine residues readily, but at higher concentrations it will also modify tryptophan and tyrosine residues, whereas ONOO⁻ will also readily oxidize these amino acids, at a lower concentration, and will also induce 3-nitrotyrosine formation (Winyard et al. 2011).

Lipids are another class of macromolecule that may be affected by ROS through the formation of radicals in unsaturated fatty acids, such as membrane phospholipids (Guéraud et al. 2010). These lipid radicals can propagate, damaging other surrounding lipids, in a process known as lipid peroxidation. These lipid peroxidation products include malondialdehyde (MDA) and 4-hydroxynonenal (4-HNE) and are able to modify both DNA and proteins or act as secondary messengers with profound biological consequences. Oxidation of the lipid moiety of low-density lipoprotein (LDL) has been demonstrated to be a key initiator of LDL antigenicity. Indeed, the uptake of oxidized LDL by macrophages is known to be key to foam cell formation and the initiation of atherosclerosis. (Witztum and Steinberg 1991) (Hörkkö et al. 1996).

These modifications may activate or inactivate proteins or cause misfolding and aggregation. H_2O_2 has been shown to inactivate protein tyrosine phosphatases (e.g., PTEN) and activate protein kinases or soluble guanylate cyclase. Oxidative modifications have also been shown to increase the antigenicity of both proteins and DNA, influencing the production of autoantibodies against these oxidatively modified molecules. Examples of this neo-epitope formation include type II collagen in RA and LDL in atherosclerosis (Witztum and Steinberg 1991; Bashir et al. 1993; Nissim et al. 2005; Eggleton et al. 2008, 2013).

Compounds which elicit ROS production are able to initiate cell proliferation or induce cell death, depending on the type of oxidant, concentration, and cell type. For example, mitochondrial complex I inhibitors, such as MPP^+ and rotenone, which are associated with Parkinson's disease, are able to disturb mitochondrial bioenergetics, causing cell death and pathology (Ryan et al. 2013; Taylor et al. 2013).

Modulation by Drugs

The generation of oxidants also diminishes the antioxidant pool in the cell, thus increasing the susceptibility to damage by further oxidant generation. A number of both intracellular and extracellular antioxidant systems are present in humans to limit the deleterious effects of excess oxidants generated *in vivo*. Two of the most studied enzymes involved in this process are SOD, which catalyzes the dismutation of $O_2^{\cdot-}$ to H_2O_2, and catalase, which catalyzes the reduction of H_2O_2 to H_2O and O_2.

One of the major antioxidant systems *in vivo* is thought to be the glutathione (GSH) system (Mari et al. 2009; Aquilano et al. 2014). Glutathione is a tripeptide (Glu-Cys-Gly) with a free thiol on the cysteine residue, which can react with a wide range of reactive species such as H_2O_2, $ONOO^-$, or lipid peroxides. These interactions, catalyzed by glutathione peroxidase, can form cysteine oxidation products (as described above) or result in the formation of a glutathione dimer, which can subsequently be reduced by glutathione reductase. The ratio of reduced to oxidized glutathione is often used as a marker of the redox status of a cell/tissue and reduced GSH/GSSG ratios are observed in numerous diseases (Sian et al. 1994; Shimizu et al. 2004).

Nuclear factor (erythroid-derived 2)-like 2 (NRF2) is a transcription factor that responds to an increased oxidative burden in cells (Itoh et al. 2004). NRF2 is usually associated with Keap1, which rapidly targets NRF2 for proteasomal degradation. However, in response to increased oxidative stress, this complex becomes dissociated and NRF2 translocates to the nucleus and binds to antioxidant response elements (ARE) in promoters of a wide range of antioxidant genes, resulting in transcription (Kansanen et al. 2013). The dissociation of the NRF2-Keap1 complex is promoted by naturally occurring compounds such as sulforaphane, curcumin, dimethyl fumarate, and lipoic acid.

Soluble antioxidant compounds, such as water-soluble ascorbic acid (vitamin C) and lipid-soluble tocopherols (vitamin E), are responsible for the scavenging of reactive species present in the cytosol and extracellularly and in lipid membranes, respectively. Despite these well-documented antioxidant functions, studies dosing various patient groups with vitamins C and E, as well as other antioxidants, such as carotenoids and flavonoids, have shown mixed results in diseases such as RA, with no clear overall benefit (Pattison and Winyard 2008). Several chelating agents (e.g., desferrioxamine, EDTA, phytic acid, etc.) have been used to attempt the inhibition of transition metal ion-dependent oxidative damage through Fenton reactions. Mitochondria-targeted antioxidants or redox modulators, such as MitoQ, accumulate in the mitochondrial matrix and sequester ROS generated by electron leak from the mitochondrial electron transport chain (Murphy 2008).

Alongside the use of drugs that limit the levels of reactive species, it is possible to find therapeutic compounds which increase the level of a particular reactive molecule. For example, some pathological conditions, such as hypertension or angina, are characterized by limited production/availability of ˙NO and have been historically treated with ˙NO donor molecules, such as inorganic or organic nitrates and nitroglycerin.

References

Aquilano, K., Sara et al. (2014) Glutathione: new roles in redox signaling for an old antioxidant. *Front Pharmacol*, 5:196. PMID PMC4144092

Bashir, S., et al. (1993). Oxidative DNA damage and cellular sensitivity to oxidative stress in human autoimmune diseases. *Annals of the Rheumatic Diseases, 52*(9), 659–666.

Bottini, N., & Firestein, G. S. (2013). Duality of fibroblast-like synoviocytes in RA: Passive responders and imprinted aggressors. *Nature Reviews. Rheumatology, 9*(1), 24–33.

Chan, E. D., & Riches, D. W. (2001). IFN-gamma + LPS induction of iNOS is modulated by ERK, JNK/SAPK, and p38(mapk) in a mouse macrophage cell line. *American Journal of Physiology. Cell Physiology, 280*(3), C441–C450.

Cooke, M. S., et al. (2003). Oxidative DNA damage: Mechanisms, mutation, and disease. *The FASEB Journal, 17*(10), 1195–1214.

Crabtree, M. J., & Channon, K. M. (2011). Synthesis and recycling of tetrahydrobiopterin in endothelial function and vascular disease. *Nitric Oxide, 25*(2), 81–88.

Eggleton, P., Haigh, R., & Winyard, P. G. (2008). Consequence of neo-antigenicity of the 'altered self'. *Rheumatology (Oxford), 47*(5), 567–571.

Eggleton, P., et al. (2013). Detection and isolation of human serum autoantibodies that recognize oxidatively modified autoantigens. *Free Radical Biology & Medicine, 57*, 79–91.

Guéraud, F., et al. (2010). Chemistry and biochemistry of lipid peroxidation products. *Free Radical Research, 44*(10), 1098–1124.

Hörkkö, S., et al. (1996). Antiphospholipid antibodies are directed against epitopes of oxidized phospholipids. Recognition of cardiolipin by monoclonal antibodies to epitopes of oxidized low density lipoprotein. *J Clin Invest, 98*(3), 815–825.

Itoh, K., Tong, K. I., & Yamamoto, M. (2004). Molecular mechanism activating nrf2–keap1 pathway in regulation of adaptive response to electrophiles. *Free Radical Biology and Medicine, 36*(10), 1208–1213.

Junn, E., et al. (2000). Requirement of hydrogen peroxide generation in TGF-β1 signal transduction in human lung fibroblast cells: Involvement of hydrogen peroxide and Ca2+ in TGF-β1-Induced IL-6 expression. *The Journal of Immunology, 165*(4), 2190–2197.

Kansanen, E., et al. (2013). The Keap1-Nrf2 pathway: Mechanisms of activation and dysregulation in cancer. *Redox Biology, 1*(1), 45–49.

Marí, M., et al. (2009) Mitochondrial Glutathione, a Key Survival Antioxidant. *Antioxid Redox Signal, 11*(11): 2685–2700. PMID PMC2821140.

Murphy, M. P. (2008). Targeting lipophilic cations to mitochondria. *Biochimica et Biophysica Acta (BBA) – Bioenergetics, 1777*(7–8), 1028–1031.

Nissim, A., et al. (2005). Generation of neoantigenic epitopes after posttranslational modification of type II collagen by factors present within the inflamed joint. *Arthritis and Rheumatism, 52*(12), 3829–3838.

Palmer, R. M. J., Ferrige, A. G., & Moncada, S. (1987). Nitric oxide release accounts for the biological activity of endothelium-derived relaxing factor. *Nature, 327*(6122), 524–526.

Pattison, D. J., & Winyard, P. G. (2008). Dietary antioxidants in inflammatory arthritis: Do they have any role in etiology or therapy? *Nature Clinical Practice Rheumatology, 4*(11), 590–596.

Quinlan, C. L., et al. (2012). Native rates of superoxide production from multiple sites in isolated mitochondria measured using endogenous reporters. *Free Radical Biology and Medicine, 53*(9), 1807–1817.

Ryan, B. J., et al. (2013). alpha-Synuclein and mitochondrial bioenergetics regulate tetrahydrobiopterin levels in a human dopaminergic model of Parkinson disease. *Free Radical Biology & Medicine, 67C*, 58–68.

Ryan, B. J., et al. (2015). Wade-Martins, R. Mitochondrial dysfunction and mitophagy in Parkinson's: from familial to sporadic disease. *Trends Biochem Sci, 40* (4), 200–210. PMID 25757399

Shimizu, H., et al. (2004). Relationship between plasma glutathione levels and cardiovascular disease in a defined population: The Hisayama study. *Stroke, 35*(9), 2072–2077.

Sian, J., et al. (1994). Alterations in glutathione levels in Parkinson's disease and other neurodegenerative disorders affecting basal ganglia. *Annals of Neurology, 36*(3), 348–355.

Sundaresan, M., et al. (1995). Requirement for generation of H2O2 for platelet-derived growth factor signal transduction. *Science, 270*(5234), 296–299.

Szabó-Taylor, K., et al. (2013). Oxidative stress in rheumatoid arthritis. In M. J. Alcaraz, O. Gualillo, & O. Sánchez-Pernaute (Eds.), *Studies on arthritis and joint disorders* (pp. 145–167). New York: Springer.

Tarr, J. M., et al. (2010). Extracellular calreticulin is present in the joints of patients with rheumatoid arthritis and inhibits FasL (CD95L)-mediated apoptosis of T cells. *Arthritis and Rheumatism, 62*(10), 2919–2929.

Taylor, T. N., et al. (2013). Region-specific deficits in dopamine, but not norepinephrine, signaling in a novel A30P alpha-synuclein BAC transgenic mouse. *Neurobiology of Disease, 62C*, 193–207.

Winyard, P. G., et al. (2011). Measurement and meaning of markers of reactive species of oxygen, nitrogen and sulfur in healthy human subjects and patients with inflammatory joint disease. *Biochemical Society Transactions, 39*(5), 1226–1232.

Witztum, J. L., & Steinberg, D. (1991). Role of oxidized low density lipoprotein in atherogenesis. *Journal of Clinical Investigation, 88*(6), 1785–1792.

Wood, Z. A., Poole, L. B., & Karplus, P. A. (2003). Peroxiredoxin evolution and the regulation of hydrogen peroxide signaling. *Science, 300*(5619), 650–653.

Rheumatic Fever

William B. Moskowitz
Pediatric Cardiac Catheterization Laboratory,
The Children's Hospital of Richmond at VCU,
Richmond, VA, USA
The Children's Hospital of Richmond at VCU,
Richmond, VA, USA

Synonyms

Acute articular rheumatism; Acute rheumatic fever; Acute rheumatism

Definition

Acute rheumatic fever (ARF) is a delayed multisystem postinfectious, nonsuppurative disease with high heritability but unpredictable severity, which occurs as a sequela of untreated or incompletely treated tonsillopharyngitis with *Streptococcus pyogenes* or group Aβ-hemolytic *Streptococcus* (GAS).

Epidemiology and Genetics

Epidemiology

The incidence of ARF varies greatly between countries. In developing countries of the world, ARF and rheumatic heart disease (RHD) are estimated to affect nearly 15 million people and are the leading causes of cardiovascular death in individuals under 50 years of age. ARF most commonly occurs in patients between 5 and 15 years of age. It is rare before the age of 3 years with 92 % of cases occurring by 18 years of age. Worldwide, there are approximately 470,000 new cases of ARF and 233,000 deaths attributable to ARF or RHD yearly (Carapetis et al. 2005).

During epidemics occurring early in the last century, 2–3 % of previously healthy patients with untreated GAS pharyngitis developed ARF; in endemic infections, the incidence of ARF is significantly less. The annual incidence of ARF is 100–200 times greater in developing countries (average 19 cases/100,000) than that observed in industrialized countries. The reported incidence of ARF is decreasing in all World Health Organization (WHO) regions except for the Americas where it appears to be increasing slightly and the Western Pacific, where it appears to be steadily increasing (Seckeler and Hoke 2011). The worldwide decreasing trend in ARF incidence, with the noted exceptions, is likely multifactorial and is attributable to improved living conditions, the use of antibiotics to treat GAS pharyngitis, and possibly shifts in GAS serotypes. The increasing trend in the Western Pacific is more likely as result of improved recognition and reporting of ARF cases from the many small island nations with isolated populations.

It is important to remember that ARF is a recurrent disease, requiring continuous secondary prophylaxis to prevent new or worsening RHD with each new bout of GAS pharyngitis. Regional ARF recurrence rates range from 8 % to 34 %, with Europe having significantly fewer recurrences than the Americas, Eastern Mediterranean, Western Pacific, and Southeast Asia (Seckeler and Hoke 2011).

While the incidence of ARF has been decreasing in most of the world, the prevalence of RHD seems to be increasing worldwide. There are two major reasons for this. First, due to advances in medical and surgical therapies for RHD, more individuals are surviving with chronic RHD. Second, with the use of echocardiography and standardization of RHD diagnostic echocardiographic criteria (Remenyi et al. 2012), "subclinical carditis" has been found at rates up to 10 times higher than that diagnosed by examination alone (Bhaya et al. 2010).

Genetics

Streptococcus pyogenes

GAS is a strict human pathogen, with no other known reservoir or species being affected by diseases unique to the organism. The complete sequence of its genome and the resulting initial analysis have revealed numerous encoded virulence factors (Ferretti et al. 2001). Additional genes have been identified that encode proteins likely associated with microbial molecular mimicry of host characteristics involved in the production of ARF, RHD, or acute glomerulonephritis. Further, the complete or partial sequences of four different bacteriophage genomes are present within the GAS genome that contains genes for super antigen-like proteins. The presence of these prophage-associated genes adds credence to the hypothesis that bacteriophages play a role in horizontal gene transfer and offers a mechanism for generating new strains of GAS with increased potential for causing disease.

Host Genetic Predisposition

The idea that ARF may be a result of GAS infection in individuals with some genetic predisposition has existed for over a century. ARF occurs more frequently within families and among monozygotic twins. One study showed that the risk of ARF in a monozygotic twin with a history of ARF in the co-twin is increased by more than six times compared to that of dizygotic twins (Engel et al. 2011). The heritability estimate was 60 %, which confirms the importance of genetic factors in ARF. However, a cosegregation study of 22 families of different ethnic background with multiple individuals with RHD supports an inheritance pattern that is dominant but with variable penetrance (Gerbase-DeLima et al. 1994).

The major histocompatibility complex (MHC) contains a multitude of genes involved in the immune response. Specific HLA antigens within the context of various immune diseases have led to investigations for such antigens in ARF. An increased frequency of MHC class II alleles, HLA-D8/17 and HLA-DR7 types are the most represented in the literature among patients with ARF and RHD. HLA-DR4 and HLA-DR2 have been noted in Caucasian and black patients with RHD (Ayoub et al. 1986). The variability in reported HLA alleles may be due to genetic differences in the populations studied or differences in local streptococcal strains.

The fact that different ethnic groups with ARF/RHD may have different HLA associations suggests that cross-reactive peptides may bind to several different HLA alleles that have structural homology in the peptide-binding groove (Bryant et al. 2009). Most researchers, however, have concluded that the association between HLA and ARF/RHD is through linkage disequilibrium and that an ARF susceptibility gene exists that is mapped within or near the HLA complex. However, although significant associations have been found between certain genetic factors and ARF/RHD, study results often conflict with each other.

Tumor necrosis factor-α (TNFA), located in the MHC class III region, has been associated with several autoimmune diseases. TNFA, a cytokine with a broad spectrum of inflammatory and immunomodulatory activities, has been implicated in several cardiac diseases, including congestive heart failure, myocarditis, and dilated cardiomyopathy, and in ARF (Ramasawmy et al. 2007). The clinical spectrum of ARF or RHD, from severe

RHD with mitral valve lesions or aortic valve lesions to milder carditis, may reflect genetic differences. For example, the TNFA -308G/A polymorphism may be associated with the development of mitral valve lesions, while the TNFA -238 G/A may be associated either a milder carditis or RHD with aortic lesions. The ability to predict which ARF patients are at highest risk for RHD has significant implications.

Pathophysiology

The close link between *Streptococcus pyogenes* and ARF is well established, but the exact pathogenesis of ARF and RHD is still not fully understood even after the association was first made at the turn of the nineteenth century. Unquestionably streptococcal tonsillopharyngeal infection is required, and genetic susceptibility may be a key factor. While the vast majority of "rheumatogenic strains" are GAS strains, in certain areas of the world such as the Eastern Caribbean, it has been suggested that ARF may be due to non-GAS (group C and group G) strains that have obtained the necessary virulence factors through phage transfection. The pharyngeal site for the streptococcal infection may be the requisite site (due to the presence of lymphoid tissue) for the prolific humoral response to those cross-reacting antigens with host target organs.

As a result of molecular mimicry, antibodies directed against GAS antigens cross-react with host antigens. Streptococcal M protein and *N*-acetyl-beta-D-glucosamine (NABG, the immunodominant carbohydrate antigen of GAS) share epitopes with myosin (Dale and Beachey 1985). Monoclonal antibodies isolated from patients with GAS pharyngitis cross-react with myosin and other proteins, and monoclonal antibodies isolated from patients with RHD are directed against myosin and NABG. Additional reactivity with an extracellular matrix protein laminin may explain the reactivity against cardiac valves (Galvin et al. 2000). Molecular mimicry is also likely involved with the development of Sydenham chorea.

Rheumatic Fever, Table 1 Jones criteria for the diagnosis of rheumatic fever

Criterion	Major	Minor
Clinical	Carditis	
	Polyarthritis	
	Chorea	Fever
	Erythema marginatum	Arthralgia
	Subcutaneous nodules	
Laboratory		Elevated acute-phase reactants (ESR, CRP)
		Prolonged PR interval (ECG)

Evidence of antecedent GAS infection: positive culture for GAS elevated or rising streptococcal antibody titer

Clinical Presentation

The clinical features of ARF are listed under Jones criteria in Table 1. These were first assembled in 1944 and have since been modified three times, aiming at minimizing over diagnosis. The diagnosis of ARF is strongly suggested when two major criteria or one major and two minor criteria are met in a patient with evidence of a recent streptococcal infection. However, a positive throat culture for GAS is inconsistently found, and the absence of a positive culture does not exclude the diagnosis. Tests for various antibodies such as antistreptolysin O (ASO) and anti-DNase B are useful though 20 % of cases of ARF are not accompanied by raised antibody titers. These cases are usually those who present with chorea alone, several months after the acute infection by GAS and after acute-phase reactant levels have waned.

The last updated revision in 1992 highlighted the exceptions to the diagnostic criteria, recognizing that the risks of underdiagnosis would be higher against the requirement of strict adherence in three situations: indolent carditis, chorea as the sole manifestation, and recurrence (Dajani et al. 1992). A high index of suspicion of recurrent ARF should be maintained in patients with a history of prior ARF attacks who present with symptoms or signs of ARF. It is particularly difficult to establish evidence of new, acute carditis in patients with preexisting RHD. Therefore, a presumptive diagnosis of recurrent ARF may be made when a single major or several minor criteria are present in a patient with a history of ARF or RHD, provided there is supporting evidence of a recent GAS infection. On the other hand, it is equally important not to use an isolated clinical finding (e.g., monoarthritis, fever, arthralgia, or elevation of acute-phase reactants) as a sole criterion for a diagnosis of ARF in this patient population.

Arthritis: Polyarthritis with fever (usually at least 39°) is the usual initial presentation. Arthritis is the most common major clinical sign and occurs in 75 % of patients. The knees, ankles, elbows, and wrists are most commonly affected, with the hips and small joints of the distal extremities rarely affected. The arthritis is usually migratory in nature and more likely bilateral and symmetrical but inconsistently so. The exquisitely rapid resolution of the arthritis (there are no long-term joint sequelae resulting from bouts of arthritis) with salicylates should always bring into question the veracity of the diagnosis when rapid resolution is not observed.

Carditis: Carditis is the most serious manifestation of ARF and occurs within 3 weeks of onset. It is present clinically in over 50 % of the cases and in up to 70–90 % of cases by echocardiography. Carditis may occur during an initial bout of ARF and certainly during recurrences. While it is an inflammatory pancarditis, significant pericarditis with a large effusion is very uncommon. Congestive heart failure may result from the inflammatory process but is thought to be secondary to severe valvular regurgitation rather than to myocarditis. Endocarditis as manifested as valvulitis (mitral valve most commonly, followed by mitral and aortic valve, and then aortic valve alone, though all valves may be affected) is the most common long-term sequela from the immune response to GAS and leads to chronic RHD. A typical murmur of mitral regurgitation may be heard at the apex into the axilla, and/or an early diastolic murmur of aortic insufficiency may be heard at the base with radiation to the apex. A wide pulse pressure should be sought to estimate the hemodynamic significance of aortic insufficiency. Serial examinations are required to follow the valvulitis with therapy and long term. Echocardiography should be obtained early and serially to aid in following the disease progression or resolution. An ECG should be obtained on all patients to look for PR interval prolongation (first-degree heart block) which may progress to complete heart block but usually resolves over time with appropriate therapy.

Chorea: Sydenham's chorea or St. Vitus' dance is a delayed sign of GAS infection, occurring as late as 6 months after onset. The movements are characterized as rapid and uncontrolled varying from loss of fine motor control (worsening penmanship), fasciculation of the tongue and speech abnormalities, facial grimacing to gait abnormalities, ballismus, clonus, and generalized purposeless movements. There can be worsening of school performance, behavioral changes including psychiatric findings of obsession and compulsion, tics, psychoses, and emotional lability. Overall, chorea is part of ARF presentation in only 10 % of cases, with a peak age at 8 years. About 20 % of patients diagnosed with Sydenham's chorea experience a recurrence, usually within 2 years of the first episode. Chorea occurs twice as frequently in females. Most women who develop Sydenham's during pregnancy have a history of ARF in childhood or of using birth control pills containing estrogen. There is a body of evidence which suggests that chorea is post-GAS immune mediated with regional localization to the basal ganglia.

Erythema marginatum: Erythema marginatum is a transient non-pruritic rash with central pallor and pink to red serpiginous borders found on the extensor surfaces of the trunk and extremities, generally sparing the face. The rash changes from hour to hour and may seem to appear, disappear, or move so rapidly that it can almost be seen doing so. It often involves multiple areas. It is exacerbated by heat (warm bath or shower) and fades when the patient is cool. There are usually other symptoms of ARF but it can recur intermittently over weeks or even months. Erythema marginatum is seen in less than 10 % of ARF cases. Erythema marginatum usually occurs early in the course of ARF in patients with acute carditis but may persist or recur when all other manifestations of disease have disappeared.

Subcutaneous nodules: Subcutaneous nodules are painless, flesh-colored bumps, usually found on the extensor surfaces of the arms and legs as well as on the bony prominences and tendons around joints, overlying the spine and scapula. The overlying skin is not inflamed and usually can be moved over the nodules that measure between a few millimeters to 2 cm in size. Nodules can occur in isolation or in crops of more than a dozen. Currently, subcutaneous nodules have been reported in less than 5 % of cases. They rarely occur as an isolated manifestation, appearing after the first weeks of illness and are associated with relatively severe carditis in most cases. With ARF therapy, the nodules usually resolve within a few weeks but may persist for longer than a month.

Diagnostic Points

In the absence of a "gold standard" for the diagnosis of ARF, no single specific laboratory test exists that is pathognomonic of ARF or its recurrences. Echocardiography is useful for confirming clinical findings and provides for serial assessment of valvular pathology and pathophysiology, chamber size and ventricular function, and the presence and significance of pericardial effusion. It should not be used as a major or minor criterion for establishing the diagnosis of carditis associated with ARF in the absence of clinical findings.

Though ARF most commonly affects children and young adults between 5 and 24 years of age, approximately 5 % of children diagnosed with ARF are younger than 5 years at diagnosis (Tani et al. 2003). Compared with older patients, children who present before 5 years of age are more likely to have moderate to severe carditis and to present with arthritis or erythema marginatum and are less likely to have chorea.

An important diagnostic dilemma is the entity of post streptococcal reactive arthritis (PSRA). Patients typically develop arthritis 3–14 days after GAS pharyngitis. The arthritis has a protracted course that fails to respond promptly to salicylate therapy – in direct contrast to the vast majority of patients with the polyarthritis of ARF. The fever and rash (scarlatiniform) are usually present during the acute phase of pharyngitis but are absent by the time arthritis appears (Shulman and Ayoub 2002). Some of these patients fulfill the Jones criteria (6 % may have mitral valve insufficiency) and should be diagnosed as having ARF and be treated as such. For those not fulfilling the Jones criteria, a diagnosis of PSRA should be made only after other rheumatologic diagnoses have been excluded.

Therapy

Treatment of GAS Infection

Prevention of both initial and recurrent bouts of ARF is dependent on appropriate and timely treatment of GAS tonsillopharyngitis. Primary prevention is accomplished through timely identification of a GAS throat infection and successfully completing a full course of antibiotic therapy (Gerber et al. 2009) (Table 2). By definition, all cases of ARF, primary and recurrent, are due to GAS infection or reinfection, respectively. Therefore, all patients should be treated for GAS pharyngitis when they present with ARF, even if the throat culture is negative. Intramuscular penicillin G should be considered over oral penicillin V and amoxicillin when compliance with a complete 10-day course of oral therapy is in question. Patients with personal or family histories of ARF or RHD or environmental factors that place them at increased risk for ARF should also receive parenteral penicillin. While anaphylaxis is rare in children, a careful history of allergic reactions to penicillin should always be obtained.

No antibiotic regimen eradicates GAS from the pharynx in 100 % of treated patients (10 % rate of carrier state has been observed). Incomplete eradication of GAS from the pharynx is more frequently seen after oral administration of penicillin than after intramuscular administration. Symptomatic patients who remain GAS positive can be retreated, intramuscularly, if compliance was an issue during the initial treatment. A second course of therapy in asymptomatic GAS-positive patients should only be considered for patients who have had ARF themselves or in members of their family (Gerber et al. 2009).

Rheumatic Fever, Table 2 Primary prevention of rheumatic fever (treatment of streptococcal tonsillopharyngitis)

Agent	Dose	Mode	Duration
Penicillins			
Penicillin V (phenoxymethylpenicillin)	Children: 250 mg 2–3 times/day for ≤27 kg; children >27 kg, adolescents, and adults: 500 mg 2–3 times/day	Oral	10 days
	or		
Amoxicillin	50 mg/kg once daily (maximum 1 g)	Oral	10 days
	or		
Benzathine penicillin G	600,000 U for patients ≤27 kg; 1,200,000 U for patients > 27 kg	Intramuscular	Once
For individuals allergic to penicillin			
Narrow-spectrum cephalosporin[a] (cephalexin, cefadroxil)	Variable	Oral	10 days
	or		
Clindamycin	20 mg/kg/day divided in three doses (maximum 1.8 g/day)	Oral	10 days
	or		
Azithromycin	12 mg/kg once daily (maximum 500 mg)	Oral	5 days
	or		
Clarithromycin	15 mg/kg/day divided in two doses (maximum 250 mg twice daily)	Oral	10 days

The following are not acceptable: sulfonamides, trimethoprim, tetracyclines, and fluoroquinolones (Adapted from Gerber et al. (2009))

[a]To be avoided in those with immediate (type I) hypersensitivity to penicillin

Posttreatment throat cultures should only be obtained after completion of therapy in individuals who remain symptomatic or have recurrence of symptoms or who have had ARF and are at high risk for recurrence.

Secondary Prevention

Prevention of recurrent bouts of ARF (secondary prevention) will prevent severe RHD that results from worsening of existing RHD from repeated inflammatory attacks. GAS tonsillopharyngitis need not be symptomatic to result in recurrence, and an ARF recurrence can occur despite successful completion of a symptomatic GAS infection. Therefore, secondary prevention must be continuous prophylaxis. The duration of prophylaxis in a given patient is dependent on the presence of RHD, which increases the risk of further carditis with recurrent bouts of ARF, increasing the severity of the valvular damage (Table 3). In addition, individuals with increased exposure to GAS (school teachers, health care providers) are at risk for acquiring GAS pharyngitis and should also consider prolonged prophylaxis. Individuals who have had ARF without carditis and have no evidence of RHD on follow-up are not immune to having carditis during recurrences of ARF, but this risk is lower.

Rheumatic Fever, Table 3 Duration of secondary rheumatic fever prophylaxis

Category	Duration after last attack
Rheumatic fever with carditis and residual heart disease (persistent valvular disease[a])	10 years or until 40 years of age (whichever is longer), sometimes lifelong prophylaxis
Rheumatic fever with carditis but no residual heart disease (no valvular disease[a])	10 years or until 21 years of age (whichever is longer)
Rheumatic fever without carditis	5 years or until 21 years of age (whichever Is longer)

[a]Clinical or echocardiographic evidence (Adapted from Gerber et al. (2009))

Successful prophylaxis is heavily dependent on patient compliance; most failures occur in noncompliant patients. Even with high compliance, the risk of recurrence is higher with oral prophylaxis than intramuscularly administer

Rheumatic Fever, Table 4 Secondary prevention of rheumatic fever (prevention of recurrent attacks)

Agent	Dose	Mode
Benzathine penicillin G	600,000 U for children ≤27 kg, 1,200,000 U for those >27 kg every 4 weeks[a]	Intramuscular
Penicillin V	250 mg twice daily	Oral
Sulfadiazine	0.5 g once daily for patients ≤27 kg, 1.0 g one daily for patients >27 kg	Oral
For individuals allergic to penicillin and sulfadiazine		
Macrolide or azalide	Variable	Oral

[a]In high-risk situations, administration every 3 weeks is justified and recommended. See discussion of high-risk situations in the text (Adapted from Gerber et al. (2009))

penicillin. Intramuscular long-acting penicillin every 4 weeks is the recommended regimen for secondary prevention for most individuals in the United States (Table 4). This is particularly important for the patient at high risk for recurrence and those with RHD. Oral agents may be more appropriate for patients at lower risk of recurrence. Once the patient has reached late adolescence, has demonstrated the required maturity and responsibility, and has been free of recurrence for at least 5 years, they may be switched to an oral regimen. The long-term demonstrated benefits of intramuscular prophylaxis have been shown to far outweigh the risk of serious allergic reaction (Markowitz et al. 1991).

General Measures and Anti-Inflammatory Treatment

All patients with suspected ARF (initial bout and recurrences) should be hospitalized. Rarely, when a diagnosis has already been established and the patient is not unwell (e.g., mild recurrent chorea in an otherwise well child without other signs or symptoms), outpatient management may be appropriate. Rest is individualized depending on symptoms. For arthritis, rest for 2 weeks should be adequate. Carditis without congestive heart failure (CHF) requires 4–6 weeks of rest. When CHF is present and under therapy, rest should continue until CHF is controlled, at least for 4–6 weeks. All patients and their families should receive GAS and ARF education, have a good understanding of the cause of ARF, and need to have pharyngitis assessed and treated early. They should understand the reason for secondary prophylaxis and the possible consequences of failing to strictly adhere to prevention measures. If carditis results in valve damage with mitral and/or aortic insufficiency, patients and families must also be made aware of the importance of antibiotic prophylaxis for dental and other dirty procedures to protect against infective endocarditis *in addition* to secondary prophylaxis.

Aspirin is the mainstay of anti-inflammatory therapy with a starting dose of 100 mg/kg/day for the initial 2–3 weeks, tapering to 60–70 mg/kg/day once symptoms have resolved. Older children may require lower doses. Naproxen at 10–20 mg/kg/day may be used in aspirin intolerant patients. The total duration of anti-inflammatory treatment should be continued for 12 weeks. Aspirin may cause tinnitus at high serum levels and obtaining blood levels may be helpful. Proton-pump inhibitors should also be prescribed with all anti-inflammatory drug regimens to prevent gastritis. If no significant response is seen to aspirin or naproxen in 4 days, steroid treatment should be instituted.

Steroid treatment in the form of oral prednisolone at a dose of 2 mg/kg/day (maximum dose 80 mg/day) is instituted from the outset of treatment for moderate to severe carditis and is maintained at the initial dose until the ESR and other acute-phase reactants normalize – usually 2 weeks. If no response to oral steroid therapy is observed, then IV methyl prednisolone 30 mg/kg/day for 3 days may be used. Tapering of oral steroid therapy is done over 2–4 weeks by reducing the daily dose by 2.5–5 mg every third day while simultaneously starting aspirin at 50–75 mg/kg/day to prevent rebound of symptoms once steroids have been weaned. A total duration of therapy again should be 12 weeks

(Saxena et al. 2008). A Cochrane database review found no significant difference in the risk of chronic RHD at 1 year following ARF between corticosteroid-treated and aspirin-treated patients (Cilliers et al. 2012).

Chorea

No treatment is usually required except for rest and a quiet environment. Sedatives may be of some benefit. For more significant chorea, haloperidol (0.25–0.5 mg/kg/day) or valproic acid (15 mg/kg/day) or carbamazepine (7–30 mg/kg/day) may be used. Treatment should be continued for 2–4 weeks after clinical improvement is observed. Importantly, if acute-phase reactants are elevated, anti-inflammatory therapy must also be administered.

Carditis and CHF

Physical activity should be restricted as previously described, strictly so when carditis is associated with CHF. As carditis is an active process with aggressive potential destruction of mitral, aortic, or both valves (all four valves are at risk), serial frequent cardiovascular examinations, both clinical and echocardiographic, are mandatory. Signs of heart failure, pulmonary congestion from severe mitral insufficiency, wide pulse pressure from progressive aortic insufficiency, and overall progressive low cardiac output from poor myocardial function should be sought out and treated. Daily weights, fluid balance, vital signs focusing on tachycardia, and pulse pressure must be closely monitored. Diuretics and salt restriction, digoxin, and afterload-reducing agents such as angiotensin-converting enzyme inhibitors or angiotensin receptor blockers are the mainstays of CHF therapy. For severe carditis and myocardial dysfunction, intravenous inotropic agents such as dopamine and dobutamine and afterload-reducing agents such as milrinone may be required. Surgical intervention for unremitting CHF as a result of severe mitral regurgitation is rarely required, but surgical therapy is absolutely indicated when chordal rupture is the cause.

Outcome

The prognosis of ARF is known to be directly related to the severity of the cardiac involvement in the initial attack. The presence of arthritis as well as the presence of chorea in the initial bout of ARF is related to a better prognosis in relation to significant chronic RHD. Despite being associated with subclinical carditis, chorea is considered as a protective factor for significant valvular heart disease. The main prognostic factors for severity of valvular heart disease are moderate or severe carditis on initial presentation, recurrences of ARF, and low maternal education (Meira et al. 2005). In patients with mild carditis, clinical manifestations disappear in the chronic phase in approximately 80 % patients, while in patients with moderate or severe carditis, chronic RHD will be the most probable outcome.

Valvular lesions and murmurs resolve in approximately one-third of the patients with acute carditis within 5–10 years (Vasan and Selvaraj 1999). In another study, improved cardiac auscultation was also observed in 33 % of the patients, however, without a corresponding Doppler echocardiographic improvement (Araujo et al. 2012). Although a 10 % reduction in the severity of significant valvulitis was observed after 2 years of onset, Doppler echocardiography became normal in only four (3.4 %) patients who had had moderate valvulitis in the first episode of ARF and in 33 (24 %) of those with mild lesions. No severe valvular lesion resolved completely, although auscultation had become normal in 13 % of the patients with severe valvulitis on the initial assessment (subclinical chronic RHD). At follow-up almost 10 years after the initial ARF episode, 69 % of children whose first episode occurred prior to age 5 years have RHD (Tani et al. 2003). Subclinical echocardiographically detected valvular abnormalities were found at both at presentation (33 %) and at follow-up (55 % of those with initial carditis).

Valvular scarring occurs beyond 15 years and is associated with recurrences of ARF. Mitral stenosis and mixed lesions of stenosis and regurgitation (as well as aortic stenosis and insufficiency) become more predominant. Atrial fibrillation may accompany severe mitral valvular disease and is poorly tolerated. Patients who suffer from severe valvular disease may require surgical intervention though children with RHD in the United States uncommonly require valve surgery. Current valve repair techniques can be performed with good long-term functional results in childhood on both mitral and aortic valves that allow for annular growth (Hillman et al. 2004). Long-term surveillance of children and adults with RHD is necessary because of the possible need for late valve replacement and the risk for atrial fibrillation.

Cross-References

▶ Autoinflammatory Syndromes
▶ Corticosteroids
▶ Genetic Susceptibility to Inflammatory Diseases
▶ Salicylates
▶ Tumor Necrosis Factor Alpha (TNFalpha)

References

Araújo, F. D. R., Goulart, E. M. A., & Meira, Z. M. A. (2012). Prognostic value of clinical and Doppler echocardiographic findings in children and adolescents with significant rheumatic valvular disease. *Annals of Pediatric Cardiology, 5*, 120–126.

Ayoub, E. M., Barrett, D. J., Maclaren, N. K., & Krischer, J. P. (1986). Association of class II human histocompatibility leukocyte antigens with rheumatic fever. *Journal of Clinical Investigation, 77*, 2019–2026.

Bhaya, M., Panwar, S., Beniwal, R., & Panwar, R. B. (2010). High prevalence of rheumatic heart disease detected by echocardiography in school children. *Echocardiography, 27*, 448–453.

Bryant, P. A., Robins-Browne, R., Carapetis, J. R., & Curtis, N. (2009). Some of the people, some of the time. Susceptibility to acute rheumatic fever. *Circulation, 119*, 742–753.

Carapetis, J. R., Steer, A. C., Mulholland, E. K., & Weber, M. (2005). The global burden of group A streptococcal diseases. *The Lancet Infectious Diseases, 5*, 685–694.

Cilliers, A., Manyemba, J., Adler, A.J., Saloojee, H. (2012). Anti-inflammatory treatment for carditis in acute rheumatic fever. *Cochrane Database of Systematic Reviews, 6*, 1–53. Art. No.: CD003176. doi:10.1002/14651858.CD003176.pub2.

Dajani, A. S., Ayoub, E., Bierman, F. Z., Bisno, A. L., Denny, F. W., Durack, D. T., et al. (1992). Guidelines for the diagnosis of rheumatic fever: Jones Criteria, 1992 update. *JAMA, 268*, 2069–2073.

Dale, J. B., & Beachey, E. H. (1985). Multiple heart cross reactive epitopes of streptococcal M proteins. *Journal of Experimental Medicine, 161*, 113–122.

Engel, M. E., Stander, R., Vogel, J., Adeyemo, A. A., & Mayosi, B. M. (2011). Genetic susceptibility to acute rheumatic fever: A systematic review and meta-analysis of twin studies. *PLoS ONE, 6*(9), e25326. doi:10.1371/journal.pone.0025326. Accessed 4 Sept 2012.

Ferretti, J. J., McShan, W. M., Ajdic, D., Savic, D. J., Savic, G., Lyon, K., et al. (2001). Complete genome of an M1 strain of *Streptococcus pyogenes*. *Proceedings of the National Academy of Sciences, 8*, 4658–4663.

Galvin, J. E., Hemric, M. E., Ward, K., & Cunningham, M. W. (2000). Cytotoxic Ab from rheumatic carditis recognizes heart valves and laminin. *Journal of Clinical Investigation, 106*, 217–224.

Gerbase-DeLima, M., Scala, L. C., Temin, J., Santos, D. V., & Otto, P. A. (1994). Rheumatic fever and the HLA complex: A cosegregation study. *Circulation, 89*, 138–141.

Gerber, M. A., Baltimore, R. S., Eaton, C. B., Gewitz, M., Rowley, A. H., Shulman, S. T., et al. (2009). Prevention of rheumatic fever and diagnosis and treatment of acute Streptococcal pharyngitis. A scientific statement from the American Heart Association Rheumatic Fever, Endocarditis, and Kawasaki Disease Committee of the Council on Cardiovascular Disease in the Young, the Interdisciplinary Council on Functional Genomics and Translational Biology, and the Interdisciplinary Council on Quality of Care and Outcomes Research. *Circulation, 119*, 1541–1551.

Hillman, N. D., Tani, L. Y., Veasy, L. G., Lambert, L. L., Di Russo, G. B., Doty, D. B., et al. (2004). Current status of surgery for rheumatic carditis in children. *The Annals of Thoracic Surgery, 78*, 1403–1408.

Markowitz, M., Kaplan, E., Cuttica, R., Berrios, X., Huang, Z., Rao, X., et al. (1991). Allergic reactions to long-term benzathine penicillin prophylaxis for rheumatic fever. *Lancet, 337*, 1308–1310.

Meira, Z. M., Goulart, E. M., Colosimo, E. A., & Mota, C. C. (2005). Long term follow up of rheumatic fever and predictors of severe rheumatic valvar disease in Brazilian children and adolescents. *Heart, 91*, 1019–1022.

Ramasawmy, R., Fae, K. C., Spina, G., Victora, G. D., Tanaka, A. C., Palacios, S. A., et al. (2007). Association of polymorphisms within the promoter region of the tumor necrosis factor-α with clinical outcomes of rheumatic fever. *Molecular Immunology, 44*, 1873–1878.

Remenyi, B., Wilson, N., Steer, A., Ferreira, B., Kado, J., Kumar, K., et al. (2012). World Health Federation criteria for echocardiographic diagnosis of rheumatic heart disease-an evidenced-based guideline. *National Reviews Cardiology, 9*, 297–309.

Saxena, A., Kumar, R. K., Gera, R. P. K., Radhakrishnan, S., Mishra, S., Ahmed, Z., et al. (2008). Consensus guidelines on pediatric acute rheumatic fever and rheumatic heart disease. *Indian Pediatrics, 45*, 565–573.

Seckeler, M. D., & Hoke, T. R. (2011). The worldwide epidemiology of acute rheumatic fever and rheumatic heart disease. *Clinical Epidemiology, 3*, 67–84.

Shulman, S. T., & Ayoub, E. M. (2002). Poststreptococcal reactive arthritis. *Current Opinion in Rheumatology, 14*, 562–565.

Tani, L. Y., Veasy, L. G., Minich, L. L., & Shaddy, R. E. (2003). Rheumatic fever in children younger than 5 years: Is the presentation different? *Pediatrics, 112*, 1065–1068.

Vasan, R. S., & Selvaraj, N. (1999). Natural history of acute rheumatic fever. In J. Narula, R. Virmani, K. S. Reddy, & R. Tandon (Eds.), *Rheumatic fever* (pp. 347–358). Washington, DC: American Registry of Pathology.

Rheumatoid Arthritis

Tangada Sudha Rao[1] and Laila Rahbar[2]
[1]Department of Rheumatology, McGuire VA Medical Center, Richmond, VA, USA
[2]Arthritis and Rheumatic Diseases PC, Richmond, VA, USA

Synonyms

Inflammatory arthritis; RA; Rheumatoid

Definition

Rheumatoid arthritis (RA) is a chronic, systemic, inflammatory autoimmune disease characterized by symmetric, small joint, inflammatory arthritis. It has a peak incidence in the fourth or fifth decade of life and affects all ethnic groups worldwide with 2.5 times higher prevalence in females compared to males. The onset appears to be more frequent in winter months compared to summer months.

Epidemiology and Genetics

The overall general estimated prevalence of RA ranges from 0.2–1.1 % with higher rates seen in European and North American populations (0.5–1.1 %) and lower rates in Asia, Middle East, and South America (0.2–0.4 %). Some Native American populations including Pima, Chippewa, and Yakima groups have particularly high prevalence rates, greater than 5 %.

The *pathogenesis of RA* is thought to be multifactorial with both environment and genetic factors playing pivotal roles. The risk of developing RA is 1.5-fold higher in first degree relatives compared to the general population. Compared to the general population, siblings of individuals with RA have a two- to fourfold higher likelihood of developing the disease. The overall heritability is about 50–60 % as estimated from twin studies.

As with other autoimmune diseases, most genetic research has been centered around *major histocompatibility complex* (*MHC*) *alleles* located on chromosome 6. This region is the most dense gene area of the mammalian genome, playing an important role in autoimmunity, reproductive success, and the immune system. Certain MHC alleles convey risk for the development of RA. The most strongly associated allele is HLA-DR4, which was first recognized in 1978. However, HLA-DR4 positivity occurs in 20–30 % of the general population, thus other factors must also be present in order for the disease to develop. The association has since been redefined to alleles containing shared epitopes of the HLA-DRB1 molecule; this locus has been associated with patients who are positive for rheumatoid factor (RF) and anti-citrullinated peptide (ACPA). Alleles that contain a shared epitope within the HLA-DRB1 region convey susceptibility. This suggests that some T cell repertoire selection, antigen presentation, or alteration in peptide affinity has a role in promoting autoreactive adaptive immune processes. Moreover, genetic variation only accounts for approximately 30 % of disease susceptibility. Over the years, researchers have also found non-HLA risk associated alleles.

Nongenetic Risk Factors

Nongenetic risk factors also play an important role in RA susceptibility. As mentioned earlier, women are two times as likely to develop RA compared to males. This suggests the importance of hormonal and reproductive factors. For example, estrogen potentially could decrease apoptosis of B cells, thus, leading to selection of autoreactive clones. Null parity has been suggested by several case–control studies to be a risk factor for RA. There have also been a few studies which showed breastfeeding could decrease the risk for development of RA.

Other Environmental Factors

Other environmental factors have been found to increase the risk for RA. Cigarette smoking, identified over a decade ago, is the strongest known environmental risk factor for the development of seropositive RA. Studies have shown that smoking or other forms of bronchial stress (silica exposure) increase risk of RA in individuals with susceptibility associated HLADR4 alleles. Similarly, smoking and HLA-DRB1 alleles increase risk of having APC-A. Conversely, alcohol consumption may lower the risk. Studies on nutritional effects and RA have been conflicting. Some studies have suggested olive oil and fish oil are protective against RA. Effects of vitamin D levels and red meat/protein intake have shown variable results. Observations have long suggested that *an infectious agent* could be responsible for the disease in a genetically susceptible host. Although a unifying mechanism is unknown, plausible biologic mechanisms have been suggested by results from animal models, likely via some form of molecular mimicry. An infection and the immune complexes that form could trigger the induction of RF. A number of potential infectious agents have been implicated as potential triggers: *Epstein-Barr virus*, *human parvovirus B19*, *Cytomegalovirus*, retroviruses, Proteus, *Mycoplasma*, and *Mycobacteria*. More recently, the association between RA and periodontal disease has been emphasized, specifically the presence of *Porphyromonas gingivalis* which expresses PADI4 and is capable of citrullinating mammalian proteins. Lastly, the gastrointestinal microbiome has also been newly implicated in the development of autoimmunity in articular models.

Pathophysiology

The cause of rheumatoid arthritis is unknown. Environmental triggers such as smoking, gut microbiome, or periodontitis can promote posttranslational citrullination of arginine in a susceptible person. Citrullinated residues act as neoepitopes and lead to formation of antibodies such as anticyclic citrullinated peptide (CCP) antibodies and rheumatoid factor. These antibodies have been found several months prior to disease onset. Factors which lead to the development of arthritis and localization of the inflammatory process in the joint are poorly understood.

Inflammation in a joint, known as synovitis, occurs as a result of migrating leukocytes infiltrating the synovial compartment. Repeated inflammation leads to destruction of normal synovium architecture, cartilage damage, and bone erosion. Numerous processes are involved in this process including endothelial activation in synovial microvessels, increased expression of adhesion molecules (integrins, selectins) and chemokines, increased cytokines, and fibroblast activation.

Systemic inflammation in rheumatoid arthritis can lead to various extra-articular manifestations involving the cardiovascular, pulmonary, and skeletal systems.

Adaptive Immune System

The synovium contains abundant T cells, myeloid cells, and plasmacytoid dendritic cells, the latter expressing cytokines (IL 12, 15, 18, and 23), HLA Class II molecules, and costimulatory molecules which are necessary for T cell activation and antigen presentation. RA is believed to be a disease mediated mostly by type 1 helper T cells; however, more recent research has also shown involvement of type 17 helper T cells. Differentiation of Th17 occurs via IL 1β, 6, 21, 23, and transforming growth factor β (TGFβ) expressed by macrophage-derived and dendritic cell derived cells. These cytokines promote differentiation of Th17 and suppress differentiation of regulatory T cells. These Th17 cells produce ILs 17A, 17 F, 21, 22, and tumor necrosis factor α (TNFα). The activity of regulatory T cells is then further blocked by TNFα. Interleukin 17A along with TNF α promotes activation of fibroblasts and chondrocytes. Humoral adaptive immunity also plays a large role in pathogenesis of RA. Synovial B cells are largely found in T cell-B cell aggregates.

Smoking, silica, other stressors of pulmonary, and other barrier tissues can promote citrullination of arginine by the enzyme peptidyl arginine deiminase 4, in the mucosal proteins. These neoepitopes elicit an anti-CCP response. Several citrullinated self-proteins have been identified in anti-CCP assays, including alpha enolase, keratin, fibrinogen, fibronectin, collagen, and vimentin.

Innate Immune System

Macrophages, mast cells, and natural killer cells are found in the synovial membrane of RA patients. Macrophages are the central effectors of synovitis. They release cytokines (IL 1, 6, 12, 18, and 23), reactive oxygen intermediates, and nitrogen. They produce prostanoids and matrix degradation enzymes and are involved in phagocytosis and antigen presentation. Macrophages are activated via Toll-like receptors (TLRs), cytokines, interactions with T cells, immune complexes, lipoprotein molecules, and liver x-receptor agonists (oxysterols, LDL, HDL). Mast cells also produce high levels of cytokines, chemokines, and proteases.

Neutrophils are found predominately in the synovial fluid. They are involved in producing prostaglandins, proteases, and reactive oxygen intermediates which further enhance synovitis.

Cytokines and Intracellular Signaling Pathways

Cytokines produced by synovial cells are central to the pathogenesis of rheumatoid arthritis. These include the IL 1 family of cytokines, IL-4, IL-6, IL-13, IL-15, and TNFα. Current therapies are targeted to block the effects of these cytokines.

Clinical Presentation

Obtaining a good history is key to making a diagnosis of RA. Most commonly, individuals will complain of symmetric, small joint pain, swelling, and stiffness. Although RA is a symmetric, polyarticular inflammatory arthritis, in the beginning stages it can present with monoarticular involvement only. The development of joint symptoms may occur overnight or be slowly progressive over many months. To diagnose RA, symptoms must be present for at least 6 weeks.

Joint Symptoms

Pain in the joints is a universal feature of RA, with the typical pattern including: proximal interphalangeal joints (PIPs), metacarpophalangeal joints (MCPs), wrists, elbows, shoulders, hips, knees, ankles, and metatarsophalangeal joints (MTPs). Usually symptoms begin with the smaller joints being affected but as the disease progresses, larger joints can become involved. The hands are most commonly affected and individuals will complain of swelling over the knuckles, decreased grip strength, and inability to make a fist. Additionally, they report increased stiffness in joints in the mornings or after prolonged periods of rest. Classically, these people will have positive RF and/or ACPA. In about 50 % of cases, individuals can have positive autoantibodies for years prior to development of symptoms.

Individuals with suspected RA should undergo a thorough physical examination. Symmetric swelling and tenderness and synovitis of the joints are almost always observed. Earlier in the disease process, it is not uncommon to see fusiform swelling of PIPs but as RA progresses, deformities of the hands such as ulnar deviation of the fingers, dorsal subluxation of MCP joints, hyperextension of PIPs or swan neck, and hyperflexion of PIPs or boutonniere can develop. Swelling and decreased range of motion of the elbows and wrists occurs frequently and can be easily detected on examination. Larger, deep joints such as the shoulders are more difficult to ascertain synovitis. Hip involvement occurs in about 20 % of RA patients. Knee involvement is quite common and effusions are easily detected on examination. Large knee effusions can also result in an enlarged popliteal cyst, also known as Baker's cyst. These can rupture and cause calf pain, pitting edema, and bruising around the ankle. Synovitis can also occur in the ankle joint, either causing inflammation in the talotibial joint (which moderates flexion/extension) and/or hindfoot (with moderated eversion/inversion). Tenosynovitis and rupture of the posterior tibial tendon is also a common feature in RA. Earlier in the disease process, a wider, puffy forefoot also known as splaying of the toes can be seen along with tenderness of MTPs. As the disease progresses, dorsal subluxation of MTPs resulting in cock-up toes deformities and bunions become common. Lastly, the cervical spine is also involved in RA, mostly through tenosynovitis of the transverse ligament of C1, which stabilizes the odontoid process of C2. Persistent inflammation can lead to erosion of the odontoid process and/or rupture of the transverse ligament resulting in cervical myelopathy.

Extra-articular Manifestations

RA is not merely confined to the joints. In fact, up to 50 % of RA patients experience some type of extra-articular manifestation throughout their disease course. The most common extra-articular manifestation in RA is Sjogren's syndrome, characterized by dry mouth (xerostomia) and dry eyes (keratoconjunctivitis sicca), occurring in about 35 % of patients. Firm, nontender nodules, known as RA nodules, occur in about 25 % of individuals. These usually form over pressure areas such as elbows, Achilles tendons, ischial tuberosities, fingers, and scalp. Lung disease is common with up to 30 % developing parenchymal lung disease (either asymptomatic pulmonary nodule or interstitial lung disease) and pleural effusion/pleurisy in up to 25 %. Pericarditis is the most common cardiac manifestation of RA, but is generally asymptomatic and usually found on autopsy. However, compared to the general population, individuals with RA have a higher incidence of myocardial infarctions and strokes. As might be suspected, anemia of chronic disease is the most common hematological finding in this patient population. Lastly, small vessel vasculitis, generally occurring in fingers/nailfolds, is an uncommon manifestation. Nevertheless, it can result in peripheral neuropathy and mononeuritis multiplex causing wrist or foot drop.

In most cases, RA is a progressive disease and if left untreated can cause significant joint destruction and immobility.

Therapy of Rheumatoid Arthritis

Therapy of rheumatoid arthritis (RA) has changed considerably over the last 25 years. Treatment strategies used in the early part of twentieth century provided symptomatic relief only, consisting of salicylates, NSAIDS and physical measures such as bed rest, splinting, and physical therapy (See also entry on "▶ Non-steroidal Anti-inflammatory Drugs: Overview"). Infection was thought to play a role in the development of RA and some of the earlier therapies used were based on this principle, e.g., gold salts (Forestier 1932) and sulfasalazine (Amos 1995). In 1949, steroids were first shown to have a dramatic clinical benefit in patients with RA (Hench et al. 1949).

Other medications such as hydroxychloroquine, D-penicillamine, azathioprine, cyclophosphamide, and cyclosporine were used with varying clinical responses and these were termed Disease Modifying Agents of Rheumatic Diseases (DMARDS). The recommendation in the early 1980s was to start with the least toxic therapy and advance if needed; this was termed the pyramidal approach. This approach led to improvement in some patients with decreased disease activity. The development and application of tools to assess outcomes of therapeutic interventions helped to recognize the fact that many of these drugs did not modify the course of the disease. These studies also revealed that structural damage often occurred within the first 2 years of disease onset.

In the mid-1980s, methotrexate (MTX) was shown to be safe and effective in RA (Weinblatt et al. 1985; Kremer and Lee 1986), was approved by the FDA for RA therapy in 1988, and a reversal of the pyramidal approach was recommended. By the 1990s, MTX had become the initial drug of choice to treat RA. Not all patients responded to MTX, some could not tolerate the effective dose, and it was contraindicated in young women contemplating pregnancy and in regular alcohol users. Because of the above limitations and as MTX alone did not result in drug-free remission in all patients, combination therapies were studied and shown to be effective and safe. Combinations used included MTX with steroids and other DMARDS.

Research during the 1980s and 1990s elucidated the role of small molecular mediators of inflammation such as arachidonic acid metabolites, cytokines, growth factors, chemokines, adhesion molecules, and metalloproteinases. In 1994, the results of the first RCT using tumor necrosis factor (TNF) inhibition with infliximab was published which changed the therapeutic landscape of RA. Based on this research information, agents targeting other cytokines, cell receptors, and intracellular signaling molecules were and continue to be studied and shown to be beneficial in the treatment of RA.

With advances in therapies over the last 25 years, it is possible to slow or halt disease progression and prevent irreversible joint damage. Aggressive therapy of RA has been shown to be beneficial in decreasing mortality in RA from associated diseases such as cardiovascular diseases. The treatment goal for RA, therefore, is no longer symptom control or low-disease activity but disease remission.

Disease-Modifying Antirheumatic Agents

For a more detailed discussion of this class of drugs, the reader is referred to the entry on "▶ Disease-Modifying Antirheumatic Drugs: Overview".

Gold Salts

Forestier started the use of gold salts in 1929 and studies since have confirmed the effectiveness of gold in controlling RA, given commonly as weekly intramuscular injections. An oral form is also available. Delay in progression of erosions was demonstrated in patients started on gold therapy within a year of disease onset.

The mechanism of action is thought to include: partial blockade of B cell maturation, inhibition of fibroblast proliferation, neutrophil chemotaxis, and PGE2 synthesis.

Adverse events occur in up to 40 % of patients, including: vasomotor nitroid postinjection reactions, dermatitis, stomatitis, nephrotic syndrome, and cytopenias. The reader is referred to the entry on "▶ Gold Complexes" for a more detailed discussion.

Sulfasalazine

Sulfasalazine was developed by Nanna Svartz, a Swedish physician in 1938, as an azo-conjugate of salicylic acid (antiarthritic) and sulfapyridine (antibacterial) for the treatment of RA, in the belief that infection was the cause of RA.

The mechanism of action includes: inhibition of production of prostanoids and B cell activation, altered distribution of lymphocyte subsets in intestinal mucosa, and reduced production of IL-1 and TNFα.

Studies showed sulfasalazine (SSZ) to be superior to placebo, as effective as gold and penicillamine, more effective than antimalarials but less effective than methotrexate. A 2-year double-blind trial in RA showed that a combination of MTX, SSZ, and HCQ was significantly more effective than MTX alone or than the combination of SSZ and HCQ.

Adverse events include nausea, upper abdominal discomfort, hepatitis, leukopenia, neutropenia, skin rash, and rare pulmonary infiltrates. The reader is referred to the entry on "▶ Sulfasalazine and Related Drugs" for a more detailed discussion.

Glucocorticoids

In 1949, Philip Hench demonstrated the dramatic clinical effects of cortisone and ACTH in the therapy of RA. Initial studies with prednisolone compared to ASA showed similar clinical benefits but decreased radiographic progression in patients treated with cortisone.

Adverse events associated with long-term use led to recommendations for use of low-dose steroids initially with tapering off after disease control. More recently, several studies have shown that treatment regimens, including low-dose steroids given early in RA, slow x-ray progression.

The reader is referred to the entry on "▶ Corticosteroids" for a more detailed discussion.

D-Penicillamine

D-penicillamine is a derivative of penicillin, obtained by acid hydrolysis. It is approved for therapy of RA and its highest use was between 1975 and 1985; use declined in the USA after that. Efficacy was first established in a UK multicenter trial. It was particularly useful in some RA patients with extra-articular manifestations such as vasculitis, Felty's syndrome, amyloidosis, lung disease, and nodulosis.

The mechanism of action of D-penicillamine in RA is unknown.

Adverse events include leukopenia, thrombocytopenia, aplastic anemia, membranous type nephropathy, skin rash, and rare hepatotoxicity and pulmonary complications.

Hydroxychloroquine

Hydroxychloroquine is the most commonly prescribed antimalarial medication in the United States.

Several double-blind, placebo-controlled trials have confirmed the benefits of HCQ in treating RA.

With regard to mechanism of action, HCQ interferes with intracellular functions dependent on an acidic microenvironment, stabilizes lysosomal membranes, inhibits PMN chemotaxis and phagocytosis, inhibits cytokine production, and may decrease autoantibody production.

Adverse events include GI distress, skin rash, neuromyopathy, and retinal toxicity including blurred vision, corneal deposits, and retinopathy. The reader is referred to the entry on "▶ Antimalarial Drugs" for a more detailed discussion.

Methotrexate

The first clinical study on the use of MTX in RA was published in 1985 and the drug was approved by the FDA for use in RA in 1988. It was shown to improve symptoms, disease activity and movement of patients, and inhibit radiographic progression to some extent. Long-term prospective studies have demonstrated sustained effects on disease activity and radiographic progression. The mechanism of action of MTX involves interference with the ability of folate to act as a cofactor for a variety of enzymes that are essential for cell replication, accumulation in the cell in the polyglutamated form, which has high affinity for folate dependent enzymes, and blockade of AICAR transformylase leading to the accumulation of adenosine, which binds to receptors on inflammatory cells and dampens inflammation.

MTX use in RA is associated with a decrease in extra-articular manifestations such as vasculitis, Felty's syndrome, and amyloidosis. MTX reduced overall mortality in RA patients by 60 % through its reduction in CVD mortality. MTX is considered the anchor drug among DMARDS and is internationally accepted as the first line choice in treatment of RA.

Adverse events include anorexia, nausea, vomiting, diarrhea, stomatitis, hematologic toxicity, hepatotoxicity, pulmonary toxicity, and it can increase the risk of EBV associated lymphomas. The reader is referred to the entry on "▶ Methotrexate" for a more detailed discussion.

Leflunomide

Leflunomide is a synthetic isoxazole derivative with immunosuppressive and antiproliferative properties. It acts through its active metabolite and was approved by the FDA for use in RA in 1998.

The mechanism of action of the active metabolite involves inhibition of the enzyme dihydroorotate dehydrogenase, important in pyrimidine synthesis, and it thereby inhibits T cell proliferation in response to inflammatory cytokines. It has a long half life of approximately 2 weeks.

In randomized controlled trials, response rates to leflunomide and MTX have been similar, greater than those of placebo and comparable to those of SSZ. Leflunomide reduced radiographic progression.

Adverse events include: diarrhea, alopecia, rash, and liver enzyme elevation. Marked liver enzyme elevation will require stopping the drug and cholestyramine administration to decrease drug levels. It is teratogenic.

Biologic Response Modifiers

Tumor Necrosis Factor Inhibitors
Since the first study using infliximab for RA therapy, four other agents have been developed and approved for treatment of RA (See also entry on "▶ Tumor Necrosis Factor (TNF) Inhibitors").

Infliximab – Chimeric antibody, binds to soluble and membrane-bound TNF alpha
Etanercept – TNF receptor-IgG1 Fc fusion protein that binds and inactivates both TNF alpha and beta
Adalimumab – Recombinant human IgG1 monoclonal antibody specific for TNF alpha
Certolizumab Pegol – Pegylated Fab fragment of a humanized anti-TNF antibody
Golimumab – Human monoclonal antibody targeted against TNF alpha

All these agents have well-established efficacy demonstrated in large randomized controlled trials. Treatment with these agents improves disease activity and may slow or arrest disease progression assessed by clinical, radiographic, and patient-reported outcome measures.

All the agents are effective as monotherapies, and studies have shown improved efficacy in combination with MTX.

Long-term postmarketing surveys and national registries have confirmed their acceptable safety profile. A Cochrane review showed that patients on these agents were likely to experience adverse events more frequently than those in the control group.

Patients need to be screened for latent tuberculosis prior to starting the medication. Unusual infections such as histoplasmosis, coccidioidomycosis, pneumocystis, listeria, and legionella can be seen in patients treated with TNF inhibitors. These agents are contraindicated in patients with chronic hepatitis B infections; no significant worsening is noted in patients with chronic hepatitis C infections. They may worsen demyelinating diseases and congestive heart failure (CHF), with unclear evidence for a risk of increased malignancy, although nonmelanoma skin cancers have been shown to be more common with this treatment.

Interleukin 1 Inhibition
Interleukin 1 receptor antagonist (IL1Ra) is a naturally occurring molecule that binds to IL 1 receptor and prevents binding of IL1 to the receptor (See also entry on "▶ Interleukin-1 (IL-1) Inhibitors: Anakinra, Rilonacept, and Canakinumab").

Anakinra is a recombinant human IL1 Ra. It is administered subcutaneously daily, and studies have demonstrated its safety and efficacy in active rheumatoid arthritis. Long-term follow-up showed the efficacy to be inferior to that of other biologics.

Interleukin 6 Inhibition
Tocilizumab is a recombinant humanized monoclonal antibody targeting the IL6 receptor, which blocks IL6 binding and inhibits its inflammatory effects (See also entry on "▶ Interleukin-6 Inhibitor: Tocilizumab").

The efficacy of tocilizumab in inadequate MTX responders was studied in three international, double-blind, placebo-controlled phase III and one multicenter US study, all of which showed a significantly higher proportion of patients receiving tocilizumab achieving primary end points compared to patients receiving placebo.

Tocilizumab is given as a monthly IV infusion and has been shown to be effective as monotherapy in the treatment of RA. Some side effects with IL 6 inhibition include reversible neutropenia, elevated lipids, transient elevation of liver enzymes, and GI perforation.

B Cell Inhibition
Rituximab is a chimeric monoclonal antibody targeting the B cell surface antigen CD 20. It can cause B cell depletion by antibody-dependent cytotoxicity, complement-dependent cytotoxicity, or apoptosis due to inhibition of Bcl2 expression (see also entry on "▶ CD20 Inhibitors: Rituximab").

Two large randomized multicenter studies have confirmed the efficacy of rituximab in RA patients who are inadequate responders to MTX and those who failed TNF inhibitor therapy. Studies also suggest that there is improvement in efficacy with subsequent cycles of rituximab, when given in early RA and when given at regular intervals.

Adverse events: There is an increased risk of infections, delayed onset neutropenia, reactivation of hepatitis B, and rare cases of progressive multifocal leuko-encephalopathy (PML) have been reported. Infusion reactions are common; premedication with IV steroids and antihistamines reduces the severity of infusion reactions. Immunoglobulin levels are less affected.

T Cell Costimulation Inhibition
Activation of T cells requires binding of the T cell receptor to the antigen-MHC complex on the antigen-presenting cell and costimulation including binding of CD28 on the T cell to CD80/86 on the antigen-presenting cell. The inhibitory coreceptor, CTLA4, on the T cell has high affinity for CD80/86 and inhibits T cell costimulation and activation (see also entry on "▶ CD80/86 Inhibitors: Abatacept").

Abatacept consists of the Fc portion of Ig attached to the extracellular domain of CTLA4. It is given as monthly IV injections, after a loading dose. Subcutaneous injections are also available. Clinical trials have demonstrated its efficacy in RA patients responding inadequately to MTX, other DMARDS, and anti-TNF agents.

Adverse events: Cochrane review has suggested that abatacept had a better safety profile than most other biologic agents. Infusion reactions are seen.

Kinase Inhibitors
Tofacitinib is an oral Janus kinase inhibitor that preferentially inhibits JAK1 and JAK3 and blocks signaling pathways for several cytokines such as ILs 2, 4, 7, 9, 15, and 21 which are important in lymphocyte function (see also entry on "▶ Janus Kinase Inhibitors").

Studies have shown it to be effective in the therapy of RA as monotherapy (34) and in combination with MTX in MTX and TNFI incomplete responders. It was also shown to inhibit progression of structural damage and improve disease activity in patients with RA who are receiving MTX.

Adverse events: neutropenia, anemia, elevated liver enzymes, dyslipidemia, elevated creatinine, and infections.

Tofacitinib is FDA approved for the treatment of rheumatoid arthritis.

Outcomes in Rheumatoid Arthritis

Goals of RA therapy include preventing joint damage, minimizing disability, and improving mortality. With recent advances in the treatment of RA, it is possible to aim for remission in some patients. A number of measures have been developed to measure disease activity, some require testing of acute phase reactants, but others do not.

ACR 20/50/70 scores are useful for measuring change in disease activity from past visits, while others such as DAS28, HAQ, Simplified Disease Activity Index (SDAI), Clinical Disease Activity Index (CDAI), and Routine Assessment of Patient Index Data (RAPID) are useful to assess disease activity at a single point. Use of these measures to guide aggressive treatment to reach a defined goal is an effective strategy.

The feasibility of attaining remission led to the development of ACR/EULAR 2011 remission criteria: at any time point the patient must satisfy all of the following: total joint count of <1, swollen joint count of <1, CRP <1 mg/dl and patient global assessment <1 (0–10 scale) *or* a SDAI of <3.3 at any time point. The role of imaging in the definition of remission remains to be determined.

Treating rheumatoid arthritis with the therapeutic target of remission or low-disease activity has significantly improved outcomes. Guidelines for treatment have been published by the American College of Rheumatology and European League Against Rheumatism.

Several studies have shown the benefit of a combination of DMARDS in achieving remission or low-disease activity. The FIN-RACo trial showed that combination therapy with low-dose prednisone achieved remission or low-disease activity in a third of the study patients compared to monotherapy (Mottonen et al. 1999). Improvement in disease activity, radiographic disease progression, physical function, and quality of life have been achieved using a strategy of tight control in an outpatient setting – TICORA study (Grigor et al. 2004).

Triple therapy with methotrexate, sulfasalazine, and hydroxychloroquine was shown to be noninferior to etanercept and methotrexate in patients with rheumatoid arthritis who had active disease despite methotrexate therapy (O'Dell et al. 2013). Biologic DMARDs are currently recommended in patients who have failed synthetic DMARDS. In patients with poor prognostic factors such as presence of autoantibodies in high titer, high-disease activity as measured by composite indices and early occurrence of erosions should receive biologic DMARD if initial DMARD failed. TNF inhibitors, abatacept, rituximab, and tocilizumab can be used as treatment choices in these patients.

The 2014 update of Treat to Target recommendations by an international task force proposes four overarching principles and ten recommendations based on strong evidence (Smolen et al. 2016).

Cross-References

▶ Antimalarial Drugs
▶ Corticosteroids
▶ Disease-Modifying Antirheumatic Drugs: Overview
▶ Gold Complexes
▶ Interleukin-6 Inhibitor: Tocilizumab
▶ Janus Kinase Inhibitors
▶ Methotrexate
▶ Non-steroidal Anti-inflammatory Drugs: Overview
▶ Sulfasalazine and Related Drugs
▶ Tumor Necrosis Factor (TNF) Inhibitors

References

Amos, R. S. (1995). The history of use of sulphasalazine in rheumatology. *British Journal of Rheumatology, 34* (Suppl 2), 2–6.

Forestier, J. (1932). The treatment of rheumatoid arthritis with gold salts. *Lancet, 219*, 441–444.

Grigor, C., Capell, H., Stirling, A., McMahon, A. D., Lock, P., Vallance, R., et al. (2004). Effect of a treatment strategy of tight control for rheumatoid arthritis (the TICORA study): A single blind randomized controlled trial. *Lancet, 364*, 263–269.

Hench, P. S., Kendall, E. C., et al. (1949). The effect of a hormone of the adrenal cortex (17-hydroxy-11-dehydrocorticosterone; compound E) and of pituitary adrenocorticotropic hormone on rheumatoid arthritis. *Proceedings of the Staff Meetings. Mayo Clinic, 24*, 181–197.

Kremer, J. M., & Lee, J. K. (1986). The safety and efficacy of the use of methotrexate in long-term therapy for rheumatoid arthritis. *Arthritis Rheum, 29*, 822–831.

Mottonen, T., Hannonen, P., Leirisalo-Repo, M., Nissila, M., Kautiainen, H., Korpela, M., et al. (1999). Comparison of combination therapy with single-drug therapy in early rheumatoid arthritis: A randomized trial. FIN-RACo trial group. *Lancet, 353*, 1568–1573.

O'Dell, J. R., Mikuls, T. R., Taylor, T., Ahluwalia, V., Brophy, M., Warren, S. R., et al. (2013). Therapies for active rheumatoid arthritis after methotrexate failure. *New England Journal of Medicine, 369*, 307–318.

Smolen, J. S., Breedveld, F. C., Burmester, G. R., Bykerk, V., Dougados, M., Emery, P., et al. (2016). Treating rheumatoid arthritis to target: 2014 update of the recommendations of an international task force. *Annals of the Rheumatic Diseases, 75*(1), 3–15. doi:10.1136/annrheumdis-2015-207524.

Weinblatt, M. E., Coblyn, J. S., Fox, D. A., et al. (1985). Efficacy of low dose methotrexate in rheumatoid arthritis. *New England Journal of Medicine, 312*, 818–822.

Salicylates

Garry G. Graham[1,2] and Richard O. Day[1]
[1]Department of Pharmacology, School of Medical Sciences, University of NSW, Sydney, NSW, Australia
[2]Department of Clinical Pharmacology and Toxicology, St Vincent's Hospital, Sydney, NSW, Australia

Definition

The salicylates are a group of nonsteroidal anti-inflammatory drugs (NSAIDs) which are derivatives of salicylate or metabolized to salicylate. Specifically, salicylate is the ionized, anionic form of salicylic acid. The general properties of the NSAIDs are described elsewhere in this encyclopedia.

Chemistry, Metabolism, and Pharmacokinetics

Aspirin (Acetylsalicylic Acid)

Aspirin (*acetylsalicylic acid*) is a synthetic compound which is metabolized to salicylate (Fig. 1). It has the pharmacological properties of salicylate together with its own activity due to acetylation and irreversible inhibition of cyclooxygenase 1 (COX-1) and COX-2 (Fig. 1). Aspirin is absorbed well in the gastrointestinal tract, but only about 70 % is absorbed intact because of the first-pass hydrolysis by esterases in the liver (Rowland et al. 1972). The total absorption, aspirin and salicylate together, is complete. Once in the circulation, aspirin is hydrolyzed rapidly, and its half-life is only about 10 min.

Salsalate (salicylsalicylic acid) is a synthetic compound which is hydrolyzed largely to salicylate. The half-life of salsalate is approximately 1 h with some salsalate metabolized and excreted as its glucuronide. As some salsalate is metabolized directly before hydrolysis, salsalate treatment yields plasma salicylate concentrations which are 15–25 % lower than those achieved by aspirin or salicylate salts. Unlike aspirin, salsalate is well tolerated by the upper gastrointestinal tract (Rainsford 2004).

Salicin is a naturally occurring compound which is present in white willow. It is hydrolyzed and oxidized to salicylate. It has been used since ancient times, but the modern use of salicin commenced with its rediscovery by Reverend Edward Stone in the eighteenth century. Subsequent chemical work led to the development of aspirin.

Methyl salicylate is a naturally occurring compound (oil of wintergreen) which is hydrolyzed to salicylate. It is used widely in liniments. The resulting concentrations of salicylate in plasma are low, typically less than 10 mg/L (compared to approximately 40 mg/L after an oral dose of 650 mg aspirin) (Rainsford 2004).

Salicylates, Fig. 1 Structures and metabolism of the salicylates. Aspirin (acetylsalicylic acid), salsalate, and salicylate are acids, but their structures are shown in the ionized forms which are the major species in the circulation. Aspirin inhibits cyclooxygenases by its acetyl group binding covalently to a serine residue in the enzymes (see section "Pharmacological Activities" below)

Salicylaldehyde is a naturally occurring compound which is present in meadowsweet and several other plants. It is oxidized to salicylate.

Salicylate is administered orally as its sodium or choline magnesium salts. It is largely metabolized by conjugation to the glycine derivative (salicyluric acid) and to glucuronides. Small amounts of gentisic acid (2,5-dihydroxy benzoic acid) are also produced. Some unchanged salicylate is also excreted in urine. Alkalinization of urine by the administration of sodium bicarbonate increases the urinary excretion and decreases the plasma concentrations of salicylate. The elimination of salicylate is saturable with its half-life ranging from about 2 h at low doses to 15 h at high doses. During continued high dosage, the metabolism of salicylate is autoinduced, and anti-inflammatory plasma concentrations decrease (Rainsford 2004).

Aspirin and salicylate bind to plasma proteins. The unbound fraction of salicylate ranges from about 3 % at low doses to 23 % at anti-inflammatory doses of aspirin (up to about 4.8 g daily). Estimates of the therapeutic plasma concentrations (total and unbound) of aspirin and salicylate are shown in Table 1.

Pharmacological Activities

Aspirin inhibits the synthesis of prostaglandins by COX-1 and COX-2 by the transfer of the acetyl group to a serine residue of COX-1 and COX-2 (Fig. 1). The pharmacological actions of aspirin are also due, in part, to its hydrolysis to salicylate.

The mechanism of action of salicylate has been unclear. Recent work indicates, however, that it inhibits prostaglandin synthesis but under the specific conditions of low peroxide tone (Aronoff et al. 2003). It should be noted that COX-1 and COX-2 are bifunctional enzymes with both cyclooxygenase and peroxidase functions (nonsteroidal anti-inflammatory drugs (NSAIDs)), and the mode of action of salicylate is probably on the peroxidase activity (Aronoff et al. 2003).

Overall, the clinical effects and pharmacological activities of salicylate are very similar to those of the selective COX-2 inhibitors (Table 2) (See ► Coxibs). In summary, this apparent COX-2 selectivity of salicylate salts is indicated by the therapeutic effects (analgesia, antipyresis, and anti-inflammatory effects), good gastrointestinal tolerance of salicylate salts. their lack of effect

Salicylates, Table 1 Approximate peak plasma concentrations of aspirin and salicylate

Dose	Aspirin[a]	Salicylate
Antiplatelet – 100 mg aspirin	Total 2 mg/L (0.01 mM)	Total 4 mg/L (0.03 mM)
	Unbound 0.8 mg/L (0.004 mM)	Unbound 0.1 mg/L (0.0007 mM)
Analgesic – 650 mg aspirin	Total 11 mg/L (0.06 mM)	Total 40 mg/L (0.3 mM)
	Unbound 4.5 mg/L (0.025 mM)	Unbound 2 mg/L (0.015 mM)
Anti-inflammatory – 4.8 g daily	Total 20 mg/L (0.1 mM)	Total up to 300 mg/L (2.2 mM)
	Unbound 8 mg/L (0.04 mM)	Unbound up to 70 mg/L (0.5 mM)

[a]Approximate peak concentrations after rapid release or soluble tablets. Lower peak concentrations after dosage with slow release or enteric-coated tablets

Salicylates, Table 2 Comparative basic and clinical pharmacological activities of salicylate and aspirin

Activity	References
Salicylate inhibits of COX-1 and COX-2 at low levels of hydroperoxides. Aspirin inhibits both COX-1 and COX-2	(Aronoff et al. 2003)
Salicylate is a potent inhibitor of the production of PGE_2 synthesis by epithelial cell line when incubated with interleukin 1b(IC_{50} ~5 mg/L; 36 µM). Effect decreased by increasing concentrations of arachidonic acid	(Mitchell et al. 1997)
Aspirin and salicylate decrease upregulation of COX-2 in some cells in response to inflammatory mediators	(Xu et al. 1999)
Both aspirin and salicylate are anti-inflammatory and analgesic in rheumatoid arthritis	(Preston et al. 1989)
Salicylate decreases prostaglandin production in inflammatory exudates in rats – similar to aspirin and nonselective NSAIDs	(Whittle et al. 1980)
Aspirin and salicylate are equipotent in reducing carrageenin-induced paw edema of the rat	(Whittle et al. 1980)
Gastrointestinal tolerance of salicylate much superior to aspirin and nonselective NSAIDs in humans and experimental animals	(Grossman et al. 1961; Whittle et al. 1980)
Salicylate has no effect on prostaglandin levels in gastric mucosa, while aspirin decreases prostaglandin concentrations	(Whittle et al. 1980)
Salicylate does not inhibit platelet aggregation. Aspirin active	(Vargaftig 1978)
Aspirin precipitates asthma in up to 20 % of adults with asthma. Salicylate inactive	(Jenkins et al. 2004; Simon 2004)

on platelet aggregation, and their inability to provoke aspirin-induced asthma (Table 2).

The potent inhibition of salicylate on myeloperoxidase (IC_{50} ~10 µM) (Kettle and Winterbourn 1991) indicates a possible use of salicylate in atherosclerosis as myeloperoxidase is expressed highly in atherosclerotic plaques and its product, hypochlorous acid (HOCl), may contribute to the development of atherosclerotic plaques. Mechanistically, inhibition of myeloperoxidase is consistent with the proposed effect of salicylate on peroxidase function of COX-2.

Many actions of salicylate, other than inhibition of peroxidase functions, have been observed in studies in vitro. However, the concentrations are generally supratherapeutic, and their relevance to the clinical effects of salicylate is very doubtful (Rainsford 2004).

Two exceptions to this general finding should be noted. Firstly, as discussed below, salicylates, other NSAIDs, and paracetamol have antidiabetic actions (Goldfine et al. 2013). The mechanism may be the activation of adenosine monophosphate kinase (AMPK). AMPK is a sensor of cellular energy (adenosine triphosphate, ATP). At low levels of ATP, AMPK phosphorylates enzymes leading to inhibition of functions involved in anabolic processes that utilize ATP while activating catabolic enzymes that generate ATP. Activation of AMPK has been suggested as the mechanism of the antidiabetic action of metformin and the similar effect of salicylate. High doses of salicylate increase the hepatic levels of

AMPK in mice, but direct activation of AMPK only occurs at supratherapeutic concentrations of salicylate (3 mM) (Hawley et al. 2012), making a direct effect of salicylate on AMPK unlikely. Possibly, an antidiabetic effect of salicylate results from inhibition of prostaglandin synthesis as this antidiabetic action is shown by other NSAIDs and paracetamol (Graham et al. 2013).

A second exception is that salicylate uncouples the oxidative phosphorylation of mitochondria (Rainsford 2004). This effect is produced at high concentrations of salicylate. While this effect may not be related to the therapeutic effect of salicylate, the uncoupling may be responsible for the hyperthermia produced by overdoses of aspirin and salicylate salts (see below).

Clinical Uses

Aspirin has been widely used for its analgesic (dose 650 mg), antipyretic (dose 650 mg), and anti-inflammatory (dose 4–5 g daily) actions. It is now largely used at low doses (about 100 mg once daily) to prevent the recurrence of myocardial infarction or stroke or their primary prevention in at risk patients. Despite its short half-life of about 10 min, aspirin has a prolonged antiplatelet action due to its irreversible inhibition (acetylation) of COX-1 in platelets.

Salicylate has been used as its sodium and choline magnesium salts which have analgesic and antipyretic effects at low doses with the addition of anti-inflammatory actions at the higher doses which are useful in the treatment of rheumatoid arthritis and rheumatic fever. Methylsalicylate is widely used in liniments for the relief of musculoskeletal pain.

The activity of salsalate in the treatment of arthritic diseases and pain is due to its metabolic product, salicylate (Fig. 1). As outlined above and in the general section on NSAIDs, clinical trials on salsalate have indicated that it lowers blood glucose in diabetic patients. However, further clinical trials are required for a more complete examination of the utility of salicylates in the treatment of T2DM.

Aspirin, like celecoxib (a selective COX-2 inhibitor), is being widely examined for its prevention of colorectal cancers. Such prevention has been shown in carcinogen-induced colon cancers in experimental animals and in clinical studies in man. Even low-dose aspirin (80–100 mg daily) may be preventative (Thun et al. 2012). This is a surprising finding as low-dose aspirin is considered to have its primary actions on COX-1-mediated synthesis of prostaglandins and related compounds, particularly the synthesis of thromboxane A_2 in platelets. Aspirin may also protect against some ovarian cancers. The widespread routine use of aspirin to prevent cancers is limited by its considerable toxic effects which are outlined below.

Adverse Effects

- Aspirin and salicylate have many adverse reactions. These include:
- Damage to the upper gastrointestinal tract (stomach and duodenum) may lead to bleeding, ulceration, and perforation. This is a major problem which is, in part, responsible for its now limited use as an analgesic and antipyretic drug. Even at low antiplatelet doses (80–100 mg daily), aspirin increases the risk of gastrointestinal damage (Musumba et al. 2012). By comparison, salicylate salts are well tolerated by the upper gastrointestinal tract.
- Reye's syndrome (a frequently fatal disease with damaging effects to many organs, particularly the brain and liver). The possible association between aspirin and Reye's syndrome has contributed to the reduced use of aspirin in children.
- Aspirin-induced asthma. Aspirin produces asthma in up to about 20 % of adult asthmatics and 5 % of child asthmatics. Cross-reactivity occurs with all nonselective NSAIDs but not selective COX-2 inhibitors, such as celecoxib or etoricoxib (Jenkins et al. 2004).
- Tinnitus and deafness. Salicylate causes tinnitus and deafness which become more severe with increasing doses of aspirin or salicylate

- and is proportional to the plasma concentrations of salicylate (Day et al. 1989).
- Hyperthermia and perturbations in acid/base balance. These effects are produced by both aspirin and salicylate salts, indicating that the toxicity is due to salicylate.

Drug Interactions

The antiplatelet action of aspirin is blocked by practically all nonselective NSAIDs such as ibuprofen, but not by selective COX-2 inhibitors, such as celecoxib. Diclofenac is the only nonselective NSAID that can be taken with aspirin, provided that the daily dose of aspirin is taken 2 h before a dose of diclofenac. In large doses, salicylate decreases the binding of several other drugs to plasma proteins (e.g., phenytoin), but the efficacy of the displaced drugs is not increased because the unbound concentrations are unaltered.

Salicylate displaces phenytoin from binding to plasma proteins, but the unbound concentration of phenytoin is unaltered. The unbound concentration and efficacy of phenytoin is unaltered.

Cross-References

▶ Asthma
▶ Coxibs
▶ Gout
▶ Non-steroidal Anti-inflammatory Drugs: Overview
▶ Osteoarthritis
▶ Prostanoids
▶ Rheumatoid arthritis

References

Aronoff, D. M., Boutaud, O., Marnett, L. J., & Oates, J. A. (2003). Inhibition of prostaglandin H2 synthases by salicylate is dependent on the oxidative state of the enzymes. *Journal of Pharmacology and Experimental Therapeutics, 304*(2), 589–595.

Day, R. O., Graham, G. G., Bieri, D., Brown, M., Cairns, D., Harris, G., et al. (1989). Concentration-response relationships for salicylate-induced ototoxicity in normal volunteers. *British Journal of Clinical Pharmacology, 28*(6), 695–702.

Goldfine, A. B., Fonseca, V., Jablonski, K. A., Chen, Y. D. A., Tipton, L., Staten, M. A., et al. (2013). Salicylate (salsalate) in patients with type 2 diabetes. *Annals of Internal Medicine, 159*(1), 1–12.

Graham, G. G., Davies, M. J., Day, R. O., Mohamudally, A., & Scott, K. F. (2013). The modern pharmacology of paracetamol: Therapeutic actions, mechanism of action, metabolism, toxicity and recent pharmacological findings. *Inflammopharmacology, 21*(3), 201–232.

Grossman, M. I., Matsumoto, K. K., & Lichter, R. J. (1961). Fecal blood loss produced by oral and intravenous administration of various salicylates. *Gastroenterology, 40*, 383–388.

Hawley, S. A., Fullerton, M. D., Ross, F. A., Schertzer, J. D., Chevtzoff, C., Walker, K. J., et al. (2012). The ancient drug salicylate directly activates AMP-activated protein kinase. *Science, 336*(6083), 918–922.

Jenkins, C., Costello, J., & Hodge, L. (2004). Systematic review of prevalence of aspirin induced asthma and its implications for clinical practice. *British Medical Journal, 328*(7437), 434.

Kettle, A. J., & Winterbourn, C. C. (1991). Mechanism of inhibition of myeloperoxidase by anti-inflammatory drugs. *Biochemical Pharmacology, 41*(10), 1485–1492.

Mitchell, J. A., Saunders, M., Barnes, P. J., Newton, R., & Belvisi, M. G. (1997). Sodium salicylate inhibits cyclooxygenase-2 activity independently of transcription factor (nuclear factor kappaB) activation: Role of arachidonic acid. *Molecular Pharmacology, 51*(6), 907–912.

Musumba, C., Jorgensen, A., Sutton, L., Van Eker, D., Moorcroft, J., Hopkins, M., et al. (2012). The relative contribution of NSAIDs and Helicobacter pylori to the aetiology of endoscopically-diagnosed peptic ulcer disease: Observations from a tertiary referral hospital in the UK between 2005 and 2010. *Alimentary Pharmacology & Therapeutics, 36*(1), 48–56.

Preston, S. J., Arnold, M. H., Beller, E. M., Brooks, P. M., & Buchanan, W. W. (1989). Comparative analgesic and anti-inflammatory properties of sodium salicylate and acetylsalicylic acid (aspirin) in rheumatoid arthritis. *British Journal of Clinical Pharmacology, 27*(5), 607–611.

Rainsford, K. D. (2004). Metabolic and related effects of salicylates. In K. D. Rainsford (Ed.), *Aspirin and related drugs* (pp. 355–366). London: Taylor & Francis.

Rowland, M., Riegelman, S., Harris, P. A., & Sholkoff, S. D. (1972). Absorption kinetics of aspirin in man following oral administration of an aqueous solution. *Journal of Pharmaceutical Sciences, 61*(3), 379–385.

Simon, R. A. (2004). Adverse respiratory reactions to aspirin and nonsteroidal anti-inflammatory drugs. *Current Allergy and Asthma Reports, 4*(1), 17–24.

Thun, M. J., Jacobs, E. J., & Patrono, C. (2012). The role of aspirin in cancer prevention. *Nature Reviews Clinical Oncology, 9*(5), 259–267.

Vargaftig, B. B. (1978). Salicylic acid fails to inhibit generation of thromboxane A2 activity in platelets after in vivo administration to the rat. *Journal of Pharmacy and Pharmacology, 30*(2), 101–104.

Whittle, B. J., Higgs, G. A., Eakins, K. E., Moncada, S., & Vane, J. R. (1980). Selective inhibition of prostaglandin production in inflammatory exudates and gastric mucosa. *Nature, 284*(5753), 271–273.

Xu, X. M., Sansores-Garcia, L., Chen, X. M., Matijevic-Aleksic, N., Du, M., & Wu, K. K. (1999). Suppression of inducible cyclooxygenase 2 gene transcription by aspirin and sodium salicylate. *Proceedings of the National Academy of Sciences of USA, 96*(9), 5292–5297.

Sarcoidosis

Robert P. Baughman
Department of Internal Medicine, Interstitial Lung Disease and Sarcoidosis Clinic, University of Cincinnati Medical Center, Cincinnati, OH, USA

Synonyms

Boecks sarcoidosis; Heerfordt's syndrome; Lofgren's syndrome; Lupus pernio

Definition

Sarcoidosis is a multisystemic inflammatory disease characterized by the presence of noncaseating granulomas. In order to make the diagnosis, all other causes of granulomas must be excluded and the patient must have the appropriate clinical presentation.

Epidemiology and Genetics

Sarcoidosis is a worldwide disease. However, the incidence of the disease varies, with some ethnic groups such as Nordic and African Americans have an increased rate of disease (Hall et al. 1969). In Japan, the disease is more frequent in northern climates and in Europe, Scandinavian countries have significantly more disease than more southern areas, such as Italy and Greece. In the USA, however, the southeastern area has a higher frequency than that reported in the northwest or the Rocky Mountain States.

In addition to geographic variation, there is a variation of the manifestations of the disease based on genetic background. For example, ocular disease is very common in Japan, while ocular disease occurs in less than a third of sarcoidosis patients in China, Europe, and America (Baughman et al. 2010a). Cardiac disease is also four times more frequent in Japan than elsewhere. On the other hand, advanced pulmonary disease is rare in Japan. Over 15 % of African Americans with sarcoidosis have pulmonary fibrosis. African Americans are also more likely to have bone marrow involvement and lupus pernio.

Table 1 reports the relative frequency of the disease around the world. In addition, it summarizes some of the more common manifestations. In some case, such as neurologic disease, there seems no difference for various areas. On the other hand, skin manifestations and cardiac disease vary based on ethnic background.

Pathophysiology

Sarcoidosis is a specific immunologic response characterized by granuloma formulation (Moller and Chen 2002). The trigger for the immunologic response appears to be an environmental agent which leads to granulomatous disease in a genetically susceptible host. While no specific etiology has been identified as the cause of sarcoidosis, several potential agents have been proposed. These include infectious agents such as mycobacteria (Song et al. 2005; Oswald-Richter and Drake 2010), commonly encountered bacteria such as *Propiniobacteria acnes*, and fungal agents. It is also possible that more than one antigen may trigger the immune response called sarcoidosis.

Sarcoidosis, Table 1 Role of ethnicity on frequency and organ manifestation of sarcoidosis

	African American	Caucasian American	Scandinavian	Southern Europe	Japan
Frequency	+3[a]	+1	+4	+2	+1
Organ involvement[b]					
Thoracic					
Overall	+4	+4	+4	+4	+4
Stage 1	+2	+3	+4	+3	+3
Stage 4	+2	+1	+/−	+1	+/−
Skin					
Erythema nodosum	Rare	+1	+2	+2	Rare
Lupus pernio	+2	+1	Rare	Rare	Very rare
Eye	+2	+2	+2	+1	+4
Cardiac	+1	+1	+1	+1	+3
Neurologic	+1	+1	+1	+1	+1
Bone marrow	+2	+1	+1	+1	Rare
Liver	+3	+3	+3	+3	+3
Fatigue	+3	+3	+3	+3	+3

[a]Relative score, with highest rate
[b]Relative score with thoracic involvement being +4

Sarcoidosis, Table 2 Pulmonary manifestations of sarcoidosis

	Manifestation	Frequency	
Upper airway disease			
Laryngeal	Stridor, hoarseness	Rare	Endoscopy
Trachea and large airway	Fixed obstruction/wheezing	Unusual	Bronchoscopy, CT scan
Lung			
Peribronchial thickening	Obstruction, cough	Common Observed	CT scan, endobronchial biopsy
Parenchymal nodules	Asymptomatic	Common	Subpleural on HRCT
Confluent nodules	Asymptomatic	10 %	CT, chest x-ray
Upper lobe bronchiectasis	Cough, dyspnea	5–15 %	HRCT
Aspergilloma	Hemoptysis	1–5 %	CT, chest imaging
Mediastinal			
Adenopathy	Cough, chest pain	Common	CT, chest x-ray

Adapted from Baughman et al. (2012a)

Clinical Manifestations

Lung

The lung is the most commonly affected organ in sarcoidosis (Baughman et al. 2012a). Table 2 summarizes the various manifestations of lung disease. The two most common areas of involvement are the lung parenchyma and mediastinal/hilar nodes. A scoring system for lung involvement was popularized by Scadding (1961) recognizing four patterns: Stage one is adenopathy alone (Fig. 1a), stage 2 is adenopathy plus lung involvement, stage 3 is lung involvement alone (Fig. 1b, c), and stage 4 is fibrosis (Fig. 1d). For most sarcoidosis patients, the chest x-ray can be characterized by one of these four patterns. In addition, stage 1 has a greater than 80 % chance of resolving in 2–5 years whether treated or not, while patients with stage 3 have only about 30 % chance that their chest x-ray will return to normal in 2–5 years. This rate of resolution is similar for different ethnic groups around the world.

Pulmonary function testing is often used to assess the severity of lung disease in sarcoidosis. This includes spirometry, including forced vital capacity (FVC) and force expiratory volume in 1 s

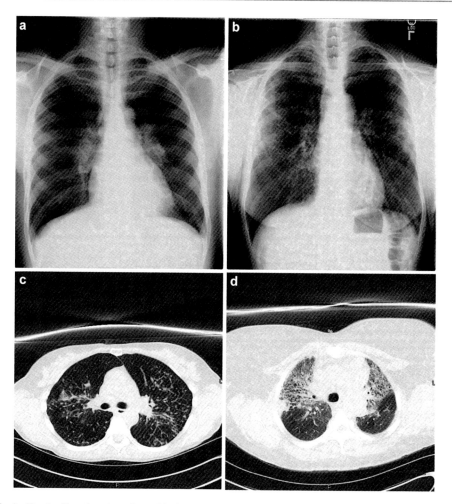

Sarcoidosis, Fig. 1 Chest imaging of sarcoidosis. (**a**) is a posterior anterior (*PA*) chest roentgenogram demonstrating bilateral hilar adenopathy and right parenchymal adenopathy termed stage 1; (**b** and **c**) demonstrate chest roentgenogram and associated computer tomograph of stage 3 disease with parenchymal nodularity and no significant adenopathy; (**d**) is a CT scan showing upper lobe fibrosis with some honeycombing in stage 4 disease

(FEV-1). The diffusion capacity may demonstrate abnormalities prior to changes in lung volumes. There is some correlation between the FVC, chest x-ray stage, and level of dyspnea reported by the patient (Yeager et al. 2005). However, there are a significant proportion of patients with advanced chest x-ray stage who still have a FVC percent predicted of greater than 80 %. Functional studies such as 6 min walk distance (Baughman et al. 2007) and cardiopulmonary exercise testing (Marcellis et al. 2013) may identify lung impairment.

Skin

Cutaneous sarcoidosis includes an array of lesions, including maculopapular lesions, lupus pernio, hypopigmentation, hyperpigmentation, and keloid formation. Also, specific lesions for sarcoidosis include *erythema nodosum* and *lupus pernio*. Figure 2 demonstrates several of these manifestations of skin lesions.

Erythema nodosum (Fig. 2a) is a very typical manifestation of sarcoidosis. It is associated with periarticular arthritis and bilateral hilar adenopathy, called Lofgren's syndrome

Sarcoidosis, Fig. 2 (**a**) shows erythema nodosum lesions on the lower leg; (**b**) demonstrates the raised purplish lesion in areas of a tattoo; (**c**) demonstrates a lesion on the cheek of a patient with *lupus pernio*

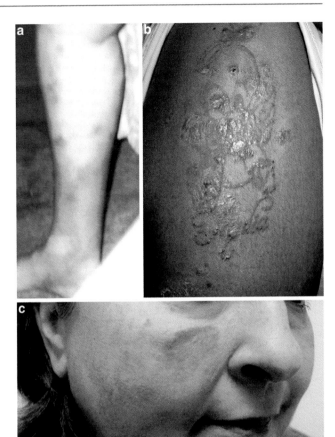

(Grunewald and Eklund 2007). Lofgren's syndrome is associated with an excellent prognosis in the majority of patients. Recent genetic studies have demonstrated that the presence of DRB1*301was associated with significantly higher rate of resolution versus those who were DRB1*301 negative (Fig. 3; Grunewald and Eklund 2009). The predictive effect of DRB*1 on prognosis from Lofgren's syndrome has been reported in other European countries (Sato et al. 2002). In patients of African descent, the presence of *erythema nodosum* is not associated with such a favorable outcome (Mana and Marcoval 2007).

Lupus pernio (Fig. 2c) is associated with a chronic course of disease (Neville et al. 1983). The lesions of *lupus pernio* include cheeks and nose (Baughman et al. 2008). The lesions are chronic and poorly respond to conventional therapy (Stagaki et al. 2009).

Eye

Ocular disease occurs to differing degrees in various ethnic groups (Table 1). There are a wide range of eye manifestations, as summarized in Table 3 (Baughman et al. 2010a). Lacrimal gland disease is perhaps the most common manifestation but often undetected. Subsequent sicca is a very common complaint.

Uveitis is the most frequent symptomatic form of sarcoidosis (Jabs and Johns 1986). Anterior uveitis is usually self-limited, while posterior uveitis is often chronic. Intermediate uveitis, with *pars planitis*, is seen in only a few conditions, including sarcoidosis and multiple sclerosis (MS) (Scott et al. 2010). In multiple sclerosis,

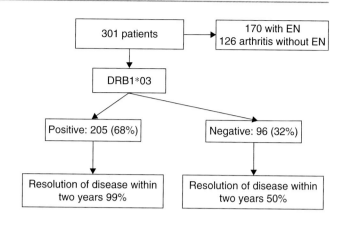

Sarcoidosis, Fig. 3 Clinical outcome of Lofgren's syndrome at the Stockholm sarcoidosis clinic (Grunewald and Eklund 2009)

Sarcoidosis, Table 3 Eye involvement in sarcoidosis

Lacrimal gland swelling
Uveitis
Optic neuritis
Adnexal nodularity
Blindness
Retinitis
Cystoid macular edema
Scleritis
Glaucoma
Cataract
Candle wax drippings
Mutton fat keratic precipitates
Iris nodules
Trabecular meshwork nodules
Snowball/string of pearls (pars planitis)
Multiple chorioretinal peripheral lesions
Optic disk nodules

pars planitis can precede other manifestations of MS by years (Prieto et al. 2001).

Optic neuritis is an unusual but important manifestation of sarcoidosis. It is a neurologic manifestation (Baughman et al. 2012b) leading to vision problems and eye pain. It is often associated with uveitis (Braswell and Kline 2007) in many of the cases. It can be a manifestation of MS, and the differential between sarcoidosis and MS can be crucial (Scott et al. 2010), since anti-inflammatory therapy is effective for acute optic neuritis. However, for chronic disease, long term anti-inflammatory therapy can reduce the rate of relapses and improve the overall prognosis in sarcoidosis but not MS (Myers et al. 2004).

Calcium Metabolism

Hypercalcemia/hypercalciuria is seen in around 10 % of sarcoidosis patients (Sharma 2000). The mechanism is usually increased 1-alpha-hydroxylase activity in the granulomas (Mason et al. 1984). This leads to increased levels of the biologically active form of vitamin D, 1,25-dihydroxy vitamin D3. Because the 1-alpha-hydroxylase enzyme can be autonomous in sarcoidosis, patients may have low 25-hydroxy-vitamin D3 levels. In one study, patients with more advanced sarcoidosis had lower 25-hydroxy-vitamin D3 and higher 1,25-dihydroxy vitamin D 3 than those with less advanced disease (Kavathia et al. 2010).

Nephrolithiasis is a complication of prolonged hypercalciuria. Sarcoidosis patients with nephrolithiasis often have chronic disease, requiring long-term treatment (Rizzato et al. 1995). Fortunately, most patients with nephrolithiasis can be treated with either low doses of prednisone and/or antimalarial agents. It is important that these patients avoid calcium and vitamin D supplements. The effectiveness of treatment can be determined by measuring 24 h calcium urine content. A value of less than 250 mg is usually achievable.

Hypercalciuria may lead to renal failure. Direct renal disease can also occur in sarcoidosis with granulomatous nephritis. The majority of these patients also have hypercalcemia (Mahevas et al. 2009). Renal failure associated with hypercalcemia can often be reversed by anti-inflammatory therapy, especially corticosteroids. Whether the improvement is simply because of reduction of serum calcium level or reversal of the granulomatous interstitial nephritis is not always clear. However, it is clear that early treatment is associated with a better prognosis (Mahevas et al. 2009).

Liver/Spleen/Bone Marrow

Sarcoidosis is often in the reticuloendothelial system. Liver involvement is common but usually asymptomatic (Kahi et al. 2006). Liver function test abnormalities are seen in sarcoidosis. The most common way it is detected is through liver function testing. Abnormalities include increased alkaline phosphatase and to a lesser extent transaminases (Table 4; Cremers et al. 2011). However, liver biopsies may detect disease in patients with normal liver function tests (Baughman et al. 1999). In some cases, hepatic sarcoidosis can be severe and lead to cirrhosis and even liver failure (Kennedy et al. 2006). In one study of 21 patients comparing liver biopsy to liver function studies, the more severe liver function testing abnormality, the more extensive the granulomatous involvement and degree of fibrosis (Cremers et al. 2011).

Spleen involvement is also rarely symptomatic (Judson 2002). However, hypersplenism can lead to pain and there is a risk for rupture (Salazar et al. 1995). Massive splenomegaly is sometimes an indication for splenectomy. However, treatment of the sarcoidosis may be sufficient to reduce symptoms and avoid surgery. Splenectomy is more commonly performed to rule out other conditions, especially lymphoma. Both liver and spleen involvement may have characteristic CT findings of nodularity (Warshauer et al. 1995).

Bone marrow involvement is also under appreciated in sarcoidosis. In one study, abnormalities of red and/or white cells were seen in more than half of sarcoidosis patients (Lower et al. 1988). The most common manifestation is lymphopenia, which may be due to sequestration of lymphocytes in areas of disease activity, such as the lung (Hunninghake and Crystal 1981). Anemia occurs in about 20 % of sarcoidosis patients, in some cases confounded by iron deficiency (Lower et al. 1988). Direct bone marrow involvement has been documented in sarcoidosis cases and may be present in up to 10 % of cases (Lower et al. 1988; Browne et al. 1978). In Americans, splenic and bone marrow involvement are more common in African Americans than Caucasians (Baughman et al. 2001).

Neurologic Disease

Neurologic involvement includes the central nervous system, cranial nerves, spine, and peripheral nerves. Central nervous involvement includes leptomeningeal disease, often at the base of brain. This can best be detected using gadolinium enhancement of MRI (Fig. 4). One can also see mass lesions and hydrocephalus. The MRI is felt to be more sensitive than CT scan to detect neurosarcoidosis (Spencer et al. 2005).

Cranial nerve involvement is common with neurosarcoidosis. The most common cranial

Sarcoidosis, Table 4 Liver function test abnormalities in hepatic sarcoidosis

	Characteristic	Number	Percent (%)
Total studied		127	
Mild	Any test 1.5–3.0 of the ULN	38	28.3
Moderate	1–2 tests ≥3.0 of the ULN	85	67
Severe	3 or 4 tests ≥3.0 of the ULN	6	4.7
Cholestasis	Alkaline phosphatase elevation predominant	55	43.3
Parenchymatous	Transaminase elevation predominant	29	22.8
Combined	Mixed alkaline phosphatase and transaminase elevation	43	33.9

ULN upper limit normal
Adapted from Cremers et al. (2011)

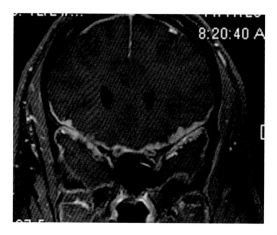

Sarcoidosis, Fig. 4 Leptomeningeal sarcoidosis demonstrated at base of brain by gadolinium enhancement (*white line*)

nerves affected are 1st (optic nerve) and 7th (Zajicek et al. 1999; Lower et al. 1997). Seventh nerve involvement can lead to paralysis which can be misdiagnosed as Bell's palsy. Seventh nerve involvement in sarcoidosis is often self-limited, resolving with or without short-term corticosteroid therapy (Lower et al. 1997). The patient and physician may not associate the paralysis with sarcoidosis.

Spinal cord involvement is a rare but serious complication of neurosarcoidosis (Cohen-Aubart et al. 2010; Bradley et al. 2006). The presenting symptoms are usually paresthesias, but untreated it can progress to paralysis. Early recognition and treatment prior to onset of paralysis is associated with a good long-term prognosis (Bradley et al. 2006).

Peripheral nerve involvement has been reported in some patients with sarcoidosis. This includes small fiber neuropathy (SFN). Small fiber neuropathy can cause significant morbidity (Tavee and Zhou 2009). It has been associated with fatigue in sarcoidosis patient (Hoitsma et al. 2002). The diagnosis of small fiber neuropathy can be difficult. It is associated with autonomic neuropathy problems, such as postural hypotension. Patients often lack temperature sensitivity. Screening for SFN should include questions regarding postural hypotension and lack of temperature sensitivity (Hoitsma et al. 2011).

A more definitive test is a skin biopsy, where the number of small nerve fibers is enumerated in the skin. The technique is tedious and requires a specialized laboratory since there can be variation with counting nerve fibers (Bakkers et al. 2009).

Cardiac

Heart involvement in sarcoidosis is of two general types: arrhythmias and reduced left ventricular function. Patients may have either or both manifestations. There is also the indirect effect of precapillary pulmonary hypertension leading to right ventricular failure. However, there is nothing intrinsically wrong with the heart in that situation (Baughman et al. 2010b).

Complete heart block has been reported in 20–30 % of sarcoidosis patients with bundle branch block alone reported in up to a third of cardiac sarcoidosis case. Ventricular tachycardia has been reported in a quarter of cardiac sarcoidosis. Sudden death, usually felt due to ventricular arrhythmias, has been reported in 25–65 % of cases. Congestive heart failure has been reported in 25–75 % of cases (Sekhri et al. 2011). Patients with ventricular arrhythmias and reduced ejection fraction have the worse prognosis in cardiac sarcoidosis (Yazaki et al. 2001).

Given the risk for sudden death, there has been interest in the detection of cardiac disease. Figure 5 summarizes the sensitivity and specificity of the individual features of history of cardiac symptoms, electrocardiogram (ECG), 24 h continuous monitoring (Holter), and echocardiogram. None of the individual variables was found in more than a quarter of patients. However, one or more abnormalities were found in half of the patients. The presence of two or more abnormalities had a specificity for cardiac sarcoidosis of greater than 95 %, but the sensitivity was quite low. In this study of 62 patients, no patient had all four abnormalities (Mehta et al. 2008).

Cardiac sarcoidosis is rarely detected directly because the yield for endomyocardial biopsy is quite low (Uemura et al. 1999). Cardiac imaging has been shown to detect cardiac involvement. Both PET and MRI scanning have been shown

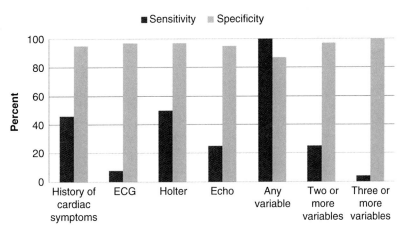

Sarcoidosis, Fig. 5 The sensitivity and specificity of individual tests, as well as the presence of one or more of the abnormalities. No patient had all four abnormalities (Mehta et al. 2008)

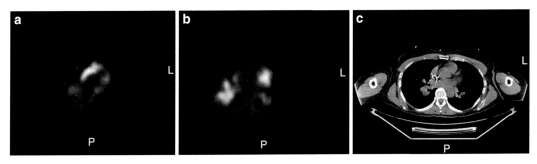

Sarcoidosis, Fig. 6 A patient with dyspnea upon exercise who was found to have complete heart block. She underwent PET/CT examination. (**a**) is a PET scan demonstrating increased activity in various part of her left ventricle; (**b**) is another picture from the PET scan demonstrating enhanced activity in the hilar lymph nodes, which were found to be enlarged on the corresponding CT scan (**c**). Endobronchial ultrasound needle biopsy of these lymph nodes demonstrated noncaseating granulomas

to detect abnormalities (Steckman et al. 2012; Ohira et al. 2011). PET scanning has the advantage of also detecting other organ involvement. Figure 6 is an example of a patient in whom both cardiac and hilar node activity were detected on PET scan.

The ventricular arrhythmias associated with cardiac sarcoidosis may lead to sudden death. Therefore, the placement of an intra-cardiac defibrillator (ICD) has been increasingly applied to sarcoidosis patients. In a retrospective study of 112 cardiac sarcoidosis patients with ICD devices, 44 % had therapy (either pacing or firing of device). Thirty six (33 % of total) patients had appropriate therapy, including sixteen who were treated for ventricular tachycardia storm. Ventricular storm was defined as three or more appropriate ICD therapies in 24 h. Risk factors for ventricular storm include left or right ventricular dysfunction (Schuller et al. 2012).

Therapy

Treatment decisions in sarcoidosis depend on the symptoms of the patient. Evidence-based recommendations have been developed for pulmonary disease (Baughman and Nunes 2012; Paramothayan et al. 2003, 2005). These are mostly based on clinical series, although there are now several double-blind randomized placebo controlled trials of glucocorticosteroids (Pietinalho et al. 1999; Gibson et al. 1996), methotrexate (Baughman et al. 2000), chloroquine (Baltzan et al. 1999), pentoxifylline (Park et al. 2009), and

Sarcoidosis, Table 5 Treatment of sarcoidosis

Manifestation/ Symptoms	Initial therapy	Maintenance	Alternatives
No symptoms	None		
Cough	Inhaled steroids		Glucocorticoids Cytotoxic drugs
Dyspnea	High dose Glucocorticoids	Low dose glucocorticoids Cytotoxic drugs	Antimalarial drugs Anti-TNF antibodies Rituximab Pentoxyfylline
Skin lesions	Topical glucocorticoids Low dose glucocorticoids	Antimalarial drugs Cytotoxic agents	Anti-TNF antibodies
Lupus pernio	Low dose glucocorticoids Anti-TNF antibodies	Low dose glucocorticoids Anti-TNF antibodies Cytotoxic drugs	Antimalarial drugs
Eye disease	Topical glucocorticoids	Cytotoxic drugs Low dose glucocorticoids	Anti-TNF antibodies Rituximab
Neurologic disease	Very high dose corticosteroids	Cytotoxic drugs Anti-TNF antibodies Low dose glucocorticoids	Rituximab Intravenous pulse corticosteroids
Cardiac disease	High dose corticoids[a]	Low dose corticoids Cytotoxic drugs	

[a]For cardiac sarcoidosis, usual maximal dose 30 mg daily

infliximab (Baughman et al. 2006a; Rossman et al. 2006).

The recommendations for extrapulmonary disease have been for the most part extended from those made for pulmonary disease. However, there have been several large series looking at specific extrapulmonary manifestations which compare the efficacy of individual agents. These usually provide information about relative effectiveness. In one large series of patients with *lupus pernio* (Stagaki et al. 2009), only glucocorticoids and the anti-TNF antibody infliximab were associated with a greater than 50 % rate of stability or improvement. In a series of 465 sarcoidosis patients with ocular disease included 365 patients treated with methotrexate, the range of treatment and response to agents for controlling chronic disease were similar to pulmonary studies (Baughman et al. 2002a). Other studies have reported on the utility of these agents for neurologic disease (Lower and Weiss 2008; Pawate et al. 2009; Dumas et al. 2000) and cardiac disease (Chapelon-Abric et al. 2004).

Table 5 summarizes treatment options for various manifestations of sarcoidosis. For pulmonary disease, the major symptoms are cough, chest pain, and dyspnea (Yeager et al. 2005). Cough is perhaps the most common symptom. In some cases, cough is the only manifestation requiring therapy. In that case, inhaled corticosteroids alone may be effective (Pietinalho et al. 1999; Baughman et al. 2002b). Dyspnea is multifactorial in sarcoidosis. Endobronchial disease can lead to fixed obstruction and dyspnea. It is more likely to reverse if diagnosed and treated within the first 6 months of onset of symptoms (Chambellan et al. 2005). While parenchymal disease often responds to corticosteroid therapy, there is a high rate of relapse when treatment is withdrawn within 2 years (Gottlieb et al. 1997; Baughman et al. 2006c). Cytotoxic agents such as methotrexate may be effective steroid sparing alternatives.

Sarcoidosis, Table 6 Drug options for sarcoidosis

Drug classification/ drug	Usual dosage	Side effects	Monitoring recommendations
Glucocorticoids			
Prednisone	Very high dose: 40–100 mg daily High dose: 20–40 mg daily Low dose: 5–15 mg daily	Weight gain, diabetes, hypertension, risk for infection, osteoporosis, cataracts, glaucoma, gastritis, acne, psychosis, increased risk for infection	Bone density, follow weight, blood pressure, ask about symptoms suggesting diabetes, cataracts
Intravenous pulse methylprednisolone	1000 mg daily	Weight gain, diabetes, hypertension, risk for infection, osteoporosis, cataracts, glaucoma, gastritis, acne, fluid retention, metallic taste, increased risk for infection	Follow weight, blood pressure, ask about symptoms suggesting diabetes, cataracts
Cytotoxic drugs			
Methotrexate	2.5–15 mg once a week	Nausea, vomiting, leukopenia, anemia, increased risk for infection, hepatotoxicity	Complete blood count, liver, renal function every 1–3 months
Azathioprine	50–250 mg daily	Nausea, vomiting, leukopenia, anemia, increased risk for infection, hepatotoxicity, increased rate of non-melanoma skin cancers	Complete blood count and liver function every 1–3 months
Leflunomide	10–20 mg daily	Nausea, vomiting, leukopenia, anemia, increased risk for infection, hepatotoxicity, hypertension, peripheral neuropathy	Complete blood count, liver, renal function every 1–3 months Ask about new neuropathy symptoms
Mycophenolate	500–1500 mg twice a day	Nausea, bloating, diarrhea, leukopenia, anemia, increased risk for infection, increased rate of non-melanoma skin cancers	Complete blood count every 1–3 months
Antimalarial drugs			
Chloroquine	250-500 mg daily	Nausea, ocular toxicity, rash	Eye examination every 6–12 months
Hydroxychloroquine	200–400 mg daily	Nausea, ocular toxicity, rash	Eye examination every 6–12 months
Anti-TNF antibodies			
Infliximab	3–5 mg/kg initially, 2 weeks later, then every 4–12 weeks	Allergic reactions, increased rate of infections, reactivation of tuberculosis, worsening congestive heart failure, increased rate of cancer, demyelinating disease	Testing for latent tuberculosis prior to initiating therapy, hold treatment when signs of significant infection, monitor for severe allergic reaction at time of infusion
Adalimumab	20 mg every 1–2 weeks	Allergic reactions, increased rate of infections, reactivation of tuberculosis, worsening congestive heart failure, increased rate of cancer, demyelinating disease	Testing for latent tuberculosis prior to initiating therapy, hold treatment when signs of significant infection
Others			
Rituximab	1000 mg initially, 2 weeks later, then every 3–52 weeks	Allergic reactions, increased rate of infections, reactivation of viral infections, leukopenia, progressive multifocal leukoencephalopathy	Test for hepatitis B and C infection prior to treatment. Hold treatment when signs of significant infection, monitor for severe allergic reaction at time of infusion
Pentoxifylline	200–400 mg three times a day	Nausea, vomiting, tachyarrythmias	None

TNF tumor necrosis factor

For patients with refractory disease requiring continued high dose treatment beyond 2 years, infliximab has been shown to provide significant improvement in lung function (Baughman et al. 2006a). Patients who fail to respond to anti-inflammatory treatment may have complications of their sarcoidosis as the cause of their dyspnea. One major factor is pulmonary hypertension, which has been reported to occur in half of the patients with moderately severe chronic dyspnea (Baughman et al. 2010b; Shorr et al. 2005).

Skin lesions may respond to topical therapy. For those with more extensive skin lesions, the antimalarial agents have been reported as helpful (Baughman and Lower 2007; Jones and Callen 1990). In patients with chronic cutaneous lesions, more aggressive treatment may be required. This could include cytotoxic drugs and anti-TNF agents. The anti-TNF agents have been shown to be effective in patients who have failed other treatments (Stagaki et al. 2009; Baughman and Lower 2001; Mallbris et al. 2003).

Anterior uveitis may be controlled by topical therapy. Since this is often a self-limited process, the patient may never require systemic therapy. However, for patients with intermediate or posterior uveitis or other ocular manifestations, systemic therapy may be required. Due to the risk of glaucoma and cataracts, alternative to systemic glucocorticoids are preferred in these patients. Several cytotoxic drugs have been used in uveitis in general (Galor et al. 2008; Samson et al. 2001) and sarcoidosis in particular (Baughman et al. 2002a; Bhat et al. 2009; Dev et al. 1999). Anti-TNF agents have been reported as effective in patients who have failed other treatments (Baughman et al. 2002a; Erckens et al. 2012). Rituximab has recently been reported as an alternative to infliximab in this group of patients (Lower et al. 2012).

Neurologic disease may require higher doses of corticosteroids for initial management (Lower et al. 1997; Scott et al. 2007). Most of these patients will require prolonged treatment and early institution of cytotoxic drugs as steroid sparing is common. However, some patients remain refractory, and the use of anti-TNF therapy can lead to reversal of neurologic deficits when other treatment regimens have failed (Sodhi et al. 2009; Moravan and Segal 2009).

Cardiac sarcoidosis can be a life-threatening complication of sarcoidosis and glucocorticoids is standard treatment. In one large retrospective study, there was no difference in outcome for those treated with very high doses of corticosteroids versus those begun on moderate doses (Yazaki et al. 2001). Cytotoxic drugs may prove steroid sparing and are commonly used in these patients (Schuller et al. 2012; Kron et al. 2013). Infliximab is contraindicated in patients with severe idiopathic cardiomyopathy (Chung et al. 2003). Infliximab has been used successfully in sarcoidosis cardiomyopathy (Barnabe et al. 2008; Uthman et al. 2007), but its use remains controversial. A major cause of death from cardiac sarcoidosis is ventricular tachycardia. The use of the implanted cardiac defibrillator (ICD) has become standard in patients with cardiac sarcoidosis (Schuller et al. 2012; Kron et al. 2013).

The individual agents described in Table 5 are summarized in Table 6. In this table, the doses and common side effects are provided. Recommendations for monitoring the nonsteroidal immunosuppressives have been made by a task force of the American College of Chest Physicians (Baughman et al. 2012c). These are summarized in Table 6.

Outcome

The outcome of sarcoidosis is quite variable. Overall about half of the patients never require systemic therapy for their disease (Baughman et al. 2006b). Of those who do require initial therapy, about half require long-term treatment. There are ethnic differences regarding outcome. African Americans have been reported to have a higher frequency of requiring long-term treatment (Gottlieb et al. 1997) than Caucasians in the USA (Hunninghake et al. 1994) or Europe (Rizzato et al. 1998). Certain manifestations such as lupus pernio, cardiac disease, and pulmonary fibrosis are associated with chronic disease (Neville et al. 1983).

The overall mortality from sarcoidosis has been reported to be around 5 % (Baughman et al. 1997). Recent studies suggest that the mortality and rate of hospitalization from sarcoidosis have risen over the past few decades (Swigris et al. 2011; Gerke et al. 2012). The most common cause of death is respiratory failure. This can be from advanced fibrosis of lungs (Nardi et al. 2011). Pulmonary hypertension is a common complication of fibrotic sarcoidosis (Shorr et al. 2005) and is a risk factor for mortality (Shorr et al. 2003). The 1 year mortality from precapillary pulmonary hypertension associated with sarcoidosis is approximately 30 % (Baughman et al. 2010b).

References

Bakkers, M., Merkies, I. S., Lauria, G., Devigili, G., Penza, P., Lombardi, R., et al. (2009). Intraepidermal nerve fiber density and its application in sarcoidosis. *Neurology, 73*(14), 1142–1148.

Baltzan, M., Mehta, S., Kirkham, T. H., & Cosio, M. G. (1999). Randomized trial of prolonged chloroquine therapy in advanced pulmonary sarcoidosis. *American Journal of Respiratory and Critical Care Medicine, 160*(1), 192–197.

Barnabe, C., McMeekin, J., Howarth, A., & Martin, L. (2008). Successful treatment of cardiac sarcoidosis with infliximab. *Journal of Rheumatology, 35*(8), 1686–1687.

Baughman, R. P., & Lower, E. E. (2001). Infliximab for refractory sarcoidosis. *Sarcoidosis, Vasculitis, and Diffuse Lung Diseases, 18*, 70–74.

Baughman, R. P., & Lower, E. E. (2007). Evidence-based therapy for cutaneous sarcoidosis. *Clinics in Dermatology, 25*(3), 334–340.

Baughman, R. P., & Nunes, H. (2012). Therapy for sarcoidosis: Evidence-based recommendations. *Expert Review of Clinical Immunology, 8*(1), 95–103.

Baughman, R. P., Winget, D. B., Bowen, E. H., & Lower, E. E. (1997). Predicting respiratory failure in sarcoidosis patients. *Sarcoidosis, 14*, 154–158.

Baughman, R. P., Weber, F. L., Bejarano, P. B., Koehler, A., & Lower, E. E. (1999). Methotrexate for chronic sarcoidosis: Hepatotoxicity assessed by liver biopsy. *American Journal of Respiratory and Critical Care Medicine, 159*, A342.

Baughman, R. P., Winget, D. B., & Lower, E. E. (2000). Methotrexate is steroid sparing in acute sarcoidosis: Results of a double blind, randomized trial. *Sarcoidosis, Vasculitis, and Diffuse Lung Diseases, 17*, 60–66.

Baughman, R. P., Teirstein, A. S., Judson, M. A., Rossman, M. D., Yeager, H. J., Bresnitz, E. A., et al. (2001). Clinical characteristics of patients in a case control study of sarcoidosis. *American Journal of Respiratory and Critical Care Medicine, 164*, 1885–1889.

Baughman, R. P., Lower, E. E., Ingledue, R., Kaufman, A. H. (2002a). Management of ocular sarcoidosis. *Sarcoidosis Vasculitis Diffuse Lung Diseases, 23*, 543–548.

Baughman, R. P., Iannuzzi, M. C., Lower, E. E., Moller, D. R., Balkissoon, R., Winget, D. B., et al. (2002b). Use of fluticasone for acute symptomatic pulmonary sarcoidosis. *Sarcoidosis, Vasculitis, and Diffuse Lung Diseases, 19*, 198–204.

Baughman, R. P., Drent, M., Kavuru, M., Judson, M. A., Costabel, U., Du, B. R., et al. (2006a). Infliximab therapy in patients with chronic sarcoidosis and pulmonary involvement. *American Journal of Respiratory and Critical Care Medicine, 174*(7), 795–802.

Baughman, R. P., Judson, M. A., Teirstein, A., Yeager, H., Rossman, M., Knatterud, G. L., et al. (2006b). Presenting characteristics as predictors of duration of treatment in sarcoidosis. *QJM, 99*(5), 307–315.

Baughman, R. P., Judson, M. A., Teirstein, A. S., Yeager, H., Rossman, M., Knatterud, G. L., et al. (2006c). Presenting characteristics as predictors of duration of treatment for sarcoidosis. *Chest, 99*, 307–315.

Baughman, R. P., Sparkman, B. K., & Lower, E. E. (2007). Six-minute walk test and health status assessment in sarcoidosis. *Chest, 132*(1), 207–213.

Baughman, R. P., Judson, M. A., Teirstein, A., Lower, E. E., Lo, K., Schlenker-Herceg, R., et al. (2008). Chronic facial sarcoidosis including lupus pernio: Clinical description and proposed scoring systems. *American Journal of Clinical Dermatology, 9*(3), 155–161.

Baughman, R. P., Lower, E. E., & Kaufman, A. H. (2010a). Ocular sarcoidosis. *Seminars in Respiratory and Critical Care Medicine, 31*(4), 452–462.

Baughman, R. P., Engel, P. J., Taylor, L., & Lower, E. E. (2010b). Survival in sarcoidosis associated pulmonary hypertension: The importance of hemodynamic evaluation. *Chest, 138*, 1078–1085.

Baughman, R. P., Lower, E. E., & Gibson, K. (2012a). Pulmonary manifestations of sarcoidosis. *Presse Médicale, 41*, e289–e302.

Baughman, R. P., Weiss, K. L., & Golnik, K. C. (2012b). Neuro-ophthalmic sarcoidosis. *Eye and Brain, 4*, 13–25.

Baughman, R. P., Meyer, K. C., Nathanson, I., Angel, L., Bhorade, S. M., Chan, K. M., et al. (2012c). Monitoring of nonsteroidal immunosuppressive drugs in patients with lung disease and lung transplant recipients: American college of chest physicians evidence-based clinical practice guidelines. *Chest, 142*(5), e1S–e111S.

Bhat, P., Cervantes-Castaneda, R. A., Doctor, P. P., Anzaar, F., & Foster, C. S. (2009). Mycophenolate mofetil therapy for sarcoidosis-associated uveitis. *Ocular Immunology and Inflammation, 17*(3), 185–190.

Bradley, D. A., Lower, E. E., & Baughman, R. P. (2006). Diagnosis and management of spinal cord sarcoidosis.

Sarcoidosis, Vasculitis, and Diffuse Lung Diseases, 23(1), 58–65.

Braswell, R. A., & Kline, L. B. (2007). Neuro-ophthalmologic manifestations of sarcoidosis. *International Ophthalmology Clinics, 47*(4), 67–77, ix.

Browne, P. M., Sharma, O. P., & Salkin, D. (1978). Bone marrow sarcoidosis. *JAMA, 240*, 43–50.

Chambellan, A., Turbie, P., Nunes, H., Brauner, M., Battesti, J. P., & Valeyre, D. (2005). Endoluminal stenosis of proximal bronchi in sarcoidosis: Bronchoscopy, function, and evolution. *Chest, 127*(2), 472–481.

Chapelon-Abric, C., de Zuttere, D., Duhaut, P., Veyssier, P., Wechsler, B., Huong, D. L., et al. (2004). Cardiac sarcoidosis: A retrospective study of 41 cases. *Medicine (Baltimore), 83*(6), 315–334.

Chung, E. S., Packer, M., Lo, K. H., Fasanmade, A. A., & Willerson, J. T. (2003). Randomized, double-blind, placebo-controlled, pilot trial of infliximab, a chimeric monoclonal antibody to tumor necrosis factor-alpha, in patients with moderate-to-severe heart failure: Results of the anti-TNF Therapy Against Congestive Heart Failure (ATTACH) trial. *Circulation, 107*(25), 3133–3140.

Cohen-Aubart, F., Galanaud, D., Grabli, D., Haroche, J., Amoura, Z., Chapelon-Abric, C., et al. (2010). Spinal cord sarcoidosis: Clinical and laboratory profile and outcome of 31 patients in a case–control study. *Medicine (Baltimore), 89*(2), 133–140.

Cremers, J., Drent, M., Driessen, A., Nieman, F., Wijnen, P., Baughman, R., et al. Liver-test abnormalities in sarcoidosis. *European Journal of Gastroenterology & Hepatology.* 2011.

Dev, S., McCallum, R. M., & Jaffe, G. J. (1999). Methotrexate for sarcoid-associated panuveitis. *Ophthalmology, 106*, 111–118.

Dumas, J. L., Valeyre, D., Chapelon-Abric, C., Belin, C., Piette, J. C., Tandjaoui-Lambiotte, H., et al. (2000). Central nervous system sarcoidosis: Follow-up at MR imaging during steroid therapy. *Radiology, 214*(2), 411–420.

Erckens, R. J., Mostard, R. L., Wijnen, P. A., Schouten, J. S., & Drent, M. (2012). Adalimumab successful in sarcoidosis patients with refractory chronic non-infectious uveitis. *Graefes Archive for Clinical and Experimental Ophthalmology, 250*, 713–720.

Galor, A., Jabs, D. A., Leder, H. A., Kedhar, S. R., Dunn, J. P., Peters, G. B., III, et al. (2008). Comparison of antimetabolite drugs as corticosteroid-sparing therapy for noninfectious ocular inflammation. *Ophthalmology, 115*(10), 1826–1832.

Gerke, A. K., Yang, M., Tang, F., Cavanaugh, J. E., & Polgreen, P. M. (2012). Increased hospitalizations among sarcoidosis patients from 1998 to 2008: A population-based cohort study. *BMC Pulmonary Medicine, 12*, 19. doi:10.1186/1471-2466-12-19.:19-12.

Gibson, G. J., Prescott, R. J., Muers, M. F., Middleton, W. G., Mitchell, D. N., Connolly, C. K., et al. (1996). British Thoracic Society Sarcoidosis study: Effects of long term corticosteroid treatment. *Thorax, 51*(3), 238–247.

Gottlieb, J. E., Israel, H. L., Steiner, R. M., Triolo, J., & Patrick, H. (1997). Outcome in sarcoidosis. The relationship of relapse to corticosteroid therapy. *Chest, 111*(3), 623–631.

Grunewald, J., & Eklund, A. (2007). Sex-specific manifestations of Lofgren's syndrome. *American Journal of Respiratory and Critical Care Medicine, 175*(1), 40–44.

Grunewald, J., & Eklund, A. (2009). Lofgren's syndrome: Human leukocyte antigen strongly influences the disease course. *American Journal of Respiratory and Critical Care Medicine, 179*(4), 307–312.

Hall, G., Naish, P., Sharma, O. P., Doe, W., & James, D. G. (1969). The epidemiology of sarcoidosis. *Postgraduate Medical Journal, 45*, 241–250.

Hoitsma, E., Marziniak, M., Faber, C. G., Reulen, J. P., Sommer, C., De Baets, M., et al. (2002). Small fibre neuropathy in sarcoidosis. *Lancet, 359*(9323), 2085–2086.

Hoitsma, E., De, V. J., & Drent, M. (2011). The small fiber neuropathy screening list: Construction and cross-validation in sarcoidosis. *Respiratory Medicine, 105*(1), 95–100.

Hunninghake, G. W., & Crystal, R. G. (1981). Pulmonary sarcoidosis: A disorder mediated by excess helper T-lymphocyte activity at sites of disease activity. *The New England Journal of Medicine, 305*, 429–432.

Hunninghake, G. W., Gilbert, S., Pueringer, R., Dayton, C., Floerchinger, C., Helmers, R., et al. (1994). Outcome of the treatment for sarcoidosis. *American Journal of Respiratory and Critical Care Medicine, 149*(4 Pt 1), 893–898.

Jabs, D. A., & Johns, C. A. (1986). Ocular involvement in chronic sarcoidosis. *American Journal of Ophthalmology, 102*, 297–301.

Jones, E., & Callen, J. P. (1990). Hydroxychloroquine is effective therapy for control of cutaneous sarcoidal granulomas. *Journal of the American Academy of Dermatology, 23*(3 Pt 1), 487–489.

Judson, M. A. (2002). Hepatc, splenic, and gastrointestinal involvement with sarcoidosis. *Seminars in Respiratory and Critical Care Medicine, 23*, 529–543.

Kahi, C. J., Saxena, R., Temkit, M., Canlas, K., Roberts, S., Knox, K., et al. (2006). Hepatobiliary disease in sarcoidosis. *Sarcoidosis, Vasculitis, and Diffuse Lung Diseases, 23*(2), 117–123.

Kavathia, D., Buckley, J. D., Rao, D., Rybicki, B., & Burke, R. (2010). Elevated 1, 25-dihydroxyvitamin D levels are associated with protracted treatment in sarcoidosis. *Respiratory Medicine, 104*(4), 564–570.

Kennedy, P. T., Zakaria, N., Modawi, S. B., Papadopoulou, A. M., Murray-Lyon, I., du Bois, R. M., et al. (2006). Natural history of hepatic sarcoidosis and its response to treatment. *European Journal of Gastroenterology and Hepatology, 18*(7), 721–726.

Kron, J., Sauer, W., Schuller, J., Bogun, F., Crawford, T., Sarsam, S., et al. (2013). Efficacy and safety of implantable cardiac defibrillators for treatment of ventricular arrhythmias in patients with cardiac sarcoidosis. *Europace, 15*, 347–354.

Lower, E. E., & Weiss, K. L. (2008). Neurosarcoidosis. *Clinics in Chest Medicine, 29*(3), 475–492.

Lower, E. E., Smith, J. T., Martelo, O. J., & Baughman, R. P. (1988). The anemia of sarcoidosis. *Sarcoidosis, 5*, 51–55.

Lower, E. E., Broderick, J. P., Brott, T. G., & Baughman, R. P. (1997). Diagnosis and management of neurologic sarcoidosis. *Archives of Internal Medicine, 157*, 1864–1868.

Lower, E. E., Baughman, R. P., & Kaufman, A. H. (2012). Rituximab for refractory granulomatous eye disease. *Clinical Ophthalmology, 6*, 1613–1618. doi:10.2147/OPTH.S35521.

Mahevas, M., Lescure, F. X., Boffa, J. J., Delastour, V., Belenfant, X., Chapelon, C., et al. (2009). Renal sarcoidosis: Clinical, laboratory, and histologic presentation and outcome in 47 patients. *Medicine (Baltimore), 88*(2), 98–106.

Mallbris, L., Ljungberg, A., Hedblad, M. A., Larsson, P., & Stahle-Backdahl, M. (2003). Progressive cutaneous sarcoidosis responding to anti-tumor necrosis factor-alpha therapy. *Journal of the American Academy of Dermatology, 48*(2), 290–293.

Mana, J., & Marcoval, J. (2007). Erythema nodosum. *Clinics in Dermatology, 25*(3), 288–294.

Marcellis, R. G., Lenssen, A. F., de Vries, G. J., Baughman, R. P., van der Grinten, C. P., Verschakelen, J. A., et al. (2013). Is there an added value of cardiopulmonary exercise testing in sarcoidosis patients? *Lung, 191*, 43–52.

Mason, R. S., Frankel, T., Chan, Y. L., Lissner, D., & Solomon, P. (1984). Vitamin D conversion by sarcoid lymph node homogenate. *Annals of Internal Medicine, 100*, 59–61.

Mehta, D., Lubitz, S. A., Frankel, Z., Wisnivesky, J. P., Einstein, A. J., Goldman, M., et al. (2008). Cardiac involvement in patients with sarcoidosis: Diagnostic and prognostic value of outpatient testing. *Chest, 133*(6), 1426–1435.

Moller, D. R., & Chen, E. S. (2002). Genetic basis of remitting sarcoidosis: Triumph of the trimolecular complex? *American Journal of Respiratory Cell and Molecular Biology, 27*(4), 391–395.

Moravan, M., & Segal, B. M. (2009). Treatment of CNS sarcoidosis with infliximab and mycophenolate mofetil. *Neurology, 72*(4), 337–340.

Myers, T. D., Smith, J. R., Wertheim, M. S., Egan, R. A., Shults, W. T., & Rosenbaum, J. T. (2004). Use of corticosteroid sparing systemic immunosuppression for treatment of corticosteroid dependent optic neuritis not associated with demyelinating disease. *British Journal of Ophthalmology, 88*(5), 673–680.

Nardi, A., Brillet, P. Y., Letoumelin, P., Girard, F., Brauner, M., Uzunhan, Y., et al. (2011). Stage IV sarcoidosis: comparison of survival with the general population and causes of death. *European Respiratory Journal, 38*(6), 1368–1373.

Neville, E., Walker, A. N., & James, D. G. (1983). Prognostic factors predicting the outcome of sarcoidosis: An analysis of 818 patients. *Quarterly Journal of Medicine, 208*, 525–533.

Ohira, H., Tsujino, I., Sato, T., Yoshinaga, K., Manabe, O., Oyama, N., et al. (2011). Early detection of cardiac sarcoid lesions with (18)F-fluoro-2-deoxyglucose positron emission tomography. *Internal Medicine, 50*(11), 1207–1209.

Oswald-Richter, K. A., & Drake, W. P. (2010). The etiologic role of infectious antigens in sarcoidosis pathogenesis. *Seminars in Respiratory and Critical Care Medicine, 31*(4), 375–379.

Paramothayan, S., Lasserson, T., & Walters, E. H. (2003). Immunosuppressive and cytotoxic therapy for pulmonary sarcoidosis. *Cochrane Database of Systematic Reviews, 3*, CD003536.

Paramothayan, N. S., Lasserson, T. J., & Jones, P. W. (2005). Corticosteroids for pulmonary sarcoidosis. *Cochrane Database of Systematic Reviews, 2*, CD001114.

Park, M. K., Fontana, J. R., Babaali, H., Gilbert-McClain, L. I., Joo, J., Moss, J., et al. (2009). Steroid sparing effects of pentoxifylline in pulmonary sarcoidosis. *Sarcoidosis, Vasculitis, and Diffuse Lung Diseases, 26*, 121–131.

Pawate, S., Moses, H., & Sriram, S. (2009). Presentations and outcomes of neurosarcoidosis: A study of 54 cases. *QJM, 102*, 449–460.

Pietinalho, A., Tukiainen, P., Haahtela, T., Persson, T., Selroos, O., & Finnish Pulmonary Sarcoidosis Study Group. (1999). Oral prednisolone followed by inhaled budesonide in newly diagnosed pulmonary sarcoidosis: A double-blind, placebo-controlled, multicenter study. *Chest, 116*, 424–431.

Prieto, J. F., Dios, E., Gutierrez, J. M., Mayo, A., Calonge, M., & Herreras, J. M. (2001). Pars planitis: Epidemiology, treatment, and association with multiple sclerosis. *Ocular Immunology and Inflammation, 9*(2), 93–102.

Rizzato, G., Fraioli, P., & Montemurro, L. (1995). Nephrolithiasis as a presenting feature of chronic sarcoidosis. *Thorax, 50*(5), 555–559.

Rizzato, G., Montemurro, L., & Colombo, P. (1998). The late follow-up of chronic sarcoid patients previously treated with corticosteroids. *Sarcoidosis, 15*, 52–58.

Rossman, M. D., Newman, L. S., Baughman, R. P., Teirstein, A., Weinberger, S. E., Miller, W. J., et al. (2006). A double-blind, randomized, placebo-controlled trial of infliximab in patients with active pulmonary sarcoidosis. *Sarcoidosis, Vasculitis, and Diffuse Lung Diseases, 23*, 201–208.

Salazar, A., Mana, J., Corbella, X., Albareda, J. M., & Pujol, R. (1995). Splenomegaly in sarcoidosis: A report of 16 cases. *Sarcoidosis, 12*, 131–134.

Samson, C. M., Waheed, N., Baltatzis, S., & Foster, C. S. (2001). Methotrexate therapy for chronic nonifectious uveitis: Analysis of a case series of 160 patients. *Ophthalmology, 108*, 1134–1139.

Sato, H., Grutters, J. C., Pantelidis, P., Mizzon, A. N., Ahmad, T., Van Houte, A. J., et al. (2002). HLA-DQB1*0201: A marker for good prognosis in British and Dutch patients with sarcoidosis. *American Journal of Respiratory Cell and Molecular Biology, 27*(4), 406–412.

Scadding, J. G. (1961). Prognosis of intrathoracic sarcoidosis in England. *British Medical Journal, 4*, 1165–1172.

Schuller, J. L., Zipse, M., Crawford, T., Bogun, F., Beshai, J., Patel, A. R., et al. (2012). Implantable cardioverter defibrillator therapy in patients with cardiac sarcoidosis. *Journal of Cardiovascular Electrophysiology, 23*(9), 925–929.

Scott, T. F., Yandora, K., Valeri, A., Chieffe, C., & Schramke, C. (2007). Aggressive therapy for neurosarcoidosis: Long-term follow-up of 48 treated patients. *Archives of Neurology, 64*(5), 691–696.

Scott, T. F., Yandora, K., Kunschner, L. J., & Schramke, C. (2010). Neurosarcoidosis mimicry of multiple sclerosis: Clinical, laboratory, and imaging characteristics. *The Neurologist, 16*(6), 386–389.

Sekhri, V., Sanal, S., Delorenzo, L. J., Aronow, W. S., & Maguire, G. P. (2011). Cardiac sarcoidosis: A comprehensive review. *Archives of Medical Science, 7*(4), 546–554.

Sharma, O. P. (2000). Hypercalcemia in granulomatous disorders: A clinical review. *Current Opinion in Pulmonary Medicine, 6*(5), 442–447.

Shorr, A. F., Davies, D. B., & Nathan, S. D. (2003). Predicting mortality in patients with sarcoidosis awaiting lung transplantation. *Chest, 124*(3), 922–928.

Shorr, A. F., Helman, D. L., Davies, D. B., & Nathan, S. D. (2005). Pulmonary hypertension in advanced sarcoidosis: Epidemiology and clinical characteristics. *European Respiratory Journal, 25*(5), 783–788.

Sodhi, M., Pearson, K., White, E. S., & Culver, D. A. (2009). Infliximab therapy rescues cyclophosphamide failure in severe central nervous system sarcoidosis. *Respiratory Medicine, 103*(2), 268–273.

Song, Z., Marzilli, L., Greenlee, B. M., Chen, E. S., Silver, R. F., Askin, F. B., et al. (2005). Mycobacterial catalase-peroxidase is a tissue antigen and target of the adaptive immune response in systemic sarcoidosis. *Journal of Experimental Medicine, 201*(5), 755–767.

Spencer, T. S., Campellone, J. V., Maldonado, I., Huang, N., Usmani, Q., & Reginato, A. J. (2005). Clinical and magnetic resonance imaging manifestations of neurosarcoidosis. *Seminars in Arthritis and Rheumatism, 34*(4), 649–661.

Stagaki, E., Mountford, W. K., Lackland, D. T., & Judson, M. A. (2009). The treatment of lupus pernio: Results of 116 treatment courses in 54 patients. *Chest, 135*(2), 468–476.

Steckman, D. A., Schneider, P. M., Schuller, J. L., Aleong, R. G., Nguyen, D. T., Sinagra, G., et al. (2012). Utility of cardiac magnetic resonance imaging to differentiate cardiac sarcoidosis from arrhythmogenic right ventricular cardiomyopathy. *American Journal of Cardiology, 110*(4), 575–579.

Swigris, J. J., Olson, A. L., Huie, T. J., Fernandez-Perez, E. R., Solomon, J., Sprunger, D., et al. (2011). Sarcoidosis-related mortality in the United States from 1988 to 2007. *American Journal of Respiratory and Critical Care Medicine, 183*(11), 1524–1530.

Tavee, J., & Zhou, L. (2009). Small fiber neuropathy: A burning problem. *Cleveland Clinic Journal of Medicine, 76*(5), 297–305.

Uemura, A., Morimoto, S., Hiramitsu, S., Kato, Y., Ito, T., & Hishida, H. (1999). Histologic diagnostic rate of cardiac sarcoidosis: Evaluation of endomyocardial biopsies. *American Heart Journal, 138*(2 Pt 1), 299–302.

Uthman, I., Touma, Z., & Khoury, M. (2007). Cardiac sarcoidosis responding to monotherapy with infliximab. *Clinical Rheumatology, 26*(11), 2001–2003.

Warshauer, D. M., Molina, P. L., Hamman, S. M., Koehler, R. E., Paulson, E. K., Bechtold, R. E., et al. (1995). Nodular sarcoidosis of the liver and spleen: Analysis of 32 cases. *Radiology, 195*(3), 757–762.

Yazaki, Y., Isobe, M., Hiroe, M., Morimoto, S., Hiramitsu, S., Nakano, T., et al. (2001). Prognostic determinants of long-term survival in Japanese patients with cardiac sarcoidosis treated with prednisone. *American Journal of Cardiology, 88*, 1006–1010.

Yeager, H., Rossman, M. D., Baughman, R. P., Teirstein, A. S., Judson, M. A., Rabin, D. L., et al. (2005). Pulmonary and psychosocial findings at enrollment in the ACCESS study. *Sarcoidosis, Vasculitis, and Diffuse Lung Diseases, 22*(2), 147–153.

Zajicek, J. P., Scolding, N. J., Foster, O., Rovaris, M., Evanson, J., Moseley, I. F., et al. (1999). Central nervous system sarcoidosis – Diagnosis and management. *QJM, 92*(Feb), 103–117.

Sepsis

Sina M. Coldewey[1,2] and Michael Bauer[1,2]
[1]Department for Anesthesiology and Intensive Care, Jena University Hospital, Jena, Germany
[2]Center for Sepsis Control and Care, Jena University Hospital, Jena, Germany

Synonyms

Blood poisoning; Infectious systemic inflammatory response syndrome (SIRS)

Definition

Sepsis is defined as an inappropriate host response to infection reflected in the development of organ dysfunction secondary to suspected, probable, or proven infection.

Epidemiology and Genetics

Sepsis, the inappropriate response of the host to an invading pathogen may occur in patients presenting with community-acquired (prototypically pneumonia) as well as with nosocomial infections. The underlying infectious trigger might involve a variety of pathogens, primarily bacteria, and to a lesser extent viruses, fungi as well as parasites. Sepsis is not restricted to multiresistant bacteria and both community and hospital-acquired infections contribute to a comparable degree to the burden of disease. In any case, sepsis is an underestimated and silently growing problem, inseparably associated with increasing invasiveness of procedures, such as major surgery and chemotherapy in cancer patients with inherently impaired immune function. Sepsis reflects the leading cause of death in contemporary intensive care units (ICUs). Additional factors, including demographic change and advent of multidrug resistance in particular of Gram-negative bacteria, further contribute to the increasing burden of disease.

Studies from high-income countries report a rising incidence but falling case fatality, while roughly a doubling of the incidence of sepsis has been estimated over the last decade. However, studies using administrative data, such as "International Statistical Classification of Diseases and Related Health Problems" (ICD) codes to estimate incidence are inherently limited by the quality of the chart records. Comparing data from 20 % of US hospitals and four methods to estimate sepsis incidence from 2004 to 2009, the incidence varied substantially (300–1,000 cases per 100,000 population per year) (Gaieski et al. 2013). A one-day prospective, point prevalence study conducted on 14,000 patients in 1,265 participating ICUs from 75 countries in May 2007 revealed that approximately half of all patients treated on an ICU were infected and ICU (25 % vs. 11 %) as well as hospital (30 % vs. 15 %) mortality rates were twice as high compared to noninfected patients (Vincent et al. 2009). The excess mortality in the infected patients can be attributed largely to the development of sepsis. A recent meta-analysis of mortality rates in the control arm of sepsis trials over two decades, at least, confirmed that the overall trend observed in administrative data, i.e., a decreasing mortality rate corrected for disease severity, is associated with recent attempts to improve care for septic patients, an example being the "Surviving Sepsis Campaign" (Stevenson et al. 2014). It is noteworthy that sepsis is a primary cause of chronic critical illness and long-term cognitive and physical disabilities in patients surviving the acute phase of the disease (Iwashyna et al. 2010; Kahn et al. 2015).

Although there is solid evidence to suggest a genetic predisposition for sepsis (Sørensen et al. 1988), the identification of genetic loci and markers underlying the inappropriate host response to infection has proved to be challenging. Despite the description of multiple genetic variations that predict survival in sepsis, a recent genome-wide association study for survival from sepsis identified only one "single nucleotide polymorphism" (SNP) as being associated with poor outcome in pneumonia (Rautanen et al. 2015). Candidate gene approaches have shown that genes for the pathogen-sensing mechanisms are important susceptibility loci, and genetic studies suggest that there is a strong genetic influence on the risk of dying from infection (Petersen et al. 2010). However, genomic variability alone is insufficient to explain differences in the host response to infection (Lambeck et al. 2012). For instance, epigenetic processes (DNA methylation, histone modifications, regulatory RNA influence) that result in different phenotypes, despite equal genomic sequence, are also proposed to affect the host response to infection in preclinical models (Cheng et al. 2014), potentially contributing to development of sepsis and postsepsis immunosuppression (Carson et al. 2011). Nevertheless, definitive proof from clinical studies is currently pending but is supported by the strong association of sepsis with an aged immune system (immunosenescence).

Pathophysiology

It is now accepted that infection triggers a complex and prolonged host response, in which both inflammatory and immunosuppressive mechanisms occur simultaneously. These processes are required for clearance of infection and tissue recovery but may also mediate tissue injury and secondary opportunistic infections (Angus and van der Poll 2013). For instance, granulocytes or macrophages are critical for the elimination of invading pathogens but are also held responsible for concomitant tissue damage and subsequent organ failure. The anti-inflammatory or immunosuppressive response, seemingly limiting local and systemic tissue injury, is important for the resolution of inflammation but can also enhance susceptibility to secondary infections.

The early host response is initiated by pathogen recognition receptors, of which four main classes – toll-like receptors, C-type lectin receptors, retinoic acid inducible gene 1-like receptors, and nucleotide-binding oligomerization domain-like receptors – have been identified, that in part act in highly organized protein complexes, for instance in inflammasomes (Takeuchi and Akira 2010). These receptors recognize conserved structures (pathogen-associated molecular patterns, PAMPs) leading to local and systemic inflammatory responses. Their ability to sense in parallel endogenous molecules derived from damaged host tissue (damage-associated molecular patterns, DAMPs) leads to diagnostic uncertainty regarding the presence of an infectious source of systemic inflammation in critically ill, such as postsurgical patients (Zhang et al. 2010). This challenge to the discrimination of systemic inflammation associated with tissue damage from severe sepsis contributes to antibiotic overuse in the ICU. Although both pro- as well as anti-inflammatory branches of the immune system are activated simultaneously, immunosuppression, such as that reflected by reduced expression of human leukocyte antigen (HLA)-DR or reduced lymphocyte populations, seems to be more sustained and to dominate the late course of disease. As a consequence, long-term survivors of sepsis frequently have persisting infectious foci despite antimicrobial therapy and develop opportunistic infections as well as reactivation of latent viruses (Limaye et al. 2008).

PAMP- and DAMP-associated innate immune responses initiate a series of closely related downstream effector mechanisms, including production of reactive oxygen species, lipid mediators, cytokines, and lysosomal enzymes. These mechanisms and effector molecules may initiate a vicious circle of immune, hormonal, and autonomous neurological alterations, microvascular dysfunction, coagulatory activation, metabolic changes, mitochondrial as well as endothelial and epithelial dysfunction. This is clinically reflected in hyper- or hypothermia, vasoplegia, coagulopathy, and capillary leak syndrome.

In systemic inflammatory states, excess activation of the immune system is linked to local and systemic coagulation abnormalities through protease-activated receptors. Systemic activation of tissue factor on a multitude of cells may initiate disseminated intravascular coagulation which will be aggravated by parallel impaired anticoagulant mechanisms via consumption or downregulation of protein C, antithrombin, and fibrinolysis systems (Fourrier 2012). These processes along with depressed cardiac function and concomitant edema formation lead to macro- and microhemodynamic alterations characteristic for distributive shock with impaired tissue oxygenation (Marik 2014). Impaired oxygen availability to the cell is further aggravated by mitochondrial damage and dysfunction, impairing cellular oxygen use (Carre et al. 2010).

All mechanisms described above promote (multiple) organ dysfunction and – if persistent – failure. Variable organ systems might be affected to a differing extent (e.g., depending on preexistent comorbidities), and irrespective of individually involved organ systems, the duration, extent, and number of failing organs predicts mortality. The pattern and severity of organ failure can be assessed by scoring systems such as the Sequential Organ Failure Score (SOFA) (Vincent et al. 1998).

Clinical Presentation

Sepsis is an important differential diagnosis for many diseases as its manifestations are variable and depend on the pathogen, the site of infection, and patterns of systemic immune response and organ dysfunction. Comorbidities such as preexisting lung, kidney, or liver disease reflect risk factors for sepsis and the development of sepsis typically leads to deterioration of chronic organ dysfunction in these patients, as in the case of spontaneous bacterial peritonitis with acute-on-chronic liver failure. Thus, clinical presentation might vary considerably, in particular, with respect to the pattern of organ dysfunction. As a result, international consensus guidelines provide a long list of warning signs for early stages of sepsis ("the golden hour of sepsis"), when treatment needs to be initiated to prevent multiple organ failure (Levy et al. 2003). Clinical signs are typically present which indicate impaired cognitive (obtundation or delirium), cardiovascular (hypotension), and respiratory functions (tachypnea). Confusion, respiratory compromise with hypoxemia, hypotension or elevated serum lactate reflect key diagnostic features of incipient sepsis in a patient with proven or suspected infection. Typically, hypotension might persist, despite volume expansion, and signs of myocardial dysfunction might be present upon clinical examination, sonography, or chest X-ray. Acute kidney injury is manifested as decreasing urine output and increasing serum creatinine and the need to initiate renal replacement therapy might indicate a septic complication in the critically ill. Paralytic ileus, altered glycemic control, and thrombocytopenia are other common features in patients developing sepsis on an ICU and should prompt a search for – and if confirmed, removal of – an infectious focus. Importantly, acute organ dysfunction may affect all organs, including those considered to fail late or those rarely based on conventional tests, such as bilirubin levels for the liver (Recknagel et al. 2012). Thus, sepsis-associated organ dysfunction reflects primarily a widespread disturbance of critical cellular functions, such as mitochondrial respiration or signal transduction.

Therapy

There is no specific treatment for sepsis. The initial management of the patient requires early source control (e.g., drainage of an abscess), initiation of broad-spectrum antibiotics, and cardiorespiratory resuscitation to mitigate the systemic consequences of uncontrolled infection. Inappropriate or delayed antibiotic treatment as well as lack of provision of early supportive cardiorespiratory care is associated with increased mortality. Resuscitation should be achieved by intravenous fluid administration (balanced crystalloids, e.g. Ringer's with or without addition of albumin). Persistend hypotension and impaired hemodynamics due to septic cardiomyopathy should be treated with vasopressors and/or inotropes, respectively. In these cases, venous oxygen saturation and kinetics of plasma lactate can serve to guide hemodynamic management; the value of (invasive) monitoring of the cardiovascular system, e.g., with pulmonary artery catheters, is not supported by data from clinical studies (Dellinger et al. 2013). Oxygen should be given to normalize saturation and mechanical ventilation is initiated if necessary, for instance in the case of the "adult respiratory distress syndrome." There is evidence to support the concept that in addition to increasing oxygen delivery, in cases of septic shock with impaired oxygen availability, reducing oxygen consumption (through control of fever or sedation) might be beneficial, primarily in cases of impaired oxygen utilization by peripheral tissues (Schortgen et al. 2012).

Although many issues regarding the management of early sepsis have been reasons for debate of late, a standardized pragmatic approach to achieve source control and hemodynamic/metabolic stabilization, based on the recommendations of the "Surviving Sepsis Campaign" currently in its third iteration (Dellinger et al. 2013), seems to be associated with an improved outcome.

These measures are easy to achieve and can be applied to patients treated in an Emergency Department or on regular wards. Nevertheless, if available, the patient should be admitted to an ICU allowing monitoring and support of organ function. In any case, "aggressive" management, as advocated in the past, including preemptive initiation of organ support (mechanical ventilation, prophylactic renal replacement therapy) or liberal use of antibiotics including their combinations, must be avoided and available data suggest de-escalation of care whenever possible. For instance, strategies involving liberal use of hemofiltration to remove nonspecifically toxic compounds from the circulation, although intuitively compelling, have been documented as harmful (Payen et al. 2009). De-escalation of initial broad-spectrum antibiotic therapy, which is attempted far too rarely, prevents emergence of resistant organisms, minimizes significant drug toxicity, while reducing costs, and has been documented as safe in noninferiority trials (Heenen et al. 2012).

Outcome

As outlined above, there is an overall decrease in case fatality rates of sepsis that varies with the specific health care system and is estimated to lie in the range of 15–40 % in high-income countries. The main factor that has favorably affected mortality due to sepsis over the past decade is increased awareness, in particular among emergency department and ICU physicians, leading to a higher share of patients that receive appropriate care, including antibiotics and volume resuscitation, during the first "golden" hours of the disease. Improved acute care, though, has unraveled a paradoxical problem, i.e., occurrence of chronic critical illness, with a significant proportion of patients requiring prolonged care in qualified facilities. Simultaneously, persistent health-related outcomes, such as cognitive and neuromuscular impairment, increased cardiovascular morbidity and mortality, and probably an increased risk to develop recurrent (septic) infection have been identified, along with comparably higher mortality rates, for years after discharge from the ICU (Mayr et al. 2014).

References

Angus, D. C., & van der Poll, T. (2013). Severe sepsis and septic shock. *The New England Journal of Medicine, 369*(9), 840–851. doi:10.1056/NEJMra1208623.

Carre, J. E., Orban, J. C., Re, L., Felsmann, K., Iffert, W., Bauer, M., et al. (2010). Survival in critical illness is associated with early activation of mitochondrial biogenesis. *American Journal of Respiratory and Critical Care Medicine, 182*(6), 745–751. doi:10.1164/rccm.201003-0326OC.

Carson, W. F., Cavassani, K. A., Dou, Y., & Kunkel, S. L. (2011). Epigenetic regulation of immune cell functions during post-septic immunosuppression. *Epigenetics, 6*(3), 273–283.

Cheng, S. C., Quintin, J., Cramer, R. A., Shepardson, K. M., Saeed, S., Kumar, V., et al. (2014). mTOR- and HIF-1alpha-mediated aerobic glycolysis as metabolic basis for trained immunity. *Science, 345*(6204), 1250684. doi:10.1126/science.1250684.

Dellinger, R. P., Levy, M. M., Rhodes, A., Annane, D., Gerlach, H., Opal, S. M., et al. (2013). Surviving sepsis campaign: International guidelines for management of severe sepsis and septic shock: 2012. *Critical Care Medicine, 41*(2), 580–637. doi:10.1097/CCM.0b013e31827e83af.

Fourrier, F. (2012). Severe sepsis, coagulation, and fibrinolysis: Dead end or one way? *Critical Care Medicine, 40*(9), 2704–2708. doi:10.1097/CCM.0b013e318258ff30.

Gaieski, D. F., Edwards, J. M., Kallan, M. J., & Carr, B. G. (2013). Benchmarking the incidence and mortality of severe sepsis in the United States. *Critical Care Medicine, 41*(5), 1167–1174. doi:10.1097/CCM.0b013e31827c09f8.

Heenen, S., Jacobs, F., & Vincent, J. L. (2012). Antibiotic strategies in severe nosocomial sepsis: Why do we not de-escalate more often? *Critical Care Medicine, 40*(5), 1404–1409. doi:10.1097/CCM.0b013e3182416ecf.

Iwashyna, T. J., Ely, E. W., Smith, D. M., & Langa, K. M. (2010). Long-term cognitive impairment and functional disability among survivors of severe sepsis. *JAMA, 304*(16), 1787–1794. doi:10.1001/jama.2010.1553.

Kahn, J. M., Le, T., Angus, D. C., Cox, C. E., Hough, C. L., White, D. B., et al. (2015). The epidemiology of chronic critical illness in the United States. *Critical Care Medicine, 43*(2), 282–287. doi:10.1097/CCM.0000000000000710.

Lambeck, S., Weber, M., Gonnert, F. A., Mrowka, R., & Bauer, M. (2012). Comparison of sepsis-induced transcriptomic changes in a murine model to clinical blood samples identifies common response patterns.

Frontiers in Microbiology, 3, 284. doi:10.3389/fmicb.2012.00284.

Levy, M. M., Fink, M. P., Marshall, J. C., Abraham, E., Angus, D., Cook, D., et al. (2003). 2001 SCCM/ESICM/ACCP/ATS/SIS International Sepsis Definitions Conference. *Critical Care Medicine, 31*(4), 1250–1256. doi:10.1097/01.CCM.0000050454.01978.3B.

Limaye, A. P., Kirby, K. A., Rubenfeld, G. D., Leisenring, W. M., Bulger, E. M., Neff, M. J., et al. (2008). Cytomegalovirus reactivation in critically ill immunocompetent patients. *JAMA, 300*(4), 413–422. doi:10.1001/jama.300.4.413.

Marik, P. E. (2014). Early management of severe sepsis: Concepts and controversies. *Chest, 145*(6), 1407–1418. doi:10.1378/chest.13-2104.

Mayr, F. B., Yende, S., & Angus, D. C. (2014). Epidemiology of severe sepsis. *Virulence, 5*(1), 4–11. doi:10.4161/viru.27372.

Payen, D., Mateo, J., Cavaillon, J. M., Fraisse, F., Floriot, C., & Vicaut, E. (2009). Impact of continuous venovenous hemofiltration on organ failure during the early phase of severe sepsis: A randomized controlled trial. *Critical Care Medicine, 37*(3), 803–810. doi:10.1097/CCM.0b013e3181962316.

Petersen, L., Andersen, P. K., & Sorensen, T. I. (2010). Genetic influences on incidence and case-fatality of infectious disease. *PLoS One, 5*(5), e10603. doi:10.1371/journal.pone.0010603.

Rautanen, A., Mills, T. C., Gordon, A. C., Hutton, P., Steffens, M., Nuamah, R., et al. (2015). Genome-wide association study of survival from sepsis due to pneumonia: an observational cohort study. *Lancet. Respiratory Medicine, 3*(1), 53–60. doi:10.1016/S2213-2600(14)70290-5.

Recknagel, P., Gonnert, F. A., Westermann, M., Lambeck, S., Lupp, A., Rudiger, A., et al. (2012). Liver dysfunction and phosphatidylinositol-3-kinase signalling in early sepsis: Experimental studies in rodent models of peritonitis. *PLoS Medicine, 9*(11), e1001338. doi:10.1371/journal.pmed.1001338.

Schortgen, F., Clabault, K., Katsahian, S., Devaquet, J., Mercat, A., Deye, N., et al. (2012). Fever control using external cooling in septic shock: A randomized controlled trial. *American Journal of Respiratory and Critical Care Medicine, 185*(10), 1088–1095. doi:10.1164/rccm.201110-1820OC.

Sørensen, T. I., Nielsen, G. G., Andersen, P. K., & Teasdale, T. W. (1988). Genetic and environmental influences on premature death in adult adoptees. *The New England Journal of Medicine, 318*(12), 727–732. doi:10.1056/NEJM198803243181202.

Stevenson, E. K., Rubenstein, A. R., Radin, G. T., Wiener, R. S., & Walkey, A. J. (2014). Two decades of mortality trends among patients with severe sepsis: A comparative meta-analysis. *Critical Care Medicine, 42*(3), 625–631. doi:10.1097/CCM.0000000000000026.

Takeuchi, O., & Akira, S. (2010). Pattern recognition receptors and inflammation. *Cell, 140*(6), 805–820. doi:10.1016/j.cell.2010.01.022.

Vincent, J. L., de Mendonca, A., Cantraine, F., Moreno, R., Takala, J., Suter, P. M., et al. (1998). Use of the SOFA score to assess the incidence of organ dysfunction/failure in intensive care units: results of a multicenter, prospective study. Working group on "sepsis-related problems" of the European Society of Intensive Care Medicine. *Critical Care Medicine, 26*(11), 1793–1800.

Vincent, J. L., Rello, J., Marshall, J., Silva, E., Anzueto, A., Martin, C. D., et al. (2009). International study of the prevalence and outcomes of infection in intensive care units. *JAMA, 302*(21), 2323–2329. doi:10.1001/jama.2009.1754.

Zhang, Q., Raoof, M., Chen, Y., Sumi, Y., Sursal, T., Junger, W., et al. (2010). Circulating mitochondrial DAMPs cause inflammatory responses to injury. *Nature, 464*(7285), 104–107. doi:10.1038/nature08780.

Sepsis Models in Animals

Aurélie Gouel-Chéron[1,2,3] and Philippe Montravers[3]
[1]Institut Pasteur, Department of Immunology, Unit of Antibodies in Therapy and Pathology, INSERM, U1222, Paris, France
[2]Université Pierre et Marie Curie, Paris, France
[3]Département d'anesthésie-réanimation, Hôpital Bichat-Claude-Bernard, Hôpitaux de Paris, Université Paris-VII, Paris, France

Synonyms

Bacteremia; CLP model; Endotoxemia; Sepsis; Septic shock

Definition

Sepsis is a complex disease that can be considered to be an immuno-inflammatory dysfunction following uncontained infection (Fig. 1). Sepsis represents a major health problem in the intensive care unit (ICU) with a growing incidence and high mortality rates of up to 28 % (Angus et al. 2001). Several levels of inflammation and infection have been described. Severe sepsis and septic shock represent the most severe steps of these host-inflammation-pathogen interactions.

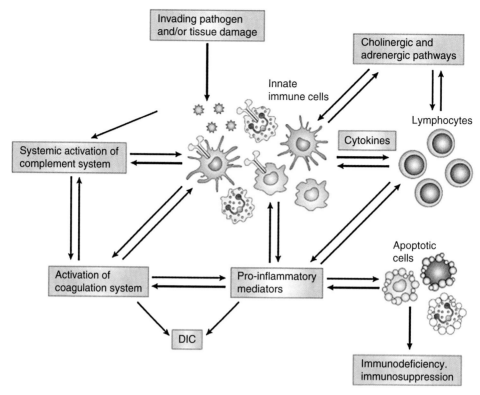

Sepsis Models in Animals, Fig. 1 The pathophysiology of sepsis. In response to pathogen invasion, a large number of type cells and inflammatory systems, involving "crosstalk," are involved to counterbalance the reactions (Adapted by permission from Macmillan Publishers Ltd: [Nature Reviews Immunology] Rittirsch et al. 2008)

Many definitions were used and induced a lot of confusion in the literature.

In 1992, a Consensus Conference defined "sepsis" as SIRS (systemic inflammatory response syndrome) with infection; "severe sepsis" as sepsis associated with organ dysfunction, hypoperfusion, or hypotension; and "septic shock" as sepsis with arterial hypotension despite "adequate" fluid resuscitation (American College of Chest Physicians and Society of Critical Care Medicine 1992). These definitions do not allow precise staging of sepsis, as they are based on highly subjective and nonspecific clinical signs. However, such artificial boundaries allow a better understanding of the dynamic process of sepsis. There are no clinical symptoms indicative of immune status and no signs suggestive of a shift from a pro-inflammatory state to an anti-inflammatory immunosuppressive state even if such changes occur during the development of the syndrome. Finally, the sequential stages of sepsis are difficult to characterize due to the absence of adequate staging biomarkers.

Inflammatory deregulation affects multiple organs via effects on endothelial, epithelial, and immune cell types that lead to irreversible damage. The first phase of sepsis, SIRS, is an overwhelming syndrome leading to the release of major pro-inflammatory cytokines via activation of the innate immune system. Compensatory anti-inflammatory response syndrome (CARS) refers to the counterregulatory mechanisms initiated to limit the inflammatory process, leading to immunosuppression and increased susceptibility to nosocomial infections (Bone 1996). Imbalance or exacerbation of one of these events results in host damage.

Structure and Function: Pathophysiology of Sepsis

For most scientists, the pathophysiology of sepsis, a complex series of cellular and humoral responses, remains a mystery (Hotchkiss and Karl 2003).

The Innate Immune System

The innate immune system is activated by bacterial cell wall products, such as lipopolysaccharide (LPS) from Gram-negative organisms, binding to host receptors, including Toll-like receptors (TLRs). These TLRs widely found on leukocytes and macrophages and on some endothelial cells have specificity for different bacterial, fungal, or viral products. Binding to these receptors leads to the production of pro-inflammatory cytokines such as tumor necrosis factor (TNF)-α and interleukins (IL) 1 and 6, which in turn activate immune cells (Fig. 1). Toxic downstream mediators, including prostaglandins, leukotrienes, platelet-activating factor, and phospholipase A2, are released. These mediators damage the endothelial lining, leading to increased capillary leakage. Furthermore, these cytokines lead to the production of adhesion molecules on endothelial cells and neutrophils. Neutrophilic endothelial interaction leads to further endothelial injury through the release of the neutrophil components. Finally, activated neutrophils release nitric oxide (NO). NO is a potent vasodilator which increased capillary permeability and has been implicated in sepsis-induced mitochondrial dysfunction.

The complement system is activated and mediates activation of leukocytes, attracting them to the site of infection where they can directly attack the organism (phagocytes, cytotoxic T lymphocytes), identify it for attack by others (antigen presenting cells, B lymphocytes), and cause the increased production and chemotaxis of more T helper cells.

The Endothelium and Coagulation System

Activated endothelium not only allows the adhesion and migration of stimulated immune cells but becomes porous to large molecules such as proteins, resulting in the pericapillary edema.

Alterations in the coagulation systems include an increase in procoagulant factors, such as plasminogen activator inhibitor type I, a potent inhibitor of fibrinolysis, and tissue factor and reduced circulating levels of natural anticoagulants, including antithrombin III and activated protein C (MacConmara et al. 2006), which also carry anti-inflammatory and modulatory roles.

The combination of these two factors leads to impairment of tissue oxygen delivery exacerbated by edema. This means that oxygen has to diffuse a greater distance to reach target cells. There is a reduction of capillary diameter due to mural edema and the procoagulant state results in capillary microthrombus formation.

Inflammation and Organ Dysfunction

Vasodilatation and increased capillary permeability are the initial hemodynamic disturbances observed in clinical practice leading to relative and absolute reductions in circulating volume. Relative and absolute hypovolemia are compounded by reduced left ventricular contractility to produce hypotension. Perfusion pressure is preserved initially through an increase in heart rate and cardiac output. As these compensatory mechanisms become exhausted, hypoperfusion and shock may result, leading to disordered blood flow through capillary beds and consequently lactic acidosis, cellular dysfunction, and multiorgan failure.

CD4 lymphocytes play a key role in the inflammatory response seen in sepsis. Early in the sepsis process, these cells assume a TH1 phenotype, where they produce large amounts of the pro-inflammatory mediators, including interferon gamma, TNF-α, and IL-2. CD4 lymphocytes may evolve over time to a Th2 phenotype, whereby the CD4 lymphocytes produce anti-inflammatory cytokines, including IL-10, IL-4, and IL-13. This is often driven by the release of stress hormones, such as catecholamines and corticosteroids. These cytokines dampen the immune response and can lead to the deactivation of monocytes. Additionally, TNF released early can cause apoptosis of lymphocytes in the gut, leading to further immunosuppression.

Pathological Relevance: Animal Models of Sepsis

In view of the impossibility to predict sepsis and the difficulty of conducting clinical trials in humans, animal models have been established to mimic human sepsis and reproduce the pathophysiology of sepsis. The main goal of animal models is to faithfully reproduce the characteristics of human sepsis.

Rodents are the mammals most commonly used in laboratory models, as they are inexpensive and easy to manage and procure. Their metabolism, physiology, and biochemistry pathways are well known and similar to those reported in humans. These biochemical similarities allow the use of human molecules or drugs in many instances. In addition, rodents are the species in which biomarkers have been most extensively investigated. Finally, rats and mice are animal models in which commercially available antibodies and biomarkers have been developed.

In humans, the early pathophysiologic response during SIRS is referred to as the hyperdynamic phase of sepsis. Subsequently, the hypodynamic state of sepsis reflects the impaired cardiac function observed during CARS. Animal models that do not reproduce this biphasic course are less relevant.

The major limitations of animal models are the time-course of sepsis development, the lack of supportive therapeutic interventions, and the study population. Human sepsis often occurs in elderly patients with comorbidities, whereas animals are carefully selected to be young adults with similar genetic background, age, gender, weight, and nutritional status. In humans, the onset and development of sepsis occur over a period of several days, versus several minutes or hours in animal models. Human patients receive various supportive and etiologic therapies (vasoactive support fluid loading, antibiotic therapy, oxygen, mechanical ventilation, etc.), making it difficult to extrapolate the results observed in animal models to humans.

Although many therapeutic agents have been shown to be beneficial in animal trials, they often fail in human trials (Riedemann et al. 2003). Animal models allow useful preliminary testing of therapeutic agents, especially at the screening phase of development. However, their advantages and caveats must be taken into account in order to accurately analyze the results of studies. Some models are devoted to general sepsis, while others have been developed to study the host response to a pathogen invading a particular organ, such as models of endocarditis, meningitis, or pneumonia. This second class of models is often devoted to the study of antimicrobial treatment. The objectives of this review are to describe the most commonly used animal models of general sepsis together with their respective advantages and limitations.

Sepsis models can be divided into three categories: (1) exogenous intravascular administration of a toxin (such as lipopolysaccharide, LPS), (2) exogenous administration of viable bacteria, and (3) alteration of the animal's endogenous protective barrier.

Exogenous Administration of a Toxin such as LPS/Endotoxemia

Burden et al., in 1951, suggested that "the administration of contaminated blood produces a characteristic clinical syndrome that may be incorrectly interpreted," corresponding to the first description of endotoxemic shock. LPS, the major component of the Gram-negative bacteria cell wall, is now the most common endotoxin used to reproduce endotoxemic shock in animal models. However, a clinical state of shock can also be induced by stimulation of TLR2 (Toll-like receptor) and TLR-9 (Freise et al. 2001).

Animal Selection in Endotoxemia Models

Sensitivity to endotoxins varies widely between species and strains. Most laboratory animals, such as rats, mice, cats, and dogs, are relatively resistant to endotoxins. The dose required to induce typical signs may differ by several orders of magnitude from that measured in humans. Presensitization with D-galactosamine enhances the animal's susceptibility to the endotoxin. However, these high doses represent the first limitation of this model. Indeed, high levels of circulating endotoxin induce side effects, and the results need to be interpreted cautiously. The reactions observed may not exactly represent the process observed in humans and may have only limited clinical relevance.

Methods of Endotoxemia

Depending on the study objectives, the endotoxin can be injected with or without anesthesia. Some anesthetic drugs, such as opiates or hypnotics, can induce immune modulation, hence the value of conscious animal models. LPS can be obtained from various strains of bacteria and from different serotypes of selected bacteria, leading to different endotoxemia models. The Gram-negative bacteria most commonly used are *Escherichia coli*. LPS can be administered to animals in several ways: intravenously and intraperitoneally, which is the most common route. The desired LPS dose is carefully injected, after stirring the solution. When appropriate, the animals are then monitored to ensure complete recovery from anesthesia.

Contribution and Limitations of the Endotoxemia Model

Endotoxemia sepsis models represent an easily controlled single-variable model in both humans and animals, making it an attractive experimental model. Unfortunately, endotoxin has a controversial role in the mechanism of human sepsis. The role of a chemical mediator may not accurately reflect the complex pathophysiologic reactions observed during septic shock. The type of Gram-negative bacteria used must be carefully selected, as the strain of LPS and its pathogenicity may vary according to the bacterial species. In addition, the endotoxemia model fails to explore Gram-positive and fungal sepsis, which represent one half of all cases of sepsis and septic shock.

As a result of these drawbacks, LPS injection is now considered to be more a model of endotoxemic shock than septic shock (Fink and Heard 1990). Despite these limitations, the endotoxemia model has provided significant contributions to sepsis research by providing knowledge about the pathways activated by pathogenic TLR agonists during the host response to infection.

Intravascular Infusion of Live Bacteria/Bacteremia

It has been proposed for a long time that bacteremia plays an important role in human sepsis mortality. Models of bacteremia have been proposed to study the conditions in which bacterial infection leads to sepsis and to assess the efficacy of sepsis therapies. Various bacterial species have been studied, mainly *Escherichia coli* (the most common one) and *Pseudomonas aeruginosa* (Fink and Heard 1990). Several confounding factors can modify interpretation of the results: the infecting bacterial load, virulence genes, the differences between human and animal pathophysiologic responses, and the choice of bacterial strain and its specificity. For example, *Salmonella typhi* does not cause systemic inflammation in rodents, while it causes typhoid fever and death in humans.

Doses of Bacteria Required

In animal models, very high doses of bacteria are required to reproduce sepsis and elicit mortality. They do not mimic the typical host responses to infection, as these bacteria do not colonize and significantly replicate following challenge, due to complement-mediated lysis. Some authors have proposed the injection of mixed bacterial flora (instead of single-species infection) to reduce the dose required. However, this may lead to difficult interpretation of the results. Moreover, a monobacterial injection does not take into account certain possible mechanisms of bacterial synergy, often demonstrated in human sepsis.

Hemodynamic Responses After Experimental Bacteremia

The hemodynamic and pulmonary changes induced by a bacterial injection are dependent on the bacterial species used. *Staphylococcus aureus* induces minimal systemic changes, whereas *Escherichia coli* and *Pseudomonas aeruginosa* both result in shock and acute respiratory failure (Dehring et al. 1983). In order to faithfully reproduce the human response, these two species are therefore mainly used, as they induce this type of immune failure.

During bacteremia, acute IV injection of live bacteria results in immediate cardiovascular collapse and early death. This pro-inflammatory reaction is more acute and more intense than the immune reaction observed after local infection. The bacteremia model can be considered to generate "pro-endotoxins" rather than a model of infective microorganisms, emphasizing why bacteremia models can be of interest in the study of early therapeutic interventions, such as fluid resuscitation.

Modalities of Bacteria Administration

Different routes of infection lead to different types of reactions. Whereas intraperitoneal injection initiates inflammatory activation and immune cell migration, blood priming induced by bacteremia induces activation of the endothelium and vascular systems. This can be monitored by cytokine concentrations as a marker of production of mediators (Zanetti et al. 1992).

To overcome the problem of early death and as many human patients develop intermittent sepsis, fractional administration of the challenge can be performed in order to create a more sustained immune reaction (Fink and Heard 1990). When the investigators performed aggressive resuscitation, the infusion can lead to a hyperdynamic state and metabolic and hormonal changes that mimic human patterns (Shaw and Wolfe 1984). These models can therefore mimic the clinical situation of patients with overwhelming infection, such as pneumococcal bacteremia in splenectomized patients, meningococcemia, or Gram-negative bacteremia during granulocytopenia (Fink and Heard 1990) and allows analysis of the early stage of sepsis and early therapies. However, these models are not commonly used, as they are highly time and cost-demanding.

Host-Barrier Disruption Models: Experimental Peritonitis

Gold Standard: The CLP Model

Host-barrier disruption models of sepsis are based on the principles of breaking the normal protective barriers between sterile compartments and pathogens, to allow the spread of infection by the host's mixed bacterial flora. The most commonly used methods are disruption of the intestinal barrier and include the cecal ligation and puncture (CLP) and colon ascendant stent peritonitis (CASP) models.

The CLP model is considered to be the gold standard for sepsis research and has been used to characterize the inflammatory stages of early sepsis (Wichterman et al. 1980). This model is highly reproducible and cost-effective and can be performed on small and large animals. The predominant species used are rodents (mice or rats) because of the relatively lower cost of performing experiments on a large numbers of animals and the availability of genetically deficient mouse strains, allowing the study of specific genes during the sepsis response (Buras et al. 2005).

CLP is a simple procedure, and survival rates are controllable and reproducible. It mimics human infection similar to that of perforated appendicitis or diverticulitis. It has become popular because of the similarity to human disease progression, reproducing the biphasic hemodynamic pattern (an early hyperdynamic phase followed by a hypodynamic phase), the metabolic phases, and the immunologic and apoptotic responses (Wichterman et al. 1980; Ayala and Chaudry 1996).

CLP Methods

Animals are not maintained on a special diet and can feed ad libitum. This is particularly important, as modifying the animal's diet results in modification of the gut flora and consequently modification of the infection observed.

After general anesthesia, the technique involves midline laparotomy, exteriorization of the cecum, ligation of the cecum distal to the ileocecal valve, and puncture of the ligated cecum (Fig. 2; Belikoff et al. 2008). This technique creates a bowel perforation with leakage of fecal contents into the peritoneum, which induces infection with mixed bacterial flora and provides an inflammatory source of necrotic tissue (Fig. 2; Wichterman et al. 1980; Ayala et al. 2000). After recovery from anesthesia, animals can receive immediate treatment, such as antibiotics or fluid replacement, to reproduce human management.

In general, the first death occurs 24 h after the procedure, eliminating any iatrogenic effects, such as hemorrhage. In the more severe CLP models, deaths can occur between 24 and 36 h after CLP with mortality typically greater than 70 % within 5 days. In the less severe CLP models, most deaths occur between 48 and 72 h after CLP with a 30–40 % range of mortality after 5 days.

Caecal ligation and puncture (CLP)

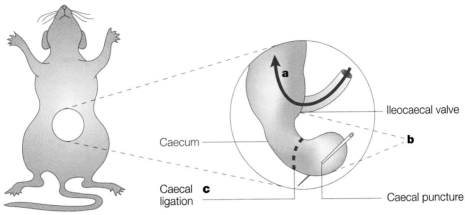

Colon ascendens stent peritonitis (CASP)

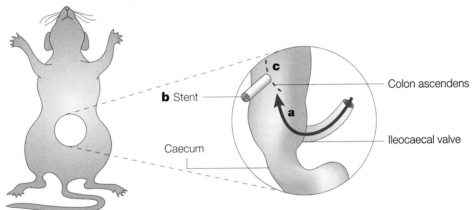

Sepsis Models in Animals, Fig. 2 Experimental models of peritonitis: cecal ligation and puncture (CLP) and colon ascendens stent peritonitis (CASP) (Adapted by permission from Macmillan Publishers Ltd: [Nat Rev Drug Discov] Buras et al. 2005)

Advantages of the CLP Model

Compared to endotoxemia, CLP allows evaluation of the mechanisms of distant organ colonization via the bloodstream and also allows control for severity of disease, as assessed by (1) mortality, (2) increasing the number of punctures, (3) puncture needle size, or (4) percentage of cecum ligated (larger amount of necrotic, stool-filled tissue creating a greater infectious and inflammatory insult) (Wichterman et al. 1980). All of these variables must be reported in subsequent publications, and their reproducibly between experiments and between operators must be verified.

Another major advantage of the CLP model is its ability to provide a control group, i.e., rodents which a sham laparotomy under anesthesia without manipulation of the gut. The CLP model also allows the possibility of surgical repair to reproduce usual human management such as appendicitis or diverticulitis.

Limitations of the CLP Model

The main disadvantage of the CLP model of sepsis is the host's capacity to minimize the development of infection. Local defense mechanisms can generate one or several abscesses around the septic focus finally preventing dissemination of the pathogen and successfully containing the infection. Some animals successfully contain the infection, do not progress to septic shock, and fully recover. This aspect may interfere with interpretation of the experimental results with respect to sepsis therapies. Variability within the CLP model itself may be underestimated, which is highlighted by comparison of control mortality rates and time to death between laboratories reporting the same experimental techniques. In addition, the load of fecal material that leaks from the ligated cecum is difficult to compare between different studies and could account for variability in the severity of disease.

Rodent bacterial flora differs from human bowel flora, directly linked to chow, as rodent bowel contains mainly Gram-positive bacteria such as *Streptococcus*, *Enterococcus*, *Clostridium*, and *Lactobacillus* and is devoid of *Enterobacteria* and strict anaerobic bacteria (Raibaud et al. 1966).

This drawback has led several authors to use calibrated inocula to assess specific issues (Montravers et al. 1994). Some authors have also proposed humanizing the rodent's bacterial flora by dietary modification.

A New Peritonitis Model: Colon Ascendens Stent Peritonitis

Fewer data are available concerning the CASP model (Maier et al. 2004), which was recently developed to improve several flaws of the CLP model. In the CASP model, after anesthesia and midline laparotomy, a stent of defined diameter is implanted into the ascending colon (Fig. 2). The stent allows the development of diffuse peritonitis due to persistent leakage of intestinal bacteria into the peritoneal cavity. This model appears to be less prone to abscess formation than the original CLP model (Maier et al. 2004). Changing the stent diameter controls the severity of sepsis. The stent can be removed after defined time periods, thereby mimicking surgical interventions.

Although CLP and CASP are both peritonitis models, the CASP model seems to better reproduce disseminated peritonitis (Maier et al. 2004). Unlike CLP mice, bacteremia in CASP mice constantly increases significantly and is associated with the development of multiple organ failure (Maier et al. 2004). The hemodynamic response to CASP has not been clearly characterized, and the disadvantages of this model are relatively unknown, as it represents a relatively new model system for acute polymicrobial septic peritonitis (Buras et al. 2005).

Interaction with Other Processes and Drugs

The goal of research is to improve the understanding of the pathophysiologic mechanisms and to develop new treatment strategies. Some reviews are already dedicated to the comparison of therapeutic interventions for sepsis that include animal experiments for explicit appraisals of the risk of bias and clinical relevance (Lamontagne et al. 2010).

Conclusion

Successful development and use of sepsis models in research require a good understanding of the advantages and caveats of each model. Animal models of sepsis do not completely reproduce human disease and/or require management similar to that delivered to human septic patients. However, they contribute to a better understanding of sepsis staging, which would facilitate comparison of animal models and allow better analysis of human sepsis. Some new treatments that failed to demonstrate their efficacy should be more carefully analyzed in light of timing of their administration at the pro- or anti-inflammatory stage, as they could be more harmful when administered at the inappropriate stage. Similarly, a better staging of animal and human sepsis could contribute to the development of new therapies and allow extrapolation of results from animal trials to humans.

Cross-References

▶ Antibiotics as Antiinflammatory Drugs
▶ Leukocyte Recruitment
▶ Platelets, Endothelium, and Inflammation
▶ Sepsis

References

American College of Chest Physicians and Society of Critical Care Medicine. (1992). Consensus conference: Definitions for sepsis and organ failure and guidelines for the use of innovative therapies in sepsis. *Critical Care Medicine, 20*, 864–874.

Angus, D. C., Linde-Zwirble, W. T., et al. (2001). Epidemiology of severe sepsis in the United States: Analysis of incidence, outcome, and associated costs of care. *Critical Care Medicine, 29*(7), 1303–1310.

Ayala, A., & Chaudry, I. H. (1996). Immune dysfunction in murine polymicrobial sepsis: Mediators, macrophages, lymphocytes and apoptosis. *Shock, 6*(Suppl 1), S27–38.

Ayala, A., Song, G. Y., et al. (2000). Immune depression in polymicrobial sepsis: The role of necrotic (injured) tissue and endotoxin. *Critical Care Medicine, 28*(8), 2949–2955.

Belikoff, B., Buras, J. A., et al. (2008). A practical approach to animal models of sepsis. In *Sourcebook of models for biomedical research* (pp. 473–482). Totowa: Humana Press.

Bone, R. C. (1996). Sir Isaac Newton, sepsis, SIRS, and CARS. *Critical Care Medicine, 24*(7), 1125–1128.

Buras, J. A., Holzmann, B., et al. (2005). Animal models of sepsis: Setting the stage. *Nature Reviews Drug Discovery, 4*(10), 854–865.

Dehring, D. J., Crocker, S. H., et al. (1983). Comparison of live bacteria infusions in a porcine model of acute respiratory failure. *Journal of Surgical Research, 34*(2), 151–158.

Fink, M. P., & Heard, S. O. (1990). Laboratory models of sepsis and septic shock. *Journal of Surgical Research, 49*(2), 186–196.

Freise, H., Bruckner, U. B., et al. (2001). Animal models of sepsis. *Journal of Investigative Surgery, 14*(4), 195–212.

Hotchkiss, R. S., & Karl, I. E. (2003). The pathophysiology and treatment of sepsis. *The New England Journal of Medicine, 348*(2), 138–150.

Lamontagne, F., Briel, M., et al. (2010). Systematic review of reviews including animal studies addressing therapeutic interventions for sepsis. *Critical Care Medicine, 38*(12), 2401–2408. doi:10.1097/CCM.2400b2013e3181fa0468.

MacConmara, M. P., Maung, A. A., et al. (2006). Increased $CD4^+$ $CD25^+$ T regulatory cell activity in trauma patients depresses protective Th1 immunity. *Annals of Surgery, 244*(4), 514–523.

Maier, S., Traeger, T., et al. (2004). Cecal ligation and puncture versus colon ascendens stent peritonitis: Two distinct animal models for polymicrobial sepsis. *Shock, 21*(6), 505–512.

Montravers, P., Andremont, A., Massias, L., Carbon, C. (1994) Investigation of the potential role of Enterococcus faecalis in the pathophysiology of experimental peritonitis. *J Infect Dis, 169*(4), 821–30.

Raibaud, P., Dickinson, A. B., et al. (1966). Microflora of the rat digestive tract. II. Quantitative study of the different microbial genera in the stomach and intestines of conventional rats. Individual quantitative variations and as a function of age. *Annales de l'Institut Pasteur (Paris), 110*(6), 861–876.

Riedemann, N. C., Guo, R. F., et al. (2003). The enigma of sepsis. *The Journal of Clinical Investigation, 112*(4), 460–467.

Rittirsch, D., Flierl, M. A., et al. (2008). Harmful molecular mechanisms in sepsis. *Nature Reviews Immunology, 8*(10), 776–787.

Shaw, J. H., & Wolfe, R. R. (1984). A conscious septic dog model with hemodynamic and metabolic responses similar to responses of humans. *Surgery, 95*(5), 553–561.

Wichterman, K. A., Baue, A. E., et al. (1980). Sepsis and septic shock – A review of laboratory models and a proposal. *Journal of Surgical Research, 29*(2), 189–201.

Zanetti, G., Heumann, D., et al. (1992). Cytokine production after intravenous or peritoneal gram-negative bacterial challenge in mice. Comparative protective efficacy of antibodies to tumor necrosis factor-alpha and to lipopolysaccharide. *The Journal of Immunology, 148*(6), 1890–1897.

Skin Inflammation Models in Animals

Stefan F. Martin
Allergy Research Group, Department of Dermatology, Medical Center - University of Freiburg, Freiburg, Germany

Synonyms

Mouse model; Skin inflammation

Definition

Inflammatory skin diseases are caused by dysregulation of tissue homeostasis due to external or internal trigger factors. Animal models are being used to study the underlying pathomechanisms.

Animal Models for Inflammatory Skin Disease

Skin diseases can be classified according to coding in the International Statistical Classification of

Diseases and Related Health Problems (ICD) published by the WHO in its latest version in 2010 (ICD-10 Chapter XII, L00-L99, www.who.int/classifications/icd/). They have many different causes and are associated with a disruption of tissue homeostasis. Tissue homeostasis is an active process that is tightly regulated by genetic and epigenetic programs and physiological processes. A tightly balanced interaction of the body surfaces such as the skin, gut, and lung with their specific – locally and individually different – microbiota is an important element of homeostasis. The skin microbiome comprises a huge community of commensal microorganisms which are not only found on the skin but also in hair follicles, sweat glands, and sebaceous glands. Some of the skin resident microorganisms are beneficial for skin function but are also potential pathogens such as S. aureus or P. acnes. These microorganisms are kept in check, for example, by the acidic skin pH and by natural antibiotics, families of antimicrobial peptides (AMP), produced in the skin. Abnormal AMP production is an important factor contributing to inflammatory skin diseases such as atopic dermatitis and psoriasis (Grice and Segre 2011; Gallo and Hooper 2012).

Any disturbance of the steady state is sensed by tissues by different means and results in processes that aim at the restoration of homeostasis. The surface area of the human skin comprises about 2 m^2. It is an important mechanical and immunological barrier to the outer world and is constantly exposed to the environment. Therefore, it has to deal with all kinds of assaults such as infectious agents, environmental noxae, chemicals, or UV irradiation. Disturbed skin homeostasis often results in breakdown of barrier function and inflammation, and a lack of restoration promotes the development of often chronic inflammatory skin diseases including infectious diseases and autoinflammatory and autoimmune diseases. Genetic defects such as null mutations of the gene for filaggrin that critically regulates skin barrier function also result in or predispose to inflammatory skin diseases (De Benedetto et al. 2012; Irvine et al. 2011). Prototypic sensors for infection but also for sterile inflammatory processes are the receptors of the innate immune system. These include the cell-associated Toll-like receptors (TLRs), cytosolic NOD-like receptors (NLRs), and RIG-I-like receptors (RLRs) but also soluble receptors in the extracellular space such as complement components, pentraxins, collectins, and ficolins (Bottazzi et al. 2010).

The common theme in inflammatory diseases of the skin and other tissues seems to be the disturbance of tissue homeostasis and barrier function leading to sterile inflammation that may be amplified by altered microbiota-tissue interaction.

This overview focuses on mouse models of inflammatory skin diseases. Inbred mouse strains have the advantage of defined and well-characterized genetics, and they allow studying the highly complex in vivo processes that are often not or only partially reproducible in in vitro systems, in part due to significant gaps of knowledge in mechanistic understanding. These gaps of knowledge can be closed by in vivo studies. Constitutive or inducible gene knockout/knockin and transgenic gene (over) expression methods as well as depletion of specific cell types with monoclonal antibodies or by modern genetic approaches and transfer of cells from wild-type, transgenic, or knockout mice enable the dissection of the molecular and cellular pathomechanisms and to design disease models. Therefore, mouse models are valuable tools for basic and preclinical research serving the understanding of human diseases, with all known limitations. They not only help to identify new drug targets and to develop new treatments, but they are also important to establish mechanistically defined endpoints which can be implemented for the development of in vitro assays for diagnostics and immunotoxicology. This should allow a stepwise reduction and eventually replacement of animal-based test systems by defined in vitro, in silico, and in chemico test systems that faithfully reproduce crucial steps of the disease pathogenesis.

Several spontaneous, inducible, transgenic, and knockout mouse models for inflammatory skin diseases exist (Table 1). A useful technology is the flanking of genes of interest with loxP sites,

Skin Inflammation Models in Animals, Table 1 Types of animal models for inflammatory skin diseases

Mouse model	Disease occurrence
Wild-type mouse strains	Induced
Mutant mouse strains	Spontaneous
Genetically engineered mouse models	**Disease occurrence**
(Inducible) knockout/transgenic mouse strains	Spontaneous/induced
(Inducible) cell type-depleting transgenic mouse strains	Spontaneous/induced

recognition sites for the adenoviral CreDNA recombinase. Cre recombinase transgenic mice are then mated with a strain expressing the loxP-flanked (floxed) gene. When Cre is expressed, the floxed gene is eliminated by a recombination process. Cre is often expressed under the control of a cell type-specific promoter and can even be made inducible. This allows, for example, a targeted and timed gene knockout. A common technique is the fusion of the Cre gene to a mutant estrogen receptor ligand binding domain that requires tamoxifen for activation. Cre is then only activated when mice receive tamoxifen (Belizário et al. 2010).

Some examples for each of the different mouse model categories will be given in the following chapters that illustrate the basic principles of the different models and the types of studies conducted.

Dysregulation of Epidermal Homeostasis Induces Skin Inflammation in Animal Models

Skin homeostasis is regulated by many factors. Among those, the transcription factor NF-κB plays a very important role. NF-κB is involved in different signaling pathways including those that regulate inflammation and apoptosis (Hoffmann and Baltimore 2006). NF-κB signaling must be tightly regulated. Therefore, NF-κB activation is negatively regulated by the inhibitory protein IκB. Upon stimulation of receptors such as TLRs, IκB is inactivated, and NF-κB translocates into the nucleus to bind DNA binding sites in promoter regions of NF-κB-regulated genes enabling their transcription. The IκB kinase beta (IKKβ or IKK2) is part of the so-called canonical NF-κB signaling pathway. IKKβ phosphorylates IκB which results in its degradation by the proteasome. Mice with a deletion of the negative regulator for NF-κB, IkBα, show strongly increased NF-κB activation and strong generalized inflammation by day 3 after birth. This includes skin inflammation characterized by erythematous thickening and scaling. The animals die by day 7 or 8. Keratinocyte-specific IκB-α-knockout (keratin 5 (K5)-Cre2) mice exhibit acanthosis and hyperkeratosis but fail to develop spontaneous epidermal inflammation. The skin disease improves progressively. Inflammation only develops when IκBα is also deleted in T cells. This underlines the role of T cells for initiation and maintenance of the inflammatory skin phenotype observed in the complete knockout mice (Rebholz et al. 2007).

Mice lacking IKKβ in keratinocytes (K5-Cre) develop inflammatory disease including spontaneous skin inflammation with infiltrating macrophages and other CD45$^+$ immune cells. The chronic inflammation is T- and B-cell independent and resembles interface dermatitis with epidermal basal cell injury, increased inflammatory cytokine production, and upregulation of genes typically associated with inflammation (Page et al. 2010).

Human polymorphisms in the gene for the ubiquitin-editing enzyme A20 (TNFAIP3) are associated with inflammatory disorders such as psoriasis. A20 negatively regulates NF-κB signaling and is involved in skin homeostasis and epidermal appendage development. Mice with a keratinocyte-specific A20 deletion (K14-Cre) develop keratinocyte hyperplasia but no spontaneous inflammation or scaling of the skin. Ectodermal appendages are abnormal: the mice have thinner, more fragile, and curly hair, longer nails, and sebocyte hyperplasia. The A20-deficient mice have a phenotype that copies the phenotype of transgenic mice overexpressing either the TNF family member ectodysplasin A (EDA) or the TNF family receptor ectodysplasin A receptor A1 (EDAR) which regulates NF-κB levels. Thus, overactivation of the EDA pathway or lack of negative regulating of EDA-induced NF-κB levels severely impacts skin homeostasis (Lippens et al. 2011).

Animal Models for Autoimmune Blistering Diseases

Autoantibody production to protein components of cellular adhesion complexes in the epidermis is associated with the loss of cell-cell adhesion between keratinocytes or loss of keratinocyte basement membrane adhesion and formation of skin blisters in locations that are characteristic for the target autoantigen and the disease. Typical diseases are pemphigus vulgaris (PV), epidermolysis bullosa acquisita (EBA), and bullous pemphigoid (BP). Two principal models exist: *Passive mouse models* are established by the injection of autoantibodies from patients or other species. *Active mouse models* make use of autoantigens injected into mice (Amagai 2008; Leighty et al. 2007; Sitaru 2007; Ganeshan et al. 2010).

In PV desmogleins (Dsg) 1 and 3, constituents of desmosomes that mediate cell-cell adhesion of keratinocytes, are the target autoantigens. Autoantibody production leads to acantholysis, the loss of intercellular keratinocyte adhesion. Passive models using injection of antibodies to Dsg failed to induce spontaneous blistering. A more successful approach makes use of Dsg3-deficient mice immunized with Dsg3 (Aoki-Ota et al. 2004). This leads to the induction of Dsg3-specific B and T cells which induce pemphigoid-like blistering disease when transferred into Rag-2-deficient recipient mice which lack B and T cells. Dsg3-deficient mice do not develop B- and T-cell tolerance to Dsg3 and can therefore be efficiently immunized. The transferred B and T cells then are activated by Dsg3 in the Rag-2 recipient, and the B cells produce pathogenic anti-Dsg3 IgG over a sustained period of time.

Collagen VII, a main constituent of anchoring fibrils of hemidesmosomes, adhesive complexes connecting epidermis and dermis, is the autoantigen in EBA. Autoantibodies to collagen VII and complement component C3 can be found deposited in the skin and are involved in the separation of the dermal-epidermal junction and subepidermal blister formation. The passive EBA mouse model makes use of anti-collagen VII antibodies (IgG) generated by immunization of rabbits with mouse or human collagen VII. Repeated injection of the purified IgG into susceptible mouse strains leads to blister formation but does not recapitulate other aspects of EBA. The injection of patient autoantibodies is of limited success. The active mouse model of EBA is based on the repeated subcutaneous injection of recombinant mouse collagen VII with adjuvants. This induces autoantibodies to collagen VII and blistering disease in susceptible mouse strains.

In BP the main autoantigens are collagen XVII (BP180) and BP230. They are both constituents of hemidesmosomes. Also here passive and active mouse models have been established. In the passive model blister formation is not induced spontaneously by injection of patient antibodies but only after shear stress applied to the skin. Injection of rabbit antibodies specific for collagen XVII induces subepidermal blister formation in mice. Recently, active models have been established that again use the injection of recombinant collagen XVII or fragments thereof. Repeated autoantigen injections can break tolerance and induce BP (Hirose et al. 2011).

The passive BP model is a prime example of the use of mouse models to dissect pathomechanisms at a cellular and molecular level. Thus, a series of publications used mast cell-deficient mice or depletion of neutrophils or macrophages, knockout mice lacking FcγRIII, T-/B-cell-deficient mice, or knockout mice lacking the complement component C5, the C5a receptor or the neutrophil effector molecules gelatinase B or elastase (Leighty et al. 2007; Heimbach et al. 2011). The results demonstrated an important role for macrophages and mast cells and of complement C5 in the recruitment of neutrophils and mast cell activation, as well as a role for FcγRIII, gelatinase, and elastase in neutrophil activation and effector function. A further study with humanized BP180 mice demonstrated a role for autoantibody-mediated activation of innate immune players such as complement, mast cells, and neutrophils which required the presence of the

Fc portion of IgG. A recent study supported the role or neutrophils as important innate effector cells in BP using a novel active mouse model (Oswald et al. 2012). In this model mice were pre-immunized against rabbit IgG and then repeatedly injected with murine collagen XVII-specific rabbit IgG leading to the formation of immune complexes. Deposition of C3 and IgG at the dermoepidermal junction was observed. The mice developed sustained subepidermal blistering disease resembling human BP. Here, a role for FcγR-dependent immune complex-triggered neutrophil activation and respiratory burst was demonstrated. Neutrophil depletion partially inhibited blister formation.

An elegant approach to study blistering disease is the generation of the so-called humanized mice (Nishie et al. 2007, 2009; Ujiie et al. 2010). This is achieved, for example, by expression of human collagen XVII under a keratinocyte-specific promoter in mice. Transplantation of the skin from transgenic mice onto wild-type mice results in autoantibody production, blister formation, and eventually rejection of the skin graft. Injection of patient antibodies into collagen XVII-humanized mice induces blister formation indicating that human collagen is efficiently targeted by the autoantibodies. Moreover, newborn human collagen XVII transgenic mice develop disease following transfer of maternal anti-collagen XVII antibodies (Nishie et al. 2009). Immunization of wild-type mice to human collagen XVII via skin transplants from the humanized mice followed by transfer of their spleen cells into Rag-2-deficient human collagen XVII transgenic mice results in sustained autoantibody production and disease.

Psoriasis

Psoriasis is a T-cell-dependent inflammatory autoimmune skin disease. The culprit autoantigens are not yet known (Perera et al. 2012). The pathogenesis depends on IFN-γ secreting T helper (Th)1 cells, but more recently a crucial contribution by IL-17 and/or IL-22 secreting $CD4^+$ Th cells (Th17, Th22 cells) has been demonstrated. For the differentiation of Th1 and Th17 cells, the cytokines TNF-α, IL-12, and IL-23, respectively, which are produced by DCs and macrophages are very important. The current rather successful immunotherapy is based on injection of a monoclonal antibody that binds and neutralizes the p40 subunit that is common to these heterodimeric cytokines. In vivo models of psoriasis also use mutant and genetically engineered mice (Danilenko 2008; Jean and Pouliot 2010). All of these mouse models only partially recapitulate the human condition. Two common spontaneous mouse models are the flaky skin mouse and the SHARPIN mouse. The flaky skin mouse develops focal parakeratosis which, in sharp contrast to human psoriasis, is however T-cell independent. The SHARPIN mouse also develops a T-cell-dependent focal parakeratosis, but other than the Th1-/Th17-mediated human disease, this phenotype is Th2 mediated. A series of transgenic mouse models have been established that use, for example, overexpression of adhesion molecules (CD18) and cytokines (IL-12/IL-23p40, TGF-b) often targeted by a keratinocyte-specific promoter to epidermal keratinocytes. Also in these models only some aspects of psoriasis are reproduced.

KC-Tie2 mice express the angiopoietin receptor 2 (Tie2) as a transgene in keratinocytes (KC) (Ward et al. 2011). These mice develop psoriasiform skin disease with infiltrating abundant macrophages in lesional skin as observed in humans. In this model macrophages were depleted by injection of liposomes containing clodronate. Liposome uptake by macrophages and clodronate release into the cell kills them. This treatment reversed acanthosis and dermal angiogenesis and resulted in normalized T-cell numbers and reduction of crucial pathogenesis driving cytokines such as IL-6 and IL-23. These findings emphasize the importance of macrophages in inflammatory skin disease.

Atopic Dermatitis

Atopic dermatitis (AD) collectively describes inflammatory hypersensitivity reactions elicited by usually harmless substances that come in contact with the skin (Boguniewicz and Leung 2011). Several mouse models are being used to study the pathomechanims of AD (Jin et al. 2009; Tanaka et al. 2012). Inbred *Nc/Nga mice* represent a spontaneous model for human atopic dermatitis (AD). They harbor a mutation on chromosome 9 that under conventional housing conditions is associated with the spontaneous development of skin barrier defects and subsequently typical signs of AD such as hyperparakeratosis, hyperplasia, and spongiosis. Another interesting spontaneous model is the *DS-Nh mouse*. These mice also develop AD under conventional housing conditions and, most interestingly, show heavy colonization of lesional skin with *S. aureus* found in human AD as one of the most important exacerbating factors. *Cby.Aly-aly mice* also develop spontaneous dermatitis with *S. aureus* infection of lesional skin.

Other approaches to study AD use the transgenic overexpression of factors associated with the immune response in the disease. One of these factors is thymic stromal lymphopoietin (TSLP) which is highly expressed in keratinocytes of lesional skin of AD patients. TSLP promotes Th2 responses by acting on DCs. TSLP transgenic mice which overexpress TSLP under the K5 promoter specifically in keratinocytes have been generated: These mice spontaneously develop a Th2-mediated inflammatory AD-like skin disease (Yoo et al. 2005). Among various other factors that upon transgenic overexpression in the epidermis lead to AD-like symptoms in mice are T-cell cytokines such as IL-4 or IL-31.

One of the most important findings in recent years was the association of AD with loss of function mutations of the protein filaggrin (FLG), a component of the outer layers of the epidermis (Irvine et al. 2011). FLG regulates hydration, surface pH, and terminal keratinocyte differentiation. The barrier function of the skin is significantly reduced in FLG-deficient spontaneous mouse mutants (Fallon et al. 2009). Homozygous flaky tail (*ft*) mice develop a spontaneous AD-like skin disease already at 3 days of age that is observed as flaking tail skin. It was recently discovered that these mice have a null mutation in the *flg* gene. Allergen application to the skin of these mice is sufficient to induce a strong inflammatory infiltrate.

Atopic dermatitis-like disease can be induced by removal of the upper skin layers by tape-stripping of shaved mouse skin and subsequent epicutaneous challenge with protein antigens such as chicken ovalbumin (OVA) or house dust mite allergen. In wild-type mice it was demonstrated that repeated application of the house dust mite allergen Der f can lead to atopy-like symptoms (Hennino et al. 2007). Interestingly, evidence was provided that not only $CD4^+$ T cells are involved in the pathogenesis but that $CD8^+$ T cells are participating. The observed early $CD8^+$ T-cell infiltration suggested a pioneering role for $CD8^+$ T cells as early initiators of the inflammatory response and a later dominance of $CD4^+$ T cells (Hennino et al. 2011). At early time points of patch test reactions, $CD8^+$ T-cell infiltration was also observed in the human skin. In this model it was even possible to establish a successful sublingual immunotherapy (SLIT) protocol (Vanbervliet et al. 2012).

Contact Dermatitis

Irritant contact dermatitis (ICD) and allergic contact dermatitis (ACD) are inflammatory skin diseases that are triggered by chemicals (Martin et al. 2011; Vocanson et al. 2009). Eczematous skin reactions are the result. While ICD is mediated by the toxic irritant effects of chemicals without involvement of the adaptive immune system, i.e., B and T cells, ACD is initiated by chemical-induced activation of the innate immune system followed by activation of chemical-specific T cells. The T cells are of the Th1/Tc1 and Th17/Tc17 type and are the main effector cells in ACD. They cause spongiosis in the skin due to cytotoxic activity. About 4,000 organic chemicals and metal ions such as nickel and cobalt are known as contact sensitizers. The mouse model for ICD and ACD is the contact hypersensitivity (CHS) model. In this model many aspects of the immune response to allergenic chemicals are analyzed that are often paralleled by adverse immune responses to drugs (Martin 2012). In the CHS protocol a single epicutaneous application of the contact sensitizer onto the shaved abdominal mouse skin leads to sensitization by activation of skin cells due to the chemical-induced innate inflammatory response. Activated dendritic cells (DCs) migrate to the skin draining lymph nodes presenting contact sensitizers in the context of MHC molecules to T cells. The contact sensitizer-specific T cells are then primed and expand. Circulating effector T cells are recruited into the inflamed skin where they exert effector functions including cytotoxicity and cytokine production after a second application of the same contact sensitizer to the ear skin, for example, on day 5 after sensitization. Within 24–48 h the ear skin shows an edematous swelling reaction which is measured using an engineer's micrometer. This is the classic mouse ear swelling test (MEST).

Metal ions such as nickel and cobalt cannot induce CHS by themselves. The reason for that is their inability to bind and activate mouse Toll-like receptor 4 (TLR4), an essential trigger for skin inflammation in CHS. They readily do that in humans since the binding sites for these metal ions are only present in human but absent in mouse TLR4. This problem can be circumvented by mixing nickel or cobalt ions with a TLR4-activating agent, the natural TLR4 ligand lipopolysaccharide (LPS) from gram-negative bacteria which in that case acts as a potent adjuvant. The case of these metal ions is a good example for the limitations of animal models, here due to species-specific differences. On the other hand, in vitro assays for DC activation with organic chemical allergens such as 2,4,6-trinitrochlorobenzene (TNCB) only yield partial activation of human and mouse DCs. This is due to the fact that signals from the extracellular matrix of the skin, i.e., degradation of hyaluronic acid to low molecular weight fragments that trigger TLR2 and TLR4 and therefore enable full DC activation, resulting especially activation of NF-κB signaling and pro-inflammatory cytokine production, are missing in vitro (see also below).

This example impressively demonstrates the limitations of in vitro models. In conclusion, it is important to understand in detail molecular and cellular mechanisms in order to define the specific limitations of each in vivo and in vitro model. Here, in vitro and in vivo studies complement each other.

Weak contact allergens such as fragrances which cause many problems with human ACD can now also be tested in the mouse model. It has been recognized that CHS does not occur unless these weak contact allergens are applied repeatedly, e.g., once daily on three consecutive days for sensitization in mice that have been pretreated with antibodies against CD4. The reason for that is the presence of CD4 expressing regulatory cell populations such as conventional regulatory cells and NKT cells which actively prevent CHS responses to weak contact allergens and can be considered as a brake for immune responses. Strong contact allergens induce enough inflammation to overcome the action of these regulatory cells and release the brake.

The CHS model is also used to study the induction of tolerance to contact allergens. In man, repeated exposure to weak contact allergens such as fragrances may induce T-cell tolerance instead of ACD. In the CHS model repeated application of very low doses of strong contact sensitizers that are 100–1000-fold lower than the dose used for sensitization leads to the so-called low zone tolerance (LZT). In this model the role of regulatory T cells and DCs in the induction of contact sensitizer-specific tolerance can be studied in great detail. Here, depletion or adoptive transfer of specific cell types is used to study the role of these cell types and their interaction with other cell types. Mice lacking cytokine receptors or other molecules involved in LZT are used to identify important molecular players.

The constitutive or inducible deletion of specific cell types is nowadays achieved by generating transgenic mice which express the diphtheria toxin A (DTA) subunit which is toxic for the respective cell type and leads to constitutive depletion or mice expressing a human or rat diphtheria toxin receptor (DTR) under the control of a cell type-specific locus. These DTR are much more sensitive than the mouse DTR and therefore allow killing the DTR+ cell type by injection of diphtheria toxin (DT) at the time point of choice. Using this technique it was shown in the CHS model that regulatory T-cell depletion leads to stronger and prolonged CHS responses (Lehtimäki et al. 2012) or that the absence or depletion of mast cells almost completely abrogated CHS (Dudeck et al. 2011). A nice strategy in the design of targeting constructs for such vectors is the use of constructs with an internal ribosomal entry site (IRES) that allows expression of two transgenes simultaneously from the same promoter. A common strategy is using a fluorochrome tag such as green fluorescent protein as the second transgene. This allows to easily identify, for example, the DTR-positive cells since they have green fluorescence. This technique is ideal to monitor the cell population expressing the transgene of interest in vivo, for example, by confocal microscopy or in vitro by immunofluorescence or flow cytometry.

Chronic application of some experimental contact sensitizers leads to a Th2-type immune response that is often described as a model for chronic contact dermatitis but is also considered to represent an inducible AD-like mouse model.

Similar to the methods described for the BP model, the pathomechanisms of CHS can also be studied using monoclonal antibody depletion, cell transfer, and transgenic and knockout mice. By using knockout mice that lack different receptors associated with innate inflammatory responses, it was demonstrated that contact sensitizers trigger an innate immune response that is very similar to the one triggered by pathogens (Martin et al. 2011). It was shown that organic chemicals such as the strong contact sensitizer TNCB induce rapid production of reactive oxygen species (ROS) that initiate the degradation of hyaluronic acid (HA), a component of the extracellular matrix of the skin (Esser et al. 2012). Moreover, ROS activate and amplify signaling pathways that promote inflammation and upregulate hyaluronidases which then further degrade HA. Low molecular weight HA fragments then trigger TLR2 and TLR4 activation leading to NF-κB signaling and production of pro-inflammatory cytokines such as IL-6, IL-12, TNF-α, and immature pro-IL-1β and pro-IL-18

Furthermore, cell stress and damage as caused by contact sensitizer leads to the release of ATP. Extracellular ATP is a danger signal that is perceived by purinergic receptors on cells. The P2X7 receptor (P2X7R) binds ATP and contributes to the activation of the NLRP3 inflammasome, a cytosolic protein complex consisting of the NLR NLRP3, and the adaptor protein ASC. The inflammasome activates caspase-1 which then cleaves immature pro-IL-1β and pro-IL-18 to their mature and secreted forms. These cytokines, especially IL-1β, are crucial mediators of skin inflammation in CHS. This very detailed picture of the sterile skin inflammation caused by contact sensitizers was revealed by studies in knockout mice. Mice lacking TLR2 and TLR4 or NLRP3 or ASC or P2X7R or IL-1 receptor are all resistant to sensitization and CHS induction. By adoptive transfer of contact sensitizer-presenting and T-cell-activating DCs from wild-type or knockout mice into wild-type or knockout recipient mice to it was then also shown that it is sufficient for sensitization that these TLRs, P2X7R, and NLRP3 or ASC are functional in DCs. Interference with the identified pathways was shown to prevent sensitization or even elicitation of CHS. These data will now lead to clinical studies and hopefully to the development of causative treatment strategies for ACD.

Summary

Mouse models have proven to be very useful to gain insights into cellular and molecular pathomechanisms of human disease in a complex in vivo setting. The limitations of each model such as species-specific differences have to be carefully defined as is the case for in vitro models. In vivo and in vitro models therefore complement each other and profit from clinical studies. Altogether the data generated can be used for the development of new treatment strategies by the identification of drug targets and for the improvement of existing and development of new in vitro assays for diagnostics and immunotoxicology.

Cross-References

▶ Allergic Disorders
▶ Atopic Dermatitis
▶ Dendritic Cells
▶ Mast Cells
▶ Neutrophil Oxidative Burst

References

Amagai, M. (2008). Pemphigus vulgaris and its active disease mouse model. *Current Directions in Autoimmunity, 10*, 167–181.

Aoki-Ota, M., Tsunoda, K., Ota, T., Iwasaki, T., Koyasu, S., Amagai, M., & Nishikawa, T. (2004). A mouse model of pemphigus vulgaris by adoptive transfer of naive splenocytes from desmoglein 3 knockout mice. *The British Journal of Dermatology, 151*, 346–354.

Belizário, J. E., Akamini, P., Wolf, P., Strauss, B., & Xavier-Neto, J. (2010). New routes for transgenesis of the mouse. *Journal of Applied Genetics, 53*, 295–315.

Boguniewicz, M., & Leung, D. Y. (2011). Atopic dermatitis: A disease of altered skin barrier and immune dysregulation. *Immunological Reviews, 242*, 233–246.

Bottazzi, B., Doni, A., Garlanda, C., & Mantovani, A. (2010). An integrated view of humoral innate immunity: Pentraxins as a paradigm. *Annual Review of Immunology, 28*, 157–183.

Danilenko, D. M. (2008). Preclinical models of psoriasis. *Veterinary Pathology Online, 45*, 563.

De Benedetto, A., Kubo, A., & Beck, L. A. (2012). Skin barrier disruption: A requirement for allergen sensitization? *Journal of Investigative Dermatology, 132*, 949–963.

Dudeck, A., Dudeck, J., Scholten, J., Petzold, A., Surianarayanan, S., Köhler, A., et al. (2011). Mast cells are key promoters of contact allergy that mediate the adjuvant effects of haptens. *Immunity, 34*, 973–984.

Esser, P. R., Wölfle, U., Dürr, C., von Loewenich, F. D., Schempp, C. M., Freudenberg, M. A., et al. (2012). Contact sensitizers induce skin inflammation via ROS production and hyaluronic acid degradation. *PLoS One, 7*, e41340.

Fallon, P. G., Sasaki, T., Sandilands, A., Campbell, L. E., Saunders, S. P., Mangan, N. E., et al. (2009). A homozygous frameshift mutation in the mouse Flg gene facilitates enhanced percutaneous allergen priming. *Nature Genetics, 41*, 602–608.

Gallo, R. L., & Hooper, L. V. (2012). Epithelial antimicrobial defence of the skin and intestine. *Nature Reviews Immunology, 12*, 503–516.

Ganeshan, R., Chen, J., & Koch, P. J. (2010). Mouse models for blistering disorders. *Dermatology Research Practice*, 584353, 7 p

Grice, E. A., & Segre, J. A. (2011). The skin microbiome. *Nature Reviews Microbiology, 9*, 244–253.

Heimbach, L., Li, Z., Berkowitz, P., Zhao, M., Li, N., Rubenstein, D. S., et al. (2011). The C5a receptor on mast cells is critical for the autoimmune skin-blistering disease bullous pemphigoid. *The Journal of Biological Chemistry, 286*, 15003–15009.

Hennino, A., Vocanson, M., Toussaint, Y., Rodet, K., Benetière, J., Schmitt, A. M., et al. (2007). Skin-infiltrating $CD8^+$ T cells initiate atopic dermatitis lesions. *The Journal of Immunology, 178*, 5571–5577.

Hennino, A., Jean-Decoster, C., Giordano-Labadie, F., Debeer, S., Vanbervliet, B., Rozières, A., et al. (2011). $CD8^+$ T cells are recruited early to allergen exposure sites in atopy patch test reactions in human atopic dermatitis. *The Journal of Allergy and Clinical Immunology, 127*, 1064–1067.

Hirose, M., Recke, A., Beckmann, T., Shimizu, A., Ishiko, A., Bieber, K., et al. (2011). Repetitive immunization breaks tolerance to type XVII collagen and leads to bullous pemphigoid in mice. *Journal of Immunology, 187*, 1176–1183.

Hoffmann, A., & Baltimore, D. (2006). Circuitry of nuclear factor kappaB signaling. *Immunological Reviews, 210*, 171–186.

Irvine, A. D., McLean, W. H., & Leung, D. Y. (2011). Filaggrin mutations associated with skin and allergic diseases. *The New England Journal of Medicine, 365*, 1315–1327.

Jean, J., & Pouliot, R. (2010). In vivo and in vitro models of psoriasis. In D. Eberli (Ed.), *Tissue engineering* (pp. 359–382). Rijeka: Intech.

Jin, H., He, R., & Geha, R. S. (2009). Animal models of atopic dermatitis. *The Journal of Investigative Dermatology, 129*, 31–40.

Lehtimäki, S., Savinko, T., Lahl, K., Sparwasser, T., Wolff, H., Lauerma, A., et al. (2012). The temporal and spatial dynamics of $Foxp3^+$ Treg cell-mediated suppression during contact hypersensitivity responses in a murine model. *The Journal of Investigative Dermatology, 132*, 2744–2751 [Epub ahead of print].

Leighty, L., Li, N., Diaz, L. A., & Liu, Z. (2007). Experimental models for autoimmune and inflammatory blistering disease, bullous pemphigoid. *Archives of Dermatological Research, 299*, 417–422.

Lippens, S., Lefebvre, S., Gilbert, B., Sze, M., Devos, M., Verhelst, K., et al. (2011). Keratinocyte-specific ablation of the NF-κB regulatory protein A20 (TNFAIP3) reveals a role in the control of epidermal homeostasis. *Cell Death and Differentiation, 18*, 1845–1853.

Martin, S. F. (2012). Allergic contact dermatitis: Xenoinflammation of the skin. *Current Opinion in Immunology, 24*, 720–729.

Martin, S. F., Esser, P. R., Weber, F. C., Jakob, T., Freudenberg, M. A., Schmidt, M., et al. (2011). Mechanisms of chemical-induced innate immunity in allergic contact dermatitis. *Allergy, 66*, 1152–1163.

Nishie, W., Sawamura, D., Goto, M., Ito, K., Shibaki, A., McMillan, J. R., et al. (2007). Humanization of autoantigen. *Nature Medicine, 13*, 378–383.

Nishie, W., Sawamura, D., Natsuga, K., Shinkuma, S., Goto, M., Shibaki, A., et al. (2009). A novel humanized neonatal autoimmune blistering skin disease model induced by maternally transferred antibodies. *The Journal of Immunology, 183*, 4088–4093.

Oswald, E., Sesarman, A., Franzke, C. W., Wölfle, U., Bruckner-Tuderman, L., Jakob, T., et al. (2012). The flavonoid luteolin inhibits Fcγ-dependent respiratory burst in granulocytes, but not skin blistering in a new model of pemphigoid in adult mice. *PLoS One, 7*, e31066.

Page, A., Navarro, M., Garin, M., Perez, P., Casanova, M. L., Moreno, R., et al. (2010). IKKbeta leads to aninflammatory skin disease resembling interface dermatitis. *The Journal of Investigative Dermatology, 130*, 1598–1610.

Perera, G. K., Di Meglio, P., & Nestle, F. O. (2012). Psoriasis. *Annual Review Pathology, 7*, 385–422.

Rebholz, B., Haase, I., Eckelt, B., Paxian, S., Flaig, M. J., Ghoreschi, K., et al. (2007). Crosstalk between keratinocytes and adaptive immune cells in an IkappaBalpha protein-mediated inflammatory disease of the skin. *Immunity, 27*, 296–307.

Sitaru, C. (2007). Experimental models of epidermolysis bullosa acquisita. *European Journal of Immunology, 16*, 520–531.

Tanaka, A., Amagai, Y., Oida, K., & Matsuda, H. (2012). Recent findings in mouse models for human atopic dermatitis. *Experimental Animals, 61*, 77–84.

Ujiie, H., Shibaki, A., Nishie, W., & Shimizu, H. (2010). What's new in bullous pemphigoid. *The Journal of Dermatology, 37*, 194–204.

Vanbervliet, B., Tourdot, S., Mascarell, L., Rouzaire, P., Vocanson, M., Rozières, A., et al. (2012). SLIT prevents the development of eczema in percutaneous allergen-sensitized mice. *The Journal of Investigative Dermatology, 132*, 244–246.

Vocanson, M., Hennino, A., Rozières, A., Poyet, G., & Nicolas, J. F. (2009). Effector and regulatory mechanisms in allergic contact dermatitis. *Allergy, 64*, 1699–1714.

Ward, N. L., Loyd, C. M., Wolfram, J. A., Diaconu, D., Michaels, C. M., & McCormick, T. S. (2011). Depletion of antigen-presenting cells by clodronate liposomes reverses the psoriatic skin phenotype in KC-Tie2 mice. *The British Journal of Dermatology, 164*, 750–758.

Yoo, J., Omori, M., Gyarmati, D., Zhou, B., Aye, T., Brewer, A., et al. (2005). Spontaneous atopic dermatitis in mice expressing an inducible thymic stromal lymphopoietin transgene specifically in the skin. *The Journal of Experimental Medicine, 202*, 541–549.

Spondyloarthritis

Fernando A. Sommefleck,
Emilce E. Schneeberger and Gustavo Citera
Section of Rheumatology, Instituto de Rehabilitación Psicofísica, Buenos Aires, Argentina

Synonyms

Ankylosing spondylitis; Arthritis associated with inflammatory bowel disease; Juvenile spondyloarthritis; Psoriatic arthritis; Spondyloarthropathies; Undifferentiated spondyloarthritis

Definition

Spondyloarthritis (SpA) is a term that covers a group of interrelated heterogeneous diseases that share the characteristic of axial engagement, asymmetric oligoarthritis, and enthesitis and are seronegative for rheumatoid factor (Gran and Skomsvoll 1997; van der Linden and van der Heijde 1998). The most relevant immunogenetic feature is the remarkable familiar aggregation and the association with the histocompatibility antigen, HLA-B27 (González-Roces and Alvarez 1996).

The concept of SpA has been gradually developed throughout the last 30 years. At present, it includes different conditions such as ankylosing spondylitis (AS), reactive arthritis (ReA), psoriatic arthritis (PsA), arthritis associated with inflammatory bowel diseases (IBD), and the juvenile forms of each of these conditions, as well as a group of less definite conditions called undifferentiated SpA (Dougados et al. 1991).

SpA is more common in males and usually starts mainly in the second and third decades of life, with an estimated mean age at onset of 25 years; its onset after age 50 is unusual. Without an appropriate treatment, SpA can produce a negative impact on the function and quality of life of the individual, leading to work disability and impairment of social relationships

Spondyloarthritis, Table 1 ASAS classification criteria for axial spondyloarthritis (axSpA) (in patient with ≥3 months of back pain and age at onset less than 45 years)

Sacroiliitis on imaging plus ≥ SpA feature	OR	HLA B27 plus ≥2 other SpA features
SpA features		
		Inflammatory back pain
		Arthritis
		Enthesitis
		Uveitis
		Dactylitis
		Psoriasis
		Crohn/ulcerative colitis
		Good response to NSAIDs
		Family history for SpA
		HLA-B27
		Elevated C-reactive protein
ASAS classification criteria for peripheral spondyloarthritis (SpA)		
Arthritis or enthesitis or dactylitis		
plus		
≥1 SpA feature		
		Uveitis
		Psoriasis
		Crohn/ulcerative colitis
		Preceding infection
		HLA B27
		Sacroiliitis on imaging
OR		
≥2 SpA feature		
		Arthritis
		Enthesitis
		Dactylitis
		Inflammatory back pain
		Family history for SpA

(Günaydin et al. 2009). The ASsessment in Ankylosing Spondylitis (ASAS) group recently established classification criteria for axial (axSpA) and peripheral SpA (Rudwaleit et al. 2009a, 2011), which are detailed in Table 1.

These criteria have reduced significantly the delay to diagnosis in SpA and lead to the appearance of two axSpA subgroups, non-radiographic (nr-axSpA) and radiographic (AS) axSpA. The nr-axSpA corresponds to patients suffering from axial SpA but in the absence of radiological sacroiliitis (Sieper and van der Heidje 2013; Fianyo et al. 2014).

Epidemiology and Genetics

SpA covers the inflammatory chronic rheumatic diseases most frequently observed in young adult males. These diseases show considerable differences among ethnic groups and populations. The prevalence is variable, ranging from 0.3 % to 1.6 %, according to different epidemiological studies, and it is directly related to the prevalence of HLA B27 in populations (González-Roces and Alvarez 1996; Scholnsstein et al. 1973; Brewerton et al. 1973; Dean et al. 2014). The prevalence of HLA-B27 is highest (near 40 %) in some tribes of North America, Canada, New Guinea, and Russia and in Scandinavian countries (Boyer et al. 1988) and is observed in 3–10 % of the population of Western Europe. HLA-B27 positivity is uncommon in the Arab countries and Japan and is nearly absent among South American, Australian, and African aborigines (Brown et al. 1997a; López-Larrea et al. 1995). In Argentina, a group of AS patients from Buenos Aires was studied, and an HLA-B27 prevalence of 90.4 % was observed compared to 5.4 % of a group of non-related persons from the general population (Citera et al. 2009). It is estimated that between 1 % and 5 % of carriers of this marker have the disease.

Relatives of patients with ankylosing spondylitis which are $B27^+$ have a 5–16-fold higher risk to develop the disease than a person who is $B27^+$ in the general population.

There is a great heterogeneity within the HLA-B27 alleles referred to as subtypes. HLA-27*05 is the most common subtype, which is found in almost all populations around the world. Other subtypes have only been reported in a limited number of persons and definitive disease association is lacking. Apart from B*2705, other common subtypes like B*2702 and B*2704 have been associated with AS in other populations. Two subtypes (B*2706 and B*2709) were unassociated or weakly associated with AS (Nasution and Marjuadi 1997).

The role of genetic factors is highlighted by demonstrating disease concordance in 75 % of monozygotic twins compared to 13 % of nonidentical twins (Van de Linden et al. 1984).

Genes other than B27 have also been implicated in the pathogenesis of the disease. HLA B60 and HLA DR1 may act as independent genes in patients with AS (Boyer et al. 1988; Brown et al. 1997b). CARD15 has been associated with Crohn's disease and psoriatic arthritis (Kim et al. 2004; Rahman et al. 2003). Tumor necrosis factor (TNF), interleukin 1 beta (IL1beta), and matrix metalloproteinase-3 polymorphisms have also been studied as possible candidate genes in patients with SpA. We have observed in our population from Buenos Aires that TNF-simple nucleotide polymorphism (SNP) -308 GA genotype was significantly more frequent in AS patients (94 %) versus controls (81 %), OR 3.96 (CI 95 % 1.14–13.70, $p = 0.02$), while SNP −238 GA genotype had a protective effect and was more frequently present in controls (76 %) versus patients (53 %), OR 0.19 (CI 95 % 0.1–0.37, $p > 0.0001$). IL1β gene -511C allele was associated with a higher susceptibility to AS, and though the CC genotype evidenced a tendency to a higher susceptibility, it did not reach statistical significance. IL1β +3954 genotype had no effect on susceptibility or a protective effect against the disease. Genome-wide association studies (GWAS), using simple nucleotide polymorphism, allowed the identification of non-HLA, new candidate genes in AS. Genes that are consistently associated are *ERAP1*, *IL23R*, and desert genes such as 2p15/21q22 (Brown 2010). Large well-controlled studies are underway to elucidate this aspect in patients with SpA.

The most common age of onset is between 15 and 30 years, and 10 % of patients begin their disease before age 15 and only 5 % after 50 years (Lau et al. 1998).

SpA has historically been a disease seen more frequently in men than women with a ratio of 3:1 or 2:1. Currently, it is recognized that these estimations are exaggerated. Males have more severe disease than women, measured by loss of spinal mobility and radiographic vertebral disease. This can produce a misdiagnosis of the disease in women. It is estimated that this relationship is probably closer to 1.5:1. In some early SpA cohorts, it was observed that patients with pre-radiographic SpA were more frequently women than men (Feldtkeller et al. 2000).

Pathophysiology

The exact mechanism by which HLA-B27 participates in the pathogenesis remains unknown; several theories exist that seek to explain the association.

Infectious Theory

The classical theory is of an arthritogenic peptide, in which some environmental factor, probably infectious, triggers the disease that has structural similarity to peptides that bind the HLA-B27, and this may result in a mechanism of molecular mimicry, with loss of tolerance and the development of autoimmunity. The role of infection as a triggering factor has been proposed by several authors (Yu 1989) with different degrees of evidence among the subgroups of SpA. Clear evidence exists in relation to reactive arthritis and the involvement is less probable in AS and PsA. It was recently shown that prolonged combination antibiotic therapies are highly effective in patients with Chlamydia-induced reactive arthritis. The influence of an environmental factor is critical for the development of the disease. The relationship between SpA and intestinal and genitourinary infections is unquestionable. It has been observed that there is some structural similarity between bacterial peptides and the hypervariable portion of the HLA-B27 and that regardless of the number of copies, in a transgenic animal possessing HLA B27, the disease does not develop in a germ-free environment. Perhaps the most involved microorganism is *Klebsiella*, especially *K. pneumoniae*. A higher titer of anti-*Klebsiella* IgA antibodies was detected in patients with EA together with cross-reactivity between antigens of *Klebsiella* and HLA-B27 (Russel and Suárez-Almanzor 1992; Ebringer 1990, 1992).

According to this infection theory, the trigger factor would be a bacterial stress and the development of autoimmunity. However, several authors question the involvement of autoimmune phenomena in SpA. Unlike rheumatoid arthritis, in SpA, there is no female predominance, no evidence of circulating autoantibodies, and no benefit of therapies directed against T or B cells. New studies on the behavior of HLA B27 in patients with SpA and participation of other intracellular proteins encoded by genes involved in the pathogenesis of the SpA, such as ERAP1, have raised a second theory.

Protein Misfolding Response

Patients with Spa and especially those with AS would be more likely to have a misfolding of the heavy chains of HLA-B27 in the endoplasmic reticulum. This would generate an inflammatory response at the level of the cell, known as "misfolding protein response." This inflammatory response increases the production of IL-23, which in turn is a major stimulus for the expansion of Th17 cells with a consequent increase in the production of IL17 and other proinflammatory cytokines. According to this theory, the main stimulus would be a mechanical stress, probably at the level of the enthesis, and the mechanism would not be autoimmune, but autoinflammatory (Colbert et al. 2010; García-Medel et al. 2012).

In SpA, in contrast to RA, the synovium is not the main site of inflammation. Most immunopathological findings can be explained by inflammation at the bony sites of tendons, ligaments, and capsules, which is called "enthesitis." However, some authors have discussed whether all aspects of the clinical and pathologic findings can be explained by enthesitis and whether it is part of a more general picture (Appel and Sieper 2008).

New Bone Formation in AS

AS, in particular, is characterized by two pathological findings: (A) inflammation of the spine and (B) new bone formation. Inflammation is prominent in early phases and induces pain and stiffness characteristic of the disease. Since new bone formation is responsible for ankylosis and the resulting disability in later stages, there is considerable evidence suggesting a negative correlation between inflammation and new bone formation (Lories et al. 2007).

Histopathological findings suggest that inflammation itself inhibits osteoproliferation, resulting in an uncoupling of inflammation and bone formation in AS. TNF alpha blocking agents may reduce inflammation at the enthesis but did not inhibit new cartilage and bone formation in a mouse model of SpA. These findings were also observed in patients with AS treated with anti-TNF therapy (van der Heijde et al. 2008).

Clinical Presentation

Clinical manifestations in the group of SpA have remarkable similarities. Age at onset of symptoms, sex, and especially disease duration are the most influential factors in clinical expression of the disease (Lau et al. 1998). Likewise, the presence of HLA B27 influences the patient's clinical condition.

Clinical Manifestations

The clinical manifestations can be divided into the extra-articular and articular, and the latter can be subdivided into axial and peripheral involvement. The axial involvement is the relevant finding in SpA. It is more frequent in AS, where 100 % of patients have axial involvement, and is less common in PsA being approximately 30 % (Pérez Alamino et al. 2011).

Inflammatory back pain is the cardinal symptom of the group. It is important to differentiate inflammatory back pain as opposed to mechanical back pain, which is highly prevalent in the general population. Inflammatory back pain is estimated to represent only 5 % of low back pain observed in the general population. Several definitions or criteria have been proposed for inflammatory back pain (Table 2). Basically inflammatory back pain is a chronic low back pain lasting more than 3 months, with a history of insidious onset that does not subside with rest, may awake the patient at night, improve with physical activity, is associated with morning stiffness, and improves with NSAIDs (Sieper et al. 2009; Braun et al. 2011).

The *peripheral involvement* is characterized by asymmetric oligoarthritis involving large joints of the lower limbs, shoulders, or less frequently affecting small joints. Symmetric polyarthritis may develop later during the disease. The frequency varies depending on the type of SpA. It can be observed in 30 % of AS and IBD patients and up to 90 % PsA and ARe. In PsA peripheral

Spondyloarthritis, Table 2 Inflammatory back pain criteria

Calin et al.	Rudwaleit et al.	ASAS group
1. Chronic low back pain of more than 3 months	1. Alternating buttock pain	1. Improvement with exercise
2. Insidious onset	2. Awakening at second half of the night because of pain	2. Insidious onset
3. Morning stiffness		3. Pain at night
4. Age at onset less than 40 years	3. Morning stiffness more than 30 min	4. Age at onset less than 40 years
5. Improve the pain with exercise	4. Improve with exercise, not with rest	5. No improvement with rest
Inflammatory back pain if 4/5 are present	**Inflammatory back pain if 2/4 are present**	**Inflammatory back pain if 4/5 are present**

joint involvement can take different forms. The asymmetric oligoarthritis is the most common form of initial symptoms, while in its evolution many patients may have polyarticular involvement, either symmetrical or asymmetrical. Some patients can have involvement of the distal interphalangeal joints, associated with the clinical picture. Dactylitis affecting hands and feet is characteristic of the SpA and especially of psoriatic arthritis. Diffuse swelling of the fingers, which adopt a "sausage aspect," is highly suggestive of SpA (Moll and Wright 1973).

Enthesitis is one of the most distinctive and characteristic lesions of SpA and occurs in 25–40 % of the patients. As mentioned in pathogenesis, enthesitis is inflammation at the bony sites of tendons, ligaments, and capsules, which often generate severe pain and swelling and after several months can produce typical radiological changes. The most commonly affected sites are the Achilles tendon, the plantar fascia, the trochanter, the iliac crest, the anterior tibial tuberosity, and the costosternal union. Objective evaluation of enthesitis can be performed using the Spondyloarthritis Research Consortium of Canada Enthesitis Index or the Maastricht Ankylosing Spondylitis Enthesitis Score (Heuft-Dorenbosch et al. 2003).

Extra-articular manifestations are very common in SpA. The main extra-articular manifestations affect the eye, skin, intestine, heart, kidney, and neurological system.

Acute anterior uveitis occurs in 20–40 % of patient and can be the first presenting symptom of the disease. The most common ocular manifestation is anterior uveitis, which is typically recurrent, unilateral, often associated with HLA B27, and is usually mild and self-limiting (Rosenbaum 1992). Osteoporosis frequently occurs and because of the rigidity of the spine, minor trauma can cause vertebral fractures (Gratacós et al. 1999). Cardiac involvement includes aortitis, aortic valve incompetence, and conduction abnormalities (Bergfeldt et al. 1982; Dik et al. 2010).

Pulmonary complications are rare and can be caused by rigidity of the chest wall and occasionally by pulmonary fibrosis, which affects the upper lobes.

Intestinal involvement is very frequent; two-thirds of patients with SpA have acute or chronic inflammatory changes in the mucosa and submucosa of the colon and terminal ileum, even without associated symptoms (Mielans et al. 1988). In patients with SpA, and inflammatory bowel disease, gastrointestinal symptoms dominate the clinical picture.

Neurological involvement is also rare and may include atlantoaxial subluxation and cauda equine. This is an infrequent complication of long-standing AS and the true prevalence is unknown. Symptoms result in severe pain, muscle wasting, sensory loss in lower limbs, and sphincter dysfunction. Anecdotal improvement with anti-TNF therapy was recently reported (Cornec et al. 2009).

Disease Assessment

Because of the chronic and progressive nature of SpA, it is crucial to make a proper assessment of the patient on the first visit as well as periodic assessments that allow the physician to judge and clearly document whether the patient is improving or worsening from baseline to last visit.

In contrast to rheumatoid arthritis, the values of acute phase reactants are poorly correlated with disease activity, especially in the axial compromise.

Various instruments have been developed to assess the severity of the disease.

On the first visit, it is recommended at least to assess the activity, function, structural damage, prognosis, and response to treatments. This assessment should include not only a medical history and complete physical examination at each visit but also other laboratory and radiological tests and assessment tools standardized and validated as in the Bath Ankylosing Spondylitis Disease Index (BASDAI), Bath Ankylosing Spondylitis Metrology Index (BASMI), Bath Ankylosing Spondylitis Functional Index (BASFI), Health Assessment Questionnaire (HAQ), Ankylosing Spondylitis Disease Activity Score (ASDAS), Ankylosing Spondylitis Quality of Life (ASQol), Psoriatic Arthritis Quality of Life (PsAQoL), and Bath Ankylosing Spondylitis Radiology Index (BASRI) (Garrett et al. 1994; Calin et al. 1994; Doward et al. 2003) Table 3.

The ASsessment in Ankylosing Spondylitis (ASAS) group has developed a composite index, ASDAS, in order to get a more objective measure, including a lab parameter. It is a composite index made up of three BASDAI questions, patient global assessment using visual analogue scale (VAS), and an objective laboratory variable such as erythrocyte sedimentation rate (ESR) or C-reactive protein (CRP) (Lukas et al. 2009).

Several studies showed that the ASDAS is reliable and reproducible and in some cases better than BASDAI; however, it is difficult to calculate without using a calculator and therefore arduous to use during daily clinical practice (Pedersen et al. 2010; Nas et al. 2010; Machado et al. 2011). For this reason, we recently developed a simplified version of the ASDAS, "the simplified" ASDAS (SASDAS). The SASDAS has the advantage of being the simple linear sum of ASDAS components and could be a useful tool in clinical practice for easy calculation (Sommerfleck et al. 2012; Salaffi et al. 2014; Solmaz et al. 2015).

Traditionally, the structural damage of SpA is assessed using conventional radiographs from sacroiliac joints and spine. Classical New York criteria for ankylosing spondylitis require the presence of sacroiliitis, grade 2 or more for diagnosis. It is important to recognize that this does not allow an early diagnosis in patients in whom structural changes have not yet occurred. However, the classical radiography remains a major tool in the evaluation of these patients. The presence of erosions, subchondral sclerosis, or bony bridges at the level of the sacroiliac joints allows the diagnosis and evaluation of these patients. In the spine, especially the lumbar region, quadrature or marginal sclerosis in the vertebrae are frequent changes. In subsequent stages, syndesmophytes joining the vertebrae together give a characteristic picture of this group of diseases.

The BASRI is a composite index that combines radiographic changes at the cervical and lumbar spine, hips, and sacroiliac joints.

More recently, magnetic resonance imaging (MRI) and ultrasonography (US) have been increasingly used in patients with SpA. MRI can identify both active inflammation and chronic structural damage of the spine and sacroiliac joints. The most valuable aspect of the MRI is that it can detect early changes of the sacroiliac joint, allowing an early diagnosis. Several scoring methods have been developed to quantify MRI

Spondyloarthritis, Table 3 Instruments to assess outcome in patients with SpA

Instrument	Type of evaluation	No of items	Range
BASDAI	Disease activity	6	1–10
BASFI	Functional capacity	10	0–10
BASMI	Metrology	5	0–10
BASRI	X-ray damage	4	0–16
ASQoL	Quality of life	18	0–18
PsAQoL	Quality of life	20	0–20

BASDAI Bath Ankylosing Spondylitis Disease Index, *BASFI* Bath Ankylosing Spondylitis Functional Index, *BASMI* Bath Ankylosing Spondylitis Metrology Index, *BASRI* Bath Ankylosing Spondylitis Radiology Index, *ASQoL* Ankylosing Spondylitis Quality of Life, *PsAQoL* Psoriatic Arthritis Quality of Life

changes of the sacroiliac joints and the spine (Rudwaleit et al. 2009b).

Ultrasonography is a noninvasive and highly sensitive tool which allows the detection and quantification of musculoskeletal disorders frequently associated with seronegative spondyloarthropathies, such as the presence of enthesitis, bone erosions, synovitis, bursitis, tenosynovitis, and calcifications (D'Agostino 2010).

In order to evaluate the response to treatment, the ASAS group has developed different response criteria as measures of therapeutic improvement, especially for measuring outcomes in therapeutic trials. They have established improvement criteria of 20 and 40 (ASAS 20, ASAS 40). These components are described in Table 4.

The ASAS 5/6 is similar to the ASAS 20 with the difference that it contains two more components, the lumbar lateral flexion and CRP. This level of efficiency is the most useful for assessing clinical efficacy in the evaluation of therapeutic agents.

Therapy

The objectives of treatment are to relieve pain, improve stiffness and fatigue, and consequently reduce disease activity and restore functional capacity and quality of life.

Therapy can be divided according to axial or peripheral involvement, also depending on the presence or not of extra-articular manifestations.

The use of nonsteroidal anti-inflammatory drugs (NSAIDs) and exercises have been the standard treatments for spinal symptoms in AS for the last decades. NSAIDs have proven to be fast and effective in controlling symptoms, when used in full doses and for an adequate period of time. More recently, some authors have shown that continuous use of NSAIDs may reduce progression of radiographic damage in the spine (Wanders et al. 2005).

Regarding exercise, those promoting extension of the muscles of the spine are especially suitable. Group and supervised activities are associated with better adherence and outcomes than those that are made individually (van Tubergen et al. 2001).

The use of steroids and disease-modifying antirheumatic drugs (DMARDs) has not been useful in the treatment of axial involvement, and these drugs are used only in cases of peripheral joint involvement (Dougados et al. 2002).

Sulfasalazine has been used in patients with peripheral arthritis in a dose of 2 g per day, with some benefit. Methotrexate and leflunomide are commonly used in rheumatoid arthritis with good results, but the results were disappointing in AS (Ferraz et al. 1990; Dougados et al. 1995; Clegg et al. 1996).

Some studies have shown benefit in patients with psoriatic arthritis, mainly for leflunomide. A multinational, double-blind, placebo-controlled study of the efficacy of leflunomide in psoriasis and psoriatic arthritis showed a significant effect on peripheral joint involvement, patient and physician's global assessment of diseases activity, and a significant improvement in Psoriatic Arthritis Response Criteria compared to placebo (59 % vs. 30 %) (Kaltwasser et al. 2004). Recently, a double-blind placebo-controlled study comparing methotrexate vs. placebo in patients with psoriatic arthritis showed no significant benefit in improving peripheral involvement (Kingsley et al. 2012).

Spondyloarthritis, Table 4 ASAS improvement criteria

ASAS 20 improvement criteria
Improvement of $\geq$ to 20 % and $\geq$1 unit in at least 3 of this 4 domains
1. Patient global (VAS)
2. Lumbar pain (VAS)
3. Function (BASFI)
4. Inflammation (BASDAI)
ASAS 5/6 improvement criteria
Improvement of $\geq$ to 20 in at least 5 of this 6 domains
1. Patient global (VAS)
2. Lumbar pain (VAS)
3. Function (BASFI)
4. Inflammation (BASDAI)
5. CRP
6. Spinal mobility (BASMI)

VAS visual analogue scale, *BASFI* Bath Ankylosing Spondylitis Functional Index, *BASDAI* Bath Ankylosing Spondylitis Disease Index, *CRP* C-reactive protein, *BASMI* Bath Ankylosing Spondylitis Metrology Index

Introduction of tumor necrosis factor (TNF) blockers has been one of the most important advances in the treatment of patients with SpA These agents have demonstrated efficacy both for controlling the axial and peripheral involvement (Gorman et al. 2002; Braun et al. 2002a).

Four anti-TNF agents have been approved for the treatment of SpA, infliximab, etanercept, adalimumab, and more recently golimumab. The four agents exhibit excellent response and improvement in disease activity and quality of life, with similar efficacy and adverse events including those similar to the adverse events observed with the same drugs in RA (Davis et al. 2003; Braun et al. 2002b).

According to an expert consensus, patient candidates for this treatment would be those who, after at least two courses of different NSAIDs at full doses or a single full-dose of NSAIDs for a period of 4 weeks, continue with active disease as measured by BASDAI higher or equal to 4 (van der Heijde et al. 2011). Recently, it was shown that treatment with anti-TNF agents might benefit patients with early pre-radiographic disease (Sieper et al. 2013). In RA, these agents produce a reduction in radiographic progression, but this has not been demonstrated in patients with SpA. Furthermore, some studies have shown that progression of radiological damage in the spine could continue despite anti-TNF therapy.

The safety profile of these agents, in patients with SpA, is comparable to that observed in patients with RA. The same safety precautions before treatment should be applied in patients with SpA.

Other biological agents, such as abatacept or rituximab, which proved to be effective in patients with RA, are not very useful in patients with SpA. In case of failure of an anti-TNF agent, switching to another agent of this family has proven effective in 50–80 % of cases (Glintborg et al. 2013).

Inhibitors of other inflammatory pathways are being investigated. Oral treatments, including phosphodiesterase 4 inhibitors, such as apremilast, and new biologics targeting interleukin-17, such as brodalumab, secukinumab, and ixekizumab, have shown encouraging clinical results in the treatment of PsA (Patel et al. 2013; Papp et al. 2012; McInnes et al. 2014) and AS (Baeten et al. 2015).

Surgery plays an important role in patients with severe hip involvement and in those with spinal deformities that limit horizontal posture.

Outcome

The clinical course and natural history of the disease is highly variable from one patient to another, making it difficult to establish an individual prognosis for each patient in the early stages of the disease (van der Linden and van der Heijde 1998). Some patients follow a relatively benign course (approximately 30–40 %), with spontaneous remissions and exacerbations, compatible with a normal family and social life. In other patients, the disease tends to be more severe and disabling, with more rapid progression.

Patients with mild disease during the first 10 years of evolution are less likely to develop extended spine ankylosis. Male gender, younger age at onset, peripheral involvement, mainly of the hips, and positive HLA-B27 are factors that have been associated with worse prognosis (Poddubnyy et al. 2012a). Other markers of disease severity are increased sedimentation rate (more than 30 mm/h), poor response to NSAIDs, smoking, dactylitis, oligoarthritis, and radiological involvement at baseline (Poddubnyy et al. 2012b; Brandt et al. 2004). SpA may have a significant social and economic impact because the disease affects young individuals, in the most active period of life. The loss of functional capacity always has a corresponding loss of quality of life and psychological health. The disease causes functional disability, decreased productivity, absenteeism, unemployment, and early retirement (Marengo et al. 2008).

Life expectancy is decreased in patients with uncontrolled SpA, and this decrease is primarily due to increased risks of cardiovascular and pulmonary diseases.

Early diagnosis by recognizing the signs and symptoms of the disease has led to the early establishment of appropriate therapeutic measures, and this has led to a less aggressive and disabling disease. There is an unacceptable delay in diagnosis as a result of the insidious onset of the disease. It is necessary to alert primary care physicians, orthopedics, and internists to the early symptoms of the disease and in this way encourage early referral to the rheumatologist (Poddubnyy et al. 2011).

Cross-References

▶ Disease-Modifying Antirheumatic Drugs: Overview
▶ Non-steroidal Anti-inflammatory Drugs: Overview
▶ Rheumatoid Arthritis
▶ Tumor Necrosis Factor (TNF) Inhibitors

References

Appel, H., & Sieper, J. (2008). Spondyloarthritis and the crossroads of imaging, pathology, and structural damage in the era of biologics. *Current Rheumatology Reports, 10*, 356–363.

Baeten D, Sieper J, Braun J, Baraliakos X, Dougados M, Emery P., et al. (2015 Dec 24) MEASURE 1 Study Group; MEASURE 2 Study Group. Secukinumab, an Interleukin-17A Inhibitor, in Ankylosing Spondylitis. *N Engl J Med, 373*(26):2534-48.

Bergfeldt, L., Edhag, O., Vedin, L., & Vallin, H. (1982). Ankylosing spondylitis: An important cause of severe disturbances of the cardiac conduction system. Prevalence among 223 pacemaker-treated men. *American Journal of Medicine, 73*(2), 187–191.

Boyer, G. S., Lanier, A. P., & Templin, D. W. (1988). Prevalence rates of spondyloarthropathies, rheumatoid arthritis, and other rheumatic disorders in an Alaskan Inupiat Eskimo population. *Journal of Rheumatology, 15*(4), 678–683.

Brandt, J., Listing, J., Sieper, J., et al. (2004). Development and preselection of criteria for short term improvement after anti-TNF alpha treatment in ankylosing spondylitis. *Annals of the Rheumatic Diseases, 63*(11), 1438–1444.

Braun, J., Sieper, J., Breban, M., Collantes-Estevez, E., Davis, J., Inman, R., et al. (2002a). Anti-tumour necrosis factor alpha therapy for ankylosing spondylitis: International experience. *Annals of the Rheumatic Diseases, 61*(Suppl 3), iii51–iii60.

Braun, J., Brandt, J., Listing, J., Zink, A., Alten, R., Golder, W., et al. (2002b). Treatment of active ankylosing spondylitis with infliximab: A randomized controlled multicenter trial. *Lancet, 359*, 1187–1193.

Braun, A., Saracbasi, E., Grifka, J., et al. (2011). Identifying patients with axial spondyloarthritis in primary care: How useful are items indicative of inflammatory back pain? *Annals of the Rheumatic Diseases, 70*, 1782–1787.

Brewerton, D. A., Cafrey, M., Hart, F. D., Jamei, D. C. O., Nichols, A., & Sturrock, R. D. (1973). Ankylosing spondylitis and HLA-B27. *Lancet, 1*, 904–990.

Brown, M. A. (2010). Genetics of ankylosing spondylitis. *Current Opinion in Rheumatology, 22*(2), 126–132.

Brown, M. A., Jepson, A., Joung, A., et al. (1997a). Ankylosing spondylitis in west africans. Evidence for a non HLA-B27 protective effect. *Annals of the Rheumatic Diseases, 56*, 68–70.

Brown, M. A., Kennedy, L. G., MacGregor, A. J., Darge, C., Duncan, E., Shatford, J. L., et al. (1997b). Susceptibility to ankylosing spondylitis in twins: The role of genes, HLA, and the environment. *Arthritis and Rheumatism, 40*, 1746–1748.

Calin, A., Garrett, S. L., Whitelock, H., Kennedy, L. G., O'Hea, J., Mallorie, P., et al. (1994). A new approach to functional ability in ankylosing spondylitis: The bath ankylosing spondylitis functional index. *Journal of Rheumatology, 21*, 2281–2285.

Clegg, D. O., Reda, D. J., Weisman, M. H., Blackburn, W. D., Cush, J. J., Cannon, G. W., et al. (1996). Comparison of sulfasalazine and placebo in the treatment of ankylosing spondylitis. A Department of Veterans Affairs Cooperative Study. *Arthritis and Rheumatism, 39*, 2004–2012.

Colbert, R., Delay, M., Klenk, E., & Layh-Schmitt, G. (2010). From HLA-B27 to spondyloarthritis: A journey through the ER. *Immunological Reviews, 233*, 181–202.

Cornec, D., Devauchelle Pensec, V., Joulin, S. J., & Saraux, A. (2009). Dramatic efficacy of infliximab in cauda equina syndrome complicating ankylosing spondylitis. *Arthritis and Rheumatism, 60*(6), 1657–1660.

D'Agostino, M. A. (2010). Ultrasound imaging in spondyloarthropathies. *Best Practice & Research. Clinical Rheumatology, 24*(5), 693–700.

Davis, J. C., Jr., van der Heijde, D., Braun, J., Dougados, M., Cush, J., Clegg, D. O., et al. (2003). Recombinant human tumor necrosis factor receptor (etanercept) for treating ankylosing spondylitis: A randomized, controlled trial. *Arthritis and Rheumatism, 48*, 3230–3236.

Dean, L. E., Jones, G. T., MacDonald, A. G., Downham, C., Sturrock, R. D., & Macfarlane, G. J. (2014). Global prevalence of ankylosing spondylitis. *Rheumatology (Oxford), 53*(4), 650–657.

Dik, V. K., Peters, M. J., & Dijkmans, P. A. (2010). The relationship between disease-related characteristics and conduction disturbances in ankylosing spondylitis. *Scandinavian Journal of Rheumatology, 39*(1), 38–41.

Dougados, M., et al. (1991). The European Spondylarthropathy Study Group preliminary criteria for the classification of spondylarthropathy. *Arthritis and Rheumatism, 34*, 1218–1227.

Dougados, M., van der Linden, S., Leirisalo-Repo, M., Huitfeldt, B., Juhlin, R., Veys, E., et al. (1995).

Sulfasalazine in the treatment of spondyloarthropathy. A randomized, multicenter, double-blind, placebo-controlled study. *Arthritis and Rheumatism, 38*, 618–627.

Dougados, M., Dijkmans, B., Khan, M., Maksymowych, W., van der Linden, S., & Brandt, J. (2002). Conventional treatments for ankylosing spondylitis. *Annals of the Rheumatic Diseases, 61*(Suppl 3), iii40–iii50.

Doward, L., Spoorerg, A., Cook, S., et al. (2003). Development of the ASQoL: A quality of life instrument specific to ankylosing spondylitis. *Annals of the Rheumatic Diseases, 62*, 20–26.

Ebringer, A. (1990). Theoretical models to explain the association of HLA-B27 with ankylosing spondylitis. *Scandinavian Journal of Rheumatology, 87*(Suppl 8t), 151–163.

Ebringer, A. (1992). Ankylosing spondylitis is caused by klebsiella: Evidence from immunogenetic, microbiologic and serologic studies. *Rheumatic Diseases Clinics of North America, 18*(1), 105–121.

Feldtkeller, E., Bruckel, J., & Khan, M. A. (2000). Scientific contributions of ankylosing spondylitis patient advocacy groups. *Current Opinion in Rheumatology, 12*, 239–247.

Ferraz, M. B., Tugwell, P., Goldsmith, C. H., & Atra, E. (1990). Meta-analysis of sulfasalazine in ankylosing spondylitis. *Journal of Rheumatology, 17*, 1482–1486.

Fianyo, E., Wendling, D., Poulain, C., Farrenq, V., & Claudepierre, P. (2014). Non-radiographic axial spondyloarthritis: What is it? *Clinical and Experimental Rheumatology, 32*(1), 1–4.

0.01w?>García-Medel, N., Sanz-Bravo, A., Van Nguyen, D., Galocha, B., Gómez-Molina, P., Martín-Esteban, A., et al. (2012). Functional interaction of the ankylosing spondylitis-associated endoplasmic reticulum aminopeptidase 1 polymorphism and HLA-B27 in vivo. *Molecular and Cellular Proteomics, 11*(11), 1416–1429.

Garrett, S., Jenkinson, T., Kennedy, L. G., Whitelock, H., Gaisford, P., & Calin, A. (1994). A new approach to defining disease status in ankylosing spondylitis: The bath ankylosing spondylitis disease activity index. *Journal of Rheumatology, 21*, 2286–2291.

Glintborg, B., Ostergaard, M., Krogh, N. S., et al. (2013). Clinical response, drug survival and predictors thereof in 432 ankylosing spondylitis patients after switching tumour necrosis factor α inhibitor therapy: Results from the Danish nationwide DANBIO registry. *Annals of the Rheumatic Diseases, 72*, 1149–1155.

González-Roces, S., & Alvarez, V. (1996). HLA-B27 structure, function, and disease association. *Current Opinion in Rheumatology, 8*(4), 296–308.

Gorman, J. D., Sack, K. E., & Davis, J. C., Jr. (2002). Treatment of ankylosing spondylitis by inhibition of tumor necrosis factor alpha. *New England Journal of Medicine, 346*, 1349–1356.

Gran, J. T., & Skomsvoll, J. F. (1997). The outcome of ankylosing spondylitis: A study of 100 patients. *British Journal of Rheumatology, 24*, 908–911.

Gratacós, J., Collado, A., Osaba, M., Moyá, Sanmartí, R., Roqué, M., et al. (1999). Significant loss of bone mass in patients with early active ankylosing spondylitis: A follow study. *Arthritis and Rheumatism, 42*, 2319–2324.

Günaydin, R., Göksel Karatepe, A., Cesmeli, N., & Kaya, T. (2009). Fatigue in patients with ankylosing spondylitis: Relationships with disease-specific variables, depression and sleep disturbance. *Clinical Rheumatology, 28*(9), 1045–1051.

Heuft-Dorenbosch, L., Spooremberg, A., Van Tubergen, R., et al. (2003). Assessment of enthesitis in ankylosing spondylitis. *Annals of the Rheumatic Diseases, 62*, 127–132.

Kaltwasser, J. P., Nash, P., Gladman, D., et al. (2004). Efficacy and safety of leflunomide in the treatment of psoriatic arthritis and psoriasis. A multinational double bind randomized placebo controlled clinical trial. *Arthritis and Rheumatism, 50*, 1939–1950.

Kim, T. H., Rahman, P., Jun, J. B., et al. (2004). Analysis of CARD15 polymorphisms in Korean patients with ankylosing spondylitis reveals absence of common variants seen in western populations. *Journal of Rheumatology, 31*, 1959–1961.

Kingsley, G. H., Kowalczyk, A., Taylor, H., et al. (2012). A randomized placebo-controlled trial of methotrexate in psoriatic arthritis. *Rheumatology (Oxford), 51*(8), 1368–1377.

Lau, C. S., Burgos-Vargas, R., Louthrenoo, W., Mok, M. Y., Wordsworth, P., & Zeng, Q. Y. (1998). Features of spondyloarthropathies around the world. *Rheumatic Diseases Clinics of North America, 24*, 753–770.

López-Larrea, C., Sujirachato, K., Mehra, N. K., et al. (1995). HLAB27 subtypes in Asian patients with AS. Evidence for new associations. *Tissue Antigens, 45*, 169–176.

Lories, R. J. U., Derese, I., Bari, C. D., & Luyten, F. R. (2007). Evidence for uncoupling of inflammation and joint remodeling in a mouse model of spondyloarthritis. *Arthritis and Rheumatism, 56*, 489–497.

Lukas, C., Landewé, R., Sieper, J., et al. (2009). Development of an ASAS-endorsed disease activity score (ASDAS) in patient with ankylosing spondylitis. *Annals of the Rheumatic Diseases, 68*, 18–24.

Machado, P., Landewé, R., Lie, E., et al. (2011). Ankylosing Spondylitis Disease Activity Score (ASDAS): Defining cut-off values for disease activity states and improvement scores. *Annals of the Rheumatic Diseases, 70*, 47–53.

Marengo, M. F., Citera, G., Schneeberger, E. E., & Maldonado Cocco, J. A. (2008). Work status among patients with ankylosing spondylitis in Argentina. *Journal of Clinical Rheumatology, 14*, 273–277.

McInnes, I. B., Sieper, J., Braun, J., et al. (2014). Efficacy and safety of secukinumab, a fully human anti-interleukin-17A monoclonal antibody, in patients with moderate-to-severe psoriatic arthritis: A 24-week, randomised, double-blind, placebo-controlled, phase

II proof-of-concept trial. *Annals of the Rheumatic Diseases, 73*, 349–356.

Mielans, H., Em, V., Cuvelier, C., et al. (1988). Ileocolonoscopic findings in seronegative spondyloarthropathies. *British Journal of Rheumatology, 27*, 95–105.

Moll, J., & Wright, V. (1973). Psoriatic arthritis. *Seminars in Arthritis and Rheumatism, 3*, 55–78.

Nas, K., Yildirim, K., Cevik, R., et al. (2010). Discrimination ability of ASDAS estimating disease activity status in patient with ankylosing spondylitis. *International Journal of Rheumatic Diseases, 13*, 240–245.

Nasution, A. R., Marjuadi, A., Kunmartini, S., et al. (1997). HLA-B27 subtypes positively and negatively associated with spondyloarthropathy. *Journal of Rheumatology, 24*, 1111–1114.

Papp, K., Cather, J., Rosoph, L., et al. (2012). The efficacy of apremilast, a phosphodiesterase-4 inhibitor, in the treatment of moderate to severe psoriasis: results of a phase 2 randomised study. *Lancet, 380*, 738–746.

Patel, D. D., Lee, D. M., Kolbinger, F., et al. (2013). Effect of IL-17A blockade with secukinumab in autoimmune diseases. *Annals of the Rheumatic Diseases, 72*(Suppl 2), ii116–ii123.

Pedersen, S. J., Sorensen, I. J., Hermann, K. G., et al. (2010). Responsiveness of the Ankylosing Spondylitis Disease Activity Score (ASDAS) and clinical and MRI measures of disease activity in a 1-year follow-up study of patients with axial spondyloarthritis treated with tumour necrosis factor alpha inhibitors. *Annals of the Rheumatic Diseases, 69*, 1065–1071.

Pérez Alamino, R., Maldonado Cocco, J. A., Citera, G., et al. (2011). Differential features between primary ankylosing spondylitis and spondylitis associated with psoriasis and inflammatory bowel disease. *Journal of Rheumatology, 38*(8), 1656–1660.

Poddubnyy, D., Vahldiek, J., Spiller, I., et al. (2011). Evaluation of 2 screening strategies for early identification of patients with axial spondyloarthritis in primary care. *Journal of Rheumatology, 38*(11), 2452–2460.

Poddubnyy, D., Rudwaleit, M., Haibel, H., et al. (2012a). Effect of non-steroidal anti-inflammatory drugs on radiographic spinal progression in patients with axial spondyloarthritis: Results from the German Spondyloarthritis Inception Cohort. *Annals of the Rheumatic Diseases, 71*, 1616–1622.

Poddubnyy, D., Haibel, H., Listing, J., et al. (2012b). Baseline radiographic damage, elevated acute-phase reactant levels, and cigarette smoking status predict spinal radiographic progression in early axial spondylarthritis. *Arthritis and Rheumatism, 64*(5), 1388–1398.

Rahman, P., Bartlett, S., Siannis, F., Pellett, F. J., Farewell, V. T., Peddle, L., et al. (2003). CARD15: A pleiotropic autoimmune gene that confers susceptibility to psoriatic arthritis. *American Journal of Human Genetics, 73*, 677–681.

Rosenbaum, J. T. (1992). Acute anterior uveitis and spondyloarthropathies. *Rheumatic Diseases Clinics of North America, 18*(1), 143–151.

Rudwaleit, M., Landewé, R., van der Heijde, D., et al. (2009a). The development of assessment of spondyloarthritis international society classification criteria for axial spondyloarthritis (part I): Classification of paper patients by expert opinion including uncertainty appraisal. *Annals of the Rheumatic Diseases, 68*(6), 770–776.

Rudwaleit, M., Jurik, A. G., Hermann, K. G., et al. (2009b). Defining active sacroiliitis on magnetic resonance imaging (MRI) for classification of axial spondyloarthritis: A consensual approach by the ASAS/OMERACT MRI group. *Annals of the Rheumatic Diseases, 68*(10), 1520–1527.

Rudwaleit, M., van der Heijde, D., Landewé, R., et al. (2011). The assessment of spondyloarthritis international society classification criteria for peripheral spondyloarthritis and for spondyloarthritis in general. *Annals of the Rheumatic Diseases, 70*(1), 25–31.

Russel, A. S., & Suárez-Almanzor, M. E. (1992). Ankylosing spondylitis is not caused by klebsiella. *Rheumatic Diseases Clinics of North America, 18*, 95–104.

Salaffi F, Ciapetti A, Carotti M, Gasparini S, Citera G, Gutierrez M. (2014). Construct validity and responsiveness of the simplified version of Ankylosing Spondylitis Disease Activity Score (SASDAS) for the evaluation of disease activity in axial spondyloarthritis. *Health Qual Life Outcomes 22*(12):129.

Scholnsstein, L., Terasaki, P. I., Bluestone, R., et al. (1973). High association of an HLA antigen B27 with ankylosing spondylitis. *New England Journal of Medicine, 288*, 704–706.

Sieper, J., & van der Heidje, D. (2013). Nonradiographic spondyloarthritis new definition of an old disease? *Arthritis and Rheumatism, 65*(3), 543–551.

Sieper, J., van der Heijde, D., Landewé, R., et al. (2009). New criteria for inflammatory back pain in patients with chronic back pain: a real patient exercise by experts from the Assessment of Spondyloarthritis international Society (ASAS). *Annals of the Rheumatic Diseases, 68*, 784–788.

Sieper, J., van der Heijde, D., Dougados, M., et al. (2013). Efficacy and safety of adalimumab in patients with non-radiographic axial spondyloarthritis: Results of a randomised placebo-controlled trial (ABILITY-1). *Annals of the Rheumatic Diseases, 72*, 815–822.

Solmaz D, Yildirim T, Avci O, Tomas N, Akar S. (2015 Dec 16). Performance characteristics of the simplified version of ankylosing spondylitis disease activity score (SASDAS). Clin Rheumatol. [Epub ahead of print].

Sommerfleck, F. A., Schneeberger, E. E., Buschiazzo, E. E., Maldonado Cocco, J. A., Citera, G., et al. (2012). A simplified version of Ankylosing Spondylitis Disease Activity Score (ASDAS) in patients with ankylosing spondylitis. *Clinical Rheumatology, 31*(11), 1599–1603.

Van de Linden, S., Valkenburg, H. A., De Jongh, B. M., et al. (1984). The risk of developing ankylosing spondylitis in HLA-B27 positive individuals: A comparison of relatives of spondylitis patients with the general population. *Arthritis and Rheumatism, 27*, 241–249.

van der Heijde, D. M., Landewe, R., Einstein, P., Ory, P., Vosse, D., Ni, L., et al. (2008). Two-year etanercept therapy does not inhibit radiographic progression in patients with ankylosing spondylitis. *Arthritis and Rheumatism, 58*, 1324–1331.

van der Heijde, D., Sieper, J., Maksymowych, W. P., et al. (2011). 2010 Update of the international ASAS recommendations for the use of anti-TNF agents in patients with axial spondyloarthritis. *Annals of the Rheumatic Diseases, 70*, 905–908.

van der Linden, S., & van der Heijde, D. (1998). Ankylosing spondylitis. Clinical features. *Rheumatic Diseases Clinics of North America, 24*, 663–673.

van Tubergen, A., Landewé, R., van der Heijde, D., et al. (2001). Combined spa-exercise therapy is effective in patients with ankylosing spondylitis: A randomized controlled trial. *Arthritis and Rheumatism, 45*(5), 430–438.

Wanders, A., Heijde, D., Landewé, R., et al. (2005). Nonsteroidal antiinflammatory drugs reduce radiographic progression in patients with ankylosing spondylitis: A randomized clinical trial. *Arthritis and Rheumatism, 52*(6), 1756–1765.

Yu, D. T. (1989). Molecular mimicry in HLAB27-related arthritis. *Annals of Internal Medicine, 111*, 581–591.

Substance P in Inflammation

Jennifer V. Bodkin[1], Gabor Pozsgai[2], Claire Sand[1], Rufino J. Klug[3], Thiago A. F. Ferro[3] and Elizabeth S. Fernandes[1,3]
[1]Cardiovascular Division, King's College London, London, UK
[2]Department of Pharmacology and Pharmacotherapy, Faculty of Medicine, University of Pécs, Pécs, Hungary
[3]Programa de Pós-Graduação, Universidade Ceuma, São Luís, MA, Brazil

Synonyms

Neuropeptide; Tachykinin

Definition

Substance P (SP) is an 11-amino acid neuropeptide and the prototypic member of the tachykinin (TK) family. TKs are widely distributed in central, peripheral, and enteric nervous systems and additionally expressed on some non-neuronal cells. They exert their actions through activation of three G protein-coupled, neurokinin (NK) receptors, known as NK1, NK2, and NK3. As with all TKs, SP is a promiscuous agonist of all NK receptors due to their conserved carboxy termini. It is now established that SP exhibits higher affinity for the NK1 receptor which signals predominantly via Gq-mediated calcium mobilization.

Biosynthesis and Release

TKs are encoded by three genes called TAC-1, TAC-3, and TAC-4, which are differentially transcribed, translated, and spliced, giving rise to several protein products. SP is generated from transcription of exon 3 of the TAC-1 gene, which is found in all TAC-1 splice variants (α, δ, β, and γ). Translation of the mRNA by ribosomes produces pre-pro-tachykinin, the precursor of SP which is further cleaved in the Golgi to form mature SP. Completed protein is then stored in intracellular vesicles until a further stimulus initiates its release into the extracellular space. It is well known now that SP release can occur following activation of several neuronal and non-neuronal cell surface receptors, classically including transient receptor potential (TRP) channels such as the TRP vanilloid 1 (TRPV1) and ankyrin 1 (TRPA1). Activation of these receptors produces a cytoplasmic Ca^{+2} influx initiating neuropeptide vesicle release. This is best characterized in neuronal cells.

Biological Activities

SP/NK1-mediated physiological effects include edema, microvasculature vasodilation, and pain signaling, which together form the key symptoms of neurogenic inflammation. The first documented biological activities of SP were its vasodilatory effect and ability to increase capillary permeability in response to SP released from sensory nerves. These phenomena are thought to contribute to the inflammatory "triple response," as defined by Lewis (reddening, flare, then wheal), and are mediated by endothelial NK1 receptors. Additionally, SP can directly mediate degranulation of mast cells, due to the basic nature of the protein. For several vascular beds, this raises the question as to whether the majority of vascular permeability is due to SP's ability to release mast cell mediators or by directly activation of endothelial NK1 receptors.

SP has a direct role in immune responses by driving cellular infiltration and cytokine release. Participation of NK1 receptors has been observed in viral, bacterial, helminthic, and protozoal infectious diseases (Douglas and Leeman 2011). Furthermore, SP is increasingly being implicated in inflammatory pathophysiology. This drives continued development of NK1 antagonists, of which only a few have made it to market. This is further discussed in the "Modulation by Drugs" section. Indeed, participation of SP in various inflammatory processes (e.g., pancreatitis, colitis, arthritis, and airway inflammation) is demonstrated by the efficacy of NK1 receptor antagonists in animal models of these pathologies. In addition to its inflammatory role, SP also mediates other effects in the body. Activation of NK1 receptors in the nucleus tractus solitarii induces nausea and vomiting, along with other effects on the central nervous system including anxiety and depression (Horii 2009). Elevated serum and cerebrospinal fluid SP levels have been reported in patients suffering from anxiety or depression, and numerous NK1 antagonists have been demonstrated to relieve symptoms in animal models (Horii 2009; Lecci and Maggi 2012). There is also evidence that SP lowers seizure threshold of experimental animals and mice lacking the TAC1 gene develop milder seizures in response to kainate (Zachrisson et al. 1998). Edema formation evoked by SP release at sites of peripheral nerve and brain injury might also contribute to cerebral ischemia and neuronal loss. On the other hand, SP has also demonstrated some beneficial effects in the nervous system, such as neurotrophic and neuroprotective effects (Pantaleo et al. 2010).

Other biological activities of NK1 activation have also been described, such as roles in hematopoiesis, tumorigenesis, and angiogenesis. These findings may pave the way for the use of NK1 receptor antagonists in antitumor therapy.

Pathophysiological Relevance

SP is increasingly being implicated in inflammatory pathophysiology, and this will be discussed in the context of some highly prevalent pathologies. The details are summarized in Fig. 1.

Sepsis

The SP/NK_1 interaction has been implicated in the progression of sepsis, largely via its key role in mediating neurogenic inflammation. Excessive stimulation or inappropriate activation of sensory neurons can result in clinical symptoms including hypotension, tissue edema, leukocyte recruitment, and the induction of inflammatory mediator production, all of which are prevalent in septicemia. Sepsis is a clinical condition comprising of an initial overwhelming and systemic pro-inflammatory response to local or systemic microorganism components, followed by an equally damaging immunosuppressive phase. In the Western world sepsis is a significant cause of mortality but has experienced little medicinal advancement in recent decades. The severity of sepsis symptoms varies between individuals but commonly includes fever, tachycardia, and hemodynamic shock, often occurring alongside high levels of pro-inflammatory mediators in the plasma. These features suggest a link with neurogenic inflammation and the actions of SP. Furthermore, SP levels in plasma are known to be elevated in both human sepsis and animal models, with strong evidence for its detrimental role provided by clinical endpoints and several landmark experimental animal studies. These studies show mice lacking TAC1 gene present with a less severe sepsis and lower mortality compared to WT mice challenged by either intestinal bacterial or the Gram-negative bacteria outer membrane component, lipopolysaccharide. Mice lacking the TAC1 gene showed reduced levels of tissue and circulating chemokines (including monocyte chemotactic protein 1, MCP-1, and macrophage inflammatory protein 2, MIP-2),

Substance P in Inflammation, Fig. 1 Substance P (*SP*) role in sepsis, arthritis, and airway inflammation. NK1 receptor antagonism has been shown to be beneficial in experimental arthritis and airway inflammation models in rodents, although these effects have not been translated into clinical practice. The benefits of blocking NK1 receptors in sepsis will depend on whether the condition is at an early or late stage. *NK1*, Neurokinin receptor 1; *TNF*, Tumor necrosis factor; *OA*, Osteoarthritis; *RA*, Rheumatoid arthritis

which correlated with reduced inflammatory cell infiltration to key organs such as the lung, liver, and kidneys. Reduced levels of lung edema, a readout of lung damage, were also noted (Puneet et al. 2006). The role for SP in sepsis was later proven to be NK1 mediated using mice genetically lacking this receptor (Hegde et al. 2007). Cytokine levels are also modulated by NK1 activation during sepsis (Hedge et al. 2007, 2010b), with a central role for IL1-β in the pro-inflammatory network, alongside a striking increase in plasma and lung levels of the endogenous IL1 receptor antagonist. Inflammatory mediator release profile changes are regulated by NK1 receptors via induction of transcription factors such as nuclear factor-κB (NF-κB), activator protein-1 (AP-1), and extracellular signal-regulated kinase (ERK) (Hedge et al. 2010a). Similar findings were not seen with the use of a NK_2 receptor antagonist, a closely related receptor which binds SP with a lower affinity than that of NK_1 and is largely expressed in the central nervous system (Hedge et al. 2010b). These findings support a modulatory role for SP in driving the pro-inflammatory phase of sepsis. A common theme emerges, suggesting SP/NK1 acts to increase leukocyte recruitment in the early stages of disease, leading to increased inflammatory mediator production and organ damage in the latter stages.

Arthritis

Rheumatoid arthritis (RA) and osteoarthritis (OA) are the most prevalent diseases involving joint inflammation in humans and greatly contribute to the difficulties experienced by aging populations. Both diseases are progressive, with decreased range of movement, structural damage of the affected joint, and ongoing pain, leading to deterioration of patients' quality of life. RA pathophysiology is driven by an ongoing autoimmune response, leading to chronic joint inflammation and damage. The etiology of OA is still unclear, but susceptibility might be determined by genetic factors. Both conditions are highly progressive. Decreased range of movement, structural damage of the affected joint, and ongoing pain lead to deterioration of life quality. Current pharmacotherapy for both RA and OA is aimed at slowing disease progression but is often associated with intolerable side effects, making novel approaches to treatment an active field of research.

SP is expressed in a subpopulation of sensory nerve fibers which innervate the synovial tissue and chondrocytes of joints. The NK1 receptor is also present in the joint, expressed on chondrocytes and epithelial cells and also on recruited immune cells such as memory T cells, suggesting the potential of SP to be immunomodulatory (Fernandes et al. 2009). Additionally, NK1 receptors expressed on neurons modulate pain pathways. Participation of SP and NK1 receptors in RA has been extensively investigated. Increased levels of SP have been detected in serum and synovial tissue of RA patients (Larsson et al. 1991). SP also has the ability to perpetuate the inflammatory response, due to its ability to induce TNF secretion from monocytes (Lavagno et al. 2001). Interestingly, circulating SP levels are reduced by treatment with the TNF inhibitor, etanercept (Origuchi et al. 2011), suggesting that SP may be playing an active role in the pro-inflammatory feedback driving RA. Data collected from animal models of RA have shown NK1 antagonists to ameliorate hyperalgesia and reduce the severity of joint inflammation (Pintér et al. 2013).

The participation of SP and NK1 receptors in OA is much less defined. While a decreased density of SP-positive sensory fibers in the OA synovium has been shown (Weidler et al. 2005), increased levels of SP have been reported in the bone tissue and joint capsule of patients. Increased SP is also found in the joints and dorsal root ganglia of rodent models of OA (Ogino et al. 2009; Ahmed et al. 2012) and returns to normal following treatment to reduce symptoms and joint destruction.

Despite showing some benefits in animal models of arthritis, NK1 receptor antagonists failed to translate into clinical practice as analgesics or as a therapy for arthritis. This may be due to differences in splice variants and distribution between human and rodent NK1 receptors. Recently, a role for the tachykinin, hemokinin-1 (HK-1), has been suggested in RA (Borbély et al. 2013). HK-1 is encoded by gene TAC-4 gene and has high affinity to NK1 receptors (Lecci and Maggi 2012); thus it has been suggested that HK-1 may be responsible for some, if not all, of the effects previously attributed to SP. In this context, it is important to highlight that HK-1 binding site on NK1 receptors differs from that of SP. This knowledge may be useful to generate antagonists and/or antibodies able to distinguish between the effects of SP and HK-1, making the role of each mediator clearer. However, this is a novel avenue of research and the potential role of HK-1 in arthritis is currently under investigation.

Airway Inflammation

SP has been implicated in a number of inflammatory airway disorders, including asthma, cough, and chronic obstructive pulmonary disease. Via actions on smooth muscle, epithelial, endothelial, and inflammatory cells, it promotes neurogenic inflammation and may contribute to the pathogenesis of several such inflammatory conditions.

SP can be released from airway sensory nerves following activation by stimuli such as cold air, inhaled irritants, or cigarette smoke and can also be produced by inflammatory cells and epithelial and endothelial cells (Maggi 1997). In the airways, SP mediates smooth muscle constriction, epithelial chloride secretion, enhanced ciliary movement and mucus hypersecretion, as well as stimulating leukocyte recruitment and increasing microvascular permeability.

Although SP-positive innervation is relatively sparse in normal human airways compared to rodent airways, several studies using rodent models have suggested that inflammatory conditions such as chronic cough are associated with increased SP-positive nerve growth. This has not been borne out in studies of asthmatic patients and remains a somewhat contentious topic. However, raised levels of SP are consistently found in the bronchoalveolar lavage of inflammatory airway patients, suggesting that SP may have a role in the inflammatory process and that it may originate from non-neuronal sources of production (Joos 2001).

An exaggerated contractile response to SP has been established both in vitro and in patients and animal models with airway inflammation. This effect was shown in vitro to be mediated by prostanoids in smaller airways and by direct activation of smooth muscle receptors and inositol phosphate release in medium-sized airways (Joos 2001). Interestingly, the in vivo bronchoconstrictor effects of inhaled SP can be blocked with cromoglycate or nedocromil sodium suggesting an indirect mechanism of action that may involve stimulation of acetylcholine release from postganglionic cholinergic airway nerves (Joos 2001).

In addition to its effects on airway smooth muscle, SP also stimulates mast cell degranulation and TNF production (Ansel et al. 1993; Fewtrell et al. 1982), as well as edema formation via its action on post-capillary endothelial cells. Other aspects of asthmatic pathophysiology that may involve SP include the formation of new blood vessels observed in asthmatic airways and the proliferation and chemotaxis of lung fibroblasts underlying asthmatic fibrosis (Harrison et al. 1995).

Pharmacological blockade of NK1 has been postulated as a potential treatment for a number of inflammatory airway conditions. NK1 antagonism prevents airway edema and coughing in animal models of airway inflammation (Bolser et al. 1997), and deletion of the NK1 gene confers protection from airway emphysema and inflammatory cell infiltration in mice exposed to cigarette smoke (De Swert et al. 2009). Preclinical studies have also revealed antitussive effects of NK1 antagonists, and in asthmatic patients, they have been shown to attenuate bronchoconstriction and exercise-induced asthma (Ichinose et al. 1996). The latter was found to occur independently of changes in maximal bronchoconstriction, indicating a vascular rather than respiratory mechanism of action, although the beneficial effects of NK1 antagonism have not been reproducible in other clinical studies. In addition to blockade of the NK1 receptor, aerosol administration of recombinant enzymes that metabolize and inactivate SP attenuates airway hyperreactivity in animal models (Kohrogi et al. 1989) and may represent a potential new strategy in the treatment of airway inflammation.

Despite evidence of the role of SP-mediated neurogenic inflammation in rodent models of airway disorders, its significance in human airways is less clear. Observations of increased SP-positive nerve innervation in asthmatic patients have not been reproducible across all studies, and clinical trials of tachykinin receptor antagonists have produced conflicting results. While elevated levels of SP may play a pathophysiological role in inflammatory airways (particularly if there is impaired breakdown of SP or increased expression or sensitivity of NK1 receptors), the significance of this role and its therapeutic potential in humans remain to be established.

Modulation by Drugs

Several antagonists have been developed to pharmacologically antagonize the NK1 receptor for therapeutic value. In a recent essay, Lecci and Maggi (2012) outlined a series of more than 30 selective and nonselective NK1 antagonists which are able to block NK1 signaling with varying affinities and efficacies across species. As discussed here and summarized in Fig. 1, evidence suggests that blocking NK1 signaling could be beneficial in several disorders which are prevalent in the western world. Poor translation of these findings from animal models (especially rodents) into human disease has been a common difficulty, likely due to the existence of divergent residues in NK1 receptor sequences between species. From animal and in vitro models of disease, NK1 antagonists have shown potential therapeutic effects in treating pain, CNS-related disorders (such as depression and anxiety), autoimmune diseases (such as RA), airway diseases, emesis, urinary incontinence, and infection, among others. On the other hand, NK1 antagonists have so far only succeeded in the clinics to treat chemotherapy-induced nausea and vomiting (aprepitant, fosaprepitant) (FDA 2010). Another NK1 antagonist, maropitant, was licensed in 2007 for the treatment of kinetosis-induced nausea and vomiting in dogs and cats (FDA 2007). There are currently promising clinical reports for the use of aprepitant to treat overactive bladder (Lecci and Maggi 2012). Therefore, although a role for SP and NK1 has been convincingly established for several clinical diseases, much further investigation will be needed to translate these findings into useful therapeutics for the future.

Cross-References

▶ Osteoarthritis
▶ Rheumatoid Arthritis
▶ Sepsis

References

Ahmed, A. S., Li, J., Erlandsson-Harris, H., Stark, A., Bakalkin, G., & Ahmed, M. (2012). Suppression of pain and joint destruction by inhibition of the proteasome system in experimental osteoarthritis. *Pain, 153*, 18–26.

Ansel, J. C., Brown, J. R., Payan, D. G., & Brown, M. A. (1993). Substance P selectively activates TNF-alpha gene expression in murine mast cells. *The Journal of Immunology, 150*, 4478–4485.

Bolser, D. C., DeGennaro, F. C., O'Reilly, S., McLeod, R. L., & Hey, J. A. (1997). Central antitussive activity of the NK1 and NK2 tachykinin receptor antagonists, CP-99,994 and SR 48968, in the guinea-pig and cat. *British Journal of Pharmacology, 121*, 165–170.

Borbély, E., Hajna, Z., Sándor, K., Kereskai, L., Tóth, I., Pintér, E., et al. (2013). Role of tachykinin 1 and 4 gene-derived neuropeptides and the neurokinin 1 receptor in adjuvant-induced chronic arthritis of the mouse. *PLoS One, 8*, e61684.

De Swert, K. O., Bracke, K. R., Demoor, T., Brusselle, G. G., & Joos, G. F. (2009). Role of the tachykinin NK1 receptor in a murine model of cigarette smoke-induced pulmonary inflammation. *Respiratory Research, 10*, 37.

Douglas, S. D., & Leeman, S. E. (2011). Neurokinin-1 receptor: Functional significance in the immune system in reference to selected infections and inflammation. *Annals of the New York Academy of Sciences, 1217*, 83–95.

Food and Drug Administration (FDA). (2007). Freedom of information summary: Original new animal drug application. NADA 141–263. CERENIA (maropitant citrate) injectable solution for the prevention and treatment of acute vomiting in dogs. http://www.fda.gov/downloads/animalveterinary/products/approved animaldrugproducts/foiadrugsummaries/ucm062313.pdf

Food and Drug Administration (FDA). (2010). Highlights of prescribing information: EMEND. http://www.accessdata.fda.gov/drugsatfda_docs/label/2010/021549s017lbl.pdf

Fernandes, E. S., Schmidhuber, S. M., & Brain, S. D. (2009). Sensory-nerve-derived neuropeptides: Possible therapeutic targets. *Handbook of Experimental Pharmacology, 194*, 393–416.

Fewtrell, C. M., Foreman, J. C., Jordan, C. C., Oehme, P., Renner, H., & Stewart, J. M. (1982). The effects of substance P on histamine and 5-hydroxytryptamine release in the rat. *The Journal of Physiology, 330*, 393–411.

Harrison, N. K., Dawes, K. E., Kwon, O. J., Barnes, P. J., Laurent, G. J., & Chung, K. F. (1995). Effects of neuropeptides on human lung fibroblast proliferation and chemotaxis. *American Journal of Physiology, 268*, L278–L283.

Hegde, A., Zhang, H., Moochhala, S. M., & Bhatia, M. (2007). Neurokinin-1 receptor antagonist treatment protects mice against lung injury in polymicrobial sepsis. *Journal of Leukocyte Biology, 82*, 678–685.

Hegde, A., Koh, Y. H., Moochhala, S. M., & Bhatia, M. (2010a). Neurokinin-1 receptor antagonist treatment in polymicrobial sepsis: Molecular insights. *International Journal of Inflammation, 2010*, 601098.

Hegde, A., Tamizhselvi, R., Manikandan, J., Melendez, A. J., Moochhala, S. M., & Bhatia, M. (2010b). Substance P in polymicrobial sepsis: Molecular fingerprint of lung injury in preprotachykinin-A-/- mice. *Molecular Medicine, 16*, 188–198.

Horii, A. (2009). Anti-motion sickness drugs. In M. D. Binder, N. Hirokawa, & U. Windhorst (Eds.), *Encyclopedia of neuroscience*. Berlin/Heidelber/New York: Springer. doi:10.1007/SpringerReference_113838.

Ichinose, M., Miura, M., Yamauchi, H., Kageyama, N., Tomaki, M., Oyake, T., et al. (1996). A neurokinin 1-receptor antagonist improves exercise-induced airway narrowing in asthmatic patients. *American Journal of Respiratory and Critical Care Medicine, 153*, 936–941.

Joos, G. F. (2001). The role of neuroeffector mechanisms in the pathogenesis of asthma. *Current Allergy and Asthma Reports, 1*, 134–143.

Kohrogi, H., Nadel, J. A., Malfroy, B., Gorman, C., Bridenbaugh, R., Patton, J. S., et al. (1989). Recombinant human enkephalinase (neutral endopeptidase) prevents cough induced by tachykinins in awake guinea pigs. *Journal of Clinical Investigation, 84*, 781–786.

Larsson, J., Ekblom, A., Henriksson, K., Lundeberg, T., & Theodorsson, E. (1991). Concentration of substance P, neurokinin A, calcitonin gene-related peptide, neuropeptide Y and vasoactive intestinal polypeptide in synovial fluid from knee joints in patients suffering from rheumatoid arthritis. *Scandinavian Journal of Rheumatology, 20*, 326–335.

Lavagno, L., Bordin, G., Colangelo, D., Viano, I., & Brunelleschi, S. (2001). Tachykinin activation of human monocytes from patients with rheumatoid arthritis: In vitro and ex-vivo effects of cyclosporin A. *Neuropeptides, 35*, 92–99.

Lecci, A., & Maggi, C. A. (2012). Tachykinins and their receptors. In S. Offermanns & W. Rosenthal (Eds.), *Encyclopedia of molecular pharmacology*. Berlin/Heidelberg/New York: Springer. doi:10.1007/SpringerReference_137970.

Maggi, C. A. (1997). The effects of tachykinins on inflammatory and immune cells. *Regulatory Peptides, 70*, 75–90.

Ogino, S., Sasho, T., Nakagawa, K., Suzuki, M., Yamaguchi, S., Higashi, M., et al. (2009). Detection of pain-related molecules in the subchondral bone of osteoarthritic knees. *Clinical Rheumatology, 28*, 1395–1402.

Origuchi, T., Iwamoto, N., Kawashiri, S. Y., Fujikawa, K., Aramaki, T., Tamai, M., et al. (2011). Reduction in serum levels of substance P in patients with rheumatoid arthritis by etanercept, a tumor necrosis factor inhibitor. *Modern Rheumatology, 21*, 244–250.

Pantaleo, N., Chadwick, W., Park, S. S., Wang, L., Zhou, Y., Martin, B., et al. (2010). The mammalian tachykinin ligand-receptor system: an emerging target for central neurological disorders. *CNS & Neurological Disorders Drug Targets, 9*, 627–635.

Pintér, E., Pozsgai, G., Hajna, Z., Helyes, Z., & Szolcsányi, J. (2013). Neuropeptide receptors as potential drug targets in the treatment of inflammatory conditions. *British Journal of Clinical Pharmacology*. doi:10.1111/bcp.12097.

Puneet, P., Hegde, A., Ng, S. W., Lau, H. Y., Lu, J., Moochhala, S. M., et al. (2006). Preprotachykinin-A gene products are key mediators of lung injury in polymicrobial sepsis. *The Journal of Immunology, 176*, 3813–3820.

Weidler, C., Holzer, C., Harbuz, M., Hofbauer, R., Angele, P., Schölmerich, J., et al. (2005). Low density of sympathetic nerve fibres and increased density of brain derived neurotrophic factor positive cells in RA synovium. *Annals of the Rheumatic Diseases, 64*, 13–20.

Zachrisson, O., Lindefors, N., & Brené, S. (1998). A tachykinin NK1 receptor antagonist, CP-122,721-1, attenuates kainic acid-induced seizure activity. *Brain Research. Molecular Brain Research, 60*, 291–295.

Sulfasalazine and Related Drugs

Garry G. Graham[1,2] and Kevin D. Pile[3]
[1]Department of Pharmacology, School of Medical Sciences, University of New South Wales, Sydney, NSW, Australia
[2]Department of Clinical Pharmacology and Toxicology, St Vincent's Hospital, Sydney, NSW, Australia
[3]Campbelltown Hospital, School of Medicine, University of Western Sydney, Campbelltown, NSW, Australia

Synonyms

Salazopyrin; Salazosulfapyridine; Sulfasalazine

Definition

Sulfasalazine is a drug which is useful in the treatment of ulcerative colitis and is also a member of a group of drugs known as disease-modifying antirheumatic drugs (DMARDs) or slow-acting antirheumatic drugs (SAARDs).

Chemical Structures and Properties

Sulfasalazine consists of mesalazine (aminosalicylate) and a sulfonamide, sulfapyridine, linked by an azo bond which is split in the small intestine (Fig. 1). Olsalazine is a related drug which is a dimer of mesalazine and also split in the large intestine. All compounds are acids with varying logP and pKa values (Table 1). (LogP is the logarithm (log10) of distribution coefficient of the drug between octanol and water.) Their chemical properties promote the absorption and distribution of the drugs within the body.

Metabolism and Pharmacokinetics

- Sulfasalazine is generally administered as enteric-coated tablets which release the drug in the small intestine.

Sulfasalazine and Related Drugs, Fig. 1 Initial metabolism of sulfasalazine and olsalazine. Both mesalazine and sulfapyridine are *N*-acetylated. The ionized forms of the drugs and metabolites are shown. These are the major forms in plasma. Key: pyr is the pyridine residue

Sulfasalazine and Related Drugs, Table 1 logP and pKa values of sulfasalazine and their metabolites

Drug	logP	pKa
Sulfasalazine	2.9	3.3
Mesalazine	0.75	2.02
Sulfasalazine	0.84	6.24
Olsalazine	2.77	2.93

- Some sulfasalazine is absorbed in the small intestine but is limited due to transport back to the intestine by breast cancer resistance protein (ATP-binding cassette subfamily G member 2).
- On reaching the large bowel, the azo bond of sulfasalazine is reduced by colonic bacteria to yield sulfapyridine and mesalazine (aminosalicylate) (Fig. 1).
- Unchanged sulfasalazine achieves substantial concentrations in plasma. During daily treatment with 2 g sulfasalazine, the average concentrations of the unchanged drug are about 4.2 mg/L (10 μM) (Taggart et al. 1992) (Table 2).
- Sulfapyridine is absorbed from the large intestine, and the appearance of sulfapyridine in blood (about 4 h) has been used experimentally to measure the transit time from the stomach to the large intestine.
- The plasma concentrations of sulfapyridine are higher in slow acetylators than in fast acetylators (Table 2).
- The azo bond of olsalazoline is split in the large intestine to yield two molecules of mesalazine from each molecule of olsalazine (Fig. 1).
- Mesalazine is poorly absorbed although low concentrations of the acetyl mesalazine are found in plasma (Table 2).
- Sulfasalazine is not dialyzable, but major metabolic products, sulfapyridine and acetyl mesalazine (Fig. 1), are dialyzable

Pharmacological Activities

This drug was synthesized in the late 1930s and originally developed and used on the basis of a belief in an infectious cause of RA (Svartz 1942). The mode of action of sulfasalazine is unclear although the unchanged drug has been reported to inhibit 5-aminoimidazole-4-carboxamidoribonucleotide (AICAR) transformylase, an enzyme involved in the biosynthesis of purines. Inhibition of this enzyme is widely considered to be the mechanism of action of methotrexate. Therefore, sulfasalazine may act in a similar fashion to methotrexate by increasing the concentrations of adenosine at sites of inflammation (Gadangi et al. 1996). Mesalazine has been evaluated in vitro for its reaction with free radicals, but the relevance of these observations to its antirheumatic actions is not known.

Clinical Uses and Efficacy

Sulfasalazine is widely used and effective in the treatment of rheumatoid arthritis as well as spondylarthropathies, ulcerative colitis, and Crohn's disease. Sulfasalazine slows the joint destruction of rheumatoid arthritis (van der Heijde et al. 1990). Oddly, both sulfasalazine and its

Sulfasalazine and Related Drugs, Table 2 Pharmacokinetics of sulfasalazine and metabolites in patients with rheumatoid arthritis (Taggart et al. 1992)

Group		Young		Half-life (hours)	
Acetylator status		Fast	Slow	Fast	Slow
Mean plasma concentrations (mg/l)	Sulfasalazine	4.2	4.4	4.6	6.7
	Acetyl mesalazine	0.41	0.63	1.2	0.95
	Sulfapyridine	11	26	11	35
Half-life (h)	Sulfasalazine	6.5	5.8	13.7	20.3
	Acetyl mesalazine	26	25	47	33
	Sulfapyridine	10	15	12	27

metabolites, sulfapyridine and mesalazine, are therapeutically active in rheumatoid arthritis (Neumann et al. 1986). On the other hand, both sulfasalazine and olsalazine appear active in the treatment of spondylarthropathies (Ferraz et al. 1990). As the only metabolite of olsalazine is mesalazine, sulfapyridine is not necessary for the activity of olsalazine in spondylarthropathies, at least. It is of note that sulfasalazine and olsalazine are used in the treatment of both rheumatoid arthritis and inflammatory bowel diseases. Furthermore, many patients with bowel diseases also develop arthritic states of varying severity. These similarities indicate a commonality in the cause of the diseases and a common mode of action of sulfasalazine and olsalazine, although the reason for these parallel diseases and treatments is unclear.

In the treatment of rheumatoid arthritis, sulfasalazine is usually started at 500 mg daily and is increased weekly to 2–3 g daily.

Adverse Effects

Although sulfasalazine can produce a wide range of side effects, it is among the best tolerated slow-acting antirheumatic drugs. Toxicity is most frequent in the first 2–3 months of usage, but its likelihood can be reduced by gradually increasing the dosage. Serious side effects are rare, and most adverse effects are eliminated if the dose is reduced from the usual 1 g twice a day.

Adverse effects include:

- Nausea and upper abdominal discomfort. These are the most frequent side effects at the start of the therapy.
- Leukopenia is very uncommon but can develop rapidly. Its occurrence is most likely in the first 6 months of therapy but can develop later.
- The metabolite, sulfapyridine, can produce hemolysis in patients with a deficiency of glucose-6-phosphate dehydrogenase. Patients known to have such a deficiency should not take sulfasalazine.
- No teratogenicity has been reported for sulfasalazine although it is generally considered that sulfasalazine should be avoided at around the time of conception and early pregnancy.
- Severe dermatological diseases with systemic aspects have been reported rarely. These include Steven-Johnson disease and DRESS (drug rash with eosinophilia and systemic symptoms).

Monitoring for adverse hematological effects is mandatory. It is recommended that a complete blood cell count be performed before commencement of treatment and then every 2–4 weeks for the first 3 months and at greater intervals subsequently. The hemolytic anemia in patients with a deficiency of glucose-6-phosphate dehydrogenase will be detected in blood counts. Baseline measurement of hepatic transaminases is advised in patients with known or suspected liver disease.

Drug Interactions

- Combinations of sulfasalazine with other SAARDs methotrexate, antimalarials, or leflunomide provide additional antirheumatic activity to that of sulfasalazine alone.
- Sulfasalazine may be used in combination with corticosteroids and/or nonsteroidal

anti-inflammatory drugs (NSAIDs), particularly when the antirheumatic effects of sulfasalazine are developing. Ultimately, it may be possible to withdraw treatment with corticosteroids and/or NSAIDs if the patients respond well to sulfasalazine or its combinations with other SAARDs.

- Sulfasalazine may potentiate oral anticoagulants (warfarin), sulfonylurea hypoglycemic agents, and thiopurines. Care should be taken when sulfasalazine is commenced in patients stabilized on these drugs.

Cross-References

▶ Antimalarial Drugs
▶ Corticosteroids
▶ Disease-Modifying Antirheumatic Drugs: Overview
▶ Inflammatory Bowel Disease
▶ Leflunomide
▶ Methotrexate
▶ Rheumatoid Arthritis
▶ Spondyloarthritis

References

Ferraz, M. B., Tugwell, P., Goldsmith, C. H., & Atra, E. (1990). Meta-analysis of sulfasalazine in ankylosing spondylitis. *Journal of Rheumatology, 17*(11), 1482–1486.

Gadangi, P., Longaker, M., Naime, D., Levin, R. I., Recht, P. A., & Montesinos, M. C. (1996). The anti-inflammatory mechanism of sulfasalazine is related to adenosine release at inflamed sites. *Journal of Immunology, 156*(5), 1937–1941.

Neumann, V. C., Taggart, A. J., Le Gallez, P., Astbury, C., Hill, J., & Bird, H. A. (1986). A study to determine the active moiety of sulphasalazine in rheumatoid arthritis. *Journal of Rheumatology, 13*(2), 285–287.

Svartz, N. (1942). Salazopyrin, a new sulfanilamide preparation: A. therapeutic results in rheumatic polyarthritis; B. therapeutic results in ulcerative colitis; C. toxic manifestations in treatment with sulfanilamide preparations. *Acta Medica Scandinavica, 110*, 577–598.

Taggart, A. J., McDermott, B. J., & Roberts, S. D. (1992). The effect of age and acetylator phenotype on the pharmacokinetics of sulfasalazine in patients with rheumatoid arthritis. *Clinical Pharmacokinetics, 23*(4), 311–320.

van der Heijde, D. M., van Riel, P. L., Nuver-Zwart, I. H., & van de Putte, L. B. (1990). Sulphasalazine versus hydroxychloroquine in rheumatoid arthritis: 3-year follow-up. *Lancet, 335*(8688), 539.

Tetracyclines

Garry G. Graham[1,2] and Kevin D. Pile[3]
[1]Department of Pharmacology, School of Medical Sciences, University of New South Wales, Sydney, NSW, Australia
[2]Department of Clinical Pharmacology and Toxicology, St Vincent's Hospital, Sydney, NSW, Australia
[3]Campbelltown Hospital, School of Medicine, University of Western Sydney, Campbelltown, NSW, Australia

Synonyms

The tetracyclines are a group of antibiotics including minocycline, doxycycline, and an individual antibiotic which is termed tetracycline.

Definition

Tetracyclines are a group of tetracyclic antibiotics which also have anti-inflammatory and antirheumatic effects. Thus, the tetracyclines are members of a group of drugs known as disease-modifying antirheumatic drugs (DMARDs) or slow-acting antirheumatic drug (SAARDs).

Chemical Structures and Properties

The tetracyclines are organic bases with pKa values of about 8 and are, therefore, largely present as the cationic (ionized forms) in blood (Fig. 1). They are usually administered as their hydrochloride salts. Many chemical analogues of the tetracyclines have been synthesized. These are termed chemically modified tetracyclines (CMTs or COLs). The structure of one such CMT (CMT-3) is shown (Fig. 1). It is a neutral unionized compound.

Metabolism and Pharmacokinetics

The tetracyclines are well absorbed and have similar ranges of elimination half-lives but different modes of elimination (Table 1). The dosage of doxycycline should be reduced because of their significant renal excretion. CMT-3 has a long half-life (about 57 h) which allows dosage every day or every second day (Gu et al. 2011).

Pharmacological Activities

The tetracyclines are well-known antibiotics which produce their antimicrobial actions through inhibition of protein synthesis in bacteria.

Tetracyclines, Fig. 1 Structures of three tetracyclines and a chemically modified tetracycline (CMT-3 or COL-3) which have been tested for the treatment of inflammatory conditions. The ionized forms of the tetracyclines are shown. These are the major forms in plasma. CMT-3 is a neutral substance

Tetracyclines, Table 1 Pharmacokinetic properties and urinary excretion of tetracyclines

	Tetracycline	Doxycycline	Minocycline
Half-life (hours)	6–12	18–22	11–22
% excretion unchanged in urine	60	40	10

The development of sulfasalazine in the late 1930s was based on the concept of rheumatoid arthritis being an infectious disease (Svartz 1942). Evidence of organisms in the joints of some patients; differences in the bowel flora of rheumatoid arthritis patients, with or without erosions; and the ability of sulfasalazine to alter bowel flora remain intriguing links between rheumatoid arthritis and infection (O'Dell et al. 1999). The possible link between mycoplasma-like organisms and rheumatoid arthritis, together with the activity of tetracyclines in mycoplasma pneumonia, prompted the testing of tetracyclines in the treatment of this disease, with trials reported since the late 1960s.

The anti-inflammatory actions of the tetracyclines and their non-antibiotic CMTs have been examined in detail in a variety of experimental systems:

- The tetracyclines show anti-inflammatory effects in experimental animals with activities in vitro consistent with their anti-inflammatory effects. For example, doxycycline and minocycline decrease the nociception produced by the injection of formalin into the hind paw of the mouse, the carrageenan-induced edema of the rat paw, and leukocyte migration into the mice peritoneal cavity (Leite et al. 2011).
- The tetracyclines inhibit several enzymes, particularly matrix metalloproteinases (MMPs), and also have interlinked immunomodulatory effects. Inhibition of MMPs is considered to lead to decreased inflammation because these enzymes may cleave some cytokines, such as vascular endothelial growth factor (VEGF) and transforming growth factor beta (TGF-β), into their active forms (Monk et al. 2011). MMPs are considered to be important in tumor invasion and angiogenesis which is necessary to provide a blood supply for the growing tumor (Richards et al. 2011) and inflammatory

tissues, in wound healing, and in the development of rosacea. Inhibition of experimental angiogenesis in a mouse model has been reported (Korting and Schöllmann 2009; Monk et al. 2011). Overall, inhibition of MMPs has provided a theoretical basis for testing tetracyclines in inflammation, cancer, and rosacea.

- Tetracyclines often reduce prostaglandin synthesis in experimental systems although they are probably not direct inhibitors of cyclooxygenase I or II, but their inhibition of prostaglandin synthesis probably results from inhibition of secreted phospholipase A_2 or from immunological effects. The tetracyclines have been shown to have a variety of immunomodulatory activities including downregulation of inducible nitric oxide synthase, inflammatory cytokines such as tumor necrosis factor and interleukin 1β, and lymphocyte proliferation (Korting and Schöllmann 2009).
- Mostly, tetracyclines reduce formation of reactive oxygen species by neutrophils or directly scavenge free radicals in vitro (Korting and Schöllmann 2009; Monk et al. 2011) although increased formation under some experimental conditions has also been reported (Monk et al. 2011). The general reduced production of reactive oxygen species has suggested that the tetracyclines may be useful in the treatment of ischemia/reperfusion injury. Consistent with this prediction, minocycline reduces experimental stroke due to ischemia in rats even when administered 4 h after the induction of the ischemia/reperfusion (Yrjanheikki et al. 1999).
- Several actions of tetracyclines are consistent with their beneficial action in asthma. For example, in asthmatic patients, serum concentrations of immunoglobulin E decrease during treatment with minocycline without effects on immunoglobulins A, G, and M (Joks and Durkin 2011). In isolated rat mast cells, doxycycline reduces histamine release induced by compound 48/80. Furthermore, doxycycline reduces the production of interleukin-8 and TNF by a human mast cell line (Joks and Durkin 2011).
- The pharmacological effects of chemically modified tetracyclines have been examined extensively, particularly CMT-3 (Fig. 1). The CMTs do not have antibiotic activities, but several have retained their inhibitory actions on MMPs and other proteinases. CMT-3 inhibits MMPs and human leukocyte elastase by two mechanisms: firstly by direct inhibition of the enzymes and secondly by decreasing the breakdown of endogenous inhibitors of these enzymes (α_1-proteinase inhibitor and tissue inhibitors of MMPs) (Gu et al. 2011). A notable effect of CMT-3 in vivo is that it blocks an experimental respiratory distress syndrome in the pig (Roy et al. 2012).

Overall, the tetracyclines and the non-antibiotic CMTs have been claimed to possess multiple anti-inflammatory and immunomodulatory effects which may make them useful in several severe chronic diseases. However, many of the enzymatic and immunomodulatory effects have been observed in experimental studies conducted only in vitro, and further work on correlating in vitro and in vivo actions of tetracyclines is required.

Clinical Uses and Efficacy

The tetracyclines and their CMT relatives have been tested in several disease states.

Dermatological Diseases

The tetracyclines are useful in treating inflammatory lesions of rosacea and acne such as erythema, papules, pustules, and blepharitis but not sebaceous changes that do not appear to be inflammatory (Monk et al. 2011). Doses of doxycycline that are low (typically 20 mg twice daily) and considered insufficient for antibiotic activity appear active in the treatment of rosacea and acne. Several other uncommon dermatological diseases have also responded to tetracyclines. These include bullous disorders, chronic wounds (inability of wounds to heal), and granulomatous disorders (including sarcoidosis) (Monk et al. 2011).

Periodontal Disease

Low-dose doxycycline has been approved for the treatment of periodontitis (pyorrhea), a disease in which there is excessive inflammation due to bacteria adhering to the teeth (Monk et al. 2011).

Rheumatoid Arthritis

Currently, minocycline is the most widely used tetracycline in the treatment of rheumatoid arthritis although its treatment of this disease is still uncommon. The usefulness of the tetracyclines is indicated from a meta-analysis of ten studies, although only three studies were assessed to be high quality (Stone et al. 2003). The small subject numbers and short duration of treatment in most studies do not allow a conclusion on the effects of the tetracyclines on the radiographic progression the disease. However, two studies lasted for 48 weeks but showed no significant reduction in erosions or joint space narrowing (Stone et al. 2003). In summary, minocycline and doxycycline are effective in the treatment of rheumatoid arthritis with a better response seen in early onset seropositive disease. The cost of treatment and monitoring is low. However, alternative antirheumatic drugs are available, and it is unlikely that the tetracyclines will achieve widespread use in rheumatological diseases.

Osteoarthritis

Doxycycline slows the progression of osteoarthritis to a small degree, but there is no reduction in pain or disability (da Costa et al. 2012; Snijders et al. 2011). However, some patients may respond to a greater degree than others. A documented example is the slowed degeneration of the knee in obese women during treatment with doxycycline. This effect was not seen in the women with malalignment of the knee (Mazzuca et al. 2010).

Respiratory Diseases

Minocycline has been trialled in recent years for the treatment of asthma where it has allowed reduction in the dose of corticosteroids (Joks and Durkin 2011). As outlined above, a recent experimental study has shown that CMT-3 prevents respiratory distress syndrome in pigs after the induction of sepsis and ischemia/reperfusion injury (Roy et al. 2012).

Cancer

The activity of tetracyclines and CMT-3 has been tested in phase I and II clinical trials for its anticancer effects with limited positive results despite their inhibition of angiogenesis and MMPs which have suggested potential anticancer activity. No tetracycline or CMT has progressed to phase III trials (i.e., a large-scale clinical trials, generally where the potentially active drug is compared against placebo). However, one clinical trial has indicated beneficial actions of CMT-3 in Kaposi's sarcoma which is seen most commonly in patients with AIDs (Monk et al. 2011). The activity of CMT-3 has considerable activity as an antiangiogenic agent, and its activity in Kaposi's sarcoma may be due to the highly vascular nature of this sarcoma (Richards et al. 2011).

Adverse Effects

The use of tetracyclines has been limited because of concerns about both their adverse effects and development of resistant organisms resulting from their antibiotic activity. This is not a problem with the non-antibiotic CMTs which would appear to have lesser adverse effects. A variety of adverse effects have been associated with tetracycline treatment:

- Continued therapy for more than 2 years is associated with hyperpigmentation of skin and nails in 10–20 % of patients and may take a year to resolve.
- Up to age of about 8 years, the tetracyclines may cause discoloration of teeth, and hypoplasia of dental enamel and their use should be avoided.
- Tetracyclines cause discoloration of babies' teeth in the second half of pregnancy, and their use should be avoided at this time.
- Tetracyclines may depress prothrombin activity, and the dosage of warfarin should be decreased accordingly.
- Iron salts are chelated by tetracyclines leading to inactivation of the tetracyclines.
- Capsules of tetracyclines have caused ulceration of the esophagus due to their retention in

the esophagus. They should be taken with milk or food to prevent this.
- The tetracyclines may cause supersensitivity to ultraviolet light (excessive sunburn). This adverse effect is also seen with some CMTs.
- Minocycline is associated with autoimmune reactions, a lupus-like syndrome and hepatitis, which are not shown by other tetracyclines. This tetracycline is metabolized by hepatic cytochrome P450 systems and the neutrophil enzyme, myeloperoxidase (which produces hypochlorous acid), to reactive products which may cause adverse effects specific to minocycline (Mannargudi et al. 2009). Appropriate monitoring is therefore indicated in patients early in their polyarthritis when diagnostic uncertainty may still exist.

Drug Interactions

Several interactions between tetracyclines and other drugs have been reported (Hansten and Horn 2012). These include the following:

- Tetracyclines may be used with corticosteroids and nonsteroidal anti-inflammatory drugs (NSAIDs).
- Doxycycline reduces the renal clearance of methotrexate by about 79 %. A similar effect on methotrexate is assumed. The significance of this interaction is unclear but requires further analysis.
- The absorption of tetracyclines is reduced by complexation with iron salts and by both calcium and bismuth antacids. The combinations should be avoided. Alternatively, a 3 h delay between the administration of the tetracycline and the interacting compounds should decrease the extent of the interactions.
- The absorption of tetracycline, but not doxycycline, is reduced by zinc sulfate supplements. Again, a significant delay between tetracycline and zinc administration should decrease the extent of the interaction.
- The efficacy of oral anticoagulants (warfarin) may be increased and oral contraceptives may become ineffective, both interactions occurring because of changes in the gut flora.

Cross-References

▶ Antibiotics as Antiinflammatory Drugs
▶ Asthma
▶ Corticosteroids
▶ Non-steroidal Anti-inflammatory Drugs: Overview
▶ Osteoarthritis
▶ Prostanoids
▶ Rheumatoid Arthritis
▶ Tumor Necrosis Factor Alpha (TNFalpha)

References

da Costa, B. R., Nüesch, E., Reichenbach, S., Jüni, P., & Rutjes, A. W. S. (2012). Doxycycline for osteoarthritis of the knee or hip. *Cochrane Database of Systematic Reviews, 11*, CD007323.

Gu, Y., Lee, H. M., Simon, S. R., & Golub, L. M. (2011). Chemically modified tetracycline-3 (CMT-3): A novel inhibitor of the serine proteinase, elastase. *Pharmacological Research, 64*(6), 595–601.

Hansten, P. D., & Horn, J. T. (2012). *Drug interactions*. St Loouis: Wolters Kluwer Health.

Joks, R., & Durkin, H. G. (2011). Non-antibiotic properties of tetracyclines as anti-allergy and asthma drugs. *Pharmacological Research, 64*(6), 602–609.

Korting, H. C., & Schöllmann, C. (2009). Tetracycline actions relevant to rosacea treatment. *Skin Pharmacology and Physiology, 22*(6), 287–294.

Leite, L. M., Carvalho, A. G. G., Ferreira, P. L., Pessoa, I. X., Goncalves, D. O., Lopes, A. D. A., et al. (2011). Antiinflammatory properties of doxycycline and minocycline in experimental models: An in vivo and in vitro comparative study. *Inflammopharmacology, 19*(2), 99–110.

Mannargudi, B., McNally, D., Reynolds, W., & Uetrecht, J. (2009). Bioactivation of minocycline to reactive intermediates by myeloperoxidase, horse peroxidase, and hepatic microsomes: Implications for minocycline-induced lupus and hepatitis. *Drug Metabolism and Disposition, 37*(9), 1806–1818.

Mazzuca, S. A., Brandt, K. D., Chakr, R., & Lane, K. A. (2010). Varus malalignment negates the structure-modifying benefits of doxycycline in obese women with knee osteoarthritis. *Osteoarthritis and Cartilage, 18*(18), 1008–1011.

Monk, E., Shalita, A., & Siegel, D. M. (2011). Clinical applications of non-antimicrobial tetracyclines in dermatology. *Pharmacological Research, 63*(2), 130–145.

O'Dell, J. R., Paulsen, G., Haire, C. E., Blakely, K., Palmer, W., Wees, S., et al. (1999). Treatment of early seropositive rheumatoid arthritis with minocycline: Four-year followup of a double-blind, placebo-controlled trial. *Arthritis and Rheumatism, 42*(8), 1691–1695.

Richards, C., Pantanowitz, L., & Dezubea, B. J. (2011). Antimicrobial and non-antimicrobial tetracyclines in human cancer trials. *Pharmacological Research, 63*(2), 151–156.

Roy, S. K., Kubiak, B. D., Albert, S. P., Vieau, C. J., Gatto, L., Golub, L., et al. (2012). Chemically modified tetracycline 3 prevents acute respiratory distress syndrome in a porcine model of sepsis + ischemia/reperfusion-induced lung injury. *Shock, 37*(4), 424–432.

Snijders, G. F., van den Ende, C. H., van Riel, P. L., van den Hoogen, F. H., & den Broeder, A. A. (2011). The effects of doxycycline on reducing symptoms in knee osteoarthritis: Results from a triple-blinded randomised controlled trial. *Annals of the Rheumatic Diseases, 70*(7), 1191–1196.

Stone, M., Fortin, P. R., Pacheco-Tena, C., & Inman, R. D. (2003). Should tetracycline treatment be used more extensively for rheumatoid arthritis? Metaanalysis demonstrates clinical benefit with reduction in disease activity. *Journal of Rheumatology, 30*(10), 2112–2122.

Svartz, N. (1942). Salazopyrin, a new sulfanilamide preparation: A. therapeutic results in rheumatic polyarthritis; B. therapeutic results in ulcerative colitis; C. toxic manifestations in treatment with sulfanilamide preparations. *Acta Medica Scandinavica, 110*(6), 577–598.

Yrjanheikki, J., Tikka, T., Keinanen, R., Goldsteins, G., Chan, P. H., & Koistinaho, J. (1999). A tetracycline derivative, minocycline, reduces inflammation and protects against focal cerebral ischemia with a wide therapeutic window. *Proceedings of the National Academy of Sciences of the United States of America, 96*(23), 13496–13500.

TGF beta Superfamily Cytokine MIC-1/GDF15 in Health and Inflammatory Diseases

Samuel N. Breit and David A. Brown
St Vincent's Centre for Applied Medical Research, St Vincent's Hospital and University of New South Wales, Sydney, NSW, Australia

Synonyms

GDF15; Macrophage inhibitory cytokine -1; MIC-1; MIC-1/GDF15; NAG-1; PL74; PLAB

Definition

MIC-1/GDF15 is cytokine belonging to the TGF-b superfamily.

Biosynthesis and Release

Macrophage inhibitory cytokine-1, now officially known as growth differentiation factor 15 (MIC-1/GDF15), is a TGF-b superfamily cytokine, which was first cloned in the 1990s based on its increased expression with macrophage activation (Bootcov et al. 1997). It is the product of a two-exon gene on human chromosome 19 (Bootcov et al. 1997). There are several SNPs that are predominantly outside its coding regions, which are associated with its altered regulation (Wang et al. 2010). However, it has one important coding region polymorphism, which results in alteration of a histidine (H, aromatic, and neutral), to an aspartic acid (D, acidic) residue at position 6 of the mature protein (H6D polymorphism) (Bootcov et al. 1997; Brown et al. 2002a). The properties of these two amino acids are very different, suggesting that they may code for differences in behavior of the protein. Supporting this are some genetic association studies that link the H6D polymorphism to cancer (Brown et al. 2003; Lindmark et al. 2004; Hayes et al. 2006), rheumatoid arthritis (Brown et al. 2007; Arlestig and Rantapaa-Dahlqvist 2012), and thromboembolic disease (Arlestig and Rantapaa-Dahlqvist 2012). The MIC-1/GDF15 gene transcript codes for a 308 amino acid peptide which dimerizes and is then usually processed and secreted as a mature dimer of about 25 kDa, comprising two 112 amino acid peptides linked by a single interchain disulphide bond (Bootcov et al. 1997; Fairlie et al. 1999). Under some circumstances, MIC-1/GDF15 is secreted as an unprocessed full-length dimer of about 60 kDa, which binds to extracellular matrix, by a heparan sulphate binding region in the propeptide (Bauskin et al. 2005, 2010). This tissue-localized protein may serve as a local reservoir, which on processing slowly releases freely diffusible, bioactive mature protein.

Unusually for a cytokine, once secreted, MIC-1/GDF15 appears in the blood of all individuals in substantial amounts, with a normal range of about 200–1,150 pg/ml (Brown et al. 2003). As discussed later, its serum levels can be dramatically increased in disease states. Although there is little tissue expression of MIC-1/GDF15 under normal physiological states other than pregnancy, it seems likely that the majority of circulating MIC-1/GDF15 comes from the liver.

To exert a biological effect, MIC-1/GDF15 must activate its cognate receptor complex, which to this point has not been definitively identified. Its receptor chains almost certainly belong to the highly conserved heterodimeric TGF-b receptor (TBR) superfamily. There is some evidence for involvement of TBRII (Johnen et al. 2007; de Jager et al. 2011), but as yet, there is no genetic or direct biochemical data to confirm this. Some functions of MIC-1/GDF15 might also be independent of receptor activation. For example, there is evidence to suggest it may bind to and inhibit connective tissue growth factor protein 2 (CCN2). This may result in decreased CCN2-induced activation of focal adhesion kinase, decreased $\alpha V\beta 3$ integrin clustering, and reduced in vitro endothelial tube formation (Whitson et al. 2012). It has also been suggested that impaired clearance of MIC-1/GDF15 by liver sinusoid endothelial cells may contribute to glomerular and hepatic fibrosis seen in mice bearing deletions of the stabilin 1 and 2 receptors (Schledzewski et al. 2011).

Biological Activities

Introduction

As discussed earlier, serum levels of MIC-1/GDF15 are elevated in a broad range of inflammatory and other diseases. An important question is whether MIC-1/GDF15 participates in disease or is just "the canary in the coal mine." There are four possible interpretations: (i) MIC-1/GDF15 may be produced as part of the disease process, but it may not be involved in its pathogenesis; (ii) it may have a harmful effect because it mediates aspects of disease processes; (iii) it may be produced in a frustrated attempt to bring disease processes under control and thus have an overall beneficial effect; and (iv) MIC-1/GDF15 might have both beneficial and detrimental effects on disease processes, in a context-dependent manner, like TGF-b itself. The current evidence, discussed below, favors one of the last two options, but with dominantly a disease-ameliorating role.

Atherosclerosis

Association studies, discussed above, very clearly link MIC-1/GDF15 serum levels to cardiovascular disease and its progression. It is expressed in atherosclerotic arteries or plaques from monkeys (Eyster et al. 2011), mice (de Jager et al. 2011), and man (Schlittenhardt et al. 2004). One mechanism by which MIC-1/GDF15 might modify cardiovascular disease is to influence the development of atherosclerosis. In the ApoE$^{-/-}$ model of atherosclerosis, mice concurrently overexpressing MIC-1/GDF15 in myelomonocytic cells had smaller atherosclerotic lesions after 6 months on a high-fat diet (Johnen et al. 2012). Another study utilized LDL receptor knockout mice which were transplanted with bone marrow from either MIC-1/GDF15 gene knockout or wild-type mice (Preusch et al. 2013). While they could not identify differences in lesion size, MIC-1/GDF15 gene knockout mice had features of plaque destabilization including increased macrophage infiltration, increased staining for ICAM1, and thinning of the fibrous cap, suggesting a beneficial effect for MIC-1/GDF15. One mechanism by which MIC-1/GDF15 might mediate these effects is by increasing macrophage cholesterol efflux by upregulation of ATP-binding cassette transporter A1, which could reduce plaque formation (Wu et al. 2014).

While the above studies clearly identified an overall protective role for this cytokine, this is contradicted by two other studies. Mice with germline deletion of both the LDL receptor and MIC-1/GDF15 had increased macrophage accumulation but, counterintuitively, decreased plaque necrotic cores (de Jager et al. 2011). Further, ApoE$^{-/-}$ mice that also had a germline deletion of MIC-1/GDF15, due to a knockin of LacZ into its locus, displayed decreased luminal stenosis and vessel wall inflammation (Bonaterra et al. 2012). However, paradoxically, the

MIC-1/GDF15 gene deleted mice had a greater vessel wall cell density and more cells in this location expressing CD68, MIF, COX2, and Mo-Ma2, but less IL6 or markers of apoptosis (Bonaterra et al. 2012). These studies vary substantially in methodology including the type of disease model, the assessment of disease activity/extent, and the type of transgenic modification (MIC-1/GDF15 deletion versus overexpression). Further studies will be required to try to directly resolve the role played by MIC-1/GDF15 in the pathogenesis of atherosclerosis.

Ischemia Reperfusion-Injury and Infarction

Myocardial infarction is a common end result of atherosclerosis, and MIC-1/GDF15 plays a protective role in this process. It is expressed at the infarct site, limiting cell damage and cell death at the infarct border (Kempf et al. 2006) and decreasing neutrophil migration into the inflamed tissue by interfering with chemokine signaling and integrin activation (Kempf et al. 2011). Further, it may promote neovasculogenesis (Song et al. 2012). In ischemia-reperfusion injury, brain MIC-1/GDF15 expression is induced, but no effect on cerebral infarct size could be demonstrated in studies comparing wild-type and MIC-1/GDF15 gene knockout mice (Schindowski et al. 2011). This of course does not exclude effects on neuronal death or recovery and functional outcomes in stroke. Indeed, several lines of evidence suggest that MIC-1/GDF15 is neurotrophic and neuroprotective (Strelau et al. 2000, 2009; Subramaniam et al. 2003).

Hemostasis and Thrombosis

Platelets are a key component of blood that play an important role in vascular biology, hemostasis/thrombosis, and inflammation. Recent evidence indicates that MIC-1/GDF15 inhibits platelet aggregation and other related functions by inhibiting activation of integrin $\alpha_{11b}\beta_3$ (GPIIb/IIIa), which is responsible for platelet adhesion, spreading, and aggregation. As a consequence, MIC-1/GDF15 knockout mice have a reduced bleeding time and an increased capacity to develop venous thromboemboli or arterial thrombosis. This is also supported by human studies indicating that MIC-1/GDF15 SNPs are associated with altered risks of these events (Arlestig and Rantapaa-Dahlqvist 2012). Vascular thrombosis plays a key role in the development of heart attacks and strokes, and these findings would be consistent with a protective role for MIC-1/GDF15 in these diseases (Rossaint et al. 2013).

Rheumatoid Arthritis

Rheumatoid arthritis is the archetypal autoinflammatory disease in which MIC-1/GDF15 serum levels relate to disease activity and predict outcome. Its direct involvement in the pathogenesis of RA is also suggested by studies in animal models. Overexpressing MIC-1^{fms} mice have markedly reduced incidence and severity of arthritis in the collagen arthritis model, which depends on both innate and adaptive immunity. Increased MIC-1/GDF15 expression also affords protection from arthritis in the KBxN serum transfer arthritis model. As this model is largely independent of adaptive immunity, a direct action of MIC-1/GDF15 on the innate immune response is implicated in mediating this protection.

Vascular Biology

Angiogenesis plays an important role in many biological processes including inflammation, but as in atherosclerosis, there are contradictory views as to the role played by MIC-1/GDF15 in endothelial cell function. MIC-1/GDF15 increases in vitro human umbilical vein endothelial cell (HUVEC) proliferation by enhancing AP-1 and E2F-dependent expression of G(1) cyclin (Jin et al. 2012) and stimulates directional vessel development in the chick chorioallantoic membrane and matrigel plug assays (Huh et al. 2010). Pretreating hypoxic UVEC with MIC-1/GDF15 promoted their in vitro angiogenesis by enhancing hypoxia-inducible factor-1a (HIF-1a) expression and nuclear translocation as well as increasing the expression of vascular endothelial growth factor (VEGF) (Song et al. 2012). Complicating interpretation of these data are two publications that express a contrary view. In these, MIC-1/GDF15 inhibited endothelial cell growth, migration and invasion in vitro, and angiogenesis in vivo (Ferrari et al. 2005). It has also been suggested that it inhibits in vitro HUVEC tube formation by binding to and inhibiting the angiogenic actions of connective tissue growth factor 2 (Whitson et al. 2012).

Diabetes

Epidemiological studies link MIC-1/GDF15 to diabetes and insulin resistance. MIC-1/GDF15 is expressed by adipocytes and strong data identifies it as an anorexigenic protein. This action, mediated by its direct effect on appetite regulatory circuits in the CNS, leads to reduced body weight and visceral fat composition (Johnen et al. 2007; Tsai et al. 2013). While these effects suggest a role in diabetes, immune and inflammatory mechanisms also play a part in both type I and type II diabetes development and might be modulated by MIC-1/GDF15. Mice overexpressing MIC-1/GDF15 are reported to have reduced macrophage infiltration of white adipose tissue and may also have reduced macrophage inflammasome activation (Wang et al. 2013). However, this situation is complicated by reduction in leptin serum levels due to reduced total fat that might also alter inflammatory indices (Kim et al. 2013).

Despite treatment, renal damage is a common and important long-term complication of diabetes and evidence indicates that MIC-1/GDF15 plays a role in limiting this process. In both induced and spontaneous diabetes in mice, germline MIC-1/GDF15 gene deletions are associated with worse renal disease due to effects on the tubular and interstitial compartments (Mazagova et al. 2013). Further, in type II diabetic patients with associated proximal tubular damage, MIC-1/GDF15 urinary levels are increased (Simonson et al. 2012). This and the finding that MIC-1/GDF15 directly participates in the early proliferation of tubular cells (Duong Van Huyen et al. 2008) suggest a role for it in injury and repair of the kidney.

CNS Injury and Repair

MIC-1/GDF15 is constitutively expressed by the CNS choroid plexus and periventricular ependymal cells (Schober et al. 2001). It is also found in the CSF and may enter the CNS, from the systemic circulation via regions of the brainstem and hypothalamus with a semipermeable blood brain barrier (Johnen et al. 2007). As in other locations, nervous system injury is associated with increased local expression of MIC-1/GDF15 (Schober et al. 2001; Schindowski et al. 2011; Mensching et al. 2012; Charalambous et al. 2013), and it may also enter the brain from the vascular compartment due to injury-associated disruption of the blood brain barrier. On the one hand, MIC-1/GDF-15-deficient mice display progressive postnatal loss of spinal, facial, and trigeminal motoneurons (Strelau et al. 2009), and on the other, MIC-1/GDF15 speeds the repair of injured rodent sciatic nerves (Mensching et al. 2012). Additionally, direct intracerebral injection of MIC-1/GDF15 protected rats from 6-hydroxydopamine (6-OHDA)-induced Parkinson's disease like injury (Strelau et al. 2000). However, while MIC-1/GDF15 was upregulated in rodent retinas following crush injury to the optic nerve, and upregulated the expression of galanin, its gene deletion did not modify the course or severity of retinal ganglion cell death (Charalambous et al. 2013).

Others

While MIC-1/GDF15 expression is consistently induced at sites of injury and inflammation, the consequences of this expression are not always clear. It is expressed in the lungs of patients with scleroderma interstitial lung disease and the bleomycin-induced lung injury model. However, its gene deletion does not modify lung fibrosis, the ultimate outcome in the bleomycin model (Lambrecht et al. 2013). MIC-1/GDF15 expression is also strongly induced in the liver injury following bile duct ligation (Koniaris et al. 2003) or administration of toxic chemicals (Hsiao et al. 2000; Zimmers et al. 2006). However, MIC-1/GDF15 gene deletion does not noticeably modify the extent of carbon tetrachloride-induced liver damage. These data again raise the question of whether MIC-1/GDF15 directly participates in pathology or merely marks its presence.

Pathophysiological Relevance

Introduction

Unlike most cytokines, substantial amounts of MIC-1/GDF15 circulate in the blood of all normal individuals with a range of about 200–1,150 pg/ml (Brown et al. 2003) and a mean of about 450 pg/ml. Serum levels rise with increasing age, various anthropomorphic measurements, environmental agents, and smoking. Serum levels can increase with extreme exercise (Tchou et al. 2009), rise substantially in early pregnancy (Moore et al. 2000), and are modestly increased by use of NSAIDs (Brown et al. 2012b). MIC-1/GDF15 serum levels can also rise further in a wide-range disease states especially those involving cancer and inflammation, and its measurement in disease is being developed for diagnosis and management of these conditions (reviewed in (Bauskin et al. 2006; Breit et al. 2011; Brown et al. 2012a). Despite this, MIC-1/GDF15 is not a classic acute phase reactant, its serum levels are not modified by corticosteroids (Brown et al. 2007), and its expression is dominantly regulated independently of the NFkB pathway. Its usefulness as an independent diagnostic marker may derive from p53 and egr-1 transcriptional pathway regulation of MIC-1/GDF15 expression, which is not commonly sampled by other biomarkers.

Cancer

Clinical studies correlating serum MIC-1/GDF15 levels with diseases or their outcome have provided important clues to its role in inflammation. MIC-1/GDF15 is overexpressed in most common cancers leading to increased serum levels in proportion to stage and extent of disease (Welsh et al. 2003), (Brown et al. 2003; Koopmann et al. 2004). In advanced cancers (Johnen et al. 2007), chronic renal (Johnen et al. 2007; Breit et al. 2012), cardiac failure (Kempf et al. 2007b), and thalassemia (Tanno et al. 2007), these serum levels can rise dramatically. Up to 100-fold elevations can be seen in advanced cancer, which can lead to anorexia mediated by actions of MIC-1/GDF15 on brain feeding centers. This action is thought to be a major cause of the cancer anorexia/cachexia syndrome. A more detailed discussion of the role of MIC-1/GDF15 in cancer can be found in the following publications (Bauskin et al. 2006; Breit et al. 2011; Brown et al. 2012a).

Cardiovascular Diseases

Clinical associations of serum MIC-1/GDF15 levels with chronic inflammatory processes have been studied extensively in the area of cardiovascular disease. Involvement in this disease process was first suggested in a nested case control study, which indicated that increased serum MIC-1/GDF15 levels were an independent risk factor for the development of acute cardiovascular events including myocardial infarction and strokes (Brown et al. 2002b). This finding was followed by a series of studies that demonstrated that serum MIC-1/GDF15 levels were useful in predicting progression of cardiovascular disease and selection of patients for interventions (Wollert et al. 2007a, b; Eggers et al. 2008; Schopfer et al. 2014). In particular, MIC-1/GDF15 is significantly and independently related to the severity of cardiac failure (Kempf et al. 2007a; Wang et al. 2012) and appears to have differential expression profiles dependent on the preservation or reduction of the cardiac ejection fraction (Santhanakrishnan et al. 2012; Izumiya et al. 2014). These properties may be due to various factors, but prime among them appears to be the upregulation of MIC-1/GDF15 expression in the context of pressure loading (Xu et al. 2006).

MIC-1 levels increased in primary pulmonary hypertension (Nickel et al. 2008) and stroke. Indeed, the initial serum levels of MIC-1/GDF15, drawn as soon as possible after hospital presentation, are related to stroke severity and predicted outcome at 90 days (Groschel et al. 2012). Another consequence of cardiovascular disease in the elderly is cognitive decline, which also is caused by other inflammatory conditions such as Alzheimer's disease. Rising MIC-1/GDF15 serum levels are associated with reduced cognitive performance and predict cognitive decline (Fuchs et al. 2013). In a nondemented elderly cohort, a serum level of MIC-1/GDF15 exceeding about 2,750 pg/ml is associated with a 20 % chance of decline from normal to mild cognitive impairment (Fuchs et al. 2013).

Diabetes

Diabetes is an important predisposing factor for cardiovascular diseases, and epidemiological evidence suggests that MIC-1/GDF15 may be involved in its development and/or complication. Glucose may directly stimulate MIC-1/GDF15 production by islet cells (Shalev et al. 2002) or vascular endothelium, (Li et al. 2013) and its serum levels independently predict the presence of insulin resistance (Kempf et al. 2012) and progress to type II diabetes (Dostalova et al. 2009; Carstensen et al. 2010; Herder et al. 2013). Further, MIC-1/GDF15 serum levels independently predict worsening of microalbuminuria (Hellemons et al. 2012), diabetic nephropathy (Lajer et al. 2010), and the outcome of chronic renal failure patients on dialysis (Breit et al. 2012), whose demise are often associated with acute vascular events.

Autoinflammatory Diseases

MIC-1/GDF15 also plays a part in the autoinflammatory diseases RA and scleroderma. Its serum levels correlate with disease extent and severity in scleroderma (Meadows et al. 2011; Yanaba et al. 2012; Lambrecht et al. 2014) and are also elevated in RA where they independently predict disease activity. Further, serum MIC-1/GDF15 levels were highest in a subgroup of patients whose RA was severe enough to require autologous stem cell transplantation therapy. In this subgroup, serum levels dropped dramatically 3 months after successful transplantation (Brown et al. 2007). Lastly, the presence of the D allelic variant of MIC-1/GDF15 was associated with earlier erosive disease and severe treatment-resistant RA (Brown et al. 2007).

All-Cause Mortality

Serum MIC-1/GDF15 levels independently predict outcome of a wide range of serious disease, including cancer and cardiovascular diseases, with higher levels generally being associated with worse outcomes. Examination of serum MIC-1/GDF15 levels at the wider population level has extended these findings and demonstrated that serum levels are an independent marker of all-cause mortality in a normal aging population (Wiklund et al. 2010; Daniels et al. 2011; Eggers et al. 2013). Over a 14-year follow-up period of a normal all-male cohort, serum levels of MIC-1/GDF15 predicted all-cause mortality with adjusted odds ratio of 3.4 (Wiklund et al. 2010). This finding was validated in an elderly mixed sex twin cohort, which was also used to demonstrate that this effect was independent of genetic background (Wiklund et al. 2010). Further validation came from independent studies of elderly community-dwelling populations where serum MIC-1/GDF15 predicted all-cause mortality with similar odds ratios (Daniels et al. 2011; Eggers et al. 2013). Thus, serum levels of MIC-1/GDF15 broadly reflect an unfavorable outcome, independently of other markers, perhaps because it is a global reflection of the activity of p53 and egr-1 transcription factor pathways (Wiklund et al. 2010), in a conceptually similar manner to CRP levels, which broadly reflect activity of NFkB pathways.

Conclusion

The data outlined above suggest a prominent role for MIC-1/GDF15 in inflammation. A wealth of epidemiological data indicates that MIC-1/GDF15 at the very least marks, if not participates, in basic biological mechanisms mediating inflammatory diseases. Many unanswered questions remain regarding its actions in the multitude of disease pathologies with which it has been associated. The major gap in knowledge that prevents us from answering many of these questions is the identity of the MIC-1/GDF15 receptor complex.

References

Arlestig, L., & Rantapaa-Dahlqvist, S. (2012). Polymorphisms of the genes encoding CD40 and growth differentiation factor 15 and in the 9p21.3 region in patients with rheumatoid arthritis and cardiovascular disease. *Journal of Rheumatology, 39*, 939–945.

Bauskin, A. R., Brown, D. A., Junankar, S., Rasiah, K. K., Eggleton, S., Hunter, M., et al. (2005). The propeptide mediates formation of stromal stores of PROMIC-1: Role in determining prostate cancer outcome. *Cancer Research, 65*, 2330–2336.

Bauskin, A. R., Brown, D. A., Kuffner, T., Johnen, H., Luo, X. W., Hunter, M., et al. (2006). Role of macrophage inhibitory cytokine-1 in tumorigenesis and diagnosis of cancer. *Cancer Research, 66*, 4983–4986.

Bauskin, A. R., Jiang, L., Luo, X. W., Wu, L., Brown, D. A., & Breit, S. N. (2010). The TGF-beta superfamily cytokine MIC-1/GDF15: Secretory mechanisms facilitate creation of latent stromal stores. *Journal of Interferon and Cytokine Research, 30*, 389–397.

Bonaterra, G. A., Zugel, S., Thogersen, J., Walter, S. A., Haberkorn, U., Strelau, J., et al. (2012). Growth differentiation factor-15 deficiency inhibits atherosclerosis progression by regulating interleukin-6-dependent inflammatory response to vascular injury. *Journal of the American Heart Association, 1*, e002550.

Bootcov, M. R., Bauskin, A. R., Valenzuela, S. M., Moore, A. G., Bansal, M., He, X. Y., et al. (1997). MIC-1, a novel macrophage inhibitory cytokine, is a divergent member of the TGF-beta superfamily. *Proceedings of the National Academy of Sciences of the United States of America, 94*, 11514–11519.

Breit, S. N., Johnen, H., Cook, A. D., Tsai, V. W., Mohammad, M. G., Kuffner, T., et al. (2011). The TGF-beta superfamily cytokine, MIC-1/GDF15: A pleotrophic cytokine with roles in inflammation, cancer and metabolism. *Growth Factors, 29*, 187–195.

Breit, S. N., Carrero, J. J., Tsai, V. W., Yagoutifam, N., Luo, W., Kuffner, T., et al. (2012). Macrophage inhibitory cytokine-1 (MIC-1/GDF15) and mortality in end-stage renal disease. *Nephrology, Dialysis, Transplantation, 27*, 70–75.

Brown, D. A., Bauskin, A. R., Fairlie, W. D., Smith, M. D., Liu, T., Xu, N., et al. (2002a). Antibody-based approach to high-volume genotyping for MIC-1 polymorphism. *Biotechniques, 33*(1), 118–20. 122, 124 passim.

Brown, D. A., Breit, S. N., Buring, J., Fairlie, W. D., Bauskin, A. R., Liu, T., et al. (2002b). Concentration in plasma of macrophage inhibitory cytokine-1 and risk of cardiovascular events in women: A nested case-control study. *Lancet, 359*, 2159–2163.

Brown, D. A., Ward, R. L., Buckhaults, P., Liu, T., Romans, K. E., Hawkins, N. J., et al. (2003). MIC-1 serum level and genotype: Associations with progress and prognosis of colorectal carcinoma. *Clinical Cancer Research, 9*, 2642–2650.

Brown, D. A., Moore, J., Johnen, H., Smeets, T. J., Bauskin, A. R., Kuffner, T., et al. (2007). Serum macrophage inhibitory cytokine 1 in rheumatoid arthritis: A potential marker of erosive joint destruction. *Arthritis and Rheumatism, 56*, 753–764.

Brown, D. A., Bauskin, A. R., & Breit, S. N. (2012a). *MIC-1. Encyclopedia of cancer.*

Brown, D. A., Hance, K. W., Rogers, C. J., Sansbury, L. B., Albert, P. S., Murphy, G., et al. (2012b). Serum macrophage inhibitory cytokine-1 (MIC-1/GDF15): A potential screening tool for the prevention of colon cancer? *Cancer Epidemiology, Biomarkers and Prevention, 21*, 337–346.

Carstensen, M., Herder, C., Brunner, E. J., Strassburger, K., Tabak, A. G., Roden, M., et al. (2010). Macrophage inhibitory cytokine-1 is increased in individuals before type 2 diabetes diagnosis but is not an independent predictor of type 2 diabetes: The Whitehall II study. *European Journal of Endocrinology, 162*, 913–917.

Charalambous, P., Wang, X., Thanos, S., Schober, A., & Unsicker, K. (2013). Regulation and effects of GDF-15 in the retina following optic nerve crush. *Cell and Tissue Research, 353*, 1–8.

Daniels, L. B., Clopton, P., Laughlin, G. A., Maisel, A. S., & Barrett-Connor, E. (2011). Growth-differentiation factor-15 is a robust, independent predictor of 11-year mortality risk in community-dwelling older adults: The Rancho Bernardo study. *Circulation, 123*, 2101–2110.

de Jager, S. C., Bermudez, B., Bot, I., Koenen, R. R., Bot, M., Kavelaars, A., et al. (2011). Growth differentiation factor 15 deficiency protects against atherosclerosis by attenuating CCR2-mediated macrophage chemotaxis. *Journal of Experimental Medicine, 208*, 217–225.

Dostalova, I., Roubicek, T., Bartlova, M., Mraz, M., Lacinova, Z., Haluzikova, D., et al. (2009). Increased serum concentrations of macrophage inhibitory cytokine-1 in patients with obesity and type 2 diabetes mellitus: The influence of very low calorie diet. *European Journal of Endocrinology, 161*, 397–404.

Duong Van Huyen, J. P., Cheval, L., Bloch-Faure, M., Belair, M. F., Heudes, D., Bruneval, P., et al. (2008). GDF15 triggers homeostatic proliferation of acid-secreting collecting duct cells. *Journal of the American Society of Nephrology, 19*, 1965–1974.

Eggers, K. M., Kempf, T., Allhoff, T., Lindahl, B., Wallentin, L., & Wollert, K. C. (2008). Growth-differentiation factor-15 for early risk stratification in patients with acute chest pain. *European Heart Journal, 29*, 2327–2335.

Eggers, K. M., Kempf, T., Wallentin, L., Wollert, K. C., & Lind, L. (2013). Change in growth differentiation factor 15 concentrations over time independently predicts mortality in community-dwelling elderly individuals. *Clinical Chemistry, 59*, 1091–1098.

Eyster, K. M., Appt, S. E., Mark-Kappeler, C. J., Chalpe, A., Register, T. C., & Clarkson, T. B. (2011). Gene expression signatures differ with extent of atherosclerosis in monkey iliac artery. *Menopause, 18*, 1087–1095.

Fairlie, W. D., Moore, A. G., Bauskin, A. R., Russell, P. K., Zhang, H. P., & Breit, S. N. (1999). MIC-1 is a novel TGF-beta superfamily cytokine associated with macrophage activation. *Journal of Leukocyte Biology, 65*(1), 2–5.

Ferrari, N., Pfeffer, U., Dell'eva, R., Ambrosini, C., Noonan, D. M., & Albini, A. (2005). The transforming growth factor-{beta} family members bone morphogenetic protein-2 and macrophage inhibitory cytokine-1 as mediators of the antiangiogenic activity of N-(4-hydroxyphenyl)retinamide. *Clinical Cancer Research, 11*(12), 4610–4619.

Fuchs, T., Trollor, J. N., Crawford, J., Brown, D., Baune, B. T., Samaras, K., et al. (2013). Macrophage inhibitory cytokine-1 is associated with cognitive impairment and predicts cognitive decline – the Sydney Memory and Ageing Study. *Aging Cell, 12*, 882–889.

Groschel, K., Schnaudigel, S., Edelmann, F., Niehaus, C. F., Weber-Kruger, M., Haase, B., et al. (2012). Growth-differentiation factor-15 and functional outcome after acute ischemic stroke. *Journal of Neurology, 259*, 1574–1579.

Hayes, V. M., Severi, G., Southey, M. C., Padilla, E. J., English, D. R., Hopper, J. L., et al. (2006). Macrophage inhibitory cytokine-1 H6D polymorphism, prostate cancer risk, and survival. *Cancer Epidemiology, Biomarkers and Prevention, 15*, 1223–1225.

Hellemons, M. E., Mazagova, M., Gansevoort, R. T., Henning, R. H., de Zeeuw, D., Bakker, S. J., et al. (2012). Growth-differentiation factor 15 predicts worsening of albuminuria in patients with type 2 diabetes. *Diabetes Care, 35*, 2340–2346.

Herder, C., Carstensen, M., & Ouwens, D. M. (2013). Anti-inflammatory cytokines and risk of type 2 diabetes. *Diabetes, Obesity & Metabolism, 15*(Suppl 3), 39–50.

Hsiao, E. C., Koniaris, L. G., Zimmers-Koniaris, T., Sebald, S. M., Huynh, T. V., & Lee, S. J. (2000). Characterization of growth-differentiation factor 15, a transforming growth factor beta superfamily member induced following liver injury. *Molecular and Cellular Biology, 20*(10), 3742–3751.

Huh, S. J., Chung, C. Y., Sharma, A., & Robertson, G. P. (2010). Macrophage inhibitory cytokine-1 regulates melanoma vascular development. *American Journal of Pathology, 176*, 2948–2957.

Izumiya, Y., Hanatani, S., Kimura, Y., Takashio, S., Yamamoto, E., Kusaka, H., et al. (2014). Growth differentiation factor-15 is a useful prognostic marker in patients with heart failure with preserved ejection fraction. *Canadian Journal of Cardiology, 30*, 338–344.

Jin, Y. J., Lee, J. H., Kim, Y. M., Oh, G. T., & Lee, H. (2012). Macrophage inhibitory cytokine-1 stimulates proliferation of human umbilical vein endothelial cells by up-regulating cyclins D1 and E through the PI3K/Akt-, ERK-, and JNK-dependent AP-1 and E2F activation signaling pathways. *Cellular Signalling, 24*, 1485–1495.

Johnen, H., Lin, S., Kuffner, T., Brown, D. A., Tsai, V. W., Bauskin, A. R., et al. (2007). Tumor-induced anorexia and weight loss are mediated by the TGF-beta superfamily cytokine MIC-1. *Nature Medicine, 13*, 1333–1340.

Johnen, H., Kuffner, T., Brown, D. A., Wu, B. J., Stocker, R., & Breit, S. N. (2012). Increased expression of the TGF-b superfamily cytokine MIC-1/GDF15 protects ApoE(-/-) mice from the development of atherosclerosis. *Cardiovascular Pathology, 21*, 499–505.

Kempf, T., Eden, M., Strelau, J., Naguib, M., Willenbockel, C., Tongers, J., et al. (2006). The transforming growth factor-beta superfamily member growth-differentiation factor-15 protects the heart from ischemia/reperfusion injury. *Circulation Research, 98*, 351–360.

Kempf, T., Horn-Wichmann, R., Brabant, G., Peter, T., Allhoff, T., Klein, G., et al. (2007a). Circulating concentrations of growth-differentiation factor 15 in apparently healthy elderly individuals and patients with chronic heart failure as assessed by a new immunoradiometric sandwich assay. *Clinical Chemistry, 53*, 284–291.

Kempf, T., von Haehling, S., Peter, T., Allhoff, T., Cicoira, M., Doehner, W., et al. (2007b). Prognostic utility of growth differentiation factor-15 in patients with chronic heart failure. *Journal of the American College of Cardiology, 50*, 1054–1060.

Kempf, T., Zarbock, A., Widera, C., Butz, S., Stadtmann, A., Rossaint, J., et al. (2011). GDF-15 is an inhibitor of leukocyte integrin activation required for survival after myocardial infarction in mice. *Nature Medicine, 17*, 581–588.

Kempf, T., Guba-Quint, A., Torgerson, J., Magnone, M. C., Haefliger, C., Bobadilla, M., et al. (2012). Growth differentiation factor 15 predicts future insulin resistance and impaired glucose control in obese nondiabetic individuals: Results from the XENDOS trial. *European Journal of Endocrinology, 167*, 671–678.

Kim, J. M., Kosak, J. P., Kim, J. K., Kissling, G., Germolec, D. R., Zeldin, D. C., et al. (2013). NAG-1/GDF15 transgenic mouse has less white adipose tissue and a reduced inflammatory response. *Mediators of Inflammation, 2013*, 641851.

Koniaris, L., Hsiao, E., Esquela, A., & Zimmers, T. (2003). Induction of MIC-1/GDF-15 following bile duct injury. *Journal of Gastrointestinal Surgery, 7*(2), 303–304.

Koopmann, J., Buckhaults, P., Brown, D. A., Zahurak, M. L., Sato, N., Fukushima, N., et al. (2004). Serum macrophage inhibitory cytokine 1 as a marker of pancreatic and other periampullary cancers. *Clinical Cancer Research, 10*, 2386–2392.

Lajer, M., Jorsal, A., Tarnow, L., Parving, H. H., & Rossing, P. (2010). Plasma growth differentiation factor-15 independently predicts all-cause and cardiovascular mortality as well as deterioration of kidney function in type 1 diabetic patients with nephropathy. *Diabetes Care, 33*, 1567–1572.

Lambrecht, S., Smith, V., De Wilde, K., Coudenys, J., Decuman, S., Deforce, D., et al. (2014). GDF15, a marker of lung involvement in systemic sclerosis, is involved in fibrosis development but does not impair fibrosis development. *Arthritis and Rheumatology, 66*, 418–427.

Li, J., Yang, L., Qin, W., Zhang, G., Yuan, J., & Wang, F. (2013). Adaptive induction of growth differentiation factor 15 attenuates endothelial cell apoptosis in response to high glucose stimulus. *PloS One, 8*, e65549.

Lindmark, F., Zheng, S. L., Wiklund, F., Bensen, J., Balter, K. A., Chang, B., et al. (2004). H6D polymorphism in macrophage-inhibitory cytokine-1 gene associated with prostate cancer. *Journal of the National Cancer Institute, 96*(16), 1248–1254.

Mazagova, M., Buikema, H., van Buiten, A., Duin, M., Goris, M., Sandovici, M., et al. (2013). Genetic deletion of growth differentiation factor 15 augments renal damage in both type 1 and type 2 models of diabetes. *American Journal of Physiology-Renal Physiology, 305*, F1249–F1264.

Meadows, C. A., Risbano, M. G., Zhang, L., Geraci, M. W., Tuder, R. M., Collier, D. H., et al. (2011). Increased expression of growth differentiation factor-15 in systemic sclerosis-associated pulmonary arterial hypertension. *Chest, 139*, 994–1002.

Mensching, L., Borger, A. K., Wang, X., Charalambous, P., Unsicker, K., & Haastert-Talini, K. (2012). Local substitution of GDF-15 improves axonal and sensory recovery after peripheral nerve injury. *Cell and Tissue Research, 350*, 225–238.

Moore, A. G., Brown, D. A., Fairlie, W. D., Bauskin, A. R., Brown, P. K., Munier, M. L., et al. (2000). The transforming growth factor-ss superfamily cytokine macrophage inhibitory cytokine-1 is present in high concentrations in the serum of pregnant women. *Journal of Clinical Endocrinology and Metabolism, 85*, 4781–4788.

Nickel, N., Kempf, T., Tapken, H., Tongers, J., Laenger, F., Lehmann, U., et al. (2008). Growth differentiation factor-15 in idiopathic pulmonary arterial hypertension. *American Journal of Respiratory and Critical Care Medicine, 178*, 534–541.

Preusch, M. R., Baeuerle, M., Albrecht, C., Blessing, E., Bischof, M., Katus, H. A., et al. (2013). GDF-15 protects from macrophage accumulation in a mouse model of advanced atherosclerosis. *European Journal of Medical Research, 18*, 19.

Rossaint, J., Vestweber, D., & Zarbock, A. (2013). GDF-15 prevents platelet integrin activation and thrombus formation. *Journal of Thrombosis and Haemostasis, 11*, 335–344.

Santhanakrishnan, R., Chong, J. P., Ng, T. P., Ling, L. H., Sim, D., Leong, K. T., et al. (2012). Growth differentiation factor 15, ST2, high-sensitivity troponin T, and N-terminal pro brain natriuretic peptide in heart failure with preserved vs. reduced ejection fraction. *European Journal of Heart Failure, 14*, 1338–1347.

Schindowski, K., von Bohlen und Halbach, O., Strelau, J., Ridder, D. A., Herrmann, O., Schober, A., et al. (2011). Regulation of GDF-15, a distant TGF-beta superfamily member, in a mouse model of cerebral ischemia. *Cell and Tissue Research, 343*, 399–409.

Schledzewski, K., Geraud, C., Arnold, B., Wang, S., Grone, H. J., Kempf, T., et al. (2011). Deficiency of liver sinusoidal scavenger receptors stabilin-1 and -2 in mice causes glomerulofibrotic nephropathy via impaired hepatic clearance of noxious blood factors. *Journal of Clinical Investigation, 121*, 703–714.

Schlittenhardt, D., Schober, A., Strelau, J., Bonaterra, G. A., Schmiedt, W., Unsicker, K., et al. (2004). Involvement of growth differentiation factor-15/ macrophage inhibitory cytokine-1 (GDF-15/MIC-1) in oxLDL-induced apoptosis of human macrophages in vitro and in arteriosclerotic lesions. *Cell and Tissue Research, 318*, 325–333.

Schober, A., Bottner, M., Strelau, J., Kinscherf, R., Bonaterra, G. A., Barth, M., et al. (2001). Expression of growth differentiation factor-15/ macrophage inhibitory cytokine-1 (GDF-15/MIC-1) in the perinatal, adult, and injured rat brain. *Journal of Comparative Neurology, 439*(1), 32–45.

Schopfer, D. W., Ku, I. A., Regan, M., & Whooley, M. A. (2014). Growth differentiation factor 15 and cardiovascular events in patients with stable ischemic heart disease (The Heart and Soul Study). *American Heart Journal, 167*, 186–192.e1.

Shalev, A., Pise-Masison, C. A., Radonovich, M., Hoffmann, S. C., Hirshberg, B., Brady, J. N., et al. (2002). Oligonucleotide microarray analysis of intact human pancreatic islets: Identification of glucose-responsive genes and a highly regulated TGFbeta signaling pathway. *Endocrinology, 143*(9), 3695–3698.

Simonson, M. S., Tiktin, M., Debanne, S. M., Rahman, M., Berger, B., Hricik, D., et al. (2012). The renal transcriptome of db/db mice identifies putative urinary biomarker proteins in patients with type 2 diabetes: A pilot study. *American Journal of Physiology. Renal Physiology, 302*, F820–F829.

Song, H., Yin, D., & Liu, Z. (2012). GDF-15 promotes angiogenesis through modulating p53/HIF-1alpha signaling pathway in hypoxic human umbilical vein endothelial cells. *Molecular Biology Reports, 39*, 4017–4022.

Strelau, J., Sullivan, A., Bottner, M., Lingor, P., Falkenstein, E., Suter-Crazzolara, C., et al. (2000). Growth/differentiation factor-15/macrophage inhibitory cytokine-1 is a novel trophic factor for midbrain dopaminergic neurons in vivo. *Journal of Neuroscience, 20*(23), 8597–8603.

Strelau, J., Strzelczyk, A., Rusu, P., Bendner, G., Wiese, S., Diella, F., et al. (2009). Progressive postnatal motoneuron loss in mice lacking GDF-15. *Journal of Neuroscience, 29*, 13640–13648.

Subramaniam, S., Strelau, J., & Unsicker, K. (2003). Growth differentiation factor-15 prevents low potassium-induced cell death of cerebellar granule neurons by differential regulation of Akt and ERK pathways. *Journal of Biological Chemistry, 278*(11), 8904–8912.

Tanno, T., Bhanu, N. V., Oneal, P. A., Goh, S. H., Staker, P., Lee, Y. T., et al. (2007). High levels of GDF15 in thalassemia suppress expression of the iron regulatory protein hepcidin. *Nature Medicine, 13*, 1096–1101.

Tchou, I., Margeli, A., Tsironi, M., Skenderi, K., Barnet, M., Kanaka-Gantenbein, C., et al. (2009). Growth-differentiation factor-15, endoglin and N-terminal pro-brain natriuretic peptide induction in athletes participating in an ultramarathon foot race. *Biomarkers, 14*, 418–422.

Tsai, V. W., Macia, L., Johnen, H., Kuffner, T., Manadhar, R., Jorgensen, S. B., et al. (2013). TGF-b superfamily cytokine MIC-1/GDF15 is a physiological appetite and body weight regulator. *PloS One, 8*, e55174.

Wang, X., Yang, X., Sun, K., Chen, J., Song, X., Wang, H., et al. (2010). The haplotype of the growth-differentiation factor 15 gene is associated with left ventricular hypertrophy in human essential hypertension. *Clinical Science (London), 118*, 137–145.

Wang, T. J., Wollert, K. C., Larson, M. G., Coglianese, E., McCabe, E. L., Cheng, S., et al. (2012). Prognostic utility of novel biomarkers of cardiovascular stress: The Framingham Heart Study. *Circulation, 126*, 1596–1604.

Wang, X., Chrysovergis, K., Kosak, J., & Eling, T. E. (2013). Lower NLRP3 inflammasome activity in NAG-1 transgenic mice is linked to a resistance to obesity and increased insulin sensitivity. *Obesity (Silver Spring), 22*, 1256–1263.

Welsh, J. B., Sapinoso, L. M., Kern, S. G., Brown, D. A., Liu, T., Bauskin, A. R., et al. (2003). Large-scale delineation of secreted protein biomarkers overexpressed in cancer tissue and serum. *Proceedings of the National Academy of Sciences of the United States of America, 100*, 3410–3415.

Whitson, R. J., Lucia, M. S., & Lambert, J. R. (2012). Growth differentiation factor-15 (GDF-15) suppresses in vitro angiogenesis through a novel interaction with connective tissue growth factor (CCN2). *Journal of Cellular Biochemistry, 114*, 1424–1433.

Wiklund, F. E., Bennet, A. M., Magnusson, P. K., Eriksson, U. K., Lindmark, F., Wu, L., et al. (2010). Macrophage inhibitory cytokine-1 (MIC-1/GDF15): A new marker of all-cause mortality. *Aging Cell, 9*, 1057–1064.

Wollert, K. C., Kempf, T., Lagerqvist, B., Lindahl, B., Olofsson, S., Allhoff, T., et al. (2007a). Growth differentiation factor 15 for risk stratification and selection of an invasive treatment strategy in non ST-elevation acute coronary syndrome. *Circulation, 116*, 1540–1548.

Wollert, K. C., Kempf, T., Peter, T., Olofsson, S., James, S., Johnston, N., et al. (2007b). Prognostic value of growth-differentiation factor-15 in patients with non-ST-elevation acute coronary syndrome. *Circulation, 115*, 962–971.

Wu, J. F., Wang, Y., Zhang, M., Tang, Y. Y., Wang, B., He, P. P., et al. (2014). Growth Differentiation Factor-15 Induces Expression of ATP-Binding Cassette Transporter A1 Through PI3K/PKCzeta/SP1 Pathway in THP-1 Macrophages. *Biochemical and Biophysical Research Communications, 444*, 325–331.

Xu, J., Kimball, T. R., Lorenz, J. N., Brown, D. A., Bauskin, A. R., Klevitsky, R., et al. (2006). GDF15/MIC-1 functions as a protective and antihypertrophic factor released from the myocardium in association with SMAD protein activation. *Circulation Research, 98*, 342–350.

Yanaba, K., Asano, Y., Tada, Y., Sugaya, M., Kadono, T., & Sato, S. (2012). Clinical significance of serum growth differentiation factor-15 levels in systemic sclerosis: Association with disease severity. *Modern Rheumatology, 22*, 668–675.

Zimmers, T. A., Jin, X., Hsiao, E. C., Perez, E. A., Pierce, R. H., Chavin, K. D., et al. (2006). Growth differentiation factor-15: Induction in liver injury through p53 and tumor necrosis factor- independent mechanisms. *Journal of Surgical Research, 130*, 45–51.

Th17 Response

Lorenzo Cosmi, Francesco Liotta and Francesco Annunziato
Department of Experimental and Clinical Medicine and Denothe Center, University of Florence, Florence, Italy

Synonyms

Autoimmunity; IL-17; IFN-γ; Inflammatory disorders; T helper lymphocytes

Definition

The Th1-Th2 paradigm: CD4+ T helper (Th) lymphocytes represent a heterogeneous population of cells, playing an essential role in adaptive immunity. These cells include effector cells, devoted to the protection against pathogens, and regulatory T cells (Treg), which defend the body from effector responses to autoantigens and also from exogenous antigens when they become dangerous for the host. Effector CD4+ Th cells are heterogeneous with regard to their protective function, enabling a type of response which is different according to the nature of the invading microorganism. Over 20 years ago, two main subsets of CD4+ Th cells with different functions and patterns of cytokine secretion have been identified in both mice and humans and named as type

1 Th (Th1) and type 2 Th (Th2) lymphocytes, respectively. Th1 cells produce high levels of interferon (IFN)-γ and are responsible for both phagocyte activation and the production of opsonizing and complement-fixing antibodies, thus playing an important role in the protection against intracellular pathogens. Th2 cells produce interleukin (IL)-4, IL-5, IL-9, and IL-13 and are mainly involved in the protection against parasitic helminths (Annunziato et al. 2014). In addition to their protective functions against invading pathogens, Th1 and Th2 cells responses can contribute to the development of human disorders: Th1 cells have been thought to be involved in the pathogenesis of organ-specific autoimmune diseases, as well as other chronic inflammatory disorders, such as Crohn's disease (CD), sarcoidosis, and atherosclerosis; Th2 cells certainly play a central role in the development of allergic disorders (Cosmi et al. 2014).

Th17 lymphocytes: The Th1 – Th2 paradigm was maintained until some years ago when a third subset of CD4+ effector Th cells, named Th17 cells, was identified. Although the existence of IL-17 as a product of activated CD4+ T cells has been known from several years, the existence of Th17 as a distinct subset of T lymphocytes was only subsequently recognized, and the main features of this population in human beings have been described (Annunziato et al. 2007).

Th17 lymphocytes produce the distinctive cytokines IL-17A and IL-17F. IL-17 family of cytokines comprehends five members, designated IL-17A–F. IL-17A is disulfide-linked homodimeric glycoprotein, consisting of 155 amino acids (Yao et al. 1995), sharing great homology with IL-17F (55 %). IL-17A and IL-17F can either exist as IL-17A homodimers and IL-17F homodimers or as IL-17A-IL-17F heterodimers. The other IL-17 family members, IL-17B, IL-17C, and IL-17D, are produced by a non-T cell source. Th17 cells play a critical role in the recruitment, activation, and migration of neutrophil granulocytes, both directly through IL-8 production and indirectly by inducing, via IL-17, the production of colony stimulatory factors (CSF) and CXCL8 in tissue resident cells (Pelletier et al. 2010). Moreover Th17 lymphocytes are also able to stimulate CXCL chemokines production and to increase mRNA and protein for the mucins, MUC5AC and MUC5B, in primary human bronchial epithelial cells in vitro, as well as the expression of human beta defensin-2 and CCL20 in lung epithelial cells (Cosmi et al. 2014). Th17 lymphocytes produce also cytokines different from IL-17, such as IL-21 and IL-22, that contribute to the activation of mononuclear and/or resident cells and therefore may induce and/or maintain a chronic inflammatory process. However, because of their unique ability to recruit neutrophils, the main protective function of Th17 cells appears to be the clearance of extracellular pathogens, including fungi (Annunziato et al. 2014).

Epidemiology and Genetics

In the last years, several phenotypic and functional features of human Th17 lymphocytes have been reported allowing a better ex vivo and in vitro identification of this cell subset. Th17 cells express the transcription factor RORC, the IL-23 receptor (IL-23R), the chemokine receptor CCR6 (Annunziato et al. 2007), and the lectin receptor CD161 (Cosmi et al. 2008). The origin of Th17 lymphocytes has been a matter of intense debate in the last years. Some years ago, we found that the simultaneous presence of IL-1β and IL-23 was required to induce the development of Th17 lymphocytes from a small subset of CD161+ CD4 + T cell precursors detectable in both human umbilical cord blood and thymus (Cosmi et al. 2008). The combination of these two cytokines also induces T-bet and IL-12Rβ2 expression, and the differentiation of Th1 cells, suggesting a possible developmental relationship between human Th17 and Th1 cells (Cosmi et al. 2014). Moreover, it has been reported that IL-1β is essential for inducing IL-17/IFN-γ double producing cells, the so-called Th17/Th1 subset, a phenotype that is frequently observed in pathological conditions (Zielinski et al. 2012).

Regarding the controversial role of TGF-β signaling for the development of human Th17 lymphocytes, it is now generally accepted that it does not have a direct effect, but it can play an indirect permissive role by suppressing the Th1 cell differentiation. Indeed, umbilical cord blood naïve CD161(+) CD4(+) T cells, which contain the precursors of human Th17 cells, differentiated into IL-17A-producing cells in response to IL-1β plus IL-23, even in the absence of TGF-β. TGF-β does not have a direct effect on the genesis of human Th17 cells, but it can indirectly favor their development by suppressing both T-bet expressions (Santarlasci et al. 2009). All these data are in keeping with the observation that patients with mutations in TGFB1, TGFBR1, and TGFBR2 (Camurati-Engelmann disease and Marfan-like syndromes) did not exhibit any difference in the number of IL-17A-producing T cells compared with healthy controls. The dispensability of TGF-β1 from Th17 cell development has been also confirmed in mice by the observation that Rorγt-expressing CD4+ lamina propria T cells were found in mice lacking TGF-β receptor subunit I in their T cells, whereas numbers of Foxp3+ CD4+ Treg cells were dramatically reduced (Ghoreschi et al. 2010).

Pathophysiology

The protective role of Th17 lymphocytes is confirmed by the finding that in autosomal dominant hyper-IgE syndrome (HIES or "Job's syndrome"), IL-17 production and Th17 differentiation are impaired, due to mutations in the gene encoding STAT3 (Ma et al. 2008). This results in recurrent and often severe pulmonary infections, pneumatoceles, eczema, staphylococcal abscesses, and mucocutaneous candidiasis. These data are consistent with a crucial role for STAT3 signaling in the generation of Th17 cells and further strengthened the key role of this subset in the clearance of fungal and extracellular bacterial infections (Milner et al. 2008). Also the autosomic recessive HIES caused by mutations in DOCK8 is characterized by candidiasis derived by defects in Th17 cell differentiation. Moreover heterozygous mutations in STAT1 have been found to be involved in chronic mucocutaneous candidiasis. These patients revealed defects in *Candida*-stimulated production of IFN-γ, IL-17, and IL-22, probably due to defective functioning of IL-12 and IL-23 signaling pathways (McDonald 2012).

Beyond their protective role in the clearance of extracellular pathogens, Th17 lymphocytes have been described to play a role in the pathogenesis of several autoimmune and inflammatory diseases, such as multiple sclerosis, rheumatoid arthritis (RA), inflammatory bowel disease (IBD), but also in psoriasis and contact dermatitis (Cosmi et al. 2014). Nevertheless, Th17 cells are very difficult to detect in the inflamed tissues of patients suffering from these diseases, and this may be explained at least by two reasons: their limited expansion in response to T cell receptor (TCR) triggering and their tendency to shift to a Th1 phenotype in the presence of IL-12. The limited proliferative capacity of Th17 lymphocytes is due to the high expression of IL-4-induced gene 1 (IL4I1) that impairs their expansion in response to T cell receptor (TCR) triggering (Santarlasci et al. 2012) and to Tob1, a member of the Tob/BTG antiproliferative protein family, which prevents cell cycle progression (Santarlasci et al. 2014).

The second important reason for explaining the rarity of Th17 cells in the inflammatory sites is their high plasticity, which allows these cells to produce IFN-γ and then rapidly shift to the Th1 phenotype. Some years ago, examining the cytokine profile of T cells derived from the inflamed mucosa of patients with Crohn's diseases, we mainly found the presence of Th1 cells, but a few Th17 cells, and appreciable levels of cells producing both IL-17 and IFN-γ that we named as Th17/Th1 cells (Annunziato et al. 2007). Interestingly, the acquisition of IFN-γ-producing potential by Th17 lymphocytes has been found to have a crucial role in pathophysiology, inasmuch as Th17 cells can induce type 1 insulin-dependent diabetes mellitus (IDDM) in lymphopenic mice only after their shift into Th1 cells (Martin-Orozco et al. 2009), and in a model of IDDM induced by the transfer of Th17 in NOD/SCID recipient mice, the onset of the disease was prevented by treatment with an anti-IFN-γ, but not an anti-IL17A, neutralizing antibody (Bending et al. 2009). Th17-derived Th1, named as non-classic Th1, can be easily distinguished from classic Th1 cells on the basis of CD161 expression as well as the consistent expression of RORC, IL-17 receptor E, CCR6, and IL-4-induced gene 1 (Maggi et al. 2012), which are all virtually absent in classic Th1 cells. The strict relationship between the Th17 phenotype and non-classic Th1 cells is further strengthened at an epigenetic level. Epigenetic modifications are commonly utilized to regulate transcription factor and polymerase access to transcriptional regulatory elements in chromatin. We demonstrated that non-classic Th1 cells, like Th17 cells, have a marked RORC2 and IL17A demethylation, while classic Th1 cells exhibit a complete methylation of these genes, thus supporting the concept that non-classic Th1 cells and Th17 cells have a common origin (Mazzoni et al. 2015).

It is not fully understood how often such plasticity occurs in the course of physiologic responses to pathogens and what is its importance in protective immunity. However, it has been reported that human Th17 cells may exhibit a different cytokine profile according to their antigen specificity. *Candida albicans*-specific Th17 cells could produce IL-17 and IFN-γ, but no IL-10, whereas *Staphylococcus aureus*-specific Th17 cells produced IL-17 and could produce IL-10 upon restimulation. Importantly, IL-1β was reported to be indispensable for the induction of IL-17/IFN-γ double producing cells and for the inhibition of the IL-10-producing capacity of differentiating Th17 cells (Zielinski et al. 2012).

Clinical Presentation

In humans, the pathogenic role of Th17 cells has been suggested in several inflammatory conditions, mainly in rheumatologic disorders, such as psoriatic arthritis (PsA), rheumatoid arthritis (RA), and juvenile idiopathic arthritis (JIA); in multiple sclerosis (MS); in inflammatory bowel diseases (IBD); and also in bronchial asthma.

Th17 lymphocytes possess an intrinsic functional plasticity that allows them to shift both to Th1-like and to Th2-like cells, in response to precise environmental signals. In inflammatory conditions Th17 lymphocytes that have shifted towards a Th1 or Th2 phenotype, acquiring the ability to produce, respectively, IFN-γ or IL-4, seem to be particularly aggressive and more pathogenic than the unshifted cells (Cosmi et al. 2014). Indeed, Th17 plasticity is not limited to the Th1 axis, but allows these cells to be shifted also towards the Th2 phenotype, in the presence of appropriate signals (Cosmi et al. 2010). Human circulating memory CD4+ T cells that produce both IL-17A and IL-4, the so-called Th17/Th2 lymphocytes, are more represented, even if very rare, in the circulation of patients with allergic asthma than in healthy donors, suggesting a possible role in the pathogenesis of the disease. The ability of Th17 cytokines to recruit neutrophils to the airway, to induce mucous cell metaplasia, and to act on airway smooth muscle provoking airway narrowing has been reported in patients with severe asthma that comprise a small percentage of asthmatics but with the highest health-care costs among asthma phenotypes.

Th17 cells are not an end stage of effector T cell differentiation, because a substantial proportion of human Th17 cells acquire or upregulate Th1-associated markers, including expression of IFN-γ, CXCR3, and T-bet, still maintaining Th17-associated markers such as CD161, IL1-R1, IL-23R, IL-17RE, RORC2, IL4I1, Tob1, and CCR6. The notion that Th17 cells are precursors that differentiate into non-classic Th1 effector progeny in a progressive, linear fashion in inflamed tissues is suggested by the ratio of these subsets found in autoimmune target organs. Indeed, Th17-derived Th1 cells, but not Th17, as well as classic Th1 cells constitute the majority of tissue-infiltrating CD4+ T cells in the joints of patients with rheumatoid arthritis (Nistala et al. 2010; Cosmi et al. 2011) and the affected gut of patients with Crohn's disease (Annunziato et al. 2007; Kleinschek et al. 2009; Maggi et al. 2013). Recently, it has been reported (Piper et al. 2014) that an enrichment of GM-CSF-producing Th cells in the joints of patients with juvenile idiopathic arthritis and, notably, the frequency of these cells were directly correlated with levels of GM-CSF protein in the joint and of serum markers of disease activity. More importantly, the same authors also demonstrated that GM-CSF-expressing T cells in human autoimmune disease have a phenotype associated with ex-Th17 cells (non-classic Th1 cells), expressing RORC2, CD161, and IFN-γ, but not IL-17A, and that the cytokine driving the polarization of Th17 cells towards the non-classic Th1 phenotype, IL-12, is also involved in the induction of GM-CSF production. Additional evidence on the pathogenic role of non-classic Th1 cells comes from a recent study (Ramesh et al. 2014) in which authors showed that pro-inflammatory human Th17 cells are restricted to a subset of CCR6+ CXCR3hi CCR4lo CCR10− CD161+ cells that express c-Kit transiently and P-glycoprotein (P-gp)/multidrug resistance type 1 (MDR1) stably. MDR1+ Th17 cells produce both Th17 (IL-17A, IL-17F, GM-CSF, and IL-22) and Th1 (IFN-γ) cytokines upon TCR stimulation.

In conclusion, non-classic Th1 cells play a more important role than classic Th1, and even Th17 themselves, in the maintenance of chronic inflammation in several autoimmune disorders. Fortunately, classic and non-classic Th1 cells exhibit not only different functional features, but they have also started to become distinguishable because of distinct phenotypic markers, which may allow their recognition and provide help for identifying the best targets for the novel immunotherapeutic strategies of chronic inflammatory, including autoimmune, disorders (Annunziato et al. 2014) (Fig. 1).

Therapy

On the basis of the observations that IL-17 plays a key role in the pathogenesis of several chronic inflammatory diseases, both in mouse models and in humans, Th17 cells have been considered as potential targets for the treatment of such disorders (Mioscsec and Kolls 2012). Among these diseases, encouraging results from clinical studies have been obtained in psoriasis. Skin biopsy samples taken from patients with psoriasis showed high expression of IL-17 together with high expression of IL-23, IL-22, and IFN-γ (Annunziato et al. 2007). Accordingly, patients with chronic moderate-to-severe plaque psoriasis treated with ixekizumab, a humanized IgG4 monoclonal antibody that neutralizes interleukin-17A, had significant improvement in clinical measures during the 12-week treatment period (Leonardi et al. 2012). Similarly, brodalumab a humanized anti-IL-17RA monoclonal antibody that antagonizes the interleukin-17 pathway, binding with high affinity to human interleukin-17RA, showed short-term efficacy in reducing clinical score and safety in patients with moderate-to-severe plaque psoriasis (Papp et al. 2012). Nevertheless, the blocking of IL-17 pathway was not efficacious in other chronic inflammatory conditions such as rheumatoid arthritis (Martin et al. 2013), or Crohn's disease (CD), where secukinumab, a fully human-selective anti-IL-17A monoclonal antibody, was not only ineffective in controlling symptoms but also provoked severe adverse events (Hueber et al. 2012).

Interestingly, it is well known since many years that the combined neutralization of IL-12 and IL-23, thanks to ustekinumab, a mAb that binds with high affinity to the p40 subunit of both cytokines, results in symptoms improvement in both psoriasis and CD (Krueger et al. 2007; Sandborn et al. 2008). These observations strengthened the concept that both IL-12 and IL-23, and consequently both Th1 and Th17 subsets, may contribute to the pathogenesis of these diseases. The therapeutic efficacy of mAbs that bind to the p40 subunit is of particular interest also considering that IL-12 not only plays a pivotal role in the induction of the Th1 response but is also involved in the shifting of memory Th17 lymphocytes towards the Th17/Th1 and the non-classic Th1 subset (Annunziato et al. 2007). Since it has been suggested that such a shifting may be associated with disease progression in JIA and in CD (Cosmi et al. 2014), we speculate that the inhibition of the polarization towards the Th1 phenotype could represent a novel and intriguing therapeutic approach for these disorders. The observation that TNF-α contributes to the shifting of Th17 lymphocytes towards the Th1 subset (Maggi et al. 2014) indirectly confirms this

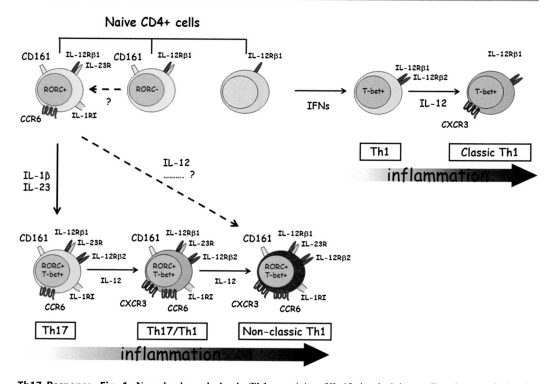

Th17 Response, Fig. 1 Non-classic and classic Th1 cell development. Developmental stages of UCB CD4+ CD161+ and CD4+ CD161- cells differentiation towards non-classic and classic Th1 phenotypes, respectively, are represented. *Continuous arrows* indicate differentiation stages; *dotted arrows* suggest hypothesized and unknown maturation processes. Among UCB CD4+ CD161+ cell fraction, RORC+ cells develop into Th17 but then can be rapidly shifted towards the non-classic Th1 subset under the activity of IL-12 signals. It is actually unknown whether the Th17 step is mandatory for the final acquisition of the non-classic Th1 status or whether UCB CD4+ CD161+ RORC+ cells can be immediately directed towards this phenotype. UCB CD4+ CD161+ RORC- cells may represent an immature population, which can acquire RORC2 expression upon proper stimulation. UCB CD4+ CD161- RORC- cells instead directly acquire a classic Th1 phenotype when in the presence of appropriate stimuli

hypothesis, since inhibitors of TNF-α are successfully used in the treatment of JIA, CD, and several other inflammatory diseases.

The possibility to target Th17 lymphocytes can also be taken in consideration in those disorders in which they coexist with Th2 cells, such as allergic asthma. Brodalumab, whose efficacy has been proven in psoriasis (Papp et al. 2012), acts by blocking the biologic activity of both IL-17 and the Th2-related cytokine, IL-25 (IL17-E), that share with IL-17 in the IL-17RA receptor subunit. This ability to interfere with both the Th2 and the Th17 pathway candidates brodalumab as a possible therapeutic approach in those forms of allergic asthma, poorly controlled by conventional therapy, that are characterized by a mixed, eosinophilic and neutrophilic, mediated inflammation.

Outcome

The discovery of a third subset of CD4+ Th cells in addition to Th1 and Th2 cells, named as Th17, has revealed the existence of a more complex pathway of adaptive cell-mediated effector immunity, which overlies the classic Th1/Th2 paradigm. However, Th17 has appeared to be a very flexible phenotype, as a result of the easy shifting of these cells in the inflammatory sites to the Th1 phenotype. Therefore, it is now clear that there are two different types of Th1 cells, which have been defined as classic and non-classic (Th17-derived), respectively. The consequence of this previously unknown dichotomy is the present awareness that in the inflammatory sites there is a mixture of the two Th1 cell types and it is therefore difficult to understand whether all the previously described pathogenic effects by Th1 cells are the result of the activity of the classic Th1 subset or they represent the result of the damage initiated by Th17 cells but maintained, or even amplified, by their non-classic Th1 progeny. The initial studies in this field seem to suggest that non-classic Th1 cells play a more important role than classic Th1, and even Th17 themselves, in the maintenance of chronic inflammation in several autoimmune disorders. Fortunately, classic and non-classic Th1 cells exhibit not only different functional features, but they have also started to become distinguishable because of distinct phenotypic markers, which may allow their recognition and provide help for identifying the best targets for the novel immunotherapeutic strategies of chronic inflammatory, including autoimmune, disorders.

References

Annunziato, F., Cosmi, L., Santarlasci, V., Maggi, L., Liotta, F., Mazzinghi, B., et al. (2007). Phenotypic and functional features of human Th17 cells. *The Journal of Experimental Medicine, 204*, 1849–1861.

Annunziato, F., Cosmi, L., Liotta, F., Maggi, E., & Romagnani, S. (2014). Human Th1 dichotomy: Origin, phenotype and biologic activities. *Immunology*. doi:10.1111/imm.12399.

Bending, D., De la Peña, H., Veldhoen, M., Phillips, J. M., Uyttenhove, C., Stockinger, B., et al. (2009). Highly purified Th17 cells from BDC2.5NOD mice convert into Th1-like cells in NOD/SCID recipient mice. *The Journal of Clinical Investigation, 119*, 565–572.

Cosmi, L., De Palma, R., Santarlasci, V., Maggi, L., Capone, M., Frosali, F., et al. (2008). Human interleukin 17-producing cells originate from a CD161+ CD4+ T cell precursor. *The Journal of Experimental Medicine, 205*, 1903–1916.

Cosmi, L., Maggi, L., Santarlasci, V., Capone, M., Cardilicchia, E., Frosali, F., et al. (2010). Identification of a novel subset of human circulating memory CD4+ T cells that produce both IL-17A and IL-4. *Journal of Allergy and Clinical Immunology, 125*, 222–230.

Cosmi, L., Cimaz, R., Maggi, L., Santarlasci, V., Capone, M., Borriello, F., et al. (2011). Evidence of the transient nature of the Th17 phenotype of CD4+ CD161+ T cells in the synovial fluid of patients with juvenile idiopathic arthritis. *Arthritis and Rheumatism, 63*, 2504–2515.

Cosmi, L., Liotta, F., Maggi, E., Romagnani, S., & Annunziato, F. (2014). Th17 and non-classic Th1 cells in chronic inflammatory disorders: Two sides of the same coin. *International Archives of Allergy and Immunology, 164*, 171–177.

Ghoreschi, K., Laurence, A., Yang, X. P., Tato, C. M., McGeachy, M. J., Konkel, J. E., et al. (2010). Generation of pathogenic T(H)17 cells in the absence of TGF-b signaling. *Nature, 467*, 967–971.

Hueber, W., Sands, B. E., Lewitzky, S., Vandemeulebroecke, M., Reinisch, W., Higgins, P. D., et al. (2012). Secukinumab, a human anti-IL-17A monoclonal antibody, for moderate to severe Crohn's disease: Unexpected results of a randomised, double-blind placebo-controlled trial. *Gut, 61*, 1693–1700.

Kleinschek, M. A., Boniface, K., Sadekova, S., Grein, J., Murphy, E. E., Turner, S. P., et al. (2009). Circulating and gut-resident human Th17 cells express CD161 and promote intestinal inflammation. *The Journal of Experimental Medicine, 206*, 525–534.

Krueger, G. G., Langley, R. G., Leonardi, C., Yeilding, N., Guzzo, C., Wang, Y., et al. (2007). A human interleukin-12/23 monoclonal antibody for the treatment of psoriasis. *The New England Journal of Medicine, 356*, 580–592.

Leonardi, C., Matheson, R., Zachariae, C., Cameron, G., Li, L., Edson-Heredia, E., et al. (2012). Anti-interleukin-17 monoclonal antibody ixekizumab in chronic plaque psoriasis. *The New England Journal of Medicine, 366*, 1190–1199.

Ma, C. S., Chew, G. Y., Simpson, N., Priyadarshi, A., Wong, M., Grimbacher, B., et al. (2008). Deficiency of Th17 cells in hyper IgE syndrome due to mutations in STAT3. *The Journal of Experimental Medicine, 205*, 1551–1557.

Maggi, L., Santarlasci, V., Capone, M., Rossi, M. C., Querci, V., Mazzoni, A., et al. (2012). Distinctive features of classic and non-classic (Th17 derived) human Th1 cells. *European Journal of Immunology, 42*, 3180–3188.

Maggi, L., Capone, M., Giudici, F., Santarlasci, V., Querci, V., Liotta, F., et al. (2013). CD4+ CD161+ T lymphocytes infiltrate Crohn's disease-associated perianal fistulas and are reduced by anti-TNF-α local therapy. *International Archives of Allergy and Immunology, 161*, 81–86.

Maggi, L., Cimaz, R., Capone, M., Santarlasci, V., Querci, V., Simonini, G., et al. (2014). Brief report: Etanercept inhibits the tumor necrosis factor α-driven shift of Th17 lymphocytes toward a nonclassic Th1 phenotype in juvenile idiopathic arthritis. *Arthritis & Rheumatology, 66*, 1372–1377.

Martin, D. A., Churchill, M., Flores-Suarez, L., Cardiel, M. H., Wallace, D., Martin, R., et al. (2013). A phase Ib multiple ascending dose study evaluating safety, pharmacokinetics, and early clinical response of brodalumab, a human anti-IL-17R antibody, in methotrexate-resistant rheumatoid arthritis. *Arthritis Research and Therapy, 15*, R164.

Martin-Orozco, N., Chung, Y., Chang, S. H., Wang, Y. H., & Dong, C. (2009). Th17 cells promote pancreatic inflammation but only induce diabetes efficiently in lymphopenic hosts after conversion into Th1 cells. *European Journal of Immunology, 39*, 216–224.

Mazzoni, A., Santarlasci, V., Maggi, L., Capone, M., Rossi, M. C., Querci, V., et al. (2015). Demethylation of the RORC2 and IL17A in human CD4+ T lymphocytes defines Th17 origin of non-classic Th1 cells. *The Journal of Immunology, 194*, 3116–3126.

McDonald, D. R. (2012). TH17 deficiency in human disease. *The Journal of Allergy and Clinical Immunology, 129*, 1429–1435.

Milner, J. D., Brenchley, J. M., Laurence, A., Freeman, A. F., Hill, B. J., Elias, K. M., et al. (2008). Impaired T(H)17 cell differentiation in subjects with autosomal dominant hyper-IgE syndrome. *Nature, 452*, 773–776.

Miossec, P., & Kolls, J. K. (2012). Targeting IL-17 and TH17 cells in chronic inflammation. *Nature Reviews Drug Discovery, 11*, 763–776.

Nistala, K., Adams, S., Cambrook, H., Ursu, S., Olivito, B., de Jager, W., et al. (2010). Th17 plasticity in human autoimmune arthritis is driven by the inflammatory environment. *Proceedings of the National Academy of Sciences of the United States of America, 107*, 14751–14756.

Papp, K. A., Leonardi, C., Menter, A., Ortonne, J. P., Krueger, J. G., Kricorian, G., et al. (2012). Brodalumab, an anti-interleukin-17-receptor antibody for psoriasis. *The New England Journal of Medicine, 366*, 1181–1189.

Pelletier, M., Maggi, L., Micheletti, A., Lazzeri, E., Tamassia, N., Costantini, C., et al. (2010). Evidence for a cross-talk between human neutrophils and Th17 cells. *Blood, 115*, 335–343.

Piper, C., Pesenacker, A. M., Bending, D., Thirugnanabalan, B., Varsani, H., Wedderburn, L. R., et al. (2014). T cell expression of granulocyte macrophage colony stimulating factor in juvenile arthritis is contingent upon Th17 plasticity. *Arthritis & Rheumatology, 66*, 1955–1960.

Ramesh, R., Kozhaya, L., McKevitt, K., Djuretic, I. M., Carlson, T. J., Quintero, M. A., et al. (2014). Pro-inflammatory human Th17 cells selectively express P-glycoprotein and are refractory to glucocorticoids. *The Journal of Experimental Medicine, 211*, 89–104.

Sandborn, W. J., Feagan, B. G., Fedorak, R. N., Scherl, E., Fleisher, M. R., Katz, S., et al. (2008). A randomized trial of Ustekinumab, a human interleukin-12/23 monoclonal antibody, in patients with moderate-to-severe Crohn's disease. *Gastroenterology, 135*, 1130–1141.

Santarlasci, V., Maggi, L., Capone, M., Frosali, F., Querci, V., De Palma, R., et al. (2009). TGF-beta indirectly favors the development of human Th17 cells by inhibiting Th1 cells. *European Journal of Immunology, 39*, 207–215.

Santarlasci, V., Maggi, L., Capone, M., Querci, V., Beltrame, L., Cavalieri, D., et al. (2012). Rarity of human T helper 17 cells is due to retinoic acid orphan receptor-dependent mechanisms that limit their expansion. *Immunity, 36*, 201–214.

Santarlasci, V., Maggi, L., Mazzoni, A., Capone, M., Querci, V., Rossi, M. C., et al. (2014). IL-4-induced gene 1 maintains high Tob1 expression which contributes to TCR unresponsiveness in human T helper 17 cells. *European Journal of Immunology, 44*, 654–661.

Yao, Z., Fanslow, W. C., Seldin, M. F., Rousseau, A. M., Painter, S. L., Comeau, M. R., et al. (1995). Herpesvirus Saimiri encodes a new cytokine, IL-17, which binds to a novel cytokine receptor. *Immunity, 3*, 811–821.

Zielinski, C. E., Mele, F., Aschenbrenner, D., Jarrossay, D., Ronchi, F., Gattorno, M., et al. (2012). Pathogen-induced human T(H)17 cells produce IFN-gamma or IL-10 and are regulated by IL-1beta. *Nature, 484*, 514–518.

Theophylline

Bernadette Prentice[1], Adam Jaffe[2,3] and Paul S. Thomas[4,5]
[1]Department of Paediatric Respiratory and Sleep Medicine, Sydney Children's Hospital, Randwick, NSW, Australia
[2]Department of Respiratory Medicine, Sydney Children's Hospital, Randwick, NSW, Australia
[3]School of Women's and Children's Health, UNSW Medicine, University of New South Wales, Sydney, NSW, Australia
[4]Inflammation and Infection Research Centre, Faculty of Medicine, University of New South Wales, Sydney, NSW, Australia
[5]Department of Respiratory Medicine, Prince of Wales Hospital, Sydney, NSW, Australia

Synonyms

Aminophylline; Choline theophyllinate; Dimethylxanthine; Oxtriphylline

Definition

Theophylline is a smooth muscle relaxant and weak diuretic with central nervous system and cardiac stimulatory effects.

Chemical Structures and Properties

Theophylline is dimethylxanthine which is closely related chemically to caffeine (Fig. 1). Like caffeine, theophylline is sparingly soluble in water (7 g/l). The logP value of theophylline is –0.02 giving the unionized form sufficient lipid solubility to diffuse passively through cell membranes. Theophylline is a weak acid (pKa 8.8). Consequently, both the unionized and ionized forms are present in plasma with the majority as the ionized species.

Theophylline is present in tea and coffee but at much lower concentrations than caffeine.

Metabolism, Pharmacokinetics, and Dosage

Controlled release formulations are the major form of administration of theophylline while, the salt, aminophylline is given intravenously.

Theophylline is well absorbed from the gastrointestinal tract although its rate of absorption can be erratic. Food is particularly important in theophylline dosing, as its rate of absorption depends upon whether or not it is taken with food. There is also the risk of "dose dumping," where there may be a sudden rise in plasma concentrations of theophylline during treatment with once daily sustained-release formulation (Steffensen and Pedersen 1986).

Theophylline, Fig. 1 Structures of theophylline and caffeine

After oral administration, the majority of the dose is absorbed with peak concentrations occurring at 0.5–2hours (Olgivie 1978). Binding to plasma proteins is extensive but varies depending upon the age of the patient. Neonates and people with liver disease reduced binding to protein (Olgivie 1978) and are thus at increased risk of toxicity. Theophylline is metabolized by the liver, specifically by several P450 enzymes including CYP1A2 (Barnes 2003) and metabolites are excreted in urine (Olgivie 1978). Factors affecting the half-life include gender, smoking, heart failure, hypoxia, age, and liver disease (Kirsten et al. 1998). Clearance can also be affected by high carbohydrate-low protein diet, or ingestion of caffeinated products (Olgivie 1978). Conversely, the risk of sub-therapeutic levels is caused by the induction of the hepatic enzymes.

Theophylline has a narrow therapeutic window and, as toxicity relates directly to plasma concentration, it is important to measure and monitor drug levels closely during long-term treatment. Therapeutic plasma levels >10mg/L, and allow for maximal bronchodilatation, but significant side effects occur >20mg/L (Kirsten et al.1998).

Pharmacological Effects

The pharmacological effects of theophylline are very similar to those of caffeine. Theophylline has bronchodilator and diuretic activities which are due to its smooth muscle relaxant action. It also has stimulatory actions on the central nervous system and heart. The mode of action of theophylline in asthma is unclear, but several mechanisms have been investigated (Barnes 2013; Cazzola et al. 2012).

- Theophylline may increase cAMP concentration in smooth muscle cells by inhibiting phosphodiesterases, in particular phosphodiesterase 3, thus causing smooth muscle relaxation and bronchodilatation (Barnes 2013). cAMP is responsible for relaxing smooth muscle and also for mitigating the response of several immune and inflammatory cells by inhibiting

the release of cytokines and chemokines and thus cell activation and migration (Barnes 2013). An effect on cAMP is, however, doubtful because inhibition of phosphodiesterases is seen at supratherapeutic concentrations.
- Theophylline binds to the adenosine A_1 and A_2 receptors and consequently blocks adenosine-mediated bronchoconstriction which may be mediated by the release of histamine and leukotrienes (Barnes 2013).
- In inflammatory states, theophylline activates histone deacetylase to prevent transcription of inflammatory genes that require the acetylation of histones for transcription to begin. The activity of histone deacetylase is reduced by oxidative stress and in corticosteroid resistance (Barnes 2013; Muller and Jacobson 2011). Thus, activation of histone deacetylase may be a mechanism by which theophylline reduces corticosteroid resistance. Unlike other suggested mechanisms of the therapeutic action of theophylline, increased histone deacetylase activity has been detected at therapeutic concentrations.

Several other xanthines (analogs of theophylline) and drugs with similar actions to theophylline have been synthesized and investigated but are not used widely (Cazzola et al. 2012).

Clinical Use

The use of theophylline has decreased greatly in recent years and is considered to be a third-line treatment for asthma (Barnes 2013; see general chapter, ▶ Anti-Asthmatic Drugs). In the long-term treatment of asthma, theophylline may be used if there is persistent poor control in adults. Its major use is therefore as a controller for chronic, persistent asthma (Tilley 2011). As outlined above, the plasma concentrations of theophylline should be checked during long-term therapy. An alternative treatment if control is inadequate is a leukotriene receptor antagonist in children and adults. Its use in acute asthma is also contentious.

Adverse Effects

The metabolism of theophylline is induced by other drugs resulting in low, even sub-therapeutic plasma concentrations. To overcome the influence of inducing agents, the dosage of theophylline may be increased. However, cessation of treatment with inducing agents increases the plasma concentrations and plasma concentrtations should be checked in order to prevent toxicity. Smoking induces the metabolism of theophylline, but deinduction may be slow in patients who have stopped the habit.

Cross-References

▶ Acute Exacerbations of Airway Inflammation
▶ Allergic Disorders
▶ Anti-asthma Drugs, Overview
▶ Antibiotics as Antiinflammatory Drugs
▶ Asthma
▶ Corticosteroids
▶ Histamine
▶ Leukotrienes
▶ Non-steroidal Anti-inflammatory Drugs: Overview

References

Barnes, P. J. (2003). Theophylline: New Perspectives for an old drug. *American Journal of Respiratory Critical Care medicine, 167*, 813–818.
Barnes, P. J. (2013). Theophylline. *American Journal of Critical Care Medicine, 188*(8), 901–906.
Cazzola, M., Page, C. P., Calzetta, L., & Matera, M. G. (2012). Pharmacology and therapeutics of bronchodilators. *Pharmacological Reviews, 64*(3), 450–504.
Kirsten, R., Nelson, K., Kirsten, D., & Heintz, B. (1998). Clincial pharmacokinetics of Vasodilators (Part 2). *Clinical Pharmacokinetics, 35*(1), 9–36.
Muller, C. E., & Jacobson, K. A. (2011). Xanthines as adenosine receptor antagonists. *Handbook of Experimental Pharmacology, 200*, 151–199.
Olgivie, R. J. (1978). Clinical Pharmacokinetics of Theophylline. *Clinical Pharmacokinetics, 3*, 267–293.
Steffensen, G., & Pedersen, S. (1986). Food induced changes in theophylline absorption from a once-a-day theophylline product. *British Journal of Clinical Pharmacology, 22*(5), 571–577.
Tilley, S. L. (2011). Methylxanthines in asthma. *Handbook of Experimental Pharmacology, 200*, 439–456.

Thiopurines: Azathioprine, Mercaptopurine, and Thioguanine

Antony B. Friedman, Miles P. Sparrow and Peter R. Gibson
Department of Gastroenterology, The Alfred Hospital and Monash University, Melbourne, VIC, Australia

Synonyms

Thiopurines

Definition

The thiopurines comprise of azathioprine (AZA), mercaptopurine (6-mercaptopurine (6MP)), and thioguanine (6-thioguanine (6TG)). In this review, the discussions of the thiopurines are concentrated on their uses in chronic inflammatory states, such as the inflammatory bowel diseases (IBD) and rheumatoid arthritis (RA). Their uses in cancer, leukemias, and organ transplantation are not reviewed in detail.

Chemical Structures and Properties

Azathioprine is a prodrug of 6MP, containing an additional imidazole group (Fig. 1). Azathioprine is a weak acid but more than 60 % is unionized at physiological pH values as indicated by its pKa of 7.9. Its sodium salt is water soluble and is used in injectable solutions. Both mercaptopurine and thioguanine are neutral substances which are lipid soluble but sparingly soluble in water and are not used in injections. The thiopurines' physicochemical properties favor the passive diffusion through cellular membranes although they are also substrates for membrane transporters.

Methodology of Assays

The thiopurine metabolites, 6-thioguanine nucleotides (6TGN) and 6-methylmercaptopurine (6MMP) in blood, are quantified during treatment using high-performance liquid chromatography. Most laboratories measure the red blood cell concentrations of 6TGN and 6MMP. Values are expressed in pmol/8×10^8 red blood cells. The clinical significance of the measurement of the thiopurines and their metabolites derived from leucocytes is greater than from red blood cells. However, assays of leucocytes are rarely performed as their purification is tedious and requires greater volumes of blood.

Metabolism and Pharmacokinetics

The pharmacokinetics and metabolism of the thiopurines are considered in two sections. Firstly, AZA and 6MP are discussed.

AZA and 6MP are well absorbed, but with a high first-pass metabolism, their bioavailability is low at about 15 %. Time to peak serum concentration is only 1–2 h for the parent drugs. AZA and 6MP have half-lives of 12 min and 1–3 h, respectively (Chouchana et al. 2012).

The metabolic pathways of thiopurines are complex (Fig. 2). The major metabolites of the thiopurines together with their full names are shown in Fig. 2.

AZA is a prodrug which is converted rapidly to 6MP, predominantly in the liver. Secondary, smaller sites of bioactivation include the small intestine, kidney, and adrenal glands (Eklund et al. 2006). 6MP is then transported intracellularly by nucleoside transporters and is metabolized to form its end metabolites 6TGN, 6MMP, and 6TU (Fig. 2) (Chouchana et al. 2012).

Metabolism of 6MP via the HPRT pathway leads to the production of TXM which is converted to the monophosphate (6TGM) which is, in turn, then metabolized sequentially to the diphosphate (6TGDP) and triphosphate (6TGTP). Collectively, the phosphates are known as 6-thioguanine nucleotides (6TGN). They are the active metabolites responsible for the efficacy of AZA and 6MP, as well as 6TG (see below). These nucleotides are also potentially myelotoxic at supratherapeutic levels. Other pathways lead to 6MMP and also to their ribotides known collectively as 6MMP(R) (Fig. 2).

Thiopurines: Azathioprine, Mercaptopurine, and Thioguanine, Fig. 1 Molecular structures of the thiopurines, azathioprine, mercaptopurine, and thioguanine and outline of their initial metabolism. Key enzymes: *GMPS* guanosine monophosphate synthetase, *HPRT* hypoxanthine phosphoribosyltransferase, *IMPDH* inosine-5-monophosphate dehydrogenase, *TPMT* thiopurine methyltransferase, *XOR* xanthine oxidoreductase. Metabolite groups: *TGN* 6TG nucleotides; *6MMP(R)* 6MMP ribonucleotides

The key enzyme, IMPDH, is thought to be the rate-limiting step in 6TGN production (Fig. 2). It competes with TPMT for the substrate 6TIMP, the first metabolite of 6MP. Metabolism of 6TIMP by IMPDH adds to 6TGN production, whereas metabolism of 6TIMP by TPMT is a secondary pathway leading to the formation of 6MMP(R).

Thiopurines: Azathioprine, Mercaptopurine, and Thioguanine, Fig. 2 Molecular structures of the thiopurines

6TG has a far longer time to peak serum concentration of approximately 8 h, but a similar half-life to 6MP of 2 h (Jharap et al. 2011). 6TG has a simpler metabolic pathway than AZA and 6MP. 6TG is directly converted by HPRT into 6TGMP and subsequent intracellular kinase activity producing 6TGDP and 6TGTP. Unlike the metabolism of AZA and 6MP, there is no significant production of 6MMP or 6TU.

The end metabolite group, 6TGN, has a half-life of between 3 and 13 days, and 6TGN steady state is only achieved after 4–5 weeks of therapy with wide interindividual variations. While metabolites are excreted via the urine, end-stage renal disease only slightly prolongs metabolite clearance (Chan et al. 1990). While 6TGN levels take approximately 4 weeks to reach steady state in red blood cells, optimal clinical efficacy is achieved much later and can take up to 12–17 weeks to occur (Prefontaine et al. 2010). It is postulated that this delayed onset of efficacy is due to the full apoptotic and antiproliferative effects of 6TGN on activated T lymphocytes taking far longer to occur. There is no correlation between the oral dose of thiopurines and concentrations of 6TGN and 6MMP in serum.

Pharmacological Activities

6TGN has several mechanisms of action (Chouchana et al. 2012).

- As a purine analogue, it triggers apoptosis and arrests the cell cycle by being incorporated into DNA in place of adenosine and guanine, leading to chromatid damage and arresting DNA replication.
- 6TGN-incorporated base pairs show reduced stability, causing small changes in local DNA structure, and increased levels of methylation, activating the DNA mismatch repair system.
- Most importantly, 6TGTP is a direct antagonist of Rac1, which blocks the activation of Vav to dampen the inflammatory cascade involving NF-κB and STAT-3.

These three mechanisms lead to apoptosis and prevent activation and proliferation of T lymphocytes implicated in the pathogenesis of autoimmune inflammatory diseases.

Clinical Uses and Efficacy

Thiopurines have an established place in the management strategies of many chronic inflammatory conditions. As shown in Table 1, their dosing and place has varied, but they share common aims – to induce and maintain disease remission. The thiopurines are often used together with corticosteroids with the aim of reducing the doses of the corticosteroid. While efficacious in many

Thiopurines: Azathioprine, Mercaptopurine, and Thioguanine, Table 1 Clinical use and efficacy of standard doses of thiopurines

Disease	Dose (per day)	Clinical use	Efficacy
Crohn's disease	AZA 2.0–2.5 mg/kg 6MP 1.0–1.5 mg/kg 6TG 20–40 mg	1st line: Mild-moderate disease – monotherapy and steroid sparing 1st line: Moderate-severe disease – in combination with TNFα antagonists	30–70 % achieve steroid free clinical remission (Colombel et al. 2010)
Ulcerative colitis	AZA 2.0–2.5 mg/kg 6MP 1.0–1.5 mg/kg 6TG 20–40 mg	2nd line: For patients failing 5ASAs and steroid sparing	50–66 % achieve steroid free clinical remission (Timmer et al. 2012)
Inflammatory myopathies (dermatomyositis and polymyositis)	AZA 2.5–3.0 mg/kg	1st line: In combination with extended steroid taper	Requires 4–6 months for full efficacy (Bunch et al. 1980)
Systemic vasculitides [granulomatosis with polyangiitis (Wegener's) and microscopic polyangiitis]	AZA 2 mg/kg	1st line: Maintain remission after induction therapy	As effective as methotrexate (Pagnoux et al. 2008) Superior to mycophenolate mofetil (Hiemstra et al. 2010)
Behcet's disease	AZA 2.5 mg/kg	1st line: Major organ involvement	Number to treat for prevention of uveitis = 2 (Hatemi et al. 2008)
Rheumatoid arthritis	AZA 150 mg	3rd line: Only if failed or intolerant of multiple other agents	Superior to placebo (Woodland et al. 1981) Inferior to 15 mg methotrexate (Jeurissen et al. 1991)
Refractory psoriasis and psoriatic arthropathy	AZA 3 mg/kg 6TG 160 mg thrice weekly	2nd line: Refractory cases	Pulsed 6TG superior to AZA with 78 % achieving complete clearance (Menter et al. 2009)

inflammatory diseases, there are wide interpatient differences in the responses to the thiopurines.

The role of 6TG in management of IBD is controversial. After initial reports that 6TG caused severe liver injuries (nodular regenerative hyperplasia and veno-occlusive disease) in both IBD and hematological malignancy patients, 6TG was no longer considered an alternative to conventional thiopurines. However, with the advent of thiopurine metabolite testing, it was discovered that these complications are probably related to the high dose of 6TG used (in adults 40–100 mg/day), leading to toxic 6TGN levels in excess of 1,000 pmol $\times 10^8$ RBCs (Dubinsky et al. 2003) well above the therapeutic window of 235–450 pmol $\times 10^8$ RBCs. In patients using a lower dose of 20 mg of 6TG and titrating to a 6TGN level of approximately 600 pmol $\times 10^8$ RBCs, long-term follow-up shows no evidence of liver injury (de Boer et al. 2008). Therefore, low-dose 6TG with dose titration to achieve therapeutic 6GTN levels remains an option for patients who have failed methotrexate and are intolerant to both thiopurines and still require immunomodulator therapy for management of their IBD.

Thiopurine Dosing Regimens

Fixed Dosing Regimen

Low-dose thiopurines at a fixed dose has only been applied in conjunction with steroids for the treatment of autoimmune hepatitis. However, both the American Association for the Study of Liver Diseases and the British Society of Gastroenterology now recommend initial treatment for autoimmune hepatitis with 30 mg/day of prednisolone together with a low, but weight-based, AZA at 1 mg/kg/day (Manns et al. 2010).

Weight-Based Dosing Regimen

It has been conventional to determine the target dose of thiopurine on the basis of the patient's weight, but, as is evident in Table 1, this has not been uniform across disease groups. Most use 1.0–1.5 mg/kg/day for 6MP and 2.0–2.5 mg/kg/day for AZA. Response rates in patients with IBD with such dosing vary between 42 % and 75 %, while flares are prevented in up to 75 % of patients with lupus nephritis, and joint swelling is reduced by at least 50 % in 33 % of patients with RA (Urowitz et al. 1973). AZA is also efficacious in the treatment of ANCA-associated vasculitis and polyarteritis nodosa.

Metabolite-Directed Dosing

The proportion of patients who achieve adequate efficacy using weight-based approaches has been far from ideal. The use of thiopurine metabolites to individualize therapy has enabled optimization of outcomes for patients and is now, at least in IBD, the preferred approach.

Correlation between 6TGN and clinical remission was demonstrated in a 1996 cohort of 25 Canadian adolescent IBD patients receiving 6MP for more than 4 months. There was a significant inverse relationship between disease activity and 6TGN levels (Cuffari et al. 1996). The follow-up study showed that higher 6TGN levels were observed in responders than in nonresponders. A secondary analysis found that if 6TGN levels were greater than 235 pmol $\times 10^8$ RBCs, then patients had an odds ratio (OR) of 5.0 of being a responder. There was also no difference in the weight-based dose for responders than in nonresponders with median dosage of 6MP in both groups being 1.25 mg/kg. The dose of 6MP correlated poorly with 6TGN levels (Dubinsky et al. 2000). In a 2006 pooled analysis of twelve studies (including these two), a 6TGN level above 230–260 pmol $\times 10^8$ RBCs had a pooled OR for remission of 3.27 (Osterman et al. 2006). Using a 6TGN threshold of 230 pmol $\times 10^8$ RBCs, an updated meta-analysis published in 2013 including 20 studies of 2,234 IBD patients found the pooled OR was 2.09 for remission (Moreau et al. 2013).

The other way that 6TGN levels have been shown to relate to clinical efficacy has been via dose escalation. In patients with a subtherapeutic concentration (below 230) and active IBD, clinical remission can be achieved in over 75 % with dose escalation. It is now accepted that for IBD patients with active disease, to achieve maximal efficacy with thiopurines, a patient's dose should be escalated until their 6TGN is in excess of 260.

Recent evidence also suggests that 6TGN measurement is beneficial in autoimmune hepatitis. There is wide variability in 6TGN levels despite identical weight-based dosing of 2 mg/kg/day with no correlation between 6TGN levels and dose. Patients in remission (ALT < 33 IU/mL) have higher 6TGN levels than those with active disease (Dhaliwal et al. 2012), but further studies are still required to establish the therapeutic window for 6TGN levels in autoimmune hepatitis.

There is a paucity of research investigating the measurement of thiopurine metabolites in rheumatology, dermatology, transplantation medicine, and hematology.

Clusters of Clinical Response and Thiopurine Metabolites

The measurement of thiopurine metabolites generates five patient clusters in the treatment of IBD (Haines et al. 2011; Kennedy et al. 2013) (Table 2):

- *Nonadherent*, where this can be as high as 9 %. Patient education and other strategies are indicated.
- *Underdosed/rapid metabolizers* where 6TGN levels are subtherapeutic. This occurs in between 29 % and 43 % of patients with active disease. This can be readily corrected by increasing the dose.
- *Thiopurine shunting*, where the 6MMP:6TGN ratio is high (>20). This occurs in about 10 % of patients. Experience has shown that increasing the dose of the drug increases 6MMP levels usually without the desired increase in 6TGN levels. Two strategies have been described for shunters. The first is splitting the dose of thiopurine to twice daily (Shih et al. 2012). This had a small impact on the ratio in one

Thiopurines: Azathioprine, Mercaptopurine, and Thioguanine, Table 2 Thiopurine metabolite results, interpretation, recommended action, and prevalence (Chouchana et al. 2012)

	Metabolite result[a]	Interpretation	Action recommended	Approximate prevalence
Group 1	No/very low 6TGN (<50) No/very low 6MMP (<50)	Nonadherence	Educate	10 %
Group 2	Low 6TGN (<260) Low 6MMP (<5,700)	Underdosed Rapid metabolizers	Dose escalate	30 %
Group 3	Low 6TGN (<260) High 6MMP (>5,700)	Thiopurine shunter	Add allopurinol and dose reduced to 25 % of original thiopurine dose	10 %
Group 4	Therapeutic 6TGN (260–450) Low or high 6MMP	Refractory	Change therapy	40 %
Group 5	High 6TGN (>450) Low or high 6MMP	Overdosed or refractory	Consider dose reduction or change in therapy	10 %

[a] All values in pmol/8×10^8 RBCs

study, but there was also a modest fall in 6TGN levels which would be unwanted if the patient has active disease. The second strategy is the concomitant use of allopurinol (Sparrow et al. 2005). This has been shown to increase the response to AZA and 6MP even while allowing a reduction in the response to the thiopurines (Sparrow et al. 2007). In the IBD setting, the utility of AZA/allopurinol combination therapy was first described in 2005. Myelosuppression is a potential complication, but counts recover and remain within normal range with a temporary drug cessation and subsequently reduced thiopurine dose. Unfortunately, all publications are retrospective analyses of prospectively collected data, and a range of allopurinol dosages (50–300 mg/day) and a variety of thiopurine dose reduction strategies were used.

- *Refractory*, where levels are therapeutic. This occurs in only about 40 %–60 % of patients who have an inadequate response to thiopurine therapy. Other therapeutic approaches are required in such patients.
- *Overdosed*, where levels are supratherapeutic. This has been observed in 20 % of patients. Whether dose reduction in these patients is indicated has not been evaluated, but drivers for such an approach include the association of higher levels with adverse effects, potential for occult hepatic toxicity (as experience with 6TG therapy discussed above) and the lack of evidence of a therapeutic benefit above 450.

Thus, metabolite-directed optimization of thiopurine usage can lead to improved outcomes in up to 85 % of IBD patients and avoid inappropriate treatment escalation in 25 % (Haines et al. 2011). There is now a compelling case for the application of thiopurine metabolite testing as standard of care in order to individualize therapy and achieve better outcomes. Similar research is yet to be undertaken in other areas of medicine.

Initiation of Therapy

There is no useful data comparing AZA or 6MP as first choice of therapy. Pretreatment assessment of

TPMT activity to guide the initial dose and to avoid life-threatening myelosuppression from TPMT deficiency is valid. Higher doses can be initiated if TPMT activity is normal. However, it must be remembered that TPMT activity is not a perfect guide to thiopurine dosage and outcomes of metabolite results and does not replace the need for regular blood monitoring. If TPMT testing is not available prior to commencement of treatment, an escalating dosage strategy is recommended. A low dose (50 mg/day for AZA or 25 mg/day for 6MP) should be initiated, increasing the dose fortnightly by a similar dose, provided leukocyte count and liver function tests remain within normal limits, until the target dose is reached. It is possible that dose escalation can improve tolerance by reducing the rates of side effects such as gastrointestinal disturbance, fatigue, myalgias, and arthralgias, but no studies have compared a dose-escalation protocol to initial full-dose thiopurine.

Adverse Effects and Monitoring

Acute and Idiosyncratic Adverse Effects

Idiosyncratic or dose-independent reactions to thiopurines such as nausea, vomiting, pancreatitis, myalgias, arthralgias, fatigue, fevers, and a flu-like illness can affect between 25 % and 40 % of patients (Hindorf et al. 2006a; Ansari et al. 2008). Pancreatitis is a rare idiosyncratic event (seen in less than 5 % of patients) that usually occurs within 3 weeks of commencement of a thiopurine. Thiopurine-induced pancreatitis is defined as clinical symptoms consistent with pancreatitis and a concomitant rise in serum lipase and/or amylase at least four times the upper limit of normal (Ansari et al. 2008). Fortunately, it resolves rapidly when the drug is withdrawn without sequelae. Another very rare idiosyncratic adverse event is acute hepatic cholestasis which can occur within 4 weeks of commencement of a thiopurine. It is unrelated to high 6MMP levels.

Dose-Dependent Reactions

There are three classes of dose-dependent adverse reactions: myelosuppression, hepatotoxicity, and nonspecific symptoms such as nausea, vomiting, myalgias, arthralgias, and fatigue.

Correlation Between 6TGN Levels and Myelosuppression

Myelosuppression, in particular leukopenia, tends to occur later than other thiopurine side effects. In non-TPMT-deficient patients, myelosuppression can occur as early as 3 months after the commencement of therapy (Ansari et al. 2008) but can be as late as 18 months (Hindorf et al. 2006a). An upper 6TGN limit of 450 pmol/ 8×10^8 RBCs was identified in the IBD literature, above which the risk of leukopenia is significant, and there is a correlation between higher 6TGN levels and leukopenia (Dubinsky et al. 2000). High 6TGN levels have also been associated with an increased risk of any adverse event with approximately 40 % of patients with a 6TGN above 400 pmol/8×10^8 RBCs may experience an adverse event, including myelotoxicity and gastrointestinal disturbances (Hindorf et al. 2006b).

It is now accepted that levels of 6MMP in excess of 5,000 pmol/8×10^8 RBCs are associated with hepatotoxicity in the form of elevated levels of hepatic transaminases. There is no correlation with 6MMP levels and therapeutic response or 6MP dose. There is also no correlation between 6TGN levels and hepatotoxicity (Dubinsky et al. 2000).

Patients who preferentially produce 6MMP rather than 6TGN are known as thiopurine shunters (see below). They are characterized by having a 6MMP to 6TGN ratio in excess of 20. It is this group that is at risk for hepatotoxicity and likely to be refractory to standard thiopurine therapy.

Long-Term Sequelae of Thiopurines

Severe liver injury due to nodular regenerative hyperplasia and veno-occlusive disease, while most commonly associated with 6TG as detailed above, also occurs in patients exposed to AZA or 6MP with a 10-year cumulative risk of up to 1.25 % (Vernier-Massouille et al. 2007).

Another long-term risk of thiopurines is lymphoproliferative disorders, particularly non-Hodgkin's lymphoma. It is estimated that for IBD patients receiving thiopurines, their risk of non-Hodgkin's lymphoma is between four and five times the general population (Beaugerie et al. (2009). There are no large-scale studies evaluating thiopurines and lymphoproliferative disorders in rheumatological or dermatological diseases.

Monitoring During Maintenance Therapy

Skin Cancer

The risk of nonmelanoma skin cancer is up to six times higher in patients on thiopurines compared with people who have never taken thiopurines. There is a persistent fourfold increased risk in patients who have been previously exposed and subsequently ceased thiopurines. As such, it is recommended that patients protect themselves from ultraviolet radiation and undergo lifelong dermatological surveillance (Peyrin-Biroulet et al. 2011).

Myeloid and Hepatic Effects

Once the target dose has been reached, 3-monthly full blood examination and liver function tests are considered mandatory due to the risk of late myelosuppression and hepatic injury (Lichtenstein et al. 2009).

Infection Monitoring

As thiopurines affect T-cell function, patients receiving thiopurines should theoretically suffer from added infections which are more severe. No specific monitoring or prophylaxis for occult infection is required with thiopurines, but it is recommended that should a patient develop a systemic infection while on thiopurines, the thiopurine should be ceased until the patient has recovered (Rahier et al. 2009).

Managing Thiopurine-Induced Adverse Effects

A major reason for failure of thiopurine therapy is the occurrence of adverse effects leading to cessation of treatment. Most episodes of myelosuppression and hepatotoxicity relate to elevated 6TGN and 6MMP levels, respectively. As such, should either occur, thiopurine metabolites should be measured and a reduced dose with or without the addition of allopurinol is indicated.

A newer development, however, is the management of idiosyncratic reactions to thiopurines such as nausea, vomiting, myalgias, arthralgias, fatigue, fevers, and a flu-like illness, which can affect up to 25 % of patients (Hindorf et al. 2006a). In IBD in particular, where the therapeutic options are more limited, labeling the patient with a "thiopurine allergy" is undesirable. On the basis that the adverse events are due to the primary drug used and not its metabolites, up to 50 % of patients who develop an idiosyncratic reaction on AZA can be safely switched to 6MP without recurrence of the adverse event (Hindorf et al. 2006a) perhaps because of AZA's imidazole group (McGovern et al. 2002). Low-dose allopurinol with dose-reduced thiopurine (irrespective of shunter status) can also be used overcome these side effects on the premise that adverse events might be related to concentration of the primary drug or its methylated metabolites. This maneuver can be effective in up to 86 % of patients (Ansari et al. 2010; Smith et al. 2012).

Thus, if an adverse event occurs on AZA, it is worthwhile to have a trial of 6MP (initially at low dose) and, if that fails, then the addition of low-dose allopurinol with 6MP, but only if a recurrence of the adverse event would be tolerated by the patient. If the adverse event occurs on 6MP as the initial drug, anecdotal experience suggests a trial of AZA may also be worthwhile, followed by combination therapy if unsuccessful.

Thiopurine Methyltransferase (TPMT): Monitoring and Toxicity

Multiple TPMT polymorphisms result in decreased TPMT activity and cause early myelosuppression from thiopurine therapy (Black et al. 1998). The prevalence of low or absent TPMT activity is approximately 1 in 300 patients who, if treated with full-dose thiopurines, will suffer life-threatening myelosuppression (Lennard et al. 1989). Between 4 % and 11 % of individuals are heterozygous for a variant and have intermediate enzyme activity. Activity testing should identify most patients at risk, but there is no consensus as to whether TPMT genotyping or phenotyping (activity testing) is the preferred test.

In theory, pre-treatment TPMT measurement modifies initiation dosage and mitigates myelosuppression, especially in patients with TPMT genetic polymorphisms. Utilisation of genotype testing for the three most common variants in TPMT (TPMT*2, TPMT*3A, and TPMT*3C) to detect patients with decreased activity leads to a ten-fold reduction in the risk of leucopenia in these patients, providing the initial thiopurine dose is reduced to 50 % of the standard weight-based dose. It is, therefore, recommended to perform TPMT testing prior to the commencement of therapy. However, it must not replace frequent blood count monitoring during commencement of therapy, as the vast majority of patients who develop leucopenia have normal TPMT levels (Coenen et al. 2015).

Drug Interactions

There are two major drug interactions with thiopurines that have direct relevance to metabolite testing. The first is with allopurinol, a potent inhibitor of xanthine oxidase, one of the critical enzymes involved in thiopurine metabolism. Allopurinol is the mainstay of treatment for gout. Traditional teaching has dictated that, because the combination of allopurinol and thiopurines causes profound myelosuppression, the two drugs should never be given in combination. However, its effect on the metabolism of thiopurines is now being used to advantage (see above). The downside of such combination therapy is that the patient is exposed to potential adverse effects of two drugs. Allopurinol is generally very well tolerated in the long term. However, rash and rare severe adverse effects such as Stevens-Johnson syndrome can occur. Studies to elucidate allopurinol's beneficial action in combination with thiopurines are needed.

The second interaction is with 5-aminosalicylates (balsalazide, mesalamine, olsalazine, or sulfasalazine) which are used frequently in IBD patients and sometimes in rheumatological conditions. Sulfasalazine and olsalazine inhibit TPMT in vitro, indicating that these drugs may increase 6TGN levels and potentially lead to myelosuppression (Lewis et al. 1997; Szumlanski et al. 1995). This effect has also been confirmed as clinically significant, but not dose dependent. The addition of mesalamine increases 6TGN levels by 50 % in a dose-independent fashion. A similar decrease in 6MMP levels can be seen, but this seems to be dose dependent. There is also a favorable improvement (i.e., fall) in the ratio of 6MMP:6TGN (de Graaf et al. 2010). This clinical effect does not appear to occur with balsalazide (Lowry et al. 2001).

Summary

Thiopurines have been the mainstay of treatment for many autoimmune conditions, transplantation medicine, and in management of hematological malignancies for over half a century. It is their end group of metabolites 6TGN that causes apoptosis and prevents activation and proliferation of T lymphocytes, thereby exerting a clinical effect. AZA and 6MP remain the most commonly used thiopurines, with 6TG reserved for exceptional circumstances. With the ability to measure thiopurine metabolites, important strides have been made in the IBD world to improve efficacy and optimize dosing of thiopurines, including in combination with low-dose allopurinol. In IBD, a therapeutic window of 235–450 pmol/8×10^8 RBCs has been established. Above this level, there are significantly increased risks of side effects, including myelotoxicity, without any gain in efficacy. In IBD, over 30 % of patients who would previously have been declared refractory or intolerant to thiopurines are now otherwise able to remain on monotherapy with improved clinical outcomes. Much of this work has yet to be undertaken within other areas of medicine. While the upper limit of 6TGN is a relevant threshold that has been established in other diseases due to the risk of universal side effects, the minimum effective 6TGN level is yet to be determined in other conditions. The addition of allopurinol should also improve thiopurine metabolic profiles in patients who are thiopurine shunters. It is prudent for patients failing thiopurines to have their metabolites checked prior to drug cessation.

References

Ansari, A., Arenas, M., Greenfield, S. M., Morris, D., Lindsay, J., Gilshenan, K., et al. (2008). Prospective evaluation of the pharmacogenetics of azathioprine in the treatment of inflammatory bowel disease. *Alimentary Pharmacology & Therapeutics, 28*, 973–983.

Ansari, A., Patel, N., Sanderson, J., O'Donohue, J., Duley, J. A., Florin, T. H., et al. (2010). Low-dose azathioprine or mercaptopurine in combination with allopurinol can bypass many adverse drug reactions in patients with inflammatory bowel disease. *Alimentary Pharmacology & Therapeutics, 31*, 640–647.

Beaugerie, L., Brousse, N., Bouvier, A. M., Colombel, J. F., Lemann, M., Cosnes, J., et al. (2009). Lymphoproliferative disorders in patients receiving thiopurines for inflammatory bowel disease: A prospective observational cohort study. *Lancet, 374*, 1617–1625.

Black, A. J., McLeod, H. L., Capell, H. A., Powrie, R. H., Matowe, L. K., Pritchard, S. C., et al. (1998). Thiopurine methyltransferase genotype predicts therapy-limiting severe toxicity from azathioprine. *Annals of Internal Medicine, 129*, 716–718.

Bunch, T. W., Worthington, J. W., Combs, J. J., Ilstrup, D. M., & Engel, A. G. (1980). Azathioprine with prednisone for polymyositis. A controlled, clinical trial. *Annals of Internal Medicine, 92*, 365–369.

Chan, G. L., Erdmann, G. R., Gruber, S. A., Matas, A. J., & Canafax, D. M. (1990). Azathioprine metabolism: Pharmacokinetics of 6-mercaptopurine, 6-thiouric acid and 6-thioguanine nucleotides in renal transplant patients. *Journal of Clinical Pharmacology, 30*, 358–363.

Chouchana, L., Narjoz, C., Beaune, P., Loriot, M. A., & Roblin, X. (2012). Review article: The benefits of pharmacogenetics for improving thiopurine therapy in inflammatory bowel disease. *Alimentary Pharmacology & Therapeutics, 35*, 15–36.

Coenen, M. J., de Jong, D. J., van Marrewijk, C. J., Derijks, L. J., Vermeulen, S. H., et al. (2015). Identification of patients With variants in TPMT and dose reduction reduces hematologic events during thiopurine treatment of inflammatory bowel disease. *Gastroenterology 149*, 907–917.

Colombel, J. F., Sandborn, W. J., Reinisch, W., Mantzaris, G. J., Kornbluth, A., et al. (2010). Infliximab, azathioprine, or combination therapy for Crohn's disease. *New England Journal of Medicine, 362*, 1383–1395.

Cuffari, C., Theoret, Y., Latour, S., & Seidman, G. (1996). 6-Mercaptopurine metabolism in Crohn's disease: Correlation with efficacy and toxicity. *Gut, 39*, 401–406.

de Boer, N. K., Zondervan, P. E., Gilissen, L. P., den Hartog, G., Westerveld, B. D., Derijks, L. J., et al. (2008). Absence of nodular regenerative hyperplasia after low-dose 6-thioguanine maintenance therapy in inflammatory bowel disease patients. *Digestive and Liver Disease, 40*, 108–113.

de Graaf, P., de Boer, N. K., Wong, D. R., Karner, S., Jharap, B., Hooymans, P. M., et al. (2010). Influence of 5-aminosalicylic acid on 6-thioguanosine phosphate metabolite levels: A prospective study in patients under steady thiopurine therapy. *British Journal of Pharmacology, 160*, 1083–1091.

Dhaliwal, H. K., Anderson, R., Thornhill, E. L., Schneider, S., McFarlane, E., Gleeson, D., et al. (2012). Clinical significance of azathioprine metabolites for the maintenance of remission in autoimmune hepatitis. *Hepatology, 56*, 1401–1408.

Dubinsky, M. C., Lamothe, S., Yang, H. Y., Targan, S. R., Sinnett, D., Theoret, Y., et al. (2000). Pharmacogenomics and metabolite measurement for 6-mercaptopurine therapy in inflammatory bowel disease. *Gastroenterology, 118*, 705–713.

Dubinsky, M. C., Vasiliauskas, E. A., Singh, H., Abreu, M. T., Papadakis, K. A., Tran, T., et al. (2003). 6-thioguanine can cause serious liver injury in inflammatory bowel disease patients. *Gastroenterology, 125*, 298–303.

Eklund, B.I., Moberg, M., Bergquist, J., Mannervik, B. (2006). Divergent activities of human glutathione transferases in the bioactivation of azathioprine. *Mol Pharmacol, 70*, 747–754.

Haines, M. L., Ajlouni, Y., Irving, P. M., Sparrow, M. P., Rose, R., Gearry, R. B., et al. (2011). Clinical usefulness of therapeutic drug monitoring of thiopurines in patients with inadequately controlled inflammatory bowel disease. *Inflammatory Bowel Diseases, 17*, 1301–1307.

Hatemi, G., Silman, A., Bang, D., Bodaghi, B., Chamberlain, A. M., Gul, A., et al. (2008). EULAR recommendations for the management of Behcet disease. *Annals of the Rheumatic Diseases, 67*, 1656–1662.

Hiemstra, T. F., Walsh, M., Mahr, A., Savage, C. O., de Groot, K., Harper, L., et al. (2010). Mycophenolate mofetil vs azathioprine for remission maintenance in antineutrophil cytoplasmic antibody-associated vasculitis: A randomized controlled trial. *JAMA, 304*, 2381–2388.

Hindorf, U., Lindqvist, M., Hildebrand, H., Fagerberg, U., & Almer, S. (2006a). Adverse events leading to modification of therapy in a large cohort of patients with inflammatory bowel disease. *Alimentary Pharmacology & Therapeutics, 24*, 331–342.

Hindorf, U., Lindqvist, M., Peterson, C., Soderkvist, P., Strom, M., Hjortswang, H., et al. (2006b). Pharmacogenetics during standardised initiation of thiopurine treatment in inflammatory bowel disease. *Gut, 55*, 1423–1431.

Jeurissen, M. E., Boerbooms, A. M., van de Putte, L. B., Doesburg, W. H., Mulder, J., Rasker, J. J., et al. (1991). Methotrexate versus azathioprine in the treatment of rheumatoid arthritis. A forty-eight-week randomized, double-blind trial. *Arthritis and Rheumatism, 34*, 961–972.

Jharap, B., de Boer, N., Vos, R., Smid, K., Zwiers, A., & Peters, G. (2011). Biotransformation of 6-thioguanine in inflammatory bowel disease patients: A comparison of oral and intravenous administration of

6-thioguanine. *British Journal of Pharmacology, 163*, 722–731.

Kennedy, N. A., Asser, T. L., Mountifield, R. E., Doogue, M. P., Andrews, J. M., & Bampton, P. A. (2013). Thiopurine metabolite measurement leads to changes in management of inflammatory bowel disease. *Internal Medicine Journal, 43*, 278–286.

Lennard, L., Van Loon, J. A., & Weinshilboum, R. M. (1989). Pharmacogenetics of acute azathioprine toxicity: Relationship to thiopurine methyltransferase genetic polymorphism. *Clinical Pharmacology & Therapeutics, 46*, 149–154.

Lewis, L. D., Benin, A., Szumlanski, C. L., Otterness, D. M., Lennard, L., Weinshilboum, R. M., et al. (1997). Olsalazine and 6-mercaptopurine-related bone marrow suppression: A possible drug-drug interaction. *Clinical Pharmacology & Therapeutics, 62*, 464–475.

Lichtenstein, G. R., Hanauer, S. B., Sandborn, W. J., & Practice Parameters Committee of American College of Gastroenterology. (2009). Management of Crohn's disease in adults. *American Journal of Gastroenterology, 104*, 465–483; quiz 464, 484.

Lowry, P. W., Franklin, C. L., Weaver, A. L., Szumlanski, C. L., Mays, D. C., Loftus, E. V., et al. (2001). Leucopenia resulting from a drug interaction between azathioprine or 6-mercaptopurine and mesalamine, sulphasalazine, or balsalazide. *Gut, 49*, 656–664.

Manns, M. P., Czaja, A. J., Gorham, J. D., Krawitt, E. L., Mieli-Vergani, G., Vergani, D., et al. (2010). Diagnosis and management of autoimmune hepatitis. *Hepatology, 51*, 2193–2213.

McGovern, D. P., Travis, S. P., Duley, J., el Shobowale-Bakre, M., & Dalton, H. R. (2002). Azathioprine intolerance in patients with IBD may be imidazole-related and is independent of TPMT activity. *Gastroenterology, 122*, 838–839.

Menter, A., Korman, N. J., Elmets, C. A., Feldman, S. R., Gelfand, J. M., Gordon, K. B., et al. (2009). Guidelines of care for the management of psoriasis and psoriatic arthritis: Section 4. Guidelines of care for the management and treatment of psoriasis with traditional systemic agents. *Journal of the American Academy of Dermatology, 61*, 451–485.

Moreau, A. C., Laporte, S., Del Tedesco, E., Rinaudo-gaujous, M., Phelip, J. M., Paul, S., et al. (2013). Association between thiopurines metabolites levels and clinical remission in IBD patients: An updated meta-analysis. *Gastroenterology, 144*, S-92.

Osterman, M. T., Kundu, R., Lichtenstein, G. R., & Lewis, J. D. (2006). Association of 6-thioguanine nucleotide levels and inflammatory bowel disease activity: A meta-analysis. *Gastroenterology, 130*, 1047–1053.

Pagnoux, C., Mahr, A., Hamidou, M. A., Boffa, J. J., Ruivard, M., Ducroix, J. P., et al. (2008). Azathioprine or methotrexate maintenance for ANCA-associated vasculitis. *New England Journal of Medicine, 359*, 2790–2803.

Peyrin-Biroulet, L., Khosrotehrani, K., Carrat, F., Bouvier, A. M., Chevaux, J. B., Tabassome, S., et al. (2011). Increased risk for nonmelanoma skin cancers in patients who receive thiopurines for inflammatory bowel disease. *Gastroenterology, 141*, 1621–1628.

Prefontaine, E., Macdonald, J.K., & Sutherland L.R. (2010). Azathioprine or 6-mercaptopurine for induction of remission in Crohn's disease. *Cochrane Database of Systematic Reviews, 16*(6):CD000545.

Rahier, J. F., Ben-Horin, S., Chowers, Y., Conlon, C., De Munter, P., D'Haens, G., et al. (2009). European evidence-based Consensus on the prevention, diagnosis and management of opportunistic infections in inflammatory bowel disease. *Journal of Crohn's and Colitis, 3*, 47–91.

Shih, D. Q., Nguyen, M., Zheng, L., Ibanez, P., Mei, L., Kwan, L. Y., et al. (2012). Split-dose administration of thiopurine drugs: A novel and effective strategy for managing preferential 6-MMP metabolism. *Alimentary Pharmacology and Therapeutics, 5*, 449–458.

Smith, M. A., Blaker, P., Marinaki, A. M., Anderson, S. H., Irving, P. M., & Sanderson, J. D. (2012). Optimising outcome on thiopurines in inflammatory bowel disease by co-prescription of allopurinol. *Journal of Crohn's and Colitis, 6*, 905–912.

Sparrow, M. P., Hande, S. A., Friedman, S., Lim, W. C., Reddy, S. I., Cao, D., et al. (2005). Allopurinol safely and effectively optimizes thioguanine metabolites in inflammatory bowel disease patients not responding to azathioprine and mercaptopurine. *Alimentary Pharmacology and Therapeutics, 5*, 441–446.

Sparrow, M. P., Hande, S. A., Friedman, S., Cao, D., & Hanauer, S. B. (2007). Effect of allopurinol on clinical outcomes in inflammatory bowel disease nonresponders to azathioprine or 6-mercaptopurine. *Clinical Gastroenterology and Hepatology, 2*, 209–214.

Szumlanski, C. L., & Weinshilboum, R. M. (1995). Sulphasalazine inhibition of thiopurine methyltransferase: Possible mechanism for interaction with 6-mercaptopurine and azathioprine. *British Journal of Clinical Pharmacology, 39*, 456–459.

Timmer, A., McDonald, J. W., Tsoulis, D. J., & Macdonald, J. K. (2012). Azathioprine and 6-mercaptopurine for maintenance of remission in ulcerative colitis. *Cochrane Database of Systematic Reviews, 9*, CD000478.

Urowitz, M. B., Gordon, D. A., Smythe, H. A., Pruzanski, W., & Ogryzio, M. A. (1973). Azathioprine in rheumatoid arthritis. A double-blind, cross over study. *Arthritis and Rheumatism, 16*, 411–418.

Vernier-Massouille, G., Cosnes, J., Lemann, M., Marteau, P., Reinisch, W., Laharie, D., et al. (2007). Nodular regenerative hyperplasia in patients with inflammatory bowel disease treated with azathioprine. *Gut, 56*, 1404–1409.

Woodland, J., Chaput de Saintonge, D. M., Evans, S. J., Sharman, V. L., & Currey, H. L. (1981). Azathioprine in rheumatoid arthritis: Double-blind study of full versus half doses versus placebo. *Annals of the Rheumatic Diseases, 40*, 355–359.

Toll-Like Receptors

Elizabeth Brint[1] and Philana Fernandes[2]
[1]Department of Pathology, Clinical Sciences Building, Cork University Hospital, University College Cork, National University of Ireland, Cork, Ireland
[2]Cork Cancer Research Centre, 5th Floor Biosciences Institute Rm 5.27, University College Cork, National University of Ireland, Cork, Ireland

Synonyms

TLR

Definitions

Toll-like receptors (TLRs) are an essential component of the innate immune system and can be considered as a first line of defense against microbial invasion. They are a well-established family of pattern recognition receptors (PRRs), which detect pathogen-associated molecular patterns (PAMPs). PAMPs are highly conserved molecules associated with microbes which are present during infection. In addition to pathogen-derived danger signals, some TLRs are also known to recognize endogenous danger signals such as those released following cell death and tissue damage. Thus, TLRs recognize both endogenous and exogenous molecules. Upon recognition of these molecules, TLRs become activated and trigger intracellular signaling cascades, the result of which is the activation of the inflammatory response and activation of both innate and adaptive immune systems. The ultimate aim of TLR activation is, therefore, to facilitate host protection through either the destruction of the invading microorganism or through initiation of host repair mechanisms to restore homeostasis.

Structure and Function

The protein Toll was originally identified in the fruit fly *Drosophila melanogaster* as a plasma membrane receptor. While activation of the Toll signaling pathway governs the differentiation of dorsal and ventral structures in the developing fly embryo, it was also observed that a Toll-deficient fly succumbed easily to fungal infections (Lemaitre et al. 1996). This established a novel role for this protein in innate immunity and host defense. In 1997, the first human homologue of Toll was identified, TLR4. This receptor was also shown to be involved in immune responses and, indeed, to be the long-sought-after receptor for lipopolysaccharide (LPS) (Medzhitov et al. 1997; Poltorak et al. 1998). Since this initial discovery, an explosion of research carried out in the TLR field has firmly established these PRRs as key initiators of the innate immune response in vertebrates. To date there are 13 known mammalian TLR genes with ten of these expressed in humans and 12 in mice. TLRs are widely expressed throughout the body, with high expression found in tissues involved in the immune response such as the spleen and peripheral blood leukocytes and with expression located on sentinel immune cells such as macrophages and dendritic cells. Expression is also detectable in non-hematopoietic cells such as the epithelial cells of the lung and the gastrointestinal tract (Takeuchi and Akira 2010).

Structure

TLRs are type I transmembrane domain proteins with a tripartite structure consisting of an amino (N)-terminal extracellular domain (ECD), a single transmembrane spanning region typically containing a stretch of approximately 20 uncharged, mostly hydrophobic, residues, and a carboxyl (C)-terminal globular cytoplasmic domain. The ECD contains leucine-rich repeats (LRRs) appearing in an xLxxLxLxx motif within a stretch of 550–800 amino acid residues and is responsible for ligand recognition.

TLRs are part of a larger superfamily of receptors which includes the IL-1 Receptor family members. All members of this family are characterized by a 200-residue intracellular domain termed the Toll-IL-1 receptor (TIR) domain. This TIR domain has been shown to be critical in the activation of signaling pathways from these receptors. Sequence conservation of the TIR domain is primarily confined to three short motifs (Boxes 1, 2, and 3) located at amino acid residue positions 10 (Box 1), 60 (Box 2), and 170 (Box 3) and with consensus sequences: Box 1, FDAFISY; Box 2, GYKLC–RD–PG; and Box 3, a conserved W surrounded by basic residues. These three regions map to the hydrophobic core of the domain structure as well as a long "BB" loop which had been proposed as the primary protein–protein interaction site. While mutations in all three lead to a loss of surface expression, only mutations in Box 1 and Box 2 cause a direct loss of signaling activity. Initial indications that the TIR domain was critical for TLR signaling were realized when a proline to histidine point mutation within the domain at position 712 of the polypeptide chain was shown to confer lipopolysaccharide (LPS) unresponsiveness to C3H/HeJ mice (Jin and Lee 2008).

Cell-surface TLRs are monomeric but form active homo- or heterodimers when exposed to their ligands. Ligand-induced dimerization of extracellular domains brings the two intracellular TIR domains into close proximity and initiates downstream signaling by providing a platform for recruitment of adaptor proteins. The structures of these TLR-ligand complexes have been extensively studied by X-ray crystallography. Crystallographic studies on TLR1–TLR2, TLR3, and TLR4 have confirmed the dimeric nature of the ligand-bound receptor. Hydrophobic interactions are important in the interaction of TLR2 with TLR1 and TLR6 following lipopeptide binding. TLR4 similarly requires a hydrophobic pocket in order to bind the lipid chains of LPS which it itself lacks but is present in the accessory protein, MD-2. Conversely, hydrophilic interactions play major roles in ligand recognition by TLR3 and TLR5. Despite the differences in their ligand interactions, the overall shape of the TLR-ligand complexes that have been solved by X-ray crystallography is very similar and supports the concept that ligand activation of TLRs brings the BB-loop structure of the two or more TIR domains into close proximity, stabilizing the receptor complex and forming a scaffold for downstream signal transduction (Gay et al. 2014).

Ligand Recognition

TLRs can be broadly subdivided into two groups based on their localization either to the plasma membrane or to acidified endolysosomal compartments. TLRs 1, 2, 4, 5, 6, and 10 are predominantly expressed on the cell surface, whereas TLRs 3, 7, 8, and 9 are mainly expressed on the surfaces of endosomes, lysosomes, and endoplasmic reticulum. Because of the cellular distribution of these TLRs, pathogen recognition can be initiated from a variety of cellular locations depending on the receptor activated. For example, the plasma membrane-associated TLRs respond to components from the microbial surface such as microbial membrane lipids or bacterial proteins, whereas the endosomal TLRs recognize various microbial nucleic acids. Compartmentalization of the nucleic acid-sensing TLRs is thought to be important for preventing autoimmune responses to self-nucleic acids. Expression of several of the cell-surface TLRs is also often detected intracellularly following ligand binding, reflecting receptor trafficking events.

Specific ligands for TLRs have been elucidated through use of *in vitro* overexpression studies, *ex vivo* studies, and knockout mice. It is known, for example, that TLR2 responds to peptidoglycan, lipopeptides, lipoteichoic acid, lipoarabinomannan, GPI anchors, phenol-soluble modulin, zymosan, and glycolipids. TLR2 seems unable to mediate an immune response through recognition of PAMPs independently but rather heterodimerizes with either TLR1 or TLR6 in order to recognize triacylated lipoproteins and diacylated lipoproteins, respectively. TLR4 recognizes and binds to LPS and lipoteichoic acid from gram-positive bacteria as well as a few viral proteins and self-antigens, e.g., fibronectin. Flagellin, found in bacterial flagella, binds to and activates TLR5. TLR3 recognizes viral dsRNA, small interfering RNAs, and self-RNAs derived from damaged cells. TLR7 is predominantly expressed in plasmacytoid DCs (pDCs) and recognizes ssRNA from viruses, and TLR8 recognizes single-stranded RNA (ssRNA) from viruses and bacterial RNA. TLR9 recognizes unmethylated CpG oligodeoxynucleotide DNA from DNA viruses and bacteria (De Nardo

Toll-Like Receptors, Table 1 **TLRs and their ligands.** TLRs: location, respective ligands, and origin of ligands

Receptor	Location	Ligand	Ligand origin
TLR1	Cell surface	Multiple triacyl lipopeptides	Bacteria
TLR2	Cell surface	Multiple glycolipids Multiple lipoproteins HSP70 Zymosan	Bacteria Bacteria Host Fungi
TLR3	Endosomal	Double-stranded RNA	Viruses
TLR4	Cell surface	Lipopolysaccharide Heat-shock proteins Fibrinogen	Gram-negative bacteria Bacterial and host cells Host
TLR5	Cell surface	Flagellin	Bacteria
TLR6	Cell surface	Multiple diacyl lipopeptides	Mycoplasma
TLR7	Endosomal	Single-stranded RNA	RNA viruses
TLR8	Endosomal	Single-stranded RNA	RNA viruses/bacteria
TLR9	Endosomal	Unmethylated CpG Oligodeoxynucleotide DNA	Bacteria, DNA viruses
TLR10	Cell surface/endosomal	?/Unidentified *L. monocytogenes* ligand	?/Bacteria
TLR11	Endosomal	Profilin-like protein	Uropathogenic bacteria and *Toxoplasma gondii*
TLR12	Endosomal	Profilin-like protein	*Toxoplasma gondii*
TLR13	Endosomal	23S ribosomal RNA	Bacteria

2015). TLR10 has recently been shown to recognize as yet unknown ligands from both influenza virus and *Listeria monocytogenes* (Lee et al. 2014; Regan et al. 2013). TLR11 exists only in a truncated form in humans and is believed to be inactive; however, in mice it has been shown to recognize uropathogenic *Escherichia coli* and a profilin-like protein on *Toxoplasma gondii*. Similarly TLRs 12 and 13 are also only expressed in mice. TLR12 has also been shown to recognize the ligand profilin, and TLR13 has been shown to recognize 23S ribosomal RNA (Yarovinsky 2014). A table showing the location of TLRs and their identified ligands is shown in Table 1.

Signal Transduction

After recognition of their respective ligand/PAMP, TLRs activate signaling pathways. The activation of these pathways provides the host with a specific immunological response tailored to the microbe expressing that PAMP. This specific response is determined, in the first instance, by recruitment of a single or a combination of TIR-domain-containing adaptor proteins. The four main adaptor proteins which bind to TLRs and facilitate signal transduction are myeloid differentiation factor 88 (MyD88), MyD88 adaptor-like (MAL), TIR-domain-containing adapter-inducing interferon-β (TRIF), and TRIF-related adaptor molecule (TRAM) (O'Neill and Bowie 2007). TLR signaling is roughly divided into two distinct pathways depending on the usage of the distinct adaptor molecules MyD88 and TRIF.

MyD88-Dependent Signaling by TLRs

MyD88 is involved in mediating signal transduction for all TLRs (apart from TLR3) as well as several IL-1 receptor family members. MyD88 is a modular protein and, along with a TIR domain, it contains a death domain (DD) by which it recruits further downstream signaling components to the receptor complex via homotypic DD interactions. Both TLR4 and TLR2 also utilize MAL (also known as TIRAP) as an additional adaptor on this pathway. MAL localizes to the plasma membrane where it can interact with the TIR domain of activated TLR2 or TLR4 and recruit MyD88. After the MyD88 TIR domain complexes with the receptor, or with the bridging adaptor protein MAL, the DD can be released from a repressed state to enable the recruitment of the IL-1R-associated kinase (IRAK) family of proteins in the next step of the signaling cascade (Akira 2006).

In vitro, the MyD88 DD forms a heterogeneous mixture of dimers and higher-order oligomers, but in the presence of IRAK4, these assemble into a discrete heterocomplex, coined the Myddosome (Gay et al. 2011). The Myd88/IRAK4 interaction promotes IRAK4 autophosphorylation and subsequent death domain-mediated recruitment of IRAK1 and IRAK2. IRAK4 subsequently phosphorylates IRAK1, leading to its activation. Activation of IRAK1 allows for the recruitment of the E3 ubiquitin ligase TRAF6 to the receptor complex followed by its subsequent activation and release into the cytosol. TRAF6 is the activator of the canonical NF-κB pathway. TRAF6 is ubiquitinated at residue K63, and this modification allows TRAF6 to activate the next component in the pathway, TGF-β-activated kinase-1 (TAK1). TAK1 then activates the IKK complex leading to the subsequent phosphorylation of the inhibitory κB (IκB) kinase protein. Under homeostatic conditions, IκB retains NFκB in an inactive state in the cytoplasm of the cell. Upon phosphorylation however, IκB is degraded by the ubiquitin-proteasome system leading to the release of NF-κB, allowing NF-κB to translocate to the cytoplasm and initiate gene transcription. TLR-mediated activation of the MAP kinase cascade also occurs via this pathway until the level of TAK-1. Activation of the MAP kinases results in activation of the AP-1 transcription factor and subsequent activation of AP-1-dependent gene expression (Akira 2006).

TRIF-Dependent Signaling by TLRs

In response to binding of its ligand, dsRNA, TLR3 recruits the adaptor protein TRIF. TLR3 has an alanine rather than a proline in the TIR domain BB loop, and this gives TLR3 its specificity for TRIF-dependent ligation and signaling. Mutation of this one residue in TLR3 to proline causes a switch in adaptor protein specificity from TRIF to MyD88 resulting in attenuated IRF3-dependent signaling and enhanced NFκB activation (Verstak et al. 2013). TLR4 triggers both MyD88- and TRIF-dependent signaling pathways. Relative to MyD88, TRIF has a more complex multimodular structure of 712 amino acids and contains the TIR domain, TRAF-binding domains, and a receptor-interacting protein (RIP) homotypic interaction motif (RHIM) domain.

TRIF is recruited via TIR–TIR interaction to TLR3. TLR4, but not TLR3, requires an additional adaptor TRAM to activate TRIF. Once activated, TRIF can then associate with TRAF3 through TRAF-binding motifs contained within its N-terminal portion. Recruitment of TRAF3 then results in activation of TBK-1 and the IκB kinases IKKε/i which induces phosphorylation and subsequent activation and dimerization of IRF-3 and IRF-7 which migrate to the nucleus. This results in an antiviral response by induction of type-1 interferons. Both TLR3 and TLR4 can also activate NFκB in a TRIF-dependent manner. This requires the binding of RIP1 to the TRIF RHIM domain causing the degradation of the inhibitor of NF-κB (IκB) kinase complex.

The localization of TLR4 is critical in determining whether the MAL-MyD88 pathway or the TRAM-TRIF pathway is activated. For example, TLR4, when located on the plasma membrane, engages with MAL, which subsequently recruits MyD88. TLR4 will subsequently traffic to endosomes, where TRAM is engaged. TRAM subsequently recruits TRIF, resulting in the activation of viral immune response gene transcription factors, the interferon regulatory factors (IRFs) (Kawai and Akira 2010).

A figure demonstrating the most common signaling pathways emanating from TLR ligation is shown in Fig. 1.

Pathological References

Due to their integral role as sentinels of the innate immune response, TLR-mediated inflammation is of paramount importance in resolving infection and mediating tissue healing.

However, due to these profound proinflammatory effects, TLRs have been linked to a

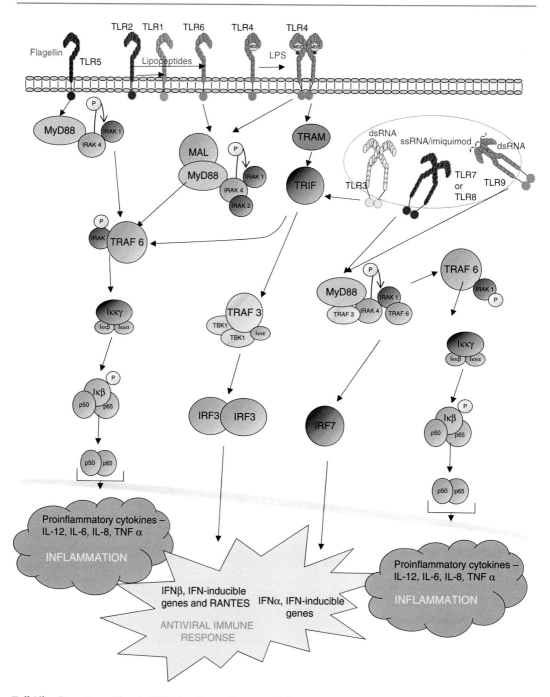

Toll-Like Receptors, Fig. 1 TLR-signaling pathways: Following ligand recognition, each TLR initiates a downstream signaling cascade using adaptor molecules. All TLRs signal through MyD88 with the exception of TLR3 which requires TRIF. Recruitment of MyD88 to the TLR receptor initiates downstream activation of TRAF6, which in turn activates NF-κB resulting in the transcription of inflammatory cytokines. Alternatively, if TLR4 is trafficked to the endosome upon ligand recognition, it recruits TRAM, which recruits TRIF. TLR3 also recruits TRIF upon ligand binding. This initiates downstream activation of TRAF3 and subsequently activates the IRF transcription factors which induce production of the antiviral interferon genes

plethora of inflammatory and immune diseases. Through the use of *in vitro* studies, *in vivo* murine studies and human studies, clear roles for TLRs in the pathogenesis of such conditions as rheumatoid arthritis, sepsis, systemic lupus erythematosus, inflammatory bowel disease, respiratory disease, liver disease, vascular disease and type I diabetes have been demonstrated. Either an aberrant expression pattern of TLRs, over-activation of TLRs, or a dysregulation of the negative TLR regulatory mechanisms may lead to the development of these diseases through the enhanced production of proinflammatory cytokines (Cook et al. 2004). Variation in the sequences of TLR genes and also many of the TLR-signaling intermediates have also been reported in the human population with certain variants increasing the risk of disease. For example, the relatively common D299G polymorphism in the TLR4 receptor increases the risk of gram-negative infections and is associated with an increased incidence of systemic inflammatory syndrome (Netea et al. 2012). These findings have further validated the role of TLRs in the inflammatory process that leads to pathology.

As chronic inflammation may be a favorable environment for tumor initiation and progression, the role of TLRs in cancer has also been investigated. In several premalignant and malignant conditions, the expression of certain TLRs has been shown to be increased, signaling through which can lead to the upregulation of immunosuppressive agents such as vascular endothelial growth factor and transforming growth factor beta within the tumor microenvironment. These modifications can lead to increased angiogenic and metastatic potential of tumors. In this context, TLR4 and TLR2 have been shown to have tumor promoting effects in the colon, liver, pancreas, skin, and stomach (Rakoff-Nahoum and Medzhitov 2009).

Interactions with Other Processes and Drugs

Interactions Between TLRs and Other PRRs

Although TLRs play a central role in the initiation of immune responses against a number of pathogens, it has become apparent that PRRs other than TLRs are also involved in recognition of PAMPs and the control of innate immunity. These include the membrane-bound C-type lectin receptors (CLRs), cytosolic proteins such as NOD-like receptors (NLRs) and RIG-I-like receptors (RLRs), and intracellular proteins that mediate sensing of cytosolic DNA or retrovirus infection. As a single pathogen can express many ligands that can be recognized by this array of PRRs, it is clear that cross talk must exist between TLRs and these other PRRs in order to facilitate an effective, coordinated, and regulated immune response.

Interaction is known to exist between TLRs and NLRs. NLRs, like TLRs, recognize bacterial PAMPs or DAMPs (Caruso et al. 2014). While TLRs are associated with plasma or endosomal membranes, NLRs are usually located in the cytosol. TLRs are thought to interact with a subset of NLRs, the NOD subfamily, in order to augment the production of cytokines. For example, iEDAP, the NOD1 agonist, and MDP, the NOD2 agonist, both synergize with LPS, the TLR4 ligand, to simulate cytokine production in human monocytes and dendritic cells (Fritz et al. 2005). However, since others have demonstrated negative cross talk between TLRs and NODS, it is likely that the interaction is stimuli and cell type specific.

Interaction between TLRs and another subset of NLRs, namely, NALP1, NALP2, NALP3, and IPAF, has also been demonstrated. These proteins are involved in the assembly of cytosolic multimeric protein complexes known as inflammasomes. Inflammasomes consist of an inflammasome sensor molecule, the adaptor protein, apoptosis-associated speck-like protein containing a caspase recruitment domain (ASC), and the protein, caspase-1. Upon inflammasome activation, the cysteine protease caspase-1 is cleaved and activated. Active caspase-1 then cleaves and activates the pro-forms of IL-1β and IL-18 into their proinflammatory cytokine counterparts (Guo et al. 2015). TLR signaling is known to be important in the first steps of inflammasome activation, also known as the inflammasome "priming step." TLR ligation leads to NFκB activation and the subsequent transcription of pro-IL-1β mRNA. Activated caspase-1 subsequently cleaves pro-IL-1β into the biologically active form.

TLR-mediated NF-κB activation is also the traditional priming signal for the transcription of the NLRP3 gene, which forms the basis of the most widely studied inflammasome. Composed of NLRP3, ASC, and procaspase-1, the NLRP3 inflammasome recognizes a wide variety of microbes, including *S. aureus*, *E. coli*, and *C. albicans*, and NLRP3 activation is critical for the resolution of infection by these pathogens. Conversely abnormal activation of the NLRP3 inflammasome is involved in many chronic inflammatory conditions such as atherosclerosis, metabolic syndrome, age-related macular degeneration (AMD), Alzheimer's disease, and gout (Guo et al. 2015).

As both certain TLRs and RLRs are capable of detecting viral infection, cooperation between these has been observed in the resolution of viral infection. For example, hepatitis C virus is recognized by both TLR3 and RIG-I (a member of the RLR superfamily), and the activation of the type I interferon-mediated antiviral response to this infectious microorganism is coordinated between these two receptors. A similar cooperation has also been observed between certain TLRs and members of the CLR superfamily, in order to mount an effective antifungal immune response (Kawai and Akira 2011).

Interactions Between TLRs and the Adaptive Immune Response

The innate immune system is an ancient and diverse collection of defenses, including the recognition of pathogens through the use of germ line-encoded PRRs. The adaptive immune system, by contrast, encompasses T and B cells which utilize somatically recombined antigen receptor genes to recognize virtually any antigen. B and T cells that have encountered antigen persist over the long term within an organism and provide rapid and specific responses to reinfection. Therefore, the adaptive immune response adds both flexibility and memory to the overall immune response.

The activation of the innate immune response, as mediated through TLR activation, plays a critical role in shaping the adaptive immune response. Innate signaling not only precedes but is also known to be essential for the generation of T-cell and B-cell responses. Central to this process are the dendritic cells (DCs), a heterogeneous family of leukocytes that integrate innate information and convey it to lymphocytes. TLRs are mainly expressed on antigen-presenting cells such as DCs, macrophages, and B cells. The engagement of TLRs on DCs leads to increased expression of MHC–peptide complexes and co-stimulatory molecules, as well as the production of immunomodulatory cytokines, all of which have a profound effect on T-cell activation, priming, and differentiation. As activation of T cells leads to subsequent activation of B cells and expansion of antibody-producing plasma cell populations, the importance of the initiating TLR signal in this upregulation of the adaptive immune response is evident (Seledtsov and Seledtsova 2012).

Interactions Between TLRs and Drugs

TLRs represent attractive drug targets for the modulation of the immune response and hold promising applications for the treatment of infection and inflammation. In addition the agonists of TLRs, particularly TLRs 7, 8, and 9, have been actively pursued for their adjuvant effects having potent antiviral and antitumor effects in multiple models of infection (Savva and Roger 2013).

Cross-References

▶ Cancer and Inflammation
▶ IkappaB
▶ Interleukin 18
▶ MAP Kinase Pathways
▶ NFkappaB
▶ Pathogen-Associated Molecular Patterns (PAMPs)
▶ Toll-Like Receptors
▶ Type I Interferons

References

Akira, S. (2006). TLR signaling. *Current Topics in Microbiology and Immunology, 311*, 1–16.

Caruso, R., Warner, N., Inohara, N., & Nunez, G. (2014). NOD1 and NOD2: Signaling, host defense, and inflammatory disease. *Immunity, 41*(6), 898–908. doi:10.1016/j.immuni.2014.12.010.

Cook, D. N., Pisetsky, D. S., & Schwartz, D. A. (2004). Toll-like receptors in the pathogenesis of human disease. *Nature Immunology, 5*(10), 975–979. doi:10.1038/ni1116.

De Nardo, D. (2015). Toll-like receptors: Activation, signalling and transcriptional modulation. *Cytokine, 74*(2), 181–189. doi:10.1016/j.cyto.2015.02.025.

Fritz, J. H., Girardin, S. E., Fitting, C., Werts, C., Mengin-Lecreulx, D., Caroff, M., et al. (2005). Synergistic stimulation of human monocytes and dendritic cells by Toll-like receptor 4 and NOD1- and NOD2-activating agonists. *European Journal of Immunology, 35*(8), 2459–2470. doi: 10.1002/eji.200526286.

Gay, N. J., Gangloff, M., & O'Neill, L. A. (2011). What the Myddosome structure tells us about the initiation of innate immunity. *Trends in Immunology, 32*(3), 104–109. doi:10.1016/j.it.2010.12.005.

Gay, N. J., Symmons, M. F., Gangloff, M., & Bryant, C. E. (2014). Assembly and localization of Toll-like receptor signalling complexes. *Nature Reviews Immunology, 14*(8), 546–558. doi:10.1038/nri3713.

Guo, H., Callaway, J. B., & Ting, J. P. (2015). Inflammasomes: Mechanism of action, role in disease, and therapeutics. *Nature Medicine, 21*(7), 677–687. doi:10.1038/nm.3893.

Jin, M. S., & Lee, J. O. (2008). Structures of the toll-like receptor family and its ligand complexes. *Immunity, 29*(2), 182–191. doi:10.1016/j.immuni.2008.07.007.

Kawai, T., & Akira, S. (2010). The role of pattern-recognition receptors in innate immunity: Update on Toll-like receptors. *Nature Immunology, 11*(5), 373–384. doi:10.1038/ni.1863.

Kawai, T., & Akira, S. (2011). Toll-like receptors and their crosstalk with other innate receptors in infection and immunity. *Immunity, 34*(5), 637–650. doi:10.1016/j.immuni.2011.05.006.

Lee, S. M., Kok, K. H., Jaume, M., Cheung, T. K., Yip, T. F., Lai, J. C., et al. (2014). Toll-like receptor 10 is involved in induction of innate immune responses to influenza virus infection. *Proceedings of the National Academy of Sciences of the United States of America, 111*(10), 3793–3798. doi: 10.1073/pnas.1324266111.

Lemaitre, B., Nicolas, E., Michaut, L., Reichhart, J. M., & Hoffmann, J. A. (1996). The dorsoventral regulatory gene cassette spatzle/Toll/cactus controls the potent antifungal response in Drosophila adults. *Cell, 86*(6), 973–983.

Medzhitov, R., Preston-Hurlburt, P., & Janeway, C. A., Jr. (1997). A human homologue of the *Drosophila* Toll protein signals activation of adaptive immunity. *Nature, 388*(6640), 394–397. doi:10.1038/41131.

Netea, M. G., Wijmenga, C., & O'Neill, L. A. (2012). Genetic variation in Toll-like receptors and disease susceptibility. *Nature Immunology, 13*(6), 535–542. doi:10.1038/ni.2284.

O'Neill, L. A., & Bowie, A. G. (2007). The family of five: TIR-domain-containing adaptors in Toll-like receptor signalling. *Nature Reviews Immunology, 7*(5), 353–364. doi:10.1038/nri2079.

Poltorak, A., He, X., Smirnova, I., Liu, M. Y., Van Huffel, C., Du, X., et al. (1998). Defective LPS signaling in C3H/HeJ and C57BL/10ScCr mice: mutations in Tlr4 gene. *Science, 282*(5396), 2085–2088.

Rakoff-Nahoum, S., & Medzhitov, R. (2009). Toll-like receptors and cancer. *Nature Reviews Cancer, 9*(1), 57–63. doi:10.1038/nrc2541.

Regan, T., Nally, K., Carmody, R., Houston, A., Shanahan, F., Macsharry, J., et al. (2013). Identification of TLR10 as a key mediator of the inflammatory response to Listeria monocytogenes in intestinal epithelial cells and macrophages. *Journal of Immunology, 191*(12), 6084–6092. doi:10.4049/jimmunol.1203245.

Savva, A., & Roger, T. (2013). Targeting toll-like receptors: Promising therapeutic strategies for the management of sepsis-associated pathology and infectious diseases. *Frontiers in Immunology, 4*, 387. doi:10.3389/fimmu.2013.00387.

Seledtsov, V. I., & Seledtsova, G. V. (2012). A balance between tissue-destructive and tissue-protective immunities: A role of toll-like receptors in regulation of adaptive immunity. *Immunobiology, 217*(4), 430–435. doi:10.1016/j.imbio.2011.10.011.

Takeuchi, O., & Akira, S. (2010). Pattern recognition receptors and inflammation. *Cell, 140*(6), 805–820. doi:10.1016/j.cell.2010.01.022.

Verstak, B., Arnot, C. J., & Gay, N. J. (2013). An alanine-to-proline mutation in the BB-loop of TLR3 Toll/IL-1R domain switches signalling adaptor specificity from TRIF to MyD88. *Journal of Immunology, 191*(12), 6101–6109. doi:10.4049/jimmunol.1300849.

Yarovinsky, F. (2014). Innate immunity to *Toxoplasma gondii* infection. *Nature Reviews Immunology, 14*(2), 109–121. doi:10.1038/nri3598.

Tumor Necrosis Factor Alpha (TNFalpha)

David Wallach and Andrew Kovalenko
Department of Biological Chemistry, Weizmann Institute of Science, Rehovot, Israel

Synonyms

Cachectin; Cytotoxin; Lymphotoxins; Necrosin

Definition

Tumor necrosis factor (TNF), like other cytokines, is a small protein that is produced by a variety of cells and acts through specific cell surface receptors to affect a variety of cellular functions, most of which contribute to immune defense. It belongs to a group of structurally related cytokines called the "TNF family," all of which share a common receptor-binding motif and act in trimeric form (in the case of TNF itself and most other members of the family, as homotrimers). All TNF ligand family members operate by triggering signaling in members of a family of single transmembrane-domain receptors. (Some also bind to other receptors as well.) TNF acts by binding to two such receptors, TNFR1 (TNFRSF1A) and TNFR2 (TNFRSF1B; see http://www.genenames.org/genefamilies/TNFSF and http://www.genenames.org/genefamilies/TNFRSF for nomenclature of the various TNF family ligands and receptors, respectively). These two receptors have distinct intracellular domains that trigger distinct signaling mechanisms and initiate distinct sets of functional changes.

Biosynthesis and Release

The Cell-Bound and Soluble Forms of TNF

Like most other members of the TNF ligand family (the only exception being lymphotoxin-α, LT-α, LTA, and a particular splice variant of Fas ligand), TNF is produced as a single-transmembrane type II protein (a transmembrane protein whose C-terminus is directed extracellularly). The receptor-binding motif of TNF occurs at its C-terminus, allowing the transmembrane form of TNF to bind receptors and trigger signaling in them. Such activation, which is evidently restricted to target cells adjacent to the TNF-expressing cell, is termed "juxtacrine."

The cell-bound form of TNF also has an intracellular domain (the N-terminal part of the molecule), which in humans is 29 amino acid long. Association of the cell-bound TNF molecules with TNF receptors not only triggers signaling by the intracellular domains of the these receptors within the TNF target cells but also initiates signaling by the intracellular domain of TNF itself within the TNF-expressing cells. The latter activity, so-called reverse signaling by TNF, was found to have several functional consequences including arrest of TNF synthesis (Eissner et al. 2004).

TNF Shedding

As well as occurring as a cell-bound molecule, TNF also occurs in a soluble form enabling it to reach, via the circulation, cells remote from those that produced it. The soluble form is derived from the cell-bound form by proteolytic cleavage, which is mediated by TACE/ADAM17, a metalloproteinase of the adamalysin family (Black et al. 1997).

Cells Producing TNF

The cells that produce TNF most effectively are the mononuclear phagocytes (Bradley 2008). They are the main source of TNF at times when its levels in the circulation are pathologically high. At lower levels, however, TNF is also produced by a variety of other cells including mast cells, T and B lymphocytes, natural killer (NK) cells, neutrophils, endothelial cells, smooth and cardiac muscle cells, fibroblasts, and osteoclasts (Bradley 2008).

Regulation of TNF Production

Generation of TNF is strictly regulated. Other than in a small number of known situations (see below), it does not occur spontaneously but only upon exposure to pathogenic stimuli – foreign antigens; various pathogens that include bacteria, viruses, and eukaryotic parasites; specific components of pathogens (pathogen-associated molecular patterns, PAMPs) such as bacterial endotoxin or viral double-stranded RNA; and danger-associated molecular patterns (DAMPs). Cells of different types generate TNF in response to different spectra of inducers. The main inducers of TNF are triggers of innate immunity, but TNF is also induced as part of the adaptive immune response. Triggering of antigen receptors induces TNF synthesis in lymphocytes, alongside that of LT-α (Falvo et al. 2010). Moreover, activated T lymphocytes initiate pronounced generation of

TNF in neighboring mononuclear monocytes (e.g., Sebbag et al. 1997).

The various inducers of TNF affect its generation on several different mechanistic levels: transcription of its gene, stabilization of the message, translocation of TNF to the cell surface together with the intracellular membranes to which it is anchored, and shedding of the soluble form of TNF.

In most cells, induction of TNF release is coupled either to the induction of both its synthesis and its shedding or of its shedding alone. An exception is in mast cells, where shed TNF molecules accumulate within intracellular granules and are released by exocytosis in response to specific inducing agents (Walsh et al. 1991).

As mentioned above, in a few cells and situations, TNF is also produced not as an outcome of immune triggering, but as a spontaneously generated occurrence at certain stages of embryonic development (Ohsawa and Natori 1989) as well as in the brain (Clark et al. 2010). The initiators of this homeostatic generation of the cytokine are not yet known.

Biological Activities

TNF is a highly multifunctional ("pleiotropic") cytokine. It affects the growth, survival, motility, differentiation, aging, and death of many cells and modulates a variety of different cellular activities. Its effect on any cell at any particular moment depends on the cell's prior state, and the nature of the effect in different cellular states might be different, even opposite. Most of its known effects serve to facilitate immune defense. As one of the major known mediators of inflammation, TNF plays a particularly prominent role in controlling innate immunity. However, it also has effects that contribute to the regulation of adaptive immunity.

TNF as a Major Coordinator of Cellular Activities that Cooperatively Mediate Stages of Local Inflammation

As described by Valy Menkin more than 80 years ago (Menkin 1948), local inflammation is characterized by various stages that serve distinct roles in the defense against the irritant (see diagrammatic representation in Fig. 1). The cellular changes occurring at the different stages vary and might even be antagonistic in character. TNF mediates functional changes associated with each of these stages and can consequently induce opposing kinds of changes in cells, depending on the stage of the inflammatory process.

Recruitment of Cellular and Humoral Blood Components to the Infection/Injury Site

The visual manifestations of the first stage of inflammation correspond to the classical signs by which inflammation has been defined: heat, reddening, and swelling. TNF contributes pivotally to the cellular changes accounting for this stage via a wide range of effects. These include induced expression of adhesion proteins in endothelial cells and in the leukocytes that adhere to them; reduced adhesion of the endothelial cells to each other (Waters et al. 2013); generation of chemokines (Waters et al. 2013); arrest of wound healing (Mori et al. 2002); enhanced dissolution of bone (Boyce et al. 2005), of cartilage (Saklatvala 1986), and of extracellular matrix components (Sorokin 2010); and other effects.

Walling Off the Inflamed Area

This occurs at the second stage of inflammation and, in practice, opposes the increase in accessibility of the inflamed site that characterized the first stage. It serves to withhold spreading of the pathogen that initiated the inflammation and of injured cell components and to restrict the action of blood components to the site to which they were recruited at the first stage. Several different kinds of TNF-induced functional changes contribute to the walling off, including enhanced coagulation (Waters et al. 2013), disassembly of capillaries (Carswell et al. 1975), cooperation between histiocytes and other cells in the generation of granulomas (Kindler et al. 1989), and others.

Beyond the local restriction of TNF function dictated by the walling off of the inflamed region, the occurrence of TNF in a cell-bound form allows for further local restriction of its function to cells that can respond to the ligand-expressing cells in a juxtacrine manner. Studies in transgenic

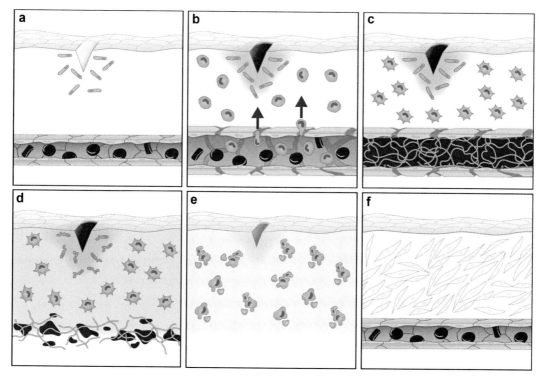

Tumor Necrosis Factor Alpha (TNFalpha), Fig. 1 Local inflammation: diagrammatic representation of its distinct stages. (**a**, **b**) Increased accessibility of the site of infection/injury by capillary dilation and recruitment of cellular and humoral blood components to the site. (**c**, **d**) Walling off: clogging and disassembly of vasculature. (**d**) Killing of pathogens and of injured/infected cells. (**e**) Death of recruited leukocytes. (**f**) Resolution and healing: regrowth of cells at the site and angiogenesis. At all of these stages, TNF is a major inducer of functional changes. The changes it induces at some stages are antagonistic to those induced in others

mice engineered to express exclusively either the soluble or the cell-bound form of TNF revealed distinct contributions of these two forms to different aspects of immune defense (Horiuchi et al. 2010).

Effects of TNF on the Infected and Injured Cells

The processes that in the optimal scenario are restricted to the interior of an injured region walled off as described above are mediated in part through direct effects of TNF on the infected and injured cells and in part less directly through a wide range of effects of TNF on activities of various cells, particularly leukocytes (Wallach et al. 2014). Notably, at least some of those effects of TNF that are considered to be direct might not be observed at all unless the cell is exposed to some additional stimuli. In most cells the induction of cell death by TNF, for example, is largely dependent on exposure of the cells to additional "sensitizing" agents. In such cases TNF might serve rather as a potentiator of the cellular response to pathogen-generated or injury-induced signals.

Cytotoxic and anti-cytotoxic effects of TNF: Among the various direct effects of TNF on infected and injured cells, the most thoroughly documented are its cytotoxic and anti-cytotoxic effects. Two distinct modes of TNF cytotoxicity are known. It can induce apoptotic death through the so-called extrinsic cell-death pathway, which yields activation of proteases of the caspase family (Ashkenazi and Salvesen 2014). Alternatively, it can induce necrotic cell death through a signaling pathway involving activation of the protein kinases RIPK1 and RIPK3 and the pseudokinase

MLKL (Ashkenazi and Salvesen 2014). Whereas induction of apoptotic cell death by TNF can contribute to arrest of the inflammatory process, necrotic cell-death induction by TNF is believed to augment inflammation through the effects of DAMPs released by the dying cells (Ashkenazi and Salvesen 2014).

TNF can also induce in cells resistance to death (Hahn et al. 1985), mainly through NF-κB-mediated activation of genes that code for some protective proteins (Ashkenazi and Salvesen 2014; Beg and Baltimore 1996). Cells infected by certain pathogens, including viruses and mycobacteria, are particularly vulnerable to the cytotoxic effect of TNF, perhaps owing to the pathogen-induced suppression of such protective cellular mechanisms (Waters et al. 2013).

Examples of other direct effects of TNF on infected and injured cells: TNF has been found to suppress virus replication in its target cells. Its antiviral effect occurs in part through TNF-induced expression of interferon (IFN)-β via TNF-dependent upregulation of IFN regulatory factor 1 (IRF1) and in part independently of IFN generation and synergistically with effects of exogenously applied IFN (Yarilina and Ivashkiv 2010). Another example of a direct effect that TNF reportedly has on infected cells is stimulation of autophagy with ensuing enhanced destruction of intracellular pathogens (Harris 2011).

Healing
TNF has a variety of effects that are opposite in character to its destructive activities. These "constructive" effects include increased resistance to cell death (Wallach 1984), stimulation of cell growth (Sugarman et al. 1985), enhanced generation of extracellular matrix components (Distler et al. 2008), angiogenesis (Baluk et al. 2009), wound epithelialization (Frank et al. 2003), and others. These various effects cooperate to promote the healing of the tissues destroyed during earlier phases of the inflammatory process.

Contribution of TNF to systemic inflammation: Proteolytic release of the cell-bound TNF into the circulation allows this cytokine to exert effects on tissues remote from its producing cells, thereby contributing to systemic inflammation. The major known systemic effects of TNF are:

- Effects on brain functions: for example, induction of fever (through activation of cyclooxygenase-2, production of prostaglandin E2 (PGE2), and activation of hypothalamic PGE2 receptors (Waters et al. 2013)), malaise (Jiang et al. 2008), anorexia (Tracey and Cerami 1990), and sleep enhancement (Zielinski and Krueger 2011)
- Effects on the liver: mainly the induced synthesis of acute-phase proteins (Waters et al. 2013)
- Suppression and enhancement of hematopoiesis: through effects on the bone marrow (Waters et al. 2013)
- Metabolic changes: suppression of lipid metabolism (Beutler et al. 1985) and initiation of a catabolic state (Tracey and Cerami 1990)

Contribution of TNF to adaptive immunity: Studies, mainly in mice, have revealed a wide range of TNF-induced effects on adaptive immunity. It facilitates humoral B-cell-mediated adaptive immune responses and is required for follicular B-cell localization and for organization of the marginal zone. Its autocrine production by B cells is indispensable for the development of follicular dendritic cell networks in the spleen and in peripheral lymph nodes (reviewed in Kruglov et al. 2008). In T lymphocytes, stimulation of TNFR2 promotes T-cell division, survival, and effector functions, while stimulation of TNFR1 can stimulate both cell growth and cell death. According to some reports TNFR2 preferentially promotes the activity of suppressor cells (Chatzidakis and Mamalaki 2010; Croft 2014).

Pathophysiological Activities

Like other cytokines, TNF is believed to have been destined by nature to facilitate immune defense. As detailed below, deficiency of TNF indeed has various pathophysiological consequences. However, since immune defense mechanisms constitute deviations from normal homeostasis, they usually also have the potential for turning out, when exerted to inappropriate extents or at unsuitable times and locations, to be deleterious. Among the known cytokines, TNF is exceptional in the extent to which its deregulated action can become deleterious and in the range of pathologies in which it does so. Functionally, therefore, it is often likened metaphorically to a double-edged sword.

Beneficial Effects of TNF in Diseases

Protective Roles of TNF in Infectious Diseases
In mice, knockout of genes encoding TNF or its receptors, or blocking of TNF function by repetitive injection of antibodies, was found to compromise the ability of the host to withstand infection by various pathogens, including extracellular bacteria such as *Salmonella*, intracellular bacteria such as *Listeria monocytogenes* and *Mycobacterium tuberculosis*, fungi like *Aspergillus fumigatus*, and others (Ellerin et al. 2003; Waters et al. 2013). TNF also plays a pivotal role in immune defense against multicellular parasites such as the malaria parasite, *Plasmodium* (Waters et al. 2013).

In humans, blocking of TNF by serial injections of antibodies or soluble TNF receptors was likewise found to increase vulnerability to infection. Most extensively documented is the increase in vulnerability to *Mycobacterium tuberculosis*, but there is also evidence for increased vulnerability to various other bacteria, various fungi, viruses, and eukaryotic parasites (Ellerin et al. 2003).

TNF as an Anticancer Agent
The discovery of TNF as a protein capable of inducing hemorrhagic necrosis in some experimental mouse models of tumors (Carswell et al. 1975) raised the possibility that TNF might play a role in immune surveillance against tumors. Accordingly, there were concerns that blocking of TNF in humans might increase the frequency of cancer. Currently, however, there seems to be no solid evidence to support this suspicion (Balkwill 2009).

Detrimental Effects of TNF in Diseases

Deleterious Effects of TNF in Infectious Diseases
Studies by Anthony Cerami, Bruce Beutler, Kevin Tracey, and their colleagues; by Pierre Vassalli, Georges Grau, and their colleagues; and by others examined the effects of anti-TNF antibodies on experimental animal models of infection as well as the effects of injection of TNF itself. With regard to infection the findings, taken together, pointed to a critical involvement of TNF in various early pathological consequences of exposure to bacterial endotoxin, including acute cardiovascular collapse and shock (Tracey et al. 1987). Also reported was a pivotal role for TNF in the pathology of cerebral malaria in mice (Gimenez et al. 2003).

Some of the known cellular activities of TNF suggest that it might also be crucially involved in the excessive loss of weight resulting from the catabolic state and anorexia in certain chronic diseases (Bradley 2008). Some support for this notion came from studies showing that chronic generation of TNF by tumor cells implanted in mice can indeed result in extreme weight loss (Tracey and Cerami 1990).

As yet, however, little is known about the relevance of the above findings to diseases in humans. The contribution of TNF to the pathology of bacterial infections in humans gained some support from observed correlations between the serum level of TNF and the pathology of meningococcal disease (Waage et al. 1987), sepsis (Cohen and Abraham 1999), and malaria (Bradley 2008), as well as from a limited beneficial effect observed soon after infusion of anti-TNF antibodies into some patients with septic shock. In the rest of the tested septic patients, these antibodies were somewhat deleterious, perhaps reflecting a role of TNF in withholding bacterial infections (Ellerin et al. 2003). A linkage between TNF and the pathology of malaria in humans appeared to be indicated both by evidence for increased vulnerability to cerebral malaria in patients with a genetic predisposition to increased generation of TNF and by correlations between the range of pathology in malaria patients and their cellular levels of both TNF and its soluble receptors (Gimenez et al. 2003). However, a beneficial effect of anti-TNF antibodies, as would be expected if TNF were indeed a crucial mediator of the pathology, could not be observed in malaria patients (Gimenez et al. 2003).

TNF as a Tumor-Promoting Mediator

Compelling evidence supports a contribution of chronic inflammation to the development of various tumors. TNF, as a major coordinator of inflammatory processes, is believed to mediate some of the functional changes underlying such a contribution (Balkwill 2009). Strong support for this possibility comes from several experimental models of tumors in mice (Balkwill 2009).

TNF as a Key Mediator of Pathologies of Chronic Inflammatory Diseases

Increased generation of TNF has been observed in a variety of autoimmune and chronic inflammatory diseases (Bradley 2008; Sfikakis 2010). Studies in mice showed that mere transgenic overexpression of TNF suffices to trigger the pathology of rheumatoid arthritis and that blocking of the chronic generation of TNF in experimental models for rheumatoid arthritis, inflammatory bowel disease, or multiple sclerosis (Apostolaki et al. 2010) significantly alleviates their respective pathologies.

Those findings pointed to the possibility that an increase in TNF plays a pivotal role in the pathologies of such diseases, a notion that has gained extensive support from the therapeutic effects of anti-TNF drugs observed in a number of chronic inflammatory conditions in humans. Such effects have been documented in detail, for example, for rheumatoid arthritis, psoriasis, Crohn's disease, ulcerative colitis, and ankylosing spondylitis (Bradley 2008; Sfikakis 2010). In addition, there is sporadic evidence for therapeutic effects of TNF-blocking agents in a number of other pathological conditions associated with chronic inflammatory states, including asthma, inflammatory myopathies, Behçet's disease, sarcoidosis, and inflammatory eye diseases (Bradley 2008; Sfikakis 2010).

Modulation by Drugs

Approaches for Therapeutic Application of TNF Function

In view of the effective and selective destructive effects of TNF found on certain experimental tumors in mice (Carswell et al. 1975), attempts were initiated to apply TNF as a tumor-destroying agent in humans. These attempts were promptly withheld when TNF was found to have life-threatening toxic effects on the patients (Balkwill 2009). However, subsequent attempts to restrict exposure of only the affected internal organs to the applied TNF (the "limb perfusion approach") yielded successful destruction of certain tumors in humans without deleterious consequences (Balkwill 2009). There have also been attempts to generate recombinant forms of TNF that would target TNF exclusively to certain tumor cells by fusing it to antibody fragments (Scherf et al. 1996) or ligand molecules (Yuan et al. 2009) that bind specifically to those cells. However, the utility of such recombinant molecules for therapy in humans has yet to be adequately explored.

Approaches for Therapeutic Arrest of TNF Function

The dramatic therapeutic effects of blocking the activity of TNF in animal models of diseases from which numerous patients suffer prompted joint attempts by scientific and industrial researchers to develop effective ways to block it (Sfikakis 2010). Currently applied means are derived either from antibodies against TNF or from the ligand-binding regions in TNF receptors. The latter approach employs receptors in the soluble form that occurs physiologically in body fluids (soluble receptors), and whose increased amounts in inflammatory conditions raise the possibility that these soluble receptors serve as a physiological means of restraining excessive TNF function. With regard to both kinds of means, attempts have been made to increase their effectiveness by modifying these molecules structurally. A form of soluble TNF receptors now widely applied for therapy is a fusion protein comprised of two chains of the extracellular ligand-binding portion of the human TNFR2 linked to the Fc portion of human IgG1. Modified forms of anti-TNF antibodies being applied nowadays to therapy include "humanized antibodies," antibody molecules in which various parts of the molecule other than the antigen-recognizing hypervariable region have been exchanged with the corresponding region in the human immunoglobulin molecule. Another modified form is a derivative of a humanized anti-TNF antibody in which the Fc portion of the antibody molecule has been deleted and the Fab portion was fused instead to polyethylene glycol (Sfikakis 2010).

Current attempts to generate more sophisticated antagonists to TNF are aimed at the following goals:

- Selective suppression of the function of the soluble form of TNF while sparing the function of the cell-bound form. This goal is sought in view of evidence that the cell-bound form of TNF contributes less than the soluble form to pathological inflammatory processes, owing in part to the preferential ability of the former to activate not only TNFR1 but also TNFR2. The techniques tried so far have employed a mutant form of TNF with a selective dominant negative effect on the function of soluble TNF (Zalevsky et al. 2007) or inhibitors of TACE, the metalloproteinase that generates the soluble form of TNF by cleaving the cell-bound form (Barbosa et al. 2011).
- Generation of small-molecule blockers of TNF binding to its receptors (Hasegawa et al. 2001).
- Chemical inhibition of signaling for generation of TNF, for example, via an inhibitor of the protein kinase Tpl2 which, through activation of ERK, signals for the nucleocytoplasmic transport of the mRNA of TNF and for the shedding of the cell-bound TNF precursor (Barbosa et al. 2011; Hall et al. 2007).
- Chemical inhibition of specific signaling pathways that TNF activates. Examples are inhibitors of the so-called canonical NF-κB pathway whose activation by TNF serves a pivotal role in the pro-inflammatory functions of this cytokine (Barbosa et al. 2011).

At the time of writing, none of these more sophisticated TNF antagonists have yet reached the market.

Cross-References

▶ Cytokines
▶ Tumor Necrosis Factor (TNF) Inhibitors

References

Apostolaki, M., Armaka, M., Victoratos, P., & Kollias, G. (2010). Cellular mechanisms of TNF function in models of inflammation and autoimmunity. *Current Directions in Autoimmunity, 11*, 1–26. doi:10.1159/000289195

Ashkenazi, A., & Salvesen, G. (2014). Regulated cell death: Signaling and mechanisms. *Annual Review of Cell and Developmental Biology, 30*, 337–356. doi:10.1146/annurev-cellbio-100913-013226.

Balkwill, F. (2009). Tumour necrosis factor and cancer. *Nature Reviews Cancer, 9*(5), 361–371. doi:10.1038/nrc2628.

Baluk, P., Yao, L. C., Feng, J., Romano, T., Jung, S. S., Schreiter, J. L., et al. (2009). TNF-alpha drives remodeling of blood vessels and lymphatics in sustained airway inflammation in mice. *Journal of Clinical Investigation, 119*(10), 2954–2964. doi:10.1172/JCI37626.

Barbosa, M. L. C., Fumian, M. M., Miranda, A. L. P., Barreiro, E. J., & Lima, L. M. (2011). Therapeutic approaches for tumor necrosis factor inhibition. *Brazilian Journal of Pharmaceutical Sciences, 47*(3), 427–446.

Beg, A. A., & Baltimore, D. (1996). An essential role for NF-kB in preventing TNF-a-induced cell death. *Science, 274*, 782–784.

Beutler, B., Mahoney, J., Le Trang, N., Pekala, P., & Cerami, A. (1985). Purification of cachectin, a lipoprotein lipase-suppressing hormone secreted by endotoxin-induced RAW 264.7 cells. *Journal of Experimental Medicine, 161*(5), 984–995.

Black, R. A., Rauch, C. T., Kozlosky, C. J., Peschon, J. J., Slack, J. L., Wolfson, M. F., et al. (1997). A metalloproteinase disintegrin that releases tumour-necrosis factor-alpha from cells. *Nature, 385*, 729–733.

Boyce, B. F., Li, P., Yao, Z., Zhang, Q., Badell, I. R., Schwarz, E. M., et al. (2005). TNF-alpha and pathologic bone resorption. *The Keio Journal of Medicine, 54*(3), 127–131.

Bradley, J. R. (2008). TNF-mediated inflammatory disease. *Journal of Pathology, 214*(2), 149–160. doi:10.1002/path.2287.

Carswell, E. A., Old, L. J., Fiore, N., & Schwartz, M. K. (1975). An endotoxin-induced serum factor that causes necrosis of tumors. *Proceedings of the National Academy of Sciences of the United States of America, 72*, 3666–3670.

Chatzidakis, I., & Mamalaki, C. (2010). T cells as sources and targets of TNF: Implications for immunity and autoimmunity. *Current Directions in Autoimmunity, 11*, 105–118. doi:10.1159/000289200.

Clark, I. A., Alleva, L. M., & Vissel, B. (2010). The roles of TNF in brain dysfunction and disease. *Pharmacology and Therapeutics, 128*(3), 519–548. doi:10.1016/j.pharmthera.2010.08.007.

Cohen, J., & Abraham, E. (1999). Microbiologic findings and correlations with serum tumor necrosis factor-alpha in patients with severe sepsis and septic shock. *Journal of Infectious Diseases, 180*(1), 116–121. doi:10.1086/314839.

Croft, M. (2014). The TNF family in T cell differentiation and function–unanswered questions and future directions. *Seminars in Immunology, 26*(3), 183–190. doi:10.1016/j.smim.2014.02.005.

Distler, J. H., Schett, G., Gay, S., & Distler, O. (2008). The controversial role of tumor necrosis factor alpha in fibrotic diseases. *Arthritis and Rheumatism, 58*(8), 2228–2235. doi:10.1002/art.23645.

Eissner, G., Kolch, W., & Scheurich, P. (2004). Ligands working as receptors: Reverse signaling by members of the TNF superfamily enhance the plasticity of the immune system. *Cytokine and Growth Factor Reviews, 15*(5), 353–366. doi:10.1016/j.cytogfr.2004.03.011.

Ellerin, T., Rubin, R. H., & Weinblatt, M. E. (2003). Infections and anti-tumor necrosis factor alpha therapy. *Arthritis and Rheumatism, 48*(11), 3013–3022. doi:10.1002/art.11301.

Falvo, J. V., Tsytsykova, A. V., & Goldfeld, A. E. (2010). Transcriptional control of the TNF gene. *Current Directions in Autoimmunity, 11*, 27–60. doi:10.1159/000289196.

Frank, J., Born, K., Barker, J. H., & Marzi, I. (2003). In vivo effect of tumor necrosis factor alpha on wound angiogenesis and epithelialization. *European Journal of Trauma, 29*, 208–219.

Gimenez, F., Barraud de Lagerie, S., Fernandez, C., Pino, P., & Mazier, D. (2003). Tumor necrosis factor alpha in the pathogenesis of cerebral malaria. *Cellular and Molecular Life Sciences, 60*(8), 1623–1635. doi:10.1007/s00018-003-2347-x.

Hahn, T., Toker, L., Budilovsky, S., Aderka, D., Eshhar, Z., & Wallach, D. (1985). Use of monoclonal antibodies to a human cytotoxin for its isolation and for examining the self-induction of resistance to this protein. *Proceedings of the National Academy of Sciences of the United States of America, 82*(11), 3814–3818.

Hall, J. P., Kurdi, Y., Hsu, S., Cuozzo, J., Liu, J., Telliez, J. B., et al. (2007). Pharmacologic inhibition of tpl2 blocks inflammatory responses in primary human monocytes, synoviocytes, and blood. *Journal of Biological Chemistry, 282*(46), 33295–33304. doi:10.1074/jbc.M703694200.

Harris, J. (2011). Autophagy and cytokines. *Cytokine, 56*(2), 140–144. doi:10.1016/j.cyto.2011.08.022.

Hasegawa, A., Takasaki, W., Greene, M. I., & Murali, R. (2001). Modifying TNFalpha for therapeutic use: A perspective on the TNF receptor system. *Mini Reviews in Medicinal Chemistry, 1*(1), 5–16.

Horiuchi, T., Mitoma, H., Harashima, S., Tsukamoto, H., & Shimoda, T. (2010). Transmembrane TNF-alpha: Structure, function and interaction with anti-TNF agents. *Rheumatology (Oxford), 49*(7), 1215–1228. doi:10.1093/rheumatology/keq031.

Jiang, Y., Deacon, R., Anthony, D. C., & Campbell, S. J. (2008). Inhibition of peripheral TNF can block the malaise associated with CNS inflammatory diseases. *Neurobiology of Disease, 32*(1), 125–132. doi:10.1016/j.nbd.2008.06.017.

Kindler, V., Sappino, A. P., Grau, G. E., Piguet, P. F., & Vassalli, P. (1989). The inducing role of tumor necrosis factor in the development of bactericidal granulomas during BCG infection. *Cell, 56*(5), 731–740.

Kruglov, A. A., Kuchmiy, A., Grivennikov, S. I., Tumanov, A. V., Kuprash, D. V., & Nedospasov, S. A. (2008). Physiological functions of tumor necrosis factor and the consequences of its pathologic overexpression or blockade: Mouse models. *Cytokine and Growth Factor Reviews, 19*(3–4), 231–244. doi:10.1016/j.cytogfr.2008.04.010.

Menkin, V. (1948). Newer Concepts of Inflammation, Springfield, Charles C Thomas.

Mori, R., Kondo, T., Ohshima, T., Ishida, Y., & Mukaida, N. (2002). Accelerated wound healing in tumor necrosis factor receptor p55-deficient mice with reduced leukocyte infiltration. *FASEB Journal, 16*(9), 963–974. doi:10.1096/fj.01-0776com.

Ohsawa, T., & Natori, S. (1989). Expression of tumor necrosis factor at a specific developmental stage of mouse embryos. *Developmental Biology, 135*(2), 459–461.

Saklatvala, J. (1986). Tumour necrosis factor alpha stimulates resorption and inhibits synthesis of proteoglycan in cartilage. *Nature, 322*(6079), 547–549. doi:10.1038/322547a0.

Scherf, U., Benhar, I., Webber, K. O., Pastan, I., & Brinkmann, U. (1996). Cytotoxic and antitumor activity of a recombinant tumor necrosis factor-B1(Fv) fusion protein on LeY antigen-expressing human cancer cells. *Clinical Cancer Research, 2*(9), 1523–1531.

Sebbag, M., Parry, S. L., Brennan, F. M., & Feldmann, M. (1997). Cytokine stimulation of T lymphocytes regulates their capacity to induce monocyte production of tumor necrosis factor-alpha, but not interleukin-10: Possible relevance to pathophysiology of rheumatoid arthritis. *European Journal of Immunology, 27*(3), 624–632. doi:10.1002/eji.1830270308.

Sfikakis, P. P. (2010). The first decade of biologic TNF antagonists in clinical practice: Lessons learned, unresolved issues and future directions. *Current Directions in Autoimmunity, 11*, 180–210. doi:10.1159/000289205.

Sorokin, L. (2010). The impact of the extracellular matrix on inflammation. *Nature Reviews Immunology, 10*(10), 712–723. doi:10.1038/nri2852.

Sugarman, B. J., Aggarwal, B. B., Hass, P. E., Figari, I. S., Palladino, M. A., Jr., & Shepard, H. M. (1985). Recombinant human tumor necrosis factor-alpha: Effects on proliferation of normal and transformed cells in vitro. *Science, 230*(4728), 943–945.

Tracey, K. J., & Cerami, A. (1990). Metabolic responses to cachectin/TNF. A brief review. *Annals of the New York Academy of Sciences, 587*, 325–331.

Tracey, K. J., Fong, Y., Hesse, D. G., Manogue, K. R., Lee, A. T., Kuo, G. C., et al. (1987). Anti-cachectin/TNF monoclonal antibodies prevent septic shock during lethal bacteraemia. *Nature, 330*(6149), 662–664. doi:10.1038/330662a0.

Waage, A., Halstensen, A., & Espevik, T. (1987). Association between tumour necrosis factor in serum and fatal outcome in patients with meningococcal disease. *Lancet, 1*(8529), 355–357.

Wallach, D. (1984). Preparations of lymphotoxin induce resistance to their own cytotoxic effect. *Journal of Immunology, 132*, 2464–2469.

Wallach, D., Kang, T. B., & Kovalenko, A. (2014). Concepts of tissue injury and cell death in inflammation: A historical perspective. *Nature Reviews Immunology, 14*(1), 51–59. doi:10.1038/nri3561.

Walsh, L. J., Trinchieri, G., Waldorf, H. A., Whitaker, D., & Murphy, G. F. (1991). Human dermal mast cells contain and release tumor necrosis factor alpha, which induces endothelial leukocyte adhesion molecule 1. *Proceedings of the National Academy of Sciences of the United States of America, 88*(10), 4220–4224.

Waters, J. P., Pober, J. S., & Bradley, J. R. (2013). Tumour necrosis factor in infectious disease. *Journal of Pathology, 230*(2), 132–147. doi:10.1002/path.4187.

Yarilina, A., & Ivashkiv, L. B. (2010). Type I interferon: A new player in TNF signaling. *Current Directions in Autoimmunity, 11*, 94–104. doi:10.1159/000289199.

Yuan, X., Lin, X., Manorek, G., Kanatani, I., Cheung, L. H., Rosenblum, M. G., et al. (2009). Recombinant CPE fused to tumor necrosis factor targets human ovarian cancer cells expressing the claudin-3 and claudin-4 receptors. *Molecular Cancer Therapeutics, 8*(7), 1906–1915. doi:10.1158/1535-7163.MCT-09-0106.

Zalevsky, J., Secher, T., Ezhevsky, S. A., Janot, L., Steed, P. M., O'Brien, C., et al. (2007). Dominant-negative inhibitors of soluble TNF attenuate experimental arthritis without suppressing innate immunity to infection. *Journal of Immunology, 179*(3), 1872–1883.

Zielinski, M. R., & Krueger, J. M. (2011). Sleep and innate immunity. *Frontiers in Bioscience (Scholar Edition), 3*, 632–642.

Tumor Necrosis Factor (TNF) Inhibitors

Kevin D. Pile[1], Garry G. Graham[2,3] and Stephen M. Mahler[4]

[1]Campbelltown Hospital, School of Medicine, University of Western Sydney, Campbelltown, NSW, Australia

[2]Department of Pharmacology, School of Medical Sciences, University of New South Wales, Sydney, NSW, Australia

[3]Department of Clinical Pharmacology and Toxicology, St Vincent's Hospital, Sydney, NSW, Australia

[4]Australian Institute for Bioengineering and Nanotechnology, University of Queensland, Brisbane, QLD, Australia

Synonyms

Anti-TNF antibodies; Anti-TNFs; TNF blockers; TNF inhibitors

Definition

The tumor necrosis factor (TNF) inhibitors are protein constructs which bind and inactivate TNF. They are members of a group of proteins commonly termed biological response modifiers (BRMs) or biological disease-modifying antirheumatic drugs (bDMARDs). They are not included in the group, slow-acting antirheumatic drug (SAARDs), because they are considered to have more specific actions and because their therapeutic actions are produced rapidly. The TNF inhibitors are active in the treatment of rheumatoid arthritis, while most also improve psoriatic arthritis, ankylosing spondylitis, ulcerative colitis, and Crohn's disease.

Chemical Structures and Properties

Five TNF inhibitors are used clinically (Fig. 1). Four bind TNF through their antibody functions, while etanercept can be regarded as a soluble receptor. The structures are summarized below, while details of their production are shown in the Appendix.

- Infliximab is a chimeric (mouse-human) monoclonal antibody in which the variable domains of mouse antibody to human TNF are fused to the constant domains of a human antibody. A biosimilar of infliximab is now available in Europe.
- Adalimumab and golimumab are fully humanized monoclonal antibodies.
- Certolizumab pegol is a humanized Fab fragment covalently bound to polyethylene glycol. Thus, it does not have an Fc domain of an immunoglobulin.
- Etanercept contains an Fc immunoglobulin backbone and a dimer of the extracellular portion of the human TNF receptor.

The TNF inhibitors have large molecular masses (approximately 150 kDa) and are water soluble. These properties prevent any passive transport through cell membranes.

Metabolism, Pharmacokinetics, and Dosage

The TNF inhibitors have immunoglobulin-like structures and, as a result, are handled in a similar fashion to immunoglobulins. For example, immunoglobulin G is absorbed in neonates, but this gastrointestinal transport is quickly lost (Lobo et al. 2004). Not unexpectedly in clinical practice, the TNF inhibitors are not absorbed orally and are only administered by injection. Intravenous infusion (infliximab) allows immediate distribution to the vasculature of inflamed tissue, but the other TNF inhibitors are administered by subcutaneous injection (adalimumab, golimumab, certolizumab pegol, etanercept). The absorption of all the subcutaneous injections is slow with peak plasma concentrations being achieved at 2–3 days after dosage. Subcutaneous injections are more convenient for the patients as intravenous infusion is time-consuming. However, as discussed below, subcutaneous injection is more likely to lead to the production of antibodies to the injected TNF inhibitors.

There is considerable interpatient variation in the pharmacokinetics of TNF inhibitors. This interpatient variation may have considerable clinical significance as marked interpatient differences in response may result (Ordás et al. 2012; Ramiro et al. 2010). Individualization of dosage according to the plasma concentrations (therapeutic drug monitoring) may improve treatment. This is a principle of the clinical pharmacology of conventional low molecular mass drugs but also appears relevant for TNF inhibitors. Alternatively, change to another TNF inhibitor or another class of antirheumatic biological drug may be considered if the response to one TNF inhibitor is inadequate.

The TNF inhibitors are cleared by proteolytic catabolism in the reticuloendothelial system. Elderly patients often have impaired clearance and longer half-lives of elimination of conventional small molecular mass drugs than younger patients. This is not the case with the TNF inhibitors, and no reduction in dosage is required in older patients. Furthermore, they are not excreted renally or dialyzable, and their dosage is therefore not altered in patients with impaired renal function or who are undergoing dialysis.

Like the native immunoglobulins, the anti-TNF antibodies have long half-lives of elimination which allow relatively infrequent dosage (see Dosage below). Loading doses of some TNF inhibitors are used (i.e., more frequent dosage at the commencement of treatment in order to achieve therapeutic levels more rapidly). The dosage and half-lives of the TNF inhibitors and some relevant pharmacokinetic details include:

- Infliximab. The most common dosage for rheumatoid arthritis is 3 mg/kg given by intravenous infusion over 2 h. This dose is repeated at 2 and 4 weeks after the first dose and thereafter every 8 weeks. Infliximab shows biphasic elimination with an initial half-life approximately 4.5 days and terminal half-life of 19 days. The steady-state dosage interval (8 weeks) is thus considerably longer

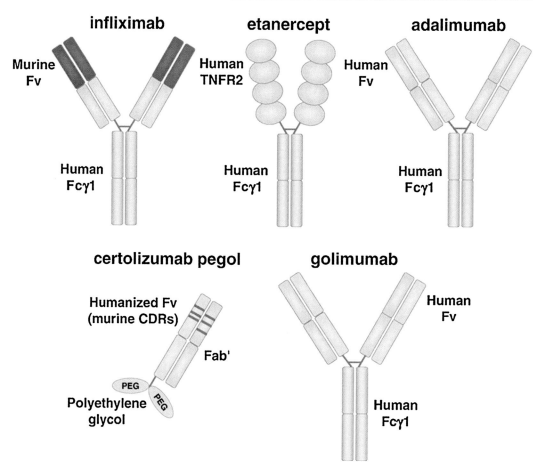

Tumor Necrosis Factor (TNF) Inhibitors, Fig. 1 Diagrammatic structures of the TNF inhibitors derived from immunoglobulin G showing the (1) variable chains (Fv) in the TNF antibodies infliximab, adalimumab, golimumab, and certolizumab pegol. The Fv regions are from mice in infliximab but humanized in adalimumab and golimumab. Fv chains from light and heavy chains make up the V domain. (2) The complementarity determining regions (*CDRs*) within the Fv regions. These are humanized in adalimumab and golimumab but are of mouse origin (murine) in infliximab and certolizumab pegol. (3) Constant chains (Fcγ1) of etanercept, infliximab, adalimumab, and golimumab which are all human. (4) Immunoglobulin fragment (Fab') fragment of a humanized anti-TNF antibody in certolizumab pegol is bound to polyethylene glycol. (5) Human TNF receptor 2 (*TNFR2*) chains in etanercept

than the half-lives, the result being large swings in the plasma concentrations over a dosage interval; the average fluctuation being 2–150 mg/L but with considerable interpatient variation. It is of note that trough concentrations below 1 mg/L are associated with a poor response to treatment (Mulleman et al. 2010). Higher doses of 5 mg/kg are recommended in the treatment of ankylosing spondylitis, psoriatic arthritis, psoriasis, Crohn's disease, and ulcerative arthritis.

The dosing frequency can be reduced to every 6 weeks for ankylosing spondylitis. The variable pharmacokinetics and response to infliximab and other TNF inhibitors may, in part, be due to interactions with antibodies which are formed to the injected TNF inhibitor. The immunogenicity of the TNF inhibitors is discussed below. Interestingly, the initial half-life decreases with increasing severity of the rheumatic disease, possibly due to higher concentrations of TNF in plasma and tissues and the formation of antibody complexes with TNF.

- Adalimumab. The terminal half-life is approximately 14 days (Nestorov 2005). Very small amounts (mostly <10 % dose) of adalimumab are excreted renally (Roberts et al. 2013), and it is likely that the renal excretion of other TNF inhibitors is also very low. No loading dose is used for rheumatoid arthritis, and the recommended dose is 40 mg subcutaneously every 2 weeks. At this dosage, the plasma concentrations range from about 4 to 8 mg/L over a dosage interval. However, plasma concentrations less than 5 mg/L have correlated with an inadequate response to adalimumab (Krieckart et al. 2013). In the treatment of ulcerative colitis and Crohn's disease, a loading dose of 160 mg is given on the first day (or 80 mg on each of the first 2 days of treatment). This is followed by 80 mg after 2 weeks then 40 mg every 2 weeks. In the treatment of psoriasis, the recommended first dose is 80 mg followed by the usual dose and 40 mg every 2 weeks.
- Golimumab. The recommended dosage is 50 mg by subcutaneous injection once monthly without a loading dose for the treatment of rheumatoid arthritis, psoriatic arthritis, and ankylosing spondylitis. The bioavailability is about 53 % (Zhuang et al. 2012). Like other TNF inhibitors administered by subcutaneous injection, the absorption of golimumab is slow with peak concentrations occurring 3 days after dosage. The half-life of golimumab is approximately 13 days during multiple dosage, resulting in the plasma concentrations swinging from about 0.5 to 3.5 mg/L over the dosage interval of 4 weeks (Zhuang et al. 2012).
- Certolizumab pegol. The recommended dosage regimen for rheumatoid arthritis is 400 mg subcutaneously every 2 weeks for the first 4 weeks then 200 mg every 2 weeks or 400 mg every 4 weeks. Absorption is slow and the bioavailability is approximately 80 %. The half-life of elimination is approximately 20 days. The half-life was extended to this level by covalent binding to polyethylene glycol (pegylation) (Smith et al. 2010).
- Etanercept. The usual dosage for rheumatoid arthritis is 50 mg subcutaneously once weekly or 25 mg twice a week. Absorption is slow with plasma concentrations peaking at approximately 2 days after dosage. Absorption is also incomplete (bioavailability about 58 %). The half-life of elimination is approximately 4 days which is shorter than those of other TNF inhibitors and responsible for the dosage interval being shorter than the antibody TNF inhibitors (Zhou 2005). The plasma concentrations fluctuate little over the usual dosage intervals, and the mean plasma concentrations are about 1.7 mg/L. The dosage for severe psoriasis is generally the same as for rheumatoid arthritis, but the initial dosage may be increased to 50 mg twice weekly for up to 12 weeks. The dosage is then reduced to the usual levels.

Pharmacological Effects

All the TNF inhibitors bind strongly to soluble and membrane-bound TNF which has a key role within the inflammatory network as well as heading an inflammatory cascade (Tracey et al. 2008). TNF activates monocytes and neutrophils, prostaglandin synthesis, apoptosis of neutrophils, and T cells and increases the expression of leukocyte and endothelial adhesion molecules, matrix metalloprotein production, and the release of fatty acids from adipocytes. TNF is present in synovial tissues in patients with rheumatoid arthritis, the major source of TNF being activated macrophages (Tracey et al. 2008). For example, overexpression of TNF in transgenic mice causes a rheumatoid-like disease while blocking TNF lessens collagen-induced arthritis, also in mice (Tracey et al. 2008). Overexpression of TNF is now considered to be a major factor in the development of several human inflammatory diseases, such as rheumatoid arthritis, psoriasis, psoriatic arthritis, and ankylosing spondylitis. The general activity of the TNF inhibitors supports the significance of TNF in these inflammatory diseases. However, the level of circulating TNF does not correlate with the response to infliximab (Pachot et al. 2007). Apart from etanercept, the TNF inhibitors are also effective in the treatment of granulomatous diseases (see below).

In rheumatoid and experimental arthritis, the TNF inhibitors quickly suppress leukocyte migration, phagocytosis, and degranulation and promote recovery of regulatory T-cell function and phenotype. Also, they mediate antibody-dependent cellular cytotoxicity (ADCC) and complement-mediated cytotoxicity (CDC) of cells expressing membrane-bound TNF (mTNF) in vitro (Stuhlmuller et al. 2010).

Etanercept may have some contrasting pharmacological effects to the four antibody TNF inhibitors, infliximab, adalimumab, golimumab, and certolizumab pegol. For example, the antibodies decrease cytokine production by monocytes under a variety of in vitro conditions, but etanercept does not (Taylor 2010). On the other hand, etanercept is unique in that it can bind to members of the lymphotoxin family, the result possibly being the reason for a reduction: "- -in the numbers of memory B cells in the peripheral blood of patients with RA as well as follicular dendritic cell networks and germinal center structures in tonsil biopsies - -" (Taylor 2010).

Clinical Use

Rheumatoid Arthritis

TNF inhibitors are in the first line of treatments of rheumatoid arthritis in those patients who have not responded to methotrexate alone or combination (cs) DMARDs. Features of the response to TNF inhibitors include:

- Not all patients respond. The 60:40:20 rule is applicable to most bDMARD therapy in RA and infers approximately 60 % of patients achieve at least an ACR20 response, 40 % at least an ACR50 response, and 20 % an ACR70 response.
- The response to TNF inhibitors has shown varying responses depending on the pretreatment severity of the rheumatoid arthritis. Poorer responses have been reported at both higher (Kievit et al. 2007) and a high level of inflammation (Hyrich et al. 2006).
- Despite the fact that the TNF inhibitors antagonize TNF, the level of circulating TNF does not correlate with the response to infliximab in the treatment of rheumatoid arthritis (Pachot et al. 2007)
- Many of the apparent failures may be due to low plasma concentrations of the anti-TNF inhibitor (see Pharmacokinetics and metabolism above).
- Introductory data indicates that the antirheumatic response to adalimumab increases with increasing expression of the mRNA of CD11c (Stuhlmuller et al. 2010). CD11c is a gene leading to the synthesis of integrins, proteins which are involved in the adherence of neutrophils and monocytes to endothelial cells. This clinical correlation, if confirmed, may indicate a further mode of action of the anti-TNF antibodies.
- The combination of methotrexate and a TNF inhibitor provides a greater improvement than either drug alone (Hyrich et al. 2006; Taylor 2010). In part, the improvement is due to the decreased incidence of neutralizing anti-drug antibodies when methotrexate is taken concomitantly (see section "Immunogenicity" below) (Taylor 2010).
- Concurrent treatment with a nonsteroidal anti-inflammatory drug (NSAID) increases the response to a TNF inhibitor (Hyrich et al. 2006). Long-term treatment with NSAIDs is not recommended because of possible adverse effects including exacerbation of hypertension and cardiac failure as well as increased risk of stroke and myocardial infarction. Furthermore, NSAIDs are associated with an increased risk of gastrointestinal damage although this can be decreased by low doses and concomitant dosage with a proton pump inhibitor (Day and Graham 2013).
- In apparent failure to one TNF inhibitor, increasing the dose of the TNF inhibitor, switching to another TNF inhibitor, or changing to another class of biological antirheumatic agent may be considered. Changing to rituximab is slightly more efficacious than

switching to another TNF inhibitor (Soliman et al. 2012).
- Treatment with TNF inhibitors may decrease the development of insulin resistance and type 2 diabetes mellitus in patients treated for rheumatoid arthritis (Antohe et al. 2012).

Other Inflammatory Diseases

The anti-TNF agents all appear active in the treatment of psoriasis, psoriatic arthritis, and ankylosing spondylitis but have weak activity in the treatment of sarcoidosis. Although etanercept has approximately equal activity to the other anti-TNF agents in the treatment of rheumatoid arthritis, it appears considerably less efficacious in the treatment of granulomatous diseases, such as Crohn's disease, ulcerative colitis, granulomatosis with polyangiitis (Wegener's granulomatosis), and sarcoidosis (Taylor 2010; Tracey et al. 2008).

Adverse Effects

Adverse effects of the TNF inhibitors include:

- An increased incidence of infections has been observed during trials and in post-marketing surveillance, presumably related to the immunosuppressive effects of these agents, with particular emphasis on the reactivation of granulomatous infections, such as tuberculosis and histoplasmosis. Etanercept appears less active (Taylor 2010). Meta-analysis showed that infliximab and adalimumab doubled the risk of serious infection with a tendency towards increased infections at higher doses (Askling and Dixon 2008). As anti-TNF agents can reactivate latent tuberculosis, screening for tuberculosis is required before treatment. There have also been reported increases in intracellular or opportunistic pathogens including *Legionella*, *Listeria*, and *Salmonella*, with caution recommended on the consumption of raw eggs and unpasteurized milk. In general, anti-TNF agents should generally be avoided in patients with active or recurrent infections.
- Concomitant therapy with corticosteroids may increase further the susceptibility to infections.
- Combined treatment with interleukin-1 (such as anakinra, rilonacept, or canakinumab) and a TNF inhibitor increases the risk of infection. These combinations are contraindicated.
- All the TNF inhibitors except infliximab are administered by subcutaneous injection and have caused local reactions such as erythema, itching, and swelling at the site of the injection. These reactions generally do not require treatment.
- Acute allergic-type reactions have been reported with all TNF inhibitors and may be associated with the presence of antibodies to the injected TNF inhibitor. In the case of infusion of infliximab, the intravenous infusion can be stopped if acute reactions develop.
- As a consequence of its structure, certolizumab pegol does not bind to transmembrane TNF. It was thought that this would prevent some of the toxic effects of the other TNF blockers, such as the activation of latent tuberculosis. Unfortunately, certolizumab has led to reactivation of tuberculosis.
- It has been postulated that inhibitors of TNF could reduce immune surveillance and lead to the development of tumors. A population-based study linking three rheumatoid arthritis cohorts to the Swedish cancer registry found the rheumatoid arthritis cohort was only at a marginally elevated overall risk of solid cancers. Not surprisingly, given that smoking is associated with rheumatoid arthritis, smoking-related cancers were increased by 20–50 %, and the risk for nonmelanoma skin cancer increased by 70 %. The good side was a 25 % decreased risk for breast and colorectal cancer, with the cancer pattern in patients treated with anti-TNF agents mirroring historic rheumatoid cohorts. The same authors observed a twofold increased risk of lymphoma and leukemia, but not myeloma. Among rheumatoid patients treated with anti-TNF agents, the risk of lymphoma was tripled, but was not higher than in other cohorts of patients with rheumatoid arthritis (Askling et al. 2005). Present recommendations are that biological agents, including TNF inhibitors, can be used in patients with any solid malignancy or nonmelanoma skin

cancer which has occurred more than 5 years in the past (Singh et al. 2012).

- In children, post-marketing surveillance of TNF inhibitors indicates an increased risk of malignancy, with infliximab having a higher reporting rate for lymphoma and all malignancies and etanercept having a higher reporting rate for lymphoma only (Diak et al. 2010) However, a clear causal link could not be established due to confounding due to the underlying illness and concomitant immunosuppressives.
- Another concern is the development of autoimmune diseases such as systemic lupus-like and demyelinating syndromes. In clinical practice, however, the risk of these diseases from treatment with TNF inhibitors appears low (Tracey et al. 2008).

Immunogenicity

Although the immunogenicity of the TNF inhibitors is potentially clinically significant, there is much uncertainty about this field. All TNF inhibitors have protein domains and have the potential to stimulate the production of antibodies with consequent interactions between the antigen (the TNF inhibitor) and the antibody. As discussed above, most TNF inhibitors are injected subcutaneously which, in general, is associated with a higher incidence of antibodies than when administered intravenously (Ordás et al. 2012). Substantial levels of antibodies to infliximab have also been detected although it is administered intravenously. The antibody production to infliximab has been associated with the mouse component, and the development of humanized antibodies was aimed to reduce the extent of antibody production to the TNF inhibitors. However, clinical findings are inconsistent with this aim.

It is difficult to compare the incidence of antibodies to TNF inhibitors because of the differing methodology in the various studies (Anderson 2005; van Schouwenburg et al. 2013). More studies have been conducted on infliximab and adalimumab which both have been associated with over a 40 % incidence of antibodies (van Schouwenburg et al. 2013). Antibodies to infliximab and adalimumab appear neutralizing (i.e., binding the agent to the antibody reduces the response the TNF inhibitor) and may be sufficient to decrease the half-life of elimination of the TNF inhibitor. Antibodies have also been associated with an increased incidence of acute reactions. Antibodies to certolizumab pegol have been detected in some patients and appear neutralizing, but there is insufficient data on golimumab antibodies to indicate the type of interaction. There are conflicting findings about the detection of antibodies to etanercept, being found by some but not other investigators. However, the detected antibodies to etanercept do not appear to be neutralizing (van Schouwenburg et al. 2013).

Drug Interactions

As outlined above, two combinations are of note:

- TNF inhibitors should not used with live vaccines or IL-1β inhibitors (anakinra, rilonacept, or canakinumab) because of the risk of infections. In children born to mothers on anti-TNF therapy, live vaccines should not be given in the first 6 months.
- TNF inhibitors are administered with methotrexate in order to produce a greater therapeutic response and potentially ameliorate the development of anti-drug antibodies.

Cross-References

▶ CD20 Inhibitors: Rituximab
▶ Corticosteroids
▶ Diabetes Type II
▶ Disease-Modifying Antirheumatic Drugs: Overview
▶ Granulomatosis with Polyangiitis (GPA)
▶ Inflammatory Bowel Disease
▶ Methotrexate
▶ Rheumatoid Arthritis
▶ Sarcoidosis
▶ Sepsis
▶ Spondyloarthritis
▶ Tumor Necrosis Factor (TNF) Inhibitors

Appendix: Structures and Preparation of TNF Inhibitors

Infliximab: A Mouse-Human Chimeric Antibody

The original mouse anti-TNF antibody was isolated from a hybridoma. The variable (V) domains of the heavy (V_H) and light (V_L) chains of the mouse antibody were cloned and fused to the constant light and heavy domains, respectively, of a human antibody. The mouse/human chimeric antibody has around 60 % human sequence and consequently has reduced immunogenicity compared to that of the original anti-TNF mouse antibody.

Adalimumab and Golimumab: Fully Human Antibodies

Adalimumab was isolated through humanizing an existing antihuman TNF mouse antibody using a technique termed guided selection and phage display technology. The V_H and V_L genes of a mouse antibody to human TNF are separated and sequentially paired with separate human genes for V_H and V_L domains from human antibody gene libraries; i.e. the V_H of the mouse antibody are paired with a diverse range of human V_L domains. Single-chain Fv (scFv) antibody fragments consisting of the single mouse V_H and different human V_L domains joined by a peptide linker are displayed on the surface of filamentous phage (phage antibodies), and the phage antibodies are screened for binding to human TNF. On selecting the phage antibody with optimal binding to TNF, the single TNF-specific human V_L domain is subsequently paired with a library of human V_H domains, followed again by screening for optimal binding to human TNF, and results in the isolation of a fully human scFv antibody containing the V_H and V_L domains. In vitro affinity maturation to increase affinity is performed by mutating CDR regions and selecting for optimal TNF binders. Selected V_H and V_L genes are subsequently fused to a human IgG1 Fc domain. The final protein is a variant of a natural fully human antibody with variation occurring in the CDR regions.

Golimumab was isolated using genetically engineered mice with human humoral immune systems; i.e. human immunoglobulin gene loci are incorporated into the genomes of the mice, creating transgenic mice with humoral immune response and antibodies with human sequence. Repeated immunization of the transgenic mice with human TNF results in somatic hypermutation of anti-TNF antibody CDRs and the generation of high affinity, anti-TNF antibodies. Subsequent generation, selection, and isolation of a mouse hybridoma secreting a human TNF-specific human IgG1 were achieved, without the need for in vitro affinity maturation.

Certolizumab Pegol: A Humanized Antibody Fab Fragment

The original anti-TNF antibody was isolated from a mouse hybridoma. The mouse antibody was sequenced and the CDR and framework regions identified. The CDR regions for both V_H and V_L domains were inserted into a human Fab framework construct, in conjunction with other selected mouse framework residues that were necessary to conserve affinity. A 40 kDa polyethylene glycol (PEG) moiety is attached to the Fab fragment to achieve more favorable pharmacokinetics and bioavailability. Certolizumab pegol binds both soluble and transmembrane TNF.

Etanercept: A Humanized Antibody Fraction

Etanercept is a fusion protein, comprising the Fc portion of IgG1 human immunoglobulin and the p75 TNF receptor, otherwise known as the TNFR2 receptor, fused to the Fc γ1 region of a human antibody molecule (i.e. constant regions CH2, CH3 and hinge region). Etanecept is a homodimer of the TNFR2 receptor, Fc γ1 fusion protein with molecular weight of 150 kDa, and is produced in Chinese hamster ovary (CHO) cells.

References

Anderson, P. J. (2005). Tumor necrosis factor inhibitors: Clinical implications of their different immunogenicity profiles. *Seminars in Arthritis and Rheumatism, 34* (Supplement 1), 19–22.

Antohe, J. L., Bili, A., Sartorius, J. A., Kirchner, H. L., Morris, S. J., Dancea, S., et al. (2012). Diabetes mellitus risk in rheumatoid arthritis: Reduced incidence with anti-tumor necrosis factor therapy. *Arthritis Care & Research, 64*(2), 215–221.

Askling, J., & Dixon, W. (2008). The safety of anti-tumour necrosis factor therapy in rheumatoid arthritis. *Current Opinion in Rheumatology, 20*(2), 138–144.

Askling, J., Fored, C. M., Brandt, L., Baecklund, E., Bertilsson, L., & Feltelius, N. (2005). Risks of solid cancers in patients with rheumatoid arthritis and after treatment with tumour necrosis factor antagonists. *Annals of Rheumatic Diseases, 64*(10), 1421–1426.

Day, R. O., & Graham, G. G. (2013). Therapeutics. Non-steroidal anti-inflammatory drugs (NSAIDs). *British Medical Journal, 346*, f3195.

Diak, P., Siegel, J., La Grenade, L., Choi, L., Lemery, S., & McMahon, A. (2010). Tumor necrosis factor alpha blockers and malignancy in children: Forty-eight cases reported to the Food and Drug Administration. *Arthritis and Rheumatism, 62*(8), 2517–2524.

Hyrich, K. L., Watson, K. D., Silman, A. J., & Symmons, D. P. (2006). Predictors of response to anti-TNF-alpha therapy among patients with rheumatoid arthritis: Results from the British Society for Rheumatology Biologics Register. *Rheumatology, 45*(12), 1558–1565.

Kievit, W., Fransen, J., Oerlemans, A. J., Kuper, H. H., van der Laar, M. A., de Rooij, D. J., et al. (2007). The efficacy of anti-TNF in rheumatoid arthritis, a comparison between randomised controlled trials and clinical practice. *Annals of the Rheumatic Diseases, 66*(11), 1473–1478.

Krieckart, C. L. M., Nair, S. C., Nurmohamed, M. T., van Dongen, C. J. J., Lems, W. F., Lafeber, F. P. J. G., et al. (2013). Personalised treatment using serum drug levels of adalimumab in patients with rheumatoid arthritis: An evaluation of costs and effects. *Annals of the Rheumatic Diseases, 74*, 361–368.

Lobo, E. D., Hansen, R. J., & Balthasar, J. P. (2004). Antibody pharmacokinetics and pharmacodynamics. *Journal of Pharmaceutical Sciences, 93*(11), 2645–2668.

Mulleman, D., Lin, D. C. M., Ducourau, E., Emond, P., Ternant, D., Magdelaine-Beuzelin, C., et al. (2010). Trough infliximab concentrations predict efficacy and sustained control of disease activity in rheumatoid arthritis. *Therapeutic Drug Monitoring, 32*(2), 232–236.

Nestorov, I. (2005). Clinical pharmacokinetics of TNF antagonists: How do they differ? *Seminars in Arthritis and Rheumatism, 34*(5 Supplement 1), 12–18.

Ordás, I., Mould, D. R., Feagan, B. G., & Sandborn, W. J. (2012). Anti-TNF monoclonal antibodies in inflammatory bowel disease: Pharmacokinetics-based dosing paradigms. *Clinical Pharmacology & Therapeutics, 91*(4), 635–646.

Pachot, A., Arnaud, B., Marrote, H., Cazalis, M. A., Diasparra, J., Gouraud, A., et al. (2007). Increased tumor necrosis factor-alpha mRNA expression in whole blood from patients with rheumatoid arthritis: Reduction after infliximab treatment does not predict response. *Journal of Rheumatology, 34*(11), 2158–2161.

Ramiro, S., Machado, P., Singh, J. A., Landewe, R. B., & da Silva, J. A. P. (2010). Applying science in practice: The optimization of biological therapy in rheumatoid arthritis. *Arthritis Research and Therapy, 12*, 220.

Roberts, B. V., Susano, I., Gipson, D. S., Trachtman, H., & Joy, M. S. (2013). Contribution of renal and non-renal clearance on increased total clearance of adalimumab in glomerular disease. *Journal of Clinical Pharmacology, 53*(9), 919–924.

Singh, J. A., Furst, D. E., Bharat, A., Curtis, J. R., Kavanaugh, A. F., Kremer, J. M., et al. (2012). 2012 update of the 2008 American College of Rheumatology recommendations for the use of disease-modifying antirheumatic drugs and biologic agents in the treatment of rheumatoid arthritis. *Arthritis Care & Research, 64*(5), 625–639.

Smith, L. S., Nelson, M., & Dolder, C. S. (2010). Certolizumab pegol: A TNF-a antagonist for the treatment of moderate-to-severe Crohn's disease. *Annals of Pharmacotherapy, 44*(2), 333–342.

Soliman, M. M., Hyrich, K. L., Lunt, M. D., Symmons, D. P., & Ashcroft, D. M. (2012). Rituximab or a second anti-tumor necrosis factor therapy for rheumatoid arthritis patients who have failed their first anti-tumor necrosis factor therapy? Comparative analysis from the British Society for Rheumatology Biologics Register. *Arthritis Care & Research, 64*(8), 1108–1115.

Stuhlmuller, B., Haupl, T., Hernandez, M. M., Grutzkau, A., Kuban, R. J., Tandon, N., et al. (2010). CD11c as a transcriptional biomarker to predict response to anti-TNF monotherapy with adalimumab in patients with rheumatoid arthritis. *Clinical Pharmacology & Therapeutics, 87*(3), 311–321.

Taylor, P. C. (2010). Pharmacology of TNF blockade in rheumatoid arthritis and other chronic inflammatory diseases. *Current Opinion in Pharmacology, 10*(3), 308–315.

Tracey, D., Klareskog, L., Sasso, E. H., Salfeld, J. G., & Tak, P. P. (2008). Tumor necrosis factor antagonist mechanisms of action: A comprehensive review. *Pharmacology & Therapeutics, 117*(2), 244–279.

van Schouwenburg, P. A., Rispens, T., & Wolbink, G. J. (2013). Immunogenicity of anti-TNF biologic therapies for rheumatoid arthritis. *Nature Reviews. Rheumatology, 9*(3), 164–172.

Zhou, H. (2005). Clinical pharmacokinetics of etanercept: A fully humanized soluble recombinant tumor necrosis factor receptor fusion protein. *Journal of Clinical Pharmacology, 45*, 490–497.

Zhuang, Y., Xu, Z., Frederick, B., de Vries, D. E., Ford, J. A., Keen, M., et al. (2012). Golimumab pharmacokinetics after repeated subcutaneous and intravenous administrations in patients with rheumatoid arthritis and the effect of concomitant methotrexate: An open-label, randomized study. *Clinical Therapeutics, 34*(1), 77–90.

Type I Interferons

Theresa K. Resch[1], Gabriele Reichmann[2] and Zoe Waibler[1]
[1]Junior Research Group "Novel Vaccination Strategies and Early Immune Responses", Paul-Ehrlich-Institut, Langen, Germany
[2]Department of Immunology, Paul-Ehrlich-Institut, Langen, Germany

Synonyms

Interferon-α/interferon-β

Definition

Type I interferons (IFNs) were identified more than 50 years ago by Isaacs and Lindenmann as a soluble factor that can interfere with viral replication and is able to block infection. Meanwhile, type I IFNs are broadly recognized for their antiviral, antitumoral, and immunomodulatory capacities. Even though virtually any cell type is able to produce type I IFNs, they are secreted mainly by plasmacytoid dendritic cells (pDC). The family of type I IFNs consists of 12 different IFN-α subtypes in man (13 subtypes in the mouse) and one subtype each of IFN-β, IFN-ω, IFN-ε, and IFN-κ. IFN-δ is expressed in pigs only, IFN-ζ in rodents, and IFN-τ in ruminants (Yoneyama and Fujita 2010).

Biosynthesis and Release

Upon appropriate stimulation of a host, cells respond within hours with the secretion of type I IFN. Hence, type I IFNs are described as first line of defense upon infections. The induction of type I IFNs is initiated via the detection of pathogen-associated molecular patterns (PAMPs), such as derived from viruses and bacteria, or danger-associated molecular patterns (DAMPs), such as components associated with damage or malfunction of host cells. Toll-like receptors (TLRs) are the most prominent pattern recognition receptors (PRRs) either localized at the cell surface or at the endosomal membrane. Cytosolic PRRs include members of the NOD-like receptor (NLR) family and members of the retinoic acid-inducible gene I (RIG-I)-like receptor (RLR) family (RIG-I and melanoma differentiation-associated gene 5 (MDA5)) (Pitha 2007; Yoneyama and Fujita 2010).

The signal transduction after ligation of PAMPs to their respective PRR is mediated by different adaptor molecules resulting in specific cellular responses. TLR adaptor molecules include the myeloid differentiation primary response gene (88) (MyD88), TIR-domain-containing adapter-inducing interferon-β (TRIF, also called TICAM1), and TRIF-related adapter molecule (TRAM). RLRs recruit the mitochondrial antiviral-signaling protein (MAVS, also known as VISA, Cardif, or IPS-1) located at the outer membrane of mitochondria (Yoneyama and Fujita 2010). Downstream of those different adaptor molecules, multi-protein complexes are formed containing IRF3, NF-κB, or ATF-2/cJun. These complexes are translocated to the nucleus leading to the induction of type I IFNs and other pro-inflammatory cytokines. After secretion, type I IFNs bind to their cellular receptor, the type I IFN receptor (IFNAR), in an auto and paracrine manner. The IFNAR is a heterodimeric transmembrane receptor consisting of an IFNAR1 and an IFNAR2 chain, expressed on almost all nucleated cells. Binding leads to the induction of the JAK/STAT signaling pathway resulting in an enhanced production of all type I IFN subtypes as well as the regulation of interferon-stimulated genes (ISGs) (Pitha 2007).

Biological Activities

Up to now, more than 2000 ISGs have been identified (de Weerd et al. 2013). Classical antiviral ISGs include the cytoplasmic protein kinase R (PKR), the 2′-5′-oligoadenylate synthetase (OAS), and Mx GTPases. PKR is activated by double-stranded (ds)RNA as derived from viral replication, and phosphorylates eukaryotic initiation factor 2 (eIF2α), preventing translation of (viral) RNA. Upon binding of dsRNA, OAS initiates synthesis of polymeric ATP whose only known function is to activate latent RNase L to degrade (viral) RNA. In addition, disruption of the RNA machinery can lead to apoptosis of the infected cell. Moreover, cleaved RNA fragments can bind to cytosolic RNA sensors such as RIG-I or MDA5 leading to an enhanced type I IFN production. In man, two Mx proteins, MxA and MxB, are induced by type I IFNs. The antiviral function of these proteins is not fully understood yet, but it could be shown that Mx GTPases interfere with viral trafficking and activation of viral polymerases (Pitha 2007).

In addition to direct anti-pathogenic effects via the induction of ISGs, type I IFNs impact adaptive immune cell functions and hence act at the interface between innate and adaptive immunity. Direct effects of type I IFNs on T-cell expansion and effector functions have been demonstrated (Frenz et al. 2010), and additionally, type I IFNs have been reported to shape antibody responses by enhancing the humoral immune response and promoting isotype switching (Le Bon et al. 2001).

Pathophysiological Relevance and Modulation by Drugs

Having the pleiotropic antiviral effects in mind, it is no surprise that type I IFNs were licensed for the treatment of severe acute viral infections such as SARS and are, along with Ribavirin, currently the means of choice for the treatment of chronic hepatitis B/C virus infections. Type I IFNs are also extensively used for the treatment of several types of cancer including hematological malignancies (e.g., hairy cell leukemia, follicular lymphoma, cutaneous T-cell lymphoma) and solid tumors (e.g., melanoma, renal cell carcinoma, Kaposi's sarcoma). Besides having a direct antiproliferative activity on tumor cells (i.e., induction of cell cycle arrest or apoptosis), type I IFNs inhibit angiogenesis and can act as an adjuvant in cancer vaccines.

Given the potency of type I IFNs as antiviral and immune-activating cytokines, their expression has to be regulated tightly in order to avoid severe cell damage. Hence, it is not surprising that deregulated type I IFN expression has been associated with a number of diseases characterized by uncontrolled inflammatory processes.

Systemic lupus erythematosus (SLE) is an inflammatory autoimmune disorder affecting the skin, kidney, musculoskeletal, and hematologic systems. Several lines of evidences indicate a central role for IFN-α in the pathogenesis of SLE. It has been shown that 70–80 % of adult patients and close to 100 % of pediatric patients overexpress type I IFN-induced genes (IFN signature) in both blood cells and peripheral tissues. The level of type I IFN target gene expression correlates with disease severity and activity. In line with this, children usually show a more aggressive course of disease than adult patients. In addition to an IFN signature, SLE patients have elevated serum type I IFN levels, particularly those patients with higher disease activity. Interestingly, several studies report the presence of anti-IFN-α antibodies in sera of SLE patients. The origin and mechanisms of induction of these autoantibodies are not clear yet. However, the majority of SLE patients positive for anti-IFN-α antibodies show significantly lower levels of serum type I IFN bioactivity, a reduced IFN signature, and lower disease activity (Lauwerys et al. 2013).

Another line of evidence indicating the central role for IFN-α in the pathology of SLE comes from long-term type I IFN-treated patients with malignant or viral diseases. The continuous administration of IFN-α to these patients can induce lupus-like symptoms indistinguishable from spontaneous SLE. Interestingly, those lupus-like symptoms were completely resolved when type I IFN administration was discontinued (Niewold and Swedler 2005).

Experimental evidences from murine models further underline the central role of type I IFNs for the etiopathology of SLE. Type I IFNs accelerate the SLE-like syndrome in lupus-prone mice, and deletion of the IFNAR protects those animals (Agrawal et al. 2009). In line with this, mice deficient for SOCS (suppressor of cytokine signaling proteins), involved in inhibiting the JAK/STAT signaling pathway, develop the

SLE-like syndrome (Banchereau and Pascual 2006). Also, mice deficient for negative regulators of TLR signaling spontaneously develop inflammatory disorders by aberrant type I IFN production (Marshak-Rothstein 2006).

Multiple mechanisms can lead to the uncontrolled type I IFN induction in SLE. Most probably, a combination of viral infection(s) and a (genetic) predisposition of the patient such as genetic variations in TLR pathway components, IRFs, or RLR pathway components culminate in aberrant type I IFN secretion. Those elevated IFN-α levels contribute to SLE. Type I IFNs were shown to induce differentiation of monocytes into DC, maturation of DC, and priming of antigen-presenting cells. Furthermore, they improve T-cell-mediated responses to (auto)antigens. With respect to (auto)antibody production, type I IFNs were shown to promote strong IgG secretion and to potentiate signaling through the B-cell receptor (Frenz et al. 2010; Lauwerys et al. 2013; Waibler et al. 2009). Hence, the various modes of action of type I IFNs, which are beneficial in an antiviral immune response (see above), are deleterious in the etiopathology of SLE.

Given the prominent role of type I IFNs for the development of SLE, a number of clinical trials targeting the type I IFN system are ongoing in SLE patients. Two IFN-α neutralizing monoclonal antibodies (mAbs) are tested, namely, Sifalimumab and Rontalizumab, one mAb blocking the IFNAR (MEDI-546), and one IFN-α vaccine (IFN-α kinoid) aiming at the transient induction of polyclonal IFN-α neutralizing antibodies (summarized in Table 1). Due to the early stage of development of all therapeutics (primarily phase II), conclusions on efficacy and safety of the treatment are limited yet. For Sifalimumab, a dose-dependent decrease in the expression of type I IFN-induced genes was reported ($n = 33$) when compared to the placebo group ($n = 17$) and a trend toward disease improvement could be observed. For Rontalizumab, in a subset of all treated patients ($n = 48$), reduction of the IFN signature was documented as well (Lauwerys et al. 2013). Whether one of the abovementioned therapeutics will prove beneficial for SLE patients will need to be demonstrated in phase III trials.

Another example for type I IFNs as central mediators of inflammation is psoriasis. Here, keratinocytes proliferate abnormally in response to a chronic inflammatory reaction. In particular at an early stage of disease, pDC are recruited to pre-plaque areas, where they become activated and secrete large quantities of type I IFNs (Nestle et al. 2005). Consequently, as reported for SLE, an IFN signature is induced in keratinocytes within psoriatic plaques (van der Fits et al. 2004). What exactly drives pDC activation in this early disease stage is not clear yet, but the involvement of TLR ligands has been postulated (Nestle et al. 2005). In a xenograft mouse model of human psoriasis, Nestle et al. demonstrated that blocking the IFNAR or inhibiting the ability of pDC to produce IFN-α prevented the development of psoriasis (Nestle et al. 2005). In line with this, it has been shown that continuous excessive type I IFN signaling in IRF2-deficient mice causes an inflammatory skin disease resembling psoriasis (Hida et al. 2000).

As for SLE, reports indicate an association between the developments of psoriatic eruptions in hepatitis C patients while receiving IFN-α therapy. Of note, cessation of the IFN-α therapy led to resolution of psoriatic-like symptoms in 93 % of the cases (Afshar et al. 2013).

Type I Interferons, Table 1 Anti-type I IFN therapeutic concepts in clinical trials

Target	Therapeutic concept	Product	Indication	Stage of clinical trial
IFN-α	Neutralizing mAb	Sifalimumab (MEDI-545)	SLE	Phase II
			Psoriasis	Phase I
IFN-α	Neutralizing mAb	Rontalizumab	SLE	Phase II
IFNAR	Blocking mAb	MEDI-546	SLE	Phase II
IFN-α	Vaccine	IFN-α kinoid	SLE	Phase I–II

All trials in this table are listed in the ClinicalTrials.gov registry

Given the central role of type I IFNs in psoriasis, a clinical trial has been conducted to evaluate the safety and tolerability of single-dose IFN-α-neutralizing Sifalimumab application in adult patients with chronic plaque psoriasis. However, the phase I clinical trial indicated no significant inhibition of the type I IFN signature and no clinical activity of the mAb (Bissonnette et al. 2010). This might be related to the fact that only initial stages of disease development seem to be type I IFN-dependent, whereas late (chronic) stages are rather depending on the activity of tumor necrosis factor (Banchereau and Pascual 2006).

In addition to SLE and psoriasis, there are a number of experimental models and inflammatory disorders for which a central role of type I IFNs in etiopathology has been proposed. In Sjögren's syndrome, foci of inflammation in salivary glands are positive for IFN-α, serum type I IFN levels are elevated, and affected organs show an IFN signature. Elevated type I IFN levels and/or an IFN signature were also prominent in dermatomyositis, Aicardi-Goutières syndrome, Cree encephalitis, and type 1 diabetes (Bronson et al. 2012). In a murine model, the development of Lyme arthritis after Borrelia burgdorferi infection is associated with an IFN signature in affected joints and blocking type I IFN signaling reduced both the IFN signature and the development of arthritis (Miller et al. 2008).

Collectively, data derived from both experimental animal models and clinical investigations pinpoint toward a crucial role of type I IFNs in numerous inflammatory diseases and underline the central role of type I IFNs as mediators of inflammation. However, type I IFNs also mediate anti-inflammatory effects, although the data regarding the anti-inflammatory and disease-limiting capacity of treatment with type I IFNs are rather inconsistent and distinct roles for IFN-α and IFN-β have been proposed.

The most prominent example for the anti-inflammatory capacity of type I IFNs is IFN-β treatment of patients suffering from multiple sclerosis (MS), an inflammatory demyelinating disease of the central nervous system (CNS). Long-term treatment with IFN-β of patients with relapsing-remitting MS attenuates the course and severity of the disease and can reduce the frequency of relapses in about one third of patients (McCormack and Scott 2004). The precise mechanism of action of IFN-β is still elusive, but it seems to act immunomodulatory at various set points of the disease. MS can be mediated by Th1 and Th17 cells, respectively. IFN-β can induce the production of IL-27 which is a potent inhibitor of Th17 cells, and in general, type I IFNs can reduce the number of effector T cells. Moreover, upon IFN-β treatment, leucocytes are retained within the lymph nodes and hence do not infiltrate the CNS. In addition, IFN-β treatment of microglia cells reduces their ability to present antigens, limiting the risk of being detected and eliminated by cytotoxic T cells or NK cells (Yong et al. 1998).

Even though IFN-β can be a powerful tool in treatment of MS, 10–50 % of relapsing-remitting MS patients do not respond to treatment (Rio et al. 2006). In some patients, treatment with IFN-β actually induces exacerbations of the disease (Shimizu et al. 2008). In line with this, acute MS-like demyelinating disease with extensive lesions throughout the brain and the thoracic spinal cord has been reported secondary to IFN-α treatment of hepatitis C patients (Matsuo and Takabatake 2002).

Rheumatoid arthritis (RA) is a chronic inflammatory disorder mainly characterized by the progressive damage of synovial joints and variable extra-articular manifestations. Several observations indicate an inflammatory role for type I IFN in RA. First, as for SLE and other indications (see above), an IFN signature could be detected in a subset of RA patients. Second, RA patients collectively display increased plasma type I IFN activity relative to levels in healthy controls. Finally, IFN-β is present in RA synovial membranes. However, IFN-β reduces synoviocyte proliferation in vitro which was considered as an anti-inflammatory feature. Anti-inflammatory and disease-limiting properties of type I IFNs were also demonstrated in a collagen-induced arthritis model and an adjuvant arthritis model. Here, the injection of IFN-β resulted in reduced disease activity and the inhibition of cartilage and bone destruction. Of note, independent clinical trials indicated that IFN-β is not effective in RA with regard to clinical or radiographic scores (Axtell et al. 2011; Crow 2010).

Inflammatory bowel disease (IBD) is an intestinal inflammation with two major clinical entities, namely, Crohn's disease and ulcerative colitis (UC). Most likely, IBD results from a disturbed interplay between the intestinal mucosa and gut microbiota. Experimental data indicated a beneficial role for type I IFN in limiting disease. First, it was shown that TLR3 and TLR9 ligand-induced production of type I IFNs reduced the severity of colonic injury and inflammation in murine models of experimental colitis and neutralizing type I IFN abolished this beneficial effect. Next, compared to their wild-type counterparts, IFNAR-deficient mice are highly susceptible to dextran sulfate sodium (DSS)-induced colitis. Finally, administration of IFN-β to DSS-treated mice decreased the disease activity and the histological scores of DSS-treated mice.

Clinical data from type I IFN-treated IBD patients are highly variable. While some studies indicated beneficial effects of IFN-α and IFN-β treatment of UC patients, other studies failed to demonstrate disease-limiting capacities of this treatment. Moreover, cases are reported of exacerbation of UC during the course of IFN-α therapy of hepatitis C virus infection and induction of UC in the context of IFN-β treatment of MS (Gonzalez-Navajas et al. 2012; Rauch et al. 2013).

Taken together, type I IFNs are potent pleiotropic cytokines involved in numerous immunological processes. They are licensed for treatment in several indications including viral infections, cancer, and MS. Given the pro-inflammatory role of type I IFNs in SLE and psoriasis, clinical trials were initiated and are ongoing, respectively, aiming at inhibiting type I IFN signaling. We summarized some (but by no means all) experimental and clinical evidence indicating highly ambivalent and even contrary effects upon both exogenous application of type I IFNs and the inhibition of type I IFN signaling. Currently, no reliable biomarkers are available predicting the outcome of (anti) type I IFN treatment.

Several attempts are discussed aiming at resolving this conundrum. Most authors agree on the distinct properties of type I IFN subtypes, particularly of IFN-α subtypes vs. IFN-β, in terms of binding to the IFNAR and subsequent target gene expression. Indeed, binding capacities of type I IFN subtypes to the IFNAR can vary (Crow 2010). Moreover, de Weerd et al. recently identified a unique IFN-β/IFNAR1 complex that signals independently of the IFNAR2 chain and induces a distinct and IFN-β-specific set of genes and functional responses (de Weerd et al. 2013). Whether this IFN-β/IFNAR1 signaling is involved in inflammatory disorders may be a matter of future investigations.

Also the severity of inflammation and the overall cytokine milieu prior to type I IFN treatment may decide between adverse or protective effects of the treatment (Axtell et al. 2011; Rauch et al. 2013). The discrepant roles of type I IFNs in two different murine sepsis models and a DSS colitis model, varying in the severity of the inflammatory response, pinpoint toward this direction (Rauch et al. 2013).

Axtell et al. argue that the T helper cell subset involved in the etiology of the respective disease determines the outcome of a type I IFN treatment: IFN-β shall inhibit symptoms in diseases with a Th1 bias but promote pathology in diseases with a Th17 bias (Axtell et al. 2011).

Without doubt, the genetic predisposition of a patient has a major impact in the onset and cause of an inflammatory disease. For SLE, multiple genetic loci associated with disease susceptibility have been described. Particularly, cytokine-induced signaling components seem to be affected (e.g., Kariuki et al. 2010). Exemplarily, genetic variations in genes encoding for TLR7, IRFs, MDA5, and IPS-1 have been associated with SLE susceptibility and with altered activation of the type I IFN pathway in SLE patients (Shrivastav and Niewold 2013). Unbiased genome-wide meta-analyses are recently increasing the number of loci, associated with a particular disease. For example, more than 100 such risk loci have been described for RA (e.g., Eyre et al. 2012). For sure, no single cause for the opposing effects mediated by type I IFNs in

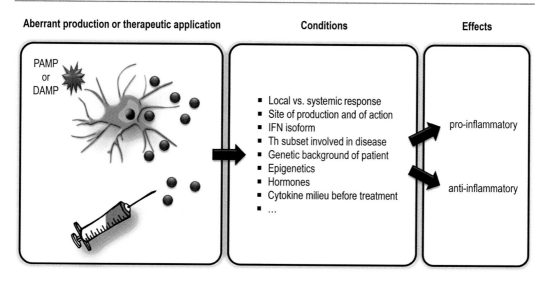

Type I Interferons, Fig. 1 Pro- and anti-inflammatory effects of type I IFNs. Aberrant production or therapeutic application of type I IFNs can mediate pro- or anti-inflammatory effects depending on patient-specific condition

inflammatory processes can be identified – in Fig. 1 conditions contributing to either pro- or anti-inflammatory properties of type I IFNs are summarized.

Cross-References

- Antiviral Responses
- Autoinflammatory Syndromes
- Cytokines
- Dendritic Cells
- Inflammatory Bowel Disease
- Inflammatory Bowel Disease Models in Animals
- Janus Kinases (Jaks)/STAT Pathway
- Pathogen-Associated Molecular Patterns (PAMPs)
- Rheumatoid Arthritis
- Toll-Like Receptors

References

Afshar, M., Martinez, A. D., Gallo, R. L., & Hata, T. R. (2013). Induction and exacerbation of psoriasis with Interferon-alpha therapy for hepatitis C: A review and analysis of 36 cases. *Journal of the European Academy of Dermatology and Venereology, 27*, 771–778.

Agrawal, H., Jacob, N., Carreras, E., Bajana, S., Putterman, C., Turner, S., et al. (2009). Deficiency of type I IFN receptor in lupus-prone New Zealand mixed 2328 mice decreases dendritic cell numbers and activation and protects from disease. *Journal of Immunology, 183*, 6021–6029.

Axtell, R. C., Raman, C., & Steinman, L. (2011). Interferon-beta exacerbates Th17-mediated inflammatory disease. *Trends in Immunology, 32*, 272–277.

Banchereau, J., & Pascual, V. (2006). Type I interferon in systemic lupus erythematosus and other autoimmune diseases. *Immunity, 25*, 383–392.

Bissonnette, R., Papp, K., Maari, C., Yao, Y., Robbie, G., White, W. I., et al. (2010). A randomized, double-blind, placebo-controlled, phase I study of MEDI-545, an anti-interferon-alfa monoclonal antibody, in subjects with chronic psoriasis. *Journal of the American Academy of Dermatology, 62*, 427–436.

Bronson, P. G., Chaivorapol, C., Ortmann, W., Behrens, T. W., & Graham, R. R. (2012). The genetics of type I interferon in systemic lupus erythematosus. *Current Opinion in Immunology, 24*, 530–537.

Crow, M. K. (2010). Type I interferon in organ-targeted autoimmune and inflammatory diseases. *Arthritis Research & Therapy, 12*(Suppl 1), S5.

de Weerd, N. A., Vivian, J. P., Nguyen, T. K., Mangan, N. E., Gould, J. A., Braniff, S. J., et al. (2013). Structural basis of a unique interferon-beta signaling axis mediated via the receptor IFNAR1. *Nature Immunology, 14*, 901–907.

Eyre, S., Bowes, J., Diogo, D., Lee, A., Barton, A., Martin, P., et al. (2012). High-density genetic mapping identifies new susceptibility loci for rheumatoid arthritis. *Nature Genetics, 44*, 1336–1340.

Frenz, T., Waibler, Z., Hofmann, J., Hamdorf, M., Lantermann, M., Reizis, B., et al. (2010). Concomitant type I IFN receptor-triggering of T cells and of DC is required to promote maximal modified vaccinia virus Ankara-induced T-cell expansion. *European Journal of Immunology, 40*, 2769–2777.

Gonzalez-Navajas, J. M., Lee, J., David, M., & Raz, E. (2012). Immunomodulatory functions of type I interferons. *Nature Reviews Immunology, 12*, 125–135.

Hida, S., Ogasawara, K., Sato, K., Abe, M., Takayanagi, H., Yokochi, T., et al. (2000). CD8(+) T cell-mediated skin disease in mice lacking IRF-2, the transcriptional attenuator of interferon-alpha/beta signaling. *Immunity, 13*, 643–655.

Kariuki, S. N., Franek, B. S., Kumar, A. A., Arrington, J., Mikolaitis, R. A., Utset, T. O., et al. (2010). Trait-stratified genome-wide association study identifies novel and diverse genetic associations with serologic and cytokine phenotypes in systemic lupus erythematosus. *Arthritis Research & Therapy, 12*, R151.

Lauwerys, B. R., Ducreux, J., Houssiau, F. A. (2013). Type I interferon blockade in systemic lupus erythematosus: Where do we stand? *Rheumatology. (Oxford), 53* (8):1369–76.

Le Bon, A., Schiavoni, G., D'Agostino, G., Gresser, I., Belardelli, F., & Tough, D. F. (2001). Type i interferons potently enhance humoral immunity and can promote isotype switching by stimulating dendritic cells in vivo. *Immunity, 14*, 461–470.

Marshak-Rothstein, A. (2006). Toll-like receptors in systemic autoimmune disease. *Nature Reviews Immunology, 6*, 823–835.

Matsuo, T., & Takabatake, R. (2002). Multiple sclerosis-like disease secondary to alpha interferon. *Ocular Immunology and Inflammation, 10*, 299–304.

McCormack, P. L., & Scott, L. J. (2004). Interferon-beta-1b: A review of its use in relapsing-remitting and secondary progressive multiple sclerosis. *CNS Drugs, 18*, 521–546.

Miller, J. C., Ma, Y., Bian, J., Sheehan, K. C., Zachary, J. F., Weis, J. H., et al. (2008). A critical role for type I IFN in arthritis development following Borrelia burgdorferi infection of mice. *Journal of Immunology, 181*, 8492–8503.

Nestle, F. O., Conrad, C., Tun-Kyi, A., Homey, B., Gombert, M., Boyman, O., et al. (2005). Plasmacytoid predendritic cells initiate psoriasis through interferon-alpha production. *Journal of Experimental Medicine, 202*, 135–143.

Niewold, T. B., & Swedler, W. I. (2005). Systemic lupus erythematosus arising during interferon-alpha therapy for cryoglobulinemic vasculitis associated with hepatitis C. *Clinical Rheumatology, 24*, 178–181.

Pitha, P. M. (2007). *Interferon: The 50th anniversary*. Berlin: Springer.

Rauch, I., Muller, M., & Decker, T. (2013). The regulation of inflammation by interferons and their STATs. *JAKSTAT, 2*, e23820.

Rio, J., Nos, C., Tintore, M., Tellez, N., Galan, I., Pelayo, R., et al. (2006). Defining the response to interferon-beta in relapsing-remitting multiple sclerosis patients. *Annals of Neurology, 59*, 344–352.

Shimizu, Y., Yokoyama, K., Misu, T., Takahashi, T., Fujihara, K., Kikuchi, S., et al. (2008). Development of extensive brain lesions following interferon beta therapy in relapsing neuromyelitis optica and longitudinally extensive myelitis. *Journal of Neurology, 255*, 305–307.

Shrivastav, M., & Niewold, T. B. (2013). Nucleic acid sensors and type I interferon production in systemic lupus erythematosus. *Frontiers in Immunology, 4*, 319.

van der Fits, L., van der Wel, L. I., Laman, J. D., Prens, E. P., & Verschuren, M. C. (2004). In psoriasis lesional skin the type I interferon signaling pathway is activated, whereas interferon-alpha sensitivity is unaltered. *The Journal of Investigative Dermatology, 122*, 51–60.

Waibler, Z., Anzaghe, M., Frenz, T., Schwantes, A., Pohlmann, C., Ludwig, H., et al. (2009). Vaccinia virus-mediated inhibition of type I interferon responses is a multifactorial process involving the soluble type I interferon receptor B18 and intracellular components. *Journal of Virology, 83*, 1563–1571.

Yoneyama, M., & Fujita, T. (2010). Recognition of viral nucleic acids in innate immunity. *Reviews in Medical Virology, 20*, 4–22.

Yong, V. W., Chabot, S., Stuve, O., & Williams, G. (1998). Interferon beta in the treatment of multiple sclerosis: Mechanisms of action. *Neurology, 51*, 682–689.

List of Entries

A
ACKR3
Acute Exacerbations of Airway Inflammation
Alarmins
Allergic Disorders
Alzheimer's disease
Anaphylaxis (Immediate Hypersensitivity): From Old to New Mechanisms
Angiogenesis Inhibitors
Anti-asthma Drugs, Overview
Antibacterial Host Defense Peptides
Antibiotics as Antiinflammatory Drugs
Antihistamines
Antimalarial Drugs
Antimuscarinics
Antiphospholipid Syndrome
Antiviral Responses
Asthma
Atherosclerosis
Atopic Dermatitis
Autoinflammatory Syndromes
Autophagy and Inflammation

B
Bacterial Lipopolysaccharide
Basophils
Behçet's Disease
Beta$_2$ Receptor Agonists
Biofilm
Bone Morphogenetic Proteins in Inflammation
Bronchiolitis Obliterans

C
Cancer and Inflammation
CCR1
CD20 Inhibitors: Rituximab
CD80/86 Inhibitors: Abatacept
Cell Signaling in Neutrophils
Cell Therapy in Autoimmune Disease
Ceramide 1-Phosphate: A Mediator of Inflammatory Responses
Chemokine CCL14
Chemokine CCL15
Chemokine CCL18
Chromones: Cromoglycate and Nedocromil
Cogan's Syndrome
Complement C5a Receptors
Complement System
Corticosteroids
Costimulatory Receptors
Coxibs
CXCL4 und CXCL4L1
CXCR1 and CXCR2 and Ligands
CXCR3 and Its Ligands
Cytokines
Cytolytic Granules

D
Dendritic Cells
Diabetes Type I
Diabetes Type II
Diffuse Panbronchiolitis
Disease-Modifying Antirheumatic Drugs: Overview

F
Fenamates
Flavonoids as Anti-inflammatory Agents
FOXP3

G
Gaucher Disease
Genetic Susceptibility to Inflammatory Diseases
GILZ-Related Regulation of Inflammation
Glial Cells
Gold Complexes
Gout
Granulomatosis with Polyangiitis (GPA)
Graves' Disease

H
Heat Shock Proteins
Henoch-Schönlein Purpura
Histamine

I
Idiopathic Thrombocytopenic Purpura
IkappaB
Immunoglobulin Receptors and Inflammation
Inflammasomes
Inflammatory Bowel Disease
Inflammatory Bowel Disease Models in Animals
Interferon gamma
Interleukin-1 (IL-1) Inhibitors: Anakinra, Rilonacept, and Canakinumab
Interleukin 2
Interleukin 4 and the Related Cytokines (Interleukin 5 and Interleukin 13)
Interleukin 6
Interleukin-6 Inhibitor: Tocilizumab
Interleukin 9
Interleukin 10
Interleukin 12
Interleukin-12/23 Inhibitors: Ustekinumab
Interleukin 17
Interleukin 18
Interleukin-18 Binding Protein
Interleukin 22
Interleukin 23
Interleukin 27
Interleukin 32
Interleukin-33
Interleukin 36 Cytokines

J
Janus Kinase Inhibitors
Janus Kinases (JAKs)/STAT Pathway

K
Kawasaki Disease
Kinins

L
Leflunomide
Leukocyte Recruitment
Leukotrienes
Lymphocyte Homing and Trafficking

M
Macrophage Heterogeneity During Inflammation
Mammalian Target of Rapamycin (mTOR)
MAP Kinase Pathways
Mast Cells
Mechanisms of Macrophage Migration in 3-Dimensional Environments
Medicinal Fatty Acids
Methotrexate
Microvascular Responses to Inflammation
Modulation of Inflammation by Key Nutrients

N
Natural Killer Cells
Neutrophil Extracellular Traps
Neutrophil Oxidative Burst
NFkappaB
Non-steroidal Anti-inflammatory Drugs: Overview
Nuclear Receptor Signaling in the Control of Inflammation

O
Obesity and Inflammation
Osteoarthritis
Osteoarthritis Genetics
Osteoclasts in Inflammation

P
Pathogen-Associated Molecular Patterns (PAMPs)
Pentraxins
Phosphodiesterase 4 Inhibitors: Apremilast and Roflumilast

Platelets, Endothelium, and Inflammation
Polymyositis and Dermatomyositis
Propionic Acid Derivative Drugs (Profens)
Prostanoids
Protease-Activated Receptors

R
Reactive Oxygen Species
Rheumatic Fever
Rheumatoid Arthritis

S
Salicylates
Sarcoidosis
Sepsis
Sepsis Models in Animals
Skin Inflammation Models in Animals

Spondyloarthritis
Substance P in Inflammation
Sulfasalazine and Related Drugs

T
Tetracyclines
TGF beta Superfamily Cytokine MIC-1/GDF15 in Health and Inflammatory Diseases
Th17 Response
Theophylline
Thiopurines: Azathioprine, Mercaptopurine, and Thioguanine
Toll-Like Receptors
Tumor Necrosis Factor Alpha (TNFalpha)
Tumor Necrosis Factor (TNF) Inhibitors
Type I Interferons

This encyclopedia includes no entries for U, V, W, X, Y and Z.

Book Erratum to the Compendium of Inflammatory Diseases

Michael J. Parnham (ed.)

DOI 10.1007/978-3-7643-8550-7, © Springer International Publishing AG 2016

By mistake the book was published under the trade name of Birkhäuser. The correct imprint is Springer.
As the publisher Springer Basel merged with Springer International Publishing AG, the copyright in the work shall be vested in © Springer International Publishing AG 2016.

This Springer imprint is published by Springer Nature
The registered company is Springer International Publishing AG
The registered company address is: Gewerbestrasse 11, 6330 Cham, Switzerland

---

The updated online version of the original book can be found under DOI 10.1007/978-3-7643-8550-7

© Springer International Publishing AG 2017
M.J. Parnham (ed.), *Compendium of Inflammatory Diseases*,
DOI 10.1007/978-3-7643-8550-7_601